FORMULAS FROM GEOMETRY

Triangle
$h = a \sin \theta$
Area $= \dfrac{1}{2} bh$
(Law of Cosines)
$c^2 = a^2 + b^2 - 2ab \cos \theta$

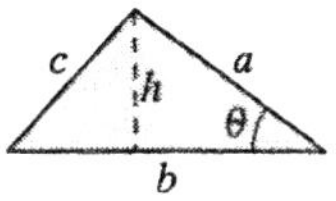

Right Triangle
(Pythagorean Theorem)
$c^2 = a^2 + b^2$

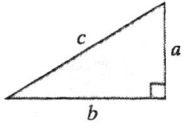

Equilateral Triangle
$h = \dfrac{\sqrt{3}\,s}{2}$
Area $= \dfrac{\sqrt{3}\,s^2}{4}$

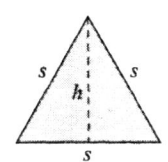

Parallelogram
Area $= bh$

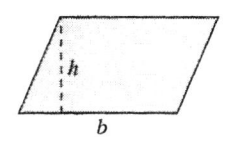

Trapezoid
Area $= \dfrac{h}{2}(a + b)$

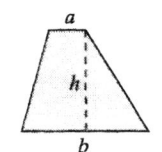

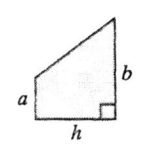

Circle
Area $= \pi r^2$
Circumference $= 2\pi r$

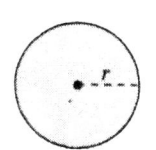

Sector of Circle
(θ in radians)
Area $= \dfrac{\theta r^2}{2}$
$s = r\theta$

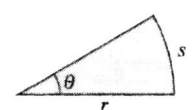

Circular Ring
(p = average radius,
w = width of ring)
Area $= \pi(R^2 - r^2)$
$\quad\quad = 2\pi p w$

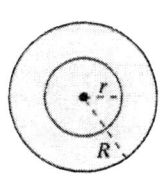

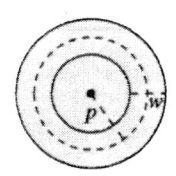

Sector of Circular Ring
(p = average radius,
w = width of ring,
θ in radians)
Area $= \theta p w$

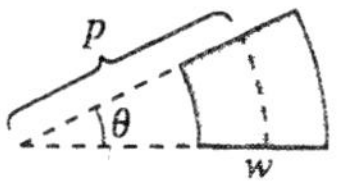

Ellipse
Area $= \pi a b$
Circumference $\approx 2\pi \sqrt{\dfrac{a^2 + b^2}{2}}$

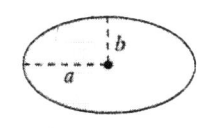

Cone
(A = area of base)
Volume $= \dfrac{Ah}{3}$

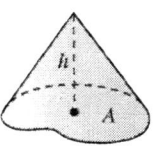

Right Circular Cone
Volume $= \dfrac{\pi r^2 h}{3}$
Lateral Surface Area $= \pi r \sqrt{r^2 + h^2}$

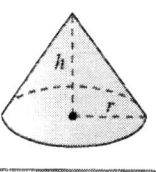

Frustum of Right Circular Cone
Volume $= \dfrac{\pi(r^2 + rR + R^2)h}{3}$
Lateral Surface Area $= \pi s(R + r)$

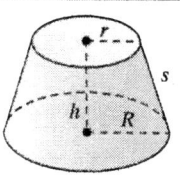

Right Circular Cylinder
Volume $= \pi r^2 h$
Lateral Surface Area $= 2\pi r h$

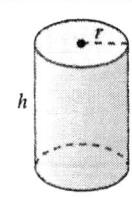

Sphere
Volume $= \dfrac{4}{3}\pi r^3$
Surface Area $= 4\pi r^2$

Wedge
(A = area of upper face,
B = area of base)
$A = B \sec \theta$

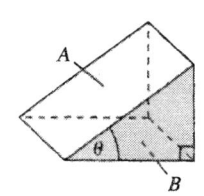

SEVENTH EDITION

Calculus
VOLUME THREE

CHAPTERS 9-14

Ron Larson
Robert P. Hostetler
The Pennsylvania State University
The Behrend College

Bruce H. Edwards
University of Florida

With the assistance of David E. Heyd
The Pennsylvania State University
The Behrend College

Houghton Mifflin Company Boston New York

Editor-in-Chief, Mathematics: Jack Shira
Managing Editor: Cathy Cantin
Development Manager: Maureen Ross
Development Editor: Laura Wheel
Assistant Editor: Rosalind Horn
Supervising Editor: Karen Carter
Project Editor: Patty Bergin
Editorial Assistant: Lindsey Gulden
Production Technology Supervisor: Gary Crespo
Senior Marketing Manager: Michael Busnach
Marketing Assistant: Nicole Mollica

We have included examples and exercises that use real-life data as well as technology output from a variety of software. This would not have been possible without the help of many people and organizations. Our wholehearted thanks goes to all for their time and effort.

Trademark Acknolwedgements: TI is a registered trademark of Texas Instruments, Inc. Mathcad is a registered trademark of MathSoft, Inc. Windows, Microsoft, and MS-DOS are registered trademarks of Microsoft, Inc. Mathematica is a registered trademark of Wolfram Research, Inc. DERIVE is a registered trademark of Texas Instruments, Inc. IBM is a registered trademark of International Business Machines Corporation. Maple is a registered trademark of Waterloo Maple, Inc. HMClassPrep is a trademark of Houghton Mifflin Company.

Custom Publishing Editor: Sheila Ellis
Custom Publishing Production Manager: Kathleen McCourt
Project Coordinator: Kim Gavrilles

Cover Designer: Galen B. Murphy

Copyright © 2002 by Houghton Mifflin Company. 2002 Impression. All rights reserved.

No part of this work may be reproduced or transmitted in any form or by any means, electronic or mechanical, including photocopying and recording, or by any information storage or retrieval system without the prior written permission of Houghton Mifflin Company unless such copying is expressly permitted by federal copyright law. Address inquiries to College Permissions, Houghton Mifflin Company, 222 Berkeley Street, Boston, MA 02116-3764.

Printed in the United States of America.

ISBN: 0-618-30133-X
N01781
1 2 3 4 5 6 7 8 9 - DMI - 04 03 02

Houghton Mifflin
Custom Publishing

222 Berkeley Street • Boston, MA 02116

Address all correspondence and order information to the above address.

Contents

A Word from the Authors (Preface) ix
Features xi

Chapter 9 — Conics, Parametric Equations, and Polar Coordinates 648

Exploring New Planets 648
9.1 Conics and Calculus 650
9.2 Plane Curves and Parametric Equations 665
Section Project: Cycloids 674
9.3 Parametric Equations and Calculus 675
9.4 Polar Coordinates and Polar Graphs 684
Section Project: Anamorphic Art 693
9.5 Area and Arc Length in Polar Coordinates 694
9.6 Polar Equations of Conics and Kepler's Laws 702
Review Exercises 709
P.S. Problem Solving 712

Chapter 10 — Vectors and the Geometry of Space 714

Suspension Bridges 714
10.1 Vectors in the Plane 716
10.2 Space Coordinates and Vectors in Space 727
10.3 The Dot Product of Two Vectors 735
10.4 The Cross Product of Two Vectors in Space 744
10.5 Lines and Planes in Space 752
Section Project: Distances in Space 762
10.6 Surfaces in Space 763
10.7 Cylindrical and Spherical Coordinates 773
Review Exercises 780
P.S. Problem Solving 782

Chapter 11 — Vector-Valued Functions 784

Race Car Cornering 784
11.1 Vector-Valued Functions 786
Section Project: Witch of Agnesi 793
11.2 Differentiation and Integration of Vector-Valued Functions 794
11.3 Velocity and Acceleration 802
11.4 Tangent Vectors and Normal Vectors 811
11.5 Arc Length and Curvature 820
 Review Exercises 832
P.S. Problem Solving 834

Chapter 12 — Functions of Several Variables 836

Satellite Receiving Dish 836
12.1 Introduction to Functions of Several Variables 838
12.2 Limits and Continuity 850
12.3 Partial Derivatives 859
Section Project: Moiré Fringes 868
12.4 Differentials 869
12.5 Chain Rules for Functions of Several Variables 876
12.6 Directional Derivatives and Gradients 884
12.7 Tangent Planes and Normal Lines 896
Section Project: Wildflowers 904
12.8 Extrema of Functions of Two Variables 905
12.9 Applications of Extrema of Functions of Two Variables 913
Section Project: Building a Pipeline 920
12.10 Lagrange Multipliers 921
 Review Exercises 929
P.S. Problem Solving 932

Chapter 13 — Multiple Integration 934

Hyperthermia Treatments for Tumors 934
13.1 Iterated Integrals and Area in the Plane 936
13.2 Double Integrals and Volume 944
13.3 Change of Variables: Polar Coordinates 955
13.4 Center of Mass and Moments of Inertia 963
Section Project: Center of Pressure on a Sail 970
13.5 Surface Area 971

Section Project: Capillary Action   977
13.6 Triple Integrals and Applications   978
13.7 Triple Integrals in Cylindrical and Spherical Coordinates   988
Section Project: Wrinkled and Bumpy Spheres   994
13.8 Change of Variables: Jacobians   995
    Review Exercises   1001
P.S. Problem Solving   1004

Chapter 14   Vector Analysis   1006

Mathematical Sculpture   1006
14.1 Vector Fields   1008
14.2 Line Integrals   1019
14.3 Conservative Vector Fields and Independence of Path   1032
14.4 Green's Theorem   1042
Section Project: Hyperbolic and Trigonometric Functions   1050
14.5 Parametric Surfaces   1051
14.6 Surface Integrals   1061
Section Project: Hyperboloid of One Sheet   1072
14.7 Divergence Theorem   1073
14.8 Stokes's Theorem   1081
    Review Exercises   1087
Section Project: The Planimeter   1089
P.S. Problem Solving   1090

Appendices   A1

Appendix A   *Additional Topics in Differential Equations*   A2
Appendix B   *Proofs of Selected Theorems*   A9
Appendix C   *Integration Tables*   A27
Appendix D   *Precalculus Review*   CD*
Appendix E   *Rotation and the General Second-Degree Equation*   CD*
Appendix F   *Complex Numbers*   CD*
Appendix G   *Business and Economic Applications*   CD*

    Answers to Odd-Numbered Exercises   A33
    Index of Applications   A167
    Index   A171

---

*Available in *e-solutions Calculus Learning Tools Student CD-ROM* and at the text-specific website at *college.hmco.com.*

A Word from the Authors

Welcome to *Calculus with Analytic Geometry*, Seventh Edition. Much has changed since we wrote the first edition—nearly 25 years ago. With each edition, we have listened to you, our users, and have tried to incorporate your suggestions for improvement.

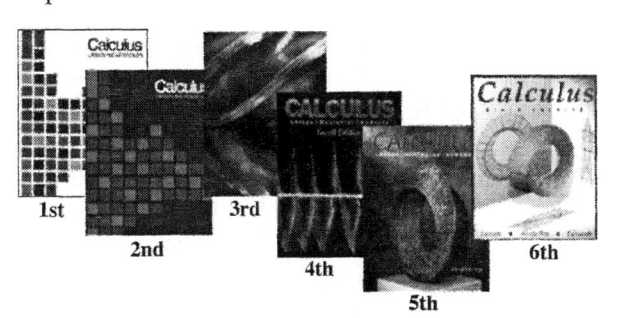

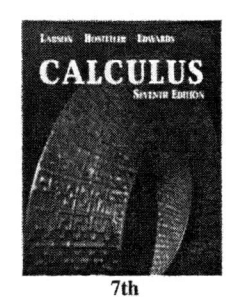

A Text Formed by Its Users

Through your support and suggestions, the text has evolved over seven editions to include these extensive enhancements:

- Expanded exercise sets containing a greater variety of tasks such as skill building, applications, explorations, writing, critical thinking, and theoretical problems
- Additional applications that more accurately represent the diverse uses of calculus in the world
- Many more open-ended activities and investigations
- Clearer, less cluttered text, full annotations and labels—carefully planned page layout
- Additional art, composed with more color, accuracy, and realism
- A more comprehensive and more mathematically rigorous text, particularly the third semester of the Seventh Edition, which is quite different when compared with the First Edition
- Increased technology use, as both a problem-solving tool and an investigative tool
- References to the history of calculus and to the mathematicians who developed it
- Updated references to current mathematical journals
- Considerably more help in the supplements package for both students and instructors
- Alternatives to the traditional print medium, particularly in the CD-ROM version
- Five different volumes from which to choose your preferred teaching approach— a great development in flexibility from the single volume in the First Edition (see page xx)

What's New and Different in the Seventh Edition

In the Seventh Edition, we continue to offer instructors and students a text that is pedagogically sound, mathematically precise, and comprehensible. There are many minor changes in the mathematics, prose, art, and design. The more significant changes are noted here.

- *New P.S. Problem Solving*   At the end of each chapter, we have included a two-page collection of new applied and theoretical exercises. These exercises offer problems that have some unusual characteristics that set them apart from exercises in a regular exercise set.
- *New Getting at the Concept*   Midway through each section exercise set we have added a set of problems that check a student's understanding of the basic concepts presented in the section.
- *New Section Objectives*   Each section in the Seventh Edition begins with a list of learning objectives. These enable students to identify and focus on the key points of the section.
- *New Downloadable Graphs*   Many exercise sets contain problems in which students are asked to draw on the graph that is provided. Because this is not feasible in the actual text, we now provide printable enlargements of these graphs on the website *www.mathgraphs.com*.
- *New Journal Articles on the Web*   The Seventh Edition contains over 60 references to articles from mathematics journals noted in the feature *For Further Information*. In order to make the articles easily accessible to instructors and students, they are now available on the website *www.matharticles.com*.
- *Revised Chapter Openers*   The chapter openers have been redesigned as two-page spreads in the Seventh Edition. Included in the chapter openers is a real-world application designed to motivate the calculus topics of the chapter.
- *Revised Review Exercises*   In order to provide a more effective study tool, we have grouped the Review Exercises by text section. This reorganization allows students to target specific concepts that may require additional study and review.
- *Exercise Sets*   Approximately 20 percent of the exercises in the Seventh Edition are new. The new exercises include skill, concept, applied, and theoretical problems.
- *Table of Contents*   Although the organization of the table of contents is much the same as in the Sixth Edition, some notable changes are as follows. In an effort to cut back on the length of the text, we have moved Section 3.10 *Business and Economic Applications* (Appendix G in the Seventh Edition), Chapter 15 *Differential Equations*, Appendix A *Precalculus Review* (Appendix D in the Seventh Edition), Appendix E *Rotation and the General Second-Degree Equation*, and Appendix F *Complex Numbers* to the text-specific website at *college.hmco.com*. We removed Appendix C *Basic Differentiation Rules for Elementary Functions* from the text; however, that material appears on the inside front cover of the text. Although Chapter 15 has been moved from the text, some of the differential equations topics have been retained and other topics have been expanded in a new Appendix A (*Additional Topics in Differential Equations*). Coverage includes slope fields, Euler's Method, and first-order linear differential equations.

Although we carefully and thoroughly revised the text by enhancing the usefulness of some features and topics and by adding others, we did not change many of the things that our colleagues and the two million students who have used this book have told us work for them. We still offer comprehensive coverage of the material required by students in a three-semester or four-quarter calculus course, including carefully stated theories and proofs.

We hope you will enjoy the Seventh Edition. We are proud to have it as our first calculus book to be published in the twenty-first century.

Ron Larson   Robert P. Hostetler   Bruce H. Edwards

Features

Chapter Openers

Each chapter opens with a real-world application designed to motivate the calculus concepts covered in the chapter. Following a brief introduction, open-ended questions guide students through an introduction to the main themes of the chapter. In addition, photographs and interesting facts related to the application are included in the chapter opener.

Section Objectives

Every section begins with a list of learning objectives that outline the key concepts of the section. This list helps instructors with class planning and provides students a study guide for the section.

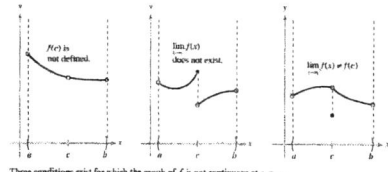

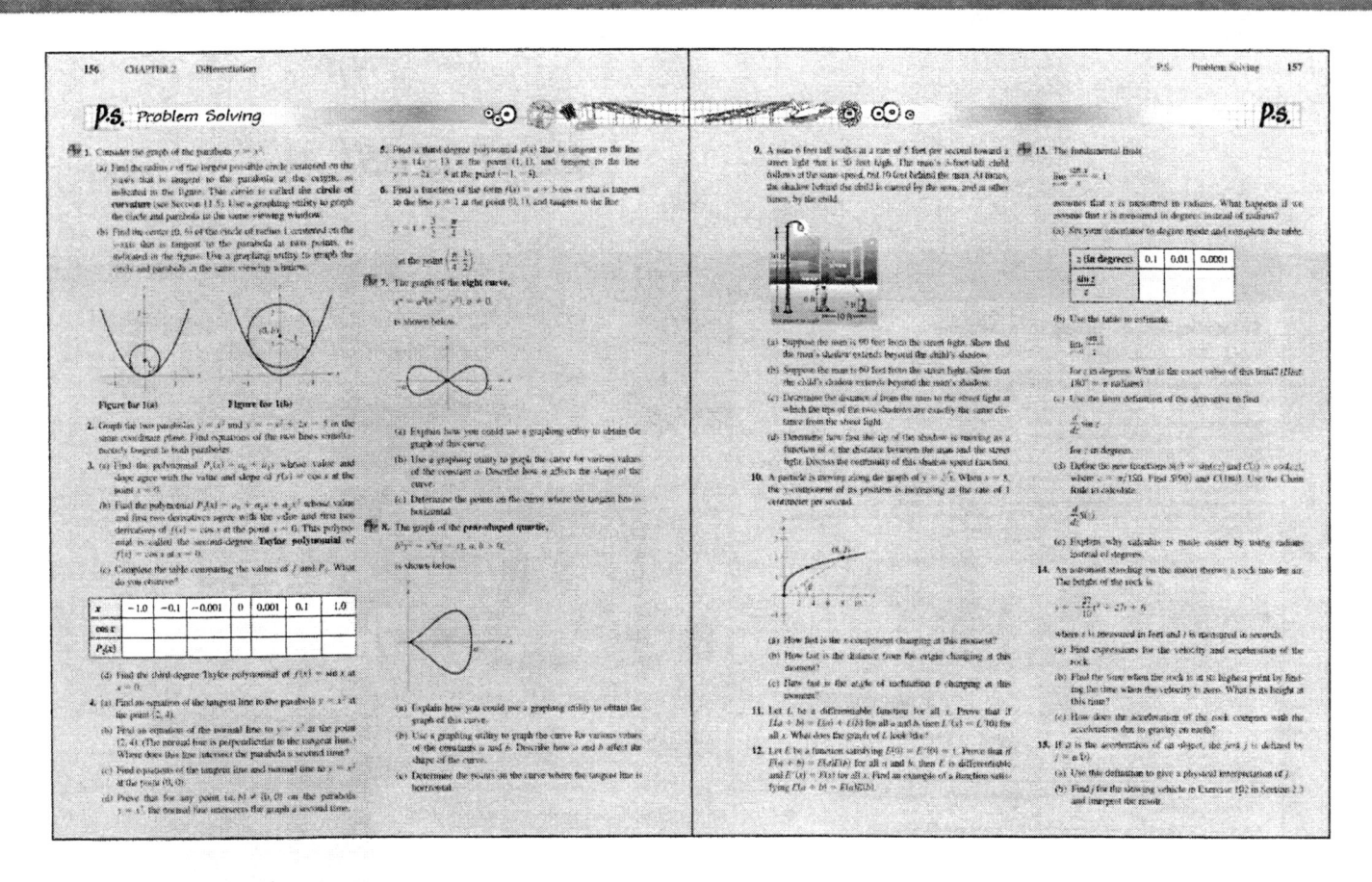

New! P.S. Problem Solving

Each chapter concludes with a collection of thought-provoking and challenging exercises that further explore and expand upon the concepts of the chapter. These exercises have unusual characteristics that set them apart from traditional calculus exercises.

Review Exercises

A set of *Review Exercises* is included at the end of each chapter. In order to provide students with a more useful study tool, these exercises are grouped by section. This organization allows students to identify specific problem types related to chapter concepts for study and review.

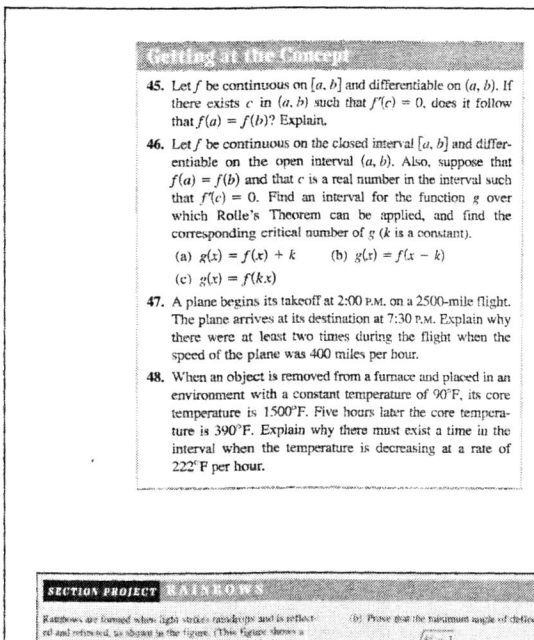

Getting at the Concept

These exercises contain questions that check a student's understanding of the basic concepts of the section. They are generally located midway through the section exercise sets and are boxed and titled for easy reference.

Section Projects

Appearing at the end of selected exercise sets, the *Section Projects* contain extended applications, which can be assigned as an individual or group activity.

Open Explorations

The *Interactive* CD-ROM version of this text contains open explorations, which further investigate selected examples throughout the text using computer algebra systems (*Maple*, *Mathematica*, *Derive*, and *Mathcad*). The icon identifies an example for which an open exploration exists.

Additional Features

Additional teaching and learning resources can be found throughout the text. These resources include explorations, technology notes, historical vignettes, study tips, journal references, lab series, and notes. For a complete description of these resources, go to the text-specific website at *college.hmco.com*.

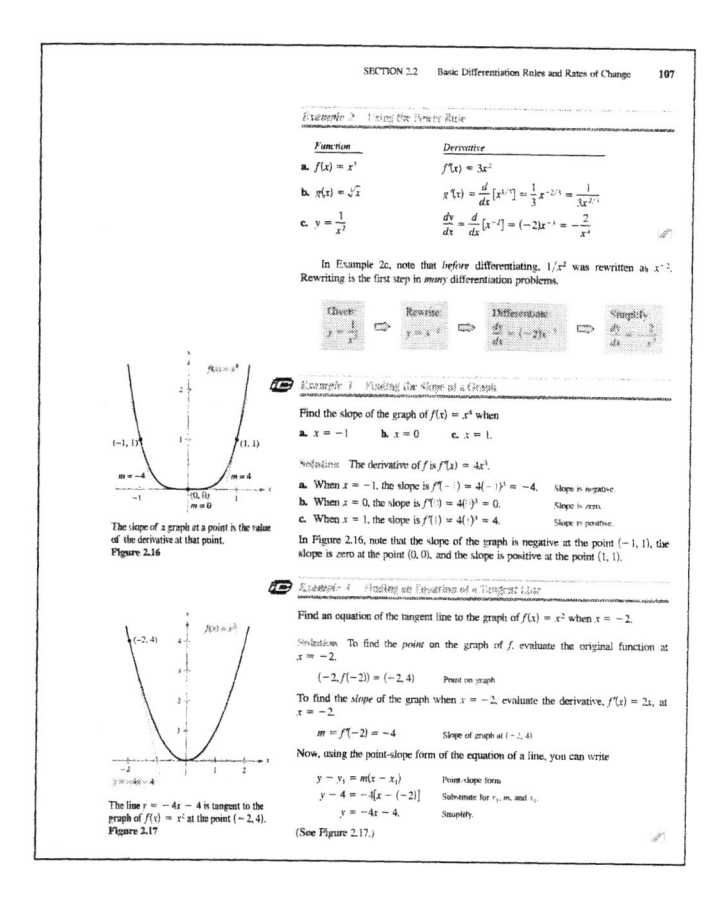

Acknowledgments

We would like to thank the many people who have helped us at various stages of this project during the past 25 years. Their encouragement, criticisms, and suggestions have been invaluable to us.

Seventh Edition Reviewers

Raymond Badalian
Los Angeles City College

John Santomas
Villanova University

Gordon Melrose
Old Dominion University

Dane R. Camp
New Trier High School, IL

Anthony Thomas
University of Wisconsin–Platteville

Eleanor Palais
Belmont High School, MA

Kathy Hoke
University of Richmond

Beth Long
Pellissippi State Technical College

Christopher Butler
Case Western Reserve University

Lynn Smith
Gloucester County College

Larry Norris
North Carolina State University

Barbara Cortzen
DePaul University

Charles Wheeler
Montgomery College

Lila Roberts
Georgia Southern University

Previous Editions' Reviewers

Dennis Alber, *Palm Beach Junior College*; James Angelos, *Central Michigan University*; Kerry D. Bailey, *Laramie County Community College*; Harry L. Baldwin, Jr., *San Diego City College*; Homer F. Bechtell, *University of New Hampshire*; Keith Bergeron, *United States Air Force Academy*; Norman Birenes, *University of Regina*; Brian Blank, *Washington University*; Andrew A. Bulleri, *Howard Community College*; Paula Castagna, *Fresno City College*; Jack Ceder, *University of California–Santa Barbara*; Charles L. Cope, *Morehouse College*; Jorge Cossio, *Miami-Dade Community College*; Jack Courtney, *Michigan State University*; James Daniels, *Palomar College*; Kathy Davis, *University of Texas*; Paul W. Davis, *Worcester Polytechnic Institute*; Luz M. DeAlba, *Drake University*; Nicolae Dinculeanu, *University of Florida*; Rosario Diprizio, *Oakton Community College*; Garret J. Etgen, *University of Houston*; Russell Euler, *Northwest Missouri State University*; Phillip A. Ferguson, *Fresno City College*; Li Fong, *Johnson County Community College*; Michael Frantz, *University of La Verne*; William R. Fuller, *Purdue University*; Dewey Furness, *Ricks College*; Javier Garza, *Tarleton State University*; K. Elayn Gay, *University of New Orleans*; Thomas M. Green, *Contra Costa College*; Ali Hajjafar, *University of Akron*; Ruth A. Hartman, *Black Hawk College*; Irvin Roy Hentzel, *Iowa State University*; Howard E. Holcomb, *Monroe Community College*; Eric R. Immel, *Georgia Institute of Technology*; Arnold J. Insel, *Illinois State University*; Elgin Johnston, *Iowa State University*; Hideaki Kaneko, *Old Dominion University*; Toni Kasper, *Borough of Manhattan Community College*; William J. Keane, *Boston College*; Timothy J. Kearns, *Boston College*;

Ronnie Khuri, *University of Florida*; Frank T. Kocher, Jr., *Pennsylvania State University*; Robert Kowalczyk, *University of Massachusetts–Dartmouth*; Joseph F. Krebs, *Boston College*; David C. Lantz, *Colgate University*; Norbert Lerner, *State University of New York at Cortland*; Maita Levine, *University of Cincinnati*; Murray Lieb, *New Jersey Institute of Technology*; Ransom Van B. Lynch, *Phillips Exeter Academy*; Bennet Manvel, *Colorado State University*; Mauricio Marroquin, *Los Angeles Valley College*; Robert L. Maynard, *Tidewater Community College*; Robert McMaster, *John Abbott College*; Darrell Minor, *Columbus State Community College*; Maurice Monahan, *South Dakota State University*; Michael Montaño, *Riverside Community College*; Philip Montgomery, *University of Kansas*; David C. Morency, *University of Vermont*; Gerald Mueller, *Columbus State Community College*; Duff A. Muir, *United States Air Force Academy*; Charlotte J. Newsom, *Tidewater Community College*; Terry J. Newton, *United States Air Force Academy*; Donna E. Nordstrom, *Pasadena City College*; Robert A. Nowlan, *Southern Connecticut State University*; Luis Ortiz-Franco, *Chapman University*; Barbara L. Osofsky, *Rutgers University*; Judith A. Palagallo, *University of Akron*; Wayne J. Peeples, *University of Texas*; Jorge A. Perez, *LaGuardia Community College*; Darrell J. Peterson, *Santa Monica College*; Donald Poulson, *Mesa Community College*; Jean L. Rubin, *Purdue University*; Barry Sarnacki, *United States Air Force Academy*; N. James Schoonmaker, *University of Vermont*; George W. Schultz, *St. Petersburg Junior College*; Richard E. Shermoen, *Washburn University*; Thomas W. Shilgalis, *Illinois State University*; J. Philip Smith, *Southern Connecticut State University*; Frank Soler, *De Anza College*; Enid Steinbart, *University of New Orleans*; Michael Steuer, *Nassau Community College*; Mark Stevenson, *Oakland Community College*; Lawrence A. Trivieri, *Mohawk Valley Community College*; John Tweed, *Old Dominion University*; Carol Urban, *College of DuPage*; Marjorie Valentine, *North Side ISD, San Antonio*; Robert J. Vojack, *Ridgewood High School, NJ*; Bert K. Waits, *Ohio State University*; Florence A. Warfel, *University of Pittsburgh*; John R. Watret, *Embry-Riddle Aeronautical University*; Carroll G. Wells, *Western Kentucky University*; Jay Wiestling, *Palomar College*; Paul D. Zahn, *Borough of Manhattan Community College*; August J. Zarcone, *College of DuPage*

During the past four years, several users of the Sixth Edition wrote to us with suggestions. We considered each and every one of them when preparing the manuscript for the Seventh Edition. We would like to extend a special thanks to Mikhail Ostrovskii of the Catholic University of America for the many thoughtful suggestions he sent to us. The time and care he invested in several correspondences was quite extraordinary.

We would like to thank the staff at Larson Texts, Inc., and the staff of Meridian Creative Group, who assisted with proofreading the manuscript, preparing and proofreading the art package, and checking and typesetting the supplements.

A special note of thanks goes to the instructors who responded to our survey and to the over 2 million students who have used earlier editions of the text.

On a personal level, we are grateful to our wives, Deanna Gilbert Larson, Eloise Hostetler, and Consuelo Edwards, for their love, patience, and support. Also, a special note of thanks goes to R. Scott O'Neil.

If you have suggestions for improving this text, please feel free to write to us. Over the past 25 years we have received many useful comments from both instructors and students, and we value these very much.

Ron Larson
Robert P. Hostetler
Bruce H. Edwards

Supplements

Resources

Website (*college.hmco.com*)

Many additional text-specific study and interactive features for students and instructors can be found at the Houghton Mifflin website.

For the Student

Study and Solutions Guide, Volumes I and II by Bruce H. Edwards (University of Florida)

Graphing Technology Guide for Precalculus and Calculus by Benjamin N. Levy and Laurel Technical Services

Graphing Calculator Videotape by Dana Mosely

Calculus, 7E, *Videotapes* by Dana Mosely

For the Instructor

Complete Solutions Guide, Volumes I, II, and III by Bruce H. Edwards (University of Florida)

Test Item File by Ann Rutledge Kraus (The Pennsylvania State University, The Behrend College)

Instructor's Resource Guide by Ann Rutledge Kraus (The Pennsylvania State University, The Behrend College)

Computerized Testing (WIN, Macintosh)

HMClassPrep™ (Instructor's CD-ROM)

New exciting study aids make the best supplements package for Calculus even better.

New! Text-Specific Video Series (available in VHS and DVD formats)

Tied directly to the Larson/Hostetler/Edwards *Calculus*, Seventh Edition, textbook, these videos created by Dana Mosely provide lecture-style instruction, review of key concepts, real-life data examples, and more. Ideal for students who want extra guidance or who have missed a class, the videos cover select material from Chapters P–9.

Announcing a whole new suite of electronic study tools for calculus: The Larson *eSolutions* for Calculus.

Calculus Learning Tools Student **CD-ROM** Contains Computer Algebra System Explorations, rotatable 3-D art, printable MathGraphs and MathArticles referenced throughout the text, as well as MathBios, labs, and more.

Companion Website Includes rotatable 3-D art and other student and instructor resources. Visit **www.college.hmco.com/mathematics.**

Interactive and *Internet Calculus 3.0* These two products are comprehensive multimedia courses in calculus. To provide you with a choice, we offer *Interactive Calculus 3.0* on CD-ROM and *Internet Calculus 3.0* online. Both contain the complete text of *Calculus*, Seventh Edition, as well as other exciting features such as solutions to odd-numbered exercises, rotatable 3-D graphs, editable 2-D graphs, Open Explorations using one of four computer algebra systems, animations, videos, simulations, Try Its for every example, and more.

CalcChat.com website An on-line resource where students can access, discuss, *and* help each other with step-by-step solutions to all the odd-numbered exercises in the Larson *Calculus* series.

EduSpace On-Line Learning Environment Instructors can easily assign, deliver, and grade homework and other assignments based on the even-numbered exercises in the text via Houghton Mifflin's new *EduSpace* platform.

Live, on-line tutoring from SMARTHINKING.COM

Houghton Mifflin has partnered with SMARTHINKING.com to give students the most advanced on-line tutoring possible. SMARTHINKING is a virtual learning assistance center created in conjunction with 31 schools. It provides qualified tutors (e-structors) and independent study resources for core courses and skills. Students can access tutors and resources at home, school, or anywhere else they have an Internet connection.

Instructors

To get copies of these supplements or for more information on packaging them at significant discounts with any Larson *Calculus*, Seventh Edition, textbook, please contact your Houghton Mifflin sales representative or call 1-800-733-1717. AP Instructors call McDougal Littell at 1-800-462-6595.

Students

For details on how you can order these exciting new learning aids, visit our website at *http://www.hmco.com* or email us at *college_math@hmco.com*.

Interactive Calculus 3.0 CD-ROM and Internet Calculus 3.0

To accommodate a wide variety of teaching and learning styles, *Calculus* is also available as *Interactive Calculus 3.0* on an interactive CD-ROM and *Internet Calculus 3.0*. These versions incorporate live mathematics throughout the entire program. Live mathematics helps students visualize and explore—leading to a deeper understanding of calculus concepts than has ever before been possible.

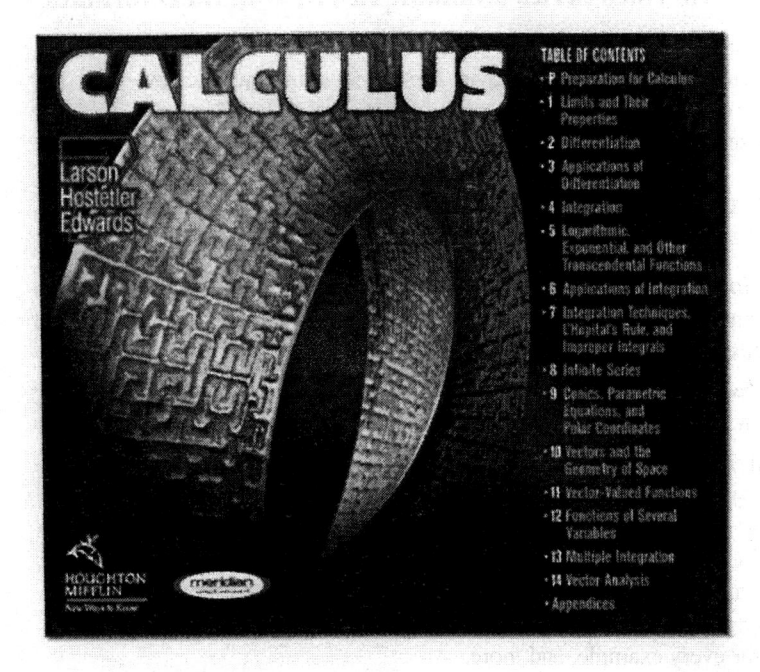

Live Mathematics Throughout

- Open Explorations give students the opportunity to explore using computer algebra systems.
- Section Quizzes require students to enter free-response answers and to click-and-drag answers into place.
- Editable two-dimensional graphs, featured throughout the entire program, provide additional opportunities to explore and investigate.
- Rotatable three-dimensional graphs allow for a whole new level of visualization.
- New and enhanced explorations, simulations, and animations make concepts come alive.

Classroom Management Tool and Syllabus Builder

All of the content of the Seventh Edition text—a wealth of applications, exercises, worked-out examples, and detailed explanations—is included in *Interactive Calculus 3.0* on CD-ROM and *Internet Calculus 3.0*. Instructors have the flexibility of customizing content and interactive features for students as desired. Instructors may simply add dates to a default syllabus or may modify the order of topics. Either way, a customized syllabus is easy to distribute electronically and update instantly. This tool is particularly useful for managing distance learning courses.

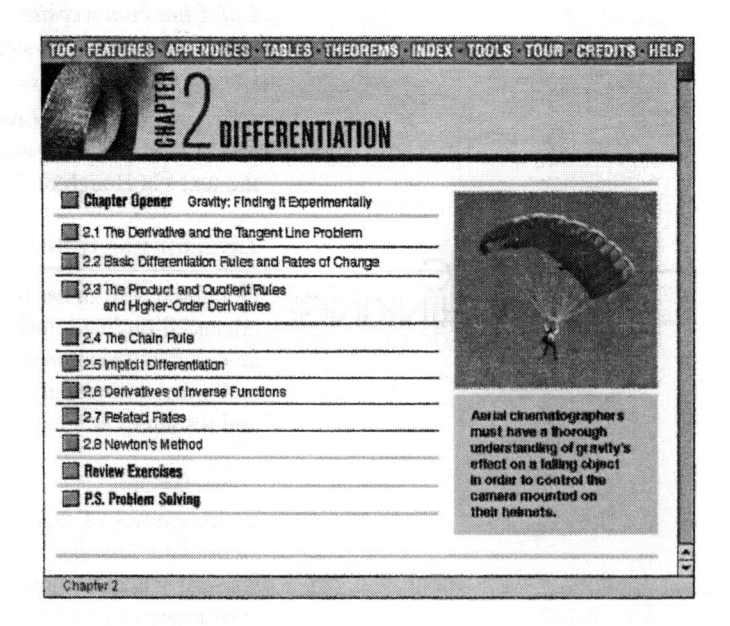

Features

Exercises with solutions to all odd exercises provide immediate feedback for students.

Try Its allow students to try problems similar to the examples and to check their work using the worked-out solutions provided.

Quizzes with responses require students to enter free responses, click-and-drag answers, and choose multiple choice answers.

Editable Graphs encourage students to explore concepts by graphing "editable" graphs as well as to change the viewing window and to use *zoom* and *trace* features.

Rotatable Graphs allow students to view three-dimensional graphs as they rotate, greatly enhancing visualization.

Simulations encourage exploration and hands-on interaction with mathematical concepts.

Animations, which use motion and sound to explain concepts, can be played and replayed, or viewed one step at a time.

Complete searchable text-specific **Content, Index, Theorem Index,** and **Features Index** facilitate cross-referencing.

Video Clips engage student interest and show connections between mathematics and other disciplines.

Syllabus Builder enables instructors to save administrative time and to convey important information online.

Bookmarking capability provides fast, efficient navigation of the site.

Other special features include:

Articles • Connections • History • Look Ahead • Math Trends • Section Projects • Technology

Conics, Parametric Equations, and Polar Coordinates

9

In April 2001, Geoffrey Marcy and Paul Butler were awarded the Henry Draper Medal by the National Academy of Sciences "for their pioneering investigations of planets orbiting other stars via high-precision radial velocities." With their colleagues, Marcy and Butler have found 38 of 53 known extra-solar planets since 1995.

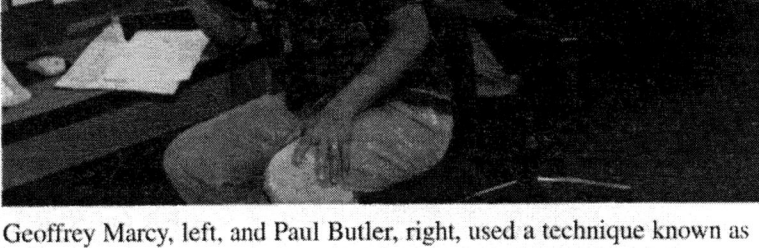

Geoffrey Marcy, left, and Paul Butler, right, used a technique known as the Doppler effect to identify the new planet 70 Vir B.

FOR FURTHER INFORMATION For more information on the discovery of the new planet 70 Vir B, see the article "Searching for Other Worlds" in *Time*. To view this article, go to the website *www.matharticles.com*.

Section 9.1 Conics and Calculus

- Understand the definition of a conic section.
- Analyze and write equations of parabolas using properties of parabolas.
- Analyze and write equations of ellipses using properties of ellipses.
- Analyze and write equations of hyperbolas using properties of hyperbolas.

Conic Sections

Each **conic section** (or simply **conic**) can be described as the intersection of a plane and a double-napped cone. Notice in Figure 9.1 that for the four basic conics, the intersecting plane does not pass through the vertex of the cone. When the plane passes through the vertex, the resulting figure is a **degenerate conic,** as shown in Figure 9.2.

Circle   Parabola   Ellipse   Hyperbola

Conic sections
Figure 9.1

Point   Line   Two intersecting lines

Degenerate conics
Figure 9.2

HYPATIA (370–415 A.D.)

The Greeks discovered conic sections sometime between 600 and 300 B.C. By the beginning of the Alexandrian period, enough was known about conics for Apollonius (262–190 B.C.) to produce an eight-volume work on the subject. Later, toward the end of the Alexandrian period, Hypatia wrote a textbook entitled *On the Conics of Apollonius*. Her death marked the end of major mathematical discoveries in Europe for several hundred years.

The early Greeks were largely concerned with the geometric properties of conics. It was not until 1900 years later, in the early seventeenth century, that the broader applicability of conics became apparent. Conics then played a prominent role in the development of calculus.

There are several ways to study conics. You could begin as the Greeks did by defining the conics in terms of the intersections of planes and cones, or you could define them algebraically in terms of the general second-degree equation

$$Ax^2 + Bxy + Cy^2 + Dx + Ey + F = 0.$$ General second-degree equation

However, a third approach, in which each of the conics is defined as a **locus** (collection) of points satisfying a certain geometric property, suits our needs best. For example, a circle can be defined as the collection of all points (x, y) that are equidistant from a fixed point (h, k). This locus definition easily produces the standard equation of a circle,

$$(x - h)^2 + (y - k)^2 = r^2.$$ Standard equation of a circle

FOR FURTHER INFORMATION To learn more about the mathematical activities of Hypatia, see the article "Hypatia and Her Mathematics" by Michael A. B. Deakin in *The American Mathematical Monthly*. To view this article, go to the website *www.matharticles.com*.

Parabolas

A **parabola** is the set of all points (x, y) that are equidistant from a fixed line called the **directrix** and a fixed point called the **focus** not on the line. The midpoint between the focus and the directrix is the **vertex**, and the line passing through the focus and the vertex is the **axis** of the parabola. Note in Figure 9.3 that a parabola is symmetric with respect to its axis.

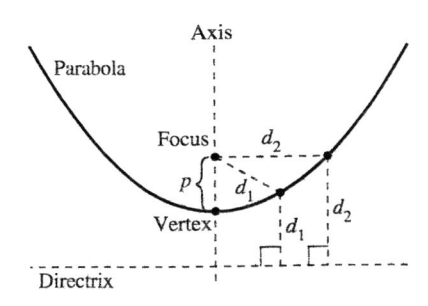

Figure 9.3

THEOREM 9.1 Standard Equation of a Parabola

The **standard form** of the equation of a parabola with vertex (h, k) and directrix $y = k - p$ is

$$(x - h)^2 = 4p(y - k). \qquad \text{Vertical axis}$$

For directrix $x = h - p$, the equation is

$$(y - k)^2 = 4p(x - h). \qquad \text{Horizontal axis}$$

The focus lies on the axis p units (*directed distance*) from the vertex. The coordinates of the focus are as follows.

$(h, k + p)$ Vertical axis

$(h + p, k)$ Horizontal axis

Example 1 Finding the Focus of a Parabola

Find the focus of the parabola given by $y = -\frac{1}{2}x^2 - x + \frac{1}{2}$.

Solution To find the focus, convert to standard form by completing the square.

$$y = \frac{1}{2} - x - \frac{1}{2}x^2 \qquad \text{Write original equation.}$$

$$y = \frac{1}{2}(1 - 2x - x^2) \qquad \text{Factor out } \tfrac{1}{2}.$$

$$2y = 1 - 2x - x^2 \qquad \text{Multiply each side by 2.}$$

$$2y = 1 - (x^2 + 2x) \qquad \text{Group terms.}$$

$$2y = 2 - (x^2 + 2x + 1) \qquad \text{Add and subtract 1 on right side.}$$

$$x^2 + 2x + 1 = -2y + 2$$

$$(x + 1)^2 = -2(y - 1) \qquad \text{Standard form}$$

Comparing this equation with $(x - h)^2 = 4p(y - k)$, you can conclude that

$$h = -1, \quad k = 1, \quad \text{and} \quad p = -\frac{1}{2}.$$

Because p is negative, the parabola opens downward, as shown in Figure 9.4. Therefore, the focus of the parabola is p units from the vertex, or

$$(h, k + p) = \left(-1, \frac{1}{2}\right). \qquad \text{Focus}$$

Parabola with a vertical axis, $p < 0$
Figure 9.4

A line segment that passes through the focus of a parabola and has endpoints on the parabola is called a **focal chord**. The specific focal chord perpendicular to the axis of the parabola is the **latus rectum**. The next example shows how to determine the length of the latus rectum and the length of the corresponding intercepted arc.

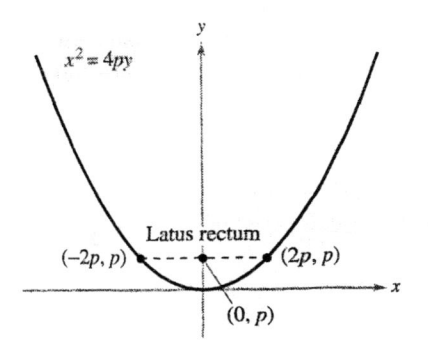

Length of latus rectum: $4p$
Arc length: $4.59p$
Figure 9.5

Example 2 Focal Chord Length and Arc Length

Find the length of the latus rectum of the parabola given by
$$x^2 = 4py.$$
Then find the length of the parabolic arc intercepted by the latus rectum.

Solution Because the latus rectum passes through the focus $(0, p)$ and is perpendicular to the y-axis, the coordinates of its endpoints are $(-x, p)$ and (x, p). Substituting p for y in the equation of the parabola produces
$$x^2 = 4p(p) \quad\Longrightarrow\quad x = \pm 2p.$$
So, the endpoints of the latus rectum are $(-2p, p)$ and $(2p, p)$, and you can conclude that its length is $4p$, as shown in Figure 9.5. In contrast, the length of the intercepted arc is given by the following.

$$\begin{aligned}
s &= \int_{-2p}^{2p} \sqrt{1 + (y')^2}\, dx & &\text{Use arc length formula.} \\
&= 2\int_0^{2p} \sqrt{1 + \left(\frac{x}{2p}\right)^2}\, dx & &y = \frac{x^2}{4p} \Longrightarrow y' = \frac{x}{2p} \\
&= \frac{1}{p}\int_0^{2p} \sqrt{4p^2 + x^2}\, dx & &\text{Simplify.} \\
&= \frac{1}{2p}\left[x\sqrt{4p^2 + x^2} + 4p^2 \ln\left|x + \sqrt{4p^2 + x^2}\right|\right]_0^{2p} & &\text{Theorem 7.2} \\
&= \frac{1}{2p}\left[2p\sqrt{8p^2} + 4p^2 \ln\left(2p + \sqrt{8p^2}\right) - 4p^2 \ln(2p)\right] \\
&= 2p\left[\sqrt{2} + \ln\left(1 + \sqrt{2}\right)\right] \\
&\approx 4.59p
\end{aligned}$$

One widely used property of a parabola is its reflective property. In physics, a surface is called **reflective** if the tangent line at any point on the surface makes equal angles with an incoming ray and the resulting outgoing ray. The angle corresponding to the incoming ray is the **angle of incidence,** and the angle corresponding to the outgoing ray is the **angle of reflection.** One example of a reflective surface is a flat mirror.

Another type of reflective surface is that formed by revolving a parabola about its axis. A special property of parabolic reflectors is that they allow us to direct all incoming rays parallel to the axis through the focus of the parabola—this is the principle behind the design of the parabolic mirrors used in reflecting telescopes. Conversely, all light rays emanating from the focus of a parabolic reflector used in a flashlight are parallel, as shown in Figure 9.6.

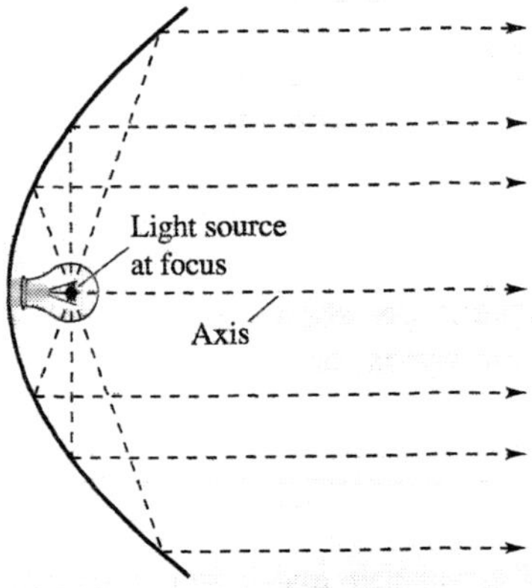

THEOREM 9.2 Reflective Property of a Parabola

Let P be a point on a parabola. The tangent line to the parabola at the point P makes equal angles with the following two lines.

1. The line passing through P and the focus
2. The line passing through P parallel to the axis of the parabola

---

Parabolic reflector: light is reflected in parallel rays.
Figure 9.6

indicates that in the *Interactive 3.0 CD-ROM* and *Internet 3.0* versions of this text (available at college.hmco.com) you will find an *Open Exploration*, which further explores this example using the computer algebra systems Maple, Mathcad, Mathematica, *and* Derive.

Ellipses

More than a thousand years after the close of the Alexandrian period of Greek mathematics, Western civilization finally began a Renaissance of mathematical and scientific discovery. One of the principal figures in this rebirth was the Polish astronomer Nicolaus Copernicus. In his work *On the Revolutions of the Heavenly Spheres*, Copernicus claimed that all of the planets, including earth, revolved about the sun in circular orbits. Although some of Copernicus's claims were invalid, the controversy set off by his heliocentric theory motivated astronomers to search for a mathematical model to explain the observed movements of the sun and planets. The first to find the correct model was the German astronomer Johannes Kepler (1571–1630). Kepler discovered that the planets move about the sun in elliptical orbits, with the sun not as the center but as a focal point of the orbit.

The use of ellipses to explain the movement of the planets is only one of many practical and aesthetic uses. As with parabolas, we begin our study of this second type of conic by defining it as a locus of points. Now, however, we use *two* focal points rather than one.

An **ellipse** is the set of all points (x, y) the sum of whose distances from two distinct fixed points called **foci** is constant. (See Figure 9.7.) The line through the foci intersects the ellipse at two points, called the **vertices**. The chord joining the vertices is the **major axis**, and its midpoint is the **center** of the ellipse. The chord perpendicular to the major axis at the center is the **minor axis** of the ellipse.

Figure 9.7

NICOLAUS COPERNICUS (1473–1543)

Copernicus began to study planetary motion when asked to revise the calendar. At that time, the exact length of the year could not be accurately predicted using the theory that earth was the center of the universe.

FOR FURTHER INFORMATION To learn about how an ellipse may be "exploded" into a parabola, see the article "Exploding the Ellipse" by Arnold Good in *Mathematics Teacher*. To view this article, go to the website *www.matharticles.com*.

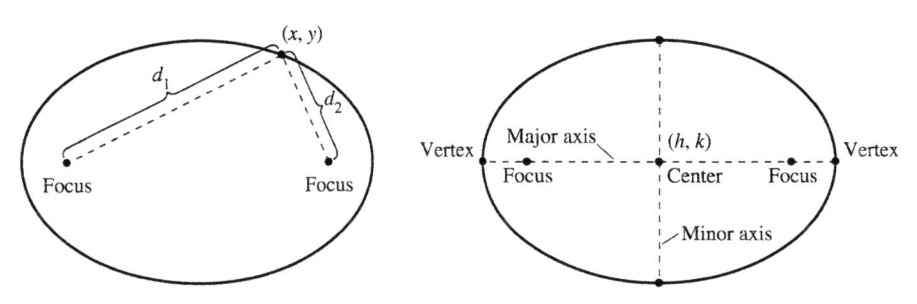

THEOREM 9.3 Standard Equation of an Ellipse

The standard form of the equation of an ellipse with center (h, k) and major and minor axes of lengths $2a$ and $2b$, where $a > b$, is

$$\frac{(x-h)^2}{a^2} + \frac{(y-k)^2}{b^2} = 1 \quad \text{Major axis is horizontal.}$$

or

$$\frac{(x-h)^2}{b^2} + \frac{(y-k)^2}{a^2} = 1. \quad \text{Major axis is vertical.}$$

The foci lie on the major axis, c units from the center, with $c^2 = a^2 - b^2$.

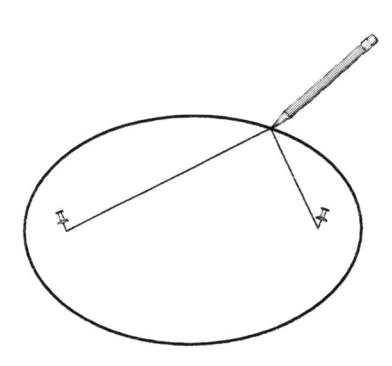

NOTE You can visualize the definition of an ellipse by imagining two thumbtacks placed at the foci, as shown in Figure 9.8. If the ends of a fixed length of string are fastened to the thumbtacks and the string is drawn taut with a pencil, the path traced by the pencil will be an ellipse.

Figure 9.8

$$\frac{(x-1)^2}{4} + \frac{(y+2)^2}{16} = 1$$

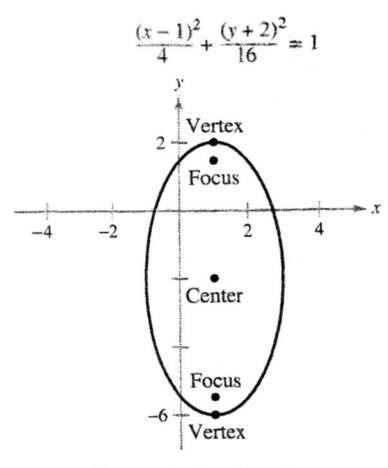

Ellipse with a vertical major axis
Figure 9.9

Example 3 Completing the Square

Find the center, vertices, and foci of the ellipse given by

$$4x^2 + y^2 - 8x + 4y - 8 = 0.$$

Solution By completing the square, you can write the given equation in standard form.

$$4x^2 + y^2 - 8x + 4y - 8 = 0 \qquad \text{Write original equation.}$$
$$4x^2 - 8x + y^2 + 4y = 8$$
$$4(x^2 - 2x + 1) + (y^2 + 4y + 4) = 8 + 4 + 4$$
$$4(x - 1)^2 + (y + 2)^2 = 16$$
$$\frac{(x-1)^2}{4} + \frac{(y+2)^2}{16} = 1 \qquad \text{Standard form}$$

So, the major axis is parallel to the y-axis, where $h = 1$, $k = -2$, $a = 4$, $b = 2$, and $c = \sqrt{16 - 4} = 2\sqrt{3}$. Therefore, you obtain the following.

Center:	$(1, -2)$	(h, k)
Vertices:	$(1, -6)$ and $(1, 2)$	$(h, k \pm a)$
Foci:	$\left(1, -2 - 2\sqrt{3}\right)$ and $\left(1, -2 + 2\sqrt{3}\right)$	$(h, k \pm c)$

The graph of the ellipse is shown in Figure 9.9.

NOTE If the constant term $F = -8$ in the equation in Example 3 had been greater than or equal to 8, you would have obtained one of the following degenerate cases.

1. $F = 8$, single point, $(1, -2)$: $\dfrac{(x-1)^2}{4} + \dfrac{(y+2)^2}{16} = 0$

2. $F > 8$, no solution points: $\dfrac{(x-1)^2}{4} + \dfrac{(y+2)^2}{16} < 0$

Example 4 The Orbit of the Moon

The moon orbits earth in an elliptical path with the center of earth at one focus, as shown in Figure 9.10. The major and minor axes of the orbit have lengths of 768,806 kilometers and 767,746 kilometers. Find the greatest and least distances (the apogee and perigee) from earth's center to the moon's center.

Solution Begin by solving for a and b.

$2a = 768{,}806$	Length of major axis
$a = 384{,}403$	Solve for a.
$2b = 767{,}746$	Length of minor axis
$b = 383{,}873$	Solve for b.

Now, using these values, you can solve for c as follows.

$$c = \sqrt{a^2 - b^2} \approx 20{,}179$$

The greatest distance between the center of earth and the center of the moon is $a + c \approx 404{,}582$ kilometers, and the least distance is $a - c \approx 364{,}224$ kilometers.

Figure 9.10

FOR FURTHER INFORMATION For more information on some uses of the reflective properties of conics, see the article "Parabolic Mirrors, Elliptic and Hyperbolic Lenses" by Mohsen Maesumi in *The American Mathematical Monthly*. Also see the article "The Geometry of Microwave Antennas" by William R. Parzynski in *Mathematics Teacher*. To view these articles, go to the website *www.matharticles.com*.

Theorem 9.2 presented a reflective property of parabolas. Ellipses have a similar reflective property. You are asked to prove the following theorem in Exercise 110.

THEOREM 9.4 Reflective Property of an Ellipse

Let P be a point on an ellipse. The tangent line to the ellipse at point P makes equal angles with the lines through P and the foci.

One of the reasons that astronomers had difficulty in detecting that the orbits of the planets are ellipses is that the foci of the planetary orbits are relatively close to the center of the sun, making the orbits nearly circular. To measure the ovalness of an ellipse, we use the concept of **eccentricity.**

Definition of Eccentricity of an Ellipse

The **eccentricity** e of an ellipse is given by the ratio

$$e = \frac{c}{a}.$$

To see how this ratio is used to describe the shape of an ellipse, note that because the foci of an ellipse are located along the major axis between the vertices and the center, it follows that

$$0 < c < a.$$

For an ellipse that is nearly circular, the foci are close to the center and the ratio c/a is small, and for an elongated ellipse, the foci are close to the vertices and the ratio is close to 1, as shown in Figure 9.11. Note that $0 < e < 1$ for every ellipse.

The orbit of the moon has an eccentricity of $e = 0.0549$, and the eccentricities of the nine planetary orbits are as follows.

Mercury:	$e = 0.2056$		Saturn:	$e = 0.0543$
Venus:	$e = 0.0068$		Uranus:	$e = 0.0460$
Earth:	$e = 0.0167$		Neptune:	$e = 0.0082$
Mars:	$e = 0.0934$		Pluto:	$e = 0.2481$
Jupiter:	$e = 0.0484$			

You can use integration to show that the area of an ellipse is $A = \pi ab$. For instance, the area of the ellipse

$$\frac{x^2}{a^2} + \frac{y^2}{b^2} = 1$$

is given by

$$A = 4\int_0^a \frac{b}{a}\sqrt{a^2 - x^2}\, dx$$
$$= \frac{4b}{a}\int_0^{\pi/2} a^2 \cos^2\theta\, d\theta. \quad \text{Trigonometric substitution } x = a\sin\theta.$$

However, it is not so simple to find the *circumference* of an ellipse. The next example shows how to use eccentricity to set up an "elliptic integral" for the circumference of an ellipse.

(a) $\frac{c}{a}$ is small.

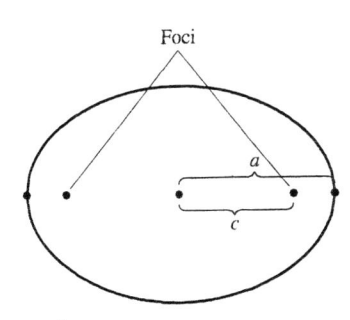

(b) $\frac{c}{a}$ is close to 1.

Eccentricity is the ratio $\frac{c}{a}$.

Figure 9.11

Example 5 Finding the Circumference of an Ellipse

Show that the circumference of the ellipse $(x^2/a^2) + (y^2/b^2) = 1$ is

$$4a \int_0^{\pi/2} \sqrt{1 - e^2 \sin^2 \theta}\, d\theta. \qquad e = \frac{c}{a}$$

Solution Because the given ellipse is symmetric with respect to both the x-axis and the y-axis, you know that its circumference C is four times the arc length of $y = (b/a)\sqrt{a^2 - x^2}$ in the first quadrant. The function y is differentiable for all x in the interval $[0, a]$ except at $x = a$. So, the circumference is given by the improper integral

$$C = \lim_{d \to a} 4 \int_0^d \sqrt{1 + (y')^2}\, dx = 4 \int_0^a \sqrt{1 + (y')^2}\, dx = 4 \int_0^a \sqrt{1 + \frac{b^2 x^2}{a^2(a^2 - x^2)}}\, dx.$$

Using the trigonometric substitution $x = a \sin \theta$, you obtain

$$C = 4 \int_0^{\pi/2} \sqrt{1 + \frac{b^2 \sin^2 \theta}{a^2 \cos^2 \theta}}\, (a \cos \theta)\, d\theta$$

$$= 4 \int_0^{\pi/2} \sqrt{a^2 \cos^2 \theta + b^2 \sin^2 \theta}\, d\theta$$

$$= 4 \int_0^{\pi/2} \sqrt{a^2(1 - \sin^2 \theta) + b^2 \sin^2 \theta}\, d\theta$$

$$= 4 \int_0^{\pi/2} \sqrt{a^2 - (a^2 - b^2)\sin^2 \theta}\, d\theta.$$

Because $e^2 = c^2/a^2 = (a^2 - b^2)/a^2$, you can rewrite this integral as

$$C = 4a \int_0^{\pi/2} \sqrt{1 - e^2 \sin^2 \theta}\, d\theta.$$

A great deal of time has been devoted to the study of elliptic integrals. Such integrals generally do not have elementary antiderivatives. To find the circumference of an ellipse, you must usually resort to an approximation technique.

AREA AND CIRCUMFERENCE OF AN ELLIPSE

In his work with elliptic orbits in the early 1600's, Johannes Kepler successfully developed a formula for the area of an ellipse, $A = \pi ab$. He was less successful in developing a formula for the circumference of an ellipse, however; the best he could do was to give the approximate formula $C = \pi(a + b)$.

Example 6 Approximating the Value of an Elliptic Integral

Use the elliptic integral in Example 5 to approximate the circumference of the ellipse

$$\frac{x^2}{25} + \frac{y^2}{16} = 1.$$

Solution Because $e^2 = c^2/a^2 = (a^2 - b^2)/a^2 = 9/25$, you have

$$C = (4)(5) \int_0^{\pi/2} \sqrt{1 - \frac{9 \sin^2 \theta}{25}}\, d\theta.$$

Applying Simpson's Rule with $n = 4$ produces

$$C \approx 20 \left(\frac{\pi}{6}\right)\left(\frac{1}{4}\right)[1 + 4(0.9733) + 2(0.9055) + 4(0.8323) + 0.8]$$

$$\approx 28.36.$$

So, the ellipse has a circumference of about 28.36 units, as shown in Figure 9.12.

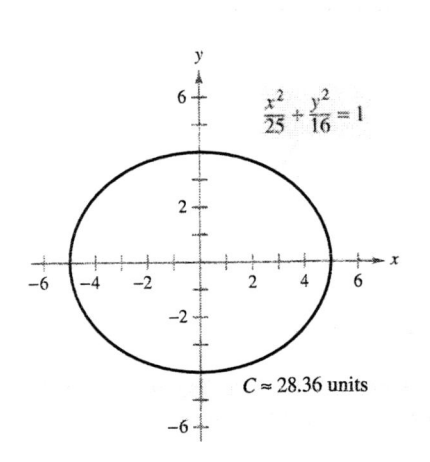

$C \approx 28.36$ units

Figure 9.12

Hyperbolas

The definition of a hyperbola is similar to that of an ellipse. For an ellipse, the *sum* of the distances between the foci and a point on the ellipse is fixed, whereas for a hyperbola, the absolute value of the *difference* between these distances is fixed.

A **hyperbola** is the set of all points (x, y) for which the absolute value of the difference between the distances from two distinct fixed points called **foci** is constant. (See Figure 9.13.) The line through the two foci intersects a hyperbola at two points called the **vertices**. The line segment connecting the vertices is the **transverse axis**, and the midpoint of the transverse axis is the **center** of the hyperbola. One distinguishing feature of a hyperbola is that its graph has two separate *branches*.

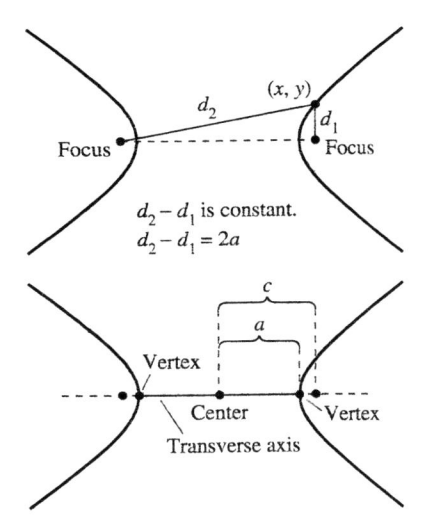

Figure 9.13

THEOREM 9.5 Standard Equation of a Hyperbola

The standard form of the equation of a hyperbola with center at (h, k) is

$$\frac{(x - h)^2}{a^2} - \frac{(y - k)^2}{b^2} = 1 \quad \text{Transverse axis is horizontal.}$$

or

$$\frac{(y - k)^2}{a^2} - \frac{(x - h)^2}{b^2} = 1. \quad \text{Transverse axis is vertical.}$$

The vertices are a units from the center, and the foci are c units from the center. Moreover, $c^2 = a^2 + b^2$.

NOTE The constants a, b, and c do not have the same relationship for hyperbolas as they do for ellipses. For hyperbolas, $c^2 = a^2 + b^2$, but for ellipses, $c^2 = a^2 - b^2$.

An important aid in sketching the graph of a hyperbola is the determination of its **asymptotes**, as shown in Figure 9.14. Each hyperbola has two asymptotes that intersect at the center of the hyperbola. The asymptotes pass through the vertices of a rectangle of dimensions $2a$ by $2b$, with its center at (h, k). The line segment of length $2b$ joining $(h, k + b)$ and $(h, k - b)$ is referred to as the **conjugate axis** of the hyperbola.

THEOREM 9.6 Asymptotes of a Hyperbola

For a *horizontal* transverse axis, the equations of the asymptotes are

$$y = k + \frac{b}{a}(x - h) \quad \text{and} \quad y = k - \frac{b}{a}(x - h).$$

For a *vertical* transverse axis, the equations of the asymptotes are

$$y = k + \frac{a}{b}(x - h) \quad \text{and} \quad y = k - \frac{a}{b}(x - h).$$

In Figure 9.14 you can see that the asymptotes coincide with the diagonals of the rectangle with dimensions $2a$ and $2b$, centered at (h, k). This provides you with a quick means of sketching the asymptotes, which in turn aids in sketching the hyperbola.

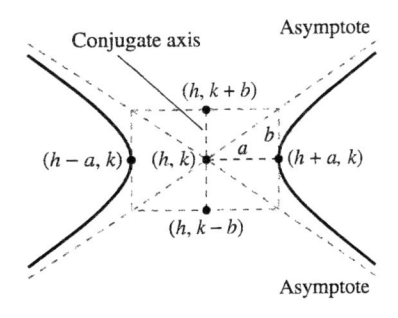

Figure 9.14

Example 7 Using Asymptotes to Sketch a Hyperbola

Sketch the graph of the hyperbola whose equation is $4x^2 - y^2 = 16$.

Solution Begin by rewriting the equation in standard form.

$$\frac{x^2}{4} - \frac{y^2}{16} = 1$$

The transverse axis is horizontal and the vertices occur at $(-2, 0)$ and $(2, 0)$. The ends of the conjugate axis occur at $(0, -4)$ and $(0, 4)$. Using these four points, you can sketch the rectangle shown in Figure 9.15(a). By drawing the asymptotes through the corners of this rectangle, you can complete the sketch as shown in Figure 9.15(b).

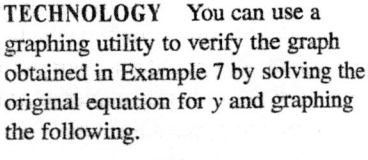

TECHNOLOGY You can use a graphing utility to verify the graph obtained in Example 7 by solving the original equation for y and graphing the following.

$$y_1 = \sqrt{4x^2 - 16}$$
$$y_2 = -\sqrt{4x^2 - 16}$$

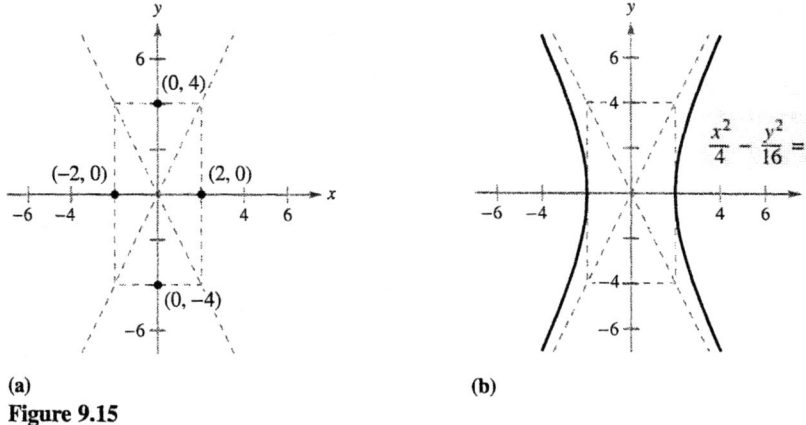

Figure 9.15

Definition of Eccentricity of a Hyperbola

The **eccentricity** e of a hyperbola is given by the ratio

$$e = \frac{c}{a}.$$

As with an ellipse, the **eccentricity** of a hyperbola is $e = c/a$. Because $c > a$ for hyperbolas, it follows that $e > 1$ for hyperbolas. If the eccentricity is large, the branches of the hyperbola are nearly flat. If the eccentricity is close to 1, the branches of the hyperbola are more pointed, as shown in Figure 9.16.

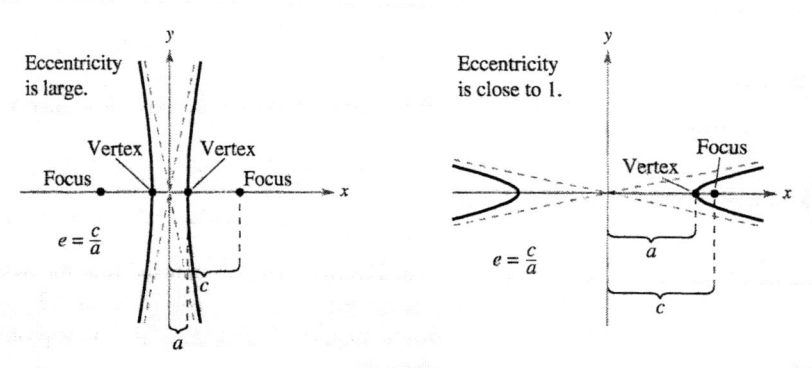

Figure 9.16

The following application was developed during World War II. It shows how the properties of hyperbolas can be used in radar and other detection systems.

Example 8 A Hyperbolic Detection System

Two microphones, 1 mile apart, record an explosion. Microphone A receives the sound 2 seconds before microphone B. Where was the explosion?

Solution Assuming that sound travels at 1100 feet per second, you know that the explosion took place 2200 feet farther from B than from A, as shown in Figure 9.17. The locus of all points that are 2200 feet closer to A than to B is one branch of the hyperbola $(x^2/a^2) - (y^2/b^2) = 1$, where

$$c = \frac{1 \text{ mile}}{2} = \frac{5280 \text{ ft}}{2} = 2640 \text{ ft}$$

and

$$a = \frac{2200 \text{ ft}}{2} = 1100 \text{ ft}.$$

Because $c^2 = a^2 + b^2$, it follows that

$$b^2 = c^2 - a^2$$
$$= 5{,}759{,}600$$

and you can conclude that the explosion occurred somewhere on the right branch of the hyperbola given by

$$\frac{x^2}{1{,}210{,}000} - \frac{y^2}{5{,}759{,}600} = 1.$$

$2c = 5280$
$d_2 - d_1 = 2a = 2200$
Figure 9.17

In Example 8, you were able to determine only the hyperbola on which the explosion occurred, but not the exact location of the explosion. If, however, you had received the sound at a third position C, then two other hyperbolas would be determined. The exact location of the explosion would be the point at which these three hyperbolas intersect.

Another interesting application of conics involves the orbits of comets in our solar system. Of the 610 comets identified prior to 1970, 245 have elliptical orbits, 295 have parabolic orbits, and 70 have hyperbolic orbits. The center of the sun is a focus of each orbit, and each orbit has a vertex at the point at which the comet is closest to the sun. Undoubtedly, many comets with parabolic or hyperbolic orbits have not been identified—such comets pass through our solar system once. Only comets with elliptical orbits such as Halley's comet remain in our solar system.

The type of orbit for a comet can be determined as follows.

1. Ellipse: $v < \sqrt{2GM/p}$
2. Parabola: $v = \sqrt{2GM/p}$
3. Hyperbola: $v > \sqrt{2GM/p}$

In these three formulas, p is the distance between one vertex and one focus of the comet's orbit (in meters), v is the velocity of the comet at the vertex (in meters per second), $M \approx 1.991 \times 10^{30}$ kilograms is the mass of the sun, and $G \approx 6.67 \times 10^{-11}$ cubic meters per kilogram-second squared is the gravitational constant.

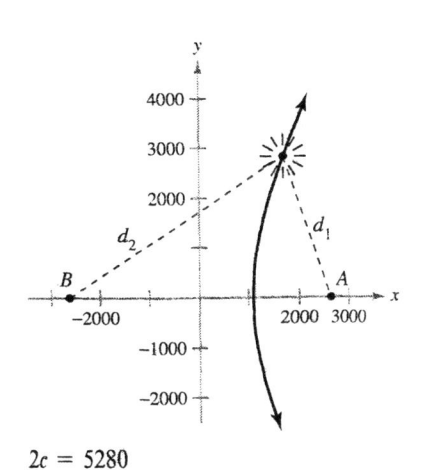

CAROLINE HERSCHEL (1750–1848)

The first woman to be credited with detecting a new comet was the English astronomer Caroline Herschel. During her life, Caroline Herschel discovered a total of eight new comets.

EXERCISES FOR SECTION 9.1

In Exercises 1–8, match the equation with its graph. [The graphs are labeled (a), (b), (c), (d), (e), (f), (g), and (h).]

(a) (b) 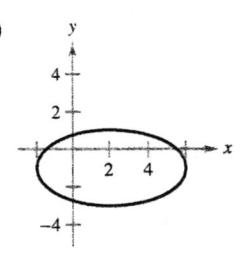 (c) 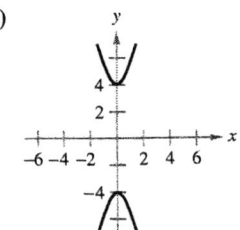 (d) 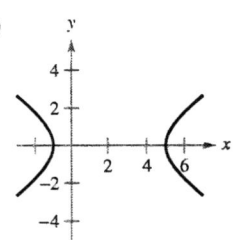 (e) 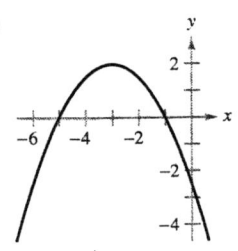 (f) 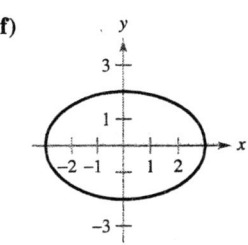 (g) 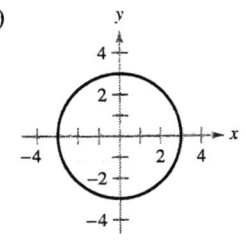 (h)

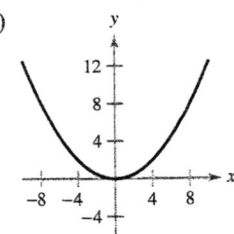

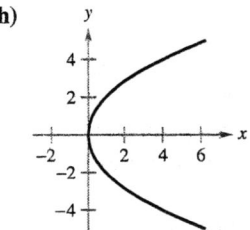

1. $y^2 = 4x$
2. $x^2 = 8y$
3. $(x + 3)^2 = -2(y - 2)$
4. $\dfrac{(x-2)^2}{16} + \dfrac{(y+1)^2}{4} = 1$
5. $\dfrac{x^2}{9} + \dfrac{y^2}{4} = 1$
6. $\dfrac{x^2}{9} + \dfrac{y^2}{9} = 1$
7. $\dfrac{y^2}{16} - \dfrac{x^2}{1} = 1$
8. $\dfrac{(x-2)^2}{9} - \dfrac{y^2}{4} = 1$

In Exercises 9–16, find the vertex, focus, and directrix of the parabola, and sketch its graph.

9. $y^2 = -6x$
10. $x^2 + 8y = 0$
11. $(x + 3) + (y - 2)^2 = 0$
12. $(x - 1)^2 + 8(y + 2) = 0$
13. $y^2 - 4y - 4x = 0$
14. $y^2 + 6y + 8x + 25 = 0$
15. $x^2 + 4x + 4y - 4 = 0$
16. $y^2 + 4y + 8x - 12 = 0$

In Exercises 17–20, find the vertex, focus, and directrix of the parabola. Then use a graphing utility to graph the parabola.

17. $y^2 + x + y = 0$
18. $y = -\frac{1}{6}(x^2 - 8x + 6)$
19. $y^2 - 4x - 4 = 0$
20. $x^2 - 2x + 8y + 9 = 0$

In Exercises 21–28, find an equation of the parabola.

21. Vertex: $(3, 2)$
    Focus: $(1, 2)$
22. Vertex: $(-1, 2)$
    Focus: $(-1, 0)$
23. Vertex: $(0, 4)$
    Directrix: $y = -2$
24. Focus: $(2, 2)$
    Directrix: $x = -2$

25. 26.

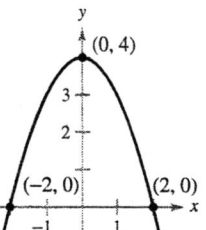

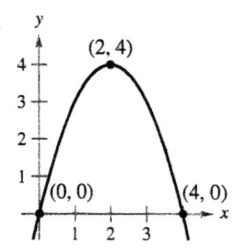

27. Axis is parallel to y-axis; graph passes through $(0, 3)$, $(3, 4)$, and $(4, 11)$.
28. Directrix: $y = -2$; endpoints of latus rectum are $(0, 2)$ and $(8, 2)$.

In Exercises 29–34, find the center, foci, vertices, and eccentricity of the ellipse, and sketch its graph.

29. $x^2 + 4y^2 = 4$
30. $5x^2 + 7y^2 = 70$
31. $\dfrac{(x-1)^2}{9} + \dfrac{(y-5)^2}{25} = 1$
32. $(x + 2)^2 + \dfrac{(y+4)^2}{1/4} = 1$
33. $9x^2 + 4y^2 + 36x - 24y + 36 = 0$
34. $16x^2 + 25y^2 - 64x + 150y + 279 = 0$

In Exercises 35–38, find the center, foci, and vertices of the ellipse. Use a graphing utility to graph the ellipse.

35. $12x^2 + 20y^2 - 12x + 40y - 37 = 0$
36. $36x^2 + 9y^2 + 48x - 36y + 43 = 0$
37. $x^2 + 2y^2 - 3x + 4y + 0.25 = 0$
38. $2x^2 + y^2 + 4.8x - 6.4y + 3.12 = 0$

In Exercises 39–44, find an equation of the ellipse.

39. Center: $(0, 0)$
    Focus: $(2, 0)$
    Vertex: $(3, 0)$
40. Vertices: $(0, 2), (4, 2)$
    Eccentricity: $\frac{1}{2}$

41. Vertices: (3, 1), (3, 9)
    Minor axis length: 6
43. Center: (0, 0)
    Major axis: horizontal
    Points on the ellipse:
    (3, 1), (4, 0)

42. Foci: (0, ±5)
    Major axis length: 14
44. Center: (1, 2)
    Major axis: vertical
    Points on the ellipse:
    (1, 6), (3, 2)

70. $25x^2 - 10x - 200y - 119 = 0$
71. $4x^2 + 4y^2 - 16y + 15 = 0$
72. $y^2 - 4y = x + 5$
73. $9x^2 + 9y^2 - 36x + 6y + 34 = 0$
74. $2x(x - y) = y(3 - y - 2x)$
75. $3(x - 1)^2 = 6 + 2(y + 1)^2$
76. $9(x + 3)^2 = 36 - 4(y - 2)^2$

In Exercises 45–52, find the center, foci, and vertices of the hyperbola, and sketch its graph using asymptotes as an aid.

45. $y^2 - \dfrac{x^2}{4} = 1$
46. $\dfrac{x^2}{25} - \dfrac{y^2}{9} = 1$
47. $\dfrac{(x - 1)^2}{4} - \dfrac{(y + 2)^2}{1} = 1$
48. $\dfrac{(y + 1)^2}{144} - \dfrac{(x - 4)^2}{25} = 1$
49. $9x^2 - y^2 - 36x - 6y + 18 = 0$
50. $y^2 - 9x^2 + 36x - 72 = 0$
51. $x^2 - 9y^2 + 2x - 54y - 80 = 0$
52. $9x^2 - 4y^2 + 54x + 8y + 78 = 0$

Getting at the Concept

77. (a) Give the definition of a parabola.
    (b) Give the standard forms of a parabola with vertex at (h, k).
    (c) In your own words, state the reflective property of a parabola.
78. (a) Give the definition of an ellipse.
    (b) Give the standard forms of an ellipse with center at (h, k).
79. (a) Give the definition of a hyperbola.
    (b) Give the standard forms of a hyperbola with center at (h, k).
    (c) Write equations for the asymptotes of a hyperbola.
80. Define the eccentricity of an ellipse. In your own words, describe how changes in the eccentricity affect the ellipse.

In Exercises 53–56, find the center, foci, and vertices of the hyperbola. Use a graphing utility to graph the hyperbola and its asymptotes.

53. $9y^2 - x^2 + 2x + 54y + 62 = 0$
54. $9x^2 - y^2 + 54x + 10y + 55 = 0$
55. $3x^2 - 2y^2 - 6x - 12y - 27 = 0$
56. $3y^2 - x^2 + 6x - 12y = 0$

In Exercises 57–64, find an equation of the hyperbola.

57. Vertices: (±1, 0)
    Asymptotes: $y = ±3x$
58. Vertices: (0, ±3)
    Asymptotes: $y = ±3x$
59. Vertices: (2, ±3)
    Point on graph: (0, 5)
60. Vertices: (2, ±3)
    Foci: (2, ±5)
61. Center: (0, 0)
    Vertex: (0, 2)
    Focus: (0, 4)
62. Center: (0, 0)
    Vertex: (3, 0)
    Focus: (5, 0)
63. Vertices: (0, 2), (6, 2)
    Asymptotes: $y = \tfrac{2}{3}x$
    $y = 4 - \tfrac{2}{3}x$
64. Focus: (10, 0)
    Asymptotes: $y = ±\tfrac{3}{4}x$

81. **Solar Collector** A solar collector for heating water is constructed with a sheet of stainless steel that is formed into the shape of a parabola (see figure). The water will flow through a pipe that is located at the focus of the parabola. At what distance from the vertex is the pipe?

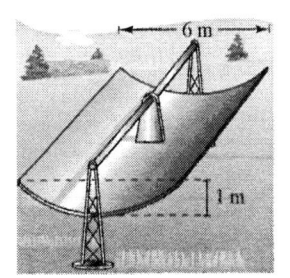

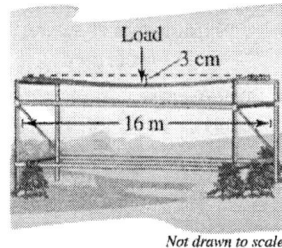

Figure for 81   Figure for 82

82. **Beam Deflection** A simply supported beam that is 16 meters long has a load concentrated at the center (see figure). The deflection of the beam at its center is 3 centimeters. Assume that the shape of the deflected beam is parabolic.
    (a) Find an equation of the parabola. (Assume that the origin is at the center of the beam.)
    (b) How far from the center of the beam is the deflection 1 centimeter?

In Exercises 65 and 66, find equations for (a) the tangent and (b) the normal lines to the hyperbola for the given value of x.

65. $\dfrac{x^2}{9} - y^2 = 1$, $x = 6$
66. $\dfrac{y^2}{4} - \dfrac{x^2}{2} = 1$, $x = 4$

In Exercises 67–76, classify the graph of the equation as a circle, a parabola, an ellipse, or a hyperbola.

67. $x^2 + 4y^2 - 6x + 16y + 21 = 0$
68. $4x^2 - y^2 - 4x - 3 = 0$
69. $y^2 - 4y - 4x = 0$

83. Find an equation of the tangent line to the parabola $y = ax^2$ at $x = x_0$. Prove that the x-intercept of this tangent line is $(x_0/2, 0)$.

84. (a) Prove that any two distinct tangent lines to a parabola intersect.

(b) Demonstrate the result in part (a) by finding the point of intersection of the tangent lines to the parabola $x^2 - 4x - 4y = 0$ at the points $(0, 0)$ and $(6, 3)$.

85. (a) Prove that if any two tangent lines to a parabola intersect at right angles, their point of intersection must lie on the directrix.

(b) Demonstrate the result in part (a) by proving that the tangent lines to the parabola $x^2 - 4x - 4y + 8 = 0$ at the points $(-2, 5)$ and $\left(3, \frac{5}{4}\right)$ intersect at right angles, and that the point of intersection lies on the directrix.

86. Find the point on the graph of $x^2 = 8y$ that is closest to the focus of the parabola.

87. *Radio and Television Reception* In mountainous areas, reception of radio and television is sometimes poor. Consider an idealized case where a hill is represented by the graph of the parabola $y = x - x^2$, a transmitter is located at the point $(-1, 1)$, and a receiver is located on the other side of the hill at the point $(x_0, 0)$. What is the closest the receiver can be to the hill so that the reception is unobstructed?

88. *Modeling Data* The per capita consumption C (in pounds) of commercially produced fruits in the United States for selected years is given in the table. (Source: U.S. Department of Agriculture)

Year	1980	1985	1990	1995	1996	1997
C	262.4	269.4	273.5	285.4	289.8	294.7

(a) Use the regression capabilities of a graphing utility to find a quadratic model for the data, where t is the time in years, with $t = 0$ corresponding to 1980.

(b) Use a graphing utility to plot the data and graph the model.

(c) Find dC/dt and sketch its graph for $0 \leq t \leq 17$. What information about the consumption of fruits is given by the graph of the derivative?

89. *Architecture* A church window is bounded on top by a parabola and below by the arc of a circle (see figure). Find the surface area of the window.

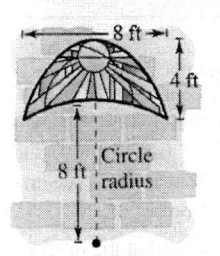

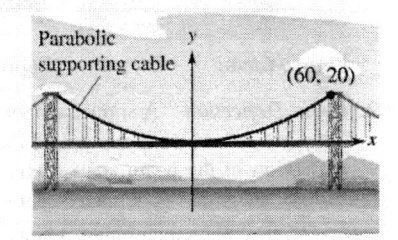

Figure for 89

Figure for 91

90. *Arc Length* Find the arc length of the parabola $4x - y^2 = 0$ over the interval $0 \leq y \leq 4$.

91. *Bridge Design* A cable of a suspension bridge is suspended (in the shape of a parabola) between two towers that are 120 meters apart and 20 meters above the roadway (see figure). The cables touch the roadway midway between the towers.

(a) Find an equation for the parabolic shape of each cable.

(b) Find the length of the parabolic supporting cable.

92. *Surface Area* A satellite-signal receiving dish is formed by revolving the parabola given by the graph of

$$x^2 = 20y$$

about the y-axis. If the radius of the dish is r feet, verify that the surface area of the dish is given by

$$2\pi \int_0^r x \sqrt{1 + \left(\frac{x}{10}\right)^2}\, dx = \frac{\pi}{15}[(100 + r^2)^{3/2} - 1000].$$

93. *Investigation* Sketch the graphs of $x^2 = 4py$ for $p = \frac{1}{4}, \frac{1}{2}, 1, \frac{3}{2}$, and 2 on the same coordinate axes. Discuss the change in the graphs as p increases.

94. *Area* Find a formula for the area of the shaded region in the figure.

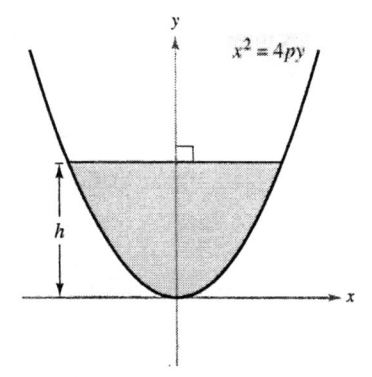

95. Sketch the ellipse that consists of all points (x, y) such that the sum of the distances between (x, y) and two fixed points is 16 units, and the foci are located at the centers of the two sets of concentric circles in the figure. To print an enlarged copy of the graph, go to the website *www.mathgraphs.com*.

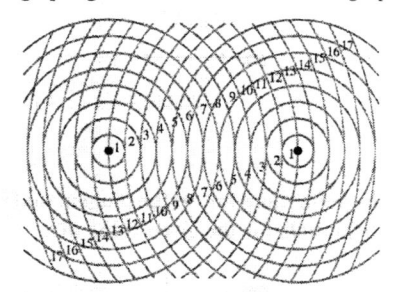

96. *Writing* On page 653, it was noted that an ellipse can be drawn using two thumbtacks, a string of fixed length (greater than the distance between the tacks), and a pencil. If the ends of the string are fastened at the tacks and the string is drawn taut with a pencil, the path traced by the pencil will be an ellipse.

(a) What is the length of the string in terms of a?

(b) Explain why the path is an ellipse.

97. Construction of a Semielliptical Arch A fireplace arch is to be constructed in the shape of a semiellipse. The opening is to have a height of 2 feet at the center and a width of 5 feet along the base (see figure). The contractor draws the outline of the ellipse by the method shown in Exercise 96. Where should the tacks be placed and what should be the length of the piece of string?

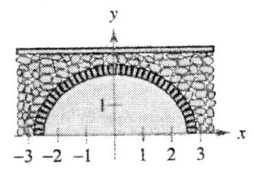

98. Orbit of the Earth Earth moves in an elliptical orbit with the sun at one of the foci. The length of half of the major axis is 149,570,000 kilometers, and the eccentricity is 0.0167. Find the minimum distance (*perihelion*) and the maximum distance (*aphelion*) of earth from the sun.

99. Satellite Orbit If the apogee and the perigee of an elliptical orbit of an earth satellite are given by A and P, show that the eccentricity of the orbit is

$$e = \frac{A - P}{A + P}.$$

100. Explorer 18 On November 26, 1963, the United States launched Explorer 18. Its low and high points above the surface of earth were 119 miles and 122,000 miles. Find the eccentricity of its elliptical orbit.

101. Halley's Comet Probably the most famous of all comets, Halley's comet, has an elliptical orbit with the sun at the focus. Its maximum distance from the sun is approximately 35.34 AU (astronomical unit $\approx 92.956 \times 10^6$ miles), and its minimum distance is approximately 0.59 AU. Find the eccentricity of the orbit.

102. The equation of an ellipse with its center at the origin can be written as

$$\frac{x^2}{a^2} + \frac{y^2}{a^2(1-e^2)} = 1.$$

Show that as $e \to 0$, with a remaining fixed, the ellipse approaches a circle.

103. Consider a particle traveling clockwise on the elliptical path $x^2/100 + y^2/25 = 1$. The particle leaves the orbit at the point $(-8, 3)$ and travels in a straight line tangent to the ellipse. At what point will the particle cross the y-axis?

104. Volume The water tank on a fire truck is 16 feet long, and its cross sections are ellipses. Find the volume of water in the partially filled tank as shown in the figure.

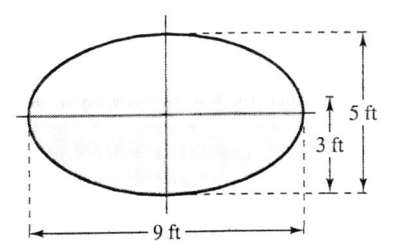

In Exercises 105 and 106, determine the points at which dy/dx is zero or does not exist to locate the endpoints of the major and minor axes of the ellipse.

105. $16x^2 + 9y^2 + 96x + 36y + 36 = 0$

106. $9x^2 + 4y^2 + 36x - 24y + 36 = 0$

Area and Volume In Exercises 107 and 108, find (a) the area of the region bounded by the ellipse, (b) the volume and surface area of the solid generated by revolving the region about its major axis (prolate spheroid), and (c) the volume and surface area of the solid generated by revolving the region about its minor axis (oblate spheroid).

107. $\dfrac{x^2}{4} + \dfrac{y^2}{1} = 1$

108. $\dfrac{x^2}{16} + \dfrac{y^2}{9} = 1$

109. Arc Length Use the integration capabilities of a graphing utility to approximate to two-decimal-place accuracy the elliptical integral representing the circumference of the ellipse.

$$\frac{x^2}{25} + \frac{y^2}{49} = 1$$

110. Prove that the tangent line to an ellipse at a point P makes equal angles with lines through P and the foci (see figure). [*Hint:* (1) Find the slope of the tangent line at P, (2) find the slopes of the lines through P and each focus, and (3) use the formula for the tangent of the angle between two lines.]

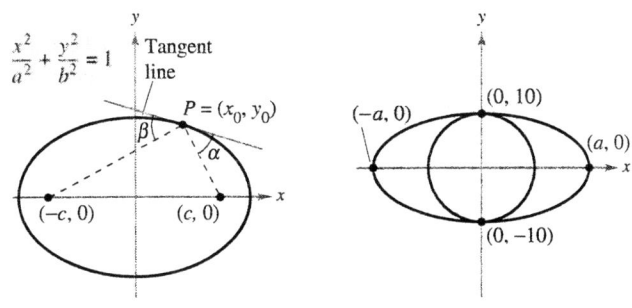

Figure for 110

Figure for 111

111. Geometry The area of the ellipse in the figure is twice the area of the circle. What is the length of the major axis?

112. Conjecture

(a) Show that the equation of an ellipse can be written as

$$\frac{(x-h)^2}{a^2} + \frac{(y-k)^2}{a^2(1-e^2)} = 1.$$

(b) Use a graphing utility to graph the ellipse

$$\frac{(x-2)^2}{4} + \frac{(y-3)^2}{4(1-e^2)} = 1$$

for $e = 0.95$, $e = 0.75$, $e = 0.5$, $e = 0.25$, and $e = 0$.

(c) Use the results in part (b) to make a conjecture about the change in the shape of the ellipse as e approaches 0.

113. Find an equation of the hyperbola such that for any point on the hyperbola, the difference between its distance from the points (2, 2) and (10, 2) is 6.

114. Find an equation of the hyperbola such that for any point on the hyperbola, the difference between its distances from the points $(-3, 0)$ and $(-3, 3)$ is 2.

115. Sketch the hyperbola that consists of all points (x, y) such that the difference of the distances between (x, y) and two fixed points is 10 units, and the foci are located at the centers of the two sets of concentric circles in the figure. To print an enlarged copy of the graph, go to the website *www.mathgraphs.com*.

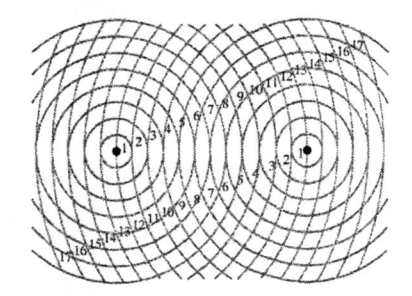

116. Consider a hyperbola centered at the origin with a horizontal transverse axis. Use the definition of a hyperbola to derive its standard form:

$$\frac{x^2}{a^2} - \frac{y^2}{b^2} = 1.$$

117. **Sound Location** A rifle positioned at point $(-c, 0)$ is fired at a target positioned at point $(c, 0)$. A person hears the sound of the rifle and the sound of the bullet hitting the target at the same time. Prove that the person is positioned on one branch of the hyperbola given by

$$\frac{x^2}{c^2 v_s^2 / v_m^2} - \frac{y^2}{c^2(v_m^2 - v_s^2)/v_m^2} = 1$$

where v_m is the muzzle velocity of the rifle and v_s is the speed of sound, which is about 1100 feet per second.

118. **Navigation** LORAN (long distance radio navigation) for aircraft and ships uses synchronized pulses transmitted by widely separated transmitting stations. These pulses travel at the speed of light (186,000 miles per second). The difference in the times of arrival of these pulses at an aircraft or ship is constant on a hyperbola having the transmitting stations as foci. Assume that two stations, 300 miles apart, are positioned on the rectangular coordinate system at $(-150, 0)$ and $(150, 0)$ and that a ship is traveling on a path with coordinates $(x, 75)$ (see figure). Find the x-coordinate of the position of the ship if the time difference between the pulses from the transmitting stations is 1000 microseconds (0.001 second).

119. **Hyperbolic Mirror** A hyperbolic mirror (used in some telescopes) has the property that a light ray directed at the focus will be reflected to the other focus. The mirror in the figure has the equation $(x^2/36) - (y^2/64) = 1$. At which point on the mirror will light from the point $(0, 10)$ be reflected to the other focus?

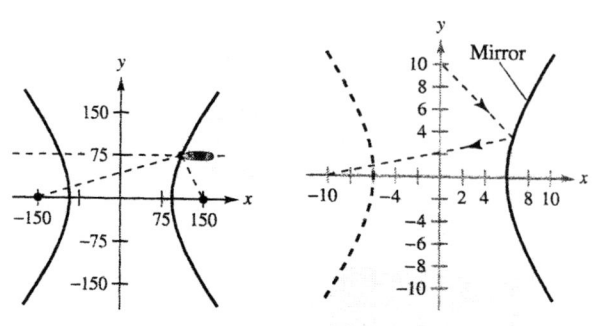

Figure for 118

Figure for 119

120. Show that the equation of the tangent line to

$$\frac{x^2}{a^2} - \frac{y^2}{b^2} = 1$$

at the point (x_0, y_0) is $(x_0/a^2)x - (y_0/b^2)y = 1$.

121. Show that the graphs of the equations intersect at right angles:

$$\frac{x^2}{a^2} + \frac{2y^2}{b^2} = 1 \quad \text{and} \quad \frac{x^2}{a^2 - b^2} - \frac{2y^2}{b^2} = 1.$$

122. Prove that the graph of the equation

$$Ax^2 + Cy^2 + Dx + Ey + F = 0$$

is one of the following (except in degenerate cases).

Conic	Condition
(a) Circle	$A = C$
(b) Parabola	$A = 0$ or $C = 0$ (but not both)
(c) Ellipse	$AC > 0$
(d) Hyperbola	$AC < 0$

True or False? In Exercises 123–129, determine whether the statement is true or false. If it is false, explain why or give an example that shows it is false.

123. It is possible for a parabola to intersect its directrix.

124. The point on a parabola closest to its focus is its vertex.

125. If C is the circumference of the ellipse

$$\frac{x^2}{a^2} + \frac{y^2}{b^2} = 1, \quad b < a,$$

then $2\pi b \leq C \leq 2\pi a$.

126. The graph of $(x^2/4) + y^4 = 1$ is an ellipse.

127. If $D \neq 0$ or $E \neq 0$, then the graph of

$$y^2 - x^2 + Dx + Ey = 0$$

is a hyperbola.

128. If the asymptotes of the hyperbola $(x^2/a^2) - (y^2/b^2) = 1$ intersect at right angles, then $a = b$.

129. Every tangent line to a hyperbola intersects the hyperbola only at the point of tangency.

Section 9.2 Plane Curves and Parametric Equations

- Sketch the graph of a curve given by a set of parametric equations.
- Eliminate the parameter in a set of parametric equations.
- Find a set of parametric equations to represent a curve.
- Understand two classic calculus problems, the tautochrone and brachistochrone problems.

Plane Curves and Parametric Equations

Until now, we have been representing a graph by a single equation involving *two* variables. In this section you will study situations in which *three* variables are used to represent a curve in the plane.

Consider the path followed by an object that is propelled into the air at an angle of 45°. If the initial velocity of the object is 48 feet per second, the object travels the parabolic path given by

$$y = -\frac{x^2}{72} + x \qquad \text{Rectangular equation}$$

as shown in Figure 9.18. However, this equation does not tell the whole story. Although it does tell you *where* the object has been, it doesn't tell you *when* the object was at a given point (x, y). To determine this time, you can introduce a third variable t, called a **parameter**. By writing both x and y as functions of t, you obtain the **parametric equations**

$$x = 24\sqrt{2}\, t \qquad \text{Parametric equation for } x$$

and

$$y = -16t^2 + 24\sqrt{2}\, t. \qquad \text{Parametric equation for } y$$

From this set of equations, you can determine that at time $t = 0$, the object is at the point $(0, 0)$. Similarly, at time $t = 1$, the object is at the point $(24\sqrt{2}, 24\sqrt{2} - 16)$, and so on. (We will discuss a method for determining this particular set of parametric equations—the equations of motion—later, in Section 11.3.)

For this particular motion problem, x and y are continuous functions of t, and the resulting path is called a **plane curve**.

Rectangular equation:
$y = -\frac{x^2}{72} + x$

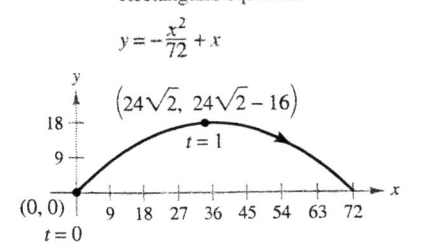

Parametric equations:
$x = 24\sqrt{2}\, t$
$y = -16t^2 + 24\sqrt{2}\, t$

Curvilinear motion: two variables for position, one variable for time
Figure 9.18

Definition of a Plane Curve

If f and g are continuous functions of t on an interval I, then the equations

$$x = f(t) \quad \text{and} \quad y = g(t)$$

are called **parametric equations** and t is called the **parameter**. The set of points (x, y) obtained as t varies over the interval I is called the **graph** of the parametric equations. Taken together, the parametric equations and the graph are called a **plane curve**, denoted by C.

NOTE At times it is important to distinguish between a graph (the set of points) and a curve (the points together with their defining parametric equations). When it is important, we will make the distinction explicit. When it is not important, we will use C to represent the graph or the curve.

When sketching (by hand) a curve represented by a pair of parametric equations, you can plot points in the xy-plane. Each set of coordinates (x, y) is determined from a value chosen for the parameter t. By plotting the resulting points in order of *increasing* values of t, the curve is traced out in a specific direction. This is called the **orientation** of the curve.

Example 1 Sketching a Curve

Sketch the curve described by the parametric equations

$$x = t^2 - 4 \quad \text{and} \quad y = \frac{t}{2}, \quad -2 \le t \le 3.$$

Solution For values of t on the given interval, the parametric equations yield the points (x, y) shown in the table.

t	-2	-1	0	1	2	3
x	0	-3	-4	-3	0	5
y	-1	$-\frac{1}{2}$	0	$\frac{1}{2}$	1	$\frac{3}{2}$

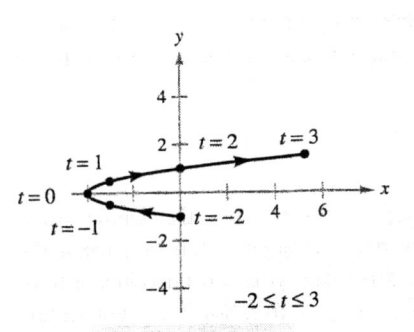

By plotting these points in order of increasing t and using the continuity of f and g, you obtain the curve C shown in Figure 9.19. Note that the arrows on the curve indicate its orientation as t increases from -2 to 3.

Parametric equations:
$x = t^2 - 4$ and $y = \frac{t}{2}$

Figure 9.19

NOTE From the vertical line test, you can see that the graph shown in Figure 9.19 does not define y as a function of x. This points out one benefit of parametric equations—they can be used to represent graphs that are more general than graphs of functions.

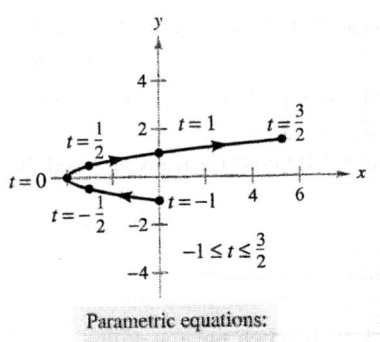

It often happens that two different sets of parametric equations have the same graph. For example, the set of parametric equations

$$x = 4t^2 - 4 \quad \text{and} \quad y = t, \quad -1 \le t \le \frac{3}{2}$$

has the same graph as the set given in Example 1. However, comparing the values of t in Figures 9.19 and 9.20, you can see that the second graph is traced out more *rapidly* (considering t as time) than the first graph. So, in applications, different parametric representations can be used to represent various *speeds* at which objects travel along a given path.

Parametric equations:
$x = 4t^2 - 4$ and $y = t$

Figure 9.20

TECHNOLOGY Most graphing utilities have a parametric graphing mode. If you have access to such a utility, try using it to confirm the graphs shown in Figures 9.19 and 9.20. Does the curve given by

$$x = 4t^2 - 8t \quad \text{and} \quad y = 1 - t, \quad -\tfrac{1}{2} \le t \le 2$$

represent the same graph as that shown in Figures 9.19 and 9.20? What do you notice about the *orientation* of this curve?

Eliminating the Parameter

Finding a rectangular equation that represents the graph of a set of parametric equations is called **eliminating the parameter**. For instance, you can eliminate the parameter from the set of parametric equations in Example 1 as follows.

Parametric equations	$\Rightarrow$	Solve for t in one equation.	$\Rightarrow$	Substitute into second equation.	$\Rightarrow$	Rectangular equation
$x = t^2 - 4$ $y = t/2$		$t = 2y$		$x = (2y)^2 - 4$		$x = 4y^2 - 4$

Once you have eliminated the parameter, you can recognize that the equation $x = 4y^2 - 4$ represents a parabola with a horizontal axis and vertex at $(-4, 0)$, as shown in Figure 9.19.

The range of x and y implied by the parametric equations may be altered by the change to rectangular form. In such instances the domain of the rectangular equation must be adjusted so that its graph matches the graph of the parametric equations. Such a situation is demonstrated in the next example.

Example 2   Adjusting the Domain After Eliminating the Parameter

Sketch the curve represented by the equations

$$x = \frac{1}{\sqrt{t+1}} \quad \text{and} \quad y = \frac{t}{t+1}, \quad t > -1$$

by eliminating the parameter and adjusting the domain of the resulting rectangular equation.

Solution   Begin by solving one of the parametric equations for t. For instance, you can solve the first equation for t as follows.

$$x = \frac{1}{\sqrt{t+1}} \quad \text{Parametric equation for } x$$

$$x^2 = \frac{1}{t+1} \quad \text{Square both sides.}$$

$$t + 1 = \frac{1}{x^2}$$

$$t = \frac{1}{x^2} - 1 = \frac{1-x^2}{x^2} \quad \text{Solve for } t.$$

Now, substituting into the parametric equation for y produces the following.

$$y = \frac{t}{t+1} \quad \text{Parametric equation for } y$$

$$y = \frac{(1-x^2)/x^2}{[(1-x^2)/x^2] + 1} \quad \text{Substitute } (1-x^2)/x^2 \text{ for } t.$$

$$y = 1 - x^2 \quad \text{Simplify.}$$

The rectangular equation, $y = 1 - x^2$, is defined for all values of x, but from the parametric equation for x you can see that the curve is defined only when $t > -1$. This implies that you should restrict the domain of x to positive values, as shown in Figure 9.21.

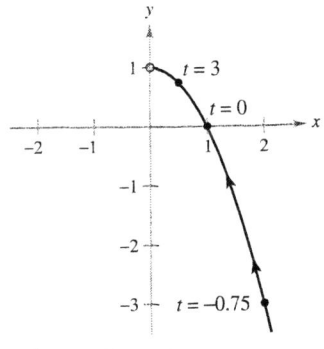

Parametric equations:
$x = \frac{1}{\sqrt{t+1}}, \ y = \frac{t}{t+1}, t > -1$

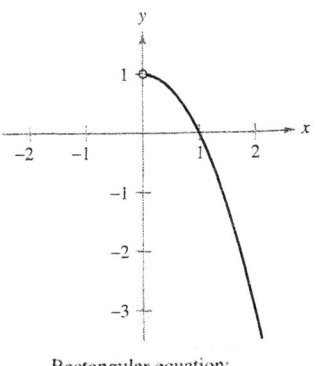

Rectangular equation:
$y = 1 - x^2, \ x > 0$

Figure 9.21

It is not necessary for the parameter in a set of parametric equations to represent time. The next example uses an *angle* as the parameter.

Example 3 Using Trigonometry to Eliminate a Parameter

Sketch the curve represented by

$$x = 3 \cos \theta \quad \text{and} \quad y = 4 \sin \theta, \quad 0 \le \theta \le 2\pi$$

by eliminating the parameter and finding the corresponding rectangular equation.

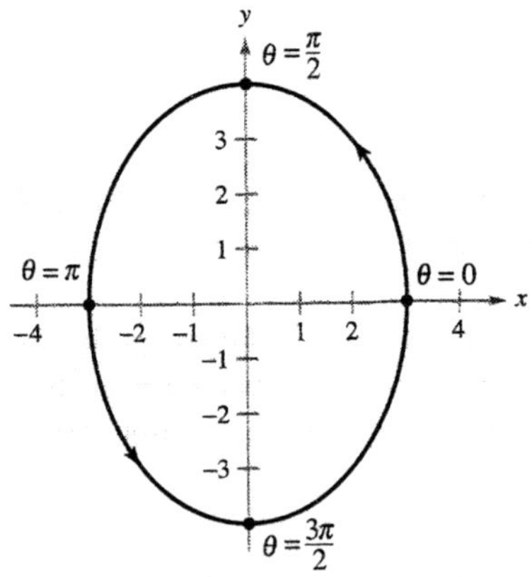

Solution Begin by solving for $\cos \theta$ and $\sin \theta$ in the given equations.

$$\cos \theta = \frac{x}{3} \quad \text{and} \quad \sin \theta = \frac{y}{4} \qquad \text{Solve for } \cos \theta \text{ and } \sin \theta.$$

Next, make use of the identity $\sin^2 \theta + \cos^2 \theta = 1$ to form an equation involving only x and y.

$$\cos^2 \theta + \sin^2 \theta = 1 \qquad \text{Trigonometric identity}$$

$$\left(\frac{x}{3}\right)^2 + \left(\frac{y}{4}\right)^2 = 1 \qquad \text{Substitute.}$$

$$\frac{x^2}{9} + \frac{y^2}{16} = 1 \qquad \text{Rectangular equation}$$

From this rectangular equation you can see that the graph is an ellipse centered at $(0, 0)$, with vertices at $(0, 4)$ and $(0, -4)$ and minor axis of length $2b = 6$, as shown in Figure 9.22. Note that the elliptic curve is traced out *counterclockwise* as θ varies from 0 to 2π.

Parametric equations:
$x = 3 \cos \theta$, $y = 4 \sin \theta$
Rectangular equation:
$\frac{x^2}{9} + \frac{y^2}{16} = 1$

Figure 9.22

Using the technique shown in Example 3, you can conclude that the graph of the parametric equations

$$x = h + a \cos \theta \quad \text{and} \quad y = k + b \sin \theta, \quad 0 \le \theta \le 2\pi$$

is the ellipse (traced counterclockwise) given by

$$\frac{(x - h)^2}{a^2} + \frac{(y - k)^2}{b^2} = 1.$$

The graph of the parametric equations

$$x = h + a \sin \theta \quad \text{and} \quad y = k + b \cos \theta, \quad 0 \le \theta \le 2\pi$$

is also the ellipse (traced clockwise) given by

$$\frac{(x - h)^2}{a^2} + \frac{(y - k)^2}{b^2} = 1.$$

Try using a graphing utility in parametric mode to sketch several ellipses.

In Examples 2 and 3, it is important to realize that eliminating the parameter is primarily an *aid to curve sketching*. If the parametric equations represent the path of a moving object, the graph alone is not sufficient to describe the object's motion. You still need the parametric equations to tell you the *position*, *direction*, and *speed* at a given time.

Finding Parametric Equations

The first three examples in this section illustrated techniques for sketching the graph represented by a set of parametric equations. We now look at the reverse problem. How can you determine a set of parametric equations for a given graph or a given physical description? From the discussion following Example 1, you know that such a representation is not unique. This is demonstrated further in the following example, which finds two different parametric representations for a given graph.

Example 4   Finding Parametric Equations for a Given Graph

Find a set of parametric equations to represent the graph of $y = 1 - x^2$, using each of the following parameters.

a. $t = x$ **b.** the slope $m = \dfrac{dy}{dx}$ at the point (x, y)

Solution

a. Letting $x = t$ produces the parametric equations

$$x = t \quad \text{and} \quad y = 1 - x^2 = 1 - t^2.$$

b. To express x and y in terms of the parameter m, you can proceed as follows.

$$m = \dfrac{dy}{dx} = -2x \qquad \text{Differentiate } y = 1 - x^2.$$

$$x = -\dfrac{m}{2} \qquad \text{Solve for } x.$$

This produces a parametric equation for x. To obtain a parametric equation for y, substitute $-m/2$ for x in the original equation.

$$y = 1 - x^2 \qquad \text{Write original rectangular equation.}$$

$$y = 1 - \left(-\dfrac{m}{2}\right)^2 \qquad \text{Substitute } -m/2 \text{ for } x.$$

$$y = 1 - \dfrac{m^2}{4} \qquad \text{Simplify.}$$

So, the parametric equations are

$$x = -\dfrac{m}{2} \quad \text{and} \quad y = 1 - \dfrac{m^2}{4}.$$

In Figure 9.23, note that the resulting curve has a right-to-left orientation as determined by the direction of increasing values of slope m. For part (a), the curve would have the opposite orientation.

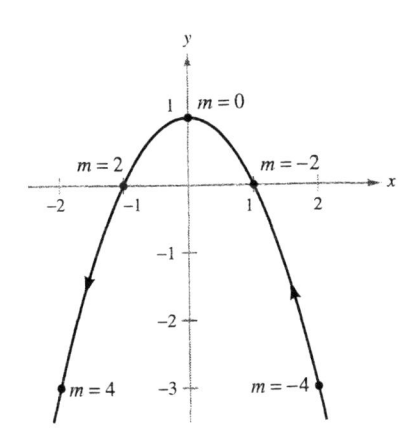

Rectangular equation: $y = 1 - x^2$
Parametric equations:

$$x = -\dfrac{m}{2}, \; y = 1 - \dfrac{m^2}{4}$$

Figure 9.23

TECHNOLOGY To be efficient at using a graphing utility, it is important that you develop skill in representing a graph by a set of parametric equations. The reason for this is that many graphing utilities have only three graphing modes—(1) functions, (2) parametric equations, and (3) polar equations. Most graphing utilities are not programmed to sketch the graph of a general equation. For instance, suppose you want to sketch the graph of the hyperbola $x^2 - y^2 = 1$. To sketch the graph in function mode, you need two equations: $y = \sqrt{x^2 - 1}$ and $y = -\sqrt{x^2 - 1}$. In parametric mode, you can represent the graph by $x = \sec t$ and $y = \tan t$.

670 CHAPTER 9 Conics, Parametric Equations, and Polar Coordinates

CYCLOIDS

Galileo first called attention to the cycloid, once recommending that it be used for the arches of bridges. Pascal once spent 8 days attempting to solve many of the problems of cycloids, such as finding the area under one arch, and the volume of the solid of revolution formed by revolving the curve about a line. The cycloid has so many interesting properties and has caused so many quarrels among mathematicians that it has been called "the Helen of geometry" and "the apple of discord."

FOR FURTHER INFORMATION For more information on cycloids, see the article "The Geometry of Rolling Curves" by John Bloom and Lee Whitt in *The American Mathematical Monthly*. To view this article, go to the website *www.matharticles.com*.

Example 5 **Parametric Equations for a Cycloid**

Determine the curve traced by a point P on the circumference of a circle of radius a rolling along a straight line in a plane. Such a curve is called a **cycloid**.

Solution Let the parameter θ be the measure of the circle's rotation, and let the point $P = (x, y)$ begin at the origin. When $\theta = 0$, P is at the origin. When $\theta = \pi$, P is at a maximum point $(\pi a, 2a)$. When $\theta = 2\pi$, P is back on the x-axis at $(2\pi a, 0)$. From Figure 9.24, you can see that $\angle APC = 180° - \theta$. Hence,

$$\sin \theta = \sin(180° - \theta) = \sin(\angle APC) = \frac{AC}{a} = \frac{BD}{a}$$

$$\cos \theta = -\cos(180° - \theta) = -\cos(\angle APC) = \frac{AP}{-a}$$

which implies that

$$AP = -a \cos \theta \quad \text{and} \quad BD = a \sin \theta.$$

Because the circle rolls along the x-axis, you know that $OD = \overset{\frown}{PD} = a\theta$. Furthermore, because $BA = DC = a$, you have

$$x = OD - BD = a\theta - a \sin \theta$$
$$y = BA + AP = a - a \cos \theta.$$

Therefore, the parametric equations are

$$x = a(\theta - \sin \theta) \quad \text{and} \quad y = a(1 - \cos \theta).$$

Cycloid:
$x = a(\theta - \sin \theta)$
$y = a(1 - \cos \theta)$

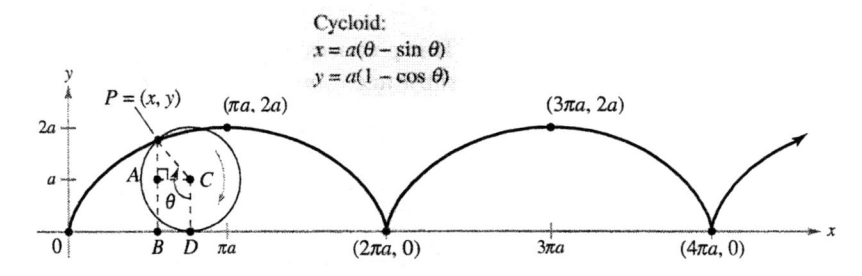

Figure 9.24

TECHNOLOGY Some graphing utilities allow you to simulate the motion of an object that is moving in the plane or in space. If you have access to such a utility, try using it to trace out the path of the cycloid shown in Figure 9.24.

The cycloid in Figure 9.24 has sharp corners at the values $x = 2n\pi a$. Notice that the derivatives $x'(\theta)$ and $y'(\theta)$ are both zero at the points for which $\theta = 2n\pi$.

$$x(\theta) = a(\theta - \sin \theta) \qquad y(\theta) = a(1 - \cos \theta)$$
$$x'(\theta) = a - a \cos \theta \qquad y'(\theta) = a \sin \theta$$
$$x'(2n\pi) = 0 \qquad y'(2n\pi) = 0$$

Between these points, the cycloid is called **smooth**.

Definition of a Smooth Curve

A curve C represented by $x = f(t)$ and $y = g(t)$ on an interval I is called **smooth** if f' and g' are continuous on I and not simultaneously 0, except possibly at the endpoints of I. The curve C is called **piecewise smooth** if it is smooth on each subinterval of some partition of I.

The Tautochrone and Brachistochrone Problems

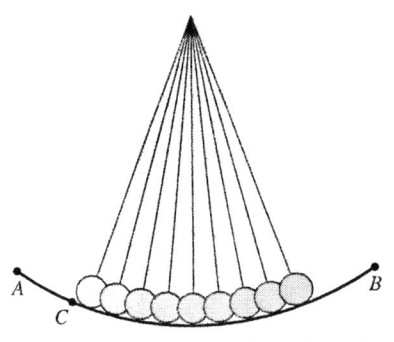

The type of curve described in Example 5 is related to one of the most famous pairs of problems in the history of calculus. The first problem (called the **tautochrone problem**) began with Galileo's discovery that the time required to complete a full swing of a given pendulum is *approximately* the same whether it makes a large movement at high speeds or a small movement at lower speeds (see Figure 9.25). Late in his life, Galileo (1564–1642) realized that he could use this principle to construct a clock. However, he was not able to conquer the mechanics of actual construction. Christian Huygens (1629–1695) was the first to design and construct a working model. In his work with pendulums, Huygens realized that a pendulum does not take *exactly* the same time to complete swings of varying lengths. (This doesn't affect a pendulum clock, because the length of the circular arc is kept constant by giving the pendulum a slight boost each time it passes its lowest point.) But, in studying the problem, Huygens discovered that a ball rolling back and forth on an inverted cycloid does complete each cycle in exactly the same time.

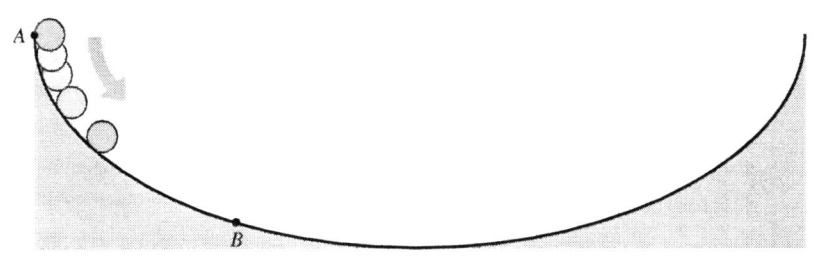

The time required to complete a full swing of the pendulum when starting from point C is only approximately the same as when starting from point A.
Figure 9.25

An inverted cycloid is the path down which a ball will roll in the shortest time.
Figure 9.26

The second problem, posed by John Bernoulli in 1696, is called the **brachistochrone problem**—in Greek, *brachys* means *short* and *chronos* means *time*. The problem was to determine the path down which a particle will slide from point A to point B in the *shortest time*. Several mathematicians took up the challenge, and the following year the problem was solved by Newton, Leibniz, L'Hôpital, John Bernoulli, and James Bernoulli. As it turns out, the solution is not a straight line from A to B, but an inverted cycloid passing through the points A and B, as shown in Figure 9.26. The amazing part of the solution is that a particle starting at rest at *any* other point C of the cycloid between A and B will take exactly the same time to reach B, as indicated in Figure 9.27.

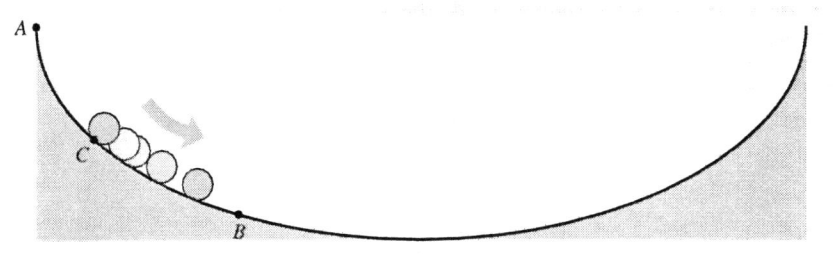

A ball starting at point C takes the same time to reach point B as one that starts at point A.
Figure 9.27

JAMES BERNOULLI (1654–1705)

James Bernoulli, also called Jacques, was the older brother of John. He was one of several accomplished mathematicians of the Swiss Bernoulli family. James's mathematical accomplishments have given him a prominent place in the early development of calculus.

FOR FURTHER INFORMATION To see a proof of the famous brachistochrone problem, see the article "A New Minimization Proof for the Brachistochrone" by Gary Lawlor in *The American Mathematical Monthly*. To view this article, go to the website *www.matharticles.com*.

EXERCISES FOR SECTION 9.2

1. Consider the parametric equations $x = \sqrt{t}$ and $y = 1 - t$.
   (a) Complete the table.

t	0	1	2	3	4
x					
y					

   (b) Plot the points (x, y) generated in the table, and sketch a graph of the parametric equations. Indicate the orientation of the graph.
   (c) Use a graphing utility to confirm your graph in part (b).
   (d) Find the rectangular equation by eliminating the parameter. Compare the graph in part (b) with the graph of the rectangular equation.

2. Consider the parametric equations $x = 4\cos^2\theta$ and $y = 2\sin\theta$.
   (a) Complete the table.

θ	$-\frac{\pi}{2}$	$-\frac{\pi}{4}$	0	$\frac{\pi}{4}$	$\frac{\pi}{2}$
x					
y					

   (b) Plot the points (x, y) generated in the table, and sketch a graph of the parametric equations. Indicate the orientation of the graph.
   (c) Use a graphing utility to confirm your graph in part (b).
   (d) Find the rectangular equation by eliminating the parameter. Compare the graph in part (b) with the graph of the rectangular equation.
   (e) If values of θ were selected from the interval $[\pi/2, 3\pi/2]$ for the table in part (a), would the graph in part (b) be different? Explain.

In Exercises 3–20, sketch the curve represented by the parametric equations (indicate the orientation of the curve), and write the corresponding rectangular equation by eliminating the parameter.

3. $x = 3t - 1$, $y = 2t + 1$
4. $x = 3 - 2t$, $y = 2 + 3t$
5. $x = t + 1$, $y = t^2$
6. $x = 2t^2$, $y = t^4 + 1$
7. $x = t^3$, $y = \dfrac{t^2}{2}$
8. $x = t^2 + t$, $y = t^2 - t$
9. $x = \sqrt{t}$, $y = t - 2$
10. $x = \sqrt[4]{t}$, $y = 3 - t$
11. $x = t - 1$, $y = \dfrac{t}{t - 1}$
12. $x = 1 + \dfrac{1}{t}$, $y = t - 1$
13. $x = 2t$, $y = |t - 2|$
14. $x = |t - 1|$, $y = t + 2$
15. $x = e^t$, $y = e^{3t} + 1$
16. $x = e^{-t}$, $y = e^{2t} - 1$
17. $x = \sec\theta$, $y = \cos\theta$, $0 \le \theta < \pi/2$, $\pi/2 < \theta \le \pi$
18. $x = \tan^2\theta$, $y = \sec^2\theta$
19. $x = 3\cos\theta$, $y = 3\sin\theta$
20. $x = 2\cos\theta$, $y = 6\sin\theta$

In Exercises 21–32, use a graphing utility to sketch the curve represented by the parametric equations (indicate the orientation of the curve). Eliminate the parameter and write the corresponding rectangular equation.

21. $x = 4\sin 2\theta$, $y = 2\cos 2\theta$
22. $x = \cos\theta$, $y = 2\sin 2\theta$
23. $x = 4 + 2\cos\theta$
    $y = -1 + \sin\theta$
24. $x = 4 + 2\cos\theta$
    $y = -1 + 2\sin\theta$
25. $x = 4 + 2\cos\theta$
    $y = -1 + 4\sin\theta$
26. $x = \sec\theta$
    $y = \tan\theta$
27. $x = 4\sec\theta$, $y = 3\tan\theta$
28. $x = \cos^3\theta$, $y = \sin^3\theta$
29. $x = t^3$, $y = 3\ln t$
30. $x = \ln 2t$, $y = t^2$
31. $x = e^{-t}$, $y = e^{3t}$
32. $x = e^{2t}$, $y = e^t$

Comparing Plane Curves In Exercises 33–36, determine any differences between the curves of the parametric equations. Are the graphs the same? Are the orientations the same? Are the curves smooth?

33. (a) $x = t$
      $y = 2t + 1$
    (b) $x = \cos\theta$
      $y = 2\cos\theta + 1$
    (c) $x = e^{-t}$
      $y = 2e^{-t} + 1$
    (d) $x = e^t$
      $y = 2e^t + 1$

34. (a) $x = 2\cos\theta$
      $y = 2\sin\theta$
    (b) $x = \sqrt{4t^2 - 1}/|t|$
      $y = 1/t$
    (c) $x = \sqrt{t}$
      $y = \sqrt{4 - t}$
    (d) $x = -\sqrt{4 - e^{2t}}$
      $y = e^t$

35. (a) $x = \cos\theta$
      $y = 2\sin^2\theta$
      $0 < \theta < \pi$
    (b) $x = \cos(-\theta)$
      $y = 2\sin^2(-\theta)$
      $0 < \theta < \pi$

36. (a) $x = t + 1$, $y = t^3$
    (b) $x = -t + 1$, $y = (-t)^3$

37. *Conjecture*
    (a) Use a graphing utility to sketch the curves represented by the two sets of parametric equations.

    $x = 4\cos t$ $\quad x = 4\cos(-t)$
    $y = 3\sin t$ $\quad y = 3\sin(-t)$

    (b) Describe the change in the graph when the sign of the parameter is changed.
    (c) Make a conjecture about the change in the graph of parametric equations when the sign of the parameter is changed.
    (d) Test your conjecture with another set of parametric equations.

38. *Writing* Review Exercises 33–36 and write a short paragraph describing how the graphs of curves represented by different sets of parametric equations can differ even though eliminating the parameter from each yields the same rectangular equation.

In Exercises 39–42, eliminate the parameter and obtain the standard form of the rectangular equation.

39. Line through (x_1, y_1) and (x_2, y_2):
$x = x_1 + t(x_2 - x_1)$, $y = y_1 + t(y_2 - y_1)$
40. Circle: $x = h + r\cos\theta$, $y = k + r\sin\theta$
41. Ellipse: $x = h + a\cos\theta$, $y = k + b\sin\theta$
42. Hyperbola: $x = h + a\sec\theta$, $y = k + b\tan\theta$

In Exercises 43–50, use the results of Exercises 39–42 to find a set of parametric equations for the line or conic.

43. Line: Passes through $(0, 0)$ and $(5, -2)$
44. Line: Passes through $(1, 4)$ and $(5, -2)$
45. Circle: Center: $(2, 1)$; Radius: 4
46. Circle: Center: $(-3, 1)$; Radius: 3
47. Ellipse: Vertices: $(\pm 5, 0)$; Foci: $(\pm 4, 0)$
48. Ellipse: Vertices: $(4, 7)$, $(4, -3)$; Foci: $(4, 5)$, $(4, -1)$
49. Hyperbola: Vertices: $(\pm 4, 0)$; Foci: $(\pm 5, 0)$
50. Hyperbola: Vertices: $(0, \pm 1)$; Foci: $(0, \pm 2)$

In Exercises 51–54, find two different sets of parametric equations for the given rectangular equation.

51. $y = 3x - 2$
52. $y = \dfrac{2}{x - 1}$
53. $y = x^3$
54. $y = x^2$

In Exercises 55–62, use a graphing utility to graph the curve represented by the parametric equations. Indicate the direction of the curve. Identify any points at which the curve is not smooth.

55. Cycloid: $x = 2(\theta - \sin\theta)$, $y = 2(1 - \cos\theta)$
56. Cycloid: $x = \theta + \sin\theta$, $y = 1 - \cos\theta$
57. Prolate cycloid: $x = \theta - \frac{3}{2}\sin\theta$, $y = 1 - \frac{3}{2}\cos\theta$
58. Prolate cycloid: $x = 2\theta - 4\sin\theta$, $y = 2 - 4\cos\theta$
59. Hypocycloid: $x = 3\cos^3\theta$, $y = 3\sin^3\theta$
60. Curtate cycloid: $x = 2\theta - \sin\theta$, $y = 2 - \cos\theta$
61. Witch of Agnesi: $x = 2\cot\theta$, $y = 2\sin^2\theta$
62. Folium of Descartes: $x = 3t/(1 + t^3)$, $y = 3t^2/(1 + t^3)$

Getting at the Concept

63. State the definition of a plane curve given by parametric equations.
64. Explain the process of sketching a plane curve given by parametric equations. What is meant by the orientation of the curve?
65. State the definition of a smooth curve.

Getting at the Concept (continued)

66. Match each graph with a set of parametric equations. Explain your reasoning.

    (i) $x = t^2 - 1$
    $y = t + 2$

    (ii) $x = \sin^2\theta - 1$
    $y = \sin\theta + 2$

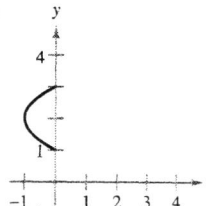

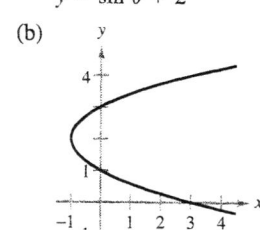

    (a)
    (b)

In Exercises 67–70, match the set of parametric equations with the correct graph. [The graphs are labeled (a), (b), (c), and (d).]

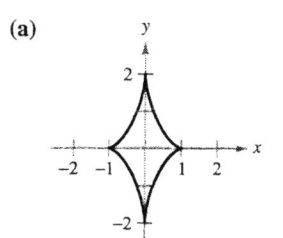

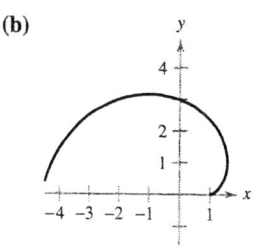

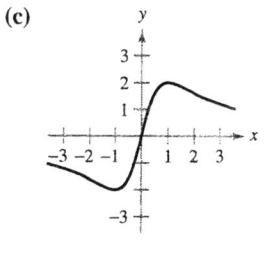

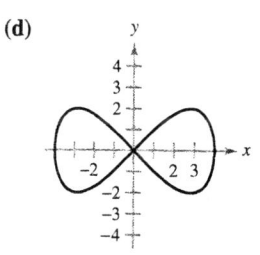

(a)
(b)
(c)
(d)

67. Lissajous curve: $x = 4\cos\theta$, $y = 2\sin 2\theta$
68. Evolute of ellipse: $x = \cos^3\theta$, $y = 2\sin^3\theta$
69. Involute of circle: $x = \cos\theta + \theta\sin\theta$, $y = \sin\theta - \theta\cos\theta$
70. Serpentine curve: $x = \cot\theta$, $y = 4\sin\theta\cos\theta$

71. **Curtate Cycloid** A wheel of radius a rolls along a line without slipping. The curve traced by a point P that is b units from the center ($b < a$) is called a **curtate cycloid** (see figure). Use the angle θ to find a set of parametric equations for this curve.

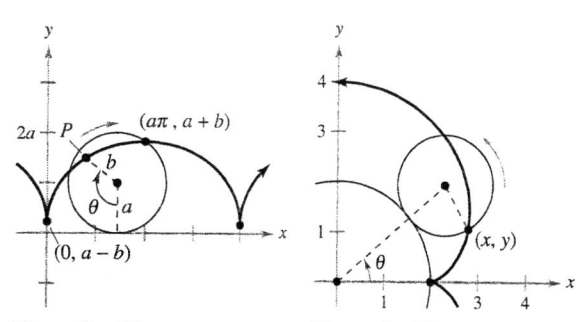

Figure for 71

Figure for 72

72. *Epicycloid* A circle of radius 1 rolls around the outside of a circle of radius 2 without slipping. The curve traced by a point on the circumference of the smaller circle is called an **epicycloid** (see figure on page 673). Use the angle θ to find a set of parametric equations for this curve.

True or False? In Exercises 73 and 74, determine whether the statement is true or false. If it is false, explain why or give an example that shows it is false.

73. The graph of the parametric equations $x = t^2$ and $y = t^2$ is the line $y = x$.

74. If y is a function of t and x is a function of t, then y is a function of x.

Projectile Motion In Exercises 75 and 76, consider a projectile launched at a height h feet above the ground and at an angle θ with the horizontal. If the initial velocity is v_0 feet per second, the path of the projectile is modeled by the parametric equations

$$x = (v_0 \cos \theta)t \quad \text{and} \quad y = h + (v_0 \sin \theta)t - 16t^2.$$

75. *Baseball* The center field fence in a ballpark is 10 feet high and 400 feet from home plate. The ball is hit 3 feet above the ground. It leaves the bat at an angle of θ degrees with the horizontal at a speed of 100 miles per hour (see figure).

(a) Write a set of parametric equations for the path of the ball.

(b) Use a graphing utility to graph the path of the ball if $\theta = 15°$. Is the hit a home run?

(c) Use a graphing utility to graph the path of the ball if $\theta = 23°$. Is the hit a home run?

(d) Find the minimum angle for the ball to leave the bat in order for the hit to be a home run.

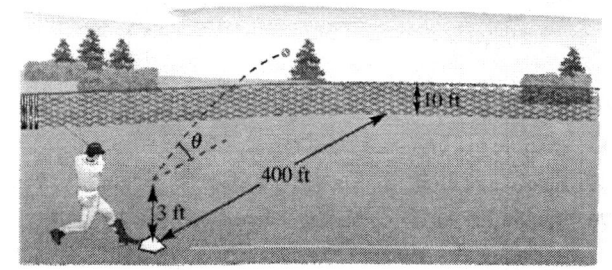

76. A rectangular equation for the path of a projectile is

$$y = 5 + x - 0.005x^2.$$

(a) Eliminate the parameter t from the position function for the motion of a projectile to show that the rectangular equation is

$$y = -\frac{16 \sec^2 \theta}{v_0^2} x^2 + (\tan \theta)x + h.$$

(b) Use the result in part (a) to find h, v_0, and θ. Find the parametric equations of the path.

(c) Use a graphing utility to graph the rectangular equation for the path of the projectile. Confirm your answer in part (b) by sketching the curve represented by the parametric equations.

(d) Use a graphing utility to approximate the maximum height of the projectile and its range.

SECTION PROJECT: CYCLOIDS

In Greek, the word *cycloid* means *wheel*, the word *hypocycloid* means *under the wheel*, and the word *epicycloid* means *upon the wheel*. Match the hypocycloid or epicycloid with its graph. [The graphs are labeled (a), (b), (c), (d), (e), and (f).]

Hypocycloid, H(A, B)

Path traced by a fixed point on a circle of radius B as it rolls around the *inside* of a circle of radius A.

$$x = (A - B) \cos t + B \cos\left(\frac{A - B}{B}\right)t$$

$$y = (A - B) \sin t - B \sin\left(\frac{A - B}{B}\right)t$$

Epicycloid, E(A, B)

Path traced by a fixed point on a circle of radius B as it rolls around the *outside* of a circle of radius A.

$$x = (A + B) \cos t - B \cos\left(\frac{A + B}{B}\right)t$$

$$y = (A + B) \sin t - B \sin\left(\frac{A + B}{B}\right)t$$

I. H(8, 3)   II. E(8, 3)
III. H(8, 7)   IV. E(24, 3)
V. H(24, 7)   VI. E(24, 7)

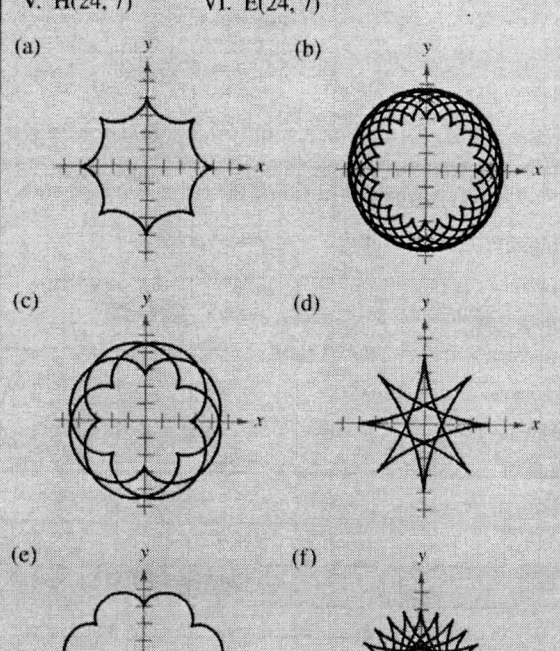

Exercises based on "Mathematical Discovery via Computer Graphics: Hypocycloids and Epicycloids" by Florence S. Gordon and Sheldon P. Gordon, *The College Mathematics Journal*, November 1984, p. 441. Used by permission of the authors.

Section 9.3 Parametric Equations and Calculus

- Find the slope of a tangent line to a curve given by a set of parametric equations.
- Find the arc length of a curve given by a set of parametric equations.
- Find the area of a surface of revolution (parametric form).

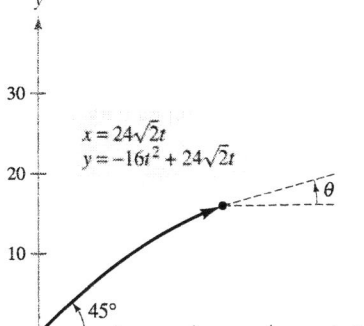

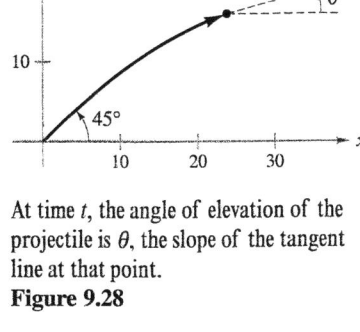

Slope and Tangent Lines

Now that you can represent a graph in the plane by a set of parametric equations, it is natural to ask how to use calculus to study plane curves. To begin, let's take another look at the projectile represented by the parametric equations

$$x = 24\sqrt{2}\,t \quad \text{and} \quad y = -16t^2 + 24\sqrt{2}\,t$$

as shown in Figure 9.28. From Section 9.2, you know that these equations enable you to locate the position of the projectile at a given time. You also know that the object is initially projected at an angle of 45°. But how can you find the angle θ representing the object's direction at some other time t? The following theorem answers this question by giving a formula for the slope of the tangent line as a function of t.

At time t, the angle of elevation of the projectile is θ, the slope of the tangent line at that point.
Figure 9.28

THEOREM 9.7 Parametric Form of the Derivative

If a smooth curve C is given by the equations $x = f(t)$ and $y = g(t)$, then the slope of C at (x, y) is

$$\frac{dy}{dx} = \frac{dy/dt}{dx/dt}, \quad \frac{dx}{dt} \neq 0.$$

Proof In Figure 9.29, consider $\Delta t > 0$ and let

$$\Delta y = g(t + \Delta t) - g(t) \quad \text{and} \quad \Delta x = f(t + \Delta t) - f(t).$$

Because $\Delta x \to 0$ as $\Delta t \to 0$, you can write

$$\frac{dy}{dx} = \lim_{\Delta x \to 0} \frac{\Delta y}{\Delta x}$$

$$= \lim_{\Delta t \to 0} \frac{g(t + \Delta t) - g(t)}{f(t + \Delta t) - f(t)}.$$

Dividing both the numerator and denominator by Δt, you can use the differentiability of f and g to conclude that

$$\frac{dy}{dx} = \lim_{\Delta t \to 0} \frac{[g(t + \Delta t) - g(t)]/\Delta t}{[f(t + \Delta t) - f(t)]/\Delta t}$$

$$= \frac{\displaystyle\lim_{\Delta t \to 0} \frac{g(t + \Delta t) - g(t)}{\Delta t}}{\displaystyle\lim_{\Delta t \to 0} \frac{f(t + \Delta t) - f(t)}{\Delta t}}$$

$$= \frac{g'(t)}{f'(t)}$$

$$= \frac{dy/dt}{dx/dt}.$$

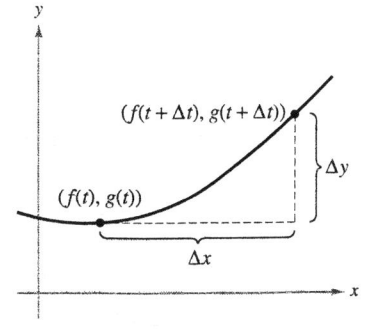

The slope of the secant line through the points $(f(t), g(t))$ and $(f(t + \Delta t), g(t + \Delta t))$ is $\Delta y/\Delta x$.
Figure 9.29

Example 1 Differentiation and Parametric Form

Find dy/dx for the curve given by $x = \sin t$ and $y = \cos t$.

STUDY TIP The curve traced out in Example 1 is a circle. Use the formula

$$\frac{dy}{dx} = -\tan t$$

to find the slope at the points $(1, 0)$ and $(0, 1)$.

Solution

$$\frac{dy}{dx} = \frac{dy/dt}{dx/dt} = \frac{-\sin t}{\cos t} = -\tan t$$

Because dy/dx is a function of t, you can use Theorem 9.7 repeatedly to find *higher-order* derivatives. For instance,

$$\frac{d^2y}{dx^2} = \frac{d}{dx}\left[\frac{dy}{dx}\right] = \frac{\frac{d}{dt}\left[\frac{dy}{dx}\right]}{dx/dt} \qquad \text{Second derivative}$$

$$\frac{d^3y}{dx^3} = \frac{d}{dx}\left[\frac{d^2y}{dx^2}\right] = \frac{\frac{d}{dt}\left[\frac{d^2y}{dx^2}\right]}{dx/dt}. \qquad \text{Third derivative}$$

Example 2 Finding Slope and Concavity

For the curve given by

$$x = \sqrt{t} \quad \text{and} \quad y = \frac{1}{4}(t^2 - 4), \quad t \geq 0$$

find the slope and concavity at the point $(2, 3)$.

Solution Because

$$\frac{dy}{dx} = \frac{dy/dt}{dx/dt} = \frac{(1/2)t}{(1/2)t^{-1/2}} = t^{3/2} \qquad \text{Parametric form of first derivative}$$

you can find the second derivative to be

$$\frac{d^2y}{dx^2} = \frac{\frac{d}{dt}[dy/dx]}{dx/dt} = \frac{\frac{d}{dt}[t^{3/2}]}{dx/dt} = \frac{(3/2)t^{1/2}}{(1/2)t^{-1/2}} = 3t. \qquad \text{Parametric form of second derivative}$$

At $(x, y) = (2, 3)$, it follows that $t = 4$, and the slope is

$$\frac{dy}{dx} = (4)^{3/2} = 8.$$

Moreover, when $t = 4$, the second derivative is

$$\frac{d^2y}{dx^2} = 3(4) = 12 > 0$$

and you can conclude that the graph is concave upward at $(2, 3)$, as shown in Figure 9.30.

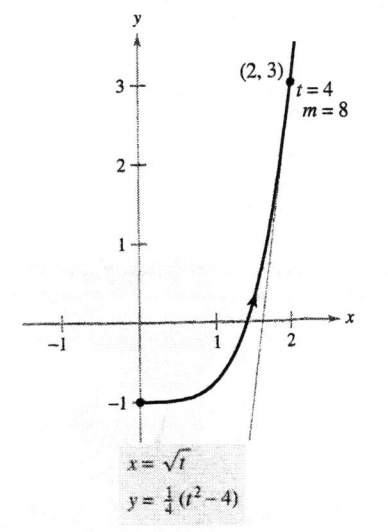

$x = \sqrt{t}$
$y = \frac{1}{4}(t^2 - 4)$

The graph is concave upward at $(2, 3)$, when $t = 4$.
Figure 9.30

Because the parametric equations $x = f(t)$ and $y = g(t)$ need not define y as a function of x, it follows that a plane curve can loop and cross itself. At such points the curve may have more than one tangent line, as shown in the next example.

Example 3 A Curve with Two Tangent Lines at a Point

The **prolate cycloid** given by

$$x = 2t - \pi \sin t \quad \text{and} \quad y = 2 - \pi \cos t$$

crosses itself at the point $(0, 2)$, as shown in Figure 9.31. Find the equations of both tangent lines at this point.

Solution Because $x = 0$ and $y = 2$ when $t = \pm \pi/2$, and

$$\frac{dy}{dx} = \frac{dy/dt}{dx/dt} = \frac{\pi \sin t}{2 - \pi \cos t}$$

you have $dy/dx = -\pi/2$ when $t = -\pi/2$ and $dy/dx = \pi/2$ when $t = \pi/2$. Therefore, the two tangent lines at $(0, 2)$ are

$$y - 2 = -\left(\frac{\pi}{2}\right)x \qquad \text{Tangent line when } t = -\frac{\pi}{2}$$

$$y - 2 = \left(\frac{\pi}{2}\right)x. \qquad \text{Tangent line when } t = \frac{\pi}{2}$$

$x = 2t - \pi \sin t$
$y = 2 - \pi \cos t$

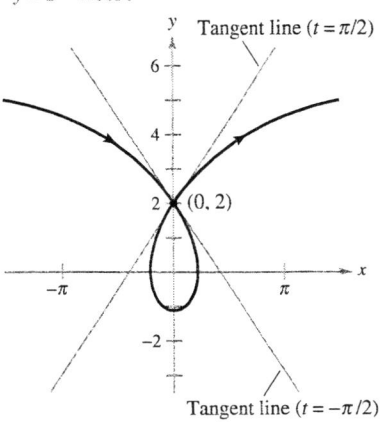

This prolate cycloid has two tangent lines at the point $(0, 2)$.
Figure 9.31

If $dy/dt = 0$ and $dx/dt \ne 0$ when $t = t_0$, the curve represented by $x = f(t)$ and $y = g(t)$ has a *horizontal* tangent at $(f(t_0), g(t_0))$. For instance, in Example 3, the given curve has a horizontal tangent at the point $(0, 2 - \pi)$ (when $t = 0$). Similarly, if $dx/dt = 0$ and $dy/dt \ne 0$ when $t = t_0$, the curve represented by $x = f(t)$ and $y = g(t)$ has a *vertical* tangent at $(f(t_0), g(t_0))$.

Arc Length

You have seen how parametric equations can be used to describe the path of a particle moving in the plane. We now develop a formula for determining the *distance* traveled by the particle along its path.

Recall from Section 6.4 that the formula for the arc length of a curve C given by $y = h(x)$ over the interval $[x_0, x_1]$ is

$$s = \int_{x_0}^{x_1} \sqrt{1 + [h'(x)]^2}\, dx$$

$$= \int_{x_0}^{x_1} \sqrt{1 + \left(\frac{dy}{dx}\right)^2}\, dx.$$

If C is represented by the parametric equations $x = f(t)$ and $y = g(t)$, $a \le t \le b$, and if $dx/dt = f'(t) > 0$, you can write

$$s = \int_{x_0}^{x_1} \sqrt{1 + \left(\frac{dy}{dx}\right)^2}\, dx = \int_{x_0}^{x_1} \sqrt{1 + \left(\frac{dy/dt}{dx/dt}\right)^2}\, dx$$

$$= \int_a^b \sqrt{\frac{(dx/dt)^2 + (dy/dt)^2}{(dx/dt)^2}}\, \frac{dx}{dt}\, dt$$

$$= \int_a^b \sqrt{\left(\frac{dx}{dt}\right)^2 + \left(\frac{dy}{dt}\right)^2}\, dt$$

$$= \int_a^b \sqrt{[f'(t)]^2 + [g'(t)]^2}\, dt.$$

NOTE When applying the arc length formula to a curve, be sure that the curve is traced out only once on the interval of integration. For instance, the circle given by $x = \cos t$ and $y = \sin t$ is traced out once on the interval $0 \le t \le 2\pi$, but is traced out twice on the interval $0 \le t \le 4\pi$.

THEOREM 9.8 Arc Length in Parametric Form

If a smooth curve C is given by $x = f(t)$ and $y = g(t)$ such that C does not intersect itself on the interval $a \le t \le b$ (except possibly at the endpoints), then the arc length of C over the interval is given by

$$s = \int_a^b \sqrt{\left(\frac{dx}{dt}\right)^2 + \left(\frac{dy}{dt}\right)^2}\, dt = \int_a^b \sqrt{[f'(t)]^2 + [g'(t)]^2}\, dt.$$

In the preceding section you saw that if a circle rolls along a line, a point on its circumference will trace a path called a cycloid. If the circle rolls around the circumference of another circle, the path of the point is an **epicycloid**. The next example shows how to find the arc length of an epicycloid.

ARCH OF A CYCLOID

The arc length of an arch of a cycloid was first calculated in 1658 by British architect and mathematician Christopher Wren, famous for rebuilding many buildings and churches in London, including St. Paul's Cathedral.

Example 4 **Finding Arc Length**

A circle of radius 1 rolls around the circumference of a larger circle of radius 4, as shown in Figure 9.32. The epicycloid traced by a point on the circumference of the smaller circle is given by

$$x = 5\cos t - \cos 5t$$

and

$$y = 5\sin t - \sin 5t.$$

Find the distance traveled by the point in one complete trip about the larger circle.

Solution Before applying Theorem 9.8, note in Figure 9.32 that the curve has sharp points when $t = 0$ and $t = \pi/2$. Between these two points, dx/dt and dy/dt are not simultaneously 0. So, the portion of the curve generated from $t = 0$ to $t = \pi/2$ is smooth. To find the total distance traveled by the point, you can find the arc length of that portion lying in the first quadrant and multiply by 4.

$$\begin{aligned}
s &= 4\int_0^{\pi/2} \sqrt{\left(\frac{dx}{dt}\right)^2 + \left(\frac{dy}{dt}\right)^2}\, dt && \text{Parametric form for arc length}\\
&= 4\int_0^{\pi/2} \sqrt{(-5\sin t + 5\sin 5t)^2 + (5\cos t - 5\cos 5t)^2}\, dt\\
&= 20\int_0^{\pi/2} \sqrt{2 - 2\sin t \sin 5t - 2\cos t \cos 5t}\, dt\\
&= 20\int_0^{\pi/2} \sqrt{2 - 2\cos 4t}\, dt\\
&= 20\int_0^{\pi/2} \sqrt{4\sin^2 2t}\, dt && \text{Trigonometric identity}\\
&= 40\int_0^{\pi/2} \sin 2t\, dt\\
&= -20\Big[\cos 2t\Big]_0^{\pi/2}\\
&= 40
\end{aligned}$$

For the epicycloid shown in Figure 9.32, an arc length of 40 seems about right because the circumference of a circle of radius 6 is $2\pi r = 12\pi \approx 37.7$.

$x = 5\cos t - \cos 5t$
$y = 5\sin t - \sin 5t$

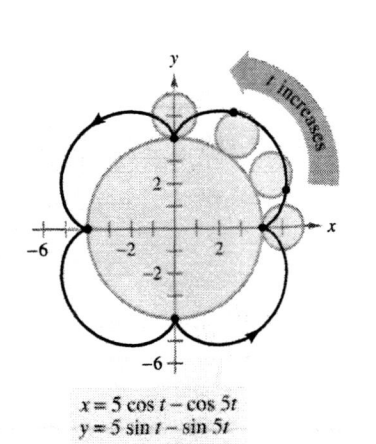

An epicycloid is traced by a point on the smaller circle as it rolls around the larger circle.
Figure 9.32

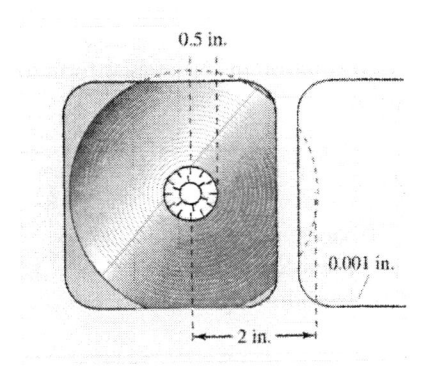

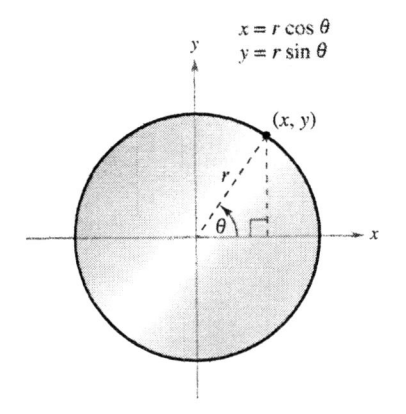

Example 5 Length of a Recording Tape

A recording tape 0.001 inch thick is wound around a reel whose inner radius is 0.5 inch and outer radius is 2 inches, as shown in Figure 9.33. How much tape is required to fill the reel?

Solution To create a model for this problem, assume that as the tape is wound around the reel its distance r from the center increases linearly at a rate of 0.001 inch per revolution, or

$$r = (0.001)\frac{\theta}{2\pi} = \frac{\theta}{2000\pi}, \qquad 1000\pi \le \theta \le 4000\pi$$

where θ is measured in radians. You can determine the coordinates of the point (x, y) corresponding to a given radius to be

$$x = r \cos \theta$$

and

$$y = r \sin \theta.$$

Substituting for r, you obtain the parametric equations

$$x = \left(\frac{\theta}{2000\pi}\right) \cos \theta \quad \text{and} \quad y = \left(\frac{\theta}{2000\pi}\right) \sin \theta.$$

You can use the arc length formula to determine the total length of the tape to be

$$s = \int_{1000\pi}^{4000\pi} \sqrt{\left(\frac{dx}{d\theta}\right)^2 + \left(\frac{dy}{d\theta}\right)^2} \, d\theta$$

$$= \frac{1}{2000\pi} \int_{1000\pi}^{4000\pi} \sqrt{(-\theta \sin \theta + \cos \theta)^2 + (\theta \cos \theta + \sin \theta)^2} \, d\theta$$

$$= \frac{1}{2000\pi} \int_{1000\pi}^{4000\pi} \sqrt{\theta^2 + 1} \, d\theta$$

$$= \frac{1}{2000\pi} \left(\frac{1}{2}\right) \left[\theta\sqrt{\theta^2 + 1} + \ln\left|\theta + \sqrt{\theta^2 + 1}\right|\right]_{1000\pi}^{4000\pi} \qquad \text{Integration tables (Appendix C), Formula 26}$$

$$\approx 11{,}781 \text{ in.}$$

$$\approx 982 \text{ ft.}$$

It takes approximately 982 feet of tape to fill the reel.
Figure 9.33

NOTE The graph of $r = a\theta$ is called the **spiral of Archimedes.** The graph of $r = \theta/2000\pi$ (in Example 5) is of this form.

FOR FURTHER INFORMATION For more information on the mathematics of recording tape, see "Tape Counters" by Richard L. Roth in *The American Mathematical Monthly*. To view this article, go to the website *www.matharticles.com*.

The length of the tape in Example 5 can be approximated by adding the circumferences of circular pieces of tape. The smallest circle has a radius of 0.501 and the largest has a radius of 2.

$$s \approx 2\pi(0.501) + 2\pi(0.502) + 2\pi(0.503) + \cdots + 2\pi(2.000)$$

$$= \sum_{i=1}^{1500} 2\pi(0.5 + 0.001i)$$

$$= 2\pi[1500(0.5) + 0.001(1500)(1501)/2]$$

$$\approx 11{,}786 \text{ in.}$$

Area of a Surface of Revolution

You can use the formula for the area of a surface of revolution in rectangular form to develop a formula for surface area in parametric form.

THEOREM 9.9   Area of a Surface of Revolution

If a smooth curve C given by $x = f(t)$ and $y = g(t)$ does not cross itself on an interval $a \leq t \leq b$, then the area S of the surface of revolution formed by revolving C about the coordinate axes is given by the following.

1. $S = 2\pi \int_a^b g(t) \sqrt{\left(\dfrac{dx}{dt}\right)^2 + \left(\dfrac{dy}{dt}\right)^2}\, dt$ Revolution about the x-axis: $g(t) \geq 0$

2. $S = 2\pi \int_a^b f(t) \sqrt{\left(\dfrac{dx}{dt}\right)^2 + \left(\dfrac{dy}{dt}\right)^2}\, dt$ Revolution about the y-axis: $f(t) \geq 0$

These formulas are easy to remember if you think of the differential of arc length as

$$ds = \sqrt{\left(\dfrac{dx}{dt}\right)^2 + \left(\dfrac{dy}{dt}\right)^2}\, dt.$$

Then the formulas are written as follows.

1. $S = 2\pi \displaystyle\int_a^b g(t)\, ds$ 2. $S = 2\pi \displaystyle\int_a^b f(t)\, ds$

Example 6   **Finding the Area of a Surface of Revolution**

Let C be the arc of the circle

$$x^2 + y^2 = 9$$

from $(3, 0)$ to $(3/2, 3\sqrt{3}/2)$, as shown in Figure 9.34. Find the area of the surface formed by revolving C about the x-axis.

Solution You can represent C parametrically by the equations

$$x = 3 \cos t \quad \text{and} \quad y = 3 \sin t, \quad 0 \leq t \leq \pi/3.$$

(Note that you can determine the interval for t by observing that $t = 0$ when $x = 3$ and $t = \pi/3$ when $x = 3/2$.) On this interval, C is smooth and y is nonnegative, and you can apply Theorem 9.9 to obtain a surface area of

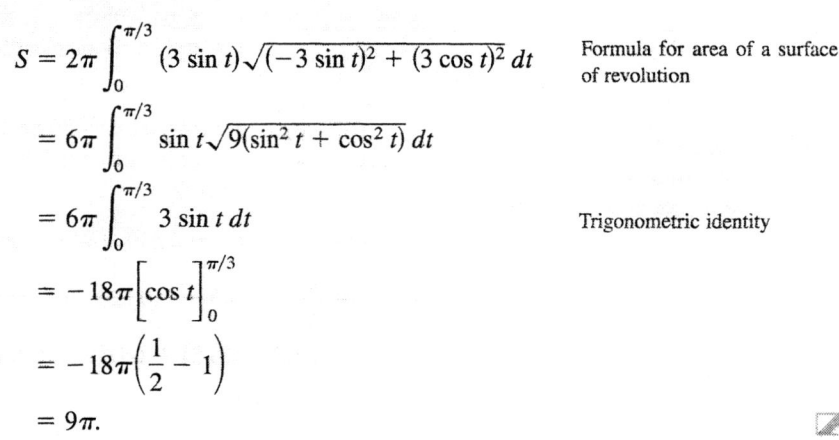

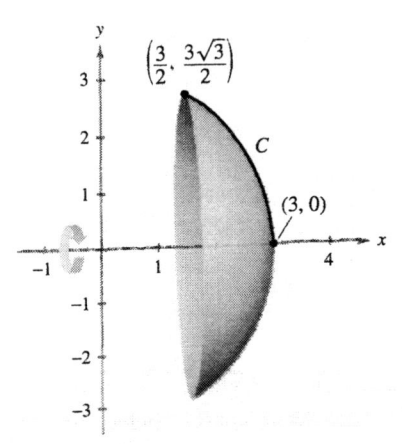

$$\begin{aligned}
S &= 2\pi \int_0^{\pi/3} (3 \sin t)\sqrt{(-3 \sin t)^2 + (3 \cos t)^2}\, dt &&\text{Formula for area of a surface of revolution}\\
&= 6\pi \int_0^{\pi/3} \sin t \sqrt{9(\sin^2 t + \cos^2 t)}\, dt \\
&= 6\pi \int_0^{\pi/3} 3 \sin t\, dt &&\text{Trigonometric identity}\\
&= -18\pi \Big[\cos t\Big]_0^{\pi/3} \\
&= -18\pi\left(\dfrac{1}{2} - 1\right) \\
&= 9\pi.
\end{aligned}$$

This surface of revolution has a surface area of 9π.
Figure 9.34

EXERCISES FOR SECTION 9.3

In Exercises 1–4, find dy/dx.

1. $x = t^2$, $y = 5 - 4t$
2. $x = \sqrt[3]{t}$, $y = 4 - t$
3. $x = \sin^2 \theta$, $y = \cos^2 \theta$
4. $x = 2e^\theta$, $y = e^{-\theta/2}$

In Exercises 5–14, find dy/dx and d^2y/dx^2, and find the slope and concavity (if possible) at the indicated value of the parameter.

	Parametric Equations	Point
5.	$x = 2t$, $y = 3t - 1$	$t = 3$
6.	$x = \sqrt{t}$, $y = 3t - 1$	$t = 1$
7.	$x = t + 1$, $y = t^2 + 3t$	$t = -1$
8.	$x = t^2 + 3t + 2$, $y = 2t$	$t = 0$
9.	$x = 2\cos\theta$, $y = 2\sin\theta$	$\theta = \dfrac{\pi}{4}$
10.	$x = \cos\theta$, $y = 3\sin\theta$	$\theta = 0$
11.	$x = 2 + \sec\theta$, $y = 1 + 2\tan\theta$	$\theta = \dfrac{\pi}{6}$
12.	$x = \sqrt{t}$, $y = \sqrt{t-1}$	$t = 2$
13.	$x = \cos^3\theta$, $y = \sin^3\theta$	$\theta = \dfrac{\pi}{4}$
14.	$x = \theta - \sin\theta$, $y = 1 - \cos\theta$	$\theta = \pi$

In Exercises 15 and 16, find an equation of the tangent line at the indicated points on the curve.

15. $x = 2\cot\theta$
    $y = 2\sin^2\theta$

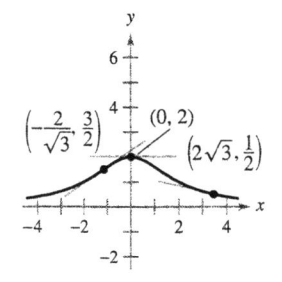

16. $x = 2 - 3\cos\theta$
    $y = 3 + 2\sin\theta$

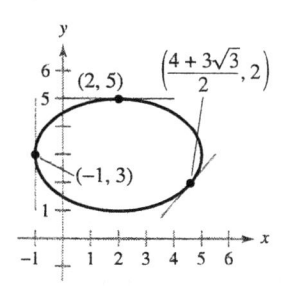

In Exercises 17–20, (a) use a graphing utility to graph the curve represented by the parametric equations, (b) use a graphing utility to find dx/dt, dy/dt, and dy/dx at the indicated value of the parameter, (c) find an equation of the tangent line to the curve at the indicated value of the parameter, and (d) confirm the result in part (c) by using a graphing utility to graph the tangent line.

	Parametric Equations	Parameter
17.	$x = 2t$, $y = t^2 - 1$	$t = 2$
18.	$x = t - 1$, $y = \dfrac{1}{t} + 1$	$t = 1$
19.	$x = t^2 - t + 2$, $y = t^3 - 3t$	$t = -1$
20.	$x = 4\cos\theta$, $y = 3\sin\theta$	$\theta = \dfrac{3\pi}{4}$

In Exercises 21 and 22, find the equations of the tangent lines at the point where the curve crosses itself.

21. $x = 2\sin 2t$, $y = 3\sin t$
22. $x = t^2 - t$, $y = t^3 - 3t - 1$

In Exercises 23 and 24, find all points (if any) of horizontal and vertical tangency to the portion of the curve shown.

23. Involute of a circle:
    $x = \cos\theta + \theta\sin\theta$
    $y = \sin\theta - \theta\cos\theta$

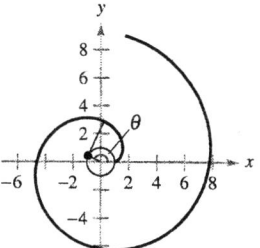

24. $x = 2\theta$
    $y = 2(1 - \cos\theta)$

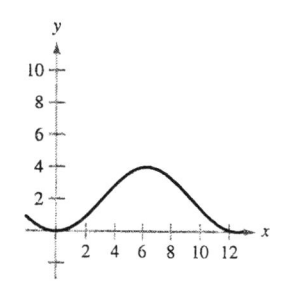

In Exercises 25–34, find all points (if any) of horizontal and vertical tangency to the curve. Use a graphing utility to confirm your results.

25. $x = 1 - t$, $y = t^2$
26. $x = t + 1$, $y = t^2 + 3t$
27. $x = 1 - t$, $y = t^3 - 3t$
28. $x = t^2 - t + 2$, $y = t^3 - 3t$
29. $x = 3\cos\theta$, $y = 3\sin\theta$
30. $x = \cos\theta$, $y = 2\sin 2\theta$
31. $x = 4 + 2\cos\theta$, $y = -1 + \sin\theta$
32. $x = 4\cos^2\theta$, $y = 2\sin\theta$
33. $x = \sec\theta$, $y = \tan\theta$
34. $x = \cos^2\theta$, $y = \cos\theta$

Arc Length In Exercises 35–40, find the arc length of the given curve on the indicated interval.

	Parametric Equations	Interval
35.	$x = t^2$, $y = 2t$	$0 \leq t \leq 2$
36.	$x = t^2 + 1$, $y = 4t^3 + 3$	$-1 \leq t \leq 0$
37.	$x = e^{-t}\cos t$, $y = e^{-t}\sin t$	$0 \leq t \leq \dfrac{\pi}{2}$
38.	$x = \arcsin t$, $y = \ln\sqrt{1-t^2}$	$0 \leq t \leq \frac{1}{2}$
39.	$x = \sqrt{t}$, $y = 3t - 1$	$0 \leq t \leq 1$
40.	$x = t$, $y = \dfrac{t^5}{10} + \dfrac{1}{6t^3}$	$1 \leq t \leq 2$

Arc Length In Exercises 41–44, find the arc length of the curve on the interval $[0, 2\pi]$.

41. Hypocycloid perimeter: $x = a\cos^3\theta$, $y = a\sin^3\theta$
42. Circle circumference: $x = a\cos\theta$, $y = a\sin\theta$
43. Cycloid arch: $x = a(\theta - \sin\theta)$, $y = a(1 - \cos\theta)$
44. Involute of a circle: $x = \cos\theta + \theta\sin\theta$, $y = \sin\theta - \theta\cos\theta$

45. *Path of a Projectile* The path of a projectile is modeled by the parametric equations

$$x = (90\cos 30°)t \quad \text{and} \quad y = (90\sin 30°)t - 16t^2$$

where x and y are measured in feet. Use a graphing utility to perform the following.

(a) Graph the path of the projectile.

(b) Approximate the range of the projectile.

(c) Use the integration capabilities of the graphing utility to approximate the arc length of the path. Compare this result with the range of the projectile.

(d) If the projectile is launched at an angle θ with the horizontal, its parametric equations are

$$x = (90\cos\theta)t \quad \text{and} \quad y = (90\sin\theta)t - 16t^2.$$

What angle maximizes its range? What angle maximizes the arc length of the trajectory?

46. *Folium of Descartes* Given the parametric equations

$$x = \frac{4t}{1 + t^3} \quad \text{and} \quad y = \frac{4t^2}{1 + t^3}$$

use a graphing utility to perform the following.

(a) Sketch the curve described by the parametric equations.

(b) Find the points of horizontal tangency to the curve.

(c) Use the integration capabilities of the graphing utility to approximate the arc length of the closed loop. (*Hint:* Use symmetry and integrate over the interval $0 \le t \le 1$.)

47. *Writing*

(a) Use a graphing utility to graph each set of parametric equations.

$x = t - \sin t$ $x = 2t - \sin(2t)$
$y = 1 - \cos t$ $y = 1 - \cos(2t)$
$0 \le t \le 2\pi$ $0 \le t \le \pi$

(b) Compare the graphs of the two sets of parametric equations in part (a). If the curve represents the motion of a particle and t is time, what can you infer about the average speed of the particle on the paths represented by the two sets of parametric equations?

(c) Without graphing the curve, determine the time required for a particle to traverse the same path as in parts (a) and (b) if the path is modeled by

$$x = \tfrac{1}{2}t - \sin(\tfrac{1}{2}t) \quad \text{and} \quad y = 1 - \cos(\tfrac{1}{2}t).$$

48. *Circumference of an Ellipse* Use the integration capabilities of a graphing utility to approximate the circumference of the ellipse given by the parametric equations $x = 3\cos\theta$ and $y = 4\sin\theta$.

Surface Area In Exercises 49–54, find the area of the surface generated by revolving the curve about the given axis.

49. $x = t$, $y = 2t$, $0 \le t \le 4$,   (a) x-axis   (b) y-axis
50. $x = t$, $y = 4 - 2t$, $0 \le t \le 2$,   (a) x-axis   (b) y-axis
51. $x = 4\cos\theta$, $y = 4\sin\theta$, $0 \le \theta \le \dfrac{\pi}{2}$, y-axis
52. $x = \tfrac{1}{3}t^3$, $y = t + 1$, $1 \le t \le 2$, y-axis
53. $x = a\cos^3\theta$, $y = a\sin^3\theta$, $0 \le \theta \le \pi$, x-axis
54. $x = a\cos\theta$, $y = b\sin\theta$, $0 \le \theta \le 2\pi$,
    (a) x-axis   (b) y-axis

Getting at the Concept

55. Give the parametric form of the derivative.

56. Mentally determine dy/dx.

(a) $x = t$ (b) $x = t$
    $y = 4$ $y = 4t - 3$

57. Sketch a graph of a curve defined by the parametric equations $x = g(t)$ and $y = f(t)$ such that $dx/dt > 0$ and $dy/dt < 0$ for all real numbers t.

58. Sketch a graph of a curve defined by the parametric equations $x = g(t)$ and $y = f(t)$ such that $dx/dt < 0$ and $dy/dt < 0$ for all real numbers t.

59. Give the integral formula for arc length in parametric form.

60. Give the integral formulas for the area of a surface of revolution formed when a smooth curve C is revolved about (a) the x-axis and (b) the y-axis.

---

61. *Surface Area* A portion of a sphere of radius r is removed by cutting out a circular cone with its vertex at the center of the sphere. Find the surface area removed from the sphere if the vertex of the cone forms an angle of 2θ.

62. Use integration by substitution to show that if y is a continuous function of x on the interval $a \le x \le b$, where $x = f(t)$ and $y = g(t)$, then

$$\int_a^b y \, dx = \int_{t_1}^{t_2} g(t) f'(t) \, dt,$$

where $f(t_1) = a$, $f(t_2) = b$, and both g and f' are continuous on $[t_1, t_2]$.

Centroid In Exercises 63 and 64, find the centroid of the region bounded by the graph of the parametric equations and the coordinate axes. (Use the result in Exercise 62.)

63. $x = \sqrt{t}$, $y = 4 - t$ 64. $x = \sqrt{4 - t}$, $y = \sqrt{t}$

Volume In Exercises 65 and 66, find the volume of the solid formed by revolving the region bounded by the graphs of the given equations about the *x*-axis. (Use the result in Exercise 62.)

65. $x = 3\cos\theta$, $y = 3\sin\theta$
66. $x = \cos\theta$, $y = 3\sin\theta$

Area In Exercises 67 and 68, find the area of the region. (Use the result in Exercise 62.)

67. $x = 2\sin^2\theta$
    $y = 2\sin^2\theta\tan\theta$
    $0 \le \theta < \dfrac{\pi}{2}$

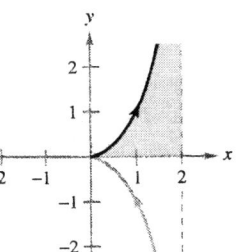

68. $x = 2\cot\theta$
    $y = 2\sin^2\theta$
    $0 < \theta < \pi$

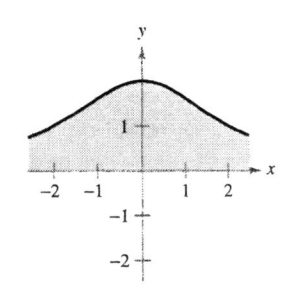

Areas of Simple Closed Curves In Exercises 69–74, use a computer algebra system and the result in Exercise 62 to match the closed curve with its area. (These exercises were adapted from the article "The Surveyor's Area Formula" by Bart Braden in the September 1986 issue of *The College Mathematics Journal*. Used by permission of the author.)

(a) $\dfrac{8}{3}ab$ (b) $\dfrac{3}{8}\pi a^2$ (c) $2\pi a^2$
(d) πab (e) $2\pi ab$ (f) $6\pi a^2$

69. Ellipse: $(0 \le t \le 2\pi)$
    $x = b\cos t$
    $y = a\sin t$

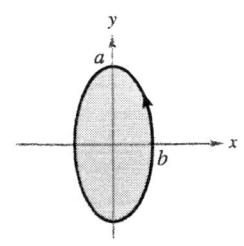

70. Asteroid: $(0 \le t \le 2\pi)$
    $x = a\cos^3 t$
    $y = a\sin^3 t$

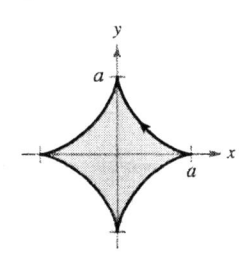

71. Cardioid: $(0 \le t \le 2\pi)$
    $x = 2a\cos t - a\cos 2t$
    $y = 2a\sin t - a\sin 2t$

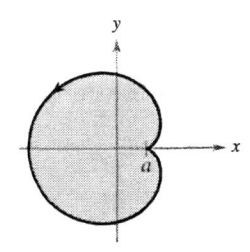

72. Deltoid: $(0 \le t \le 2\pi)$
    $x = 2a\cos t + a\cos 2t$
    $y = 2a\sin t - a\sin 2t$

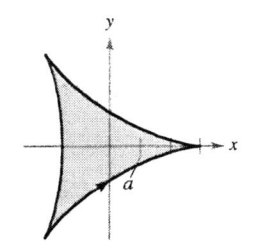

73. Hourglass: $(0 \le t \le 2\pi)$
    $x = a\sin 2t$
    $y = b\sin t$

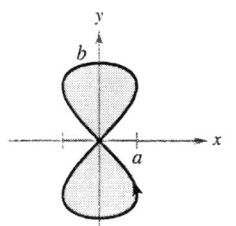

74. Teardrop: $(0 \le t \le 2\pi)$
    $x = 2a\cos t - a\sin 2t$
    $y = b\sin t$

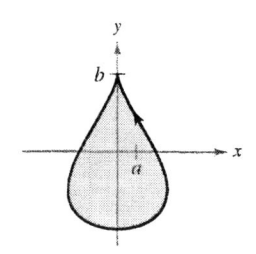

75. Use a graphing utility to graph the curve given by
$$x = \dfrac{1-t^2}{1+t^2}, \quad y = \dfrac{2t}{1+t^2}, \quad -20 \le t \le 20.$$

(a) Describe the graph and confirm your result analytically.

(b) Discuss the speed at which the curve is traced as *t* increases from -20 to 20.

76. **Tractrix** A person moves from the origin along the positive *y*-axis pulling a weight at the end of a 12-meter rope. Initially, the weight is located at the point $(12, 0)$.

(a) In Exercise 75 of Section 7.4, it was shown that the path of the weight is modeled by the rectangular equation
$$y = -12\ln\left(\dfrac{12 - \sqrt{144 - x^2}}{x}\right) - \sqrt{144 - x^2}$$
where $0 < x \le 12$. Use a graphing utility to graph the rectangular equation.

(b) Use a graphing utility to graph the parametric equations
$$x = 12\operatorname{sech}\dfrac{t}{12} \quad \text{and} \quad y = t - 12\tanh\dfrac{t}{12}$$
where $t \ge 0$. How does this graph compare with the graph in part (a)? Which graph (if either) do you think is a better representation of the path?

(c) Use the parametric equations for the tractrix to verify that the distance from the *y*-intercept of the tangent line to the point of tangency is independent of the location of the point of tangency.

True or False? In Exercises 77 and 78, determine whether the statement is true or false. If it is false, explain why or give an example that shows it is false.

77. If $x = f(t)$ and $y = g(t)$, then $d^2y/dx^2 = g''(t)/f''(t)$.

78. The curve given by $x = t^3$, $y = t^2$ has a horizontal tangent at the origin because $dy/dt = 0$ when $t = 0$.

684 CHAPTER 9 Conics, Parametric Equations, and Polar Coordinates

Section 9.4 Polar Coordinates and Polar Graphs

- Understand the polar coordinate system.
- Rewrite rectangular equations in polar form and vice versa.
- Sketch the graph of an equation given in polar form.
- Find the slope of a tangent line to a polar graph.
- Identify several types of special polar graphs.

Polar Coordinates

So far, we have been representing graphs as collections of points (x, y) on the rectangular coordinate system. The corresponding equations for these graphs have been in either rectangular or parametric form. In this section we introduce a coordinate system called the **polar coordinate system.**

To form the polar coordinate system in the plane, we fix a point O, called the **pole** (or **origin**), and construct from O an initial ray called the **polar axis,** as shown in Figure 9.35. Then each point P in the plane can be assigned **polar coordinates** (r, θ), as follows.

$r =$ *directed distance* from O to P

$\theta =$ *directed angle*, counterclockwise from polar axis to segment $\overline{OP}$

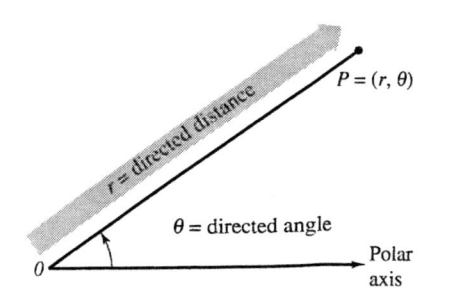

Polar coordinates
Figure 9.35

Figure 9.36 shows three points on the polar coordinate system. Notice that in this system, it is convenient to locate points with respect to a grid of concentric circles intersected by **radial lines** through the pole.

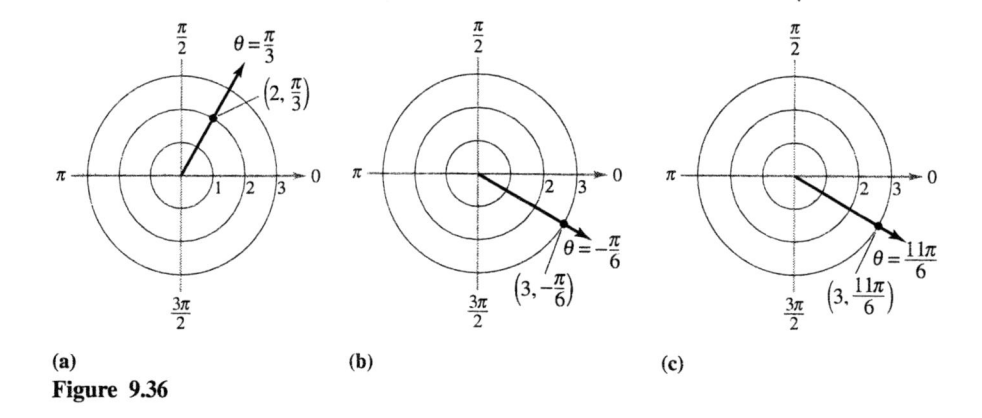

(a)   (b)   (c)
Figure 9.36

With rectangular coordinates, each point (x, y) has a unique representation. This is not true with polar coordinates. For instance, the coordinates (r, θ) and $(r, 2\pi + \theta)$ represent the same point [see parts (b) and (c) in Figure 9.36]. Also, because r is a *directed distance*, the coordinates (r, θ) and $(-r, \theta + \pi)$ represent the same point. In general, the point (r, θ) can be written as

$$(r, \theta) = (r, \theta + 2n\pi)$$

or

$$(r, \theta) = (-r, \theta + (2n + 1)\pi)$$

where n is any integer. Moreover, the pole is represented by $(0, \theta)$, where θ is any angle.

POLAR COORDINATES

The mathematician credited with first using polar coordinates was James Bernoulli, who introduced them in 1691. However, there is some evidence that it may have been Isaac Newton who first used them.

Coordinate Conversion

To establish the relationship between polar and rectangular coordinates, let the polar axis coincide with the positive x-axis and the pole with the origin, as shown in Figure 9.37. Because (x, y) lies on a circle of radius r, it follows that $r^2 = x^2 + y^2$. Moreover, for $r > 0$, the definition of the trigonometric functions implies that

$$\tan \theta = \frac{y}{x}, \quad \cos \theta = \frac{x}{r}, \quad \text{and} \quad \sin \theta = \frac{y}{r}.$$

If $r < 0$, you can show that the same relationships hold.

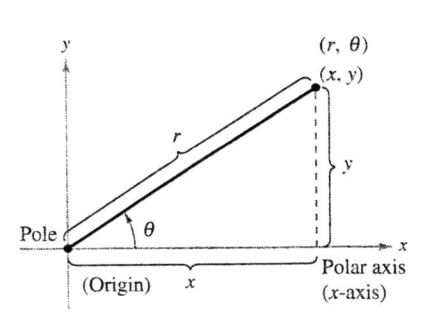

Relating polar and rectangular coordinates
Figure 9.37

THEOREM 9.10 Coordinate Conversion

The polar coordinates (r, θ) of a point are related to the rectangular coordinates (x, y) of the point as follows.

1. $x = r \cos \theta$ 2. $\tan \theta = \dfrac{y}{x}$

    $y = r \sin \theta$ $r^2 = x^2 + y^2$

Example 1   Polar-to-Rectangular Conversion

a. For the point $(r, \theta) = (2, \pi)$,

$$x = r \cos \theta = 2 \cos \pi = -2 \quad \text{and} \quad y = r \sin \theta = 2 \sin \pi = 0.$$

So, the rectangular coordinates are $(x, y) = (-2, 0)$.

b. For the point $(r, \theta) = (\sqrt{3}, \pi/6)$,

$$x = \sqrt{3} \cos \frac{\pi}{6} = \frac{3}{2} \quad \text{and} \quad y = \sqrt{3} \sin \frac{\pi}{6} = \frac{\sqrt{3}}{2}.$$

So, the rectangular coordinates are $(x, y) = (3/2, \sqrt{3}/2)$.

(See Figure 9.38.)

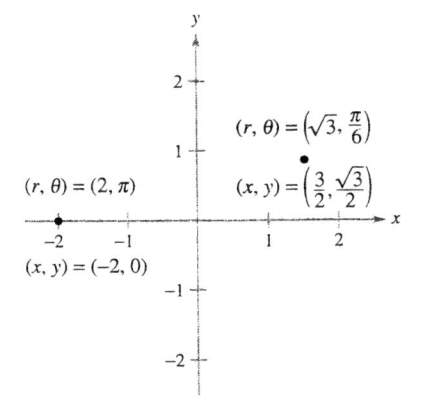

To convert from polar to rectangular coordinates, let $x = r \cos \theta$ and $y = r \sin \theta$.
Figure 9.38

Example 2   Rectangular-to-Polar Conversion

a. For the second quadrant point $(x, y) = (-1, 1)$,

$$\tan \theta = \frac{y}{x} = -1 \quad \Longrightarrow \quad \theta = \frac{3\pi}{4}.$$

Because θ was chosen to be in the same quadrant as (x, y), you should use a positive value of r.

$$r = \sqrt{x^2 + y^2}$$
$$= \sqrt{(-1)^2 + (1)^2}$$
$$= \sqrt{2}$$

This implies that *one* set of polar coordinates is $(r, \theta) = (\sqrt{2}, 3\pi/4)$.

b. Because the point $(x, y) = (0, 2)$ lies on the positive y-axis, we choose $\theta = \pi/2$ and $r = 2$, and one set of polar coordinates is $(r, \theta) = (2, \pi/2)$.

(See Figure 9.39.)

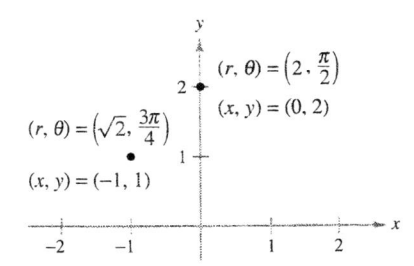

To convert from rectangular to polar coordinates, let $\tan \theta = y/x$ and $r = \sqrt{x^2 + y^2}$.
Figure 9.39

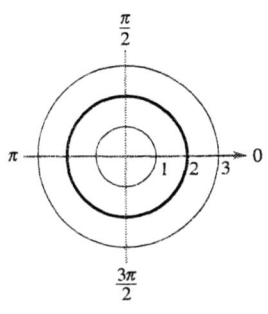

(a) Circle: $r = 2$

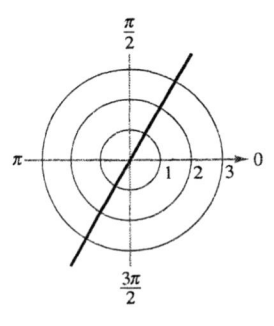

(b) Radial line: $\theta = \dfrac{\pi}{3}$

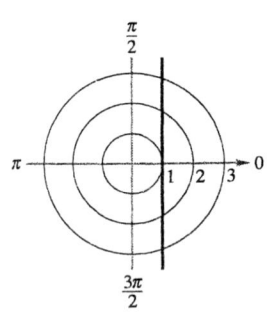

(c) Vertical line: $r = \sec\theta$

Figure 9.40

Polar Graphs

One way to sketch the graph of a polar equation is to convert to rectangular coordinates and then sketch the graph of the rectangular equation.

Example 3 Graphing Polar Equations

Describe the graph of each polar equation. Confirm each description by converting to a rectangular equation.

a. $r = 2$ **b.** $\theta = \dfrac{\pi}{3}$ **c.** $r = \sec\theta$

Solution

a. The graph of the polar equation $r = 2$ consists of all points that are two units from the pole. In other words, this graph is a circle centered at the origin with a radius of 2. (See Figure 9.40a.) You can confirm this by using the relationship $r^2 = x^2 + y^2$ to obtain the rectangular equation

$$x^2 + y^2 = 2^2. \quad \text{Rectangular equation}$$

b. The graph of the polar equation $\theta = \pi/3$ consists of all points on the line that makes an angle of $\pi/3$ with the positive x-axis. (See Figure 9.40b.) You can confirm this by using the relationship $\tan\theta = y/x$ to obtain the rectangular equation

$$y = \sqrt{3}\, x. \quad \text{Rectangular equation}$$

c. The graph of the polar equation $r = \sec\theta$ is not evident by simple inspection, so you can begin by converting to rectangular form using the relationship $r\cos\theta = x$.

$$r = \sec\theta \quad \text{Polar equation}$$
$$r\cos\theta = 1$$
$$x = 1 \quad \text{Rectangular equation}$$

From the rectangular equation, you can see that the graph is a vertical line. (See Figure 9.40c.)

TECHNOLOGY Sketching the graphs of complicated polar equations *by hand* can be tedious. With technology, however, the task is not difficult. If your graphing utility has a polar mode, try using it to sketch the graphs in the exercise set. If your graphing utility doesn't have a polar mode, but does have a parametric mode, you can sketch the graph of $r = f(\theta)$ by writing the equation as

$$x = f(\theta)\cos\theta$$
$$y = f(\theta)\sin\theta.$$

For instance, the graph of $r = \tfrac{1}{2}\theta$ shown in Figure 9.41 was produced with a graphing calculator in parametric mode. To sketch the graph, we entered the parametric equations

$$x = \dfrac{1}{2}\theta\cos\theta$$
$$y = \dfrac{1}{2}\theta\sin\theta$$

and let the values of θ vary from -4π to 4π. This curve is of the form $r = a\theta$ and is called a **spiral of Archimedes**.

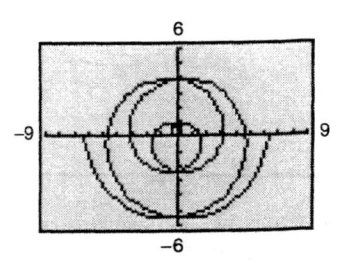

Spiral of Archimedes
Figure 9.41

Example 4 Sketching a Polar Graph

Sketch the graph of $r = 2 \cos 3\theta$.

Solution Begin by writing the polar equation in parametric form.

$$x = 2 \cos 3\theta \cos \theta \quad \text{and} \quad y = 2 \cos 3\theta \sin \theta$$

After some experimentation, you will find that the entire curve, which is called a **rose curve**, can be sketched by letting θ vary from 0 to π, as shown in Figure 9.42. If you try duplicating this graph with a graphing utility, you will find that by letting θ vary from 0 to 2π, you will actually trace the entire curve *twice*.

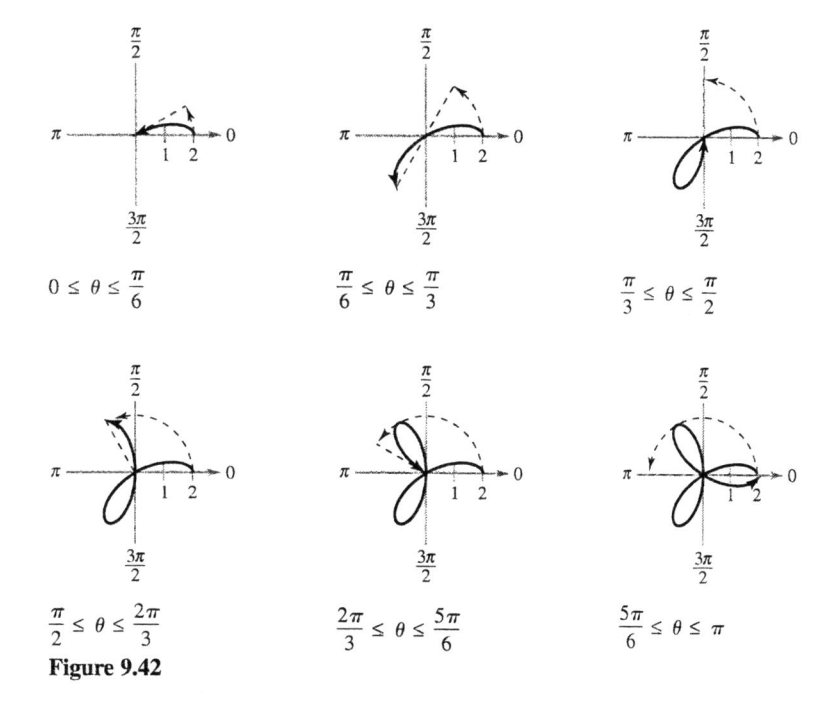

NOTE One way to sketch the graph of $r = 2 \cos 3\theta$ by hand is to make a table of values.

θ	0	$\frac{\pi}{6}$	$\frac{\pi}{3}$	$\frac{\pi}{2}$	$\frac{2\pi}{3}$
r	2	0	-2	0	2

By extending the table and plotting the points, you will obtain the curve shown in Example 4.

$0 \leq \theta \leq \frac{\pi}{6}$ $\quad$ $\frac{\pi}{6} \leq \theta \leq \frac{\pi}{3}$ $\quad$ $\frac{\pi}{3} \leq \theta \leq \frac{\pi}{2}$

$\frac{\pi}{2} \leq \theta \leq \frac{2\pi}{3}$ $\quad$ $\frac{2\pi}{3} \leq \theta \leq \frac{5\pi}{6}$ $\quad$ $\frac{5\pi}{6} \leq \theta \leq \pi$

Figure 9.42

Try using a graphing utility to experiment with other rose curves (they are of the form $r = a \cos n\theta$ or $r = a \sin n\theta$). For instance, Figure 9.43 shows the graphs of two other rose curves.

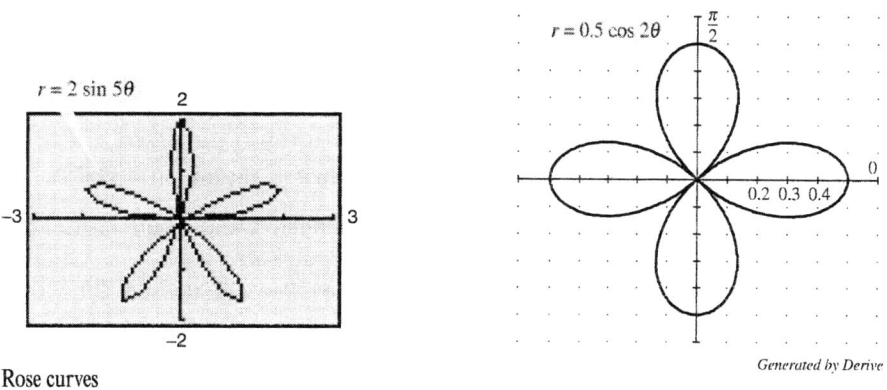

$r = 2 \sin 5\theta$

$r = 0.5 \cos 2\theta$

Generated by Derive

Rose curves
Figure 9.43

Slope and Tangent Lines

To find the slope of a tangent line to a polar graph, consider a differentiable function given by $r = f(\theta)$. To find the slope in polar form, use the parametric equations

$$x = r \cos \theta = f(\theta) \cos \theta \quad \text{and} \quad y = r \sin \theta = f(\theta) \sin \theta.$$

Using the parametric form of dy/dx given in Theorem 9.7, you have

$$\frac{dy}{dx} = \frac{dy/d\theta}{dx/d\theta}$$

$$= \frac{f(\theta) \cos \theta + f'(\theta) \sin \theta}{-f(\theta) \sin \theta + f'(\theta) \cos \theta}$$

which establishes the following theorem.

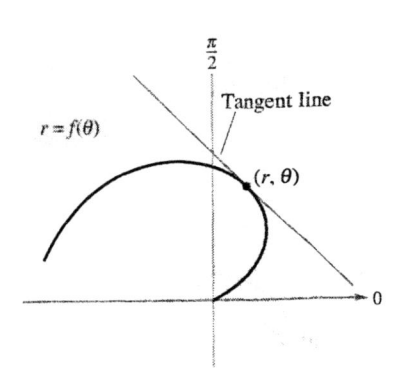

THEOREM 9.11 Slope in Polar Form

If f is a differentiable function of θ, then the *slope* of the tangent line to the graph of $r = f(\theta)$ at the point (r, θ) is

$$\frac{dy}{dx} = \frac{dy/d\theta}{dx/d\theta} = \frac{f(\theta) \cos \theta + f'(\theta) \sin \theta}{-f(\theta) \sin \theta + f'(\theta) \cos \theta}$$

provided that $dx/d\theta \neq 0$ at (r, θ). (See Figure 9.44.)

Tangent line to polar curve
Figure 9.44

From Theorem 9.11, you can make the following observations.

1. Solutions to $\dfrac{dy}{d\theta} = 0$ yield horizontal tangents, provided that $\dfrac{dx}{d\theta} \neq 0$.

2. Solutions to $\dfrac{dx}{d\theta} = 0$ yield vertical tangents, provided that $\dfrac{dy}{d\theta} \neq 0$.

If $dy/d\theta$ and $dx/d\theta$ are *simultaneously* 0, no conclusion can be drawn about tangent lines.

Example 5 **Finding Horizontal and Vertical Tangent Lines**

Find the horizontal and vertical tangent lines of $r = \sin \theta$, $0 \leq \theta \leq \pi$.

Solution Begin by writing the equation in parametric form.

$$x = r \cos \theta = \sin \theta \cos \theta$$

and

$$y = r \sin \theta = \sin \theta \sin \theta = \sin^2 \theta$$

Next, differentiate x and y with respect to θ and set each derivative equal to 0.

$$\frac{dx}{d\theta} = \cos^2 \theta - \sin^2 \theta = \cos 2\theta = 0 \quad \Longrightarrow \quad \theta = \frac{\pi}{4}, \frac{3\pi}{4}$$

$$\frac{dy}{d\theta} = 2 \sin \theta \cos \theta = \sin 2\theta = 0 \quad \Longrightarrow \quad \theta = 0, \frac{\pi}{2}$$

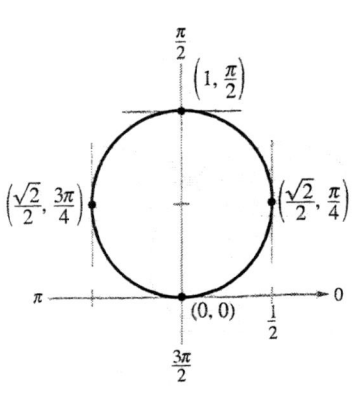

So, the graph has vertical tangent lines at $(\sqrt{2}/2, \pi/4)$ and $(\sqrt{2}/2, 3\pi/4)$, and it has horizontal tangent lines at $(0, 0)$ and $(1, \pi/2)$, as shown in Figure 9.45.

Horizontal and vertical tangent lines of $r = \sin \theta$
Figure 9.45

Example 6 Finding Horizontal and Vertical Tangent Lines

Find the horizontal and vertical tangents to the graph of $r = 2(1 - \cos \theta)$.

Solution Using $y = r \sin \theta$, differentiate and set $dy/d\theta$ equal to 0.

$$y = r \sin \theta = 2(1 - \cos \theta) \sin \theta$$

$$\frac{dy}{d\theta} = 2[(1 - \cos \theta)(\cos \theta) + \sin \theta(\sin \theta)]$$

$$= -2(2 \cos \theta + 1)(\cos \theta - 1) = 0$$

So, $\cos \theta = -\frac{1}{2}$ and $\cos \theta = 1$, and you can conclude that $dy/d\theta = 0$ when $\theta = 2\pi/3, 4\pi/3$, and 0. Similarly, using $x = r \cos \theta$, you have

$$x = r \cos \theta = 2 \cos \theta - 2 \cos^2 \theta$$

$$\frac{dx}{d\theta} = -2 \sin \theta + 4 \cos \theta \sin \theta = 2 \sin \theta(2 \cos \theta - 1) = 0.$$

So, $\sin \theta = 0$ or $\cos \theta = \frac{1}{2}$, and you can conclude that $dx/d\theta = 0$ when $\theta = 0$, π, $\pi/3$, and $5\pi/3$. From these results, and from the graph shown in Figure 9.46, you can conclude that the graph has horizontal tangents at $(3, 2\pi/3)$ and $(3, 4\pi/3)$, and has vertical tangents at $(1, \pi/3)$, $(1, 5\pi/3)$, and $(4, \pi)$. This graph is called a **cardioid**. Note that both derivatives $(dy/d\theta$ and $dx/d\theta)$ are 0 when $\theta = 0$. Using this information alone, you don't know whether the graph has a horizontal or vertical tangent line at the pole. From Figure 9.46, however, you can see that the graph has a cusp at the pole.

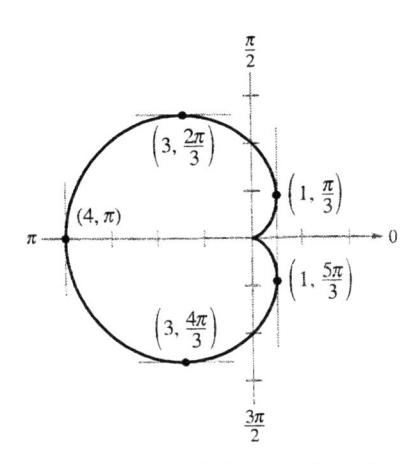

Horizontal and vertical tangent lines of $r = 2(1 - \cos \theta)$
Figure 9.46

Theorem 9.11 has an important consequence. Suppose the graph of $r = f(\theta)$ passes through the pole when $\theta = \alpha$ and $f'(\alpha) \neq 0$. Then the formula for dy/dx simplifies as follows.

$$\frac{dy}{dx} = \frac{f'(\alpha) \sin \alpha + f(\alpha) \cos \alpha}{f'(\alpha) \cos \alpha - f(\alpha) \sin \alpha} = \frac{f'(\alpha) \sin \alpha + 0}{f'(\alpha) \cos \alpha - 0} = \frac{\sin \alpha}{\cos \alpha} = \tan \alpha$$

So, the line $\theta = \alpha$ is tangent to the graph at the pole, $(0, \alpha)$.

THEOREM 9.12 Tangent Lines at the Pole

If $f(\alpha) = 0$ and $f'(\alpha) \neq 0$, then the line $\theta = \alpha$ is tangent at the pole to the graph of $r = f(\theta)$.

Theorem 9.12 is useful because it states that the zeros of $r = f(\theta)$ can be used to find the tangent lines at the pole. Note that because a polar curve can cross the pole more than once, it can have more than one tangent line at the pole. For example, the rose curve

$$f(\theta) = 2 \cos 3\theta$$

$f(\theta) = 2 \cos 3\theta$

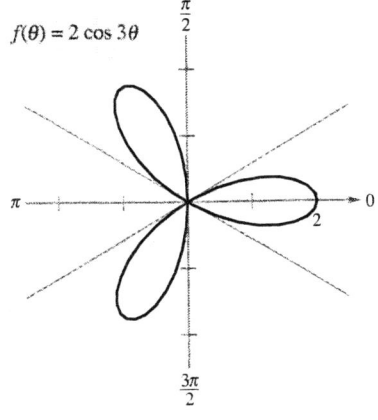

This rose curve has three tangent lines $(\theta = \pi/6, \theta = \pi/2,$ and $\theta = 5\pi/6)$ at the pole.
Figure 9.47

has three tangent lines at the pole, as shown in Figure 9.47. For this curve, $f(\theta) = 2 \cos 3\theta$ is 0 when θ is $\pi/6$, $\pi/2$, and $5\pi/6$. Moreover, the derivative $f'(\theta) = -6 \sin 3\theta$ is not 0 for these values of θ.

Special Polar Graphs

Several important types of graphs have equations that are simpler in polar form than in rectangular form. For example, the polar equation of a circle having a radius of a and centered at the origin is simply $r = a$. Later in the text you will come to appreciate this benefit. For now, we summarize some other types of graphs that have simpler equations in polar form. (Conics are considered in Section 9.6.)

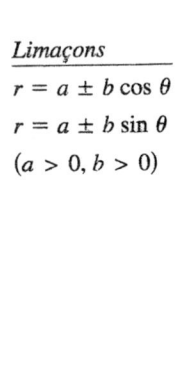

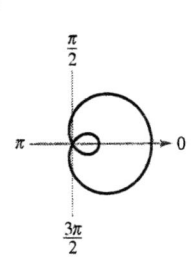

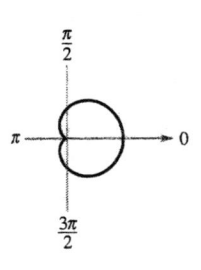

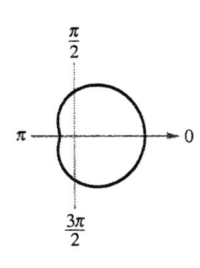

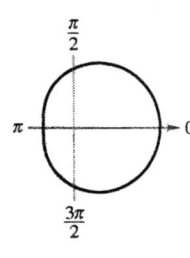

Limaçons
$r = a \pm b \cos \theta$
$r = a \pm b \sin \theta$
$(a > 0, b > 0)$

$\dfrac{a}{b} < 1$	$\dfrac{a}{b} = 1$	$1 < \dfrac{a}{b} < 2$	$\dfrac{a}{b} \geq 2$
Limaçon with inner loop	Cardioid (heart-shaped)	Dimpled limaçon	Convex limaçon

Rose Curves

n petals if n is odd
$2n$ petals if n is even
$(n \geq 2)$

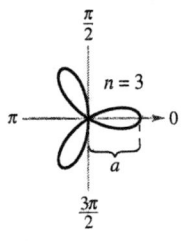

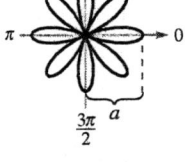

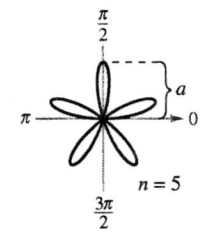

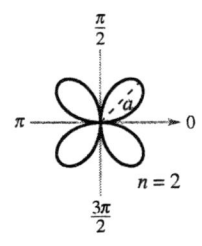

$r = a \cos n\theta$	$r = a \cos n\theta$	$r = a \sin n\theta$	$r = a \sin n\theta$
Rose curve ($n=3$)	Rose curve ($n=4$)	Rose curve ($n=5$)	Rose curve ($n=2$)

Circles and Lemniscates

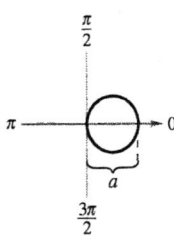

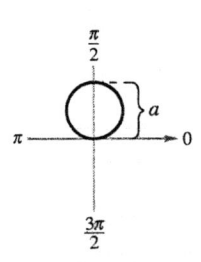

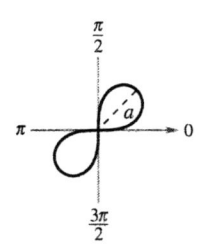

$r = a \cos \theta$	$r = a \sin \theta$	$r^2 = a^2 \sin 2\theta$	$r^2 = a^2 \cos 2\theta$
Circle	Circle	Lemniscate	Lemniscate

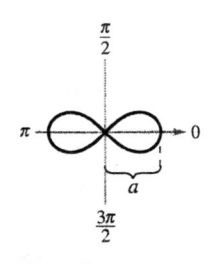

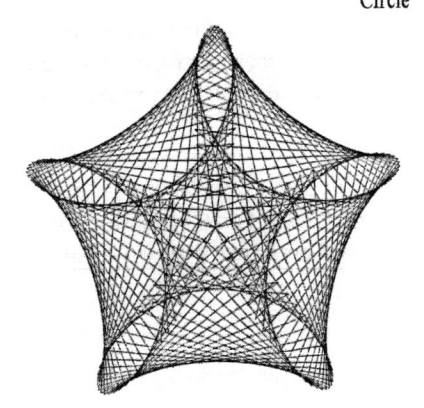

TECHNOLOGY The rose curves described above are of the form $r = a \cos n\theta$ or $r = a \sin n\theta$, where n is a positive integer that is greater than or equal to 2. Try using a graphing utility to sketch the graph of $r = a \cos n\theta$ or $r = a \sin n\theta$ for some noninteger values of n. Are these graphs also rose curves? For example, try sketching the graph of $r = \cos \frac{2}{3}\theta$, $0 \leq \theta \leq 6\pi$.

FOR FURTHER INFORMATION For more information on rose curves and related curves, see the article "A Rose is a Rose . . ." by Peter M. Maurer in *The American Mathematical Monthly*. To view this article, go to the website *www.matharticles.com*. (The computer-generated graph at the left is the result of an algorithm that Maurer calls "The Rose.")

EXERCISES FOR SECTION 9.4

In Exercises 1–6, plot the point in polar coordinates and find the corresponding rectangular coordinates for the point.

1. $(4, 3\pi/6)$
2. $(-2, 7\pi/4)$
3. $(-4, -\pi/3)$
4. $(0, -7\pi/6)$
5. $(\sqrt{2}, 2.36)$
6. $(-3, -1.57)$

In Exercises 7–10, use the *angle* feature of a graphing utility to find the rectangular coordinates for the point given in polar coordinates. Plot the point.

7. $(5, 3\pi/4)$
8. $(-2, 11\pi/6)$
9. $(-3.5, 2.5)$
10. $(8.25, 1.3)$

In Exercises 11–14, the rectangular coordinates of a point are given. Plot the point and find *two* sets of polar coordinates for the point for $0 \le \theta < 2\pi$.

11. $(1, 1)$
12. $(0, -5)$
13. $(-3, 4)$
14. $(4, -2)$

In Exercises 15–18, use the *angle* feature of a graphing utility to find one set of polar coordinates for the point given in rectangular coordinates.

15. $(3, -2)$
16. $(3\sqrt{2}, 3\sqrt{2})$
17. $\left(\frac{5}{2}, \frac{4}{3}\right)$
18. $(0, -5)$

19. Plot the point $(4, 3.5)$ if the point is given in (a) rectangular coordinates and (b) polar coordinates.

20. *Graphical Reasoning*

    (a) Set the window format of a graphing utility to rectangular coordinates and locate the cursor at any position off the coordinate axes. Move the cursor horizontally and describe any changes in the displayed coordinates of the points. Repeat the process moving the cursor vertically.

    (b) Set the window format of a graphing utility to polar coordinates and locate the cursor at any position off the coordinate axes. Move the cursor horizontally and describe any changes in the displayed coordinates of the points. Repeat the process moving the cursor vertically.

    (c) Why are the results in parts (a) and (b) different?

In Exercises 21–28, convert the rectangular equation to polar form and sketch its graph.

21. $x^2 + y^2 = a^2$
22. $x^2 + y^2 - 2ax = 0$
23. $y = 4$
24. $x = 10$
25. $3x - y + 2 = 0$
26. $xy = 4$
27. $y^2 = 9x$
28. $(x^2 + y^2)^2 - 9(x^2 - y^2) = 0$

In Exercises 29–36, convert the polar equation to rectangular form and sketch its graph.

29. $r = 3$
30. $r = -2$
31. $r = \sin\theta$
32. $r = 5\cos\theta$
33. $r = \theta$
34. $\theta = \frac{5\pi}{6}$
35. $r = 3\sec\theta$
36. $r = 2\csc\theta$

In Exercises 37–46, use a graphing utility to graph the polar equation. Find an interval for θ over which the graph is traced only once.

37. $r = 3 - 4\cos\theta$
38. $r = 5(1 - 2\sin\theta)$
39. $r = 2 + \sin\theta$
40. $r = 4 + 3\cos\theta$
41. $r = \dfrac{2}{1 + \cos\theta}$
42. $r = \dfrac{2}{4 - 3\sin\theta}$
43. $r = 2\cos\left(\dfrac{3\theta}{2}\right)$
44. $r = 3\sin\left(\dfrac{5\theta}{2}\right)$
45. $r^2 = 4\sin 2\theta$
46. $r^2 = \dfrac{1}{\theta}$

47. Convert the equation

    $r = 2(h\cos\theta + k\sin\theta)$

    to rectangular form and verify that it is the equation of a circle. Find the radius and the rectangular coordinates of the center of the circle.

48. *Distance Formula*

    (a) Verify that the Distance Formula for the distance between the two points (r_1, θ_1) and (r_2, θ_2) in polar coordinates is

    $d = \sqrt{r_1^2 + r_2^2 - 2r_1 r_2 \cos(\theta_1 - \theta_2)}.$

    (b) Describe the position of the points relative to each other if $\theta_1 = \theta_2$. Simplify the Distance Formula for this case. Is the simplification what you expected? Explain.

    (c) Simplify the Distance Formula if $\theta_1 - \theta_2 = 90°$. Is the simplification what you expected? Explain.

    (d) Choose two points on the polar coordinate system and find the distance between them. Then choose different polar representations of the same two points and apply the Distance Formula again. Discuss the result.

In Exercises 49–52, use the result of Exercise 48 to approximate the distance between the two points in polar coordinates.

49. $\left(4, \dfrac{2\pi}{3}\right), \left(2, \dfrac{\pi}{6}\right)$
50. $\left(10, \dfrac{7\pi}{6}\right), (3, \pi)$
51. $(2, 0.5), (7, 1.2)$
52. $(4, 2.5), (12, 1)$

692 CHAPTER 9 Conics, Parametric Equations, and Polar Coordinates

In Exercises 53 and 54, find dy/dx and the slope of the tangent lines shown on the graph of the polar equation.

53. $r = 2 + 3 \sin \theta$

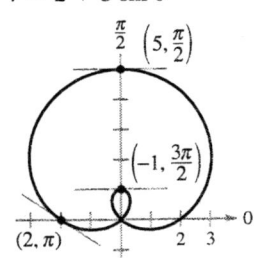

54. $r = 2(1 - \sin \theta)$

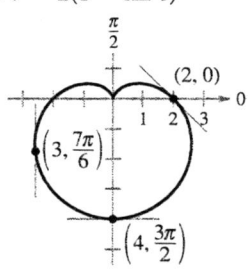

In Exercises 55–58, use a graphing utility to (a) graph the polar equation, (b) draw the tangent line at the given value of θ, and (c) find dy/dx at the given value of θ. (*Hint:* Let the increment between the values of θ equal $\pi/24$.)

55. $r = 3(1 - \cos \theta)$, $\theta = \dfrac{\pi}{2}$

56. $r = 3 - 2 \cos \theta$, $\theta = 0$

57. $r = 3 \sin \theta$, $\theta = \dfrac{\pi}{3}$

58. $r = 4$, $\theta = \dfrac{\pi}{4}$

In Exercises 59 and 60, find the points of horizontal and vertical tangency (if any) to the polar curve.

59. $r = 1 - \sin \theta$

60. $r = a \sin \theta$

In Exercises 61 and 62, find the points of horizontal tangency (if any) to the polar curve.

61. $r = 2 \csc \theta + 3$

62. $r = a \sin \theta \cos^2 \theta$

In Exercises 63–66, use a graphing utility to graph the polar equation and find all points of horizontal tangency.

63. $r = 4 \sin \theta \cos^2 \theta$

64. $r = 3 \cos 2\theta \sec \theta$

65. $r = 2 \csc \theta + 5$

66. $r = 2 \cos(3\theta - 2)$

In Exercises 67–74, sketch the graph of the polar equation and find the tangents at the pole.

67. $r = 3 \sin \theta$

68. $r = 3 \cos \theta$

69. $r = 2(1 - \sin \theta)$

70. $r = 3(1 - \cos \theta)$

71. $r = 2 \cos 3\theta$

72. $r = -\sin 5\theta$

73. $r = 3 \sin 2\theta$

74. $r = 3 \cos 2\theta$

In Exercises 75–86, sketch the graph of the polar equation.

75. $r = 5$

76. $r = 2$

77. $r = 4(1 + \cos \theta)$

78. $r = 1 + \sin \theta$

79. $r = 3 - 2 \cos \theta$

80. $r = 5 - 4 \sin \theta$

81. $r = 3 \csc \theta$

82. $r = \dfrac{6}{2 \sin \theta - 3 \cos \theta}$

83. $r = 2\theta$

84. $r = \dfrac{1}{\theta}$

85. $r^2 = 4 \cos 2\theta$

86. $r^2 = 4 \sin \theta$

In Exercises 87–90, use a graphing utility to graph the equation and show that the indicated line is an asymptote of the graph.

Name of Graph	Polar Equation	Asymptote
87. Conchoid	$r = 2 - \sec \theta$	$x = -1$
88. Conchoid	$r = 2 + \csc \theta$	$y = 1$
89. Hyperbolic spiral	$r = 2/\theta$	$y = 2$
90. Strophoid	$r = 2 \cos 2\theta \sec \theta$	$x = -2$

Getting at the Concept

91. In your own words, describe the differences between the rectangular coordinate system and the polar coordinate system.

92. Give the equations for the coordinate conversion from rectangular to polar coordinates and vice versa.

93. For constants a and b, describe the graphs of the equations $r = a$ and $\theta = b$ in polar coordinates.

94. How are the slopes of tangent lines determined in polar coordinates? What are tangent lines at the pole and how are they determined?

In Exercises 95–98, match the graph with its polar equation. [The graphs are labeled (a), (b), (c), and (d).]

(a) (b) (c) 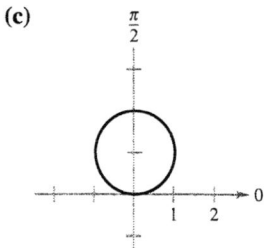 (d)

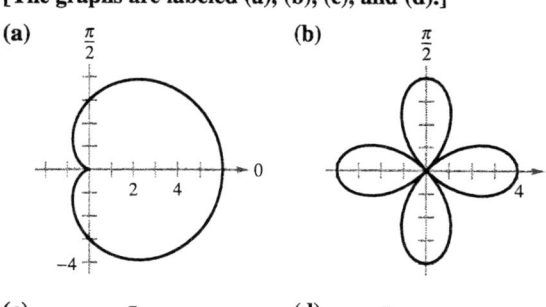

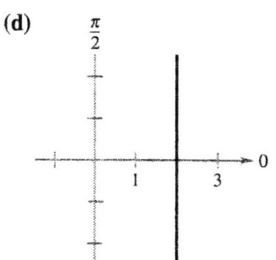

95. $r = 2 \sin \theta$

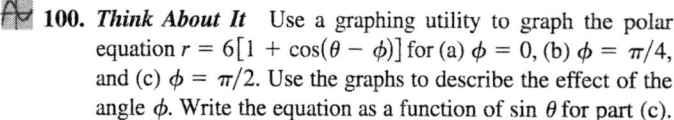

96. $r = 4 \cos 2\theta$

97. $r = 3(1 + \cos \theta)$

98. $r = 2 \sec \theta$

99. Sketch the graph of $r = 4 \sin \theta$ over each interval.

(a) $0 \leq \theta \leq \dfrac{\pi}{2}$ (b) $\dfrac{\pi}{2} \leq \theta \leq \pi$ (c) $-\dfrac{\pi}{2} \leq \theta \leq \dfrac{\pi}{2}$

100. **Think About It** Use a graphing utility to graph the polar equation $r = 6[1 + \cos(\theta - \phi)]$ for (a) $\phi = 0$, (b) $\phi = \pi/4$, and (c) $\phi = \pi/2$. Use the graphs to describe the effect of the angle ϕ. Write the equation as a function of $\sin \theta$ for part (c).

101. Verify that if the curve whose polar equation is $r = f(\theta)$ is rotated about the pole through an angle ϕ, then an equation for the rotated curve is $r = f(\theta - \phi)$.

102. The polar form of an equation for a curve is $r = f(\sin \theta)$. Show that the form becomes
    (a) $r = f(-\cos \theta)$ if the curve is rotated counterclockwise $\pi/2$ radians about the pole.
    (b) $r = f(-\sin \theta)$ if the curve is rotated counterclockwise π radians about the pole.
    (c) $r = f(\cos \theta)$ if the curve is rotated counterclockwise $3\pi/2$ radians about the pole.

In Exercises 103–106, use the results of Exercises 101 and 102.

103. Write an equation for the limaçon $r = 2 - \sin \theta$ after it has been rotated by the given amount. Verify the results by using a graphing utility to graph the rotated limaçon.
    (a) $\dfrac{\pi}{4}$ (b) $\dfrac{\pi}{2}$ (c) π (d) $\dfrac{3\pi}{2}$

104. Write an equation for the rose curve $r = 2 \sin 2\theta$ after it has been rotated by the given amount. Verify the results by using a graphing utility to graph the rotated rose curve.
    (a) $\dfrac{\pi}{6}$ (b) $\dfrac{\pi}{2}$ (c) $\dfrac{2\pi}{3}$ (d) π

105. Sketch the graph of each equation.
    (a) $r = 1 - \sin \theta$ (b) $r = 1 - \sin\left(\theta - \dfrac{\pi}{4}\right)$

106. Prove that the tangent of the angle ψ ($0 \le \psi \le \pi/2$) between the radial line and the tangent line at the point (r, θ) on the graph of $r = f(\theta)$ (see figure) is given by $\tan \psi = |r/(dr/d\theta)|$.

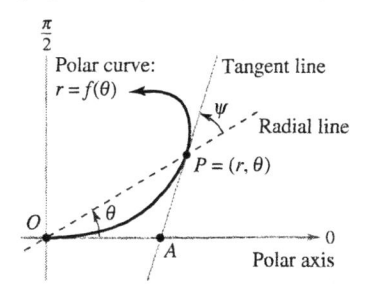

In Exercises 107–112, use the result of Exercise 106 to find the angle ψ between the radial and tangent lines to the graph for the indicated value of θ. Use a graphing utility to graph the polar equation, the radial line, and the tangent line for the indicated value of θ. Identify the angle ψ.

Polar Equation	Value of θ
107. $r = 2(1 - \cos \theta)$	$\theta = \pi$
108. $r = 3(1 - \cos \theta)$	$\theta = 3\pi/4$
109. $r = 2 \cos 3\theta$	$\theta = \pi/6$
110. $r = 4 \sin 2\theta$	$\theta = \pi/6$
111. $r = \dfrac{6}{1 - \cos \theta}$	$\theta = 2\pi/3$
112. $r = 5$	$\theta = \pi/6$

True or False? **In Exercises 113–116, determine whether the statement is true or false. If it is false, explain why or give an example that shows it is false.**

113. If (r_1, θ_1) and (r_2, θ_2) represent the same point on the polar coordinate system, then $|r_1| = |r_2|$.

114. If (r, θ_1) and (r, θ_2) represent the same point on the polar coordinate system, then $\theta_1 = \theta_2 + 2\pi n$ for some integer n.

115. If $x > 0$, then the point (x, y) on the rectangular coordinate system can be represented by (r, θ) on the polar coordinate system, where $r = \sqrt{x^2 + y^2}$ and $\theta = \arctan(y/x)$.

116. The polar equations $r = \sin 2\theta$ and $r = -\sin 2\theta$ have the same graph.

SECTION PROJECT: ANAMORPHIC ART

Use the anamorphic transformations

$$r = y + 16 \quad \text{and} \quad \theta = -\dfrac{\pi}{8}x, \quad -\dfrac{3\pi}{4} \le \theta \le \dfrac{3\pi}{4}$$

to sketch the transformed polar image of the rectangular graph. When the reflection (in a cylindrical mirror centered at the pole) of each polar image is viewed from the polar axis, the viewer will see the original rectangular image.
(a) $y = 3$ (b) $x = 2$
(c) $y = x + 5$ (d) $x^2 + (y - 5)^2 = 5^2$

This example of anamorphic art is from the Museum of Science and Industry in Manchester, England. When the reflection of the transformed "polar painting" is viewed in the mirror, the viewer sees faces.

FOR FURTHER INFORMATION For more information on anamorphic art, see the article "Anamorphisms" by Philip Hickin in the *Mathematical Gazette*. To view this article, go to the website *www.matharticles.com*.

Section 9.5 Area and Arc Length in Polar Coordinates

- Find the area of a region bounded by a polar graph.
- Find the points of intersection of two polar graphs.
- Find the arc length of a polar graph.
- Find the area of a surface of revolution (polar form).

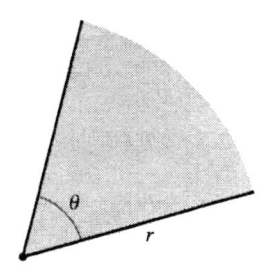

Area of a Polar Region

The development of a formula for the area of a polar region parallels that for the area of a region on the rectangular coordinate system, but uses *sectors* of a circle instead of rectangles as the basic element of area. In Figure 9.48, note that the area of a circular sector of radius r is given by $\frac{1}{2}\theta r^2$, provided θ is measured in radians.

Consider the function given by $r = f(\theta)$, where f is continuous and nonnegative in the interval given by $\alpha \leq \theta \leq \beta$. The region bounded by the graph of f and the radial lines $\theta = \alpha$ and $\theta = \beta$ is shown in Figure 9.49. To find the area of this region, partition the interval $[\alpha, \beta]$ into n equal subintervals,

$$\alpha = \theta_0 < \theta_1 < \theta_2 < \cdots < \theta_{n-1} < \theta_n = \beta.$$

Then, approximate the area of the region by the sum of the areas of the n sectors.

Radius of ith sector $= f(\theta_i)$

Central angle of ith sector $= \dfrac{\beta - \alpha}{n} = \Delta\theta$

$$A \approx \sum_{i=1}^{n} \left(\frac{1}{2}\right) \Delta\theta [f(\theta_i)]^2$$

Taking the limit as $n \to \infty$ produces

$$A = \lim_{n \to \infty} \frac{1}{2} \sum_{i=1}^{n} [f(\theta_i)]^2 \Delta\theta$$

$$= \frac{1}{2} \int_{\alpha}^{\beta} [f(\theta)]^2 \, d\theta$$

which leads to the following theorem.

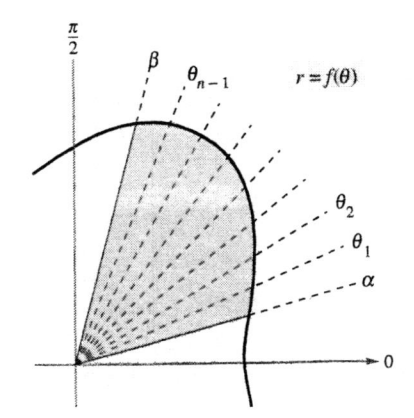

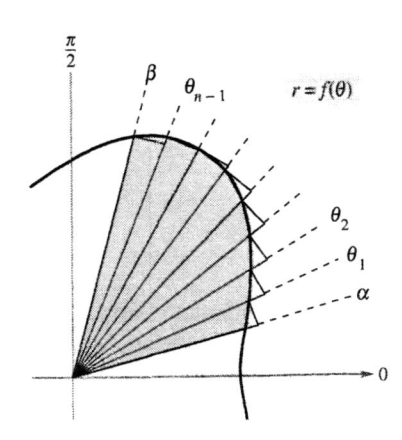

The area of a sector of a circle is $A = \frac{1}{2}\theta r^2$.
Figure 9.48

Figure 9.49

THEOREM 9.13 Area in Polar Coordinates

If f is continuous and nonnegative on the interval $[\alpha, \beta]$, $0 < \beta - \alpha \leq 2\pi$, then the area of the region bounded by the graph of $r = f(\theta)$ between the radial lines $\theta = \alpha$ and $\theta = \beta$ is given by

$$A = \frac{1}{2} \int_{\alpha}^{\beta} [f(\theta)]^2 \, d\theta$$

$$= \frac{1}{2} \int_{\alpha}^{\beta} r^2 \, d\theta. \qquad 0 < \beta - \alpha \leq 2\pi$$

NOTE You can use the same formula to find the area of a region bounded by the graph of a continuous *nonpositive* function. However, the formula is not necessarily valid if f takes on both positive *and* negative values in the interval $[\alpha, \beta]$.

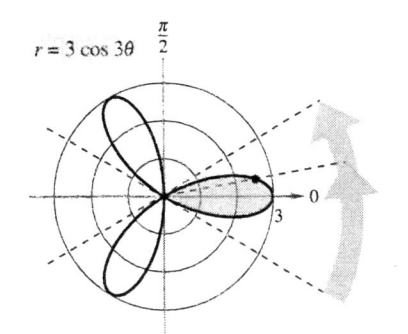

SECTION 9.5 Area and Arc Length in Polar Coordinates 695

Example 1 Finding the Area of a Polar Region

Find the area of *one petal* of the rose curve given by $r = 3\cos 3\theta$.

Solution In Figure 9.50, you can see that the right petal is traced as θ increases from $-\pi/6$ to $\pi/6$. So, the area is

$$A = \frac{1}{2}\int_\alpha^\beta r^2\,d\theta = \frac{1}{2}\int_{-\pi/6}^{\pi/6}(3\cos 3\theta)^2\,d\theta \qquad \text{Formula for area in polar coordinates}$$

$$= \frac{9}{2}\int_{-\pi/6}^{\pi/6}\frac{1+\cos 6\theta}{2}\,d\theta \qquad \text{Trigonometric identity}$$

$$= \frac{9}{4}\left[\theta + \frac{\sin 6\theta}{6}\right]_{-\pi/6}^{\pi/6}$$

$$= \frac{9}{4}\left(\frac{\pi}{6} + \frac{\pi}{6}\right)$$

$$= \frac{3\pi}{4}.$$

$r = 3\cos 3\theta$

The area of one petal of the rose curve that lies between the radial lines $\theta = -\pi/6$ and $\theta = \pi/6$ is $3\pi/4$.
Figure 9.50

NOTE: To find the area of the region lying inside all three petals of the rose curve in Example 1, you could not simply integrate between 0 and 2π. In doing this you would obtain $9\pi/2$, which is twice the area of the three petals—the duplication occurs because the rose curve is traced *twice* as θ increases from 0 to 2π.

Example 2 Finding the Area Bounded by a Single Curve

Find the area of the region lying between the inner and outer loops of the limaçon $r = 1 - 2\sin\theta$.

Solution In Figure 9.51, note that the inner loop is traced as θ increases from $\pi/6$ to $5\pi/6$. So, the area inside the *inner loop* is

$$A_1 = \frac{1}{2}\int_\alpha^\beta r^2\,d\theta = \frac{1}{2}\int_{\pi/6}^{5\pi/6}(1-2\sin\theta)^2\,d\theta \qquad \text{Formula for area in polar coordinates}$$

$$= \frac{1}{2}\int_{\pi/6}^{5\pi/6}(1-4\sin\theta+4\sin^2\theta)\,d\theta$$

$$= \frac{1}{2}\int_{\pi/6}^{5\pi/6}\left[1-4\sin\theta+4\left(\frac{1-\cos 2\theta}{2}\right)\right]d\theta \qquad \text{Trigonometric identity}$$

$$= \frac{1}{2}\int_{\pi/6}^{5\pi/6}(3-4\sin\theta-2\cos 2\theta)\,d\theta \qquad \text{Simplify.}$$

$$= \frac{1}{2}\Big[3\theta + 4\cos\theta - \sin 2\theta\Big]_{\pi/6}^{5\pi/6}$$

$$= \frac{1}{2}(2\pi - 3\sqrt{3})$$

$$= \pi - \frac{3\sqrt{3}}{2}.$$

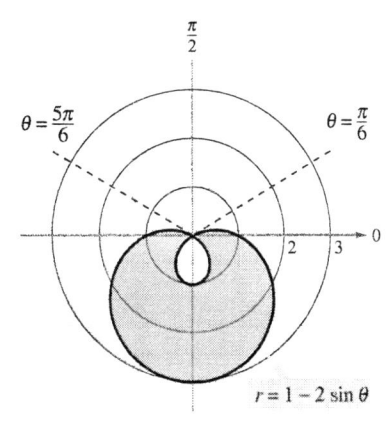

$\theta = \frac{5\pi}{6}$, $\theta = \frac{\pi}{6}$

$r = 1 - 2\sin\theta$

The area between the inner and outer loops is approximately 8.34.
Figure 9.51

In a similar way, you can integrate from $5\pi/6$ to $13\pi/6$ to find that the area of the region lying inside the *outer loop* is $A_2 = 2\pi + (3\sqrt{3}/2)$. The area of the region lying between the two loops is the difference of A_2 and A_1.

$$A = A_2 - A_1 = \left(2\pi + \frac{3\sqrt{3}}{2}\right) - \left(\pi - \frac{3\sqrt{3}}{2}\right) = \pi + 3\sqrt{3} \approx 8.34$$

Points of Intersection of Polar Graphs

Because a point may be represented in different ways in polar coordinates, care must be taken in determining the points of intersection of two polar graphs. For example, consider the points of intersection of the graphs of

$$r = 1 - 2\cos\theta \quad \text{and} \quad r = 1$$

as shown in Figure 9.52. If, as with rectangular equations, you attempted to find the points of intersection by solving the two equations simultaneously, you would obtain the following.

$r = 1 - 2\cos\theta$ First equation
$1 = 1 - 2\cos\theta$ Substitute $r = 1$ from 2nd equation into 1st equation.
$\cos\theta = 0$ Simplify.
$\theta = \dfrac{\pi}{2}, \dfrac{3\pi}{2}$ Solve for θ.

The corresponding points of intersection are $(1, \pi/2)$ and $(1, 3\pi/2)$. However, from Figure 9.52 you can see that there is a *third* point of intersection that did not show up when the two polar equations were solved simultaneously. (This is one reason we stress sketching a graph when finding the area of a polar region.) The reason the third point was not found is that it does not occur with the same coordinates in the two graphs. On the graph of $r = 1$, the point occurs with coordinates $(1, \pi)$, but on the graph of $r = 1 - 2\cos\theta$, the point occurs with coordinates $(-1, 0)$.

You can compare the problem of finding points of intersection of two polar graphs with that of finding collision points of two satellites in intersecting orbits about earth, as shown in Figure 9.53. The satellites will not collide as long as they reach the points of intersection at different times (θ-values). A collision will occur only at the points of intersection that are "simultaneous points"—those reached at the same time (θ-value).

NOTE   Because the pole can be represented by $(0, \theta)$, where θ is *any* angle, you should check separately for the pole when hunting for points of intersection.

FOR FURTHER INFORMATION For more information on using technology to find points of intersection, see the article "Finding Points of Intersection of Polar-Coordinate Graphs" by Warren W. Esty in *Mathematics Teacher*. To view this article, go to the website *www.matharticles.com*.

Limaçon: $r = 1 - 2\cos\theta$
Circle: $r = 1$

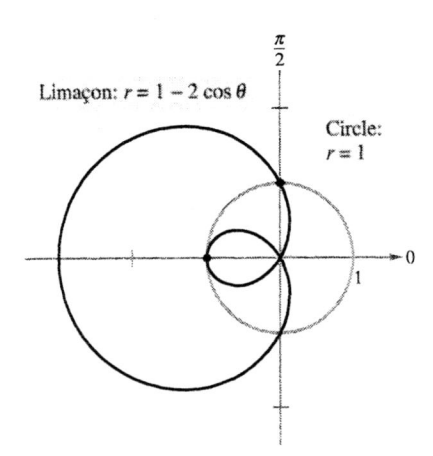

Three points of intersection: $(1, \pi/2)$, $(-1, 0), (1, 3\pi/2)$
Figure 9.52

The paths of satellites can cross without causing a collision.
Figure 9.53

Example 3 Finding the Area of a Region Between Two Curves

Find the area of the region common to the two regions bounded by the following curves.

$r = -6 \cos \theta$ Circle

$r = 2 - 2 \cos \theta$ Cardioid

Solution Because both curves are symmetric with respect to the x-axis, you can work with the upper half-plane, as shown in Figure 9.54. The gray shaded region lies between the circle and the radial line $\theta = 2\pi/3$. Because the circle has coordinates $(0, \pi/2)$ at the pole, you can integrate between $\pi/2$ and $2\pi/3$ to obtain the area of this region. The region that is shaded red is bounded by the radial lines $\theta = 2\pi/3$ and $\theta = \pi$ and the cardioid. So, you can find the area of this second region by integrating between $2\pi/3$ and π. The sum of these two integrals gives the area of the common region lying *above* the radial line $\theta = \pi$.

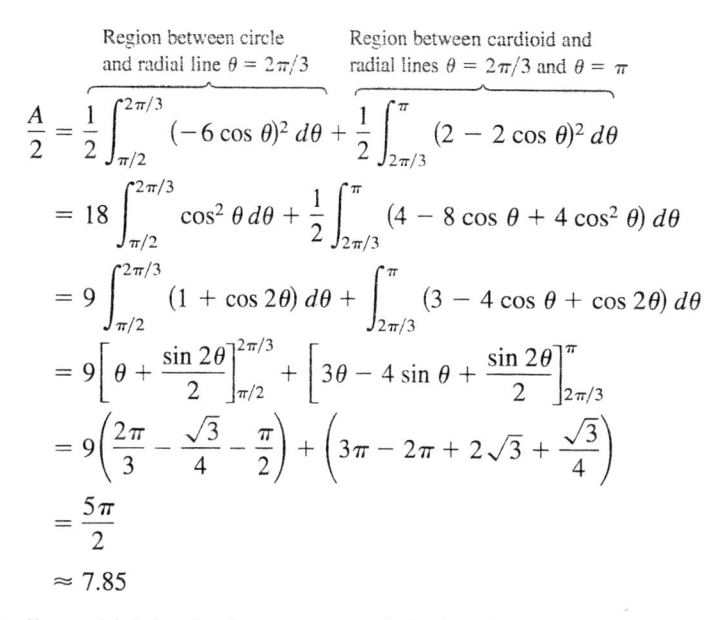

$$\frac{A}{2} = \overbrace{\frac{1}{2}\int_{\pi/2}^{2\pi/3} (-6 \cos \theta)^2 \, d\theta}^{\text{Region between circle and radial line } \theta = 2\pi/3} + \overbrace{\frac{1}{2}\int_{2\pi/3}^{\pi} (2 - 2 \cos \theta)^2 \, d\theta}^{\text{Region between cardioid and radial lines } \theta = 2\pi/3 \text{ and } \theta = \pi}$$

$$= 18 \int_{\pi/2}^{2\pi/3} \cos^2 \theta \, d\theta + \frac{1}{2}\int_{2\pi/3}^{\pi} (4 - 8 \cos \theta + 4 \cos^2 \theta) \, d\theta$$

$$= 9 \int_{\pi/2}^{2\pi/3} (1 + \cos 2\theta) \, d\theta + \int_{2\pi/3}^{\pi} (3 - 4 \cos \theta + \cos 2\theta) \, d\theta$$

$$= 9 \left[\theta + \frac{\sin 2\theta}{2} \right]_{\pi/2}^{2\pi/3} + \left[3\theta - 4 \sin \theta + \frac{\sin 2\theta}{2} \right]_{2\pi/3}^{\pi}$$

$$= 9 \left(\frac{2\pi}{3} - \frac{\sqrt{3}}{4} - \frac{\pi}{2} \right) + \left(3\pi - 2\pi + 2\sqrt{3} + \frac{\sqrt{3}}{4} \right)$$

$$= \frac{5\pi}{2}$$

$$\approx 7.85$$

Finally, multiplying by 2, you can conclude that the total area is 5π.

NOTE To check the reasonableness of the result obtained in Example 3, note that the area of the circular region is $\pi r^2 = 9\pi$. So, it seems reasonable that the area of the region lying inside the circle and the cardioid is 5π.

To see the benefit of polar coordinates for finding the area in Example 3, consider the following integral, which gives the comparable area in rectangular coordinates.

$$\frac{A}{2} = \int_{-4}^{-3/2} \sqrt{2\sqrt{1 - 2x} - x^2 - 2x + 2} \, dx + \int_{-3/2}^{0} \sqrt{-x^2 - 6x} \, dx$$

Try using the integration capabilities of a graphing utility to show that you obtain the same area as that found in Example 3.

Circle: $r = -6 \cos \theta$

Cardioid: $r = 2 - 2 \cos \theta$

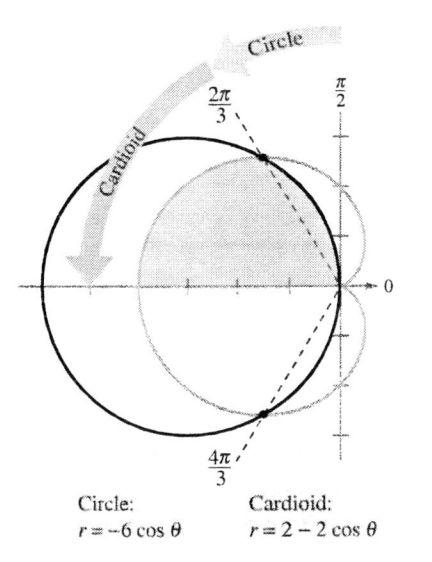

Figure 9.54

Arc Length in Polar Form

The formula for the length of a polar arc can be obtained from the arc length formula for a curve described by parametric equations. (See Exercise 65.)

THEOREM 9.14 Arc Length of a Polar Curve

Let f be a function whose derivative is continuous on an interval $\alpha \leq \theta \leq \beta$. The length of the graph of $r = f(\theta)$ from $\theta = \alpha$ to $\theta = \beta$ is

$$s = \int_\alpha^\beta \sqrt{[f(\theta)]^2 + [f'(\theta)]^2}\, d\theta = \int_\alpha^\beta \sqrt{r^2 + \left(\frac{dr}{d\theta}\right)^2}\, d\theta.$$

NOTE When applying the arc length formula to a polar curve, be sure that the curve is traced out only once on the interval of integration. For instance, the rose curve given by $r = \cos 3\theta$ is traced out once on the interval $0 \leq \theta \leq \pi$, but is traced out twice on the interval $0 \leq \theta \leq 2\pi$.

Example 4 Finding the Length of a Polar Curve

Find the length of the arc from $\theta = 0$ to $\theta = 2\pi$ for the cardioid

$$r = f(\theta) = 2 - 2\cos\theta$$

as shown in Figure 9.55.

Solution Because $f'(\theta) = 2\sin\theta$, you can find the arc length as follows.

$$s = \int_\alpha^\beta \sqrt{[f(\theta)]^2 + [f'(\theta)]^2}\, d\theta \qquad \text{Formula for arc length of a polar curve}$$

$$= \int_0^{2\pi} \sqrt{(2 - 2\cos\theta)^2 + (2\sin\theta)^2}\, d\theta$$

$$= 2\sqrt{2} \int_0^{2\pi} \sqrt{1 - \cos\theta}\, d\theta \qquad \text{Simplify.}$$

$$= 2\sqrt{2} \int_0^{2\pi} \sqrt{2\sin^2\frac{\theta}{2}}\, d\theta \qquad \text{Trigonometric identity}$$

$$= 4 \int_0^{2\pi} \sin\frac{\theta}{2}\, d\theta \qquad \sin\frac{\theta}{2} \geq 0 \text{ for } 0 \leq \theta \leq 2\pi$$

$$= 8\left[-\cos\frac{\theta}{2}\right]_0^{2\pi}$$

$$= 8(1 + 1)$$

$$= 16$$

In the fifth step of the solution, it is legitimate to write

$$\sqrt{2\sin^2(\theta/2)} = \sqrt{2}\sin(\theta/2)$$

rather than

$$\sqrt{2\sin^2(\theta/2)} = \sqrt{2}|\sin(\theta/2)|$$

because $\sin(\theta/2) \geq 0$ for $0 \leq \theta \leq 2\pi$.

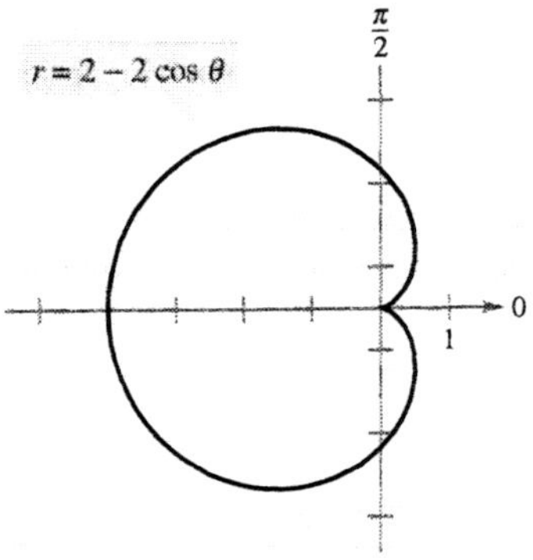

NOTE Using Figure 9.55, you can determine the reasonableness of this answer by comparing it with the circumference of a circle. For example, a circle of radius $\frac{5}{2}$ has a circumference of $5\pi \approx 15.7$.

$r = 2 - 2\cos\theta$

The arc length of this cardioid is 16.
Figure 9.55

Area of a Surface of Revolution

The polar coordinate version of the formulas for the area of a surface of revolution can be obtained from the parametric versions given in Theorem 9.9, using the equations $x = r\cos\theta$ and $y = r\sin\theta$.

THEOREM 9.15 Area of a Surface of Revolution

Let f be a function whose derivative is continuous on an interval $\alpha \leq \theta \leq \beta$. The area of the surface formed by revolving the graph of $r = f(\theta)$ from $\theta = \alpha$ to $\theta = \beta$ about the indicated line is as follows.

1. $S = 2\pi \int_\alpha^\beta f(\theta) \sin\theta \sqrt{[f(\theta)]^2 + [f'(\theta)]^2}\, d\theta \qquad$ About the polar axis

2. $S = 2\pi \int_\alpha^\beta f(\theta) \cos\theta \sqrt{[f(\theta)]^2 + [f'(\theta)]^2}\, d\theta \qquad$ About the line $\theta = \dfrac{\pi}{2}$

NOTE When using Theorem 9.15, check to see that the graph of $r = f(\theta)$ is traced only once on the interval $\alpha \leq \theta \leq \beta$. For example, the circle given by $r = \cos\theta$ is traced once on the interval $0 \leq \theta \leq \pi$.

Example 5   Finding the Area of a Surface of Revolution

Find the area of the surface formed by revolving the circle $r = f(\theta) = \cos\theta$ about the line $\theta = \pi/2$, as shown in Figure 9.56.

(a)

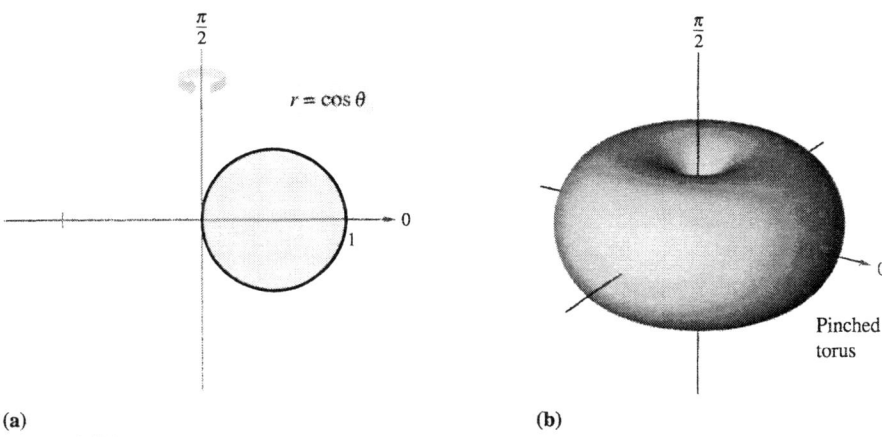

(b) Pinched torus

Figure 9.56

Solution You can use the second formula given in Theorem 9.15 with $f'(\theta) = -\sin\theta$. Because the circle is traced once as θ increases from 0 to π, we have

$$S = 2\pi \int_\alpha^\beta f(\theta) \cos\theta \sqrt{[f(\theta)]^2 + [f'(\theta)]^2}\, d\theta \qquad \text{Formula for area of a surface of revolution}$$

$$= 2\pi \int_0^\pi \cos\theta(\cos\theta)\sqrt{\cos^2\theta + \sin^2\theta}\, d\theta$$

$$= 2\pi \int_0^\pi \cos^2\theta\, d\theta \qquad \text{Trigonometric identity}$$

$$= \pi \int_0^\pi (1 + \cos 2\theta)\, d\theta \qquad \text{Trigonometric identity}$$

$$= \pi\left[\theta + \frac{\sin 2\theta}{2}\right]_0^\pi = \pi^2.$$

EXERCISES FOR SECTION 9.5

In Exercises 1 and 2, find the area of the region bounded by the graph of the polar equation using (a) a geometric formula, and (b) integration.

1. $r = 8 \sin \theta$
2. $r = 3 \cos \theta$

In Exercises 3–8, find the area of the region.

3. One petal of $r = 2 \cos 3\theta$
4. One petal of $r = 6 \sin 2\theta$
5. One petal of $r = \cos 2\theta$
6. One petal of $r = \cos 5\theta$
7. Interior of $r = 1 - \sin \theta$
8. Interior of $r = 1 - \sin \theta$ (above the polar axis)

In Exercises 9–12, use a graphing utility to graph the polar equation and find the area of the indicated region.

9. Inner loop of $r = 1 + 2 \cos \theta$
10. Inner loop of $r = 4 - 6 \sin \theta$
11. Between the loops of $r = 1 + 2 \cos \theta$
12. Between the loops of $r = 2(1 + 2 \sin \theta)$

In Exercises 13–22, find the points of intersection of the graphs of the equations.

13. $r = 1 + \cos \theta$
    $r = 1 - \cos \theta$

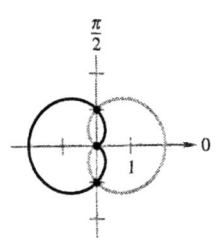

14. $r = 3(1 + \sin \theta)$
    $r = 3(1 - \sin \theta)$

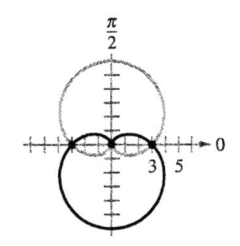

15. $r = 1 + \cos \theta$
    $r = 1 - \sin \theta$

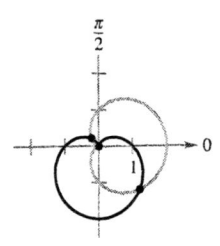

16. $r = 2 - 3 \cos \theta$
    $r = \cos \theta$

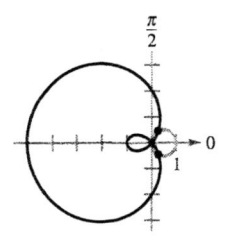

17. $r = 4 - 5 \sin \theta$
    $r = 3 \sin \theta$

18. $r = 1 + \cos \theta$
    $r = 3 \cos \theta$

19. $r = \dfrac{\theta}{2}$
    $r = 2$

20. $\theta = \dfrac{\pi}{4}$
    $r = 2$

21. $r = 4 \sin 2\theta$
    $r = 2$

22. $r = 3 + \sin \theta$
    $r = 2 \csc \theta$

In Exercises 23 and 24, use a graphing utility to approximate the points of intersection of the graphs of the polar equations. Confirm your results analytically.

23. $r = 2 + 3 \cos \theta$
    $r = \dfrac{\sec \theta}{2}$

24. $r = 3(1 - \cos \theta)$
    $r = \dfrac{6}{1 - \cos \theta}$

Writing In Exercises 25 and 26, use a graphing utility to find the points of intersection of the graphs of the polar equations. Watch the graphs as they are traced in the viewing window. Explain why the pole is not a point of intersection obtained by solving the equations simultaneously.

25. $r = \cos \theta$
    $r = 2 - 3 \sin \theta$

26. $r = 4 \sin \theta$
    $r = 2(1 + \sin \theta)$

In Exercises 27–32, use a graphing utility to graph the polar equations and find the area of the indicated region.

27. Common interior of $r = 4 \sin 2\theta$ and $r = 2$
28. Common interior of $r = 3(1 + \sin \theta)$ and $r = 3(1 - \sin \theta)$
29. Common interior of $r = 3 - 2 \sin \theta$ and $r = -3 + 2 \sin \theta$
30. Common interior of $r = 5 - 3 \sin \theta$ and $r = 5 - 3 \cos \theta$
31. Common interior of $r = 4 \sin \theta$ and $r = 2$
32. Inside $r = 3 \sin \theta$ and outside $r = 2 - \sin \theta$

In Exercises 33–36, find the area of the region.

33. Inside $r = a(1 + \cos \theta)$ and outside $r = a \cos \theta$
34. Inside $r = 2a \cos \theta$ and outside $r = a$
35. Common interior of $r = a(1 + \cos \theta)$ and $r = a \sin \theta$
36. Common interior of $r = a \cos \theta$ and $r = a \sin \theta$ where $a > 0$.

37. **Antenna Radiation** The radiation from a transmitting antenna is not uniform in all directions. The intensity from a particular antenna is modeled by

    $r = a \cos^2 \theta.$

    (a) Convert the polar equation to rectangular form.
    (b) Use a graphing utility to graph the model for $a = 4$ and $a = 6$.
    (c) Find the area of the geographical region between the two curves in part (b).

38. **Area** The area inside one or more of the three interlocking circles

    $r = 2a \cos \theta, \quad r = 2a \sin \theta, \quad \text{and} \quad r = a$

    is divided into seven regions. Find the area of each region.

39. **Conjecture** Find the area of the region enclosed by $r = a \cos(n\theta)$ for $n = 1, 2, 3, \ldots$. Use the results to make a conjecture about the area enclosed by the function if n is even and if n is odd.

40. Area Sketch the strophoid

$$r = \sec\theta - 2\cos\theta, \quad -\frac{\pi}{2} < \theta < \frac{\pi}{2}.$$

Convert this equation to rectangular coordinates. Find the area enclosed by the loop.

In Exercises 41–44, find the length of the curve over the indicated interval.

Polar Equation	Interval
41. $r = a$	$0 \le \theta \le 2\pi$
42. $r = 2a\cos\theta$	$-\frac{\pi}{2} \le \theta \le \frac{\pi}{2}$
43. $r = 1 + \sin\theta$	$0 \le \theta \le 2\pi$
44. $r = 8(1 + \cos\theta)$	$0 \le \theta \le 2\pi$

In Exercises 45–50, use a graphing utility to graph the polar equation over the indicated interval. Use the integration capabilities of the graphing utility to approximate the length of the curve accurate to two decimal places.

Polar Equation	Interval
45. $r = 2\theta$	$0 \le \theta \le \frac{\pi}{2}$
46. $r = \sec\theta$	$0 \le \theta \le \frac{\pi}{3}$
47. $r = \frac{1}{\theta}$	$\pi \le \theta \le 2\pi$
48. $r = e^\theta$	$0 \le \theta \le \pi$
49. $r = \sin(3\cos\theta)$	$0 \le \theta \le \pi$
50. $r = 2\sin(2\cos\theta)$	$0 \le \theta \le \pi$

In Exercises 51–54, find the area of the surface formed by revolving the curve about the given line.

Polar Equation	Interval	Axis of Revolution
51. $r = 6\cos\theta$	$0 \le \theta \le \frac{\pi}{2}$	Polar axis
52. $r = a\cos\theta$	$0 \le \theta \le \frac{\pi}{2}$	$\theta = \frac{\pi}{2}$
53. $r = e^{a\theta}$	$0 \le \theta \le \frac{\pi}{2}$	$\theta = \frac{\pi}{2}$
54. $r = a(1 + \cos\theta)$	$0 \le \theta \le \pi$	Polar axis

In Exercises 55 and 56, use the integration capabilities of a graphing utility to approximate to two decimal places the area of the surface formed by revolving the curve about the polar axis.

Polar Equation	Interval
55. $r = 4\cos 2\theta$	$0 \le \theta \le \frac{\pi}{4}$
56. $r = \theta$	$0 \le \theta \le \pi$

Getting at the Concept

57. Give the integral formulas for area and arc length in polar coordinates.

58. Explain why finding points of intersection of polar graphs may require further analysis beyond solving two equations simultaneously.

59. Which integral yields the arc length of $r = 3(1 - \cos 2\theta)$? State why the other integrals are incorrect.

(a) $3\int_0^{2\pi} \sqrt{(1 - \cos 2\theta)^2 + 4\sin^2 2\theta}\, d\theta$

(b) $12\int_0^{\pi/4} \sqrt{(1 - \cos 2\theta)^2 + 4\sin^2 2\theta}\, d\theta$

(c) $3\int_0^{\pi} \sqrt{(1 - \cos 2\theta)^2 + 4\sin^2 2\theta}\, d\theta$

(d) $6\int_0^{\pi/2} \sqrt{(1 - \cos 2\theta)^2 + 4\sin^2 2\theta}\, d\theta$

60. Give the integral formulas for the area of the surface of revolution formed when the graph of $r = f(\theta)$ is revolved about (a) the x-axis and (b) the y-axis.

61. Surface Area of a Torus Find the surface area of the torus generated by revolving the circle given by $r = a$ about the line $r = b\sec\theta$, where $0 < a < b$.

62. Approximating Area Consider the circle $r = 8\cos\theta$.

(a) Find the area of the circle.

(b) Complete the table giving the areas A of the sectors of the circle between $\theta = 0$ and the values of θ in the table.

θ	0.2	0.4	0.6	0.8	1.0	1.2	1.4
A							

(c) Use the table in part (b) to approximate the values of θ for which the sector of the circle composes $\frac{1}{4}$, $\frac{1}{2}$, and $\frac{3}{4}$ of the total area of the circle.

(d) Use a graphing utility to approximate to two-decimal-place accuracy the angles θ for which the sector of the circle composes $\frac{1}{4}$, $\frac{1}{2}$, and $\frac{3}{4}$ of the total area of the circle.

(e) Do the results in part (d) depend on the radius of the circle? Explain.

True or False? In Exercises 63 and 64, determine whether the statement is true or false. If it is false, explain why or give an example that shows it is false.

63. If $f(\theta) > 0$ for all θ and $g(\theta) < 0$ for all θ, then the graphs of $r = f(\theta)$ and $r = g(\theta)$ do not intersect.

64. If $f(\theta) = g(\theta)$ for $\theta = 0$, $\pi/2$, and $3\pi/2$, then the graphs of $r = f(\theta)$ and $r = g(\theta)$ have at least four points of intersection.

65. Use the formula for the arc length of a curve in parametric form to derive the formula for the arc length of a polar curve.

Section 9.6 Polar Equations of Conics and Kepler's Laws

- Analyze and write polar equations of conics.
- Understand and use Kepler's Laws of planetary motion.

Polar Equations of Conics

In this chapter you have seen that the rectangular equations of ellipses and hyperbolas take simple forms when the origin lies at their *centers*. As it happens, there are many important applications of conics in which it is more convenient to use one of the *foci* as the reference point (the origin) for the coordinate system. For example, the sun lies at a focus of earth's orbit. Similarly, the light source of a parabolic reflector lies at its focus. In this section you will see that polar equations of conics take simple forms if one of the foci lies at the pole.

The following theorem uses the concept of *eccentricity*, as defined in Section 9.1, to classify the three basic types of conics. A proof of this theorem is given in Appendix B.

EXPLORATION

Graphing Conics Set a graphing utility to polar mode and enter polar equations of the form

$$r = \frac{a}{1 \pm b \cos \theta}$$

or

$$r = \frac{a}{1 \pm b \sin \theta}.$$

As long as $a \neq 0$, the graph should be a conic. Describe the values of a and b that produce parabolas. What values produce ellipses? What values produce hyperbolas?

THEOREM 9.16 Classification of Conics by Eccentricity

Let F be a fixed point (*focus*) and D be a fixed line (*directrix*) in the plane. Let P be another point in the plane and let e (*eccentricity*) be the ratio of the distance between P and F to the distance between P and D. The collection of all points P with a given eccentricity is a conic.

1. The conic is an ellipse if $0 < e < 1$.
2. The conic is a parabola if $e = 1$.
3. The conic is a hyperbola if $e > 1$.

Ellipse: $0 < e < 1$
$$\frac{PF}{PQ} < 1$$

Parabola: $e = 1$
$$PF = PQ$$

Hyperbola: $e > 1$
$$\frac{PF}{PQ} = \frac{P'F}{P'Q'} > 1$$

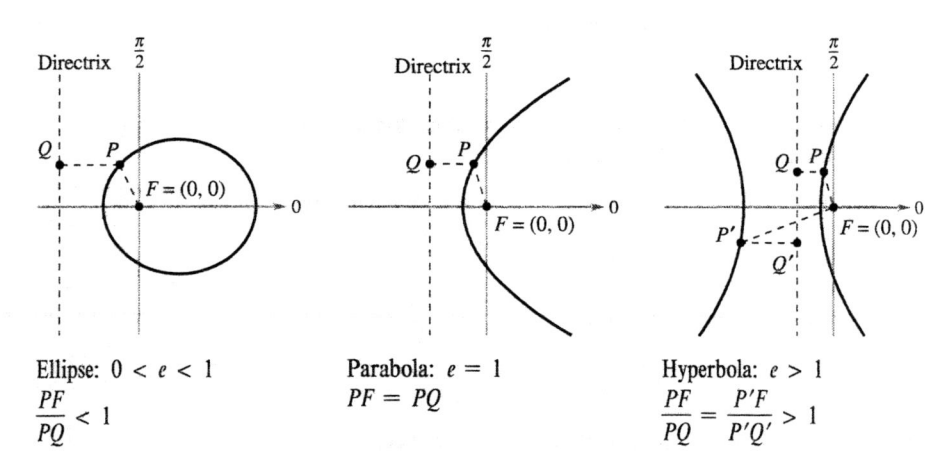

Figure 9.57

In Figure 9.57, note that for each type of conic the pole corresponds to the fixed point (focus) given in the definition. The benefit of this location can be seen in the proof of the following theorem.

THEOREM 9.17 Polar Equations of Conics

The graph of a polar equation of the form

$$r = \frac{ed}{1 \pm e \cos \theta} \quad \text{or} \quad r = \frac{ed}{1 \pm e \sin \theta}$$

is a conic, where $e > 0$ is the eccentricity and $|d|$ is the distance between the focus at the pole and its corresponding directrix.

Proof We give a proof for $r = ed/(1 + e \cos \theta)$ with $d > 0$. In Figure 9.58, consider a vertical directrix d units to the right of the focus $F = (0, 0)$. If $P = (r, \theta)$ is a point on the graph of $r = ed/(1 + e \cos \theta)$, the distance between P and the directrix can be shown to be

$$PQ = |d - x| = |d - r \cos \theta| = \left| \frac{r(1 + e \cos \theta)}{e} - r \cos \theta \right| = \left| \frac{r}{e} \right|.$$

Because the distance between P and the pole is simply $PF = |r|$, the ratio of PF to PQ is $PF/PQ = |r|/|r/e| = |e| = e$ and, by Theorem 9.16, the graph of the equation must be a conic. The proofs of the other cases are similar.

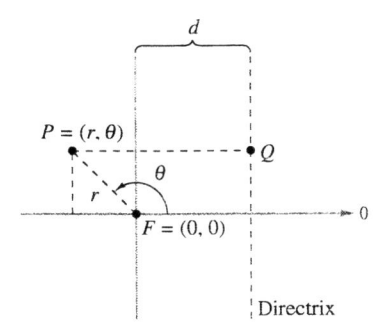

Figure 9.58

The four types of equations indicated in Theorem 9.17 can be classified as follows, where $d > 0$.

a. Horizontal directrix above the pole: $\quad r = \dfrac{ed}{1 + e \sin \theta}$

b. Horizontal directrix below the pole: $\quad r = \dfrac{ed}{1 - e \sin \theta}$

c. Vertical directrix to the right of the pole: $\quad r = \dfrac{ed}{1 + e \cos \theta}$

d. Vertical directrix to the left of the pole: $\quad r = \dfrac{ed}{1 - e \cos \theta}$

Figure 9.59 illustrates these four possibilities for a parabola.

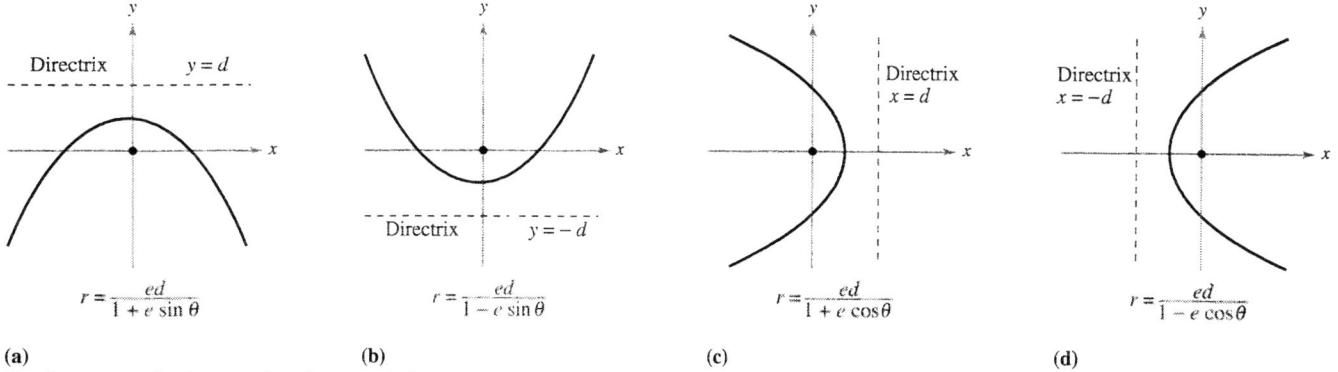

(a) $r = \dfrac{ed}{1 + e \sin \theta}$

(b) $r = \dfrac{ed}{1 - e \sin \theta}$

(c) $r = \dfrac{ed}{1 + e \cos \theta}$

(d) $r = \dfrac{ed}{1 - e \cos \theta}$

The four types of polar equations for a parabola
Figure 9.59

704 CHAPTER 9 Conics, Parametric Equations, and Polar Coordinates

The graph of the conic is an ellipse with $e = \frac{2}{3}$.
Figure 9.60

Example 1 Determining a Conic from Its Equation

Sketch the graph of the conic given by $r = \dfrac{15}{3 - 2\cos\theta}$.

Solution To determine the type of conic, rewrite the equation as

$$r = \frac{15}{3 - 2\cos\theta}$$

$$= \frac{5}{1 - (2/3)\cos\theta}. \qquad \text{Divide numerator and denominator by 3.}$$

So, the graph is an ellipse with $e = \frac{2}{3}$. You can sketch the upper half of the ellipse by plotting points from $\theta = 0$ to $\theta = \pi$, as shown in Figure 9.60. Then, using symmetry with respect to the polar axis, you can sketch the lower half.

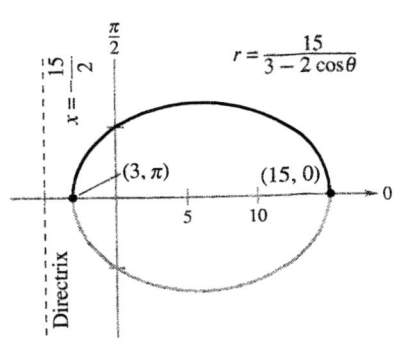

For the ellipse in Figure 9.60, the major axis is horizontal and the vertices lie at $(15, 0)$ and $(3, \pi)$. So, the length of the *major* axis is $2a = 18$. To find the length of the *minor* axis, you can use the equations $e = c/a$ and $b^2 = a^2 - c^2$ to conclude

$$b^2 = a^2 - c^2 = a^2 - (ea)^2 = a^2(1 - e^2). \qquad \text{Ellipse}$$

Because $e = \frac{2}{3}$, you have

$$b^2 = 9^2\left[1 - \left(\frac{2}{3}\right)^2\right] = 45$$

which implies that $b = \sqrt{45} = 3\sqrt{5}$. So, the length of the minor axis is $2b = 6\sqrt{5}$. A similar analysis for hyperbolas yields

$$b^2 = c^2 - a^2 = (ea)^2 - a^2 = a^2(e^2 - 1). \qquad \text{Hyperbola}$$

Example 2 Sketching a Conic from Its Polar Equation

Sketch the graph of the polar equation $r = \dfrac{32}{3 + 5\sin\theta}$.

Solution Dividing the numerator and denominator by 3 produces

$$r = \frac{32/3}{1 + (5/3)\sin\theta}.$$

Because $e = \frac{5}{3} > 1$, the graph is a hyperbola. Because $d = \frac{32}{5}$, the directrix is the line $y = \frac{32}{5}$. The transverse axis of the hyperbola lies on the line $\theta = \pi/2$, and the vertices occur at

$$(r, \theta) = \left(4, \frac{\pi}{2}\right) \quad \text{and} \quad (r, \theta) = \left(-16, \frac{3\pi}{2}\right).$$

Because the length of the transverse axis is 12, you can see that $a = 6$. To find b, write

$$b^2 = a^2(e^2 - 1) = 6^2\left[\left(\frac{5}{3}\right)^2 - 1\right] = 64.$$

Therefore, $b = 8$. Finally, you can use a and b to determine the asymptotes of the hyperbola and obtain the sketch shown in Figure 9.61.

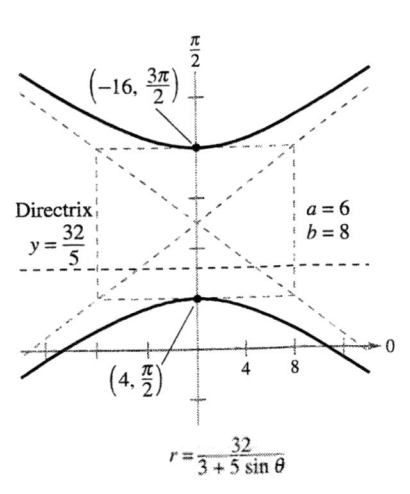

The graph of the conic is a hyperbola with $e = \frac{5}{3}$.
Figure 9.61

Kepler's Laws

Kepler's Laws, named after the German astronomer Johannes Kepler, can be used to describe the orbits of the planets about the sun.

1. Each planet moves in an elliptical orbit with the sun as a focus.
2. The ray from the sun to the planet sweeps out equal areas of the ellipse in equal times.
3. The square of the period is proportional to the cube of the mean distance between the planet and the sun.*

Although Kepler derived these laws empirically, they were later validated by Newton. In fact, Newton was able to show that each law can be deduced from a set of universal laws of motion and gravitation that govern the movement of all heavenly bodies, including comets and satellites. This is illustrated in the next example, involving the comet named after the English mathematician and physicist Edmund Halley (1656–1742).

Example 3 Halley's Comet

Halley's comet has an elliptical orbit with an eccentricity of $e \approx 0.97$. The length of the major axis of the orbit is approximately 36.18 astronomical units. (An astronomical unit is defined to be the mean distance between earth and the sun, 93 million miles.) Find a polar equation for the orbit. How close does Halley's comet come to the sun?

Solution Using a vertical axis, you can choose an equation of the form

$$r = \frac{ed}{(1 + e \sin \theta)}.$$

Because the vertices of the ellipse occur when $\theta = \pi/2$ and $\theta = 3\pi/2$, you can determine the length of the major axis to be the sum of the r-values of the vertices, as shown in Figure 9.62. That is,

$$2a = \frac{0.97d}{1 + 0.97} + \frac{0.97d}{1 - 0.97}$$

$$36.18 \approx 32.83d. \qquad 2a \approx 36.18$$

So, $d \approx 1.102$ and $ed \approx (0.97)(1.102) \approx 1.069$. Using this value in the equation produces

$$r = \frac{1.069}{1 + 0.97 \sin \theta}$$

where r is measured in astronomical units. To find the closest point to the sun (the focus), you can write $c = ea \approx (0.97)(18.09) \approx 17.55$. Because c is the distance between the focus and the center, the closest point is

$$a - c \approx 18.09 - 17.55$$
$$\approx 0.54 \text{ AU}$$
$$\approx 50{,}000{,}000 \text{ miles}$$

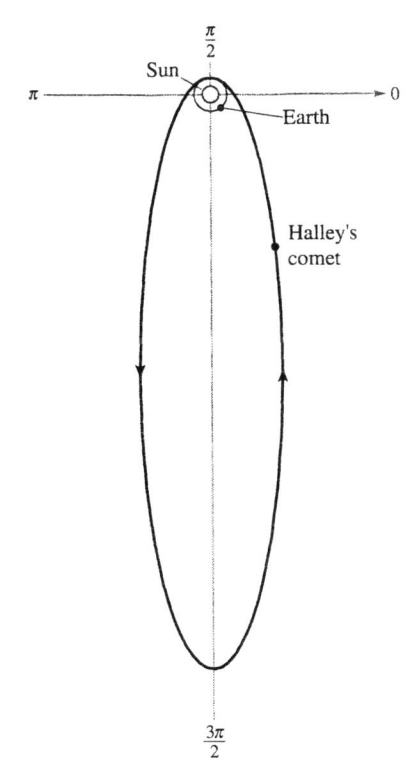

Figure 9.62

---

*If earth is used as a reference with a period of 1 year and a distance of 1 astronomical unit, the proportionality constant is 1. For example, because Mars has a mean distance to the sun of $D = 1.523$ AU, its period P is given by $D^3 = P^2$. So, the period for Mars is $P = 1.88$.

Kepler's Second Law states that as a planet moves about the sun, a ray from the sun to the planet sweeps out equal areas in equal times. This law can also be applied to comets or asteroids with elliptical orbits. For example, Figure 9.63 shows the orbit of the asteroid Apollo about the sun. Applying Kepler's Second Law to this asteroid, you know that the closer it is to the sun, the greater its velocity, because a short ray must be moving quickly to sweep out as much area as a long ray.

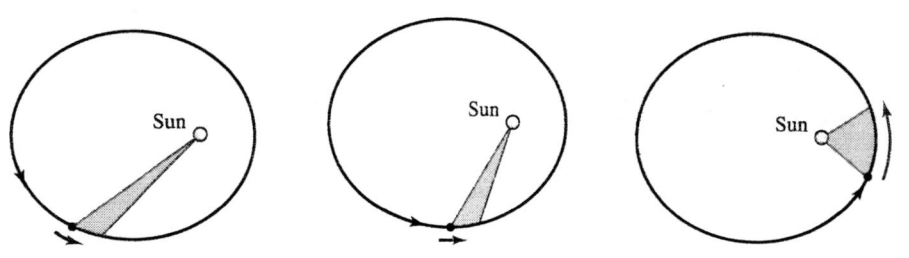

A ray from the sun to the asteroid sweeps out equal areas in equal times.
Figure 9.63

Example 4 The Asteroid Apollo

The asteroid Apollo has a period of 478 earth days, and its orbit is approximated by the ellipse

$$r = \frac{1}{1 + (5/9) \cos \theta} = \frac{9}{9 + 5 \cos \theta}$$

where r is measured in astronomical units. How long does it take Apollo to move from the position given by $\theta = -\pi/2$ to $\theta = \pi/2$, as shown in Figure 9.64?

Solution Begin by finding the area swept out as θ increases from $-\pi/2$ to $\pi/2$.

$$A = \frac{1}{2}\int_\alpha^\beta r^2 \, d\theta \qquad \text{Formula for area of a polar graph}$$

$$= \frac{1}{2}\int_{-\pi/2}^{\pi/2} \left(\frac{9}{9 + 5 \cos \theta}\right)^2 d\theta$$

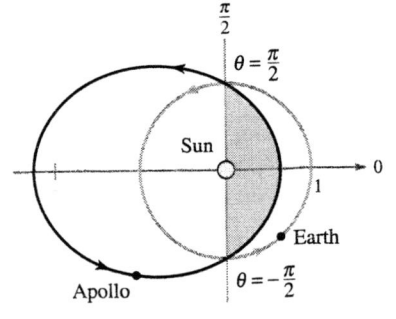

Figure 9.64

Using the substitution $u = \tan(\theta/2)$, as discussed in Section 7.6, you obtain

$$A = \frac{81}{112}\left[\frac{-5 \sin \theta}{9 + 5 \cos \theta} + \frac{18}{\sqrt{56}} \arctan \frac{\sqrt{56} \tan(\theta/2)}{14}\right]_{-\pi/2}^{\pi/2} \approx 0.90429.$$

Because the major axis of the ellipse has length $2a = 81/28$ and the eccentricity is $e = 5/9$, you can determine that $b = a\sqrt{1 - e^2} = 9/\sqrt{56}$. So, the area of the ellipse is

$$\text{Area of ellipse} = \pi ab = \pi\left(\frac{81}{56}\right)\left(\frac{9}{\sqrt{56}}\right) \approx 5.46507.$$

Because the time required to complete the orbit is 478 days, you can apply Kepler's Second Law to conclude that the time t required to move from the position $\theta = -\pi/2$ to $\theta = \pi/2$ is given by

$$\frac{t}{478} = \frac{\text{area of elliptical segment}}{\text{area of ellipse}} \approx \frac{0.90429}{5.46507}$$

which implies that $t \approx 79$ days.

EXERCISES FOR SECTION 9.6

Graphical Reasoning In Exercises 1–4, use a graphing utility to graph the polar equation when (a) $e = 1$, (b) $e = 0.5$, and (c) $e = 1.5$. Identify the conic.

1. $r = \dfrac{2e}{1 + e \cos \theta}$
2. $r = \dfrac{2e}{1 - e \cos \theta}$
3. $r = \dfrac{2e}{1 - e \sin \theta}$
4. $r = \dfrac{2e}{1 + e \sin \theta}$

5. Consider the polar equation
$$r = \dfrac{4}{1 + e \sin \theta}.$$

   (a) Use a graphing utility to graph the equation for $e = 0.1$, $e = 0.25$, $e = 0.5$, $e = 0.75$, and $e = 0.9$. Identify the conic and discuss the change in its shape as $e \to 1^-$ and $e \to 0^+$.

   (b) Use a graphing utility to graph the equation for $e = 1$. Identify the conic.

   (c) Use a graphing utility to graph the equation for $e = 1.1$, $e = 1.5$, and $e = 2$. Identify the conic and discuss the change in its shape as $e \to 1^+$ and $e \to \infty$.

6. Consider the polar equation
$$r = \dfrac{4}{1 - 0.4 \cos \theta}.$$

   (a) Identify the conic without graphing the equation.

   (b) Without graphing the following polar equations, describe how each differs from the polar equation above.
   $$r = \dfrac{4}{1 + 0.4 \cos \theta}, \quad r = \dfrac{4}{1 - 0.4 \sin \theta}$$

   (c) Verify the results in part (b) graphically.

In Exercises 7–12, match the polar equation with the correct graph. [The graphs are labeled (a), (b), (c), (d), (e), and (f).]

(a) (b) 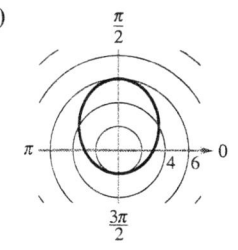 (c) 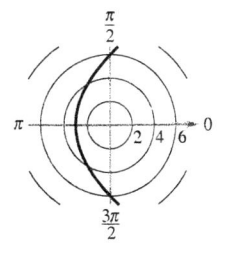 (d) 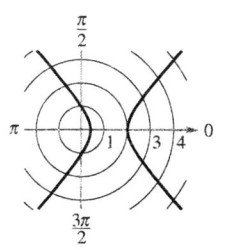 (e) 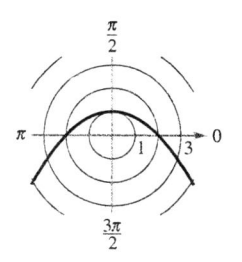 (f)

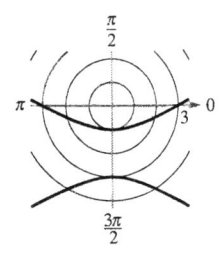

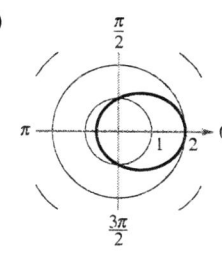

7. $r = \dfrac{6}{1 - \cos \theta}$
8. $r = \dfrac{2}{2 - \cos \theta}$
9. $r = \dfrac{3}{1 - 2 \sin \theta}$
10. $r = \dfrac{2}{1 + \sin \theta}$
11. $r = \dfrac{6}{2 - \sin \theta}$
12. $r = \dfrac{2}{2 + 3 \cos \theta}$

In Exercises 13–22, sketch and identify the graph. Use a graphing utility to confirm your results.

13. $r = \dfrac{-1}{1 - \sin \theta}$
14. $r = \dfrac{6}{1 + \cos \theta}$
15. $r = \dfrac{6}{2 + \cos \theta}$
16. $r = \dfrac{5}{5 + 3 \sin \theta}$
17. $r(2 + \sin \theta) = 4$
18. $r(3 - 2 \cos \theta) = 6$
19. $r = \dfrac{5}{-1 + 2 \cos \theta}$
20. $r = \dfrac{-6}{3 + 7 \sin \theta}$
21. $r = \dfrac{3}{2 + 6 \sin \theta}$
22. $r = \dfrac{4}{1 + 2 \cos \theta}$

In Exercises 23–26, use a graphing utility to graph the polar equation. Identify the graph.

23. $r = \dfrac{3}{-4 + 2 \sin \theta}$
24. $r = \dfrac{-3}{2 + 4 \sin \theta}$
25. $r = \dfrac{-1}{1 - \cos \theta}$
26. $r = \dfrac{2}{2 + 3 \sin \theta}$

In Exercises 27–30, use a graphing utility to graph the conic. Describe how the graph differs from that in the indicated exercise.

27. $r = \dfrac{-1}{1 - \sin(\theta - \pi/4)}$ (See Exercise 13.)
28. $r = \dfrac{6}{1 + \cos(\theta - \pi/3)}$ (See Exercise 14.)
29. $r = \dfrac{6}{2 + \cos(\theta + \pi/6)}$ (See Exercise 15.)
30. $r = \dfrac{-6}{3 + 7 \sin(\theta + 2\pi/3)}$ (See Exercise 20.)

31. Write the equation for the ellipse rotated $\pi/4$ radians clockwise from the ellipse $r = 5/(5 + 3 \cos\theta)$.

32. Write the equation for the parabola rotated $\pi/6$ radians counterclockwise from the parabola $r = 2/(1 + \sin \theta)$.

In Exercises 33–44, find a polar equation for the conic with its focus at the pole. (For convenience, the equation for the directrix is given in rectangular form.)

	Conic	Eccentricity	Directrix
33.	Parabola	$e = 1$	$x = -1$
34.	Parabola	$e = 1$	$y = 1$
35.	Ellipse	$e = \frac{1}{2}$	$y = 1$
36.	Ellipse	$e = \frac{3}{4}$	$y = -2$
37.	Hyperbola	$e = 2$	$x = 1$
38.	Hyperbola	$e = \frac{3}{2}$	$x = -1$

	Conic	Vertex or Vertices
39.	Parabola	$(1, -\pi/2)$
40.	Parabola	$(5, \pi)$
41.	Ellipse	$(2, 0), (8, \pi)$
42.	Ellipse	$(2, \pi/2), (4, 3\pi/2)$
43.	Hyperbola	$(1, 3\pi/2), (9, 3\pi/2)$
44.	Hyperbola	$(2, 0), (10, 0)$

Getting at the Concept

45. Classify the conics by their eccentricities.

46. Explain how the graph of each conic differs from the graph of $r = \dfrac{4}{1 + \sin \theta}$.

    (a) $r = \dfrac{4}{1 - \cos \theta}$ (b) $r = \dfrac{4}{1 - \sin \theta}$

    (c) $r = \dfrac{4}{1 + \cos \theta}$ (d) $r = \dfrac{4}{1 - \sin(\theta - \pi/4)}$

47. Identify the conic.

    (a) $r = \dfrac{5}{1 - 2\cos \theta}$ (b) $r = \dfrac{5}{10 - \sin \theta}$

    (c) $r = \dfrac{5}{3 - 3\cos \theta}$ (d) $r = \dfrac{5}{1 - 3\sin(\theta - \pi/4)}$

48. (a) Show that the polar equation for $(x^2/a^2) + (y^2/b^2) = 1$ is
$$r^2 = \dfrac{b^2}{1 - e^2 \cos^2 \theta}. \quad \text{Ellipse}$$

    (b) Show that the polar equation for $(x^2/a^2) - (y^2/b^2) = 1$ is
$$r^2 = \dfrac{-b^2}{1 - e^2 \cos^2 \theta}. \quad \text{Hyperbola}$$

In Exercises 49–52, use the results of Exercise 48 to write the polar form of the equation of the conic.

49. Ellipse: Focus at $(4, 0)$; Vertices at $(5, 0), (5, \pi)$

50. Hyperbola: Focus at $(5, 0)$; Vertices at $(4, 0), (4, \pi)$

51. $\dfrac{x^2}{9} - \dfrac{y^2}{16} = 1$ 52. $\dfrac{x^2}{4} + y^2 = 1$

In Exercises 53 and 54, use the integration capabilities of a graphing utility to approximate to two decimal places the area of the region bounded by the graph of the polar equation.

53. $r = \dfrac{3}{2 - \cos \theta}$ 54. $r = \dfrac{2}{3 - 2 \sin \theta}$

55. **Explorer 18** On November 26, 1963, the United States launched Explorer 18. Its low and high points above the surface of earth were 119 miles and 122,000 miles (see figure). The center of earth is the focus of the orbit. Find the polar equation for the orbit and find the distance between the surface of earth and the satellite when $\theta = 60°$. (Assume that the radius of earth is 4000 miles.)

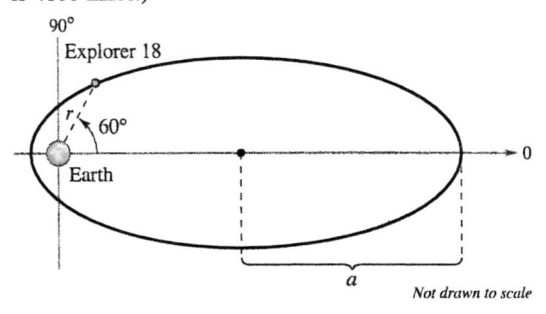

56. **Planetary Motion** The planets travel in elliptical orbits with the sun as a focus, as shown in the figure.

    (a) Show that the polar equation of the orbit is given by
$$r = \dfrac{(1 - e^2)a}{1 - e \cos \theta}$$
    where e is the eccentricity.

    (b) Show that the minimum distance (*perihelion distance*) from the sun to the planet is $r = a(1 - e)$ and the maximum distance (*aphelion distance*) is $r = a(1 + e)$.

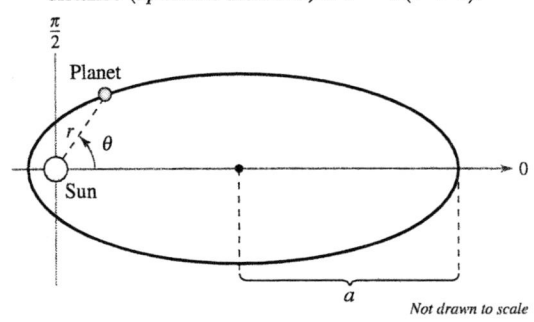

In Exercises 57–60, use Exercise 56 to find the polar equation of the elliptical orbit of the planet, and the perihelion and aphelion distances.

57. Earth $a = 92.957 \times 10^6$ miles
              $e = 0.0167$

58. Saturn $a = 1.427 \times 10^9$ kilometers
              $e = 0.0543$

59. Pluto $a = 5.900 \times 10^9$ kilometers
              $e = 0.2481$

60. Mercury $a = 36.0 \times 10^6$ miles
              $e = 0.206$

61. Planetary Motion In Exercise 59, the polar equation for the elliptical orbit of Pluto was found. Use the equation and a computer algebra system to perform each of the following.

(a) Approximate the area swept out by a ray from the sun to the planet as θ increases from 0 to $\pi/9$. Use this result to determine the number of years for the planet to move through this arc if the period of one revolution around the sun is 248 years.

(b) By trial and error, approximate the angle α such that the area swept out by a ray from the sun to the planet as θ increases from π to α equals the area found in part (a) (see figure). Does the ray sweep through a larger or smaller angle than in part (a) to generate the same area? Why is this the case?

(c) Approximate the distances the planet traveled in parts (a) and (b). Use these distances to approximate the average number of kilometers per year the planet traveled in the two cases.

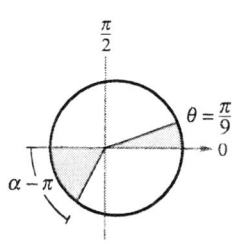

Figure for 61

62. What conic section does the following polar equation represent?
$$r = a \sin \theta + b \cos \theta$$

63. Show that the graphs of the following equations intersect at right angles.
$$r = \frac{ed}{1 + \sin \theta} \quad \text{and} \quad r = \frac{ed}{1 - \sin \theta}$$

REVIEW EXERCISES FOR CHAPTER 9

9.1 In Exercises 1–4, match the equation with the correct graph. [The graphs are labeled (a), (b), (c), and (d).]

(a) (b) 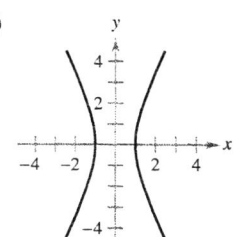 (c) 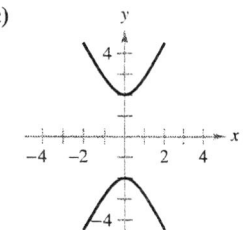 (d)

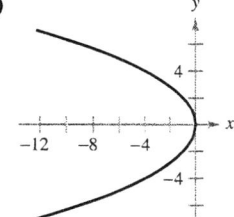

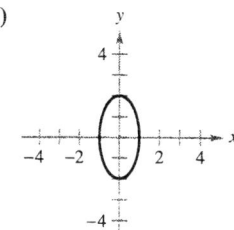

1. $4x^2 + y^2 = 4$
2. $4x^2 - y^2 = 4$
3. $y^2 = -4x$
4. $y^2 - 4x^2 = 4$

In Exercises 5–10, analyze each equation and sketch its graph. Use a graphing utility to confirm your results.

5. $16x^2 + 16y^2 - 16x + 24y - 3 = 0$
6. $y^2 - 12y - 8x + 20 = 0$
7. $3x^2 - 2y^2 + 24x + 12y + 24 = 0$
8. $4x^2 + y^2 - 16x + 15 = 0$
9. $3x^2 + 2y^2 - 12x + 12y + 29 = 0$
10. $4x^2 - 4y^2 - 4x + 8y - 11 = 0$

In Exercises 11 and 12, find an equation of the parabola.

11. Vertex: $(0, 2)$; Directrix: $x = -3$
12. Vertex: $(4, 2)$; Focus: $(4, 0)$

In Exercises 13 and 14, find an equation of the ellipse.

13. Vertices: $(-3, 0)$, $(7, 0)$; Foci: $(0, 0)$, $(4, 0)$
14. Center: $(0, 0)$; Solution points: $(1, 2)$, $(2, 0)$

In Exercises 15 and 16, find an equation of the hyperbola.

15. Vertices: $(\pm 4, 0)$; Foci: $(\pm 6, 0)$
16. Foci: $(0, \pm 8)$; Asymptotes: $y = \pm 4x$

In Exercises 17 and 18, use a graphing utility to approximate the perimeter of the ellipse.

17. $\dfrac{x^2}{9} + \dfrac{y^2}{4} = 1$
18. $\dfrac{x^2}{4} + \dfrac{y^2}{25} = 1$

19. A line is tangent to the parabola $y = x^2 - 2x + 2$ and perpendicular to the line $y = x - 2$. Find the equation of the line.

20. **Satellite Antenna** A cross section of a large parabolic antenna is modeled by the graph of $y = x^2/200$, $-100 \leq x \leq 100$. The receiving and transmitting equipment is positioned at the focus.

(a) Find the coordinates of the focus.
(b) Find the surface area of the antenna.

21. Consider a fire truck with a water tank 16 feet long whose vertical cross sections are ellipses modeled by the equation $x^2/16 + y^2/9 = 1$.

(a) Find the volume of the tank.
(b) Find the force on the end of the tank when it is full of water. (The density of water is 62.4 pounds per cubic foot.)
(c) Find the depth of the water in the tank if it is $\frac{3}{4}$ full (by volume) and the truck is on level ground.
(d) Approximate the tank's surface area.

22. Consider the region bounded by the ellipse

$$\frac{x^2}{a^2} + \frac{y^2}{b^2} = 1,$$

eccentricity $e = c/a$.

(a) Show that the area of the region is πab.

(b) Show that the solid (oblate spheroid) generated by revolving the region about the minor axis of the ellipse has a volume of $V = 4\pi a^2 b/3$ and a surface area of

$$S = 2\pi a^2 + \pi\left(\frac{b^2}{e}\right)\ln\left(\frac{1+e}{1-e}\right).$$

(c) Show that the solid (prolate spheroid) generated by revolving the region about the major axis of the ellipse has a volume of $V = 4\pi ab^2/3$ and a surface area of

$$S = 2\pi b^2 + 2\pi\left(\frac{ab}{e}\right)\arcsin e.$$

9.2 In Exercises 23–28, sketch the curve represented by the parametric equations (indicate the orientation of the curve), and write the corresponding rectangular equation by eliminating the parameter.

23. $x = 1 + 4t, \ y = 2 - 3t$

24. $x = t + 4, \ y = t^2$

25. $x = 6\cos\theta, \ y = 6\sin\theta$

26. $x = 3 + 3\cos\theta, \ y = 2 + 5\sin\theta$

27. $x = 2 + \sec\theta, \ y = 3 + \tan\theta$

28. $x = 5\sin^3\theta, \ y = 5\cos^3\theta$

In Exercises 29–32, find a parametric representation of the line or conic.

29. Line: Passes through $(-2, 6)$ and $(3, 2)$

30. Circle: Center at $(5, 3)$; Radius 2

31. Ellipse: Center at $(-3, 4)$; Horizontal major axis of length 8 and minor axis of length 6

32. Hyperbola: Vertices at $(0, \pm 4)$; Foci at $(0, \pm 5)$

33. *Rotary Engine* The rotary engine was developed by Felix Wankel in the 1950s (see Chapter 4). It features a rotor, which is a modified equilateral triangle. The rotor moves in a chamber that, in two dimensions, is an epitrochoid. Use a graphing utility to graph the chamber modeled by the parametric equations.

$x = \cos 3\theta + 5\cos\theta$

and

$y = \sin 3\theta + 5\sin\theta.$

34. *Hypocycloids* A hypocycloid has the parametric equations

$$x = (a-b)\cos t + b\cos\left(\frac{a-b}{b}t\right) \quad \text{and}$$

$$y = (a-b)\sin t - b\sin\left(\frac{a-b}{b}t\right).$$

Use a graphing utility to graph the hypocycloid for each of the following values of a and b.

(a) $a = 2, \ b = 1$ (b) $a = 3, \ b = 1$ (c) $a = 4, \ b = 1$
(d) $a = 10, \ b = 1$ (e) $a = 3, \ b = 2$ (f) $a = 4, \ b = 3$

35. *Serpentine Curve* Consider the parametric equations $x = 2\cot\theta$ and $y = 4\sin\theta\cos\theta, \ 0 < \theta < \pi$.

(a) Use a graphing utility to sketch the curve.

(b) Eliminate the parameter to show that the rectangular equation of the serpentine curve is $(4 + x^2)y = 8x$.

36. *Involute of a Circle* The involute of a circle is described by the endpoint P of a string that is held taut as it is unwound from a spool that does not turn (see figure). Show that a parametric representation of the involute is

$x = r(\cos\theta + \theta\sin\theta)$ and $y = r(\sin\theta - \theta\cos\theta).$

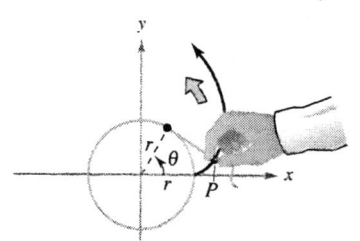

9.3 In Exercises 37–46, (a) find dy/dx and all points of horizontal tangency, (b) eliminate the parameter where possible, and (c) sketch the curve represented by the parametric equations.

37. $x = 1 + 4t, \ y = 2 - 3t$

38. $x = t + 4, \ y = t^2$

39. $x = \dfrac{1}{t}, \ y = 2t + 3$

40. $x = \dfrac{1}{t}, \ y = t^2$

41. $x = \dfrac{1}{2t+1}$

$y = \dfrac{1}{t^2 - 2t}$

42. $x = 2t - 1$

$y = \dfrac{1}{t^2 - 2t}$

43. $x = 3 + 2\cos\theta$

$y = 2 + 5\sin\theta$

44. $x = 6\cos\theta$

$y = 6\sin\theta$

45. $x = \cos^3\theta$

$y = 4\sin^3\theta$

46. $x = e^t$

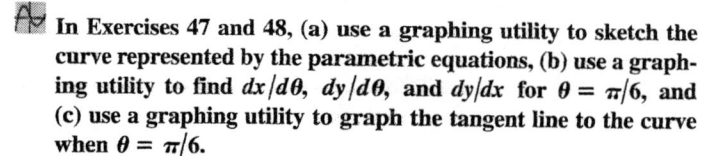

$y = e^{-t}$

In Exercises 47 and 48, (a) use a graphing utility to sketch the curve represented by the parametric equations, (b) use a graphing utility to find $dx/d\theta$, $dy/d\theta$, and dy/dx for $\theta = \pi/6$, and (c) use a graphing utility to graph the tangent line to the curve when $\theta = \pi/6$.

47. $x = \cot\theta$

$y = \sin 2\theta$

48. $x = 2\theta - \sin\theta$

$y = 2 - \cos\theta$

In Exercises 49 and 50, find the length of the curve represented by the parametric equations over the given interval.

49. $x = r(\cos\theta + \theta\sin\theta)$
$y = r(\sin\theta - \theta\cos\theta)$
$0 \leq \theta \leq \pi$

50. $x = 6\cos\theta$
$y = 6\sin\theta$
$0 \leq \theta \leq \pi$

9.4 In Exercises 51 and 52, the rectangular coordinates of a point are given. Plot the point and find two sets of polar coordinates for the point for $0 \leq \theta \leq 2\pi$.

51. $(4, -4)$
52. $(-1, 3)$

In Exercises 53–60, convert the polar equation to rectangular form.

53. $r = 3\cos\theta$
54. $r = 10$
55. $r = -2(1 + \cos\theta)$
56. $r = \dfrac{1}{2 - \cos\theta}$
57. $r^2 = \cos 2\theta$
58. $r = 4\sec\left(\theta - \dfrac{\pi}{3}\right)$
59. $r = 4\cos 2\theta \sec\theta$
60. $\theta = \dfrac{3\pi}{4}$

In Exercises 61–64, convert the rectangular equation to polar form.

61. $(x^2 + y^2)^2 = ax^2 y$
62. $x^2 + y^2 - 4x = 0$
63. $x^2 + y^2 = a^2\left(\arctan\dfrac{y}{x}\right)^2$
64. $(x^2 + y^2)\left(\arctan\dfrac{y}{x}\right)^2 = a^2$

In Exercises 65–76, sketch a graph of the polar equation.

65. $r = 4$
66. $\theta = \dfrac{\pi}{12}$
67. $r = -\sec\theta$
68. $r = 3\csc\theta$
69. $r = -2(1 + \cos\theta)$
70. $r = 3 - 4\cos\theta$
71. $r = 4 - 3\cos\theta$
72. $r = 2\theta$
73. $r = -3\cos 2\theta$
74. $r = \cos 5\theta$
75. $r^2 = 4\sin^2 2\theta$
76. $r^2 = \cos 2\theta$

In Exercises 77–80, use a graphing utility to graph the polar equation.

77. $r = \dfrac{3}{\cos(\theta - \pi/4)}$
78. $r = 2\sin\theta \cos^2\theta$
79. $r = 4\cos 2\theta \sec\theta$
80. $r = 4(\sec\theta - \cos\theta)$

In Exercises 81 and 82, (a) find the tangents at the pole, (b) find all points of horizontal and vertical tangency, and (c) use a graphing utility to graph the polar equation and draw a tangent line to the graph for $\theta = \pi/6$.

81. $r = 1 - 2\cos\theta$
82. $r^2 = 4\sin 2\theta$

83. Find the angle between the circle $r = 3\sin\theta$ and the limaçon $r = 4 - 5\sin\theta$ at the point of intersection $(3/2, \pi/6)$.

84. *True or False?* There is a unique polar coordinate representation for each point in the plane. Explain.

9.5 In Exercises 85 and 86, show that the graphs of the polar equations are orthogonal at the points of intersection. Use a graphing utility to confirm your results graphically.

85. $r = 1 + \cos\theta$
$r = 1 - \cos\theta$
86. $r = a\sin\theta$
$r = a\cos\theta$

In Exercises 87–94, use a graphing utility to graph the polar equation. Set up an integral for finding the area of the indicated region and use the integration capabilities of a graphing utility to approximate the integral accurate to two decimal places.

87. Interior of $r = 2 + \cos\theta$
88. Interior of $r = 5(1 - \sin\theta)$
89. Interior of $r = \sin\theta \cos^2\theta$
90. Interior of $r = 4\sin 3\theta$
91. Interior of $r^2 = 4\sin 2\theta$
92. Common interior of $r = 3$ and $r^2 = 18\sin 2\theta$
93. Common interior of $r = 4\cos\theta$ and $r = 2$
94. Region bounded by the polar axis and $r = e^\theta$ for $0 \leq \theta \leq \pi$

In Exercises 95 and 96, find the perimeter of the curve.

95. $r = a(1 - \cos\theta)$
96. $r = a\cos 2\theta$

9.6 In Exercises 97–102, sketch and identify the graph. Use a graphing utility to confirm your results.

97. $r = \dfrac{2}{1 - \sin\theta}$
98. $r = \dfrac{2}{1 + \cos\theta}$
99. $r = \dfrac{6}{3 + 2\cos\theta}$
100. $r = \dfrac{4}{5 - 3\sin\theta}$
101. $r = \dfrac{4}{2 - 3\sin\theta}$
102. $r = \dfrac{8}{2 - 5\cos\theta}$

In Exercises 103–108, find a polar equation for the line or conic.

103. Circle
Center: $(5, \pi/2)$
Solution point: $(0, 0)$

104. Line
Solution point: $(0, 0)$
Slope: $\sqrt{3}$

105. Parabola
Vertex: $(2, \pi)$
Focus: $(0, 0)$

106. Parabola
Vertex: $(2, \pi/2)$
Focus: $(0, 0)$

107. Ellipse
Vertices: $(5, 0), (1, \pi)$
One focus: $(0, 0)$

108. Hyperbola
Vertices: $(1, 0), (7, 0)$
One focus: $(0, 0)$

P.S. Problem Solving

1. Consider the parabola $x^2 = 4y$ and the focal chord $y = \frac{3}{4}x + 1$.
   (a) Sketch the graph of the parabola and the focal chord.
   (b) Show that the tangent lines to the parabola at the endpoints of the focal chord intersect at right angles.
   (c) Show that the tangent lines to the parabola at the endpoints of the focal chord intersect on the directrix of the parabola.

2. Consider the parabola $x^2 = 4py$ and one of its focal chords.
   (a) Show that the tangent lines to the parabola at the endpoints of the focal chord intersect at right angles.
   (b) Show that the tangent lines to the parabola at the endpoints of the focal chord intersect on the directrix of the parabola.

3. Prove Theorem 9.2, the Reflective Property of a Parabola, as illustrated in the figure.

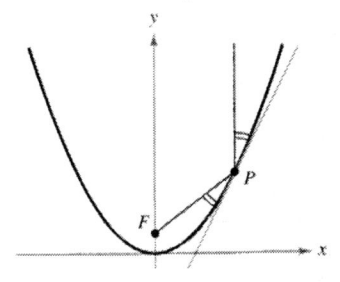

4. Consider the hyperbola
   $$\frac{x^2}{a^2} - \frac{y^2}{b^2} = 1$$
   with foci F_1 and F_2, as indicated in the figure. Let T be the tangent line at a point M on the hyperbola. Show that incoming rays of light aimed at one focus are reflected by a hyperbolic mirror toward the other focus.

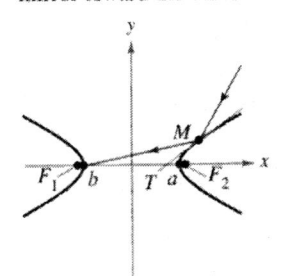

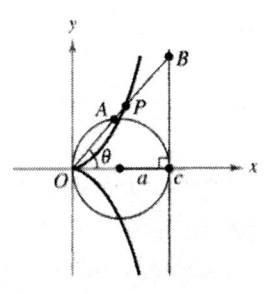

   Figure for 4   **Figure for 5**

5. Consider a circle of radius a tangent to the y-axis and the line $x = 2a$, as indicated in the figure. Let A be the point where the segment OB intersects the circle. The **cissoid of Diocles** consists of all points P such that $OP = AB$.
   (a) Find a polar equation of the cissoid.
   (b) Find a set of parametric equations for the cissoid that does not contain trigonometric functions.
   (c) Find a rectangular equation of the cissoid.

6. The curve given by the parametric equations
   $$x(t) = \frac{1 - t^2}{1 + t^2} \quad \text{and} \quad y(t) = \frac{t(1 - t^2)}{1 + t^2}$$
   is called a **strophoid.**
   (a) Find a rectangular equation of the strophoid.
   (b) Find a polar equation of the strophoid.
   (c) Sketch a graph of the strophoid.
   (d) Find the equations of the two tangent lines at the origin.
   (e) Find the points on the graph where the tangent lines are horizontal.

7. Find the rectangular equation of the portion of the cycloid given by the parametric equations $x = a(\theta - \sin \theta)$ and $y = a(1 - \cos \theta)$, $0 \leq \theta \leq \pi$, as indicated in the figure.

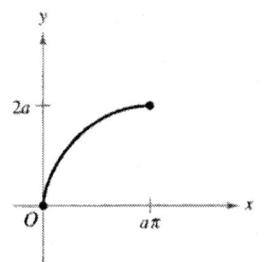

8. Consider the **cornu spiral** given by
   $$x(t) = \int_0^t \cos\left(\frac{\pi u^2}{2}\right) du \quad \text{and} \quad y(t) = \int_0^t \sin\left(\frac{\pi u^2}{2}\right) du.$$
   (a) Use a graphing utility to graph the spiral over the interval $-\pi \leq t \leq \pi$.
   (b) Show that the cornu spiral is symmetric with respect to the origin.
   (c) Find the length of the cornu spiral from $t = 0$ to $t = a$. What is the length of the spiral from $t = -\pi$ to $t = \pi$?

9. A particle is moving along the path described by the parametric equations
   $$x = \frac{1}{t} \quad \text{and} \quad y = \frac{\sin t}{t}, \quad 1 \leq t < \infty,$$
   as indicated in the figure. Find the length of this path.

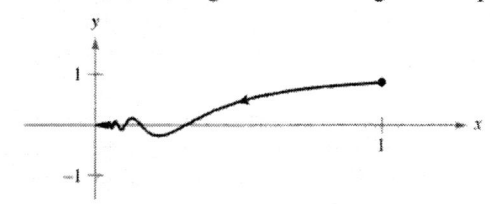

10. Let a and b be positive constants. Find the area of the region in the first quadrant bounded by the graph of the polar equation

$$r = \frac{ab}{(a\sin\theta + b\cos\theta)}, \quad 0 \leq \theta \leq \frac{\pi}{2}.$$

11. Consider the right triangle in the figure.

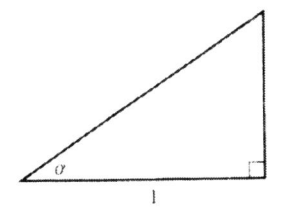

    (a) Show that the area of the triangle is

    $$A(\alpha) = \frac{1}{2}\int_0^\alpha \sec^2\theta\, d\theta.$$

    (b) Show that $\tan\alpha = \int_0^\alpha \sec^2\theta\, d\theta.$

    (c) Use part (b) to derive the formula for the derivative of the tangent function.

12. Determine the polar equation of the set of all points (r, θ), the product of whose distances from the points $(1, 0)$ and $(-1, 0)$ is equal to 1, as indicated in the figure.

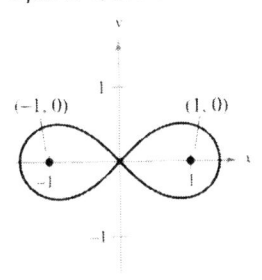

13. Four dogs are located at the corners of a square with sides of length d. The dogs all move counterclockwise at the same speed directly toward the next dog, as indicated in the figure. Find the polar equation of a dog's path as it spirals toward the center of the square.

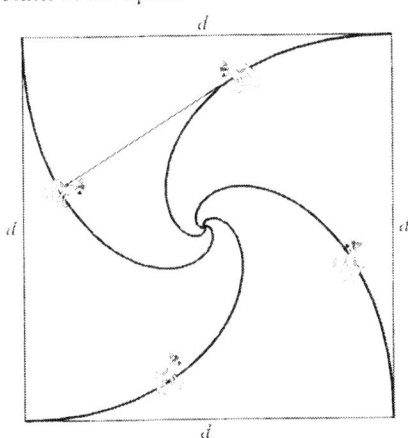

14. Use a graphing utility to graph the polar equation $r = 2 + k\cos\theta$ for $k = 0, 1, 2,$ and 3. Identify each graph.

15. A controller spots two planes at the same altitude flying toward each other (see figure). Their flight paths are S 20° W and S 45° E. One plane is 150 miles from point P with a speed of 375 miles per hour. The other is 190 miles from point P with a speed of 450 miles per hour.

    (a) Find parametric equations for the path of each plane where t is the time in hours, with $t = 0$ corresponding to the time at which the air traffic controller spots the planes.

    (b) Use the result in part (a) to write the distance between the planes as a function of t.

    (c) Use a graphing utility to graph the function in part (b). When will the distance between the planes be minimum? If the planes must keep a separation of at least 3 miles, is the requirement met?

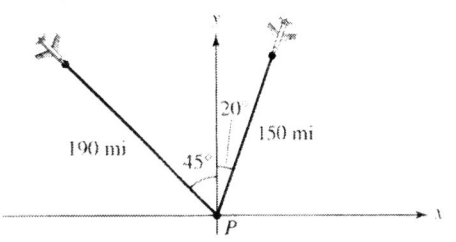

16. Use a graphing utility to produce the curve shown below. The curve is given by

$$r = e^{\cos\theta} - 2\cos 4\theta + \sin^5\frac{\theta}{12}.$$

Over what interval must θ vary to produce the curve?

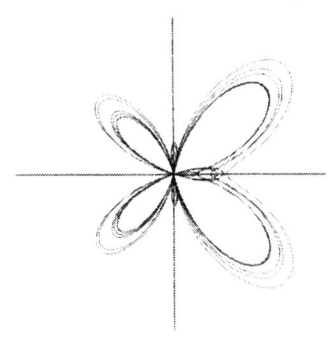

FOR FURTHER INFORMATION For more information on this curve, see the article "A Study in Step Size" by Temple H. Fay in *Mathematics Magazine*. To view this article, go to the website *www.matharticles.com*.

17. Use a graphing utility to graph the polar equation

$$r = \cos 5\theta + n\cos\theta$$

for $0 \leq \theta < \pi$ for the integers $n = -5$ to $n = 5$. What values of n produce the "heart" portion of the curve? What values of n produce the "bell" portion? (This curve, created by Michael W. Chamberlin, appeared in *The College Mathematics Journal*.)

Suspension Bridges

Bridges have been around since primitive people first threw a tree trunk across a stream. The oldest known bridge was constructed of stone slabs. Since primitive times, engineers have strived to construct bridges that are longer, sturdier, and more aesthetically pleasing than their predecessors.

One of the most commonly used bridge designs is the suspension bridge. This type of bridge is used in situations that require long single spans. The roadway is hung from cables that are supported by stationary towers. Well-designed suspension bridges, such as the Golden Gate Bridge (shown on the facing page) and the Brooklyn Bridge, can be functional for many years.

On the other hand, a poor design can result in tragedy. For a suspension bridge to be stable, the forces acting on its main cables must be in equilibrium. This equilibrium occurs when the main cables are in the shape of parabolas.

The forces acting on the main cable are shown in the diagram at the left below as directed line segments, indicating both the magnitude and the direction of the forces. At the center of the parabolic cable, the tension force $\mathbf{T}_0$ is horizontal. $\mathbf{T}$ is the tension force at point D, and is directed along the tangent at point D. The uniformly distributed load supported by the section CD of the cable is represented by $\mathbf{W}$.

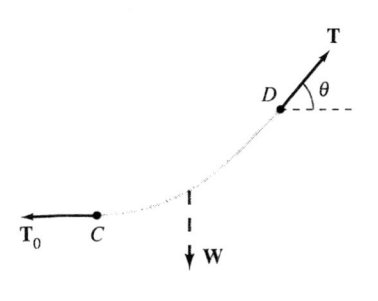

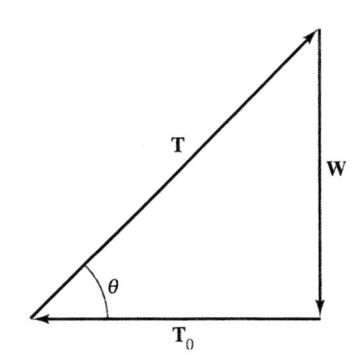

QUESTIONS

1. The forces acting on the cable are related by a "force triangle," as shown at the right above. Use trigonometric functions to relate the magnitudes $\|\mathbf{T}_0\|$, $\|\mathbf{T}\|$, and $\|\mathbf{W}\|$ of the vectors $\mathbf{T}_0$, $\mathbf{T}$, and $\mathbf{W}$.

2. The main cable of the Golden Gate Bridge is suspended from towers 520 feet above the roadway at either end of a 4200-foot span. The low point in the center of the cable is 6 feet above the roadway. Given that the cable hangs in the shape of a parabola, find an equation describing the shape of the cable.

3. Use the equation you found in Question 2 to find the acute angle θ of the force $\mathbf{T}$ at a point located 400 feet horizontally from the center of the span. Express the magnitude of $\mathbf{T}$ in terms of $\mathbf{T}_0$ and $\mathbf{W}$.

4. Find the angle and magnitude of $\mathbf{T}$ at a point located 200 feet horizontally from the center of the span.

5. In general, where in a suspension cable would you expect the magnitude of $\mathbf{T}$ to be the greatest? Where would you expect the magnitude of $\mathbf{T}$ to be the least? Explain.

The concepts presented here will be explored further in this chapter. For an extension of this application, see Lab 14 in the lab series that accompanies this text at college.hmco.com.

Vectors and the Geometry of Space

10

According to the National Science Foundation, about 250,000 bridges in the United States need to be fixed or replaced due to corrosion. With an estimated price tag of $50 billion, it is no wonder the NSF is funding research to find materials to replace steel and concrete. The best options are glass, plastic, and carbon fiber, as these materials are not only lighter but also more durable.

In 1930, Joseph Strauss designed the Golden Gate Bridge in San Francisco, still considered to be one of the world's greatest civil engineering masterpieces.

Section 10.1 Vectors in the Plane

- Write the component form of a vector.
- Perform vector operations and interpret the results geometrically.
- Write a vector as a linear combination of standard unit vectors.
- Use vectors to solve problems involving force or velocity.

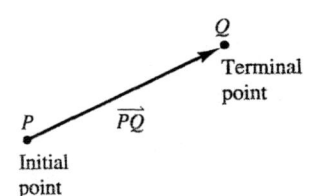

Component Form of a Vector

Many quantities in geometry and physics, such as area, volume, temperature, mass, and time, can be characterized by single real numbers scaled to appropriate units of measure. We call these **scalar quantities,** and the real number associated with each is called a **scalar.**

Other quantities, such as force, velocity, and acceleration, involve both magnitude and direction and cannot be characterized completely by single real numbers. A **directed line segment** is used to represent such a quantity, as shown in Figure 10.1. The directed line segment $\overrightarrow{PQ}$ has **initial point** P and **terminal point** Q, and its **length** (or **magnitude**) is denoted by $\|\overrightarrow{PQ}\|$. Directed line segments that have the same length and direction are **equivalent,** as shown in Figure 10.2. The set of all directed line segments that are equivalent to a given directed line segment $\overrightarrow{PQ}$ is a **vector in the plane** and is denoted by $\mathbf{v} = \overrightarrow{PQ}$. In typeset material, vectors are usually denoted by lowercase, boldface letters such as **u, v,** and **w.** When written by hand, however, vectors are often denoted by letters with arrows above them, such as $\vec{u}$, $\vec{v}$, and $\vec{w}$.

Be sure you see that a vector in the plane can be represented by many different directed line segments—all pointing in the same direction and all of the same length.

Example 1 Vector Representation by Directed Line Segments

Let **v** be represented by the directed line segment from $(0, 0)$ to $(3, 2)$, and let **u** be represented by the directed line segment from $(1, 2)$ to $(4, 4)$. Show that **v** and **u** are equivalent.

Solution Let $P(0, 0)$ and $Q(3, 2)$ be the initial and terminal points of **v**, and let $R(1, 2)$ and $S(4, 4)$ be the initial and terminal points of **u**, as shown in Figure 10.3. You can use the Distance Formula to show that $\overrightarrow{PQ}$ and $\overrightarrow{RS}$ have the *same length*.

$$\|\overrightarrow{PQ}\| = \sqrt{(3 - 0)^2 + (2 - 0)^2} = \sqrt{13} \qquad \text{Length of } \overrightarrow{PQ}$$
$$\|\overrightarrow{RS}\| = \sqrt{(4 - 1)^2 + (4 - 2)^2} = \sqrt{13} \qquad \text{Length of } \overrightarrow{RS}$$

Both line segments have the *same direction*, because they both are directed toward the upper right on lines having the same slope.

$$\text{Slope of } \overrightarrow{PQ} = \frac{2 - 0}{3 - 0} = \frac{2}{3}$$

and

$$\text{Slope of } \overrightarrow{RS} = \frac{4 - 2}{4 - 1} = \frac{2}{3}$$

Because $\overrightarrow{PQ}$ and $\overrightarrow{RS}$ have the same length and direction, you can conclude that the two vectors are equivalent. That is, **v** and **u** are equivalent.

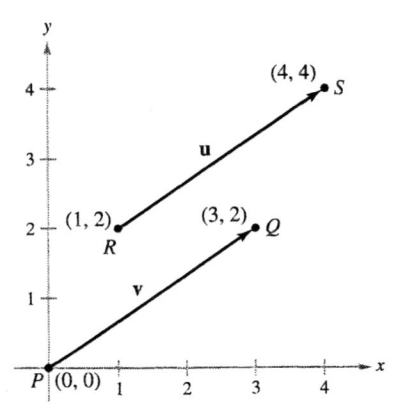

A directed line segment
Figure 10.1

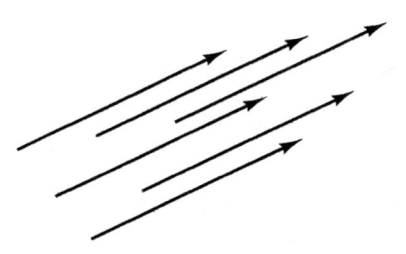

Equivalent directed line segments
Figure 10.2

The vectors **u** and **v** are equivalent.
Figure 10.3

The directed line segment whose initial point is the origin is often the most convenient representative of a set of equivalent directed line segments such as those shown in Figure 10.3. This representation of **v** is said to be in **standard position**. A directed line segment whose initial point is at the origin can be uniquely represented by the coordinates of its terminal point $Q(v_1, v_2)$, as shown in Figure 10.4.

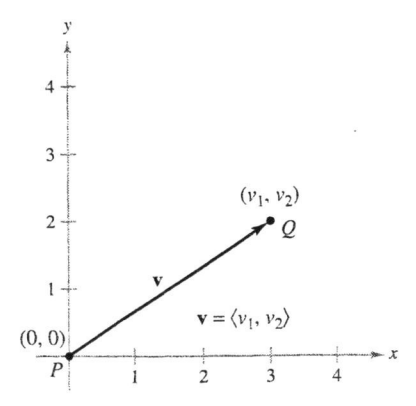

The standard position of a vector
Figure 10.4

Definition of Component Form of a Vector in the Plane

If **v** is a vector in the plane whose initial point is the origin and whose terminal point is (v_1, v_2), then the **component form of v** is given by

$$\mathbf{v} = \langle v_1, v_2 \rangle.$$

The coordinates v_1 and v_2 are called the **components of v**. If both the initial point and the terminal point lie at the origin, then **v** is called the **zero vector** and is denoted by $\mathbf{0} = \langle 0, 0 \rangle$.

This definition implies that two vectors $\mathbf{u} = \langle u_1, u_2 \rangle$ and $\mathbf{v} = \langle v_1, v_2 \rangle$ are **equal** if and only if $u_1 = v_1$ and $u_2 = v_2$.

The following procedures can be used to convert directed line segments to component form or vice versa.

NOTE It is important to understand that a vector represents a *set* of directed line segments (each having the same length and direction). In practice, however, it is common not to distinguish between a vector and one of its representatives.

1. If $P(p_1, p_2)$ and $Q(q_1, q_2)$ are the initial and terminal points of a directed line segment, the component form of the vector **v** represented by $\overrightarrow{PQ}$ is $\langle v_1, v_2 \rangle = \langle q_1 - p_1, q_2 - p_2 \rangle$. Moreover, the **length** (or **magnitude**) **of v** is

$$\|\mathbf{v}\| = \sqrt{(q_1 - p_1)^2 + (q_2 - p_2)^2} \qquad \text{Length of a vector}$$
$$= \sqrt{v_1^2 + v_2^2}.$$

2. If $\mathbf{v} = \langle v_1, v_2 \rangle$, **v** can be represented by the directed line segment, in standard position, from $P(0, 0)$ to $Q(v_1, v_2)$.

The length of **v** is also called the **norm of v**. If $\|\mathbf{v}\| = 1$, **v** is a **unit vector**. Moreover, $\|\mathbf{v}\| = 0$ if and only if **v** is the zero vector **0**.

Example 2 Finding the Component Form and Length of a Vector

Find the component form and length of the vector **v** that has initial point $(3, -7)$ and terminal point $(-2, 5)$.

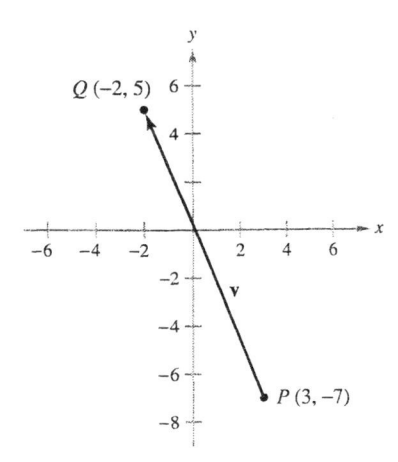

Solution Let $P(3, -7) = (p_1, p_2)$ and $Q(-2, 5) = (q_1, q_2)$. Then the components of $\mathbf{v} = \langle v_1, v_2 \rangle$ are

$$v_1 = q_1 - p_1 = -2 - 3 = -5$$
$$v_2 = q_2 - p_2 = 5 - (-7) = 12.$$

So, as shown in Figure 10.5, $\mathbf{v} = \langle -5, 12 \rangle$, and the length of **v** is

$$\|\mathbf{v}\| = \sqrt{(-5)^2 + 12^2}$$
$$= \sqrt{169}$$
$$= 13.$$

Component form of **v**: $\mathbf{v} = \langle -5, 12 \rangle$
Figure 10.5

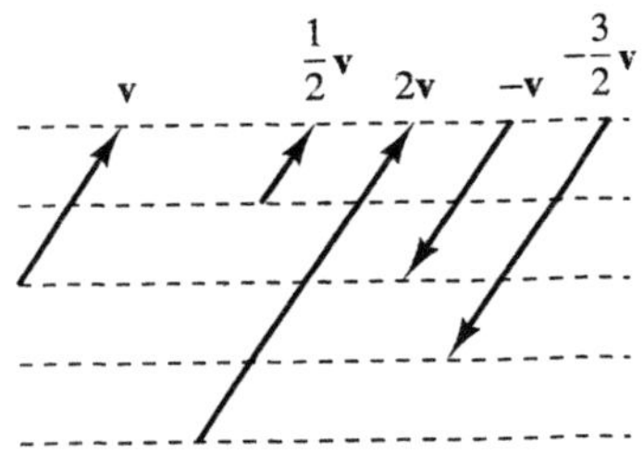

The scalar multiplication of **v**
Figure 10.6

ISAAC WILLIAM ROWAN HAMILTON (1805–1865)

Some of the earliest work with vectors was done by the Irish mathematician William Rowan Hamilton. Hamilton spent many years developing a system of vector-like quantities called *quaternions*. Although Hamilton was convinced of the benefits of quaternions, the operations he defined did not produce good models for physical phenomena. It wasn't until the latter half of the nineteenth century that the Scottish physicist James Maxwell (1831–1879) restructured Hamilton's quaternions in a form useful for representing physical quantities such as force, velocity, and acceleration.

Vector Operations

Definitions of Vector Addition and Scalar Multiplication

Let $\mathbf{u} = \langle u_1, u_2 \rangle$ and $\mathbf{v} = \langle v_1, v_2 \rangle$ be vectors and let c be a scalar.

1. The **vector sum** of $\mathbf{u}$ and $\mathbf{v}$ is the vector
$$\mathbf{u} + \mathbf{v} = \langle u_1 + v_1, u_2 + v_2 \rangle.$$

2. The **scalar multiple** of c and $\mathbf{u}$ is the vector
$$c\mathbf{u} = \langle cu_1, cu_2 \rangle.$$

3. The **negative** of $\mathbf{v}$ is the vector
$$-\mathbf{v} = (-1)\mathbf{v} = \langle -v_1, -v_2 \rangle.$$

4. The **difference** of $\mathbf{u}$ and $\mathbf{v}$ is
$$\mathbf{u} - \mathbf{v} = \mathbf{u} + (-\mathbf{v}) = \langle u_1 - v_1, u_2 - v_2 \rangle.$$

Geometrically, the scalar multiple of a vector $\mathbf{v}$ and a scalar c is the vector that is $|c|$ times as long as $\mathbf{v}$, as shown in Figure 10.6. If c is positive, $c\mathbf{v}$ has the same direction as $\mathbf{v}$. If c is negative, $c\mathbf{v}$ has the opposite direction.

The sum of two vectors can be represented geometrically by positioning the vectors (without changing their magnitudes or directions) so that the initial point of one coincides with the terminal point of the other, as shown in Figure 10.7. The vector $\mathbf{u} + \mathbf{v}$, called the **resultant vector,** is the diagonal of a parallelogram having $\mathbf{u}$ and $\mathbf{v}$ as its adjacent sides.

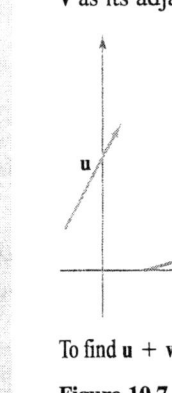

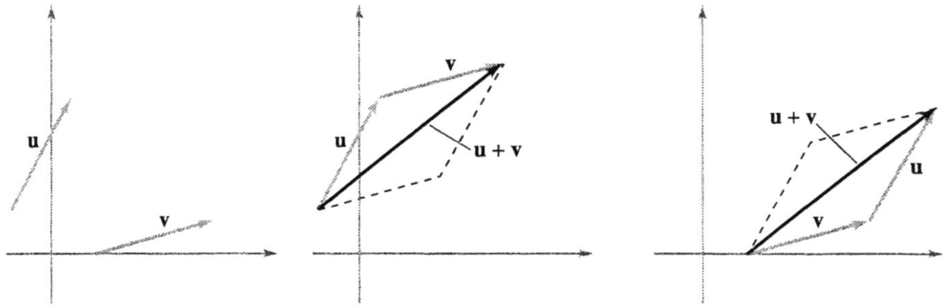

To find $\mathbf{u} + \mathbf{v}$,
Figure 10.7

(1) move the initial point of $\mathbf{v}$ to the terminal point of $\mathbf{u}$, or

(2) move the initial point of $\mathbf{u}$ to the terminal point of $\mathbf{v}$.

Figure 10.8 shows the equivalence of the geometric and algebraic definitions of vector addition and scalar multiplication, and presents (at far right) a geometric interpretation of $\mathbf{u} - \mathbf{v}$.

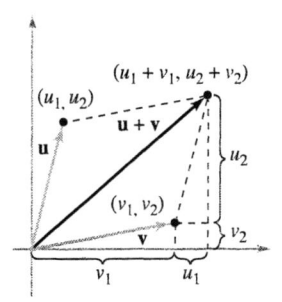

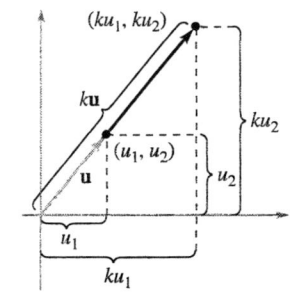

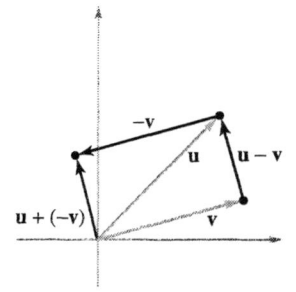

Vector addition   Scalar multiplication   Vector subtraction
Figure 10.8

Example 3 Vector Operations

Given $\mathbf{v} = \langle -2, 5 \rangle$ and $\mathbf{w} = \langle 3, 4 \rangle$, find each of the vectors.

a. $\frac{1}{2}\mathbf{v}$ **b.** $\mathbf{w} - \mathbf{v}$ **c.** $\mathbf{v} + 2\mathbf{w}$

Solution

a. $\frac{1}{2}\mathbf{v} = \langle \frac{1}{2}(-2), \frac{1}{2}(5) \rangle = \langle -1, \frac{5}{2} \rangle$

b. $\mathbf{w} - \mathbf{v} = \langle w_1 - v_1, w_2 - v_2 \rangle = \langle 3 - (-2), 4 - 5 \rangle = \langle 5, -1 \rangle$

c. Using $2\mathbf{w} = \langle 6, 8 \rangle$, you have

$$\mathbf{v} + 2\mathbf{w} = \langle -2, 5 \rangle + \langle 6, 8 \rangle$$
$$= \langle -2 + 6, 5 + 8 \rangle$$
$$= \langle 4, 13 \rangle.$$

Vector addition and scalar multiplication share many properties of ordinary arithmetic, as shown in the following theorem.

THEOREM 10.1 Properties of Vector Operations

Let $\mathbf{u}$, $\mathbf{v}$, and $\mathbf{w}$ be vectors in the plane, and let c and d be scalars.

1. $\mathbf{u} + \mathbf{v} = \mathbf{v} + \mathbf{u}$ Commutative property
2. $(\mathbf{u} + \mathbf{v}) + \mathbf{w} = \mathbf{u} + (\mathbf{v} + \mathbf{w})$ Associative property
3. $\mathbf{u} + \mathbf{0} = \mathbf{u}$ Additive identity property
4. $\mathbf{u} + (-\mathbf{u}) = \mathbf{0}$ Additive inverse property
5. $c(d\mathbf{u}) = (cd)\mathbf{u}$
6. $(c + d)\mathbf{u} = c\mathbf{u} + d\mathbf{u}$ Distributive property
7. $c(\mathbf{u} + \mathbf{v}) = c\mathbf{u} + c\mathbf{v}$ Distributive property
8. $1(\mathbf{u}) = \mathbf{u}, 0(\mathbf{u}) = \mathbf{0}$

Proof The proof of the *associative property* of vector addition uses the associative property of addition of real numbers.

$$(\mathbf{u} + \mathbf{v}) + \mathbf{w} = [\langle u_1, u_2 \rangle + \langle v_1, v_2 \rangle] + \langle w_1, w_2 \rangle$$
$$= \langle u_1 + v_1, u_2 + v_2 \rangle + \langle w_1, w_2 \rangle$$
$$= \langle (u_1 + v_1) + w_1, (u_2 + v_2) + w_2 \rangle$$
$$= \langle u_1 + (v_1 + w_1), u_2 + (v_2 + w_2) \rangle$$
$$= \langle u_1, u_2 \rangle + \langle v_1 + w_1, v_2 + w_2 \rangle = \mathbf{u} + (\mathbf{v} + \mathbf{w})$$

Similarly, the proof of the *distributive property* depends on the distributive property of real numbers.

$$(c + d)\mathbf{u} = (c + d)\langle u_1, u_2 \rangle$$
$$= \langle (c + d)u_1, (c + d)u_2 \rangle$$
$$= \langle cu_1 + du_1, cu_2 + du_2 \rangle$$
$$= \langle cu_1, cu_2 \rangle + \langle du_1, du_2 \rangle = c\mathbf{u} + d\mathbf{u}$$

The other properties can be proved in a similar manner.

Any set of vectors (with an accompanying set of scalars) that satisfies the eight properties given in Theorem 10.1 is a **vector space.**[*] The eight properties are the *vector space axioms*. So, this theorem states that the set of vectors in the plane (with the set of real numbers) forms a vector space.

THEOREM 10.2 Length of a Scalar Multiple

Let $\mathbf{v}$ be a vector and c be a scalar. Then

$$\|c\mathbf{v}\| = |c|\|\mathbf{v}\|. \qquad |c| \text{ is the absolute value of } c.$$

Proof Because $c\mathbf{v} = \langle cv_1, cv_2 \rangle$, it follows that

$$\begin{aligned}
\|c\mathbf{v}\| = \|\langle cv_1, cv_2 \rangle\| &= \sqrt{(cv_1)^2 + (cv_2)^2} \\
&= \sqrt{c^2 v_1^2 + c^2 v_2^2} \\
&= \sqrt{c^2(v_1^2 + v_2^2)} \\
&= |c|\sqrt{v_1^2 + v_2^2} \\
&= |c|\|\mathbf{v}\|.
\end{aligned}$$

In many applications of vectors, it is useful to find a unit vector that has the same direction as a given vector. The following theorem gives a procedure for doing this.

THEOREM 10.3 Unit Vector in the Direction of v

If $\mathbf{v}$ is a nonzero vector in the plane, then the vector

$$\mathbf{u} = \frac{\mathbf{v}}{\|\mathbf{v}\|} = \frac{1}{\|\mathbf{v}\|}\mathbf{v}$$

has length 1 and the same direction as $\mathbf{v}$.

Proof Because $1/\|\mathbf{v}\|$ is positive and $\mathbf{u} = (1/\|\mathbf{v}\|)\mathbf{v}$, you can conclude that $\mathbf{u}$ has the same direction as $\mathbf{v}$. To see that $\|\mathbf{u}\| = 1$, note that

$$\begin{aligned}
\|\mathbf{u}\| &= \left\|\left(\frac{1}{\|\mathbf{v}\|}\right)\mathbf{v}\right\| \\
&= \left|\frac{1}{\|\mathbf{v}\|}\right| \|\mathbf{v}\| \\
&= \frac{1}{\|\mathbf{v}\|}\|\mathbf{v}\| \\
&= 1.
\end{aligned}$$

So, $\mathbf{u}$ has length 1 and the same direction as $\mathbf{v}$.

In Theorem 10.3, $\mathbf{u}$ is called a **unit vector in the direction of v.** The process of multiplying $\mathbf{v}$ by $1/\|\mathbf{v}\|$ to get a unit vector is called **normalization of v.**

EMMY NOETHER (1882–1935)

One person who contributed to our knowledge of axiomatic systems was the German mathematician Emmy Noether. Noether is generally recognized as the leading woman mathematician in recent history.

FOR FURTHER INFORMATION For more information on Emmy Noether, see the article "Emmy Noether, Greatest Woman Mathematician" by Clark Kimberling in *The Mathematics Teacher*. To view this article, go to the website *www.matharticles.com*.

---

[*] *For more information about vector spaces, see* Elementary Linear Algebra, *Fourth Edition, by Larson and Edwards (Boston: Houghton Mifflin Company, 2000).*

Example 4 Finding a Unit Vector

Find a unit vector in the direction of $\mathbf{v} = \langle -2, 5 \rangle$ and verify that it has length 1.

Solution From Theorem 10.3, the unit vector in the direction of $\mathbf{v}$ is

$$\frac{\mathbf{v}}{\|\mathbf{v}\|} = \frac{\langle -2, 5 \rangle}{\sqrt{(-2)^2 + (5)^2}} = \frac{1}{\sqrt{29}} \langle -2, 5 \rangle = \left\langle \frac{-2}{\sqrt{29}}, \frac{5}{\sqrt{29}} \right\rangle.$$

This vector has length 1, because

$$\sqrt{\left(\frac{-2}{\sqrt{29}}\right)^2 + \left(\frac{5}{\sqrt{29}}\right)^2} = \sqrt{\frac{4}{29} + \frac{25}{29}} = \sqrt{\frac{29}{29}} = 1.$$

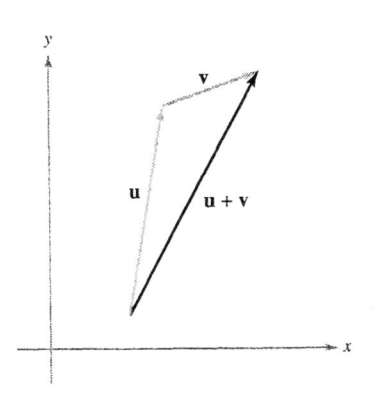

Generally, the length of the sum of two vectors is not equal to the sum of their lengths. To see this, consider the vectors $\mathbf{u}$ and $\mathbf{v}$ as shown in Figure 10.9. By considering $\mathbf{u}$ and $\mathbf{v}$ as two sides of a triangle, you can see that the length of the third side is $\|\mathbf{u} + \mathbf{v}\|$, and you have

$$\|\mathbf{u} + \mathbf{v}\| \leq \|\mathbf{u}\| + \|\mathbf{v}\|.$$

Equality occurs only if the vectors $\mathbf{u}$ and $\mathbf{v}$ have the *same direction*. This result is called the **triangle inequality** for vectors. (You are asked to prove this in Exercise 81, Section 10.3.)

Triangle inequality
Figure 10.9

Standard Unit Vectors

The unit vectors $\langle 1, 0 \rangle$ and $\langle 0, 1 \rangle$ are called the **standard unit vectors** in the plane and are denoted by

$$\mathbf{i} = \langle 1, 0 \rangle \quad \text{and} \quad \mathbf{j} = \langle 0, 1 \rangle \qquad \text{Standard unit vectors}$$

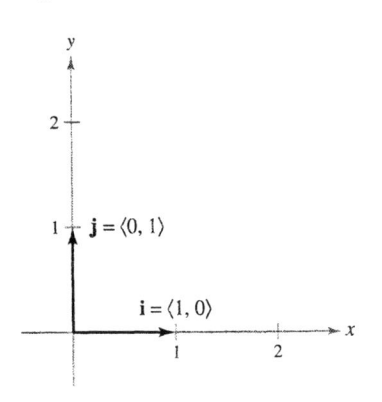

as shown in Figure 10.10. These vectors can be used to uniquely represent any vector, as follows.

$$\mathbf{v} = \langle v_1, v_2 \rangle = \langle v_1, 0 \rangle + \langle 0, v_2 \rangle = v_1 \langle 1, 0 \rangle + v_2 \langle 0, 1 \rangle = v_1 \mathbf{i} + v_2 \mathbf{j}$$

The vector $\mathbf{v} = v_1 \mathbf{i} + v_2 \mathbf{j}$ is called a **linear combination** of $\mathbf{i}$ and $\mathbf{j}$. The scalars v_1 and v_2 are called the **horizontal** and **vertical components of v.**

Standard unit vectors **i** and **j**
Figure 10.10

Example 5 Writing a Vector as a Linear Combination of Unit Vectors

Let $\mathbf{u}$ be the vector with initial point $(2, -5)$ and terminal point $(-1, 3)$, and let $\mathbf{v} = 2\mathbf{i} - \mathbf{j}$. Write each of the vectors as a linear combination of $\mathbf{i}$ and $\mathbf{j}$.

a. $\mathbf{u}$ **b.** $\mathbf{w} = 2\mathbf{u} - 3\mathbf{v}$

Solution

a. $\mathbf{u} = \langle q_1 - p_1, q_2 - p_2 \rangle = \langle -1 - 2, 3 - (-5) \rangle = \langle -3, 8 \rangle = -3\mathbf{i} + 8\mathbf{j}$

b. $\mathbf{w} = 2\mathbf{u} - 3\mathbf{v} = 2(-3\mathbf{i} + 8\mathbf{j}) - 3(2\mathbf{i} - \mathbf{j})$

$= -6\mathbf{i} + 16\mathbf{j} - 6\mathbf{i} + 3\mathbf{j}$

$= -12\mathbf{i} + 19\mathbf{j}$

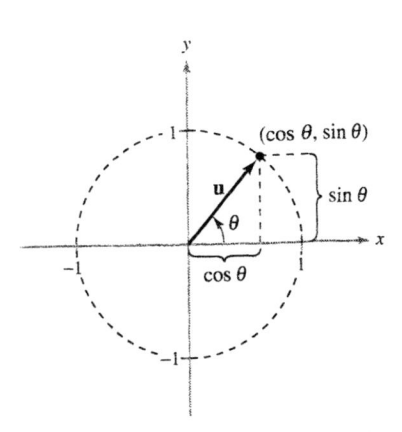

If **u** is a unit vector such that θ is the angle (measured counterclockwise) from the positive x-axis to **u**, then the terminal point of **u** lies on the unit circle, and you have

$$\mathbf{u} = \langle \cos\theta, \sin\theta \rangle = \cos\theta\,\mathbf{i} + \sin\theta\,\mathbf{j} \qquad \text{Unit vector}$$

as shown in Figure 10.11. Moreover, it follows that any other nonzero vector **v** making an angle θ with the positive x-axis has the same direction as **u**, and you can write

$$\mathbf{v} = \|\mathbf{v}\|\langle \cos\theta, \sin\theta \rangle = \|\mathbf{v}\|\cos\theta\,\mathbf{i} + \|\mathbf{v}\|\sin\theta\,\mathbf{j}.$$

The angle θ from the positive x-axis to the vector **u**
Figure 10.11

Example 6 Writing a Vector of Given Length and Direction

The vector **v** has a length of 3 and makes an angle of $30° = \pi/6$ with the positive x-axis. Write **v** as a linear combination of the unit vectors **i** and **j**.

Solution Because the angle between **v** and the positive x-axis is $\theta = \pi/6$, you can write the following.

$$\mathbf{v} = \|\mathbf{v}\|\cos\theta\,\mathbf{i} + \|\mathbf{v}\|\sin\theta\,\mathbf{j}$$
$$= 3\cos\frac{\pi}{6}\mathbf{i} + 3\sin\frac{\pi}{6}\mathbf{j}$$
$$= \frac{3\sqrt{3}}{2}\mathbf{i} + \frac{3}{2}\mathbf{j}$$

Applications of Vectors

There are many applications of vectors in physics and engineering. One example is force. A vector can be used to represent force because force has both magnitude and direction. If two or more forces are acting on an object, then the **resultant force** on the object is the vector sum of the vector forces.

Example 7 Finding the Resultant Force

Two tugboats are pushing an ocean liner, as shown in Figure 10.12. Each boat is exerting a force of 400 pounds. What is the resultant force on the ocean liner?

Solution Using Figure 10.12, you can represent the forces exerted by the first and second tugboats as

$$\mathbf{F}_1 = 400\langle \cos 20°, \sin 20° \rangle$$
$$= 400\cos(20°)\mathbf{i} + 400\sin(20°)\mathbf{j}$$
$$\mathbf{F}_2 = 400\langle \cos(-20°), \sin(-20°) \rangle$$
$$= 400\cos(20°)\mathbf{i} - 400\sin(20°)\mathbf{j}.$$

The resultant force on the ocean liner is

$$\mathbf{F} = \mathbf{F}_1 + \mathbf{F}_2$$
$$= [400\cos(20°)\mathbf{i} + 400\sin(20°)\mathbf{j}] + [400\cos(20°)\mathbf{i} - 400\sin(20°)\mathbf{j}]$$
$$= 800\cos(20°)\mathbf{i}$$
$$\approx 752\mathbf{i}.$$

So, the resultant force on the ocean liner is approximately 752 pounds in the direction of the positive x-axis.

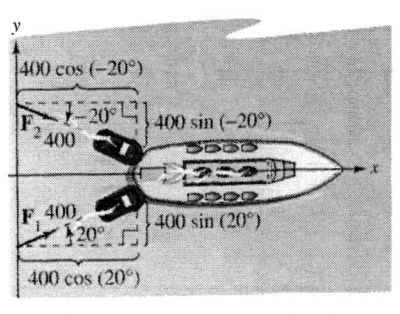

The resultant force on the ocean liner is approximately 752 pounds in the direction of the positive x-axis.
Figure 10.12

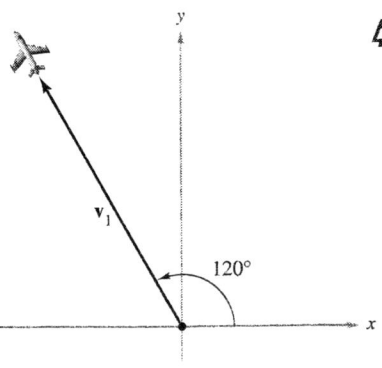

Example 8 Finding a Velocity

An airplane is traveling at a fixed altitude with a negligible wind factor. The plane is headed N 30° W (30° west of north) at a speed of 500 miles per hour, as shown in Figure 10.13. As the plane reaches a certain point, it encounters wind with a velocity of 70 miles per hour in the direction E 45° N. What are the resultant speed and direction of the plane?

Solution Using Figure 10.13, you can represent the velocity of the plane as

$$\mathbf{v}_1 = 500 \cos(120°)\mathbf{i} + 500 \sin(120°)\mathbf{j}.$$

The velocity of the wind is represented by the vector

$$\mathbf{v}_2 = 70 \cos(45°)\mathbf{i} + 70 \sin(45°)\mathbf{j}.$$

The resultant velocity of the plane is

$$\begin{aligned}\mathbf{v} &= \mathbf{v}_1 + \mathbf{v}_2 \\ &= 500 \cos(120°)\mathbf{i} + 500 \sin(120°)\mathbf{j} + 70 \cos(45°)\mathbf{i} + 70 \sin(45°)\mathbf{j} \\ &\approx -200.5\mathbf{i} + 482.5\mathbf{j}.\end{aligned}$$

To find the speed and direction, write $\mathbf{v} = \|\mathbf{v}\|(\cos\theta\,\mathbf{i} + \sin\theta\,\mathbf{j})$. Because $\|\mathbf{v}\| \approx \sqrt{(-200.5)^2 + (482.5)^2} \approx 522.5$ you can write

$$\mathbf{v} \approx 522.5\left(\frac{-200.5}{522.5}\mathbf{i} + \frac{482.5}{522.5}\mathbf{j}\right) \approx 522.5[\cos(112.6°)\mathbf{i} + \sin(112.6°)\mathbf{j}].$$

The new speed of the plane, as altered by the wind, is approximately 522.5 miles per hour in a path that makes an angle of 112.6° with the positive x-axis.

Direction without wind

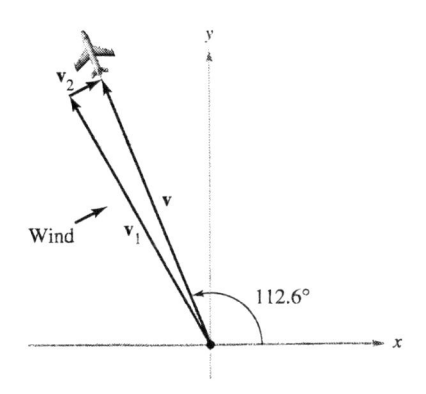

Direction with wind
Figure 10.13

EXERCISES FOR SECTION 10.1

In Exercises 1–4, (a) find the component form of the vector **v** and (b) sketch the vector with its initial point at the origin.

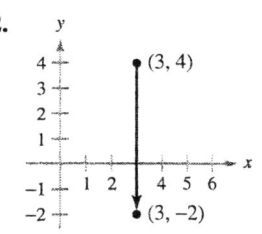

1. 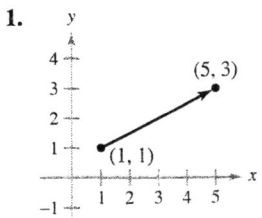 (1, 1) to (5, 3)
2. (3, 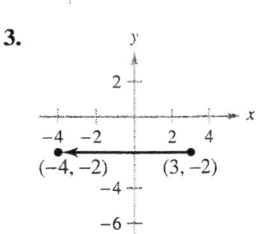 −2) to (3, 4) 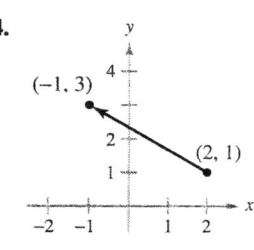
3. (−4, −2) to (3, −2)
4. (2, 1) to (−1, 3)

In Exercises 5–8, find the vectors **u** and **v** whose initial and terminal points are given. Show that **u** and **v** are equivalent.

5. **u**: (3, 2), (5, 6)
   v: (−1, 4), (1, 8)
6. **u**: (−4, 0), (1, 8)
   v: (2, −1), (7, 7)
7. **u**: (0, 3), (6, −2)
   v: (3, 10), (9, 5)
8. **u**: (−4, −1), (11, −4)
   v: (10, 13), (25, 10)

In Exercises 9–16, the initial and terminal points of a vector **v** are given. (a) Sketch the given directed line segment, (b) write the vector in component form, and (c) sketch the vector with its initial point at the origin.

	Initial Point	Terminal Point		Initial Point	Terminal Point
9.	(1, 2)	(5, 5)	10.	(2, −6)	(3, 6)
11.	(10, 2)	(6, −1)	12.	(0, −4)	(−5, −1)

indicates that in the Interactive 3.0 *CD-ROM and* Internet 3.0 *versions of this text (available at* college.hmco.com) *you will find an Open Exploration, which further explores this example using the computer algebra systems* Maple, Mathcad, Mathematica, *and* Derive.

	Initial Point	Terminal Point		Initial Point	Terminal Point
13.	(6, 2)	(6, 6)	14.	(7, −1)	(−3, −1)
15.	$\left(\frac{3}{2}, \frac{4}{3}\right)$	$\left(\frac{1}{2}, 3\right)$	16.	(0.12, 0.60)	(0.84, 1.25)

In Exercises 17 and 18, sketch each scalar multiple of v.

17. $\mathbf{v} = \langle 2, 3 \rangle$
    (a) $2\mathbf{v}$ (b) $-3\mathbf{v}$ (c) $\frac{7}{2}\mathbf{v}$ (d) $\frac{2}{3}\mathbf{v}$

18. $\mathbf{v} = \langle -1, 5 \rangle$
    (a) $4\mathbf{v}$ (b) $-\frac{1}{2}\mathbf{v}$ (c) $0\mathbf{v}$ (d) $-6\mathbf{v}$

In Exercises 19–22, use the figure to sketch a graph of the indicated vector. To print an enlarged copy of the graph, go to the website www.mathgraphs.com.

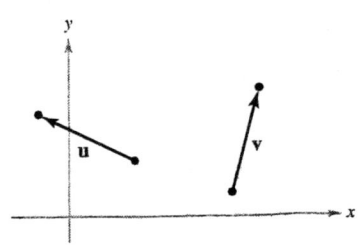

19. $-\mathbf{u}$
20. $2\mathbf{u}$
21. $\mathbf{u} - \mathbf{v}$
22. $\mathbf{u} + 2\mathbf{v}$

In Exercises 23 and 24, find (a) $\frac{2}{3}\mathbf{u}$, (b) $\mathbf{v} - \mathbf{u}$, and (c) $2\mathbf{u} + 5\mathbf{v}$.

23. $\mathbf{u} = \langle 4, 9 \rangle$
    $\mathbf{v} = \langle 2, -5 \rangle$
24. $\mathbf{u} = \langle -3, -8 \rangle$
    $\mathbf{v} = \langle 8, 25 \rangle$

In Exercises 25–28, find the vector v where $\mathbf{u} = \langle 2, -1 \rangle$ and $\mathbf{w} = \langle 1, 2 \rangle$. Illustrate the vector operations geometrically.

25. $\mathbf{v} = \frac{3}{2}\mathbf{u}$
26. $\mathbf{v} = \mathbf{u} + \mathbf{w}$
27. $\mathbf{v} = \mathbf{u} + 2\mathbf{w}$
28. $\mathbf{v} = 5\mathbf{u} - 3\mathbf{w}$

In Exercises 29 and 30, the vector v and its initial point are given. Find the terminal point.

29. $\mathbf{v} = \langle -1, 3 \rangle$, initial point (4, 2)
30. $\mathbf{v} = \langle 4, -9 \rangle$, initial point (3, 2)

In Exercises 31–36, find the magnitude of v.

31. $\mathbf{v} = \langle 4, 3 \rangle$
32. $\mathbf{v} = \langle 12, -5 \rangle$
33. $\mathbf{v} = 6\mathbf{i} - 5\mathbf{j}$
34. $\mathbf{v} = -10\mathbf{i} + 3\mathbf{j}$
35. $\mathbf{v} = 4\mathbf{j}$
36. $\mathbf{v} = \mathbf{i} - \mathbf{j}$

In Exercises 37–40, find the unit vector in the direction of u and verify that it has length 1.

37. $\mathbf{u} = \langle 3, 12 \rangle$
38. $\mathbf{u} = \langle 5, 15 \rangle$
39. $\mathbf{u} = \left\langle \frac{3}{2}, \frac{5}{2} \right\rangle$
40. $\mathbf{u} = \langle -6.2, 3.4 \rangle$

In Exercises 41–44, find the following.

(a) $\|\mathbf{u}\|$ (b) $\|\mathbf{v}\|$ (c) $\|\mathbf{u} + \mathbf{v}\|$
(d) $\left\|\dfrac{\mathbf{u}}{\|\mathbf{u}\|}\right\|$ (e) $\left\|\dfrac{\mathbf{v}}{\|\mathbf{v}\|}\right\|$ (f) $\left\|\dfrac{\mathbf{u} + \mathbf{v}}{\|\mathbf{u} + \mathbf{v}\|}\right\|$

41. $\mathbf{u} = \langle 1, -1 \rangle$
    $\mathbf{v} = \langle -1, 2 \rangle$
42. $\mathbf{u} = \langle 0, 1 \rangle$
    $\mathbf{v} = \langle 3, -3 \rangle$
43. $\mathbf{u} = \left\langle 1, \frac{1}{2} \right\rangle$
    $\mathbf{v} = \langle 2, 3 \rangle$
44. $\mathbf{u} = \langle 2, -4 \rangle$
    $\mathbf{v} = \langle 5, 5 \rangle$

In Exercises 45 and 46, demonstrate the triangle inequality using the vectors u and v.

45. $\mathbf{u} = \langle 2, 1 \rangle$
    $\mathbf{v} = \langle 5, 4 \rangle$
46. $\mathbf{u} = \langle -3, 2 \rangle$
    $\mathbf{v} = \langle 1, -2 \rangle$

In Exercises 47–50, find the vector v with the given magnitude and the same direction as u.

	Magnitude	Direction
47.	$\|\mathbf{v}\| = 4$	$\mathbf{u} = \langle 1, 1 \rangle$
48.	$\|\mathbf{v}\| = 4$	$\mathbf{u} = \langle -1, 1 \rangle$
49.	$\|\mathbf{v}\| = 2$	$\mathbf{u} = \langle \sqrt{3}, 3 \rangle$
50.	$\|\mathbf{v}\| = 3$	$\mathbf{u} = \langle 0, 3 \rangle$

In Exercises 51–54, find the component form of v given its magnitude and the angle it makes with the positive x-axis.

51. $\|\mathbf{v}\| = 3, \ \theta = 0°$
52. $\|\mathbf{v}\| = 5, \ \theta = 120°$
53. $\|\mathbf{v}\| = 2, \ \theta = 150°$
54. $\|\mathbf{v}\| = 1, \ \theta = 3.5°$

In Exercises 55–58, find the component form of u + v given the magnitudes of u and v and the angles that u and v make with the positive x-axis.

55. $\|\mathbf{u}\| = 1, \ \theta_{\mathbf{u}} = 0°$
    $\|\mathbf{v}\| = 3, \ \theta_{\mathbf{v}} = 45°$
56. $\|\mathbf{u}\| = 4, \ \theta_{\mathbf{u}} = 0°$
    $\|\mathbf{v}\| = 2, \ \theta_{\mathbf{v}} = 60°$
57. $\|\mathbf{u}\| = 2, \ \theta_{\mathbf{u}} = 4$
    $\|\mathbf{v}\| = 1, \ \theta_{\mathbf{v}} = 2$
58. $\|\mathbf{u}\| = 5, \ \theta_{\mathbf{u}} = -0.5$
    $\|\mathbf{v}\| = 5, \ \theta_{\mathbf{v}} = 0.5$

Getting at the Concept

59. In your own words, state the difference between a scalar and a vector. Give examples of each.
60. Give geometric descriptions of the operations of addition of vectors and multiplication of a vector by a scalar.
61. What is meant by the normalization of a vector?
62. State the eight vector space axioms.

In Exercises 63–68, find a and b such that $\mathbf{v} = a\mathbf{u} + b\mathbf{w}$, where $\mathbf{u} = \langle 1, 2 \rangle$ and $\mathbf{w} = \langle 1, -1 \rangle$.

63. $\mathbf{v} = \langle 2, 1 \rangle$
64. $\mathbf{v} = \langle 0, 3 \rangle$
65. $\mathbf{v} = \langle 3, 0 \rangle$
66. $\mathbf{v} = \langle 3, 3 \rangle$
67. $\mathbf{v} = \langle 1, 1 \rangle$
68. $\mathbf{v} = \langle -1, 7 \rangle$

In Exercises 69–72, find a unit vector (a) parallel to and (b) normal to the graph of $f(x)$ at the indicated point.

Function	Point
69. $f(x) = x^3$	$(1, 1)$
70. $f(x) = x^3$	$(-2, -8)$
71. $f(x) = \sqrt{25 - x^2}$	$(3, 4)$
72. $f(x) = \tan x$	$\left(\dfrac{\pi}{4}, 1\right)$

In Exercises 73 and 74, find the component form of $\mathbf{v}$ given the magnitudes of $\mathbf{u}$ and $\mathbf{u} + \mathbf{v}$ and the angles that $\mathbf{u}$ and $\mathbf{u} + \mathbf{v}$ make with the positive x-axis.

73. $\|\mathbf{u}\| = 1, \theta = 45°$
    $\|\mathbf{u} + \mathbf{v}\| = \sqrt{2}, \theta = 90°$

74. $\|\mathbf{u}\| = 4, \theta = 30°$
    $\|\mathbf{u} + \mathbf{v}\| = 6, \theta = 120°$

75. *Programming* You are given the magnitudes of $\mathbf{u}$ and $\mathbf{v}$ and the angles $\mathbf{u}$ and $\mathbf{v}$ make with the positive x-axis. Write a program for a graphing utility in which the output is the following.
    (a) $\mathbf{u} + \mathbf{v}$
    (b) $\|\mathbf{u} + \mathbf{v}\|$
    (c) The angle $\mathbf{u} + \mathbf{v}$ makes with the positive x-axis

76. Use the program of Exercise 75 to find the magnitude and direction of the resultant of the vectors.

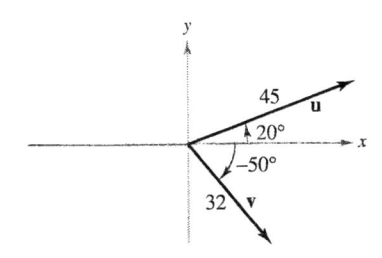

In Exercises 77 and 78, use a graphing utility to find the magnitude and direction of the resultant of the vectors.

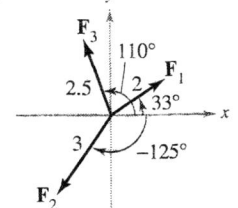

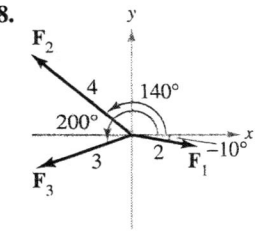

77.

78.

79. *Numerical and Graphical Analysis* Forces with magnitudes of 180 newtons and 275 newtons act on a hook (see figure). The angle between the two forces is θ degrees.
    (a) If $\theta = 30°$, find the direction and magnitude of the resultant force.
    (b) Express the magnitude M and direction α of the resultant force as functions of θ, where $0° \leq \theta \leq 180°$.
    (c) Use a graphing utility to complete the table.

θ	0°	30°	60°	90°	120°	150°	180°
M							
α							

    (d) Use a graphing utility to graph the two functions M and α.
    (e) Explain why one of the functions decreases for increasing values of θ whereas the other does not.

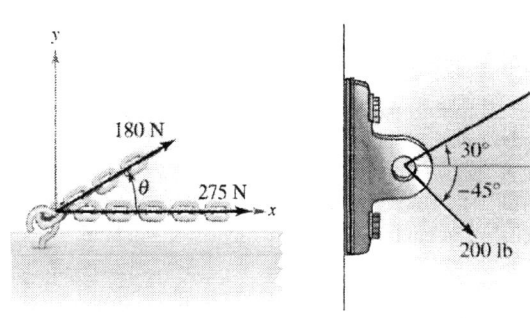

Figure for 79   **Figure for 80**

80. *Resultant Force* Forces with magnitudes of 500 pounds and 200 pounds act on a machine part at angles of 30° and −45° with the x-axis (see figure). Find the direction and magnitude of the resultant force.

81. *Resultant Force* Three forces with magnitudes of 75 pounds, 100 pounds, and 125 pounds act on an object at angles of 30°, 45°, and 120° with the positive x-axis. Find the direction and magnitude of the resultant force.

82. *Resultant Force* Three forces with magnitudes of 400 newtons, 280 newtons, and 350 newtons act on an object at angles of −30°, 45°, and 135° with the positive x-axis. Find the direction and magnitude of the resultant force.

83. *Think About It* Consider two forces of equal magnitude acting on a point.
    (a) If the magnitude of the resultant is the sum of the magnitudes of the two forces, make a conjecture about the angle between the forces.
    (b) If the resultant of the forces is **0**, make a conjecture about the angle between the forces.
    (c) Can the magnitude of the resultant be greater than the sum of the magnitudes of the two forces? Explain.

84. Graphical Reasoning Consider two forces $\mathbf{F}_1 = \langle 20, 0 \rangle$ and $\mathbf{F}_2 = 10\langle \cos\theta, \sin\theta \rangle$.
(a) Find $\|\mathbf{F}_1 + \mathbf{F}_2\|$.
(b) Determine the magnitude of the resultant as a function of θ. Use a graphing utility to graph the function for $0 \leq \theta < 2\pi$.
(c) Use the graph in part (b) to determine the range of the function. What is its maximum and for what value of θ does it occur? What is its minimum and for what value of θ does it occur?
(d) Explain why the magnitude of the resultant is never 0.

85. Three vertices of a parallelogram are $(1, 2)$, $(3, 1)$, and $(8, 4)$. Find the three possible fourth vertices (see figure).

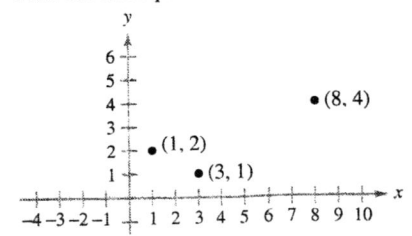

86. Use vectors to find the points of trisection of the line segment with endpoints $(1, 2)$ and $(7, 5)$.

Cable Tension In Exercises 87 and 88, use the figure to determine the tension in each cable supporting the given load.

87.

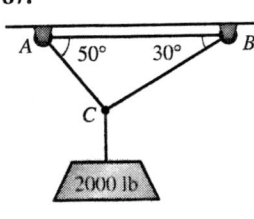

88.

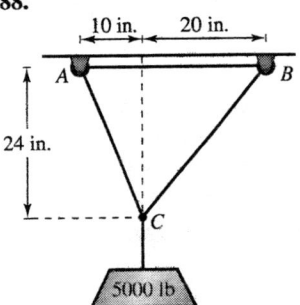

89. Projectile Motion A gun with a muzzle velocity of 1200 feet per second is fired at an angle of 6° above the horizontal. Find the vertical and horizontal components of the velocity.

90. Shared Load To carry a 100-pound cylindrical weight, two workers lift on the ends of short ropes tied to an eyelet on the top center of the cylinder. One rope makes a 20° angle away from the vertical and the other a 30° angle (see figure).
(a) Find each rope's tension if the resultant force is vertical.
(b) Find the vertical component of each worker's force.

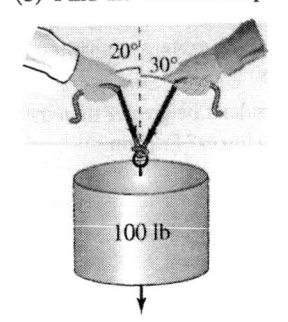

91. Navigation An airplane is headed W 32° N. Its speed with respect to the air is 900 kilometers per hour. The wind at the plane's altitude is from the southwest at 100 kilometers per hour (see figure). What is the true direction of the plane, and what is its speed with respect to the ground?

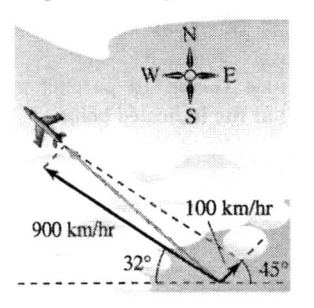

92. Navigation A plane flies at a constant groundspeed of 400 miles per hour due east and encounters a 50-mile-per-hour wind from the northwest. Find the airspeed and compass direction that will allow the plane to maintain its groundspeed and eastward direction.

93. If $\mathbf{F}_1 + \mathbf{F}_2 + \mathbf{F}_3 = \mathbf{0}$, find T_2 and T_3.
$$\mathbf{F}_1 = -3600\mathbf{j}, \quad \mathbf{F}_2 = T_2(\cos 35°\mathbf{i} - \sin 35°\mathbf{j}),$$
$$\mathbf{F}_3 = T_3(\cos 92°\mathbf{i} + \sin 92°\mathbf{j})$$

94. Prove that $\mathbf{u} = (\cos\theta)\mathbf{i} - (\sin\theta)\mathbf{j}$ and $\mathbf{v} = (\sin\theta)\mathbf{i} + (\cos\theta)\mathbf{j}$ are unit vectors for any angle θ.

95. Geometry Using vectors, prove that the line segment joining the midpoints of two sides of a triangle is parallel to, and one half the length of, the third side.

96. Geometry Using vectors, prove that the diagonals of a parallelogram bisect each other.

97. Prove that the vector $\mathbf{w} = \|\mathbf{u}\|\mathbf{v} + \|\mathbf{v}\|\mathbf{u}$ bisects the angle between $\mathbf{u}$ and $\mathbf{v}$.

98. Consider the vector $\mathbf{u} = \langle x, y \rangle$. Describe the set of all points (x, y) such that $\|\mathbf{u}\| = 5$.

True or False? In Exercises 99–104, determine whether the statement is true or false. If it is false, explain why or give an example that shows it is false.

99. If $\mathbf{u}$ and $\mathbf{v}$ have the same magnitude and direction, then $\mathbf{u}$ and $\mathbf{v}$ are equivalent.

100. If $\mathbf{u}$ is a unit vector in the direction of $\mathbf{v}$, then $\mathbf{v} = \|\mathbf{v}\|\mathbf{u}$.

101. If $\mathbf{u} = a\mathbf{i} + b\mathbf{j}$ is a unit vector, then $a^2 + b^2 = 1$.

102. If $\mathbf{v} = a\mathbf{i} + b\mathbf{j} = \mathbf{0}$, then $a = -b$.

103. If $a = b$, then $\|a\mathbf{i} + b\mathbf{j}\| = \sqrt{2}a$.

104. If $\mathbf{u}$ and $\mathbf{v}$ have the same magnitude but opposite directions, then $\mathbf{u} + \mathbf{v} = \mathbf{0}$.

Section 10.2 Space Coordinates and Vectors in Space

- Understand the three-dimensional rectangular coordinate system.
- Analyze vectors in space.
- Use three-dimensional vectors to solve real-life problems.

Coordinates in Space

Up to this point in the text, you have been primarily concerned with the two-dimensional coordinate system. Much of the remaining part of your study of calculus will involve the three-dimensional coordinate system.

Before extending the concept of a vector to three dimensions, we introduce the **three-dimensional coordinate system.** You can construct this system by passing a z-axis perpendicular to both the x- and y-axes at the origin. Figure 10.14 shows the positive portion of each coordinate axis. Taken as pairs, the axes determine three **coordinate planes:** the *xy*-plane, the *xz*-plane, and the *yz*-plane. These three coordinate planes separate three-space into eight **octants.** The first octant is the one for which all three coordinates are positive. In this three-dimensional system, a point P in space is determined by an ordered triple (x, y, z) where x, y, and z are as follows.

The three-dimensional coordinate system
Figure 10.14

$x = $ directed distance from yz-plane to P

$y = $ directed distance from xz-plane to P

$z = $ directed distance from xy-plane to P

Several points are shown in Figure 10.15.

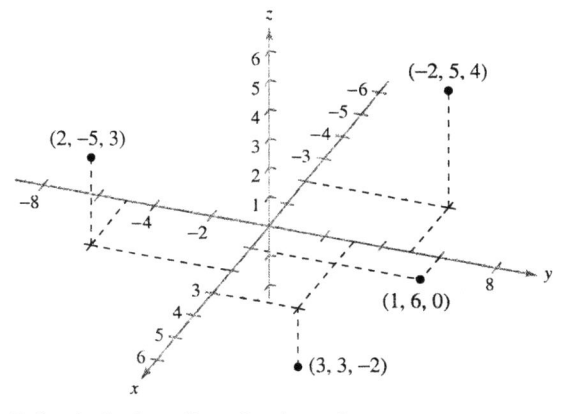

Points in the three-dimensional coordinate system are represented by ordered triples.
Figure 10.15

A three-dimensional coordinate system can have either a **left-handed** or a **right-handed** orientation. To determine the orientation of a system, imagine that you are standing at the origin, with your arms pointing in the direction of the positive x- and y-axes, and with the z-axis pointing up, as shown in Figure 10.16. The system is right-handed or left-handed depending on which hand points along the x-axis. In this text we work exclusively with the right-handed system.

Right-handed system   Left-handed system
Figure 10.16

NOTE To help visualize points or objects in a three-dimensional system, try using the CD-ROM three-dimensional software that accompanies this text.

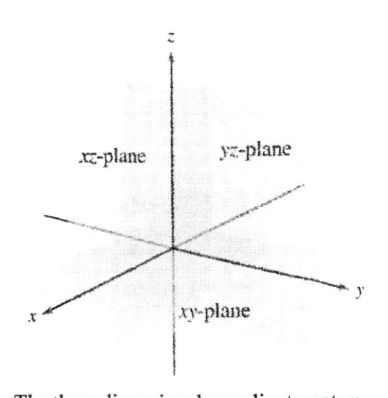

Many of the formulas established for the two-dimensional coordinate system can be extended to three dimensions. For example, to find the distance between two points in space, you can use the Pythagorean Theorem twice, as shown in Figure 10.17. By doing this, you will obtain the formula for the distance between the points (x_1, y_1, z_1) and (x_2, y_2, z_2).

$$d = \sqrt{(x_2 - x_1)^2 + (y_2 - y_1)^2 + (z_2 - z_1)^2} \quad \text{Distance Formula}$$

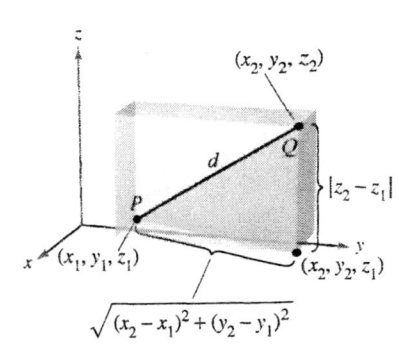

The distance between two points in space
Figure 10.17

Example 1 Finding the Distance Between Two Points in Space

The distance between the points $(2, -1, 3)$ and $(1, 0, -2)$ is

$$\begin{aligned} d &= \sqrt{(1 - 2)^2 + (0 + 1)^2 + (-2 - 3)^2} \quad \text{Distance Formula} \\ &= \sqrt{1 + 1 + 25} \\ &= \sqrt{27} \\ &= 3\sqrt{3}. \end{aligned}$$

A **sphere** with center at (x_0, y_0, z_0) and radius r is defined to be the set of all points (x, y, z) such that the distance between (x, y, z) and (x_0, y_0, z_0) is r. You can use the Distance Formula to find the **standard equation of a sphere** of radius r, centered at (x_0, y_0, z_0). If (x, y, z) is an arbitrary point on the sphere, the equation of the sphere is

$$(x - x_0)^2 + (y - y_0)^2 + (z - z_0)^2 = r^2 \quad \text{Equation of sphere}$$

as shown in Figure 10.18. Moreover, the midpoint of the line segment joining the points (x_1, y_1, z_1) and (x_2, y_2, z_2) has coordinates

$$\left(\frac{x_1 + x_2}{2}, \frac{y_1 + y_2}{2}, \frac{z_1 + z_2}{2} \right). \quad \text{Midpoint Rule}$$

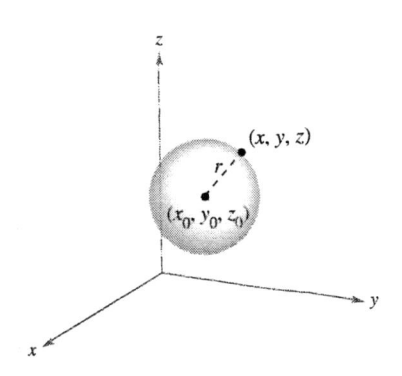

Figure 10.18

Example 2 Finding the Equation of a Sphere

Find the standard equation of the sphere that has the points $(5, -2, 3)$ and $(0, 4, -3)$ as endpoints of a diameter.

Solution By the Midpoint Rule, the center of the sphere is

$$\left(\frac{5 + 0}{2}, \frac{-2 + 4}{2}, \frac{3 - 3}{2} \right) = \left(\frac{5}{2}, 1, 0 \right). \quad \text{Midpoint Rule}$$

By the Distance Formula, the radius is

$$r = \sqrt{\left(0 - \frac{5}{2} \right)^2 + (4 - 1)^2 + (-3 - 0)^2} = \sqrt{\frac{97}{4}} = \frac{\sqrt{97}}{2}.$$

Therefore, the standard equation of the sphere is

$$\left(x - \frac{5}{2} \right)^2 + (y - 1)^2 + (z - 0)^2 = \frac{97}{4}. \quad \text{Equation of sphere}$$

Vectors in Space

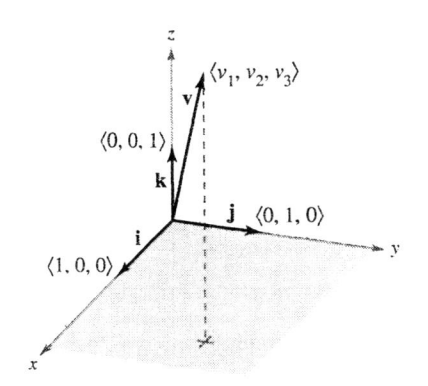

In space, vectors are denoted by ordered triples $\mathbf{v} = \langle v_1, v_2, v_3 \rangle$. The **zero vector** is denoted by $\mathbf{0} = \langle 0, 0, 0 \rangle$. Using the unit vectors $\mathbf{i} = \langle 1, 0, 0 \rangle$, $\mathbf{j} = \langle 0, 1, 0 \rangle$, and $\mathbf{k} = \langle 0, 0, 1 \rangle$ in the direction of the positive z-axis, the **standard unit vector notation** for $\mathbf{v}$ is

$$\mathbf{v} = v_1 \mathbf{i} + v_2 \mathbf{j} + v_3 \mathbf{k}$$

as shown in Figure 10.19. If $\mathbf{v}$ is represented by the directed line segment from $P(p_1, p_2, p_3)$ to $Q(q_1, q_2, q_3)$, as shown in Figure 10.20, the component form of $\mathbf{v}$ is given by subtracting the coordinates of the initial point from the coordinates of the terminal point, as follows.

$$\mathbf{v} = \langle v_1, v_2, v_3 \rangle = \langle q_1 - p_1, q_2 - p_2, q_3 - p_3 \rangle$$

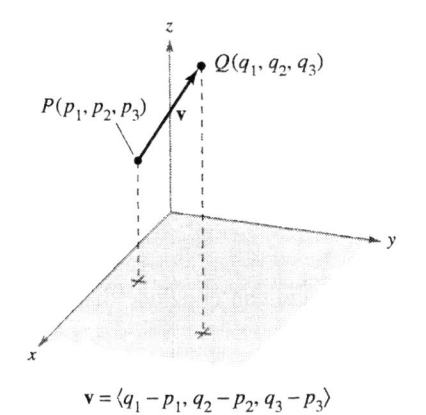

The standard unit vectors in space
Figure 10.19

$\mathbf{v} = \langle q_1 - p_1, q_2 - p_2, q_3 - p_3 \rangle$
Figure 10.20

Vectors in Space

Let $\mathbf{u} = \langle u_1, u_2, u_3 \rangle$ and $\mathbf{v} = \langle v_1, v_2, v_3 \rangle$ be vectors in space and let c be a scalar.

1. *Equality of Vectors:* $\mathbf{u} = \mathbf{v}$ if and only if $u_1 = v_1$, $u_2 = v_2$, and $u_3 = v_3$.
2. *Component Form:* If $\mathbf{v}$ is represented by the directed line segment from $P(p_1, p_2, p_3)$ to $Q(q_1, q_2, q_3)$, then
$$\mathbf{v} = \langle v_1, v_2, v_3 \rangle = \langle q_1 - p_1, q_2 - p_2, q_3 - p_3 \rangle.$$
3. *Length:* $\|\mathbf{v}\| = \sqrt{v_1^2 + v_2^2 + v_3^2}$
4. *Unit Vector in the Direction of $\mathbf{v}$:* $\dfrac{\mathbf{v}}{\|\mathbf{v}\|} = \left(\dfrac{1}{\|\mathbf{v}\|}\right) \langle v_1, v_2, v_3 \rangle, \quad \mathbf{v} \neq \mathbf{0}$
5. *Vector Addition:* $\mathbf{v} + \mathbf{u} = \langle v_1 + u_1, v_2 + u_2, v_3 + u_3 \rangle$
6. *Scalar Multiplication:* $c\mathbf{v} = \langle cv_1, cv_2, cv_3 \rangle$

NOTE The properties of vector addition and scalar multiplication given in Theorem 10.1 are also valid for vectors in space.

Example 3 Finding the Component Form of a Vector in Space

Find the component form and length of the vector $\mathbf{v}$ having initial point $(-2, 3, 1)$ and terminal point $(0, -4, 4)$. Then find a unit vector in the direction of $\mathbf{v}$.

Solution The component form of $\mathbf{v}$ is

$$\mathbf{v} = \langle q_1 - p_1, q_2 - p_2, q_3 - p_3 \rangle = \langle 0 - (-2), -4 - 3, 4 - 1 \rangle$$
$$= \langle 2, -7, 3 \rangle$$

which implies that its length is

$$\|\mathbf{v}\| = \sqrt{(2)^2 + (-7)^2 + (3)^2} = \sqrt{62}.$$

The unit vector in the direction of $\mathbf{v}$ is.

$$\mathbf{u} = \frac{\mathbf{v}}{\|\mathbf{v}\|} = \frac{1}{\sqrt{62}} \langle 2, -7, 3 \rangle.$$

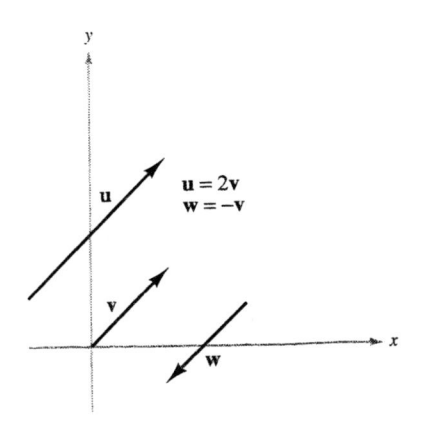

Recall from the definition of scalar multiplication that positive scalar multiples of a nonzero vector **v** have the same direction as **v**, whereas negative multiples have the direction opposite of **v**. In general, two nonzero vectors **u** and **v** are **parallel** if there is some scalar c such that $\mathbf{u} = c\mathbf{v}$.

Definition of Parallel Vectors

Two nonzero vectors **u** and **v** are **parallel** if there is some scalar c such that $\mathbf{u} = c\mathbf{v}$.

For example, in Figure 10.21, the vectors **u, v,** and **w** are parallel because $\mathbf{u} = 2\mathbf{v}$ and $\mathbf{w} = -\mathbf{v}$.

Parallel vectors
Figure 10.21

Example 4   Parallel Vectors

Vector **w** has initial point $(2, -1, 3)$ and terminal point $(-4, 7, 5)$. Which of the following vectors is parallel to **w**?

a. $\mathbf{u} = \langle 3, -4, -1 \rangle$ **b.** $\mathbf{v} = \langle 12, -16, 4 \rangle$

Solution   Begin by writing **w** in component form.

$$\mathbf{w} = \langle -4 - 2, 7 - (-1), 5 - 3 \rangle = \langle -6, 8, 2 \rangle$$

a. Because $\mathbf{u} = \langle 3, -4, -1 \rangle = -\frac{1}{2}\langle -6, 8, 2 \rangle = -\frac{1}{2}\mathbf{w}$, you can conclude that **u** is parallel to **w**.

b. In this case, you want to find a scalar c such that

$$\langle 12, -16, 4 \rangle = c\langle -6, 8, 2 \rangle.$$
$$12 = -6c \rightarrow c = -2$$
$$-16 = 8c \rightarrow c = -2$$
$$4 = 2c \rightarrow c = 2$$

Because there is no c for which the equation has a solution, the vectors are *not* parallel.

Example 5   Using Vectors to Determine Collinear Points

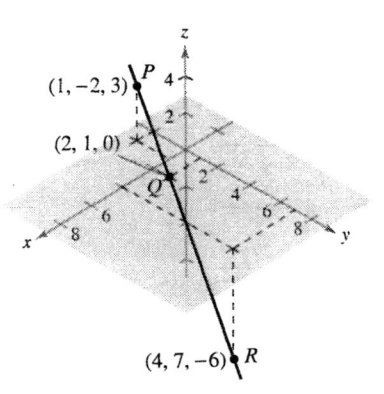

Determine whether the points $P(1, -2, 3)$, $Q(2, 1, 0)$, and $R(4, 7, -6)$ lie on the same line.

Solution   The component forms of $\overrightarrow{PQ}$ and $\overrightarrow{PR}$ are

$$\overrightarrow{PQ} = \langle 2 - 1, 1 - (-2), 0 - 3 \rangle = \langle 1, 3, -3 \rangle$$

and

$$\overrightarrow{PR} = \langle 4 - 1, 7 - (-2), -6 - 3 \rangle = \langle 3, 9, -9 \rangle.$$

These two vectors have a common initial point. Hence, P, Q, and R lie on the same line if and only if $\overrightarrow{PQ}$ and $\overrightarrow{PR}$ are parallel—which they are because $\overrightarrow{PR} = 3\overrightarrow{PQ}$, as shown in Figure 10.22.

The points P, Q, and R lie on the same line.
Figure 10.22

Example 6 Standard Unit Vector Notation

a. Write the vector $\mathbf{v} = 4\mathbf{i} - 5\mathbf{k}$ in component form.
b. Find the terminal point of the vector $\mathbf{v} = 7\mathbf{i} - \mathbf{j} + 3\mathbf{k}$, given that the initial point is $P(-2, 3, 5)$.

Solution

a. Because $\mathbf{j}$ is missing, its component is 0 and
$$\mathbf{v} = 4\mathbf{i} - 5\mathbf{k} = \langle 4, 0, -5 \rangle.$$

b. You need to find $Q(q_1, q_2, q_3)$ such that $\mathbf{v} = \overrightarrow{PQ} = 7\mathbf{i} - \mathbf{j} + 3\mathbf{k}$. This implies that $q_1 - (-2) = 7$, $q_2 - 3 = -1$, and $q_3 - 5 = 3$. The solution of these three equations is $q_1 = 5$, $q_2 = 2$, and $q_3 = 8$. Therefore, Q is $(5, 2, 8)$.

Application

Example 7 Measuring Force

A television camera weighing 120 pounds is supported by a tripod, as shown in Figure 10.23. Represent the force exerted on each leg of the tripod as a vector.

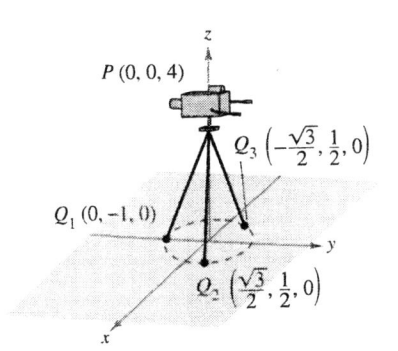

Solution Let the vectors $\mathbf{F}_1$, $\mathbf{F}_2$, and $\mathbf{F}_3$ represent the forces exerted on the three legs. From Figure 10.23, you can determine the directions of $\mathbf{F}_1$, $\mathbf{F}_2$, and $\mathbf{F}_3$ to be as follows.

$$\overrightarrow{PQ_1} = \langle 0 - 0, -1 - 0, 0 - 4 \rangle = \langle 0, -1, -4 \rangle$$

$$\overrightarrow{PQ_2} = \left\langle \frac{\sqrt{3}}{2} - 0, \frac{1}{2} - 0, 0 - 4 \right\rangle = \left\langle \frac{\sqrt{3}}{2}, \frac{1}{2}, -4 \right\rangle$$

$$\overrightarrow{PQ_3} = \left\langle -\frac{\sqrt{3}}{2} - 0, \frac{1}{2} - 0, 0 - 4 \right\rangle = \left\langle -\frac{\sqrt{3}}{2}, \frac{1}{2}, -4 \right\rangle$$

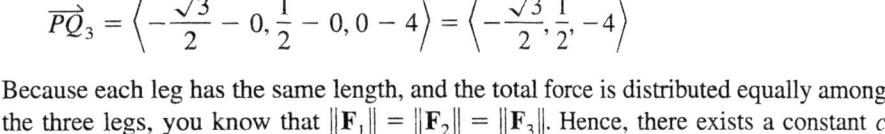

Because each leg has the same length, and the total force is distributed equally among the three legs, you know that $\|\mathbf{F}_1\| = \|\mathbf{F}_2\| = \|\mathbf{F}_3\|$. Hence, there exists a constant c such that

$$\mathbf{F}_1 = c\langle 0, -1, -4 \rangle, \quad \mathbf{F}_2 = c\left\langle \frac{\sqrt{3}}{2}, \frac{1}{2}, -4 \right\rangle, \quad \text{and} \quad \mathbf{F}_3 = c\left\langle -\frac{\sqrt{3}}{2}, \frac{1}{2}, -4 \right\rangle.$$

Let the total force exerted by the object be given by $\mathbf{F} = -120\mathbf{k}$. Then, using the fact that

$$\mathbf{F} = \mathbf{F}_1 + \mathbf{F}_2 + \mathbf{F}_3$$

you can conclude that $\mathbf{F}_1$, $\mathbf{F}_2$, and $\mathbf{F}_3$ all have a vertical component of -40. This implies that $c(-4) = -40$ and $c = 10$. Therefore, the forces exerted on the legs can be represented by

$$\mathbf{F}_1 = \langle 0, -10, -40 \rangle$$
$$\mathbf{F}_2 = \langle 5\sqrt{3}, 5, -40 \rangle$$
$$\mathbf{F}_3 = \langle -5\sqrt{3}, 5, -40 \rangle.$$

Figure 10.23

EXERCISES FOR SECTION 10.2

In Exercises 1–4, plot the points on the same three-dimensional coordinate system.

1. (a) (2, 1, 3) (b) (−1, 2, 1)
2. (a) (3, −2, 5) (b) $(\frac{3}{2}, 4, -2)$
3. (a) (5, −2, 2) (b) (5, −2, −2)
4. (a) (0, 4, −5) (b) (4, 0, 5)

In Exercises 5 and 6, approximate the coordinates of the points.

5.

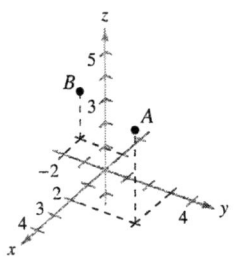

6.

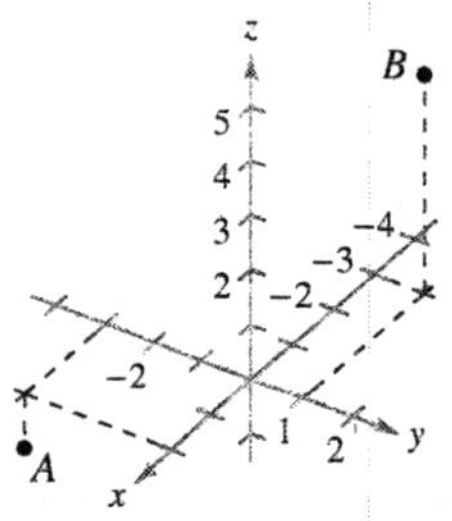

In Exercises 7–10, find the coordinates of the point.

7. The point is located 3 units behind the yz-plane, 4 units to the right of the xz-plane, and 5 units above the xy-plane.
8. The point is located 7 units in front of the yz-plane, 2 units to the left of the xz-plane, and 1 unit below the xy-plane.
9. The point is located on the x-axis, 10 units in front of the yz-plane.
10. The point is located in the yz-plane, 3 units to the right of the xz-plane, and 2 units above the xy-plane.
11. *Think About It* What is the z-coordinate of any point in the xy-plane?
12. *Think About It* What is the x-coordinate of any point in the yz-plane?

In Exercises 13–24, determine the location of a point (x, y, z) that satisfies the condition(s).

13. $z = 6$
14. $y = 2$
15. $x = 4$
16. $z = -3$
17. $y < 0$
18. $x < 0$
19. $|y| \le 3$
20. $|x| > 4$
21. $xy > 0, z = -3$
22. $xy < 0, z = 4$
23. $xyz < 0$
24. $xyz > 0$

In Exercises 25–28, find the distance between the points.

25. (0, 0, 0), (5, 2, 6)
26. (−2, 3, 2), (2, −5, −2)
27. (1, −2, 4), (6, −2, −2)
28. (2, 2, 3), (4, −5, 6)

In Exercises 29–32, find the lengths of the sides of the triangle with the indicated vertices, and determine whether the triangle is a right triangle, an isosceles triangle, or neither.

29. (0, 0, 0), (2, 2, 1), (2, −4, 4)
30. (5, 3, 4), (7, 1, 3), (3, 5, 3)
31. (1, −3, −2), (5, −1, 2), (−1, 1, 2)
32. (5, 0, 0), (0, 2, 0), (0, 0, −3)

33. *Think About It* The triangle in Exercise 29 is translated 5 units upward along the z-axis. Determine the coordinates of the translated triangle.
34. *Think About It* The triangle in Exercise 30 is translated 3 units to the right along the y-axis. Determine the coordinates of the translated triangle.

In Exercises 35 and 36, find the coordinates of the midpoint of the line segment joining the points.

35. (5, −9, 7), (−2, 3, 3)
36. (4, 0, −6), (8, 8, 20)

In Exercises 37–40, find the standard form of the equation of the sphere.

37. Center: (0, 2, 5)
    Radius: 2
38. Center: (4, −1, 1)
    Radius: 5
39. Endpoints of a diameter: (2, 0, 0), (0, 6, 0)
40. Center: (−3, 2, 4), tangent to the yz-plane

In Exercises 41–44, complete the square to write the equation of the sphere in standard form. Find the center and radius.

41. $x^2 + y^2 + z^2 - 2x + 6y + 8z + 1 = 0$
42. $x^2 + y^2 + z^2 + 9x - 2y + 10z + 19 = 0$
43. $9x^2 + 9y^2 + 9z^2 - 6x + 18y + 1 = 0$
44. $4x^2 + 4y^2 + 4z^2 - 4x - 32y + 8z + 33 = 0$

In Exercises 45 and 46, describe the solid satisfying the condition.

45. $x^2 + y^2 + z^2 \le 36$
46. $x^2 + y^2 + z^2 < 4x - 6y + 8z - 13$

In Exercises 47–50, (a) find the component form of the vector **v** and (b) sketch the vector with its initial point at the origin.

47.

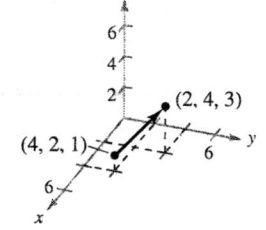

48.

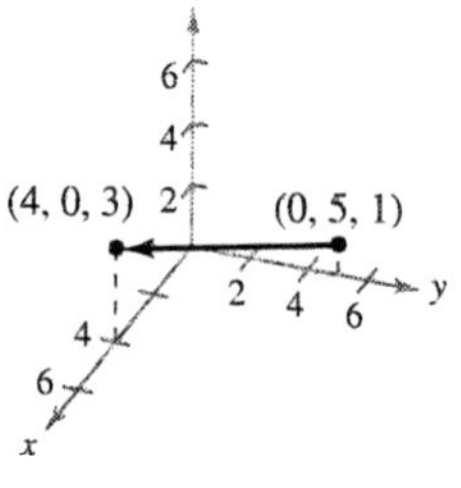

49.

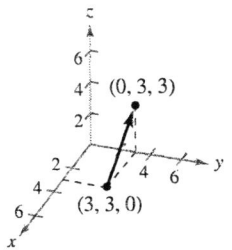

50.

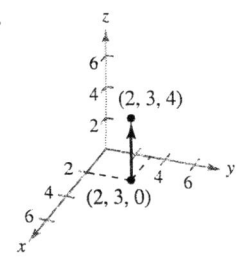

In Exercises 51–54, find the component form and length of the vector u with the given initial and terminal points. Then find the unit vector in the direction of u.

	Initial Point	Terminal Point
51.	(3, 2, 0)	(4, 1, 6)
52.	(4, −5, 2)	(−1, 7, −3)
53.	(−4, 3, 1)	(−5, 3, 0)
54.	(1, −2, 4)	(2, 4, −2)

In Exercises 55 and 56, the initial and terminal points of a vector v are given. (a) Sketch the directed line segment, (b) find the component form of the vector, and (c) sketch the vector with its initial point at the origin.

55. Initial point: (−1, 2, 3)
Terminal point: (3, 3, 4)

56. Initial point: (2, −1, −2)
Terminal point: (−4, 3, 7)

In Exercises 57 and 58, the vector v and its initial point are given. Find the terminal point.

57. $\mathbf{v} = \langle 3, -5, 6 \rangle$
Initial point: (0, 6, 2)

58. $\mathbf{v} = \langle 1, -\frac{2}{3}, \frac{1}{2} \rangle$
Initial point: $(0, 2, \frac{5}{2})$

In Exercises 59 and 60, sketch each scalar multiple of v.

59. $\mathbf{v} = \langle 1, 2, 2 \rangle$
(a) $2\mathbf{v}$ (b) $-\mathbf{v}$ (c) $\frac{3}{2}\mathbf{v}$ (d) $0\mathbf{v}$

60. $\mathbf{v} = \langle 2, -2, 1 \rangle$
(a) $-\mathbf{v}$ (b) $2\mathbf{v}$ (c) $\frac{1}{2}\mathbf{v}$ (d) $\frac{5}{2}\mathbf{v}$

In Exercises 61–66, find the vector z, given that $\mathbf{u} = \langle 1, 2, 3 \rangle$, $\mathbf{v} = \langle 2, 2, -1 \rangle$, and $\mathbf{w} = \langle 4, 0, -4 \rangle$.

61. $\mathbf{z} = \mathbf{u} - \mathbf{v}$
62. $\mathbf{z} = \mathbf{u} - \mathbf{v} + 2\mathbf{w}$
63. $\mathbf{z} = 2\mathbf{u} + 4\mathbf{v} - \mathbf{w}$
64. $\mathbf{z} = 5\mathbf{u} - 3\mathbf{v} - \frac{1}{2}\mathbf{w}$
65. $2\mathbf{z} - 3\mathbf{u} = \mathbf{w}$
66. $2\mathbf{u} + \mathbf{v} - \mathbf{w} + 3\mathbf{z} = \mathbf{0}$

In Exercises 67–70, determine which of the vectors are parallel to z. Use a graphing utility to confirm your results.

67. $\mathbf{z} = \langle 3, 2, -5 \rangle$
(a) $\langle -6, -4, 10 \rangle$ (b) $\langle 2, \frac{4}{3}, -\frac{10}{3} \rangle$
(c) $\langle 6, 4, 10 \rangle$ (d) $\langle 1, -4, 2 \rangle$

68. $\mathbf{z} = \frac{1}{2}\mathbf{i} - \frac{2}{3}\mathbf{j} + \frac{3}{4}\mathbf{k}$
(a) $6\mathbf{i} - 4\mathbf{j} + 9\mathbf{k}$ (b) $-\mathbf{i} + \frac{4}{3}\mathbf{j} - \frac{3}{2}\mathbf{k}$
(c) $12\mathbf{i} + 9\mathbf{k}$ (d) $\frac{3}{4}\mathbf{i} - \mathbf{j} + \frac{9}{8}\mathbf{k}$

69. z has initial point (1, −1, 3) and terminal point (−2, 3, 5).
(a) $-6\mathbf{i} + 8\mathbf{j} + 4\mathbf{k}$ (b) $4\mathbf{j} + 2\mathbf{k}$

70. z has initial point (5, 4, 1) and terminal point (−2, −4, 4).
(a) $\langle 7, 6, 2 \rangle$ (b) $\langle 14, 16, -6 \rangle$

In Exercises 71–74, use vectors to determine whether the points lie in a straight line.

71. (0, −2, −5), (3, 4, 4), (2, 2, 1)
72. (4, −2, 7), (−2, 0, 3), (7, −3, 9)
73. (1, 2, 4), (2, 5, 0), (0, 1, 5)
74. (0, 0, 0), (1, 3, −2), (2, −6, 4)

In Exercises 75 and 76, use vectors to show that the points form the vertices of a parallelogram.

75. (2, 9, 1), (3, 11, 4), (0, 10, 2), (1, 12, 5)
76. (1, 1, 3), (9, −1, −2), (11, 2, −9), (3, 4, −4)

In Exercises 77–82, find the magnitude of v.

77. $\mathbf{v} = \langle 0, 0, 0 \rangle$
78. $\mathbf{v} = \langle 1, 0, 3 \rangle$
79. $\mathbf{v} = \mathbf{i} - 2\mathbf{j} - 3\mathbf{k}$
80. $\mathbf{v} = -4\mathbf{i} + 3\mathbf{j} + 7\mathbf{k}$
81. Initial point of v: (1, −3, 4)
Terminal point of v: (1, 0, −1)
82. Initial point of v: (0, −1, 0)
Terminal point of v: (1, 2, −2)

In Exercises 83–86, find a unit vector (a) in the direction of u and (b) in the direction opposite of u.

83. $\mathbf{u} = \langle 2, -1, 2 \rangle$
84. $\mathbf{u} = \langle 6, 0, 8 \rangle$
85. $\mathbf{u} = \langle 3, 2, -5 \rangle$
86. $\mathbf{u} = \langle 8, 0, 0 \rangle$

87. *Programming* You are given the component forms of the vectors **u** and **v**. Write a program for a graphing utility in which the output is (a) the component form of $\mathbf{u} + \mathbf{v}$, (b) $\|\mathbf{u} + \mathbf{v}\|$, (c) $\|\mathbf{u}\|$, and (d) $\|\mathbf{v}\|$.

88. Run the program you wrote in Exercise 87 for the vectors $\mathbf{u} = \langle -1, 3, 4 \rangle$ and $\mathbf{v} = \langle 5, 4.5, -6 \rangle$.

In Exercises 89 and 90, determine the values of c that satisfy the equation. Let $\mathbf{u} = \mathbf{i} + 2\mathbf{j} + 3\mathbf{k}$ and $\mathbf{v} = 2\mathbf{i} + 2\mathbf{j} - \mathbf{k}$.

89. $\|c\mathbf{v}\| = 5$
90. $\|c\mathbf{u}\| = 3$

In Exercises 91–94, find the vector v with the given magnitude and direction u.

	Magnitude	Direction
91.	10	$\mathbf{u} = \langle 0, 3, 3 \rangle$
92.	3	$\mathbf{u} = \langle 1, 1, 1 \rangle$
93.	$\frac{3}{2}$	$\mathbf{u} = \langle 2, -2, 1 \rangle$
94.	$\sqrt{5}$	$\mathbf{u} = \langle -4, 6, 2 \rangle$

In Exercises 95 and 96, sketch the vector v and write its component form.

95. **v** lies in the yz-plane, has magnitude 2, and makes an angle of 30° with the positive y-axis.

96. **v** lies in the xz-plane, has magnitude 5, and makes an angle of 45° with the positive z-axis.

In Exercises 97 and 98, use vectors to find the point that lies two-thirds of the way from P to Q.

97. $P(4, 3, 0)$, $Q(1, -3, 3)$
98. $P(1, 2, 5)$, $Q(6, 8, 2)$

99. Let $\mathbf{u} = \mathbf{i} + \mathbf{j}$, $\mathbf{v} = \mathbf{j} + \mathbf{k}$, and $\mathbf{w} = a\mathbf{u} + b\mathbf{v}$.
    (a) Sketch **u** and **v**.
    (b) If $\mathbf{w} = \mathbf{0}$, show that a and b must both be zero.
    (c) Find a and b such that $\mathbf{w} = \mathbf{i} + 2\mathbf{j} + \mathbf{k}$.
    (d) Show that no choice of a and b yields $\mathbf{w} = \mathbf{i} + 2\mathbf{j} + 3\mathbf{k}$.

Getting at the Concept

100. A point in the three-dimensional coordinate system has coordinates (x_0, y_0, z_0). Describe what each coordinate measures.

101. Give the formula for the distance between the points (x_1, y_1, z_1) and (x_2, y_2, z_2).

102. Give the standard equation of a sphere of radius r, centered at (x_0, y_0, z_0).

103. State the definition of parallel vectors.

104. The initial and terminal points of the vector **v** are (x_1, y_1, z_1) and (x, y, z). Describe the set of all points (x, y, z) such that $\|\mathbf{v}\| = 4$.

105. *Numerical, Graphical, and Analytic Analysis* The lights in an auditorium are 24-pound discs of radius 18 inches. Each disc is supported by three equally spaced cables that are L inches long (see figure).

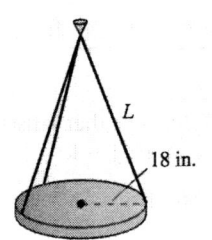

    (a) Write the tension T in each cable as a function of L. Determine the domain of the function.
    (b) Use a graphing utility and the model in part (a) to complete the table.

L	20	25	30	35	40	45	50
T							

(c) Use a graphing utility to graph the model in part (a). Determine the asymptotes of the graph.
(d) Confirm the asymptotes of the graph in part (c) analytically.
(e) Determine the minimum length of each cable if a cable is designed to carry a maximum load of 10 pounds.

106. *Think About It* Suppose the length of each cable in Exercise 105 has a fixed length $L = a$, and the radius of each disc is r_0 inches. Make a conjecture about the limit
$$\lim_{r_0 \to a^-} T$$
and give a reason for your answer.

107. *Diagonal of a Cube* Find the component form of the unit vector **v** in the direction of the diagonal of the cube shown in the figure.

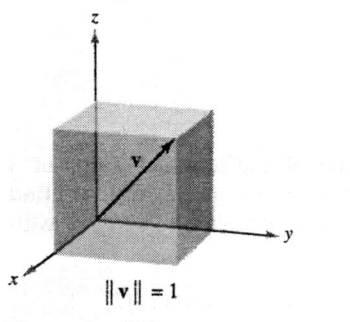

Figure for 107 **Figure for 108**

108. *Tower Guy Wire* The guy wire to a 100-foot tower has a tension of 550 pounds. Using the distances shown in the figure, write the component form of the vector **F** representing the tension in the wire.

109. *Load Supports* Find the tension in each of the supporting cables in the figure if the weight of the crate is 500 newtons.

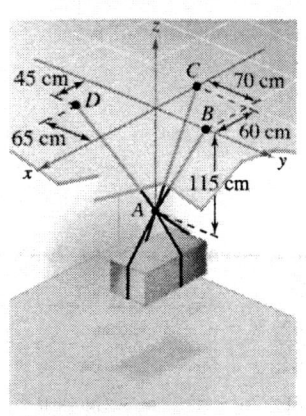

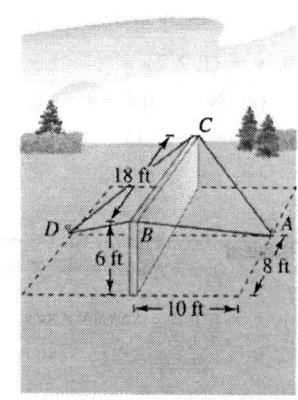

Figure for 109 **Figure for 110**

110. *Building Construction* A precast concrete wall is temporarily kept in its vertical position by ropes (see figure). Find the total force exerted on the pin at position A if the tensions in AB and AC are 420 pounds and 650 pounds.

111. Write an equation whose graph consists of the set of points $P(x, y, z)$ that are twice as far from $A(0, -1, 1)$ as from $B(1, 2, 0)$.

Section 10.3 — The Dot Product of Two Vectors

- Use properties of the dot product of two vectors.
- Find the angle between two vectors using the dot product.
- Find the direction cosines of a vector in space.
- Find the projection of a vector onto another vector.
- Use vectors to find the work done by a constant force.

EXPLORATION

Interpreting a Dot Product Several vectors are shown below on the unit circle. Find the dot products of several pairs of vectors. Then find the angle between each pair that you used. Make a conjecture about the relationship between the dot product of two vectors and the angle between the vectors.

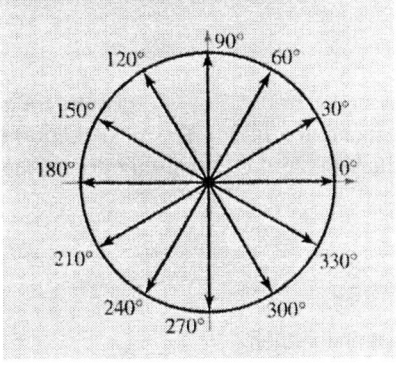

The Dot Product

So far you have studied two operations with vectors—vector addition and multiplication by a scalar—each of which yields another vector. In this section you will study a third vector operation, called the **dot product.** This product yields a scalar, rather than a vector.

Definition of Dot Product

The **dot product** of $\mathbf{u} = \langle u_1, u_2 \rangle$ and $\mathbf{v} = \langle v_1, v_2 \rangle$ is

$$\mathbf{u} \cdot \mathbf{v} = u_1 v_1 + u_2 v_2.$$

The **dot product** of $\mathbf{u} = \langle u_1, u_2, u_3 \rangle$ and $\mathbf{v} = \langle v_1, v_2, v_3 \rangle$ is

$$\mathbf{u} \cdot \mathbf{v} = u_1 v_1 + u_2 v_2 + u_3 v_3.$$

NOTE Because the dot product of two vectors yields a scalar, it is also called the **inner product** (or **scalar product**) of the two vectors.

THEOREM 10.4 Properties of the Dot Product

Let $\mathbf{u}$, $\mathbf{v}$, and $\mathbf{w}$ be vectors in the plane or in space and let c be a scalar.

1. $\mathbf{u} \cdot \mathbf{v} = \mathbf{v} \cdot \mathbf{u}$ Commutative property
2. $\mathbf{u} \cdot (\mathbf{v} + \mathbf{w}) = \mathbf{u} \cdot \mathbf{v} + \mathbf{u} \cdot \mathbf{w}$ Distributive property
3. $c(\mathbf{u} \cdot \mathbf{v}) = c\mathbf{u} \cdot \mathbf{v} = \mathbf{u} \cdot c\mathbf{v}$
4. $\mathbf{0} \cdot \mathbf{v} = 0$
5. $\mathbf{v} \cdot \mathbf{v} = \|\mathbf{v}\|^2$

Proof To prove the first property, let $\mathbf{u} = \langle u_1, u_2, u_3 \rangle$ and $\mathbf{v} = \langle v_1, v_2, v_3 \rangle$. Then

$$\mathbf{u} \cdot \mathbf{v} = u_1 v_1 + u_2 v_2 + u_3 v_3$$
$$= v_1 u_1 + v_2 u_2 + v_3 u_3$$
$$= \mathbf{v} \cdot \mathbf{u}.$$

For the fifth property, let $\mathbf{v} = \langle v_1, v_2, v_3 \rangle$. Then

$$\mathbf{v} \cdot \mathbf{v} = v_1^2 + v_2^2 + v_3^2$$
$$= \left(\sqrt{v_1^2 + v_2^2 + v_3^2}\right)^2$$
$$= \|\mathbf{v}\|^2.$$

Proofs of the other properties are left to you.

Example 1 **Finding Dot Products**

Given $\mathbf{u} = \langle 2, -2 \rangle$, $\mathbf{v} = \langle 5, 8 \rangle$, and $\mathbf{w} = \langle -4, 3 \rangle$, find each of the following.

a. $\mathbf{u} \cdot \mathbf{v}$ **b.** $(\mathbf{u} \cdot \mathbf{v})\mathbf{w}$ **c.** $\mathbf{u} \cdot (2\mathbf{v})$ **d.** $\|\mathbf{w}\|^2$

Solution

a. $\mathbf{u} \cdot \mathbf{v} = \langle 2, -2 \rangle \cdot \langle 5, 8 \rangle = 2(5) + (-2)(8) = -6$

b. $(\mathbf{u} \cdot \mathbf{v})\mathbf{w} = -6\langle -4, 3 \rangle = \langle 24, -18 \rangle$

c. $\mathbf{u} \cdot (2\mathbf{v}) = 2(\mathbf{u} \cdot \mathbf{v}) = 2(-6) = -12$

d. $\|\mathbf{w}\|^2 = \mathbf{w} \cdot \mathbf{w}$ Theorem 10.4

$\qquad\quad = \langle -4, 3 \rangle \cdot \langle -4, 3 \rangle$ Substitute $\langle -4, 3 \rangle$ for $\mathbf{w}$.

$\qquad\quad = (-4)(-4) + (3)(3)$ Definition of dot product

$\qquad\quad = 25$ Simplify.

Notice that the result of part (b) is a *vector* quantity, whereas the results of the other three parts are *scalar* quantities.

Angle Between Two Vectors

The **angle between two nonzero vectors** is the angle θ, $0 \leq \theta \leq \pi$, between their respective standard position vectors, as shown in Figure 10.24. The next theorem shows how to find this angle using the dot product. (Note that we do not define the angle between the zero vector and another vector.)

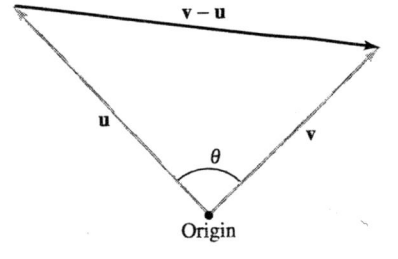

THEOREM 10.5 Angle Between Two Vectors

If θ is the angle between two nonzero vectors $\mathbf{u}$ and $\mathbf{v}$, then

$$\cos \theta = \frac{\mathbf{u} \cdot \mathbf{v}}{\|\mathbf{u}\| \|\mathbf{v}\|}.$$

The angle between two vectors
Figure 10.24

Proof Consider the triangle determined by vectors $\mathbf{u}$, $\mathbf{v}$, and $\mathbf{v} - \mathbf{u}$, as shown in Figure 10.24. By the Law of Cosines, you can write

$$\|\mathbf{v} - \mathbf{u}\|^2 = \|\mathbf{u}\|^2 + \|\mathbf{v}\|^2 - 2\|\mathbf{u}\| \|\mathbf{v}\| \cos \theta.$$

Using the properties of the dot product, the left side can be rewritten as

$$\begin{aligned}\|\mathbf{v} - \mathbf{u}\|^2 &= (\mathbf{v} - \mathbf{u}) \cdot (\mathbf{v} - \mathbf{u}) \\ &= (\mathbf{v} - \mathbf{u}) \cdot \mathbf{v} - (\mathbf{v} - \mathbf{u}) \cdot \mathbf{u} \\ &= \mathbf{v} \cdot \mathbf{v} - \mathbf{u} \cdot \mathbf{v} - \mathbf{v} \cdot \mathbf{u} + \mathbf{u} \cdot \mathbf{u} \\ &= \|\mathbf{v}\|^2 - 2\mathbf{u} \cdot \mathbf{v} + \|\mathbf{u}\|^2\end{aligned}$$

and substitution back into the Law of Cosines yields

$$\|\mathbf{v}\|^2 - 2\mathbf{u} \cdot \mathbf{v} + \|\mathbf{u}\|^2 = \|\mathbf{u}\|^2 + \|\mathbf{v}\|^2 - 2\|\mathbf{u}\| \|\mathbf{v}\| \cos \theta$$

$$-2\mathbf{u} \cdot \mathbf{v} = -2\|\mathbf{u}\| \|\mathbf{v}\| \cos \theta$$

$$\cos \theta = \frac{\mathbf{u} \cdot \mathbf{v}}{\|\mathbf{u}\| \|\mathbf{v}\|}.$$

If the angle between two vectors is known, rewriting Theorem 10.5 in the form

$$\mathbf{u} \cdot \mathbf{v} = \|\mathbf{u}\| \|\mathbf{v}\| \cos \theta \qquad \text{Alternative form of dot product}$$

produces an alternative way to calculate the dot product. From this form, you can see that because $\|\mathbf{u}\|$ and $\|\mathbf{v}\|$ are always positive, $\mathbf{u} \cdot \mathbf{v}$ and $\cos \theta$ will always have the same sign. Figure 10.25 shows the possible orientations of two vectors.

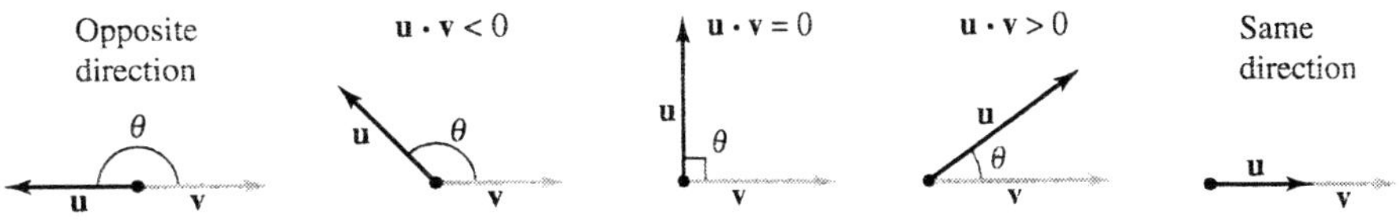

Opposite direction	$\mathbf{u} \cdot \mathbf{v} < 0$	$\mathbf{u} \cdot \mathbf{v} = 0$	$\mathbf{u} \cdot \mathbf{v} > 0$	Same direction
$\theta = \pi$	$\pi/2 < \theta < \pi$	$\theta = \pi/2$	$0 < \theta < \pi/2$	$\theta = 0$
$\cos \theta = -1$	$-1 < \cos \theta < 0$	$\cos \theta = 0$	$0 < \cos \theta < 1$	$\cos \theta = 1$

Figure 10.25

From Figure 10.25, you can see that two nonzero vectors meet at a right angle if and only if their dot product is zero. Two such vectors are said to be **orthogonal**.

Definition of Orthogonal Vectors

The vectors $\mathbf{u}$ and $\mathbf{v}$ are orthogonal if $\mathbf{u} \cdot \mathbf{v} = 0$.

NOTE The terms "perpendicular," "orthogonal," and "normal" all mean essentially the same thing—meeting at right angles. However, we usually say that two vectors are *orthogonal*, two lines or planes are *perpendicular*, and a vector is *normal* to a given line or plane.

From this definition, it follows that the zero vector is orthogonal to every vector $\mathbf{u}$, because $\mathbf{0} \cdot \mathbf{u} = 0$. Moreover, for $0 \le \theta \le \pi$, you know that $\cos \theta = 0$ if and only if $\theta = \pi/2$. So, you can use Theorem 10.5 to conclude that two *nonzero* vectors are orthogonal if and only if the angle between them is $\pi/2$.

Example 2 Finding the Angle Between Two Vectors

For $\mathbf{u} = \langle 3, -1, 2 \rangle$, $\mathbf{v} = \langle -4, 0, 2 \rangle$, $\mathbf{w} = \langle 1, -1, -2 \rangle$, and $\mathbf{z} = \langle 2, 0, -1 \rangle$, find the angle between each pair of vectors.

a. $\mathbf{u}$ and $\mathbf{v}$ **b.** $\mathbf{u}$ and $\mathbf{w}$ **c.** $\mathbf{v}$ and $\mathbf{z}$

Solution

a. $\cos \theta = \dfrac{\mathbf{u} \cdot \mathbf{v}}{\|\mathbf{u}\| \|\mathbf{v}\|} = \dfrac{-12 + 4}{\sqrt{14}\sqrt{20}} = \dfrac{-8}{2\sqrt{14}\sqrt{5}} = \dfrac{-4}{\sqrt{70}}$

Because $\mathbf{u} \cdot \mathbf{v} < 0$, $\theta = \arccos \dfrac{-4}{\sqrt{70}} \approx 2.069$ radians.

b. $\cos \theta = \dfrac{\mathbf{u} \cdot \mathbf{w}}{\|\mathbf{u}\| \|\mathbf{w}\|} = \dfrac{3 + 1 - 4}{\sqrt{14}\sqrt{6}} = \dfrac{0}{\sqrt{84}} = 0$

Because $\mathbf{u} \cdot \mathbf{w} = 0$, $\mathbf{u}$ and $\mathbf{w}$ are *orthogonal*. Thus, $\theta = \pi/2$.

c. $\cos \theta = \dfrac{\mathbf{v} \cdot \mathbf{z}}{\|\mathbf{v}\| \|\mathbf{z}\|} = \dfrac{-8 + 0 - 2}{\sqrt{20}\sqrt{5}} = \dfrac{-10}{\sqrt{100}} = -1$

Consequently, $\theta = \pi$. Note that $\mathbf{v}$ and $\mathbf{z}$ are parallel, with $\mathbf{v} = -2\mathbf{z}$.

Direction Cosines

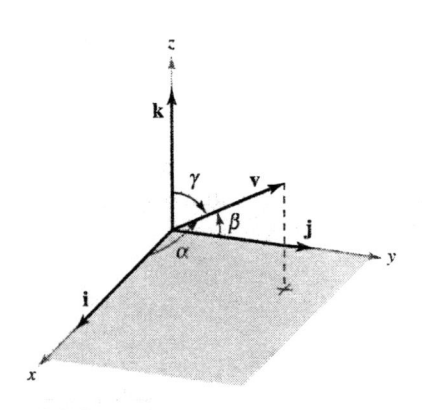

For a vector in the plane, you have seen that it is convenient to measure direction in terms of the angle, measured counterclockwise, *from* the positive x-axis *to* the vector. In space it is more convenient to measure direction in terms of the angles *between* the nonzero vector **v** and the three unit vectors **i**, **j**, and **k**, as shown in Figure 10.26. The angles α, β, and γ are the **direction angles of v**, and $\cos \alpha$, $\cos \beta$, and $\cos \gamma$ are the **direction cosines of v**. Because

$$\mathbf{v} \cdot \mathbf{i} = \|\mathbf{v}\| \|\mathbf{i}\| \cos \alpha = \|\mathbf{v}\| \cos \alpha$$

and

$$\mathbf{v} \cdot \mathbf{i} = \langle v_1, v_2, v_3 \rangle \cdot \langle 1, 0, 0 \rangle = v_1$$

it follows that $\cos \alpha = v_1/\|\mathbf{v}\|$. By similar reasoning with the unit vectors **j** and **k**, you have

$$\cos \alpha = \frac{v_1}{\|\mathbf{v}\|} \qquad \text{α is the angle between \textbf{v} and \textbf{i}.}$$

$$\cos \beta = \frac{v_2}{\|\mathbf{v}\|} \qquad \text{β is the angle between \textbf{v} and \textbf{j}.}$$

$$\cos \gamma = \frac{v_3}{\|\mathbf{v}\|}. \qquad \text{γ is the angle between \textbf{v} and \textbf{k}.}$$

Consequently, any nonzero vector **v** in space has the normalized form

$$\frac{\mathbf{v}}{\|\mathbf{v}\|} = \frac{v_1}{\|\mathbf{v}\|}\mathbf{i} + \frac{v_2}{\|\mathbf{v}\|}\mathbf{j} + \frac{v_3}{\|\mathbf{v}\|}\mathbf{k} = \cos \alpha \, \mathbf{i} + \cos \beta \, \mathbf{j} + \cos \gamma \, \mathbf{k}$$

and because $\mathbf{v}/\|\mathbf{v}\|$ is a unit vector, it follows that

$$\cos^2 \alpha + \cos^2 \beta + \cos^2 \gamma = 1.$$

Direction angles
Figure 10.26

α = angle between **v** and **i**
β = angle between **v** and **j**
γ = angle between **v** and **k**

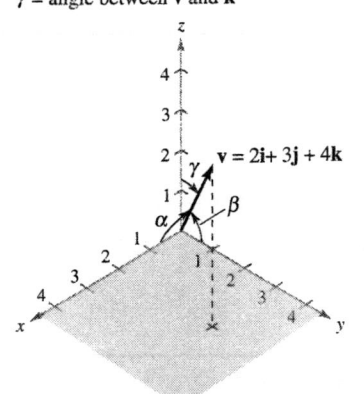

The direction angles of **v**
Figure 10.27

Example 3 Finding Direction Angles

Find the direction cosines and angles for the vector $\mathbf{v} = 2\mathbf{i} + 3\mathbf{j} + 4\mathbf{k}$, and show that $\cos^2 \alpha + \cos^2 \beta + \cos^2 \gamma = 1$.

Solution Because $\|\mathbf{v}\| = \sqrt{2^2 + 3^2 + 4^2} = \sqrt{29}$, you can write the following.

$$\cos \alpha = \frac{v_1}{\|\mathbf{v}\|} = \frac{2}{\sqrt{29}} \implies \alpha \approx 68.2° \qquad \text{Angle between \textbf{v} and \textbf{i}}$$

$$\cos \beta = \frac{v_2}{\|\mathbf{v}\|} = \frac{3}{\sqrt{29}} \implies \beta \approx 56.1° \qquad \text{Angle between \textbf{v} and \textbf{j}}$$

$$\cos \gamma = \frac{v_3}{\|\mathbf{v}\|} = \frac{4}{\sqrt{29}} \implies \gamma \approx 42.0° \qquad \text{Angle between \textbf{v} and \textbf{k}}$$

Furthermore, the sum of the squares of the direction cosines is

$$\cos^2 \alpha + \cos^2 \beta + \cos^2 \gamma = \frac{4}{29} + \frac{9}{29} + \frac{16}{29}$$

$$= \frac{29}{29}$$

$$= 1.$$

(See Figure 10.27.)

Projections and Vector Components

You have already seen applications in which two vectors are added to produce a resultant vector. Many applications in physics and engineering pose the reverse problem—decomposing a given vector into the sum of two **vector components.** To see the usefulness of this procedure, we look at a physical example.

Consider a boat on an inclined ramp, as shown in Figure 10.28. The force **F** due to gravity pulls the boat *down* the ramp and *against* the ramp. These two forces, $\mathbf{w}_1$ and $\mathbf{w}_2$, are orthogonal—they are called the vector components of **F**.

$$\mathbf{F} = \mathbf{w}_1 + \mathbf{w}_2 \qquad \text{Vector components of } \mathbf{F}$$

The forces $\mathbf{w}_1$ and $\mathbf{w}_2$ help you analyze the effect of gravity on the boat. For example, $\mathbf{w}_1$ indicates the force necessary to keep the boat from rolling down the ramp, whereas $\mathbf{w}_2$ indicates the force that the tires must withstand.

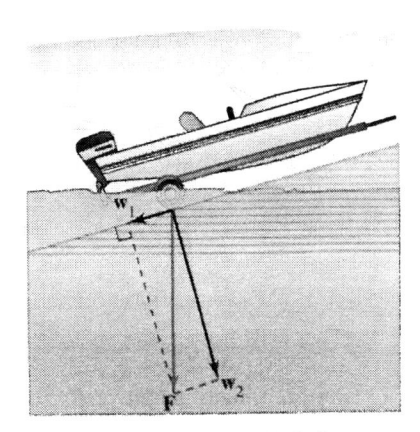

The force due to gravity pulls the boat against the ramp and down the ramp.
Figure 10.28

Definition of Projection and Vector Components

Let **u** and **v** be nonzero vectors. Moreover, let $\mathbf{u} = \mathbf{w}_1 + \mathbf{w}_2$, where $\mathbf{w}_1$ is parallel to **v** and $\mathbf{w}_2$ is orthogonal to **v**, as shown in Figure 10.29.

1. $\mathbf{w}_1$ is called the **projection of u onto v** or the **vector component of u along v**, and is denoted by $\mathbf{w}_1 = \text{proj}_\mathbf{v}\mathbf{u}$.
2. $\mathbf{w}_2 = \mathbf{u} - \mathbf{w}_1$ is called the **vector component of u orthogonal to v.**

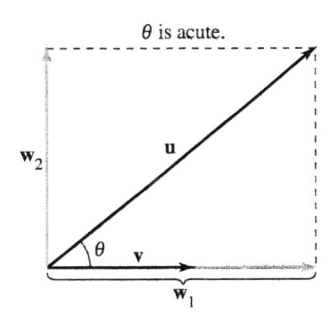

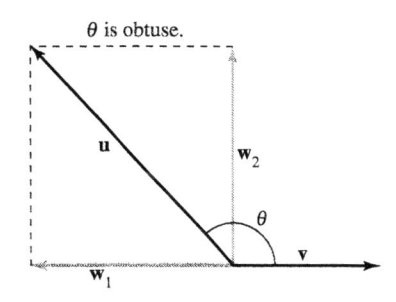

$\mathbf{w}_1 = \text{proj}_\mathbf{v}\mathbf{u} = $ projection of **u** onto **v** = vector component of **u** along **v**
$\mathbf{w}_2 = $ vector component of **u** orthogonal to **v**
Figure 10.29

Example 4 Finding a Vector Component of **u** Orthogonal to **v**

Find the vector component of $\mathbf{u} = \langle 7, 4 \rangle$ that is orthogonal to $\mathbf{v} = \langle 2, 3 \rangle$, given that $\mathbf{w}_1 = \text{proj}_\mathbf{v}\mathbf{u} = \langle 4, 6 \rangle$ and

$$\mathbf{u} = \langle 7, 4 \rangle = \mathbf{w}_1 + \mathbf{w}_2.$$

Solution Because $\mathbf{u} = \mathbf{w}_1 + \mathbf{w}_2$, where $\mathbf{w}_1$ is parallel to **v**, it follows that $\mathbf{w}_2$ is the vector component of **u** orthogonal to **v**. So, you have

$$\begin{aligned} \mathbf{w}_2 &= \mathbf{u} - \mathbf{w}_1 \\ &= \langle 7, 4 \rangle - \langle 4, 6 \rangle \\ &= \langle 3, -2 \rangle. \end{aligned}$$

Check to see that $\mathbf{w}_2$ is orthogonal to **v**, as shown in Figure 10.30.

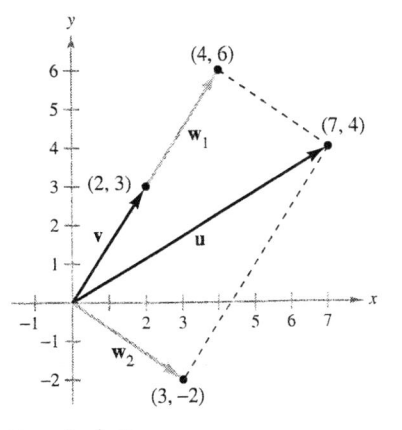

$\mathbf{u} = \mathbf{w}_1 + \mathbf{w}_2$
Figure 10.30

NOTE Note the distinction between the terms "component" and "vector component." For example, using the standard unit vectors with $\mathbf{u} = u_1\mathbf{i} + u_2\mathbf{j}$, u_1 is the *component* of $\mathbf{u}$ in the direction of $\mathbf{i}$ and $u_1\mathbf{i}$ is the *vector component* in the direction of $\mathbf{i}$.

From Example 4, you can see that it is easy to find the vector component $\mathbf{w}_2$ once you have found the projection, $\mathbf{w}_1$, of $\mathbf{u}$ onto $\mathbf{v}$. To find this projection, use the dot product given in the theorem below, which is proven in Exercise 82.

THEOREM 10.6 Projection Using the Dot Product

If $\mathbf{u}$ and $\mathbf{v}$ are nonzero vectors, then the projection of $\mathbf{u}$ onto $\mathbf{v}$ is given by

$$\text{proj}_\mathbf{v}\mathbf{u} = \left(\frac{\mathbf{u} \cdot \mathbf{v}}{\|\mathbf{v}\|^2}\right)\mathbf{v}.$$

The projection of $\mathbf{u}$ onto $\mathbf{v}$ can be written as a scalar multiple of a unit vector in the direction of $\mathbf{v}$. That is,

$$\left(\frac{\mathbf{u} \cdot \mathbf{v}}{\|\mathbf{v}\|^2}\right)\mathbf{v} = \left(\frac{\mathbf{u} \cdot \mathbf{v}}{\|\mathbf{v}\|}\right)\frac{\mathbf{v}}{\|\mathbf{v}\|} = (k)\frac{\mathbf{v}}{\|\mathbf{v}\|} \implies k = \frac{\mathbf{u} \cdot \mathbf{v}}{\|\mathbf{v}\|} = \|\mathbf{u}\|\cos\theta.$$

The scalar k is called the **component of u in the direction of v**.

Example 5 **Decomposing a Vector into Vector Components**

Find the projection of $\mathbf{u}$ onto $\mathbf{v}$ and the vector component of $\mathbf{u}$ orthogonal to $\mathbf{v}$ for the vectors $\mathbf{u} = 3\mathbf{i} - 5\mathbf{j} + 2\mathbf{k}$ and $\mathbf{v} = 7\mathbf{i} + \mathbf{j} - 2\mathbf{k}$. (See Figure 10.31.)

Solution The projection of $\mathbf{u}$ onto $\mathbf{v}$ is

$$\mathbf{w}_1 = \left(\frac{\mathbf{u} \cdot \mathbf{v}}{\|\mathbf{v}\|^2}\right)\mathbf{v} = \left(\frac{12}{54}\right)(7\mathbf{i} + \mathbf{j} - 2\mathbf{k}) = \frac{14}{9}\mathbf{i} + \frac{2}{9}\mathbf{j} - \frac{4}{9}\mathbf{k}.$$

The vector component of $\mathbf{u}$ orthogonal to $\mathbf{v}$ is the vector

$$\mathbf{w}_2 = \mathbf{u} - \mathbf{w}_1 = (3\mathbf{i} - 5\mathbf{j} + 2\mathbf{k}) - \left(\frac{14}{9}\mathbf{i} + \frac{2}{9}\mathbf{j} - \frac{4}{9}\mathbf{k}\right) = \frac{13}{9}\mathbf{i} - \frac{47}{9}\mathbf{j} + \frac{22}{9}\mathbf{k}.$$

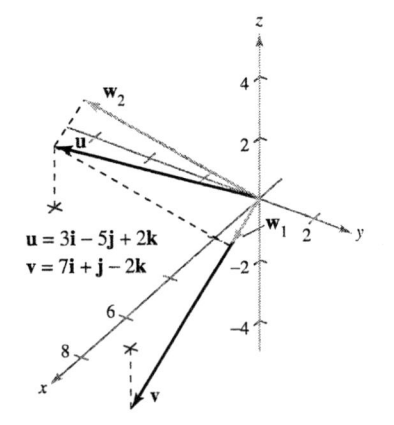

$\mathbf{u} = 3\mathbf{i} - 5\mathbf{j} + 2\mathbf{k}$
$\mathbf{v} = 7\mathbf{i} + \mathbf{j} - 2\mathbf{k}$

$\mathbf{u} = \mathbf{w}_1 + \mathbf{w}_2$
Figure 10.31

Example 6 **Finding a Force**

A 600-pound boat sits on a ramp inclined at 30°, as shown in Figure 10.32. What force is required to keep the boat from rolling down the ramp?

Solution Because the force due to gravity is vertical and downward, you can represent the gravitational force by the vector $\mathbf{F} = -600\mathbf{j}$. To find the force required to keep the boat from rolling down the ramp, we project $\mathbf{F}$ onto a unit vector $\mathbf{v}$ in the direction of the ramp, as follows.

$$\mathbf{v} = \cos 30°\mathbf{i} + \sin 30°\mathbf{j} = \frac{\sqrt{3}}{2}\mathbf{i} + \frac{1}{2}\mathbf{j} \qquad \text{Unit vector along ramp}$$

Therefore, the projection of $\mathbf{F}$ onto $\mathbf{v}$ is given by

$$\mathbf{w}_1 = \text{proj}_\mathbf{v}\mathbf{F} = \left(\frac{\mathbf{F} \cdot \mathbf{v}}{\|\mathbf{v}\|^2}\right)\mathbf{v} = (\mathbf{F} \cdot \mathbf{v})\mathbf{v} = (-600)\left(\frac{1}{2}\right)\mathbf{v} = -300\left(\frac{\sqrt{3}}{2}\mathbf{i} + \frac{1}{2}\mathbf{j}\right).$$

The magnitude of this force is 300, and therefore a force of 300 pounds is required to keep the boat from rolling down the ramp.

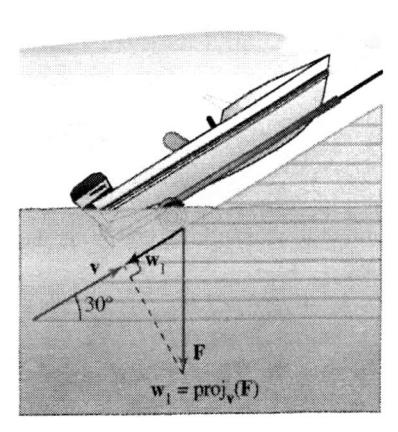

Figure 10.32

Work

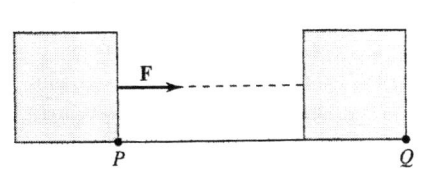

The work W done by the constant force $\mathbf{F}$ acting along the line of motion of an object is given by

$$W = (\text{magnitude of force})(\text{distance}) = \|\mathbf{F}\|\,\|\overrightarrow{PQ}\|$$

as shown in Figure 10.33a. If the constant force $\mathbf{F}$ is not directed along the line of motion, you can see from Figure 10.33b that the work W done by the force is

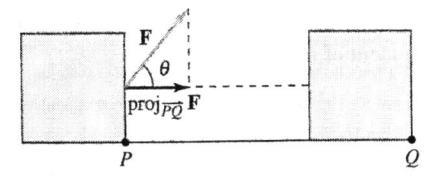

$$W = \|\text{proj}_{\overrightarrow{PQ}}\mathbf{F}\|\,\|\overrightarrow{PQ}\| = (\cos\theta)\|\mathbf{F}\|\,\|\overrightarrow{PQ}\| = \mathbf{F} \cdot \overrightarrow{PQ}.$$

We summarize this notion of work in the following definition.

(a) Force acts along the line of motion.

(b) Force acts at angle θ with the line of motion.

Figure 10.33

Definition of Work

The work W done by a constant force $\mathbf{F}$ as its point of application moves along the vector $\overrightarrow{PQ}$ is given by either of the following.

1. $W = \|\text{proj}_{\overrightarrow{PQ}}\mathbf{F}\|\,\|\overrightarrow{PQ}\|$ Projection form
2. $W = \mathbf{F} \cdot \overrightarrow{PQ}$ Dot product form

Example 7 Finding Work

To close a sliding door, a person pulls on a rope with a constant force of 50 pounds at a constant angle of 60°, as shown in Figure 10.34. Find the work done in moving the door 12 feet to its closed position.

Solution Using a projection, you can calculate the work as follows.

$$\begin{aligned}
W &= \|\text{proj}_{\overrightarrow{PQ}}\mathbf{F}\|\,\|\overrightarrow{PQ}\| \quad \text{Projection form for work}\\
&= \cos(60°)\|\mathbf{F}\|\,\|\overrightarrow{PQ}\|\\
&= \tfrac{1}{2}(50)(12)\\
&= 300 \text{ foot-pounds}
\end{aligned}$$

Figure 10.34

EXERCISES FOR SECTION 10.3

In Exercises 1–6, find (a) $\mathbf{u}\cdot\mathbf{v}$, (b) $\mathbf{u}\cdot\mathbf{u}$, (c) $\|\mathbf{u}\|^2$, (d) $(\mathbf{u}\cdot\mathbf{v})\mathbf{v}$, and (e) $\mathbf{u}\cdot(2\mathbf{v})$.

1. $\mathbf{u} = \langle 3, 4\rangle$
   $\mathbf{v} = \langle 2, -3\rangle$
2. $\mathbf{u} = \langle 4, 10\rangle$
   $\mathbf{v} = \langle -2, 3\rangle$
3. $\mathbf{u} = \langle 2, -3, 4\rangle$
   $\mathbf{v} = \langle 0, 6, 5\rangle$
4. $\mathbf{u} = \mathbf{i}$
   $\mathbf{v} = \mathbf{i}$
5. $\mathbf{u} = 2\mathbf{i} - \mathbf{j} + \mathbf{k}$
   $\mathbf{v} = \mathbf{i} - \mathbf{k}$
6. $\mathbf{u} = 2\mathbf{i} + \mathbf{j} - 2\mathbf{k}$
   $\mathbf{v} = \mathbf{i} - 3\mathbf{j} + 2\mathbf{k}$

7. *Revenue* The vector $\mathbf{u} = \langle 3240, 1450, 2235\rangle$ gives the numbers of units for products X, Y, and Z. The vector $\mathbf{v} = \langle 2.22, 1.85, 3.25\rangle$ gives the price (in dollars) per unit for each product. Find the dot product $\mathbf{u}\cdot\mathbf{v}$, and explain what information it gives.

8. *Revenue* Repeat Exercise 7 after increasing prices by 4%. Identify the vector operation used to increase prices by 4%.

In Exercises 9 and 10, find $\mathbf{u}\cdot\mathbf{v}$.

9. $\|\mathbf{u}\| = 8$, $\|\mathbf{v}\| = 5$, and the angle between $\mathbf{u}$ and $\mathbf{v}$ is $\pi/3$.
10. $\|\mathbf{u}\| = 40$, $\|\mathbf{v}\| = 25$, and the angle between $\mathbf{u}$ and $\mathbf{v}$ is $5\pi/6$.

In Exercises 11–18, find the angle θ between the vectors.

11. $\mathbf{u} = \langle 1, 1\rangle, \mathbf{v} = \langle 2, -2\rangle$
12. $\mathbf{u} = \langle 3, 1\rangle, \mathbf{v} = \langle 2, -1\rangle$
13. $\mathbf{u} = 3\mathbf{i} + \mathbf{j}, \mathbf{v} = -2\mathbf{i} + 4\mathbf{j}$
14. $\mathbf{u} = \cos\left(\dfrac{\pi}{6}\right)\mathbf{i} + \sin\left(\dfrac{\pi}{6}\right)\mathbf{j}$
    $\mathbf{v} = \cos\left(\dfrac{3\pi}{4}\right)\mathbf{i} + \sin\left(\dfrac{3\pi}{4}\right)\mathbf{j}$
15. $\mathbf{u} = \langle 1, 1, 1\rangle$
    $\mathbf{v} = \langle 2, 1, -1\rangle$
16. $\mathbf{u} = 3\mathbf{i} + 2\mathbf{j} + \mathbf{k}$
    $\mathbf{v} = 2\mathbf{i} - 3\mathbf{j}$
17. $\mathbf{u} = 3\mathbf{i} + 4\mathbf{j}$
    $\mathbf{v} = -2\mathbf{j} + 3\mathbf{k}$
18. $\mathbf{u} = 2\mathbf{i} - 3\mathbf{j} + \mathbf{k}$
    $\mathbf{v} = \mathbf{i} - 2\mathbf{j} + \mathbf{k}$

In Exercises 19–26, determine whether **u** and **v** are orthogonal, parallel, or neither.

19. $\mathbf{u} = \langle 4, 0 \rangle$
    $\mathbf{v} = \langle 1, 1 \rangle$
20. $\mathbf{u} = \langle 2, 18 \rangle$
    $\mathbf{v} = \langle \frac{3}{2}, -\frac{1}{6} \rangle$
21. $\mathbf{u} = \langle 4, 3 \rangle$
    $\mathbf{v} = \langle \frac{1}{2}, -\frac{2}{3} \rangle$
22. $\mathbf{u} = -\frac{1}{3}(\mathbf{i} - 2\mathbf{j})$
    $\mathbf{v} = 2\mathbf{i} - 4\mathbf{j}$
23. $\mathbf{u} = \mathbf{j} + 6\mathbf{k}$
    $\mathbf{v} = \mathbf{i} - 2\mathbf{j} - \mathbf{k}$
24. $\mathbf{u} = -2\mathbf{i} + 3\mathbf{j} - \mathbf{k}$
    $\mathbf{v} = 2\mathbf{i} + \mathbf{j} - \mathbf{k}$
25. $\mathbf{u} = \langle 2, -3, 1 \rangle$
    $\mathbf{v} = \langle -1, -1, -1 \rangle$
26. $\mathbf{u} = \langle \cos\theta, \sin\theta, -1 \rangle$
    $\mathbf{v} = \langle \sin\theta, -\cos\theta, 0 \rangle$

In Exercises 27–30, find the direction cosines of **u** and demonstrate that the sum of the squares of the direction cosines is 1.

27. $\mathbf{u} = \mathbf{i} + 2\mathbf{j} + 2\mathbf{k}$
28. $\mathbf{u} = 5\mathbf{i} + 3\mathbf{j} - \mathbf{k}$
29. $\mathbf{u} = \langle 0, 6, -4 \rangle$
30. $\mathbf{u} = \langle a, b, c \rangle$

In Exercises 31–34, find the direction angles of the vector.

31. $\mathbf{u} = 3\mathbf{i} + 2\mathbf{j} - 2\mathbf{k}$
32. $\mathbf{u} = -4\mathbf{i} + 3\mathbf{j} + 5\mathbf{k}$
33. $\mathbf{u} = \langle -1, 5, 2 \rangle$
34. $\mathbf{u} = \langle -2, 6, 1 \rangle$

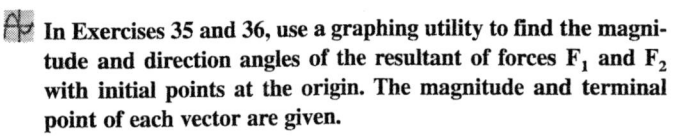

In Exercises 35 and 36, use a graphing utility to find the magnitude and direction angles of the resultant of forces F_1 and F_2 with initial points at the origin. The magnitude and terminal point of each vector are given.

Vector	Magnitude	Terminal Point
35. F_1	50 lb	(10, 5, 3)
F_2	80 lb	(12, 7, −5)
36. F_1	300 N	(−20, −10, 5)
F_2	100 N	(5, 15, 0)

37. Find the angle between a cube's diagonal and one of its edges.
38. Find the angle between the diagonal of a cube and the diagonal of one of its sides.
39. *Load-Supporting Cables* A load is supported by three cables, as shown in the figure. Find the direction angles of the load-supporting cable *OA*.

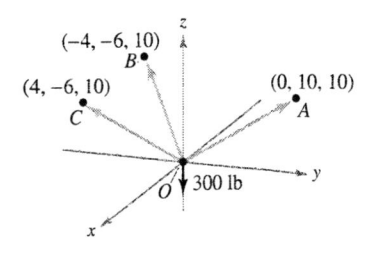

40. *Load-Supporting Cables* Determine the weight of the load if the tension in the cable *OA* in Exercise 39 is 200 newtons.

In Exercises 41–44, find the component of **u** that is orthogonal to **v**, given $\mathbf{w}_1 = \text{proj}_\mathbf{v}\mathbf{u}$.

41. $\mathbf{u} = \langle 6, 7 \rangle$
    $\mathbf{v} = \langle 1, 4 \rangle$
    $\text{proj}_\mathbf{v}\mathbf{u} = \langle 2, 8 \rangle$
42. $\mathbf{u} = \langle 9, 7 \rangle$
    $\mathbf{v} = \langle 1, 3 \rangle$
    $\text{proj}_\mathbf{v}\mathbf{u} = \langle 3, 9 \rangle$
43. $\mathbf{u} = \langle 0, 3, 3 \rangle$
    $\mathbf{v} = \langle -1, 1, 1 \rangle$
    $\text{proj}_\mathbf{v}\mathbf{u} = \langle -2, 2, 2 \rangle$
44. $\mathbf{u} = \langle 8, 2, 0 \rangle$
    $\mathbf{v} = \langle 2, 1, -1 \rangle$
    $\text{proj}_\mathbf{v}\mathbf{u} = \langle 6, 3, -3 \rangle$

In Exercises 45–48, (a) find the projection of **u** onto **v**, and (b) find the vector component of **u** orthogonal to **v**.

45. $\mathbf{u} = \langle 2, 3 \rangle, \mathbf{v} = \langle 5, 1 \rangle$
46. $\mathbf{u} = \langle 2, -3 \rangle, \mathbf{v} = \langle 3, 2 \rangle$
47. $\mathbf{u} = \langle 2, 1, 2 \rangle$
    $\mathbf{v} = \langle 0, 3, 4 \rangle$
48. $\mathbf{u} = \langle 1, 0, 4 \rangle$
    $\mathbf{v} = \langle 3, 0, 2 \rangle$

Getting at the Concept

49. Define the dot product of vectors **u** and **v**.
50. State the definition of orthogonal vectors. If vectors are neither parallel nor orthogonal, how do you find the angle between them?
51. What is known about θ, the angle between two nonzero vectors **u** and **v**, if
    (a) $\mathbf{u} \cdot \mathbf{v} = 0$? (b) $\mathbf{u} \cdot \mathbf{v} > 0$? (c) $\mathbf{u} \cdot \mathbf{v} < 0$?
52. Determine which of the following are defined for nonzero vectors **u**, **v**, and **w**.
    (a) $\mathbf{u} \cdot (\mathbf{v} + \mathbf{w})$ (b) $(\mathbf{u} \cdot \mathbf{v})\mathbf{w}$
    (c) $\mathbf{u} \cdot \mathbf{v} + \mathbf{w}$ (d) $\|\mathbf{u}\| \cdot (\mathbf{v} + \mathbf{w})$
53. Describe direction cosines and direction angles of a vector **v**.
54. Give a geometric description of the projection of **u** onto **v**.
55. What can be said about the vectors **u** and **v** if (a) the projection of **u** onto **v** equals **u** and (b) the projection of **u** onto **v** equals **0**?
56. If the projection of **u** onto **v** has the same magnitude as the projection of **v** onto **u**, can we conclude that $\|\mathbf{u}\| = \|\mathbf{v}\|$? Explain.

57. *Programming* Given vectors **u** and **v** in component form, write a program for a graphing utility in which the output is (a) $\|\mathbf{u}\|$, (b) $\|\mathbf{v}\|$, and (c) the angle between **u** and **v**.
58. Use Exercise 57 to find the angle between the vectors.
    (a) $\mathbf{u} = \langle 3, 4 \rangle, \mathbf{v} = \langle -7, 5 \rangle$
    (b) $\mathbf{u} = \langle 8, -4, 2 \rangle, \mathbf{v} = \langle 2, 5, 2 \rangle$
59. *Programming* Given vectors **u** and **v** in component form, write a program for a graphing utility in which the output is the component form of the projection of **u** onto **v**.
60. Use the program you wrote in Exercise 59 to find the projection of **u** onto **v**.
    (a) $\mathbf{u} = \langle 3, 4 \rangle, \mathbf{v} = \langle 8, 2 \rangle$
    (b) $\mathbf{u} = \langle 5, 6, 2 \rangle, \mathbf{v} = \langle -1, 3, 4 \rangle$

Think About It In Exercises 61 and 62, use the figure to mentally determine the projection of u onto v. (The coordinates of the terminal points of the vectors in standard position are given.) Verify your results analytically.

61.

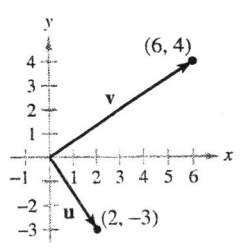

62.

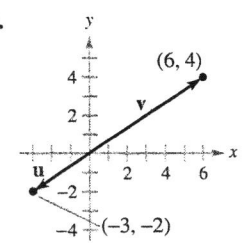

In Exercises 63–66, find two vectors in opposite directions that are orthogonal to the vector u. (The answers are not unique.)

63. $\mathbf{u} = \frac{1}{2}\mathbf{i} - \frac{2}{3}\mathbf{j}$
64. $\mathbf{u} = -8\mathbf{i} + 3\mathbf{j}$
65. $\mathbf{u} = \langle 3, 1, -2 \rangle$
66. $\mathbf{u} = \langle 0, -3, 6 \rangle$

67. *Braking Load* A 48,000-pound truck is parked on a 10° slope (see figure). Assume the only force to overcome is that due to gravity. Find (a) the force required to keep the truck from rolling down the hill, and (b) the force perpendicular to the hill.

Figure for 67

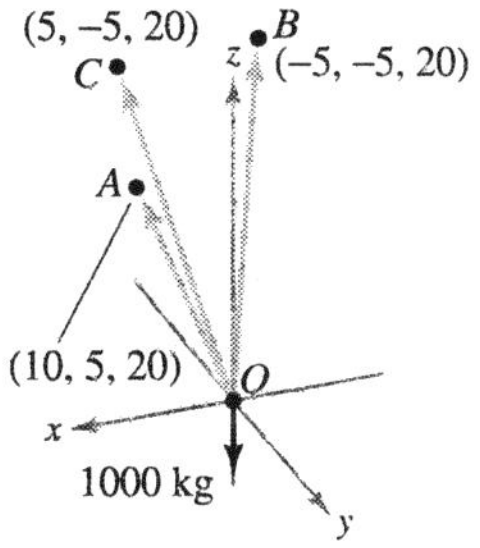

Figure for 68

68. *Load-Supporting Cables* Find the magnitude of the projection of the load-supporting cable OA onto the positive z-axis as shown in the figure.

69. *Work* An object is pulled 10 feet across a floor, using a force of 85 pounds. Find the work done if the direction of the force is 60° above the horizontal (see figure).

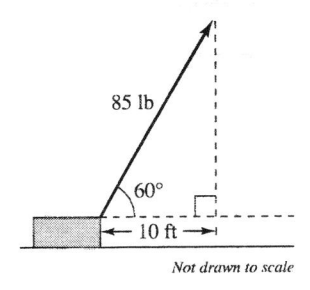

70. *Work* A toy wagon is pulled by exerting a force of 25 pounds on a handle that makes a 20° angle with the horizontal (see figure). Find the work done in pulling the wagon 50 feet.

Work In Exercises 71 and 72, find the work done in moving a particle from P to Q if the magnitude and direction of the force are given by v.

71. $P(0, 0, 0)$, $Q(4, 7, 5)$, $\mathbf{v} = \langle 1, 4, 8 \rangle$
72. $P(1, 3, 0)$, $Q(-3, 5, 10)$, $\mathbf{v} = -2\mathbf{i} + 3\mathbf{j} + 6\mathbf{k}$

True or False? In Exercises 73 and 74, determine whether the statement is true or false. If it is false, explain why or give an example that shows it is false.

73. If $\mathbf{u} \cdot \mathbf{v} = \mathbf{u} \cdot \mathbf{w}$ and $\mathbf{u} \neq \mathbf{0}$, then $\mathbf{v} = \mathbf{w}$.

74. If $\mathbf{u}$ and $\mathbf{v}$ are orthogonal to $\mathbf{w}$, then $\mathbf{u} + \mathbf{v}$ is orthogonal to $\mathbf{w}$.

75. Use vectors to prove that the diagonals of a rhombus are perpendicular.

76. *Bond Angle* Consider a regular tetrahedron with vertices $(0, 0, 0)$, $(k, k, 0)$, $(k, 0, k)$, and $(0, k, k)$, where k is a positive real number.
   (a) Sketch the graph of the tetrahedron.
   (b) Find the length of each edge.
   (c) Find the angle between any two edges.
   (d) Find the angle between the line segments from the centroid $(k/2, k/2, k/2)$ to two vertices. This is the *bond angle* for a molecule such as CH_4 or $PbCl_4$, where the structure of the molecule is a tetrahedron.

77. Consider the vectors $\mathbf{u} = \langle \cos \alpha, \sin \alpha, 0 \rangle$ and $\mathbf{v} = \langle \cos \beta, \sin \beta, 0 \rangle$, where $\alpha > \beta$. Find the dot product of the vectors and use the result to prove the identity
$$\cos(\alpha - \beta) = \cos \alpha \cos \beta + \sin \alpha \sin \beta.$$

78. Consider the two curves $y_1 = x^2$ and $y_2 = x^{1/3}$. Find the unit tangent vectors to each curve at their points of intersection. Find the angles between the curves at their points of intersection.

79. Prove that $\|\mathbf{u} - \mathbf{v}\|^2 = \|\mathbf{u}\|^2 + \|\mathbf{v}\|^2 - 2\mathbf{u} \cdot \mathbf{v}$.

80. Prove the **Cauchy-Schwarz Inequality** $|\mathbf{u} \cdot \mathbf{v}| \leq \|\mathbf{u}\| \|\mathbf{v}\|$.

81. Prove the triangle inequality $\|\mathbf{u} + \mathbf{v}\| \leq \|\mathbf{u}\| + \|\mathbf{v}\|$.

82. Prove Theorem 10.6.

Section 10.4 The Cross Product of Two Vectors in Space

- Find the cross product of two vectors in space.
- Use the triple scalar product of three vectors in space.

EXPLORATION

Geometric Property of the Cross Product Three pairs of vectors are shown below. Use the definition to find the cross product of each pair. Sketch all three vectors in a three-dimensional system. Describe any relationships among the three vectors. Use your description to write a conjecture about $\mathbf{u}$, $\mathbf{v}$, and $\mathbf{u} \times \mathbf{v}$.

a. $\mathbf{u} = \langle 3, 0, 3 \rangle$, $\mathbf{v} = \langle 3, 0, -3 \rangle$

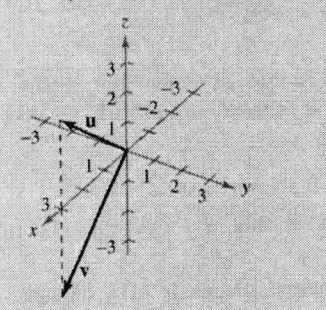

b. $\mathbf{u} = \langle 0, 3, 3 \rangle$, $\mathbf{v} = \langle 0, -3, 3 \rangle$

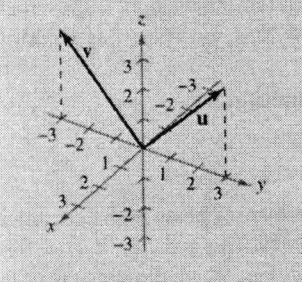

c. $\mathbf{u} = \langle 3, 3, 0 \rangle$, $\mathbf{v} = \langle 3, -3, 0 \rangle$

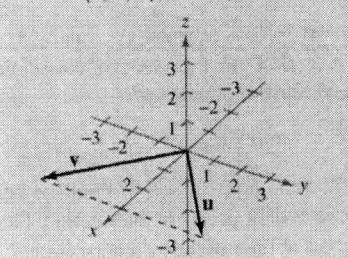

The Cross Product

Many applications in physics, engineering, and geometry involve finding a vector in space that is orthogonal to two given vectors. In this section you will study a product that will yield such a vector. It is called the **cross product**, and it is most conveniently defined and calculated using the standard unit vector form. Because the cross product yields a vector, it is also called the **vector product.**

Definition of Cross Product of Two Vectors in Space

Let $\mathbf{u} = u_1\mathbf{i} + u_2\mathbf{j} + u_3\mathbf{k}$ and $\mathbf{v} = v_1\mathbf{i} + v_2\mathbf{j} + v_3\mathbf{k}$ be vectors in space. The **cross product** of $\mathbf{u}$ and $\mathbf{v}$ is the vector

$$\mathbf{u} \times \mathbf{v} = (u_2v_3 - u_3v_2)\mathbf{i} - (u_1v_3 - u_3v_1)\mathbf{j} + (u_1v_2 - u_2v_1)\mathbf{k}.$$

NOTE Be sure you see that this definition applies only to three-dimensional vectors. The cross product is not defined for two-dimensional vectors.

A convenient way to calculate $\mathbf{u} \times \mathbf{v}$ is to use the following *determinant form* with cofactor expansion. (This 3×3 determinant form is used simply to help remember the formula for the cross product—it is technically not a determinant because the entries of the corresponding matrix are not all real numbers.)

$$\mathbf{u} \times \mathbf{v} = \begin{vmatrix} \mathbf{i} & \mathbf{j} & \mathbf{k} \\ u_1 & u_2 & u_3 \\ v_1 & v_2 & v_3 \end{vmatrix} \quad \begin{array}{l} \leftarrow \text{Put ``}\mathbf{u}\text{'' in Row 2.} \\ \leftarrow \text{Put ``}\mathbf{v}\text{'' in Row 3.} \end{array}$$

$$= \begin{vmatrix} \mathbf{i} & \mathbf{j} & \mathbf{k} \\ u_1 & u_2 & u_3 \\ v_1 & v_2 & v_3 \end{vmatrix}\mathbf{i} - \begin{vmatrix} \mathbf{i} & \mathbf{j} & \mathbf{k} \\ u_1 & u_2 & u_3 \\ v_1 & v_2 & v_3 \end{vmatrix}\mathbf{j} + \begin{vmatrix} \mathbf{i} & \mathbf{j} & \mathbf{k} \\ u_1 & u_2 & u_3 \\ v_1 & v_2 & v_3 \end{vmatrix}\mathbf{k}$$

$$= \begin{vmatrix} u_2 & u_3 \\ v_2 & v_3 \end{vmatrix}\mathbf{i} - \begin{vmatrix} u_1 & u_3 \\ v_1 & v_3 \end{vmatrix}\mathbf{j} + \begin{vmatrix} u_1 & u_2 \\ v_1 & v_2 \end{vmatrix}\mathbf{k}$$

$$= (u_2v_3 - u_3v_2)\mathbf{i} - (u_1v_3 - u_3v_1)\mathbf{j} + (u_1v_2 - u_2v_1)\mathbf{k}$$

Note the minus sign in front of the **j**-component. Each of the three 2×2 determinants can be evaluated by using the following diagonal pattern.

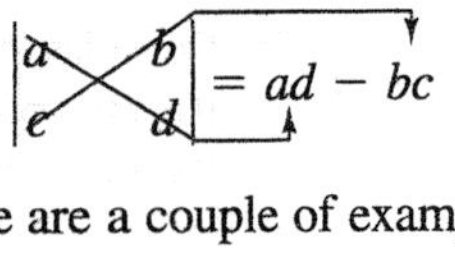

$$\begin{vmatrix} a & b \\ c & d \end{vmatrix} = ad - bc$$

Here are a couple of examples.

$$\begin{vmatrix} 2 & 4 \\ 3 & -1 \end{vmatrix} = (2)(-1) - (4)(3) = -2 - 12 = -14$$

$$\begin{vmatrix} 4 & 0 \\ -6 & 3 \end{vmatrix} = (4)(3) - (0)(-6) = 12$$

NOTATION FOR DOT AND CROSS PRODUCTS

The notation for the dot product and cross product of vectors was first introduced by the American physicist Josiah Willard Gibbs (1839–1903). In the early 1880s, Gibbs built a system to represent physical quantities called "vector analysis." The system was a departure from Hamilton's theory of quaternions.

Example 1 Finding the Cross Product

Given $\mathbf{u} = \mathbf{i} - 2\mathbf{j} + \mathbf{k}$ and $\mathbf{v} = 3\mathbf{i} + \mathbf{j} - 2\mathbf{k}$, find each of the following.

a. $\mathbf{u} \times \mathbf{v}$ **b.** $\mathbf{v} \times \mathbf{u}$ **c.** $\mathbf{v} \times \mathbf{v}$

Solution

a. $\mathbf{u} \times \mathbf{v} = \begin{vmatrix} \mathbf{i} & \mathbf{j} & \mathbf{k} \\ 1 & -2 & 1 \\ 3 & 1 & -2 \end{vmatrix} = \begin{vmatrix} -2 & 1 \\ 1 & -2 \end{vmatrix}\mathbf{i} - \begin{vmatrix} 1 & 1 \\ 3 & -2 \end{vmatrix}\mathbf{j} + \begin{vmatrix} 1 & -2 \\ 3 & 1 \end{vmatrix}\mathbf{k}$

$= (4 - 1)\mathbf{i} - (-2 - 3)\mathbf{j} + (1 + 6)\mathbf{k}$

$= 3\mathbf{i} + 5\mathbf{j} + 7\mathbf{k}$

b. $\mathbf{v} \times \mathbf{u} = \begin{vmatrix} \mathbf{i} & \mathbf{j} & \mathbf{k} \\ 3 & 1 & -2 \\ 1 & -2 & 1 \end{vmatrix} = \begin{vmatrix} 1 & -2 \\ -2 & 1 \end{vmatrix}\mathbf{i} - \begin{vmatrix} 3 & -2 \\ 1 & 1 \end{vmatrix}\mathbf{j} + \begin{vmatrix} 3 & 1 \\ 1 & -2 \end{vmatrix}\mathbf{k}$

$= (1 - 4)\mathbf{i} - (3 + 2)\mathbf{j} + (-6 - 1)\mathbf{k}$

$= -3\mathbf{i} - 5\mathbf{j} - 7\mathbf{k}$

Note that this result is the negative of that in part (a).

c. $\mathbf{v} \times \mathbf{v} = \begin{vmatrix} \mathbf{i} & \mathbf{j} & \mathbf{k} \\ 3 & 1 & -2 \\ 3 & 1 & -2 \end{vmatrix} = \mathbf{0}$

The results obtained in Example 1 suggest some interesting *algebraic* properties of the cross product. For instance, $\mathbf{u} \times \mathbf{v} = -(\mathbf{v} \times \mathbf{u})$, and $\mathbf{v} \times \mathbf{v} = \mathbf{0}$. These properties, and several others, are summarized in the following theorem.

THEOREM 10.7 Algebraic Properties of the Cross Product

Let $\mathbf{u}$, $\mathbf{v}$, and $\mathbf{w}$ be vectors in space, and let c be a scalar.

1. $\mathbf{u} \times \mathbf{v} = -(\mathbf{v} \times \mathbf{u})$
2. $\mathbf{u} \times (\mathbf{v} + \mathbf{w}) = (\mathbf{u} \times \mathbf{v}) + (\mathbf{u} \times \mathbf{w})$
3. $c(\mathbf{u} \times \mathbf{v}) = (c\mathbf{u}) \times \mathbf{v} = \mathbf{u} \times (c\mathbf{v})$
4. $\mathbf{u} \times \mathbf{0} = \mathbf{0} \times \mathbf{u} = \mathbf{0}$
5. $\mathbf{u} \times \mathbf{u} = \mathbf{0}$
6. $\mathbf{u} \cdot (\mathbf{v} \times \mathbf{w}) = (\mathbf{u} \times \mathbf{v}) \cdot \mathbf{w}$

Proof To prove Property 1, let $\mathbf{u} = u_1\mathbf{i} + u_2\mathbf{j} + u_3\mathbf{k}$ and $\mathbf{v} = v_1\mathbf{i} + v_2\mathbf{j} + v_3\mathbf{k}$. Then,

$$\mathbf{u} \times \mathbf{v} = (u_2v_3 - u_3v_2)\mathbf{i} - (u_1v_3 - u_3v_1)\mathbf{j} + (u_1v_2 - u_2v_1)\mathbf{k}$$

and

$$\mathbf{v} \times \mathbf{u} = (v_2u_3 - v_3u_2)\mathbf{i} - (v_1u_3 - v_3u_1)\mathbf{j} + (v_1u_2 - v_2u_1)\mathbf{k}$$

which implies that $\mathbf{u} \times \mathbf{v} = -(\mathbf{v} \times \mathbf{u})$. Proofs of Properties 2, 3, 5, and 6 are left as exercises (see Exercises 57–60).

NOTE It follows from Properties 1 and 2 in Theorem 10.8 that if **n** is a unit vector orthogonal to both **u** and **v**, then

$$\mathbf{u} \times \mathbf{v} = \pm(\|\mathbf{u}\|\,\|\mathbf{v}\|\sin\theta)\mathbf{n}.$$

Note that Property 1 of Theorem 10.7 indicates that the cross product is *not commutative*. In particular, this property indicates that the vectors $\mathbf{u} \times \mathbf{v}$ and $\mathbf{v} \times \mathbf{u}$ have equal lengths but opposite directions. The following theorem lists some other *geometric* properties of the cross product of two vectors.

THEOREM 10.8 Geometric Properties of the Cross Product

Let **u** and **v** be nonzero vectors in space, and let θ be the angle between **u** and **v**.

1. $\mathbf{u} \times \mathbf{v}$ is orthogonal to both **u** and **v**.
2. $\|\mathbf{u} \times \mathbf{v}\| = \|\mathbf{u}\|\,\|\mathbf{v}\|\sin\theta$
3. $\mathbf{u} \times \mathbf{v} = \mathbf{0}$ if and only if **u** and **v** are scalar multiples of each other.
4. $\|\mathbf{u} \times \mathbf{v}\|$ = area of parallelogram having **u** and **v** as adjacent sides.

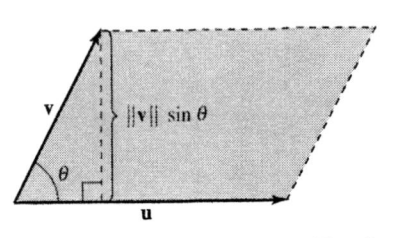

The vectors **u** and **v** form adjacent sides of a parallelogram.
Figure 10.35

Proof To prove Property 2, note that because $\cos\theta = (\mathbf{u} \cdot \mathbf{v})/(\|\mathbf{u}\|\,\|\mathbf{v}\|)$, it follows that

$$\begin{aligned}
\|\mathbf{u}\|\,\|\mathbf{v}\|\sin\theta &= \|\mathbf{u}\|\,\|\mathbf{v}\|\sqrt{1 - \cos^2\theta} \\
&= \|\mathbf{u}\|\,\|\mathbf{v}\|\sqrt{1 - \frac{(\mathbf{u}\cdot\mathbf{v})^2}{\|\mathbf{u}\|^2\|\mathbf{v}\|^2}} \\
&= \sqrt{\|\mathbf{u}\|^2\|\mathbf{v}\|^2 - (\mathbf{u}\cdot\mathbf{v})^2} \\
&= \sqrt{(u_1^2 + u_2^2 + u_3^2)(v_1^2 + v_2^2 + v_3^2) - (u_1v_1 + u_2v_2 + u_3v_3)^2} \\
&= \sqrt{(u_2v_3 - u_3v_2)^2 + (u_1v_3 - u_3v_1)^2 + (u_1v_2 - u_2v_1)^2} \\
&= \|\mathbf{u} \times \mathbf{v}\|.
\end{aligned}$$

To prove Property 4, refer to Figure 10.35, which is a parallelogram having **v** and **u** as adjacent sides. Because the height of the parallelogram is $\|\mathbf{v}\|\sin\theta$, the area is

$$\begin{aligned}
\text{Area} &= (\text{base})(\text{height}) \\
&= \|\mathbf{u}\|\,\|\mathbf{v}\|\sin\theta \\
&= \|\mathbf{u} \times \mathbf{v}\|.
\end{aligned}$$

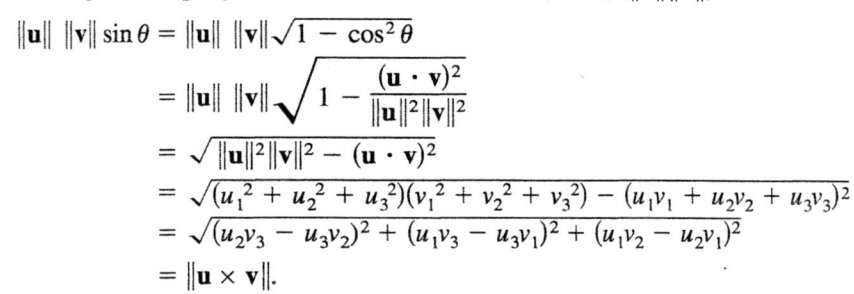

Proofs of Properties 1 and 3 are left as exercises (see Exercises 61 and 62).

Both $\mathbf{u} \times \mathbf{v}$ and $\mathbf{v} \times \mathbf{u}$ are perpendicular to the plane determined by **u** and **v**. One way to remember the orientation of the vectors **u**, **v**, and $\mathbf{u} \times \mathbf{v}$ is to compare them with the unit vectors **i**, **j**, and $\mathbf{k} = \mathbf{i} \times \mathbf{j}$, as shown in Figure 10.36. The three vectors **u**, **v**, and $\mathbf{u} \times \mathbf{v}$ form a *right-handed system*, whereas the three vectors **u**, **v**, and $\mathbf{v} \times \mathbf{u}$ form a *left-handed system*.

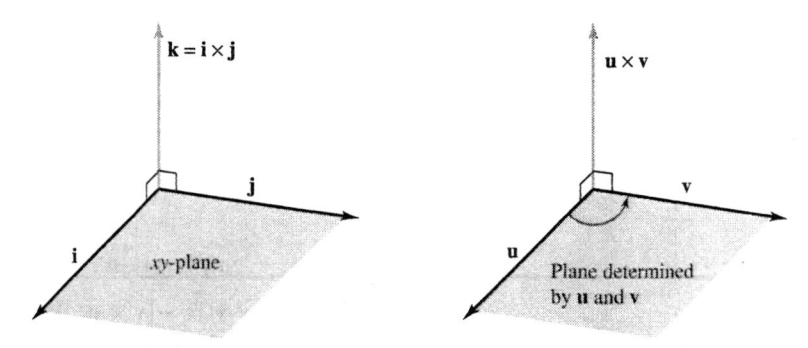

Right-handed systems
Figure 10.36

Example 2 Using the Cross Product

Find a unit vector that is orthogonal to both

$$\mathbf{u} = \mathbf{i} - 4\mathbf{j} + \mathbf{k} \quad \text{and} \quad \mathbf{v} = 2\mathbf{i} + 3\mathbf{j}.$$

Solution The cross product $\mathbf{u} \times \mathbf{v}$, as shown in Figure 10.37, is orthogonal to both $\mathbf{u}$ and $\mathbf{v}$.

$$\mathbf{u} \times \mathbf{v} = \begin{vmatrix} \mathbf{i} & \mathbf{j} & \mathbf{k} \\ 1 & -4 & 1 \\ 2 & 3 & 0 \end{vmatrix} \quad \text{Cross product}$$

$$= -3\mathbf{i} + 2\mathbf{j} + 11\mathbf{k}$$

Because

$$\|\mathbf{u} \times \mathbf{v}\| = \sqrt{(-3)^2 + 2^2 + 11^2} = \sqrt{134},$$

a unit vector orthogonal to both $\mathbf{u}$ and $\mathbf{v}$ is

$$\frac{\mathbf{u} \times \mathbf{v}}{\|\mathbf{u} \times \mathbf{v}\|} = -\frac{3}{\sqrt{134}}\mathbf{i} + \frac{2}{\sqrt{134}}\mathbf{j} + \frac{11}{\sqrt{134}}\mathbf{k}.$$

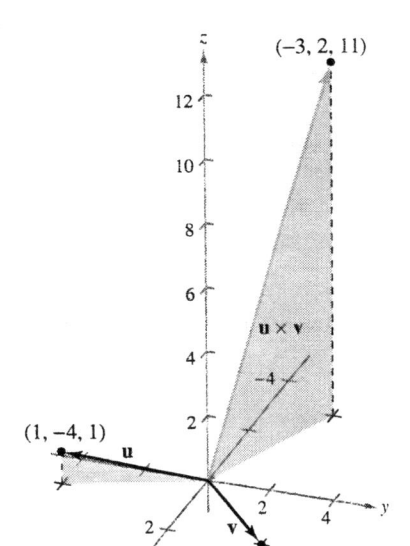

The vector $\mathbf{u} \times \mathbf{v}$ is orthogonal to both $\mathbf{u}$ and $\mathbf{v}$.
Figure 10.37

NOTE In Example 2, note that you could have used the cross product $\mathbf{v} \times \mathbf{u}$ to form a unit vector that is orthogonal to both $\mathbf{u}$ and $\mathbf{v}$. With that choice, you would have obtained the negative of the unit vector found in the example.

Example 3 Geometric Application of the Cross Product

Show that the quadrilateral with vertices at the following points is a parallelogram, and find its area.

$$A = (5, 2, 0) \quad B = (2, 6, 1)$$
$$C = (2, 4, 7) \quad D = (5, 0, 6)$$

Solution From Figure 10.38 you can see that the sides of the quadrilateral correspond to the following four vectors.

$$\overrightarrow{AB} = -3\mathbf{i} + 4\mathbf{j} + \mathbf{k} \qquad \overrightarrow{CD} = 3\mathbf{i} - 4\mathbf{j} - \mathbf{k} = -\overrightarrow{AB}$$
$$\overrightarrow{AD} = 0\mathbf{i} - 2\mathbf{j} + 6\mathbf{k} \qquad \overrightarrow{CB} = 0\mathbf{i} + 2\mathbf{j} - 6\mathbf{k} = -\overrightarrow{AD}$$

So, $\overrightarrow{AB}$ is parallel to $\overrightarrow{CD}$ and $\overrightarrow{AD}$ is parallel to $\overrightarrow{CB}$, and you can conclude that the quadrilateral is a parallelogram with $\overrightarrow{AB}$ and $\overrightarrow{AD}$ as adjacent sides. Moreover, because

$$\overrightarrow{AB} \times \overrightarrow{AD} = \begin{vmatrix} \mathbf{i} & \mathbf{j} & \mathbf{k} \\ -3 & 4 & 1 \\ 0 & -2 & 6 \end{vmatrix} \quad \text{Cross product}$$

$$= 26\mathbf{i} + 18\mathbf{j} + 6\mathbf{k}$$

the area of the parallelogram is

$$\|\overrightarrow{AB} \times \overrightarrow{AD}\| = \sqrt{1036} \approx 32.19.$$

Is the parallelogram a rectangle? You can determine whether it is by finding the angle between the vectors $\overrightarrow{AB}$ and $\overrightarrow{AD}$.

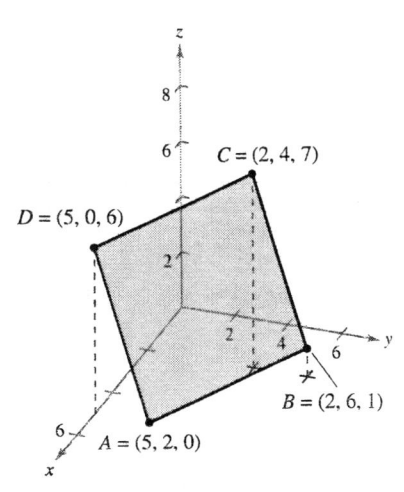

The area of the parallelogram is approximately 32.19.
Figure 10.38

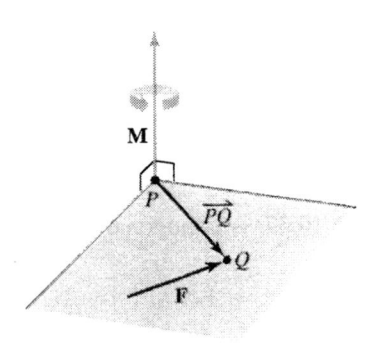

In physics, the cross product can be used to measure **torque**—the **moment M of a force F about a point** *P*, as shown in Figure 10.39. If the point of application of the force is *Q*, the moment of **F** about *P* is given by

$$\mathbf{M} = \overrightarrow{PQ} \times \mathbf{F}. \qquad \text{Moment of } \mathbf{F} \text{ about } P$$

The magnitude of the moment **M** measures the tendency of the vector $\overrightarrow{PQ}$ to rotate counterclockwise (using the right-hand rule) about an axis directed along the vector **M**.

The moment of **F** about *P*
Figure 10.39

Example 4 An Application of the Cross Product

A vertical force of 50 pounds is applied to the end of a 1-foot lever that is attached to an axle at point *P*, as shown in Figure 10.40. Find the moment of this force about the point *P* when $\theta = 60°$.

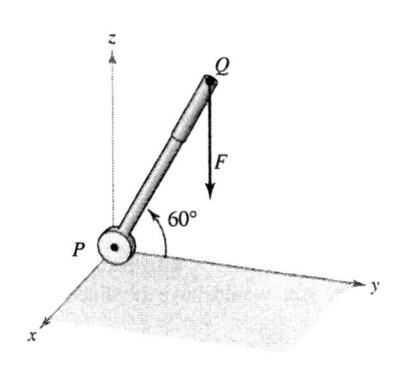

Solution If you represent the 50-pound force as $\mathbf{F} = -50\mathbf{k}$ and the lever as

$$\overrightarrow{PQ} = \cos(60°)\mathbf{j} + \sin(60°)\mathbf{k} = \frac{1}{2}\mathbf{j} + \frac{\sqrt{3}}{2}\mathbf{k}$$

the moment of **F** about *P* is given by

$$\mathbf{M} = \overrightarrow{PQ} \times \mathbf{F} = \begin{vmatrix} \mathbf{i} & \mathbf{j} & \mathbf{k} \\ 0 & \frac{1}{2} & \frac{\sqrt{3}}{2} \\ 0 & 0 & -50 \end{vmatrix} = -25\mathbf{i}. \qquad \text{Moment of } \mathbf{F} \text{ about } P$$

The magnitude of this moment is 25 foot-pounds.

A vertical force of 50 pounds is applied at point *Q*.
Figure 10.40

NOTE In Example 4, note that the moment (the tendency of the lever to rotate about its axle) is dependent on the angle θ. When $\theta = \pi/2$, the moment is **0**. The moment is greatest when $\theta = 0$.

The Triple Scalar Product

For vectors **u**, **v**, and **w** in space, the dot product of **u** and **v** × **w**

$$\mathbf{u} \cdot (\mathbf{v} \times \mathbf{w})$$

is called the **triple scalar product**, as defined in Theorem 10.9. The proof of this theorem is left as an exercise (see Exercise 56).

FOR FURTHER INFORMATION To see how the cross product is used to model the torque of the robot arm of a space shuttle, see the article "The Long Arm of Calculus" by Ethan Berkove and Rich Marchand in *The College Mathematics Journal*. To view this article, go to the website *www.matharticles.com*.

THEOREM 10.9 The Triple Scalar Product

For $\mathbf{u} = u_1\mathbf{i} + u_2\mathbf{j} + u_3\mathbf{k}$, $\mathbf{v} = v_1\mathbf{i} + v_2\mathbf{j} + v_3\mathbf{k}$, and $\mathbf{w} = w_1\mathbf{i} + w_2\mathbf{j} + w_3\mathbf{k}$, the triple scalar product is given by

$$\mathbf{u} \cdot (\mathbf{v} \times \mathbf{w}) = \begin{vmatrix} u_1 & u_2 & u_3 \\ v_1 & v_2 & v_3 \\ w_1 & w_2 & w_3 \end{vmatrix}.$$

NOTE The value of a determinant is multiplied by -1 if two rows are interchanged. After two such interchanges, the value of the determinant will be unchanged. So, the following triple scalar products are equivalent.

$$\mathbf{u} \cdot (\mathbf{v} \times \mathbf{w}) = \mathbf{v} \cdot (\mathbf{w} \times \mathbf{u}) = \mathbf{w} \cdot (\mathbf{u} \times \mathbf{v})$$

v × w

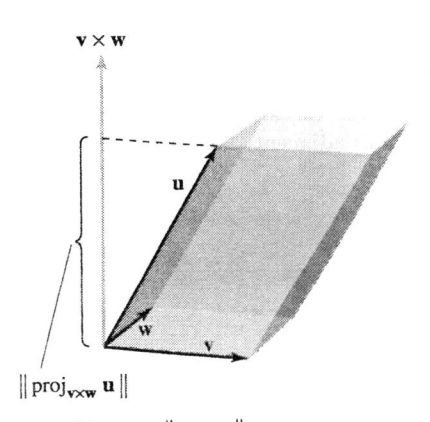

Area of base = $\|\mathbf{v} \times \mathbf{w}\|$
Volume of parallelepiped = $|\mathbf{u} \cdot (\mathbf{v} \times \mathbf{w})|$
Figure 10.41

If the vectors **u**, **v**, and **w** do not lie in the same plane, the triple scalar product $\mathbf{u} \cdot (\mathbf{v} \times \mathbf{w})$ can be used to determine the volume of the parallelepiped (a polyhedron, all of whose faces are parallelograms) with **u**, **v**, and **w** as adjacent edges, as shown in Figure 10.41. This is established in the following theorem.

THEOREM 10.10   Geometric Property of Triple Scalar Product

The volume V of a parallelepiped with vectors **u**, **v**, and **w** as adjacent edges is given by

$$V = |\mathbf{u} \cdot (\mathbf{v} \times \mathbf{w})|.$$

Proof   In Figure 10.41, note that

$$\|\mathbf{v} \times \mathbf{w}\| = \text{area of base}$$

and

$$\|\text{proj}_{\mathbf{v} \times \mathbf{w}} \mathbf{u}\| = \text{height of parallelepiped}.$$

Therefore, the volume is

$$V = (\text{height})(\text{area of base}) = \|\text{proj}_{\mathbf{v} \times \mathbf{w}} \mathbf{u}\| \|\mathbf{v} \times \mathbf{w}\|$$
$$= \left|\frac{\mathbf{u} \cdot (\mathbf{v} \times \mathbf{w})}{\|\mathbf{v} \times \mathbf{w}\|}\right| \|\mathbf{v} \times \mathbf{w}\|$$
$$= |\mathbf{u} \cdot (\mathbf{v} \times \mathbf{w})|.$$

Example 5   Volume by the Triple Scalar Product

Find the volume of the parallelepiped having $\mathbf{u} = 3\mathbf{i} - 5\mathbf{j} + \mathbf{k}$, $\mathbf{v} = 2\mathbf{j} - 2\mathbf{k}$, and $\mathbf{w} = 3\mathbf{i} + \mathbf{j} + \mathbf{k}$ as adjacent edges (see Figure 10.42).

Solution   By Theorem 10.10, you have

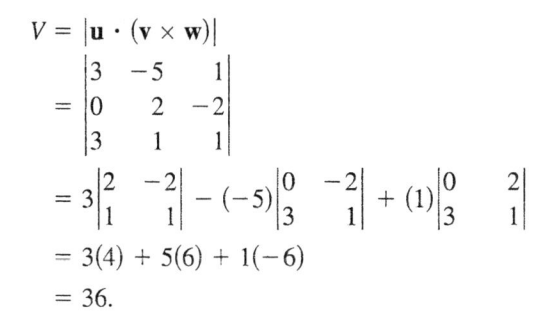

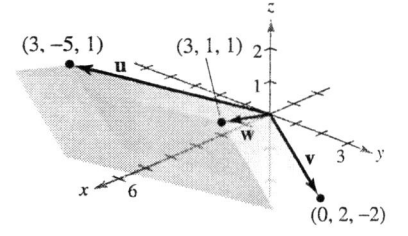

$$V = |\mathbf{u} \cdot (\mathbf{v} \times \mathbf{w})| \qquad \text{Triple scalar product}$$
$$= \begin{vmatrix} 3 & -5 & 1 \\ 0 & 2 & -2 \\ 3 & 1 & 1 \end{vmatrix}$$
$$= 3\begin{vmatrix} 2 & -2 \\ 1 & 1 \end{vmatrix} - (-5)\begin{vmatrix} 0 & -2 \\ 3 & 1 \end{vmatrix} + (1)\begin{vmatrix} 0 & 2 \\ 3 & 1 \end{vmatrix}$$
$$= 3(4) + 5(6) + 1(-6)$$
$$= 36.$$

The parallelepiped has a volume of 36.
Figure 10.42

A natural consequence of Theorem 10.10 is that the volume of the parallelepiped is 0 if and only if the three vectors are coplaner. That is, if the vectors $\mathbf{u} = \langle u_1, u_2, u_3 \rangle$, $\mathbf{v} = \langle v_1, v_2, v_3 \rangle$, and $\mathbf{w} = \langle w_1, w_2, w_3 \rangle$ have the same initial point, they lie in the same plane if and only if

$$\mathbf{u} \cdot (\mathbf{v} \times \mathbf{w}) = \begin{vmatrix} u_1 & u_2 & u_3 \\ v_1 & v_2 & v_3 \\ w_1 & w_2 & w_3 \end{vmatrix} = 0.$$

EXERCISES FOR SECTION 10.4

In Exercises 1–6, find the cross product of the unit vectors and sketch your result.

1. $\mathbf{j} \times \mathbf{i}$
2. $\mathbf{i} \times \mathbf{j}$
3. $\mathbf{j} \times \mathbf{k}$
4. $\mathbf{k} \times \mathbf{j}$
5. $\mathbf{i} \times \mathbf{k}$
6. $\mathbf{k} \times \mathbf{i}$

In Exercises 7–10, find (a) $\mathbf{u} \times \mathbf{v}$, (b) $\mathbf{v} \times \mathbf{u}$, and (c) $\mathbf{v} \times \mathbf{v}$.

7. $\mathbf{u} = -2\mathbf{i} + 3\mathbf{j} + 4\mathbf{k}$
   $\mathbf{v} = 3\mathbf{i} + 7\mathbf{j} + 2\mathbf{k}$
8. $\mathbf{u} = 3\mathbf{i} + 5\mathbf{k}$
   $\mathbf{v} = 2\mathbf{i} + 3\mathbf{j} - 2\mathbf{k}$
9. $\mathbf{u} = \langle 7, 3, 2 \rangle$
   $\mathbf{v} = \langle 1, -1, 5 \rangle$
10. $\mathbf{u} = \langle 3, -2, -2 \rangle$
    $\mathbf{v} = \langle 1, 5, 1 \rangle$

In Exercises 11–16, find $\mathbf{u} \times \mathbf{v}$ and show that it is orthogonal to both $\mathbf{u}$ and $\mathbf{v}$.

11. $\mathbf{u} = \langle 2, -3, 1 \rangle$
    $\mathbf{v} = \langle 1, -2, 1 \rangle$
12. $\mathbf{u} = \langle -1, 1, 2 \rangle$
    $\mathbf{v} = \langle 0, 1, 0 \rangle$
13. $\mathbf{u} = \langle 12, -3, 0 \rangle$
    $\mathbf{v} = \langle -2, 5, 0 \rangle$
14. $\mathbf{u} = \langle -10, 0, 6 \rangle$
    $\mathbf{v} = \langle 7, 0, 0 \rangle$
15. $\mathbf{u} = \mathbf{i} + \mathbf{j} + \mathbf{k}$
    $\mathbf{v} = 2\mathbf{i} + \mathbf{j} - \mathbf{k}$
16. $\mathbf{u} = \mathbf{i} + 6\mathbf{j}$
    $\mathbf{v} = -2\mathbf{i} + \mathbf{j} + \mathbf{k}$

Think About It In Exercises 17–20, use the vectors $\mathbf{u}$ and $\mathbf{v}$ shown in the figure to sketch a vector in the direction of the indicated cross product in a right-handed system.

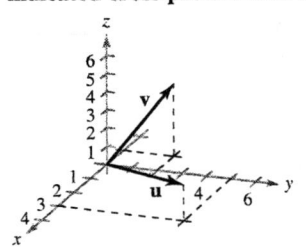

17. $\mathbf{u} \times \mathbf{v}$
18. $\mathbf{v} \times \mathbf{u}$
19. $(-\mathbf{v}) \times \mathbf{u}$
20. $\mathbf{u} \times (\mathbf{u} \times \mathbf{v})$

In Exercises 21–24, use a computer algebra system to find $\mathbf{u} \times \mathbf{v}$ and a unit vector orthogonal to $\mathbf{u}$ and $\mathbf{v}$.

21. $\mathbf{u} = \langle 4, -3.5, 7 \rangle$
    $\mathbf{v} = \langle -1, 8, 4 \rangle$
22. $\mathbf{u} = \langle -8, -6, 4 \rangle$
    $\mathbf{v} = \langle 10, -12, -2 \rangle$
23. $\mathbf{u} = -3\mathbf{i} + 2\mathbf{j} - 5\mathbf{k}$
    $\mathbf{v} = \frac{1}{2}\mathbf{i} - \frac{3}{4}\mathbf{j} + \frac{1}{10}\mathbf{k}$
24. $\mathbf{u} = \frac{2}{3}\mathbf{k}$
    $\mathbf{v} = \frac{1}{2}\mathbf{i} + 6\mathbf{k}$

25. *Programming* Given the vectors $\mathbf{u}$ and $\mathbf{v}$ in component form, write a program for a graphing utility in which the output is $\mathbf{u} \times \mathbf{v}$ and $\|\mathbf{u} \times \mathbf{v}\|$.

26. Use the program you wrote in Exercise 25 to find $\mathbf{u} \times \mathbf{v}$ and $\|\mathbf{u} \times \mathbf{v}\|$.
    (a) $\mathbf{u} = \langle 8, -4, 2 \rangle$
        $\mathbf{v} = \langle 2, 5, 2 \rangle$
    (b) $\mathbf{u} = \langle -2, 6, 10 \rangle$
        $\mathbf{v} = \langle 3, 8, 5 \rangle$

Area In Exercises 27–30, find the area of the parallelogram that has the given vectors as adjacent sides. Use a computer algebra system or a graphing utility to verify your result.

27. $\mathbf{u} = \mathbf{j}$
    $\mathbf{v} = \mathbf{j} + \mathbf{k}$
28. $\mathbf{u} = \mathbf{i} + \mathbf{j} + \mathbf{k}$
    $\mathbf{v} = \mathbf{j} + \mathbf{k}$
29. $\mathbf{u} = \langle 3, 2, -1 \rangle$
    $\mathbf{v} = \langle 1, 2, 3 \rangle$
30. $\mathbf{u} = \langle 2, -1, 0 \rangle$
    $\mathbf{v} = \langle -1, 2, 0 \rangle$

Area In Exercises 31 and 32, verify that the points are the vertices of a parallelogram, and find its area.

31. $(1, 1, 1)$, $(2, 3, 4)$, $(6, 5, 2)$, $(7, 7, 5)$
32. $(2, -3, 1)$, $(6, 5, -1)$, $(3, -6, 4)$, $(7, 2, 2)$

Area In Exercises 33–36, find the area of the triangle with the given vertices. $\left(\text{Hint: } \frac{1}{2}\|\mathbf{u} \times \mathbf{v}\|\right.$ is the area of the triangle having $\mathbf{u}$ and $\mathbf{v}$ as adjacent sides.$\left.\right)$

33. $(0, 0, 0)$, $(1, 2, 3)$, $(-3, 0, 0)$
34. $(2, -3, 4)$, $(0, 1, 2)$, $(-1, 2, 0)$
35. $(2, -7, 3)$, $(-1, 5, 8)$, $(4, 6, -1)$
36. $(1, 2, 0)$, $(-2, 1, 0)$, $(0, 0, 0)$

37. *Torque* A child applies the brakes on a bicycle by applying a downward force of 20 pounds on the pedal when the crank makes a 40° angle with the horizontal (see figure). Find the torque at P if the crank is 6 inches in length.

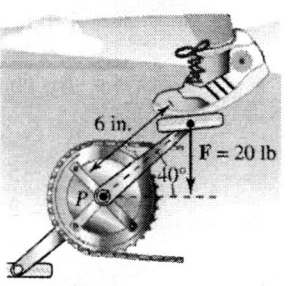

Figure for 37 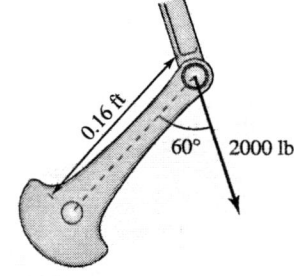   Figure for 38

38. *Torque* Both the magnitude and the direction of the force on a crankshaft change as the crankshaft rotates. Find the torque on the crankshaft using the position and data shown in the figure.

39. *Optimization* A force of 60 pounds acts on the pipe wrench shown in the figure.
    (a) Find the magnitude of the moment about O by evaluating $\|\overrightarrow{OA} \times \mathbf{F}\|$. Use a graphing utility to graph the resulting function of θ.
    (b) Use the result in part (a) to determine the magnitude of the moment when $\theta = 45°$.
    (c) Use the result in part (a) to determine the angle θ when the magnitude of the moment is maximum. Is the answer what you expected? Why or why not?

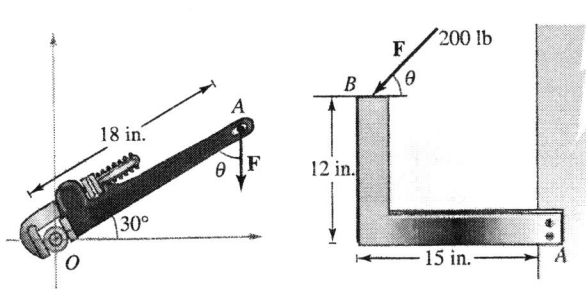

Figure for 39   **Figure for 40**

40. *Optimization* A force of 200 pounds acts on the bracket shown in the figure.

(a) Determine the vector $\overrightarrow{AB}$ and the vector **F** representing the force. (**F** will be in terms of θ.)

(b) Find the magnitude of the moment about A by evaluating $\|\overrightarrow{AB} \times \mathbf{F}\|$.

(c) Use the result in part (b) to determine the magnitude of the moment when $\theta = 30°$.

(d) Use the result in part (b) to determine the angle θ when the magnitude of the moment is maximum. At that angle, what is the relationship between the vectors **F** and $\overrightarrow{AB}$? Is it what you expected? Why or why not?

(e) Use a graphing utility to graph the function for the magnitude of the moment about A for $0° \le \theta \le 180°$. Find the zero of the function in the given domain. Interpret the meaning of the zero in the context of the problem.

In Exercises 41–44, find $\mathbf{u} \cdot (\mathbf{v} \times \mathbf{w})$.

41. $\mathbf{u} = \mathbf{i}$
    $\mathbf{v} = \mathbf{j}$
    $\mathbf{w} = \mathbf{k}$

42. $\mathbf{u} = \langle 1, 1, 1 \rangle$
    $\mathbf{v} = \langle 2, 1, 0 \rangle$
    $\mathbf{w} = \langle 0, 0, 1 \rangle$

43. $\mathbf{u} = \langle 2, 0, 1 \rangle$
    $\mathbf{v} = \langle 0, 3, 0 \rangle$
    $\mathbf{w} = \langle 0, 0, 1 \rangle$

44. $\mathbf{u} = \langle 2, 0, 0 \rangle$
    $\mathbf{v} = \langle 1, 1, 1 \rangle$
    $\mathbf{w} = \langle 0, 2, 2 \rangle$

Volume In Exercises 45 and 46, use the triple scalar product to find the volume of the parallelepiped having adjacent edges **u**, **v**, and **w**.

45. $\mathbf{u} = \mathbf{i} + \mathbf{j}$
    $\mathbf{v} = \mathbf{j} + \mathbf{k}$
    $\mathbf{w} = \mathbf{i} + \mathbf{k}$

46. $\mathbf{u} = \langle 1, 3, 1 \rangle$
    $\mathbf{v} = \langle 0, 6, 6 \rangle$
    $\mathbf{w} = \langle -4, 0, -4 \rangle$

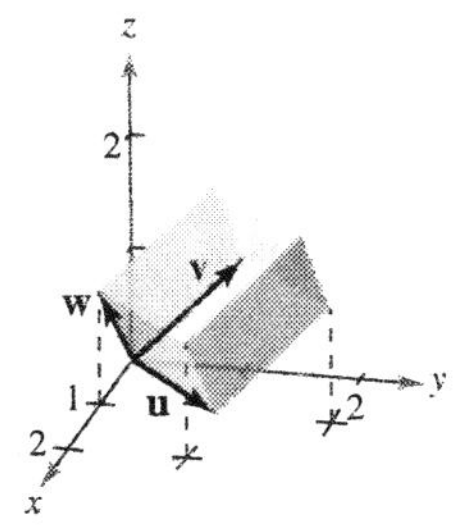

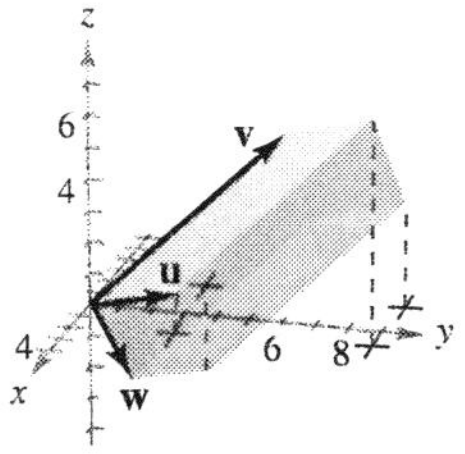

Volume In Exercises 47 and 48, find the volume of the parallelepiped with the given vertices (see figures).

47. $(0, 0, 0)$, $(3, 0, 0)$, $(0, 5, 1)$, $(3, 5, 1)$
    $(2, 0, 5)$, $(5, 0, 5)$, $(2, 5, 6)$, $(5, 5, 6)$

48. $(0, 0, 0)$, $(1, 1, 0)$, $(1, 0, 2)$, $(0, 1, 1)$
    $(2, 1, 2)$, $(1, 1, 3)$, $(1, 2, 1)$, $(2, 2, 3)$

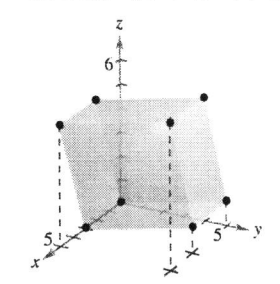

Figure for 47 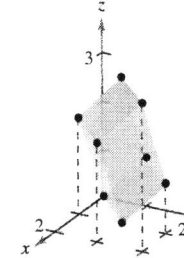   **Figure for 48**

Getting at the Concept

49. Define the cross product of vectors **u** and **v**.

50. State the geometric properties of the cross product.

51. If the magnitudes of two vectors are doubled, how will the magnitude of the cross product of the vectors change? Explain.

52. The vertices of a triangle in space are (x_1, y_1, z_1), (x_2, y_2, z_2), and (x_3, y_3, z_3). Explain how to find a vector perpendicular to the triangle.

True or False? In Exercises 53–55, determine whether the statement is true or false. If it is false, explain why or give an example that shows it is false.

53. It is possible to find the cross product of two vectors in a two-dimensional coordinate system.

54. If $\mathbf{u} \neq \mathbf{0}$ and $\mathbf{u} \times \mathbf{v} = \mathbf{u} \times \mathbf{w}$, then $\mathbf{v} = \mathbf{w}$.

55. If $\mathbf{u} \neq \mathbf{0}$, $\mathbf{u} \cdot \mathbf{v} = \mathbf{u} \cdot \mathbf{w}$, and $\mathbf{u} \times \mathbf{v} = \mathbf{u} \times \mathbf{w}$, then $\mathbf{v} = \mathbf{w}$.

56. Prove Theorem 10.9.

In Exercises 57–64, prove the property of the cross product.

57. $\mathbf{u} \times (\mathbf{v} + \mathbf{w}) = (\mathbf{u} \times \mathbf{v}) + (\mathbf{u} \times \mathbf{w})$

58. $c(\mathbf{u} \times \mathbf{v}) = (c\mathbf{u}) \times \mathbf{v} = \mathbf{u} \times (c\mathbf{v})$

59. $\mathbf{u} \times \mathbf{u} = \mathbf{0}$

60. $\mathbf{u} \cdot (\mathbf{v} \times \mathbf{w}) = (\mathbf{u} \times \mathbf{v}) \cdot \mathbf{w}$

61. $\mathbf{u} \times \mathbf{v}$ is orthogonal to both **u** and **v**.

62. $\mathbf{u} \times \mathbf{v} = \mathbf{0}$ if and only if **u** and **v** are scalar multiples of each other.

63. $\|\mathbf{u} \times \mathbf{v}\| = \|\mathbf{u}\| \|\mathbf{v}\|$ if **u** and **v** are orthogonal.

64. $\mathbf{u} \times (\mathbf{v} \times \mathbf{w}) = (\mathbf{u} \cdot \mathbf{w})\mathbf{v} - (\mathbf{u} \cdot \mathbf{v})\mathbf{w}$

Section 10.5 Lines and Planes in Space

- Write a set of parametric equations for a line in space.
- Write a linear equation to represent a plane in space.
- Sketch the plane given by a linear equation.
- Find the distance between points, planes, and lines in space.

Lines in Space

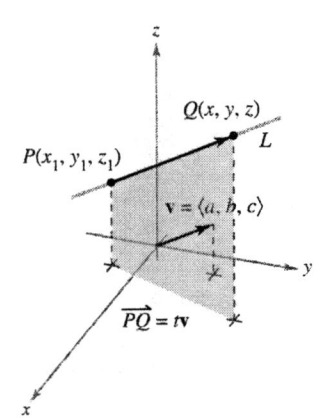

In the plane, *slope* is used to determine an equation of a line. In space, it is more convenient to use *vectors* to determine the equation of a line.

In Figure 10.43, consider the line L through the point $P(x_1, y_1, z_1)$ and parallel to the vector $\mathbf{v} = \langle a, b, c \rangle$. The vector $\mathbf{v}$ is the **direction vector** for the line L, and a, b, and c are the **direction numbers**. One way of describing the line L is to say that it consists of all points $Q(x, y, z)$ for which the vector $\overrightarrow{PQ}$ is parallel to $\mathbf{v}$. This means that $\overrightarrow{PQ}$ is a scalar multiple of $\mathbf{v}$, and you can write $\overrightarrow{PQ} = t\mathbf{v}$, where t is scalar (a real number).

$$\overrightarrow{PQ} = \langle x - x_1, y - y_1, z - z_1 \rangle = \langle at, bt, ct \rangle = t\mathbf{v}$$

By equating corresponding components, you can obtain the **parametric equations** of a line in space.

Line L and its direction vector $\mathbf{v}$
Figure 10.43

THEOREM 10.11 Parametric Equations of a Line in Space

A line L parallel to the vector $\mathbf{v} = \langle a, b, c \rangle$ and passing through the point $P(x_1, y_1, z_1)$ is represented by the **parametric equations**

$$x = x_1 + at, \quad y = y_1 + bt, \quad \text{and} \quad z = z_1 + ct.$$

If the direction numbers a, b, and c are all nonzero, you can eliminate the parameter t to obtain the **symmetric equations** of a line.

$$\frac{x - x_1}{a} = \frac{y - y_1}{b} = \frac{z - z_1}{c} \qquad \text{Symmetric equations}$$

Example 1 **Finding Parametric and Symmetric Equations**

Find parametric and symmetric equations of the line L that passes through the point $(1, -2, 4)$ and is parallel to $\mathbf{v} = \langle 2, 4, -4 \rangle$.

Solution To find a set of parametric equations of the line, use the coordinates $x_1 = 1$, $y_1 = -2$, and $z_1 = 4$ and direction numbers $a = 2$, $b = 4$, and $c = -4$ (see Figure 10.44).

$$x = 1 + 2t, \quad y = -2 + 4t, \quad z = 4 - 4t \qquad \text{Parametric equations}$$

Because a, b, and c are all nonzero, a set of symmetric equations is

$$\frac{x - 1}{2} = \frac{y + 2}{4} = \frac{z - 4}{-4}. \qquad \text{Symmetric equations}$$

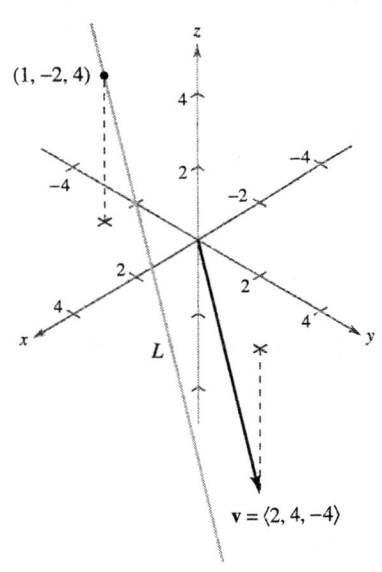

The vector $\mathbf{v}$ is parallel to the line L.
Figure 10.44

Neither the parametric equations nor the symmetric equations of a given line are unique. For instance, in Example 1, by letting $t = 1$ in the parametric equations you would obtain the point $(3, 2, 0)$. Using this point with the direction numbers $a = 2$, $b = 4$, and $c = -4$ would produce the different parametric equations

$$x = 3 + 2t, \quad y = 2 + 4t, \quad \text{and} \quad z = -4t.$$

Example 2 Parametric Equations of a Line Through Two Points

Find a set of parametric equations of the line that passes through the points $(-2, 1, 0)$ and $(1, 3, 5)$.

Solution Begin by using the points $P(-2, 1, 0)$ and $Q(1, 3, 5)$ to find a direction vector for the line passing through P and Q, given by

$$\mathbf{v} = \overrightarrow{PQ} = \langle 1 - (-2), 3 - 1, 5 - 0 \rangle = \langle 3, 2, 5 \rangle = \langle a, b, c \rangle.$$

Using the direction numbers $a = 3$, $b = 2$, and $c = 5$ with the point $P(-2, 1, 0)$, you can obtain the parametric equations

$$x = -2 + 3t, \quad y = 1 + 2t, \quad \text{and} \quad z = 5t.$$

NOTE As t varies over all real numbers, the parametric equations in Example 2 determine the points (x, y, z) on the line. In particular, note that $t = 0$ and $t = 1$ give the original points $(-2, 1, 0)$ and $(1, 3, 5)$.

Planes in Space

You have seen how an equation of a line in space can be obtained from a point on the line and a vector *parallel* to it. You will now see that an equation of a plane in space can be obtained from a point in the plane and a vector *normal* (perpendicular) to it.

Consider the plane containing the point $P(x_1, y_1, z_1)$ having a nonzero normal vector $\mathbf{n} = \langle a, b, c \rangle$, as shown in Figure 10.45. This plane consists of all points $Q(x, y, z)$ for which vector $\overrightarrow{PQ}$ is orthogonal to $\mathbf{n}$. Using the dot product, you can write the following.

$$\mathbf{n} \cdot \overrightarrow{PQ} = 0$$
$$\langle a, b, c \rangle \cdot \langle x - x_1, y - y_1, z - z_1 \rangle = 0$$
$$a(x - x_1) + b(y - y_1) + c(z - z_1) = 0$$

The third equation of the plane is said to be in **standard form.**

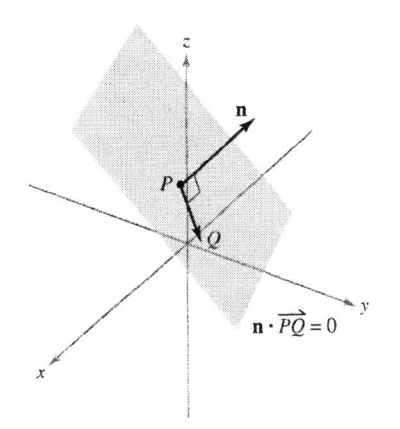

The normal vector $\mathbf{n}$ is orthogonal to each vector $\overrightarrow{PQ}$ in the plane.
Figure 10.45

THEOREM 10.12 Standard Equation of a Plane in Space

The plane containing the point (x_1, y_1, z_1) and having a normal vector $\mathbf{n} = \langle a, b, c \rangle$ can be represented, in **standard form,** by the equation

$$a(x - x_1) + b(y - y_1) + c(z - z_1) = 0.$$

By regrouping terms, you obtain the **general form** of the equation of a plane in space.

$$ax + by + cz + d = 0 \qquad \text{General form of equation of plane}$$

Given the general form of the equation of a plane, it is easy to find a normal vector to the plane. Simply use the coefficients of x, y, and z and write $\mathbf{n} = \langle a, b, c \rangle$.

Example 3 Finding an Equation of a Plane in Three-Space

Find the general equation of the plane containing the points $(2, 1, 1)$, $(0, 4, 1)$, and $(-2, 1, 4)$.

Solution To apply Theorem 10.12 you need a point in the plane and a vector that is normal to the plane. There are three choices for the point, but no normal vector is given. To obtain a normal vector, use the cross product of vectors $\mathbf{u}$ and $\mathbf{v}$ extending from the point $(2, 1, 1)$ to the points $(0, 4, 1)$ and $(-2, 1, 4)$, as shown in Figure 10.46. The component forms of $\mathbf{u}$ and $\mathbf{v}$ are

$$\mathbf{u} = \langle 0 - 2, 4 - 1, 1 - 1 \rangle = \langle -2, 3, 0 \rangle$$
$$\mathbf{v} = \langle -2 - 2, 1 - 1, 4 - 1 \rangle = \langle -4, 0, 3 \rangle$$

and it follows that

$$\mathbf{n} = \mathbf{u} \times \mathbf{v}$$
$$= \begin{vmatrix} \mathbf{i} & \mathbf{j} & \mathbf{k} \\ -2 & 3 & 0 \\ -4 & 0 & 3 \end{vmatrix}$$
$$= 9\mathbf{i} + 6\mathbf{j} + 12\mathbf{k}$$
$$= \langle a, b, c \rangle$$

is normal to the given plane. Using the direction numbers for $\mathbf{n}$ and the point $(x_1, y_1, z_1) = (2, 1, 1)$, you can determine an equation of the plane to be

$$a(x - x_1) + b(y - y_1) + c(z - z_1) = 0$$
$$9(x - 2) + 6(y - 1) + 12(z - 1) = 0 \quad \text{Standard form}$$
$$9x + 6y + 12z - 36 = 0$$
$$3x + 2y + 4z - 12 = 0. \quad \text{General form}$$

NOTE In Example 3, check to see that each of the three original points satisfies the equation $3x + 2y + 4z - 12 = 0$.

Two distinct planes in three-space either are parallel or intersect in a line. If they intersect, you can determine the angle $(0 \leq \theta \leq \pi/2)$ between them from the angle between their normal vectors, as shown in Figure 10.47. Specifically, if vectors $\mathbf{n}_1$ and $\mathbf{n}_2$ are normal to two intersecting planes, the angle θ between the normal vectors is equal to the angle between the two planes and is given by

$$\cos \theta = \frac{|\mathbf{n}_1 \cdot \mathbf{n}_2|}{\|\mathbf{n}_1\| \|\mathbf{n}_2\|}. \quad \text{Angle between two planes}$$

Consequently, two planes with normal vectors $\mathbf{n}_1$ and $\mathbf{n}_2$ are

1. *perpendicular* if $\mathbf{n}_1 \cdot \mathbf{n}_2 = 0$.
2. *parallel* if $\mathbf{n}_1$ is a scalar multiple of $\mathbf{n}_2$.

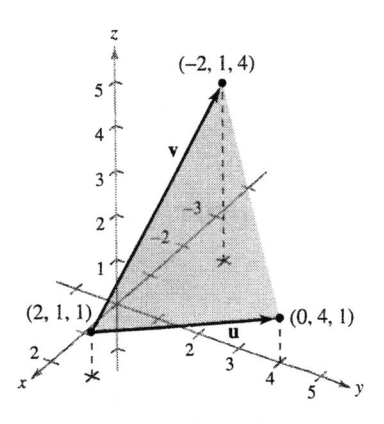

A plane determined by $\mathbf{u}$ and $\mathbf{v}$
Figure 10.46

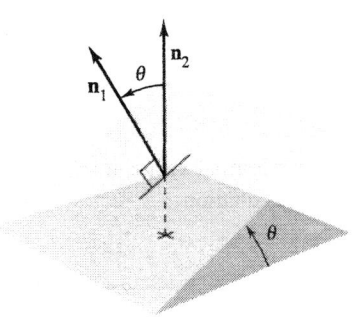

The angle θ between two planes
Figure 10.47

Example 4 Finding the Line of Intersection of Two Planes

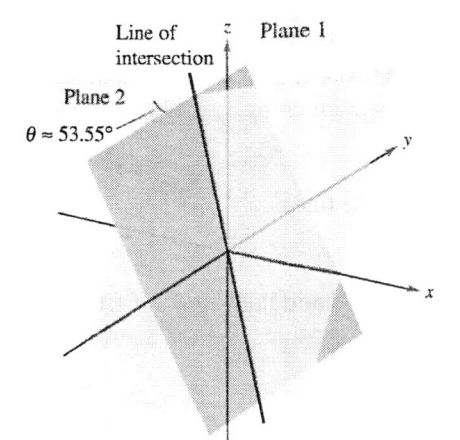

Find the angle between the two planes given by

$$x - 2y + z = 0 \qquad \text{Equation of plane 1}$$
$$2x + 3y - 2z = 0 \qquad \text{Equation of plane 2}$$

and find parametric equations of their line of intersection (see Figure 10.48).

Solution The normal vectors for the planes are $\mathbf{n}_1 = \langle 1, -2, 1 \rangle$ and $\mathbf{n}_2 = \langle 2, 3, -2 \rangle$. Consequently, the angle between the two planes is determined as follows.

$$\cos \theta = \frac{|\mathbf{n}_1 \cdot \mathbf{n}_2|}{\|\mathbf{n}_1\| \|\mathbf{n}_2\|} \qquad \text{Cosine of angle between } \mathbf{n}_1 \text{ and } \mathbf{n}_2$$

$$= \frac{|-6|}{\sqrt{6}\sqrt{17}}$$

$$= \frac{6}{\sqrt{102}}$$

$$\approx 0.59409$$

This implies that the angle between the two planes is $\theta \approx 53.55°$. You can find the line of intersection of the two planes by simultaneously solving the two linear equations representing the planes. One way to do this is to multiply the first equation by -2 and add the result to the second equation.

$$\begin{array}{l} x - 2y + z = 0 \\ 2x + 3y - 2z = 0 \end{array} \implies \begin{array}{l} -2x + 4y - 2z = 0 \\ \underline{2x + 3y - 2z = 0} \\ 7y - 4z = 0 \end{array} \implies y = \frac{4z}{7}$$

Substituting $y = 4z/7$ back into one of the original equations, you can determine that $x = z/7$. Finally, by letting $t = z/7$, you obtain the parametric equations

$$x = t, \quad y = 4t, \quad \text{and} \quad z = 7t \qquad \text{Line of intersection}$$

which indicate that 1, 4, and 7 are direction numbers for the line of intersection.

The angle between the planes is approximately 53.55°.
Figure 10.48

Note that the direction numbers in Example 4 can be obtained from the cross product of the two normal vectors as follows.

$$\mathbf{n}_1 \times \mathbf{n}_2 = \begin{vmatrix} \mathbf{i} & \mathbf{j} & \mathbf{k} \\ 1 & -2 & 1 \\ 2 & 3 & -2 \end{vmatrix}$$

$$= \begin{vmatrix} -2 & 1 \\ 3 & -2 \end{vmatrix} \mathbf{i} - \begin{vmatrix} 1 & 1 \\ 2 & -2 \end{vmatrix} \mathbf{j} + \begin{vmatrix} 1 & -2 \\ 2 & 3 \end{vmatrix} \mathbf{k}$$

$$= \mathbf{i} + 4\mathbf{j} + 7\mathbf{k}$$

This means that the line of intersection of the two planes is parallel to the cross product of their normal vectors.

NOTE The three-dimensional rotating software that accompanies this text can help you visualize surfaces such as those shown in Figure 10.48. If you have access to this software, we suggest that you use it to help your spatial intuition when studying this section and other sections in the text that deal with vectors, curves, or surfaces in space.

Sketching Planes in Space

If a plane in space intersects one of the coordinate planes, we call the line of intersection the **trace** of the given plane in the coordinate plane. To sketch a plane in space, it is helpful to find its points of intersection with the coordinate axes and its traces in the coordinate planes. For example, consider the plane given by

$$3x + 2y + 4z = 12. \qquad \text{Equation of plane}$$

We find the xy-trace by letting $z = 0$ and sketching the line

$$3x + 2y = 12 \qquad xy\text{-trace}$$

in the xy-plane. This line intersects the x-axis at $(4, 0, 0)$ and the y-axis at $(0, 6, 0)$. In Figure 10.49, we continue this process by finding the yz-trace and the xz-trace, and then shading the triangular region lying in the first octant.

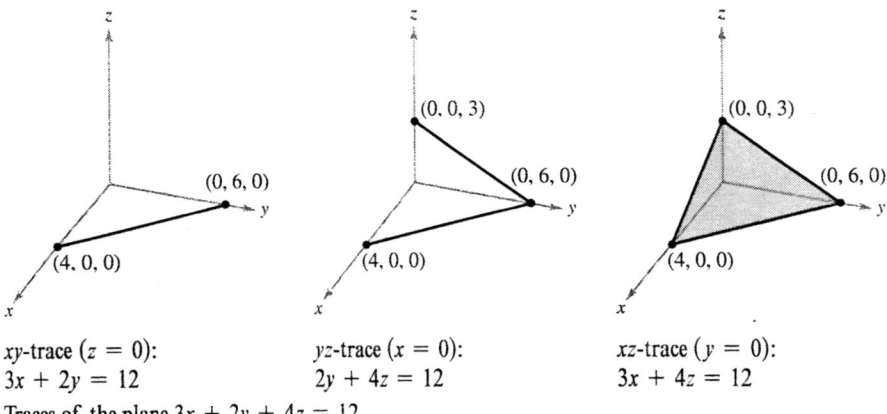

xy-trace ($z = 0$):
$3x + 2y = 12$

yz-trace ($x = 0$):
$2y + 4z = 12$

xz-trace ($y = 0$):
$3x + 4z = 12$

Traces of the plane $3x + 2y + 4z = 12$
Figure 10.49

If the equation of a plane has a missing variable, such as $2x + z = 1$, the plane must be *parallel to the axis* represented by the missing variable, as shown in Figure 10.50. If two variables are missing from the equation of a plane, it is *parallel to the coordinate plane* represented by the missing variables, as shown in Figure 10.51.

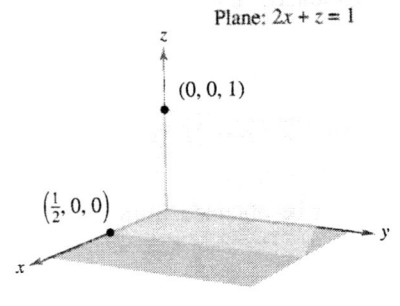

Plane $2x + z = 1$ is parallel to the y-axis.
Figure 10.50

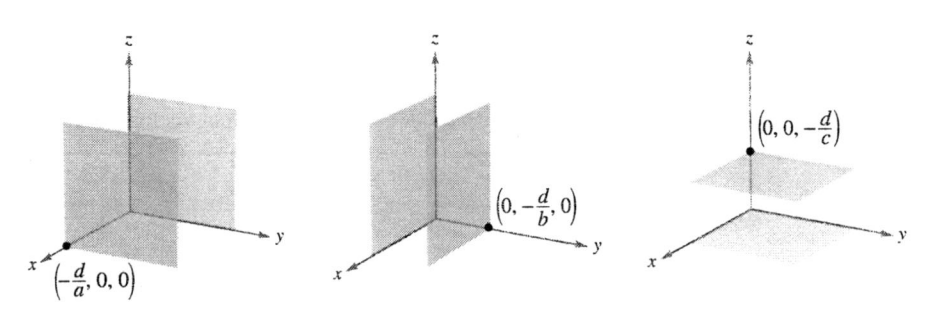

Plane $ax + d = 0$ is parallel to yz-plane.

Plane $by + d = 0$ is parallel to xz-plane.

Plane $cz + d = 0$ is parallel to xy-plane.
Figure 10.51

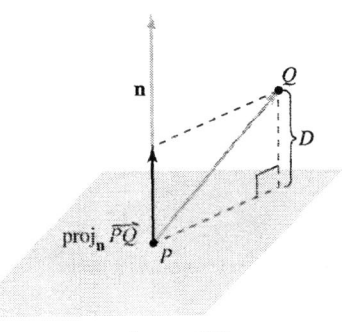

Distances Between Points, Planes, and Lines

We conclude this section with a discussion of two basic types of problems involving distance in space.

1. Finding the distance between a point and a plane
2. Finding the distance between a point and a line

The solutions of these problems illustrate the versatility and usefulness of vectors in coordinate geometry: the first problem uses the *dot product* of two vectors, and the second problem uses the *cross product*.

The distance D between a point Q and a plane is the length of the shortest line segment connecting Q to the plane, as shown in Figure 10.52. If P is *any* point in the plane, you can find this distance by projecting the vector $\overrightarrow{PQ}$ onto the normal vector $\mathbf{n}$. The length of this projection is the desired distance.

$D = \|\text{proj}_\mathbf{n} \overrightarrow{PQ}\|$

The distance between a point and a plane
Figure 10.52

THEOREM 10.13 Distance Between a Point and a Plane

The distance between a plane and a point Q (not in the plane) is

$$D = \|\text{proj}_\mathbf{n} \overrightarrow{PQ}\| = \frac{|\overrightarrow{PQ} \cdot \mathbf{n}|}{\|\mathbf{n}\|}$$

where P is a point in the plane and $\mathbf{n}$ is normal to the plane.

To find a point in the plane given by $ax + by + cz + d = 0$ ($a \neq 0$), let $y = 0$ and $z = 0$. Then, from the equation $ax + d = 0$, you can conclude that the point $(-d/a, 0, 0)$ lies in the plane.

Example 5 Finding the Distance Between a Point and a Plane

Find the distance between the point $Q(1, 5, -4)$ and the plane given by

$$3x - y + 2z = 6.$$

Solution You know that $\mathbf{n} = \langle 3, -1, 2 \rangle$ is normal to the given plane. To find a point in the plane, let $y = 0$ and $z = 0$, and obtain the point $P(2, 0, 0)$. The vector from P to Q is given by

$$\overrightarrow{PQ} = \langle 1 - 2, 5 - 0, -4 - 0 \rangle$$
$$= \langle -1, 5, -4 \rangle.$$

Using the distance formula given in Theorem 10.13 produces

$$D = \frac{|\overrightarrow{PQ} \cdot \mathbf{n}|}{\|\mathbf{n}\|} = \frac{|\langle -1, 5, -4 \rangle \cdot \langle 3, -1, 2 \rangle|}{\sqrt{9 + 1 + 4}} \quad \text{Distance between a point and a plane}$$
$$= \frac{|-3 - 5 - 8|}{\sqrt{14}}$$
$$= \frac{16}{\sqrt{14}}.$$

NOTE The choice of the point P in Example 5 is arbitrary. Try choosing a different point in the plane to verify that you obtain the same distance.

From Theorem 10.13, you can determine that the distance between the point $Q(x_0, y_0, z_0)$ and the plane given by $ax + by + cz + d = 0$ is

$$D = \frac{|a(x_0 - x_1) + b(y_0 - y_1) + c(z_0 - z_1)|}{\sqrt{a^2 + b^2 + c^2}}$$

or

$$D = \frac{|ax_0 + by_0 + cz_0 + d|}{\sqrt{a^2 + b^2 + c^2}} \qquad \text{Distance between a point and a plane}$$

where $P(x_1, y_1, z_1)$ is a point in the plane and $d = -(ax_1 + by_1 + cz_1)$.

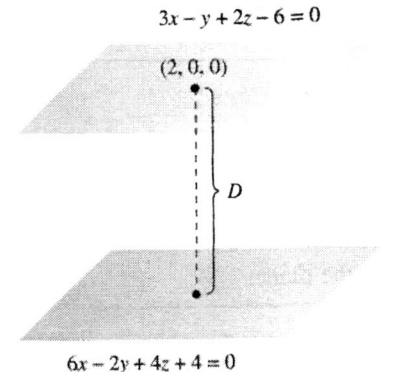

Example 6 Finding the Distance Between Two Parallel Planes

Find the distance between the two parallel planes given by

$$3x - y + 2z - 6 = 0 \quad \text{and} \quad 6x - 2y + 4z + 4 = 0.$$

Solution The two planes are shown in Figure 10.53. To find the distance between the planes, choose a point in the first plane, say $(x_0, y_0, z_0) = (2, 0, 0)$. Then, from the second plane, you can determine that $a = 6$, $b = -2$, $c = 4$, and $d = 4$, and conclude that the distance is

$$D = \frac{|ax_0 + by_0 + cz_0 + d|}{\sqrt{a^2 + b^2 + c^2}} \qquad \text{Distance between a point and a plane}$$

$$= \frac{|6(2) + (-2)(0) + (4)(0) + 4|}{\sqrt{6^2 + (-2)^2 + 4^2}}$$

$$= \frac{16}{\sqrt{56}} = \frac{8}{\sqrt{14}} \approx 2.14.$$

The distance between the parallel planes is approximately 2.14.
Figure 10.53

The formula for the distance between a point and a line in space resembles that for the distance between a point and a plane—except that you replace the dot product with the cross product and the normal vector **n** with a direction vector for the line.

THEOREM 10.14 Distance Between a Point and a Line in Space

The distance between a point Q and a line in space is given by

$$D = \frac{\|\overrightarrow{PQ} \times \mathbf{u}\|}{\|\mathbf{u}\|}$$

where **u** is the direction vector for the line and P is a point on the line.

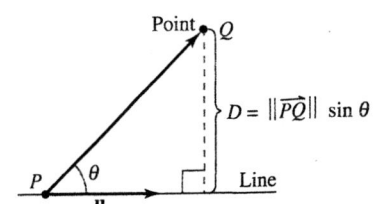

Proof In Figure 10.54, let D be the distance between the point Q and the given line. Then $D = \|\overrightarrow{PQ}\| \sin \theta$, where θ is the angle between **u** and $\overrightarrow{PQ}$. By Theorem 10.8, you have

$$\|\mathbf{u}\| \|\overrightarrow{PQ}\| \sin \theta = \|\mathbf{u} \times \overrightarrow{PQ}\| = \|\overrightarrow{PQ} \times \mathbf{u}\|.$$

Consequently,

$$D = \|\overrightarrow{PQ}\| \sin \theta = \frac{\|\overrightarrow{PQ} \times \mathbf{u}\|}{\|\mathbf{u}\|}.$$

The distance between a point and a line
Figure 10.54

Example 7 Finding the Distance Between a Point and a Line

Find the distance between the point $Q(3, -1, 4)$ and the line given by

$$x = -2 + 3t, \quad y = -2t, \quad \text{and} \quad z = 1 + 4t.$$

Solution Using the direction numbers 3, -2, and 4, you know that the direction vector for the line is

$$\mathbf{u} = \langle 3, -2, 4 \rangle. \qquad \text{Direction vector for line}$$

To find a point on the line, let $t = 0$ and obtain

$$P = (-2, 0, 1). \qquad \text{Point on the line}$$

So,

$$\vec{PQ} = \langle 3 - (-2), -1 - 0, 4 - 1 \rangle = \langle 5, -1, 3 \rangle$$

and you can form the cross product

$$\vec{PQ} \times \mathbf{u} = \begin{vmatrix} \mathbf{i} & \mathbf{j} & \mathbf{k} \\ 5 & -1 & 3 \\ 3 & -2 & 4 \end{vmatrix} = 2\mathbf{i} - 11\mathbf{j} - 7\mathbf{k} = \langle 2, -11, -7 \rangle.$$

Finally, using Theorem 10.14, you can find the distance to be

$$D = \frac{\|\vec{PQ} \times \mathbf{u}\|}{\|\mathbf{u}\|} = \frac{\sqrt{174}}{\sqrt{29}} = \sqrt{6} \approx 2.45. \qquad \text{(See Figure 10.55.)}$$

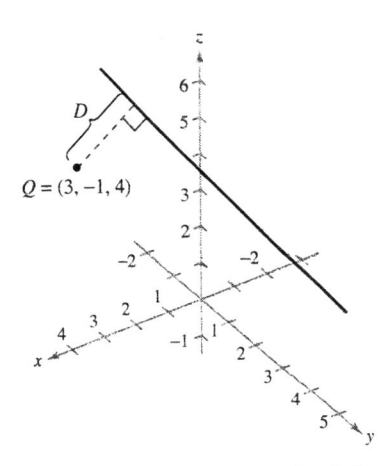

The distance between the point Q and the line is $\sqrt{6} \approx 2.45$.
Figure 10.55

EXERCISES FOR SECTION 10.5

In Exercises 1 and 2, the figure shows the graph of a line given by the parametric equations. (a) Draw an arrow on the line to indicate its orientation. To print an enlarged copy of the graph, go to the website *www.mathgraphs.com*. (b) Find the coordinates of two points, P and Q, on the line. Determine the vector $\vec{PQ}$. What is the relationship between the components of the vector and the coefficients of t in the parametric equations? Why is this true? (c) Determine the coordinates of any points of intersection with the coordinate planes. If the line does not intersect a coordinate plane, explain why.

1. $x = 1 + 3t$
   $y = 2 - t$
   $z = 2 + 5t$

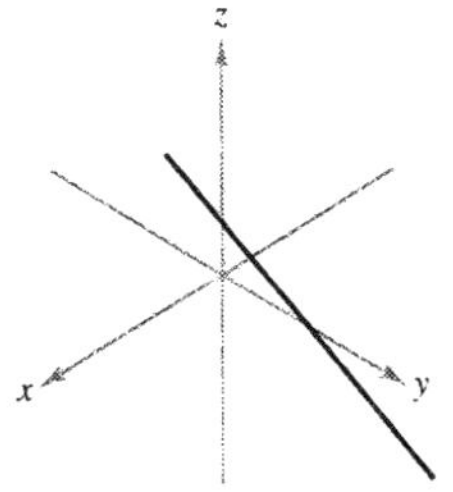

2. $x = 2 - 3t$
   $y = 2$
   $z = 1 - t$

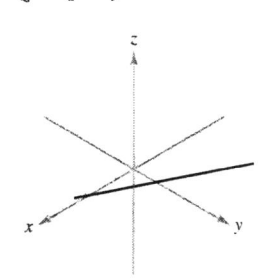

In Exercises 3–8, find sets of (a) parametric equations and (b) symmetric equations of the line through the point parallel to the indicated vector or line. (For each line, express the direction numbers as integers.)

Point	Parallel to
3. $(0, 0, 0)$	$\mathbf{v} = \langle 1, 2, 3 \rangle$
4. $(0, 0, 0)$	$\mathbf{v} = \langle -2, \frac{5}{2}, 1 \rangle$
5. $(-2, 0, 3)$	$\mathbf{v} = 2\mathbf{i} + 4\mathbf{j} - 2\mathbf{k}$
6. $(-3, 0, 2)$	$\mathbf{v} = 6\mathbf{j} + 3\mathbf{k}$
7. $(1, 0, 1)$	$x = 3 + 3t, y = 5 - 2t, z = -7 + t$
8. $(-3, 5, 4)$	$\dfrac{x-1}{3} = \dfrac{y+1}{-2} = z - 3$

In Exercises 9–12, find sets of (a) parametric equations and (b) symmetric equations of the line through the two points. (For each line, express the direction numbers as integers.)

9. $(5, -3, -2), \left(-\frac{2}{3}, \frac{2}{3}, 1\right)$
10. $(2, 0, 2), (1, 4, -3)$
11. $(2, 3, 0), (10, 8, 12)$
12. $(0, 0, 25), (10, 10, 0)$

In Exercises 13 and 14, find a set of parametric equations of the line.

13. The line passes through the point $(2, 3, 4)$ and is parallel to the xz-plane and the yz-plane.

14. The line passes through the point $(2, 3, 4)$ and is perpendicular to the plane given by $3x + 2y - z = 6$.

In Exercises 15 and 16, determine which points lie on the line L.

15. The line L passes through the point $(-2, 3, 1)$ and is parallel to the vector $\mathbf{v} = 4\mathbf{i} - \mathbf{k}$.
    (a) $(2, 3, 0)$ (b) $(-6, 3, 2)$ (c) $(2, 1, 0)$ (d) $(6, 3, -2)$

16. The line L passes through the points $(2, 0, -3)$ and $(4, 2, -2)$.
    (a) $(4, 1, -2)$ (b) $\left(\frac{5}{2}, \frac{1}{2}, -\frac{11}{4}\right)$ (c) $(-1, -3, -4)$

In Exercises 17 and 18, determine if any of the lines are parallel or identical.

17. L_1: $x = 6 - 3t,\ y = -2 + 2t,\ z = 5 + 4t$
    L_2: $x = 6t,\ y = 2 - 4t,\ z = 13 - 8t$
    L_3: $x = 10 - 6t,\ y = 3 + 4t,\ z = 7 + 8t$
    L_4: $x = -4 + 6t,\ y = 3 + 4t,\ z = 5 - 6t$

18. L_1: $\dfrac{x - 8}{4} = \dfrac{y + 5}{-2} = \dfrac{z + 9}{3}$
    L_2: $\dfrac{x + 7}{2} = \dfrac{y - 4}{1} = \dfrac{z + 6}{5}$
    L_3: $\dfrac{x + 4}{-8} = \dfrac{y - 1}{4} = \dfrac{z + 18}{-6}$
    L_4: $\dfrac{x - 2}{-2} = \dfrac{y + 3}{1} = \dfrac{z - 4}{1.5}$

In Exercises 19–22, determine whether the lines intersect, and if so, find the point of intersection and the cosine of the angle of intersection.

19. $x = 4t + 2,\ y = 3,\ z = -t + 1$
    $x = 2s + 2,\ y = 2s + 3,\ z = s + 1$

20. $x = -3t + 1,\ y = 4t + 1,\ z = 2t + 4$
    $x = 3s + 1,\ y = 2s + 4,\ z = -s + 1$

21. $\dfrac{x}{3} = \dfrac{y - 2}{-1} = z + 1,\quad \dfrac{x - 1}{4} = y + 2 = \dfrac{z + 3}{-3}$

22. $\dfrac{x - 2}{-3} = \dfrac{y - 2}{6} = z - 3,\quad \dfrac{x - 3}{2} = y + 5 = \dfrac{z + 2}{4}$

In Exercises 23 and 24, use a computer algebra system to graph the pair of intersecting lines and find the point of intersection.

23. $x = 2t + 3,\ y = 5t - 2,\ z = -t + 1$
    $x = -2s + 7,\ y = s + 8,\ z = 2s - 1$

24. $x = 2t - 1,\ y = -4t + 10,\ z = t$
    $x = -5s - 12,\ y = 3s + 11,\ z = -2s - 4$

Cross Product In Exercises 25 and 26, (a) find the coordinates of three points P, Q, and R in the plane, and determine the vectors $\overrightarrow{PQ}$ and $\overrightarrow{PR}$. (b) Find $\overrightarrow{PQ} \times \overrightarrow{PR}$. What is the relationship between the components of the cross product and the coefficients of the equation of the plane? Why is this true?

25. $4x - 3y - 6z = 6$ 26. $2x + 3y + 4z = 4$

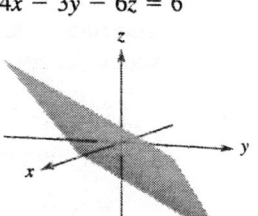

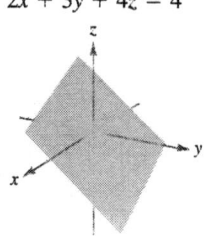

In Exercises 27–32, find an equation of the plane passing through the point perpendicular to the indicated vector or line.

Point	Perpendicular to
27. $(2, 1, 2)$	$\mathbf{n} = \mathbf{i}$
28. $(1, 0, -3)$	$\mathbf{n} = \mathbf{k}$
29. $(3, 2, 2)$	$\mathbf{n} = 2\mathbf{i} + 3\mathbf{j} - \mathbf{k}$
30. $(0, 0, 0)$	$\mathbf{n} = -3\mathbf{i} + 2\mathbf{k}$
31. $(0, 0, 6)$	$x = 1 - t,\ y = 2 + t,\ z = 4 - 2t$
32. $(3, 2, 2)$	$\dfrac{x - 1}{4} = y + 2 = \dfrac{z + 3}{-3}$

In Exercises 33–44, find an equation of the plane.

33. The plane passes through $(0, 0, 0)$, $(1, 2, 3)$, and $(-2, 3, 3)$.

34. The plane passes through $(2, 3, -2)$, $(3, 4, 2)$, and $(1, -1, 0)$.

35. The plane passes through $(1, 2, 3)$, $(3, 2, 1)$, and $(-1, -2, 2)$.

36. The plane passes through the point $(1, 2, 3)$ and is parallel to the yz-plane.

37. The plane passes through the point $(1, 2, 3)$ and is parallel to the xy-plane.

38. The plane contains the y-axis and makes an angle of $\pi/6$ with the positive x-axis.

39. The plane contains the lines given by
    $\dfrac{x - 1}{-2} = y - 4 = z\quad$ and $\quad\dfrac{x - 2}{-3} = \dfrac{y - 1}{4} = \dfrac{z - 2}{-1}$.

40. The plane passes through the point $(2, 2, 1)$ and contains the line given by
    $\dfrac{x}{2} = \dfrac{y - 4}{-1} = z$.

41. The plane passes through the points $(2, 2, 1)$ and $(-1, 1, -1)$ and is perpendicular to the plane $2x - 3y + z = 3$.

42. The plane passes through the points $(3, 2, 1)$ and $(3, 1, -5)$ and is perpendicular to the plane $6x + 7y + 2z = 10$.

43. The plane passes through the points $(1, -2, -1)$ and $(2, 5, 6)$ and is parallel to the x-axis.

44. The plane passes through the points $(4, 2, 1)$ and $(-3, 5, 7)$ and is parallel to the z-axis.

In Exercises 45–50, determine whether the planes are parallel, orthogonal, or neither. If they are neither parallel nor orthogonal, find the angle of intersection.

45. $5x - 3y + z = 4$
    $x + 4y + 7z = 1$

46. $3x + y - 4z = 3$
    $-9x - 3y + 12z = 4$

47. $x - 3y + 6z = 4$
    $5x + y - z = 4$

48. $3x + 2y - z = 7$
    $x - 4y + 2z = 0$

49. $x - 5y - z = 1$
    $5x - 25y - 5z = -3$

50. $2x - z = 1$
    $4x + y + 8z = 10$

In Exercises 51–58, mark any intercepts and sketch a graph of the plane.

51. $4x + 2y + 6z = 12$
52. $3x + 6y + 2z = 6$
53. $2x - y + 3z = 4$
54. $2x - y + z = 4$
55. $y + z = 5$
56. $x + 2y = 4$
57. $x = 5$
58. $z = 8$

In Exercises 59–62, use a computer algebra system to graph the plane.

59. $2x + y - z = 6$
60. $x - 3z = 3$
61. $-5x + 4y - 6z = -8$
62. $2.1x - 4.7y - z = -3$

In Exercises 63 and 64, determine if any of the planes are parallel or identical.

63. P_1: $3x - 2y + 5z = 10$
    P_2: $-6x + 4y - 10z = 5$
    P_3: $-3x + 2y + 5z = 8$
    P_4: $75x - 50y + 125z = 250$

64. P_1: $-60x + 90y + 30z = 27$
    P_2: $6x - 9y - 3z = 2$
    P_3: $-20x + 30y + 10z = 9$
    P_4: $12x - 18y + 6z = 5$

In Exercises 65 and 66, describe the family of planes represented by the equation, where c is any real number.

65. $x + y + z = c$
66. $cy + z = 0$

In Exercises 67 and 68, find a set of parametric equations for the line of intersection of the planes.

67. $3x + 2y - z = 7$
    $x - 4y + 2z = 0$

68. $6x - 3y + z = 5$
    $-x + y + 5z = 5$

In Exercises 69–72, find the point(s) of intersection (if any) of the plane and the line. Also determine whether the line lies in the plane.

69. $2x - 2y + z = 12$, $x - \frac{1}{2} = \frac{y + (3/2)}{-1} = \frac{z + 1}{2}$

70. $2x + 3y = -5$, $\frac{x-1}{4} = \frac{y}{2} = \frac{z-3}{6}$

71. $2x + 3y = 10$, $\frac{x-1}{3} = \frac{y+1}{-2} = z - 3$

72. $5x + 3y = 17$, $\frac{x-4}{2} = \frac{y+1}{-3} = \frac{z+2}{5}$

In Exercises 73–76, find the distance between the point and the plane.

73. $(0, 0, 0)$
    $2x + 3y + z = 12$

74. $(0, 0, 0)$
    $8x - 4y + z = 8$

75. $(2, 8, 4)$
    $2x + y + z = 5$

76. $(3, 2, 1)$
    $x - y + 2z = 4$

In Exercises 77–80, find the distance between the planes.

77. $x - 3y + 4z = 10$
    $x - 3y + 4z = 6$

78. $4x - 4y + 9z = 7$
    $4x - 4y + 9z = 18$

79. $-3x + 6y + 7z = 1$
    $6x - 12y - 14z = 25$

80. $2x - 4z = 4$
    $2x - 4z = 10$

In Exercises 81 and 82, find the distance between the point and the line given by the set of parametric equations.

81. $(1, 5, -2)$; $x = 4t - 2, y = 3, z = -t + 1$
82. $(1, -2, 4)$; $x = 2t, y = t - 3, z = 2t + 2$

Getting at the Concept

83. Give the parametric equations and the symmetric equations of a line in space. Describe what is required to find these equations.

84. Give the standard equation of a plane in space. Describe what is required to find this equation.

85. Describe a method of finding the line of intersection of two planes.

86. Describe each surface given by the equations $x = a$, $y = b$, and $z = c$.

87. (a) Describe and find an equation for the surface generated by all points (x, y, z) that are 4 units from the point $(3, -2, 5)$.
    (b) Describe and find an equation for the surface generated by all points (x, y, z) that are 4 units from the plane
    $4x - 3y + z = 10$.

88. Consider the two nonzero vectors **u** and **v**. Describe the geometric figure generated by the terminal points of the following vectors, where s and t represent all real numbers.
    (a) $t\mathbf{v}$ (b) $\mathbf{u} + t\mathbf{v}$ (c) $s\mathbf{u} + t\mathbf{v}$

89. Modeling Data Per capita consumption (in gallons) of different types of milk in the United States for selected years is shown in the table. Consumption of skim milk, reduced-fat milk, and whole milk are represented by the variables x, y, and z, respectively. *(Source: U.S. Department of Agriculture)*

Year	1980	1985	1990	1994	1995	1996	1997
x	3.1	3.2	4.9	5.8	6.2	6.4	6.6
y	6.3	7.9	9.1	8.7	8.2	8.0	7.7
z	16.5	13.9	10.2	8.8	8.4	8.4	8.2

A model for the data is given by

$1.83x + 1.09y + z = 28.7$.

(a) Complete a fourth row in the table using the model to approximate z for the given values of x and y. Compare the approximations with the actual values of z.

(b) According to this model, any increases in consumption of two types of milk will have what effect on the consumption of the third type?

(c) Because x, y, and z must be nonnegative, sketch the traces of the plane and the first octant portion of the plane.

90. Mechanical Design A chute at the top of a grain elevator of a combine funnels the grain into a bin (see figure). Find the angle between two adjacent sides.

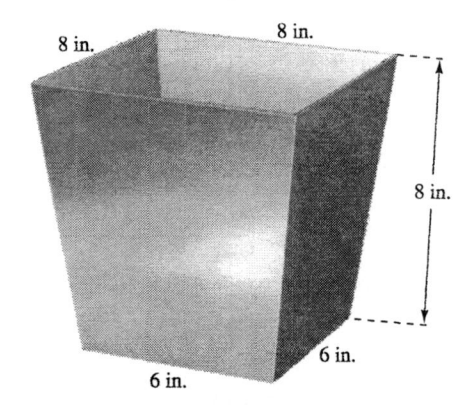

True or False? In Exercises 91 and 92, determine whether the statement is true or false. If it is false, explain why or give an example that shows it is false.

91. If $\mathbf{v} = a_1\mathbf{i} + b_1\mathbf{j} + c_1\mathbf{k}$ is any vector in the plane given by $a_2 x + b_2 y + c_2 z + d_2 = 0$, then $a_1 a_2 + b_1 b_2 + c_1 c_2 = 0$.

92. Every pair of lines in space are either intersecting or parallel.

SECTION PROJECT: DISTANCES IN SPACE

We have developed two distance formulas in this section—the distance between a point and a plane, and the distance between a point and a line. In this project you will study a third distance problem—the distance between two skew lines. Two lines in space are *skew* if they are neither parallel nor intersecting (see figure).

(a) Consider the following two lines in space.

$L_1: x = 4 + 5t, \ y = 5 + 5t, \ z = 1 - 4t$
$L_2: x = 4 + s, \ y = -6 + 8s, \ z = 7 - 3s$

 (i) Show that these lines are not parallel.

 (ii) Show that these lines do not intersect, and hence are skew lines.

 (iii) Show that the two lines lie in parallel planes.

 (iv) Find the distance between the parallel planes from part (iii). This is the distance between the original skew lines.

(b) Use the procedure in part (a) to find the distance between the lines.

$L_1: x = 2t, \ y = 4t, \ z = 6t$
$L_2: x = 1 - s, \ y = 4 + s, \ z = -1 + s$

(c) Use the procedure in part (a) to find the distance between the lines.

$L_1: x = 3t, \ y = 2 - t, \ z = -1 + t$
$L_2: x = 1 + 4s, \ y = -2 + s, \ z = -3 - 3s$

(d) Develop a formula for finding the distance between the skew lines.

$L_1: x = x_1 + a_1 t, \ y = y_1 + b_1 t, \ z = z_1 + c_1 t$
$L_2: x = x_2 + a_2 s, \ y = y_2 + b_2 s, \ z = z_2 + c_2 s$

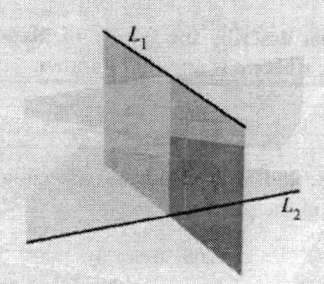

Section 10.6 Surfaces in Space

- Recognize and write equations for cylindrical surfaces.
- Recognize and write equations for quadratic surfaces.
- Recognize and write equations for surfaces of revolution.

Cylindrical Surfaces

The first five sections of this chapter contained the vector portion of the preliminary work necessary to study vector calculus and the calculus of space. In this and the next section, you will study surfaces in space and alternative coordinate systems for space. You have already studied two special types of surfaces.

1. Spheres: $(x - x_0)^2 + (y - y_0)^2 + (z - z_0)^2 = r^2$ Section 10.2
2. Planes: $ax + by + cz + d = 0$ Section 10.5

A third type of surface in space is called a **cylindrical surface,** or simply a **cylinder.** To define a cylinder, consider the familiar right circular cylinder shown in Figure 10.56. You can imagine that this cylinder is generated by a vertical line moving around the circle $x^2 + y^2 = a^2$ in the xy-plane. This circle is called a **generating curve** for the cylinder, as indicated in the following definition.

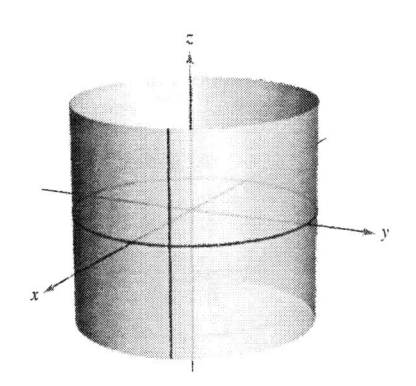

Right circular cylinder:
$x^2 + y^2 = a^2$

Rulings are parallel to z-axis.
Figure 10.56

Definition of a Cylinder

Let C be a curve in a plane and let L be a line not in a parallel plane. The set of all lines parallel to L and intersecting C is called a **cylinder.** C is called the **generating curve** (or **directrix**) of the cylinder, and the parallel lines are called **rulings.**

NOTE Without loss of generality, you can assume that C lies in one of the three coordinate planes. Moreover, in this text we restrict the discussion to *right* cylinders—cylinders whose rulings are perpendicular to the coordinate plane containing C, as shown in Figure 10.57.

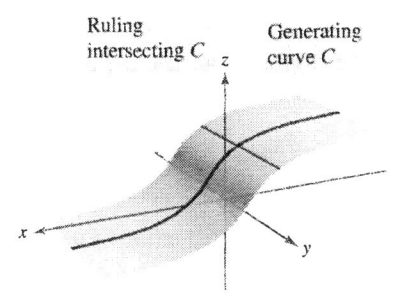

For the right circular cylinder shown in Figure 10.56, the equation of the generating curve is

$$x^2 + y^2 = a^2 \qquad \text{Equation of generating curve in } xy\text{-plane}$$

To find an equation for the cylinder, note that you can generate any one of the rulings by fixing the values of x and y and then allowing z to take on all real values. In this sense the value of z is arbitrary and is, therefore, not included in the equation. In other words, the equation of this cylinder is simply the equation of its generating curve.

Cylinder: Rulings intersect C and are parallel to the given line.
Figure 10.57

$$x^2 + y^2 = a^2 \qquad \text{Equation of cylinder in space}$$

Equations of Cylinders

The equation of a cylinder whose rulings are parallel to one of the coordinate axes contains only the variables corresponding to the other two axes.

Example 1 **Sketching a Cylinder**

Sketch the surface represented by each of the equations.

a. $z = y^2$ **b.** $z = \sin x, \quad 0 \leq x \leq 2\pi$

Solution

a. The graph is a cylinder whose generating curve, $z = y^2$, is a parabola in the yz-plane. The rulings of the cylinder are parallel to the x-axis, as shown in Figure 10.58a.

b. The graph is a cylinder generated by the sine curve in the xz-plane. The rulings are parallel to the y-axis, as shown in Figure 10.58b.

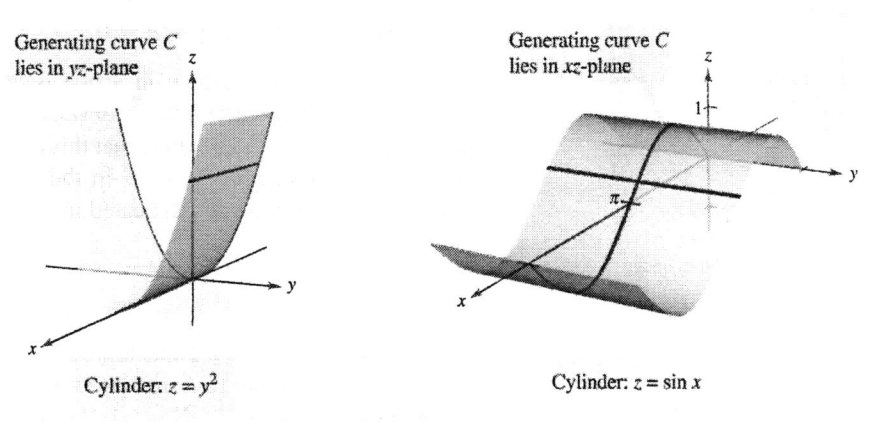

Generating curve C lies in yz-plane

Generating curve C lies in xz-plane

Cylinder: $z = y^2$

Cylinder: $z = \sin x$

(a) Rulings are parallel to x-axis.
(b) Rulings are parallel to y-axis.
Figure 10.58

Quadric Surfaces

STUDY TIP In the table on pages 765 and 766, only one of several orientations of each quadric surface is shown. If the surface is oriented along a different axis, its standard equation will change accordingly, as illustrated in Examples 2 and 3. The fact that the two types of paraboloids have one variable raised to the first power can be helpful in classifying quadric surfaces. The other four types of basic quadric surfaces have equations that are of *second degree* in all three variables.

The fourth basic type of surface in space is a **quadric surface**. Quadric surfaces are the three-dimensional analogs of conic sections.

Quadric Surface

The equation of a **quadric surface** in space is a second-degree equation of the form

$$Ax^2 + By^2 + Cz^2 + Dxy + Exz + Fyz + Gx + Hy + Iz + J = 0.$$

There are six basic types of quadric surfaces: **ellipsoid, hyperboloid of one sheet, hyperboloid of two sheets, elliptic cone, elliptic paraboloid,** and **hyperbolic paraboloid.**

The intersection of a surface with a plane is called the **trace of the surface** in the plane. To visualize a surface in space, it is helpful to determine its traces in some well-chosen planes. The traces of quadric surfaces are conics. These traces, together with the **standard form** of the equation of each quadric surface, are shown in the table on pages 765 and 766.

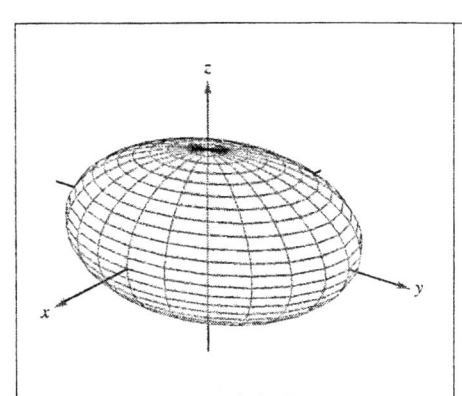

Ellipsoid

$$\frac{x^2}{a^2} + \frac{y^2}{b^2} + \frac{z^2}{c^2} = 1$$

Trace	Plane
Ellipse	Parallel to xy-plane
Ellipse	Parallel to xz-plane
Ellipse	Parallel to yz-plane

The surface is a sphere if $a = b = c \neq 0$.

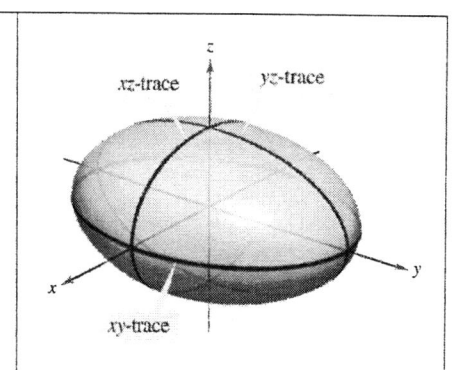

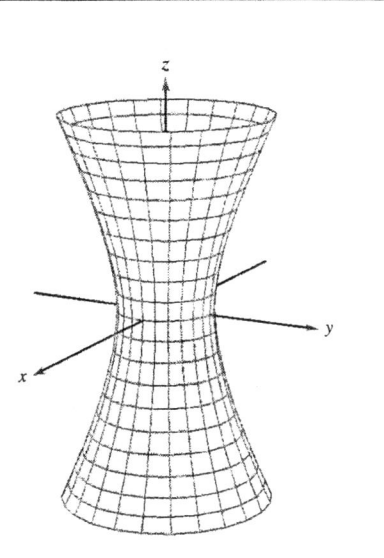

Hyperboloid of One Sheet

$$\frac{x^2}{a^2} + \frac{y^2}{b^2} - \frac{z^2}{c^2} = 1$$

Trace	Plane
Ellipse	Parallel to xy-plane
Hyperbola	Parallel to xz-plane
Hyperbola	Parallel to yz-plane

The axis of the hyperboloid corresponds to the variable whose coefficient is negative.

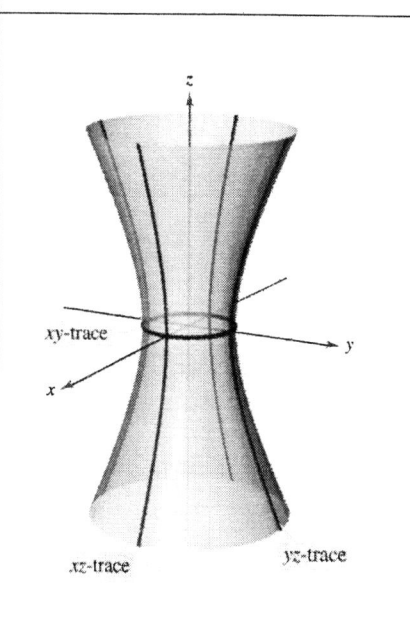

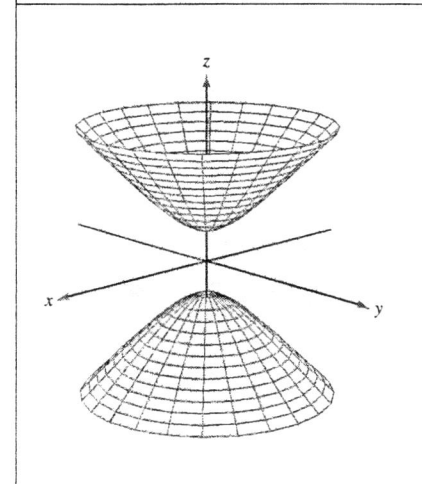

Hyperboloid of Two Sheets

$$\frac{z^2}{c^2} - \frac{x^2}{a^2} - \frac{y^2}{b^2} = 1$$

Trace	Plane
Ellipse	Parallel to xy-plane
Hyperbola	Parallel to xz-plane
Hyperbola	Parallel to yz-plane

The axis of the hyperboloid corresponds to the variable whose coefficient is positive. There is no trace in the coordinate plane perpendicular to this axis.

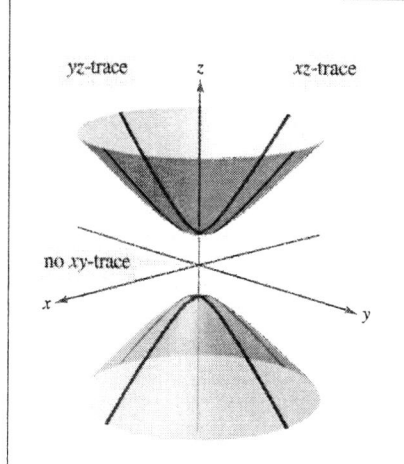

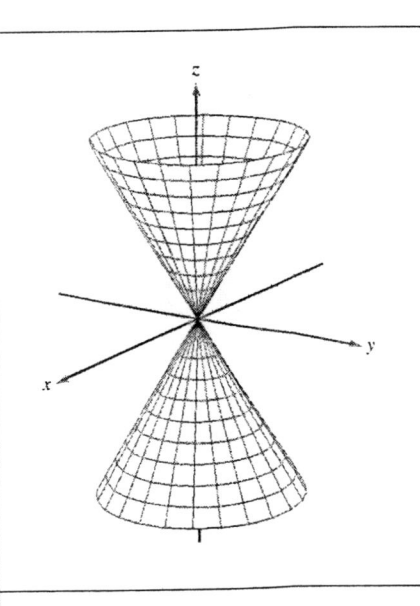

Elliptic Cone

$$\frac{x^2}{a^2} + \frac{y^2}{b^2} - \frac{z^2}{c^2} = 0$$

Trace	Plane
Ellipse	Parallel to xy-plane
Hyperbola	Parallel to xz-plane
Hyperbola	Parallel to yz-plane

The axis of the cone corresponds to the variable whose coefficient is negative. The traces in the coordinate planes parallel to this axis are intersecting lines.

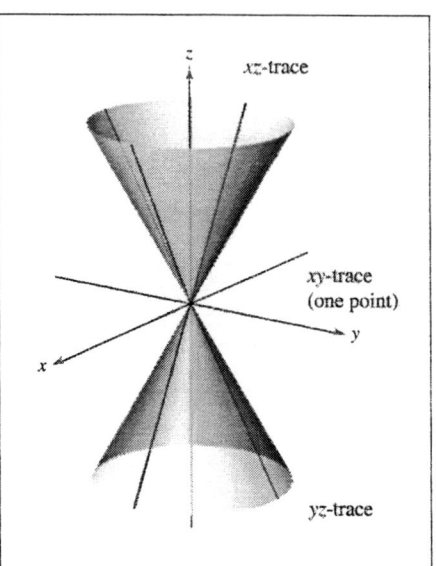

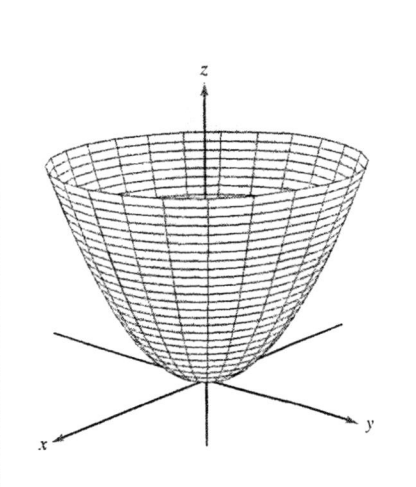

Elliptic Paraboloid

$$z = \frac{x^2}{a^2} + \frac{y^2}{b^2}$$

Trace	Plane
Ellipse	Parallel to xy-plane
Parabola	Parallel to xz-plane
Parabola	Parallel to yz-plane

The axis of the paraboloid corresponds to the variable raised to the first power.

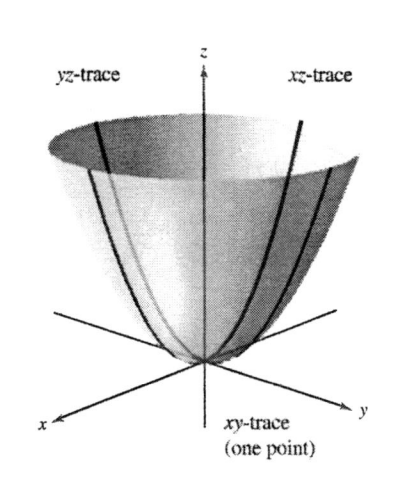

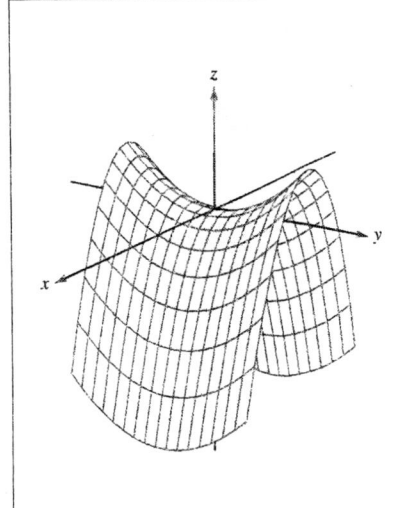

Hyperbolic Paraboloid

$$z = \frac{y^2}{b^2} - \frac{x^2}{a^2}$$

Trace	Plane
Hyperbola	Parallel to xy-plane
Parabola	Parallel to xz-plane
Parabola	Parallel to yz-plane

The axis of the paraboloid corresponds to the variable raised to the first power.

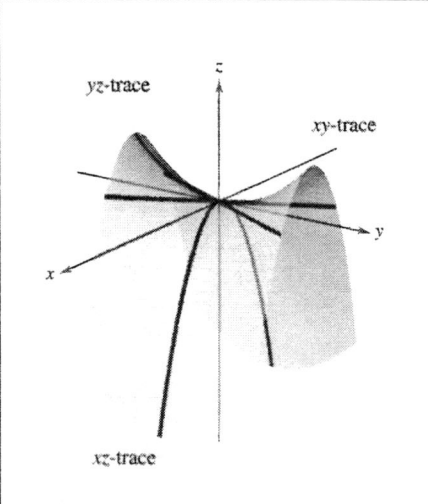

To classify a quadric surface, begin by writing the surface in standard form. Then, determine several traces taken in the coordinate planes *or* taken in planes that are parallel to the coordinate planes.

Example 2   Sketching a Quadric Surface

Classify and sketch the surface given by $4x^2 - 3y^2 + 12z^2 + 12 = 0$.

Solution   Begin by writing the equation in standard form.

$$4x^2 - 3y^2 + 12z^2 + 12 = 0 \qquad \text{Write original equation.}$$

$$\frac{x^2}{-3} + \frac{y^2}{4} - z^2 - 1 = 0 \qquad \text{Divide by } -12.$$

$$\frac{y^2}{4} - \frac{x^2}{3} - \frac{z^2}{1} = 1 \qquad \text{Standard form}$$

From the table on pages 765 and 766, you can conclude that the surface is a hyperboloid of two sheets with the y-axis as its axis. To sketch the graph of this surface, it helps to find the traces in the coordinate planes.

$$xy\text{-trace } (z = 0): \quad \frac{y^2}{4} - \frac{x^2}{3} = 1 \qquad \text{Hyperbola}$$

$$xz\text{-trace } (y = 0): \quad \frac{x^2}{3} + \frac{z^2}{1} = -1 \qquad \text{No trace}$$

$$yz\text{-trace } (x = 0): \quad \frac{y^2}{4} - \frac{z^2}{1} = 1 \qquad \text{Hyperbola}$$

The graph is shown in Figure 10.59.

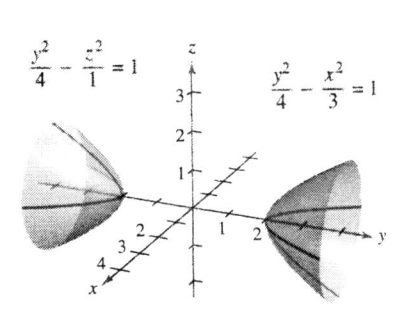

Hyperboloid of two sheets:
$$\frac{y^2}{4} - \frac{x^2}{3} - z^2 = 1$$

Figure 10.59

Example 3   Sketching a Quadric Surface

Classify and sketch the surface given by $x - y^2 - 4z^2 = 0$.

Solution   Because x is raised only to the first power, the surface is a paraboloid. The axis of the paraboloid is the x-axis. In the standard form, the equation is

$$x = y^2 + 4z^2. \qquad \text{Standard form}$$

Some convenient traces are as follows.

$$xy\text{-trace } (z = 0): \qquad x = y^2 \qquad \text{Parabola}$$
$$xz\text{-trace } (y = 0): \qquad x = 4z^2 \qquad \text{Parabola}$$
$$\text{parallel to } yz\text{-plane } (x = 4): \quad \frac{y^2}{4} + \frac{z^2}{1} = 1 \qquad \text{Ellipse}$$

The surface is an *elliptic* paraboloid, as shown in Figure 10.60.

Elliptic paraboloid:
$$x = y^2 + 4z^2$$

Figure 10.60

Some second-degree equations in x, y, and z do not represent one of the basic types of quadric surfaces. Here are two examples.

$$x^2 + y^2 + z^2 = 0 \qquad \text{Single point}$$
$$x^2 + y^2 = 1 \qquad \text{Right circular cylinder}$$

$$\frac{(x-2)^2}{4} + \frac{(y+1)^2}{2} + \frac{(z-1)^2}{4} = 1$$

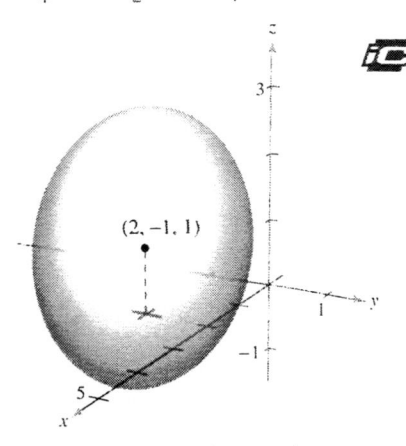

An ellipsoid centered at $(2, -1, 1)$
Figure 10.61

For a quadric surface not centered at the origin, you can form the standard equation by completing the square, as demonstrated in Example 4.

Example 4 A Quadric Surface Not Centered at the Origin

Classify and sketch the surface given by

$$x^2 + 2y^2 + z^2 - 4x + 4y - 2z + 3 = 0.$$

Solution Completing the square for each variable produces the following.

$$(x^2 - 4x +) + 2(y^2 + 2y +) + (z^2 - 2z +) = -3$$
$$(x^2 - 4x + 4) + 2(y^2 + 2y + 1) + (z^2 - 2z + 1) = -3 + 4 + 2 + 1$$
$$(x - 2)^2 + 2(y + 1)^2 + (z - 1)^2 = 4$$
$$\frac{(x-2)^2}{4} + \frac{(y+1)^2}{2} + \frac{(z-1)^2}{4} = 1$$

From this equation, you can see that the quadric surface is an ellipsoid that is centered at $(2, -1, 1)$. Its graph is shown in Figure 10.61.

TECHNOLOGY A computer algebra system can help you visualize a surface in space.* Most of these computer algebra systems create three-dimensional illusions by sketching several traces of the surface and then applying a "hidden-line" routine that blocks out portions of the surface that lie behind other portions of the surface. Two examples of figures that were generated by *Mathematica* are shown below.

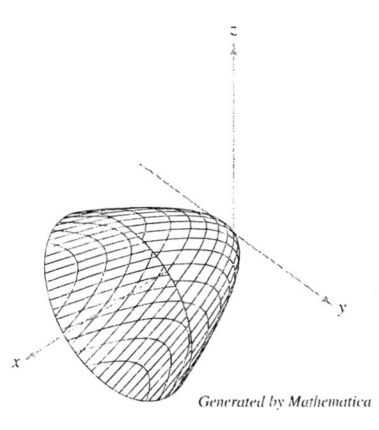

Elliptic paraboloid
$$x = \frac{y^2}{2} + \frac{z^2}{2}$$

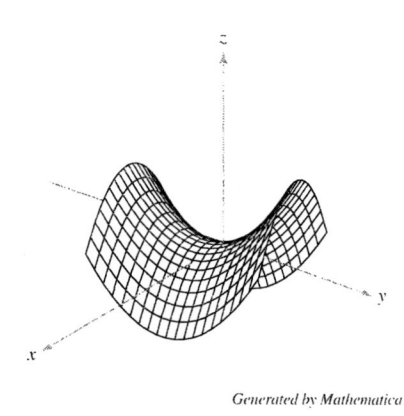

Hyperbolic paraboloid
$$z = \frac{y^2}{16} - \frac{x^2}{16}$$

Generated by Mathematica

Using a graphing utility to sketch the graph of a surface in space requires practice. For one thing, you must know enough about the surface to be able to specify a *viewing window* that gives a representative view of the surface. Also, you can often improve the view of a surface by rotating the axes. For instance, note that the elliptic paraboloid in the figure is seen from a line of sight that is "higher" than the line of sight used to view the hyperbolic paraboloid.

---

*Some 3-D graphing utilities require surfaces to be entered with parametric equations. For a discussion of this technique, see Section 14.5.

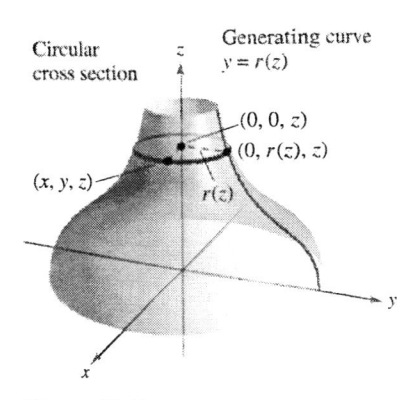

Surfaces of Revolution

The fifth special type of surface you will study is called a **surface of revolution.** In Section 6.4, you studied a method for finding the *area* of such a surface. You will now look at a procedure for finding its *equation*. Consider the graph of the **radius function**

$$y = r(z) \qquad \text{Generating curve}$$

in the yz-plane. If this graph is revolved about the z-axis, it forms a surface of revolution, as shown in Figure 10.62. The trace of the surface in the plane $z = z_0$ is a circle whose radius is $r(z_0)$ and whose equation is

$$x^2 + y^2 = [r(z_0)]^2. \qquad \text{Circular trace in plane: } z = z_0$$

Replacing z_0 with z produces an equation that is valid for all values of z. In a similar manner, we can obtain equations for surfaces of revolution for the other two axes, and we summarize the results as follows.

Surface of Revolution

If the graph of a radius function r is revolved about one of the coordinate axes, the equation of the resulting surface of revolution has one of the following forms.

1. Revolved about the x-axis: $y^2 + z^2 = [r(x)]^2$
2. Revolved about the y-axis: $x^2 + z^2 = [r(y)]^2$
3. Revolved about the z-axis: $x^2 + y^2 = [r(z)]^2$

Example 5 Finding an Equation for a Surface of Revolution

a. An equation for the surface of revolution formed by revolving the graph of

$$y = \frac{1}{z} \qquad \text{Radius function}$$

about the z-axis is

$$x^2 + y^2 = [r(z)]^2 \qquad \text{Revolved about the } z\text{-axis}$$

$$x^2 + y^2 = \left(\frac{1}{z}\right)^2. \qquad \text{Substitute } 1/z \text{ for } r(z).$$

b. To find an equation for the surface formed by revolving the graph of $9x^2 = y^3$ about the y-axis, solve for x in terms of y to obtain

$$x = \tfrac{1}{3}y^{3/2} = r(y). \qquad \text{Radius function}$$

So, the equation for this surface is

$$x^2 + z^2 = [r(y)]^2 \qquad \text{Revolved about the } y\text{-axis}$$

$$x^2 + z^2 = \left(\tfrac{1}{3}y^{3/2}\right)^2 \qquad \text{Substitute } \tfrac{1}{3}y^{3/2} \text{ for } r(y).$$

$$x^2 + z^2 = \tfrac{1}{9}y^3. \qquad \text{Equation of surface}$$

The graph is shown in Figure 10.63.

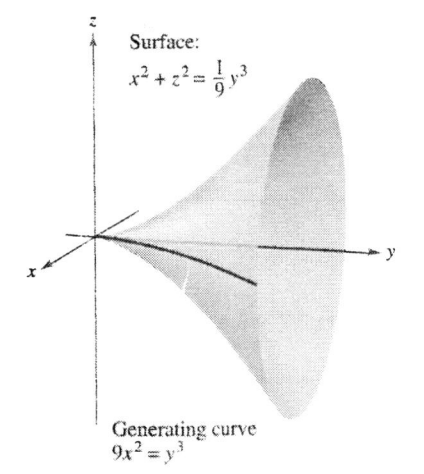

The generating curve for a surface of revolution is not unique. For instance, the surface

$$x^2 + z^2 = e^{-2y}$$

can be formed by revolving either the graph of $x = e^{-y}$ about the y-axis or the graph of $z = e^{-y}$ about the y-axis, as shown in Figure 10.64.

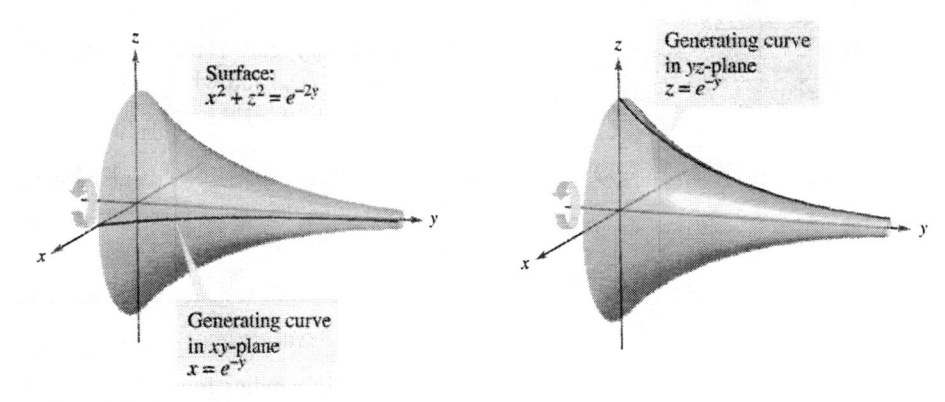

Figure 10.64

Example 6   Finding a Generating Curve for a Surface of Revolution

Find a generating curve and the axis of revolution for the surface given by

$$x^2 + 3y^2 + z^2 = 9.$$

Solution   You now know that the equation has one of the following forms.

$x^2 + y^2 = [r(z)]^2$ Revolved about z-axis
$y^2 + z^2 = [r(x)]^2$ Revolved about x-axis
$x^2 + z^2 = [r(y)]^2$ Revolved about y-axis

Because the coefficients of x^2 and z^2 are equal, you should choose the third form and write

$$x^2 + z^2 = 9 - 3y^2.$$

The y-axis is the axis of revolution. You can choose a generating curve from either of the following traces.

$x^2 = 9 - 3y^2$ Trace in xy-plane
$z^2 = 9 - 3y^2$ Trace in yz-plane

For example, using the first trace, the generating curve is the semiellipse given by

$$x = \sqrt{9 - 3y^2}.$$ Generating curve

The graph of this surface is shown in Figure 10.65.

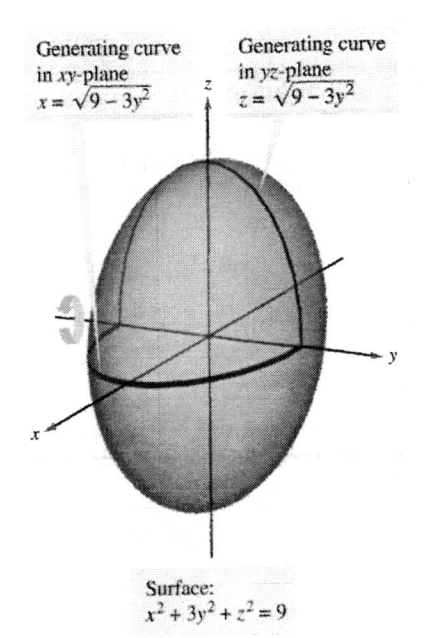

Figure 10.65

EXERCISES FOR SECTION 10.6

In Exercises 1–6, match the equation with its graph. [The graphs are labeled (a), (b), (c), (d), (e), and (f).]

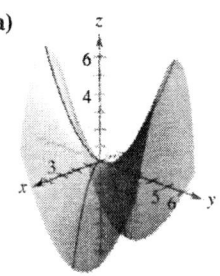

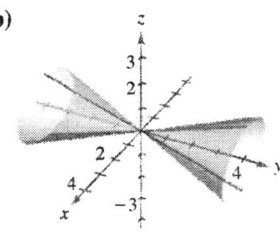

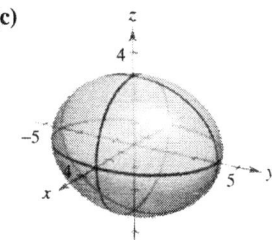

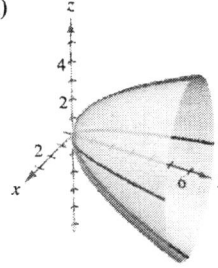

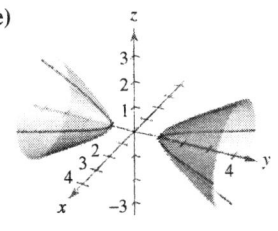

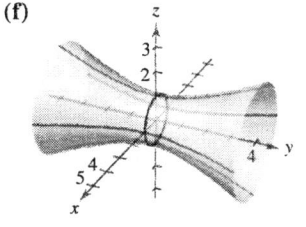

(a)

(b)

(c)

(d)

(e)

(f)

1. $\dfrac{x^2}{9} + \dfrac{y^2}{16} + \dfrac{z^2}{9} = 1$
2. $15x^2 - 4y^2 + 15z^2 = -4$
3. $4x^2 - y^2 + 4z^2 = 4$
4. $y^2 = 4x^2 + 9z^2$
5. $4x^2 - 4y + z^2 = 0$
6. $4x^2 - y^2 + 4z = 0$

In Exercises 7–16, describe and sketch the surface.

7. $z = 3$
8. $x = 4$
9. $y^2 + z^2 = 9$
10. $x^2 + z^2 = 25$
11. $x^2 - y = 0$
12. $y^2 + z = 4$
13. $4x^2 + y^2 = 4$
14. $y^2 - z^2 = 4$
15. $z - \sin y = 0$
16. $z - e^y = 0$

17. **Think About It** The four figures are graphs of the quadric surface $z = x^2 + y^2$. Match each of the four graphs with the point in space from which the paraboloid is viewed. The four points are $(0, 0, 20)$, $(0, 20, 0)$, $(20, 0, 0)$, and $(10, 10, 20)$.

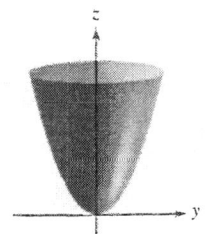

(a)

(b)

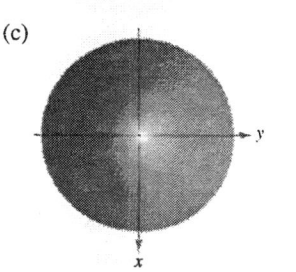

(c)

(d)

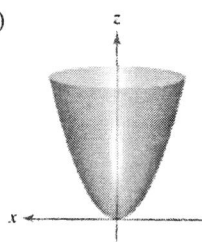

Figures for 17

18. Use a computer algebra system to sketch a view of the cylinder $y^2 + z^2 = 4$ from each of the following points.
    (a) $(10, 0, 0)$ (b) $(0, 10, 0)$ (c) $(10, 10, 10)$

In Exercises 19–30, identify and sketch the quadric surface. Use a computer algebra system to confirm your sketch.

19. $x^2 + \dfrac{y^2}{4} + z^2 = 1$
20. $\dfrac{x^2}{16} + \dfrac{y^2}{25} + \dfrac{z^2}{25} = 1$
21. $16x^2 - y^2 + 16z^2 = 4$
22. $z^2 - x^2 - \dfrac{y^2}{4} = 1$
23. $x^2 - y + z^2 = 0$
24. $z = x^2 + 4y^2$
25. $x^2 - y^2 + z = 0$
26. $3z = -y^2 + x^2$
27. $z^2 = x^2 + \dfrac{y^2}{4}$
28. $x^2 = 2y^2 + 2z^2$
29. $16x^2 + 9y^2 + 16z^2 - 32x - 36y + 36 = 0$
30. $9x^2 + y^2 - 9z^2 - 54x - 4y - 54z + 4 = 0$

In Exercises 31–40, use a computer algebra system to graph the surface. (*Hint:* It may be necessary to solve for z and acquire two equations to graph the surface.)

31. $z = 2 \sin x$
32. $z = x^2 + 0.5y^2$
33. $z^2 = x^2 + 4y^2$
34. $4y = x^2 + z^2$
35. $x^2 + y^2 = \left(\dfrac{2}{z}\right)^2$
36. $x^2 + y^2 = e^{-z}$
37. $z = 4 - \sqrt{|xy|}$
38. $z = \dfrac{-x}{8 + x^2 + y^2}$
39. $4x^2 - y^2 + 4z^2 = -16$
40. $9x^2 + 4y^2 - 8z^2 = 72$

In Exercises 41–44, sketch the region bounded by the graphs of the equations.

41. $z = 2\sqrt{x^2 + y^2}$, $z = 2$
42. $z = \sqrt{4 - x^2}$, $y = \sqrt{4 - x^2}$, $x = 0$, $y = 0$, $z = 0$
43. $x^2 + y^2 = 1$, $x + z = 2$, $z = 0$
44. $z = \sqrt{4 - x^2 - y^2}$, $y = 2z$, $z = 0$

In Exercises 45–50, find an equation for the surface of revolution generated by revolving the curve in the indicated coordinate plane about the given axis.

Equation of Curve	Coordinate Plane	Axis of Revolution
45. $z^2 = 4y$	yz-plane	y-axis
46. $z = 3y$	yz-plane	y-axis
47. $z = 2y$	yz-plane	z-axis
48. $2z = \sqrt{4 - x^2}$	xz-plane	x-axis
49. $xy = 2$	xy-plane	x-axis
50. $z = \ln y$	yz-plane	z-axis

In Exercises 51 and 52, find an equation of a generating curve given the equation of its surface of revolution.

51. $x^2 + y^2 - 2z = 0$ 52. $x^2 + z^2 = \cos^2 y$

Getting at the Concept

53. State the definition of a cylinder.
54. What is meant by the trace of a surface? How do you find a trace?
55. Identify the six quadric surfaces and give the standard form of each.
56. The graph of a radius function r is revolved about one of the coordinate axes. Give the standard form of the resulting surface of revolution for each coordinate axis.

In Exercises 57 and 58, use the shell method to find the volume of the solid below the surface of revolution and above the xy-plane.

57. The curve $z = 4x - x^2$ in the xz-plane is revolved about the z-axis.
58. The curve $z = \sin y$ $(0 \leq y \leq \pi)$ in the yz-plane is revolved about the z-axis.

In Exercises 59 and 60, analyze the trace when the surface

$$z = \tfrac{1}{2}x^2 + \tfrac{1}{4}y^2$$

is intersected by the indicated planes.

59. Find the length of the major and minor axes and the coordinates of the foci of the ellipse generated when the surface is intersected by the planes given by
    (a) $z = 2$ and   (b) $z = 8$.
60. Find the coordinates of the focus of the parabola formed when the surface is intersected by the planes given by
    (a) $y = 4$ and   (b) $x = 2$.

In Exercises 61 and 62, find an equation of the surface satisfying the conditions, and identify the surface.

61. The set of all points equidistant from the point $(0, 2, 0)$ and the plane $y = -2$

62. The set of all points equidistant from the point $(0, 0, 4)$ and the xy-plane

63. **Shape of Earth** Because of the forces caused by its rotation, earth is an oblate ellipsoid rather than a sphere. The equatorial radius is 3963 miles and the polar radius is 3942 miles. Find an equation of the ellipsoid. (Assume the center of earth is at the origin and the trace formed by the plane $z = 0$ corresponds to the equator.)

64. **Modeling Data** The table shows the amounts of public medical expenditures (in billions of dollars) for worker's compensation x, public assistance y, and Medicare z for selected years. (Source: U.S. Health Care Financing Administration)

Year	1980	1985	1990	1995	1996	1997
x	5.1	8.0	16.1	17.1	15.2	14.1
y	28.0	44.4	80.5	151.6	159.7	165.2
z	37.5	72.2	111.5	185.2	200.1	214.6

A mathematical model for the data is

$0.775x^2 - 0.007y^2 - 22.15x + 0.54y + z + 45.4 = 0$.

(a) Use the model to approximate z for the given values of x and y.
(b) Use a computer algebra system to graph the model for $5 \leq x \leq 20$ and $25 \leq y \leq 175$.
(c) Determine the concavity of the traces parallel to the xz-plane. Interpret the result in the context of the problem.
(d) Determine the concavity of the traces parallel to the yz-plane. Interpret the result in the context of the problem.

65. Determine the intersection of the hyperbolic paraboloid $z = y^2/b^2 - x^2/a^2$ with the plane $bx + ay - z = 0$. (Assume $a, b > 0$.)

66. Explain why the curve of intersection of the surfaces $x^2 + 3y^2 - 2z^2 + 2y = 4$ and $2x^2 + 6y^2 - 4z^2 - 3x = 2$ lies in a plane.

67. **Think About It** Three types of classic "topological" surfaces are shown below. The sphere and torus have both an "inside" and an "outside." Does the Klein bottle have both an inside and an outside? Explain.

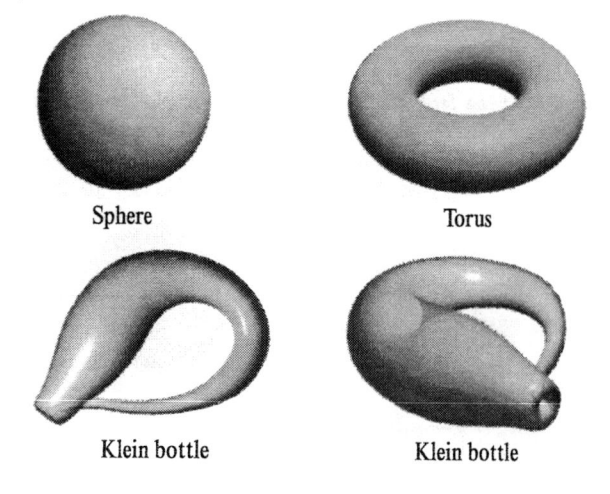

Sphere

Torus

Klein bottle

Klein bottle

Section 10.7 Cylindrical and Spherical Coordinates

- Use cylindrical coordinates to represent surfaces in space.
- Use spherical coordinates to represent surfaces in space.

Cylindrical Coordinates

You have already seen that some two-dimensional graphs are easier to represent in polar coordinates than in rectangular coordinates. A similar situation exists for surfaces in space. In this section, you will study two alternative space-coordinate systems. The first, the **cylindrical coordinate system,** is an extension of polar coordinates in the plane to three-dimensional space.

> **The Cylindrical Coordinate System**
>
> In a **cylindrical coordinate system,** a point P in space is represented by an ordered triple (r, θ, z).
>
> 1. (r, θ) is a polar representation of the projection of P in the xy-plane.
> 2. z is the directed distance from (r, θ) to P.

To convert from rectangular to cylindrical coordinates (or vice versa), use the following conversion guidelines for polar coordinates, as illustrated in Figure 10.66.

Cylindrical to rectangular:

$$x = r \cos \theta, \qquad y = r \sin \theta, \qquad z = z$$

Rectangular to cylindrical:

$$r^2 = x^2 + y^2, \qquad \tan \theta = \frac{y}{x}, \qquad z = z$$

The point $(0, 0, 0)$ is called the **pole.** Moreover, because the representation of a point in the polar coordinate system is not unique, it follows that the representation in the cylindrical coordinate system is also not unique.

Example 1 **Changing from Cylindrical to Rectangular Coordinates**

Express the point $(r, \theta, z) = (4, 5\pi/6, 3)$ in rectangular coordinates.

Solution Using the *cylindrical-to-rectangular* conversion equations produces

$$x = 4 \cos \frac{5\pi}{6} = 4\left(-\frac{\sqrt{3}}{2}\right) = -2\sqrt{3}$$

$$y = 4 \sin \frac{5\pi}{6} = 4\left(\frac{1}{2}\right) = 2$$

$$z = 3.$$

So, in rectangular coordinates, the point is $(x, y, z) = \left(-2\sqrt{3}, 2, 3\right)$, as shown in Figure 10.67.

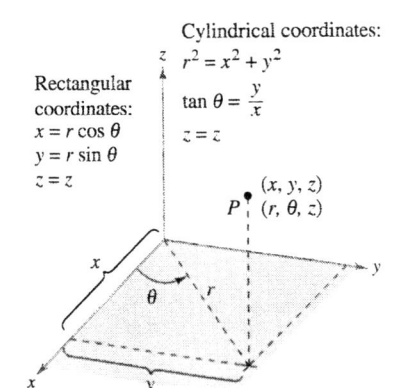

Figure 10.66

Figure 10.67

Example 2 Changing from Rectangular to Cylindrical Coordinates

Express the point $(x, y, z) = (1, \sqrt{3}, 2)$ in cylindrical coordinates.

Solution Use the *rectangular-to-cylindrical* conversion equations.

$$r = \pm\sqrt{1 + 3} = \pm 2$$

$$\tan \theta = \sqrt{3} \quad \Longrightarrow \quad \theta = \arctan\left(\sqrt{3}\right) + n\pi = \frac{\pi}{3} + n\pi$$

$$z = 2$$

You have two choices for r and infinitely many choices for θ. As shown in Figure 10.68, two convenient representations of the point are

$$\left(2, \frac{\pi}{3}, 2\right) \qquad r > 0 \text{ and } \theta \text{ in Quadrant I}$$

$$\left(-2, \frac{4\pi}{3}, 2\right). \qquad r < 0 \text{ and } \theta \text{ in Quadrant III}$$

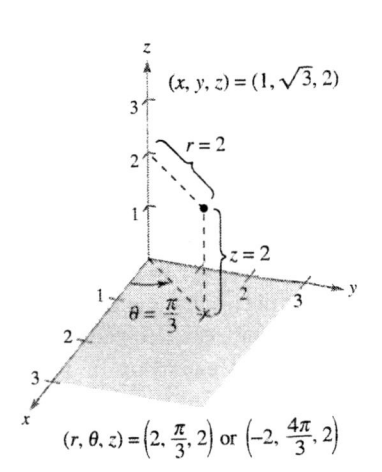

Figure 10.68

Cylindrical coordinates are especially convenient for representing cylindrical surfaces and surfaces of revolution with the z-axis as the axis of symmetry, as shown in Figure 10.69.

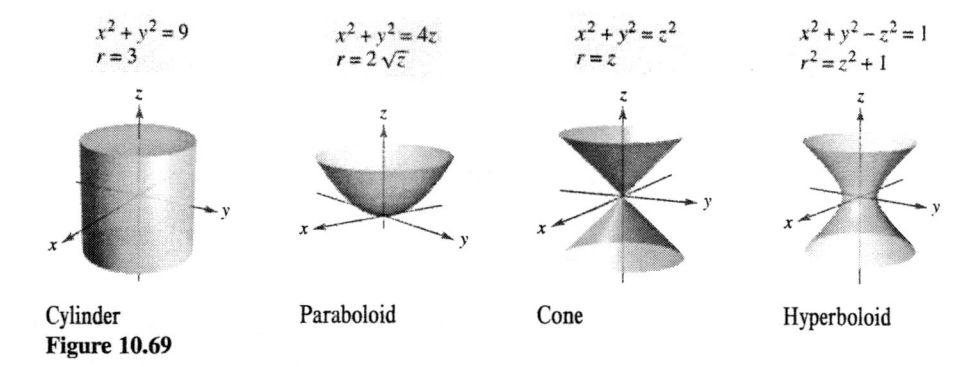

$x^2 + y^2 = 9$
$r = 3$
Cylinder

$x^2 + y^2 = 4z$
$r = 2\sqrt{z}$
Paraboloid

$x^2 + y^2 = z^2$
$r = z$
Cone

$x^2 + y^2 - z^2 = 1$
$r^2 = z^2 + 1$
Hyperboloid

Figure 10.69

Vertical planes containing the z-axis and horizontal planes also have simple cylindrical coordinate equations, as shown in Figure 10.70.

Vertical plane: $\theta = c$

Horizontal plane: $z = c$

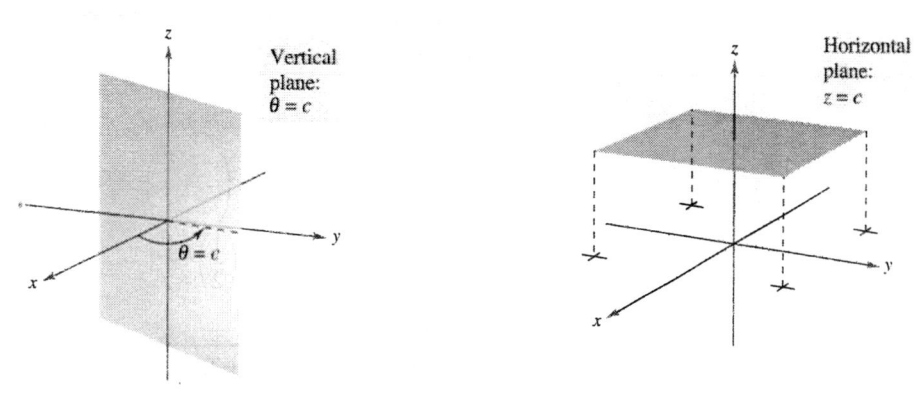

Figure 10.70

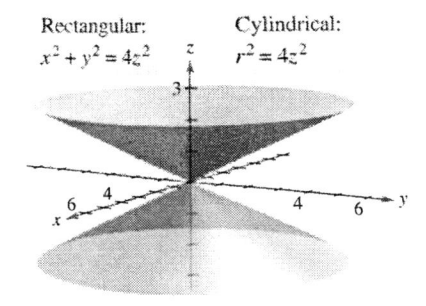

Example 3 Rectangular-to-Cylindrical Conversion

Find equations in cylindrical coordinates for the surfaces whose rectangular equations are as follows.

a. $x^2 + y^2 = 4z^2$

b. $y^2 = x$

Solution

a. From the preceding section, you know that the graph $x^2 + y^2 = 4z^2$ is a "double-napped" cone with its axis along the z-axis, as shown in Figure 10.71. If you replace $x^2 + y^2$ with r^2, the equation in cylindrical coordinates is

$$x^2 + y^2 = 4z^2 \quad \text{Rectangular equation}$$
$$r^2 = 4z^2. \quad \text{Cylindrical equation}$$

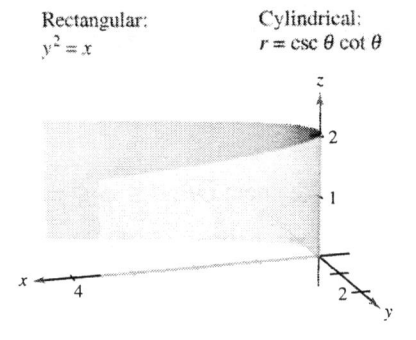

b. The graph of the surface $y^2 = x$ is a parabolic cylinder with rulings parallel to the z-axis, as shown in Figure 10.72. By replacing y^2 with $r^2 \sin^2 \theta$ and x with $r \cos \theta$, you obtain the following equation in cylindrical coordinates.

$$y^2 = x \quad \text{Rectangular equation}$$
$$r^2 \sin^2 \theta = r \cos \theta \quad \text{Substitute } r \sin \theta \text{ for } y \text{ and } r \cos \theta \text{ for } x.$$
$$r(r \sin^2 \theta - \cos \theta) = 0 \quad \text{Collect terms and factor.}$$
$$r \sin^2 \theta - \cos \theta = 0 \quad \text{Divide both sides by } r.$$
$$r = \frac{\cos \theta}{\sin^2 \theta} \quad \text{Solve for } r.$$
$$r = \csc \theta \cot \theta \quad \text{Cylindrical equation}$$

Note that this equation includes a point for which $r = 0$, so nothing was lost by dividing both sides by the factor r.

Converting from rectangular coordinates to cylindrical coordinates is more straightforward than converting from cylindrical coordinates to rectangular coordinates, as demonstrated in Example 4.

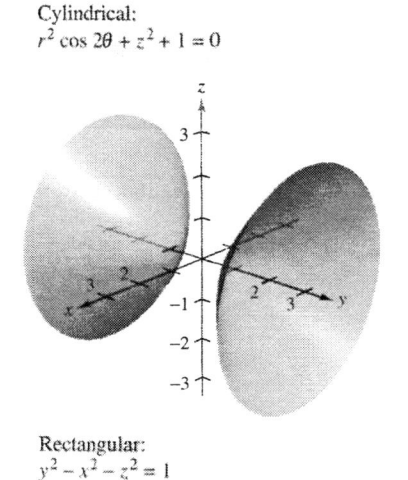

Example 4 Cylindrical-to-Rectangular Conversion

Find a rectangular equation for the graph represented by the cylindrical equation

$$r^2 \cos 2\theta + z^2 + 1 = 0.$$

Solution

$$r^2 \cos 2\theta + z^2 + 1 = 0 \quad \text{Cylindrical equation}$$
$$r^2(\cos^2 \theta - \sin^2 \theta) + z^2 + 1 = 0 \quad \text{Trigonometric identity}$$
$$r^2 \cos^2 \theta - r^2 \sin^2 \theta + z^2 = -1$$
$$x^2 - y^2 + z^2 = -1 \quad \text{Replace } r \cos \theta \text{ with } x \text{ and } r \sin \theta \text{ with } y.$$
$$y^2 - x^2 - z^2 = 1 \quad \text{Rectangular equation}$$

This is a hyperboloid of two sheets whose axis lies along the y-axis, as shown in Figure 10.73.

Spherical Coordinates

In the **spherical coordinate system,** each point is represented by an ordered triple: the first coordinate is a distance, and the second and third coordinates are angles. This system is similar to the latitude-longitude system used to identify points on the surface of earth. For example, the point on the surface of earth whose latitude is 40° North (of the equator) and whose longitude is 80° West (of the prime meridian) is shown in Figure 10.74. Assuming that the earth is spherical and has a radius of 4000 miles, you would label this point as

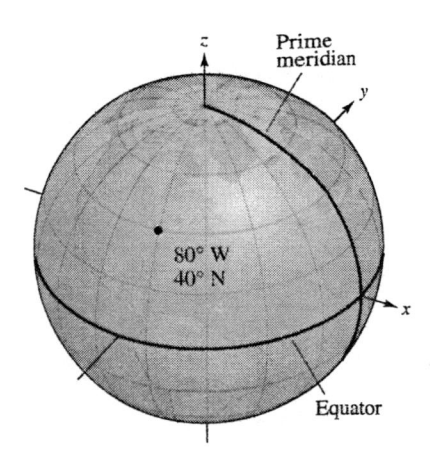

$$(4000, -80°, 50°).$$

Radius — 80° clockwise from prime meridian — 50° down from North Pole

Figure 10.74

The Spherical Coordinate System

In a **spherical coordinate system,** a point P in space is represented by an ordered triple (ρ, θ, ϕ).

1. ρ is the distance between P and the origin, $\rho \geq 0$.
2. θ is the same angle used in cylindrical coordinates for $r \geq 0$.
3. ϕ is the angle *between* the positive z-axis and the line segment $\overrightarrow{OP}$, $0 \leq \phi \leq \pi$.

Note that the first and third coordinates, ρ and ϕ, are nonnegative. ρ is the lowercase Greek letter *rho*, and ϕ is the lowercase Greek letter *phi*.

The relationship between rectangular and spherical coordinates is illustrated in Figure 10.75. To convert from one system to the other, use the following.

Spherical to rectangular:

$$x = \rho \sin \phi \cos \theta, \quad y = \rho \sin \phi \sin \theta, \quad z = \rho \cos \phi$$

Rectangular to spherical:

$$\rho^2 = x^2 + y^2 + z^2, \quad \tan \theta = \frac{y}{x}, \quad \phi = \arccos\left(\frac{z}{\sqrt{x^2 + y^2 + z^2}}\right)$$

Spherical coordinates
Figure 10.75

To change coordinates between the cylindrical and spherical systems, use the following.

Spherical to cylindrical ($r \geq 0$):

$$r^2 = \rho^2 \sin^2 \phi, \quad \theta = \theta, \quad z = \rho \cos \phi$$

Cylindrical to spherical ($r \geq 0$):

$$\rho = \sqrt{r^2 + z^2}, \quad \theta = \theta, \quad \phi = \arccos\left(\frac{z}{\sqrt{r^2 + z^2}}\right)$$

SECTION 10.7 Cylindrical and Spherical Coordinates

The spherical coordinate system is useful primarily for surfaces in space that have a *point* or *center* of symmetry. For example, Figure 10.76 shows three surfaces with simple spherical equations.

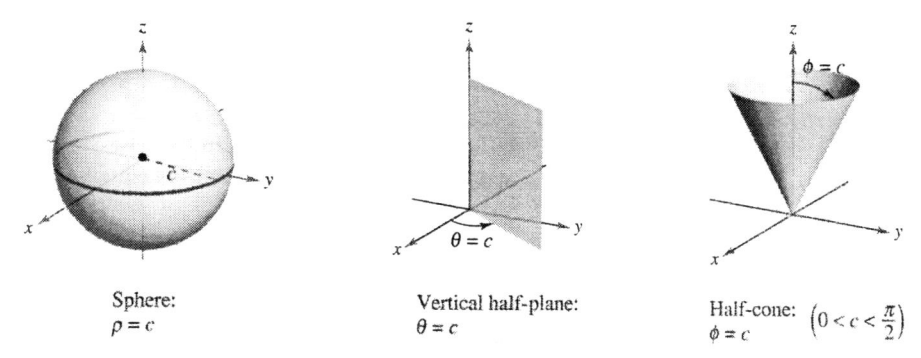

Sphere:
$\rho = c$

Vertical half-plane:
$\theta = c$

Half-cone: $\left(0 < c < \dfrac{\pi}{2}\right)$
$\phi = c$

Figure 10.76

Example 5 Rectangular-to-Spherical Conversion

Find an equation in spherical coordinates for the surface represented by each of the rectangular equations.

a. Cone: $x^2 + y^2 = z^2$
b. Sphere: $x^2 + y^2 + z^2 - 4z = 0$

Solution

a. Making the appropriate replacements for x, y, and z in the given equation yields the following.

$$x^2 + y^2 = z^2$$
$$\rho^2 \sin^2 \phi \cos^2 \theta + \rho^2 \sin^2 \phi \sin^2 \theta = \rho^2 \cos^2 \phi$$
$$\rho^2 \sin^2 \phi (\cos^2 \theta + \sin^2 \theta) = \rho^2 \cos^2 \phi$$
$$\rho^2 \sin^2 \phi = \rho^2 \cos^2 \phi$$
$$\dfrac{\sin^2 \phi}{\cos^2 \phi} = 1 \qquad \rho \geq 0$$
$$\tan^2 \phi = 1 \qquad \phi = \pi/4 \text{ or } \phi = 3\pi/4$$

The equation $\phi = \pi/4$ represents the *upper* half-cone, and the equation $\phi = 3\pi/4$ represents the *lower* half-cone.

b. Because $\rho^2 = x^2 + y^2 + z^2$ and $z = \rho \cos \phi$, the given equation has the following spherical form.

$$\rho^2 - 4\rho \cos \phi = 0 \implies \rho(\rho - 4 \cos \phi) = 0$$

Temporarily discarding the possibility that $\rho = 0$, you have the spherical equation

$$\rho - 4 \cos \phi = 0 \qquad \text{or} \qquad \rho = 4 \cos \phi.$$

Note that the solution set for this equation includes a point for which $\rho = 0$, so nothing is lost by discarding the factor ρ. The sphere represented by the equation $\rho = 4 \cos \phi$ is shown in Figure 10.77.

Rectangular:
$x^2 + y^2 + z^2 - 4z = 0$

Spherical:
$\rho = 4 \cos \phi$

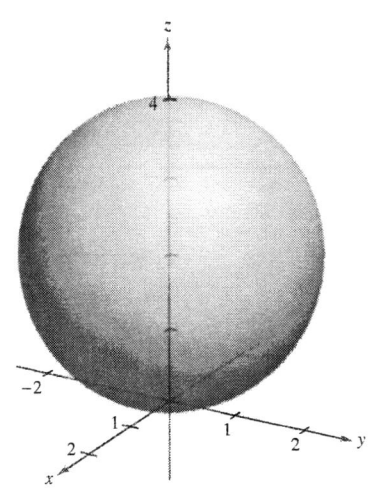

Figure 10.77

EXERCISES FOR SECTION 10.7

In Exercises 1–6, convert the point from cylindrical coordinates to rectangular coordinates.

1. $(5, 0, 2)$
2. $(4, \pi/2, -2)$
3. $(2, \pi/3, 2)$
4. $(6, -\pi/4, 2)$
5. $(4, 7\pi/6, 3)$
6. $(1, 3\pi/2, 1)$

In Exercises 7–12, convert the point from rectangular coordinates to cylindrical coordinates.

7. $(0, 5, 1)$
8. $(2\sqrt{2}, -2\sqrt{2}, 4)$
9. $(1, \sqrt{3}, 4)$
10. $(2\sqrt{3}, -2, 6)$
11. $(2, -2, -4)$
12. $(-3, 2, -1)$

In Exercises 13–16, find an equation in cylindrical coordinates for the rectangular equation.

13. $x^2 + y^2 + z^2 = 10$
14. $z = x^2 + y^2 - 2$
15. $y = x^2$
16. $x^2 + y^2 = 8x$

In Exercises 17–24, find an equation in rectangular coordinates for the equation in cylindrical coordinates, and sketch its graph.

17. $r = 2$
18. $z = 2$
19. $\theta = \pi/6$
20. $r = \frac{1}{2}z$
21. $r = 2 \sin \theta$
22. $r = 2 \cos \theta$
23. $r^2 + z^2 = 4$
24. $z = r^2 \cos^2 \theta$

In Exercises 25–30, convert the point from rectangular coordinates to spherical coordinates.

25. $(4, 0, 0)$
26. $(1, 1, 1)$
27. $(-2, 2\sqrt{3}, 4)$
28. $(2, 2, 4\sqrt{2})$
29. $(\sqrt{3}, 1, 2\sqrt{3})$
30. $(-4, 0, 0)$

In Exercises 31–36, convert the point from spherical coordinates to rectangular coordinates.

31. $(4, \pi/6, \pi/4)$
32. $(12, 3\pi/4, \pi/9)$
33. $(12, -\pi/4, 0)$
34. $(9, \pi/4, \pi)$
35. $(5, \pi/4, 3\pi/4)$
36. $(6, \pi, \pi/2)$

37. *Programming*
   (a) Write a program for a graphing utility that converts a point from rectangular coordinates to spherical coordinates.
   (b) Use the program in part (a) to convert the point $(3, -4, 2)$ from rectangular coordinates to spherical coordinates.

38. *Programming*
   (a) Write a program for a graphing utility that converts a point from spherical coordinates to rectangular coordinates.
   (b) Use the program in part (a) to convert the point $(5, 1, 0.5)$ from spherical coordinates to rectangular coordinates.

In Exercises 39–42, find an equation in spherical coordinates for the rectangular equation.

39. $x^2 + y^2 + z^2 = 36$
40. $x^2 + y^2 - 3z^2 = 0$
41. $x^2 + y^2 = 9$
42. $x = 10$

In Exercises 43–50, find an equation in rectangular coordinates for the equation in spherical coordinates, and sketch its graph.

43. $\rho = 2$
44. $\theta = \dfrac{3\pi}{4}$
45. $\phi = \dfrac{\pi}{6}$
46. $\phi = \dfrac{\pi}{2}$
47. $\rho = 4 \cos \phi$
48. $\rho = 2 \sec \phi$
49. $\rho = \csc \phi$
50. $\rho = 4 \csc \phi \sec \theta$

In Exercises 51–58, convert the point from cylindrical coordinates to spherical coordinates.

51. $(4, \pi/4, 0)$
52. $(3, -\pi/4, 0)$
53. $(4, \pi/2, 4)$
54. $(2, 2\pi/3, -2)$
55. $(4, -\pi/6, 6)$
56. $(-4, \pi/3, 4)$
57. $(12, \pi, 5)$
58. $(4, \pi/2, 3)$

In Exercises 59–66, convert the point from spherical coordinates to cylindrical coordinates.

59. $(10, \pi/6, \pi/2)$
60. $(4, \pi/18, \pi/2)$
61. $(36, \pi, \pi/2)$
62. $(18, \pi/3, \pi/3)$
63. $(6, -\pi/6, \pi/3)$
64. $(5, -5\pi/6, \pi)$
65. $(8, 7\pi/6, \pi/6)$
66. $(7, \pi/4, 3\pi/4)$

In Exercises 67–80, use a computer algebra system or graphing utility to convert the point from one system to another among the rectangular, cylindrical, and spherical coordinate systems.

	Rectangular	Cylindrical	Spherical
67.	$(4, 6, 3)$		
68.	$(6, -2, -3)$		
69.		$(5, \pi/9, 8)$	
70.		$(10, -0.75, 6)$	
71.			$(20, 2\pi/3, \pi/4)$
72.			$(7.5, 0.25, 1)$
73.	$(3, -2, 2)$		
74.	$(3\sqrt{2}, 3\sqrt{2}, -3)$		
75.	$(5/2, 4/3, -3/2)$		
76.	$(0, -5, 4)$		
77.		$(5, 3\pi/4, -5)$	
78.		$(-2, 11\pi/6, 3)$	
79.		$(-3.5, 2.5, 6)$	
80.		$(8.25, 1.3, -4)$	

In Exercises 81–86, match the equation (expressed in terms of cylindrical or spherical coordinates) with its graph. [The graphs are labeled (a), (b), (c), (d), (e), and (f).]

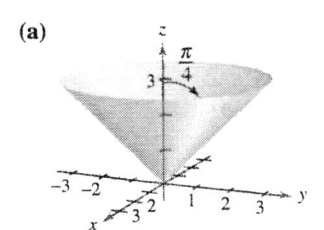

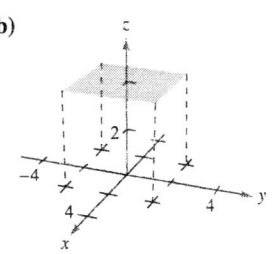

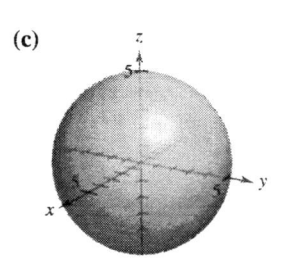

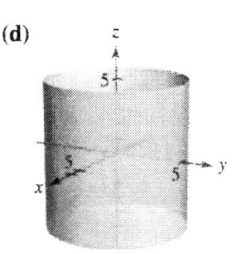

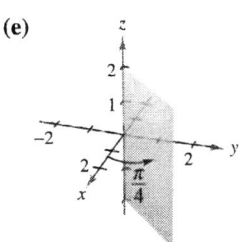

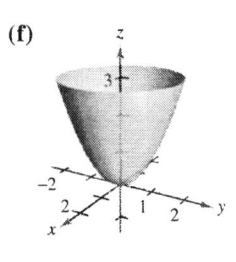

(a) (b) (c) (d) (e) (f)

81. $r = 5$
82. $\theta = \dfrac{\pi}{4}$
83. $\rho = 5$
84. $\phi = \dfrac{\pi}{4}$
85. $r^2 = z$
86. $\rho = 4 \sec \phi$

Getting at the Concept

87. Give the equations for the coordinate conversion from rectangular to cylindrical coordinates and vice versa.
88. For constants a, b, and c, describe the graphs of the equations $r = a$, $\theta = b$, and $z = c$ in cylindrical coordinates.
89. Give the equations for the coordinate conversion from rectangular to spherical coordinates and vice versa.
90. For constants a, b, and c, describe the graphs of the equations $\rho = a$, $\theta = b$, and $\phi = c$ in spherical coordinates.

In Exercises 91–98, convert the rectangular equation to an equation in (a) cylindrical coordinates and (b) spherical coordinates.

91. $x^2 + y^2 + z^2 = 16$
92. $4(x^2 + y^2) = z^2$
93. $x^2 + y^2 + z^2 - 2z = 0$
94. $x^2 + y^2 = z$
95. $x^2 + y^2 = 4y$
96. $x^2 + y^2 = 16$
97. $x^2 - y^2 = 9$
98. $y = 4$

In Exercises 99–102, sketch the solid that has the given description in cylindrical coordinates.

99. $0 \leq \theta \leq \pi/2, 0 \leq r \leq 2, 0 \leq z \leq 4$
100. $-\pi/2 \leq \theta \leq \pi/2, 0 \leq r \leq 3, 0 \leq z \leq r \cos \theta$
101. $0 \leq \theta \leq 2\pi, 0 \leq r \leq a, r \leq z \leq a$
102. $0 \leq \theta \leq 2\pi, 2 \leq r \leq 4, z^2 \leq -r^2 + 6r - 8$

In Exercises 103 and 104, sketch the solid that has the given description in spherical coordinates.

103. $0 \leq \theta \leq 2\pi, 0 \leq \phi \leq \pi/6, 0 \leq \rho \leq a \sec \phi$
104. $0 \leq \theta \leq 2\pi, \pi/4 \leq \phi \leq \pi/2, 0 \leq \rho \leq 1$

Think About It **In Exercises 105–108, find inequalities that describe the solid, and state the coordinate system used. Position the solid on the coordinate system such that the inequalities are as simple as possible.**

105. A cube with each edge 10 centimeters long
106. A cylindrical shell 8 meters long with an inside diameter of 0.75 meter and an outside diameter of 1.25 meters
107. A spherical shell with inside and outside radii of 4 inches and 6 inches
108. The solid that remains after a hole 1 inch in diameter is drilled through the center of a sphere 6 inches in diameter

109. Identify the curve of intersection of the surfaces (in cylindrical coordinates) $z = \sin \theta$ and $r = 1$.
110. Identify the curve of intersection of the surfaces (in spherical coordinates) $\rho = 2 \sec \phi$ and $\rho = 4$.

REVIEW EXERCISES FOR CHAPTER 10

10.1 In Exercises 1 and 2, let $\mathbf{u} = \vec{PQ}$ and $\mathbf{v} = \vec{PR}$, and find (a) the component forms of **u** and **v**, (b) the magnitude of **v**, and (c) $2\mathbf{u} + \mathbf{v}$.

1. $P = (1, 2), Q = (4, 1), R = (5, 4)$
2. $P = (-2, -1), Q = (5, -1), R = (2, 4)$

In Exercises 3 and 4, find the component form of the vector **v** given its magnitude and the angle it makes with the x-axis.

3. $\|\mathbf{v}\| = 8, \theta = 120°$
4. $\|\mathbf{v}\| = \frac{1}{2}, \theta = 225°$

5. *Equilibrium* A 100-pound collar slides on a frictionless vertical rod (see figure). Find the distance y for which the system is in equilibrium if the counterweight weighs 120 pounds.

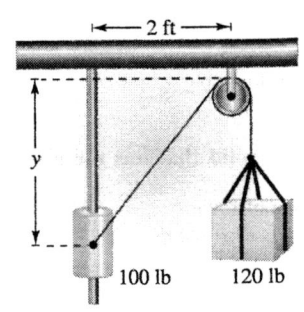

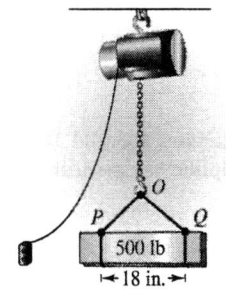

Figure for 5

Figure for 6

6. *Minimum Length* In a manufacturing process, an electric hoist lifts 500-pound ingots (see figure). The length of the cable connecting points P, O, and Q is L inches. (Assume that O is at the midpoint of the cable.)

   (a) Write the tension T in the cable as a function of L. What is the domain of the function?

   (b) Use the function in part (a) to complete the table.

L	19	20	21	22	23	24	25
T							

   (c) Use a graphing utility to graph the tension function.

   (d) Find the shortest cable connecting points P, O, and Q that can be used if the tension in the cable cannot exceed 400 pounds.

   (e) Find (if possible) $\lim_{L \to \infty} T$. Interpret the result in the context of the problem.

10.2

7. Find the coordinates of the point in the xy-plane 4 units to the right of the xz-plane and 5 units behind the yz-plane.

8. Find the coordinates of the point located on the y-axis and 7 units to the left of the xz-plane.

In Exercises 9 and 10, determine the location of the point (x, y, z) such that the given condition is satisfied.

9. $yz > 0$
10. $xy < 0$

In Exercises 11 and 12, find the standard form of the equation of the sphere.

11. Center: $(3, -2, 6)$; Diameter: 15
12. Endpoints of a diameter: $(0, 0, 4), (4, 6, 0)$

In Exercises 13 and 14, find the center and radius of the sphere and sketch its graph.

13. $x^2 + y^2 + z^2 - 4x - 6y + 4 = 0$
14. $x^2 + y^2 + z^2 - 10x + 6y - 4z + 34 = 0$

In Exercises 15 and 16, the initial and terminal point of a vector are given. Sketch the directed line segment and find the component form of the vector.

15. Initial point: $(2, -1, 3)$
    Terminal point: $(4, 4, -7)$
16. Initial point: $(6, 2, 0)$
    Terminal point: $(3, -3, 8)$

In Exercises 17 and 18, use vectors to determine whether the points lie in a straight line.

17. $(3, 4, -1), (-1, 6, 9), (5, 3, -6)$
18. $(5, -4, 7), (8, -5, 5), (11, 6, 3)$

19. Find a unit vector in the direction of $\mathbf{u} = \langle 2, 3, 5 \rangle$.
20. Find the vector **v** of magnitude 8 in the direction $\langle 6, -3, 2 \rangle$.

10.3 In Exercises 21 and 22, let $\mathbf{u} = \vec{PQ}$ and $\mathbf{v} = \vec{PR}$, and find (a) the component forms of **u** and **v**, (b) $\mathbf{u} \cdot \mathbf{v}$, and (c) $\mathbf{v} \cdot \mathbf{v}$.

21. $P = (5, 0, 0), Q = (4, 4, 0), R = (2, 0, 6)$
22. $P = (2, -1, 3), Q = (0, 5, 1), R = (5, 5, 0)$

In Exercises 23 and 24, determine whether the vectors are orthogonal, parallel, or neither.

23. $\langle 7, -2, 3 \rangle, \langle -1, 4, 5 \rangle$
24. $\langle -4, 3, -6 \rangle, \langle 16, -12, 24 \rangle$

In Exercises 25–28, find the angle θ between the vectors **u** and **v**.

25. $\mathbf{u} = 5[\cos(3\pi/4)\mathbf{i} + \sin(3\pi/4)\mathbf{j}]$
    $\mathbf{v} = 2[\cos(2\pi/3)\mathbf{i} + \sin(2\pi/3)\mathbf{j}]$
26. $\mathbf{u} = \langle 4, -1, 5 \rangle, \quad \mathbf{v} = \langle 3, 2, -2 \rangle$
27. $\mathbf{u} = \langle 10, -5, 15 \rangle, \quad \mathbf{v} = \langle -2, 1, -3 \rangle$
28. $\mathbf{u} = \langle 1, 0, -3 \rangle, \quad \mathbf{v} = \langle 2, -2, 1 \rangle$

29. Find two vectors in opposite directions that are orthogonal to the vector $\mathbf{u} = \langle 5, 6, -3 \rangle$.

30. *Work* An object is pulled 8 feet across a floor using a force of 75 pounds. Find the work done if the direction of the force is $30°$ above the horizontal.

In Exercises 31–34, let $\mathbf{u} = \langle 3, -2, 1 \rangle$, $\mathbf{v} = \langle 2, -4, -3 \rangle$, and $\mathbf{w} = \langle -1, 2, 2 \rangle$.

31. Show that $\mathbf{u} \cdot \mathbf{u} = \|\mathbf{u}\|^2$.
32. Find the angle between $\mathbf{u}$ and $\mathbf{v}$.
33. Determine the projection of $\mathbf{w}$ onto $\mathbf{u}$.
34. Find the work done in moving an object along the vector $\mathbf{u}$ if the applied force is $\mathbf{w}$.

10.4 In Exercises 35–40, let $\mathbf{u} = \langle 3, -2, 1 \rangle$, $\mathbf{v} = \langle 2, -4, -3 \rangle$, and $\mathbf{w} = \langle -1, 2, 2 \rangle$.

35. Determine a unit vector perpendicular to the plane containing $\mathbf{v}$ and $\mathbf{w}$.
36. Show that $\mathbf{u} \times \mathbf{v} = -(\mathbf{v} \times \mathbf{u})$.
37. Find the volume of the solid whose edges are $\mathbf{u}$, $\mathbf{v}$, and $\mathbf{w}$.
38. Show that $\mathbf{u} \times (\mathbf{v} + \mathbf{w}) = (\mathbf{u} \times \mathbf{v}) + (\mathbf{u} \times \mathbf{w})$.
39. Find the area of the parallelogram with adjacent sides $\mathbf{u}$ and $\mathbf{v}$.
40. Find the area of the triangle with adjacent sides $\mathbf{v}$ and $\mathbf{w}$.

41. *Torque* The specifications for a tractor state that the torque on a bolt with head size 7/8 inch cannot exceed 200 foot-pounds. Determine the maximum force $\|\mathbf{F}\|$ that can be applied to the wrench in the figure.

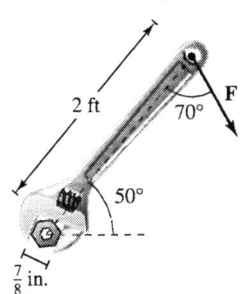

42. *Volume* Use the triple scalar product to find the volume of the parallelepiped with edges $\mathbf{u} = 2\mathbf{i} + \mathbf{j}$, $\mathbf{v} = 2\mathbf{j} + \mathbf{k}$, and $\mathbf{w} = -\mathbf{j} + 2\mathbf{k}$.

10.5 In Exercises 43–46, find (a) a set of parametric equations and (b) a set of symmetric equations for the line.

43. The line passes through the point $(1, 2, 3)$ and is perpendicular to the xz-plane.
44. The line passes through the point $(1, 2, 3)$ and is parallel to the line given by $x = y = z$.
45. The intersection of the planes $3x - 3y - 7z = -4$ and $x - y + 2z = 3$.
46. The line passes through the point $(0, 1, 4)$ and is perpendicular to $\mathbf{u} = \langle 2, -5, 1 \rangle$ and $\mathbf{v} = \langle -3, 1, 4 \rangle$.

In Exercises 47 and 48, find an equation of the plane.

47. The plane contains the lines $(x - 1)/(-2) = y = z + 1$ and $(x + 1)/(-2) = y - 1 = z - 2$.
48. The plane passes through the points $(-3, -4, 2)$, $(-3, 4, 1)$, and $(1, 1, -2)$.

49. Find the distance between the point $(1, 0, 2)$ and the plane $2x - 3y + 6z = 6$.
50. Find the distance between the planes $5x - 3y + z = 2$ and $5x - 3y + z = -3$.
51. Find the distance between the point $(3, -2, 4)$ and the plane $2x - 5y + z = 10$.
52. Find the distance between the point $(-5, 1, 3)$ and the line given by $x = 1 + t$, $y = 3 - 2t$, and $z = 5 - t$.

10.6 In Exercises 53–60, sketch the graph of the surface.

53. $x + 2y + 3z = 6$
54. $y = z^2$
55. $y = \frac{1}{2}z$
56. $y = \cos z$
57. $\dfrac{x^2}{16} + \dfrac{y^2}{9} + z^2 = 1$
58. $16x^2 + 16y^2 - 9z^2 = 0$
59. $\dfrac{x^2}{16} - \dfrac{y^2}{9} + z^2 = -1$
60. $\dfrac{x^2}{25} + \dfrac{y^2}{4} - \dfrac{z^2}{100} = 1$

61. *Machine Design* The top of a rubber bushing designed to absorb vibrations in an automobile is the surface of revolution generated by revolving the curve $z = \frac{1}{2}y^2 + 1$ $(0 \le y \le 2)$ in the yz-plane about the z-axis.
    (a) Find an equation for the surface of revolution.
    (b) If all measurements are in centimeters and the bushing is set on the xy-plane, use the shell method to find its volume.
    (c) Suppose the bushing has a hole of diameter 1 centimeter through its center and parallel to the axis of revolution. Find the volume of the rubber bushing.

62. Find an equation of the generating curve of the surface of revolution $y^2 + z^2 - 4x = 0$.

10.7 In Exercises 63 and 64, convert the point from rectangular coordinates to (a) cylindrical coordinates and (b) spherical coordinates.

63. $(-2\sqrt{2}, 2\sqrt{2}, 2)$
64. $\left(\dfrac{\sqrt{3}}{4}, \dfrac{3}{4}, \dfrac{3\sqrt{3}}{2}\right)$

In Exercises 65 and 66, convert the point from cylindrical coordinates to spherical coordinates.

65. $\left(100, -\dfrac{\pi}{6}, 50\right)$
66. $\left(81, -\dfrac{5\pi}{6}, 27\sqrt{3}\right)$

In Exercises 67 and 68, convert the point from spherical coordinates to cylindrical coordinates.

67. $\left(25, -\dfrac{\pi}{4}, \dfrac{3\pi}{4}\right)$
68. $\left(12, -\dfrac{\pi}{2}, \dfrac{2\pi}{3}\right)$

In Exercises 69 and 70, find an equation of the surface in (a) cylindrical coordinates and (b) spherical coordinates.

69. $x^2 - y^2 = 2z$
70. $x^2 + y^2 + z^2 = 16$

P.S. Problem Solving

1. Using vectors, prove the Law of Sines: If **a**, **b**, and **c** are the three sides of the triangle shown in the figure, then

   $$\frac{\sin A}{\|\mathbf{a}\|} = \frac{\sin B}{\|\mathbf{b}\|} = \frac{\sin C}{\|\mathbf{c}\|}.$$

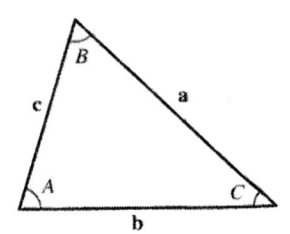

2. Consider the function $f(x) = \int_0^x \sqrt{t^4 + 1}\, dt$.

   (a) Use a graphing utility to graph the function on the interval $-2 \le x \le 2$.

   (b) Find a unit vector parallel to the graph of f at the point $(0, 0)$.

   (c) Find a unit vector perpendicular to the graph of f at the point $(0, 0)$.

   (d) Find the parametric equations of the tangent line to the graph of f at the point $(0, 0)$.

3. Using vectors, prove that the line segments joining the midpoints of the sides of a parallelogram form a parallelogram.

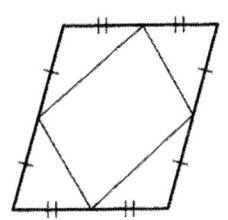

4. Using vectors, prove that the diagonals of a rhombus are perpendicular.

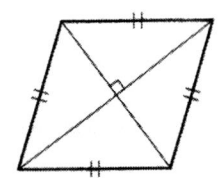

5. (a) Find the shortest distance between the point $Q(2, 0, 0)$ and the line determined by the points $P_1(0, 0, 1)$ and $P_2(0, 1, 2)$.

   (b) Find the shortest distance between the point $Q(2, 0, 0)$ and the line segment joining the points $P_1(0, 0, 1)$ and $P_2(0, 1, 2)$.

6. Let P_0 be a point in the plane with normal vector **n**. Describe the set of points P in the plane for which $(\mathbf{n} + \overrightarrow{PP_0})$ is orthogonal to $(\mathbf{n} - \overrightarrow{PP_0})$.

7. (a) Find the volume of the solid bounded below by the paraboloid $z = x^2 + y^2$ and above by the plane $z = 1$.

   (b) Find the volume of the solid bounded below by the elliptic paraboloid $z = \frac{x^2}{a^2} + \frac{y^2}{b^2}$ and above by the plane $z = k$, where $k > 0$.

   (c) Show that the volume of the solid in part (b) is equal to one-half the product of the base times the altitude, as indicated in the figure.

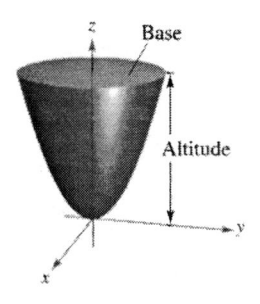

8. (a) Use the Disk Method to find the volume of the sphere $x^2 + y^2 + z^2 = r^2$.

   (b) Find the volume of the ellipsoid $\frac{x^2}{a^2} + \frac{y^2}{b^2} + \frac{z^2}{c^2} = 1$.

9. Sketch the graph of each equation given in spherical coordinates.

   (a) $\rho = 2 \sin \phi$

   (b) $\rho = 2 \cos \phi$

10. Sketch the graph of each equation given in cylindrical coordinates.

    (a) $r = 2 \cos \theta$

    (b) $z = r^2 \cos 2\theta$

11. Prove the following property of the cross product.

    $$(\mathbf{u} \times \mathbf{v}) \times (\mathbf{w} \times \mathbf{z}) = (\mathbf{u} \times \mathbf{v} \cdot \mathbf{z})\mathbf{w} - (\mathbf{u} \times \mathbf{v} \cdot \mathbf{w})\mathbf{z}$$

12. Consider the line given by the parametric equations

    $$x = -t + 3, \quad y = \tfrac{1}{2}t + 1, \quad z = 2t - 1$$

    and the point $(4, 3, s)$ for any real number s.

    (a) Write the distance between the point and the line as a function of s.

    (b) Use a graphing utility to graph the function in part (a). Use the graph to find the value of s such that the distance between the point and the line is minimum.

    (c) Use the *zoom* feature of a graphing utility to zoom out several times on the graph in part (b). Does it appear that the graph has slant asymptotes? Explain. If it appears to have slant asymptotes, find them.

13. A tetherball weighing 1 pound is pulled outward from the pole by a horizontal force **u** until the rope makes an angle of θ degrees with the pole (see figure).

    (a) Determine the resulting tension in the rope and the magnitude of **u** when $\theta = 30°$.

    (b) Write the tension T in the rope and the magnitude of **u** as functions of θ. Determine the domains of the functions.

    (c) Use a graphing utility to complete the table.

θ	0°	10°	20°	30°	40°	50°	60°
T							
$\|\mathbf{u}\|$							

    (d) Use a graphing utility to graph the two functions for $0° \leq \theta \leq 60°$.

    (e) Compare T and $\|\mathbf{u}\|$ as θ increases.

    (f) Find (if possible) $\lim_{\theta \to \pi/2} T$ and $\lim_{\theta \to \pi/2} \|\mathbf{u}\|$. Are the results what you expected? Explain.

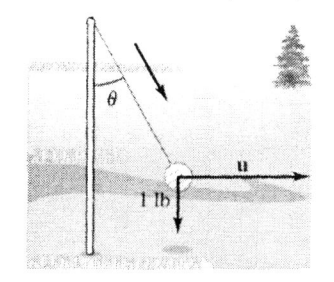

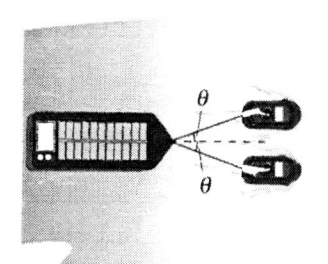

    Figure for 13   **Figure for 14**

14. A loaded barge is being towed by two tugboats, and the magnitude of the resultant is 6000 pounds directed along the axis of the barge (see figure). Each towline makes an angle of θ degrees with the axis of the barge.

    (a) Find the tension in the towlines if $\theta = 20°$.

    (b) Write the tension T of each line as a function of θ. Determine the domain of the function.

    (c) Use a graphing utility to complete the table.

θ	10°	20°	30°	40°	50°	60°
T						

    (d) Use a graphing utility to graph the tension function.

    (e) Explain why the tension increases as θ increases.

15. Consider the vectors $\mathbf{u} = \langle \cos \alpha, \sin \alpha, 0 \rangle$ and $\mathbf{v} = \langle \cos \beta, \sin \beta, 0 \rangle$, where $\alpha > \beta$. Find the cross product of the vectors and use the result to prove the identity

    $\sin(\alpha - \beta) = \sin \alpha \cos \beta - \cos \alpha \sin \beta$.

16. Los Angeles is located at 34.05° North latitude and 118.24° West longitude, and Rio de Janeiro, Brazil is located at 22.90° South latitude and 43.22° West longitude (see figure). Assume that the earth is spherical and has a radius of 4000 miles.

    (a) Find the spherical coordinates for the location of each city.

    (b) Find the rectangular coordinates for the location of each city.

    (c) Find the angle (in radians) between the vectors from the center of the earth to each city.

    (d) Find the great-circle distance s between the cities. (*Hint:* $s = r\theta$)

    (e) Repeat parts (a)–(d) for the cities of Boston, located at 42.36° North latitude and 71.06° West longitude, and Honolulu, located at 21.31° North latitude and 157.86° West longitude.

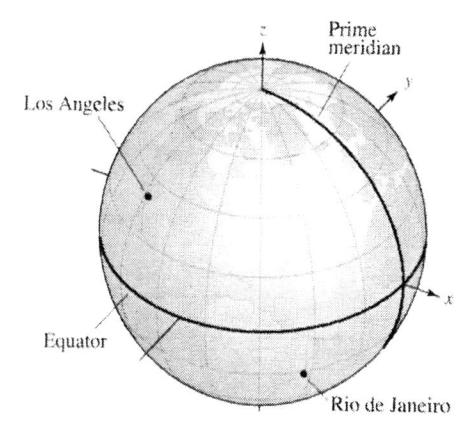

17. Consider the plane that passes through the points P, R, and S. Show that the distance from a point Q to this plane is

    $$\text{Distance} = \frac{|\mathbf{u} \cdot (\mathbf{v} \times \mathbf{w})|}{\|\mathbf{u} \times \mathbf{v}\|}$$

    where $\mathbf{u} = \overrightarrow{PR}$, $\mathbf{v} = \overrightarrow{PS}$, and $\mathbf{w} = \overrightarrow{PQ}$.

18. Show that the distance between the parallel planes $ax + by + cz + d_1 = 0$ and $ax + by + cz + d_2 = 0$ is

    $$\text{Distance} = \frac{|d_1 - d_2|}{\sqrt{a^2 + b^2 + c^2}}.$$

19. If a_1, b_1, c_1 and a_2, b_2, c_2 are two sets of direction numbers for the same line, show that there exists a scalar d such that $a_1 = a_2 d$, $b_1 = b_2 d$, and $c_1 = c_2 d$.

20. Read the article "Tooth Tables: Solution of a Dental Problem by Vector Algebra" by Gary Hosler Meisters in *Mathematics Magazine*. (To view this article, go to the website www.matharticles.com.) Then write a paragraph explaining how vectors and vector algebra can be used in the construction of dental inlays.

Race Car Cornering

Just northwest of Indianapolis, Indiana, sits the Indianapolis Motor Speedway, host to the internationally famous Indianapolis 500 Mile Race. The $2\frac{1}{2}$-mile track consists of front and back stretches that measure 3300 feet each, north and south straightaways that are each 660 feet long, and four 1320-foot-long turns.

A race car's position on a turn (see figure) could be modeled by the vector function

$$\mathbf{r}(t) = x(t)\mathbf{i} + y(t)\mathbf{j} \quad \text{Position function}$$
$$= 840(\cos 0.349t)\mathbf{i} + 840(\sin 0.349t)\mathbf{j},$$
$$0 \le t \le 4.5.$$

The velocity (comprising both speed and direction) can also be represented by a vector function. Any change in velocity, whether in speed or direction, is called acceleration. If the velocity is constant, you do not "feel" the motion and the acceleration is zero. You might say that you cannot feel motion, you can only feel *changes* in motion.

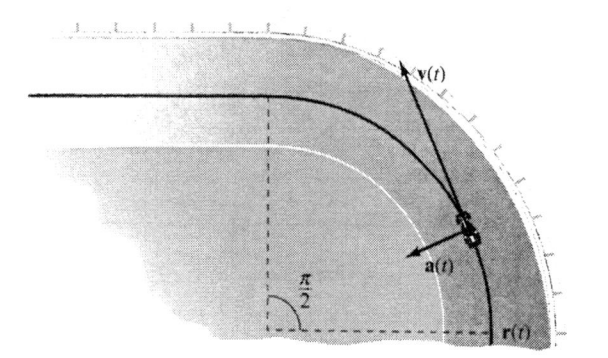

QUESTIONS

1. Use the function **r** to make a table of the car's positions for several values of t between $t = 0$ and $t = 4.5$.

t	0	0.5	1	1.5	2	2.5	3	3.5	4	4.5
$x(t)$										
$y(t)$										

   Plot your results. Describe the car's path during this interval of time.

2. Use a graphing utility to graph the car's path. Do the results agree with your conclusion in Question 1?

3. As the graphing utility plots the car's position, does the car appear to be taking the turn at a constant speed? Does it appear to be taking the turn at a constant velocity? Explain the distinction between these two questions.

4. In this chapter you will learn that the velocity of an object whose position function is **r** is given by

   $$\mathbf{v}(t) = x'(t)\mathbf{i} + y'(t)\mathbf{j}.$$

   Find the velocity vector for the car described above. Use your result to find a function that describes the speed of the car at any time. Is the speed of the car constant? Explain your reasoning.

5. The acceleration vector for the car is given by $\mathbf{a}(t) = x''(t) + y''(t)\mathbf{j}$. Describe the physical relationship between the car's velocity and acceleration. Is the driver "feeling" any acceleration? Explain your reasoning.

The concepts presented here will be explored further in this chapter. For an extension of this application, see Lab 15 in the lab series that accompanies this text at college.hmco.com.

Vector-Valued Functions 11

Lyn St. James of Daytona Beach, Florida, has qualified seven times for the Indianapolis 500. In 2000, her qualifying speed was more than 218 miles per hour.

The Indianapolis Motor Speedway, also called the Brickyard, opened in 1909. Ray Harroun won the first Indianapolis 500 in 1911 with an average speed of 74.6 miles per hour.

Section 11.1 Vector-Valued Functions

- Analyze and sketch a space curve given by a vector-valued function.
- Extend the concepts of limits and continuity to vector-valued functions.

Space Curves and Vector-Valued Functions

In Section 9.2, a *plane curve* was defined as the set of ordered pairs $(f(t), g(t))$ together with their defining parametric equations

$$x = f(t) \quad \text{and} \quad y = g(t)$$

where f and g are continuous functions of t on an interval I. This definition can be extended naturally to three-dimensional space as follows. A **space curve** C is the set of all ordered triples $(f(t), g(t), h(t))$ together with their defining parametric equations

$$x = f(t), \quad y = g(t), \quad \text{and} \quad z = h(t)$$

where f, g, and h are continuous functions of t on an interval I.

Before looking at examples of space curves, we introduce a new type of function, called a **vector-valued function**, that maps real numbers onto vectors.

Definition of a Vector-Valued Function

A function of the form

$$\mathbf{r}(t) = f(t)\mathbf{i} + g(t)\mathbf{j} \qquad \text{Plane}$$

or

$$\mathbf{r}(t) = f(t)\mathbf{i} + g(t)\mathbf{j} + h(t)\mathbf{k} \qquad \text{Space}$$

is a **vector-valued function**, where the **component functions** f, g, and h are real-valued functions of the parameter t. Vector-valued functions are sometimes denoted as $\mathbf{r}(t) = \langle f(t), g(t) \rangle$ or $\mathbf{r}(t) = \langle f(t), g(t), h(t) \rangle$.

Technically, a curve in the plane or in space consists of a collection of points *and* the defining parametric equations. Two different curves can have the same graph. For instance, each of the curves given by

$$\mathbf{r} = \sin t \, \mathbf{i} + \cos t \, \mathbf{j} \quad \text{and} \quad \mathbf{r} = \sin t^2 \, \mathbf{i} + \cos t^2 \, \mathbf{j}$$

has the unit circle as its graph, but these equations do not represent the same curve— because the circle is traced out in different ways on the graphs.

Be sure you see the distinction between the vector-valued function $\mathbf{r}$ and the real-valued functions f, g, and h. All are functions of the real variable t, but $\mathbf{r}(t)$ is a vector, whereas $f(t)$, $g(t)$, and $h(t)$ are real numbers (for each specific value of t).

Vector-valued functions serve dual roles in the representation of curves. By letting the parameter t represent time, you can use a vector-valued function to represent *motion* along a curve. Or, in the more general case, you can use a vector-valued function to *trace the graph* of a curve. In either case, the terminal point of the position vector $\mathbf{r}(t)$ coincides with the point (x, y) or (x, y, z) on the curve given by the parametric equations, as shown in Figure 11.1. The arrowhead on the curve indicates the curve's *orientation* by pointing in the direction of increasing values of t.

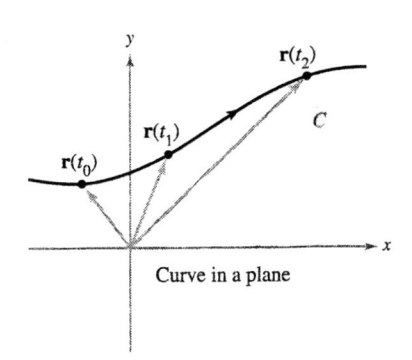

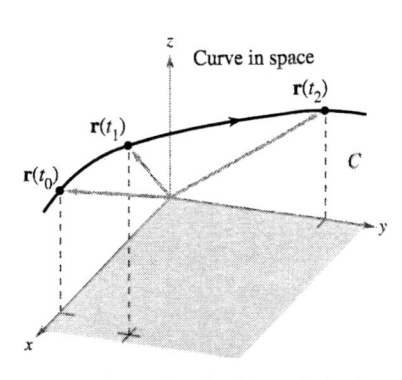

Curve C is traced out by the terminal point of position vector $\mathbf{r}(t)$.
Figure 11.1

Unless stated otherwise, the **domain** of a vector-valued function **r** is considered to be the intersection of the domains of the component functions f, g, and h. For instance, the domain of

$$\mathbf{r}(t) = (\ln t)\mathbf{i} + \sqrt{1-t}\,\mathbf{j} + t\mathbf{k}$$

is the interval $(0, 1]$.

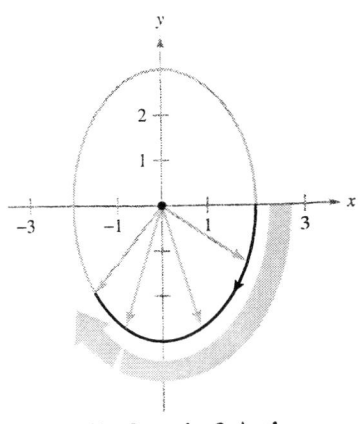

Example 1 Sketching a Plane Curve

Sketch the plane curve represented by the vector-valued function

$$\mathbf{r}(t) = 2\cos t\mathbf{i} - 3\sin t\mathbf{j}, \quad 0 \le t \le 2\pi. \quad \text{Vector-valued function}$$

Solution From the position vector $\mathbf{r}(t)$, you can write the parametric equations $x = 2\cos t$ and $y = -3\sin t$. Solving for $\cos t$ and $\sin t$ and using the identity $\cos^2 t + \sin^2 t = 1$ produces the rectangular equation

$$\frac{x^2}{2^2} + \frac{y^2}{3^2} = 1. \quad \text{Rectangular equation}$$

The graph of this rectangular equation is the ellipse shown in Figure 11.2. The curve has a *clockwise* orientation. That is, as t increases from 0 to 2π, the position vector $\mathbf{r}(t)$ moves clockwise, and its terminal point traces the ellipse.

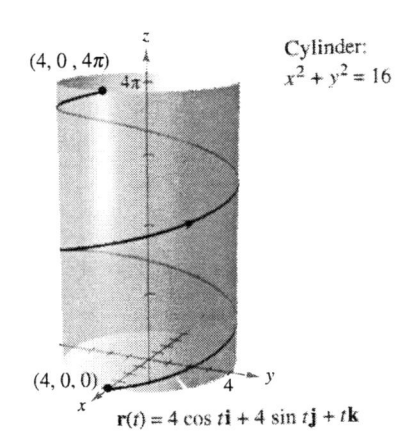

Example 2 Sketching a Space Curve

Sketch the space curve represented by the vector-valued function

$$\mathbf{r}(t) = 4\cos t\mathbf{i} + 4\sin t\mathbf{j} + t\mathbf{k}, \quad 0 \le t \le 4\pi. \quad \text{Vector-valued function}$$

Solution From the first two parametric equations $x = 4\cos t$ and $y = 4\sin t$, you can obtain

$$x^2 + y^2 = 16. \quad \text{Rectangular equation}$$

This means that the curve lies on a right circular cylinder of radius 4, centered about the z-axis. To locate the curve on this cylinder, you can use the third parametric equation $z = t$. In Figure 11.3, note that as t increases from 0 to 4π, the point (x, y, z) spirals up the cylinder to produce a **helix**. A real-life example of a helix is shown in the drawing at the lower left.

In Examples 1 and 2, you were given a vector-valued function and asked to sketch the corresponding curve. The next two examples address the reverse problem—finding a vector-valued function to represent a given graph. Of course, if the graph is described parametrically, representation by a vector-valued function is straightforward. For instance, to represent the line in space given by

$$x = 2 + t, \quad y = 3t, \quad \text{and} \quad z = 4 - t$$

you can simply use the vector-valued function given by

$$\mathbf{r}(t) = (2 + t)\mathbf{i} + 3t\mathbf{j} + (4 - t)\mathbf{k}.$$

If a set of parametric equations for the graph is not given, the problem of representing the graph by a vector-valued function boils down to finding a set of parametric equations.

$\mathbf{r}(t) = 2\cos t\mathbf{i} - 3\sin t\mathbf{j}$

The ellipse is traced clockwise as t increases from 0 to 2π.
Figure 11.2

$\mathbf{r}(t) = 4\cos t\mathbf{i} + 4\sin t\mathbf{j} + t\mathbf{k}$

As t increases from 0 to 4π, two spirals on the helix are traced out.
Figure 11.3

The DNA molecule, discovered in 1962 by Francis Crick and James D. Watson, is in the shape of a double helix.

indicates that in the Interactive 3.0 CD-ROM and Internet 3.0 versions of this text (available at college.hmco.com) you will find an Open Exploration, which further explores this example using the computer algebra systems Maple, Mathcad, Mathematica, and Derive.

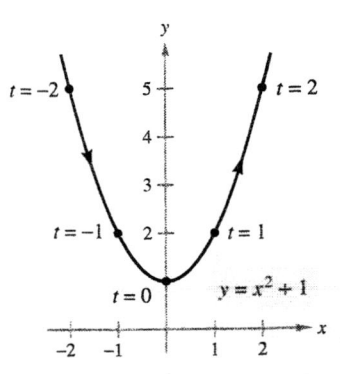

Example 3 Representing a Graph by a Vector-Valued Function

Represent the parabola given by $y = x^2 + 1$ by a vector-valued function.

Solution Although there are many ways to choose the parameter t, a natural choice is to let $x = t$. Then $y = t^2 + 1$ and you have

$$\mathbf{r}(t) = t\mathbf{i} + (t^2 + 1)\mathbf{j}.\qquad \text{Vector-valued function}$$

Note in Figure 11.4 the orientation produced by this particular choice of parameter. Had you chosen $x = -t$ as the parameter, the curve would have been oriented in the opposite direction.

There are many ways to parametrize this graph. One way is to let $x = t$.
Figure 11.4

Example 4 Representing a Graph by a Vector-Valued Function

Sketch the graph C represented by the intersection of the semiellipsoid

$$\frac{x^2}{12} + \frac{y^2}{24} + \frac{z^2}{4} = 1, \qquad z \geq 0$$

and the parabolic cylinder $y = x^2$. Then, find a vector-valued function to represent the graph.

Solution The intersection of the two surfaces is shown in Figure 11.5. As in Example 3, a natural choice of parameter is $x = t$. For this choice, you can use the given equation $y = x^2$ to obtain $y = t^2$. Then, it follows that

$$\frac{z^2}{4} = 1 - \frac{x^2}{12} - \frac{y^2}{24} = 1 - \frac{t^2}{12} - \frac{t^4}{24} = \frac{24 - 2t^2 - t^4}{24}.$$

Because the curve lies above the xy-plane, you should choose the positive square root for z and obtain the following parametric equations.

$$x = t, \qquad y = t^2, \qquad \text{and} \qquad z = \sqrt{\frac{24 - 2t^2 - t^4}{6}}$$

The resulting vector-valued function is

$$\mathbf{r}(t) = t\mathbf{i} + t^2\mathbf{j} + \sqrt{\frac{24 - 2t^2 - t^4}{6}}\,\mathbf{k}, \qquad -2 \leq t \leq 2. \qquad \text{Vector-valued function}$$

From the points $(-2, 4, 0)$ and $(2, 4, 0)$ shown in Figure 11.5, you can see that the curve is traced as t increases from -2 to 2.

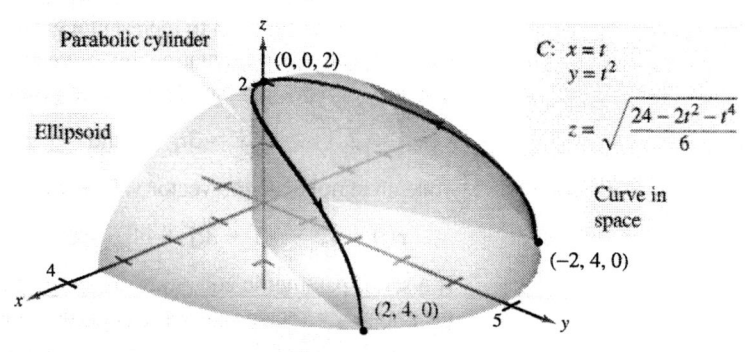

NOTE Curves in space can be specified in various ways. For instance, the curve in Example 4 is described as the intersection of two surfaces in space.

The curve C is the intersection of the semiellipsoid and the parabolic cylinder.
Figure 11.5

Limits and Continuity

Many techniques and definitions used in the calculus of real-valued functions can be applied to vector-valued functions. For instance, you can add and subtract vector-valued functions, multiply a vector-valued function by a scalar, take the limit of a vector-valued function, differentiate a vector-valued function, and so on. The basic approach is to capitalize on the linearity of vector operations by extending the definitions on a component-by-component basis. For example, to add or subtract two vector-valued functions (in the plane), you can write

$$\mathbf{r}_1(t) + \mathbf{r}_2(t) = [f_1(t)\mathbf{i} + g_1(t)\mathbf{j}] + [f_2(t)\mathbf{i} + g_2(t)\mathbf{j}] \quad \text{Sum}$$
$$= [f_1(t) + f_2(t)]\mathbf{i} + [g_1(t) + g_2(t)]\mathbf{j}$$
$$\mathbf{r}_1(t) - \mathbf{r}_2(t) = [f_1(t)\mathbf{i} + g_1(t)\mathbf{j}] - [f_2(t)\mathbf{i} + g_2(t)\mathbf{j}] \quad \text{Difference}$$
$$= [f_1(t) - f_2(t)]\mathbf{i} + [g_1(t) - g_2(t)]\mathbf{j}.$$

Similarly, to multiply and divide a vector-valued function by a scalar, you can write

$$c\mathbf{r}(t) = c[f_1(t)\mathbf{i} + g_1(t)\mathbf{j}] \quad \text{Scalar multiplication}$$
$$= cf_1(t)\mathbf{i} + cg_1(t)\mathbf{j}$$
$$\frac{\mathbf{r}(t)}{c} = \frac{[f_1(t)\mathbf{i} + g_1(t)\mathbf{j}]}{c}, \quad c \neq 0 \quad \text{Scalar division}$$
$$= \frac{f_1(t)}{c}\mathbf{i} + \frac{g_1(t)}{c}\mathbf{j}.$$

This component-by-component extension of operations with real-valued functions to vector-valued functions is further illustrated in the following definition of the limit of a vector-valued function.

Definition of the Limit of a Vector-Valued Function

1. If $\mathbf{r}$ is a vector-valued function such that $\mathbf{r}(t) = f(t)\mathbf{i} + g(t)\mathbf{j}$, then

$$\lim_{t \to a} \mathbf{r}(t) = \left[\lim_{t \to a} f(t)\right]\mathbf{i} + \left[\lim_{t \to a} g(t)\right]\mathbf{j} \quad \text{Plane}$$

provided f and g have limits as $t \to a$.

2. If $\mathbf{r}$ is a vector-valued function such that $\mathbf{r}(t) = f(t)\mathbf{i} + g(t)\mathbf{j} + h(t)\mathbf{k}$, then

$$\lim_{t \to a} \mathbf{r}(t) = \left[\lim_{t \to a} f(t)\right]\mathbf{i} + \left[\lim_{t \to a} g(t)\right]\mathbf{j} + \left[\lim_{t \to a} h(t)\right]\mathbf{k} \quad \text{Space}$$

provided f, g, and h have limits as $t \to a$.

If $\mathbf{r}(t)$ approaches the vector $\mathbf{L}$ as $t \to a$, the length of the vector $\mathbf{r}(t) - \mathbf{L}$ approaches 0. That is,

$$\|\mathbf{r}(t) - \mathbf{L}\| \to 0 \quad \text{as} \quad t \to a.$$

This is illustrated graphically in Figure 11.6. With this definition of the limit of a vector-valued function, you can develop vector versions of most of the limit theorems given in Chapter 1. For example, the limit of the sum of two vector-valued functions is the sum of their individual limits. Also, you can use the orientation of the curve $\mathbf{r}(t)$ to define one-sided limits of vector-valued functions. The next definition extends the notion of continuity to vector-valued functions.

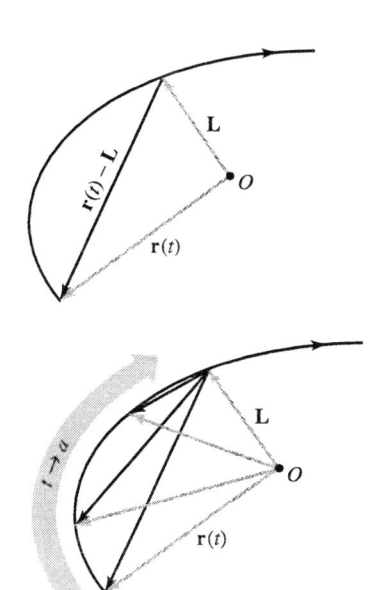

As t approaches a, $\mathbf{r}(t)$ approaches the limit $\mathbf{L}$. For the limit $\mathbf{L}$ to exist, it is not necessary that $\mathbf{r}(a)$ be defined or that $\mathbf{r}(a)$ be equal to $\mathbf{L}$.
Figure 11.6

Definition of Continuity of a Vector-Valued Function

A vector-valued function **r** is **continuous at the point** given by $t = a$ if the limit of **r**(t) exists as $t \to a$ and

$$\lim_{t \to a} \mathbf{r}(t) = \mathbf{r}(a).$$

A vector-valued function **r** is **continuous on an interval** I if it is continuous at every point in the interval.

From this definition, it follows that a vector-valued function is continuous at $t = a$ if and only if each of its component functions is continuous at $t = a$.

Example 5 Continuity of Vector-Valued Functions

Discuss the continuity of the vector-valued function given by

$$\mathbf{r}(t) = t\mathbf{i} + a\mathbf{j} + (a^2 - t^2)\mathbf{k} \qquad a \text{ is a constant.}$$

at $t = 0$.

Solution As t approaches 0, the limit is

$$\lim_{t \to 0} \mathbf{r}(t) = \left[\lim_{t \to 0} t\right]\mathbf{i} + \left[\lim_{t \to 0} a\right]\mathbf{j} + \left[\lim_{t \to 0} (a^2 - t^2)\right]\mathbf{k}$$
$$= 0\mathbf{i} + a\mathbf{j} + a^2\mathbf{k}$$
$$= a\mathbf{j} + a^2\mathbf{k}.$$

Because

$$\mathbf{r}(0) = (0)\mathbf{i} + (a)\mathbf{j} + (a^2)\mathbf{k}$$
$$= a\mathbf{j} + a^2\mathbf{k}$$

you can conclude that **r** is continuous at $t = 0$. By similar reasoning, you can conclude that the vector-valued function **r** is continuous at all real-number values of t.

For each value of a, the curve represented by the vector-valued function in Example 5,

$$\mathbf{r}(t) = t\mathbf{i} + a\mathbf{j} + (a^2 - t^2)\mathbf{k} \qquad a \text{ is a constant.}$$

is a parabola. You can think of each parabola as the intersection of the vertical plane $y = a$ and the hyperbolic paraboloid

$$y^2 - x^2 = z$$

as shown in Figure 11.7.

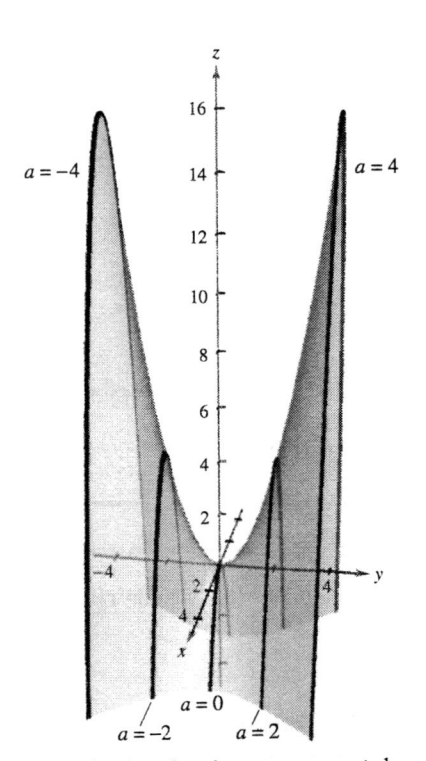

For each value of a, the curve represented by the vector-valued function
$\mathbf{r}(t) = t\mathbf{i} + a\mathbf{j} + (a^2 - t^2)\mathbf{k}$ is a parabola.
Figure 11.7

TECHNOLOGY Almost any type of three-dimensional sketch is difficult to do by hand, but sketching curves in space is especially difficult. The problem is in trying to create the illusion of three dimensions. Graphing utilities use a variety of techniques to add "three-dimensionality" to sketches of space curves: one way is to show the curve on a surface, as in Figure 11.7.

EXERCISES FOR SECTION 11.1

In Exercises 1–8, find the domain of the vector-valued function.

1. $\mathbf{r}(t) = 5t\mathbf{i} - 4t\mathbf{j} - \frac{1}{t}\mathbf{k}$
2. $\mathbf{r}(t) = \sqrt{4 - t^2}\,\mathbf{i} + t^2\mathbf{j} - 6t\mathbf{k}$
3. $\mathbf{r}(t) = \ln t\,\mathbf{i} - e^t\mathbf{j} - t\mathbf{k}$
4. $\mathbf{r}(t) = \sin t\,\mathbf{i} + 4\cos t\,\mathbf{j} + t\mathbf{k}$
5. $\mathbf{r}(t) = \mathbf{F}(t) + \mathbf{G}(t)$ where
   $\mathbf{F}(t) = \cos t\,\mathbf{i} - \sin t\,\mathbf{j} + \sqrt{t}\,\mathbf{k}$, $\mathbf{G}(t) = \cos t\,\mathbf{i} + \sin t\,\mathbf{j}$
6. $\mathbf{r}(t) = \mathbf{F}(t) - \mathbf{G}(t)$ where
   $\mathbf{F}(t) = \ln t\,\mathbf{i} + 5t\mathbf{j} - 3t^2\mathbf{k}$, $\mathbf{G}(t) = \mathbf{i} + 4t\mathbf{j} - 3t^2\mathbf{k}$
7. $\mathbf{r}(t) = \mathbf{F}(t) \times \mathbf{G}(t)$ where
   $\mathbf{F}(t) = \sin t\,\mathbf{i} + \cos t\,\mathbf{j}$, $\mathbf{G}(t) = \sin t\,\mathbf{j} + \cos t\,\mathbf{k}$
8. $\mathbf{r}(t) = \mathbf{F}(t) \times \mathbf{G}(t)$ where
   $\mathbf{F}(t) = t^3\mathbf{i} - t\mathbf{j} + t\mathbf{k}$, $\mathbf{G}(t) = \sqrt[3]{t}\,\mathbf{i} + \frac{1}{t+1}\mathbf{j} + (t+2)\mathbf{k}$

In Exercises 9–12, evaluate (if possible) the vector-valued function at the indicated value of t.

9. $\mathbf{r}(t) = \frac{1}{2}t^2\mathbf{i} - (t-1)\mathbf{j}$
   (a) $\mathbf{r}(1)$ (b) $\mathbf{r}(0)$ (c) $\mathbf{r}(s+1)$
   (d) $\mathbf{r}(2 + \Delta t) - \mathbf{r}(2)$

10. $\mathbf{r}(t) = \cos t\,\mathbf{i} + 2\sin t\,\mathbf{j}$
    (a) $\mathbf{r}(0)$ (b) $\mathbf{r}(\pi/4)$ (c) $\mathbf{r}(\theta - \pi)$
    (d) $\mathbf{r}(\pi/6 + \Delta t) - \mathbf{r}(\pi/6)$

11. $\mathbf{r}(t) = \ln t\,\mathbf{i} + \frac{1}{t}\mathbf{j} + 3t\mathbf{k}$
    (a) $\mathbf{r}(2)$ (b) $\mathbf{r}(-3)$ (c) $\mathbf{r}(t-4)$
    (d) $\mathbf{r}(1 + \Delta t) - \mathbf{r}(1)$

12. $\mathbf{r}(t) = \sqrt{t}\,\mathbf{i} + t^{3/2}\mathbf{j} + e^{-t/4}\mathbf{k}$
    (a) $\mathbf{r}(0)$ (b) $\mathbf{r}(4)$ (c) $\mathbf{r}(c+2)$
    (d) $\mathbf{r}(9 + \Delta t) - \mathbf{r}(9)$

In Exercises 13 and 14, find $\|\mathbf{r}(t)\|$.

13. $\mathbf{r}(t) = \sin 3t\,\mathbf{i} + \cos 3t\,\mathbf{j} + t\mathbf{k}$
14. $\mathbf{r}(t) = \sqrt{t}\,\mathbf{i} + 3t\mathbf{j} - 4t\mathbf{k}$

Think About It In Exercises 15 and 16, find $\mathbf{r}(t) \cdot \mathbf{u}(t)$. Is the result a vector-valued function? Explain.

15. $\mathbf{r}(t) = (3t - 1)\mathbf{i} + \frac{1}{4}t^3\mathbf{j} + 4\mathbf{k}$
    $\mathbf{u}(t) = t^2\mathbf{i} - 8\mathbf{j} + t^3\mathbf{k}$
16. $\mathbf{r}(t) = \langle 3\cos t, 2\sin t, t - 2\rangle$
    $\mathbf{u}(t) = \langle 4\sin t, -6\cos t, t^2\rangle$

In Exercises 17–20, match the equation with its graph. [The graphs are labeled (a), (b), (c), and (d).]

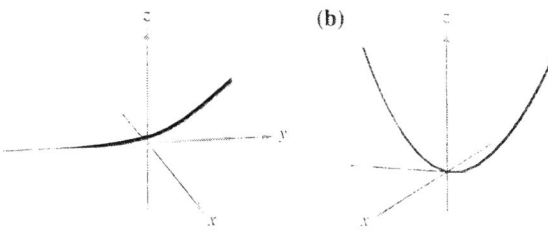

(a)

(b)

(c)

(d)

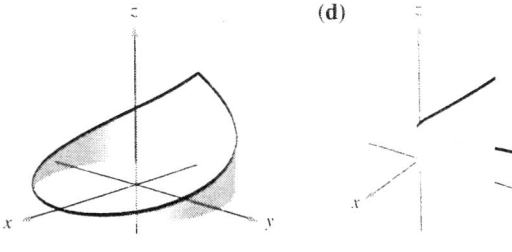

17. $\mathbf{r}(t) = t\mathbf{i} + 2t\mathbf{j} + t^2\mathbf{k}$, $-2 \leq t \leq 2$
18. $\mathbf{r}(t) = \cos(\pi t)\mathbf{i} + \sin(\pi t)\mathbf{j} + t^2\mathbf{k}$, $-1 \leq t \leq 1$
19. $\mathbf{r}(t) = t\mathbf{i} + t^2\mathbf{j} + e^{0.75t}\mathbf{k}$, $-2 \leq t \leq 2$
20. $\mathbf{r}(t) = t\mathbf{i} + \ln t\,\mathbf{j} + \frac{2t}{3}\mathbf{k}$, $0.1 \leq t \leq 5$

21. *Think About It* The four figures below are graphs of the vector-valued function

    $\mathbf{r}(t) = 4\cos t\,\mathbf{i} + 4\sin t\,\mathbf{j} + \frac{t}{4}\mathbf{k}$.

    Match the four graphs with the point in space from which the helix is viewed. The four points are $(0, 0, 20)$, $(20, 0, 0)$, $(-20, 0, 0)$, and $(10, 20, 10)$.

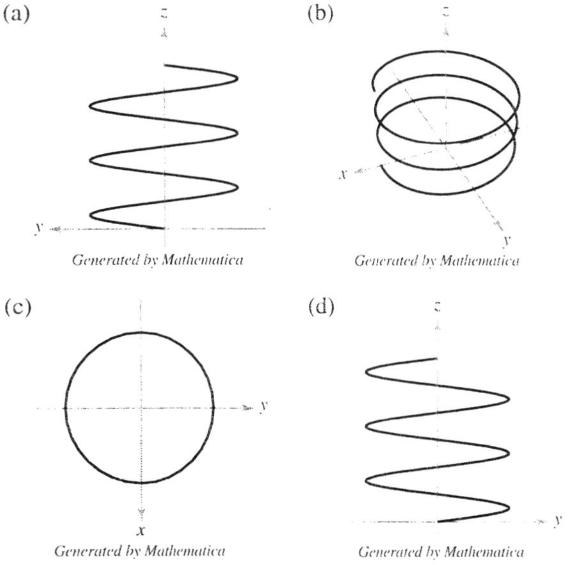

    (a)

    (b)

    Generated by Mathematica   *Generated by Mathematica*

    (c)

    (d)

    Generated by Mathematica   *Generated by Mathematica*

22. Sketch three graphs of the vector-valued function

$$r(t) = t\mathbf{i} + t\mathbf{j} + 2\mathbf{k}$$

as viewed from the points
(a) $(0, 0, 20)$ (b) $(10, 0, 0)$ (c) $(5, 5, 5)$.

In Exercises 23–38, sketch the curve represented by the vector-valued function and give the orientation of the curve.

23. $\mathbf{r}(t) = 3t\mathbf{i} + (t - 1)\mathbf{j}$
24. $\mathbf{r}(t) = (1 - t)\mathbf{i} + \sqrt{t}\mathbf{j}$
25. $\mathbf{r}(t) = t^3\mathbf{i} + t^2\mathbf{j}$
26. $\mathbf{r}(t) = (t^2 + t)\mathbf{i} + (t^2 - t)\mathbf{j}$
27. $\mathbf{r}(\theta) = \cos\theta\mathbf{i} + 3\sin\theta\mathbf{j}$
28. $\mathbf{r}(t) = 2\cos t\mathbf{i} + 2\sin t\mathbf{j}$
29. $\mathbf{r}(\theta) = 3\sec\theta\mathbf{i} + 2\tan\theta\mathbf{j}$
30. $\mathbf{r}(t) = 2\cos^3 t\mathbf{i} + 2\sin^3 t\mathbf{j}$
31. $\mathbf{r}(t) = (-t + 1)\mathbf{i} + (4t + 2)\mathbf{j} + (2t + 3)\mathbf{k}$
32. $\mathbf{r}(t) = t\mathbf{i} + (2t - 5)\mathbf{j} + 3t\mathbf{k}$
33. $\mathbf{r}(t) = 2\cos t\mathbf{i} + 2\sin t\mathbf{j} + t\mathbf{k}$
34. $\mathbf{r}(t) = 3\cos t\mathbf{i} + 4\sin t\mathbf{j} + \frac{t}{2}\mathbf{k}$
35. $\mathbf{r}(t) = 2\sin t\mathbf{i} + 2\cos t\mathbf{j} + e^{-t}\mathbf{k}$
36. $\mathbf{r}(t) = t^2\mathbf{i} + 2t\mathbf{j} + \frac{3}{2}t\mathbf{k}$
37. $\mathbf{r}(t) = \langle t, t^2, \frac{2}{3}t^3 \rangle$
38. $\mathbf{r}(t) = \langle \cos t + t\sin t, \sin t - t\cos t, t \rangle$

In Exercises 39–42, use a computer algebra system to graph the vector-valued function and identify the common curve.

39. $\mathbf{r}(t) = -\frac{1}{2}t^2\mathbf{i} + t\mathbf{j} - \frac{\sqrt{3}}{2}t^2\mathbf{k}$
40. $\mathbf{r}(t) = t\mathbf{i} - \frac{\sqrt{3}}{2}t^2\mathbf{j} + \frac{1}{2}t^2\mathbf{k}$
41. $\mathbf{r}(t) = \sin t\mathbf{i} + \left(\frac{\sqrt{3}}{2}\cos t - \frac{1}{2}t\right)\mathbf{j} + \left(\frac{1}{2}\cos t + \frac{\sqrt{3}}{2}\right)\mathbf{k}$
42. $\mathbf{r}(t) = -\sqrt{2}\sin t\mathbf{i} + 2\cos t\mathbf{j} + \sqrt{2}\sin t\mathbf{k}$

Think About It **In Exercises 43 and 44, use a computer algebra system to graph the vector-valued function r(t). For each u(t), make a conjecture about the transformation (if any) of the graph of r(t). Use a computer algebra system to check your conjecture.**

43. $\mathbf{r}(t) = 2\cos t\mathbf{i} + 2\sin t\mathbf{j} + \frac{1}{2}t\mathbf{k}$
(a) $\mathbf{u}(t) = 2(\cos t - 1)\mathbf{i} + 2\sin t\mathbf{j} + \frac{1}{2}t\mathbf{k}$
(b) $\mathbf{u}(t) = 2\cos t\mathbf{i} + 2\sin t\mathbf{j} + 2t\mathbf{k}$
(c) $\mathbf{u}(t) = 2\cos(-t)\mathbf{i} + 2\sin(-t)\mathbf{j} + \frac{1}{2}(-t)\mathbf{k}$
(d) $\mathbf{u}(t) = \frac{1}{2}t\mathbf{i} + 2\sin t\mathbf{j} + 2\cos t\mathbf{k}$
(e) $\mathbf{u}(t) = 6\cos t\mathbf{i} + 6\sin t\mathbf{j} + \frac{1}{2}t\mathbf{k}$

44. $\mathbf{r}(t) = t\mathbf{i} + t^2\mathbf{j} + \frac{1}{2}t^3\mathbf{k}$
(a) $\mathbf{u}(t) = t\mathbf{i} + (t^2 - 2)\mathbf{j} + \frac{1}{2}t^3\mathbf{k}$
(b) $\mathbf{u}(t) = t^2\mathbf{i} + t\mathbf{j} + \frac{1}{2}t^3\mathbf{k}$
(c) $\mathbf{u}(t) = t\mathbf{i} + t^2\mathbf{j} + \left(\frac{1}{2}t^3 + 4\right)\mathbf{k}$
(d) $\mathbf{u}(t) = t\mathbf{i} + t^2\mathbf{j} + \frac{1}{8}t^3\mathbf{k}$
(e) $\mathbf{u}(t) = (-t)\mathbf{i} + (-t)^2\mathbf{j} + \frac{1}{2}(-t)^3\mathbf{k}$

In Exercises 45–52, represent the plane curve by a vector-valued function. (There are many correct answers.)

45. $y = 4 - x$
46. $2x - 3y + 5 = 0$
47. $y = (x - 2)^2$
48. $y = 4 - x^2$
49. $x^2 + y^2 = 25$
50. $(x - 2)^2 + y^2 = 4$
51. $\frac{x^2}{16} - \frac{y^2}{4} = 1$
52. $\frac{x^2}{16} + \frac{y^2}{9} = 1$

53. A particle moves on a straight line path that passes through the points $(2, 3, 0)$ and $(0, 8, 8)$. Find a vector-valued function for the path. Use a computer algebra system to graph your function. (There are many correct answers.)

54. The outer edge of a playground slide is in the shape of a helix of radius 1.5 meters. The slide has a height of 2 meters and makes one complete revolution from top to bottom. Find a vector-valued function for the helix. Use a computer algebra system to graph your function. (There are many correct answers.)

In Exercises 55–58, find vector-valued functions forming the boundaries of the region in the figure. State the interval for the parameter of each function.

55.

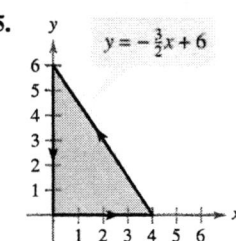

$y = -\frac{3}{2}x + 6$

56.

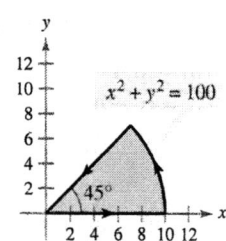

$x^2 + y^2 = 100$

$45°$

57.

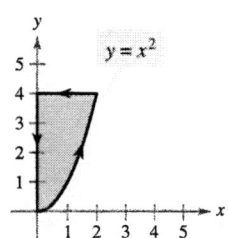

$y = x^2$

58.

$y = \sqrt{x}$

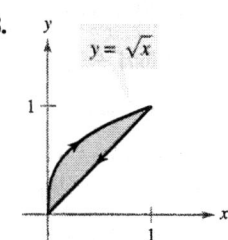

In Exercises 59–66, sketch the space curve represented by the intersection of the surfaces. Then represent the curve by a vector-valued function using the given parameter.

Surfaces	Parameter
59. $z = x^2 + y^2$, $x + y = 0$	$x = t$
60. $z = x^2 + y^2$, $z = 4$	$x = 2\cos t$
61. $x^2 + y^2 = 4$, $z = x^2$	$x = 2\sin t$
62. $4x^2 + 4y^2 + z^2 = 16$, $x = z^2$	$z = t$
63. $x^2 + y^2 + z^2 = 4$, $x + z = 2$	$x = 1 + \sin t$
64. $x^2 + y^2 + z^2 = 10$, $x + y = 4$	$x = 2 + \sin t$
65. $x^2 + z^2 = 4$, $y^2 + z^2 = 4$	$x = t$ (first octant)
66. $x^2 + y^2 + z^2 = 16$, $xy = 4$	$x = t$ (first octant)

67. Show that the vector-valued function

$\mathbf{r}(t) = t\mathbf{i} + 2t\cos t\mathbf{j} + 2t\sin t\mathbf{k}$

lies on the cone $4x^2 = y^2 + z^2$. Sketch the curve.

68. Show that the vector-valued function

$\mathbf{r}(t) = e^{-t}\cos t\mathbf{i} + e^{-t}\sin t\mathbf{j} + e^{-t}\mathbf{k}$

lies on the cone $z^2 = x^2 + y^2$. Sketch the curve.

In Exercises 69–74, evaluate the limit.

69. $\lim\limits_{t \to 2} \left(t\mathbf{i} + \dfrac{t^2 - 4}{t^2 - 2t}\mathbf{j} + \dfrac{1}{t}\mathbf{k} \right)$

70. $\lim\limits_{t \to 0} \left(e^t\mathbf{i} + \dfrac{\sin t}{t}\mathbf{j} + e^{-t}\mathbf{k} \right)$

71. $\lim\limits_{t \to 0} \left(t^2\mathbf{i} + 3t\mathbf{j} + \dfrac{1 - \cos t}{t}\mathbf{k} \right)$

72. $\lim\limits_{t \to 1} \left(\sqrt{t}\,\mathbf{i} + \dfrac{\ln t}{t^2 - 1}\mathbf{j} + 2t^2\mathbf{k} \right)$

73. $\lim\limits_{t \to 0} \left(\dfrac{1}{t}\mathbf{i} + \cos t\mathbf{j} + \sin t\mathbf{k} \right)$

74. $\lim\limits_{t \to \infty} \left(e^{-t}\mathbf{i} + \dfrac{1}{t}\mathbf{j} + \dfrac{t}{t^2 + 1}\mathbf{k} \right)$

In Exercises 75–80, determine the interval(s) on which the vector-valued function is continuous.

75. $\mathbf{r}(t) = t\mathbf{i} + \dfrac{1}{t}\mathbf{j}$

76. $\mathbf{r}(t) = \sqrt{t}\,\mathbf{i} + \sqrt{t - 1}\,\mathbf{j}$

77. $\mathbf{r}(t) = t\mathbf{i} + \arcsin t\mathbf{j} + (t - 1)\mathbf{k}$

78. $\mathbf{r}(t) = 2e^{-t}\mathbf{i} + e^{-t}\mathbf{j} + \ln(t - 1)\mathbf{k}$

79. $\mathbf{r}(t) = \langle e^{-t}, t^2, \tan t \rangle$

80. $\mathbf{r}(t) = \langle 8, \sqrt{t}, \sqrt[3]{t} \rangle$

Getting at the Concept

81. State the definition of a vector-valued function in the plane and in space.

82. If $\mathbf{r}(t)$ is a vector-valued function, is the graph of the vector-valued function $\mathbf{u}(t) = \mathbf{r}(t - 2)$ a horizontal translation of the graph of $\mathbf{r}(t)$? Explain.

83. Consider the vector-valued function

$\mathbf{r}(t) = t^2\mathbf{i} + (t - 3)\mathbf{j} + t\mathbf{k}$.

Write a vector-valued function $\mathbf{s}(t)$ that is the specified transformation of $\mathbf{r}$.

(a) A vertical translation 3 units upward

(b) A horizontal translation 2 units in the direction of the negative x-axis

(c) A horizontal translation 5 units in the direction of the positive y-axis

84. State the definition of continuity of a vector-valued function. Give an example of a vector-valued function that is defined but not continuous at $t = 2$.

85. Let $\mathbf{r}(t)$ and $\mathbf{u}(t)$ be vector-valued functions whose limits exist as $t \to c$. Prove that

$\lim\limits_{t \to c} [\mathbf{r}(t) \times \mathbf{u}(t)] = \lim\limits_{t \to c} \mathbf{r}(t) \times \lim\limits_{t \to c} \mathbf{u}(t)$.

86. Let $\mathbf{r}(t)$ and $\mathbf{u}(t)$ be vector-valued functions whose limits exist as $t \to c$. Prove that

$\lim\limits_{t \to c} [\mathbf{r}(t) \cdot \mathbf{u}(t)] = \lim\limits_{t \to c} \mathbf{r}(t) \cdot \lim\limits_{t \to c} \mathbf{u}(t)$.

87. Prove that if $\mathbf{r}$ is a vector-valued function that is continuous at c, then $\|\mathbf{r}\|$ is continuous at c.

88. Verify that the converse of Exercise 87 is *not* true by finding a vector-valued function $\mathbf{r}$ such that $\|\mathbf{r}\|$ is continuous at c but $\mathbf{r}$ is not continuous at c.

True or False? **In Exercises 89 and 90, determine whether the statement is true or false. If it is false, explain why or give an example that shows it is false.**

89. If f, g, and h are first-degree polynomial functions, then the curve given by $x = f(t)$, $y = g(t)$, and $z = h(t)$ is a line.

90. If the curve given by $x = f(t)$, $y = g(t)$, and $z = h(t)$ is a line, then f, g, and h are first-degree polynomial functions of t.

SECTION PROJECT: WITCH OF AGNESI

On page 195 in Section 3.5, you studied a famous curve called the **Witch of Agnesi**. In this project you will take a closer look at this function.

Consider a circle of radius a centered on the y-axis at $(0, a)$. Let A be a point on the horizontal line $y = 2a$, let O be the origin, and let B be the point where the segment OA intersects the circle. A point P is on the Witch of Agnesi if P lies on the horizontal line through B and on the vertical line through A.

(a) Show that the point A is traced out by the vector-valued function

$\mathbf{r}_A(\theta) = 2a \cot \theta \mathbf{i} + 2a\mathbf{j}, \quad 0 < \theta < \pi$

where θ is the angle OA makes with the positive x-axis.

(b) Show that the point B is traced out by the vector-valued function

$\mathbf{r}_B(\theta) = a \sin 2\theta \mathbf{i} + a(1 - \cos 2\theta)\mathbf{j}, \quad 0 < \theta < \pi$.

(c) Combine the results in parts (a) and (b) to find the vector-valued function $\mathbf{r}(\theta)$ for the Witch of Agnesi. Use a graphing utility to graph this curve for $a = 1$.

(d) Describe the following limits.

$\lim\limits_{\theta \to 0^+} \mathbf{r}(\theta) \qquad \lim\limits_{\theta \to \pi^-} \mathbf{r}(\theta)$

(e) Eliminate the parameter θ and determine the rectangular equation of the Witch of Agnesi. Use a graphing utility to graph this function for $a = 1$ and compare your graph with that obtained in part (c).

Section 11.2 Differentiation and Integration of Vector-Valued Functions

- Differentiate a vector-valued function.
- Integrate a vector-valued function.

Differentiation of Vector-Valued Functions

In Sections 11.3–11.5, you will study several important applications involving the calculus of vector-valued functions. In preparation for that study, this section is devoted to the mechanics of differentiation and integration of vector-valued functions.

The definition of the derivative of a vector-valued function parallels that given for real-valued functions.

Definition of the Derivative of a Vector-Valued Function

The **derivative of a vector-valued function r** is defined by

$$\mathbf{r}'(t) = \lim_{\Delta t \to 0} \frac{\mathbf{r}(t + \Delta t) - \mathbf{r}(t)}{\Delta t}$$

for all t for which the limit exists. If $\mathbf{r}'(c)$ exists, then $\mathbf{r}$ is **differentiable at c.** If $\mathbf{r}'(c)$ exists for all c in an open interval I, then $\mathbf{r}$ is **differentiable on the interval I.** Differentiability of vector-valued functions can be extended to closed intervals by considering one-sided limits.

NOTE   In addition to $\mathbf{r}'(t)$, other notations for the derivative of a vector-valued function are

$$D_t[\mathbf{r}(t)], \quad \frac{d}{dt}[\mathbf{r}(t)], \quad \text{and} \quad \frac{d\mathbf{r}}{dt}.$$

Differentiation of vector-valued functions can be done on a *component-by-component basis*. To see why this is true, consider the function given by

$$\mathbf{r}(t) = f(t)\mathbf{i} + g(t)\mathbf{j}.$$

Applying the definition of the derivative produces the following.

$$\begin{aligned}
\mathbf{r}'(t) &= \lim_{\Delta t \to 0} \frac{\mathbf{r}(t + \Delta t) - \mathbf{r}(t)}{\Delta t} \\
&= \lim_{\Delta t \to 0} \frac{f(t + \Delta t)\mathbf{i} + g(t + \Delta t)\mathbf{j} - f(t)\mathbf{i} - g(t)\mathbf{j}}{\Delta t} \\
&= \lim_{\Delta t \to 0} \left\{ \left[\frac{f(t + \Delta t) - f(t)}{\Delta t} \right]\mathbf{i} + \left[\frac{g(t + \Delta t) - g(t)}{\Delta t} \right]\mathbf{j} \right\} \\
&= \left\{ \lim_{\Delta t \to 0} \left[\frac{f(t + \Delta t) - f(t)}{\Delta t} \right] \right\}\mathbf{i} + \left\{ \lim_{\Delta t \to 0} \left[\frac{g(t + \Delta t) - g(t)}{\Delta t} \right] \right\}\mathbf{j} \\
&= f'(t)\mathbf{i} + g'(t)\mathbf{j}
\end{aligned}$$

This important result is listed in the following theorem. Note that the derivative of the vector-valued function $\mathbf{r}$ is itself a vector-valued function. You can see from Figure 11.8 that $\mathbf{r}'(t)$ is a vector tangent to the curve given by $\mathbf{r}(t)$ and pointing in the direction of increasing t-values.

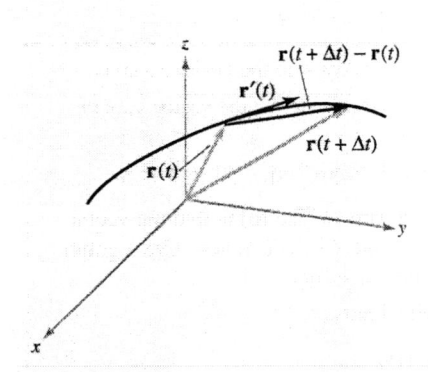

Figure 11.8

SECTION 11.2 Differentiation and Integration of Vector-Valued Functions

> **THEOREM 11.1 Differentiation of Vector-Valued Functions**
>
> 1. If $\mathbf{r}(t) = f(t)\mathbf{i} + g(t)\mathbf{j}$, where f and g are differentiable functions of t, then
> $$\mathbf{r}'(t) = f'(t)\mathbf{i} + g'(t)\mathbf{j}. \qquad \text{Plane}$$
> 2. If $\mathbf{r}(t) = f(t)\mathbf{i} + g(t)\mathbf{j} + h(t)\mathbf{k}$, where f, g, and h are differentiable functions of t, then
> $$\mathbf{r}'(t) = f'(t)\mathbf{i} + g'(t)\mathbf{j} + h'(t)\mathbf{k}. \qquad \text{Space}$$

Example 1 Differentiation of Vector-Valued Functions

Find the derivative of each vector-valued function.

a. $\mathbf{r}(t) = t^2\mathbf{i} - 4\mathbf{j}$ **b.** $\mathbf{r}(t) = \dfrac{1}{t}\mathbf{i} + \ln t\,\mathbf{j} + e^{2t}\mathbf{k}$

Solution Differentiating on a component-by-component basis produces the following.

a. $\mathbf{r}'(t) = 2t\mathbf{i} - 0\mathbf{j}$
$\qquad\quad = 2t\mathbf{i}$ \hfill Derivative

b. $\mathbf{r}'(t) = -\dfrac{1}{t^2}\mathbf{i} + \dfrac{1}{t}\mathbf{j} + 2e^{2t}\mathbf{k}$ \hfill Derivative

Higher-order derivatives of vector-valued functions are obtained by successive differentiation of each component function.

Example 2 Higher-Order Differentiation

For the vector-valued function given by $\mathbf{r}(t) = \cos t\,\mathbf{i} + \sin t\,\mathbf{j} + 2t\mathbf{k}$, find each of the following.

a. $\mathbf{r}'(t)$ **b.** $\mathbf{r}''(t)$ **c.** $\mathbf{r}'(t) \cdot \mathbf{r}''(t)$ **d.** $\mathbf{r}'(t) \times \mathbf{r}''(t)$

Solution

a. $\mathbf{r}'(t) = -\sin t\,\mathbf{i} + \cos t\,\mathbf{j} + 2\mathbf{k}$ \hfill First derivative

b. $\mathbf{r}''(t) = -\cos t\,\mathbf{i} - \sin t\,\mathbf{j} + 0\mathbf{k}$
$\qquad\quad\; = -\cos t\,\mathbf{i} - \sin t\,\mathbf{j}$ \hfill Second derivative

c. $\mathbf{r}'(t) \cdot \mathbf{r}''(t) = \sin t \cos t - \sin t \cos t = 0$ \hfill Dot product

d. $\mathbf{r}'(t) \times \mathbf{r}''(t) = \begin{vmatrix} \mathbf{i} & \mathbf{j} & \mathbf{k} \\ -\sin t & \cos t & 2 \\ -\cos t & -\sin t & 0 \end{vmatrix}$ \hfill Cross product

$\qquad\qquad\quad = \begin{vmatrix} \cos t & 2 \\ -\sin t & 0 \end{vmatrix}\mathbf{i} - \begin{vmatrix} -\sin t & 2 \\ -\cos t & 0 \end{vmatrix}\mathbf{j} + \begin{vmatrix} -\sin t & \cos t \\ -\cos t & -\sin t \end{vmatrix}\mathbf{k}$

$\qquad\qquad\quad = 2\sin t\,\mathbf{i} - 2\cos t\,\mathbf{j} + \mathbf{k}$

Note that the dot product in part (c) is a *real-valued* function, not a vector-valued function.

The parametrization of the curve represented by the vector-valued function

$$\mathbf{r}(t) = f(t)\mathbf{i} + g(t)\mathbf{j} + h(t)\mathbf{k}$$

is **smooth on an open interval** I if f', g', and h' are continuous on I and $\mathbf{r}'(t) \ne \mathbf{0}$ for any value of t in the interval I.

Example 3   Finding Intervals on Which a Curve Is Smooth

Find the intervals on which the epicycloid C given by

$$\mathbf{r}(t) = (5\cos t - \cos 5t)\mathbf{i} + (5\sin t - \sin 5t)\mathbf{j}, \quad 0 \le t \le 2\pi$$

is smooth.

Solution   The derivative of $\mathbf{r}$ is

$$\mathbf{r}'(t) = (-5\sin t + 5\sin 5t)\mathbf{i} + (5\cos t - 5\cos 5t)\mathbf{j}.$$

In the interval $[0, 2\pi]$, the only values of t for which

$$\mathbf{r}'(t) = 0\mathbf{i} + 0\mathbf{j}$$

are $t = 0, \pi/2, \pi, 3\pi/2,$ and 2π. Therefore, you can conclude that C is smooth in the intervals

$$\left(0, \frac{\pi}{2}\right), \left(\frac{\pi}{2}, \pi\right), \left(\pi, \frac{3\pi}{2}\right), \text{ and } \left(\frac{3\pi}{2}, 2\pi\right)$$

as shown in Figure 11.9.

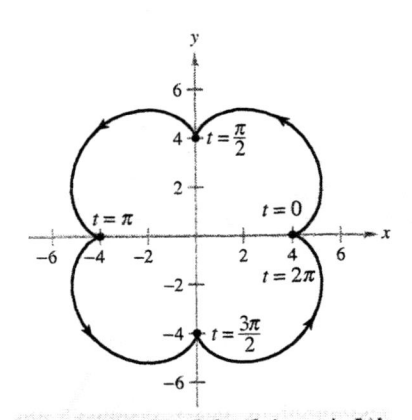

$\mathbf{r}(t) = (5\cos t - \cos 5t)\mathbf{i} + (5\sin t - \sin 5t)\mathbf{j}$

The epicycloid is not smooth at the points where it intersects the axes.
Figure 11.9

NOTE   In Figure 11.9, note that the curve is not smooth at points at which the curve makes an abrupt change in direction. Such points are called **cusps** or **nodes**.

Most of the differentiation rules in Chapter 2 have counterparts for vector-valued functions, and several are listed in the following theorem. Note that the theorem contains three versions of "product rules." Property 3 gives the derivative of the product of a real-valued function f and a vector-valued function $\mathbf{r}$, Property 4 gives the derivative of the dot product of two vector-valued functions, and Property 5 gives the derivative of the cross product of two vector-valued functions (in space). (Property 5 applies only to three-dimensional vector-valued functions, because the cross product is not defined for two-dimensional vectors.)

THEOREM 11.2   Properties of the Derivative

Let $\mathbf{r}$ and $\mathbf{u}$ be differentiable vector-valued functions of t, let f be a differentiable real-valued function of t, and let c be a scalar.

1. $D_t[c\mathbf{r}(t)] = c\mathbf{r}'(t)$
2. $D_t[\mathbf{r}(t) \pm \mathbf{u}(t)] = \mathbf{r}'(t) \pm \mathbf{u}'(t)$
3. $D_t[f(t)\mathbf{r}(t)] = f(t)\mathbf{r}'(t) + f'(t)\mathbf{r}(t)$
4. $D_t[\mathbf{r}(t) \cdot \mathbf{u}(t)] = \mathbf{r}(t) \cdot \mathbf{u}'(t) + \mathbf{r}'(t) \cdot \mathbf{u}(t)$
5. $D_t[\mathbf{r}(t) \times \mathbf{u}(t)] = \mathbf{r}(t) \times \mathbf{u}'(t) + \mathbf{r}'(t) \times \mathbf{u}(t)$
6. $D_t[\mathbf{r}(f(t))] = \mathbf{r}'(f(t))f'(t)$
7. If $\mathbf{r}(t) \cdot \mathbf{r}(t) = c$, then $\mathbf{r}(t) \cdot \mathbf{r}'(t) = 0$.

EXPLORATION

Let $\mathbf{r}(t) = \cos t\mathbf{i} + \sin t\mathbf{j}$. Sketch the graph of $\mathbf{r}(t)$. Explain why the graph is a circle of radius 1 centered at the origin. Calculate $\mathbf{r}(\pi/4)$ and $\mathbf{r}'(\pi/4)$. Position the vector $\mathbf{r}'(\pi/4)$ so that its initial point is at the terminal point of $\mathbf{r}(\pi/4)$. What do you observe? Show that $\mathbf{r}(t) \cdot \mathbf{r}(t)$ is constant and that $\mathbf{r}(t) \cdot \mathbf{r}'(t) = 0$ for all t. How does this example relate to Property 7 of Theorem 11.2?

Proof To prove Property 4, let

$$\mathbf{r}(t) = f_1(t)\mathbf{i} + g_1(t)\mathbf{j} \quad \text{and} \quad \mathbf{u}(t) = f_2(t)\mathbf{i} + g_2(t)\mathbf{j}$$

where f_1, f_2, g_1, and g_2 are differentiable functions of t. Then,

$$\mathbf{r}(t) \cdot \mathbf{u}(t) = f_1(t)f_2(t) + g_1(t)g_2(t)$$

and it follows that

$$\begin{aligned} D_t[\mathbf{r}(t) \cdot \mathbf{u}(t)] &= f_1(t)f_2'(t) + f_1'(t)f_2(t) + g_1(t)g_2'(t) + g_1'(t)g_2(t) \\ &= [f_1(t)f_2'(t) + g_1(t)g_2'(t)] + [f_1'(t)f_2(t) + g_1'(t)g_2(t)] \\ &= \mathbf{r}(t) \cdot \mathbf{u}'(t) + \mathbf{r}'(t) \cdot \mathbf{u}(t). \end{aligned}$$

Proofs of the other properties are left as exercises (see Exercises 65–69 and Exercise 72).

Example 4 Using Properties of the Derivative

For the vector-valued functions given by

$$\mathbf{r}(t) = \frac{1}{t}\mathbf{i} - \mathbf{j} + \ln t\mathbf{k} \quad \text{and} \quad \mathbf{u}(t) = t^2\mathbf{i} - 2t\mathbf{j} + \mathbf{k}$$

find

a. $D_t[\mathbf{r}(t) \cdot \mathbf{u}(t)]$ and   **b.** $D_t[\mathbf{u}(t) \times \mathbf{u}'(t)]$.

Solution

a. Because $\mathbf{r}'(t) = -\frac{1}{t^2}\mathbf{i} + \frac{1}{t}\mathbf{k}$ and $\mathbf{u}'(t) = 2t\mathbf{i} - 2\mathbf{j}$, you have

$$\begin{aligned} D_t[\mathbf{r}(t) \cdot \mathbf{u}(t)] &= \mathbf{r}(t) \cdot \mathbf{u}'(t) + \mathbf{r}'(t) \cdot \mathbf{u}(t) \\ &= \left(\frac{1}{t}\mathbf{i} - \mathbf{j} + \ln t\mathbf{k}\right) \cdot (2t\mathbf{i} - 2\mathbf{j}) \\ &\quad + \left(-\frac{1}{t^2}\mathbf{i} + \frac{1}{t}\mathbf{k}\right) \cdot (t^2\mathbf{i} - 2t\mathbf{j} + \mathbf{k}) \\ &= 2 + 2 + (-1) + \frac{1}{t} \\ &= 3 + \frac{1}{t}. \end{aligned}$$

b. Because $\mathbf{u}'(t) = 2t\mathbf{i} - 2\mathbf{j}$ and $\mathbf{u}''(t) = 2\mathbf{i}$, you have

$$\begin{aligned} D_t[\mathbf{u}(t) \times \mathbf{u}'(t)] &= [\mathbf{u}(t) \times \mathbf{u}''(t)] + [\mathbf{u}'(t) \times \mathbf{u}'(t)] \\ &= \begin{vmatrix} \mathbf{i} & \mathbf{j} & \mathbf{k} \\ t^2 & -2t & 1 \\ 2 & 0 & 0 \end{vmatrix} + \mathbf{0} \\ &= \begin{vmatrix} -2t & 1 \\ 0 & 0 \end{vmatrix}\mathbf{i} - \begin{vmatrix} t^2 & 1 \\ 2 & 0 \end{vmatrix}\mathbf{j} + \begin{vmatrix} t^2 & -2t \\ 2 & 0 \end{vmatrix}\mathbf{k} \\ &= 0\mathbf{i} - (-2)\mathbf{j} + 4t\mathbf{k} \\ &= 2\mathbf{j} + 4t\mathbf{k}. \end{aligned}$$

NOTE Try reworking parts (a) and (b) in Example 4 by first forming the dot and cross products and then differentiating to see that you obtain the same results.

Integration of Vector-Valued Functions

The following definition is a rational consequence of the definition of the derivative of a vector-valued function.

Definition of Integration of Vector-Valued Functions

1. If $\mathbf{r}(t) = f(t)\mathbf{i} + g(t)\mathbf{j}$, where f and g are continuous on $[a, b]$, then the **indefinite integral (antiderivative)** of $\mathbf{r}$ is

$$\int \mathbf{r}(t)\, dt = \left[\int f(t)\, dt\right]\mathbf{i} + \left[\int g(t)\, dt\right]\mathbf{j} \qquad \text{Plane}$$

and its **definite integral** over the interval $a \leq t \leq b$ is

$$\int_a^b \mathbf{r}(t)\, dt = \left[\int_a^b f(t)\, dt\right]\mathbf{i} + \left[\int_a^b g(t)\, dt\right]\mathbf{j}.$$

2. If $\mathbf{r}(t) = f(t)\mathbf{i} + g(t)\mathbf{j} + h(t)\mathbf{k}$, where f, g, and h are continuous on $[a, b]$, then the **indefinite integral (antiderivative)** of $\mathbf{r}$ is

$$\int \mathbf{r}(t)\, dt = \left[\int f(t)\, dt\right]\mathbf{i} + \left[\int g(t)\, dt\right]\mathbf{j} + \left[\int h(t)\, dt\right]\mathbf{k} \qquad \text{Space}$$

and its **definite integral** over the interval $a \leq t \leq b$ is

$$\int_a^b \mathbf{r}(t)\, dt = \left[\int_a^b f(t)\, dt\right]\mathbf{i} + \left[\int_a^b g(t)\, dt\right]\mathbf{j} + \left[\int_a^b h(t)\, dt\right]\mathbf{k}.$$

The antiderivative of a vector-valued function is a family of vector-valued functions all differing by a constant vector $\mathbf{C}$. For instance, if $\mathbf{r}(t)$ is a three-dimensional vector-valued function, then for the indefinite integral $\int \mathbf{r}(t)\, dt$, you obtain three constants of integration

$$\int f(t)\, dt = F(t) + C_1, \qquad \int g(t)\, dt = G(t) + C_2, \qquad \int h(t)\, dt = H(t) + C_3$$

where $F'(t) = f(t)$, $G'(t) = g(t)$, and $H'(t) = h(t)$. These three *scalar* constants produce one *vector* constant of integration,

$$\int \mathbf{r}(t)\, dt = [F(t) + C_1]\mathbf{i} + [G(t) + C_2]\mathbf{j} + [H(t) + C_3]\mathbf{k}$$
$$= [F(t)\mathbf{i} + G(t)\mathbf{j} + H(t)\mathbf{k}] + [C_1\mathbf{i} + C_2\mathbf{j} + C_3\mathbf{k}]$$
$$= \mathbf{R}(t) + \mathbf{C}$$

where $\mathbf{R}'(t) = \mathbf{r}(t)$.

Example 5 Integrating a Vector-Valued Function

Evaluate the indefinite integral

$$\int (t\mathbf{i} + 3\mathbf{j})\, dt.$$

Solution Integrating on a component-by-component basis produces

$$\int (t\mathbf{i} + 3\mathbf{j})\, dt = \frac{t^2}{2}\mathbf{i} + 3t\mathbf{j} + \mathbf{C}.$$

Example 6 shows how to evaluate the definite integral of a vector-valued function.

Example 6 Definite Integral of a Vector-Valued Function

Evaluate the integral

$$\int_0^1 \mathbf{r}(t)\, dt = \int_0^1 \left(\sqrt[3]{t}\,\mathbf{i} + \frac{1}{t+1}\mathbf{j} + e^{-t}\mathbf{k} \right) dt.$$

Solution

$$\int_0^1 \mathbf{r}(t)\, dt = \left(\int_0^1 t^{1/3}\, dt \right)\mathbf{i} + \left(\int_0^1 \frac{1}{t+1}\, dt \right)\mathbf{j} + \left(\int_0^1 e^{-t}\, dt \right)\mathbf{k}$$

$$= \left[\left(\frac{3}{4}\right) t^{4/3} \right]_0^1 \mathbf{i} + \Big[\ln|t+1| \Big]_0^1 \mathbf{j} + \Big[-e^{-t} \Big]_0^1 \mathbf{k}$$

$$= \frac{3}{4}\mathbf{i} + (\ln 2)\mathbf{j} + \left(1 - \frac{1}{e}\right)\mathbf{k}$$

As with real-valued functions, you can narrow the family of antiderivatives of a vector-valued function $\mathbf{r}'$ down to a single antiderivative by imposing an initial condition on the vector-valued function $\mathbf{r}$. This is demonstrated in the next example.

Example 7 The Antiderivative of a Vector-Valued Function

Find the antiderivative of

$$\mathbf{r}'(t) = \cos 2t\,\mathbf{i} - 2\sin t\,\mathbf{j} + \frac{1}{1+t^2}\mathbf{k}$$

that satisfies the initial condition $\mathbf{r}(0) = 3\mathbf{i} - 2\mathbf{j} + \mathbf{k}$.

Solution

$$\mathbf{r}(t) = \int \mathbf{r}'(t)\, dt$$

$$= \left(\int \cos 2t\, dt \right)\mathbf{i} + \left(\int -2\sin t\, dt \right)\mathbf{j} + \left(\int \frac{1}{1+t^2}\, dt \right)\mathbf{k}$$

$$= \left(\frac{1}{2}\sin 2t + C_1 \right)\mathbf{i} + (2\cos t + C_2)\mathbf{j} + (\arctan t + C_3)\mathbf{k}$$

Letting $t = 0$ and using the fact that $\mathbf{r}(0) = 3\mathbf{i} - 2\mathbf{j} + \mathbf{k}$, you have

$$\mathbf{r}(0) = (0 + C_1)\mathbf{i} + (2 + C_2)\mathbf{j} + (0 + C_3)\mathbf{k}$$
$$= 3\mathbf{i} + (-2)\mathbf{j} + \mathbf{k}.$$

Equating corresponding components produces

$$C_1 = 3, \quad 2 + C_2 = -2, \quad \text{and} \quad C_3 = 1.$$

So, the antiderivative that satisfies the given initial condition is

$$\mathbf{r}(t) = \left(\frac{1}{2}\sin 2t + 3 \right)\mathbf{i} + (2\cos t - 4)\mathbf{j} + (\arctan t + 1)\mathbf{k}.$$

EXERCISES FOR SECTION 11.2

In Exercises 1–4, (a) sketch the plane curve represented by the vector-valued function, and (b) sketch the vectors $\mathbf{r}(t_0)$ and $\mathbf{r}'(t_0)$ for the indicated value of t_0. Position the vectors such that the initial point of $\mathbf{r}(t_0)$ is at the origin and the initial point of $\mathbf{r}'(t_0)$ is at the terminal point of $\mathbf{r}(t_0)$. What is the relationship between $\mathbf{r}'(t_0)$ and the curve?

1. $\mathbf{r}(t) = t^2\mathbf{i} + t\mathbf{j}$ $t_0 = 2$
2. $\mathbf{r}(t) = t\mathbf{i} + t^3\mathbf{j}$ $t_0 = 1$
3. $\mathbf{r}(t) = \cos t\mathbf{i} + \sin t\mathbf{j}$ $t_0 = \dfrac{\pi}{2}$
4. $\mathbf{r}(t) = t^2\mathbf{i} + \dfrac{1}{t}\mathbf{j}$ $t_0 = 2$

5. *Investigation*   Consider the vector-valued function

$$\mathbf{r}(t) = t\mathbf{i} + t^2\mathbf{j}.$$

  (a) Sketch the graph of $\mathbf{r}(t)$. Use a graphing utility to verify your graph.

  (b) Sketch the vectors $\mathbf{r}(1/4)$, $\mathbf{r}(1/2)$, and $\mathbf{r}(1/2) - \mathbf{r}(1/4)$ on the graph in part (a).

  (c) Compare the vector $\mathbf{r}'(1/4)$ with the vector

$$\frac{\mathbf{r}(1/2) - \mathbf{r}(1/4)}{1/2 - 1/4}.$$

6. *Investigation*   Consider the vector-valued function

$$\mathbf{r}(t) = t\mathbf{i} + (4 - t^2)\mathbf{j}.$$

  (a) Sketch the graph of $\mathbf{r}(t)$. Use a graphing utility to verify your graph.

  (b) Sketch the vectors $\mathbf{r}(1)$, $\mathbf{r}(1.25)$, and $\mathbf{r}(1.25) - \mathbf{r}(1)$ on the graph in part (a).

  (c) Compare the vector $\mathbf{r}'(1)$ with the vector

$$\frac{\mathbf{r}(1.25) - \mathbf{r}(1)}{1.25 - 1}.$$

In Exercises 7 and 8, (a) sketch the space curve represented by the vector-valued function, and (b) sketch the vectors $\mathbf{r}(t_0)$ and $\mathbf{r}'(t_0)$ for the indicated value of t_0.

7. $\mathbf{r}(t) = 2\cos t\mathbf{i} + 2\sin t\mathbf{j} + t\mathbf{k}$ $t_0 = \dfrac{3\pi}{2}$
8. $\mathbf{r}(t) = t\mathbf{i} + t^2\mathbf{j} + \tfrac{3}{2}\mathbf{k}$ $t_0 = 2$

In Exercises 9–16, find $\mathbf{r}'(t)$.

9. $\mathbf{r}(t) = 6t\mathbf{i} - 7t^2\mathbf{j} + t^3\mathbf{k}$
10. $\mathbf{r}(t) = \dfrac{1}{t}\mathbf{i} + 16t\mathbf{j} + \dfrac{t^2}{2}\mathbf{k}$
11. $\mathbf{r}(t) = a\cos^3 t\mathbf{i} + a\sin^3 t\mathbf{j} + \mathbf{k}$
12. $\mathbf{r}(t) = 4\sqrt{t}\mathbf{i} + t^2\sqrt{t}\mathbf{j} + \ln t^2\mathbf{k}$
13. $\mathbf{r}(t) = e^{-t}\mathbf{i} + 4\mathbf{j}$
14. $\mathbf{r}(t) = \langle \sin t - t\cos t, \cos t + t\sin t, t^2 \rangle$
15. $\mathbf{r}(t) = \langle t\sin t, t\cos t, t \rangle$
16. $\mathbf{r}(t) = \langle \arcsin t, \arccos t, 0 \rangle$

In Exercises 17–24, find (a) $\mathbf{r}''(t)$ and (b) $\mathbf{r}'(t) \cdot \mathbf{r}''(t)$.

17. $\mathbf{r}(t) = t^3\mathbf{i} + \tfrac{1}{2}t^2\mathbf{j}$
18. $\mathbf{r}(t) = (t^2 + t)\mathbf{i} + (t^2 - t)\mathbf{j}$
19. $\mathbf{r}(t) = 4\cos t\mathbf{i} + 4\sin t\mathbf{j}$
20. $\mathbf{r}(t) = 8\cos t\mathbf{i} + 3\sin t\mathbf{j}$
21. $\mathbf{r}(t) = \tfrac{1}{2}t^2\mathbf{i} - t\mathbf{j} + \tfrac{1}{6}t^3\mathbf{k}$
22. $\mathbf{r}(t) = t\mathbf{i} + (2t + 3)\mathbf{j} + (3t - 5)\mathbf{k}$
23. $\mathbf{r}(t) = \langle \cos t + t\sin t, \sin t - t\cos t, t \rangle$
24. $\mathbf{r}(t) = \langle e^{-t}, t^2, \tan t \rangle$

In Exercises 25 and 26, a vector-valued function and its graph are given. The graph also shows the unit vectors $\mathbf{r}'(t_0)/\|\mathbf{r}'(t_0)\|$ and $\mathbf{r}''(t_0)/\|\mathbf{r}''(t_0)\|$. Find these two unit vectors and identify them on the graph.

25. $\mathbf{r}(t) = \cos(\pi t)\mathbf{i} + \sin(\pi t)\mathbf{j} + t^2\mathbf{k}$ $t_0 = -\tfrac{1}{4}$
26. $\mathbf{r}(t) = t\mathbf{i} + t^2\mathbf{j} + e^{0.75t}\mathbf{k}$ $t_0 = \tfrac{1}{4}$

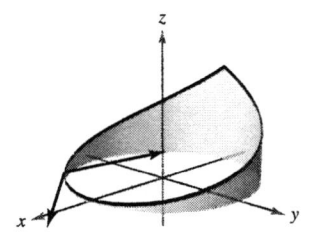

Figure for 25    Figure for 26

In Exercises 27–36, find the open interval(s) on which the curve given by the vector-valued function is smooth.

27. $\mathbf{r}(t) = t^2\mathbf{i} + t^3\mathbf{j}$
28. $\mathbf{r}(t) = \dfrac{1}{t-1}\mathbf{i} + 3t\mathbf{j}$
29. $\mathbf{r}(\theta) = 2\cos^3\theta\mathbf{i} + 3\sin^3\theta\mathbf{j}$
30. $\mathbf{r}(\theta) = (\theta + \sin\theta)\mathbf{i} + (1 - \cos\theta)\mathbf{j}$
31. $\mathbf{r}(\theta) = (\theta - 2\sin\theta)\mathbf{i} + (1 - 2\cos\theta)\mathbf{j}$
32. $\mathbf{r}(t) = \dfrac{2t}{8+t^3}\mathbf{i} + \dfrac{2t^2}{8+t^3}\mathbf{j}$
33. $\mathbf{r}(t) = (t-1)\mathbf{i} + \dfrac{1}{t}\mathbf{j} - t^2\mathbf{k}$
34. $\mathbf{r}(t) = e^t\mathbf{i} - e^{-t}\mathbf{j} + 3t\mathbf{k}$
35. $\mathbf{r}(t) = t\mathbf{i} - 3t\mathbf{j} + \tan t\mathbf{k}$
36. $\mathbf{r}(t) = \sqrt{t}\mathbf{i} + (t^2 - 1)\mathbf{j} + \tfrac{1}{4}t\mathbf{k}$

In Exercises 37 and 38, use the properties of the derivative to find the following.

(a) $\mathbf{r}'(t)$ (b) $\mathbf{r}''(t)$
(c) $D_t[\mathbf{r}(t) \cdot \mathbf{u}(t)]$ (d) $D_t[3\mathbf{r}(t) - \mathbf{u}(t)]$
(e) $D_t[\mathbf{r}(t) \times \mathbf{u}(t)]$ (f) $D_t[\|\mathbf{r}(t)\|]$, $t > 0$

37. $\mathbf{r}(t) = t\mathbf{i} + 3t\mathbf{j} + t^2\mathbf{k}$, $\mathbf{u}(t) = 4t\mathbf{i} + t^2\mathbf{j} + t^3\mathbf{k}$
38. $\mathbf{r}(t) = t\mathbf{i} + 2\sin t\mathbf{j} + 2\cos t\mathbf{k}$,
    $\mathbf{u}(t) = \frac{1}{t}\mathbf{i} + 2\sin t\mathbf{j} + 2\cos t\mathbf{k}$

In Exercises 39 and 40, find the angle θ between $\mathbf{r}(t)$ and $\mathbf{r}'(t)$ as a function of t. Use a graphing utility to graph $\theta(t)$. Use the graph to find any extrema of the function. Find any values of t at which the vectors are orthogonal.

39. $\mathbf{r}(t) = 3\sin t\mathbf{i} + 4\cos t\mathbf{j}$
40. $\mathbf{r}(t) = t^2\mathbf{i} + t\mathbf{j}$

In Exercises 41 and 42, use the definition of the derivative to find $\mathbf{r}'(t)$.

41. $\mathbf{r}(t) = (3t + 2)\mathbf{i} + (1 - t^2)\mathbf{j}$
42. $\mathbf{r}(t) = \sqrt{t}\mathbf{i} + \frac{3}{t}\mathbf{j} - 2t\mathbf{k}$

In Exercises 43–50, evaluate the indefinite integral.

43. $\int (2t\mathbf{i} + \mathbf{j} + \mathbf{k})\, dt$
44. $\int (4t^3\mathbf{i} + 6t\mathbf{j} - 4\sqrt{t}\mathbf{k})\, dt$
45. $\int \left(\frac{1}{t}\mathbf{i} + \mathbf{j} - t^{3/2}\mathbf{k}\right) dt$
46. $\int \left(\ln t\mathbf{i} + \frac{1}{t}\mathbf{j} + \mathbf{k}\right) dt$
47. $\int \left[(2t - 1)\mathbf{i} + 4t^3\mathbf{j} + 3\sqrt{t}\mathbf{k}\right] dt$
48. $\int (e^t\mathbf{i} + \sin t\mathbf{j} + \cos t\mathbf{k})\, dt$
49. $\int \left(\sec^2 t\mathbf{i} + \frac{1}{1 + t^2}\mathbf{j}\right) dt$
50. $\int (e^{-t}\sin t\mathbf{i} + e^{-t}\cos t\mathbf{j})\, dt$

In Exercises 51–54, evaluate the definite integral.

51. $\int_0^1 (8t\mathbf{i} + t\mathbf{j} - \mathbf{k})\, dt$
52. $\int_{-1}^1 (t\mathbf{i} + t^3\mathbf{j} + \sqrt[3]{t}\mathbf{k})\, dt$
53. $\int_0^{\pi/2} [(a\cos t)\mathbf{i} + (a\sin t)\mathbf{j} + \mathbf{k}]\, dt$
54. $\int_0^2 (t\mathbf{i} + e^t\mathbf{j} - te^t\mathbf{k})\, dt$

In Exercises 55–60, find $\mathbf{r}(t)$ for the given conditions.

55. $\mathbf{r}'(t) = 4e^{2t}\mathbf{i} + 3e^t\mathbf{j}$, $\mathbf{r}(0) = 2\mathbf{i}$
56. $\mathbf{r}'(t) = 3t^2\mathbf{j} + 6\sqrt{t}\mathbf{k}$, $\mathbf{r}(0) = \mathbf{i} + 2\mathbf{j}$
57. $\mathbf{r}''(t) = -32\mathbf{j}$
    $\mathbf{r}'(0) = 600\sqrt{3}\mathbf{i} + 600\mathbf{j}$, $\mathbf{r}(0) = \mathbf{0}$
58. $\mathbf{r}''(t) = -4\cos t\mathbf{j} - 3\sin t\mathbf{k}$
    $\mathbf{r}'(0) = 3\mathbf{k}$, $\mathbf{r}(0) = 4\mathbf{j}$
59. $\mathbf{r}'(t) = te^{-t^2}\mathbf{i} - e^{-t}\mathbf{j} + \mathbf{k}$, $\mathbf{r}(0) = \frac{1}{2}\mathbf{i} - \mathbf{j} + \mathbf{k}$
60. $\mathbf{r}'(t) = \frac{1}{1 + t^2}\mathbf{i} + \frac{1}{t^2}\mathbf{j} + \frac{1}{t}\mathbf{k}$, $\mathbf{r}(1) = 2\mathbf{i}$

Getting at the Concept

61. State the definition of the derivative of a vector-valued function. Describe how to find the derivative of a vector-valued function and give its geometric interpretation.
62. How do you find the integral of a vector-valued function?
63. The three components of the derivative of the vector-valued function $\mathbf{u}$ are positive at $t = t_0$. Describe the behavior of $\mathbf{u}$ at $t = t_0$.
64. The z-component of the derivative of the vector-valued function $\mathbf{u}$ is 0 for t in the domain of the function. What does this information imply about the graph of $\mathbf{u}$?

In Exercises 65–72, prove the given property. In each case, assume that $\mathbf{r}$, $\mathbf{u}$, and $\mathbf{v}$ are differentiable vector-valued functions of t, f is a differentiable real-valued function of t, and c is a scalar.

65. $D_t[c\mathbf{r}(t)] = c\mathbf{r}'(t)$
66. $D_t[\mathbf{r}(t) \pm \mathbf{u}(t)] = \mathbf{r}'(t) \pm \mathbf{u}'(t)$
67. $D_t[f(t)\mathbf{r}(t)] = f(t)\mathbf{r}'(t) + f'(t)\mathbf{r}(t)$
68. $D_t[\mathbf{r}(t) \times \mathbf{u}(t)] = \mathbf{r}(t) \times \mathbf{u}'(t) + \mathbf{r}'(t) \times \mathbf{u}(t)$
69. $D_t[\mathbf{r}(f(t))] = \mathbf{r}'(f(t))f'(t)$
70. $D_t[\mathbf{r}(t) \times \mathbf{r}'(t)] = \mathbf{r}(t) \times \mathbf{r}''(t)$
71. $D_t\{\mathbf{r}(t) \cdot [\mathbf{u}(t) \times \mathbf{v}(t)]\} = \mathbf{r}'(t) \cdot [\mathbf{u}(t) \times \mathbf{v}(t)] + \mathbf{r}(t) \cdot [\mathbf{u}'(t) \times \mathbf{v}(t)] + \mathbf{r}(t) \cdot [\mathbf{u}(t) \times \mathbf{v}'(t)]$
72. If $\mathbf{r}(t) \cdot \mathbf{r}(t)$ is a constant, then $\mathbf{r}(t) \cdot \mathbf{r}'(t) = 0$.

True or False? In Exercises 73 and 74, determine whether the statement is true or false. If it is false, explain why or give an example that shows it is false.

73. $\frac{d}{dt}[\|\mathbf{r}(t)\|] = \|\mathbf{r}'(t)\|$

74. If $\mathbf{r}$ and $\mathbf{u}$ are differentiable vector-valued functions of t, then
    $D_t[\mathbf{r}(t) \cdot \mathbf{u}(t)] = \mathbf{r}'(t) \cdot \mathbf{u}'(t)$.

Section 11.3 Velocity and Acceleration

- Describe the velocity and acceleration associated with a vector-valued function.
- Use a vector-valued function to analyze projectile motion.

Velocity and Acceleration

You are now ready to combine your study of parametric equations, curves, vectors, and vector-valued functions to form a model for motion along a curve. We begin by looking at the motion of an object in the plane. (The motion of an object in space can be developed similarly.)

As an object moves along a curve in the plane, the coordinates x and y of its center of mass are each functions of time t. Rather than using f and g to represent these two functions, it is convenient to write $x = x(t)$ and $y = y(t)$. So, the position vector $\mathbf{r}(t)$ takes the form

$$\mathbf{r}(t) = x(t)\mathbf{i} + y(t)\mathbf{j}. \qquad \text{Position vector}$$

The beauty of this vector model for representing motion is that you can use the first and second derivatives of the vector-valued function $\mathbf{r}$ to find the object's velocity and acceleration. (Recall from the preceding chapter that velocity and acceleration are both vector quantities having magnitude and direction.) To find the velocity and acceleration vectors at a given time t, consider a point $Q(x(t + \Delta t), y(t + \Delta t))$ that is approaching the point $P(x(t), y(t))$ along the curve C given by $\mathbf{r}(t) = x(t)\mathbf{i} + y(t)\mathbf{j}$, as shown in Figure 11.10. As $\Delta t \to 0$, the direction of the vector $\overrightarrow{PQ}$ (denoted by $\Delta \mathbf{r}$) approaches the *direction of motion* at time t.

$$\Delta \mathbf{r} = \mathbf{r}(t + \Delta t) - \mathbf{r}(t)$$

$$\frac{\Delta \mathbf{r}}{\Delta t} = \frac{\mathbf{r}(t + \Delta t) - \mathbf{r}(t)}{\Delta t}$$

$$\lim_{\Delta t \to 0} \frac{\Delta \mathbf{r}}{\Delta t} = \lim_{\Delta t \to 0} \frac{\mathbf{r}(t + \Delta t) - \mathbf{r}(t)}{\Delta t}$$

If this limit exists, it is defined to be the **velocity vector** or **tangent vector** to the curve at point P. Note that this is the same limit used to define $\mathbf{r}'(t)$. So, the direction of $\mathbf{r}'(t)$ gives the direction of motion at time t. Moreover, the magnitude of the vector $\mathbf{r}'(t)$

$$\|\mathbf{r}'(t)\| = \|x'(t)\mathbf{i} + y'(t)\mathbf{j}\| = \sqrt{[x'(t)]^2 + [y'(t)]^2}$$

gives the **speed** of the object at time t. Similarly, you can use $\mathbf{r}''(t)$ to represent acceleration, as indicated in the definition at the top of page 803.

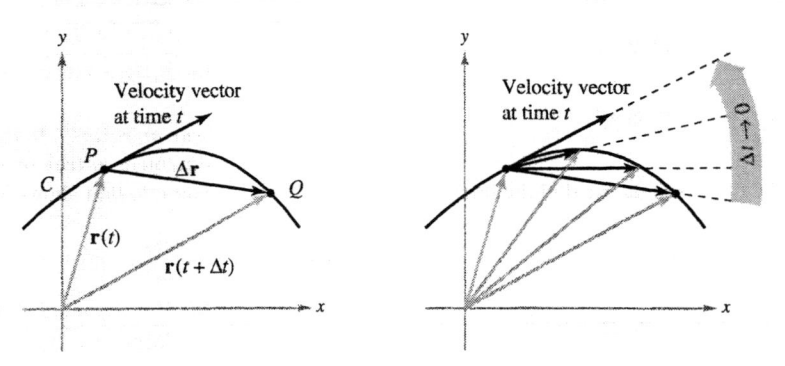

As $\Delta t \to 0$, $\dfrac{\Delta \mathbf{r}}{\Delta t}$ approaches the velocity vector.

Figure 11.10

EXPLORATION

Exploring Velocity Consider the circle given by

$$\mathbf{r}(t) = (\cos \omega t)\mathbf{i} + (\sin \omega t)\mathbf{j}.$$

Use a graphing utility in parametric mode to sketch this circle for several values of ω. How does ω affect the velocity of the terminal point as it traces out the curve? For a given value of ω, does the speed appear constant? Does the acceleration appear constant? Explain your reasoning.

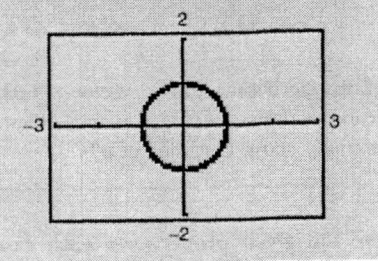

Definition of Velocity and Acceleration

If x and y are twice-differentiable functions of t, and $\mathbf{r}$ is a vector-valued function given by $\mathbf{r}(t) = x(t)\mathbf{i} + y(t)\mathbf{j}$, then the velocity vector, acceleration vector, and speed at time t are as follows.

$$\text{Velocity} = \mathbf{v}(t) = \mathbf{r}'(t) = x'(t)\mathbf{i} + y'(t)\mathbf{j}$$
$$\text{Acceleration} = \mathbf{a}(t) = \mathbf{r}''(t) = x''(t)\mathbf{i} + y''(t)\mathbf{j}$$
$$\text{Speed} = \|\mathbf{v}(t)\| = \|\mathbf{r}'(t)\| = \sqrt{[x'(t)]^2 + [y'(t)]^2}$$

For motion along a space curve, the definitions are similar. That is, if $\mathbf{r}(t) = x(t)\mathbf{i} + y(t)\mathbf{j} + z(t)\mathbf{k}$, you have

$$\text{Velocity} = \mathbf{v}(t) = \mathbf{r}'(t) = x'(t)\mathbf{i} + y'(t)\mathbf{j} + z'(t)\mathbf{k}$$
$$\text{Acceleration} = \mathbf{a}(t) = \mathbf{r}''(t) = x''(t)\mathbf{i} + y''(t)\mathbf{j} + z''(t)\mathbf{k}$$
$$\text{Speed} = \|\mathbf{v}(t)\| = \|\mathbf{r}'(t)\| = \sqrt{[x'(t)]^2 + [y'(t)]^2 + [z'(t)]^2}.$$

Example 1 Finding Velocity and Acceleration Along a Plane Curve

Find the velocity vector, speed, and acceleration vector of a particle that moves along the plane curve C described by

$$\mathbf{r}(t) = 2\sin\frac{t}{2}\mathbf{i} + 2\cos\frac{t}{2}\mathbf{j}. \qquad \text{Position vector}$$

Solution The velocity vector is

$$\mathbf{v}(t) = \mathbf{r}'(t) = \cos\frac{t}{2}\mathbf{i} - \sin\frac{t}{2}\mathbf{j}. \qquad \text{Velocity vector}$$

The speed (at any time) is

$$\|\mathbf{r}'(t)\| = \sqrt{\cos^2\frac{t}{2} + \sin^2\frac{t}{2}} = 1. \qquad \text{Speed}$$

The acceleration vector is

$$\mathbf{a}(t) = \mathbf{r}''(t) = -\frac{1}{2}\sin\frac{t}{2}\mathbf{i} - \frac{1}{2}\cos\frac{t}{2}\mathbf{j}. \qquad \text{Acceleration vector}$$

The parametric equations for the curve in Example 1 are

$$x = 2\sin\frac{t}{2} \quad \text{and} \quad y = 2\cos\frac{t}{2}.$$

By eliminating the parameter t, you can obtain the rectangular equation

$$x^2 + y^2 = 4. \qquad \text{Rectangular equation}$$

So, the curve is a circle of radius 2 centered at the origin, as shown in Figure 11.11. Because the velocity vector

$$\mathbf{v}(t) = \cos\frac{t}{2}\mathbf{i} - \sin\frac{t}{2}\mathbf{j}$$

has a constant magnitude but a changing direction as t increases, the particle moves around the circle at a constant speed.

NOTE In Example 1, note that the velocity and acceleration vectors are orthogonal at any point in time. This is characteristic of motion at a constant speed. (See Exercise 53.)

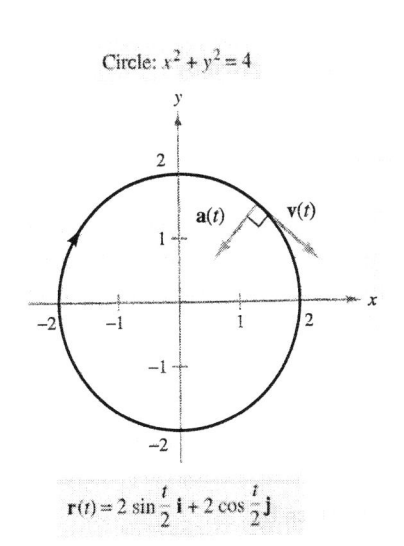

Circle: $x^2 + y^2 = 4$

$\mathbf{r}(t) = 2\sin\frac{t}{2}\mathbf{i} + 2\cos\frac{t}{2}\mathbf{j}$

The particle moves around the circle at a constant speed.
Figure 11.11

$\mathbf{r}(t) = (t^2 - 4)\mathbf{i} + t\mathbf{j}$

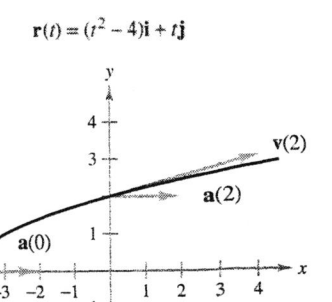

At each point on the curve, the acceleration vector points to the right.
Figure 11.12

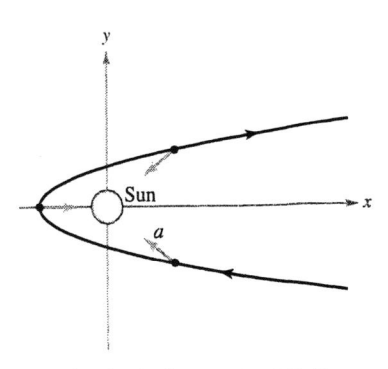

At each point in the comet's orbit, the acceleration vector points toward the sun.
Figure 11.13

Curve:
$\mathbf{r}(t) = t\mathbf{i} + t^3\mathbf{j} + 3t\mathbf{k}, \ t \geq 0$

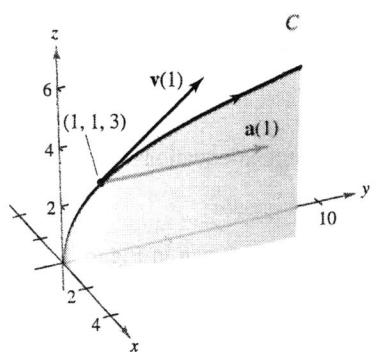

Figure 11.14

Example 2 Sketching Velocity and Acceleration Vectors in the Plane

Sketch the path of an object moving along the plane curve given by

$$\mathbf{r}(t) = (t^2 - 4)\mathbf{i} + t\mathbf{j} \qquad \text{Position vector}$$

and find the velocity and acceleration vectors when $t = 0$ and $t = 2$.

Solution Using the parametric equations $x = t^2 - 4$ and $y = t$, you can determine that the curve is a parabola given by $x = y^2 - 4$, as shown in Figure 11.12. The velocity vector (at any time) is

$$\mathbf{v}(t) = \mathbf{r}'(t) = 2t\mathbf{i} + \mathbf{j} \qquad \text{Velocity vector}$$

and the acceleration vector (at any time) is

$$\mathbf{a}(t) = \mathbf{r}''(t) = 2\mathbf{i}. \qquad \text{Acceleration vector}$$

When $t = 0$, the velocity and acceleration vectors are given by

$$\mathbf{v}(0) = 2(0)\mathbf{i} + \mathbf{j} = \mathbf{j} \quad \text{and} \quad \mathbf{a}(0) = 2\mathbf{i}.$$

When $t = 2$, the velocity and acceleration vectors are given by

$$\mathbf{v}(2) = 2(2)\mathbf{i} + \mathbf{j} = 4\mathbf{i} + \mathbf{j} \quad \text{and} \quad \mathbf{a}(2) = 2\mathbf{i}.$$

For the object moving along the path shown in Figure 11.12, note that the acceleration vector is constant (it has a magnitude of 2 and points to the right). This implies that the speed of the object is decreasing as the object moves toward the vertex of the parabola, and the speed is increasing as the object moves away from the vertex of the parabola.

This type of motion is *not* characteristic of comets that travel on parabolic paths through our solar system. For such comets, the acceleration vector always points to the origin (the sun), which implies that the comet's speed increases as it approaches the vertex of the path and decreases as it moves away from the vertex. (See Figure 11.13.)

Example 3 Sketching Velocity and Acceleration Vectors in Space

Sketch the path of an object moving along the space curve C given by

$$\mathbf{r}(t) = t\mathbf{i} + t^3\mathbf{j} + 3t\mathbf{k}, \quad t \geq 0 \qquad \text{Position vector}$$

and find the velocity and acceleration vectors when $t = 1$.

Solution Using the parametric equations $x = t$ and $y = t^3$, you can determine that the path of the object lies on the cubic cylinder given by $y = x^3$. Moreover, because $z = 3t$, the object starts at $(0, 0, 0)$ and moves upward as t increases, as shown in Figure 11.14. Because $\mathbf{r}(t) = t\mathbf{i} + t^3\mathbf{j} + 3t\mathbf{k}$, you have

$$\mathbf{v}(t) = \mathbf{r}'(t) = \mathbf{i} + 3t^2\mathbf{j} + 3\mathbf{k} \qquad \text{Velocity vector}$$

and

$$\mathbf{a}(t) = \mathbf{r}''(t) = 6t\mathbf{j}. \qquad \text{Acceleration vector}$$

When $t = 1$, the velocity and acceleration vectors are given by

$$\mathbf{v}(1) = \mathbf{r}'(1) = \mathbf{i} + 3\mathbf{j} + 3\mathbf{k} \quad \text{and} \quad \mathbf{a}(1) = \mathbf{r}''(1) = 6\mathbf{j}.$$

So far in this section, we have concentrated on finding the velocity and acceleration by differentiating the position function. Many practical applications involve the reverse problem—finding the position function for a given velocity or acceleration. This is demonstrated in the next example.

Example 4 Finding a Position Function by Integration

An object starts from rest at the point $P(1, 2, 0)$ and moves with an acceleration of

$$\mathbf{a}(t) = \mathbf{j} + 2\mathbf{k} \qquad \text{Acceleration vector}$$

where $\|\mathbf{a}(t)\|$ is measured in feet per second per second. Find the location of the object after $t = 2$ seconds.

Solution From the description of the object's motion, you can deduce the following *initial conditions*. Because the object starts from rest, you have

$$\mathbf{v}(0) = \mathbf{0}.$$

Moreover, because the object starts at the point $(x, y, z) = (1, 2, 0)$, you have

$$\mathbf{r}(0) = x(0)\mathbf{i} + y(0)\mathbf{j} + z(0)\mathbf{k}$$
$$= 1\mathbf{i} + 2\mathbf{j} + 0\mathbf{k}$$
$$= \mathbf{i} + 2\mathbf{j}.$$

To find the position function, you should integrate twice, each time using one of the initial conditions to solve for the constant of integration. The velocity vector is

$$\mathbf{v}(t) = \int \mathbf{a}(t)\, dt = \int (\mathbf{j} + 2\mathbf{k})\, dt$$
$$= t\mathbf{j} + 2t\mathbf{k} + \mathbf{C}$$

where $\mathbf{C} = C_1\mathbf{i} + C_2\mathbf{j} + C_3\mathbf{k}$. Letting $t = 0$ and applying the initial condition $\mathbf{v}(0) = \mathbf{0}$, you obtain

$$\mathbf{v}(0) = C_1\mathbf{i} + C_2\mathbf{j} + C_3\mathbf{k} = \mathbf{0} \quad \Longrightarrow \quad C_1 = C_2 = C_3 = 0.$$

So, the *velocity* at any time t is

$$\mathbf{v}(t) = t\mathbf{j} + 2t\mathbf{k}. \qquad \text{Velocity vector}$$

Integrating once more produces

$$\mathbf{r}(t) = \int \mathbf{v}(t)\, dt = \int (t\mathbf{j} + 2t\mathbf{k})\, dt$$
$$= \frac{t^2}{2}\mathbf{j} + t^2\mathbf{k} + \mathbf{C}$$

where $\mathbf{C} = C_4\mathbf{i} + C_5\mathbf{j} + C_6\mathbf{k}$. Letting $t = 0$ and applying the initial condition $\mathbf{r}(0) = \mathbf{i} + 2\mathbf{j}$, you have

$$\mathbf{r}(0) = C_4\mathbf{i} + C_5\mathbf{j} + C_6\mathbf{k} = \mathbf{i} + 2\mathbf{j} \quad \Longrightarrow \quad C_4 = 1, C_5 = 2, C_6 = 0.$$

So, the *position* vector is

$$\mathbf{r}(t) = \mathbf{i} + \left(\frac{t^2}{2} + 2\right)\mathbf{j} + t^2\mathbf{k}. \qquad \text{Position vector}$$

The location of the object after $t = 2$ seconds is given by $\mathbf{r}(2) = \mathbf{i} + 4\mathbf{j} + 4\mathbf{k}$, as shown in Figure 11.15.

Curve:
$\mathbf{r}(t) = \mathbf{i} + \left(\frac{t^2}{2} + 2\right)\mathbf{j} + t^2\mathbf{k}$

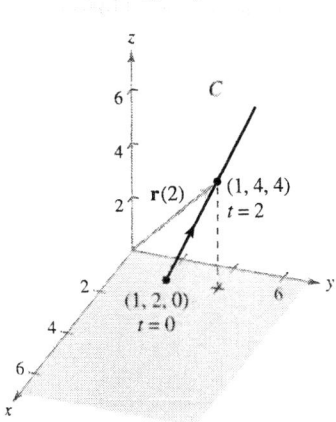

The object takes 2 seconds to move from point $(1, 2, 0)$ to point $(1, 4, 4)$ along C.
Figure 11.15

Projectile Motion

We now have the machinery to derive the parametric equations for the path of a projectile. We assume that gravity is the only force acting on the projectile after it is launched. Hence, the motion occurs in a vertical plane, which we represent by the xy-coordinate system with the origin as a point on the earth's surface, as shown in Figure 11.16. For a projectile of mass m, the force due to gravity is

$$\mathbf{F} = -mg\mathbf{j} \qquad \text{Force due to gravity}$$

where the gravitational constant is $g = 32$ feet per second per second, or 9.81 meters per second per second. By **Newton's Second Law of Motion,** this same force produces an acceleration $\mathbf{a} = \mathbf{a}(t)$, and satisfies the equation $\mathbf{F} = m\mathbf{a}$. Consequently, the acceleration of the projectile is given by $m\mathbf{a} = -mg\mathbf{j}$, which implies that

$$\mathbf{a} = -g\mathbf{j}. \qquad \text{Acceleration of projectile}$$

Example 5 Derivation of the Position Function for a Projectile

---

A projectile of mass m is launched from an initial position $\mathbf{r}_0$ with an initial velocity $\mathbf{v}_0$. Find its position vector as a function of time.

Solution Begin with the acceleration $\mathbf{a}(t) = -g\mathbf{j}$ and integrate twice.

$$\mathbf{v}(t) = \int \mathbf{a}(t)\,dt = \int -g\mathbf{j}\,dt = -gt\mathbf{j} + \mathbf{C}_1$$

$$\mathbf{r}(t) = \int \mathbf{v}(t)\,dt = \int (-gt\mathbf{j} + \mathbf{C}_1)\,dt = -\frac{1}{2}gt^2\mathbf{j} + \mathbf{C}_1 t + \mathbf{C}_2$$

You can use the facts that $\mathbf{v}(0) = \mathbf{v}_0$ and $\mathbf{r}(0) = \mathbf{r}_0$ to solve for the constant vectors $\mathbf{C}_1$ and $\mathbf{C}_2$. Doing this produces $\mathbf{C}_1 = \mathbf{v}_0$ and $\mathbf{C}_2 = \mathbf{r}_0$. Therefore, the position vector is

$$\mathbf{r}(t) = -\frac{1}{2}gt^2\mathbf{j} + t\mathbf{v}_0 + \mathbf{r}_0. \qquad \text{Position vector}$$

In many projectile problems, the constant vectors $\mathbf{r}_0$ and $\mathbf{v}_0$ are not given explicitly. Often you are given the initial height h, the initial speed v_0, and the angle θ at which the projectile is launched, as shown in Figure 11.17. From the given height, you can deduce that $\mathbf{r}_0 = h\mathbf{j}$. Because the speed gives the magnitude of the initial velocity, it follows that $v_0 = \|\mathbf{v}_0\|$ and you can write

$$\mathbf{v}_0 = x\mathbf{i} + y\mathbf{j}$$
$$= (\|\mathbf{v}_0\|\cos\theta)\mathbf{i} + (\|\mathbf{v}_0\|\sin\theta)\mathbf{j}$$
$$= v_0\cos\theta\,\mathbf{i} + v_0\sin\theta\,\mathbf{j}.$$

So, the position vector can be written in the form

$$\mathbf{r}(t) = -\frac{1}{2}gt^2\mathbf{j} + t\mathbf{v}_0 + \mathbf{r}_0 \qquad \text{Position vector}$$

$$= -\frac{1}{2}gt^2\mathbf{j} + tv_0\cos\theta\,\mathbf{i} + tv_0\sin\theta\,\mathbf{j} + h\mathbf{j}$$

$$= (v_0\cos\theta)t\,\mathbf{i} + \left[h + (v_0\sin\theta)t - \frac{1}{2}gt^2\right]\mathbf{j}.$$

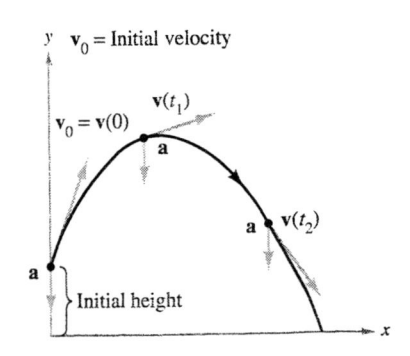

Figure 11.16

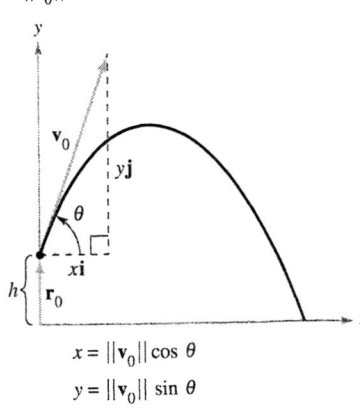

$\|\mathbf{v}_0\| = v_0 = $ initial speed
$\|\mathbf{r}_0\| = h = $ initial height

$x = \|\mathbf{v}_0\|\cos\theta$
$y = \|\mathbf{v}_0\|\sin\theta$

Figure 11.17

THEOREM 11.3 Position Function for a Projectile

Neglecting air resistance, the path of a projectile launched from an initial height h with initial speed v_0 and angle of elevation θ is described by the vector function

$$\mathbf{r}(t) = (v_0 \cos \theta)t\mathbf{i} + \left[h + (v_0 \sin \theta)t + \frac{1}{2}gt^2\right]\mathbf{j}$$

where g is the gravitational constant.

Example 6 Describing the Path of a Baseball

A baseball is hit 3 feet above ground level at 100 feet per second and at an angle of 45° with respect to the ground, as shown in Figure 11.18. Find the maximum height reached by the baseball. Will it clear a 10-foot high fence located 300 feet from home plate?

Solution You are given $h = 3$, $v_0 = 100$, and $\theta = 45°$. So, using $g = -32$ feet per second per second produces

$$\mathbf{r}(t) = \left(100 \cos \frac{\pi}{4}\right)t\mathbf{i} + \left[3 + \left(100 \sin \frac{\pi}{4}\right)t - 16t^2\right]\mathbf{j}$$
$$= (50\sqrt{2}\,t)\mathbf{i} + (3 + 50\sqrt{2}\,t - 16t^2)\mathbf{j}$$
$$\mathbf{v}(t) = \mathbf{r}'(t) = 50\sqrt{2}\,\mathbf{i} + (50\sqrt{2} - 32t)\mathbf{j}.$$

The maximum height occurs when

$$y'(t) = 50\sqrt{2} - 32t = 0$$

which implies that

$$t = \frac{25\sqrt{2}}{16} \approx 2.21 \text{ seconds.}$$

Hence, the maximum height reached by the ball is

$$y = 3 + 50\sqrt{2}\left(\frac{25\sqrt{2}}{16}\right) - 16\left(\frac{25\sqrt{2}}{16}\right)^2$$
$$= \frac{649}{8}$$
$$\approx 81 \text{ feet.} \qquad \text{Maximum height when } t \approx 2.21 \text{ seconds}$$

The ball is 300 feet from where it was hit when

$$300 = x(t) = 50\sqrt{2}\,t.$$

Solving this equation for t produces $t = 3\sqrt{2} \approx 4.24$ seconds. At this time, the height of the ball is

$$y = 3 + 50\sqrt{2}\,(3\sqrt{2}) - 16(3\sqrt{2})^2$$
$$= 303 - 288$$
$$= 15 \text{ feet.} \qquad \text{Height when } t \approx 4.24 \text{ seconds}$$

Therefore, the ball clears the 10-foot fence for a home run.

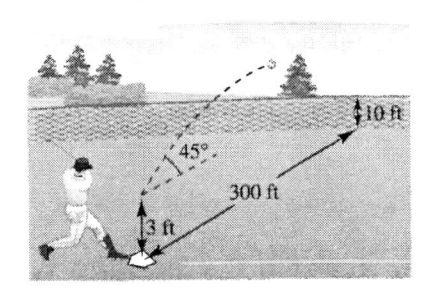

Figure 11.18

EXERCISES FOR SECTION 11.3

In Exercises 1–8, the position vector r describes the path of an object moving in the xy-plane. Sketch a graph of the path and sketch the velocity and acceleration vectors at the given point.

Position Function	Point
1. $\mathbf{r}(t) = 3t\mathbf{i} + (t-1)\mathbf{j}$	$(3, 0)$
2. $\mathbf{r}(t) = (6-t)\mathbf{i} + t\mathbf{j}$	$(3, 3)$
3. $\mathbf{r}(t) = t^2\mathbf{i} + t\mathbf{j}$	$(4, 2)$
4. $\mathbf{r}(t) = t^2\mathbf{i} + t^3\mathbf{j}$	$(1, 1)$
5. $\mathbf{r}(t) = 2\cos t\mathbf{i} + 2\sin t\mathbf{j}$	$(\sqrt{2}, \sqrt{2})$
6. $\mathbf{r}(t) = 3\cos t\mathbf{i} + 2\sin t\mathbf{j}$	$(3, 0)$
7. $\mathbf{r}(t) = \langle t - \sin t, 1 - \cos t \rangle$	$(\pi, 2)$
8. $\mathbf{r}(t) = \langle e^{-t}, e^{t} \rangle$	$(1, 1)$

In Exercises 9–16, the position vector r describes the path of an object moving in space. Find the velocity, speed, and acceleration of the object.

9. $\mathbf{r}(t) = t\mathbf{i} + (2t-5)\mathbf{j} + 3t\mathbf{k}$
10. $\mathbf{r}(t) = 4t\mathbf{i} + 4t\mathbf{j} + 2t\mathbf{k}$
11. $\mathbf{r}(t) = t\mathbf{i} + t^2\mathbf{j} + \dfrac{t^2}{2}\mathbf{k}$
12. $\mathbf{r}(t) = 3t\mathbf{i} + t\mathbf{j} + \frac{1}{4}t^2\mathbf{k}$
13. $\mathbf{r}(t) = t\mathbf{i} + t\mathbf{j} + \sqrt{9-t^2}\mathbf{k}$
14. $\mathbf{r}(t) = t^2\mathbf{i} + t\mathbf{j} + 2t^{3/2}\mathbf{k}$
15. $\mathbf{r}(t) = \langle 4t, 3\cos t, 3\sin t \rangle$
16. $\mathbf{r}(t) = \langle e^t \cos t, e^t \sin t, e^t \rangle$

Linear Approximation In Exercises 17 and 18, the graph of the vector-valued function $\mathbf{r}(t)$ and a tangent vector to the graph at $t = t_0$ are given.

(a) Find a set of parametric equations for the tangent line to the graph at $t = t_0$.

(b) Use the equations for the line to approximate $\mathbf{r}(t_0 + 0.1)$.

17. $\mathbf{r}(t) = \langle t, -t^2, \frac{1}{4}t^3 \rangle$, $t_0 = 1$
18. $\mathbf{r}(t) = \langle t, \sqrt{25-t^2}, \sqrt{25-t^2} \rangle$, $t_0 = 3$

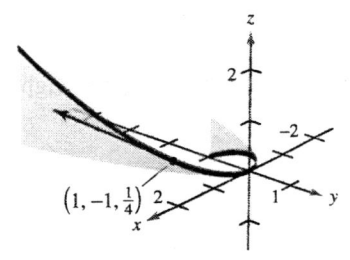

Figure for 17

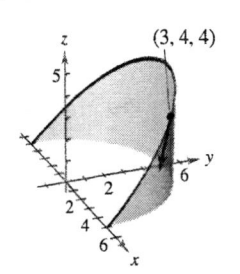

Figure for 18

In Exercises 19–22, use the given acceleration function to find the velocity and position vectors. Then find the position at time $t = 2$.

19. $\mathbf{a}(t) = \mathbf{i} + \mathbf{j} + \mathbf{k}$
    $\mathbf{v}(0) = \mathbf{0}$, $\mathbf{r}(0) = \mathbf{0}$

20. $\mathbf{a}(t) = 2\mathbf{i} + 3\mathbf{k}$
    $\mathbf{v}(0) = 4\mathbf{j}$, $\mathbf{r}(0) = \mathbf{0}$

21. $\mathbf{a}(t) = t\mathbf{j} + t\mathbf{k}$
    $\mathbf{v}(1) = 5\mathbf{j}$, $\mathbf{r}(1) = \mathbf{0}$

22. $\mathbf{a}(t) = -\cos t\mathbf{i} - \sin t\mathbf{j}$
    $\mathbf{v}(0) = \mathbf{j} + \mathbf{k}$, $\mathbf{r}(0) = \mathbf{i}$

Getting at the Concept

23. In your own words, explain the difference between the velocity of an object and its speed.

24. What is known about the speed of an object if the angle between the velocity and acceleration vectors is (a) acute and (b) obtuse?

Projectile Motion In Exercises 25–40, use the model for projectile motion, assuming there is no air resistance.

25. Find the vector-valued function for the path of a projectile launched at a height of 10 feet above the ground with an initial velocity of 88 feet per second and at an angle of 30° above the horizontal. Use a graphing utility to sketch the path of the projectile.

26. Determine the maximum height and range of a projectile fired at a height of 3 feet above the ground with an initial velocity of 900 feet per second and at an angle of 45° above the horizontal.

27. A baseball, hit 3 feet above the ground, leaves the bat at an angle of 45° and is caught by an outfielder 300 feet from home plate. What is the initial speed of the ball, and how high does it rise if it is caught 3 feet above the ground?

28. A baseball player at second base throws a ball 90 feet to the player at first base. The ball is thrown at 50 miles per hour at an angle of 15° above the horizontal. At what height does the player at first base catch the ball if the ball is thrown from a height of 5 feet?

29. Eliminate the parameter t from the position function for the motion of a projectile to show that the rectangular equation is

$$y = -\frac{16 \sec^2 \theta}{v_0^2}x^2 + (\tan \theta)x + h.$$

30. The path of a ball is given by the rectangular equation

$$y = x - 0.005x^2.$$

Use the result of Exercise 29 to find the position function. Then find the speed and direction of the ball at the point when it has traveled 60 feet horizontally.

31. *Modeling Data* After the path of a ball thrown by a baseball player is videotaped, it is analyzed on a television set with a grid covering the screen. The tape is paused three times and the positions of the ball are measured. The coordinates are approximately (0, 6.0), (15, 10.6), and (30, 13.4). (The x-coordinate measures the horizontal distance from the player in feet and the y-coordinate measures the height in feet.)

(a) Use a graphing utility to fit a quadratic model to the data.

(b) Use a graphing utility to plot the data and graph the model.

(c) Determine the maximum height of the ball.

(d) Find the initial velocity of the ball and the angle at which it was thrown.

32. A baseball is hit from a height of 2.5 feet above the ground with an initial velocity of 140 feet per second and at an angle of 22° above the horizontal. An eight-mile-per-hour wind is blowing horizontally toward the batter. Use a graphing utility to graph the path of the ball and determine whether it will clear a 10-foot-high fence located 375 feet from home plate.

33. The SkyDome in Toronto, Ontario has a center field fence that is 10 feet high and 400 feet from home plate. A ball is hit 3 feet above the ground and leaves the bat at a speed of 100 miles per hour.

(a) If the ball leaves the bat at an angle of $\theta = \theta_0$ with the horizontal, write the vector-valued function for the path of the ball.

(b) Use a graphing utility to graph the vector-valued function for $\theta_0 = 10°$, $\theta_0 = 15°$, $\theta_0 = 20°$, and $\theta_0 = 25°$. Use the graphs to approximate the minimum angle required for the hit to be a home run.

(c) Determine analytically the minimum angle required for the hit to be a home run.

34. The quarterback of a football team releases a pass at a height of 7 feet above the playing field, and the football is caught by a receiver 30 yards directly downfield at a height of 4 feet. The pass is released at an angle of 35° with the horizontal.

(a) Find the speed of the football when it is released.

(b) Find the maximum height of the football.

(c) Find the time the receiver has to reach the proper position after the quarterback releases the football.

35. A bale ejector consists of two variable-speed belts at the end of a baler. Its purpose is to toss bales into a trailing wagon. In loading the back of a wagon, a bale must be thrown to a position 8 feet above and 16 feet behind the ejector.

(a) Find the minimum initial speed of the bale and the corresponding angle at which it must be ejected from the baler.

(b) If the ejector has a fixed angle of 45°, find the initial speed required.

36. A bomber is flying at an altitude of 30,000 feet at a speed of 540 miles per hour (see figure). When should the bomb be released for it to hit the target? (Give your answer in terms of the angle of depression from the plane to the target.) What is the speed of the bomb at the time of impact?

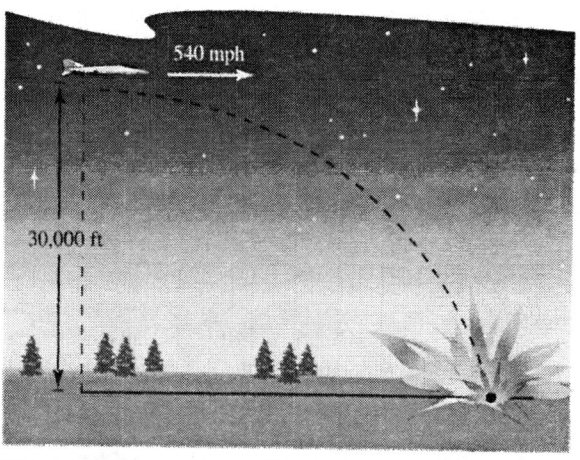

Figure for 36

37. A shot fired from a gun with a muzzle velocity of 1200 feet per second is to hit a target 3000 feet away. Determine the minimum angle of elevation of the gun.

38. A projectile is fired from ground level at an angle of 12° with the horizontal. Find the minimum initial velocity necessary if the projectile is to have a range of 150 feet.

39. Use a graphing utility to graph the paths of a projectile for the indicated values of θ and v_0. For each case, use the graph to approximate the maximum height and range of the projectile. (Assume that the projectile is launched from ground level.)

(a) $\theta = 10°$, $v_0 = 66$ ft/sec

(b) $\theta = 10°$, $v_0 = 146$ ft/sec

(c) $\theta = 45°$, $v_0 = 66$ ft/sec

(d) $\theta = 45°$, $v_0 = 146$ ft/sec

(e) $\theta = 60°$, $v_0 = 66$ ft/sec

(f) $\theta = 60°$, $v_0 = 146$ ft/sec

40. Find the angle at which an object must be thrown to obtain

(a) the maximum range and

(b) the maximum height.

Projectile Motion In Exercises 41 and 42, use the model for projectile motion, assuming there is no resistance. $[a(t) = -9.8$ meters per second per second]

41. Determine the maximum height and range of a projectile fired at a height of 1.5 meters above the ground with an initial velocity of 100 meters per second and at an angle of 30° above the horizontal.

42. A projectile is fired from ground level at an angle of 8° with the horizontal. Find the minimum velocity necessary if the projectile is to have a range of 50 meters.

Cycloidal Motion In Exercises 43 and 44, consider the motion of a point (or particle) on the circumference of a rolling circle. As the circle rolls, it generates the cycloid

$$\mathbf{r}(t) = b(\omega t - \sin \omega t)\mathbf{i} + b(1 - \cos \omega t)\mathbf{j}$$

where ω is the constant angular velocity of the circle.

43. Find the velocity and acceleration vectors of the particle. Use the results to determine the times at which the speed of the particle will be (a) zero and (b) maximized.

44. Find the maximum speed of a point on the circumference of an automobile tire of radius 1 foot when the automobile is traveling at 55 miles per hour. Compare this speed with the speed of the automobile.

Circular Motion In Exercises 45–48, consider a particle moving on a circular path of radius b described by

$$\mathbf{r}(t) = b \cos \omega t \, \mathbf{i} + b \sin \omega t \, \mathbf{j}$$

where $\omega = d\theta/dt$ is the constant angular velocity.

45. Find the velocity vector and show that it is orthogonal to $\mathbf{r}(t)$.

46. (a) Show that the speed of the particle is $b\omega$.
    (b) Use a graphing utility in parametric mode to sketch the circle for $b = 6$. Try different values of ω. Does the graphing utility draw the circle faster for greater values of ω?

47. Find the acceleration vector and show that its direction is always toward the center of the circle.

48. Show that the magnitude of the acceleration vector is $\omega^2 b$.

Circular Motion In Exercises 49 and 50, use the results of Exercises 45–48.

49. A stone weighing 1 pound is attached to a 2-foot string and is whirled horizontally (see figure). The string will break under a force of 10 pounds. Find the maximum speed the stone can attain without breaking the string. (Use $\mathbf{F} = m\mathbf{a}$, where $m = \frac{1}{32}$.)

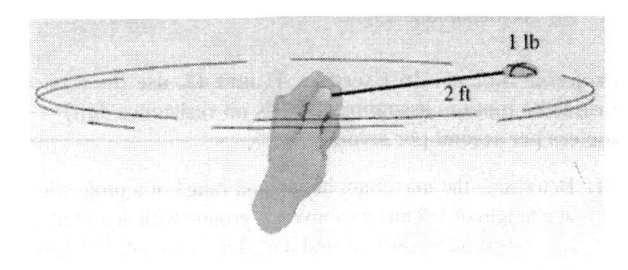

50. A 3000-pound automobile is negotiating a circular interchange of radius 300 feet at 30 miles per hour (see figure). Assuming the roadway to be level, find the force between the tires and the road such that the car stays on the circular path and does not skid. (Use $\mathbf{F} = m\mathbf{a}$, where $m = 3000/32$.) Find the angle at which the roadway should be banked so that no lateral frictional force is exerted on the tires of the automobile.

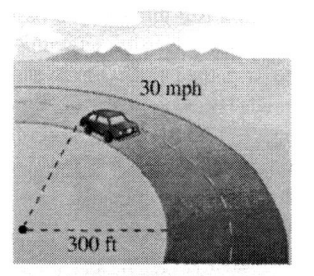

Figure for 50

51. *Shot-Put Throw* The path of a shot thrown at an angle θ is

$$\mathbf{r}(t) = (v_0 \cos \theta) t \mathbf{i} + \left[h + (v_0 \sin \theta) t - \frac{1}{2} g t^2 \right] \mathbf{j}$$

where v_0 is the initial speed, h is the initial height, t is the time in seconds, and g is the acceleration due to gravity. Verify that the shot will remain in the air for a total of

$$t = \frac{v_0 \sin \theta + \sqrt{v_0^2 \sin^2 \theta + 2gh}}{g} \text{ seconds}$$

and will travel a horizontal distance of

$$\frac{v_0^2 \cos \theta}{g} \left(\sin \theta + \sqrt{\sin^2 \theta + \frac{2gh}{v_0^2}} \right) \text{ feet.}$$

52. *Shot-Put Throw* A shot is thrown from a height of $h = 6$ feet with an initial speed of $v_0 = 45$ feet per second. Find the total time of travel and the total horizontal distance traveled if the shot is thrown at an angle of $\theta = 42.5°$ with the horizontal.

53. Prove that if an object is traveling at a constant speed, its velocity and acceleration vectors are orthogonal.

54. Prove that an object moving in a straight line at a constant speed has an acceleration of 0.

55. *Investigation* An object moves on an elliptical path given by the vector-valued function $\mathbf{r}(t) = 6 \cos t \mathbf{i} + 3 \sin t \mathbf{j}$.
    (a) Find $\mathbf{v}(t)$, $\|\mathbf{v}(t)\|$, and $\mathbf{a}(t)$.
    (b) Use a graphing utility to complete the table.

t	0	$\frac{\pi}{4}$	$\frac{\pi}{2}$	$\frac{2\pi}{3}$	π
Speed					

(c) Graph the elliptical path and the velocity and acceleration vectors at the values of t given in the table in part (b).
(d) Use the results in parts (b) and (c) to describe the geometric relationship between the velocity and acceleration vectors when the speed of the particle is increasing, and when it is decreasing.

56. *Writing* Consider a particle moving on the path

$$\mathbf{r}_1(t) = x(t)\mathbf{i} + y(t)\mathbf{j} + z(t)\mathbf{k}.$$

Discuss any changes in the position, velocity, or acceleration of the particle if its position is given by the vector-valued function $\mathbf{r}_2(t) = \mathbf{r}_1(2t)$. Generalize the results for the position function $\mathbf{r}_3(t) = \mathbf{r}_1(\omega t)$.

Section 11.4 Tangent Vectors and Normal Vectors

- Find a unit tangent vector at a point on a space curve.
- Find the tangential and normal components of acceleration.

Tangent Vectors and Normal Vectors

In the preceding section, you learned that the velocity vector points in the direction of motion. This observation leads to the following definition, which applies to any smooth curve—not just to those for which the parameter represents time.

Definition of Unit Tangent Vector

Let C be a smooth curve represented by $\mathbf{r}$ on an open interval I. The **unit tangent vector** $\mathbf{T}(t)$ at t is defined to be

$$\mathbf{T}(t) = \frac{\mathbf{r}'(t)}{\|\mathbf{r}'(t)\|}, \quad \mathbf{r}'(t) \neq \mathbf{0}.$$

Recall that a curve is *smooth* on an interval if $\mathbf{r}'$ is continuous and nonzero on the interval. So, "smoothness" is sufficient to guarantee that a curve has a unit tangent vector.

Example 1 Finding the Unit Tangent Vector

Find the unit tangent vector to the curve given by

$$\mathbf{r}(t) = t\mathbf{i} + t^2\mathbf{j}$$

when $t = 1$.

Solution The derivative of $\mathbf{r}(t)$ is

$$\mathbf{r}'(t) = \mathbf{i} + 2t\mathbf{j}. \qquad \text{Derivative of } \mathbf{r}(t)$$

So, the unit tangent vector is

$$\mathbf{T}(t) = \frac{\mathbf{r}'(t)}{\|\mathbf{r}'(t)\|} \qquad \text{Definition of } \mathbf{T}(t)$$

$$= \frac{1}{\sqrt{1 + 4t^2}}(\mathbf{i} + 2t\mathbf{j}). \qquad \text{Substitute for } \mathbf{r}'(t).$$

When $t = 1$, the unit tangent vector is

$$\mathbf{T}(1) = \frac{1}{\sqrt{5}}(\mathbf{i} + 2\mathbf{j})$$

as shown in Figure 11.19.

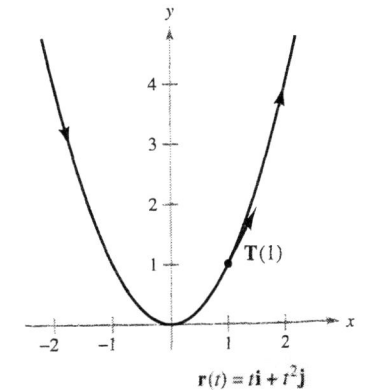

The direction of the unit tangent vector depends on the orientation of the curve.
Figure 11.19

NOTE In Example 1, note that the direction of the unit tangent vector depends on the orientation of the curve. For instance, if the parabola in Figure 11.19 were given by

$$\mathbf{r}(t) = -(t - 2)\mathbf{i} + (t - 2)^2\mathbf{j},$$

$\mathbf{T}(1)$ would still represent the unit tangent vector at the point $(1, 1)$, but it would point in the opposite direction. (Try verifying this.)

The **tangent line to a curve** at a point is the line passing through the point and parallel to the unit tangent vector. In Example 2, the unit tangent vector is used to find the tangent line at a point on a helix.

Example 2 Finding the Tangent Line at a Point on a Curve

Find $\mathbf{T}(t)$ and then find a set of parametric equations for the tangent line to the helix given by

$$\mathbf{r}(t) = 2\cos t\,\mathbf{i} + 2\sin t\,\mathbf{j} + t\mathbf{k}$$

at the point corresponding to $t = \pi/4$.

Solution The derivative of $\mathbf{r}(t)$ is $\mathbf{r}'(t) = -2\sin t\,\mathbf{i} + 2\cos t\,\mathbf{j} + \mathbf{k}$, which implies that $\|\mathbf{r}'(t)\| = \sqrt{4\sin^2 t + 4\cos^2 t + 1} = \sqrt{5}$. Therefore, the unit tangent vector is

$$\mathbf{T}(t) = \frac{\mathbf{r}'(t)}{\|\mathbf{r}'(t)\|} = \frac{1}{\sqrt{5}}(-2\sin t\,\mathbf{i} + 2\cos t\,\mathbf{j} + \mathbf{k}). \qquad \text{Unit tangent vector}$$

When $t = \pi/4$, the unit tangent vector is

$$\mathbf{T}\left(\frac{\pi}{4}\right) = \frac{1}{\sqrt{5}}\left(-2\frac{\sqrt{2}}{2}\mathbf{i} + 2\frac{\sqrt{2}}{2}\mathbf{j} + \mathbf{k}\right)$$

$$= \frac{1}{\sqrt{5}}(-\sqrt{2}\,\mathbf{i} + \sqrt{2}\,\mathbf{j} + \mathbf{k}).$$

Using the direction numbers $a = -\sqrt{2}$, $b = \sqrt{2}$, and $c = 1$, and the point $(x_1, y_1, z_1) = (\sqrt{2}, \sqrt{2}, \pi/4)$, you can obtain the following parametric equations (given with parameter s).

$$x = x_1 + as = \sqrt{2} - \sqrt{2}s$$
$$y = y_1 + bs = \sqrt{2} + \sqrt{2}s$$
$$z = z_1 + cs = \frac{\pi}{4} + s$$

This tangent line is shown in Figure 11.20.

Curve:
$\mathbf{r}(t) = 2\cos t\,\mathbf{i} + 2\sin t\,\mathbf{j} + t\mathbf{k}$

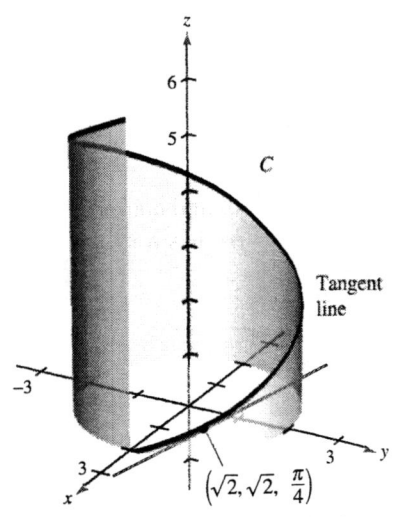

The tangent line to a curve at a point is determined by the unit tangent vector at the point.
Figure 11.20

In Example 2, there are infinitely many vectors that are orthogonal to the tangent vector $\mathbf{T}(t)$. One of these is the vector $\mathbf{T}'(t)$. This follows from Property 7 of Theorem 11.2. That is,

$$\mathbf{T}(t) \cdot \mathbf{T}(t) = \|\mathbf{T}(t)\|^2 = 1 \quad \Longrightarrow \quad \mathbf{T}(t) \cdot \mathbf{T}'(t) = 0.$$

By normalizing the vector $\mathbf{T}'(t)$, you obtain a special vector called the **principal unit normal vector**, as indicated in the following definition.

Definition of Principal Unit Normal Vector

Let C be a smooth curve represented by $\mathbf{r}$ on an open interval I. If $\mathbf{T}'(t) \neq \mathbf{0}$, then the **principal unit normal vector** at t is defined to be

$$\mathbf{N}(t) = \frac{\mathbf{T}'(t)}{\|\mathbf{T}'(t)\|}.$$

Example 3 Finding the Principal Unit Normal Vector

Find $\mathbf{N}(t)$ and $\mathbf{N}(1)$ for the curve represented by

$$\mathbf{r}(t) = 3t\mathbf{i} + 2t^2\mathbf{j}.$$

Solution By differentiating, you obtain

$$\mathbf{r}'(t) = 3\mathbf{i} + 4t\mathbf{j} \quad \text{and} \quad \|\mathbf{r}'(t)\| = \sqrt{9 + 16t^2}$$

which implies that the unit tangent vector is

$$\mathbf{T}(t) = \frac{\mathbf{r}'(t)}{\|\mathbf{r}'(t)\|}$$

$$= \frac{1}{\sqrt{9 + 16t^2}}(3\mathbf{i} + 4t\mathbf{j}). \qquad \text{Unit tangent vector}$$

Using Theorem 11.2, differentiate $\mathbf{T}(t)$ with respect to t to obtain

$$\mathbf{T}'(t) = \frac{1}{\sqrt{9 + 16t^2}}(4\mathbf{j}) - \frac{16t}{(9 + 16t^2)^{3/2}}(3\mathbf{i} + 4t\mathbf{j})$$

$$= \frac{12}{(9 + 16t^2)^{3/2}}(-4t\mathbf{i} + 3\mathbf{j})$$

$$\|\mathbf{T}'(t)\| = 12\sqrt{\frac{9 + 16t^2}{(9 + 16t^2)^3}} = \frac{12}{9 + 16t^2}.$$

Therefore, the principal unit normal vector is

$$\mathbf{N}(t) = \frac{\mathbf{T}'(t)}{\|\mathbf{T}'(t)\|}$$

$$= \frac{1}{\sqrt{9 + 16t^2}}(-4t\mathbf{i} + 3\mathbf{j}). \qquad \text{Principal unit normal vector}$$

When $t = 1$, the principal unit normal vector is

$$\mathbf{N}(1) = \frac{1}{5}(-4\mathbf{i} + 3\mathbf{j})$$

as shown in Figure 11.21.

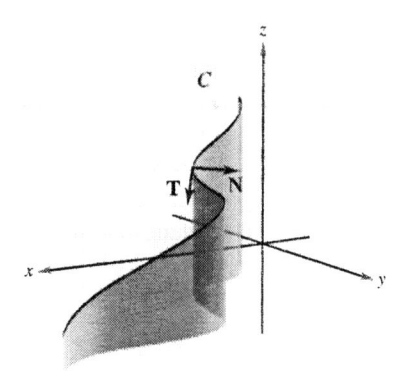

Curve: $\mathbf{r}(t) = 3t\mathbf{i} + 2t^2\mathbf{j}$

$\mathbf{N}(1) = \frac{1}{5}(-4\mathbf{i} + 3\mathbf{j})$

$\mathbf{T}(1) = \frac{1}{5}(3\mathbf{i} + 4\mathbf{j})$

The principal unit normal vector points toward the concave side of the curve.
Figure 11.21

The principal unit normal vector can be difficult to evaluate algebraically. For plane curves, you can simplify the algebra by finding

$$\mathbf{T}(t) = x(t)\mathbf{i} + y(t)\mathbf{j} \qquad \text{Unit tangent vector}$$

and observing that $\mathbf{N}(t)$ must be either

$$\mathbf{N}_1(t) = y(t)\mathbf{i} - x(t)\mathbf{j} \quad \text{or} \quad \mathbf{N}_2(t) = -y(t)\mathbf{i} + x(t)\mathbf{j}.$$

Because $\sqrt{[x(t)]^2 + [y(t)]^2} = 1$, it follows that both $\mathbf{N}_1(t)$ and $\mathbf{N}_2(t)$ are unit normal vectors. The *principal* unit normal vector $\mathbf{N}$ is the one that points toward the concave side of the curve, as indicated in Figure 11.21 (see Exercise 61). This also holds for curves in space. That is, for an object moving along a curve C in space, the vector $\mathbf{T}(t)$ points in the direction the object is moving, whereas the vector $\mathbf{N}(t)$ is orthogonal to $\mathbf{T}(t)$ and points in the direction the object is turning, as shown in Figure 11.22.

At any point on a curve, a unit normal vector is orthogonal to the unit tangent vector. The *principal* unit normal vector points in the direction the curve is turning.
Figure 11.22

Helix:
$\mathbf{r}(t) = 2\cos t\,\mathbf{i} + 2\sin t\,\mathbf{j} + t\mathbf{k}$

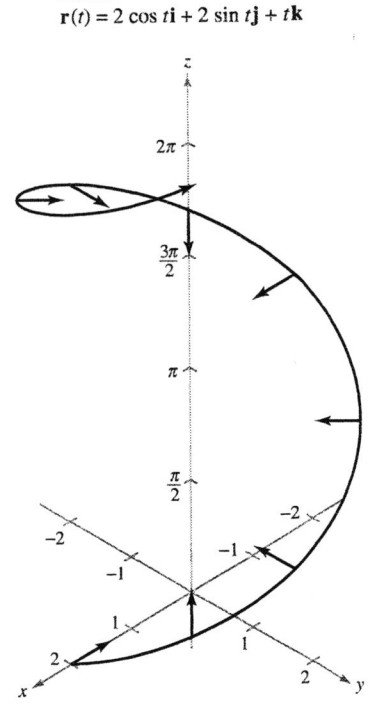

N(t) is horizontal and points toward the z-axis.
Figure 11.23

Example 4 Finding the Principal Unit Normal Vector

Find the principal unit normal vector for the helix given by

$$\mathbf{r}(t) = 2\cos t\,\mathbf{i} + 2\sin t\,\mathbf{j} + t\mathbf{k}.$$

Solution From Example 2, you know that the unit tangent vector is

$$\mathbf{T}(t) = \frac{1}{\sqrt{5}}(-2\sin t\,\mathbf{i} + 2\cos t\,\mathbf{j} + \mathbf{k}). \qquad \text{Unit tangent vector}$$

So, $\mathbf{T}'(t)$ is given by

$$\mathbf{T}'(t) = \frac{1}{\sqrt{5}}(-2\cos t\,\mathbf{i} - 2\sin t\,\mathbf{j}).$$

Because $\|\mathbf{T}'(t)\| = 2/\sqrt{5}$, it follows that the principal unit normal vector is

$$\mathbf{N}(t) = \frac{\mathbf{T}'(t)}{\|\mathbf{T}'(t)\|}$$

$$= \frac{1}{2}(-2\cos t\,\mathbf{i} - 2\sin t\,\mathbf{j})$$

$$= -\cos t\,\mathbf{i} - \sin t\,\mathbf{j}. \qquad \text{Principal unit normal vector}$$

Note that this vector is horizontal and points toward the z-axis, as shown in Figure 11.23.

Tangential and Normal Components of Acceleration

We now return to the problem of describing the motion of an object along a curve. In the preceding section, you saw that for an object traveling at a *constant speed*, the velocity and acceleration vectors are perpendicular. This seems reasonable, because the speed would not be constant if any acceleration were acting in the direction of motion. You can verify this observation by noting that

$$\mathbf{r}''(t) \cdot \mathbf{r}'(t) = 0$$

if $\|\mathbf{r}'(t)\|$ is a constant. (See Property 7 of Theorem 11.2.)

However, for an object traveling at a *variable speed*, the velocity and acceleration vectors are not necessarily perpendicular. For instance, you saw that the acceleration vector for a projectile always points down, regardless of the direction of motion.

In general, part of the acceleration (the tangential component) acts in the line of motion, and part (the normal component) acts perpendicular to the line of motion. In order to determine these two components, you can use the unit vectors $\mathbf{T}(t)$ and $\mathbf{N}(t)$, which serve in much the same way as do $\mathbf{i}$ and $\mathbf{j}$ in representing vectors in the plane. The following theorem states that the acceleration vector lies in the plane determined by $\mathbf{T}(t)$ and $\mathbf{N}(t)$.

THEOREM 11.4 Acceleration Vector

If $\mathbf{r}(t)$ is the position vector for a smooth curve C and $\mathbf{N}(t)$ exists, then the acceleration vector $\mathbf{a}(t)$ lies in the plane determined by $\mathbf{T}(t)$ and $\mathbf{N}(t)$.

Proof To simplify the notation, we write $\mathbf{T}$ for $\mathbf{T}(t)$, $\mathbf{T}'$ for $\mathbf{T}'(t)$, and so on. Because $\mathbf{T} = \mathbf{r}'/\|\mathbf{r}'\| = \mathbf{v}/\|\mathbf{v}\|$, it follows that

$$\mathbf{v} = \|\mathbf{v}\|\mathbf{T}.$$

By differentiating, you obtain

$$\mathbf{a} = \mathbf{v}' = D_t[\|\mathbf{v}\|]\mathbf{T} + \|\mathbf{v}\|\mathbf{T}' \qquad \text{Product Rule}$$

$$= D_t[\|\mathbf{v}\|]\mathbf{T} + \|\mathbf{v}\|\mathbf{T}'\left(\frac{\|\mathbf{T}'\|}{\|\mathbf{T}'\|}\right)$$

$$= D_t[\|\mathbf{v}\|]\mathbf{T} + \|\mathbf{v}\|\|\mathbf{T}'\|\mathbf{N}. \qquad \mathbf{N} = \mathbf{T}'/\|\mathbf{T}'\|$$

Because $\mathbf{a}$ is written as a linear combination of $\mathbf{T}$ and $\mathbf{N}$, it must lie in the plane determined by $\mathbf{T}$ and $\mathbf{N}$.

The coefficients of $\mathbf{T}$ and $\mathbf{N}$ in the proof of Theorem 11.4 are called the **tangential and normal components of acceleration** and are denoted by $a_\mathbf{T} = D_t[\|\mathbf{v}\|]$ and $a_\mathbf{N} = \|\mathbf{v}\|\|\mathbf{T}'\|$. So, you can write

$$\mathbf{a}(t) = a_\mathbf{T}\mathbf{T}(t) + a_\mathbf{N}\mathbf{N}(t).$$

The following theorem gives some convenient formulas for $a_\mathbf{N}$ and $a_\mathbf{T}$.

THEOREM 11.5 Tangential and Normal Components of Acceleration

If $\mathbf{r}(t)$ is the position vector for a smooth curve C [for which $\mathbf{N}(t)$ exists], then the tangential and normal components of acceleration are as follows.

$$a_\mathbf{T} = D_t[\|\mathbf{v}\|] = \mathbf{a} \cdot \mathbf{T} = \frac{\mathbf{v} \cdot \mathbf{a}}{\|\mathbf{v}\|}$$

$$a_\mathbf{N} = \|\mathbf{v}\|\|\mathbf{T}'\| = \mathbf{a} \cdot \mathbf{N} = \frac{\|\mathbf{v} \times \mathbf{a}\|}{\|\mathbf{v}\|} = \sqrt{\|\mathbf{a}\|^2 - a_\mathbf{T}^2}$$

Note that $a_\mathbf{N} \geq 0$. The normal component of acceleration is also called the **centripetal component of acceleration.**

Proof Note that $\mathbf{a}$ lies in the plane of $\mathbf{T}$ and $\mathbf{N}$. So, you can use Figure 11.24 to conclude that, for any time t, the component of the projection of the acceleration vector onto $\mathbf{T}$ is given by $a_\mathbf{T} = \mathbf{a} \cdot \mathbf{T}$, and onto $\mathbf{N}$ is given by $a_\mathbf{N} = \mathbf{a} \cdot \mathbf{N}$. Moreover, because $\mathbf{a} = \mathbf{v}'$ and $\mathbf{T} = \mathbf{v}/\|\mathbf{v}\|$, you have

$$a_\mathbf{T} = \mathbf{a} \cdot \mathbf{T}$$
$$= \mathbf{T} \cdot \mathbf{a}$$
$$= \frac{\mathbf{v}}{\|\mathbf{v}\|} \cdot \mathbf{a}$$
$$= \frac{\mathbf{v} \cdot \mathbf{a}}{\|\mathbf{v}\|}.$$

In Exercises 63 and 64, you are asked to prove the other parts of the theorem.

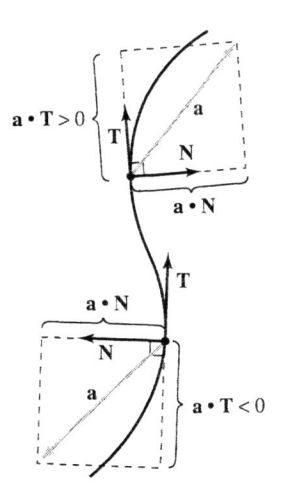

The tangential and normal components of acceleration are obtained by projecting $\mathbf{a}$ onto $\mathbf{T}$ and $\mathbf{N}$.
Figure 11.24

NOTE The formulas from Theorem 11.5, together with several other formulas from this chapter, are summarized on page 828.

Example 5 Tangential and Normal Components of Acceleration

Find the tangential and normal components of acceleration for the position vector given by $\mathbf{r}(t) = 3t\mathbf{i} - t\mathbf{j} + t^2\mathbf{k}$.

Solution Begin by finding the velocity, speed, and acceleration.

$$\mathbf{v}(t) = \mathbf{r}'(t) = 3\mathbf{i} - \mathbf{j} + 2t\mathbf{k}$$
$$\|\mathbf{v}(t)\| = \sqrt{9 + 1 + 4t^2} = \sqrt{10 + 4t^2}$$
$$\mathbf{a}(t) = \mathbf{r}''(t) = 2\mathbf{k}$$

By Theorem 11.5, the tangential component of acceleration is

$$a_\mathbf{T} = \frac{\mathbf{v} \cdot \mathbf{a}}{\|\mathbf{v}\|} = \frac{4t}{\sqrt{10 + 4t^2}} \qquad \text{Tangential component of acceleration}$$

and because

$$\mathbf{v} \times \mathbf{a} = \begin{vmatrix} \mathbf{i} & \mathbf{j} & \mathbf{k} \\ 3 & -1 & 2t \\ 0 & 0 & 2 \end{vmatrix} = -2\mathbf{i} - 6\mathbf{j}$$

the normal component of acceleration is

$$a_\mathbf{N} = \frac{\|\mathbf{v} \times \mathbf{a}\|}{\|\mathbf{v}\|} = \frac{\sqrt{4 + 36}}{\sqrt{10 + 4t^2}} = \frac{2\sqrt{10}}{\sqrt{10 + 4t^2}}. \qquad \text{Normal component of acceleration}$$

NOTE In Example 5, you could have used the alternative formula for $a_\mathbf{N}$ as follows.

$$a_\mathbf{N} = \sqrt{\|\mathbf{a}\|^2 - a_\mathbf{T}^2} = \sqrt{(2)^2 - \frac{16t^2}{10 + 4t^2}} = \frac{2\sqrt{10}}{\sqrt{10 + 4t^2}}$$

Example 6 Finding $a_\mathbf{T}$ and $a_\mathbf{N}$ for a Circular Helix

Find the tangential and normal components of acceleration for the helix given by $\mathbf{r}(t) = b\cos t\mathbf{i} + b\sin t\mathbf{j} + ct\mathbf{k}, b > 0$.

Solution

$$\mathbf{v}(t) = \mathbf{r}'(t) = -b\sin t\mathbf{i} + b\cos t\mathbf{j} + c\mathbf{k}$$
$$\|\mathbf{v}(t)\| = \sqrt{b^2\sin^2 t + b^2\cos^2 t + c^2} = \sqrt{b^2 + c^2}$$
$$\mathbf{a}(t) = \mathbf{r}''(t) = -b\cos t\mathbf{i} - b\sin t\mathbf{j}$$

By Theorem 11.5, the tangential component of acceleration is

$$a_\mathbf{T} = \frac{\mathbf{v} \cdot \mathbf{a}}{\|\mathbf{v}\|} = \frac{b^2\sin t\cos t - b^2\sin t\cos t + 0}{\sqrt{b^2 + c^2}} = 0. \qquad \text{Tangential component of acceleration}$$

Moreover, because $\|\mathbf{a}\| = \sqrt{b^2\cos^2 t + b^2\sin^2 t} = b$, you can use the alternative formula for the normal component of acceleration to obtain

$$a_\mathbf{N} = \sqrt{\|\mathbf{a}\|^2 - a_\mathbf{T}^2} = \sqrt{b^2 - 0^2} = b. \qquad \text{Normal component of acceleration}$$

Note that the normal component of acceleration is equal to the magnitude of the acceleration. In other words, because the speed is constant, the acceleration is perpendicular to the velocity. (See Figure 11.25.)

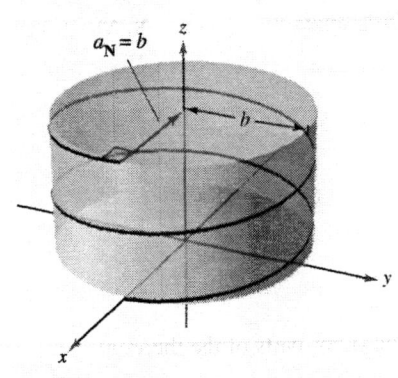

The normal component of acceleration is equal to the radius of the cylinder around which the helix is spiraling.
Figure 11.25

$\mathbf{r}(t) = (50\sqrt{2}t)\mathbf{i} + (50\sqrt{2}t - 16t^2)\mathbf{j}$

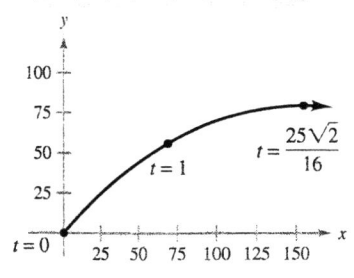

The path of a projectile
Figure 11.26

Example 7 Projectile Motion

The position vector for the projectile shown in Figure 11.26 is given by

$$\mathbf{r}(t) = (50\sqrt{2}\,t)\mathbf{i} + (50\sqrt{2}\,t - 16t^2)\mathbf{j}. \qquad \text{Position vector}$$

Find the tangential component of acceleration when $t = 0$, 1, and $25\sqrt{2}/16$.

Solution

$$\mathbf{v}(t) = 50\sqrt{2}\,\mathbf{i} + (50\sqrt{2} - 32t)\mathbf{j} \qquad \text{Velocity vector}$$
$$\|\mathbf{v}(t)\| = 2\sqrt{50^2 - 16(50)\sqrt{2}\,t + 16^2 t^2} \qquad \text{Speed}$$
$$\mathbf{a}(t) = -32\mathbf{j} \qquad \text{Acceleration vector}$$

The tangential component of acceleration is

$$a_\mathbf{T}(t) = \frac{\mathbf{v}(t) \cdot \mathbf{a}(t)}{\|\mathbf{v}(t)\|} = \frac{-32(50\sqrt{2} - 32t)}{2\sqrt{50^2 - 16(50)\sqrt{2}\,t + 16^2 t^2}}. \qquad \text{Tangential component of acceleration}$$

At the specified times, you have

$$a_\mathbf{T}(0) = \frac{-32(50\sqrt{2})}{100} = -16\sqrt{2} \approx -22.6$$

$$a_\mathbf{T}(1) = \frac{-32(50\sqrt{2} - 32)}{2\sqrt{50^2 - 16(50)\sqrt{2} + 16^2}} \approx -15.4$$

$$a_\mathbf{T}\left(\frac{25\sqrt{2}}{16}\right) = \frac{-32(50\sqrt{2} - 50\sqrt{2})}{50\sqrt{2}} = 0.$$

You can see from Figure 11.26 that, at the maximum height, when $t = 25\sqrt{2}/16$, the tangential component is 0. This is reasonable because the direction of motion is horizontal at the point and the tangential component of the acceleration is equal to the horizontal component of the acceleration.

EXERCISES FOR SECTION 11.4

In Exercises 1–4, find the unit tangent vector to the curve at the specified value of the parameter.

1. $\mathbf{r}(t) = t^2\mathbf{i} + 2t\mathbf{j}$, $t = 1$
2. $\mathbf{r}(t) = t^3\mathbf{i} + 2t^2\mathbf{j}$, $t = 1$
3. $\mathbf{r}(t) = 4\cos t\,\mathbf{i} + 4\sin t\,\mathbf{j}$, $t = \dfrac{\pi}{4}$
4. $\mathbf{r}(t) = 6\cos t\,\mathbf{i} + 2\sin t\,\mathbf{j}$, $t = \dfrac{\pi}{3}$

In Exercises 5–10, find the unit tangent vector $\mathbf{T}(t)$ and find a set of parametric equations for the line tangent to the space curve at point P.

5. $\mathbf{r}(t) = t\mathbf{i} + t^2\mathbf{j} + t\mathbf{k}$, $P(0, 0, 0)$
6. $\mathbf{r}(t) = t^2\mathbf{i} + t\mathbf{j} + \frac{4}{3}\mathbf{k}$, $P\left(1, 1, \frac{4}{3}\right)$
7. $\mathbf{r}(t) = 2\cos t\,\mathbf{i} + 2\sin t\,\mathbf{j} + t\mathbf{k}$, $P(2, 0, 0)$
8. $\mathbf{r}(t) = \langle t, t, \sqrt{4 - t^2}\rangle$, $P(1, 1, \sqrt{3})$
9. $\mathbf{r}(t) = \langle 2\cos t, 2\sin t, 4\rangle$, $P(\sqrt{2}, \sqrt{2}, 4)$
10. $\mathbf{r}(t) = \langle 2\sin t, 2\cos t, 4\sin^2 t\rangle$, $P(1, \sqrt{3}, 1)$

In Exercises 11 and 12, use a computer algebra system to graph the space curve. Then find $\mathbf{T}(t)$ and find a set of parametric equations for the line tangent to the space curve at point P. Graph the tangent line.

11. $\mathbf{r}(t) = \langle t, t^2, 2t^3/3\rangle$, $P(3, 9, 18)$
12. $\mathbf{r}(t) = 3\cos t\,\mathbf{i} + 4\sin t\,\mathbf{j} + \frac{1}{2}t\mathbf{k}$, $P(0, 4, \pi/4)$

Linear Approximation In Exercises 13 and 14, find a set of parametric equations for the tangent line to the graph at $t = t_0$ and use the equations for the line to approximate $\mathbf{r}(t_0 + 0.1)$.

13. $\mathbf{r}(t) = \langle t, \ln t, \sqrt{t}\rangle$, $t_0 = 1$
14. $\mathbf{r}(t) = \langle e^{-t}, 2\cos t, 2\sin t\rangle$, $t_0 = 0$

In Exercises 15 and 16, verify that the space curves intersect at the given values of the parameters. Find the angle between the tangent vectors to the curves at the point of intersection.

15. $\mathbf{r}(t) = \langle t - 2, t^2, \frac{1}{2}t\rangle$, $t = 4$
    $\mathbf{u}(s) = \langle \frac{1}{4}s, 2s, \sqrt[3]{s}\rangle$, $s = 8$

16. $\mathbf{r}(t) = \langle t, \cos t, \sin t \rangle, \quad t = 0$
 $\mathbf{u}(s) = \langle -\frac{1}{2}\sin^2 s - \sin s, 1 - \frac{1}{2}\sin^2 s - \sin s,$
 $\frac{1}{2}\sin s \cos s + \frac{1}{2}s \rangle, \quad s = 0$

In Exercises 17–20, find the principal unit normal vector to the curve at the specified value of the parameter.

17. $\mathbf{r}(t) = t\mathbf{i} + \frac{1}{2}t^2\mathbf{j}, \quad t = 2$

18. $\mathbf{r}(t) = t\mathbf{i} + \frac{6}{t}\mathbf{j}, \quad t = 3$

19. $\mathbf{r}(t) = 6\cos t\mathbf{i} + 6\sin t\mathbf{j} + t\mathbf{k}, \quad t = \frac{3\pi}{4}$

20. $\mathbf{r}(t) = \cos t\mathbf{i} + 2\sin t\mathbf{j} + \mathbf{k}, \quad t = -\frac{\pi}{4}$

In Exercises 21–24, find $\mathbf{v}(t)$, $\mathbf{a}(t)$, $\mathbf{T}(t)$, and $\mathbf{N}(t)$ (if it exists) for an object moving along the path given by the vector-valued function $\mathbf{r}(t)$. Use the results to determine the form of the path. Is the speed of the object constant or changing?

21. $\mathbf{r}(t) = 4t\mathbf{i}$

22. $\mathbf{r}(t) = 4t\mathbf{i} - 2t\mathbf{j}$

23. $\mathbf{r}(t) = 4t^2\mathbf{i}$

24. $\mathbf{r}(t) = t^2\mathbf{j} + \mathbf{k}$

In Exercises 25–30, find $\mathbf{T}(t)$, $\mathbf{N}(t)$, a_T, and a_N at the given time t for the plane curve $\mathbf{r}(t)$.

25. $\mathbf{r}(t) = t\mathbf{i} + \frac{1}{t}\mathbf{j}, \quad t = 1$

26. $\mathbf{r}(t) = t^2\mathbf{i} + 2t\mathbf{j}, \quad t = 1$

27. $\mathbf{r}(t) = e^t\cos t\mathbf{i} + e^t\sin t\mathbf{j}, \quad t = \frac{\pi}{2}$

28. $\mathbf{r}(t) = a\cos\omega t\mathbf{i} + b\sin\omega t\mathbf{j}, \quad t = 0$

29. $\mathbf{r}(t) = \langle \cos\omega t + \omega t\sin\omega t, \sin\omega t - \omega t\cos\omega t \rangle, \quad t = t_0$

30. $\mathbf{r}(t) = \langle \omega t - \sin\omega t, 1 - \cos\omega t \rangle, \quad t = t_0$

Circular Motion In Exercises 31–34, consider an object moving according to the position function

$\mathbf{r}(t) = a\cos\omega t\,\mathbf{i} + a\sin\omega t\,\mathbf{j}.$

31. Find $\mathbf{T}(t)$, $\mathbf{N}(t)$, a_T, and a_N.

32. Determine the directions of **T** and **N** relative to the position function **r**.

33. Determine the speed of the object at any time t and explain its value relative to the value of a_T.

34. If the angular velocity ω is halved, by what factor is a_N changed?

In Exercises 35 and 36, sketch the graph of the plane curve given by the vector-valued function, and, at the point on the curve determined by $\mathbf{r}(t_0)$, sketch the vectors T and N. Note that N points toward the concave side of the curve.

Function	Time
35. $\mathbf{r}(t) = t\mathbf{i} + \frac{1}{t}\mathbf{j}$	$t_0 = 2$
36. $\mathbf{r}(t) = 2\cos t\mathbf{i} + 2\sin t\mathbf{j}$	$t_0 = \frac{\pi}{4}$

In Exercises 37–40, find $\mathbf{T}(t)$, $\mathbf{N}(t)$, a_T, and a_N at the given time t for the space curve $\mathbf{r}(t)$. (Hint: Find $\mathbf{a}(t)$, $\mathbf{T}(t)$, and a_N. Solve for N in the equation $\mathbf{a}(t) = a_T\mathbf{T} + a_N\mathbf{N}$.)

Function	Time
37. $\mathbf{r}(t) = t\mathbf{i} + 2t\mathbf{j} - 3t\mathbf{k}$	$t = 1$
38. $\mathbf{r}(t) = 4t\mathbf{i} - 4t\mathbf{j} + 2t\mathbf{k}$	$t = 2$
39. $\mathbf{r}(t) = t\mathbf{i} + t^2\mathbf{j} + \frac{t^2}{2}\mathbf{k}$	$t = 1$
40. $\mathbf{r}(t) = e^t\sin t\mathbf{i} + e^t\cos t\mathbf{j} + e^t\mathbf{k}$	$t = 0$

In Exercises 41 and 42, use a computer algebra system to graph the space curve. Then find $\mathbf{T}(t)$, $\mathbf{N}(t)$, a_T, and a_N at the given time t. Sketch $\mathbf{T}(t)$ and $\mathbf{N}(t)$ on the space curve.

Function	Time
41. $\mathbf{r}(t) = 4t\mathbf{i} + 3\cos t\mathbf{j} + 3\sin t\mathbf{k}$	$t = \frac{\pi}{2}$
42. $\mathbf{r}(t) = t\mathbf{i} + 3t^2\mathbf{j} + \frac{t^2}{2}\mathbf{k}$	$t = 2$

Getting at the Concept

43. Define the unit tangent vector, the principle unit normal vector, and the tangential and normal components of acceleration.

44. How is the unit tangent vector related to the orientation of a curve?

45. Describe the motion of a particle if the normal component of acceleration is 0.

46. Describe the motion of a particle if the tangential component of acceleration is 0.

47. **Cycloidal Motion** The figure shows the path of a particle modeled by the vector-valued function

 $\mathbf{r}(t) = \langle \pi t - \sin\pi t, 1 - \cos\pi t \rangle.$

 The figure also shows the vectors $\mathbf{v}(t)/\|\mathbf{v}(t)\|$ and $\mathbf{a}(t)/\|\mathbf{a}(t)\|$ at the indicated values of t.

 (a) Find a_T and a_N at $t = \frac{1}{2}$, $t = 1$, and $t = \frac{3}{2}$.

 (b) Determine whether the speed of the particle is increasing or decreasing at each of the indicated values of t. Give reasons for your answers.

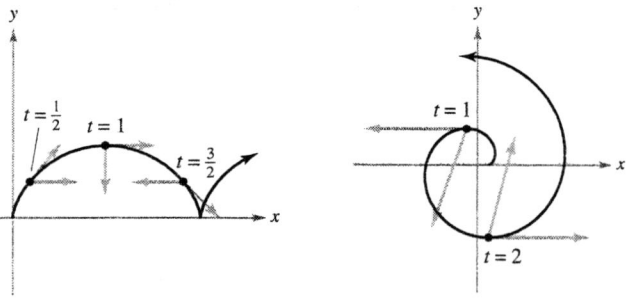

Figure for 47

Figure for 48

48. *Motion Along an Involute of a Circle* The figure shows a particle moving along a path modeled by

$$\mathbf{r}(t) = \langle \cos \pi t + \pi t \sin \pi t, \sin \pi t - \pi t \cos \pi t \rangle.$$

The figure also shows the vectors $\mathbf{v}(t)$ and $\mathbf{a}(t)$ for $t = 1$ and $t = 2$.

(a) Find a_T and a_N at $t = 1$ and $t = 2$.

(b) Determine whether the speed of the particle is increasing or decreasing at each of the indicated values of t. Give reasons for your answers.

In Exercises 49 and 50, find the vectors T and N, and the unit binormal vector $\mathbf{B} = \mathbf{T} \times \mathbf{N}$, for the vector-valued function $\mathbf{r}(t)$ at the indicated value of t.

49. $\mathbf{r}(t) = 2 \cos t \mathbf{i} + 2 \sin t \mathbf{j} + \dfrac{t}{2}\mathbf{k}$

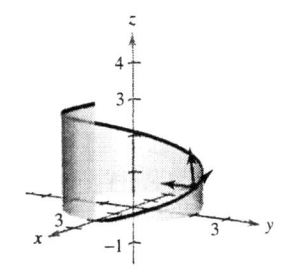

$t_0 = \dfrac{\pi}{2}$

50. $\mathbf{r}(t) = t\mathbf{i} + t^2 \mathbf{j} + \dfrac{t^3}{3}\mathbf{k}$

$t_0 = 1$

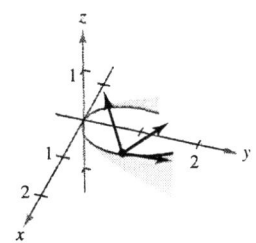

51. *Projectile Motion* Find the tangential and normal components of acceleration for a projectile fired at an angle θ with the horizontal at an initial speed of v_0. What are the components when the projectile is at its maximum height?

52. *Projectile Motion* A projectile is launched with an initial velocity of 100 feet per second at a height of 5 feet and at an angle of 30° with the horizontal.

(a) Determine the vector-valued function for the path of the projectile.

(b) Use a graphing utility to graph the path and approximate the maximum height and range of the projectile.

(c) Find $\mathbf{v}(t)$, $\|\mathbf{v}(t)\|$, and $\mathbf{a}(t)$.

(d) Use a graphing utility to complete the table.

t	0.5	1.0	1.5	2.0	2.5	3.0
Speed						

(e) Use a computer algebra system to find $\mathbf{T}(t)$, $\mathbf{N}(t)$, a_T, and a_N, and verify that $\mathbf{a} = a_T \mathbf{T} + a_N \mathbf{N}$.

(f) Use a graphing utility to graph the scalar functions a_T and a_N. How is the speed of the projectile changing when a_T and a_N have opposite signs?

53. *Air Traffic Control* Because of a storm, ground controllers instruct the pilot of a plane flying at an altitude of 4 miles to make a 90° turn and climb to an altitude of 4.2 miles. The model for the path of the plane during this maneuver is

$$\mathbf{r}(t) = \langle 10 \cos 10\pi t, 10 \sin 10\pi t, 4 + 4t \rangle, \quad 0 \leq t \leq \tfrac{1}{20}$$

where t is the time in hours and $\mathbf{r}$ is the distance in miles.

(a) Determine the speed of the plane.

(b) Use a computer algebra system to calculate a_T and a_N. Why is one of these equal to 0?

54. *Projectile Motion* A plane flying at an altitude of 36,000 feet at a speed of 600 miles per hour releases a bomb. Find the tangential and normal components of acceleration acting on the bomb.

55. *Centripetal Acceleration* An object is spinning at a constant speed on the end of a string, according to the position function given in Exercises 31–34.

(a) If the angular velocity ω is doubled, how is the centripetal component of acceleration changed?

(b) If the angular velocity is unchanged but the length of the string is halved, how is the centripetal component of acceleration changed?

56. *Centripetal Force* An object of mass m moves at a constant speed v in a circular path of radius r. The force required to produce the centripetal component of acceleration is called the *centripetal force* and is given by $F = mv^2/r$. Newton's Law of Universal Gravitation is given by $F = GMm/d^2$, where d is the distance between the centers of the two bodies of masses M and m, and G is a gravitational constant. Use this law to show that the speed required for circular motion is $v = \sqrt{GM/r}$.

Orbital Speed **In Exercises 57–60, use the result of Exercise 56 to find the speed necessary for the given circular orbit around earth. Let $GM = 9.56 \times 10^4$ cubic miles per second per second, and assume the radius of earth is 4000 miles.**

57. The orbit of a space shuttle 100 miles above the surface of earth

58. The orbit of a space shuttle 200 miles above the surface of earth

59. The orbit of a heat capacity mapping satellite 385 miles above the surface of earth

60. The orbit of a SYNCOM satellite r miles above the surface of earth that is in geosynchronous orbit [The satellite completes one orbit per sidereal day (23 hours, 56 minutes), and thus appears to remain stationary above a point on earth.]

61. Prove that the principal unit normal vector $\mathbf{N}$ points toward the concave side of a plane curve.

62. Prove that the vector $\mathbf{T}'(t)$ is $\mathbf{0}$ for an object moving in a straight line.

63. Prove that $a_N = \dfrac{\|\mathbf{v} \times \mathbf{a}\|}{\|\mathbf{v}\|}$.

64. Prove that $a_N = \sqrt{\|\mathbf{a}\|^2 - a_T^2}$.

Arc Length and Curvature

- Find the arc length of a space curve.
- Use the arc length parameter to describe a plane curve or space curve.
- Find the curvature of a curve at a point on the curve.
- Use a vector-valued function to find frictional force.

Arc Length

In Section 9.3, you saw that the arc length of a smooth *plane* curve C given by the parametric equations $x = x(t)$ and $y = y(t)$, $a \le t \le b$, is

$$s = \int_a^b \sqrt{[x'(t)]^2 + [y'(t)]^2}\, dt.$$

In vector form, where C is given by $\mathbf{r}(t) = x(t)\mathbf{i} + y(t)\mathbf{j}$, you can rewrite this equation for arc length as

$$s = \int_a^b \|\mathbf{r}'(t)\|\, dt.$$

The formula for the arc length of a plane curve has a natural extension to a smooth curve in *space*, as stated in the following theorem.

> **THEOREM 11.6 Arc Length of a Space Curve**
>
> If C is a smooth curve given by $\mathbf{r}(t) = x(t)\mathbf{i} + y(t)\mathbf{j} + z(t)\mathbf{k}$, on an interval $[a, b]$, then the arc length of C on the interval is
>
> $$s = \int_a^b \sqrt{[x'(t)]^2 + [y'(t)]^2 + [z'(t)]^2}\, dt = \int_a^b \|\mathbf{r}'(t)\|\, dt.$$

Example 1 Finding the Arc Length of a Curve in Space

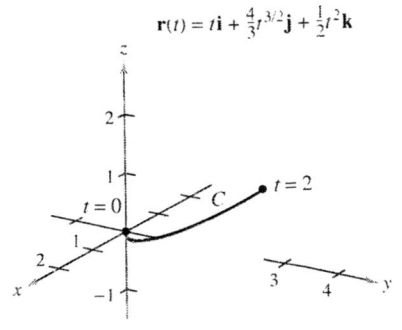

Find the arc length of the curve given by

$$\mathbf{r}(t) = t\mathbf{i} + \frac{4}{3}t^{3/2}\mathbf{j} + \frac{1}{2}t^2\mathbf{k}$$

from $t = 0$ to $t = 2$, as shown in Figure 11.27.

Solution Using $x(t) = t$, $y(t) = \frac{4}{3}t^{3/2}$, and $z(t) = \frac{1}{2}t^2$, you obtain $x'(t) = 1$, $y'(t) = 2t^{1/2}$, and $z'(t) = t$. So, the arc length from $t = 0$ to $t = 2$ is given by

$$\begin{aligned}
s &= \int_0^2 \sqrt{[x'(t)]^2 + [y'(t)]^2 + [z'(t)]^2}\, dt && \text{Formula for arc length}\\
&= \int_0^2 \sqrt{1 + 4t + t^2}\, dt \\
&= \int_0^2 \sqrt{(t+2)^2 - 3}\, dt && \text{Integration tables (Appendix C), Formula 26}\\
&= \left[\frac{t+2}{2}\sqrt{(t+2)^2 - 3} - \frac{3}{2}\ln\left|(t+2) + \sqrt{(t+2)^2 - 3}\right|\right]_0^2 \\
&= 2\sqrt{13} - \frac{3}{2}\ln(4 + \sqrt{13}) - 1 + \frac{3}{2}\ln 3 \\
&\approx 4.816.
\end{aligned}$$

EXPLORATION

Arc Length Formula The formula for the arc length of a space curve is given in terms of the parametric equations used to represent the curve. Does this mean that the arc length of the curve depends on the parameter being used? Would you want this to be true? Explain your reasoning.

Here is a different parametric representation of the curve in Example 1.

$$\mathbf{r}(t) = t^2\mathbf{i} + \frac{4}{3}t^3\mathbf{j} + \frac{1}{2}t^4\mathbf{k}$$

Find the arc length from $t = 0$ to $t = \sqrt{2}$ and compare the result with that found in Example 1.

As t increases from 0 to 2, the vector $\mathbf{r}(t)$ traces out a curve whose length is approximately 4.816.
Figure 11.27

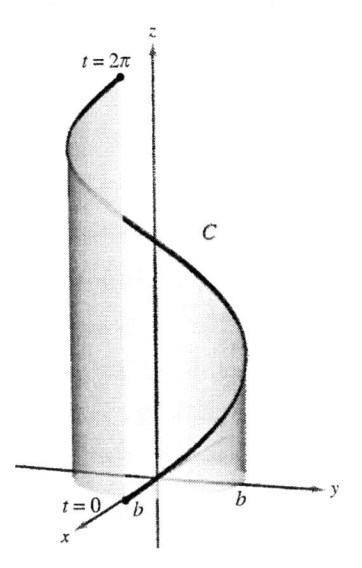

Curve:
$\mathbf{r}(t) = b\cos t\,\mathbf{i} + b\sin t\,\mathbf{j} + \sqrt{1-b^2}\,t\,\mathbf{k}$

One turn of a helix
Figure 11.28

Example 2 Finding the Arc Length of a Helix

Find the length of one turn of the helix given by

$$\mathbf{r}(t) = b\cos t\,\mathbf{i} + b\sin t\,\mathbf{j} + \sqrt{1-b^2}\,t\,\mathbf{k}$$

as shown in Figure 11.28.

Solution Begin by finding the derivative.

$\mathbf{r}(t) = b\cos t\,\mathbf{i} + b\sin t\,\mathbf{j} + \sqrt{1-b^2}\,t\,\mathbf{k}$ Write original function.

$\mathbf{r}'(t) = -b\sin t\,\mathbf{i} + b\cos t\,\mathbf{j} + \sqrt{1-b^2}\,\mathbf{k}$ Derivative

Now, using the formula for arc length, you can find the length of one turn of the helix by integrating $\|\mathbf{r}'(t)\|$ from 0 to 2π.

$$s = \int_0^{2\pi} \|\mathbf{r}'(t)\|\, dt \qquad \text{Formula for arc length}$$

$$= \int_0^{2\pi} \sqrt{b^2(\sin^2 t + \cos^2 t) + (1-b^2)}\, dt$$

$$= \int_0^{2\pi} dt$$

$$= 2\pi.$$

So, the length is 2π units.

Arc Length Parameter

You have seen that curves can be represented by vector-valued functions in different ways, depending on the choice of parameter. For *motion* along a curve, the convenient parameter is time t. However, for studying the *geometric properties* of a curve, the convenient parameter is often arc length s.

$s(t) = \int_a^t \sqrt{[x'(u)]^2 + [y'(u)]^2 + [z'(u)]^2}\, du$

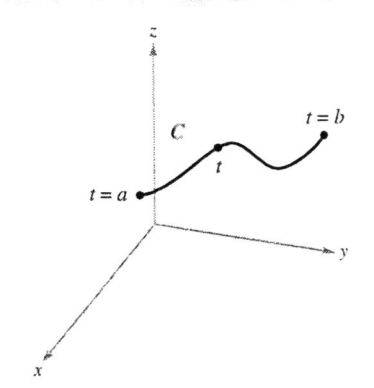

Figure 11.29

Definition of Arc Length Function

Let C be a smooth curve given by $\mathbf{r}(t)$ defined on the closed interval $[a, b]$. For $a \leq t \leq b$, the **arc length function** is given by

$$s(t) = \int_a^t \|\mathbf{r}'(u)\|\, du = \int_a^t \sqrt{[x'(u)]^2 + [y'(u)]^2 + [z'(u)]^2}\, du.$$

The arc length s is called the **arc length parameter.** (See Figure 11.29.)

NOTE The arc length function s is *nonnegative*. It measures the distance along C from the initial point $(x(a), y(a), z(a))$ to the point $(x(t), y(t), z(t))$.

Using the definition of the arc length function and the Second Fundamental Theorem of Calculus, you can conclude that

$$\frac{ds}{dt} = \|\mathbf{r}'(t)\|. \qquad \text{Derivative of arc length function}$$

In differential form, you can write

$$ds = \|\mathbf{r}'(t)\|\, dt.$$

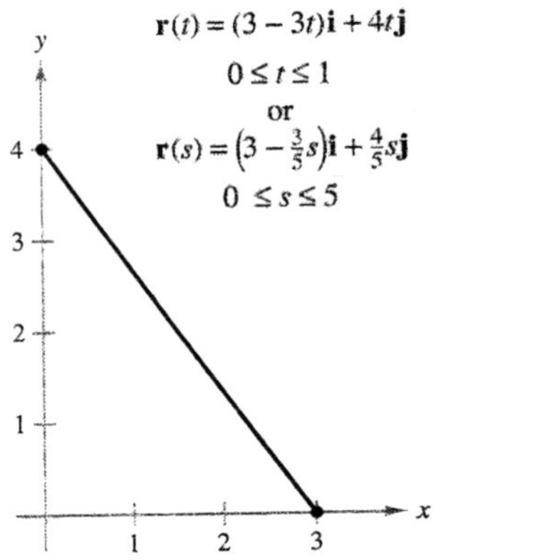

The line segment from (3, 0) to (0, 4) can be parametrized using the arc length parameter s.
Figure 11.30

Example 3 Finding the Arc Length Function for a Line

Find the arc length function $s(t)$ for the line segment given by

$$\mathbf{r}(t) = (3 - 3t)\mathbf{i} + 4t\mathbf{j}, \qquad 0 \le t \le 1$$

and express $\mathbf{r}$ as a function of the parameter s. (See Figure 11.30.)

Solution Because $\mathbf{r}'(t) = -3\mathbf{i} + 4\mathbf{j}$ and

$$\|\mathbf{r}'(t)\| = \sqrt{3^2 + 4^2} = 5$$

you have

$$\begin{aligned} s(t) &= \int_0^t \|\mathbf{r}'(u)\|\, du \\ &= \int_0^t 5\, du \\ &= 5t. \end{aligned}$$

Using $s = 5t$ (or $t = s/5$), you can rewrite $\mathbf{r}$ using the arc length parameter as follows.

$$\mathbf{r}(s) = \left(3 - \tfrac{3}{5}s\right)\mathbf{i} + \tfrac{4}{5}s\mathbf{j}, \qquad 0 \le s \le 5.$$

One of the advantages of writing a vector-valued function in terms of the arc length parameter is that $\|\mathbf{r}'(s)\| = 1$. For instance, in Example 3, you have

$$\|\mathbf{r}'(s)\| = \sqrt{\left(-\frac{3}{5}\right)^2 + \left(\frac{4}{5}\right)^2} = 1.$$

So, for a smooth curve C represented by $\mathbf{r}(s)$, where s is the arc length parameter, the arc length between a and b is

$$\begin{aligned} \text{Length of arc} &= \int_a^b \|\mathbf{r}'(s)\|\, ds \\ &= \int_a^b ds \\ &= b - a \\ &= \text{length of interval.} \end{aligned}$$

Furthermore, if t is *any* parameter such that $\|\mathbf{r}'(t)\| = 1$, then t must be the arc length parameter. These results are summarized in the following theorem, which we state without proof.

THEOREM 11.7 Arc Length Parameter

If C is a smooth curve given by

$$\mathbf{r}(s) = x(s)\mathbf{i} + y(s)\mathbf{j} \quad \text{or} \quad \mathbf{r}(s) = x(s)\mathbf{i} + y(s)\mathbf{j} + z(s)\mathbf{k}$$

where s is the arc length parameter, then

$$\|\mathbf{r}'(s)\| = 1.$$

Moreover, if t is *any* parameter for the vector-valued function $\mathbf{r}$ such that $\|\mathbf{r}'(t)\| = 1$, then t must be the arc length parameter.

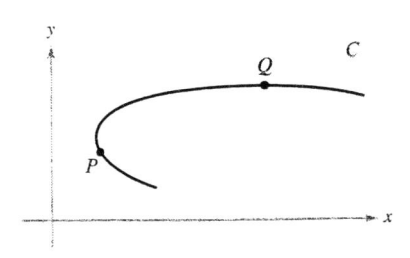

Curvature at P is greater than at Q.
Figure 11.31

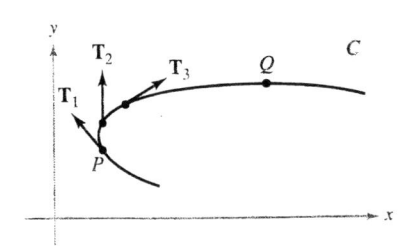

The magnitude of the rate of change of T with respect to the arc length is the curvature of a curve.
Figure 11.32

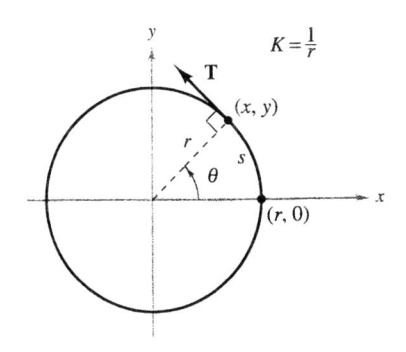

The curvature of a circle is constant.
Figure 11.33

Curvature

An important use of the arc length parameter is to find **curvature**—the measure of how sharply a curve bends. For instance, in Figure 11.31 the curve bends more sharply at P than at Q, and we say that the curvature is greater at P than at Q. You can calculate curvature by calculating the magnitude of the rate of change of the unit tangent vector **T** with respect to the arc length s, as indicated in Figure 11.32.

Definition of Curvature

Let C be a smooth curve (in the plane or in space) given by $\mathbf{r}(s)$, where s is the arc length parameter. The **curvature** at s is given by

$$K = \left\| \frac{d\mathbf{T}}{ds} \right\| = \|\mathbf{T}'(s)\|.$$

A circle has the same curvature at any point. Moreover, the curvature and the radius of the circle are inversely related. That is, a circle with a large radius has a small curvature, and a circle with a small radius has a large curvature. This inverse relationship is made explicit in the following example.

Example 4 Finding the Curvature of a Circle

Show that the curvature of a circle of radius r is $K = 1/r$.

Solution Without loss of generality you can consider the circle to be centered at the origin. Let (x, y) be any point on the circle and let s be the length of the arc from $(r, 0)$ to (x, y), as shown in Figure 11.33. By letting θ be the central angle of the circle, you can represent the circle by

$$\mathbf{r}(\theta) = r\cos\theta\,\mathbf{i} + r\sin\theta\,\mathbf{j}. \qquad \theta \text{ is the parameter.}$$

Using the formula for the length of a circular arc $s = r\theta$, you can rewrite $\mathbf{r}(\theta)$ in terms of the arc length parameter as follows.

$$\mathbf{r}(s) = r\cos\frac{s}{r}\mathbf{i} + r\sin\frac{s}{r}\mathbf{j} \qquad \text{Arc length } s \text{ is the parameter.}$$

So, $\mathbf{r}'(s) = -\sin\frac{s}{r}\mathbf{i} + \cos\frac{s}{r}\mathbf{j}$, and it follows that $\|\mathbf{r}'(s)\| = 1$, which implies that the unit tangent vector is

$$\mathbf{T}(s) = \frac{\mathbf{r}'(s)}{\|\mathbf{r}'(s)\|} = -\sin\frac{s}{r}\mathbf{i} + \cos\frac{s}{r}\mathbf{j}$$

and the curvature is given by

$$K = \|\mathbf{T}'(s)\| = \left\| -\frac{1}{r}\cos\frac{s}{r}\mathbf{i} - \frac{1}{r}\sin\frac{s}{r}\mathbf{j} \right\| = \frac{1}{r}$$

at every point on the circle.

NOTE Because a straight line doesn't "curve," its curvature is 0. Try checking this by finding the curvature of the line given by

$$\mathbf{r}(s) = \left(3 - \frac{3}{5}s\right)\mathbf{i} + \frac{4}{5}s\mathbf{j}.$$

In Example 4, the curvature was found by applying the definition directly. This requires that the curve be written in terms of the arc length parameter s. The following theorem gives two other formulas for finding the curvature of a curve written in terms of an arbitrary parameter t. We leave the proof of this theorem as an exercise [see Exercise 84, parts (a) and (b)].

THEOREM 11.8 Formulas for Curvature

If C is a smooth curve given by $\mathbf{r}(t)$, then the curvature of C at t is given by

$$K = \frac{\|\mathbf{T}'(t)\|}{\|\mathbf{r}'(t)\|} = \frac{\|\mathbf{r}'(t) \times \mathbf{r}''(t)\|}{\|\mathbf{r}'(t)\|^3}.$$

Because $\|\mathbf{r}'(t)\| = ds/dt$, the first formula implies that curvature is the ratio of the rate of change in the tangent vector $\mathbf{T}$ to the rate of change in arc length. To see that this is reasonable, let Δt be a "small number." Then,

$$\frac{\mathbf{T}'(t)}{ds/dt} \approx \frac{[\mathbf{T}(t + \Delta t) - \mathbf{T}(t)]/\Delta t}{[s(t + \Delta t) - s(t)]/\Delta t} = \frac{\mathbf{T}(t + \Delta t) - \mathbf{T}(t)}{s(t + \Delta t) - s(t)} = \frac{\Delta \mathbf{T}}{\Delta s}.$$

In other words, for a given Δs, the greater the length of $\Delta \mathbf{T}$, the more the curve bends at t, as shown in Figure 11.34.

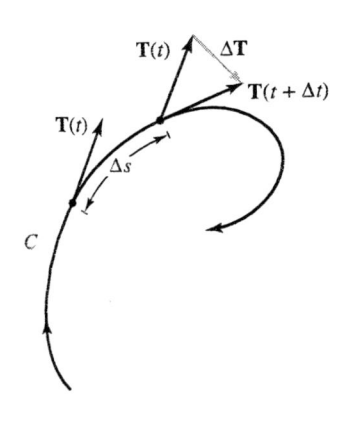

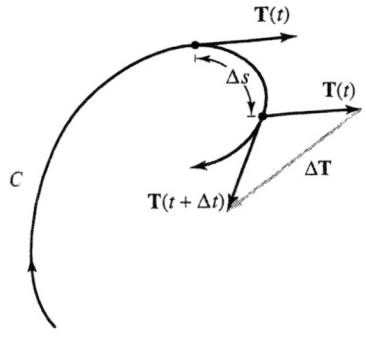

Figure 11.34

Example 5 Finding the Curvature of a Space Curve

Find the curvature of the curve given by $\mathbf{r}(t) = 2t\mathbf{i} + t^2\mathbf{j} - \frac{1}{3}t^3\mathbf{k}$.

Solution It is not apparent whether this parameter represents arc length, so you should use the formula $K = \|\mathbf{T}'(t)\|/\|\mathbf{r}'(t)\|$.

$$\mathbf{r}'(t) = 2\mathbf{i} + 2t\mathbf{j} - t^2\mathbf{k}$$

$$\|\mathbf{r}'(t)\| = \sqrt{4 + 4t^2 + t^4} = t^2 + 2 \qquad \text{Length of } \mathbf{r}'(t)$$

$$\mathbf{T}(t) = \frac{\mathbf{r}'(t)}{\|\mathbf{r}'(t)\|} = \frac{2\mathbf{i} + 2t\mathbf{j} - t^2\mathbf{k}}{t^2 + 2}$$

$$\mathbf{T}'(t) = \frac{(t^2 + 2)(2\mathbf{j} - 2t\mathbf{k}) - (2t)(2\mathbf{i} + 2t\mathbf{j} - t^2\mathbf{k})}{(t^2 + 2)^2}$$

$$= \frac{-4t\mathbf{i} + (4 - 2t^2)\mathbf{j} - 4t\mathbf{k}}{(t^2 + 2)^2}$$

$$\|\mathbf{T}'(t)\| = \frac{\sqrt{16t^2 + 16 - 16t^2 + 4t^4 + 16t^2}}{(t^2 + 2)^2}$$

$$= \frac{2(t^2 + 2)}{(t^2 + 2)^2}$$

$$= \frac{2}{t^2 + 2} \qquad \text{Length of } \mathbf{T}'(t)$$

Therefore,

$$K = \frac{\|\mathbf{T}'(t)\|}{\|\mathbf{r}'(t)\|} = \frac{2}{(t^2 + 2)^2}. \qquad \text{Curvature}$$

The following theorem presents a formula for calculating the curvature of a plane curve given by $y = f(x)$.

THEOREM 11.9 Curvature in Rectangular Coordinates

If C is the graph of a twice-differentiable function given by $y = f(x)$, then the curvature at the point (x, y) is given by

$$K = \frac{|y''|}{[1 + (y')^2]^{3/2}}.$$

Proof By representing the curve C by $\mathbf{r}(x) = x\mathbf{i} + f(x)\mathbf{j} + 0\mathbf{k}$ (where x is the parameter), you obtain $\mathbf{r}'(x) = \mathbf{i} + f'(x)\mathbf{j}$,

$$\|\mathbf{r}'(x)\| = \sqrt{1 + [f'(x)]^2}$$

and $\mathbf{r}''(x) = f''(x)\mathbf{j}$. Because $\mathbf{r}'(x) \times \mathbf{r}''(x) = f''(x)\mathbf{k}$, it follows that the curvature is

$$K = \frac{\|\mathbf{r}'(x) \times \mathbf{r}''(x)\|}{\|\mathbf{r}'(x)\|^3}$$

$$= \frac{|f''(x)|}{\{1 + [f'(x)]^2\}^{3/2}}$$

$$= \frac{|y''|}{[1 + (y')^2]^{3/2}}.$$

Let C be a curve with curvature K at point P. The circle passing through point P with radius $r = 1/K$ is called the **circle of curvature** if the circle lies on the concave side of the curve and shares a common tangent line with the curve at point P. The radius is called the **radius of curvature** at P, and the center of the circle is called the **center of curvature.**

The circle of curvature gives us a nice way to estimate graphically the curvature K at a point P on a curve. Using a compass, you can sketch a circle that snuggles up against the concave side of the curve at point P, as shown in Figure 11.35. If the circle has a radius of r, you can estimate the curvature to be $K = 1/r$.

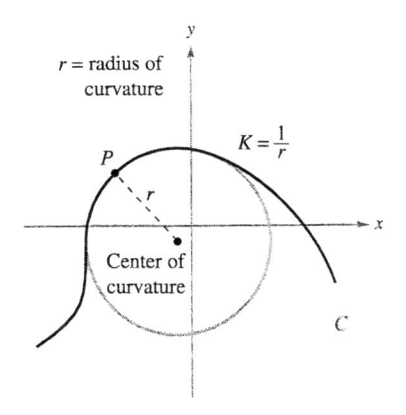

The circle of curvature
Figure 11.35

Example 6 Finding Curvature in Rectangular Coordinates

Find the curvature of the parabola given by $y = x - \frac{1}{4}x^2$ at $x = 2$. Sketch the circle of curvature at $(2, 1)$.

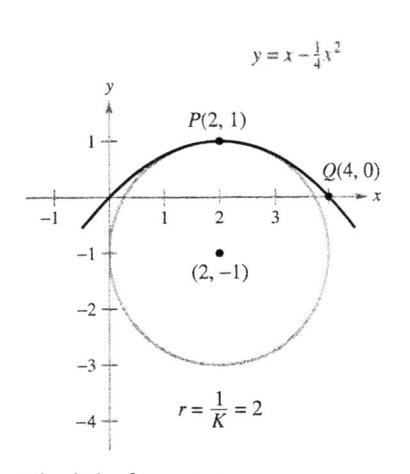

Solution The curvature at $x = 2$ is as follows.

$$y' = 1 - \frac{x}{2} \qquad y' = 0$$

$$y'' = -\frac{1}{2} \qquad y'' = -\frac{1}{2}$$

$$K = \frac{|y''|}{[1 + (y')^2]^{3/2}} \qquad K = \frac{1}{2}$$

Because the curvature at $P(2, 1)$ is $\frac{1}{2}$, it follows that the radius of the circle of curvature at that point is 2. So, the center of curvature is $(2, -1)$, as shown in Figure 11.36. [In the figure, note that the curve has the greatest curvature at P. Try showing that the curvature at $Q(4, 0)$ is $1/2^{5/2} \approx 0.177$.]

The circle of curvature
Figure 11.36

Arc length and curvature are closely related to the tangential and normal components of acceleration. The tangential component of acceleration is the rate of change of the speed, which in turn is the rate of change of the arc length. This component is negative as a moving object slows down and positive as it speeds up—regardless of whether the object is turning or traveling in a straight line. So, the tangential component is solely a function of the arc length and is independent of the curvature.

On the other hand, the normal component of acceleration is a function of **both** speed and curvature. This component measures the acceleration acting perpendicular to the direction of motion. To see why the normal component is affected by both speed and curvature, imagine that you are driving a car around a turn, as shown in Figure 11.37. If your speed is high and the turn is sharp, you feel yourself thrown against the car door. By lowering your speed *or* taking a more gentle turn, you are able to lessen this sideways thrust.

The next theorem explicitly states the relationships among speed, curvature, and the components of acceleration.

The amount of thrust felt by passengers in a car that is turning depends on two things—the speed of the car and the sharpness of the turn.
Figure 11.37

THEOREM 11.10 Acceleration, Speed, and Curvature

If $\mathbf{r}(t)$ is the position vector for a smooth curve C, then the acceleration vector is given by

$$\mathbf{a}(t) = \frac{d^2s}{dt^2}\mathbf{T} + K\left(\frac{ds}{dt}\right)^2\mathbf{N}$$

where K is the curvature of C and ds/dt is the speed.

NOTE Note that Theorem 11.10 gives additional formulas for $a_\mathbf{T}$ and $a_\mathbf{N}$.

Proof For the position vector $\mathbf{r}(t)$, you have

$$\begin{aligned}\mathbf{a}(t) &= a_\mathbf{T}\mathbf{T} + a_\mathbf{N}\mathbf{N} \\ &= D_t[\|\mathbf{v}\|]\mathbf{T} + \|\mathbf{v}\|\,\|\mathbf{T}'\|\mathbf{N} \\ &= \frac{d^2s}{dt^2}\mathbf{T} + \frac{ds}{dt}(\|\mathbf{v}\|K)\mathbf{N} \\ &= \frac{d^2s}{dt^2}\mathbf{T} + K\left(\frac{ds}{dt}\right)^2\mathbf{N}.\end{aligned}$$

Example 7 Tangential and Normal Components of Acceleration

Find $a_\mathbf{T}$ and $a_\mathbf{N}$ for the curve given by

$$\mathbf{r}(t) = 2t\mathbf{i} + t^2\mathbf{j} - \tfrac{1}{3}t^3\mathbf{k}.$$

Solution From Example 5, you know that

$$\frac{ds}{dt} = \|\mathbf{r}'(t)\| = t^2 + 2 \quad \text{and} \quad K = \frac{2}{(t^2+2)^2}.$$

Therefore,

$$a_\mathbf{T} = \frac{d^2s}{dt^2} = 2t \qquad \text{Tangential component}$$

and

$$a_\mathbf{N} = K\left(\frac{ds}{dt}\right)^2 = \frac{2}{(t^2+2)^2}(t^2+2)^2 = 2. \qquad \text{Normal component}$$

Applications

There are many applications in physics and engineering dynamics that involve the relationships among speed, arc length, curvature, and acceleration. One such application concerns frictional force.

Suppose a moving object with mass m is in contact with a stationary object. The total force required to produce an acceleration **a** along a given path is

$$\mathbf{F} = m\mathbf{a} = m\left(\frac{d^2s}{dt^2}\right)\mathbf{T} + mK\left(\frac{ds}{dt}\right)^2 \mathbf{N}$$
$$= ma_{\mathbf{T}}\mathbf{T} + ma_{\mathbf{N}}\mathbf{N}.$$

The portion of this total force that is supplied by the stationary object is called the **force of friction**. For example, if a car moving with constant speed is rounding a turn, the roadway exerts a frictional force that keeps the car from sliding off the road. If the car is not sliding, the frictional force is perpendicular to the direction of motion and has magnitude equal to the normal component of acceleration, as shown in Figure 11.38. The potential frictional force of a road around a turn can be increased by banking the roadway.

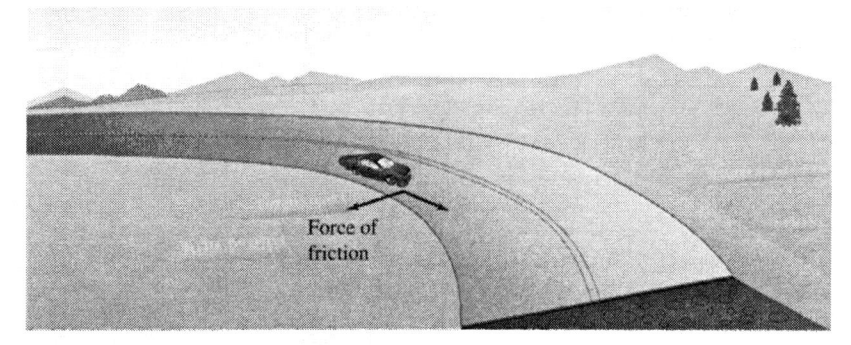

The force of friction is perpendicular to the direction of motion.
Figure 11.38

Example 8 Frictional Force

A 360-kilogram go-cart is driven at a speed of 60 kilometers per hour around a circular racetrack of radius 12 meters. To keep the cart from skidding off course, what frictional force must the track surface exert on the tires? (See Figure 11.39.)

Solution The frictional force must equal the mass times the normal component of acceleration. For this circular path, you know that the curvature is

$$K = \frac{1}{12}. \quad \text{Curvature of circular racetrack}$$

Therefore, the frictional force is

$$ma_{\mathbf{N}} = mK\left(\frac{ds}{dt}\right)^2$$
$$= (360 \text{ kg})\left(\frac{1}{12 \text{ m}}\right)\left(\frac{60{,}000 \text{ m}}{3600 \text{ sec}}\right)^2$$
$$\approx 8333 \text{ (kg)(m)/sec}^2.$$

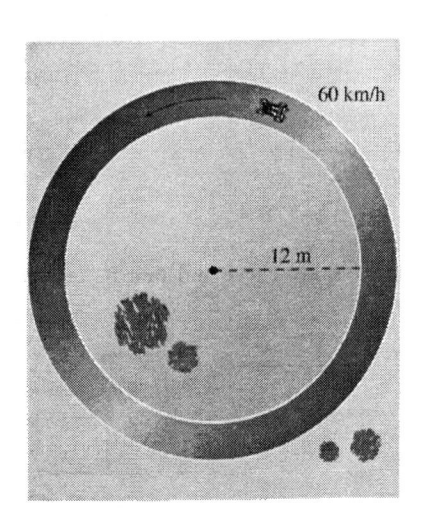

Figure 11.39

Summary of Velocity, Acceleration, and Curvature

Let C be a curve (in the plane or in space) given by the position function

$\mathbf{r}(t) = x(t)\mathbf{i} + y(t)\mathbf{j}$ Curve in the plane

$\mathbf{r}(t) = x(t)\mathbf{i} + y(t)\mathbf{j} + z(t)\mathbf{k}.$ Curve in space

Velocity vector, speed, and acceleration vector:

$\mathbf{v}(t) = \mathbf{r}'(t)$ Velocity vector

$\|\mathbf{v}(t)\| = \dfrac{ds}{dt} = \|\mathbf{r}'(t)\|$ Speed

$\mathbf{a}(t) = \mathbf{r}''(t) = a_\mathbf{T}\mathbf{T}(t) + a_\mathbf{N}\mathbf{N}(t)$ Acceleration vector

Unit tangent vector and principal unit normal vector:

$\mathbf{T}(t) = \dfrac{\mathbf{r}'(t)}{\|\mathbf{r}'(t)\|}$ and $\mathbf{N}(t) = \dfrac{\mathbf{T}'(t)}{\|\mathbf{T}'(t)\|}$

Components of acceleration:

$a_\mathbf{T} = \mathbf{a} \cdot \mathbf{T} = \dfrac{\mathbf{v} \cdot \mathbf{a}}{\|\mathbf{v}\|} = \dfrac{d^2 s}{dt^2}$

$a_\mathbf{N} = \mathbf{a} \cdot \mathbf{N} = \dfrac{\|\mathbf{v} \times \mathbf{a}\|}{\|\mathbf{v}\|} = \sqrt{\|\mathbf{a}\|^2 - a_\mathbf{T}^2} = K\left(\dfrac{ds}{dt}\right)^2$

Formulas for curvature in the plane:

$K = \dfrac{|y''|}{[1 + (y')^2]^{3/2}}$ C given by $y = f(x)$

$K = \dfrac{|x'y'' - y'x''|}{[(x')^2 + (y')^2]^{3/2}}$ C given by $x = x(t),\ y = y(t)$

Formulas for curvature in the plane or in space:

$K = \|\mathbf{T}'(s)\| = \|\mathbf{r}''(s)\|$ s is arc length parameter.

$K = \dfrac{\|\mathbf{T}'(t)\|}{\|\mathbf{r}'(t)\|} = \dfrac{\|\mathbf{r}'(t) \times \mathbf{r}''(t)\|}{\|\mathbf{r}'(t)\|^3}$ t is general parameter.

$K = \dfrac{\mathbf{a}(t) \cdot \mathbf{N}(t)}{\|\mathbf{v}(t)\|^2}$

Cross product formulas apply only to curves in space.

EXERCISES FOR SECTION 11.5

In Exercises 1–4, sketch the plane curve and find its length over the indicated interval.

Function	Interval
1. $\mathbf{r}(t) = t\mathbf{i} + 3t\mathbf{j}$	$[0, 4]$
2. $\mathbf{r}(t) = t\mathbf{i} + t^2\mathbf{k}$	$[0, 4]$
3. $\mathbf{r}(t) = a\cos^3 t\,\mathbf{i} + a\sin^3 t\,\mathbf{j}$	$[0, 2\pi]$
4. $\mathbf{r}(t) = a\cos t\,\mathbf{i} + a\sin t\,\mathbf{j}$	$[0, 2\pi]$

5. **Projectile Motion** A baseball is hit 3 feet above the ground at 100 feet per second and at an angle of 45° with respect to the ground.
   (a) Find the vector-valued function for the path of the baseball.
   (b) Find the maximum height.
   (c) Find the range.
   (d) Find the arc length of the trajectory.

6. **Projectile Motion** If an object is launched from ground level, determine the angle of the launch to obtain (a) the maximum height, (b) the maximum range, and (c) the maximum length of the trajectory. For part (c), let $v_0 = 96$ feet per second.

In Exercises 7–10, sketch the space curve and find its length over the indicated interval.

Function	Interval
7. $\mathbf{r}(t) = 2t\mathbf{i} - 3t\mathbf{j} + t\mathbf{k}$	$[0, 2]$
8. $\mathbf{r}(t) = \langle 3t, 2\cos t, 2\sin t\rangle$	$\left[0, \dfrac{\pi}{2}\right]$
9. $\mathbf{r}(t) = a\cos t\,\mathbf{i} + a\sin t\,\mathbf{j} + bt\,\mathbf{k}$	$[0, 2\pi]$
10. $\mathbf{r}(t) = \langle \cos t + t\sin t,\ \sin t - t\cos t,\ t^2\rangle$	$\left[0, \dfrac{\pi}{2}\right]$

SECTION 11.5 Arc Length and Curvature

In Exercises 11 and 12, use the integration capabilities of a graphing utility to approximate the length of the space curve over the indicated interval.

Function	Interval
11. $\mathbf{r}(t) = t^2\mathbf{i} + t\mathbf{j} + \ln t\mathbf{k}$	$1 \le t \le 3$
12. $\mathbf{r}(t) = \sin \pi t\mathbf{i} + \cos \pi t\mathbf{j} + t^3\mathbf{k}$	$0 \le t \le 2$

13. *Investigation* Consider the graph of the vector-valued function $\mathbf{r}(t) = t\mathbf{i} + (4 - t^2)\mathbf{j} + t^3\mathbf{k}$ on the interval $[0, 2]$.

(a) Approximate the length of the curve by finding the length of the line segment connecting its endpoints.

(b) Approximate the length of the curve by summing the lengths of the line segments connecting the terminal points of the vectors $\mathbf{r}(0)$, $\mathbf{r}(0.5)$, $\mathbf{r}(1)$, $\mathbf{r}(1.5)$, and $\mathbf{r}(2)$.

(c) Describe how you could obtain a more accurate approximation by continuing the processes in parts (a) and (b).

(d) Use the integration capabilities of a graphing utility to approximate the length of the curve. Compare this result with the answers in parts (a) and (b).

14. *Investigation* Repeat Exercise 13 for the vector-valued function $\mathbf{r}(t) = 6\cos(\pi t/4)\mathbf{i} + 2\sin(\pi t/4)\mathbf{j} + t\mathbf{k}$.

15. *Investigation* Consider the helix represented by the vector-valued function $\mathbf{r}(t) = \langle 2\cos t, 2\sin t, t\rangle$.

(a) Express the length of the arc s on the helix as a function of t by evaluating the integral
$$s = \int_0^t \sqrt{[x'(u)]^2 + [y'(u)]^2 + [z'(u)]^2}\, du.$$

(b) Solve for t in the relationship derived in part (a), and substitute the result into the original set of parametric equations. This yields a parametrization of the curve in terms of the arc length parameter s.

(c) Find the coordinates of the point on the helix when the length of the arc is $s = \sqrt{5}$ and $s = 4$.

(d) Verify that $\|\mathbf{r}'(s)\| = 1$.

16. *Investigation* Repeat Exercise 15 for the curve represented by the vector-valued function
$$\mathbf{r}(t) = \langle 4(\sin t - t\cos t), 4(\cos t + t\sin t), \tfrac{3}{2}t^2\rangle.$$

In Exercises 17–20, find the curvature K of the curve, where s is the arc length parameter.

17. $\mathbf{r}(s) = \left(1 + \frac{\sqrt{2}}{2}s\right)\mathbf{i} + \left(1 - \frac{\sqrt{2}}{2}s\right)\mathbf{j}$

18. $\mathbf{r}(s) = (3 + s)\mathbf{i} + \mathbf{j}$

19. Helix in Exercise 15

20. Curve in Exercise 16

In Exercises 21–24, find the curvature K of the plane curve at the indicated value of the parameter.

21. $\mathbf{r}(t) = 4t\mathbf{i} - 2t\mathbf{j},\ t = 1$
22. $\mathbf{r}(t) = t^2\mathbf{j} + \mathbf{k},\ t = 0$
23. $\mathbf{r}(t) = t\mathbf{i} + \frac{1}{t}\mathbf{j},\ t = 1$
24. $\mathbf{r}(t) = t\mathbf{i} + t^2\mathbf{j},\ t = 1$

In Exercises 25–36, find the curvature K of the curve.

25. $\mathbf{r}(t) = 4\cos 2\pi t\mathbf{i} + 4\sin 2\pi t\mathbf{j}$
26. $\mathbf{r}(t) = 2\cos \pi t\mathbf{i} + \sin \pi t\mathbf{j}$
27. $\mathbf{r}(t) = a\cos \omega t\mathbf{i} + a\sin \omega t\mathbf{j}$
28. $\mathbf{r}(t) = a\cos \omega t\mathbf{i} + b\sin \omega t\mathbf{j}$
29. $\mathbf{r}(t) = e^t\cos t\mathbf{i} + e^t\sin t\mathbf{j}$
30. $\mathbf{r}(t) = \langle a(\omega t - \sin \omega t), a(1 - \cos \omega t)\rangle$
31. $\mathbf{r}(t) = \langle \cos \omega t + \omega t\sin \omega t, \sin \omega t - \omega t\cos \omega t\rangle$
32. $\mathbf{r}(t) = 4t\mathbf{i} - 4t\mathbf{j} + 2t\mathbf{k}$
33. $\mathbf{r}(t) = t\mathbf{i} + t^2\mathbf{j} + \dfrac{t^2}{2}\mathbf{k}$
34. $\mathbf{r}(t) = 2t^2\mathbf{i} + t\mathbf{j} + \dfrac{1}{2}t^2\mathbf{k}$
35. $\mathbf{r}(t) = 4t\mathbf{i} + 3\cos t\mathbf{j} + 3\sin t\mathbf{k}$
36. $\mathbf{r}(t) = e^t\cos t\mathbf{i} + e^t\sin t\mathbf{j} + e^t\mathbf{k}$

In Exercises 37–42, find the curvature and radius of curvature of the plane curve at the indicated value of x.

Function	Point
37. $y = 3x - 2$	$x = a$
38. $y = mx + b$	$x = a$
39. $y = 2x^2 + 3$	$x = -1$
40. $y = 2x + \dfrac{4}{x}$	$x = 1$
41. $y = \sqrt{a^2 - x^2}$	$x = 0$
42. $y = \tfrac{3}{4}\sqrt{16 - x^2}$	$x = 0$

Writing **In Exercises 43 and 44, two circles of curvature to the graph of the function are given. (a) Find the equation of the smaller circle, and (b) write a short paragraph explaining why the circles have different radii.**

43. $f(x) = \sin x$

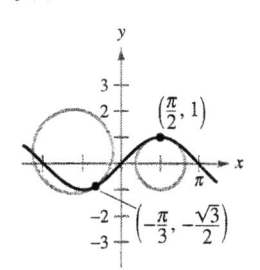

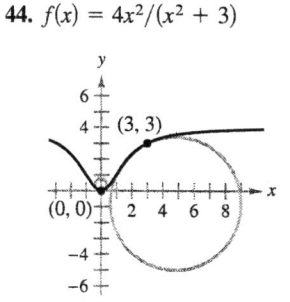

44. $f(x) = 4x^2/(x^2 + 3)$

In Exercises 45–48, use a graphing utility to graph the function. In the same viewing window, graph the circle of curvature to the graph at the indicated value of x.

45. $y = x + \dfrac{1}{x},\ x = 1$
46. $y = \ln x,\ x = 1$
47. $y = e^x,\ x = 0$
48. $y = \tfrac{1}{3}x^3,\ x = 1$

Evolute An *evolute* is the curve formed by the set of centers of curvature of a curve. In Exercises 49 and 50, a curve and its evolute are given. Use a compass to sketch the circles of curvature with centers at points *A* and *B*. To print an enlarged copy of the graph, go to the website *www.mathgraphs.com*.

49. Cycloid: $x = t - \sin t$
    $y = 1 - \cos t$
    Evolute: $x = \sin t + t$
    $y = \cos t - 1$

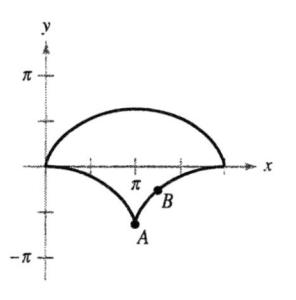

50. Ellipse: $x = 3 \cos t$
    $y = 2 \sin t$
    Evolute: $x = \frac{5}{3} \cos^3 t$
    $y = \frac{5}{2} \sin^3 t$

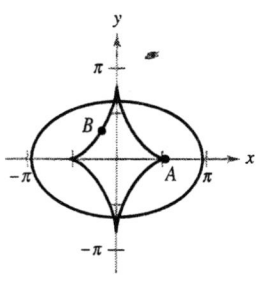

In Exercises 51–54, (a) find the point on the curve at which the curvature *K* is a maximum and (b) find the limit of *K* as $x \to \infty$.

51. $y = (x - 1)^2 + 3$
52. $y = x^3$
53. $y = x^{2/3}$
54. $y = \ln x$

In Exercises 55 and 56, find all points on the graph of the function such that the curvature is zero.

55. $y = (x - 1)^3 + 3$
56. $y = \cos x$

57. *Writing* Use the result of Exercise 55 to write a paragraph describing the points on the graph of $y = f(x)$ at which the curvature is 0.

58. Verify that the curvature at any point (x, y) on the graph of $y = \cosh x$ is $1/y^2$.

Getting at the Concept

59. Give the formula for the arc length of a smooth curve in space.

60. Give the formulas for curvature in the plane and curvature for a space curve.

61. Describe the graph of a vector-valued function for which the curvature is 0 for all values of *t* in its domain.

62. Given a twice-differentiable function $y = f(x)$, determine its curvature at a relative extremum. Can the curvature ever be greater than it is at a relative extremum? Why or why not?

63. Show that the curvature is greatest at the endpoints of the major axis and is least at the endpoints of the minor axis for the ellipse given by $x^2 + 4y^2 = 4$.

64. *Investigation* Find all *a* and *b* such that the two curves given by
$$y_1 = ax(b - x) \text{ and } y_2 = \frac{x}{x + 2}$$
intersect at only one point and have a common tangent line and equal curvature at that point. Sketch a graph for each set of values of *a* and *b*.

65. *Investigation* Consider the function $f(x) = x^4 - x^2$.
    (a) Use a computer algebra system to find the curvature *K* of the curve as a function of *x*.
    (b) Use the result of part (a) to find the circle of curvature to the graph of *f* when $x = 0$ and $x = 1$. Use a computer algebra system to graph the function and the two circles of curvature.
    (c) Graph the function $K(x)$ and compare it with the graph of $f(x)$. For example, do the extrema of *f* and *K* occur at the same critical numbers? Explain.

66. *Investigation* The surface of a goblet is formed by revolving the graph of the function
$$y = \tfrac{1}{4} x^{8/5}, \quad 0 \le x \le 5$$
about the *y*-axis. The measurements are given in centimeters.
    (a) Use a computer algebra system to graph the surface.
    (b) Find the volume of the goblet.
    (c) Find the curvature *K* of the generating curve as a function of *x*. Use a graphing utility to graph *K*.
    (d) If a spherical object is dropped into the goblet, is it possible for it to touch the bottom? Explain.

67. A sphere of radius 4 is dropped into the paraboloid given by $z = x^2 + y^2$.
    (a) How close will the sphere come to the vertex of the paraboloid?
    (b) What is the radius of the largest sphere that will touch the vertex?

68. *Speed* The smaller the curvature in a bend of a road, the faster a car can travel. Assume that the maximum speed around a turn is inversely proportional to the square root of the curvature. A car moving on the path
$$y = \tfrac{1}{3} x^3$$
(*x* and *y* are measured in miles) can safely go 30 miles per hour at $(1, \tfrac{1}{3})$. How fast can it go at $(\tfrac{3}{2}, \tfrac{9}{8})$?

69. Let *C* be given by $y = f(x)$. Show that the center of curvature for (x, y) on *C* is $(x_0, y_0) = (x - y'z, y + z)$, where
$$z = \frac{1 + (y')^2}{y''}.$$
Find the center of curvature for $y = e^x$ at the point (0, 1) on the curve.

70. A curve C is given by the polar equation $r = f(\theta)$. Show that the curvature K at the point (r, θ) is

$$K = \frac{|2(r')^2 - rr'' + r^2|}{[(r')^2 + r^2]^{3/2}}.$$

[*Hint:* Represent the curve by $\mathbf{r}(\theta) = r\cos\theta\mathbf{i} + r\sin\theta\mathbf{j}$.]

In Exercises 71–74, use the result of Exercise 70 to find the curvature of the polar curve.

71. $r = 1 + \sin\theta$ **72.** $r = \theta$

73. $r = a\sin\theta$ **74.** $r = e^\theta$

75. Given the polar curve $r = e^{a\theta}$, $a > 0$, find the curvature K and determine the limit of K as (a) $\theta \to \infty$ and (b) $a \to \infty$.

76. Show that the formula for the curvature of a polar curve $r = f(\theta)$ given in Exercise 70 reduces to

$$K = 2/|r'|$$

for the curvature *at the pole*.

In Exercises 77 and 78, use the result of Exercise 76 to find the curvature of the rose curve at the pole.

77. $r = 4\sin 2\theta$

78. $r = 6\cos 3\theta$

79. For a smooth curve given by the parametric equations $x = f(t)$ and $y = g(t)$, prove that the curvature is given by

$$K = \frac{|f'(t)g''(t) - g'(t)f''(t)|}{\{[f'(t)]^2 + [g'(t)]^2\}^{3/2}}.$$

80. Use the result of Exercise 79 to find the curvature K of the curve represented by the parametric equations $x(t) = t^3$ and $y(t) = \frac{1}{2}t^2$. Use a graphing utility to graph K and determine any horizontal asymptotes. Interpret the asymptotes in the context of the problem.

81. Use the result of Exercise 79 to find the curvature K of the cycloid represented by the parametric equations

$$x(\theta) = a(\theta - \sin\theta) \quad \text{and} \quad y(\theta) = a(1 - \cos\theta).$$

What are the minimum and maximum values of K?

82. Use Theorem 11.10 to find a_T and a_N for each curve given by the vector-valued function.

(a) $\mathbf{r}(t) = 3t^2\mathbf{i} + (3t - t^3)\mathbf{j}$

(b) $\mathbf{r}(t) = t\mathbf{i} + t^2\mathbf{j} + \frac{1}{2}t^2\mathbf{k}$

83. *Frictional Force* A 5500-pound vehicle is driven at a speed of 30 miles per hour on a circular interchange of radius 100 feet. To keep the vehicle from skidding off course, what frictional force must the road surface exert on the tires?

84. Use the definition of curvature in space, $K = \|\mathbf{T}'(s)\| = \|\mathbf{r}''(s)\|$, to verify the following three formulas.

(a) $K = \dfrac{\|\mathbf{T}'(t)\|}{\|\mathbf{r}'(t)\|}$

(b) $K = \dfrac{\|\mathbf{r}'(t) \times \mathbf{r}''(t)\|}{\|\mathbf{r}'(t)\|^3}$

(c) $K = \dfrac{\mathbf{a}(t) \cdot \mathbf{N}(t)}{\|\mathbf{v}(t)\|^2}$

Kepler's Laws **In Exercises 85–92, you are asked to verify Kepler's Laws of Planetary Motion. For these exercises, assume that each planet moves in an orbit given by the vector-valued function r. Let $r = \|\mathbf{r}\|$, let G represent the universal gravitational constant, let M represent the mass of the sun, and let m represent the mass of the planet.**

85. Prove that $\mathbf{r} \cdot \mathbf{r}' = r\dfrac{dr}{dt}$.

86. Using Newton's Second Law of Motion, $\mathbf{F} = m\mathbf{a}$, and Newton's Second Law of Gravitation, $\mathbf{F} = -(GmM/r^3)\mathbf{r}$, show that $\mathbf{a}$ and $\mathbf{r}$ are parallel, and that

$$\mathbf{r}(t) \times \mathbf{r}'(t) = \mathbf{L}$$

is a constant vector. Hence, $\mathbf{r}(t)$ moves in a *fixed plane*, orthogonal to $\mathbf{L}$.

87. Prove that $\dfrac{d}{dt}\left[\dfrac{\mathbf{r}}{r}\right] = \dfrac{1}{r^3}\{[\mathbf{r} \times \mathbf{r}'] \times \mathbf{r}\}$.

88. Show that

$$\frac{\mathbf{r}'}{GM} \times \mathbf{L} - \frac{\mathbf{r}}{r} = \mathbf{e}$$

is a constant vector.

89. Prove Kepler's First Law: Each planet moves in an elliptical orbit with the sun as a focus.

90. Assume that the elliptical orbit

$$r = \frac{ed}{1 + e\cos\theta}$$

is in the xy-plane, with $\mathbf{L}$ along the z-axis. Prove that

$$\|\mathbf{L}\| = r^2\frac{d\theta}{dt}.$$

91. Prove Kepler's Second Law: Each ray from the sun to a planet sweeps out equal areas of the ellipse in equal times.

92. Prove Kepler's Third Law: The square of the period of a planet's orbit is proportional to the cube of the mean distance between the planet and the sun.

REVIEW EXERCISES FOR CHAPTER 11

11.1 In Exercises 1–4, (a) find the domain of r and (b) determine the values (if any) of t for which the function is continuous.

1. $\mathbf{r}(t) = t\mathbf{i} + \csc t\,\mathbf{k}$
2. $\mathbf{r}(t) = \sqrt{t}\,\mathbf{i} + \dfrac{1}{t-4}\mathbf{j} + \mathbf{k}$
3. $\mathbf{r}(t) = \ln t\,\mathbf{i} + t\mathbf{j} + t\mathbf{k}$
4. $\mathbf{r}(t) = (2t+1)\mathbf{i} + t^2\mathbf{j} + t\mathbf{k}$

In Exercises 5 and 6, evaluate (if possible) the vector-valued function at the indicated value of t.

5. $\mathbf{r}(t) = (2t+1)\mathbf{i} + t^2\mathbf{j} - \tfrac{1}{3}t^3\mathbf{k}$
   (a) $\mathbf{r}(0)$ (b) $\mathbf{r}(-2)$ (c) $\mathbf{r}(c-1)$ (d) $\mathbf{r}(1+\Delta t) - \mathbf{r}(1)$

6. $\mathbf{r}(t) = 3\cos t\,\mathbf{i} + (1-\sin t)\mathbf{j} - t\mathbf{k}$
   (a) $\mathbf{r}(0)$ (b) $\mathbf{r}\!\left(\dfrac{\pi}{2}\right)$ (c) $\mathbf{r}(s-\pi)$ (d) $\mathbf{r}(\pi+\Delta t) - \mathbf{r}(\pi)$

In Exercises 7 and 8, sketch the plane curve represented by the vector-valued function and give the orientation of the curve.

7. $\mathbf{r}(t) = \langle \cos t,\, 2\sin^2 t \rangle$
8. $\mathbf{r}(t) = \langle t,\, t/(t-1) \rangle$

In Exercises 9–14, use a computer algebra system to graph the space curve represented by the vector-valued function.

9. $\mathbf{r}(t) = \mathbf{i} + t\mathbf{j} + t^2\mathbf{k}$
10. $\mathbf{r}(t) = 2t\mathbf{i} + t\mathbf{j} + t^2\mathbf{k}$
11. $\mathbf{r}(t) = \langle 1, \sin t, 1 \rangle$
12. $\mathbf{r}(t) = \langle 2\cos t, t, 2\sin t \rangle$
13. $\mathbf{r}(t) = \langle t, \ln t, \tfrac{1}{2}t^2 \rangle$
14. $\mathbf{r}(t) = \langle \tfrac{1}{2}t, \sqrt{t}, \tfrac{1}{4}t^3 \rangle$

In Exercises 15 and 16, find vector-valued functions forming the boundaries of the region in the figure.

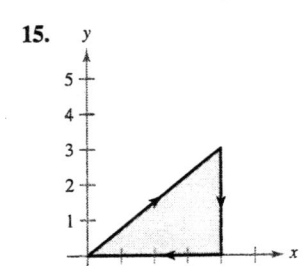

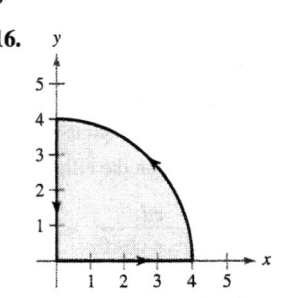

15.

16.

17. A particle moves on a straight-line path that passes through the points $(-2, -3, 8)$ and $(5, 1, -2)$. Find a vector-valued function for the path. (There are many correct answers.)

18. The outer edge of a spiral staircase is in the shape of a helix of radius 2 meters. The staircase has a height of 2 meters and is three-fourths of one complete revolution from bottom to top. Find a vector-valued function for the helix. (There are many correct answers.)

In Exercises 19 and 20, sketch the space curve represented by the intersection of the surfaces. Use the parameter $x = t$ to find a vector-valued function for the space curve.

19. $z = x^2 + y^2$, $x + y = 0$
20. $x^2 + z^2 = 4$, $x - y = 0$

In Exercises 21 and 22, evaluate the limit.

21. $\displaystyle\lim_{t \to 2^-} \left(t^2\mathbf{i} + \sqrt{4-t^2}\,\mathbf{j} + \mathbf{k} \right)$
22. $\displaystyle\lim_{t \to 0} \left(\dfrac{\sin 2t}{t}\mathbf{i} + e^{-t}\mathbf{j} + e^t\mathbf{k} \right)$

11.2 In Exercises 23 and 24, find the following.

(a) $\mathbf{r}'(t)$ (b) $\mathbf{r}''(t)$ (c) $D_t[\mathbf{r}(t) \cdot \mathbf{u}(t)]$
(d) $D_t[\mathbf{u}(t) - 2\mathbf{r}(t)]$ (e) $D_t[\|\mathbf{r}(t)\|]$, $t > 0$ (f) $D_t[\mathbf{r}(t) \times \mathbf{u}(t)]$

23. $\mathbf{r}(t) = 3t\mathbf{i} + (t-1)\mathbf{j},\quad \mathbf{u}(t) = t\mathbf{i} + t^2\mathbf{j} + \tfrac{2}{3}t^3\mathbf{k}$

24. $\mathbf{r}(t) = \sin t\,\mathbf{i} + \cos t\,\mathbf{j} + t\mathbf{k},\quad \mathbf{u}(t) = \sin t\,\mathbf{i} + \cos t\,\mathbf{j} + \dfrac{1}{t}\mathbf{k}$

25. *Writing* The x- and y-components of the derivative of the vector-valued function $\mathbf{u}$ are positive at $t = t_0$, and the z-component is negative. Describe the behavior of $\mathbf{u}$ at $t = t_0$.

26. *Writing* The x-component of the derivative of the vector-valued function $\mathbf{u}$ is 0 for t in the domain of the function. What does this information imply about the graph of $\mathbf{u}$?

In Exercises 27–30, find the indefinite integral.

27. $\displaystyle\int (\cos t\,\mathbf{i} + t\cos t\,\mathbf{j})\,dt$
28. $\displaystyle\int (\ln t\,\mathbf{i} + t\ln t\,\mathbf{j} + \mathbf{k})\,dt$
29. $\displaystyle\int \|\cos t\,\mathbf{i} + \sin t\,\mathbf{j} + t\mathbf{k}\|\,dt$
30. $\displaystyle\int (t\mathbf{j} + t^2\mathbf{k}) \times (\mathbf{i} + t\mathbf{j} + t\mathbf{k})\,dt$

In Exercises 31 and 32, find $\mathbf{r}(t)$ for the given condition.

31. $\mathbf{r}'(t) = 2t\mathbf{i} + e^t\mathbf{j} + e^{-t}\mathbf{k},\quad \mathbf{r}(0) = \mathbf{i} + 3\mathbf{j} - 5\mathbf{k}$
32. $\mathbf{r}'(t) = \sec t\,\mathbf{i} + \tan t\,\mathbf{j} + t^2\mathbf{k},\quad \mathbf{r}(0) = 3\mathbf{k}$

In Exercises 33–36, evaluate the definite integral.

33. $\displaystyle\int_{-2}^{2} (3t\mathbf{i} + 2t^2\mathbf{j} - t^3\mathbf{k})\,dt$
34. $\displaystyle\int_{0}^{1} \left(\sqrt{t}\,\mathbf{j} + t\sin t\,\mathbf{k}\right)dt$
35. $\displaystyle\int_{0}^{2} (e^{t/2}\mathbf{i} - 3t^2\mathbf{j} - \mathbf{k})\,dt$
36. $\displaystyle\int_{-1}^{1} (t^3\mathbf{i} + \arcsin t\,\mathbf{j} - t^2\mathbf{k})\,dt$

11.3 In Exercises 37 and 38, the position vector $\mathbf{r}$ describes the path of an object moving in space. Find the velocity, speed, and acceleration of the object.

37. $\mathbf{r}(t) = \langle \cos^3 t,\, \sin^3 t,\, 3t \rangle$
38. $\mathbf{r}(t) = \langle t,\, -\tan t,\, e^t \rangle$

Linear Approximation In Exercises 39 and 40, find a set of parametric equations for the tangent line to the graph of the vector-valued function at $t = t_0$. Use the equations for the line to approximate $r(t_0 + 0.1)$.

39. $\mathbf{r}(t) = \ln(t-3)\mathbf{i} + t^2\mathbf{j} + \frac{1}{2}t\mathbf{k}, \quad t_0 = 4$
40. $\mathbf{r}(t) = 3\cosh t\,\mathbf{i} + \sinh t\,\mathbf{j} - 2t\,\mathbf{k}, \quad t_0 = 0$

Projectile Motion In Exercises 41–44, use the model for projectile motion, assuming there is no air resistance. [$a(t) = -32$ ft/sec² or $a(t) = -9.8$ m/sec²]

41. A projectile is fired from ground level at an angle of 30° with the horizontal. Find the range of the projectile if the initial velocity is 75 feet per second.

42. The center of a truckbed is 6 feet below and 4 feet horizontally from the end of a horizontal conveyor that is discharging gravel (see figure). Determine the speed ds/dt at which the conveyor belt should be moving so that the gravel falls onto the center of the truck bed.

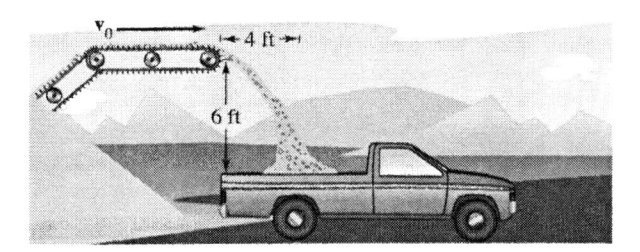

43. A projectile is fired from ground level at an angle of 20° with the horizontal. Find the minimum initial velocity if the projectile has a range of 80 meters.

44. Use a graphing utility to graph the paths of a projectile if $v_0 = 20$ meters per second and
    (a) $\theta = 30°$, (b) $\theta = 45°$, and (c) $\theta = 60°$.
    Use the graphs to approximate the maximum height and range of the projectile for each case.

11.4 In Exercises 45–52, find the velocity, speed, and acceleration at time t. Then find $\mathbf{a} \cdot \mathbf{T}$ and $\mathbf{a} \cdot \mathbf{N}$ at time t.

45. $\mathbf{r}(t) = 5t\mathbf{i}$
46. $\mathbf{r}(t) = (1+4t)\mathbf{i} + (2-3t)\mathbf{j}$
47. $\mathbf{r}(t) = t\mathbf{i} + \sqrt{t}\,\mathbf{j}$
48. $\mathbf{r}(t) = 2(t+1)\mathbf{i} + \frac{2}{t+1}\mathbf{j}$
49. $\mathbf{r}(t) = e^t\mathbf{i} + e^{-t}\mathbf{j}$
50. $\mathbf{r}(t) = t\cos t\,\mathbf{i} + t\sin t\,\mathbf{j}$
51. $\mathbf{r}(t) = t\mathbf{i} + t^2\mathbf{j} + \frac{1}{2}t^2\mathbf{k}$
52. $\mathbf{r}(t) = (t-1)\mathbf{i} + t\mathbf{j} + \frac{1}{t}\mathbf{k}$

In Exercises 53 and 54, find a set of parametric equations for the line tangent to the space curve at the indicated point.

53. $\mathbf{r}(t) = 2\cos t\,\mathbf{i} + 2\sin t\,\mathbf{j} + t\,\mathbf{k}, \quad t = \frac{3\pi}{4}$
54. $\mathbf{r}(t) = t\mathbf{i} + t^2\mathbf{j} + \frac{2}{3}t^3\mathbf{k}, \quad t = 2$

55. *Satellite Orbit* Find the speed necessary for a satellite to maintain a circular orbit 600 miles above the surface of earth.

56. *Centripetal Force* An automobile in a circular traffic exchange is traveling at twice the posted speed. By what factor is the centripetal force increased over that which would occur at the posted speed?

11.5 In Exercises 57–60, sketch the plane curve and find its length over the indicated interval.

Function	Interval
57. $\mathbf{r}(t) = 2t\mathbf{i} - 3t\mathbf{j}$	$[0, 5]$
58. $\mathbf{r}(t) = t^2\mathbf{i} + 2t\mathbf{k}$	$[0, 3]$
59. $\mathbf{r}(t) = 10\cos^3 t\,\mathbf{i} + 10\sin^3 t\,\mathbf{j}$	$[0, 2\pi]$
60. $\mathbf{r}(t) = 10\cos t\,\mathbf{i} + 10\sin t\,\mathbf{j}$	$[0, 2\pi]$

In Exercises 61–64, sketch the space curve and find its length over the indicated interval.

Function	Interval
61. $\mathbf{r}(t) = -3t\mathbf{i} + 2t\mathbf{j} + 4t\mathbf{k}$	$[0, 3]$
62. $\mathbf{r}(t) = t\mathbf{i} + t^2\mathbf{j} + 2t\mathbf{k}$	$[0, 2]$
63. $\mathbf{r}(t) = \langle 8\cos t, 8\sin t, t\rangle$	$\left[0, \frac{\pi}{2}\right]$
64. $\mathbf{r}(t) = \langle 2(\sin t - t\cos t), 2(\cos t + t\sin t), t\rangle$	$\left[0, \frac{\pi}{2}\right]$

In Exercises 65 and 66, use a computer algebra system to find the length of the space curve over the indicated interval.

65. $\mathbf{r}(t) = \frac{1}{2}t\mathbf{i} + \sin t\,\mathbf{j} + \cos t\,\mathbf{k}, \quad 0 \le t \le \pi$
66. $\mathbf{r}(t) = e^t\sin t\,\mathbf{i} + e^t\cos t\,\mathbf{k}, \quad 0 \le t \le \pi$

In Exercises 67–70, find the curvature K of the curve.

67. $\mathbf{r}(t) = 3t\mathbf{i} + 2t\mathbf{j}$
68. $\mathbf{r}(t) = 2\sqrt{t}\,\mathbf{i} + 3t\mathbf{j}$
69. $\mathbf{r}(t) = 2t\mathbf{i} + \frac{1}{2}t^2\mathbf{j} + t^2\mathbf{k}$
70. $\mathbf{r}(t) = 2t\mathbf{i} + 5\cos t\,\mathbf{j} + 5\sin t\,\mathbf{k}$

In Exercises 71–74, find the curvature and radius of curvature of the plane curve at the specified value of x.

71. $y = \frac{1}{2}x^2 + 2, \quad x = 4$
72. $y = e^{-x/2}, \quad x = 0$
73. $y = \ln x, \quad x = 1$
74. $y = \tan x, \quad x = \frac{\pi}{4}$

75. *Writing* A civil engineer designs a highway as indicated in the figure. BC is an arc of the circle. AB and CD are straight lines tangent to the circular arc. Criticize the design.

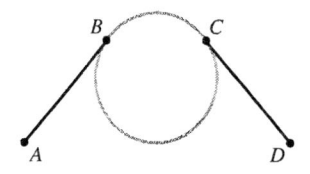

P.S. Problem Solving

1. The **cornu spiral** is given by

   $$x(t) = \int_0^t \cos\left(\frac{\pi u^2}{2}\right) du \quad \text{and} \quad y(t) = \int_0^t \sin\left(\frac{\pi u^2}{2}\right) du.$$

   The spiral shown in the figure was plotted over the interval $-\pi \leq t \leq \pi$.

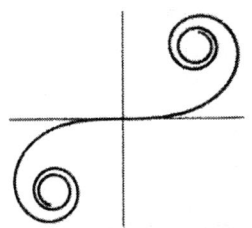

   Generated by Mathematica

   (a) Find the arc length of this curve from $t = 0$ to $t = a$.

   (b) Find the curvature of the graph when $t = a$.

   (c) The cornu spiral was discovered by James Bernoulli. He found that the spiral has an amazing relationship between curvature and arc length. What is this relationship?

2. Let T be the tangent line at the point $P(x, y)$ to the graph of the curve $x^{2/3} + y^{2/3} = a^{2/3}$, $a > 0$, as shown in the figure. Show that the radius of curvature at P is three times the distance from the origin to the tangent line T.

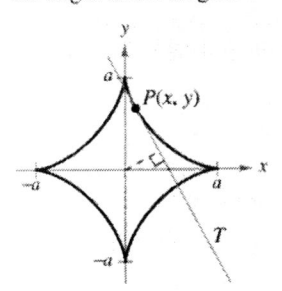

3. A bomber is flying horizontally at an altitude of 3200 feet with a velocity of 400 feet per second when it releases a bomb. A projectile is launched five seconds later from a cannon at a site facing the bomber and 5000 feet from the point beneath the original position of the bomber, as indicated in the figure. If the projectile is to intercept the bomb at an altitude of 1600 feet, determine the initial speed and angle of inclination of the projectile. (Ignore air resistance.)

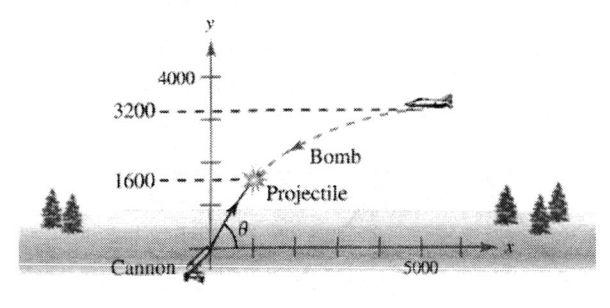

4. Repeat Exercise 3 if the bomber is facing **away** from the launch site, as indicated in the figure.

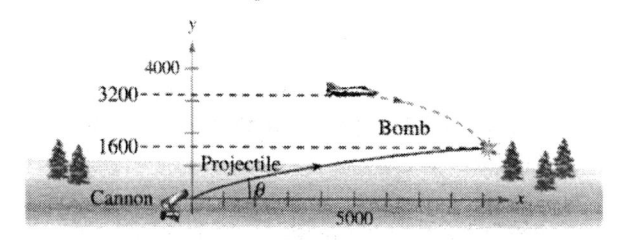

5. Consider one arch of the cycloid

   $$\mathbf{r}(\theta) = (\theta - \sin\theta)\mathbf{i} + (1 - \cos\theta)\mathbf{j}, \ 0 \leq \theta \leq 2\pi$$

   as indicated in the figure. Let $s(\theta)$ be the arc length from the highest point on the arch to the point $(x(\theta), y(\theta))$, and let $\rho(\theta) = \dfrac{1}{K}$ be the radius of curvature at the point $(x(\theta), y(\theta))$. Show that s and ρ are related by the equation $s^2 + \rho^2 = 16$. (This equation is called a *natural equation* for the curve.)

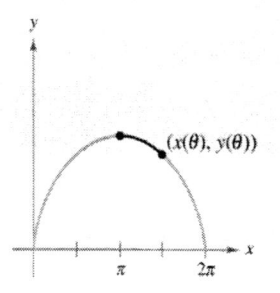

6. Consider the cardioid $r = 1 - \cos\theta$, $0 \leq \theta \leq 2\pi$, as shown in the figure. Let $s(\theta)$ be the arc length from the point $(2, \pi)$ on the cardioid to the point (r, θ), and let $\rho(\theta) = \dfrac{1}{K}$ be the radius of curvature at the point (r, θ). Show that s and ρ are related by the equation $s^2 + 9\rho^2 = 16$. (This equation is called a *natural equation* for the curve.)

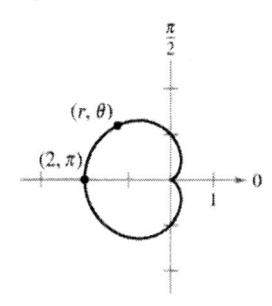

7. If $\mathbf{r}(t)$ is a nonzero differentiable function of t, prove that

   $$\frac{d}{dt}(\|\mathbf{r}(t)\|) = \frac{1}{\|\mathbf{r}(t)\|}\mathbf{r}(t) \cdot \mathbf{r}'(t).$$

8. A communications satellite moves in a circular orbit around earth at a distance of 42,000 kilometers from the center of earth. The angular velocity

$$\frac{d\theta}{dt} = \omega = \frac{\pi}{12} \text{ radians per hour}$$

is constant.

(a) Use polar coordinates to show that the acceleration vector is given by

$$\mathbf{a} = \frac{d^2\mathbf{r}}{dt^2} = \left[\frac{d^2r}{dt^2} - r\left(\frac{d\theta}{dt}\right)^2\right]\mathbf{u}_r + \left[r\frac{d^2\theta}{dt^2} + 2\frac{dr}{dt}\frac{d\theta}{dt}\right]\mathbf{u}_\theta$$

where $\mathbf{u}_r = \cos\theta\mathbf{i} + \sin\theta\mathbf{j}$ is the unit vector in the radial direction and $\mathbf{u}_\theta = -\sin\theta\mathbf{i} + \cos\theta\mathbf{j}$.

(b) Find the radial and angular components of the acceleration for the satellite.

In Exercises 9–11, use the binormal vector defined by the equation $\mathbf{B} = \mathbf{T} \times \mathbf{N}$.

9. Find the unit tangent, unit normal, and binormal vectors for the helix $\mathbf{r}(t) = 4\cos t\mathbf{i} + 4\sin t\mathbf{j} + 3t\mathbf{k}$ at $t = \frac{\pi}{2}$. Sketch the helix together with these three mutually orthogonal unit vectors.

10. Find the unit tangent, unit normal, and binormal vectors for the curve $\mathbf{r}(t) = \cos t\mathbf{i} + \sin t\mathbf{j} - \mathbf{k}$ at $t = \frac{\pi}{4}$. Sketch the curve together with these three mutually orthogonal unit vectors.

11. (a) Prove that there exists a scalar τ, called the **torsion**, such that $d\mathbf{B}/ds = -\tau\mathbf{N}$.

   (b) Prove that $\frac{d\mathbf{N}}{ds} = -K\mathbf{T} + \tau\mathbf{B}$.

   (The three equations $d\mathbf{T}/ds = K\mathbf{N}$, $d\mathbf{N}/ds = -K\mathbf{T} + \tau\mathbf{B}$, and $d\mathbf{B}/ds = -\tau\mathbf{N}$ are called the *Frenet-Serret formulas*.)

12. A highway has an exit ramp that begins at the origin of a coordinate system and follows the curve $y = \frac{1}{32}x^{5/2}$ to the point $(4, 1)$ (see figure). Then it follows a circular path whose curvature is that given by the curve at $(4, 1)$. What is the radius of the circular arc? Explain why the curve and the circular arc should have the same curvature at $(4, 1)$.

13. Consider the vector-valued function $\mathbf{r}(t) = (t\cos\pi t, t\sin\pi t)$, $0 \le t \le 2$.

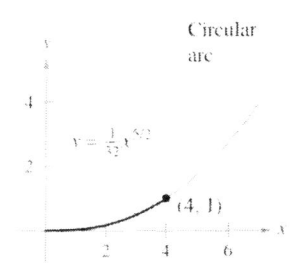

   (a) Use a graphing utility to graph the function.
   (b) Find the length of the arc in part (a).
   (c) Find the curvature K as a function of t. Find the curvature when t is 0, 1, and 2.
   (d) Use a graphing utility to graph the function K.
   (e) Find (if possible) $\lim_{t\to\infty} K$.
   (f) Using the result in part (e), make a conjecture about the graph of $\mathbf{r}$ as $t \to \infty$.

14. You want to toss an object to a friend who is riding a Ferris wheel (see figure). The following parametric equations give the path of the friend $\mathbf{r}_1(t)$ and the path of the object $\mathbf{r}_2(t)$. Distance is measured in meters and time is measured in seconds.

$$\mathbf{r}_1(t) = 15\left(\sin\frac{\pi t}{10}\right)\mathbf{i} + \left(16 - 15\cos\frac{\pi t}{10}\right)\mathbf{j}$$

$$\mathbf{r}_2(t) = [22 - 8.03(t - t_0)]\mathbf{i} + [1 + 11.47(t - t_0) - 4.9(t - t_0)^2]\mathbf{j}$$

   (a) Locate your friend's position on the Ferris wheel at time $t = 0$.
   (b) Determine the number of revolutions per minute of the Ferris wheel.
   (c) What are the speed and angle of inclination (in degrees) at which the object is thrown at time $t = t_0$?
   (d) Use a graphing utility to graph the vector-valued functions using a value of t_0 that allows your friend to be within reach of the object. (Do this by trial and error.) Explain the significance of t_0.
   (e) Find the approximate time your friend should be able to catch the object. Approximate the speeds of your friend and the object at that time.

Satellite Receiving Dish

From 1970 through today, home satellite dishes have evolved from large, unwieldy things sitting in yards to small units that can be attached to roofs or other stationary objects. The shape of these dishes, however, has remained the same. Each dish, whether large or small, has the shape of a circular paraboloid. This shape allows the dish to receive signals whose strength is only a few billionths of a watt.

To model a circular paraboloid, you can modify the elliptic paraboloid formula described in Section 10.6.

$$z = \frac{x^2}{a^2} + \frac{y^2}{a^2} \qquad \text{Circular paraboloid}$$

For this model, the paraboloid opens up and has its vertex at the origin. For a satellite dish, the axis of the paraboloid should pass through the satellite so that all incoming signals are parallel to the paraboloid's axis. When incoming rays strike the surface of the paraboloid at a point P, they reflect at the same angle as they would if they were reflecting off a plane that was tangent to the surface at point P, as shown in the left-hand figure below.

All parabolas have a special reflective property. One way to describe the property is to say that *all incoming rays that are parallel to the axis of the parabola reflect directly through the focus of the parabola*, as shown in the right-hand figure below.

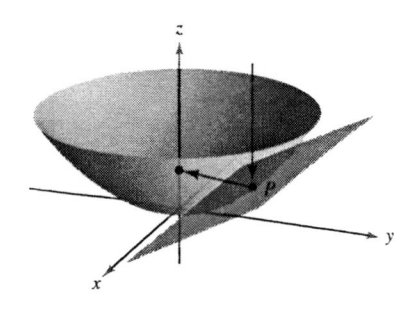

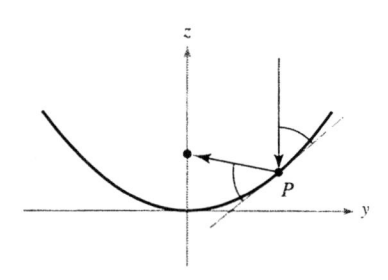

QUESTIONS

1. You are designing a satellite receiving dish. At what point should you place the receiver? Explain your reasoning.

2. Other than satellite receiving dishes, what other common objects use the reflective property of parabolas in their design?

3. Consider the circular paraboloid given by

    $$z = x^2 + y^2$$

    and the point $P(1, 1, 2)$ on this surface. You can characterize the tangent plane to the surface at this point by saying that it is the only plane whose intersection with the paraboloid consists of the single point P. Find an equation for this tangent plane. Explain your strategy.

4. Does it make sense to talk about the "slope" of the tangent plane in Question 3? If so, what is the slope of this plane? How might you use differentiation to discover the slope of this plane?

The concepts presented here will be explored further in this chapter. For an extension of this application, see Lab 17 in the lab series that accompanies this text at college.hmco.com.

Functions of Several Variables

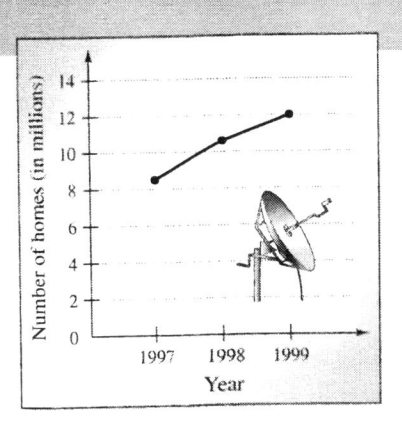

12

The number of satellite television users is growing each year. This trend is expected to continue as costs decrease and additional services, such as cellular and Internet access, become available via satellite.

Older home satellite dishes have diameters of 10 feet. Today's consumers, subscribing to a direct-broadcast satellite system, may have satellite dishes with diameters as small as $1\frac{1}{2}$ feet.

Section 12.1 Introduction to Functions of Several Variables

- Understand the notation for a function of several variables.
- Sketch the graph of a function of two variables.
- Sketch level curves for a function of two variables.
- Sketch level surfaces for a function of three variables.
- Use computer graphics to sketch the graph of a function of two variables.

Functions of Several Variables

So far in this text, we have dealt only with functions of single (independent) variables. Many familiar quantities, however, are functions of two or more variables. For instance, the work done by a force ($W = FD$) and the volume of a right circular cylinder ($V = \pi r^2 h$) are both functions of two variables. The volume of a rectangular solid ($V = lwh$) is a function of three variables. The notation for a function of two or more variables is similar to that for a function of a single variable. Here are two examples.

$$z = \underbrace{f(x, y)}_{\text{2 variables}} = x^2 + xy \qquad \text{Function of two variables}$$

and

$$w = \underbrace{f(x, y, z)}_{\text{3 variables}} = x + 2y - 3z \qquad \text{Function of three variables}$$

> **Definition of a Function of Two Variables**
>
> Let D be a set of ordered pairs of real numbers. If to each ordered pair (x, y) in D there corresponds a unique real number $f(x, y)$, then f is called a **function of x and y**. The set D is the **domain** of f, and the corresponding set of values for $f(x, y)$ is the **range** of f.

For the function given by $z = f(x, y)$, x and y are called the **independent variables** and z is called the **dependent variable**.

Similar definitions can be given to functions of three, four, or n variables, where the domains consist of ordered triples (x_1, x_2, x_3), quadruples (x_1, x_2, x_3, x_4), and n-tuples $(x_1, x_2, \ldots, x_n)$. In all cases, the range is a set of real numbers. In this chapter, we limit the discussion to functions of two or three variables.

As with functions of one variable, the most common way to describe a function of several variables is with an *equation*, and unless otherwise restricted, you can assume that the domain is the set of all points for which the equation is defined. For instance, the domain of the function given by

$$f(x, y) = x^2 + y^2$$

is assumed to be the entire xy-plane. Similarly, the domain of

$$f(x, y) = \ln xy$$

is the set of all points (x, y) in the plane for which $xy > 0$. This consists of all points in the first and third quadrants.

EXPLORATION

Comparing Dimensions Without using a graphing utility, describe the graph of each function of two variables.

a. $z = x^2 + y^2$
b. $z = x + y$
c. $z = x^2 + y$
d. $z = \sqrt{x^2 + y^2}$
e. $z = \sqrt{1 - x^2 + y^2}$

MARY FAIRFAX SOMERVILLE (1780–1872)

Somerville was interested in the problem of creating geometric models for functions of several variables. Her most well-known book, *The Mechanics of the Heavens*, was published in 1831.

Example 1 Domains of Functions of Several Variables

Find the domain of each function.

a. $f(x, y) = \dfrac{\sqrt{x^2 + y^2 - 9}}{x}$ **b.** $g(x, y, z) = \dfrac{x}{\sqrt{9 - x^2 - y^2 - z^2}}$

Solution

a. The function f is defined for all points (x, y) such that $x \neq 0$ and

$$x^2 + y^2 \geq 9.$$

So, the domain is the set of all points lying on or outside the circle $x^2 + y^2 = 9$, *except* those on the y-axis, as shown in Figure 12.1.

b. The function g is defined for all points (x, y, z) such that

$$x^2 + y^2 + z^2 < 9.$$

Consequently, the domain is the set of all points (x, y, z) lying inside a sphere of radius 3 that is centered at the origin.

Domain of

$f(x, y) = \dfrac{\sqrt{x^2 + y^2 - 9}}{x}$

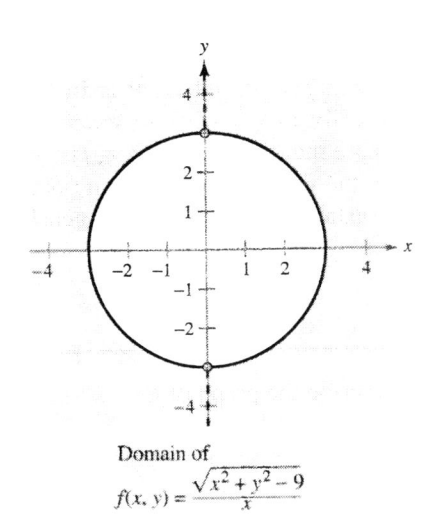

Figure 12.1

Functions of several variables can be combined in the same ways as functions of single variables. For instance, you can form the sum, difference, product, and quotient of two functions of two variables as follows.

$(f \pm g)(x, y) = f(x, y) \pm g(x, y)$ Sum or difference
$(fg)(x, y) = f(x, y)g(x, y)$ Product
$\dfrac{f}{g}(x, y) = \dfrac{f(x, y)}{g(x, y)}, \quad g(x, y) \neq 0$ Quotient

You cannot form the composite of two functions of several variables. However, if h is a function of several variables and g is a function of a single variable, you can form the **composite** function $(g \circ h)(x, y)$ as follows.

$(g \circ h)(x, y) = g(h(x, y))$ Composition

The domain of this composite function consists of all (x, y) in the domain of h such that $h(x, y)$ is in the domain of g. For example, the function given by

$$f(x, y) = \sqrt{16 - 4x^2 - y^2}$$

can be viewed as the composite of the function of two variables given by $h(x, y) = 16 - 4x^2 - y^2$ and the function of a single variable given by $g(u) = \sqrt{u}$. The domain of this function is the set of all points lying on or inside the ellipse given by $4x^2 + y^2 = 16$.

A function that can be expressed as a sum of functions of the form $cx^m y^n$ (where c is a real number and m and n are nonnegative integers) is called a **polynomial function** of two variables. For instance, the functions given by

$$f(x, y) = x^2 + y^2 - 2xy + x + 2 \quad \text{and} \quad g(x, y) = 3xy^2 + x - 2$$

are polynomial functions of two variables. A **rational function** is the quotient of two polynomial functions. Similar terminology is used for functions of more than two variables.

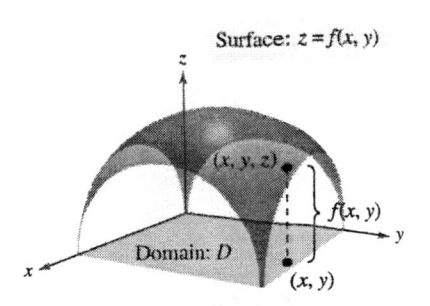

840 CHAPTER 12 Functions of Several Variables

The Graph of a Function of Two Variables

As with functions of a single variable, you can learn a lot about the behavior of a function of two variables by sketching its graph. The **graph** of a function f of two variables is the set of all points (x, y, z) for which $z = f(x, y)$ and (x, y) is in the domain of f. This graph can be interpreted geometrically as a *surface in space*, as discussed in Sections 10.5 and 10.6. In Figure 12.2, note that the graph of $z = f(x, y)$ is a surface whose projection onto the xy-plane is D, the domain of f. To each point (x, y) in D there corresponds a point (x, y, z) on the surface, and, conversely, to each point (x, y, z) on the surface there corresponds a point (x, y) in D.

Figure 12.2

Example 2 Describing the Graph of a Function of Two Variables

What is the range of $f(x, y) = \sqrt{16 - 4x^2 - y^2}$? Describe the graph of f.

Solution The domain D implied by the equation for f is the set of all points (x, y) such that $16 - 4x^2 - y^2 \geq 0$. So, D is the set of all points lying on or inside the ellipse given by

$$\frac{x^2}{4} + \frac{y^2}{16} = 1. \qquad \text{Ellipse in the } xy\text{-plane}$$

The range of f is all values $z = f(x, y)$ such that $0 \leq z \leq \sqrt{16}$ or

$$0 \leq z \leq 4. \qquad \text{Range of } f$$

A point (x, y, z) is on the graph of f if and only if

$$z = \sqrt{16 - 4x^2 - y^2}$$
$$z^2 = 16 - 4x^2 - y^2$$
$$4x^2 + y^2 + z^2 = 16$$
$$\frac{x^2}{4} + \frac{y^2}{16} + \frac{z^2}{16} = 1, \qquad 0 \leq z \leq 4.$$

From Section 10.6, you know that the graph of f is the upper half of an ellipsoid, as shown in Figure 12.3.

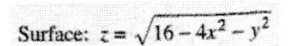

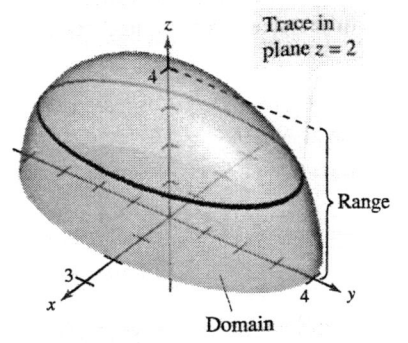

The graph of $f(x, y) = \sqrt{16 - 4x^2 - y^2}$ is the upper half of an ellipsoid.
Figure 12.3

To sketch a surface in space *by hand*, it helps to use traces in planes parallel to the coordinate planes, as shown in Figure 12.3. For example, to find the trace of the surface in the plane $z = 2$, substitute $z = 2$ in the equation $z = \sqrt{16 - 4x^2 - y^2}$ and obtain

$$2 = \sqrt{16 - 4x^2 - y^2} \quad \Longrightarrow \quad \frac{x^2}{3} + \frac{y^2}{12} = 1.$$

So, the trace is an ellipse centered at the point $(0, 0, 2)$ with major and minor axes of lengths $4\sqrt{3}$ and $2\sqrt{3}$.

Traces are also used with most three-dimensional graphing utilities. For instance, Figure 12.4 shows a computer-generated version of the surface given in Example 2. For this sketch, the computer took 25 traces parallel to the xy-plane and 12 traces in vertical planes.

If you have access to a three-dimensional graphing utility, try using it to sketch the graphs of several surfaces.

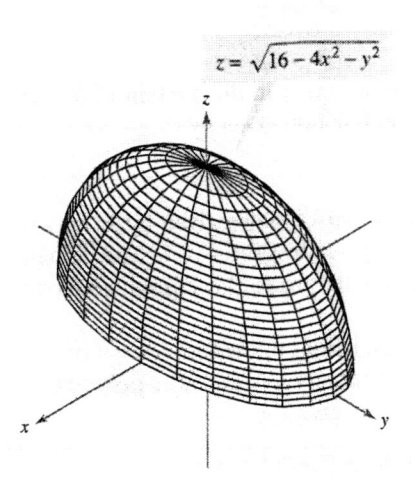

Figure 12.4

Level Curves

A second way to visualize a function of two variables is to use a **scalar field** in which the scalar $z = f(x, y)$ is assigned to the point (x, y). A scalar field can be characterized by **level curves** (or **contour lines**) along which the value of $f(x, y)$ is constant. For instance, the weather map in Figure 12.5 shows level curves of equal pressure called **isobars.** In weather maps for which the level curves represent points of equal temperature, the level curves are called **isotherms,** as shown in Figure 12.6. Another common use of level curves is in representing electric potential fields. In this type of map, the level curves are called **equipotential lines.**

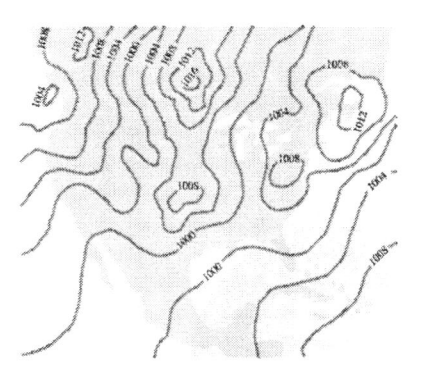

Level curves show the lines of equal pressure (isobars) measured in millibars.
Figure 12.5

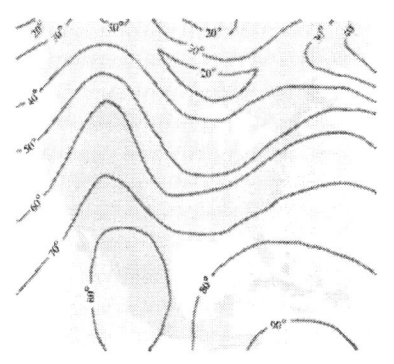

Level curves show the lines of equal temperature (isotherms) measured in degrees Fahrenheit.
Figure 12.6

Contour maps are commonly used to show regions on earth's surface, with the level curves representing the height above sea level. This type of map is called a **topographic map.** For example, the mountain shown in Figure 12.7 is represented by the topographic map in Figure 12.8.

A contour map depicts the variation of z with respect to x and y by the spacing between level curves. Much space between level curves indicates that z is changing slowly, whereas little space indicates a rapid change in z. Furthermore, to give a good three-dimensional illusion in a contour map, it is important to choose c-values that are *evenly spaced.*

Figure 12.7

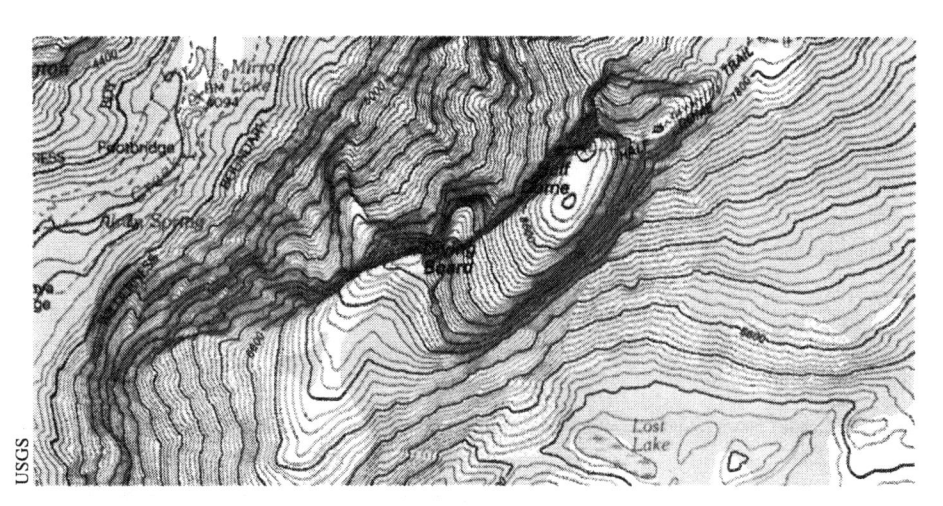

Figure 12.8

Example 3 Sketching a Contour Map

The hemisphere given by $f(x, y) = \sqrt{64 - x^2 - y^2}$ is shown in Figure 12.9. Sketch a contour map for this surface using level curves corresponding to $c = 0, 1, 2, \ldots, 8$.

Solution For each value of c, the equation given by $f(x, y) = c$ is a circle (or point) in the xy-plane. For example, when $c_1 = 0$, the level curve is

$$x^2 + y^2 = 64 \quad \text{Circle of radius 8}$$

which is a circle of radius 8. Figure 12.10 shows the nine level curves for the hemisphere.

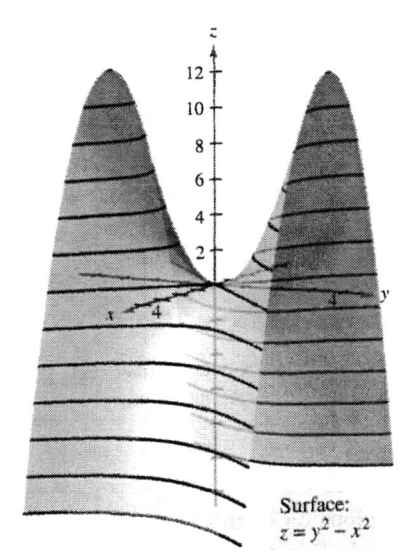

Surface:
$f(x, y) = \sqrt{64 - x^2 - y^2}$

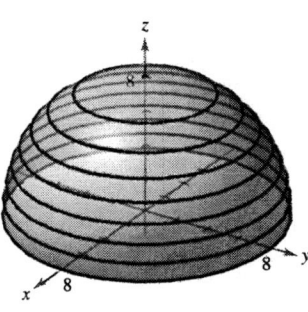

$c_1 = 0$
$c_2 = 1$
$c_3 = 2$
$c_4 = 3$
$c_5 = 4$
$c_6 = 5$
$c_7 = 6$
$c_8 = 7$
$c_9 = 8$

Level curves
Figure 12.9

Contour map
Figure 12.10

Surface:
$z = y^2 - x^2$

Hyperbolic paraboloid
Figure 12.11

Example 4 Sketching a Contour Map

The hyperbolic paraboloid given by

$$z = y^2 - x^2$$

is shown in Figure 12.11. Sketch a contour map for this surface.

Solution For each value of c, we let $f(x, y) = c$ and sketch the resulting level curve in the xy-plane. For this function, each of the level curves ($c \neq 0$) is a hyperbola whose asymptotes are the lines $y = \pm x$. If $c < 0$, the transverse axis is horizontal. For instance, the level curve for $c = -4$ is given by

$$\frac{x^2}{2^2} - \frac{y^2}{2^2} = 1. \quad \text{Hyperbola with horizontal transverse axis}$$

If $c > 0$, the transverse axis is vertical. For instance, the level curve for $c = 4$ is given by

$$\frac{y^2}{2^2} - \frac{x^2}{2^2} = 1. \quad \text{Hyperbola with vertical transverse axis}$$

If $c = 0$, the level curve is the degenerate conic representing the intersecting asymptotes, as shown in Figure 12.12.

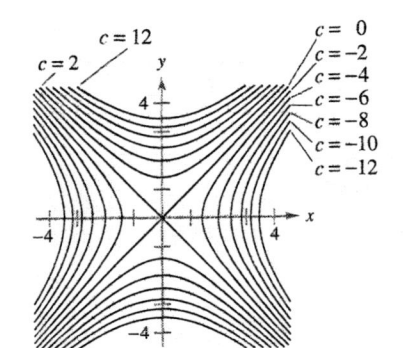

$c = 12$, $c = 2$, $c = 0$, $c = -2$, $c = -4$, $c = -6$, $c = -8$, $c = -10$, $c = -12$

Hyperbolic level curves (at increments of 2)
Figure 12.12

indicates that in the Interactive 3.0 *CD-ROM and* Internet 3.0 *versions of this text (available at* college.hmco.com) *you will find an Open Exploration, which further explores this example using the computer algebra systems* Maple, Mathcad, Mathematica, *and* Derive.

One example of a function of two variables used in economics is the **Cobb-Douglas production function.** This function is used as a model to represent the number of units produced by varying amounts of labor and capital. If x measures the units of labor and y measures the units of capital, the number of units produced is given by

$$f(x, y) = Cx^a y^{1-a}$$

where C is a constant and $0 < a < 1$.

Example 5 The Cobb-Douglas Production Function

A manufacturer estimates a production function to be $f(x, y) = 100x^{0.6}y^{0.4}$, where x is the number of units of labor and y is the number of units of capital. Compare the production level when $x = 1000$ and $y = 500$ with the production level when $x = 2000$ and $y = 1000$.

Solution When $x = 1000$ and $y = 500$, the production level is

$$f(1000, 500) = 100(1000^{0.6})(500^{0.4}) \approx 75{,}786.$$

When $x = 2000$ and $y = 1000$, the production level is

$$f(2000, 1000) = 100(2000^{0.6})(1000^{0.4}) \approx 151{,}572.$$

The level curves of $z = f(x, y)$ are shown in Figure 12.13. Note that by doubling *both* x and y, you double the production level (see Exercise 78).

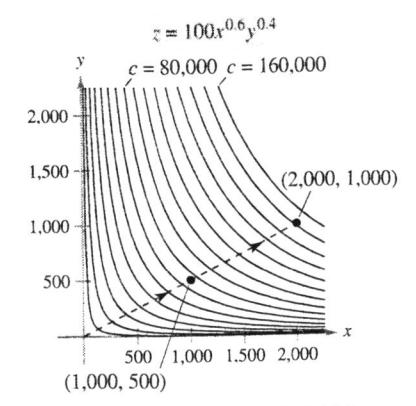

Level curves (at increments of 10,000)
Figure 12.13

Level Surfaces

The concept of a level curve can be extended by one dimension to define a **level surface.** If f is a function of three variables and c is a constant, the graph of the equation $f(x, y, z) = c$ is a **level surface** of the function f, as shown in Figure 12.14.

With computers, engineers and scientists have developed other ways to view functions of three variables. For instance, Figure 12.15 shows a computer simulation that uses color to represent the optimal strain distribution of a car door.

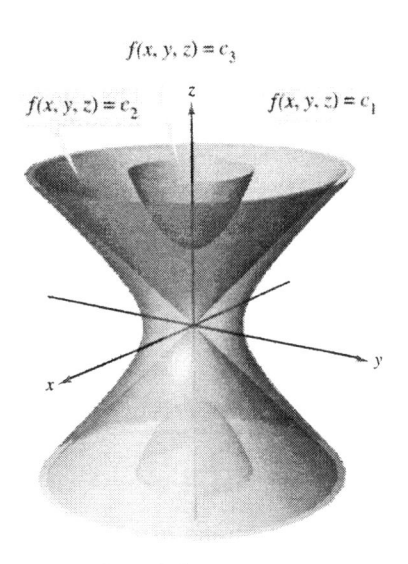

Level surfaces of f
Figure 12.14

Figure 12.15

Example 6   Level Surfaces

Describe the level surfaces of the function
$$f(x, y, z) = 4x^2 + y^2 + z^2.$$

Solution   Each level surface has an equation of the form
$$4x^2 + y^2 + z^2 = c. \quad \text{Equation of level surface}$$

Therefore, the level surfaces are ellipsoids (whose cross sections parallel to the yz-plane are circles). As c increases, the radii of the circular cross sections increase according to the square root of c. For example, the level surfaces corresponding to the values $c = 0$, $c = 4$, and $c = 16$ are as follows.

$$4x^2 + y^2 + z^2 = 0 \quad \text{Level surface for } c = 0 \text{ (single point)}$$

$$\frac{x^2}{1} + \frac{y^2}{4} + \frac{z^2}{4} = 1 \quad \text{Level surface for } c = 4 \text{ (ellipsoid)}$$

$$\frac{x^2}{4} + \frac{y^2}{16} + \frac{z^2}{16} = 1 \quad \text{Level surface for } c = 16 \text{ (ellipsoid)}$$

These level surfaces are shown in Figure 12.16.

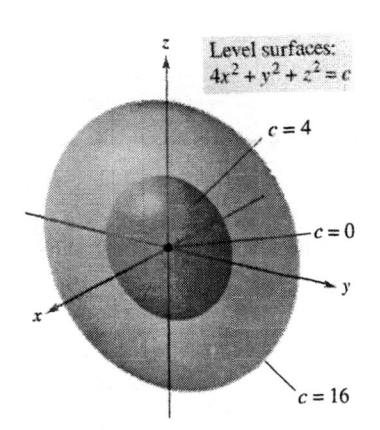

Figure 12.16

NOTE   If the function in Example 6 represented the *temperature* at the point (x, y, z), the level surfaces shown in Figure 12.16 would be called **isothermal surfaces.**

Computer Graphics

The problem of sketching the graph of a surface in space can be simplified by using a computer. Although there are several types of three-dimensional graphing utilities, most use some form of trace analysis to give the illusion of three dimensions. To use such a graphing utility, you usually need to enter the equation of the surface, the region in the xy-plane over which the surface is to be plotted, and the number of traces to be taken. For instance, to sketch the surface given by

$$f(x, y) = (x^2 + y^2)e^{1-x^2-y^2}$$

you might choose the following bounds for x, y, and z.

$-3 \leq x \leq 3$ Bounds for x

$-3 \leq y \leq 3$ Bounds for y

$0 \leq z \leq 3$ Bounds for z

Figure 12.17 shows a computer-generated sketch of this surface using 26 traces taken parallel to the yz-plane. To heighten the three-dimensional effect, the program uses a "hidden line" routine. That is, it begins by plotting the traces in the foreground (those corresponding to the largest x-values), and then, as each new trace is plotted, the program determines whether all or only part of the next trace should be shown.

The graphs on page 845 show a variety of surfaces that were plotted by computer. If you have access to a computer drawing program, try using it to reproduce these surfaces. Remember also that the three-dimensional graphics in this text can be viewed and rotated using the three-dimensional software that accompanies the text and is also incorporated into the *Interactive* CD-ROM version of this text (available at *college.hmco.com*).

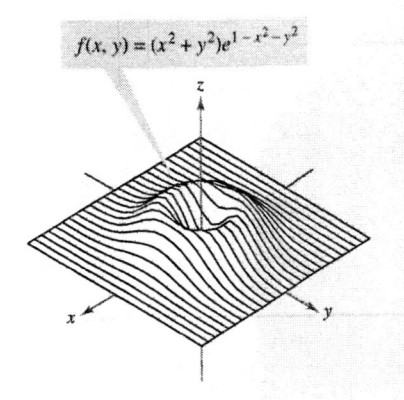

Figure 12.17

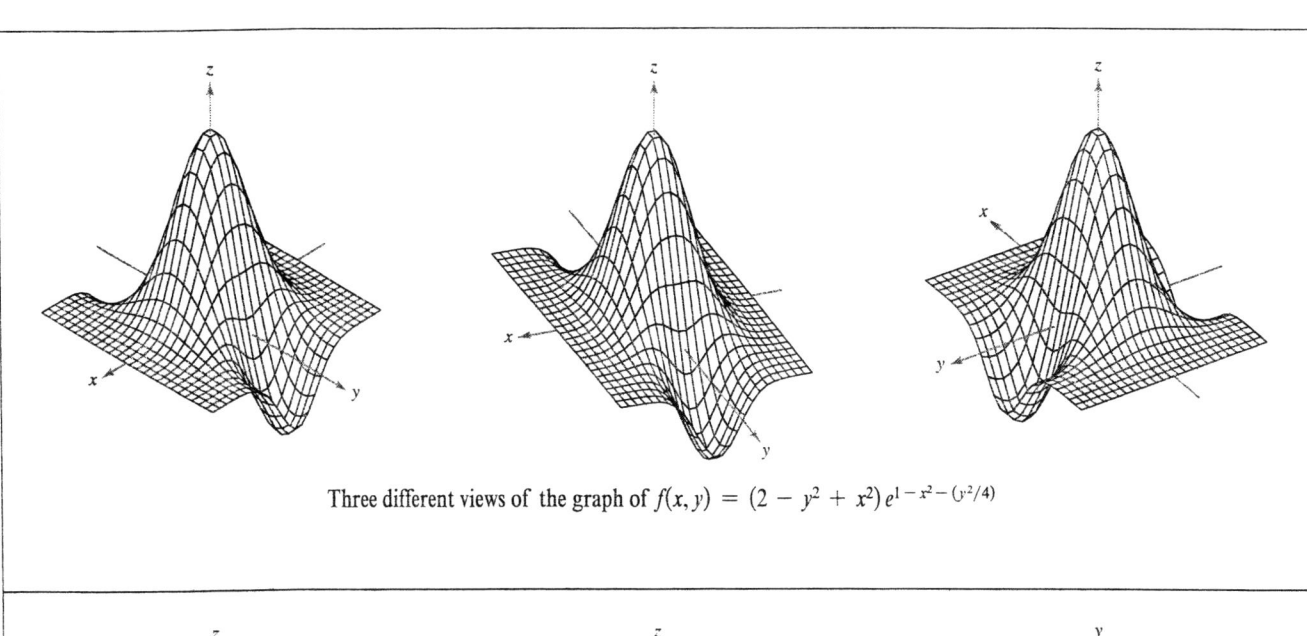

Three different views of the graph of $f(x, y) = (2 - y^2 + x^2) e^{1 - x^2 - (y^2/4)}$

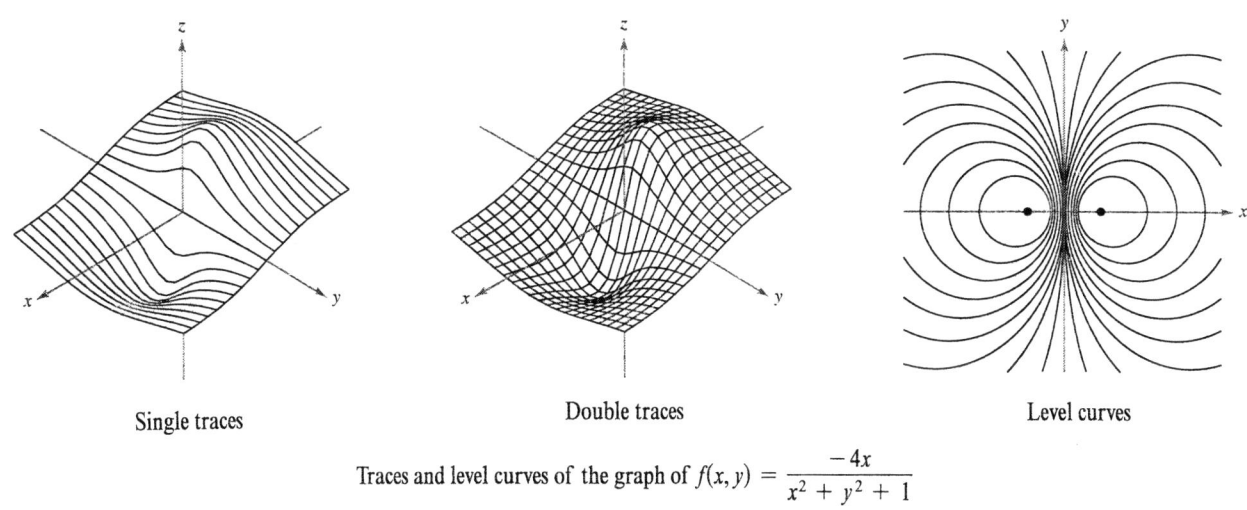

Single traces

Double traces

Level curves

Traces and level curves of the graph of $f(x, y) = \dfrac{-4x}{x^2 + y^2 + 1}$

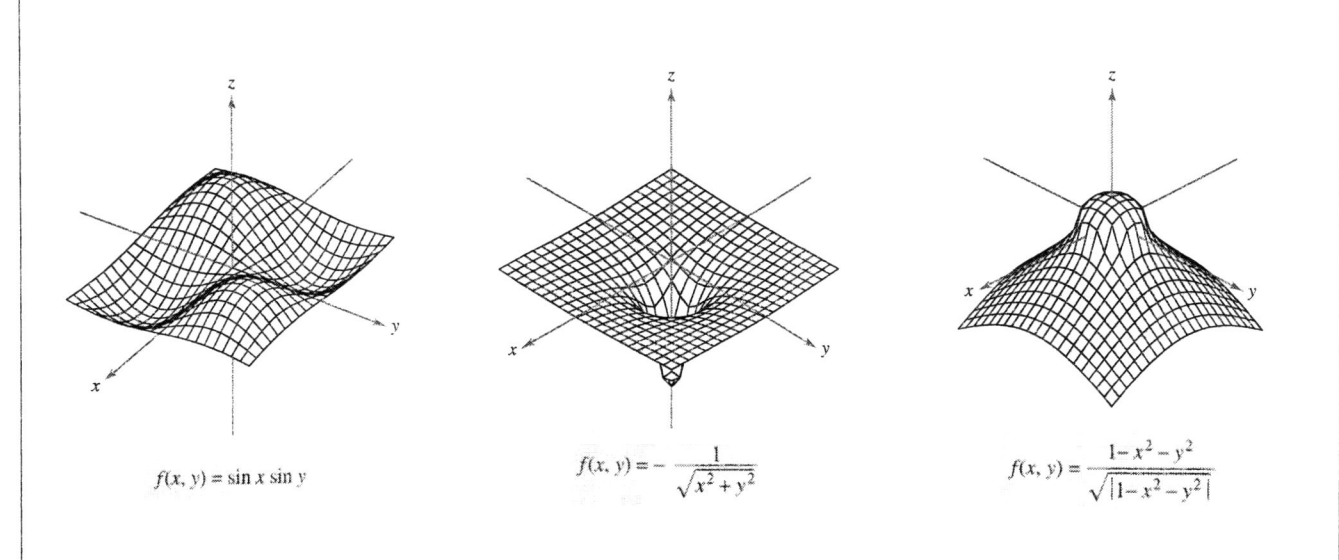

$f(x, y) = \sin x \sin y$

$f(x, y) = -\dfrac{1}{\sqrt{x^2 + y^2}}$

$f(x, y) = \dfrac{1 - x^2 - y^2}{\sqrt{|1 - x^2 - y^2|}}$

EXERCISES FOR SECTION 12.1

In Exercises 1–4, determine whether z is a function of x and y.

1. $x^2z + yz - xy = 10$
2. $xz^2 + 2xy - y^2 = 4$
3. $\dfrac{x^2}{4} + \dfrac{y^2}{9} + z^2 = 1$
4. $z + x \ln y - 8 = 0$

In Exercises 5–16, find and simplify the function values.

5. $f(x, y) = x/y$
   (a) $(3, 2)$ (b) $(-1, 4)$ (c) $(30, 5)$
   (d) $(5, y)$ (e) $(x, 2)$ (f) $(5, t)$

6. $f(x, y) = 4 - x^2 - 4y^2$
   (a) $(0, 0)$ (b) $(0, 1)$ (c) $(2, 3)$
   (d) $(1, y)$ (e) $(x, 0)$ (f) $(t, 1)$

7. $f(x, y) = xe^y$
   (a) $(5, 0)$ (b) $(3, 2)$ (c) $(2, -1)$
   (d) $(5, y)$ (e) $(x, 2)$ (f) (t, t)

8. $g(x, y) = \ln|x + y|$
   (a) $(2, 3)$ (b) $(5, 6)$ (c) $(e, 0)$
   (d) $(0, 1)$ (e) $(2, -3)$ (f) (e, e)

9. $h(x, y, z) = \dfrac{xy}{z}$
   (a) $(2, 3, 9)$ (b) $(1, 0, 1)$

10. $f(x, y, z) = \sqrt{x + y + z}$
    (a) $(0, 5, 4)$ (b) $(6, 8, -3)$

11. $f(x, y) = x \sin y$
    (a) $\left(2, \dfrac{\pi}{4}\right)$ (b) $(3, 1)$

12. $V(r, h) = \pi r^2 h$
    (a) $(3, 10)$ (b) $(5, 2)$

13. $g(x, y) = \displaystyle\int_x^y (2t - 3)\, dt$
    (a) $(0, 4)$ (b) $(1, 4)$

14. $g(x, y) = \displaystyle\int_x^y \dfrac{1}{t}\, dt$
    (a) $(4, 1)$ (b) $(6, 3)$

15. $f(x, y) = x^2 - 2y$
    (a) $\dfrac{f(x + \Delta x, y) - f(x, y)}{\Delta x}$
    (b) $\dfrac{f(x, y + \Delta y) - f(x, y)}{\Delta y}$

16. $f(x, y) = 3xy + y^2$
    (a) $\dfrac{f(x + \Delta x, y) - f(x, y)}{\Delta x}$
    (b) $\dfrac{f(x, y + \Delta y) - f(x, y)}{\Delta y}$

In Exercises 17–28, describe the domain and range of the function.

17. $f(x, y) = \sqrt{4 - x^2 - y^2}$
18. $f(x, y) = \sqrt{4 - x^2 - 4y^2}$
19. $f(x, y) = \arcsin(x + y)$
20. $f(x, y) = \arccos(y/x)$
21. $f(x, y) = \ln(4 - x - y)$
22. $f(x, y) = \ln(xy - 6)$
23. $z = \dfrac{x + y}{xy}$
24. $z = \dfrac{xy}{x - y}$
25. $f(x, y) = e^{x/y}$
26. $f(x, y) = x^2 + y^2$
27. $g(x, y) = \dfrac{1}{xy}$
28. $g(x, y) = x\sqrt{y}$

29. **Think About It** The graphs labeled (a), (b), (c), and (d) are graphs of the function $f(x, y) = -4x/(x^2 + y^2 + 1)$. Match the four graphs with the points in space from which the surface is viewed. The four points are $(20, 15, 25)$, $(-15, 10, 20)$, $(20, 20, 0)$, and $(20, 0, 0)$.

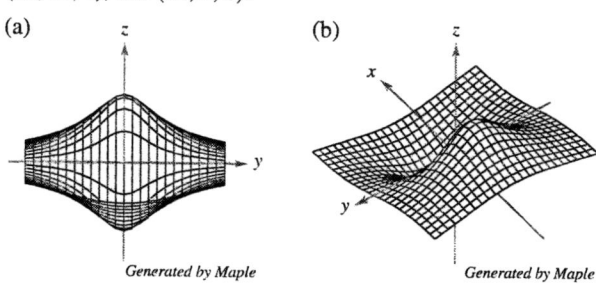

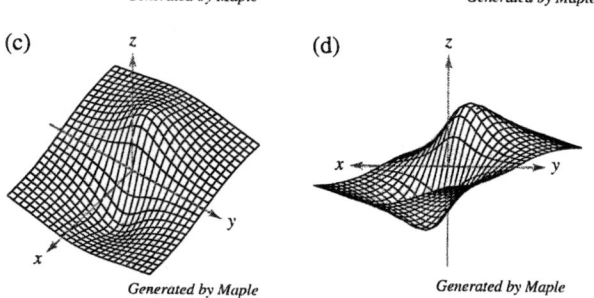

(a) (b) (c) (d)
Generated by Maple

30. **Think About It** Use the function given in Exercise 29.
    (a) Find the domain and range of the function.
    (b) Identify the points in the xy-plane where the function value is 0.
    (c) Does the surface pass through all the octants of the rectangular coordinate system? Give reasons for your answer.

In Exercises 31–38, sketch the surface given by the function.

31. $f(x, y) = 5$
32. $f(x, y) = 6 - 2x - 3y$
33. $f(x, y) = y^2$
34. $g(x, y) = \tfrac{1}{2}x$
35. $z = 4 - x^2 - y^2$
36. $z = \tfrac{1}{2}\sqrt{x^2 + y^2}$
37. $f(x, y) = e^{-x}$
38. $f(x, y) = \begin{cases} xy, & x \geq 0,\ y \geq 0 \\ 0, & x < 0 \text{ or } y < 0 \end{cases}$

In Exercises 39–42, use a computer algebra system to graph the function.

39. $z = y^2 - x^2 + 1$
40. $z = \tfrac{1}{12}\sqrt{144 - 16x^2 - 9y^2}$
41. $f(x, y) = x^2 e^{(-xy/2)}$
42. $f(x, y) = x \sin y$

43. Conjecture Consider the function $f(x, y) = x^2 + y^2$.
(a) Sketch the graph of the surface given by f.
(b) Make a conjecture about the relationship between the graphs of f and $g(x, y) = f(x, y) + 2$. Use a computer algebra system to confirm your answer.
(c) Make a conjecture about the relationship between the graphs of f and $g(x, y) = f(x, y - 2)$. Use a computer algebra system to confirm your answer.
(d) Make a conjecture about the relationship between the graphs of f and $g(x, y) = 4 - f(x, y)$. Use a computer algebra system to confirm your answer.
(e) On the surface in part (a), sketch the graphs of $z = f(1, y)$ and $z = f(x, 1)$.

44. Conjecture Consider the function $f(x, y) = xy$ for $x \geq 0$ and $y \geq 0$.
(a) Sketch the graph of the surface given by f.
(b) Make a conjecture about the relationship between the graphs of f and $g(x, y) = f(x, y) - 3$. Use a computer algebra system to confirm your answer.
(c) Make a conjecture about the relationship between the graphs of f and $g(x, y) = -f(x, y)$. Use a computer algebra system to confirm your answer.
(d) Make a conjecture about the relationship between the graphs of f and $g(x, y) = \frac{1}{2}f(x, y)$. Use a computer algebra system to confirm your answer.
(e) On the surface in part (a), sketch the graph of $z = f(x, x)$.

In Exercises 45–48, match the graph of the surface with one of the contour maps. [The contour maps are labeled (a), (b), (c), and (d).]

(a) 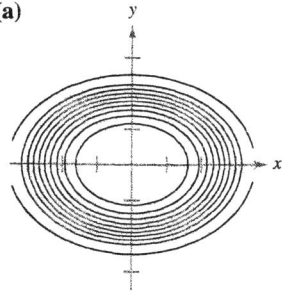
(b) 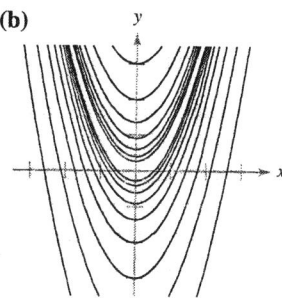
(c) 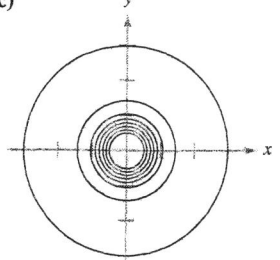
(d)

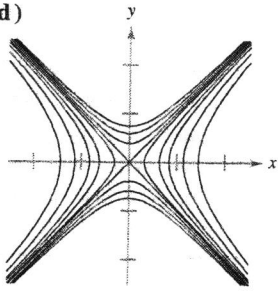

45. $f(x, y) = e^{1-x^2-y^2}$ 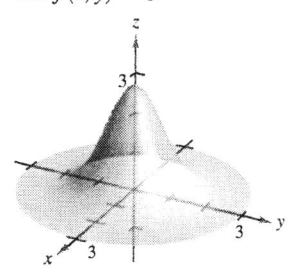
46. $f(x, y) = e^{1-x^2+y^2}$ 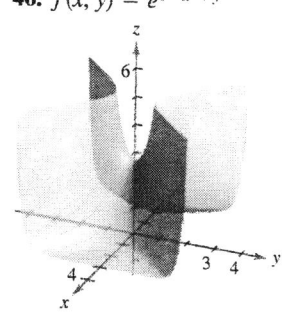
47. $f(x, y) = \ln|y - x^2|$ 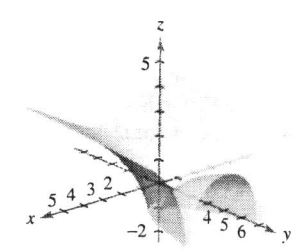
48. $f(x, y) = \cos\left(\dfrac{x^2 + 2y^2}{4}\right)$

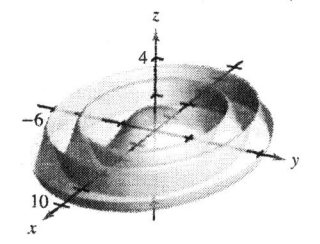

In Exercises 49–56, describe the level curves of the function. Sketch the level curves for the given c-values.

49. $z = x + y$ $c = -1, 0, 2, 4$
50. $z = 6 - 2x - 3y$ $c = 0, 2, 4, 6, 8, 10$
51. $z = \sqrt{25 - x^2 - y^2}$ $c = 0, 1, 2, 3, 4, 5$
52. $f(x, y) = x^2 + 2y^2$ $c = 0, 2, 4, 6, 8$
53. $f(x, y) = xy$ $c = \pm 1, \pm 2, \ldots, \pm 6$
54. $f(x, y) = e^{xy/2}$ $c = 2, 3, 4, \frac{1}{2}, \frac{1}{3}, \frac{1}{4}$
55. $f(x, y) = \dfrac{x}{x^2 + y^2}$ $c = \pm\frac{1}{2}, \pm 1, \pm\frac{3}{2}, \pm 2$
56. $f(x, y) = \ln(x - y)$ $c = 0, \pm\frac{1}{2}, \pm 1, \pm\frac{3}{2}, \pm 2$

In Exercises 57–60, use a graphing utility to sketch six level curves of the function.

57. $f(x, y) = x^2 - y^2 + 2$
58. $f(x, y) = |xy|$
59. $g(x, y) = \dfrac{8}{1 + x^2 + y^2}$
60. $h(x, y) = 3\sin(|x| + |y|)$

Getting at the Concept

61. Define a function of two variables.
62. What is a graph of a function of two variables? How is it interpreted geometrically? Describe level curves.
63. All of the level curves of the surface given by $z = f(x, y)$ are concentric circles. Does this imply that the graph of f is a hemisphere? Illustrate your answer with an example.
64. Construct a function whose level curves are lines passing through the origin.

Writing In Exercises 65 and 66, use the graphs of the level curves (*c*-values evenly spaced) of the function *f* to write a description of a possible graph of *f*. Is the graph of *f* unique? Explain.

65.

66.

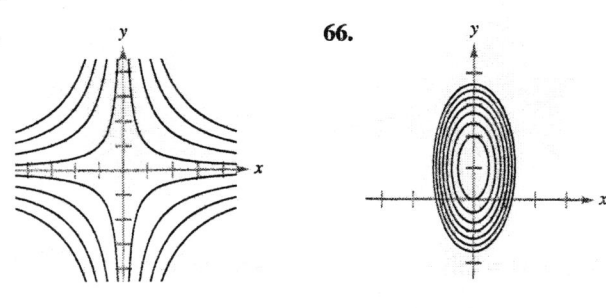

67. *Investment* In 2002, an investment of $1000 was made in a bond earning 10% compounded annually. Assume that the buyer pays tax at rate *R* and the annual rate of inflation is *I*. In the year 2012, the value *V* of the investment in constant 2002 dollars is

$$V(I, R) = 1000\left[\frac{1 + 0.10(1 - R)}{1 + I}\right]^{10}.$$

Use this function of two variables to complete the table.

	Inflation Rate		
Tax Rate	0	0.03	0.05
0			
0.28			
0.35			

68. *Investment* A principal of $1000 is deposited in a savings account that earns an interest rate of *r* (expressed as a decimal), compounded continuously. The amount $A(r, t)$ after *t* years is

$$A(r, t) = 1000e^{rt}.$$

Use this function of two variables to complete the table.

	Number of Years			
Rate	5	10	15	20
0.08				
0.10				
0.12				
0.14				

In Exercises 69–74, sketch the graph of the level surface $f(x, y, z) = c$ at the indicated value of *c*.

69. $f(x, y, z) = x - 2y + 3z$ $c = 6$
70. $f(x, y, z) = 4x + y + 2z$ $c = 4$
71. $f(x, y, z) = x^2 + y^2 + z^2$ $c = 9$
72. $f(x, y, z) = x^2 + \frac{1}{4}y^2 - z$ $c = 1$
73. $f(x, y, z) = 4x^2 + 4y^2 - z^2$ $c = 0$
74. $f(x, y, z) = \sin x - z$ $c = 0$

75. *Forestry* The **Doyle Log Rule** is one of several methods used to determine the lumber yield of a log (in board-feet) in terms of its diameter *d* (in inches) and its length *L* (in feet). The number of board-feet is

$$N(d, L) = \left(\frac{d - 4}{4}\right)^2 L.$$

(a) Find the number of board-feet of lumber in a log 22 inches in diameter and 12 feet in length.

(b) Find $N(30, 12)$.

76. *Queuing Model* The average length of time that a customer waits in line for service is

$$W(x, y) = \frac{1}{x - y}, \quad x > y$$

where *y* is the average arrival rate, expressed as the number of customers per unit of time, and *x* is the average service rate, expressed in the same units. Evaluate each of the following.

(a) $W(15, 10)$ (b) $W(12, 9)$ (c) $W(12, 6)$ (d) $W(4, 2)$

77. *Temperature Distribution* The temperature *T* (in degrees Celsius) at any point (x, y) in a circular steel plate of radius 10 meters is

$$T = 600 - 0.75x^2 - 0.75y^2$$

where *x* and *y* are measured in meters. Sketch some of the isothermal curves.

78. *Cobb-Douglas Production Function* Use the Cobb-Douglas production function (see Example 5) to show that if the number of units of labor and the number of units of capital are doubled, the production level is also doubled.

79. *Construction Cost* A rectangular box with an open top has a length of *x* feet, a width of *y* feet, and a height of *z* feet. Express the cost *C* of constructing the box as a function of *x*, *y*, and *z* if it costs $0.75 per square foot to build the base and $0.40 per square foot to build the sides.

80. *Volume* A propane tank is constructed by welding hemispheres to the ends of a right circular cylinder. Write the volume *V* of the tank as a function of *r* and *l*, where *r* is the radius of the cylinder and hemispheres and *l* is the length of the cylinder.

81. Ideal Gas Law According to the Ideal Gas Law, $PV = kT$, where P is pressure, V is volume, T is temperature (in Kelvins), and k is a constant of proportionality. A tank contains 2600 cubic inches of nitrogen at a pressure of 20 pounds per square inch and a temperature of 300 K.

(a) Determine k.

(b) Express P as a function of V and T and describe the level curves.

82. Modeling Data The table shows the net sales x (in billions of dollars), the total assets y (in billions of dollars), and the shareholder's equity z (in billions of dollars) for Wal-Mart for the years 1995 through 2000. *(Source: 2000 Annual Report)*

Year	1995	1996	1997	1998	1999	2000
x	82.5	93.6	104.9	118.0	137.6	165.0
y	32.8	37.5	39.6	45.4	50.0	70.3
z	12.7	14.8	17.1	18.5	21.1	25.8

A model for these data is

$z = f(x, y) = 0.143x + 0.024y + 0.502$.

(a) Use a graphing utility and the model to approximate z for the given values of x and y.

(b) Which of the two variables in this model has the greater influence on shareholder's equity?

(c) Simplify the expression for $f(x, 25)$ and interpret its meaning in the context of the problem.

83. Meteorology Meteorologists measure the atmospheric pressure in millibars. From these observations they create weather maps on which the curves of equal atmospheric pressure (isobars) are drawn (see figure). On the map, the closer the isobars the higher the wind speed. Match points A, B, and C with (a) highest pressure, (b) lowest pressure, and (c) highest wind velocity.

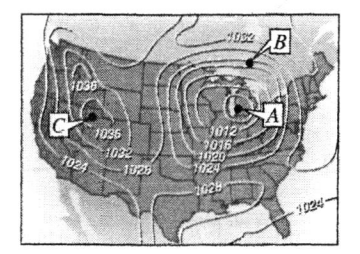

Figure for 83    Figure for 84

84. Acid Rain The acidity of rainwater is measured in units called pH. A pH of 7 is neutral, smaller values are increasingly acidic, and larger values are increasingly alkaline. The map shows the curves of equal pH and gives evidence that downwind of heavily industrialized areas the acidity has been increasing. Using the level curves on the map, determine the direction of the prevailing winds in the northeastern United States.

85. Air Conditioner Use The contour map in the figure represents the estimated annual hours of air conditioner use for an average household. *(Source: Association of Home Appliance Manufacturers)*

(a) Discuss the use of color to represent the level curves.

(b) Do the level curves correspond to equally spaced annual usage hours? Explain.

(c) Describe how to obtain a more detailed contour map.

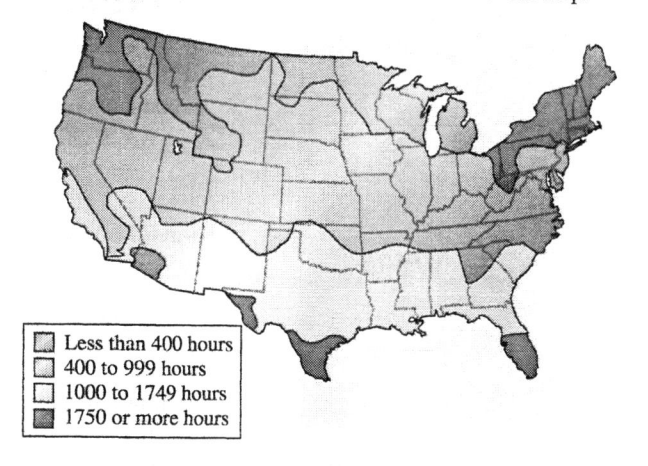

- Less than 400 hours
- 400 to 999 hours
- 1000 to 1749 hours
- 1750 or more hours

86. Annual Temperature Range The colored regions shown on the weather map of the world depict the normal annual temperature range of each region on Earth. Which two geographic features have the greatest effect on temperature range?

True or False? In Exercises 87–90, determine whether the statement is true or false. If it is false, explain why or give an example that shows it is false.

87. If $f(x_0, y_0) = f(x_1, y_1)$, then $x_0 = x_1$ and $y_0 = y_1$.

88. A vertical line can intersect the graph of $z = f(x, y)$ at most once.

89. If f is a function, then $f(ax, ay) = a^2 f(x, y)$.

90. The graph of $f(x, y) = x^2 - y^2$ is a hyperbolic paraboloid.

Section 12.2 Limits and Continuity

- Understand the definition of a neighborhood in the plane.
- Understand the definition of the limit of a function of two variables.
- Extend the concept of continuity to a function of two variables.
- Extend the concept of continuity to a function of three variables.

Neighborhoods in the Plane

In this section, you will study limits and continuity involving functions of two or three variables. The section begins with functions of two variables. At the end of the section, the concepts are extended to functions of three variables.

We begin our discussion of the limit of a function of two variables by defining a two-dimensional analog to an interval on the real line. Using the formula for the distance between two points (x, y) and (x_0, y_0) in the plane, you can define the **δ-neighborhood** about (x_0, y_0) to be the **disk** centered at (x_0, y_0) with radius $\delta > 0$

$$\{(x, y): \sqrt{(x - x_0)^2 + (y - y_0)^2} < \delta\} \qquad \text{Open disk}$$

as shown in Figure 12.18. When this formula contains the *less than* inequality, $<$, the disk is called **open**, and when it contains the *less than or equal to* inequality, $\leq$, the disk is called **closed**. This corresponds to the use of $<$ and $\leq$ to define open and closed intervals.

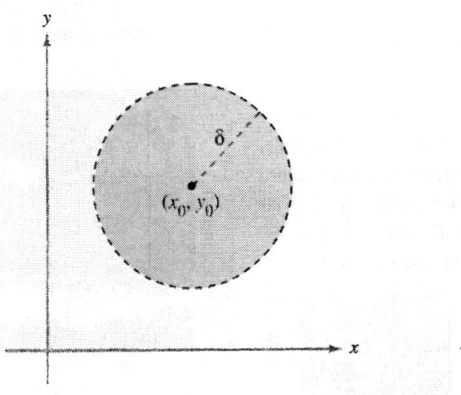

An open disk
Figure 12.18

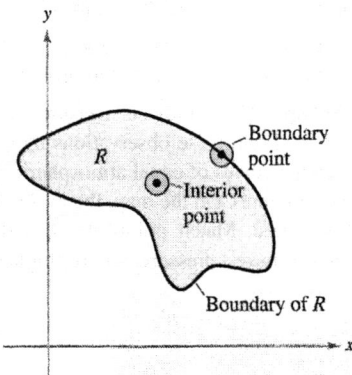

The boundary and interior points of a region R
Figure 12.19

A point (x_0, y_0) in a plane region R is an **interior point** of R if there exists a δ-neighborhood about (x_0, y_0) that lies entirely in R, as shown in Figure 12.19. If every point in R is an interior point, then R is an **open region**. A point (x_0, y_0) is a **boundary point** of R if every open disk centered at (x_0, y_0) contains points inside R *and* points outside R. By definition, a region must contain its interior points, but it need not contain its boundary points. If a region contains all its boundary points, the region is **closed**. A region that contains some but not all of its boundary points is neither open nor closed.

SONYA KOVALEVSKY (1850–1891)

Much of the terminology used to define limits and continuity of a function of two or three variables was introduced by the German mathematician Karl Weierstrass (1815–1897). Weierstrass's rigorous approach to limits and other topics in calculus gained him the reputation as the "father of modern analysis." Weierstrass was a gifted teacher. One of his best-known students was the Russian mathematician Sonya Kovalevsky, who applied many of Weierstrass's techniques to problems in mathematical physics and became one of the first women to gain acceptance as a research mathematician.

FOR FURTHER INFORMATION For more information on Sonya Kovalevsky, see the article "S. Kovalevsky: A Mathematical Lesson" by Karen D. Rappaport in *The American Mathematical Monthly*. To view this article, go to the website *www.matharticles.com*.

Limit of a Function of Two Variables

Definition of the Limit of a Function of Two Variables

Let f be a function of two variables defined, except possibly at (x_0, y_0), on an open disk centered at (x_0, y_0), and let L be a real number. Then

$$\lim_{(x, y) \to (x_0, y_0)} f(x, y) = L$$

if for each $\varepsilon > 0$ there corresponds a $\delta > 0$ such that

$$|f(x, y) - L| < \varepsilon \quad \text{whenever} \quad 0 < \sqrt{(x - x_0)^2 + (y - y_0)^2} < \delta.$$

NOTE Graphically, this definition of a limit implies that for any point $(x, y) \neq (x_0, y_0)$ in the disk of radius δ, the value $f(x, y)$ lies between $L + \varepsilon$ and $L - \varepsilon$, as shown in Figure 12.20.

The definition of the limit of a function of two variables is similar to the definition of the limit of a function of a single variable, yet there is a critical difference. To determine whether a function of a single variable possesses a limit, you need only test the approach from two directions—from the left and from the right. If the function approaches the same limit from the right and from the left, you can conclude that the limit exists. However, for a function of two variables, the statement

$$(x, y) \to (x_0, y_0)$$

means that the point (x, y) is allowed to approach (x_0, y_0) from any "direction." If the value of

$$\lim_{(x, y) \to (x_0, y_0)} f(x, y)$$

is not the same for all possible approaches, or **paths,** to (x_0, y_0), the limit does not exist.

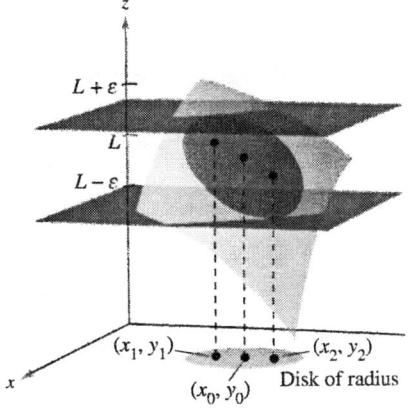

For any (x, y) in the circle of radius δ, the value $f(x, y)$ lies between $L + \varepsilon$ and $L - \varepsilon$.
Figure 12.20

Example 1 **Verifying a Limit by the Definition**

Show that

$$\lim_{(x, y) \to (a, b)} x = a.$$

Solution Let $f(x, y) = x$ and $L = a$. You need to show that for each $\varepsilon > 0$, there exists a δ-neighborhood about (a, b) such that

$$|f(x, y) - L| = |x - a| < \varepsilon$$

whenever $(x, y) \neq (a, b)$ lies in the neighborhood. You can first observe that from

$$0 < \sqrt{(x - a)^2 + (y - b)^2} < \delta$$

it follows that

$$\begin{aligned}|f(x, y) - a| &= |x - a| \\ &= \sqrt{(x - a)^2} \\ &\leq \sqrt{(x - a)^2 + (y - b)^2} \\ &< \delta.\end{aligned}$$

So, you can choose $\delta = \varepsilon$, and the limit is verified.

Limits of functions of several variables have the same properties regarding sums, differences, products, and quotients as do limits of functions of single variables. (See Theorem 1.2 in Section 1.3.) Some of these properties are used in the next example.

Example 2 Verifying a Limit

Evaluate $\lim_{(x, y) \to (1, 2)} \dfrac{5x^2y}{x^2 + y^2}$.

Solution By using the properties of limits of products and sums, you obtain

$$\lim_{(x, y) \to (1, 2)} 5x^2y = 5(1^2)(2)$$
$$= 10$$

and

$$\lim_{(x, y) \to (1, 2)} (x^2 + y^2) = (1^2 + 2^2)$$
$$= 5.$$

Because the limit of a quotient is equal to the quotient of the limits (and the denominator is not 0), you have

$$\lim_{(x, y) \to (1, 2)} \dfrac{5x^2y}{x^2 + y^2} = \dfrac{10}{5}$$
$$= 2.$$

Example 3 Verifying a Limit

Evaluate $\lim_{(x, y) \to (0, 0)} \dfrac{5x^2y}{x^2 + y^2}$.

Solution In this case, the limits of the numerator and of the denominator are both 0, and so you cannot determine the existence (or nonexistence) of a limit by taking the limits of the numerator and denominator separately and then dividing. However, from the graph of f in Figure 12.21, it seems reasonable that the limit might be 0. So, you can try applying the definition to $L = 0$. First, note that

$$|y| \leq \sqrt{x^2 + y^2} \quad \text{and} \quad \dfrac{x^2}{x^2 + y^2} \leq 1.$$

Then, in a δ-neighborhood about $(0, 0)$, you have $0 < \sqrt{x^2 + y^2} < \delta$, and it follows that, for $(x, y) \neq (0, 0)$,

$$|f(x, y) - 0| = \left|\dfrac{5x^2y}{x^2 + y^2}\right|$$
$$= 5|y|\left(\dfrac{x^2}{x^2 + y^2}\right)$$
$$\leq 5|y|$$
$$\leq 5\sqrt{x^2 + y^2}$$
$$< 5\delta.$$

So, you can choose $\delta = \varepsilon/5$ and conclude that

$$\lim_{(x, y) \to (0, 0)} \dfrac{5x^2y}{x^2 + y^2} = 0.$$

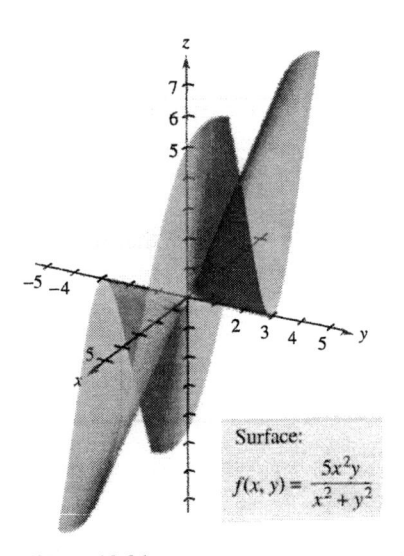

Surface: $f(x, y) = \dfrac{5x^2y}{x^2 + y^2}$

Figure 12.21

$$\lim_{(x,y)\to(0,0)} \frac{1}{x^2+y^2} \text{ does not exist.}$$
Figure 12.22

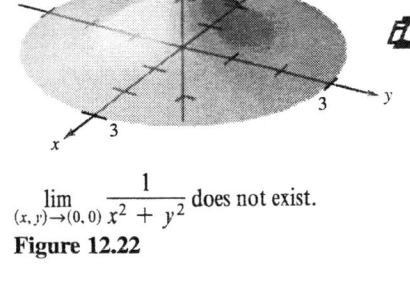

NOTE In Example 4, you could conclude that the limit does not exist because you found two approaches that produced different limits. If two approaches had produced the same limit, you still could not have concluded that the limit exists. To form such a conclusion, you must show that the limit is the same along *all* possible approaches.

For some functions, it is easy to recognize that a limit does not exist. For instance, it is clear that the limit

$$\lim_{(x,y)\to(0,0)} \frac{1}{x^2+y^2}$$

does not exist because the values of $f(x, y)$ increase without bound as (x, y) approaches $(0, 0)$ along *any* path (see Figure 12.22).

For other functions, it is not so easy to recognize that a limit does not exist. For instance, the next example describes a limit that does not exist because the function approaches different values along different paths.

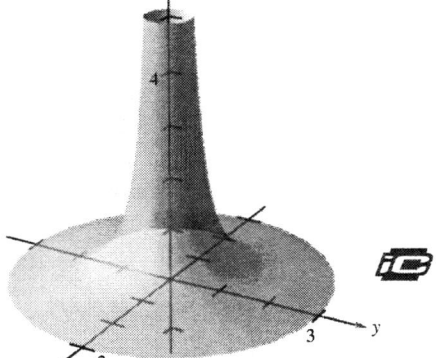

Example 4 A Limit That Does Not Exist

Show that the following limit does not exist.

$$\lim_{(x,y)\to(0,0)} \left(\frac{x^2-y^2}{x^2+y^2}\right)^2$$

Solution The domain of the function given by

$$f(x,y) = \left(\frac{x^2-y^2}{x^2+y^2}\right)^2$$

consists of all points in the *xy*-plane except for the point $(0, 0)$. To show that the limit as (x, y) approaches $(0, 0)$ does not exist, consider approaching $(0, 0)$ along two different "paths," as shown in Figure 12.23. Along the *x*-axis, every point is of the form $(x, 0)$, and the limit along this approach is

$$\lim_{(x,0)\to(0,0)} \left(\frac{x^2-0^2}{x^2+0^2}\right)^2 = \lim_{(x,0)\to(0,0)} (1)^2 = 1. \quad \text{Limit along } x\text{-axis}$$

However, if (x, y) approaches $(0, 0)$ along the line $y = x$, you obtain

$$\lim_{(x,x)\to(0,0)} \left(\frac{x^2-x^2}{x^2+x^2}\right)^2 = \lim_{(x,x)\to(0,0)} \left(\frac{0}{2x^2}\right)^2 = 0. \quad \text{Limit along line } y=x$$

This means that in any open disk centered at $(0, 0)$ there are points (x, y) at which f takes on the value 1, and other points at which f takes on the value 0. For instance, $f(x, y) = 1$ at the points

$(1, 0), (0.1, 0), (0.01, 0),$ and $(0.001, 0)$

and $f(x, y) = 0$ at the points

$(1, 1), (0.1, 0.1), (0.01, 0.01),$ and $(0.001, 0.001).$

Hence, f does not have a limit as $(x, y) \to (0, 0)$.

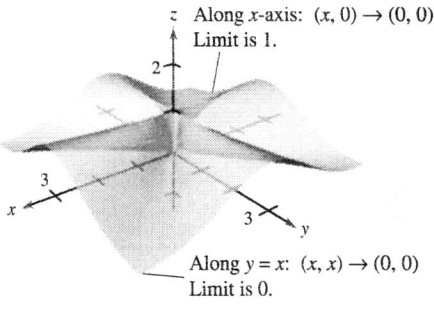

$$\lim_{(x,y)\to(0,0)} \left(\frac{x^2-y^2}{x^2+y^2}\right)^2 \text{ does not exist.}$$
Figure 12.23

Continuity of a Function of Two Variables

Notice in Example 2 that the limit of $f(x, y) = 5x^2y/(x^2 + y^2)$ as $(x, y) \to (1, 2)$ can be evaluated by direct substitution. That is, the limit is $f(1, 2) = 2$. In such cases the function f is said to be **continuous** at the point $(1, 2)$.

NOTE This definition of continuity can be extended to *boundary points* of the open region R by considering a special type of limit in which (x, y) is allowed to approach (x_0, y_0) along paths lying in the region R. This notion is similar to that of one-sided limits, as discussed in Chapter 1.

> **Definition of Continuity of a Function of Two Variables**
>
> A function f of two variables is **continuous at a point** (x_0, y_0) in an open region R if $f(x_0, y_0)$ is equal to the limit of $f(x, y)$ as (x, y) approaches (x_0, y_0). That is,
>
> $$\lim_{(x, y) \to (x_0, y_0)} f(x, y) = f(x_0, y_0).$$
>
> The function f is **continuous in the open region** R if it is continuous at every point in R.

In Example 3, it is shown that the function

$$f(x, y) = \frac{5x^2y}{x^2 + y^2}$$

is not continuous at $(0, 0)$. However, because the limit at this point exists, you can remove the discontinuity by defining f at $(0, 0)$ as being equal to its limit there. Such a discontinuity is called **removable**. In Example 4, the function

$$f(x, y) = \left(\frac{x^2 - y^2}{x^2 + y^2}\right)^2$$

is also shown not to be continuous at $(0, 0)$, but this discontinuity is **nonremovable**.

> **THEOREM 12.1 Continuous Functions of Two Variables**
>
> If k is a real number and f and g are continuous at (x_0, y_0), then the following functions are continuous at (x_0, y_0).
>
> 1. Scalar multiple: kf
> 2. Sum and difference: $f \pm g$
> 3. Product: fg
> 4. Quotient: f/g, if $g(x_0, y_0) \neq 0$

Theorem 12.1 establishes the continuity of *polynomial* and *rational* functions at every point in their domains. Furthermore, the continuity of other types of functions can be extended naturally from one to two variables. For instance, the functions whose graphs are shown in Figures 12.24 and 12.25 are continuous at every point in the plane.

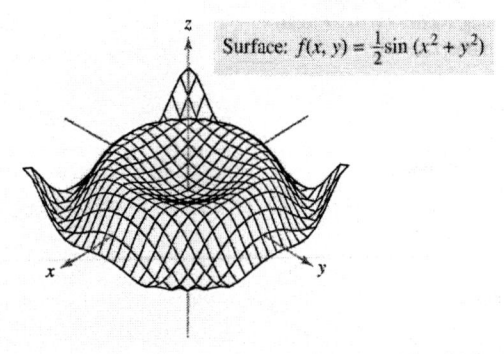

Surface: $f(x, y) = \frac{1}{2}\sin(x^2 + y^2)$

The function f is continuous at every point in the plane.
Figure 12.24

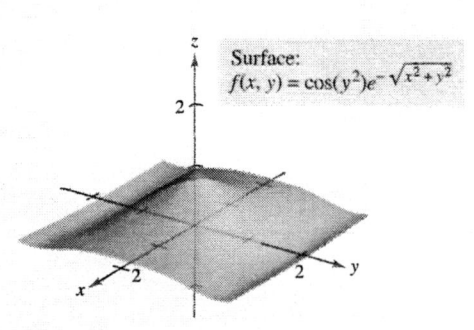

Surface: $f(x, y) = \cos(y^2)e^{-\sqrt{x^2+y^2}}$

The function f is continuous at every point in the plane.
Figure 12.25

EXPLORATION

Hold a spoon a foot or so from your eyes. Look at your image in the spoon. It should be upside down. Now, move the spoon closer and closer to one eye. At some point, your image will be right side up. Could it be that your image is being continuously deformed? Talk about this question and the general meaning of continuity with other members of your class. (This exploration was suggested by Irvin Roy Hentzel, Iowa State University.)

The next theorem states conditions under which a composite function is continuous.

THEOREM 12.2 Continuity of a Composite Function

If h is continuous at (x_0, y_0) and g is continuous at $h(x_0, y_0)$, then the composite function given by $(g \circ h)(x, y) = g(h(x, y))$ is continuous at (x_0, y_0). That is,

$$\lim_{(x, y) \to (x_0, y_0)} g(h(x, y)) = g(h(x_0, y_0)).$$

NOTE Note in Theorem 12.2 that h is a function of two variables and g is a function of one variable.

Example 5 Testing for Continuity

Discuss the continuity of each function.

a. $f(x, y) = \dfrac{x - 2y}{x^2 + y^2}$ **b.** $g(x, y) = \dfrac{2}{y - x^2}$

Solution

a. Because a rational function is continuous at every point in its domain, you can conclude that f is continuous at each point in the xy-plane except at $(0, 0)$, as shown in Figure 12.26.

b. The function given by $g(x, y) = 2/(y - x^2)$ is continuous except at the points at which the denominator is 0, $y - x^2 = 0$. So, you can conclude that the function is continuous at all points except those lying on the parabola $y = x^2$. Inside this parabola, you have $y > x^2$, and the surface represented by the function lies above the xy-plane, as shown in Figure 12.27. Outside the parabola, $y < x^2$, and the surface lies below the xy-plane.

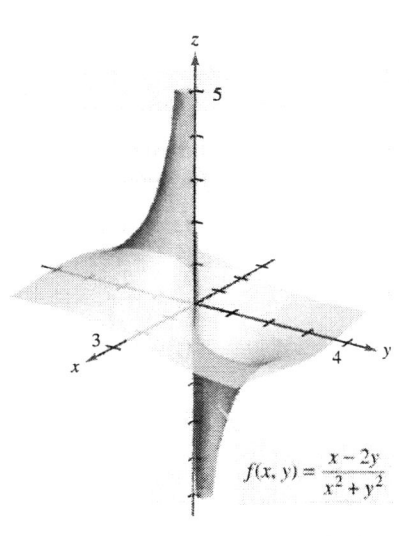

The function f is not continuous at $(0, 0)$.
Figure 12.26

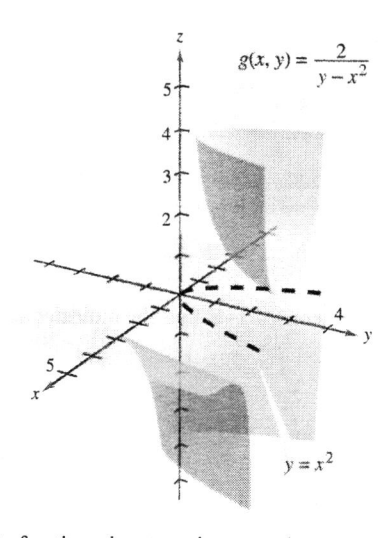

The function g is not continuous at the parabola $y = x^2$.
Figure 12.27

Continuity of a Function of Three Variables

The preceding definitions of limits and continuity can be extended to functions of three variables by considering points (x, y, z) within the *open sphere*

$$(x - x_0)^2 + (y - y_0)^2 + (z - z_0)^2 < \delta^2. \quad \text{Open sphere}$$

The radius of this sphere is δ, and the sphere is centered at (x_0, y_0, z_0), as shown in Figure 12.28. A point (x_0, y_0, z_0) in a region R in space is an **interior point** of R if there exists a δ-sphere about (x_0, y_0, z_0) that lies entirely in R. If every point in R is an interior point, then R is called **open.**

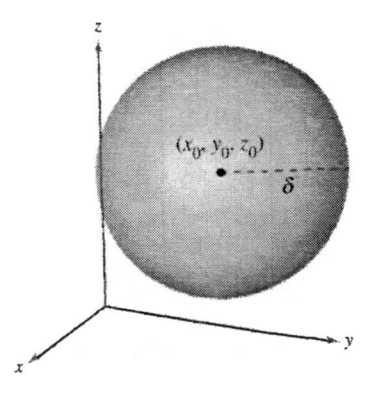

Open sphere in space
Figure 12.28

Definition of Continuity of a Function of Three Variables

A function f of three variables is **continuous at a point** (x_0, y_0, z_0) in an open region R if $f(x_0, y_0, z_0)$ is defined and is equal to the limit of $f(x, y, z)$ as (x, y, z) approaches (x_0, y_0, z_0). That is,

$$\lim_{(x, y, z) \to (x_0, y_0, z_0)} f(x, y, z) = f(x_0, y_0, z_0).$$

The function f is **continuous in the open region** R if it is continuous at every point in R.

Example 6 **Testing Continuity of a Function of Three Variables**

The function

$$f(x, y, z) = \frac{1}{x^2 + y^2 - z}$$

is continuous at each point in space except at the points on the paraboloid given by $z = x^2 + y^2$.

EXERCISES FOR SECTION 12.2

In Exercises 1 and 2, use the definition of the limit of a function of two variables to verify the limit.

1. $\lim_{(x, y) \to (a, b)} y = b$
2. $\lim_{(x, y) \to (4, -1)} x = 4$

In Exercises 3–6, find the indicated limit by using the limits

$$\lim_{(x, y) \to (a, b)} f(x, y) = 5 \text{ and } \lim_{(x, y) \to (a, b)} g(x, y) = 3.$$

3. $\lim_{(x, y) \to (a, b)} [f(x, y) - g(x, y)]$
4. $\lim_{(x, y) \to (a, b)} \left[\frac{4f(x, y)}{g(x, y)} \right]$
5. $\lim_{(x, y) \to (a, b)} [f(x, y) g(x, y)]$
6. $\lim_{(x, y) \to (a, b)} \left[\frac{f(x, y) - g(x, y)}{f(x, y)} \right]$

In Exercises 7–16, find the limit and discuss the continuity of the function.

7. $\lim_{(x, y) \to (2, 1)} (x + 3y^2)$
8. $\lim_{(x, y) \to (0, 0)} (5x + y + 1)$
9. $\lim_{(x, y) \to (2, 4)} \frac{x + y}{x - y}$
10. $\lim_{(x, y) \to (1, 1)} \frac{x}{\sqrt{x + y}}$
11. $\lim_{(x, y) \to (0, 1)} \frac{\arcsin(x/y)}{1 + xy}$
12. $\lim_{(x, y) \to (\pi/4, 2)} y \cos xy$
13. $\lim_{(x, y) \to (-1, 2)} e^{xy}$
14. $\lim_{(x, y) \to (1, 1)} \frac{xy}{x^2 + y^2}$
15. $\lim_{(x, y, z) \to (1, 2, 5)} \sqrt{x + y + z}$
16. $\lim_{(x, y, z) \to (2, 0, 1)} xe^{yz}$

In Exercises 17–20, discuss the continuity of the function and evaluate the limit of $f(x, y)$ (if it exists) as $(x, y) \to (0, 0)$.

17. $f(x, y) = e^{xy}$

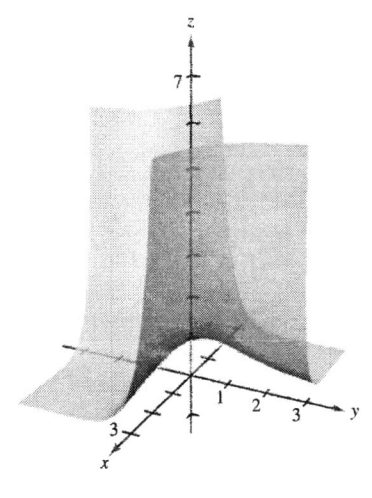

18. $f(x, y) = \dfrac{x^2}{(x^2 + 1)(y^2 + 1)}$

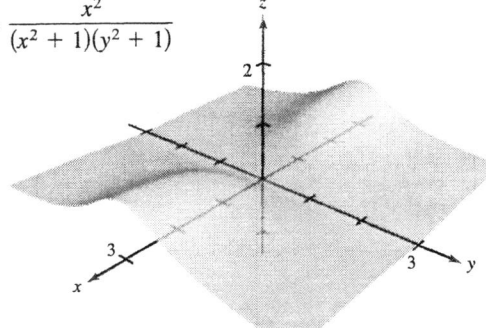

19. $f(x, y) = \ln(x^2 + y^2)$

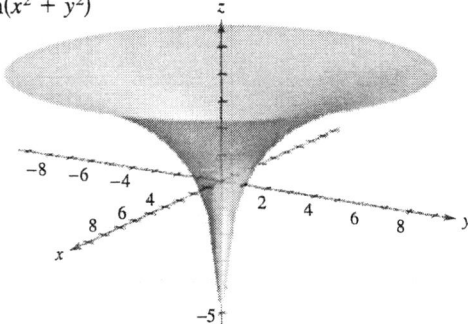

20. $f(x, y) = 1 - \dfrac{\cos(x^2 + y^2)}{x^2 + y^2}$

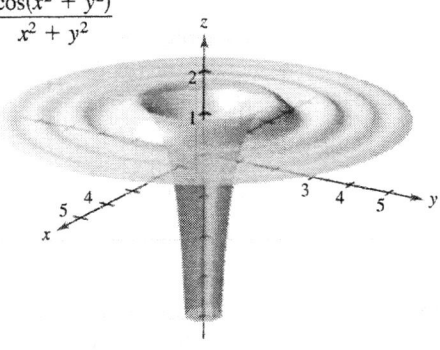

In Exercises 21–24, use a graphing utility to make a table showing the values of $f(x, y)$ at the indicated points. Use the result to make a conjecture about the limit of $f(x, y)$ as $(x, y) \to (0, 0)$. Determine whether the limit exists analytically and discuss the continuity of the function.

21. $f(x, y) = \dfrac{xy}{x^2 + y^2}$

Path: $y = 0$
Points: $(1, 0)$, $(0.5, 0)$, $(0.1, 0)$, $(0.01, 0)$, $(0.001, 0)$
Path: $y = x$
Points: $(1, 1)$, $(0.5, 0.5)$, $(0.1, 0.1)$, $(0.01, 0.01)$, $(0.001, 0.001)$

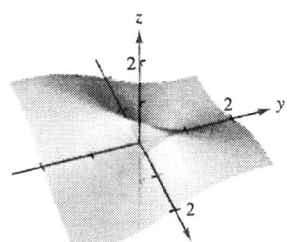

22. $f(x, y) = \dfrac{y}{x^2 + y^2}$

Path: $y = 0$
Points: $(1, 0)$, $(0.5, 0)$, $(0.1, 0)$, $(0.01, 0)$, $(0.001, 0)$
Path: $y = x$
Points: $(1, 1)$, $(0.5, 0.5)$, $(0.1, 0.1)$, $(0.01, 0.01)$, $(0.001, 0.001)$

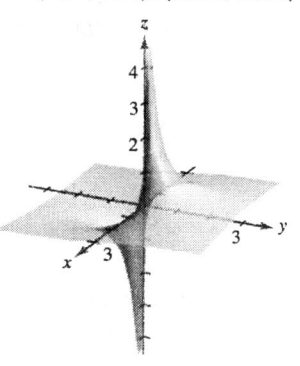

23. $f(x, y) = -\dfrac{xy^2}{x^2 + y^4}$

Path: $x = y^2$
Points: $(1, 1)$, $(0.25, 0.5)$, $(0.01, 0.1)$, $(0.0001, 0.01)$, $(0.000001, 0.001)$
Path: $x = -y^2$
Points: $(-1, 1)$, $(-0.25, 0.5)$, $(-0.01, 0.1)$, $(-0.0001, 0.01)$, $(-0.000001, 0.001)$

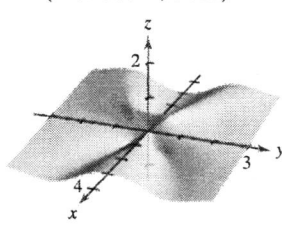

24. $f(x, y) = \dfrac{2x - y^2}{2x^2 + y}$

Path: $y = 0$
Points: $(1, 0)$, $(0.25, 0)$, $(0.01, 0)$, $(0.001, 0)$, $(0.000001, 0)$
Path: $y = x$
Points: $(1, 1)$, $(0.25, 0.25)$, $(0.01, 0.01)$, $(0.001, 0.001)$, $(0.0001, 0.0001)$

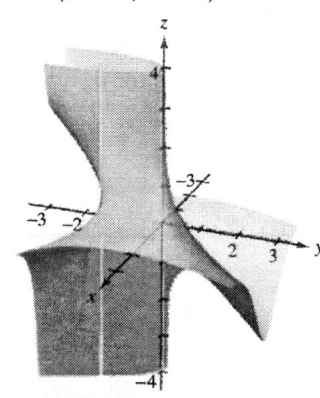

In Exercises 25 and 26, discuss the continuity of the functions f and g. Explain any differences.

25. $f(x, y) = \begin{cases} \dfrac{x^2 + 2xy^2 + y^2}{x^2 + y^2}, & (x, y) \neq (0, 0) \\ 0, & (x, y) = (0, 0) \end{cases}$

$g(x, y) = \begin{cases} \dfrac{x^2 + 2xy^2 + y^2}{x^2 + y^2}, & (x, y) \neq (0, 0) \\ 1, & (x, y) = (0, 0) \end{cases}$

26. $f(x, y) = \begin{cases} \dfrac{4x^2y^2}{x^2 + y^2}, & (x, y) \neq (0, 0) \\ 0, & (x, y) = (0, 0) \end{cases}$

$g(x, y) = \begin{cases} \dfrac{4x^2y^2}{x^2 + y^2}, & (x, y) \neq (0, 0) \\ 2, & (x, y) = (0, 0) \end{cases}$

In Exercises 27–32, use a computer algebra system to graph the function and find $\lim_{(x, y) \to (0, 0)} f(x, y)$ (if it exists).

27. $f(x, y) = \sin x + \sin y$

28. $f(x, y) = \sin \dfrac{1}{x} + \cos \dfrac{1}{x}$

29. $f(x, y) = \dfrac{x^2 y}{x^4 + 4y^2}$

30. $f(x, y) = \dfrac{x^2 + y^2}{x^2 y}$

31. $f(x, y) = \dfrac{10xy}{2x^2 + 3y^2}$

32. $f(x, y) = \dfrac{2xy}{x^2 + y^2 + 1}$

In Exercises 33–36, use polar coordinates to find the limit. [Hint: Let $x = r \cos \theta$ and $y = r \sin \theta$, and note that $(x, y) \to (0, 0)$ implies $r \to 0$.]

33. $\lim_{(x, y) \to (0, 0)} \dfrac{\sin(x^2 + y^2)}{x^2 + y^2}$

34. $\lim_{(x, y) \to (0, 0)} \dfrac{xy^2}{x^2 + y^2}$

35. $\lim_{(x, y) \to (0, 0)} \dfrac{x^3 + y^3}{x^2 + y^2}$

36. $\lim_{(x, y) \to (0, 0)} \dfrac{x^2 y^2}{x^2 + y^2}$

In Exercises 37–40, discuss the continuity of the function.

37. $f(x, y, z) = \dfrac{1}{\sqrt{x^2 + y^2 + z^2}}$

38. $f(x, y, z) = \dfrac{z}{x^2 + y^2 - 9}$

39. $f(x, y, z) = \dfrac{\sin z}{e^x + e^y}$

40. $f(x, y, z) = xy \sin z$

In Exercises 41–44, discuss the continuity of the composite function $f \circ g$.

41. $f(t) = t^2$
    $g(x, y) = 3x - 2y$

42. $f(t) = \dfrac{1}{t}$
    $g(x, y) = x^2 + y^2$

43. $f(t) = \dfrac{1}{t}$
    $g(x, y) = 3x - 2y$

44. $f(t) = \dfrac{1}{4 - t}$
    $g(x, y) = x^2 + y^2$

In Exercises 45–48, find each of the following limits.

(a) $\lim_{\Delta x \to 0} \dfrac{f(x + \Delta x, y) - f(x, y)}{\Delta x}$

(b) $\lim_{\Delta y \to 0} \dfrac{f(x, y + \Delta y) - f(x, y)}{\Delta y}$

45. $f(x, y) = x^2 - 4y$
46. $f(x, y) = x^2 + y^2$
47. $f(x, y) = 2x + xy - 3y$
48. $f(x, y) = \sqrt{y}\,(y + 1)$

Getting at the Concept

49. Define the limit of a function of two variables. Describe a method for showing that
$$\lim_{(x, y) \to (x_0, y_0)} f(x, y)$$
does not exist.

50. State the definition of continuity of a function of two variables.

51. If $f(2, 3) = 4$, can you conclude anything about
$$\lim_{(x, y) \to (2, 3)} f(x, y)?$$
Give reasons for your answer.

52. Consider $\lim_{(x, y) \to (0, 0)} \dfrac{x^2 + y^2}{xy}$.

    (a) Determine (if possible) the limit along any line of the form $y = ax$.

    (b) Determine (if possible) the limit along the parabola $y = x^2$.

    (c) Does the limit exist? Explain.

53. Prove that
$$\lim_{(x, y) \to (a, b)} [f(x, y) + g(x, y)] = L_1 + L_2$$
where $f(x, y)$ approaches L_1 and $g(x, y)$ approaches L_2 as $(x, y) \to (a, b)$.

54. Prove that if f is continuous and $f(a, b) < 0$, there exists a δ-neighborhood about (a, b) such that $f(x, y) < 0$ for every point (x, y) in the neighborhood.

True or False? In Exercises 55–58, determine whether the statement is true or false. If it is false, explain why or give an example that shows it is false.

55. If $\lim_{(x, y) \to (0, 0)} f(x, y) = 0$, then $\lim_{x \to 0} f(x, 0) = 0$.

56. If $\lim_{(x, y) \to (0, 0)} f(0, y) = 0$, then $\lim_{(x, y) \to (0, 0)} f(x, y) = 0$.

57. If f is continuous for all nonzero x and y, and $f(0, 0) = 0$, then $\lim_{(x, y) \to (0, 0)} f(x, y) = 0$.

58. If g and h are continuous functions of x and y, and $f(x, y) = g(x) + h(y)$, then f is continuous.

Partial Derivatives

Partial Derivatives of a Function of Two Variables

In applications of functions of several variables, the question often arises, "How will a function be affected by a change in one of its independent variables?" You can answer this by considering the independent variables one at a time. For example, to determine the effect of a catalyst in an experiment, a chemist could conduct the experiment several times using varying amounts of the catalyst, while keeping constant other variables such as temperature and pressure. You can use a similar procedure to determine the rate of change of a function f with respect to one of its several independent variables. This process is called **partial differentiation,** and the result is referred to as the **partial derivative** of f with respect to the chosen independent variable.

Definition of Partial Derivatives of a Function of Two Variables

If $z = f(x, y)$, then the **first partial derivatives** of f with respect to x and y are the functions f_x and f_y defined by

$$f_x(x, y) = \lim_{\Delta x \to 0} \frac{f(x + \Delta x, y) - f(x, y)}{\Delta x}$$

$$f_y(x, y) = \lim_{\Delta y \to 0} \frac{f(x, y + \Delta y) - f(x, y)}{\Delta y}$$

provided the limits exist.

This definition indicates that if $z = f(x, y)$, then to find f_x you *consider y constant* and differentiate with respect to x. Similarly, to find f_y, you *consider x constant* and differentiate with respect to y.

Example 1   Finding Partial Derivatives

Find the partial derivatives f_x and f_y for the function

$$f(x, y) = 3x - x^2y^2 + 2x^3y. \quad \text{Original function}$$

Solution   Considering y to be constant and differentiating with respect to x produces

$$f(x, y) = 3x - x^2y^2 + 2x^3y \quad \text{Write original function.}$$
$$f_x(x, y) = 3 - 2xy^2 + 6x^2y. \quad \text{Partial derivative with respect to } x$$

Considering x to be constant and differentiating with respect to y produces

$$f(x, y) = 3x - x^2y^2 + 2x^3y \quad \text{Write original function.}$$
$$f_y(x, y) = -2x^2y + 2x^3. \quad \text{Partial derivative with respect to } y$$

JEAN LE ROND D'ALEMBERT (1717–1783)

The introduction of partial derivatives followed Newton's and Leibniz's work in calculus by several years. Between 1730 and 1760, Leonhard Euler and Jean Le Rond d'Alembert separately published several papers on dynamics, in which they established much of the theory of partial derivatives. These papers used functions of two or more variables to study problems involving equilibrium, fluid motion, and vibrating strings.

Notation for First Partial Derivatives

For $z = f(x, y)$, the partial derivatives f_x and f_y are denoted by

$$\frac{\partial}{\partial x} f(x, y) = f_x(x, y) = z_x = \frac{\partial z}{\partial x}$$

and

$$\frac{\partial}{\partial y} f(x, y) = f_y(x, y) = z_y = \frac{\partial z}{\partial y}.$$

The first partials evaluated at the point (a, b) are denoted by

$$\left.\frac{\partial z}{\partial x}\right|_{(a, b)} = f_x(a, b) \quad \text{and} \quad \left.\frac{\partial z}{\partial y}\right|_{(a, b)} = f_y(a, b).$$

Example 2 **Finding and Evaluating Partial Derivatives**

For $f(x, y) = xe^{x^2 y}$, find f_x and f_y, and evaluate each at the point $(1, \ln 2)$.

Solution Because

$$f_x(x, y) = xe^{x^2 y}(2xy) + e^{x^2 y} \quad \text{Partial derivative with respect to } x$$

the partial derivative of f with respect to x at $(1, \ln 2)$ is

$$f_x(1, \ln 2) = e^{\ln 2}(2 \ln 2) + e^{\ln 2}$$
$$= 4 \ln 2 + 2.$$

Because

$$f_y(x, y) = xe^{x^2 y}(x^2)$$
$$= x^3 e^{x^2 y} \quad \text{Partial derivative with respect to } y$$

the partial derivative of f with respect to y at $(1, \ln 2)$ is

$$f_y(1, \ln 2) = e^{\ln 2}$$
$$= 2.$$

The partial derivatives of a function of two variables, $z = f(x, y)$, have a useful geometric interpretation. If $y = y_0$, then $z = f(x, y_0)$ represents the curve formed by intersecting the surface $z = f(x, y)$ with the plane $y = y_0$, as shown in Figure 12.29. Therefore,

$$f_x(x_0, y_0) = \lim_{\Delta x \to 0} \frac{f(x_0 + \Delta x, y_0) - f(x_0, y_0)}{\Delta x}$$

represents the slope of this curve at the point $(x_0, y_0, f(x_0, y_0))$. Note that both the curve and the tangent line lie in the plane $y = y_0$. Similarly,

$$f_y(x_0, y_0) = \lim_{\Delta y \to 0} \frac{f(x_0, y_0 + \Delta y) - f(x_0, y_0)}{\Delta y}$$

represents the slope of the curve given by the intersection of $z = f(x, y)$ and the plane $x = x_0$ at $(x_0, y_0, f(x_0, y_0))$, as shown in Figure 12.30.

Informally, we say that the values of $\partial f / \partial x$ and $\partial f / \partial y$ at the point (x_0, y_0, z_0) denote the **slopes of the surface in the x- and y-directions.**

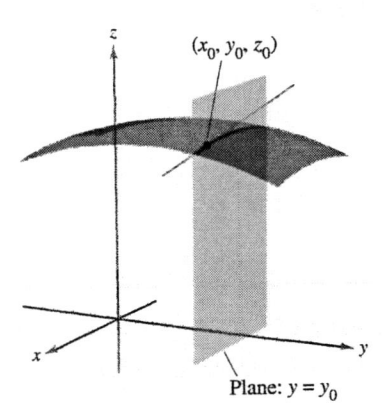

$\dfrac{\partial f}{\partial x}$ = slope in x-direction
Figure 12.29

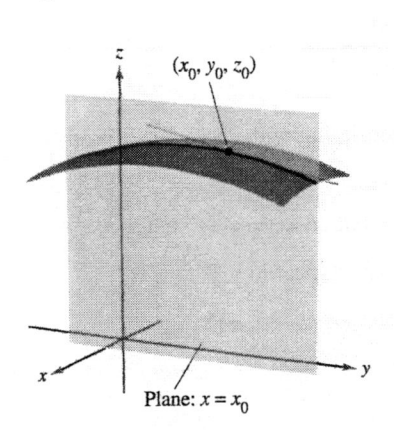

$\dfrac{\partial f}{\partial y}$ = slope in y-direction
Figure 12.30

Example 3 Finding the Slopes of a Surface in the x- and y-Directions

Find the slopes of the surface given by

$$f(x, y) = -\frac{x^2}{2} - y^2 + \frac{25}{8}$$

at the point $\left(\frac{1}{2}, 1, 2\right)$ in the x-direction and in the y-direction.

Solution The partial derivatives of f with respect to x and y are

$$f_x(x, y) = -x \quad \text{and} \quad f_y(x, y) = -2y. \qquad \textit{Partial derivatives}$$

So, in the x-direction, the slope is

$$f_x\left(\frac{1}{2}, 1\right) = -\frac{1}{2} \qquad \textit{Figure 12.31(a)}$$

and in the y-direction, the slope is

$$f_y\left(\frac{1}{2}, 1\right) = -2. \qquad \textit{Figure 12.31(b)}$$

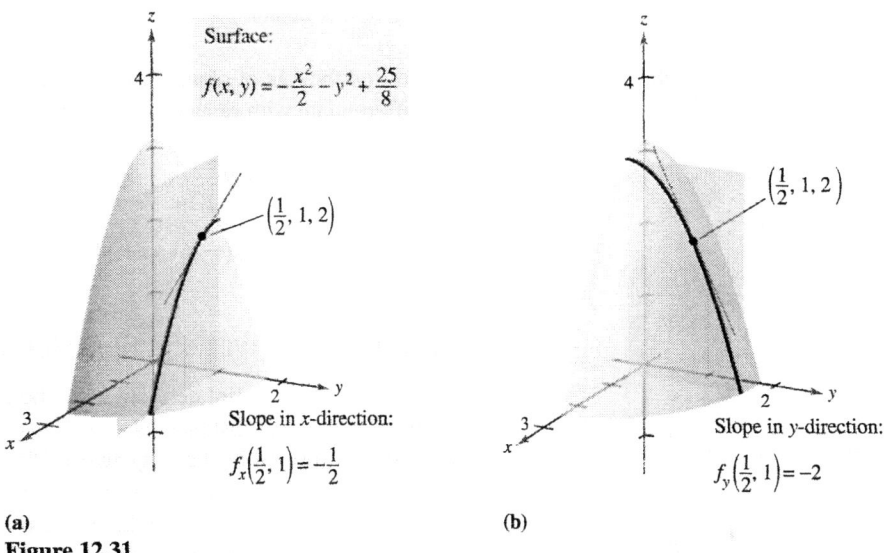

Figure 12.31

Example 4 Finding the Slopes of a Surface in the x- and y-Directions

Find the slopes of the surface given by

$$f(x, y) = 1 - (x - 1)^2 - (y - 2)^2$$

at the point $(1, 2, 1)$ in the x-direction and in the y-direction.

Solution The partial derivatives of f with respect to x and y are

$$f_x(x, y) = -2(x - 1) \quad \text{and} \quad f_y(x, y) = -2(y - 2). \qquad \textit{Partial derivatives}$$

So, at the point $(1, 2, 1)$, the slopes in the x- and y-directions are

$$f_x(1, 2) = -2(1 - 1) = 0 \quad \text{and} \quad f_y(1, 2) = -2(2 - 2) = 0$$

as indicated in Figure 12.32.

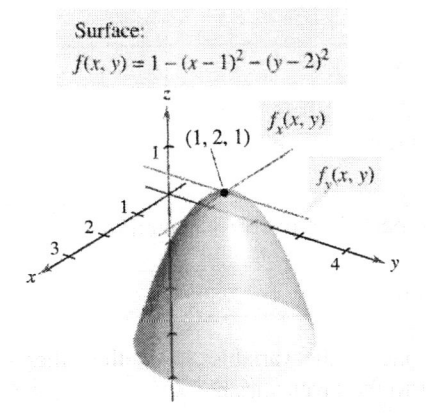

Figure 12.32

No matter how many variables are involved, partial derivatives of several variables can be interpreted as *rates of change*.

Example 5 Using Partial Derivatives to Find Rates of Change

The area of a parallelogram with adjacent sides a and b and included angle θ is given by $A = ab \sin \theta$, as shown in Figure 12.33.

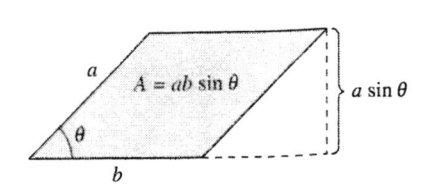

a. Find the rate of change of A with respect to a for $a = 10$, $b = 20$, and $\theta = \dfrac{\pi}{6}$.

b. Find the rate of change of A with respect to θ for $a = 10$, $b = 20$, and $\theta = \dfrac{\pi}{6}$.

Solution

a. To find the rate of change of the area with respect to a, hold b and θ constant and differentiate with respect to a to obtain

$$\frac{\partial A}{\partial a} = b \sin \theta \qquad \text{Find partial with respect to } a.$$

$$\frac{\partial A}{\partial a} = 20 \sin \frac{\pi}{6} = 10. \qquad \text{Substitute for } b \text{ and } \theta.$$

b. To find the rate of change of the area with respect to θ, hold a and b constant and differentiate with respect to θ to obtain

$$\frac{\partial A}{\partial \theta} = ab \cos \theta \qquad \text{Find partial with respect to } \theta.$$

$$\frac{\partial A}{\partial \theta} = 200 \cos(\pi/6) = 100\sqrt{3}. \qquad \text{Substitute for } a, b, \text{ and } \theta.$$

The area of the parallelogram is $ab \sin \theta$.
Figure 12.33

Partial Derivatives of a Function of Three or More Variables

The concept of a partial derivative can be extended naturally to functions of three or more variables. For instance, if $w = f(x, y, z)$, there are three partial derivatives, each of which is formed by holding two of the variables constant. That is, to define the partial derivative of w with respect to x, consider y and z to be constant and differentiate with respect to x. A similar process is used to find the derivatives of w with respect to y and with respect to z.

$$\frac{\partial w}{\partial x} = f_x(x, y, z) = \lim_{\Delta x \to 0} \frac{f(x + \Delta x, y, z) - f(x, y, z)}{\Delta x}$$

$$\frac{\partial w}{\partial y} = f_y(x, y, z) = \lim_{\Delta y \to 0} \frac{f(x, y + \Delta y, z) - f(x, y, z)}{\Delta y}$$

$$\frac{\partial w}{\partial z} = f_z(x, y, z) = \lim_{\Delta z \to 0} \frac{f(x, y, z + \Delta z) - f(x, y, z)}{\Delta z}$$

In general, if $w = f(x_1, x_2, \ldots, x_n)$, there are n partial derivatives denoted by

$$\frac{\partial w}{\partial x_k} = f_{x_k}(x_1, x_2, \ldots, x_n), \quad k = 1, 2, \ldots, n.$$

To find the partial derivative with respect to one of the variables, hold the other variables constant and differentiate with respect to the given variable.

Example 6 Finding Partial Derivatives

a. To find the partial derivative of $f(x, y, z) = xy + yz^2 + xz$ with respect to z, consider x and y to be constant and obtain

$$\frac{\partial}{\partial z}[xy + yz^2 + xz] = 2yz + x.$$

b. To find the partial derivative of $f(x, y, z) = z \sin(xy^2 + 2z)$ with respect to z, consider x and y to be constant. Then, using the Product Rule, you obtain

$$\frac{\partial}{\partial z}[z \sin(xy^2 + 2z)] = (z)\frac{\partial}{\partial z}[\sin(xy^2 + 2z)] + \sin(xy^2 + 2z)\frac{\partial}{\partial z}[z]$$
$$= (z)[\cos(xy^2 + 2z)](2) + \sin(xy^2 + 2z)$$
$$= 2z \cos(xy^2 + 2z) + \sin(xy^2 + 2z).$$

c. To find the partial derivative of $f(x, y, z, w) = (x + y + z)/w$ with respect to w, consider x, y, and z to be constant and obtain

$$\frac{\partial}{\partial w}\left[\frac{x + y + z}{w}\right] = -\frac{x + y + z}{w^2}.$$

Higher-Order Partial Derivatives

As is true for ordinary derivatives, it is possible to take second, third, and higher partial derivatives of a function of several variables, provided such derivatives exist. Higher-order derivatives are denoted by the order in which the differentiation occurs. For instance, the function $z = f(x, y)$ has the following second partial derivatives.

1. Differentiate twice with respect to x:

$$\frac{\partial}{\partial x}\left(\frac{\partial f}{\partial x}\right) = \frac{\partial^2 f}{\partial x^2} = f_{xx}.$$

2. Differentiate twice with respect to y:

$$\frac{\partial}{\partial y}\left(\frac{\partial f}{\partial y}\right) = \frac{\partial^2 f}{\partial y^2} = f_{yy}.$$

3. Differentiate first with respect to x and then with respect to y:

$$\frac{\partial}{\partial y}\left(\frac{\partial f}{\partial x}\right) = \frac{\partial^2 f}{\partial y \partial x} = f_{xy}.$$

4. Differentiate first with respect to y and then with respect to x:

$$\frac{\partial}{\partial x}\left(\frac{\partial f}{\partial y}\right) = \frac{\partial^2 f}{\partial x \partial y} = f_{yx}.$$

The third and fourth cases are called **mixed partial derivatives.**

NOTE Note that the two types of notation for mixed partials have different conventions for indicating the order of differentiation.

$$\frac{\partial}{\partial y}\left(\frac{\partial f}{\partial x}\right) = \frac{\partial^2 f}{\partial y \partial x} \quad \text{Right-to-left order}$$

$$(f_x)_y = f_{xy} \quad \text{Left-to-right order}$$

You can remember the order by observing that in both notations, you differentiate first with respect to the variable "nearest" f.

Example 7 Finding Second Partial Derivatives

Find the second partial derivatives of $f(x, y) = 3xy^2 - 2y + 5x^2y^2$, and determine the value of $f_{xy}(-1, 2)$.

Solution Begin by finding the first partial derivatives with respect to x and y.

$$f_x(x, y) = 3y^2 + 10xy^2 \quad \text{and} \quad f_y(x, y) = 6xy - 2 + 10x^2y$$

Then, differentiate each of these with respect to x and y.

$$f_{xx}(x, y) = 10y^2 \quad \text{and} \quad f_{yy}(x, y) = 6x + 10x^2$$
$$f_{xy}(x, y) = 6y + 20xy \quad \text{and} \quad f_{yx}(x, y) = 6y + 20xy$$

At $(-1, 2)$, the value of f_{xy} is $f_{xy}(-1, 2) = 12 - 40 = -28$.

NOTE Notice in Example 7 that the two mixed partials are equal. Sufficient conditions for this occurrence are given in Theorem 12.3.

THEOREM 12.3 Equality of Mixed Partial Derivatives

If f is a function of x and y such that f_{xy} and f_{yx} are continuous on an open disk R, then, for every (x, y) in R,

$$f_{xy}(x, y) = f_{yx}(x, y).$$

Theorem 12.3 also applies to a function f of *three or more variables* so long as all second partial derivatives are continuous. For example, if $w = f(x, y, z)$ and all the second partial derivatives are continuous in an open region R, then at each point in R the order of differentiation in the mixed second partial derivatives is irrelevant. If the third partial derivatives of f are also continuous, the order of differentiation of the mixed third partial derivatives is irrelevant.

Example 8 Finding Higher-Order Partial Derivatives

Show that $f_{xz} = f_{zx}$ and $f_{xzz} = f_{zxz} = f_{zzx}$ for the function given by

$$f(x, y, z) = ye^x + x \ln z.$$

Solution

First partials:

$$f_x(x, y, z) = ye^x + \ln z, \quad f_z(x, y, z) = \frac{x}{z}$$

Second partials (note that the first two are equal):

$$f_{xz}(x, y, z) = \frac{1}{z}, \quad f_{zx}(x, y, z) = \frac{1}{z}, \quad f_{zz}(x, y, z) = -\frac{x}{z^2}$$

Third partials (note that all three are equal):

$$f_{xzz}(x, y, z) = -\frac{1}{z^2}, \quad f_{zxz}(x, y, z) = -\frac{1}{z^2}, \quad f_{zzx}(x, y, z) = -\frac{1}{z^2}$$

EXERCISES FOR SECTION 12.3

Think About It In Exercises 1–4, use the graph of the surface to determine the sign of the indicated partial derivative.

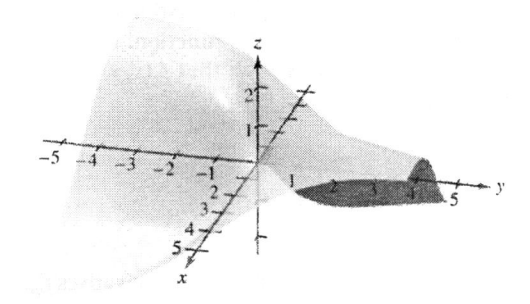

1. $f_x(4, 1)$
2. $f_y(-1, -2)$
3. $f_y(4, 1)$
4. $f_x(-1, -1)$

In Exercises 5–28, find both first partial derivatives.

5. $f(x, y) = 2x - 3y + 5$
6. $f(x, y) = x^2 - 3y^2 + 7$
7. $z = x\sqrt{y}$
8. $z = 2y^2\sqrt{x}$
9. $z = x^2 - 5xy + 3y^2$
10. $z = y^3 - 4xy^2 - 1$
11. $z = x^2 e^{2y}$
12. $z = xe^{x/y}$
13. $z = \ln(x^2 + y^2)$
14. $z = \ln\sqrt{xy}$
15. $z = \ln\dfrac{x+y}{x-y}$
16. $z = \ln(x^2 - y^2)$
17. $z = \dfrac{x^2}{2y} + \dfrac{4y^2}{x}$
18. $z = \dfrac{xy}{x^2 + y^2}$
19. $h(x, y) = e^{-(x^2+y^2)}$
20. $g(x, y) = \ln\sqrt{x^2 + y^2}$
21. $f(x, y) = \sqrt{x^2 + y^2}$
22. $f(x, y) = \sqrt{2x + y^3}$
23. $z = \tan(2x - y)$
24. $z = \sin 3x \cos 3y$
25. $z = e^y \sin xy$
26. $z = \cos(x^2 + y^2)$
27. $f(x, y) = \displaystyle\int_x^y (t^2 - 1)\, dt$
28. $f(x, y) = \displaystyle\int_x^y (2t + 1)\, dt + \displaystyle\int_y^x (2t - 1)\, dt$

In Exercises 29–32, use the limit definition of partial derivatives to find $f_x(x, y)$ and $f_y(x, y)$.

29. $f(x, y) = 2x + 3y$
30. $f(x, y) = x^2 - 2xy + y^2$
31. $f(x, y) = \sqrt{x + y}$
32. $f(x, y) = \dfrac{1}{x + y}$

In Exercises 33–36, find the slopes of the surface in the x- and y-directions at the indicated point.

33. $g(x, y) = 4 - x^2 - y^2$, $(1, 1, 2)$
34. $h(x, y) = x^2 - y^2$, $(-2, 1, 3)$
35. $z = e^{-x} \cos y$, $(0, 0, 1)$
36. $z = \cos(2x - y)$, $(\pi/4, \pi/3, \sqrt{3}/2)$

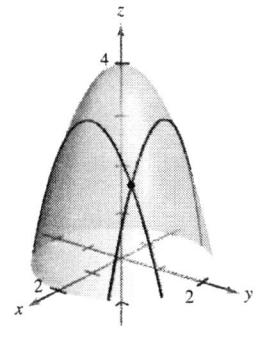

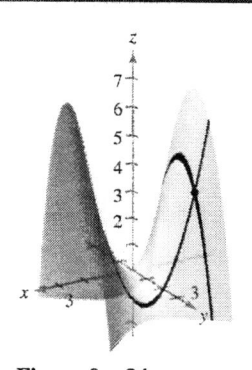

Figure for 33

Figure for 34

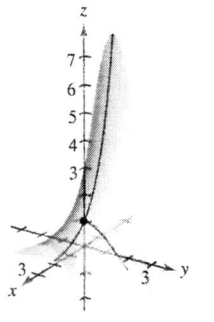

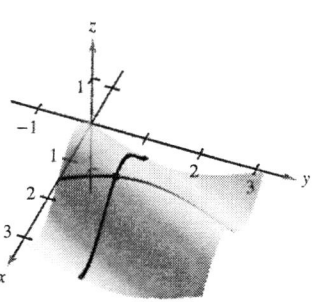

Figure for 35

Figure for 36

In Exercises 37–40, evaluate f_x and f_y at the indicated point.

Function	Point
37. $f(x, y) = \arctan\dfrac{y}{x}$	$(2, -2)$
38. $f(x, y) = \arccos xy$	$(1, 1)$
39. $f(x, y) = xy/(x - y)$	$(2, -2)$
40. $f(x, y) = \dfrac{6xy}{\sqrt{4x^2 + 5y^2}}$	$(1, 1)$

In Exercises 41–44, use a computer algebra system to graph the curve formed by the intersection of the surface and the plane. Find the slope of the curve at the given point.

Surface	Plane	Point
41. $z = \sqrt{49 - x^2 - y^2}$	$x = 2$	$(2, 3, 6)$
42. $z = x^2 + 4y^2$	$y = 1$	$(2, 1, 8)$
43. $z = 9x^2 - y^2$	$y = 3$	$(1, 3, 0)$
44. $z = 9x^2 - y^2$	$x = 1$	$(1, 3, 0)$

In Exercises 45–48, for $f(x, y)$, find all values of x and y such that $f_x(x, y) = 0$ and $f_y(x, y) = 0$ simultaneously.

45. $f(x, y) = x^2 + 4xy + y^2 - 4x + 16y + 3$
46. $f(x, y) = 3x^3 - 12xy + y^3$
47. $f(x, y) = \dfrac{1}{x} + \dfrac{1}{y} + xy$
48. $f(x, y) = \ln(x^2 + y^2 + 1)$

Think About It In Exercises 49 and 50, the graph of a function f and its two partial derivatives f_x and f_y are given. Identify f_x and f_y and give reasons for your answers.

49.

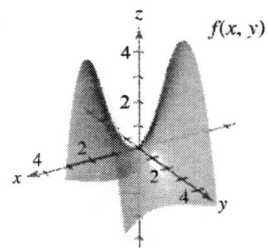

(a) 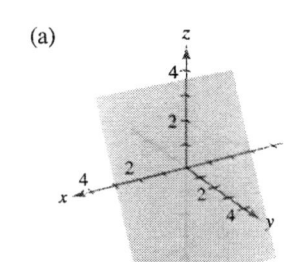 (b)

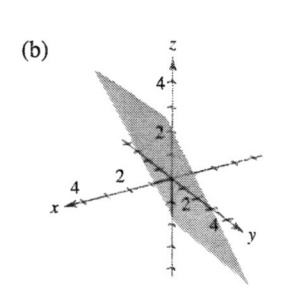

50.

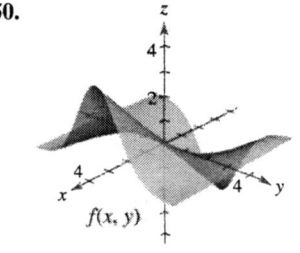

(a) 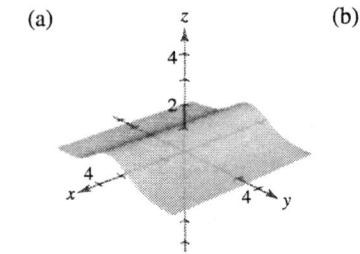 (b)

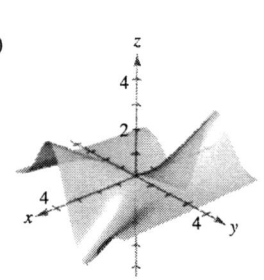

In Exercises 51–56, find the first partial derivatives with respect to x, y, and z.

51. $w = \sqrt{x^2 + y^2 + z^2}$

52. $w = \dfrac{3xz}{x + y}$

53. $F(x, y, z) = \ln\sqrt{x^2 + y^2 + z^2}$

54. $G(x, y, z) = \dfrac{1}{\sqrt{1 - x^2 - y^2 - z^2}}$

55. $H(x, y, z) = \sin(x + 2y + 3z)$

56. $f(x, y, z) = 3x^2 y - 5xyz + 10yz^2$

In Exercises 57–64, find the four second partial derivatives. Observe that the second mixed partials are equal.

57. $z = x^2 - 2xy + 3y^2$

58. $z = x^4 - 3x^2 y^2 + y^4$

59. $z = \sqrt{x^2 + y^2}$

60. $z = \ln(x - y)$

61. $z = e^x \tan y$

62. $z = 2xe^y - 3ye^{-x}$

63. $z = \arctan \dfrac{y}{x}$

64. $z = \sin(x - 2y)$

In Exercises 65–68, use a computer algebra system to find the first and second partial derivatives of the function. Determine whether there exist values of x and y such that $f_x(x, y) = 0$ and $f_y(x, y) = 0$ simultaneously.

65. $f(x, y) = x \sec y$

66. $f(x, y) = \sqrt{9 - x^2 - y^2}$

67. $f(x, y) = \ln \dfrac{x}{x^2 + y^2}$

68. $f(x, y) = \dfrac{xy}{x - y}$

In Exercises 69–72, show that the mixed partial derivatives f_{xyy}, f_{yxy}, and f_{yyx} are equal.

69. $f(x, y, z) = xyz$

70. $f(x, y, z) = x^2 - 3xy + 4yz + z^3$

71. $f(x, y, z) = e^{-x} \sin yz$

72. $f(x, y, z) = \dfrac{2z}{x + y}$

Laplace's Equation In Exercises 73–76, show that the function satisfies Laplace's equation $\partial^2 z/\partial x^2 + \partial^2 z/\partial y^2 = 0$.

73. $z = 5xy$

74. $z = \tfrac{1}{2}(e^y - e^{-y})\sin x$

75. $z = e^x \sin y$

76. $z = \arctan \dfrac{y}{x}$

Wave Equation In Exercises 77 and 78, show that the function satisfies the wave equation $\partial^2 z/\partial t^2 = c^2(\partial^2 z/\partial x^2)$.

77. $z = \sin(x - ct)$

78. $z = \sin \omega ct \sin \omega x$

Heat Equation In Exercises 79 and 80, show that the function satisfies the heat equation $\partial z/\partial t = c^2(\partial^2 z/\partial x^2)$.

79. $z = e^{-t} \cos \dfrac{x}{c}$

80. $z = e^{-t} \sin \dfrac{x}{c}$

Getting at the Concept

81. Define the first partial derivatives of a function f of two variables x and y.

82. Let f be a function of two variables x and y. Describe the procedure for finding the first partial derivatives.

83. Sketch a surface representing a function f of two variables x and y. Use the sketch to give a geometric interpretation of $\partial f/\partial x$ and $\partial f/\partial y$.

84. Sketch the graph of a function $z = f(x, y)$ whose derivative f_x is always negative and whose derivative f_y is always positive.

85. Sketch the graph of a function $z = f(x, y)$ whose derivatives f_x and f_y are always positive.

86. If f is a function of x and y such that f_{xy} and f_{yx} are continuous, what is the relationship between the mixed partial derivatives?

87. Marginal Costs A company manufactures two types of wood-burning stoves: a freestanding model and a fireplace-insert model. The cost function for producing x freestanding and y fireplace-insert stoves is

$$C = 32\sqrt{xy} + 175x + 205y + 1050.$$

(a) Find the marginal costs ($\partial C/\partial x$ and $\partial C/\partial y$) when $x = 80$ and $y = 20$.

(b) When additional production is required, which model stove results in the cost increasing at a faster rate? How can this be determined from the cost model?

88. Marginal Productivity Consider the Cobb-Douglas production function $f(x, y) = 200x^{0.7}y^{0.3}$. When $x = 1000$ and $y = 500$, find

(a) the marginal productivity of labor, $\partial f/\partial x$.

(b) the marginal productivity of capital, $\partial f/\partial y$.

89. Think About It Let N be the number of applicants to a university, p the charge for food and housing at the university, and t the tuition. Suppose that N is a function of p and t such that $\partial N/\partial p < 0$ and $\partial N/\partial t < 0$. What information is gained by noticing that both partials are negative?

90. Investment The value of an investment of $1000 earning 10% compounded annually is

$$V(I, R) = 1000\left[\frac{1 + 0.10(1 - R)}{1 + I}\right]^{10}$$

where I is the annual rate of inflation and R is the tax rate for the person making the investment. Calculate $V_I(0.03, 0.28)$ and $V_R(0.03, 0.28)$. Determine whether the tax rate or the rate of inflation is the greater "negative" factor on the growth of the investment.

91. Temperature Distribution The temperature at any point (x, y) in a steel plate is

$$T = 500 - 0.6x^2 - 1.5y^2$$

where x and y are measured in meters. At the point $(2, 3)$, find the rate of change of the temperature with respect to the distance moved along the plate in the directions of the x- and y-axes.

92. Apparent Temperature A measure of what hot weather feels like to two average persons is the Apparent Temperature Index. A model for this index is

$$A = 0.885t - 22.4h + 1.20th - 0.544,$$

where A is the apparent temperature in °C, t is the air temperature, and h is the relative humidity in decimal form. *(Source: The UMAP Journal, Fall 1984)*

(a) Find $\partial A/\partial t$ and $\partial A/\partial h$ when $t = 30°$ and $h = 0.80$.

(b) Which has a greater effect on A, air temperature or humidity? Explain.

93. Ideal Gas Law The Ideal Gas Law states that $PV = nRT$, where P is pressure, V is volume, n is the number of moles of gas, R is a fixed constant (the gas constant), and T is absolute temperature. Show that

$$\frac{\partial T}{\partial P}\frac{\partial P}{\partial V}\frac{\partial V}{\partial T} = -1.$$

94. Marginal Utility The utility function $U = f(x, y)$ is a measure of the utility (or satisfaction) derived by a person from the consumption of two products x and y. Suppose the utility function is

$$U = -5x^2 + xy - 3y^2.$$

(a) Determine the marginal utility of product x.

(b) Determine the marginal utility of product y.

(c) When $x = 2$ and $y = 3$, should a person consume one more unit of product x or one more unit of product y? Explain your reasoning.

(d) Use a computer algebra system to graph the function. Interpret the marginal utilities of products x and y graphically.

95. Modeling Data Per capita consumption (in gallons) of different types of milk in the United States for selected years is shown in the table. Consumption of skim milk, reduced-fat milk, and whole milk are represented by the variables x, y, and z, respectively. *(Source: U.S. Department of Agriculture)*

Year	1980	1985	1990	1994	1995	1996	1997
x	3.1	3.2	4.9	5.8	6.2	6.4	6.6
y	6.3	7.9	9.1	8.7	8.2	8.0	7.7
z	16.5	13.9	10.2	8.8	8.4	8.4	8.2

A model for the data is given by

$$z = -1.83x - 1.09y + 28.7.$$

(a) Find $\dfrac{\partial z}{\partial x}$ and $\dfrac{\partial z}{\partial y}$.

(b) Interpret the partial derivatives in the context of the problem.

96. Modeling Data The table shows the amount of public medical expenditures (in billions of dollars) for worker's compensation x, public assistance y, and Medicare z for selected years. *(Source: U.S. Health Care Financing Administration)*

Year	1980	1985	1990	1995	1996	1997
x	5.1	8.0	16.1	17.1	15.2	14.1
y	28.0	44.4	80.5	151.6	159.7	165.2
z	37.5	72.2	111.5	185.2	200.1	214.6

A model for the data is

$$z = -0.775x^2 + 0.007y^2 + 22.15x - 0.54y - 45.4.$$

(a) Find $\dfrac{\partial^2 z}{\partial x^2}$ and $\dfrac{\partial^2 z}{\partial y^2}$.

(b) Determine the concavity of traces parallel to the xz-plane. Interpret the result in the context of the problem.

(c) Determine the concavity of traces parallel to the yz-plane. Interpret the result in the context of the problem.

97. Consider the function defined by

$$f(x, y) = \begin{cases} \dfrac{xy(x^2 - y^2)}{x^2 + y^2}, & (x, y) \neq (0, 0) \\ 0, & (x, y) = (0, 0). \end{cases}$$

(a) Find $f_x(x, y)$ and $f_y(x, y)$ for $(x, y) \neq (0, 0)$.

(b) Use the definition of partial derivatives to find $f_x(0, 0)$ and $f_y(0, 0)$.

$$\left[\text{Hint: } f_x(0, 0) = \lim_{\Delta x \to 0} \frac{f(\Delta x, 0) - f(0, 0)}{\Delta x}.\right]$$

(c) Use the definition of partial derivatives to find $f_{xy}(0, 0)$ and $f_{yx}(0, 0)$.

(d) Using Theorem 12.3 and the result in part (c), what can be said about f_{xy} or f_{yx}?

True or False? In Exercises 98–101, determine whether the statement is true or false. If it is false, explain why or give an example that shows it is false.

98. If $z = f(x, y)$ and $\partial z/\partial x = \partial z/\partial y$, then $z = c(x + y)$.

99. If $z = f(x)g(y)$, then $(\partial z/\partial x) + (\partial z/\partial y) = f'(x)g(y) + f(x)g'(y)$.

100. If $z = e^{xy}$, then $\dfrac{\partial^2 z}{\partial y \partial x} = (xy + 1)e^{xy}$.

101. If a cylindrical surface $z = f(x, y)$ has rulings parallel to the y-axis, then $\partial z/\partial y = 0$.

102. Let $f(x, y) = \displaystyle\int_x^y \sqrt{1 + t^3}\, dt$. Find $f_x(x, y)$ and $f_y(x, y)$.

SECTION PROJECT: MOIRÉ FRINGES

Read the article "Moiré Fringes and the Conic Sections" by Mike Cullen in *The College Mathematics Journal*. (To view this article, go to the website *www.matharticles.com*.) The article describes how two families of level curves given by

$$f(x, y) = a$$

and

$$g(x, y) = b$$

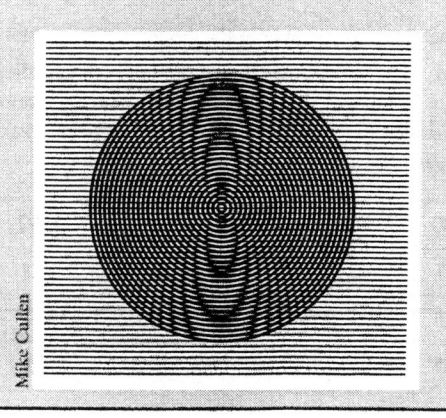

can form Moiré patterns. After reading the article, write a paper explaining how the expression

$$\frac{\partial f}{\partial x} \cdot \frac{\partial g}{\partial x} + \frac{\partial f}{\partial y} \cdot \frac{\partial g}{\partial y}$$

is related to the Moiré patterns formed by intersecting the two families of level curves. Use one of the following patterns as an example in your paper.

Section 12.4 Differentials

- Understand the concepts of increments and differentials.
- Extend the concept of differentiability to a function of two variables.
- Use a differential as an approximation.

Increments and Differentials

In this section, we generalize the concepts of increments and differentials to functions of two or more variables. Recall from Chapter 3 that for $y = f(x)$, the differential of y was defined as

$$dy = f'(x)\, dx.$$

Similar terminology is used for a function of two variables, $z = f(x, y)$. That is, Δx and Δy are the **increments of x and y**, and the **increment of z** is given by

$$\Delta z = f(x + \Delta x, y + \Delta y) - f(x, y). \qquad \text{Increment of } z$$

Definition of Total Differential

If $z = f(x, y)$ and Δx and Δy are increments of x and y, then the **differentials** of the independent variables x and y are

$$dx = \Delta x \quad \text{and} \quad dy = \Delta y$$

and the **total differential** of the dependent variable z is

$$dz = \frac{\partial z}{\partial x}\, dx + \frac{\partial z}{\partial y}\, dy = f_x(x, y)\, dx + f_y(x, y)\, dy.$$

This definition can be extended to a function of three or more variables. For instance, if $w = f(x, y, z, u)$, then $dx = \Delta x$, $dy = \Delta y$, $dz = \Delta z$, $du = \Delta u$, and the total differential of w is

$$dw = \frac{\partial w}{\partial x}\, dx + \frac{\partial w}{\partial y}\, dy + \frac{\partial w}{\partial z}\, dz + \frac{\partial w}{\partial u}\, du.$$

Example 1 Finding the Total Differential

Find the total differential for each function.

a. $z = 2x \sin y - 3x^2 y^2$ **b.** $w = x^2 + y^2 + z^2$

Solution

a. The total differential dz for $z = 2x \sin y - 3x^2 y^2$ is

$$dz = \frac{\partial z}{\partial x}\, dx + \frac{\partial z}{\partial y}\, dy \qquad \text{Total differential } dz$$

$$= (2 \sin y - 6xy^2)\, dx + (2x \cos y - 6x^2 y)\, dy.$$

b. The total differential dw for $w = x^2 + y^2 + z^2$ is

$$dw = \frac{\partial w}{\partial x}\, dx + \frac{\partial w}{\partial y}\, dy + \frac{\partial w}{\partial z}\, dz \qquad \text{Total differential } dw$$

$$= 2x\, dx + 2y\, dy + 2z\, dz.$$

Differentiability

In Chapter 3, you learned that for a *differentiable* function given by $y = f(x)$, you can use the differential $dy = f'(x)\,dx$ as an approximation (for small Δx) to the value $\Delta y = f(x + \Delta x) - f(x)$. When a similar approximation is possible for a function of two variables, the function is said to be **differentiable**. This is stated explicitly in the following definition.

Definition of Differentiability

A function f given by $z = f(x, y)$ is **differentiable** at (x_0, y_0) if Δz can be expressed in the form

$$\Delta z = f_x(x_0, y_0)\,\Delta x + f_y(x_0, y_0)\,\Delta y + \varepsilon_1 \Delta x + \varepsilon_2 \Delta y$$

where both ε_1 and $\varepsilon_2 \to 0$ as $(\Delta x, \Delta y) \to (0, 0)$. The function f is **differentiable in a region R** if it is differentiable at each point in R.

Example 2 Showing That a Function Is Differentiable

Show that the function given by

$$f(x, y) = x^2 + 3y$$

is differentiable at every point in the plane.

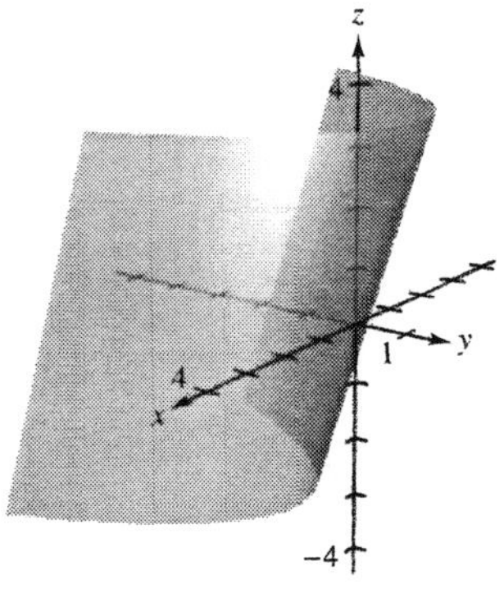

Solution Letting $z = f(x, y)$, the increment of z at an arbitrary point (x, y) in the plane is

$$\begin{aligned}
\Delta z &= f(x + \Delta x, y + \Delta y) - f(x, y) && \text{Increment of } z\\
&= (x^2 + 2x\Delta x + \Delta x^2) + 3(y + \Delta y) - (x^2 + 3y)\\
&= 2x\Delta x + \Delta x^2 + 3\Delta y\\
&= 2x(\Delta x) + 3(\Delta y) + \Delta x(\Delta x) + 0(\Delta y)\\
&= f_x(x, y)\,\Delta x + f_y(x, y)\,\Delta y + \varepsilon_1 \Delta x + \varepsilon_2 \Delta y
\end{aligned}$$

where $\varepsilon_1 = \Delta x$ and $\varepsilon_2 = 0$. Because $\varepsilon_1 \to 0$ and $\varepsilon_2 \to 0$ as $(\Delta x, \Delta y) \to (0, 0)$, it follows that f is differentiable at every point in the plane. The graph of f is shown in Figure 12.34.

Figure 12.34

Be sure you see that the term "differentiable" is used differently for functions of two variables than for functions of one variable. A function of one variable is differentiable at a point if its derivative exists at the point. However, for a function of two variables, the existence of the partial derivatives f_x and f_y does not guarantee that the function is differentiable (see Example 5). The following theorem gives a *sufficient* condition for differentiability of a function of two variables. A proof of Theroem 12.4 is given in Appendix B.

THEOREM 12.4 Sufficient Condition for Differentiability

If f is a function of x and y, where f_x and f_y are continuous in an open region R, then f is differentiable on R.

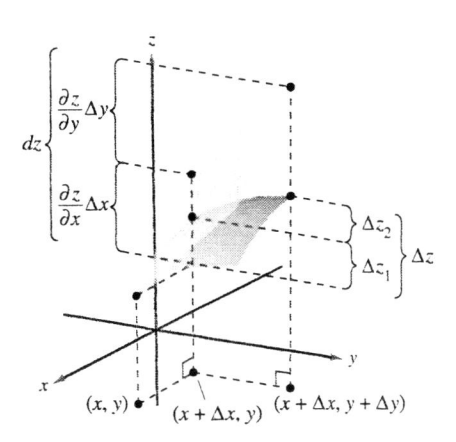

Approximation by Differentials

Theorem 12.4 tells you that you can choose $(x + \Delta x, y + \Delta y)$ close enough to (x, y) to make $\varepsilon_1 \Delta x$ and $\varepsilon_2 \Delta y$ insignificant. In other words, for small Δx and Δy, you can use the approximation

$$\Delta z \approx dz.$$

This approximation is illustrated graphically in Figure 12.35. Recall that the partial derivatives $\partial z/\partial x$ and $\partial z/\partial y$ can be interpreted as the slopes of the surface in the x- and y-directions. This means that

$$dz = \frac{\partial z}{\partial x}\Delta x + \frac{\partial z}{\partial y}\Delta y$$

represents the change in height of a plane that is tangent to the surface at the point $(x, y, f(x, y))$. Because a plane in space is represented by a linear equation in the variables x, y, and z, the approximation of Δz by dz is called a **linear approximation.** You will learn more about this geometric interpretation in Section 12.7.

The exact change in z is Δz. This change can be approximated by the differential dz.
Figure 12.35

Example 3 Using a Differential as an Approximation

Use the differential dz to approximate the change in $z = \sqrt{4 - x^2 - y^2}$ as (x, y) moves from the point $(1, 1)$ to the point $(1.01, 0.97)$. Compare this approximation with the exact change in z.

Solution Letting $(x, y) = (1, 1)$ and $(x + \Delta x, y + \Delta y) = (1.01, 0.97)$ produces $dx = \Delta x = 0.01$ and $dy = \Delta y = -0.03$. So, the change in z can be approximated by

$$\Delta z \approx dz = \frac{\partial z}{\partial x}dx + \frac{\partial z}{\partial y}dy = \frac{-x}{\sqrt{4 - x^2 - y^2}}\Delta x + \frac{-y}{\sqrt{4 - x^2 - y^2}}\Delta y.$$

When $x = 1$ and $y = 1$, you have

$$\Delta z \approx -\frac{1}{\sqrt{2}}(0.01) - \frac{1}{\sqrt{2}}(-0.03) = \frac{0.02}{\sqrt{2}} = \sqrt{2}(0.01) \approx 0.0141.$$

In Figure 12.36 you can see that the exact change corresponds to the difference in the heights of two points on the surface of a hemisphere. This difference is given by

$$\Delta z = f(1.01, 0.97) - f(1, 1)$$
$$= \sqrt{4 - (1.01)^2 - (0.97)^2} - \sqrt{4 - 1^2 - 1^2} \approx 0.0137.$$

A function of three variables $w = f(x, y, z)$ is called **differentiable** at (x, y, z) provided that

$$\Delta w = f(x + \Delta x, y + \Delta y, z + \Delta z) - f(x, y, z)$$

can be expressed in the form

$$\Delta w = f_x \Delta x + f_y \Delta y + f_z \Delta z + \varepsilon_1 \Delta x + \varepsilon_2 \Delta y + \varepsilon_3 \Delta z$$

where ε_1, ε_2, and $\varepsilon_3 \to 0$ as $(\Delta x, \Delta y, \Delta z) \to (0, 0, 0)$. With this definition of differentiability, Theorem 12.4 has the following extension for functions of three variables: If f is a function of x, y, and z, where f, f_x, f_y, and f_z are continuous in an open region R, then f is differentiable on R.

In Chapter 3, you used differentials to approximate the propagated error introduced by an error in measurement. This application of differentials is further illustrated in Example 4.

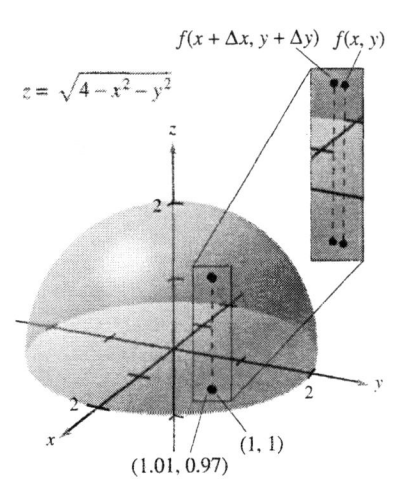

As (x, y) moves from $(1, 1)$ to the point $(1.01, 0.97)$, the value of $f(x, y)$ changes by about 0.0137.
Figure 12.36

Example 4 Error Analysis

The possible error involved in measuring each dimension of a rectangular box is ±0.1 millimeter. The dimensions of the box are $x = 50$ centimeters, $y = 20$ centimeters, and $z = 15$ centimeters, as shown in Figure 12.37. Use dV to estimate the propagated error and the relative error in the calculated volume of the box.

Solution The volume of the box is given by $V = xyz$, and thus

$$dV = \frac{\partial V}{\partial x}dx + \frac{\partial V}{\partial y}dy + \frac{\partial V}{\partial z}dz$$
$$= yz\, dx + xz\, dy + xy\, dz.$$

Using 0.1 millimeter = 0.01 centimeter, you have $dx = dy = dz = \pm 0.01$, and the propagated error is approximately

$$dV = (20)(15)(\pm 0.01) + (50)(15)(\pm 0.01) + (50)(20)(\pm 0.01)$$
$$= 300(\pm 0.01) + 750(\pm 0.01) + 1000(\pm 0.01)$$
$$= 2050(\pm 0.01) = \pm 20.5 \text{ cm}^3.$$

Because the measured volume is

$$V = (50)(20)(15) = 15{,}000 \text{ cm}^3$$

the relative error, $\Delta V/V$, is approximately

$$\frac{\Delta V}{V} \approx \frac{dV}{V} = \frac{20.5}{15{,}000} \approx 0.14\%.$$

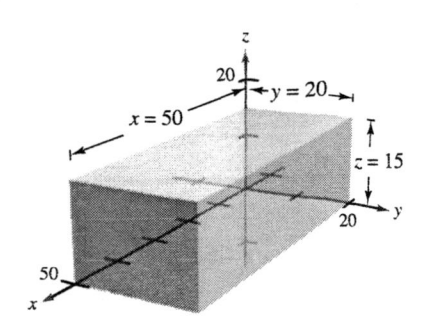

Volume = xyz
Figure 12.37

As is true for a function of a single variable, if a function in two or more variables is differentiable at a point, it is also continuous there.

THEOREM 12.5 Differentiability Implies Continuity

If a function of x and y is differentiable at (x_0, y_0), then it is continuous at (x_0, y_0).

Proof Let f be differentiable at (x_0, y_0), where $z = f(x, y)$. Then

$$\Delta z = [f_x(x_0, y_0) + \varepsilon_1]\Delta x + [f_y(x_0, y_0) + \varepsilon_2]\Delta y$$

where both ε_1 and $\varepsilon_2 \to 0$ as $(\Delta x, \Delta y) \to (0, 0)$. However, by definition, you know that Δz is given by

$$\Delta z = f(x_0 + \Delta x, y_0 + \Delta y) - f(x_0, y_0).$$

Letting $x = x_0 + \Delta x$ and $y = y_0 + \Delta y$ produces

$$f(x, y) - f(x_0, y_0) = [f_x(x_0, y_0) + \varepsilon_1]\Delta x + [f_y(x_0, y_0) + \varepsilon_2]\Delta y$$
$$= [f_x(x_0, y_0) + \varepsilon_1](x - x_0) + [f_y(x_0, y_0) + \varepsilon_2](y - y_0).$$

Taking the limit as $(x, y) \to (x_0, y_0)$, you have

$$\lim_{(x, y) \to (x_0, y_0)} f(x, y) = f(x_0, y_0)$$

which means that f is continuous at (x_0, y_0).

Remember that the existence of f_x and f_y is not sufficient to guarantee differentiability, as illustrated in the next example.

Example 5 A Function That Is Not Differentiable

Show that $f_x(0, 0)$ and $f_y(0, 0)$ both exist, but that f is not differentiable at $(0, 0)$ where f is defined as

$$f(x, y) = \begin{cases} \dfrac{-3xy}{x^2 + y^2}, & \text{if } (x, y) \neq (0, 0) \\ 0, & \text{if } (x, y) = (0, 0). \end{cases}$$

TECHNOLOGY Try using a graphing utility to graph the function given in Example 5. For instance, the graph shown below was generated by *Mathematica*.

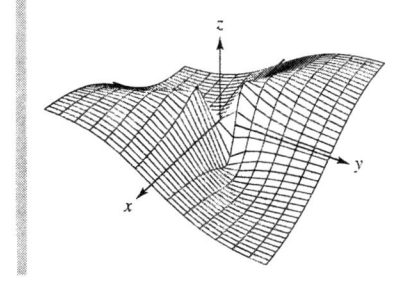

Solution You can show that f is not differentiable at $(0, 0)$ by showing that it is not continuous at this point. To see that f is not continuous at $(0, 0)$, look at the values of $f(x, y)$ along two different approaches to $(0, 0)$, as shown in Figure 12.38. Along the line $y = x$, the limit is

$$\lim_{(x, x) \to (0, 0)} f(x, y) = \lim_{(x, x) \to (0, 0)} \frac{-3x^2}{2x^2} = -\frac{3}{2}$$

whereas along $y = -x$ you have

$$\lim_{(x, -x) \to (0, 0)} f(x, y) = \lim_{(x, -x) \to (0, 0)} \frac{3x^2}{2x^2} = \frac{3}{2}.$$

So, the limit of $f(x, y)$ as $(x, y) \to (0, 0)$ does not exist, and you can conclude that f is not continuous at $(0, 0)$. Therefore, by Theorem 12.5, you know that f is not differentiable at $(0, 0)$. On the other hand, by the definition of the partial derivatives f_x and f_y, you have

$$f_x(0, 0) = \lim_{\Delta x \to 0} \frac{f(\Delta x, 0) - f(0, 0)}{\Delta x} = \lim_{\Delta x \to 0} \frac{0 - 0}{\Delta x} = 0$$

and

$$f_y(0, 0) = \lim_{\Delta y \to 0} \frac{f(0, \Delta y) - f(0, 0)}{\Delta y} = \lim_{\Delta y \to 0} \frac{0 - 0}{\Delta y} = 0.$$

So, the partial derivatives at $(0, 0)$ exist.

Along the line $y = -x$
$f(x, y)$ approaches $3/2$.

Along the line $y = x$
$f(x, y)$ approaches $-3/2$.

Figure 12.38

EXERCISES FOR SECTION 12.4

In Exercises 1–10, find the total differential.

1. $z = 3x^2y^3$
2. $z = \dfrac{x^2}{y}$
3. $z = \dfrac{-1}{x^2 + y^2}$
4. $w = \dfrac{x + y}{z - 2y}$
5. $z = x \cos y - y \cos x$
6. $z = \tfrac{1}{2}(e^{x^2+y^2} - e^{-x^2-y^2})$
7. $z = e^x \sin y$
8. $w = e^y \cos x + z^2$
9. $w = 2z^3 y \sin x$
10. $w = x^2yz^2 + \sin yz$

In Exercises 11–16, (a) evaluate $f(1, 2)$ and $f(1.05, 2.1)$ and calculate Δz, and (b) use the total differential dz to approximate Δz.

11. $f(x, y) = 9 - x^2 - y^2$
12. $f(x, y) = \sqrt{x^2 + y^2}$
13. $f(x, y) = x \sin y$
14. $f(x, y) = xe^y$
15. $f(x, y) = 3x - 4y$
16. $f(x, y) = \dfrac{x}{y}$

In Exercises 17–20, find $z = f(x, y)$ and use the total differential to approximate the quantity.

17. $\sqrt{(5.05)^2 + (3.1)^2} - \sqrt{5^2 + 3^2}$
18. $(2.03)^2(1 + 8.9)^3 - 2^2(1 + 9)^3$
19. $\dfrac{1 - (3.05)^2}{(5.95)^2} - \dfrac{1 - 3^2}{6^2}$
20. $\sin[(1.05)^2 + (0.95)^2] - \sin(1^2 + 1^2)$

Getting at the Concept

21. Define the total differential of a function of two variables.
22. Describe the change in accuracy of dz as an approximation of Δz as Δx and Δy increase.
23. What is meant by a linear approximation of $z = f(x, y)$ at the point $P(x_0, y_0)$?
24. When using differentials, what is meant by the terms *propagated error* and *relative error*?

25. **Area** The area of the shaded rectangle in the figure is $A = lh$. The possible errors in the length and height are Δl and Δh. Find dA and identify the regions in the figure whose areas are given by the terms of dA. What region represents the difference between ΔA and dA?

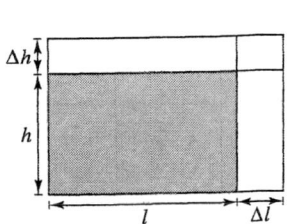

Figure for 25

Figure for 26

26. **Volume** The volume of the red right circular cylinder in the figure is $V = \pi r^2 h$. The possible errors in the radius and the height are Δr and Δh. Find dV and identify the solids in the figure whose volumes are given by the terms of dV. What solid represents the difference between ΔV and dV?

27. **Numerical Analysis** A right circular cone of height $h = 6$ and radius $r = 3$ is constructed, and in the process errors Δr and Δh are made in the radius and height. Complete the table to show the relationship between ΔV and dV for the indicated errors.

Δr	Δh	dV or dS	ΔV or ΔS	$\Delta V - dV$ or $\Delta S - dS$
0.1	0.1			
0.1	-0.1			
0.001	0.002			
-0.0001	0.0002			

28. **Numerical Analysis** The height and radius of a right circular cone are measured as $h = 20$ meters and $r = 8$ meters. In the process of measuring, errors Δr and Δh are made. If S is the lateral surface area of a cone, complete the table above to show the relationship between ΔS and dS for the indicated errors.

29. **Modeling Data** Per capita consumption (in gallons) of different types of milk in the United States for selected years is shown in the table. Consumption of skim milk, reduced-fat milk, and whole milk are represented by the variables x, y, and z, respectively. *(Source: U.S. Department of Agriculture)*

Year	1980	1985	1990	1994	1995	1996	1997
x	3.1	3.2	4.9	5.8	6.2	6.4	6.6
y	6.3	7.9	9.1	8.7	8.2	8.0	7.7
z	16.5	13.9	10.2	8.8	8.4	8.4	8.2

A model for the data is given by

$$z = -1.83x - 1.09y + 28.7.$$

(a) Find the total differential of the model.

(b) Suppose a dairy industry forecast for a certain year is that per capita consumption of skim milk will be 7.2 ± 0.25 gallons and that per capita consumption of reduced-fat milk will be 8.5 ± 0.25 gallons. Use dz to estimate the maximum possible propagated error and relative error in the prediction for the consumption of whole milk.

30. **Wind Chill** The formula for wind chill C is given by

$$C = 0.0817(3.71\sqrt{v} + 5.81 - 0.25v)(T - 91.4) + 91.4$$

where v is the wind speed in miles per hour and T is the temperature in degrees Fahrenheit. Suppose the wind speed is 23 ± 3 miles per hour and the temperature is $8° \pm 1°$. Use dC to estimate the maximum possible propagated error and relative error in calculating the wind chill.

31. Volume The radius r and height h of a right circular cylinder are measured with possible errors of 4% and 2%. Approximate the maximum possible percent error in measuring the volume.

32. Rectangular to Polar Coordinates A rectangular coordinate system is placed over a map and the coordinates of a point of interest are (8.5, 3.2). There is a possible error of 0.05 in each coordinate. Approximate the maximum possible error in measuring the polar coordinates for the point.

33. Area A triangle is measured and two adjacent sides are found to be 3 and 4 inches long, with an included angle of $\pi/4$. The possible errors in measurement are $\frac{1}{16}$ inch for the sides and 0.02 radian for the angle. Approximate the maximum possible error in the computation of the area.

34. Acceleration The centripetal acceleration of a particle moving in a circle is

$$a = \frac{v^2}{r}$$

where v is the velocity and r is the radius of the circle. Approximate the maximum percent error in measuring the acceleration due to errors of 3% in v and 2% in r.

35. Volume A trough is 16 feet long (see figure). Its cross sections are isosceles triangles; each of the two equal sides is 18 inches long. The angle between the two equal sides is θ.

(a) Express the volume of the trough as a function of θ and determine the value of θ such that the volume is a maximum.

(b) Approximate the change from the maximum volume if the maximum error in the linear measurements is one-half inch and the maximum error in the angle measure is 2°.

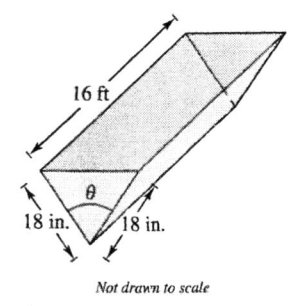

Figure for 35   **Figure for 36**

36. Baseball A baseball player in center field is playing approximately 330 feet from a television camera that is behind home plate. A batter hits a fly ball that goes to a wall 420 feet from the camera (see figure).

(a) Approximate the number of feet that the center fielder had to run to make the catch if the camera turned 9° to follow the play.

(b) Approximate the maximum possible error in the result in part (a) if the position of the center fielder could be in error by as much as 6 feet and the maximum error in measuring the rotation of the camera is 1°.

37. Power Electrical power P is given by

$$P = \frac{E^2}{R}$$

where E is voltage and R is resistance. Approximate the maximum percent error in calculating power if 200 volts is applied to a 4000-ohm resistor and the possible percent errors in measuring E and R are 2% and 3%.

38. Resistance The total resistance R of two resistors connected in parallel is

$$\frac{1}{R} = \frac{1}{R_1} + \frac{1}{R_2}.$$

Approximate the change in R as R_1 is increased from 10 ohms to 10.5 ohms and R_2 is decreased from 15 ohms to 13 ohms.

39. Inductance The inductance L (in microhenrys) of a straight nonmagnetic wire in free space is

$$L = 0.00021\left(\ln\frac{2h}{r} - 0.75\right)$$

where h is the length of the wire in millimeters and r is the radius of a circular cross section. Approximate L when $r = 2 \pm \frac{1}{16}$ millimeters and $h = 100 \pm \frac{1}{100}$ millimeters.

40. Pendulum The period T of a pendulum of length L is $T = 2\pi\sqrt{L/g}$, where g is the acceleration due to gravity. A pendulum is moved from the Canal Zone, where $g = 32.09$ feet per second per second, to Greenland, where $g = 32.24$ feet per second per second. Because of the change in temperature, the length of the pendulum changes from 2.5 feet to 2.48 feet. Approximate the change in the period of the pendulum.

In Exercises 41–44, show that the function is differentiable by finding values for ε_1 and ε_2 as designated in the definition of differentiability, and verify that both ε_1 and $\varepsilon_2 \to 0$ as $(\Delta x, \Delta y) \to (0, 0)$.

41. $f(x, y) = x^2 - 2x + y$ **42.** $f(x, y) = x^2 + y^2$
43. $f(x, y) = x^2 y$ **44.** $f(x, y) = 5x - 10y + y^3$

In Exercises 45 and 46, use the function to prove that (a) $f_x(0, 0)$ and $f_y(0, 0)$ exist, and (b) f is not differentiable at $(0, 0)$.

45. $f(x, y) = \begin{cases} \dfrac{3x^2 y}{x^4 + y^2}, & (x, y) \neq (0, 0) \\ 0, & (x, y) = (0, 0) \end{cases}$

46. $f(x, y) = \begin{cases} \dfrac{5x^2 y}{x^3 + y^3}, & (x, y) \neq (0, 0) \\ 0, & (x, y) = (0, 0) \end{cases}$

47. Interdisciplinary Problem Consider measurements and formulas you are using, or have used, in other science or engineering courses. Show how to apply differentials to these measurements and formulas to estimate possible propagated errors.

Section 12.5 Chain Rules for Functions of Several Variables

- Use the Chain Rules for functions of several variables.
- Find partial derivatives implicitly.

Chain Rules for Functions of Several Variables

Your work with differentials in the preceding section provides the basis for the extension of the Chain Rule to functions of two variables. There are two cases—the first case involves w as a function of x and y, where x and y are functions of a single independent variable t. (A proof of this theorem is given in Appendix B.)

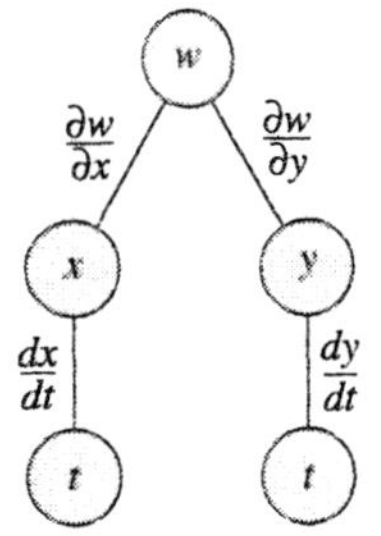

THEOREM 12.6   Chain Rule: One Independent Variable

Let $w = f(x, y)$, where f is a differentiable function of x and y. If $x = g(t)$ and $y = h(t)$, where g and h are differentiable functions of t, then w is a differentiable function of t, and

$$\frac{dw}{dt} = \frac{\partial w}{\partial x}\frac{dx}{dt} + \frac{\partial w}{\partial y}\frac{dy}{dt}. \quad \text{(See Figure 12.39.)}$$

Chain Rule: one independent variable w is a function of x and y, which are each functions of t. This diagram represents the derivative of w with respect to t.
Figure 12.39

Example 1   Using the Chain Rule with One Independent Variable

Let $w = x^2y - y^2$, where $x = \sin t$ and $y = e^t$. Find dw/dt when $t = 0$.

Solution   By the Chain Rule for one independent variable, you have

$$\frac{dw}{dt} = \frac{\partial w}{\partial x}\frac{dx}{dt} + \frac{\partial w}{\partial y}\frac{dy}{dt}$$
$$= 2xy(\cos t) + (x^2 - 2y)e^t$$
$$= 2(\sin t)(e^t)(\cos t) + (\sin^2 t - 2e^t)e^t$$
$$= 2e^t \sin t \cos t + e^t \sin^2 t - 2e^{2t}.$$

When $t = 0$, it follows that

$$\frac{dw}{dt} = -2.$$

The Chain Rules presented in this section provide alternative techniques for solving many problems in single-variable calculus. For instance, in Example 1, you could have used single-variable techniques to find dw/dt by first writing w as a function of t,

$$w = x^2y - y^2$$
$$= (\sin t)^2(e^t) - (e^t)^2$$
$$= e^t \sin^2 t - e^{2t}$$

and then differentiating as usual.

$$\frac{dw}{dt} = 2e^t \sin t \cos t + e^t \sin^2 t - 2e^{2t}$$

The Chain Rule in Theorem 12.6 can be extended to any number of variables. For example, if each x_i is a differentiable function of a single variable t, then for

$$w = f(x_1, x_2, \ldots, x_n)$$

you have

$$\frac{dw}{dt} = \frac{\partial w}{\partial x_1}\frac{dx_1}{dt} + \frac{\partial w}{\partial x_2}\frac{dx_2}{dt} + \cdots + \frac{\partial w}{\partial x_n}\frac{dx_n}{dt}.$$

Example 2 An Application of a Chain Rule to Related Rates

Two objects are traveling in elliptical paths given by the following parametric equations.

$x_1 = 4\cos t$ and $y_1 = 2\sin t$ First object
$x_2 = 2\sin 2t$ and $y_2 = 3\cos 2t$ Second object

At what rate is the distance between the two objects changing when $t = \pi$?

Solution From Figure 12.40, you can see that the distance s between the two objects is given by

$$s = \sqrt{(x_2 - x_1)^2 + (y_2 - y_1)^2}$$

and that when $t = \pi$, you have $x_1 = -4$, $y_1 = 0$, $x_2 = 0$, $y_2 = 3$, and

$$s = \sqrt{(0+4)^2 + (3-0)^2} = 5.$$

When $t = \pi$, the partial derivatives of s are as follows.

$$\frac{\partial s}{\partial x_1} = \frac{-(x_2 - x_1)}{\sqrt{(x_2 - x_1)^2 + (y_2 - y_1)^2}} = -\frac{1}{5}(0+4) = -\frac{4}{5}$$

$$\frac{\partial s}{\partial y_1} = \frac{-(y_2 - y_1)}{\sqrt{(x_2 - x_1)^2 + (y_2 - y_1)^2}} = -\frac{1}{5}(3-0) = -\frac{3}{5}$$

$$\frac{\partial s}{\partial x_2} = \frac{(x_2 - x_1)}{\sqrt{(x_2 - x_1)^2 + (y_2 - y_1)^2}} = \frac{1}{5}(0+4) = \frac{4}{5}$$

$$\frac{\partial s}{\partial y_2} = \frac{(y_2 - y_1)}{\sqrt{(x_2 - x_1)^2 + (y_2 - y_1)^2}} = \frac{1}{5}(3-0) = \frac{3}{5}$$

When $t = \pi$, the derivatives of x_1, y_1, x_2, and y_2 are

$$\frac{dx_1}{dt} = -4\sin t = 0 \qquad \frac{dy_1}{dt} = 2\cos t = -2$$

$$\frac{dx_2}{dt} = 4\cos 2t = 4 \qquad \frac{dy_2}{dt} = -6\sin 2t = 0.$$

Therefore, using the appropriate Chain Rule, you know that the distance is changing at the rate of

$$\frac{ds}{dt} = \frac{\partial s}{\partial x_1}\frac{dx_1}{dt} + \frac{\partial s}{\partial y_1}\frac{dy_1}{dt} + \frac{\partial s}{\partial x_2}\frac{dx_2}{dt} + \frac{\partial s}{\partial y_2}\frac{dy_2}{dt}$$

$$= \left(-\frac{4}{5}\right)(0) + \left(-\frac{3}{5}\right)(-2) + \left(\frac{4}{5}\right)(4) + \left(\frac{3}{5}\right)(0)$$

$$= \frac{22}{5}.$$

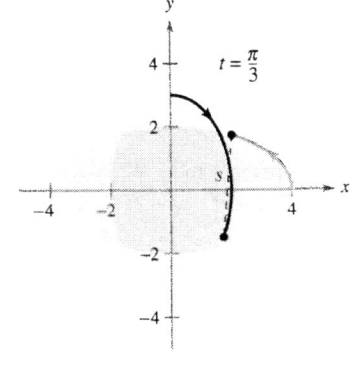

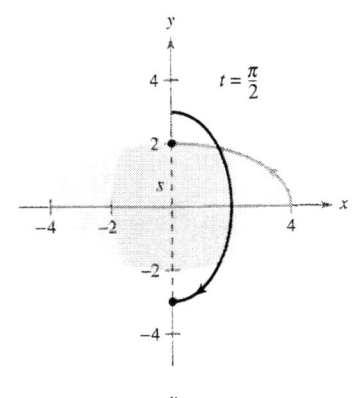

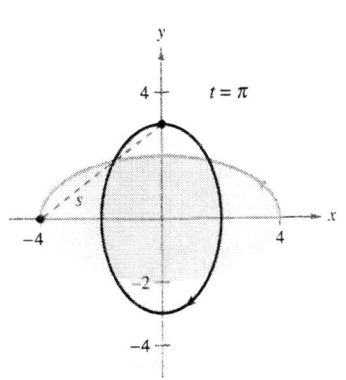

Paths of two objects traveling in elliptical orbits
Figure 12.40

In Example 2, note that s is the function of four *intermediate* variables, x_1, y_1, x_2, and y_2, each of which is a function of a single variable t. Another type of composite function is one in which the intermediate variables are themselves functions of more than one variable. For instance, if $w = f(x, y)$, where $x = g(s, t)$ and $y = h(s, t)$, it follows that w is a function of s and t, and you can consider the partial derivatives of w with respect to s and t. One way to find these partial derivatives is to write w as a function of s and t explicitly by substituting the equations $x = g(s, t)$ and $y = h(s, t)$ into the equation $w = f(x, y)$. Then you can find the partial derivatives in the usual way, as demonstrated in the next example.

Example 3 **Finding Partial Derivatives by Substitution**

Find $\partial w/\partial s$ and $\partial w/\partial t$ for $w = 2xy$, where $x = s^2 + t^2$ and $y = s/t$.

Solution Begin by substituting $x = s^2 + t^2$ and $y = s/t$ into the equation $w = 2xy$ to obtain

$$w = 2xy = 2(s^2 + t^2)\left(\frac{s}{t}\right) = 2\left(\frac{s^3}{t} + st\right).$$

Then, to find $\partial w/\partial s$, hold t constant and differentiate with respect to s.

$$\frac{\partial w}{\partial s} = 2\left(\frac{3s^2}{t} + t\right)$$

$$= \frac{6s^2 + 2t^2}{t}$$

Similarly, to find $\partial w/\partial t$, hold s constant and differentiate with respect to t to obtain

$$\frac{\partial w}{\partial t} = 2\left(-\frac{s^3}{t^2} + s\right)$$

$$= 2\left(\frac{-s^3 + st^2}{t^2}\right)$$

$$= \frac{2st^2 - 2s^3}{t^2}.$$

Theorem 12.7 gives an alternative method for finding the partial derivatives in Example 3, without explicitly writing w as a function of s and t.

THEOREM 12.7   Chain Rule: Two Independent Variables

Let $w = f(x, y)$, where f is a differentiable function of x and y. If $x = g(s, t)$ and $y = h(s, t)$ such that the first partials $\partial x/\partial s$, $\partial x/\partial t$, $\partial y/\partial s$, and $\partial y/\partial t$ all exist, then $\partial w/\partial s$ and $\partial w/\partial t$ exist and are given by

$$\frac{\partial w}{\partial s} = \frac{\partial w}{\partial x}\frac{\partial x}{\partial s} + \frac{\partial w}{\partial y}\frac{\partial y}{\partial s} \quad \text{and} \quad \frac{\partial w}{\partial t} = \frac{\partial w}{\partial x}\frac{\partial x}{\partial t} + \frac{\partial w}{\partial y}\frac{\partial y}{\partial t}.$$

Proof To obtain $\partial w/\partial s$, hold t constant and apply Theorem 12.6 to obtain the desired result. Similarly, for $\partial w/\partial t$ hold s constant and apply Theorem 12.6.

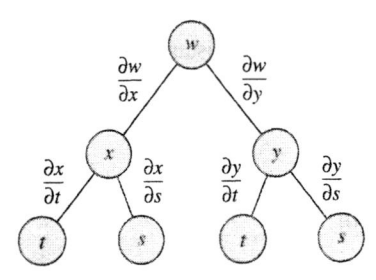

Chain Rule: two independent variables
Figure 12.41

NOTE The Chain Rule in this theorem is shown schematically in Figure 12.41.

Example 4   The Chain Rule with Two Independent Variables

Use the Chain Rule to find $\partial w/\partial s$ and $\partial w/\partial t$ for

$$w = 2xy$$

where $x = s^2 + t^2$ and $y = s/t$.

Solution (These same partials were found in Example 3.) Using Theorem 12.7, you can hold t constant and differentiate with respect to s to obtain

$$\frac{\partial w}{\partial s} = \frac{\partial w}{\partial x}\frac{\partial x}{\partial s} + \frac{\partial w}{\partial y}\frac{\partial y}{\partial s}$$

$$= 2y(2s) + 2x\left(\frac{1}{t}\right)$$

$$= 4\left(\frac{s^2}{t}\right) + \frac{2s^2 + 2t^2}{t} \quad\quad \text{Substitute } (s/t) \text{ for } y \text{ and } s^2 + t^2 \text{ for } x.$$

$$= \frac{6s^2 + 2t^2}{t}.$$

Similarly, holding s constant gives

$$\frac{\partial w}{\partial t} = \frac{\partial w}{\partial x}\frac{\partial x}{\partial t} + \frac{\partial w}{\partial y}\frac{\partial y}{\partial t}$$

$$= 2y(2t) + 2x\left(\frac{-s}{t^2}\right)$$

$$= 2\left(\frac{s}{t}\right)(2t) + 2(s^2 + t^2)\left(\frac{-s}{t^2}\right) \quad\quad \text{Substitute } (s/t) \text{ for } y \text{ and } s^2 + t^2 \text{ for } x.$$

$$= 4s - \frac{2s^3 + 2st^2}{t^2}$$

$$= \frac{4st^2 - 2s^3 - 2st^2}{t^2}$$

$$= \frac{2st^2 - 2s^3}{t^2}.$$

The Chain Rule in Theorem 12.7 can also be extended to any number of variables. For example, if w is a differentiable function of the n variables $x_1, x_2, \ldots, x_n$, where each x_i is a differentiable function of the m variables $t_1, t_2, \ldots, t_m$, then for

$$w = f(x_1, x_2, \ldots, x_n)$$

you obtain the following.

$$\frac{\partial w}{\partial t_1} = \frac{\partial w}{\partial x_1}\frac{\partial x_1}{\partial t_1} + \frac{\partial w}{\partial x_2}\frac{\partial x_2}{\partial t_1} + \cdots + \frac{\partial w}{\partial x_n}\frac{\partial x_n}{\partial t_1}$$

$$\frac{\partial w}{\partial t_2} = \frac{\partial w}{\partial x_1}\frac{\partial x_1}{\partial t_2} + \frac{\partial w}{\partial x_2}\frac{\partial x_2}{\partial t_2} + \cdots + \frac{\partial w}{\partial x_n}\frac{\partial x_n}{\partial t_2}$$

$$\vdots$$

$$\frac{\partial w}{\partial t_m} = \frac{\partial w}{\partial x_1}\frac{\partial x_1}{\partial t_m} + \frac{\partial w}{\partial x_2}\frac{\partial x_2}{\partial t_m} + \cdots + \frac{\partial w}{\partial x_n}\frac{\partial x_n}{\partial t_m}$$

Example 5 **The Chain Rule for a Function of Three Variables**

Find $\partial w/\partial s$ and $\partial w/\partial t$ when $s = 1$ and $t = 2\pi$ for the function given by

$$w = xy + yz + xz$$

where $x = s \cos t$, $y = s \sin t$, and $z = t$.

Solution By extending the result of Theorem 12.7, you have

$$\frac{\partial w}{\partial s} = \frac{\partial w}{\partial x}\frac{\partial x}{\partial s} + \frac{\partial w}{\partial y}\frac{\partial y}{\partial s} + \frac{\partial w}{\partial z}\frac{\partial z}{\partial s}$$

$$= (y + z)(\cos t) + (x + z)(\sin t) + (y + x)(0)$$

$$= (y + z)(\cos t) + (x + z)(\sin t).$$

When $s = 1$ and $t = 2\pi$, you have $x = 1$, $y = 0$, and $z = 2\pi$. Therefore, $\partial w/\partial s = 2\pi(1) + (1 + 2\pi)(0) + 0 = 2\pi$. Furthermore,

$$\frac{\partial w}{\partial t} = \frac{\partial w}{\partial x}\frac{\partial x}{\partial t} + \frac{\partial w}{\partial y}\frac{\partial y}{\partial t} + \frac{\partial w}{\partial z}\frac{\partial z}{\partial t}$$

$$= (y + z)(-s \sin t) + (x + z)(s \cos t) + (y + x)(1)$$

and for $s = 1$ and $t = 2\pi$ it follows that

$$\frac{\partial w}{\partial t} = (0 + 2\pi)(0) + (1 + 2\pi)(1) + (0 + 1)(1)$$

$$= 2 + 2\pi.$$

Implicit Partial Differentiation

We conclude this section with an application of the Chain Rule to determine the derivative of a function defined *implicitly*. Suppose that x and y are related by the equation $F(x, y) = 0$, where it is assumed that $y = f(x)$ is a differentiable function of x. To find dy/dx, you could use the techniques discussed in Section 2.5. However, you will see that the Chain Rule provides a convenient alternative. If you consider the function given by

$$w = F(x, y) = F(x, f(x))$$

you can apply Theorem 12.6 to obtain

$$\frac{dw}{dx} = F_x(x, y)\frac{dx}{dx} + F_y(x, y)\frac{dy}{dx}.$$

Because $w = F(x, y) = 0$ for all x in the domain of f, you know that $dw/dx = 0$ and you have

$$F_x(x, y)\frac{dx}{dx} + F_y(x, y)\frac{dy}{dx} = 0.$$

Now, if $F_y(x, y) \neq 0$, you can use the fact that $dx/dx = 1$ to conclude that

$$\frac{dy}{dx} = -\frac{F_x(x, y)}{F_y(x, y)}.$$

A similar procedure can be used to find the partial derivatives of functions of several variables that are defined implicitly.

THEOREM 12.8 Chain Rule: Implicit Differentiation

If the equation $F(x, y) = 0$ defines y implicitly as a differentiable function of x, then

$$\frac{dy}{dx} = -\frac{F_x(x, y)}{F_y(x, y)}, \qquad F_y(x, y) \neq 0.$$

If the equation $F(x, y, z) = 0$ defines z implicitly as a differentiable function of x and y, then

$$\frac{\partial z}{\partial x} = -\frac{F_x(x, y, z)}{F_z(x, y, z)} \quad \text{and} \quad \frac{\partial z}{\partial y} = -\frac{F_y(x, y, z)}{F_z(x, y, z)}, \qquad F_z(x, y, z) \neq 0.$$

This theorem can be extended to differentiable functions defined implicitly with any number of variables.

Example 6 Finding a Derivative Implicitly

Find dy/dx, given $y^3 + y^2 - 5y - x^2 + 4 = 0$.

Solution Begin by defining a function F as

$$F(x, y) = y^3 + y^2 - 5y - x^2 + 4.$$

Then, using Theorem 12.8, you have

$$F_x(x, y) = -2x \quad \text{and} \quad F_y(x, y) = 3y^2 + 2y - 5$$

and it follows that

$$\frac{dy}{dx} = -\frac{F_x(x, y)}{F_y(x, y)} = \frac{-(-2x)}{3y^2 + 2y - 5} = \frac{2x}{3y^2 + 2y - 5}.$$

NOTE Compare the solution of Example 6 with the solution of Example 2 in Section 2.5.

Example 7 Finding Partial Derivatives Implicitly

Find $\partial z/\partial x$ and $\partial z/\partial y$, given $3x^2z - x^2y^2 + 2z^3 + 3yz - 5 = 0$.

Solution To apply Theorem 12.8, let

$$F(x, y, z) = 3x^2z - x^2y^2 + 2z^3 + 3yz - 5.$$

Then

$$F_x(x, y, z) = 6xz - 2xy^2$$
$$F_y(x, y, z) = -2x^2y + 3z$$
$$F_z(x, y, z) = 3x^2 + 6z^2 + 3y$$

and you obtain

$$\frac{\partial z}{\partial x} = -\frac{F_x}{F_z} = \frac{2xy^2 - 6xz}{3x^2 + 6z^2 + 3y}$$
$$\frac{\partial z}{\partial y} = -\frac{F_y}{F_z} = \frac{2x^2y - 3z}{3x^2 + 6z^2 + 3y}.$$

EXERCISES FOR SECTION 12.5

In Exercises 1–4, find dw/dt using the appropriate Chain Rule.

1. $w = x^2 + y^2$
   $x = e^t, \quad y = e^{-t}$

2. $w = \sqrt{x^2 + y^2}$
   $x = \cos t, \quad y = e^t$

3. $w = x \sec y$
   $x = e^t, \quad y = \pi - t$

4. $w = \ln \dfrac{y}{x}$
   $x = \cos t, \quad y = \sin t$

In Exercises 5–10, find dw/dt (a) using the appropriate Chain Rule and (b) by converting w to a function of t before differentiating.

5. $w = xy, \quad x = 2 \sin t, \quad y = \cos t$
6. $w = \cos(x - y), \quad x = t^2, \quad y = 1$
7. $w = x^2 + y^2 + z^2, \quad x = e^t \cos t, \quad y = e^t \sin t, \quad z = e^t$
8. $w = xy \cos z, \quad x = t, \quad y = t^2, \quad z = \arccos t$
9. $w = xy + xz + yz, \quad x = t - 1, \quad y = t^2 - 1, \quad z = t$
10. $w = xyz, \quad x = t^2, \quad y = 2t, \quad z = e^{-t}$

Projectile Motion In Exercises 11 and 12, the parametric equations for the paths of two projectiles are given. At what rate is the distance between the two objects changing at the specified value of t?

11. $x_1 = 10 \cos 2t, \; y_1 = 6 \sin 2t$ First object
    $x_2 = 7 \cos t, \; y_2 = 4 \sin t$ Second object
    $t = \pi/2$

12. $x_1 = 48\sqrt{2}\,t, \; y_1 = 48\sqrt{2}\,t - 16t^2$ First object
    $x_2 = 48\sqrt{3}\,t, \; y_2 = 48t - 16t^2$ Second object
    $t = 1$

In Exercises 13 and 14, find d^2w/dt^2 using the appropriate Chain Rule. Evaluate d^2w/dt^2 at the given value of t.

13. $w = \arctan(2xy), \quad x = \cos t, \quad y = \sin t, \quad t = 0$
14. $w = \dfrac{x^2}{y}, \quad x = t^2, \quad y = t + 1, \quad t = 1$

In Exercises 15–18, find $\partial w/\partial s$ and $\partial w/\partial t$ using the appropriate Chain Rule, and evaluate each partial derivative at the indicated values of s and t.

Function	Point
15. $w = x^2 + y^2$	$s = 2, \quad t = -1$
$x = s + t, \quad y = s - t$	
16. $w = y^3 - 3x^2 y$	$s = 0, \quad t = 1$
$x = e^s, \quad y = e^t$	
17. $w = x^2 - y^2$	$s = 3, \quad t = \dfrac{\pi}{4}$
$x = s \cos t, \quad y = s \sin t$	
18. $w = \sin(2x + 3y)$	$s = 0, \quad t = \dfrac{\pi}{2}$
$x = s + t, \quad y = s - t$	

In Exercises 19–22, find $\partial w/\partial r$ and $\partial w/\partial \theta$ (a) using the appropriate Chain Rule and (b) by converting w to a function of r and θ before differentiating.

19. $w = x^2 - 2xy + y^2, \quad x = r + \theta, \quad y = r - \theta$
20. $w = \sqrt{25 - 5x^2 - 5y^2}, \quad x = r \cos \theta, \quad y = r \sin \theta$
21. $w = \arctan \dfrac{y}{x}, \quad x = r \cos \theta, \quad y = r \sin \theta$
22. $w = \dfrac{yz}{x}, \quad x = \theta^2, \quad y = r + \theta, \quad z = r - \theta$

In Exercises 23–26, find $\partial w/\partial s$ and $\partial w/\partial t$ by using the appropriate Chain Rule.

23. $w = xyz, \quad x = s + t, \quad y = s - t, \quad z = st^2$
24. $w = x \cos yz, \quad x = s^2, \quad y = t^2, \quad z = s - 2t$
25. $w = ze^{x/y}, \quad x = s - t, \quad y = s + t, \quad z = st$
26. $w = x^2 + y^2 + z^2, \quad x = t \sin s, \quad y = t \cos s, \quad z = st^2$

In Exercises 27–30, differentiate implicitly to find dy/dx.

27. $x^2 - 3xy + y^2 - 2x + y - 5 = 0$
28. $\cos x + \tan xy + 5 = 0$
29. $\ln \sqrt{x^2 + y^2} + xy = 4$
30. $\dfrac{x}{x^2 + y^2} - y^2 = 6$

In Exercises 31–38, differentiate implicitly to find the first partial derivatives of z.

31. $x^2 + y^2 + z^2 = 25$
32. $xz + yz + xy = 0$
33. $\tan(x + y) + \tan(y + z) = 1$
34. $z = e^x \sin(y + z)$
35. $x^2 + 2yz + z^2 = 1$
36. $x + \sin(y + z) = 0$
37. $e^{xz} + xy = 0$
38. $x \ln y + y^2 z + z^2 = 8$

In Exercises 39–42, differentiate implicitly to find the first partial derivatives of w.

39. $xyz + xzw - yzw + w^2 = 5$
40. $x^2 + y^2 + z^2 - 5yw + 10w^2 = 2$
41. $\cos xy + \sin yz + wz = 20$
42. $w - \sqrt{x - y} - \sqrt{y - z} = 0$

Homogeneous Functions In Exercises 43–46, the function f is homogeneous of degree n if $f(tx, ty) = t^n f(x, y)$. Determine the degree of the homogeneous function, and show that
$$xf_x(x, y) + yf_y(x, y) = nf(x, y).$$

43. $f(x, y) = \dfrac{xy}{\sqrt{x^2 + y^2}}$
44. $f(x, y) = x^3 - 3xy^2 + y^3$
45. $f(x, y) = e^{x/y}$
46. $f(x, y) = \dfrac{x^2}{\sqrt{x^2 + y^2}}$

Getting at the Concept

47. Let $w = f(x, y)$ be a function where x and y are functions of a single variable t. Give the Chain Rule for finding dw/dt.

48. Let $w = f(x, y)$ be a function where x and y are functions of two variables s and t. Give the Chain Rule for finding $\partial w/\partial s$ and $\partial w/\partial t$.

49. Describe the difference between the explicit form of a function of two variables x and y and the implicit form. Give an example of each.

50. If $f(x, y) = 0$, give the rule for finding dy/dx implicitly. If $f(x, y, z) = 0$, give the rule for finding $\partial z/\partial x$ and $\partial z/\partial y$ implicitly.

---

51. Area Let θ be the angle between equal sides of an isosceles triangle and let x be the length of these sides. If x is increasing at $\frac{1}{2}$ meter per hour and θ is increasing at $\pi/90$ radian per hour, find the rate of increase of the area when $x = 6$ and $\theta = \pi/4$.

52. Volume and Surface Area The radius of a right circular cylinder is increasing at a rate of 6 inches per minute, and the height is decreasing at a rate of 4 inches per minute. What is the rate of change of the volume and surface area when the radius is 12 inches and the height is 36 inches?

53. Volume and Surface Area Repeat Exercise 52 for a right circular cone.

54. Volume and Surface Area The two radii of the frustum of a right circular cone are increasing at a rate of 4 centimeters per minute, and the height is increasing at a rate of 12 centimeters per minute (see figure). Find the rate at which the volume and surface area are changing when the two radii are 15 centimeters and 25 centimeters, and the height is 10 centimeters.

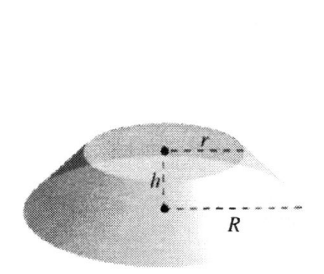

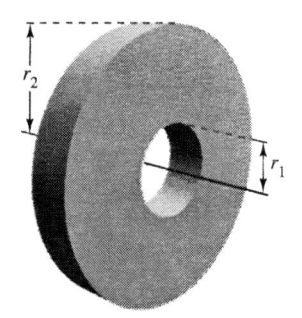

Figure for 54   **Figure for 55**

55. Moment of Inertia An annular cylinder has an inside radius of r_1 and an outside radius of r_2 (see figure). Its moment of inertia is

$$I = \tfrac{1}{2}m(r_1^2 + r_2^2)$$

where m is the mass. Find the rate at which I is changing at the instant the radii are 6 centimeters and 8 centimeters if the two radii are increasing at a rate of 2 centimeters per second.

56. Ideal Gas Law The Ideal Gas Law is $pV = mRT$, where R is a constant and m is a constant mass. If p and V are functions of time, find dT/dt, the rate at which the temperature changes with respect to time.

57. Maximum Angle A 2-foot-high picture hangs on a wall such that the bottom is 6 feet from the floor. A child whose eyes are 4 feet above the floor stands x feet from the wall (see figure).

(a) Show that $x^2 \tan \theta - 2x + 8 \tan \theta = 0$.

(b) Use implicit differentiation to find $d\theta/dx$.

(c) Find x such that θ is maximum.

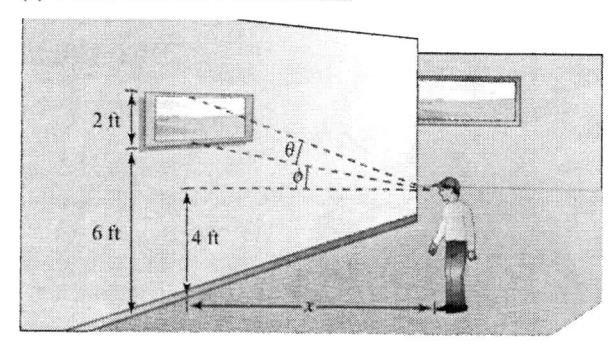

58. Show that if $f(x,y)$ is homogeneous of degree n, then

$$xf_x(x, y) + yf_y(x, y) = nf(x, y).$$

[*Hint:* Let $g(t) = f(tx, ty) = t^n f(x, y)$. Find $g'(t)$ and then let $t = 1$.]

59. Show that

$$\frac{\partial w}{\partial u} + \frac{\partial w}{\partial v} = 0$$

for $w = f(x, y)$, $x = u - v$, and $y = v - u$.

60. Demonstrate the result in Exercise 59 for

$$w = (x - y)\sin(y - x).$$

61. Consider the function $w = f(x, y)$, where $x = r\cos\theta$ and $y = r\sin\theta$. Prove each of the following.

(a) $\dfrac{\partial w}{\partial x} = \dfrac{\partial w}{\partial r}\cos\theta - \dfrac{\partial w}{\partial \theta}\dfrac{\sin\theta}{r}$

$\dfrac{\partial w}{\partial y} = \dfrac{\partial w}{\partial r}\sin\theta + \dfrac{\partial w}{\partial \theta}\dfrac{\cos\theta}{r}$

(b) $\left(\dfrac{\partial w}{\partial x}\right)^2 + \left(\dfrac{\partial w}{\partial y}\right)^2 = \left(\dfrac{\partial w}{\partial r}\right)^2 + \left(\dfrac{1}{r^2}\right)\left(\dfrac{\partial w}{\partial \theta}\right)^2$

62. Demonstrate the result in Exercise 61(b) for $w = \arctan(y/x)$.

63. Cauchy-Riemann Equations Given the functions $u(x, y)$ and $v(x, y)$, verify that the **Cauchy-Riemann differential equations**

$$\frac{\partial u}{\partial x} = \frac{\partial v}{\partial y} \quad \text{and} \quad \frac{\partial u}{\partial y} = -\frac{\partial v}{\partial x}$$

can be written in polar coordinate form as

$$\frac{\partial u}{\partial r} = \frac{1}{r}\frac{\partial v}{\partial \theta} \quad \text{and} \quad \frac{\partial v}{\partial r} = -\frac{1}{r}\frac{\partial u}{\partial \theta}.$$

64. Demonstrate the result in Exercise 63 for the functions

$$u = \ln\sqrt{x^2 + y^2} \quad \text{and} \quad v = \arctan\frac{y}{x}.$$

Section 12.6 Directional Derivatives and Gradients

- Find and use directional derivatives of a function of two variables.
- Find the gradient of a function of two variables.
- Use the gradient of a function of two variables in applications.
- Find directional derivatives and gradients for functions of three variables.

Directional Derivative

Suppose you are standing on the hillside pictured in Figure 12.42 and want to determine the hill's incline toward the z-axis. If the hill were represented by $z = f(x, y)$, you would already know how to determine the slopes in two different directions—the slope in the y-direction would be given by the partial derivative $f_y(x, y)$, and the slope in the x-direction would be given by the partial derivative $f_x(x, y)$. In this section, you will see that these two partial derivatives can be used to find the slope in *any* direction.

To determine the slope at a point on a surface, we define a new type of derivative called a **directional derivative**. We begin by letting $z = f(x, y)$ be a *surface* and $P(x_0, y_0)$ a *point* in the domain of f, as shown in Figure 12.43. The "direction" of the directional derivative is given by a unit vector

$$\mathbf{u} = \cos\theta\, \mathbf{i} + \sin\theta\, \mathbf{j}$$

where θ is the angle the vector makes with the positive x-axis. To find the desired slope, reduce the problem to two dimensions by intersecting the surface with a vertical plane passing through the point P and parallel to $\mathbf{u}$, as shown in Figure 12.44. This vertical plane intersects the surface to form a curve C. The slope of the surface at $(x_0, y_0, f(x_0, y_0))$ in the direction of $\mathbf{u}$ is defined as the slope of the curve C at that point.

Informally, you can write the slope of the curve C as a limit that looks much like those used in single-variable calculus. The vertical plane used to form C intersects the xy-plane in a line L, represented by the parametric equations

$$x = x_0 + t\cos\theta$$

and

$$y = y_0 + t\sin\theta$$

so that for any value of t, the point $Q(x, y)$ lies on the line L. For each of the points P and Q, there is a corresponding point on the surface.

$(x_0, y_0, f(x_0, y_0))$ Point above P
$(x, y, f(x, y))$ Point above Q

Moreover, because the distance between P and Q is

$$\sqrt{(x - x_0)^2 + (y - y_0)^2} = \sqrt{(t\cos\theta)^2 + (t\sin\theta)^2} = |t|$$

you can write the slope of the secant line through $(x_0, y_0, f(x_0, y_0))$ and $(x, y, f(x, y))$ as

$$\frac{f(x, y) - f(x_0, y_0)}{t} = \frac{f(x_0 + t\cos\theta, y_0 + t\sin\theta) - f(x_0, y_0)}{t}.$$

Finally, by letting t approach 0, you arrive at the following definition.

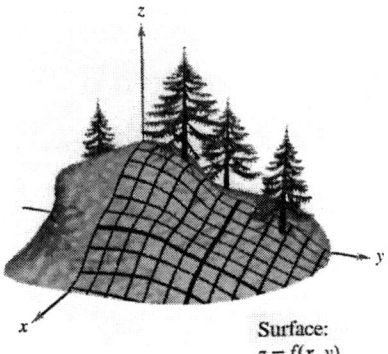

Figure 12.42

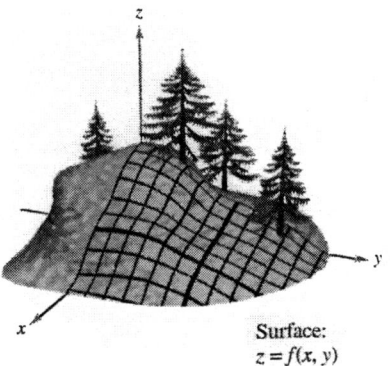

Surface: $z = f(x, y)$

Figure 12.43

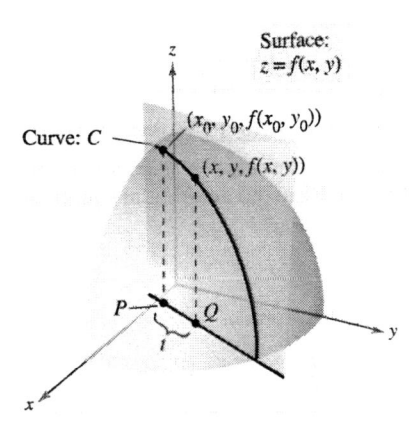

Figure 12.44

Surface: $z = f(x, y)$

Curve: C

Definition of Directional Derivative

Let f be a function of two variables x and y and let $\mathbf{u} = \cos\theta\,\mathbf{i} + \sin\theta\,\mathbf{j}$ be a unit vector. Then the **directional derivative of f in the direction of u,** denoted by $D_{\mathbf{u}}f$, is

$$D_{\mathbf{u}}f(x, y) = \lim_{t \to 0} \frac{f(x + t\cos\theta, y + t\sin\theta) - f(x, y)}{t}$$

provided this limit exists.

Calculating directional derivatives by this definition is similar to finding the derivative of a function of one variable by the limit process (given in Section 2.1). A simpler "working" formula for finding directional derivatives involves the partial derivatives f_x and f_y.

THEOREM 12.9 Directional Derivative

If f is a differentiable function of x and y, then the directional derivative of f in the direction of the unit vector $\mathbf{u} = \cos\theta\,\mathbf{i} + \sin\theta\,\mathbf{j}$ is

$$D_{\mathbf{u}}f(x, y) = f_x(x, y)\cos\theta + f_y(x, y)\sin\theta.$$

Proof For a fixed point (x_0, y_0), let $x = x_0 + t\cos\theta$ and let $y = y_0 + t\sin\theta$. Then, let $g(t) = f(x, y)$. Because f is differentiable, you can apply the Chain Rule given in Theorem 12.7 to obtain

$$g'(t) = f_x(x, y)x'(t) + f_y(x, y)y'(t) = f_x(x, y)\cos\theta + f_y(x, y)\sin\theta.$$

If $t = 0$, then $x = x_0$ and $y = y_0$, so

$$g'(0) = f_x(x_0, y_0)\cos\theta + f_y(x_0, y_0)\sin\theta.$$

By the definition of $g'(t)$, it is also true that

$$g'(0) = \lim_{t \to 0} \frac{g(t) - g(0)}{t}$$

$$= \lim_{t \to 0} \frac{f(x_0 + t\cos\theta, y_0 + t\sin\theta) - f(x_0, y_0)}{t}.$$

Consequently, $D_{\mathbf{u}}f(x_0, y_0) = f_x(x_0, y_0)\cos\theta + f_y(x_0, y_0)\sin\theta.$

There are infinitely many directional derivatives to a surface at a given point—one for each direction specified by $\mathbf{u}$, as indicated in Figure 12.45. Two of these are the partial derivatives f_x and f_y.

1. Direction of positive x-axis ($\theta = 0$): $\mathbf{u} = \cos 0\,\mathbf{i} + \sin 0\,\mathbf{j} = \mathbf{i}$

$$D_{\mathbf{i}}f(x, y) = f_x(x, y)\cos 0 + f_y(x, y)\sin 0 = f_x(x, y)$$

2. Direction of positive y-axis ($\theta = \pi/2$): $\mathbf{u} = \cos\frac{\pi}{2}\mathbf{i} + \sin\frac{\pi}{2}\mathbf{j} = \mathbf{j}$

$$D_{\mathbf{j}}f(x, y) = f_x(x, y)\cos\frac{\pi}{2} + f_y(x, y)\sin\frac{\pi}{2} = f_y(x, y)$$

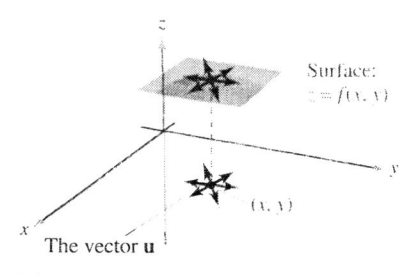

Figure 12.45 The vector $\mathbf{u}$; Surface: $z = f(x, y)$

886 CHAPTER 12 Functions of Several Variables

Example 1 Finding a Directional Derivative

Find the directional derivative of

$$f(x, y) = 4 - x^2 - \tfrac{1}{4}y^2 \qquad \text{Surface}$$

at $(1, 2)$ in the direction of

$$\mathbf{u} = \left(\cos \tfrac{\pi}{3}\right)\mathbf{i} + \left(\sin \tfrac{\pi}{3}\right)\mathbf{j}. \qquad \text{Direction}$$

Solution Because f_x and f_y are continuous, f is differentiable, and you can apply Theorem 12.9.

$$D_{\mathbf{u}}f(x, y) = f_x(x, y) \cos \theta + f_y(x, y) \sin \theta$$

$$= (-2x) \cos \theta + \left(-\tfrac{y}{2}\right) \sin \theta$$

Evaluating at $\theta = \pi/3$, $x = 1$, and $y = 2$ produces

$$D_{\mathbf{u}}f(1, 2) = (-2)\left(\tfrac{1}{2}\right) + (-1)\left(\tfrac{\sqrt{3}}{2}\right)$$

$$= -1 - \tfrac{\sqrt{3}}{2}$$

$$\approx -1.866. \qquad \text{(See Figure 12.46.)}$$

Surface:
$f(x, y) = 4 - x^2 - \tfrac{1}{4}y^2$

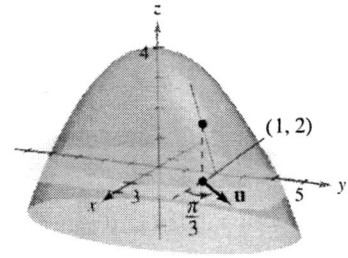

Figure 12.46

NOTE Note in Figure 12.46 that you can interpret the directional derivative as giving the slope of the surface at the point $(1, 2, 2)$ in the direction of the unit vector $\mathbf{u}$.

We have been specifying direction by a unit vector $\mathbf{u}$. If the direction is given by a vector whose length is not 1, we must normalize the vector before applying the formula in Theorem 12.9.

Example 2 Finding a Directional Derivative

Find the directional derivative of

$$f(x, y) = x^2 \sin 2y \qquad \text{Surface}$$

at $(1, \pi/2)$ in the direction of

$$\mathbf{v} = 3\mathbf{i} - 4\mathbf{j}. \qquad \text{Direction}$$

Solution Because f_x and f_y are continuous, f is differentiable, and you can apply Theorem 12.9. Begin by finding a unit vector in the direction of $\mathbf{v}$.

$$\mathbf{u} = \frac{\mathbf{v}}{\|\mathbf{v}\|} = \tfrac{3}{5}\mathbf{i} - \tfrac{4}{5}\mathbf{j} = \cos \theta \mathbf{i} + \sin \theta \mathbf{j}$$

Using this unit vector, you have

$$D_{\mathbf{u}}f(x, y) = (2x \sin 2y)(\cos \theta) + (2x^2 \cos 2y)(\sin \theta)$$

$$D_{\mathbf{u}}f\left(1, \tfrac{\pi}{2}\right) = (2 \sin \pi)\left(\tfrac{3}{5}\right) + (2 \cos \pi)\left(-\tfrac{4}{5}\right)$$

$$= (0)\left(\tfrac{3}{5}\right) + (-2)\left(-\tfrac{4}{5}\right)$$

$$= \tfrac{8}{5}. \qquad \text{(See Figure 12.47.)}$$

Surface:
$f(x, y) = x^2 \sin 2y$

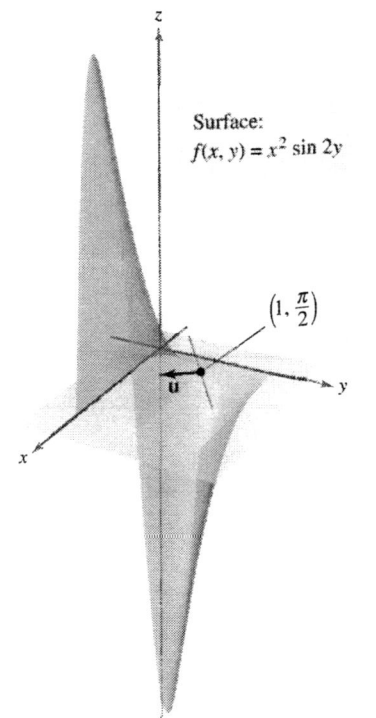

Figure 12.47

The Gradient of a Function of Two Variables

The **gradient** of a function of two variables is a vector-valued function of two variables. This function has many important uses, some of which are described later in this section.

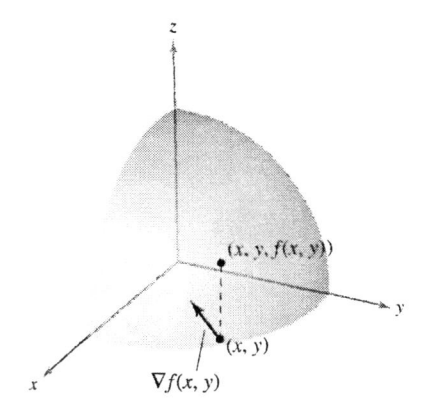

Definition of Gradient of a Function of Two Variables

Let $z = f(x, y)$ be a function of x and y such that f_x and f_y exist. Then the **gradient of f**, denoted by $\nabla f(x, y)$, is the vector

$$\nabla f(x, y) = f_x(x, y)\mathbf{i} + f_y(x, y)\mathbf{j}.$$

∇f is read as "del f." Another notation for the gradient is **grad** $f(x, y)$. In Figure 12.48, note that for each (x, y), the gradient $\nabla f(x, y)$ is a vector in the plane (not a vector in space).

The gradient of f is a vector in the xy-plane.
Figure 12.48

NOTE No value is assigned to the symbol ∇ by itself. It is an operator in the same sense that d/dx is an operator. When ∇ operates on $f(x, y)$, it produces the vector $\nabla f(x, y)$.

Example 3 Finding the Gradient of a Function

Find the gradient of $f(x, y) = y \ln x + xy^2$ at the point $(1, 2)$.

Solution Using

$$f_x(x, y) = \frac{y}{x} + y^2$$

and

$$f_y(x, y) = \ln x + 2xy$$

you have

$$\nabla f(x, y) = \left(\frac{y}{x} + y^2\right)\mathbf{i} + (\ln x + 2xy)\mathbf{j}.$$

At the point $(1, 2)$, the gradient is

$$\nabla f(1, 2) = \left(\frac{2}{1} + 2^2\right)\mathbf{i} + [\ln 1 + 2(1)(2)]\mathbf{j}$$

$$= 6\mathbf{i} + 4\mathbf{j}.$$

Because the gradient of f is a vector, you can write the directional derivative of f in the direction of $\mathbf{u}$ as

$$D_{\mathbf{u}} f(x, y) = [f_x(x, y)\mathbf{i} + f_y(x, y)\mathbf{j}] \cdot [\cos \theta \mathbf{i} + \sin \theta \mathbf{j}].$$

In other words, the directional derivative is the dot product of the gradient and the direction vector. This useful result is summarized in the following theorem.

THEOREM 12.10 Alternative Form of the Directional Derivative

If f is a differentiable function of x and y, then the directional derivative of f in the direction of the unit vector $\mathbf{u}$ is

$$D_{\mathbf{u}} f(x, y) = \nabla f(x, y) \cdot \mathbf{u}.$$

Example 4 Using $\nabla f(x, y)$ to Find a Directional Derivative

Find the directional derivative of

$$f(x, y) = 3x^2 - 2y^2$$

at $\left(-\frac{3}{4}, 0\right)$ in the direction from $P\left(-\frac{3}{4}, 0\right)$ to $Q(0, 1)$.

Solution Because the partials of f are continuous, f is differentiable and you can apply Theorem 12.10. A vector in the specified direction is

$$\overrightarrow{PQ} = \mathbf{v} = \left(0 + \frac{3}{4}\right)\mathbf{i} + (1 - 0)\mathbf{j}$$
$$= \frac{3}{4}\mathbf{i} + \mathbf{j}$$

and a unit vector in this direction is

$$\mathbf{u} = \frac{\mathbf{v}}{\|\mathbf{v}\|} = \frac{3}{5}\mathbf{i} + \frac{4}{5}\mathbf{j}. \qquad \text{Unit vector in direction of } \overrightarrow{PQ}$$

Because $\nabla f(x, y) = f_x(x, y)\mathbf{i} + f_y(x, y)\mathbf{j} = 6x\mathbf{i} - 4y\mathbf{j}$, the gradient at $\left(-\frac{3}{4}, 0\right)$ is

$$\nabla f\left(-\frac{3}{4}, 0\right) = -\frac{9}{2}\mathbf{i} + 0\mathbf{j}. \qquad \text{Gradient at } \left(-\frac{3}{4}, 0\right)$$

Consequently, at $\left(-\frac{3}{4}, 0\right)$ the directional derivative is

$$D_{\mathbf{u}}f\left(-\frac{3}{4}, 0\right) = \nabla f\left(-\frac{3}{4}, 0\right) \cdot \mathbf{u}$$
$$= \left(-\frac{9}{2}\mathbf{i} + 0\mathbf{j}\right) \cdot \left(\frac{3}{5}\mathbf{i} + \frac{4}{5}\mathbf{j}\right)$$
$$= -\frac{27}{10}. \qquad \text{Directional derivative at } \left(-\frac{3}{4}, 0\right)$$

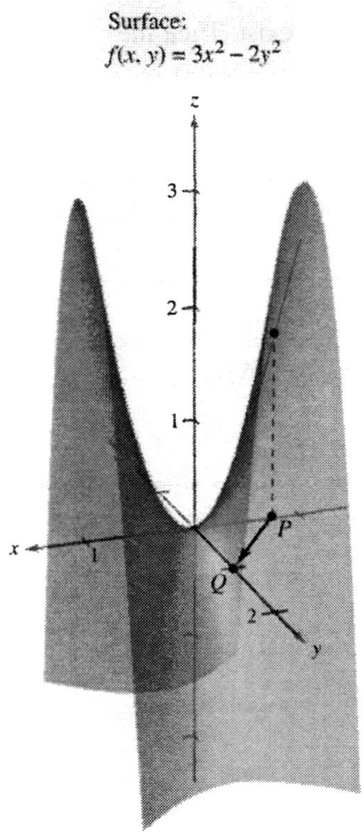

(See Figure 12.49.)

Surface:
$f(x, y) = 3x^2 - 2y^2$

Figure 12.49

Applications of the Gradient

You have already seen that there are many directional derivatives at the point (x, y) on a surface. In many applications we would like to know in which direction to move so that $f(x, y)$ increases most rapidly. This direction is called the direction of steepest ascent, and it is given by the gradient, as stated in the following theorem.

THEOREM 12.11 Properties of the Gradient

Let f be differentiable at the point (x, y).

1. If $\nabla f(x, y) = \mathbf{0}$, then $D_{\mathbf{u}}f(x, y) = 0$ for all $\mathbf{u}$.
2. The direction of *maximum* increase of f is given by $\nabla f(x, y)$. The maximum value of $D_{\mathbf{u}}f(x, y)$ is $\|\nabla f(x, y)\|$.
3. The direction of *minimum* increase of f is given by $-\nabla f(x, y)$. The minimum value of $D_{\mathbf{u}}f(x, y)$ is $-\|\nabla f(x, y)\|$.

NOTE Part 2 of Theorem 12.11 says that at the point (x, y), f increases most rapidly in the direction of the gradient, $\nabla f(x, y)$.

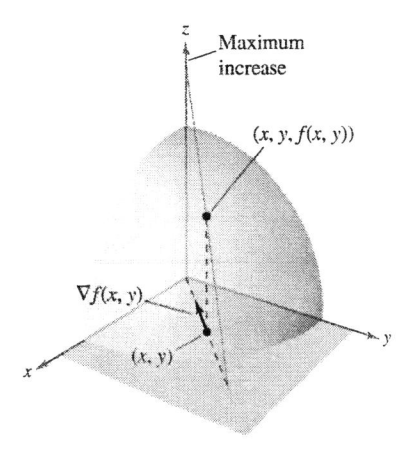

The gradient of f is a vector in the xy-plane that points in the direction of maximum increase on the surface given by $z = f(x, y)$.
Figure 12.50

Proof If $\nabla f(x, y) = \mathbf{0}$, then for any direction (any $\mathbf{u}$), you have

$$D_{\mathbf{u}} f(x, y) = \nabla f(x, y) \cdot \mathbf{u}$$
$$= (0\mathbf{i} + 0\mathbf{j}) \cdot (\cos\theta \mathbf{i} + \sin\theta \mathbf{j})$$
$$= 0.$$

If $\nabla f(x, y) \ne \mathbf{0}$, then let ϕ be the angle between $\nabla f(x, y)$ and a unit vector $\mathbf{u}$. Using the dot product, you can apply Theorem 10.5 to conclude that

$$D_{\mathbf{u}} f(x, y) = \nabla f(x, y) \cdot \mathbf{u}$$
$$= \|\nabla f(x, y)\| \, \|\mathbf{u}\| \cos\phi$$
$$= \|\nabla f(x, y)\| \cos\phi$$

and it follows that the maximum value of $D_{\mathbf{u}} f(x, y)$ will occur when $\cos\phi = 1$. So, $\phi = 0$, and the maximum value for the directional derivative occurs when $\mathbf{u}$ has the same direction as $\nabla f(x, y)$. Moreover, this largest value for $D_{\mathbf{u}} f(x, y)$ is precisely

$$\|\nabla f(x, y)\| \cos\phi = \|\nabla f(x, y)\|.$$

Similarly, the minimum value of $D_{\mathbf{u}} f(x, y)$ can be obtained by letting $\phi = \pi$ so that $\mathbf{u}$ points in the direction opposite that of $\nabla f(x, y)$, as indicated in Figure 12.50.

To visualize one of the properties of the gradient, imagine a skier coming down a mountainside. If $f(x, y)$ denotes the altitude of the skier, then $-\nabla f(x, y)$ indicates the *compass direction* the skier should take to ski the path of steepest descent. (Remember that the gradient indicates direction in the xy-plane and does not itself point up or down the mountainside.)

As another illustration of the gradient, consider the temperature $T(x, y)$ at any point (x, y) on a flat metal plate. In this case, $\nabla T(x, y)$ gives the direction of greatest temperature increase at the point (x, y), as illustrated in the next example.

Example 5 Finding the Direction of Maximum Increase

The temperature in degrees Celsius on the surface of a metal plate is

$$T(x, y) = 20 - 4x^2 - y^2$$

where x and y are measured in centimeters. In what direction from $(2, -3)$ does the temperature increase most rapidly? What is this rate of increase?

Solution The gradient is

$$\nabla T(x, y) = T_x(x, y)\mathbf{i} + T_y(x, y)\mathbf{j}$$
$$= -8x\mathbf{i} - 2y\mathbf{j}.$$

It follows that the direction of maximum increase is given by

$$\nabla T(2, -3) = -16\mathbf{i} + 6\mathbf{j}$$

as shown in Figure 12.51, and the rate of increase is

$$\|\nabla T(2, -3)\| = \sqrt{256 + 36}$$
$$= \sqrt{292}$$
$$\approx 17.09° \text{ per centimeter.}$$

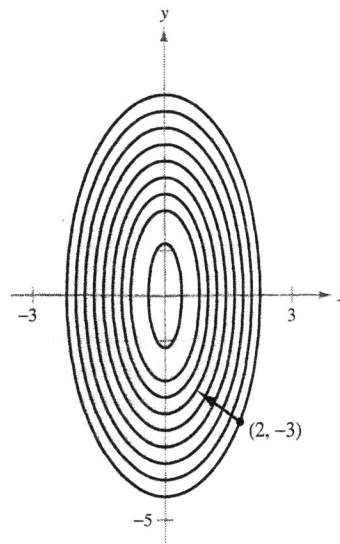

The direction of most rapid increase in temperature at $(2, -3)$ is given by $-16\mathbf{i} + 6\mathbf{j}$.
Figure 12.51

The solution presented in Example 5 can be misleading. Although the gradient points in the direction of maximum temperature increase, it does not necessarily point toward the hottest spot on the plate. In other words, the gradient provides a local solution to finding an increase relative to the temperature at the point $(2, -3)$. *Once you leave that position, the direction of maximum increase may change.*

Example 6 Finding the Path of a Heat-Seeking Particle

A heat-seeking particle is located at the point $(2, -3)$ on a metal plate whose temperature at (x, y) is

$$T(x, y) = 20 - 4x^2 - y^2.$$

Find the path of the particle as it continuously moves in the direction of maximum temperature increase.

Solution Let the path be represented by the position function

$$\mathbf{r}(t) = x(t)\mathbf{i} + y(t)\mathbf{j}.$$

A tangent vector at each point $(x(t), y(t))$ is given by

$$\mathbf{r}'(t) = \frac{dx}{dt}\mathbf{i} + \frac{dy}{dt}\mathbf{j}.$$

Because the particle seeks maximum temperature increase, the directions of $\mathbf{r}'(t)$ and $\nabla T(x, y) = -8x\mathbf{i} - 2y\mathbf{j}$ are the same at each point on the path. So,

$$-8x = k\frac{dx}{dt} \quad \text{and} \quad -2y = k\frac{dy}{dt}$$

where k depends on t. By solving each equation for dt/k and equating the results, you obtain

$$\frac{dx}{-8x} = \frac{dy}{-2y}.$$

The solution of this differential equation is $x = Cy^4$. Because the particle starts at the point $(2, -3)$, you can determine that $C = 2/81$. So, the path of the heat-seeking particle is

$$x = \frac{2}{81}y^4.$$

The path is shown in Figure 12.52.

Level curves:
$T(x, y) = 20 - 4x^2 - y^2$

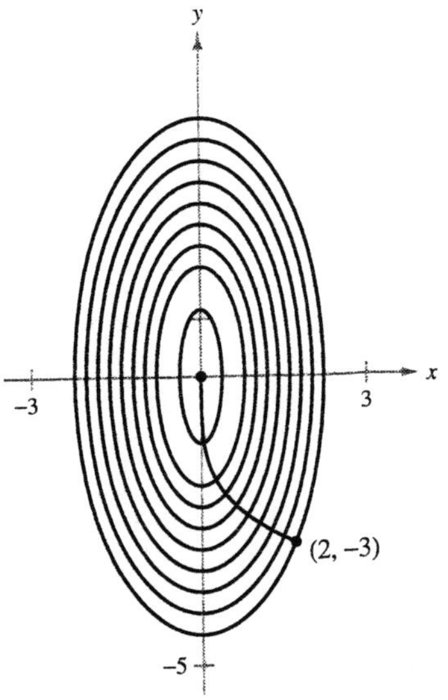

Path followed by a heat-seeking particle
Figure 12.52

In Figure 12.52, the path of the particle (determined by the gradient at each point) appears to be orthogonal to each of the level curves. This becomes clear when you consider that the temperature $T(x, y)$ is constant along a given level curve. Hence, at any point (x, y) on the curve, the rate of change of T in the direction of a unit tangent vector $\mathbf{u}$ is 0, and you can write

$$\nabla f(x, y) \cdot \mathbf{u} = D_\mathbf{u} T(x, y) = 0. \quad \text{\textbf{u} is a unit tangent vector.}$$

Because the dot product of $\nabla f(x, y)$ and $\mathbf{u}$ is 0, you can conclude that they must be orthogonal. This result is stated in the following theorem.

> **THEOREM 12.12 Gradient Is Normal to Level Curves**
>
> If f is differentiable at (x_0, y_0) and $\nabla f(x_0, y_0) \neq \mathbf{0}$, then $\nabla f(x_0, y_0)$ is normal to the level curve through (x_0, y_0).

Example 7 Finding a Normal Vector to a Level Curve

Sketch the level curve corresponding to $c = 0$ for the function given by

$$f(x, y) = y - \sin x$$

and find a normal vector at several points on the curve.

Solution The level curve for $c = 0$ is given by

$$0 = y - \sin x$$
$$y = \sin x$$

as shown in Figure 12.53(a). Because the gradient vector of f at (x, y) is

$$\nabla f(x, y) = f_x(x, y)\mathbf{i} + f_y(x, y)\mathbf{j}$$
$$= -\cos x \mathbf{i} + \mathbf{j}$$

you can use Theorem 12.12 to conclude that $\nabla f(x, y)$ is normal to the level curve at the point (x, y). Some gradient vectors are

$$\nabla f(-\pi, 0) = \mathbf{i} + \mathbf{j}$$
$$\nabla f\left(-\frac{2\pi}{3}, -\frac{\sqrt{3}}{2}\right) = \frac{1}{2}\mathbf{i} + \mathbf{j}$$
$$\nabla f\left(-\frac{\pi}{2}, -1\right) = \mathbf{j}$$
$$\nabla f\left(-\frac{\pi}{3}, -\frac{\sqrt{3}}{2}\right) = -\frac{1}{2}\mathbf{i} + \mathbf{j}$$
$$\nabla f(0, 0) = -\mathbf{i} + \mathbf{j}$$
$$\nabla f\left(\frac{\pi}{3}, \frac{\sqrt{3}}{2}\right) = -\frac{1}{2}\mathbf{i} + \mathbf{j}$$
$$\nabla f\left(\frac{\pi}{2}, 1\right) = \mathbf{j}.$$

These are shown in Figure 12.53(b).

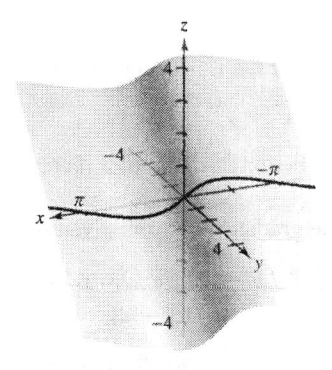

(a) The surface is given by $f(x, y) = y - \sin x$

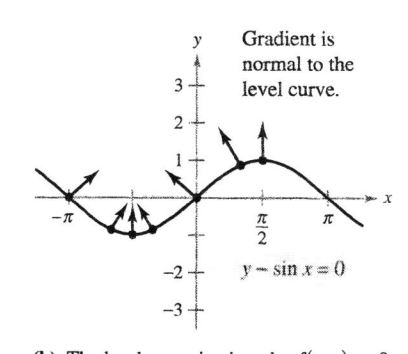

(b) The level curve is given by $f(x, y) = 0$.

Figure 12.53

Functions of Three Variables

The definitions of the directional derivative and the gradient can be extended naturally to functions of three or more variables. As often happens, some of the geometric interpretation is lost in the generalization from functions of two variables to those of three variables. For example, you cannot interpret the directional derivative of a function of three variables to represent slope.

The definitions and properties of the directional derivative and the gradient of a function of three variables are given in the following summary.

Directional Derivative and Gradient for Three Variables

Let f be a function of x, y, and z, with continuous first partial derivatives. The **directional derivative of f** in the direction of a unit vector $\mathbf{u} = a\mathbf{i} + b\mathbf{j} + c\mathbf{k}$ is given by

$$D_\mathbf{u} f(x, y, z) = af_x(x, y, z) + bf_y(x, y, z) + cf_z(x, y, z).$$

The **gradient of f** is defined to be

$$\nabla f(x, y, z) = f_x(x, y, z)\mathbf{i} + f_y(x, y, z)\mathbf{j} + f_z(x, y, z)\mathbf{k}.$$

Properties of the gradient are as follows.

1. $D_\mathbf{u} f(x, y, z) = \nabla f(x, y, z) \cdot \mathbf{u}$
2. If $\nabla f(x, y, z) = \mathbf{0}$, then $D_\mathbf{u} f(x, y, z) = 0$ for all $\mathbf{u}$.
3. The direction of *maximum* increase of f is given by $\nabla f(x, y, z)$. The maximum value of $D_\mathbf{u} f(x, y, z)$ is

    $\|\nabla f(x, y, z)\|.$ Maximum value of $D_\mathbf{u} f(x, y, z)$

4. The direction of *minimum* increase of f is given by $-\nabla f(x, y, z)$. The minimum value of $D_\mathbf{u} f(x, y, z)$ is

    $-\|\nabla f(x, y, z)\|.$ Minimum value of $D_\mathbf{u} f(x, y, z)$

NOTE You can generalize Theorem 12.12 to functions of three variables. Under suitable hypotheses,

$$\nabla f(x_0, y_0, z_0)$$

is normal to the *level surface* through (x_0, y_0, z_0).

Example 8   Finding the Gradient for a Function of Three Variables

Find $\nabla f(x, y, z)$ for the function given by

$$f(x, y, z) = x^2 + y^2 - 4z$$

and find the direction of maximum increase of f at the point $(2, -1, 1)$.

Solution The gradient vector is given by

$$\nabla f(x, y, z) = f_x(x, y, z)\mathbf{i} + f_y(x, y, z)\mathbf{j} + f_z(x, y, z)\mathbf{k}$$
$$= 2x\mathbf{i} + 2y\mathbf{j} - 4\mathbf{k}.$$

Hence, it follows that the direction of maximum increase at $(2, -1, 1)$ is

$$\nabla f(2, -1, 1) = 4\mathbf{i} - 2\mathbf{j} - 4\mathbf{k}.$$

EXERCISES FOR SECTION 12.6

In Exercises 1–12, find the directional derivative of the function at P in the direction of $\mathbf{v}$.

1. $f(x, y) = 3x - 4xy + 5y$, $P(1, 2)$, $\mathbf{v} = \frac{1}{2}(\mathbf{i} + \sqrt{3}\mathbf{j})$
2. $f(x, y) = x^3 - y^3$, $P(4, 3)$, $\mathbf{v} = \frac{\sqrt{2}}{2}(\mathbf{i} + \mathbf{j})$
3. $f(x, y) = xy$, $P(2, 3)$, $\mathbf{v} = \mathbf{i} + \mathbf{j}$
4. $f(x, y) = \frac{x}{y}$, $P(1, 1)$, $\mathbf{v} = -\mathbf{j}$
5. $g(x, y) = \sqrt{x^2 + y^2}$, $P(3, 4)$, $\mathbf{v} = 3\mathbf{i} - 4\mathbf{j}$
6. $g(x, y) = \arccos xy$, $P(1, 0)$, $\mathbf{v} = \mathbf{i} + 5\mathbf{j}$
7. $h(x, y) = e^x \sin y$, $P\left(1, \frac{\pi}{2}\right)$, $\mathbf{v} = -\mathbf{i}$
8. $h(x, y) = e^{-(x^2+y^2)}$, $P(0, 0)$, $\mathbf{v} = \mathbf{i} + \mathbf{j}$
9. $f(x, y, z) = xy + yz + xz$, $P(1, 1, 1)$, $\mathbf{v} = 2\mathbf{i} + \mathbf{j} - \mathbf{k}$
10. $f(x, y, z) = x^2 + y^2 + z^2$, $P(1, 2, -1)$, $\mathbf{v} = \mathbf{i} - 2\mathbf{j} + 3\mathbf{k}$
11. $h(x, y, z) = x \arctan yz$, $P(4, 1, 1)$, $\mathbf{v} = \langle 1, 2, -1 \rangle$
12. $h(x, y, z) = xyz$, $P(2, 1, 1)$, $\mathbf{v} = \langle 2, 1, 2 \rangle$

In Exercises 13–16, find the directional derivative of the function in the direction of $\mathbf{u} = \cos\theta\mathbf{i} + \sin\theta\mathbf{j}$.

13. $f(x, y) = x^2 + y^2$, $\theta = \frac{\pi}{4}$
14. $f(x, y) = \frac{y}{x + y}$, $\theta = -\frac{\pi}{6}$
15. $f(x, y) = \sin(2x - y)$, $\theta = -\frac{\pi}{3}$
16. $g(x, y) = xe^y$, $\theta = \frac{2\pi}{3}$

In Exercises 17–20, find the directional derivative of the function at P in the direction of Q.

17. $f(x, y) = x^2 + 4y^2$, $P(3, 1)$, $Q(1, -1)$
18. $f(x, y) = \cos(x + y)$, $P(0, \pi)$, $Q\left(\frac{\pi}{2}, 0\right)$
19. $h(x, y, z) = \ln(x + y + z)$, $P(1, 0, 0)$, $Q(4, 3, 1)$
20. $g(x, y, z) = xye^z$, $P(2, 4, 0)$, $Q(0, 0, 0)$

In Exercises 21–26, find the gradient of the function at the indicated point.

21. $f(x, y) = 3x - 5y^2 + 10$, $(2, 1)$
22. $g(x, y) = 2xe^{y/x}$, $(2, 0)$
23. $z = \cos(x^2 + y^2)$, $(3, -4)$
24. $z = \ln(x^2 - y)$, $(2, 3)$
25. $w = 3x^2y - 5yz + z^2$, $(1, 1, -2)$
26. $w = x\tan(y + z)$, $(4, 3, -1)$

In Exercises 27–30, use the gradient to find the directional derivative of the function at P in the direction of Q.

27. $g(x, y) = x^2 + y^2 + 1$, $P(1, 2)$, $Q(3, 6)$
28. $f(x, y) = 3x^2 - y^2 + 4$, $P(3, 1)$, $Q(1, 8)$
29. $f(x, y) = e^{-x}\cos y$, $P(0, 0)$, $Q(2, 1)$
30. $f(x, y) = \sin 2x \cos y$, $P(0, 0)$, $Q\left(\frac{\pi}{2}, \pi\right)$

In Exercises 31–38, find the gradient of the function and the maximum value of the directional derivative at the indicated point.

Function	Point
31. $h(x, y) = x \tan y$	$\left(2, \frac{\pi}{4}\right)$
32. $h(x, y) = y\cos(x - y)$	$\left(0, \frac{\pi}{3}\right)$
33. $g(x, y) = \ln\sqrt[3]{x^2 + y^2}$	$(1, 2)$
34. $g(x, y) = ye^{-x^2}$	$(0, 5)$
35. $f(x, y, z) = \sqrt{x^2 + y^2 + z^2}$	$(1, 4, 2)$
36. $w = \dfrac{1}{\sqrt{1 - x^2 - y^2 - z^2}}$	$(0, 0, 0)$
37. $f(x, y, z) = xe^{yz}$	$(2, 0, -4)$
38. $w = xy^2z^2$	$(2, 1, 1)$

In Exercises 39–46, use the function

$$f(x, y) = 3 - \frac{x}{3} - \frac{y}{2}.$$

39. Sketch the graph of f in the first octant and plot the point $(3, 2, 1)$ on the surface.
40. Find $D_\mathbf{u} f(3, 2)$ where $\mathbf{u} = \cos\theta\mathbf{i} + \sin\theta\mathbf{j}$.
    (a) $\theta = \frac{\pi}{4}$ (b) $\theta = \frac{2\pi}{3}$
41. Find $D_\mathbf{u} f(3, 2)$ where $\mathbf{u} = \cos\theta\mathbf{i} + \sin\theta\mathbf{j}$.
    (a) $\theta = \frac{4\pi}{3}$ (b) $\theta = -\frac{\pi}{6}$
42. Find $D_\mathbf{u} f(3, 2)$ where $\mathbf{u} = \mathbf{v}/\|\mathbf{v}\|$.
    (a) $\mathbf{v} = \mathbf{i} + \mathbf{j}$ (b) $\mathbf{v} = -3\mathbf{i} - 4\mathbf{j}$
43. Find $D_\mathbf{u} f(3, 2)$ where $\mathbf{u} = \mathbf{v}/\|\mathbf{v}\|$.
    (a) $\mathbf{v}$ is the vector from $(1, 2)$ to $(-2, 6)$.
    (b) $\mathbf{v}$ is the vector from $(3, 2)$ to $(4, 5)$.
44. Find $\nabla f(x, y)$.
45. Find the maximum value of the directional derivative at $(3, 2)$.
46. Find a unit vector $\mathbf{u}$ orthogonal to $\nabla f(3, 2)$ and calculate $D_\mathbf{u} f(3, 2)$. Discuss the geometric meaning of the result.

In Exercises 47–50, use the function

$f(x, y) = 9 - x^2 - y^2$.

47. Sketch the graph of f in the first octant and plot the point $(1, 2, 4)$ on the surface.

48. Find $D_\mathbf{u} f(1, 2)$ where $\mathbf{u} = \cos\theta \mathbf{i} + \sin\theta \mathbf{j}$.

    (a) $\theta = -\dfrac{\pi}{4}$ (b) $\theta = \dfrac{\pi}{3}$

49. Find $\nabla f(1, 2)$ and $\|\nabla f(1, 2)\|$.

50. Find a unit vector $\mathbf{u}$ orthogonal to $\nabla f(1, 2)$ and calculate $D_\mathbf{u} f(1, 2)$. Discuss the geometric meaning of the result.

Investigation In Exercises 51 and 52, (a) use the graph to estimate the components of the vector in the direction of the maximum rate of increase in the function at the indicated point. (b) Find the gradient at the point and compare it with your estimate in part (a). (c) In what direction would the function be decreasing at the greatest rate? Explain.

51. $f(x, y) = \frac{1}{10}(x^2 - 3xy + y^2)$,   52. $f(x, y) = \frac{1}{2}y\sqrt{x}$,
    $(1, 2)$ $(1, 2)$

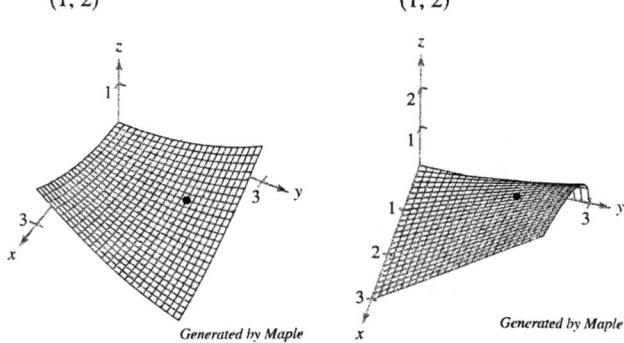

53. **Investigation** Consider the function

    $f(x, y) = x^2 - y^2$

    at the point $(4, -3, 7)$.

    (a) Use a computer algebra system to graph the surface represented by the function.

    (b) Determine the directional derivative $D_\mathbf{u} f(4, -3)$ as a function of θ where $\mathbf{u} = \cos\theta \mathbf{i} + \sin\theta \mathbf{j}$. Use a computer algebra system to graph the function on the interval $[0, 2\pi)$.

    (c) Approximate the zeros of the function in part (b) and interpret each in the context of the problem.

    (d) Approximate the critical numbers of the function in part (b) and interpret each in the context of the problem.

    (e) Find $\|\nabla f(4, -3)\|$ and explain its relationship to the answers in part (d).

    (f) Use a computer algebra system to graph the level curve of the function f at the level $c = 7$. On this curve, graph the vector in the direction of $\nabla f(4, -3)$, and state its relationship to the level curve.

54. **Investigation** The figure below shows the level curve of the function

    $f(x, y) = \dfrac{8y}{1 + x^2 + y^2}$

    at the level $c = 2$.

    (a) Analytically verify that the curve is a circle.

    (b) At the point $(\sqrt{3}, 2)$ on the level curve, sketch the vector showing the direction of the greatest rate of increase of the function. (To print an enlarged copy of the graph, go to the website *www.mathgraphs.com*.)

    (c) At the point $(\sqrt{3}, 2)$ on the level curve, sketch the vector such that the directional derivative is 0.

    (d) Use a computer algebra system to graph the surface to verify your answers in parts (a) to (c).

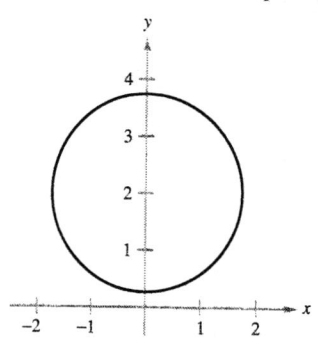

In Exercises 55–58, find a normal vector to the level curve $f(x, y) = c$ **at** P.

55. $f(x, y) = x^2 + y^2$ 56. $f(x, y) = 6 - 2x - 3y$
    $c = 25$, $P(3, 4)$ $c = 6$, $P(0, 0)$

57. $f(x, y) = \dfrac{x}{x^2 + y^2}$ 58. $f(x, y) = xy$
    $c = \frac{1}{2}$, $P(1, 1)$ $c = -3$, $P(-1, 3)$

In Exercises 59–62, use the gradient to find a unit normal vector to the graph of the equation at the indicated point. Sketch your results.

59. $4x^2 - y = 6$, $(2, 10)$ 60. $3x^2 - 2y^2 = 1$, $(1, 1)$
61. $9x^2 + 4y^2 = 40$, $(2, -1)$ 62. $xe^y - y = 5$, $(5, 0)$

63. **Temperature Distribution** The temperature at the point (x, y) on a metal plate is

    $T = \dfrac{x}{x^2 + y^2}$.

    Find the direction of greatest increase in heat from the point $(3, 4)$.

64. **Topography** The surface of a mountain is modeled by the equation $h(x, y) = 5000 - 0.001x^2 - 0.004y^2$. Suppose that a mountain climber is at the point $(500, 300, 4390)$. In what direction should the climber move in order to ascend at the greatest rate?

Getting at the Concept

65. Define the derivative of the function $z = f(x, y)$ in the direction $\mathbf{u} = \cos\theta \mathbf{i} + \sin\theta \mathbf{j}$.

66. In your own words, give a geometric description of the directional derivative of $z = f(x, y)$.

67. Write a paragraph describing the directional derivative of the function f in the direction $\mathbf{u} = \cos\theta \mathbf{i} + \sin\theta \mathbf{j}$ if (a) $\theta = 0°$ and (b) $\theta = 90°$.

68. Define the gradient of a function of two variables. State the properties of the gradient.

69. Sketch the graph of a surface and select a point P on the surface. Sketch a vector in the xy-plane giving the direction of steepest ascent on the surface at P.

70. Describe the relationship of the gradient to the level curves of a surface given by $z = f(x, y)$.

71. **Topography** The figure shows a topographic map carried by a group of hikers. Sketch the paths of steepest descent if the hikers start at point A and if they start at point B. (To print an enlarged copy of the graph, go to the website www.mathgraphs.com.)

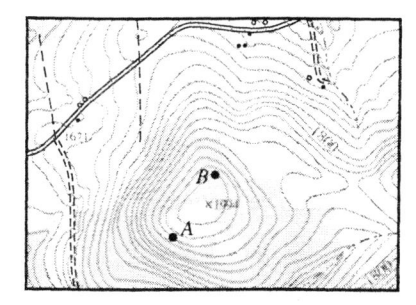

72. **Meteorology** Meteorologists measure the atmospheric pressure in units called millibars. From these observations they create weather maps on which the curves of equal atmospheric pressure (isobars) are drawn (see figure). These are level curves to the function $P(x, y)$ yielding the pressure at any point. Sketch the gradients to the isobars at the points A, B, and C. Although the magnitudes of the gradients are unknown, their lengths relative to each other can be estimated. At which of the three points is the wind speed greatest if the speed increases as the pressure gradient increases? (To print an enlarged copy of the graph, go to the website www.mathgraphs.com.)

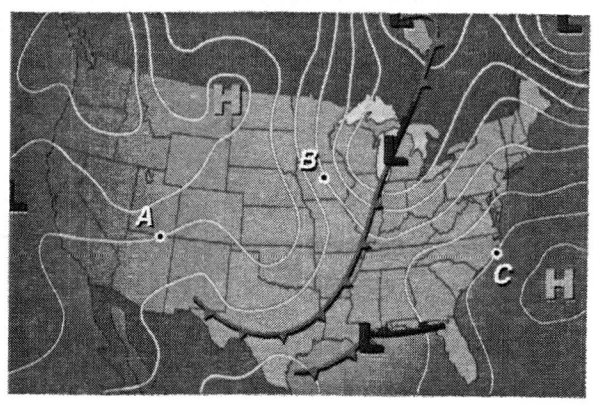

Heat-Seeking Path In Exercises 73 and 74, find the path followed by a heat-seeking particle placed at point P on a metal plate with a temperature field $T(x, y)$.

Temperature Field	Point
73. $T(x, y) = 400 - 2x^2 - y^2$	$P(10, 10)$
74. $T(x, y) = 100 - x^2 - 2y^2$	$P(4, 3)$

75. **Investigation** A team of oceanographers is mapping the ocean floor to assist in the recovery of a sunken ship. Using sonar, they develop the model

$$D = 250 + 30x^2 + 50\sin\frac{\pi y}{2}, \quad 0 \leq x \leq 2, 0 \leq y \leq 2$$

where D is the depth in meters, and x and y are the distances in kilometers.

(a) Use a computer algebra system to graph the surface.

(b) Because the graph in part (a) is showing depth, it is not a map of the ocean floor. How could the model be changed so that the graph of the ocean floor could be obtained?

(c) What is the depth of the ship if it is located at the coordinates $x = 1$ and $y = 0.5$?

(d) Determine the steepness of the ocean floor in the positive x-direction from the position of the ship.

(e) Determine the steepness of the ocean floor in the positive y-direction from the position of the ship.

(f) Determine the direction of the greatest rate of change of depth from the position of the ship.

76. **Temperature** The temperature at the point (x, y) on a metal plate is modeled by

$$T(x, y) = 400e^{-(x^2+y)/2}, \quad x \geq 0, y \geq 0.$$

(a) Use a computer algebra system to graph the temperature distribution function.

(b) Find the directions of no change in heat on the plate from the point $(3, 5)$.

(c) Find the direction of greatest increase in heat from the point $(3, 5)$.

True or False? In Exercises 77–80, determine whether the statement is true or false. If it is false, explain why or give an example that shows it is false.

77. If $f(x, y) = \sqrt{1 - x^2 - y^2}$, then $D_{\mathbf{u}} f(0, 0) = 0$ for any unit vector $\mathbf{u}$.

78. If $f(x, y) = x + y$, then $-1 \leq D_{\mathbf{u}} f(x, y) \leq 1$.

79. If $D_{\mathbf{u}} f(x, y)$ exists, then $D_{\mathbf{u}} f(x, y) = -D_{-\mathbf{u}} f(x, y)$.

80. If $D_{\mathbf{u}} f(x_0, y_0) = c$ for any unit vector $\mathbf{u}$, then $c = 0$.

81. Find a function f such that

$$\nabla f = e^x \cos y \mathbf{i} - e^x \sin y \mathbf{j} + z \mathbf{k}.$$

Section 12.7 Tangent Planes and Normal Lines

- Find equations of tangent planes and normal lines to surfaces.
- Find the angle of inclination of a plane in space.
- Compare the gradients $\nabla f(x, y)$ and $\nabla F(x, y, z)$.

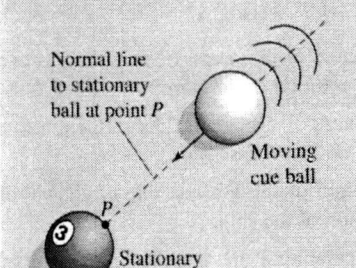

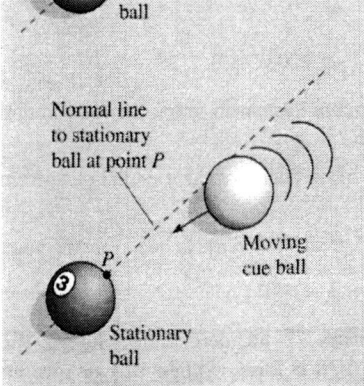

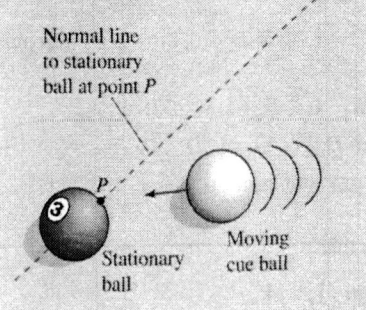

Tangent Plane and Normal Line to a Surface

So far we have represented surfaces in space primarily by equations of the form

$$z = f(x, y). \qquad \text{Equation of a surface } S$$

In the development to follow, however, it is convenient to use the more general representation $F(x, y, z) = 0$. For a surface S given by $z = f(x, y)$, you can convert to the general form by defining F as

$$F(x, y, z) = f(x, y) - z.$$

Because $f(x, y) - z = 0$, you can consider S to be the level surface of F given by

$$F(x, y, z) = 0. \qquad \text{Alternative equation of surface } S$$

Example 1 Writing an Equation of a Surface

For the function given by

$$F(x, y, z) = x^2 + y^2 + z^2 - 4,$$

describe the level surface given by $F(x, y, z) = 0$.

Solution The level surface given by $F(x, y, z) = 0$ can be written as

$$x^2 + y^2 + z^2 = 4$$

which is a sphere of radius 2 whose center is at the origin.

You have seen many examples of the usefulness of normal lines in applications involving curves. Normal lines are equally important in analyzing surfaces and solids. For example, consider the collision of two billiard balls. When a stationary ball is struck at a point P on its surface, it moves along the **line of impact** determined by P and the center of the ball. The impact can occur in *two* ways. If the cue ball is moving along the line of impact, it stops dead and imparts all of its momentum to the stationary ball, as shown in Figure 12.54. If the cue ball is not moving along the line of impact, it is deflected to one side or the other and retains part of its momentum. That part of the momentum that is transferred to the stationary ball occurs along the line of impact, *regardless* of the direction of the cue ball, as shown in Figure 12.55. This line of impact is called the **normal line** to the surface of the ball at the point P.

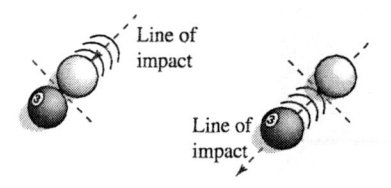

Figure 12.54

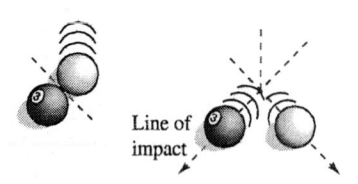

Figure 12.55

EXPLORATION

Billiard Balls and Normal Lines
In each of the three figures below, the cue ball is about to strike a stationary ball at point P. Explain how you can use the normal line to the stationary ball at point P to describe the resulting motion of each of the two balls. Assuming that each cue ball has the same speed, which stationary ball will acquire the greatest speed? Which will acquire the least? Explain your reasoning.

In the process of finding a normal line to a surface, you are also able to solve the problem of finding a **tangent plane** to the surface. Let S be a surface given by

$$F(x, y, z) = 0$$

and let $P(x_0, y_0, z_0)$ be a point on S. Let C be a curve on S through P that is defined by the vector-valued function

$$\mathbf{r}(t) = x(t)\mathbf{i} + y(t)\mathbf{j} + z(t)\mathbf{k}.$$

Then, for all t,

$$F(x(t), y(t), z(t)) = 0.$$

If F is differentiable and $x'(t)$, $y'(t)$, and $z'(t)$ all exist, it follows from the Chain Rule that

$$0 = F'(t)$$
$$= F_x(x, y, z)x'(t) + F_y(x, y, z)y'(t) + F_z(x, y, z)z'(t).$$

At (x_0, y_0, z_0), the equivalent vector form is

$$0 = \underbrace{\nabla F(x_0, y_0, z_0)}_{\text{Gradient}} \cdot \underbrace{\mathbf{r}'(t_0)}_{\substack{\text{Tangent}\\ \text{Vector}}}.$$

This result means that the gradient at P is orthogonal to the tangent vector of every curve on S through P. So, all tangent lines at S lie in a plane that is normal to $\nabla F(x_0, y_0, z_0)$ and contains P, as shown in Figure 12.56.

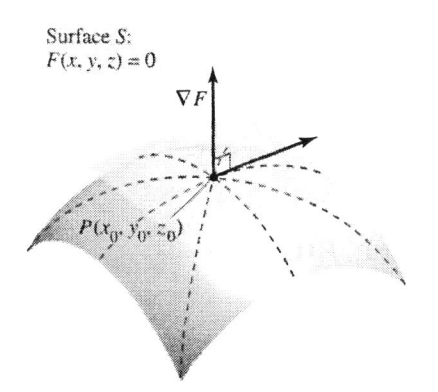

Surface S:
$F(x, y, z) = 0$

Tangent plane to surface S at P
Figure 12.56

Definition of Tangent Plane and Normal Line

Let F be differentiable at the point $P(x_0, y_0, z_0)$ on the surface S given by $F(x, y, z) = 0$ such that $\nabla F(x_0, y_0, z_0) \neq \mathbf{0}$.

1. The plane through P that is normal to $\nabla F(x_0, y_0, z_0)$ is called the **tangent plane to S at P**.
2. The line through P having the direction of $\nabla F(x_0, y_0, z_0)$ is called the **normal line to S at P**.

NOTE In the remainder of this section, we assume $\nabla F(x_0, y_0, z_0)$ to be nonzero unless stated otherwise.

To find an equation for the tangent plane to S at (x_0, y_0, z_0), let (x, y, z) be an arbitrary point in the tangent plane. Then the vector

$$\mathbf{v} = (x - x_0)\mathbf{i} + (y - y_0)\mathbf{j} + (z - z_0)\mathbf{k}$$

lies in the tangent plane. Because $\nabla F(x_0, y_0, z_0)$ is normal to the tangent plane at (x_0, y_0, z_0), it must be orthogonal to every vector in the tangent plane, and you have $\nabla F(x_0, y_0, z_0) \cdot \mathbf{v} = 0$, which leads to the following theorem.

THEOREM 12.13   Equation of Tangent Plane

If F is differentiable at (x_0, y_0, z_0), then an equation of the tangent plane to the surface given by $F(x, y, z) = 0$ at (x_0, y_0, z_0) is

$$F_x(x_0, y_0, z_0)(x - x_0) + F_y(x_0, y_0, z_0)(y - y_0) + F_z(x_0, y_0, z_0)(z - z_0) = 0.$$

Example 2 Finding an Equation of a Tangent Plane

Find an equation of the tangent plane to the hyperboloid given by

$$z^2 - 2x^2 - 2y^2 = 12$$

at the point $(1, -1, 4)$.

Solution Begin by writing the equation of the surface as

$$z^2 - 2x^2 - 2y^2 - 12 = 0.$$

Then, considering

$$F(x, y, z) = z^2 - 2x^2 - 2y^2 - 12$$

you have

$$F_x(x, y, z) = -4x, \quad F_y(x, y, z) = -4y, \quad \text{and} \quad F_z(x, y, z) = 2z.$$

At the point $(1, -1, 4)$ the partial derivatives are

$$F_x(1, -1, 4) = -4, \quad F_y(1, -1, 4) = 4, \quad \text{and} \quad F_z(1, -1, 4) = 8.$$

Therefore, an equation of the tangent plane at $(1, -1, 4)$ is

$$-4(x - 1) + 4(y + 1) + 8(z - 4) = 0$$
$$-4x + 4 + 4y + 4 + 8z - 32 = 0$$
$$-4x + 4y + 8z - 24 = 0$$
$$x - y - 2z + 6 = 0.$$

Figure 12.57 shows a portion of the hyperboloid and tangent plane.

Surface:
$z^2 - 2x^2 - 2y^2 - 12 = 0$

$\nabla F(1, -1, 4)$

Tangent plane to surface
Figure 12.57

TECHNOLOGY Some three-dimensional graphing utilities are capable of sketching tangent planes to surfaces. Two examples are shown below.

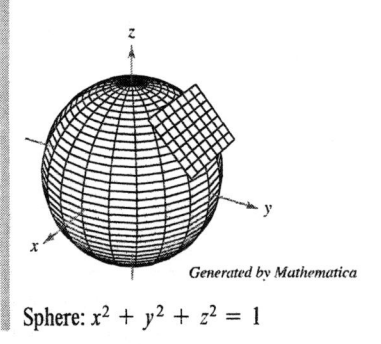

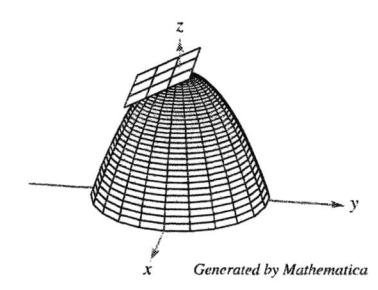

Sphere: $x^2 + y^2 + z^2 = 1$ *Generated by Mathematica*

Paraboloid: $z = 2 - x^2 - y^2$ *Generated by Mathematica*

To find the equation of the tangent plane at a point on a surface given by $z = f(x, y)$, you can define the function F by

$$F(x, y, z) = f(x, y) - z.$$

Then S is given by the level surface $F(x, y, z) = 0$, and by Theorem 12.13 an equation of the tangent plane to S at the point (x_0, y_0, z_0) is

$$f_x(x_0, y_0)(x - x_0) + f_y(x_0, y_0)(y - y_0) - (z - z_0) = 0.$$

Example 3 Finding an Equation of the Tangent Plane

Find the equation of the tangent plane to the paraboloid

$$z = 1 - \frac{1}{10}(x^2 + 4y^2)$$

at the point $\left(1, 1, \frac{1}{2}\right)$.

Solution From $z = f(x, y) = 1 - \frac{1}{10}(x^2 + 4y^2)$, you obtain

$$f_x(x, y) = -\frac{x}{5} \implies f_x(1, 1) = -\frac{1}{5}$$

and

$$f_y(x, y) = -\frac{4y}{5} \implies f_y(1, 1) = -\frac{4}{5}.$$

Therefore, an equation of the tangent plane at $\left(1, 1, \frac{1}{2}\right)$ is

$$f_x(1, 1)(x - 1) + f_y(1, 1)(y - 1) - \left(z - \tfrac{1}{2}\right) = 0$$
$$-\frac{1}{5}(x - 1) - \frac{4}{5}(y - 1) - \left(z - \tfrac{1}{2}\right) = 0$$
$$-\frac{1}{5}x - \frac{4}{5}y - z + \frac{3}{2} = 0.$$

This tangent plane is shown in Figure 12.58.

Surface:
$z = 1 - \frac{1}{10}(x^2 + 4y^2)$

Figure 12.58

The gradient $\nabla F(x, y, z)$ gives a convenient way to find equations of normal lines, as shown in Example 4.

Example 4 Finding an Equation of a Normal Line to a Surface

Find a set of symmetric equations for the normal line to the surface given by

$$xyz = 12$$

at the point $(2, -2, -3)$.

Solution Begin by letting

$$F(x, y, z) = xyz - 12.$$

Then, the gradient is given by

$$\nabla F(x, y, z) = F_x(x, y, z)\mathbf{i} + F_y(x, y, z)\mathbf{j} + F_z(x, y, z)\mathbf{k}$$
$$= yz\mathbf{i} + xz\mathbf{j} + xy\mathbf{k}$$

and at the point $(2, -2, -3)$ you have

$$\nabla F(2, -2, -3) = (-2)(-3)\mathbf{i} + (2)(-3)\mathbf{j} + (2)(-2)\mathbf{k}$$
$$= 6\mathbf{i} - 6\mathbf{j} - 4\mathbf{k}.$$

The normal line at $(2, -2, -3)$ has direction numbers 6, -6, and -4, and the corresponding set of symmetric equations is

$$\frac{x - 2}{6} = \frac{y + 2}{-6} = \frac{z + 3}{-4}.$$

(See Figure 12.59.)

Surface: $xyz = 12$

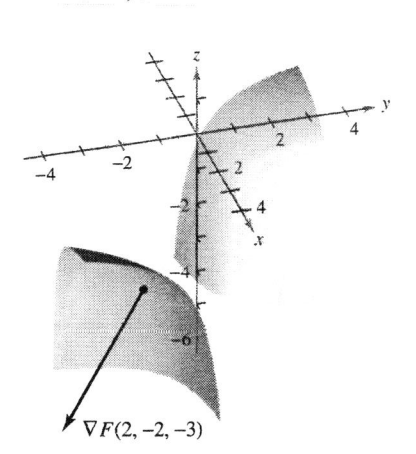

Figure 12.59

Knowing that the gradient $\nabla F(x, y, z)$ is normal to the surface given by $F(x, y, z) = 0$ allows you to solve a variety of problems dealing with surfaces and curves in space.

Example 5 Finding the Equation of a Tangent Line to a Curve

Describe the tangent line to the curve of intersection of the surfaces

$$x^2 + 2y^2 + 2z^2 = 20 \qquad \text{Ellipsoid}$$
$$x^2 + y^2 + z = 4 \qquad \text{Paraboloid}$$

at the point (0, 1, 3), as shown in Figure 12.60.

Solution Begin by finding the gradients to both surfaces at the point (0, 1, 3).

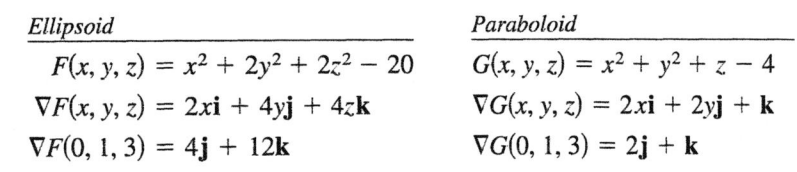

Ellipsoid	Paraboloid
$F(x, y, z) = x^2 + 2y^2 + 2z^2 - 20$	$G(x, y, z) = x^2 + y^2 + z - 4$
$\nabla F(x, y, z) = 2x\mathbf{i} + 4y\mathbf{j} + 4z\mathbf{k}$	$\nabla G(x, y, z) = 2x\mathbf{i} + 2y\mathbf{j} + \mathbf{k}$
$\nabla F(0, 1, 3) = 4\mathbf{j} + 12\mathbf{k}$	$\nabla G(0, 1, 3) = 2\mathbf{j} + \mathbf{k}$

The cross product of these two gradients is a vector that is tangent to both surfaces at the point (0, 1, 3).

$$\nabla F(0, 1, 3) \times \nabla G(0, 1, 3) = \begin{vmatrix} \mathbf{i} & \mathbf{j} & \mathbf{k} \\ 0 & 4 & 12 \\ 0 & 2 & 1 \end{vmatrix} = -20\mathbf{i}.$$

So, the tangent line to the curve of intersection of the two surfaces at the point (0, 1, 3) is a line that is parallel to the x-axis and passes through the point (0, 1, 3).

Ellipsoid: $x^2 + 2y^2 + 2z^2 = 20$

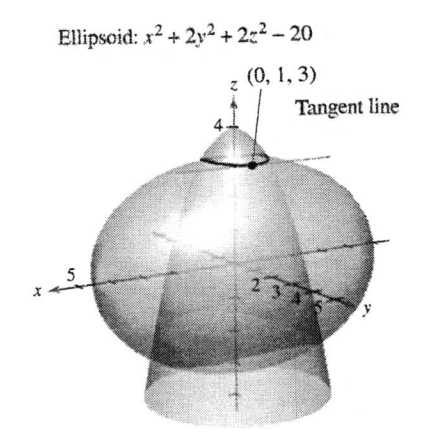

Paraboloid: $x^2 + y^2 + z = 4$

Figure 12.60

The Angle of Inclination of a Plane

Another use of the gradient $\nabla F(x, y, z)$ is to determine the angle of inclination of the tangent plane to a surface. The **angle of inclination** of a plane is defined to be the angle θ, $0 \le \theta \le \pi/2$, between the given plane and the xy-plane, as shown in Figure 12.61. (The angle of inclination of a horizontal plane is defined to be zero.) Because the vector $\mathbf{k}$ is normal to the xy-plane, you can use the formula for the cosine of the angle between two planes (given in Section 10.5) to conclude that the angle of inclination of a plane with normal vector $\mathbf{n}$ is given by

$$\cos \theta = \frac{|\mathbf{n} \cdot \mathbf{k}|}{\|\mathbf{n}\| \|\mathbf{k}\|} = \frac{|\mathbf{n} \cdot \mathbf{k}|}{\|\mathbf{n}\|}. \qquad \text{Angle of inclination of a plane}$$

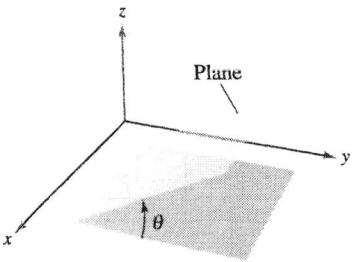

The angle of inclination
Figure 12.61

Example 6 Finding the Angle of Inclination of a Tangent Plane

Find the angle of inclination of the tangent plane to the ellipsoid given by

$$\frac{x^2}{12} + \frac{y^2}{12} + \frac{z^2}{3} = 1$$

at the point $(2, 2, 1)$.

Solution If you let

$$F(x, y, z) = \frac{x^2}{12} + \frac{y^2}{12} + \frac{z^2}{3} - 1$$

the gradient of F at the point $(2, 2, 1)$ is given by

$$\nabla F(x, y, z) = \frac{x}{6}\mathbf{i} + \frac{y}{6}\mathbf{j} + \frac{2z}{3}\mathbf{k}$$

$$\nabla F(2, 2, 1) = \frac{1}{3}\mathbf{i} + \frac{1}{3}\mathbf{j} + \frac{2}{3}\mathbf{k}.$$

Because $\nabla F(2, 2, 1)$ is normal to the tangent plane and $\mathbf{k}$ is normal to the xy-plane, it follows that the angle of inclination of the tangent plane is given by

$$\cos\theta = \frac{|\nabla F(2, 2, 1) \cdot \mathbf{k}|}{\|\nabla F(2, 2, 1)\|} = \frac{2/3}{\sqrt{(1/3)^2 + (1/3)^2 + (2/3)^2}} = \sqrt{\frac{2}{3}}$$

which implies that

$$\theta = \arccos\sqrt{\frac{2}{3}} \approx 35.3°,$$

as shown in Figure 12.62.

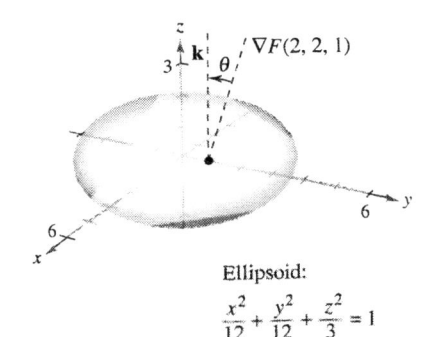

Figure 12.62

Ellipsoid:
$$\frac{x^2}{12} + \frac{y^2}{12} + \frac{z^2}{3} = 1$$

NOTE A special case of the procedure shown in Example 6 is worth noting. The angle of inclination θ of the tangent plane to the surface $z = f(x, y)$ at (x_0, y_0, z_0) is given by

$$\cos\theta = \frac{1}{\sqrt{[f_x(x_0, y_0)]^2 + [f_y(x_0, y_0)]^2 + 1}}.$$
Alternative formula for angle of inclination (See Exercise 62.)

A Comparison of the Gradients $\nabla f(x, y)$ and $\nabla F(x, y, z)$

We conclude this section with a comparison of the gradients $\nabla f(x, y)$ and $\nabla F(x, y, z)$. In the preceding section, you saw that the gradient of a function f of two variables is normal to the *level curves* of f. Specifically, Theorem 12.12 stated that if f is differentiable at (x_0, y_0) and $\nabla f(x_0, y_0) \neq \mathbf{0}$, then $\nabla f(x_0, y_0)$ is normal to the level curve through (x_0, y_0). Having developed normal lines to surfaces, you can now extend this result to a function of three variables.

THEOREM 12.14 Gradient Is Normal to Level Surfaces

If F is differentiable at (x_0, y_0, z_0) and $\nabla F(x_0, y_0, z_0) \neq \mathbf{0}$, then $\nabla F(x_0, y_0, z_0)$ is normal to the level surface through (x_0, y_0, z_0).

When working with the gradients $\nabla f(x, y)$ and $\nabla F(x, y, z)$, be sure you remember that $\nabla f(x, y)$ is a vector in the xy-plane and $\nabla F(x, y, z)$ is a vector in space.

EXERCISES FOR SECTION 12.7

In Exercises 1–4, describe the level surface $F(x, y, z) = 0$.

1. $F(x, y, z) = 3x - 5y + 3z - 15$
2. $F(x, y, z) = x^2 + y^2 + z^2 - 25$
3. $F(x, y, z) = 4x^2 + 9y^2 - 4z^2$
4. $F(x, y, z) = 16x^2 - 9y^2 + 144z$

In Exercises 5–14, find a unit normal vector to the surface at the indicated point. [*Hint*: Normalize the gradient vector $\nabla F(x, y, z)$.]

Surface	Point
5. $x + y + z = 4$	$(2, 0, 2)$
6. $x^2 + y^2 + z^2 = 11$	$(3, 1, 1)$
7. $z = \sqrt{x^2 + y^2}$	$(3, 4, 5)$
8. $z = x^3$	$(2, 1, 8)$
9. $x^2 y^4 - z = 0$	$(1, 2, 16)$
10. $x^2 + 3y + z^3 = 9$	$(2, -1, 2)$
11. $\ln\left(\dfrac{x}{y - z}\right) = 0$	$(1, 4, 3)$
12. $ze^{x^2 - y^2} - 3 = 0$	$(2, 2, 3)$
13. $z - x \sin y = 4$	$\left(6, \dfrac{\pi}{6}, 7\right)$
14. $\sin(x - y) - z = 2$	$\left(\dfrac{\pi}{3}, \dfrac{\pi}{6}, -\dfrac{3}{2}\right)$

In Exercises 15–18, find an equation of the tangent plane to the surface at the indicated point.

15. $z = 25 - x^2 - y^2$,
$(3, 1, 15)$

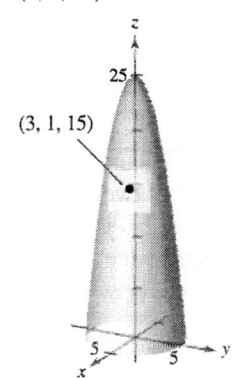

16. $f(x, y) = \dfrac{y}{x}$,
$(1, 2, 2)$

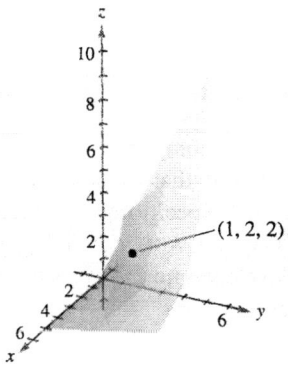

17. $z = \sqrt{x^2 + y^2}$,
$(3, 4, 5)$

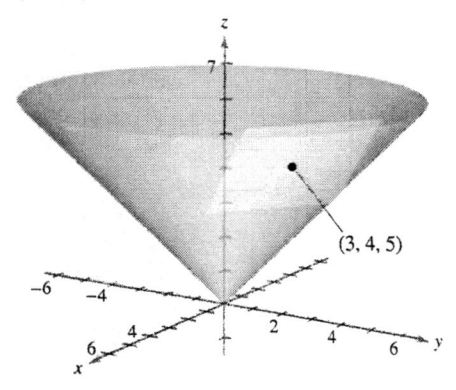

18. $g(x, y) = \arctan \dfrac{y}{x}$,
$(1, 0, 0)$

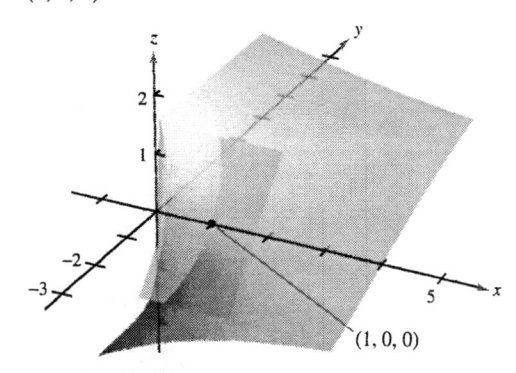

In Exercises 19–28, find an equation of the tangent plane to the surface at the indicated point.

Surface	Point
19. $g(x, y) = x^2 - y^2$	$(5, 4, 9)$
20. $f(x, y) = 2 - \frac{2}{3}x - y$	$(3, -1, 1)$
21. $z = e^x(\sin y + 1)$	$\left(0, \dfrac{\pi}{2}, 2\right)$
22. $z = x^2 - 2xy + y^2$	$(1, 2, 1)$
23. $h(x, y) = \ln \sqrt{x^2 + y^2}$	$(3, 4, \ln 5)$
24. $h(x, y) = \cos y$	$\left(5, \dfrac{\pi}{4}, \dfrac{\sqrt{2}}{2}\right)$
25. $x^2 + 4y^2 + z^2 = 36$	$(2, -2, 4)$
26. $x^2 + 2z^2 = y^2$	$(1, 3, -2)$
27. $xy^2 + 3x - z^2 = 4$	$(2, 1, -2)$
28. $x = y(2z - 3)$	$(4, 4, 2)$

In Exercises 29–34, find an equation of the tangent plane and find symmetric equations of the normal line to the surface at the indicated point.

Surface	Point
29. $x^2 + y^2 + z = 9$	$(1, 2, 4)$
30. $x^2 + y^2 + z^2 = 9$	$(1, 2, 2)$
31. $xy - z = 0$	$(-2, -3, 6)$
32. $x^2 - y^2 + z^2 = 0$	$(5, 13, -12)$
33. $z = \arctan(y/x)$	$(1, 1, \pi/4)$
34. $xyz = 10$	$(1, 2, 5)$

35. *Investigation* Consider the function

$$f(x, y) = \frac{4xy}{(x^2 + 1)(y^2 + 1)}$$

on the intervals $-2 \le x \le 2$ and $0 \le y \le 3$.

(a) Find a set of parametric equations of the normal line and an equation of the tangent plane to the surface at the point $(1, 1, 1)$.

(b) Repeat part (a) for the point $\left(-1, 2, -\frac{4}{5}\right)$.

(c) Use a computer algebra system to graph the surface, the normal lines, and the tangent planes found in parts (a) and (b).

(d) Use analytical and graphical analysis to write a brief description of the surface at the two indicated points.

Getting at the Concept

36. Consider the function $F(x, y, z) = 0$, which is differentiable at $P(x_0, y_0, z_0)$. Give the definition of the tangent plane at P and the normal line at P.

37. Give the standard form of the equation of the tangent plane to a surface given by $F(x, y, z) = 0$ at (x_0, y_0, z_0).

38. For some surfaces, the normal lines at any point pass through the same geometric object. What is the common geometric object for a sphere? What is the common geometric object for a right circular cylinder? Explain.

In Exercises 39–44, (a) find symmetric equations of the tangent line to the curve of intersection of the surfaces at the indicated point, and (b) find the cosine of the angle between the gradient vectors at this point. State whether or not the surfaces are orthogonal at the point of intersection.

Surfaces	Point
39. $x^2 + y^2 = 5$, $z = x$	$(2, 1, 2)$
40. $z = x^2 + y^2$, $z = 4 - y$	$(2, -1, 5)$
41. $x^2 + z^2 = 25$, $y^2 + z^2 = 25$	$(3, 3, 4)$
42. $z = \sqrt{x^2 + y^2}$, $5x - 2y + 3z = 22$	$(3, 4, 5)$
43. $x^2 + y^2 + z^2 = 6$, $x - y - z = 0$	$(2, 1, 1)$
44. $z = x^2 + y^2$, $x + y + 6z = 33$	$(1, 2, 5)$

45. Consider the functions

$$f(x, y) = 6 - x^2 - y^2/4 \quad \text{and} \quad g(x, y) = 2x + y.$$

(a) Find a set of parametric equations of the tangent line to the curve of intersection of the surfaces at the point $(1, 2, 4)$, and find the angle between the gradient vectors.

(b) Use a computer algebra system to graph the surfaces. Graph the tangent line found in part (a).

46. Consider the functions

$$f(x, y) = \sqrt{16 - x^2 - y^2 + 2x - 4y} \quad \text{and}$$

$$g(x, y) = \frac{\sqrt{2}}{2}\sqrt{1 - 3x^2 + y^2 + 6x + 4y}.$$

(a) Use a computer algebra system to graph the first-octant portion of the surfaces represented by f and g.

(b) Find two first-octant points on the curve of intersection and show that the surfaces are orthogonal at these points.

(c) These surfaces are orthogonal along the curve of intersection. Does part (b) prove this fact? Explain.

In Exercises 47–50, find the angle of inclination θ of the tangent plane to the given surface at the indicated point.

Surface	Point
47. $3x^2 + 2y^2 - z = 15$	$(2, 2, 5)$
48. $2xy - z^3 = 0$	$(2, 2, 2)$
49. $x^2 - y^2 + z = 0$	$(1, 2, 3)$
50. $x^2 + y^2 = 5$	$(2, 1, 3)$

In Exercises 51 and 52, find the point on the surface where the tangent plane is horizontal. Use a computer algebra system to graph the surface and the horizontal tangent plane. Describe the surface where the tangent plane is horizontal.

51. $z = 3 - x^2 - y^2 + 6y$

52. $z = 3x^2 + 2y^2 - 3x + 4y - 5$

In Exercises 53 and 54, find the path of a heat-seeking particle in the temperature field T, starting at the indicated point.

53. $T(x, y, z) = 400 - 2x^2 - y^2 - 4z^2$, $(4, 3, 10)$

54. $T(x, y, z) = 100 - 3x - y - z^2$, $(2, 2, 5)$

In Exercises 55 and 56, show that the tangent plane to the quadric surface at the point (x_0, y_0, z_0) can be written in the given form.

55. Ellipsoid: $\dfrac{x^2}{a^2} + \dfrac{y^2}{b^2} + \dfrac{z^2}{c^2} = 1$

Plane: $\dfrac{x_0 x}{a^2} + \dfrac{y_0 y}{b^2} + \dfrac{z_0 z}{c^2} = 1$

56. Hyperboloid: $\dfrac{x^2}{a^2} + \dfrac{y^2}{b^2} - \dfrac{z^2}{c^2} = 1$

Plane: $\dfrac{x_0 x}{a^2} + \dfrac{y_0 y}{b^2} - \dfrac{z_0 z}{c^2} = 1$

57. Show that any tangent plane to the cone $z^2 = a^2x^2 + b^2y^2$ passes through the origin.

58. Let f be a differentiable function and consider the surface $z = xf(y/x)$. Show that the tangent plane at any point $P(x_0, y_0, z_0)$ on the surface passes through the origin.

59. Approximation Consider the following approximations for a function $f(x, y)$ centered at $(0, 0)$.

Linear approximation:

$$P_1(x, y) = f(0, 0) + f_x(0, 0)x + f_y(0, 0)y$$

Quadratic approximation:

$$P_2(x, y) = f(0, 0) + f_x(0, 0)x + f_y(0, 0)y + \tfrac{1}{2}f_{xx}(0, 0)x^2 + f_{xy}(0, 0)xy + \tfrac{1}{2}f_{yy}(0, 0)y^2$$

[Note that the linear approximation is the tangent plane to the surface at $(0, 0, f(0, 0))$.]

(a) Find the linear approximations of $f(x, y) = e^{(x-y)}$ centered at $(0, 0)$.

(b) Find the quadratic approximation of $f(x, y) = e^{(x-y)}$ centered at $(0, 0)$.

(c) If $x = 0$ in the quadratic approximation, you obtain the second-degree Taylor polynomial for what function? Answer the same question for $y = 0$.

(d) Complete the table.

x	y	$f(x, y)$	$P_1(x, y)$	$P_2(x, y)$
0	0			
0	0.1			
0.2	0.1			
0.2	0.5			
1	0.5			

(e) Use a computer algebra system to graph the surfaces $z = f(x, y)$, $z = P_1(x, y)$, and $z = P_2(x, y)$.

60. Approximation Repeat Exercise 59 for the function $f(x, y) = \cos(x + y)$.

61. Prove Theorem 12.14.

62. Prove that the angle of inclination θ of the tangent plane to the surface $z = f(x, y)$ at the point (x_0, y_0, z_0) is given by

$$\cos\theta = \frac{1}{\sqrt{[f_x(x_0, y_0)]^2 + [f_y(x_0, y_0)]^2 + 1}}.$$

SECTION PROJECT: WILDFLOWERS

The diversity of wildflowers in a meadow can be measured by counting the number of daisies, buttercups, shooting stars, and so on. If there are n types of wildflowers, each with a proportion p_i of the total population, it follows that $p_1 + p_2 + \cdots + p_n = 1$. The measure of diversity of the population is defined as

$$H = -\sum_{i=1}^{n} p_i \log_2 p_i.$$

In this definition, it is understood that $p_i \log_2 p_i = 0$ when $p_i = 0$. The tables show proportions of wildflowers in a meadow in May, June, August, and September.

May

Flower type	1	2	3	4
Proportion	$\frac{5}{16}$	$\frac{5}{16}$	$\frac{5}{16}$	$\frac{1}{16}$

June

Flower type	1	2	3	4
Proportion	$\frac{1}{4}$	$\frac{1}{4}$	$\frac{1}{4}$	$\frac{1}{4}$

August

Flower type	1	2	3	4
Proportion	$\frac{1}{4}$	0	$\frac{1}{4}$	$\frac{1}{2}$

September

Flower type	1	2	3	4
Proportion	0	0	0	1

(a) Determine the wildflower diversity for each month. How would you interpret September's diversity? Which month had the greatest diversity?

(b) If the meadow contains ten types of wildflowers in roughly equal proportions, is the diversity of the population greater than or less than the diversity of a similar distribution of four types of flowers? What type of distribution (of ten types of wildflowers) would produce maximum diversity?

(c) Let H_n represent the maximum diversity of n types of wildflowers. Does H_n approach a limit as $n \to \infty$?

FOR FURTHER INFORMATION Biologists use the concept of diversity to measure the proportions of different types of organisms within an environment. For more information on this technique, see the article "Information Theory and Biological Diversity" by Steven Kolmes and Kevin Mitchell in the *UMAP Modules*. To view this article, go to the website *www.matharticles.com*.

Section 12.8

Extrema of Functions of Two Variables

- Find absolute and relative extrema of a function of two variables.
- Use the Second Partials Test to find relative extrema of a function of two variables.

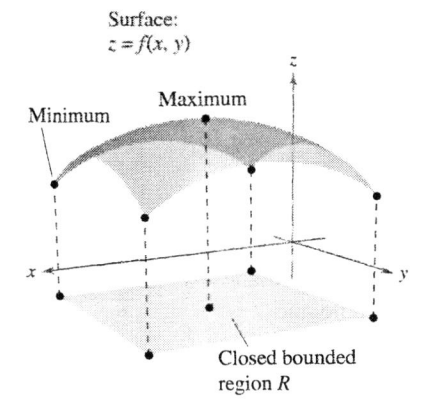

Absolute Extrema and Relative Extrema

In Chapter 3, you studied techniques for finding the extreme values of a function of a single variable. In this section, we extend these techniques to functions of two variables. For example, in Theorem 12.15 the Extreme Value Theorem for a function of a single variable is extended to a function of two variables.

Consider the continuous function f of two variables, defined on a closed bounded region R. The values $f(a, b)$ and $f(c, d)$ such that

$$f(a, b) \leq f(x, y) \leq f(c, d) \qquad (a, b) \text{ and } (c, d) \text{ are in } R.$$

for all (x, y) in R are called the **minimum** and **maximum** of f in the region R, as shown in Figure 12.63. Recall from Section 12.2 that a region in the plane is *closed* if it contains all of its boundary points. The Extreme Value Theorem deals with a region in the plane that is both closed and *bounded*. A region in the plane is called **bounded** if it is a subregion of a closed disk in the plane.

Surface:
$z = f(x, y)$

R contains point(s) at which $f(x, y)$ is a minimum and point(s) at which $f(x, y)$ is a maximum.
Figure 12.63

THEOREM 12.15 Extreme Value Theorem

Let f be a continuous function of two variables x and y defined on a closed bounded region R in the xy-plane.

1. There is at least one point in R where f takes on a minimum value.
2. There is at least one point in R where f takes on a maximum value.

A minimum is also called an **absolute minimum** and a maximum is also called an **absolute maximum**. As in single-variable calculus, we distinguish between absolute extrema and **relative extrema**.

Definition of Relative Extrema

Let f be a function defined on a region R containing (x_0, y_0).

1. The function f has a **relative minimum** at (x_0, y_0) if
   $$f(x, y) \geq f(x_0, y_0)$$
   for all (x, y) in an *open* disk containing (x_0, y_0).
2. The function f has a **relative maximum** at (x_0, y_0) if
   $$f(x, y) \leq f(x_0, y_0)$$
   for all (x, y) in an *open* disk containing (x_0, y_0).

Relative extrema
Figure 12.64

To say that f has a relative maximum at (x_0, y_0) means that the point (x_0, y_0, z_0) is at least as high as all nearby points on the graph of $z = f(x, y)$. Similarly, f has a relative minimum at (x_0, y_0) if (x_0, y_0, z_0) is at least as low as all nearby points on the graph. (See Figure 12.64.)

KARL WEIERSTRASS (1815–1897)

Although the Extreme Value Theorem had been used by earlier mathematicians, the first to provide a rigorous proof was the German mathematician Karl Weierstrass. Weierstrass also provided rigorous justifications for many other mathematical results already in common use. We are indebted to him for much of the logical foundation on which modern calculus is built.

To locate relative extrema of f, you can investigate the points at which the gradient of f is $\mathbf{0}$ or the points at which one of the partial derivatives does not exist. Such points are called **critical points** of f.

Definition of Critical Point

Let f be defined on an open region R containing (x_0, y_0). The point (x_0, y_0) is a **critical point** of f if one of the following is true.

1. $f_x(x_0, y_0) = 0$ and $f_y(x_0, y_0) = 0$
2. $f_x(x_0, y_0)$ or $f_y(x_0, y_0)$ does not exist.

Recall from Theorem 12.11 that if f is differentiable and

$$\nabla f(x_0, y_0) = f_x(x_0, y_0)\mathbf{i} + f_y(x_0, y_0)\mathbf{j}$$
$$= 0\mathbf{i} + 0\mathbf{j}$$

then every directional derivative at (x_0, y_0) must be 0. This implies that the function has a horizontal tangent plane at the point (x_0, y_0), as shown in Figure 12.65. It appears that such a point is a likely location of a relative extremum. This is confirmed by Theorem 12.16.

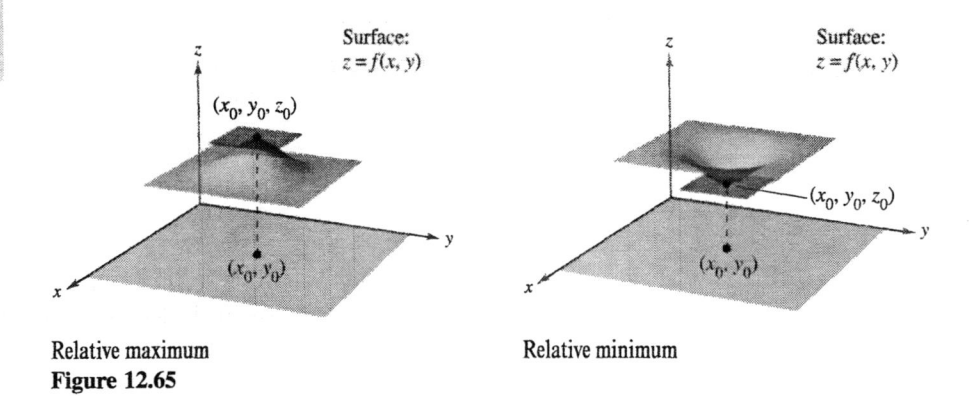

Relative maximum   Relative minimum
Figure 12.65

THEOREM 12.16 Relative Extrema Occur Only at Critical Points

If f has a relative extremum at (x_0, y_0) on an open region R, then (x_0, y_0) is a critical point of f.

EXPLORATION

Try using a graphing utility to sketch the graph of

$$z = x^3 - 3xy + y^3$$

using the bounds $0 \leq x \leq 3$, $0 \leq y \leq 3$, and $-3 \leq z \leq 3$. This view makes it appear as though the surface has an absolute minimum. But does it?

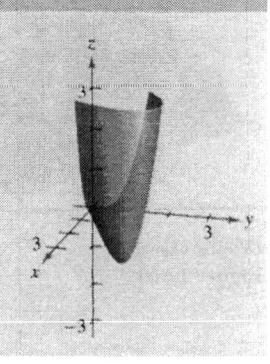

Example 1 Finding a Relative Extremum

Determine the relative extrema of

$$f(x, y) = 2x^2 + y^2 + 8x - 6y + 20.$$

Solution Begin by finding the critical points of f. Because

$$f_x(x, y) = 4x + 8 \qquad \text{Partial with respect to } x$$

and

$$f_y(x, y) = 2y - 6 \qquad \text{Partial with respect to } y$$

are defined for all x and y, the only critical points are those for which both first partial derivatives are 0. To locate these points, let $f_x(x, y)$ and $f_y(x, y)$ be 0, and solve the equations

$$4x + 8 = 0 \quad \text{and} \quad 2y - 6 = 0$$

to obtain the critical point $(-2, 3)$. By completing the square, you can conclude that for all $(x, y) \neq (-2, 3)$,

$$f(x, y) = 2(x + 2)^2 + (y - 3)^2 + 3 > 3.$$

Therefore, a relative *minimum* of f occurs at $(-2, 3)$. The value of the relative minimum is $f(-2, 3) = 3$, as shown in Figure 12.66.

Surface:
$f(x, y) = 2x^2 + y^2 + 8x - 6y + 20$

The function $z = f(x, y)$ has a relative minimum at $(-2, 3)$.
Figure 12.66

Example 1 shows a relative minimum occurring at one type of critical point—the type for which both $f_x(x, y)$ and $f_y(x, y)$ are 0. The next example concerns a relative maximum that occurs at the other type of critical point—the type for which either $f_x(x, y)$ or $f_y(x, y)$ does not exist.

Example 2 Finding a Relative Extremum

Determine the relative extrema of $f(x, y) = 1 - (x^2 + y^2)^{1/3}$.

Solution Because

$$f_x(x, y) = -\frac{2x}{3(x^2 + y^2)^{2/3}} \qquad \text{Partial with respect to } x$$

and

$$f_y(x, y) = -\frac{2y}{3(x^2 + y^2)^{2/3}} \qquad \text{Partial with respect to } y$$

it follows that both partial derivatives exist for all points in the xy-plane except for $(0, 0)$. Moreover, because the partial derivatives cannot both be 0 unless both x and y are 0, you can conclude that $(0, 0)$ is the only critical point. In Figure 12.67, note that $f(0, 0)$ is 1. For all other (x, y) it is clear that

$$f(x, y) = 1 - (x^2 + y^2)^{1/3} < 1.$$

Therefore, f has a relative *maximum* at $(0, 0)$.

Surface:
$f(x, y) = 1 - (x^2 + y^2)^{1/3}$

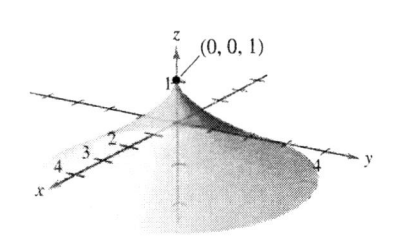

$f_x(x, y)$ and $f_y(x, y)$ are undefined at $(0, 0)$.
Figure 12.67

NOTE In Example 2, $f_x(x, y) = 0$ for every point on the y-axis other than $(0, 0)$. However, because $f_y(x, y)$ is nonzero, these are not critical points. Remember that *one* of the partials must not exist or *both* must be 0 in order to yield a critical point.

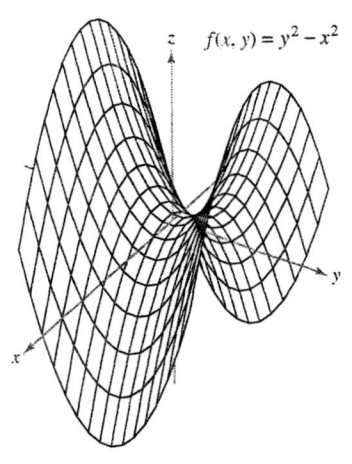

Saddle point at $(0, 0, 0)$:
$f_x(0, 0) = f_y(0, 0) = 0$
Figure 12.68

The Second Partials Test

Theorem 12.16 tells you that to find relative extrema you need only examine values of $f(x, y)$ at critical points. However, as is true for a function of one variable, the critical points of a function of two variables do not always yield relative maxima or minima. Some critical points yield **saddle points,** which are neither relative maxima nor relative minima.

As an example of a critical point that does not yield a relative extremum, consider the surface given by

$$f(x, y) = y^2 - x^2 \qquad \text{Hyperbolic paraboloid}$$

as shown in Figure 12.68. At the point $(0, 0)$, both partial derivatives are 0. The function f does not, however, have a relative extremum at this point because in any open disk centered at $(0, 0)$ the function takes on both negative values (along the x-axis) *and* positive values (along the y-axis). So, the point $(0, 0, 0)$ is a saddle point of the surface. (The name "saddle point" comes from the fact that the surface shown in Figure 12.68 resembles a saddle.)

For the functions in Examples 1 and 2, it was relatively easy to determine the relative extrema, because each function was either given, or able to be written, in completed square form. For more complicated functions, algebraic arguments are less convenient and it is better to rely on the analytical means presented in the following Second Partials Test. This is the two-variable counterpart of the Second Derivative Test for functions of one variable. The proof of this theorem is best left to a course in advanced calculus.

THEOREM 12.17   Second Partials Test

Let f have continuous second partial derivatives on an open region containing a point (a, b) for which

$$f_x(a, b) = 0 \quad \text{and} \quad f_y(a, b) = 0.$$

To test for relative extrema of f, consider the quantity

$$d = f_{xx}(a, b)f_{yy}(a, b) - [f_{xy}(a, b)]^2.$$

1. If $d > 0$ and $f_{xx}(a, b) > 0$, then f has a **relative minimum** at (a, b).
2. If $d > 0$ and $f_{xx}(a, b) < 0$, then f has a **relative maximum** at (a, b).
3. If $d < 0$, then $(a, b, f(a, b))$ is a **saddle point.**
4. The test is inconclusive if $d = 0$.

NOTE   If $d > 0$, then $f_{xx}(a, b)$ and $f_{yy}(a, b)$ must have the same sign. This means that $f_{xx}(a, b)$ can be replaced by $f_{yy}(a, b)$ in the first two parts of the test.

A convenient device for remembering the formula for d in the Second Partials Test is given by the 2×2 determinant

$$d = \begin{vmatrix} f_{xx}(a, b) & f_{xy}(a, b) \\ f_{yx}(a, b) & f_{yy}(a, b) \end{vmatrix}$$

where $f_{xy}(a, b) = f_{yx}(a, b)$ by Theorem 12.3.

Example 3 Using the Second Partials Test

Find the relative extrema of $f(x, y) = -x^3 + 4xy - 2y^2 + 1$.

Solution Begin by finding the critical points of f. Because

$$f_x(x, y) = -3x^2 + 4y \quad \text{and} \quad f_y(x, y) = 4x - 4y$$

exist for all x and y, the only critical points are those for which both first partial derivatives are 0. To locate these points, let $f_x(x, y)$ and $f_y(x, y)$ be 0 to obtain $-3x^2 + 4y = 0$ and $4x - 4y = 0$. From the second equation you know that $x = y$, and, by substitution into the first equation, you obtain two solutions: $y = x = 0$ and $y = x = \frac{4}{3}$. Because

$$f_{xx}(x, y) = -6x, \quad f_{yy}(x, y) = -4, \quad \text{and} \quad f_{xy}(x, y) = 4$$

it follows that, for the critical point $(0, 0)$,

$$d = f_{xx}(0, 0)f_{yy}(0, 0) - [f_{xy}(0, 0)]^2 = 0 - 16 < 0$$

and, by the Second Partials Test, you can conclude that $(0, 0, 1)$ is a saddle point of f. Furthermore, for the critical point $\left(\frac{4}{3}, \frac{4}{3}\right)$,

$$d = f_{xx}\left(\tfrac{4}{3}, \tfrac{4}{3}\right) f_{yy}\left(\tfrac{4}{3}, \tfrac{4}{3}\right) - \left[f_{xy}\left(\tfrac{4}{3}, \tfrac{4}{3}\right)\right]^2$$
$$= -8(-4) - 16$$
$$= 16$$
$$> 0$$

and because $f_{xx}\left(\tfrac{4}{3}, \tfrac{4}{3}\right) = -8 < 0$ you can conclude that f has a relative maximum at $\left(\tfrac{4}{3}, \tfrac{4}{3}\right)$, as shown in Figure 12.69.

$f(x, y) = -x^3 + 4xy - 2y^2 + 1$

Figure 12.69

The Second Partials Test can fail to find relative extrema in two ways. If either of the first partial derivatives does not exist, you cannot use the test. Also, if

$$d = f_{xx}(a, b)f_{yy}(a, b) - [f_{xy}(a, b)]^2 = 0$$

the test fails. In such cases, you can try a sketch or some other approach, as demonstrated in the next example.

Example 4 Failure of the Second Partials Test

Find the relative extrema of $f(x, y) = x^2y^2$.

Solution Because $f_x(x, y) = 2xy^2$ and $f_y(x, y) = 2x^2y$, you know that both partial derivatives are 0 if $x = 0$ or $y = 0$. That is, every point along the x- or y-axis is a critical point. Moreover, because

$$f_{xx}(x, y) = 2y^2, \quad f_{yy}(x, y) = 2x^2, \quad \text{and} \quad f_{xy}(x, y) = 4xy$$

you know that if either $x = 0$ or $y = 0$, then

$$d = f_{xx}(x, y)f_{yy}(x, y) - [f_{xy}(x, y)]^2$$
$$= 4x^2y^2 - 16x^2y^2 = -12x^2y^2 = 0.$$

So, the Second Partials Test fails. However, because $f(x, y) = 0$ for every point along the x- or y-axis and $f(x, y) = x^2y^2 > 0$ for all other points, you can conclude that each of these critical points yields an absolute minimum, as shown in Figure 12.70.

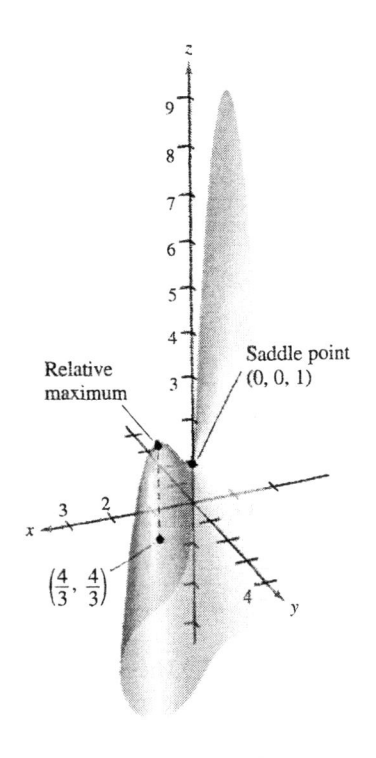

$f(x, y) = x^2y^2$

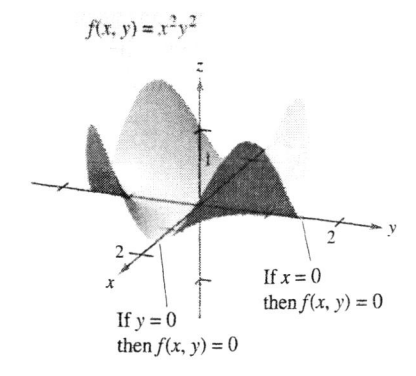

Figure 12.70

Absolute extrema of a function can occur in two ways. First, some relative extrema also happen to be absolute extrema. For instance, in Example 1, $f(-2, 3)$ is an absolute minimum of the function. (On the other hand, the relative maximum found in Example 3 is not an absolute maximum of the function.) Second, absolute extrema can occur at a boundary point of the domain. This is illustrated in Example 5.

Example 5 Finding Absolute Extrema

Find the absolute extrema of the function

$$f(x, y) = \sin xy$$

on the closed region given by $0 \leq x \leq \pi$ and $0 \leq y \leq 1$.

Solution From the partial derivatives

$$f_x(x, y) = y \cos xy \quad \text{and} \quad f_y(x, y) = x \cos xy$$

you can see that each point lying on the hyperbola given by $xy = \pi/2$ is a critical point. These points each yield the value

$$f(x, y) = \sin\left(\frac{\pi}{2}\right) = 1$$

which you know is the absolute maximum, as shown in Figure 12.71. The only other critical point of f *lying in the given region* is $(0, 0)$. It yields an absolute minimum of 0, because

$$0 \leq xy \leq \pi$$

implies that

$$0 \leq \sin xy \leq 1.$$

To hunt for other absolute extrema, you should consider the four boundaries of the region formed by taking traces with the vertical planes $x = 0$, $x = \pi$, $y = 0$, and $y = 1$. In doing this, you will find that $\sin xy = 0$ at all points on the x-axis, at all points on the y-axis, and at the point $(\pi, 1)$. Each of these points yields an absolute minimum for the surface, as shown in Figure 12.71.

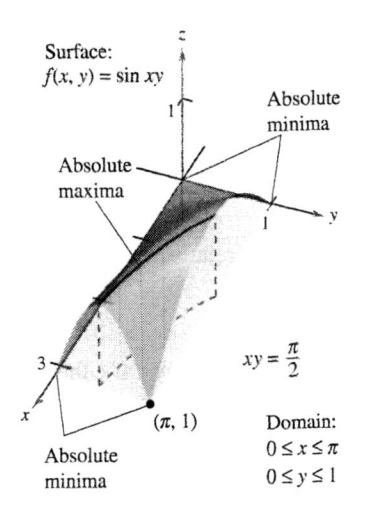

Figure 12.71

The concepts of relative extrema and critical points can be extended to functions of three or more variables. If all first partial derivatives of

$$w = f(x_1, x_2, x_3, \ldots, x_n)$$

exist, it can be shown that a relative maximum or minimum can occur at $(x_1, x_2, x_3, \ldots, x_n)$ only if every first partial derivative is 0 at that point. This means that the critical points are obtained by solving the following system of equations.

$$f_{x_1}(x_1, x_2, x_3, \ldots, x_n) = 0$$
$$f_{x_2}(x_1, x_2, x_3, \ldots, x_n) = 0$$
$$\vdots$$
$$f_{x_n}(x_1, x_2, x_3, \ldots, x_n) = 0$$

The extension of Theorem 12.17 to three or more variables is also possible, although we will not consider such an extension in this text.

EXERCISES FOR SECTION 12.8

In Exercises 1–6, identify any extrema of the function by recognizing its given form or its form after completing the square. Verify your results by using the partial derivatives to locate any critical points and test for relative extrema. Use a computer algebra system to graph the function and label any extrema.

1. $g(x, y) = (x - 1)^2 + (y - 3)^2$
2. $g(x, y) = 9 - (x - 3)^2 - (y + 2)^2$
3. $f(x, y) = \sqrt{x^2 + y^2 + 1}$
4. $f(x, y) = \sqrt{25 - (x - 2)^2 - y^2}$
5. $f(x, y) = x^2 + y^2 + 2x - 6y + 6$
6. $f(x, y) = -x^2 - y^2 + 4x + 8y - 11$

In Exercises 7–16, examine each function for relative extrema.

7. $f(x, y) = 2x^2 + 2xy + y^2 + 2x - 3$
8. $f(x, y) = -x^2 - 5y^2 + 10x - 30y - 62$
9. $f(x, y) = -5x^2 + 4xy - y^2 + 16x + 10$
10. $f(x, y) = x^2 + 6xy + 10y^2 - 4y + 4$
11. $z = 2x^2 + 3y^2 - 4x - 12y + 13$
12. $z = -3x^2 - 2y^2 + 3x - 4y + 5$
13. $f(x, y) = 2\sqrt{x^2 + y^2} + 3$
14. $h(x, y) = (x^2 + y^2)^{1/3} + 2$
15. $g(x, y) = 4 - |x| - |y|$
16. $f(x, y) = |x + y| - 2$

In Exercises 17–20, use a computer algebra system to graph the surface and locate any relative extrema and saddle points.

17. $z = \dfrac{-4x}{x^2 + y^2 + 1}$
18. $f(x, y) = y^3 - 3yx^2 - 3y^2 - 3x^2 + 1$
19. $z = (x^2 + 4y^2)e^{1 - x^2 - y^2}$
20. $z = e^{xy}$

In Exercises 21–28, examine each function for relative extrema and saddle points.

21. $h(x, y) = x^2 - y^2 - 2x - 4y - 4$
22. $g(x, y) = 120x + 120y - xy - x^2 - y^2$
23. $h(x, y) = x^2 - 3xy - y^2$
24. $g(x, y) = xy$
25. $f(x, y) = x^3 - 3xy + y^3$ 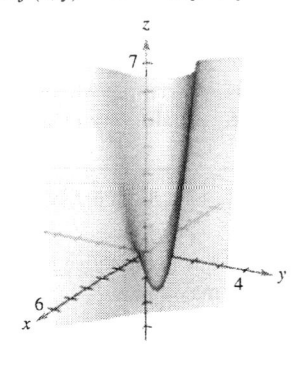
26. $f(x, y) = 2xy - \frac{1}{2}(x^4 + y^4) + 1$ 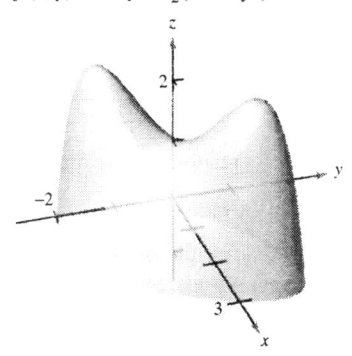
27. $z = e^{-x} \sin y$ 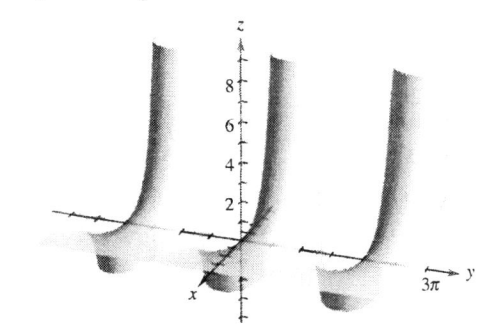
28. $z = \left(\dfrac{1}{2} - x^2 + y^2\right)e^{1 - x^2 - y^2}$

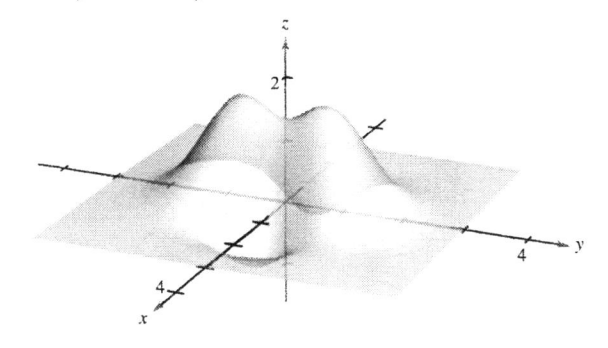

In Exercises 29 and 30, examine the function for extrema without using the derivative tests and use a computer algebra system to graph the surface. (*Hint:* By observation determine if it is possible for z to be negative. When is z equal to 0?)

29. $z = \dfrac{(x - y)^4}{x^2 + y^2}$
30. $z = \dfrac{(x^2 - y^2)^2}{x^2 + y^2}$

Think About It In Exercises 31–34, determine whether there is a relative maximum, a relative minimum, a saddle point, or insufficient information to determine the nature of the function $f(x, y)$ at the critical point (x_0, y_0).

31. $f_{xx}(x_0, y_0) = 9$, $f_{yy}(x_0, y_0) = 4$, $f_{xy}(x_0, y_0) = 6$
32. $f_{xx}(x_0, y_0) = -3$, $f_{yy}(x_0, y_0) = -8$, $f_{xy}(x_0, y_0) = 2$
33. $f_{xx}(x_0, y_0) = -9$, $f_{yy}(x_0, y_0) = 6$, $f_{xy}(x_0, y_0) = 10$
34. $f_{xx}(x_0, y_0) = 25$, $f_{yy}(x_0, y_0) = 8$, $f_{xy}(x_0, y_0) = 10$

Getting at the Concept

35. Define each of the following for a function of two variables.
  (a) Relative minimum
  (b) Relative maximum
  (c) Saddle point
  (d) Critical point

36. State the Second Partials Test for relative extrema and saddle points.

In Exercises 37–40, sketch the graph of an arbitrary function f satisfying the given conditions. State whether the function has any extrema or saddle points. (There are many correct answers.)

37. $f_x(x, y) > 0$ and $f_y(x, y) < 0$ for all (x, y).

38. All of the first and second partial derivatives of f are 0.

39. $f_x(0, 0) = 0,\ f_y(0, 0) = 0$

$f_x(x, y) \begin{cases} < 0, & x < 0 \\ > 0, & x > 0 \end{cases},\quad f_y(x, y) \begin{cases} > 0, & y < 0 \\ < 0, & y > 0 \end{cases}$

$f_{xx}(x, y) > 0, f_{yy}(x, y) < 0$, and $f_{xy}(x, y) = 0$ for all (x, y).

40. $f_x(2, 1) = 0,\quad f_y(2, 1) = 0$

$f_x(x, y) \begin{cases} > 0, & x < 2 \\ < 0, & x > 2 \end{cases},\quad f_y(x, y) \begin{cases} > 0, & y < 1 \\ < 0, & y > 1 \end{cases}$

$f_{xx}(x, y) < 0, f_{yy}(x, y) < 0$, and $f_{xy}(x, y) = 0$ for all (x, y).

41. The figure shows the level curves for an unknown function $f(x, y)$. What, if any, information can be given about f at the point A?

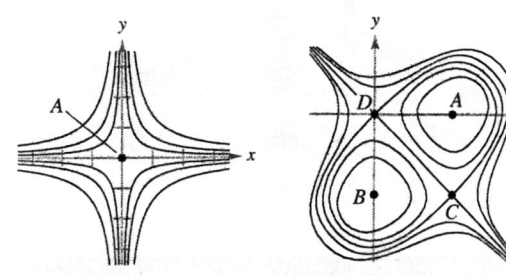

Figure for 41   **Figure for 42**

42. The figure shows the level curves for an unknown function $f(x, y)$. What, if any, information can be given about f at the points A, B, C, and D?

43. A function f has continuous second partial derivatives on an open region containing the critical point $(3, 7)$. The function has a minimum at $(3, 7)$. Determine the interval for $f_{xy}(3, 7)$ if $f_{xx}(3, 7) = 2$ and $f_{yy}(3, 7) = 8$.

44. A function f has continuous second partial derivatives on an open region containing the critical point (a, b). If $f_x(a, b)$ and $f_y(a, b)$ have opposite signs, what is implied? Explain.

In Exercises 45–50, find the critical points and test for relative extrema. List the critical points for which the Second Partials Test fails.

45. $f(x, y) = x^3 + y^3$

46. $f(x, y) = x^3 + y^3 - 6x^2 + 9y^2 + 12x + 27y + 19$

47. $f(x, y) = (x - 1)^2(y + 4)^2$

48. $f(x, y) = \sqrt{(x - 1)^2 + (y + 2)^2}$

49. $f(x, y) = x^{2/3} + y^{2/3}$ **50.** $f(x, y) = (x^2 + y^2)^{2/3}$

In Exercises 51 and 52, find the critical points of the function and, from the form of the function, determine whether each point is a relative maximum or a relative minimum.

51. $f(x, y, z) = x^2 + (y - 3)^2 + (z + 1)^2$

52. $f(x, y, z) = 4 - [x(y - 1)(z + 2)]^2$

In Exercises 53–62, find the absolute extrema of the function over the region R. (In each case, R contains the boundaries.) Use a computer algebra system to confirm your results.

53. $f(x, y) = 12 - 3x - 2y$

  R: The triangular region in the xy-plane with vertices $(2, 0)$, $(0, 1)$, and $(1, 2)$.

54. $f(x, y) = (2x - y)^2$

  R: The triangular region in the xy-plane with vertices $(2, 0)$, $(0, 1)$, and $(1, 2)$.

55. $f(x, y) = 3x^2 + 2y^2 - 4y$

  R: The region in the xy-plane bounded by the graphs of $y = x^2$ and $y = 4$.

56. $f(x, y) = 2x - 2xy + y^2$

  R: The region in the xy-plane bounded by the graphs of $y = x^2$ and $y = 1$.

57. $f(x, y) = x^2 + xy,\quad R = \{(x, y): |x| \leq 2, |y| \leq 1\}$

58. $f(x, y) = x^2 + 2xy + y^2,\quad R = \{(x, y): |x| \leq 2, |y| \leq 1\}$

59. $f(x, y) = x^2 + 2xy + y^2,\quad R = \{(x, y): x^2 + y^2 \leq 8\}$

60. $f(x, y) = x^2 - 4xy + 5$

  $R = \{(x, y): 0 \leq x \leq 4,\ 0 \leq y \leq \sqrt{x}\}$

61. $f(x, y) = \dfrac{4xy}{(x^2 + 1)(y^2 + 1)}$

  $R = \{(x, y): 0 \leq x \leq 1,\ 0 \leq y \leq 1\}$

62. $f(x, y) = \dfrac{4xy}{(x^2 + 1)(y^2 + 1)}$

  $R = \{(x, y): x \geq 0, y \geq 0, x^2 + y^2 \leq 1\}$

True or False? **In Exercises 63 and 64, determine whether the statement is true or false. If it is false, explain why or give an example that shows it is false.**

63. If f has a relative maximum at (x_0, y_0, z_0), then $f_x(x_0, y_0) = f_y(x_0, y_0) = 0$.

64. If f is continuous for all x and y and has two relative minima, then f must have at least one relative maximum.

Section 12.9 Applications of Extrema of Functions of Two Variables

- Solve optimization problems involving functions of several variables.
- Use the method of least squares.

Applied Optimization Problems

In this section, we survey a few of the many applications of extrema of functions of two (or more) variables.

Example 1 Finding Maximum Volume

A rectangular box is resting on the xy-plane with one vertex at the origin. The opposite vertex lies in the plane

$$6x + 4y + 3z = 24$$

as shown in Figure 12.72. Find the maximum volume of such a box.

Solution Let x, y, and z represent the length, width, and height of the box. Because one vertex of the box lies in the plane $6x + 4y + 3z = 24$, you know that $z = \frac{1}{3}(24 - 6x - 4y)$, and you can write the volume xyz of the box as a function of two variables.

$$V(x, y) = (x)(y)\left[\tfrac{1}{3}(24 - 6x - 4y)\right]$$
$$= \tfrac{1}{3}(24xy - 6x^2y - 4xy^2)$$

By setting the first partial derivatives equal to 0

$$V_x(x, y) = \tfrac{1}{3}(24y - 12xy - 4y^2) = \frac{y}{3}(24 - 12x - 4y) = 0$$

$$V_y(x, y) = \tfrac{1}{3}(24x - 6x^2 - 8xy) = \frac{x}{3}(24 - 6x - 8y) = 0$$

you obtain the critical points $(0, 0)$ and $\left(\tfrac{4}{3}, 2\right)$. At $(0, 0)$ the volume is 0, so that point does not yield a maximum volume. At the point $\left(\tfrac{4}{3}, 2\right)$, you can apply the Second Partials Test.

$$V_{xx}(x, y) = -4y$$
$$V_{yy}(x, y) = \frac{-8x}{3}$$
$$V_{xy}(x, y) = \tfrac{1}{3}(24 - 12x - 8y)$$

Because

$$V_{xx}\left(\tfrac{4}{3}, 2\right)V_{yy}\left(\tfrac{4}{3}, 2\right) - \left[V_{xy}\left(\tfrac{4}{3}, 2\right)\right]^2 = (-8)\left(-\tfrac{32}{9}\right) - \left(-\tfrac{8}{3}\right)^2 = \tfrac{64}{3} > 0$$

and

$$V_{xx}\left(\tfrac{4}{3}, 2\right) = -8 < 0$$

you can conclude from the Second Partials Test that the maximum volume is

$$V\left(\tfrac{4}{3}, 2\right) = \tfrac{1}{3}\left[24\left(\tfrac{4}{3}\right)(2) - 6\left(\tfrac{4}{3}\right)^2(2) - 4\left(\tfrac{4}{3}\right)(2^2)\right]$$
$$= \tfrac{64}{9} \text{ cubic units.}$$

(Note that the volume is 0 at the boundary points of the triangular domain of V.)

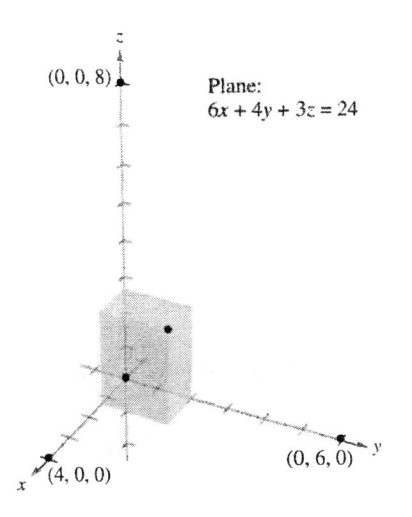

The maximum volume of the box is $\tfrac{64}{9}$ cubic units.
Figure 12.72

NOTE In many applied problems, the domain of the function to be optimized is a closed bounded region. To find minimum or maximum points, you must not only test critical points, but also consider the values of the function at points on the boundary.

Applications of extrema in economics and business often involve more than one independent variable. For instance, a company may produce several models of one type of product. The price per unit and profit per unit are usually different for each model. Moreover, the demand for each model is often a function of the prices of the other models (as well as its own price). The next example illustrates an application involving two products.

Example 2 Finding the Maximum Profit

The profit obtained by producing x units of product A and y units of product B is approximated by the model

$$P(x, y) = 8x + 10y - (0.001)(x^2 + xy + y^2) - 10{,}000.$$

Find the production level that produces a maximum profit. What is the maximum profit?

Solution The partial derivatives of the profit function are

$$P_x(x, y) = 8 - (0.001)(2x + y)$$

and

$$P_y(x, y) = 10 - (0.001)(x + 2y).$$

By setting these partial derivatives equal to 0, you obtain the following system of equations.

$$8 - (0.001)(2x + y) = 0$$
$$10 - (0.001)(x + 2y) = 0$$

After simplifying, this system of linear equations can be written as

$$2x + y = 8{,}000$$
$$x + 2y = 10{,}000.$$

Solving this system produces $x = 2000$ and $y = 4000$. The second partial derivatives of P are

$$P_{xx}(2000, 4000) = -0.002$$
$$P_{yy}(2000, 4000) = -0.002$$
$$P_{xy}(2000, 4000) = -0.001.$$

Because $P_{xx} < 0$ and

$$P_{xx}(2000, 4000)P_{yy}(2000, 4000) - [P_{xy}(2000, 4000)]^2 =$$
$$(-0.002)^2 - (-0.001)^2 > 0$$

you can conclude that the production level of $x = 2000$ units and $y = 4000$ units yields a *maximum* profit. The maximum profit is

$$P(2000, 4000) = 8(2000) + 10(4000) -$$
$$(0.001)[2000^2 + 2000(4000) - 4000^2)] - 10{,}000$$
$$= 50{,}000.$$

FOR FURTHER INFORMATION
For more information on the use of mathematics in economics, see the article "Mathematical Methods of Economics" by Joel Franklin in *The American Mathematical Monthly*. To view this article, go to the website *www.matharticles.com*.

NOTE In Example 2, we assumed that the manufacturing plant is able to produce the required number of units to yield a maximum profit. In actual practice, the production would be bounded by physical constraints. You will study such constrained optimization problems in the next section.

The Method of Least Squares

Many of the examples in this text have involved **mathematical models.** For instance, Example 2 involves a quadratic model for profit. There are several ways to develop such models; one is called the **method of least squares.**

In constructing a model to represent a particular phenomenon, the goals are simplicity and accuracy. Of course, these goals often conflict. For instance, a simple linear model for the points in Figure 12.73 is

$$y = 1.8566x - 5.0246.$$

However, Figure 12.74 shows that by choosing the slightly more complicated quadratic model*

$$y = 0.1996x^2 - 0.7281x + 1.3749$$

you can achieve greater accuracy.

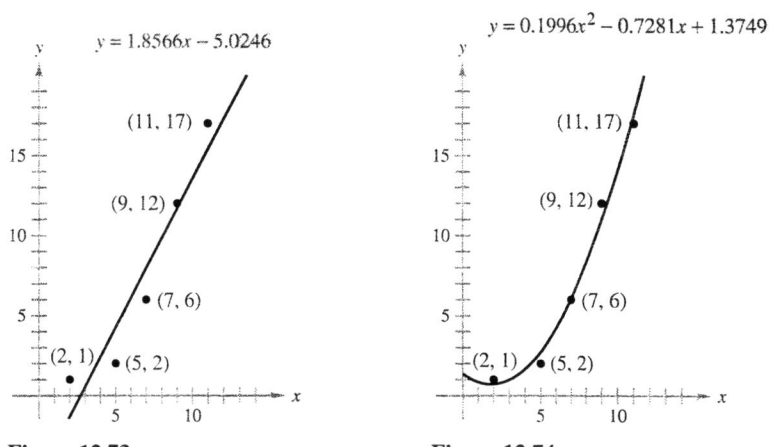

Figure 12.73

Figure 12.74

As a measure of how well the model $y = f(x)$ fits the collection of points

$$\{(x_1, y_1), (x_2, y_2), (x_3, y_3), \ldots, (x_n, y_n)\}$$

you can add the squares of the differences between the actual y-values and the values given by the model to obtain the **sum of the squared errors**

$$S = \sum_{i=1}^{n} [f(x_i) - y_i]^2. \qquad \text{Sum of the squared errors}$$

Graphically, S can be interpreted as the sum of the squares of the vertical distances between the graph of f and the given points in the plane, as shown in Figure 12.75. If the model is perfect, then $S = 0$. However, when perfection is not feasible, we settle for a model that minimizes S. Statisticians call the *linear model* that minimizes S the **least squares regression line.** The proof that this line actually minimizes S involves the minimizing of a function of two variables.

Sum of the squared errors:
$S = d_1^2 + d_2^2 + d_3^2$
Figure 12.75

---

*A method for finding the least squares quadratic model for a collection of data is described in Exercise 39.

ADRIEN-MARIE LEGENDRE (1752–1833)

The method of least squares was introduced by the French mathematician Adrien-Marie Legendre. Legendre is best known for his work in geometry. In fact, his text *Elements of Geometry* was so popular in the United States that it continued to be used for 33 editions, spanning a period of more than 100 years.

THEOREM 12.18 Least Squares Regression Line

The **least squares regression line** for $\{(x_1, y_1), (x_2, y_2), \ldots, (x_n, y_n)\}$ is given by $f(x) = ax + b$, where

$$a = \frac{n\sum_{i=1}^{n} x_i y_i - \sum_{i=1}^{n} x_i \sum_{i=1}^{n} y_i}{n\sum_{i=1}^{n} x_i^2 - \left(\sum_{i=1}^{n} x_i\right)^2} \quad \text{and} \quad b = \frac{1}{n}\left(\sum_{i=1}^{n} y_i - a\sum_{i=1}^{n} x_i\right).$$

Proof Let $S(a, b)$ represent the sum of the squared errors for the model $f(x) = ax + b$ and the given set of points. That is,

$$S(a, b) = \sum_{i=1}^{n} [f(x_i) - y_i]^2$$

$$= \sum_{i=1}^{n} (ax_i + b - y_i)^2$$

where the points (x_i, y_i) represent constants. Because S is a function of a and b, you can use the methods discussed in the preceding section to find the minimum value of S. Specifically, the first partial derivatives of S are

$$S_a(a, b) = \sum_{i=1}^{n} 2x_i(ax_i + b - y_i)$$

$$= 2a\sum_{i=1}^{n} x_i^2 + 2b\sum_{i=1}^{n} x_i - 2\sum_{i=1}^{n} x_i y_i$$

$$S_b(a, b) = \sum_{i=1}^{n} 2(ax_i + b - y_i)$$

$$= 2a\sum_{i=1}^{n} x_i + 2nb - 2\sum_{i=1}^{n} y_i.$$

By setting these two partial derivatives equal to 0, you obtain the values for a and b that are listed in the theorem. We leave it to you to apply the Second Partials Test (see Exercise 40) to verify that these values of a and b yield a minimum.

If the x-values are symmetrically spaced about the y-axis, then $\sum x_i = 0$ and the formulas for a and b simplify to

$$a = \frac{\sum_{i=1}^{n} x_i y_i}{\sum_{i=1}^{n} x_i^2}$$

and

$$b = \frac{1}{n}\sum_{i=1}^{n} y_i.$$

This simplification is often possible with a translation of the x-values. For instance, if the x-values in a data collection consist of the years 2000, 2001, 2002, 2003, and 2004, you could let 2002 be represented by 0.

Example 3 Finding the Least Squares Regression Line

Find the least squares regression line for the points $(-3, 0)$, $(-1, 1)$, $(0, 2)$, and $(2, 3)$.

Solution The table shows the calculations involved in finding the least squares regression line using $n = 4$.

TECHNOLOGY Many calculators have "built-in" least squares regression programs. If your calculator has such a program, try using it to duplicate the results of Example 3.

x	y	xy	x^2
-3	0	0	9
-1	1	-1	1
0	2	0	0
2	3	6	4
$\sum_{i=1}^{n} x_i = -2$	$\sum_{i=1}^{n} y_i = 6$	$\sum_{i=1}^{n} x_i y_i = 5$	$\sum_{i=1}^{n} x_i^2 = 14$

Applying Theorem 12.18 produces

$$a = \frac{n\sum_{i=1}^{n} x_i y_i - \sum_{i=1}^{n} x_i \sum_{i=1}^{n} y_i}{n\sum_{i=1}^{n} x_i^2 - \left(\sum_{i=1}^{n} x_i\right)^2} = \frac{4(5) - (-2)(6)}{4(14) - (-2)^2} = \frac{8}{13}$$

and

$$b = \frac{1}{n}\left(\sum_{i=1}^{n} y_i - a\sum_{i=1}^{n} x_i\right) = \frac{1}{4}\left[6 - \frac{8}{13}(-2)\right] = \frac{47}{26}.$$

The least squares regression line is $y = \frac{8}{13}x + \frac{47}{26}$, as shown in Figure 12.76.

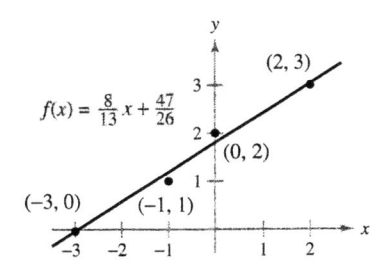

$f(x) = \frac{8}{13}x + \frac{47}{26}$

Least squares regression line
Figure 12.76

EXERCISES FOR SECTION 12.9

In Exercises 1 and 2, find the minimum distance from the point to the plane $2x + 3y + z = 12$. (*Hint:* To simplify the computations, minimize the square of the distance.)

1. $(0, 0, 0)$ **2.** $(1, 2, 3)$

In Exercises 3 and 4, find the minimum distance from the point to the paraboloid $z = x^2 + y^2$.

3. $(5, 5, 0)$ **4.** $(5, 0, 0)$

In Exercises 5–8, find three positive numbers x, y, and z that satisfy the indicated conditions.

5. The sum is 30 and the product is a maximum.

6. The sum is 32 and $P = xy^2z$ is a maximum.

7. The sum is 30 and the sum of the squares is a minimum.

8. The sum is 1 and the sum of the squares is a minimum.

9. *Volume* The sum of the length and the girth (perimeter of a cross section) of packages carried by a delivery service cannot exceed 108 inches. Find the dimensions of the rectangular package of largest volume that may be sent.

10. *Volume* The material for constructing the base of an open box costs 1.5 times as much per unit area as the material for constructing the sides. For a fixed amount of money C, find the dimensions of the box of largest volume that can be made.

11. *Volume* The volume of an ellipsoid

$$\frac{x^2}{a^2} + \frac{y^2}{b^2} + \frac{z^2}{c^2} = 1$$

is $4\pi abc/3$. For a fixed sum $a + b + c$, show that the ellipsoid of maximum volume is a sphere.

12. *Volume* Show that the rectangular box of maximum volume inscribed in a sphere of radius r is a cube.

13. **Volume and Surface Area** Show that a rectangular box of given volume and minimum surface area is a cube.

14. **Volume** Repeat Exercise 9 under the condition that the sum of the perimeters of the two cross sections shown in the figure cannot exceed 144 inches.

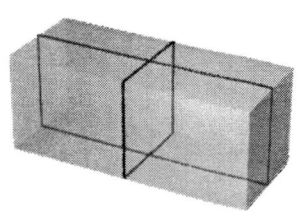

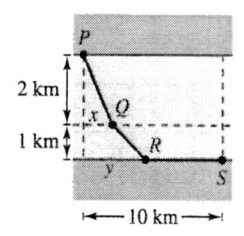

Figure for 14

Figure for 15

15. **Minimum Cost** A water line is to be built from point P to point S and must pass through regions where construction costs differ (see figure). Find x and y such that the total cost C will be minimized if the cost per kilometer in dollars is $3k$ from P to Q, $2k$ from Q to R, and k from R to S.

16. **Area** A trough with trapezoidal cross sections is formed by turning up the edges of a 30-inch-wide sheet of aluminum (see figure). Find the cross section of maximum area.

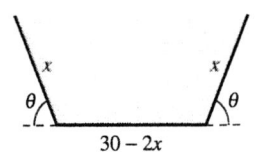

17. **Area** Repeat Exercise 16 for a sheet that is w inches wide.

18. **Hardy-Weinberg Law** Common blood types are determined genetically by three alleles A, B, and O. (An allele is any of a group of possible mutational forms of a gene.) A person whose blood type is AA, BB, or OO is homozygous. A person whose blood type is AB, AO, or BO is heterozygous. The Hardy-Weinberg Law states that the proportion P of heterozygous individuals in any given population is

$$P(p, q, r) = 2pq + 2pr + 2qr$$

where p represents the percent of allele A in the population, q represents the percent of allele B in the population, and r represents the percent of allele O in the population. Use the fact that $p + q + r = 1$ to show that the maximum proportion of heterozygous individuals in any population is $\frac{2}{3}$.

19. **Revenue** A company manufactures two products. The total revenue from x_1 units of product 1 and x_2 units of product 2 is $R = -5x_1^2 - 8x_2^2 - 2x_1x_2 + 42x_1 + 102x_2$. Find x_1 and x_2 so as to maximize the revenue.

20. **Revenue** A retail outlet sells two competitive products, the prices of which are p_1 and p_2. Find p_1 and p_2 so as to maximize total revenue, where $R = 515p_1 + 805p_2 + 1.5p_1p_2 - 1.5p_1^2 - p_2^2$.

21. **Profit** A corporation manufactures a product at two locations. The cost of producing x_1 units at location 1 is

$$C_1 = 0.02x_1^2 + 4x_1 + 500$$

and the cost of producing x_2 units at location 2 is

$$C_2 = 0.05x_2^2 + 4x_2 + 275.$$

If the product sells for $15 per unit, find the quantity that should be produced at each location to maximize the profit $P = 15(x_1 + x_2) - C_1 - C_2$.

22. **Distance** A company has retail outlets located at the points $(0, 0)$, $(2, 2)$, and $(-2, 2)$ (see figure). Management plans to build a distribution center located such that the sum of the distances S from the center to the outlets is minimum. From the symmetry of the problem it is clear that the distribution center will be located on the y-axis, and therefore S is a function of the single variable y. Using techniques presented in Chapter 3, find the required value of y.

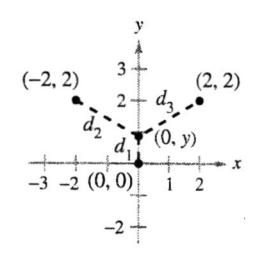

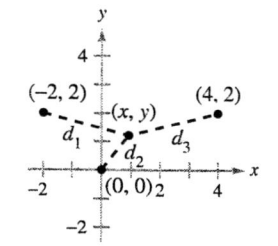

Figure for 22

Figure for 23

23. **Investigation** The retail outlets described in Exercise 22 are located at $(0, 0)$, $(4, 2)$, and $(-2, 2)$ (see figure). The location of the distribution center is (x, y), and therefore the sum of the distances S is a function of x and y.

(a) Write the expression giving the sum of the distances S. Use a computer algebra system to graph S. Does the surface have a minimum?

(b) Use a computer algebra system to obtain S_x and S_y. Observe that solving the system $S_x = 0$ and $S_y = 0$ is very difficult. Therefore, you will approximate the location of the distribution center.

(c) An initial estimate of the critical point is $(x_1, y_1) = (1, 1)$. Calculate $-\nabla S(1, 1)$ with components $-S_x(1, 1)$ and $-S_y(1, 1)$. What direction is given by the vector $-\nabla S(1, 1)$?

(d) The second estimate of the critical point is

$$(x_2, y_2) = (x_1 - S_x(x_1, y_1)t, y_1 - S_y(x_1, y_1)t).$$

If these coordinates are substituted into $S(x, y)$, then S becomes a function of the single variable t. Find the value of t that minimizes S. Use this value of t to estimate (x_2, y_2).

(e) Complete two more iterations of the process in part (d) to obtain (x_4, y_4). For this location of the distribution center, what is the sum of the distances to the retail outlets?

(f) Explain why $-\nabla S(x, y)$ was used to approximate the minimum value of S. In what types of problems would you use $\nabla S(x, y)$?

24. **Investigation** Repeat Exercise 23 for retail outlets located at the points $(-4, 0)$, $(1, 6)$, and $(12, 2)$.

Getting at the Concept

25. In your own words, state the problem-solving strategy for applied minimum and maximum problems.

26. In you own words, describe the method of least squares for finding mathematical models.

In Exercises 27–30, (a) find the least squares regression line and (b) calculate S, the sum of the squared errors. Use the regression capabilities of a graphing utility to verify your results.

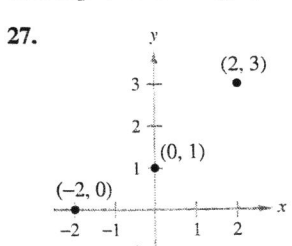

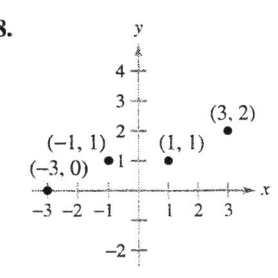

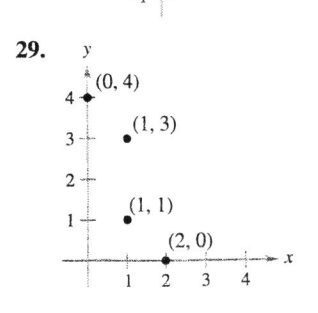

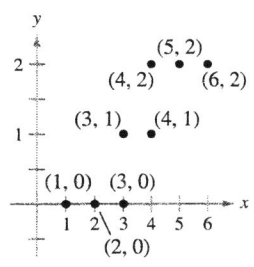

27. Points: $(-2, 0)$, $(0, 1)$, $(2, 3)$

28. Points: $(-3, 0)$, $(-1, 1)$, $(1, 1)$, $(3, 2)$

29. Points: $(0, 4)$, $(1, 3)$, $(1, 1)$, $(2, 0)$

30. Points: $(1, 0)$, $(3, 0)$, $(2, 0)$, $(3, 1)$, $(4, 1)$, $(4, 2)$, $(5, 2)$, $(6, 2)$

In Exercises 31–34, find the least squares regression line for the points. Use the regression capabilities of a graphing utility to verify your results. Use the graphing utility to plot the points and graph the regression line.

31. $(0, 0)$, $(1, 1)$, $(3, 4)$, $(4, 2)$, $(5, 5)$

32. $(1, 0)$, $(3, 3)$, $(5, 6)$

33. $(0, 6)$, $(4, 3)$, $(5, 0)$, $(8, -4)$, $(10, -5)$

34. $(6, 4)$, $(1, 2)$, $(3, 3)$, $(8, 6)$, $(11, 8)$, $(13, 8)$

35. *Modeling Data* The ages x (in years) of seven men and their systolic blood pressures y are shown in the table.

x	16	25	39	45	49	64	70
y	109	122	143	132	199	185	199

(a) Use the regression capabilities of a graphing utility to find the least squares regression line for the data.

(b) Use a graphing utility to plot the data and graph the model.

(c) Use the model to approximate the change in systolic blood pressure for each one-year increase in age.

36. *Modeling Data* A store manager wants to know the demand for a certain product as a function of price. The daily sales for three different prices of the product are shown in the table.

Price (x)	$1.00	$1.25	$1.50
Demand (y)	450	375	330

(a) Use the regression capabilities of a graphing utility to find the least squares regression line for the data.

(b) Estimate the demand when the price is $1.40.

37. *Modeling Data* An agronomist used four test plots to determine the relationship between the wheat yield (in bushels per acre) and the amount of fertilizer (in hundreds of pounds per acre). The results are shown in the table.

Fertilizer (x)	1.0	1.5	2.0	2.5
Yield (y)	32	41	48	53

Use the regression capabilities of a graphing utility to find the least squares regression line for the data, and estimate the yield for a fertilizer application of 160 pounds per acre.

38. *Modeling Data* The table shows the percent and number (in millions) of women in the work force for selected years. *(Source: Department of Labor)*

Year	1960	1970	1980	1990
Percent (x)	37.7	43.3	51.5	57.5
Number (y)	23.2	31.5	45.5	56.8

Year	1995	1996	1997	1998
Percent (x)	58.9	59.3	59.8	59.8
Number (y)	60.9	61.9	63.0	63.7

(a) Use the regression capabilities of a graphing utility to find the least squares regression line for the data.

(b) According to this model, approximately how many women enter the labor force for each one-point increase in the percent of women in the labor force?

39. Find a system of equations whose solution yields the coefficients a, b, and c for the least squares regression quadratic $y = ax^2 + bx + c$ for the points

$$(x_1, y_1), (x_2, y_2), \ldots, (x_n, y_n)$$

by minimizing the sum

$$S(a, b, c) = \sum_{i=1}^{n} (y_i - ax_i^2 - bx_i - c)^2.$$

40. Use the Second Partials Test to verify that the formulas for a and b given in Theorem 12.18 yield a minimum.

$$\left[\textit{Hint: } \text{Use the fact that } n\sum_{i=1}^{n} x_i^2 \geq \left(\sum_{i=1}^{n} x_i \right)^2. \right]$$

In Exercises 41–44, use the result of Exercise 39 to find the least squares regression quadratic for the given points. Use the regression capabilities of a graphing utility to confirm your results. Use the graphing utility to plot the points and graph the least squares regression quadratic.

41. $(-2, 0), (-1, 0), (0, 1), (1, 2), (2, 5)$
42. $(-4, 5), (-2, 6), (2, 6), (4, 2)$
43. $(0, 0), (2, 2), (3, 6), (4, 12)$
44. $(0, 10), (1, 9), (2, 6), (3, 0)$

45. *Modeling Data* After a new turbocharger for an automobile engine was developed, the following experimental data were obtained for speed in miles per hour at 2-second intervals.

Time (x)	0	2	4	6	8	10
Speed (y)	0	15	30	50	65	70

(a) Find a least squares regression quadratic for the data. Use a graphing utility to confirm your results.

(b) Use a graphing utility to plot the points and graph the model.

46. *Modeling Data* The table shows the world population (in billions) for five different years. *(Source: U.S. Bureau of the Census)*

Year (x)	1960	1970	1980	1990	2000
Population (y)	3.0	3.7	4.5	5.3	6.1

Let $x = 0$ represent the year 1960.

(a) Use the regression capabilities of a graphing utility to find the least squares regression line for the data.

(b) Use the regression capabilities of a graphing utility to find the least squares regression quadratic for the data.

(c) Use a graphing utility to plot the data and graph the models.

(d) Use both models to forecast the world population for the year 2010. How do the two models differ as you extrapolate into the future?

47. *Modeling Data* A meteorologist measures the atmospheric pressure P (in kilograms per square meter) at altitude h (in kilometers). The data are shown below.

h	0	5	10	15	20
P	10,332	5583	2376	1240	517

(a) Use the regression capabilities of a graphing utility to find a least squares regression line for the points $(h, \ln P)$.

(b) The result in part (a) is an equation of the form $\ln P = ah + b$. Write this logarithmic form in exponential form.

(c) Use a graphing utility to plot the original data and graph the exponential model in part (b).

(d) If your graphing utility can fit logarithmic models to data, use it to verify the result in part (b).

48. *Modeling Data* The endpoints of the interval over which distinct vision is possible are called the near point and far point of the eye. With increasing age, these points normally change. The table shows the approximate near point y in centimeters for various ages x.

x	10	20	30	40	50
y	7	10	14	22	40

(a) Find a rational model for the data by taking the reciprocal of the near points to generate the points $(x, 1/y)$. Use the regression capabilities of a graphing utility to find a least squares regression line for the revised data. The resulting line has the form

$$\frac{1}{y} = ax + b.$$

Solve for y.

(b) Use a graphing utility to plot the data and graph the model.

(c) Do you think the model can be used to predict the near point for a person who is 60 years old? Explain.

SECTION PROJECT BUILDING A PIPELINE

An oil company wishes to construct a pipeline from its offshore facility A to its refinery B. The offshore facility is 2 miles from shore, and the refinery is 1 mile inland. Furthermore, A and B are 5 miles apart, as indicated in the figure.

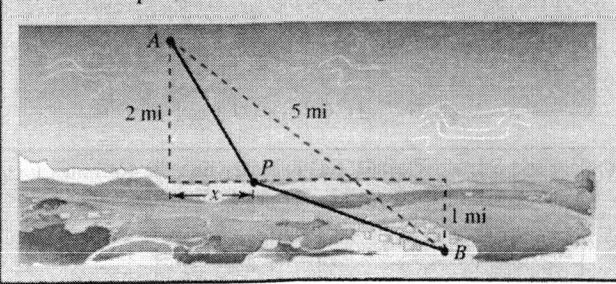

The cost of building the pipeline is $3 million per mile in the water, and $4 million per mile on land. Hence, the cost of the pipeline depends on the location of point P, where it meets the shore. What would be the most economical route of the pipeline?

Imagine that you are to write a report to the oil company about this problem. Let x be the distance indicated in the figure. Determine the cost of building the pipeline from A to P, and the cost from P to B. Analyze some sample pipeline routes and their corresponding costs. For instance, what is the cost of the most direct route? Then use calculus to determine the route of the pipeline that minimizes the cost. Explain all steps of your development and include any relevant graphs.

Section 12.10

Lagrange Multipliers

- Understand the Method of Lagrange Multipliers.
- Use Lagrange Multipliers to solve constrained optimization problems.
- Use the Method of Lagrange Multipliers with two constraints.

Lagrange Multipliers

Many optimization problems have restrictions, or **constraints**, on the values that can be used to produce the optimal solution. Such constraints tend to complicate optimization problems because the optimal solution can occur at a boundary point of the domain. In this section, you will study an ingenious technique for solving such problems. It is called the **Method of Lagrange Multipliers.**

To see how this technique works, suppose you want to find the rectangle of maximum area that can be inscribed in the ellipse given by

$$\frac{x^2}{3^2} + \frac{y^2}{4^2} = 1.$$

Let (x, y) be the vertex of the rectangle in the first quadrant, as shown in Figure 12.77. Because the rectangle has sides of lengths $2x$ and $2y$, its area is given by

$$f(x, y) = 4xy. \quad \text{Objective function}$$

You want to find x and y such that $f(x, y)$ is a maximum. Your choice of (x, y) is restricted to first-quadrant points that lie on the ellipse

$$\frac{x^2}{3^2} + \frac{y^2}{4^2} = 1. \quad \text{Constraint}$$

Now, consider the constraint equation to be a fixed level curve of

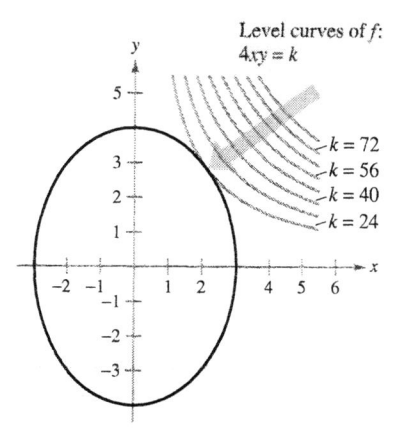

$$g(x, y) = \frac{x^2}{3^2} + \frac{y^2}{4^2}.$$

The level curves of f represent a family of hyperbolas

$$f(x, y) = 4xy = k.$$

In this family, the level curves that meet the given constraint correspond to the hyperbolas that intersect the ellipse. Moreover, to maximize $f(x, y)$, you want to find the hyperbola that just barely satisfies the constraint. The level curve that does this is the one that is *tangent* to the ellipse, as shown in Figure 12.78.

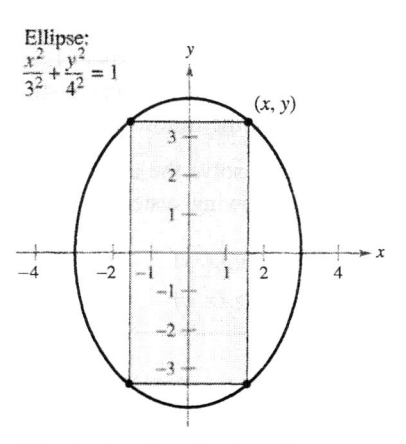

Objective function: $f(x, y) = 4xy$
Figure 12.77

Constraint: $g(x, y) = \dfrac{x^2}{3^2} + \dfrac{y^2}{4^2} = 1$
Figure 12.78

To find the appropriate hyperbola, use the fact that two curves are tangent at a point if and only if their gradient vectors are parallel. This means that $\nabla f(x, y)$ must be a scalar multiple of $\nabla g(x, y)$ at the point of tangency. In the context of constrained optimization problems, this scalar is denoted by λ (the lowercase Greek letter lambda).

$$\nabla f(x, y) = \lambda \nabla g(x, y)$$

The scalar λ is called a **Lagrange multiplier.** Theorem 12.19 gives the necessary conditions for the existence of such multipliers.

THEOREM 12.19 Lagrange's Theorem

Let f and g have continuous first partial derivatives such that f has an extremum at a point (x_0, y_0) on the smooth constraint curve $g(x, y) = c$. If $\nabla g(x_0, y_0) \neq \mathbf{0}$, then there is a real number λ such that

$$\nabla f(x_0, y_0) = \lambda \nabla g(x_0, y_0).$$

JOSEPH-LOUIS LAGRANGE (1736–1813)

The Method of Lagrange Multipliers is named after the French mathematician Joseph-Louis Lagrange. Lagrange first introduced the method in his famous paper on mechanics, written when he was just 19 years old.

Proof To begin, represent the smooth curve given by $g(x, y) = c$ by the vector-valued function

$$\mathbf{r}(t) = x(t)\mathbf{i} + y(t)\mathbf{j}, \qquad \mathbf{r}'(t) \neq \mathbf{0}$$

where x' and y' are continuous on an open interval I. Define the function h as $h(t) = f(x(t), y(t))$. Then, because $f(x_0, y_0)$ is an extreme value of f, you know that

$$h(t_0) = f(x(t_0), y(t_0)) = f(x_0, y_0)$$

is an extreme value of h. This implies that $h'(t_0) = 0$, and, by the Chain Rule,

$$h'(t_0) = f_x(x_0, y_0)x'(t_0) + f_y(x_0, y_0)y'(t_0) = \nabla f(x_0, y_0) \cdot \mathbf{r}'(t_0) = 0.$$

Therefore, $\nabla f(x_0, y_0)$ is orthogonal to $\mathbf{r}'(t_0)$. Moreover, by Theorem 12.12, $\nabla g(x_0, y_0)$ is also orthogonal to $\mathbf{r}'(t_0)$. Consequently, the gradients $\nabla f(x_0, y_0)$ and $\nabla g(x_0, y_0)$ are parallel, and there must exist a scalar λ such that

$$\nabla f(x_0, y_0) = \lambda \nabla g(x_0, y_0).$$

NOTE Lagrange's Theorem can be shown to be true for functions of three variables, using a similar argument with level surfaces and Theorem 12.14.

The Method of Lagrange Multipliers uses Theorem 12.19 to find the extreme values of a function f subject to a constraint.

Method of Lagrange Multipliers

Let f and g satisfy the hypothesis of Lagrange's Theorem, and let f have a minimum or maximum subject to the constraint $g(x, y) = c$. To find the minimum or maximum of f, use the following steps.

1. Simultaneously solve the equations $\nabla f(x, y) = \lambda \nabla g(x, y)$ and $g(x, y) = c$ by solving the following system of equations.

$$f_x(x, y) = \lambda g_x(x, y)$$
$$f_y(x, y) = \lambda g_y(x, y)$$
$$g(x, y) = c$$

2. Evaluate f at each solution point obtained in the first step. The largest value yields the maximum of f subject to the constraint $g(x, y) = c$, and the smallest value yields the minimum of f subject to the constraint $g(x, y) = c$.

NOTE As you will see in Examples 1 and 2, the Method of Lagrange Multipliers requires solving systems of nonlinear equations. This often can require some tricky algebraic manipulation.

Constrained Optimization Problems

At the beginning of this section, we described a problem in which we wanted to maximize the area of a rectangle that is inscribed in an ellipse. Example 1 shows how to use Lagrange multipliers to solve this problem.

Example 1 Using a Lagrange Multiplier with One Constraint

Find the maximum value of $f(x, y) = 4xy$ where $x > 0$ and $y > 0$, subject to the constraint $(x^2/3^2) + (y^2/4^2) = 1$.

Solution To begin, let

$$g(x, y) = \frac{x^2}{3^2} + \frac{y^2}{4^2} = 1.$$

By equating $\nabla f(x, y) = 4y\,\mathbf{i} + 4x\,\mathbf{j}$ and $\lambda \nabla g(x, y) = (2\lambda x/9)\mathbf{i} + (\lambda y/8)\mathbf{j}$, you can obtain the following system of equations.

$$4y = \frac{2}{9}\lambda x \qquad f_x(x, y) = \lambda g_x(x, y)$$

$$4x = \frac{1}{8}\lambda y \qquad f_y(x, y) = \lambda g_y(x, y)$$

$$\frac{x^2}{3^2} + \frac{y^2}{4^2} = 1 \qquad \text{Constraint}$$

From the first equation, you obtain $\lambda = 18y/x$, and substitution into the second equation produces

$$4x = \frac{1}{8}\left(\frac{18y}{x}\right)y \quad \Longrightarrow \quad x^2 = \frac{9}{16}y^2.$$

Substituting this value for x^2 into the third equation produces

$$\frac{1}{9}\left(\frac{9}{16}y^2\right) + \frac{1}{16}y^2 = 1 \quad \Longrightarrow \quad y^2 = 8.$$

So, $y = \pm 2\sqrt{2}$. Because it is required that $y > 0$, choose the positive value and find that

$$x^2 = \frac{9}{16}y^2$$

$$= \frac{9}{16}(8) = \frac{9}{2}$$

$$x = \frac{3}{\sqrt{2}}.$$

So, the maximum value of f is

$$f\left(\frac{3}{\sqrt{2}}, 2\sqrt{2}\right) = 4xy = 4\left(\frac{3}{\sqrt{2}}\right)(2\sqrt{2}) = 24.$$

Note that writing the constraint as

$$g(x, y) = \frac{x^2}{3^2} + \frac{y^2}{4^2} = 1 \quad \text{or} \quad g(x, y) = \frac{x^2}{3^2} + \frac{y^2}{4^2} - 1 = 0$$

does not affect the solution—the constant is eliminated when you form ∇g.

NOTE Example 1 can also be solved using the techniques you learned in Chapter 3. To see how, try to find the maximum value of $A = 4xy$ given that

$$\frac{x^2}{3^2} + \frac{y^2}{4^2} = 1.$$

To begin, solve the second equation for y to obtain

$$y = \tfrac{4}{3}\sqrt{9 - x^2}.$$

Then substitute into the first equation to obtain

$$A = 4x\left(\tfrac{4}{3}\sqrt{9 - x^2}\right).$$

Finally, use the techniques of Chapter 3 to maximize A.

Example 2 A Business Application

The Cobb-Douglas production function (see Example 5, Section 12.1) for a particular manufacturer is given by

$$f(x, y) = 100x^{3/4}y^{1/4} \quad \text{Objective function}$$

where x represents the units of labor (at \$150 per unit) and y represents the units of capital (at \$250 per unit). The total cost of labor and capital is limited to \$50,000. Find the maximum production level for this manufacturer.

Solution From the given function, you have

$$\nabla f(x, y) = 75x^{-1/4}y^{1/4}\mathbf{i} + 25x^{3/4}y^{-3/4}\mathbf{j}.$$

The limit on the cost of labor and capital produces the constraint

$$g(x, y) = 150x + 250y = 50,000. \quad \text{Constraint}$$

So, $\lambda \nabla g(x, y) = 150\lambda\mathbf{i} + 250\lambda\mathbf{j}$. This gives rise to the following system of equations.

$$75x^{-1/4}y^{1/4} = 150\lambda \qquad f_x(x, y) = \lambda g_x(x, y)$$
$$25x^{3/4}y^{-3/4} = 250\lambda \qquad f_y(x, y) = \lambda g_y(x, y)$$
$$150x + 250y = 50,000 \qquad \text{Constraint}$$

By solving for λ in the first equation

$$\lambda = \frac{75x^{-1/4}y^{1/4}}{150} = \frac{x^{-1/4}y^{1/4}}{2}$$

and substituting into the second equation, you obtain

$$25x^{3/4}y^{-3/4} = 250\left(\frac{x^{-1/4}y^{1/4}}{2}\right)$$

$$25x = 125y. \quad \text{Multiply by } x^{1/4}y^{3/4}.$$

So, $x = 5y$. By substituting into the third equation, you have

$$150(5y) + 250y = 50,000$$
$$1000y = 50,000$$
$$y = 50 \text{ units of capital}$$
$$x = 250 \text{ units of labor}.$$

So, the maximum production level is

$$f(250, 50) = 100(250)^{3/4}(50)^{1/4}$$
$$\approx 16,719 \text{ product units.}$$

Economists call the Lagrange multiplier obtained in a production function the **marginal productivity of money.** For instance, in Example 2 the marginal productivity of money at $x = 250$ and $y = 50$ is

$$\lambda = \frac{x^{-1/4}y^{1/4}}{2} = \frac{(250)^{-1/4}(50)^{1/4}}{2} \approx 0.334$$

which means that for each additional dollar spent on production, 0.334 additional units of the product can be produced.

FOR FURTHER INFORMATION For more information on the use of Lagrange multipliers in economics, see the article "Lagrange Multiplier Problems in Economics" by John V. Baxley and John C. Moorhouse in *The American Mathematical Monthly*. To view this article, go to the website www.matharticles.com.

Example 3 Lagrange Multipliers and Three Variables

Find the minimum value of

$$f(x, y, z) = 2x^2 + y^2 + 3z^2 \qquad \text{Objective function}$$

subject to the constraint $2x - 3y - 4z = 49$.

Solution Let $g(x, y, z) = 2x - 3y - 4z = 49$. Then, because

$$\nabla f(x, y, z) = 4x\mathbf{i} + 2y\mathbf{j} + 6z\mathbf{k} \qquad \text{and} \qquad \lambda \nabla g(x, y, z) = 2\lambda\mathbf{i} - 3\lambda\mathbf{j} - 4\lambda\mathbf{k}$$

you obtain the following system of equations.

$$\begin{aligned} 4x &= 2\lambda & f_x(x, y, z) &= \lambda g_x(x, y, z) \\ 2y &= -3\lambda & f_y(x, y, z) &= \lambda g_y(x, y, z) \\ 6z &= -4\lambda & f_z(x, y, z) &= \lambda g_z(x, y, z) \\ 2x - 3y - 4z &= 49 & \text{Constraint} \end{aligned}$$

The solution of this system is $x = 3$, $y = -9$, and $z = -4$. Therefore, the optimum value of f is

$$\begin{aligned} f(3, -9, -4) &= 2(3)^2 + (-9)^2 + 3(-4)^2 \\ &= 147. \end{aligned}$$

From the original function and constraint, it is clear that $f(x, y, z)$ has no maximum. So, the optimum value of f determined above is a minimum.

At the beginning of this section, we gave a graphical interpretation of constrained optimization problems in two variables. In three variables, the interpretation is similar, except that we use level surfaces instead of level curves. For instance, in Example 3, the level surfaces of f are ellipsoids centered at the origin, and the constraint

$$2x - 3y - 4z = 49$$

is a plane. The minimum value of f is represented by the ellipsoid that is tangent to the constraint plane, as shown in Figure 12.79.

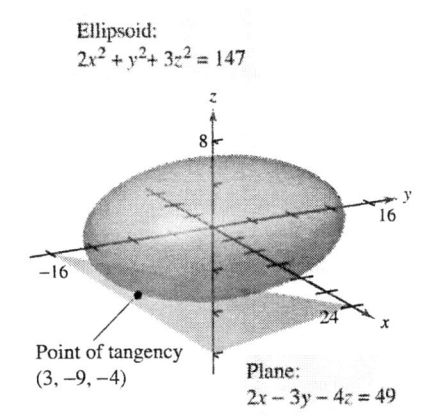

Ellipsoid: $2x^2 + y^2 + 3z^2 = 147$
Point of tangency $(3, -9, -4)$
Plane: $2x - 3y - 4z = 49$

Figure 12.79

Example 4 Optimization Inside a Region

Find the extreme values of

$$f(x, y) = x^2 + 2y^2 - 2x + 3 \qquad \text{Objective function}$$

subject to the constraint $x^2 + y^2 \leq 10$.

Solution To solve this problem, you can break the constraint into two cases.

a. For points *on the circle* $x^2 + y^2 = 10$, you can use Lagrange multipliers to find that the maximum value of $f(x, y)$ is 24—this value occurs at $(-1, 3)$ and at $(-1, -3)$. In a similar way, you can determine that the minimum value of $f(x, y)$ is approximately 6.675—this value occurs at $(\sqrt{10}, 0)$.

b. For points *inside the circle*, you can use the techniques discussed in Section 12.8 to conclude that the function has a relative minimum of 2 at the point $(1, 0)$.

By combining these two results, you can conclude that f has a maximum of 24 at $(-1, \pm 3)$ and a minimum of 2 at $(1, 0)$, as shown in Figure 12.80.

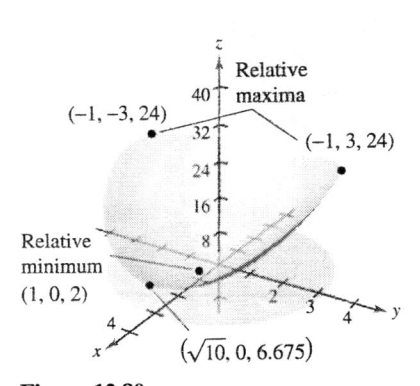

Figure 12.80

The Method of Lagrange Multipliers with Two Constraints

For optimization problems involving *two* constraint functions g and h, we introduce a second Lagrange multiplier, μ (the lowercase Greek letter mu), and solve the equation

$$\nabla f = \lambda \nabla g + \mu \nabla h$$

where the gradient vectors are not parallel, as illustrated in Example 5.

Example 5 Optimization with Two Constraints

Let $T(x, y, z) = 20 + 2x + 2y + z^2$ represent the temperature at each point on the sphere $x^2 + y^2 + z^2 = 11$. Find the extreme temperatures on the curve formed by the intersection of the plane $x + y + z = 3$ and the sphere.

Solution The two constraints are

$$g(x, y, z) = x^2 + y^2 + z^2 = 11 \quad \text{and} \quad h(x, y, z) = x + y + z = 3.$$

Using

$$\nabla T(x, y, z) = 2\mathbf{i} + 2\mathbf{j} + 2z\mathbf{k}$$
$$\lambda \nabla g(x, y, z) = 2\lambda x \mathbf{i} + 2\lambda y \mathbf{j} + 2\lambda z \mathbf{k}$$

and

$$\mu \nabla h(x, y, z) = \mu \mathbf{i} + \mu \mathbf{j} + \mu \mathbf{k}$$

you can write the following system of equations.

$$\begin{aligned}
2 &= 2\lambda x + \mu & T_x(x, y, z) &= \lambda g_x(x, y, z) + \mu h_x(x, y, z) \\
2 &= 2\lambda y + \mu & T_y(x, y, z) &= \lambda g_y(x, y, z) + \mu h_y(x, y, z) \\
2z &= 2\lambda z + \mu & T_z(x, y, z) &= \lambda g_z(x, y, z) + \mu h_z(x, y, z) \\
x^2 + y^2 + z^2 &= 11 & &\text{Constraint 1} \\
x + y + z &= 3 & &\text{Constraint 2}
\end{aligned}$$

By subtracting the second equation from the first, you can obtain the following system.

$$\begin{aligned}
\lambda(x - y) &= 0 \\
2z(1 - \lambda) - \mu &= 0 \\
x^2 + y^2 + z^2 &= 11 \\
x + y + z &= 3
\end{aligned}$$

STUDY TIP The system of equations that arises in the Method of Lagrange Multipliers is not, in general, a linear system, and the solution often requires ingenuity.

From the first equation, you can conclude that $\lambda = 0$ or $x = y$. If $\lambda = 0$, you can show that the critical points are $(3, -1, 1)$ and $(-1, 3, 1)$. (Try doing this—it takes a little work.) If $\lambda \neq 0$, then $x = y$ and you can show that the critical points occur when $x = y = (3 \pm 2\sqrt{3})/3$ and $z = (3 \mp 4\sqrt{3})/3$. Finally, to find the optimal solutions, compare the temperatures at the four critical points.

$$T(3, -1, 1) = T(-1, 3, 1) = 25$$
$$T\left(\frac{3 - 2\sqrt{3}}{3}, \frac{3 - 2\sqrt{3}}{3}, \frac{3 + 4\sqrt{3}}{3}\right) = \frac{91}{3} \approx 30.33$$
$$T\left(\frac{3 + 2\sqrt{3}}{3}, \frac{3 + 2\sqrt{3}}{3}, \frac{3 - 4\sqrt{3}}{3}\right) = \frac{91}{3} \approx 30.33$$

So, $T = 25$ is the minimum temperature and $T = \frac{91}{3}$ is the maximum temperature on the curve.

EXERCISES FOR SECTION 12.10

In Exercises 1–4, identify the constraint and level curves of the objective function in the given figure. Use the figure to approximate the indicated extrema, assuming that x and y are positive. Use Lagrange multipliers to verify your result.

1. Maximize $z = xy$
   Constraint: $x + y = 10$

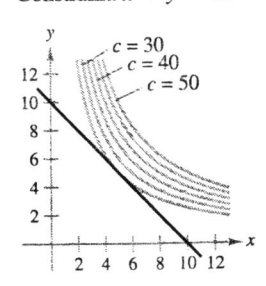

2. Maximize $z = xy$
   Constraint: $2x + y = 4$

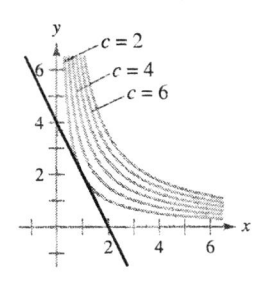

3. Minimize $z = x^2 + y^2$
   Constraint: $x + y - 4 = 0$

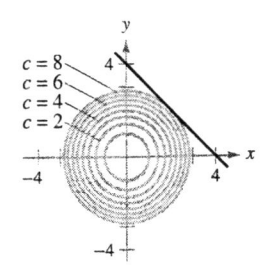

4. Minimize $z = x^2 + y^2$
   Constraint: $2x + 4y = 5$

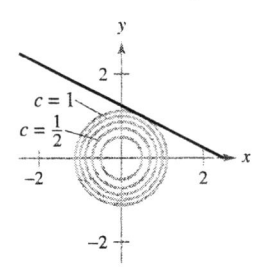

In Exercises 5–12, use Lagrange multipliers to find the indicated extrema, assuming that x and y are positive.

5. Minimize $f(x, y) = x^2 - y^2$
   Constraint: $x - 2y + 6 = 0$

6. Maximize $f(x, y) = x^2 - y^2$
   Constraint: $2y - x^2 = 0$

7. Maximize $f(x, y) = 2x + 2xy + y$
   Constraint: $2x + y = 100$

8. Minimize $f(x, y) = 3x + y + 10$
   Constraint: $x^2 y = 6$

9. Maximize $f(x, y) = \sqrt{6 - x^2 - y^2}$
   Constraint: $x + y - 2 = 0$

10. Minimize $f(x, y) = \sqrt{x^2 + y^2}$
    Constraint: $2x + 4y - 15 = 0$

11. Maximize $f(x, y) = e^{xy}$
    Constraint: $x^2 + y^2 = 8$

12. Minimize $f(x, y) = 2x + y$
    Constraint: $xy = 32$

In Exercises 13 and 14, use Lagrange multipliers to find any extrema of the function subject to the constraint $x^2 + y^2 \leq 1$.

13. $f(x, y) = x^2 + 3xy + y^2$

14. $f(x, y) = e^{-xy/4}$

In Exercises 15–18, use Lagrange multipliers to find the indicated extrema, assuming that x, y, and z are positive.

15. Minimize $f(x, y, z) = x^2 + y^2 + z^2$
    Constraint: $x + y + z - 6 = 0$

16. Maximize $f(x, y, z) = xyz$
    Constraint: $x + y + z - 6 = 0$

17. Minimize $f(x, y, z) = x^2 + y^2 + z^2$
    Constraint: $x + y + z = 1$

18. Minimize $f(x, y) = x^2 - 10x + y^2 - 14y + 70$
    Constraint: $x + y = 10$

In Exercises 19–22, use Lagrange multipliers to find the indicated extrema of f subject to two constraints. In each case, assume that x, y, and z are nonnegative.

19. Maximize $f(x, y, z) = xyz$
    Constraints: $x + y + z = 32,\quad x - y + z = 0$

20. Minimize $f(x, y, z) = x^2 + y^2 + z^2$
    Constraints: $x + 2z = 6,\quad x + y = 12$

21. Maximize $f(x, y, z) = xy + yz$
    Constraints: $x + 2y = 6,\quad x - 3z = 0$

22. Maximize $f(x, y, z) = xyz$
    Constraints: $x^2 + z^2 = 5,\quad x - 2y = 0$

In Exercises 23–26, use Lagrange multipliers to find the minimum distance from the curve or surface to the indicated point. [*Hint:* In Exercise 23, minimize $f(x, y) = x^2 + y^2$ subject to the constraint $2x + 3y = -1$.]

Curve	Point
23. Line: $2x + 3y = -1$	$(0, 0)$
24. Circle: $(x - 4)^2 + y^2 = 4$	$(0, 10)$

Surface	Point
25. Plane: $x + y + z = 1$	$(2, 1, 1)$
26. Cone: $z = \sqrt{x^2 + y^2}$	$(4, 0, 0)$

In Exercises 27 and 28, find the highest point on the curve of intersection of the surfaces.

27. Sphere: $x^2 + y^2 + z^2 = 36$,   Plane: $2x + y - z = 2$

28. Cone: $x^2 + y^2 - z^2 = 0$,   Plane: $x + 2z = 4$

Getting at the Concept

29. What is meant by constrained optimization problems?

30. Explain the Method of Lagrange Multipliers for solving constrained optimization problems.

31. Volume Use Lagrange multipliers to find the dimensions of the rectangular package of largest volume subject to the constraint that the sum of the length and the girth cannot exceed 108 inches. Compare the answer to that obtained in Exercise 9, Section 12.9.

32. Volume The material for the base of an open box costs 1.5 times as much as the material for the sides. Use Lagrange multipliers to find the dimensions of the box of largest volume that can be made for a fixed cost C. (Maximize $V = xyz$ subject to $1.5xy + 2xz + 2yz = C$.) Compare the answer to that obtained in Exercise 10, Section 12.9.

33. Cost A cargo container (in the shape of a rectangular solid) must have a volume of 480 cubic feet. Use Lagrange multipliers to find the dimensions of the container of this size that has minimum cost if the bottom will cost $5 per square foot to construct and the sides and the top will cost $3 per square foot to construct.

34. Surface Area Use Lagrange multipliers to find the dimensions of a right circular cylinder with volume V_0 cubic units and minimum surface area.

35. Volume Use Lagrange multipliers to find the dimensions of a rectangular box of maximum volume that can be inscribed (with edges parallel to the coordinate axes) in the ellipsoid
$$\frac{x^2}{a^2} + \frac{y^2}{b^2} + \frac{z^2}{c^2} = 1.$$

36. Geometric and Arithmetic Means

(a) Use Lagrange multipliers to prove that the product of three positive numbers x, y, and z, whose sum has the constant value S, is a maximum when the three numbers are equal. Use this result to prove that
$$\sqrt[3]{xyz} \leq \frac{x + y + z}{3}.$$

(b) Generalize the result in part (a) to prove that the product $x_1 x_2 x_3 \cdots x_n$ is a maximum when $x_1 = x_2 = x_3 = \cdots = x_n$, $\sum_{i=1}^{n} x_i = S$, and all $x_i \geq 0$. Use this result to prove that
$$\sqrt[n]{x_1 x_2 x_3 \cdots x_n} \leq \frac{x_1 + x_2 + x_3 + \cdots + x_n}{n}.$$

This shows that the geometric mean is never greater than the arithmetic mean.

37. Refraction of Light When light waves traveling in a transparent medium strike the surface of a second transparent medium, they tend to "bend" in order to follow the path of minimum time. This tendency is called *refraction* and is described by **Snell's Law of Refraction**,
$$\frac{\sin \theta_1}{v_1} = \frac{\sin \theta_2}{v_2}$$
where θ_1 and θ_2 are the magnitudes of the angles shown in the figure, and v_1 and v_2 are the velocities of light in the two media. Use Lagrange multipliers to derive this law using the constraint $x + y = a$.

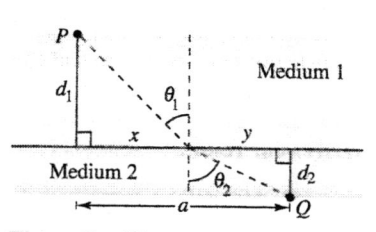

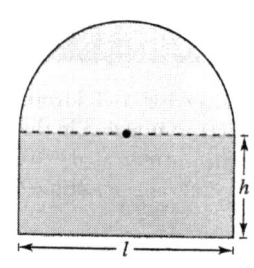

Figure for 37

Figure for 38

38. Area and Perimeter A semicircle is on top of a rectangle (see figure). If the area is fixed and the perimeter is a minimum, or if the perimeter is fixed and the area is a maximum, use Lagrange multipliers to verify that the length of the rectangle is twice its height.

39. Hardy-Weinberg Law Use Lagrange multipliers to maximize $P(p, q, r) = 2pq + 2pr + 2qr$ subject to $p + q + r = 1$. (See Exercise 18 in Section 12.9.)

40. Temperature Distribution Let $T(x, y, z) = 100 + x^2 + y^2$ represent the temperature at each point on the sphere $x^2 + y^2 + z^2 = 50$. Find the maximum temperature on the curve formed by the intersection of the sphere and the plane $x - z = 0$.

Production Level In Exercises 41 and 42, find the maximum production level P if the total cost of labor (at $48 per unit) and capital (at $36 per unit) is limited to $100,000, where x is the number of units of labor and y is the number of units of capital.

41. $P(x, y) = 100x^{0.25}y^{0.75}$ **42.** $P(x, y) = 100x^{0.4}y^{0.6}$

Cost In Exercises 43 and 44, find the minimum cost of producing 20,000 units of a product, where x is the number of units of labor (at $48 per unit) and y is the number of units of capital (at $36 per unit).

43. $P(x, y) = 100x^{0.25}y^{0.75}$ **44.** $P(x, y) = 100x^{0.6}y^{0.4}$

45. Investigation Consider the objective function $g(\alpha, \beta, \gamma) = \cos \alpha \cos \beta \cos \gamma$ subject to the constraint that α, β, and γ are the angles of a triangle.

(a) Use Lagrange multipliers to maximize g.

(b) Use the constraint to reduce the function g to a function of two independent variables. Use a computer algebra system to graph the surface represented by g. Identify the maximum values on the graph.

46. Investigation Consider the objective function $f(x, y) = ax + by$ subject to the constraint $x^2/64 + y^2/36 = 1$. Assume that x and y are positive.

(a) Use a computer algebra system to graph the constraint. If $a = 4$ and $b = 3$, use the computer algebra system to graph the level curves of the objective function. By trial and error, find the level curve that appears to be tangent to the ellipse. Use the result to approximate the maximum of f subject to the constraint.

(b) Repeat part (a) for $a = 4$ and $b = 9$.

REVIEW EXERCISES FOR CHAPTER 12

12.1 In Exercises 1 and 2, use the graph to determine whether z is a function of x and y. Explain.

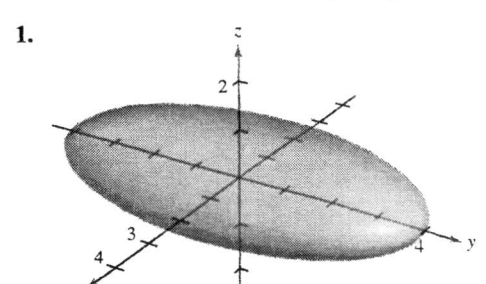

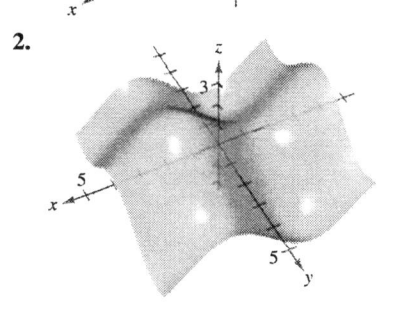

1.

2.

In Exercises 3–6, use a computer algebra system to graph several level curves for the function.

3. $f(x, y) = e^{x^2 + y^2}$
4. $f(x, y) = \ln xy$
5. $f(x, y) = x^2 - y^2$
6. $f(x, y) = \dfrac{x}{x + y}$

In Exercises 7 and 8, use a computer algebra system to graph the function.

7. $f(x, y) = e^{-(x^2 + y^2)}$
8. $g(x, y) = |y|^{1 + |x|}$

In Exercises 9 and 10, sketch the graph of the level surface $f(x, y, z) = c$ at the specified value of c.

9. $f(x, y, z) = x^2 - y + z^2, \quad c = 1$
10. $f(x, y, z) = 9x^2 - y^2 + 9z^2, \quad c = 0$

12.2 In Exercises 11–14, discuss the continuity of the function and evaluate the limit, if it exists.

11. $\lim\limits_{(x, y) \to (1, 1)} \dfrac{xy}{x^2 + y^2}$
12. $\lim\limits_{(x, y) \to (1, 1)} \dfrac{xy}{x^2 - y^2}$
13. $\lim\limits_{(x, y) \to (0, 0)} \dfrac{-4x^2 y}{x^4 + y^2}$
14. $\lim\limits_{(x, y) \to (0, 0)} \dfrac{y + xe^{-y^2}}{1 + x^2}$

12.3 In Exercises 15–24, find all first partial derivatives.

15. $f(x, y) = e^x \cos y$
16. $f(x, y) = \dfrac{xy}{x + y}$
17. $z = xe^y + ye^x$
18. $z = \ln(x^2 + y^2 + 1)$
19. $g(x, y) = \dfrac{xy}{x^2 + y^2}$
20. $w = \sqrt{x^2 + y^2 + z^2}$
21. $f(x, y, z) = z \arctan \dfrac{y}{x}$
22. $f(x, y, z) = \dfrac{1}{\sqrt{1 - x^2 - y^2 - z^2}}$
23. $u(x, t) = ce^{-n^2 t} \sin nx$
24. $u(x, t) = c \sin(akx) \cos kt$

25. *Think About It* Sketch a graph of a function $z = f(x, y)$ whose derivative f_x is always negative and whose derivative f_y is always negative.

26. Find the slopes of the surface $z = x^2 \ln(y + 1)$ in the x- and y-directions at the point $(2, 0, 0)$.

In Exercises 27–30, find all second partial derivatives and verify that the second mixed partials are equal.

27. $f(x, y) = 3x^2 - xy + 2y^3$
28. $h(x, y) = \dfrac{x}{x + y}$
29. $h(x, y) = x \sin y + y \cos x$
30. $g(x, y) = \cos(x - 2y)$

Laplace Equation In Exercises 31–34, show that the function satisfies the Laplace equation

$$\dfrac{\partial^2 z}{\partial x^2} + \dfrac{\partial^2 z}{\partial y^2} = 0.$$

31. $z = x^2 - y^2$
32. $z = x^3 - 3xy^2$
33. $z = \dfrac{y}{x^2 + y^2}$
34. $z = e^x \sin y$

12.4 In Exercises 35 and 36, find dz.

35. $z = x \sin \dfrac{y}{x}$
36. $z = \dfrac{xy}{\sqrt{x^2 + y^2}}$

37. *Error Analysis* The legs of a right triangle are measured to be 5 centimeters and 12 centimeters, with a possible error of $\frac{1}{2}$ centimeter. Approximate the maximum possible error in computing the length of the hypotenuse. Approximate the maximum percent error.

38. *Error Analysis* To determine the height of a tower, the angle of elevation to the top of the tower was measured from a point 100 feet $\pm \frac{1}{2}$ foot from the base. The angle is measured at 33°, with a possible error of 1°. Assuming that the ground is horizontal, approximate the maximum error in determining the height of the tower.

39. Error Analysis The volume of a right circular cone is $V = \frac{1}{3}\pi r^2 h$. The measured values of r and h for a cone are found to be 2 and 5 inches. Find the approximate error in the computation of the volume because of a possible error of $\frac{1}{8}$ inch in the measurements of the radius and the height.

40. Error Analysis Approximate the error in the computation of the lateral surface area of the cone in Exercise 39. (The lateral surface area is given by $A = \pi r \sqrt{r^2 + h^2}$.)

12.5 In Exercises 41–44, find the indicated derivatives (a) by the Chain Rule and (b) by substitution before differentiating.

41. $w = \ln(x^2 + y^2)$, $\dfrac{dw}{dt}$

  $x = 2t + 3$, $y = 4 - t$

42. $u = y^2 - x$, $\dfrac{du}{dt}$

  $x = \cos t$, $y = \sin t$

43. $u = x^2 + y^2 + z^2$, $\dfrac{\partial u}{\partial r}, \dfrac{\partial u}{\partial t}$

  $x = r \cos t$, $y = r \sin t$, $z = t$

44. $w = \dfrac{xy}{z}$, $\dfrac{\partial w}{\partial r}, \dfrac{\partial w}{\partial t}$

  $x = 2r + t$, $y = rt$, $z = 2r - t$

In Exercises 45 and 46, find $\partial z/\partial x$ and $\partial z/\partial y$.

45. $x^2 y - 2yz - xz - z^2 = 0$

46. $xz^2 - y \sin z = 0$

12.6 In Exercises 47–50, find the directional derivative in the direction of **v** at the indicated point.

Function	Direction	Point
47. $f(x, y) = x^2 y$	$\mathbf{v} = \mathbf{i} - \mathbf{j}$	(2, 1)
48. $f(x, y) = \frac{1}{4}y^2 - x^2$	$\mathbf{v} = 2\mathbf{i} + \mathbf{j}$	(1, 4)
49. $w = y^2 + xz$	$\mathbf{v} = 2\mathbf{i} - \mathbf{j} + 2\mathbf{k}$	(1, 2, 2)
50. $w = 6x^2 + 3xy - 4y^2 z$	$\mathbf{v} = \mathbf{i} + \mathbf{j} - \mathbf{k}$	(1, 0, 1)

In Exercises 51–54, find the gradient and the maximum value of the directional derivative of the function at the indicated point.

51. $z = \dfrac{y}{x^2 + y^2}$, $(1, 1)$

52. $z = \dfrac{x^2}{x - y}$, $(2, 1)$

53. $z = e^{-x} \cos y$, $\left(0, \dfrac{\pi}{4}\right)$

54. $z = x^2 y$, $(2, 1)$

In Exercises 55 and 56, use the gradient to find the unit normal vector to the graph of the equation at the indicated point.

55. $9x^2 - 4y^2 = 65$, $(3, 2)$

56. $4y \sin x - y^2 = 3$, $(\pi/2, 1)$

12.7 In Exercises 57–60, find an equation of the tangent plane and parametric equations of the normal line to the surface at the indicated point.

Surface	Point
57. $f(x, y) = x^2 y$	(2, 1, 4)
58. $f(x, y) = \sqrt{25 - y^2}$	(2, 3, 4)
59. $z = -9 + 4x - 6y - x^2 - y^2$	(2, -3, 4)
60. $z = \sqrt{9 - x^2 - y^2}$	(1, 2, 2)

In Exercises 61 and 62, find symmetric equations of the tangent line to the curve of intersection of the surfaces at the indicated point.

Surfaces	Point
61. $z = x^2 - y^2$, $z = 3$	(2, 1, 3)
62. $z = 25 - y^2$, $y = x$	(4, 4, 9)

63. Find the angle of inclination θ of the tangent plane to the surface $x^2 + y^2 + z^2 = 14$ at the point (2, 1, 3).

64. **Approximation** Consider the following approximations centered at (0, 0) for a function $f(x, y)$.

  Linear approximation:

  $P_1(x, y) = f(0, 0) + f_x(0, 0)x + f_y(0, 0)y$

  Quadratic approximation:

  $P_2(x, y) = f(0, 0) + f_x(0, 0)x + f_y(0, 0)y + \frac{1}{2}f_{xx}(0, 0)x^2 + f_{xy}(0, 0)xy + \frac{1}{2}f_{yy}(0, 0)y^2$

  [Note that the linear approximation is the tangent plane to the surface at $(0, 0, f(0, 0))$.]

  (a) Find the linear approximation of $f(x, y) = \cos x + \sin y$ centered at (0, 0).

  (b) Find the quadratic approximation of $f(x, y) = \cos x + \sin y$ centered at (0, 0).

  (c) If $y = 0$ in the quadratic approximation, you obtain the second-degree Taylor polynomial for what function?

  (d) Complete the table.

x	y	$f(x, y)$	$P_1(x, y)$	$P_2(x, y)$
0	0			
0	0.1			
0.2	0.1			
0.5	0.3			
1	0.5			

  (e) Use a computer algebra system to graph the surfaces $z = f(x, y)$, $z = P_1(x, y)$, and $z = P_2(x, y)$. How does the accuracy of the approximations change as the distance from (0, 0) increases?

12.8

In Exercises 65–68, locate and classify any extrema of the function. Use a computer algebra system to graph the function and confirm your analytical results.

65. $f(x, y) = x^3 - 3xy + y^2$
66. $f(x, y) = 2x^2 + 6xy + 9y^2 + 8x + 14$
67. $f(x, y) = xy + \dfrac{1}{x} + \dfrac{1}{y}$
68. $z = 50(x + y) - (0.1x^3 + 20x + 150) - (0.05y^3 + 20.6y + 125)$

Writing In Exercises 69 and 70, write a short paragraph about the surface whose level curves (c-values evenly spaced) are given. Comment on possible extrema, saddle points, the magnitude of the gradient, etc.

69.

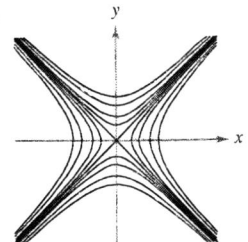

70.

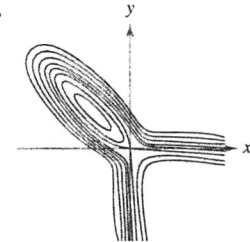

12.9

71. **Profit** A corporation manufactures a product at two locations. The cost functions for producing x_1 units at location 1 and x_2 units at location 2 are

$C_1 = 0.05x_1^2 + 15x_1 + 5400$
$C_2 = 0.03x_2^2 + 15x_2 + 6100$

and the total revenue function is

$R = [225 - 0.4(x_1 + x_2)](x_1 + x_2)$.

Find the production levels at the two locations that will maximize the profit $P(x_1, x_2) = R - C_1 - C_2$.

72. **Cost** A manufacturer has an order for 1000 units that can be produced at two locations. Let x_1 and x_2 be the numbers of units produced at the two locations. Find the number that should be produced at each to meet the order and minimize cost, if the cost function is

$C = 0.25x_1^2 + 10x_1 + 0.15x_2^2 + 12x_2$.

73. **Production Level** The production function for a manufacturer is

$f(x, y) = 4x + xy + 2y$

where x is the number of units of labor and y is the number of units of capital. Assume that the total amount available for labor and capital is $2000, and that units of labor and capital cost $20 and $4, respectively. Find the maximum production level for this manufacturer.

74. Find the minimum distance from the point $(2, 2, 0)$ to the surface $z = x^2 + y^2$.

75. **Modeling Data** The data in the table show the yield y (in milligrams) of a chemical reaction after t minutes.

t	1	2	3	4
y	1.5	7.4	10.2	13.4

t	5	6	7	8
y	15.8	16.3	18.2	18.3

(a) Use a graphing utility to plot the data. Use the graphing utility to find a linear model for the data and graph the model.

(b) Use a graphing utility to plot the points $(\ln t, y)$. Do these points appear to follow a linear pattern more closely than the plot of the given data in part (a)?

(c) Use a graphing utility to find a linear model for the data $(\ln t, y)$ and obtain the logarithmic model

$y = a + b \ln t$.

(d) Use a graphing utility to plot the data and graph the linear and logarithmic models. Which is a better model? Explain.

76. **Modeling Data** The table shows the drag force y in kilograms for a certain motor vehicle at indicated speeds x in kilometers per hour.

Speed x	25	50	75	100	125
Drag y	28	38	54	75	102

(a) Use the regression capabilities of a graphing utility to find the least squares regression quadratic for the data.

(b) Use the quadratic to estimate the total drag when the vehicle is moving at 80 kilometers per hour.

12.10

In Exercises 77 and 78, locate and classify any extrema of the function by using Lagrange multipliers.

77. $w = xy + yz + xz$

Constraint: $x + y + z = 1$

78. $z = x^2 y$

Constraint: $x + 2y = 2$

79. **Minimum Cost** A water line is to be built from point P to point S and must pass through regions where construction costs differ (see figure). Use Lagrange multipliers to find x, y, and z such that the total cost C will be minimized if the cost per kilometer in dollars is $3k$ from P to Q, $2k$ from Q to R, and k from R to S. For simplicity, let $k = 1$.

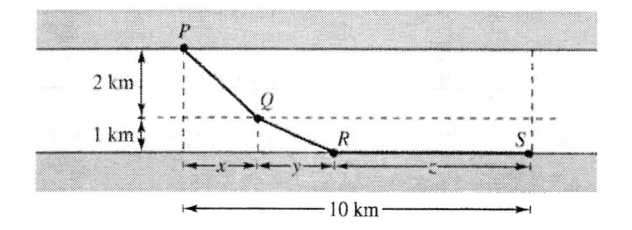

P.S. Problem Solving

1. Heron's Formula states that the area of a triangle with sides of lengths a, b, and c is given by
$$A = \sqrt{s(s-a)(s-b)(s-c)}$$
where $s = \dfrac{a+b+c}{2}$, as indicated in the figure.

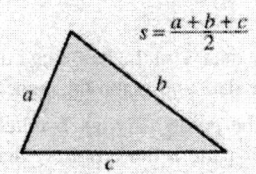

   (a) Use Heron's Formula to find the area of the triangle with vertices $(0, 0)$, $(3, 4)$, and $(6, 0)$.

   (b) Show that among all triangles having a fixed perimeter, the triangle with the largest area is an equilateral triangle.

   (c) Show that among all triangles having a fixed area, the triangle with the smallest perimeter is an equilateral triangle.

2. An industrial container is in the shape of a cylinder with hemispherical ends, as shown in the figure. If the container must hold 1000 liters of fluid, determine the radius r and length l that minimize the amount of material used in the construction of the tank.

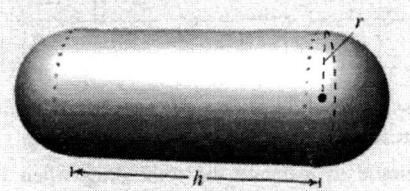

3. Let $P(x_0, y_0, z_0)$ be a point in the first octant on the surface $xyz = 1$.

   (a) Find the equation of the tangent plane to the surface at the point P.

   (b) Show that the volume of the tetrahedron formed by the three coordinate planes and the tangent plane is constant, independent of the point of tangency (see figure).

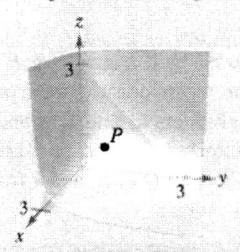

4. Use a graphing utility to graph the functions $f(x) = \sqrt[3]{x^3 - 1}$ and $g(x) = x$ in the same viewing window.

   (a) Show that
$$\lim_{x \to \infty} [f(x) - g(x)] = 0 \text{ and } \lim_{x \to -\infty} [f(x) - g(x)] = 0.$$

   (b) Find the point on the graph of f that is farthest from the graph of g.

5. Consider the function
$$f(x, y) = \begin{cases} \dfrac{4xy}{x^2 + y^2}, & x \neq 0 \\ 0, & x = 0 \end{cases}$$
and the unit vector $\mathbf{u} = \dfrac{1}{\sqrt{2}}(\mathbf{i} + \mathbf{j})$.

   Does the directional derivative of f at $P(0, 0)$ in the direction of $\mathbf{u}$ exist? If $f(0, 0)$ were defined as 2 instead of 0, would the directional derivative exist?

6. A heated storage room is shaped like a rectangular box and has a volume of 1000 cubic feet, as indicated in the figure. Because warm air rises, the heat loss per unit of area through the ceiling is five times as great as the heat loss through the floor. If the heat loss through the four walls is three times as great as the heat loss through the floor, determine the room dimensions that will minimize heat loss and thus minimize heating costs.

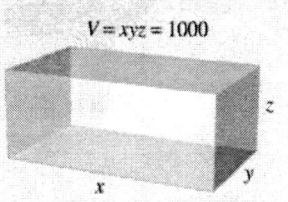

$$V = xyz = 1000$$

7. Repeat Exercise 6 assuming that the heat loss through the walls and ceiling remain the same, but the floor is insulated so that there is no heat loss through the floor.

8. Consider a circular plate of radius 1 given by $x^2 + y^2 \leq 1$, as shown in the figure. The temperature at any point $P(x, y)$ on the plate is $T(x, y) = 2x^2 + y^2 - y + 10$.

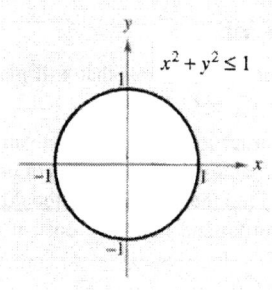

   (a) Sketch the isotherm $T(x, y) = 10$. To print an enlarged copy of the graph, go to the website www.mathgraphs.com.

   (b) Find the hottest and coldest points on the plate.

9. Consider the Cobb-Douglas production function
$$f(x, y) = Cx^a y^{1-a}, \quad 0 < a < 1.$$

   (a) Show that f satisfies the equation $x\dfrac{\partial f}{\partial x} + y\dfrac{\partial f}{\partial y} = f$.

   (b) Show that $f(tx, ty) = tf(x, y)$.

10. Rewrite Laplace's equation $\dfrac{\partial^2 u}{\partial x^2} + \dfrac{\partial^2 u}{\partial y^2} + \dfrac{\partial^2 u}{\partial z^2} = 0$ in cylindrical coordinates.

11. A projectile is launched at an angle of 45° with the horizontal and with an initial velocity of 64 feet per second. A television camera is located in the plane of the path of the projectile 50 feet behind the launch site (see figure).

    (a) Find parametric equations for the path of the projectile in terms of the parameter t representing time.

    (b) Express the angle α that the camera makes with the horizontal in terms of x and y and in terms of t.

    (c) Use the results in part (b) to find $d\alpha/dt$.

    (d) Use a graphing utility to obtain a graph of α in terms of t. Is the graph symmetric to the axis of the parabolic arch of the projectile? At what time is the rate of change of α greatest?

    (e) At what time is the angle α maximum? Does this occur when the projectile is at its greatest height?

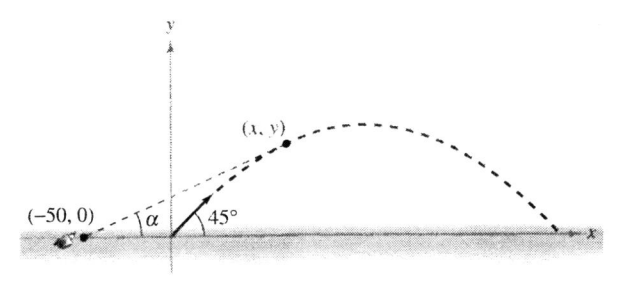

12. Consider the distance d between the launch site and the projectile in Exercise 11.

    (a) Express the distance d in terms of x and y and in terms of the parameter t.

    (b) Use the results in part (a) to find the rate of change of d.

    (c) Find the rate of change of the distance when $t = 2$.

    (d) When is the rate of change of d minimum during the flight of the projectile? Does this occur at the time when the projectile reaches its maximum height?

13. Consider the function

    $$f(x, y) = (\alpha x^2 + \beta y^2)e^{-(x^2+y^2)}, \quad 0 < |\alpha| < \beta.$$

    (a) Use a computer algebra system to graph the function for $\alpha = 1$ and $\beta = 2$, and identify any extrema or saddle points.

    (b) Use a computer algebra system to graph the function for $\alpha = -1$ and $\beta = 2$, and identify any extrema or saddle points.

    (c) Generalize the results in parts (a) and (b) for the function f.

14. Prove that if f is a differentiable function such that $\nabla f(x_0, y_0) = \mathbf{0}$, then the tangent plane at (x_0, y_0) is horizontal.

15. The figure shows a rectangle that is approximately $l = 6$ centimeters long and $h = 1$ centimeter high.

    (a) Draw a rectangular strip along the rectangular region showing a small increase in length.

    (b) Draw a rectangular strip along the rectangular region showing a small increase in height.

    (c) Use the results in parts (a) and (b) to identify the measurement that has more effect on the area A of the rectangle.

    (d) Verify your answer in part (c) analytically by comparing the value of dA when $dl = 0.01$ and when $dh = 0.01$.

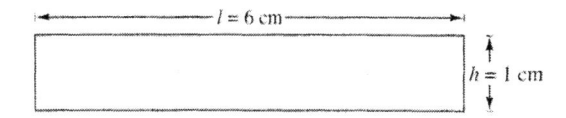

16. Consider converting a point $(5 \pm 0.05, \pi/18 \pm 0.05)$ in polar coordinates to rectangular coordinates (x, y).

    (a) Use a geometric argument to determine whether the accuracy in x is more dependent on the accuracy in r or on the accuracy in θ. Explain. Verify your answer analytically.

    (b) Use a geometric argument to determine whether the accuracy in y is more dependent on the accuracy in r or on the accuracy in θ. Explain. Verify your answer analytically.

17. Use the results of Exercises 15 and 16 to write a short paragraph discussing the importance of making accurate measurements and identifying which variables in a formula have the greater effect on the outcome when the formula is applied.

18. Show that

    $$u(x, t) = \frac{1}{2}[\sin(x - t) + \sin(x + t)]$$

    is a solution to the one-dimensional wave equation

    $$\frac{\partial^2 u}{\partial t^2} = \frac{\partial^2 u}{\partial x^2}.$$

19. Show that

    $$u(x, t) = \frac{1}{2}[f(x - ct) + f(x + ct)]$$

    is a solution to the one-dimensional wave equation

    $$\frac{\partial^2 u}{\partial t^2} = c^2 \frac{\partial^2 u}{\partial x^2}.$$

    (This equation describes the small transverse vibration of an elastic string such as those on certain musical instruments.)

Hyperthermia Treatments for Tumors

Hyperthermia treatment uses elevated temperatures to destroy malignant tissues. Microwaves are employed to raise tumor temperatures to about 107°F. The concept of heat treatment is not a new one, and doctors estimate that application of sufficient heat can increase the effectiveness of both radiation and chemotherapy by a factor of 2. Still, the technique is not in wide use today because of difficulties in focusing the energy on the tumor site.

The latest technology, Adaptive Phased Array (APA), is expected to solve the focusing problem. Originally developed to track targets in defense radar systems, APA technology was adapted by the Massachusetts Institute of Technology. In 1996, MIT licensed the technology to Celsion Corporation (formerly Cheung Laboratories), who incorporated it into a device that is able to direct a beam of energy directly into the tumor, minimizing temperature elevation in surrounding areas.

The tumor temperature during treatment is highest at the center and gradually decreases toward the edges. Regions of tissue having the same temperature, or equitherms, can be visualized as closed surfaces that are nested one inside the other. The problem of determining the portion of the tumor that has been heated to an effective temperature reduces to finding the ratio V_T/V, where V is the volume of the entire tumor and V_T is the volume of the portion of the tumor that is heated above temperature T. Using APA technology, the shapes of the equithermal surfaces are determined by the type of applicator that is used, which in turn is determined by the shape of the tumor.

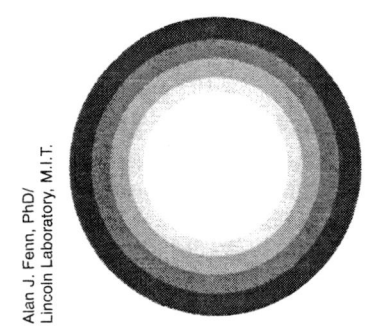

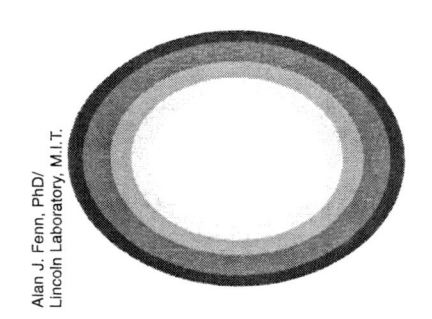

Thermal patterns determined by APA technology

QUESTIONS

1. When treating a spherical tumor, a technician uses a probe to determine that the temperature has reached an appropriate level to about half the radius of the tumor. What is the ratio V_T/V? Is it $\frac{1}{2}$? Explain your reasoning.

2. Consider an ellipsoidal tumor that can be modeled by the equation

$$\frac{x^2}{2.5} + \frac{y^2}{6.5} + \frac{z^2}{2.5} = 1.$$

Consider a sequence of five ellipsoidal equitherms whose major and minor axes increase linearly until the fifth equitherm is the entire tumor. Write an equation for each of the five equitherms. Then find the ratio V_T/V for each of the five equitherms.

The concepts presented here will be explored further in this chapter. For an extension of this application, see Lab 18 in the lab series that accompanies this text at college.hmco.com.

Multiple Integration 13

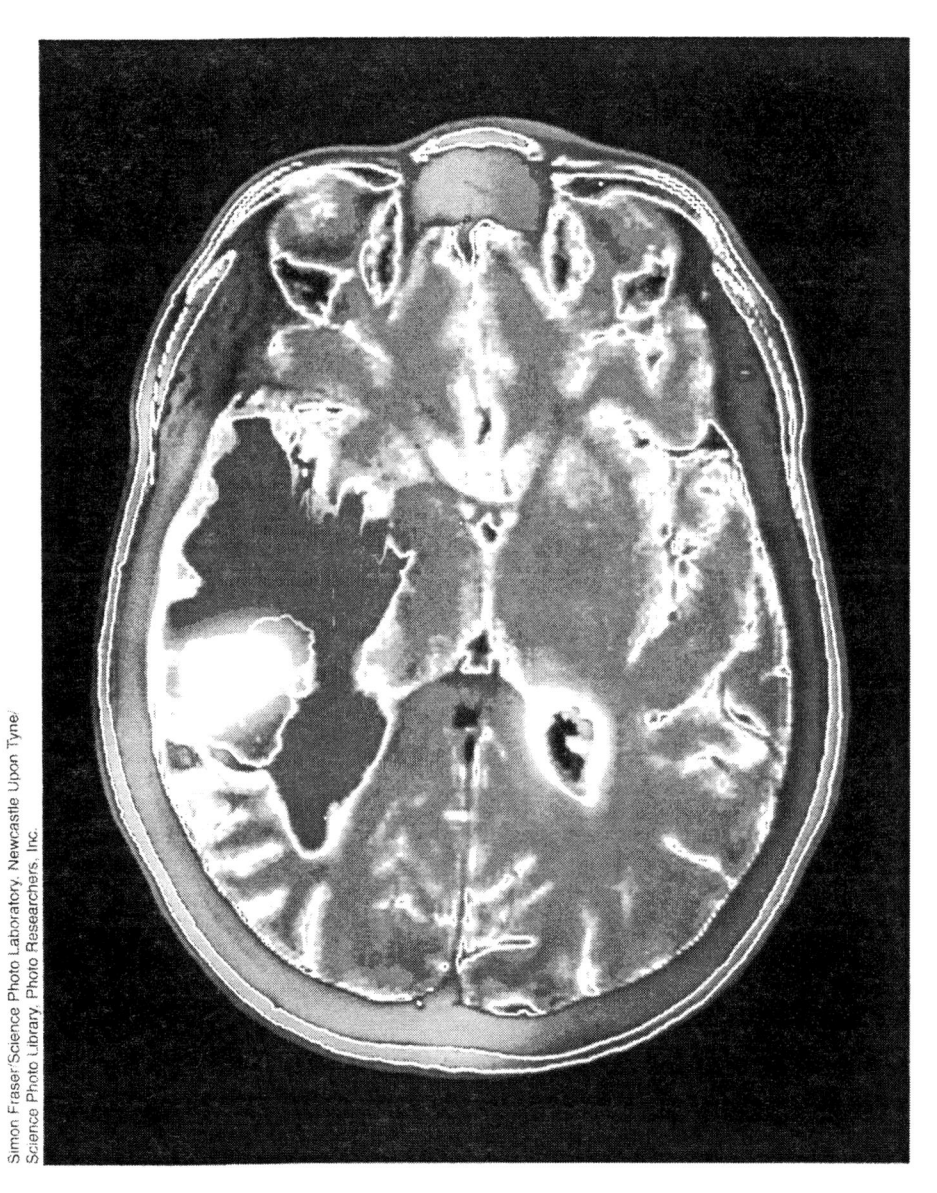

CLI's Hyperthermia Microfocus 1000 introduces a practical way to apply microwaves to tumors, and may bring heat therapy into general use.

Haim I. Bicher, M.D., a leader in hyperthermia treatment, founded the Valley Cancer Institute in 1985. Since then, it has become the largest nonprofit hyperthermic treatment facility in the world.

NOTE   Equations for other possible tumor shapes are given in the Section Project in Section 13.7.

Section 13.1 Iterated Integrals and Area in the Plane

- Evaluate an iterated integral.
- Use an iterated integral to find the area of a plane region.

NOTE In Chapters 13 and 14, you will study several applications of integration involving functions of several variables. Chapter 13 is much like Chapter 6 in that it surveys the use of integration to find plane areas, volumes, surface areas, moments, and centers of mass.

Iterated Integrals

In Chapter 12, you saw that it is meaningful to differentiate functions of several variables with respect to one variable while holding the other variables constant. You can *integrate* functions of several variables by a similar procedure. For example, if you are given the partial derivative

$$f_x(x, y) = 2xy$$

then, by considering y constant, you can integrate with respect to x to obtain

$$\begin{aligned}
f(x, y) &= \int f_x(x, y)\, dx & &\text{Integrate with respect to } x. \\
&= \int 2xy\, dx & &\text{Hold } y \text{ constant.} \\
&= y \int 2x\, dx & &\text{Factor out constant } y. \\
&= y(x^2) + C(y) & &\text{Antiderivative of } 2x \text{ is } x^2. \\
&= x^2 y + C(y). & &C(y) \text{ is a function of } y.
\end{aligned}$$

The "constant" of integration, $C(y)$, is a function of y. In other words, by integrating with respect to x, you are able to recover $f(x, y)$ only partially. The total recovery of a function of x and y from its partial derivatives is a topic you will study in Chapter 14. For now, we are more concerned with extending definite integrals to functions of several variables. For instance, by considering y constant, you can apply the Fundamental Theorem of Calculus to evaluate

$$\int_1^{2y} 2xy\, dx = x^2 y \Big]_1^{2y} = (2y)^2 y - (1)^2 y = 4y^3 - y.$$

- x is the variable of integration and y is fixed.
- Replace x by the limits of integration.
- The result is a function of y.

Similarly, you can integrate with respect to y by holding x fixed. Both procedures are summarized as follows.

$$\int_{h_1(y)}^{h_2(y)} f_x(x, y)\, dx = f(x, y) \Big]_{h_1(y)}^{h_2(y)} = f(h_2(y), y) - f(h_1(y), y) \quad \text{With respect to } x$$

$$\int_{g_1(x)}^{g_2(x)} f_y(x, y)\, dy = f(x, y) \Big]_{g_1(x)}^{g_2(x)} = f(x, g_2(x)) - f(x, g_1(x)) \quad \text{With respect to } y$$

Note that the variable of integration cannot appear in either limit of integration. For instance, it makes no sense to write

$$\int_0^x y\, dx.$$

Example 1 Integrating with Respect to y

Evaluate $\int_1^x (2x^2 y^{-2} + 2y)\, dy$.

Solution Considering x to be constant and integrating with respect to y produces

$$\int_1^x (2x^2 y^{-2} + 2y)\, dy = \left[\frac{-2x^2}{y} + y^2 \right]_1^x \quad \text{Integrate with respect to } y.$$

$$= \left(\frac{-2x^2}{x} + x^2 \right) - \left(\frac{-2x^2}{1} + 1 \right)$$

$$= 3x^2 - 2x - 1.$$

Notice in Example 1 that the integral defines a function of x and can *itself* be integrated, as shown in the next example.

Example 2 The Integral of an Integral

Evaluate $\int_1^2 \left[\int_1^x (2x^2 y^{-2} + 2y)\, dy \right] dx$.

Solution Using the result of Example 1, you have

$$\int_1^2 \left[\int_1^x (2x^2 y^{-2} + 2y)\, dy \right] dx = \int_1^2 (3x^2 - 2x - 1)\, dx$$

$$= \left[x^3 - x^2 - x \right]_1^2 \quad \text{Integrate with respect to } x.$$

$$= 2 - (-1)$$

$$= 3.$$

The integral in Example 2 is an **iterated integral**. The brackets used in Example 2 are normally not written. Instead, iterated integrals are usually written simply as

$$\int_a^b \int_{g_1(x)}^{g_2(x)} f(x, y)\, dy\, dx \quad \text{and} \quad \int_c^d \int_{h_1(y)}^{h_2(y)} f(x, y)\, dx\, dy.$$

The **inside limits of integration** can be variable with respect to the outer variable of integration. However, the **outside limits of integration** must be constant with respect to both variables of integration. After performing the inside integration, you obtain a "standard" definite integral, and the second integration produces a real number. The limits of integration for an iterated integral identify two sets of boundary intervals for the variables. For instance, in Example 2, the outside limits indicate that x lies in the interval $1 \leq x \leq 2$ and the inside limits indicate that y lies in the interval $1 \leq y \leq x$. Together, these two intervals determine the **region of integration R** of the iterated integral, as shown in Figure 13.1.

Because an iterated integral is just a special type of definite integral—one in which the integrand is also an integral—you can use the properties of definite integrals to evaluate iterated integrals.

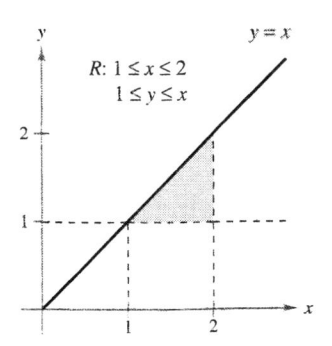

The region of integration for

$$\int_1^2 \int_1^x f(x, y)\, dy\, dx$$

Figure 13.1

Area of a Plane Region

In the remainder of this section, you will take a new look at an old problem—that of finding the area of a plane region. Consider the plane region R bounded by $a \le x \le b$ and $g_1(x) \le y \le g_2(x)$, as shown in Figure 13.2. The area of R is given by the definite integral

$$\int_a^b [g_2(x) - g_1(x)]\, dx. \qquad \text{Area of } R$$

Using the Fundamental Theorem of Calculus, you can rewrite the integrand $g_2(x) - g_1(x)$ as a definite integral. Specifically, if you consider x to be fixed and let y vary from $g_1(x)$ to $g_2(x)$, you can write

$$\int_{g_1(x)}^{g_2(x)} dy = y\Big]_{g_1(x)}^{g_2(x)} = g_2(x) - g_1(x).$$

Combining these two integrals, you can write the area of the region R as an iterated integral

$$\int_a^b \int_{g_1(x)}^{g_2(x)} dy\, dx = \int_a^b y\Big]_{g_1(x)}^{g_2(x)} dx \qquad \text{Area of } R$$

$$= \int_a^b [g_2(x) - g_1(x)]\, dx.$$

Placing a representative rectangle in the region R helps determine both the order and the limits of integration. A vertical rectangle implies the order $dy\, dx$, with the inside limits corresponding to the upper and lower bounds of the rectangle, as shown in Figure 13.2. This type of region is called **vertically simple,** because the outside limits of integration represent the vertical lines $x = a$ and $x = b$.

Similarly, a horizontal rectangle implies the order $dx\, dy$, with the inside limits determined by the left and right bounds of the rectangle, as shown in Figure 13.3. This type of region is called **horizontally simple,** because the outside limits represent the horizontal lines $y = c$ and $y = d$. The iterated integrals used for these two types of simple regions are summarized as follows.

Region is bounded by
$a \le x \le b$
$g_1(x) \le y \le g_2(x)$

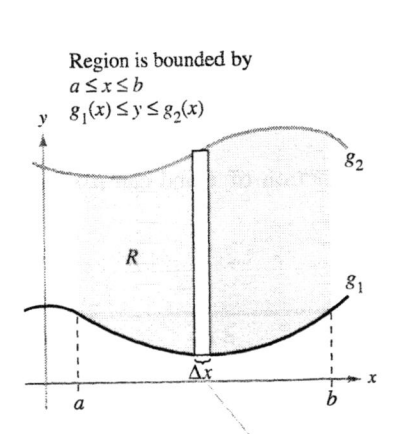

Area $= \displaystyle\int_a^b \int_{g_1(x)}^{g_2(x)} dy\, dx$

Vertically simple region
Figure 13.2

Region is bounded by
$c \le y \le d$
$h_1(y) \le x \le h_2(y)$

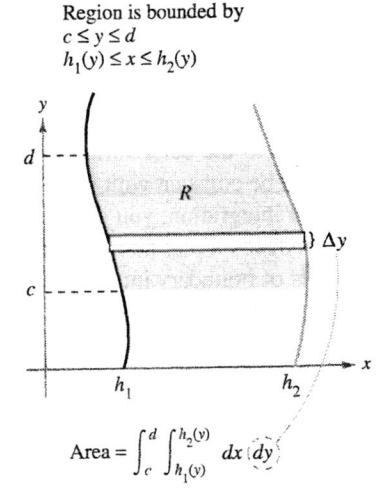

Area $= \displaystyle\int_c^d \int_{h_1(y)}^{h_2(y)} dx\, dy$

Horizontally simple region
Figure 13.3

Area of a Region in the Plane

1. If R is defined by $a \le x \le b$ and $g_1(x) \le y \le g_2(x)$, where g_1 and g_2 are continuous on $[a, b]$, then the area of R is given by

$$A = \int_a^b \int_{g_1(x)}^{g_2(x)} dy\, dx. \qquad \text{Figure 13.2 (vertically simple)}$$

2. If R is defined by $c \le y \le d$ and $h_1(y) \le x \le h_2(y)$, where h_1 and h_2 are continuous on $[c, d]$, then the area of R is given by

$$A = \int_c^d \int_{h_1(y)}^{h_2(y)} dx\, dy. \qquad \text{Figure 13.3 (horizontally simple)}$$

NOTE Be sure you see that the order of integration of these two integrals is different—the order $dy\, dx$ corresponds to a vertically simple region, and the order $dx\, dy$ corresponds to a horizontally simple region.

If all four limits of integration happen to be constants, the region of integration is rectangular, as shown in Example 3.

Example 3 The Area of a Rectangular Region

Use an iterated integral to represent the area of the rectangle shown in Figure 13.4.

Solution The region shown in Figure 13.4 is both vertically simple and horizontally simple, so you can use either order of integration. By choosing the order $dy\,dx$, you obtain the following.

$$\int_a^b \int_c^d dy\,dx = \int_a^b \Big[y \Big]_c^d dx \qquad \text{Integrate with respect to } y.$$

$$= \int_a^b (d-c)\,dx$$

$$= \Big[(d-c)x \Big]_a^b \qquad \text{Integrate with respect to } x.$$

$$= (d-c)(b-a)$$

Notice that this answer is consistent with what you know from geometry.

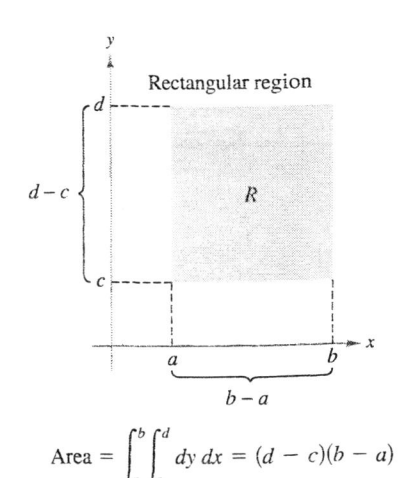

Figure 13.4

Area $= \int_a^b \int_c^d dy\,dx = (d-c)(b-a)$

Example 4 Finding Area by an Iterated Integral

Use an iterated integral to find the area of the region bounded by the graphs of

$f(x) = \sin x$ Sine curve forms upper boundary.

$g(x) = \cos x$ Cosine curve forms lower boundary.

between $x = \pi/4$ and $x = 5\pi/4$.

Solution Because f and g are given as functions of x, a vertical representative rectangle is convenient, and you can choose $dy\,dx$ as the order of integration, as shown in Figure 13.5. The outside limits of integration are $\pi/4 \leq x \leq 5\pi/4$. Moreover, because the rectangle is bounded above by $f(x) = \sin x$ and below by $g(x) = \cos x$, you have

$$\text{Area of } R = \int_{\pi/4}^{5\pi/4} \int_{\cos x}^{\sin x} dy\,dx$$

$$= \int_{\pi/4}^{5\pi/4} \Big[y \Big]_{\cos x}^{\sin x} dx \qquad \text{Integrate with respect to } y.$$

$$= \int_{\pi/4}^{5\pi/4} (\sin x - \cos x)\,dx$$

$$= \Big[-\cos x - \sin x \Big]_{\pi/4}^{5\pi/4} \qquad \text{Integrate with respect to } x.$$

$$= 2\sqrt{2}.$$

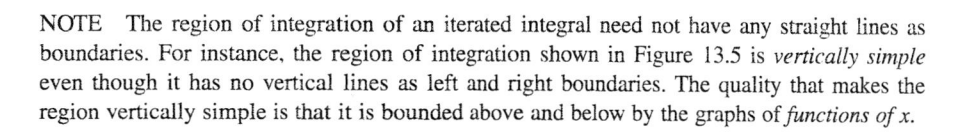

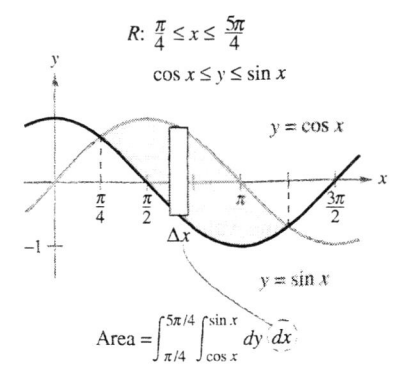

Figure 13.5

$R: \frac{\pi}{4} \leq x \leq \frac{5\pi}{4}$

$\cos x \leq y \leq \sin x$

Area $= \int_{\pi/4}^{5\pi/4} \int_{\cos x}^{\sin x} dy\,dx$

NOTE The region of integration of an iterated integral need not have any straight lines as boundaries. For instance, the region of integration shown in Figure 13.5 is *vertically simple* even though it has no vertical lines as left and right boundaries. The quality that makes the region vertically simple is that it is bounded above and below by the graphs of *functions of x*.

One order of integration will often produce a simpler integration problem than the other order. For instance, try reworking Example 4 with the order $dx\,dy$—you may be surprised to see that the task is formidable. However, if you succeed, you will see that the answer is the same. In other words, the order of integration affects the ease of integration, but not the value of the integral.

Example 5 Comparing Different Orders of Integration

Sketch the region whose area is represented by the integral

$$\int_0^2 \int_{y^2}^4 dx\,dy.$$

Then find another iterated integral using the order $dy\,dx$ to represent the same area and show that both integrals yield the same value.

Solution From the given limits of integration, you know that

$$y^2 \le x \le 4 \qquad \text{Inner limits of integration}$$

which means that the region R is bounded on the left by the parabola $x = y^2$ and on the right by the line $x = 4$. Furthermore, because

$$0 \le y \le 2 \qquad \text{Outer limits of integration}$$

you know that R is bounded below by the x-axis, as shown in Figure 13.6(a). The value of this integral is

$$\int_0^2 \int_{y^2}^4 dx\,dy = \int_0^2 \left[x \right]_{y^2}^4 dy \qquad \text{Integrate with respect to } x.$$

$$= \int_0^2 (4 - y^2)\,dy$$

$$= \left[4y - \frac{y^3}{3} \right]_0^2 = \frac{16}{3}. \qquad \text{Integrate with respect to } y.$$

To change the order of integration to $dy\,dx$, place a vertical rectangle in the region, as shown in Figure 13.6(b). From this you can see that the constant bounds $0 \le x \le 4$ serve as the outer limits of integration. By solving for y in the equation $x = y^2$, you can conclude that the inner bounds are $0 \le y \le \sqrt{x}$. Therefore, the area of the region can also be represented by

$$\int_0^4 \int_0^{\sqrt{x}} dy\,dx.$$

By evaluating this integral, you can see that it has the same value as the original integral.

$$\int_0^4 \int_0^{\sqrt{x}} dy\,dx = \int_0^4 \left[y \right]_0^{\sqrt{x}} dx \qquad \text{Integrate with respect to } y.$$

$$= \int_0^4 \sqrt{x}\,dx$$

$$= \frac{2}{3} x^{3/2} \bigg]_0^4 = \frac{16}{3} \qquad \text{Integrate with respect to } x.$$

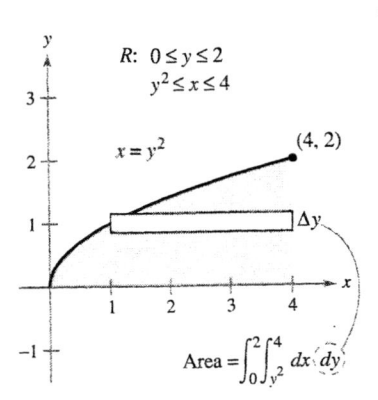

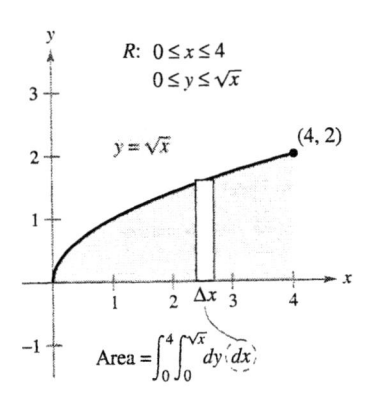

(a) $R: 0 \le y \le 2,\ y^2 \le x \le 4$; $x = y^2$; Area $= \int_0^2 \int_{y^2}^4 dx\,dy$

(b) $R: 0 \le x \le 4,\ 0 \le y \le \sqrt{x}$; $y = \sqrt{x}$; Area $= \int_0^4 \int_0^{\sqrt{x}} dy\,dx$

Figure 13.6

indicates that in the Interactive 3.0 *CD-ROM* and Internet 3.0 *versions of this text (available at* college.hmco.com) *you will find an Open Exploration, which further explores this example using the computer algebra systems* Maple, Mathcad, Mathematica, *and* Derive.

TECHNOLOGY Some computer software can perform symbolic integration for integrals such as those in Example 6. If you have access to such software, try using it to evaluate the integrals in the exercises and examples given in this section.

Sometimes it is not possible to calculate the area of a region with a single iterated integral. In these cases you can divide the region into subregions such that the area of each subregion can be calculated by an iterated integral. The total area is then the sum of the iterated integrals.

Example 6 An Area Represented by Two Iterated Integrals

Find the area of the region R that lies below the parabola

$$y = 4x - x^2 \qquad \text{Parabola forms upper boundary.}$$

above the x-axis, and above the line

$$y = -3x + 6 \qquad \text{Line and } x\text{-axis form lower boundary.}$$

as shown in Figure 13.7.

$$\text{Area} = \int_1^2 \int_{-3x+6}^{4x-x^2} dy\, dx + \int_2^4 \int_0^{4x-x^2} dy\, dx$$

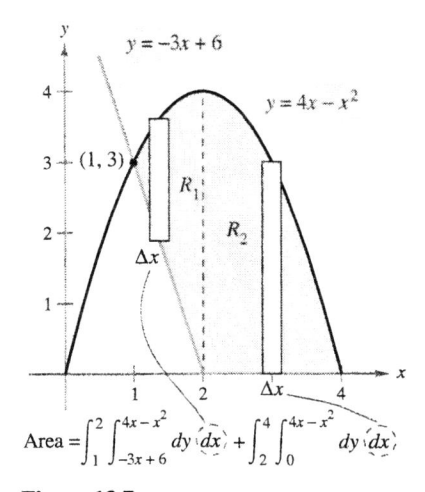

Figure 13.7

Solution Begin by dividing R into the two subregions R_1 and R_2 shown in Figure 13.7. In both regions, it is convenient to use vertical rectangles, and you have

$$\text{Area} = \int_1^2 \int_{-3x+6}^{4x-x^2} dy\, dx + \int_2^4 \int_0^{4x-x^2} dy\, dx$$

$$= \int_1^2 (4x - x^2 + 3x - 6)\, dx + \int_2^4 (4x - x^2)\, dx$$

$$= \left[\frac{7x^2}{2} - \frac{x^3}{3} - 6x \right]_1^2 + \left[2x^2 - \frac{x^3}{3} \right]_2^4$$

$$= \left(14 - \frac{8}{3} - 12 - \frac{7}{2} + \frac{1}{3} + 6 \right) + \left(32 - \frac{64}{3} - 8 + \frac{8}{3} \right) = \frac{15}{2}.$$

The area of the region is $15/2$ square units. Try checking this using the procedure for finding the area between two curves, as presented in Section 6.1.

NOTE In Examples 3 to 6, be sure you see the benefit of sketching the region of integration. We strongly recommend that you develop the habit of making sketches to help determine the limits of integration for all iterated integrals in this chapter.

At this point you may be wondering why you would need iterated integrals. After all, you already know how to use conventional integration to find the area of a region in the plane. (For instance, compare the solution of Example 4 in this section with that given in Example 3 in Section 6.1.) The need for iterated integrals will become clear in the next section. In this section, we have chosen to give primary attention to procedures for finding the limits of integration of the region of an iterated integral, and the following exercise set is designed to develop skill in this important procedure.

EXERCISES FOR SECTION 13.1

In Exercises 1–10, evaluate the integral.

1. $\int_0^x (2x - y)\, dy$

2. $\int_x^{x^2} \frac{y}{x}\, dy$

3. $\int_1^{2y} \frac{y}{x}\, dx$

4. $\int_0^{\cos y} y\, dx$

5. $\int_0^{\sqrt{4-x^2}} x^2 y\, dy$

6. $\int_{x^3}^{\sqrt{x}} (x^2 + 3y^2)\, dy$

7. $\int_{e^y}^{y} \frac{y \ln x}{x}\, dx$

8. $\int_{-\sqrt{1-y^2}}^{\sqrt{1-y^2}} (x^2 + y^2)\, dx$

9. $\int_0^{x^3} y e^{-y/x}\, dy$

10. $\int_y^{\pi/2} \sin^3 x \cos y\, dx$

In Exercises 11–22, evaluate the iterated integral.

11. $\int_0^1 \int_0^2 (x + y)\, dy\, dx$

12. $\int_{-1}^1 \int_{-2}^2 (x^2 - y^2)\, dy\, dx$

13. $\int_0^1 \int_0^x \sqrt{1 - x^2}\, dy\, dx$

14. $\int_{-4}^4 \int_0^{x^2} \sqrt{64 - x^3}\, dy\, dx$

15. $\int_1^2 \int_0^4 (x^2 - 2y^2 + 1)\, dx\, dy$

16. $\int_0^2 \int_y^{2y} (10 + 2x^2 + 2y^2)\, dx\, dy$

17. $\int_0^1 \int_0^{\sqrt{1-y^2}} (x + y)\, dx\, dy$

18. $\int_0^2 \int_{3y^2-6y}^{2y-y^2} 3y\, dx\, dy$

19. $\int_0^2 \int_0^{\sqrt{4-y^2}} \frac{2}{\sqrt{4-y^2}}\, dx\, dy$

20. $\int_0^{\pi/2} \int_0^{2\cos\theta} r\, dr\, d\theta$

21. $\int_0^{\pi/2} \int_0^{\sin\theta} \theta r\, dr\, d\theta$

22. $\int_0^{\pi/4} \int_0^{\cos\theta} 3r^2 \sin\theta\, dr\, d\theta$

In Exercises 23–26, evaluate the improper iterated integral.

23. $\int_1^{\infty} \int_0^{1/x} y\, dy\, dx$

24. $\int_0^3 \int_0^{\infty} \frac{x^2}{1 + y^2}\, dy\, dx$

25. $\int_1^{\infty} \int_1^{\infty} \frac{1}{xy}\, dx\, dy$

26. $\int_0^{\infty} \int_0^{\infty} xy e^{-(x^2+y^2)}\, dx\, dy$

In Exercises 27–32, use an iterated integral to find the area of the region.

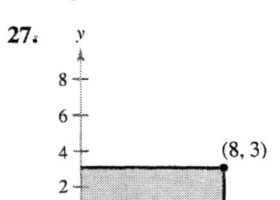

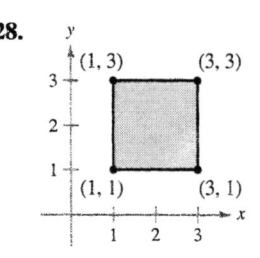

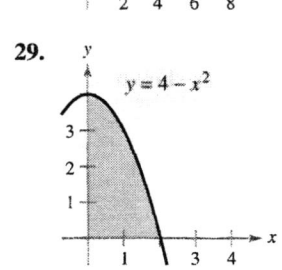

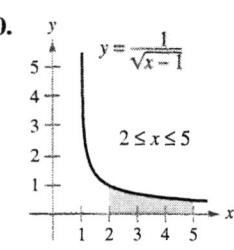

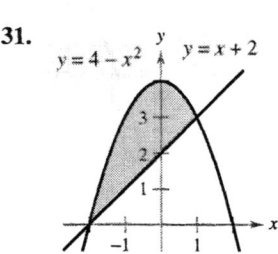

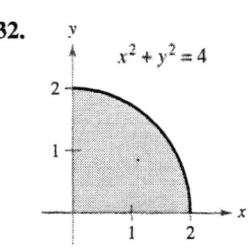

27.

28.

29.

30.

31.

32.

In Exercises 33–38, use an iterated integral to find the area of the region bounded by the graphs of the equations.

33. $\sqrt{x} + \sqrt{y} = 2, x = 0, y = 0$

34. $y = x^{3/2}, y = 2x$

35. $2x - 3y = 0, x + y = 5, y = 0$

36. $xy = 9, y = x, y = 0, x = 9$

37. $\dfrac{x^2}{a^2} + \dfrac{y^2}{b^2} = 1$

38. $y = x, y = 2x, x = 2$

In Exercises 39–46, sketch the region R of integration and switch the order of integration.

39. $\int_0^4 \int_0^y f(x, y)\, dx\, dy$

40. $\int_0^4 \int_{\sqrt{y}}^2 f(x, y)\, dx\, dy$

41. $\int_{-2}^2 \int_0^{\sqrt{4-x^2}} f(x, y)\, dy\, dx$

42. $\int_0^2 \int_0^{4-x^2} f(x, y)\, dy\, dx$

43. $\int_1^{10} \int_0^{\ln y} f(x, y)\, dx\, dy$

44. $\int_{-1}^2 \int_0^{e^{-x}} f(x, y)\, dy\, dx$

45. $\int_{-1}^1 \int_{x^2}^1 f(x, y)\, dy\, dx$

46. $\int_{-\pi/2}^{\pi/2} \int_0^{\cos x} f(x, y)\, dy\, dx$

In Exercises 47–56, sketch the region R whose area is given by the iterated integral. Then switch the order of integration and show that both orders yield the same area.

47. $\int_0^1 \int_0^2 dy\, dx$

48. $\int_1^2 \int_2^4 dx\, dy$

49. $\int_0^1 \int_{-\sqrt{1-y^2}}^{\sqrt{1-y^2}} dx\, dy$

50. $\int_{-2}^2 \int_{-\sqrt{4-x^2}}^{\sqrt{4-x^2}} dy\, dx$

51. $\int_0^2 \int_0^x dy\, dx + \int_2^4 \int_0^{4-x} dy\, dx$

52. $\int_0^4 \int_0^{x/2} dy\, dx + \int_4^6 \int_0^{6-x} dy\, dx$

53. $\int_0^2 \int_{x/2}^1 dy\, dx$

54. $\int_0^9 \int_{\sqrt{x}}^3 dy\, dx$

55. $\int_0^1 \int_{y^2}^{\sqrt[3]{y}} dx\, dy$

56. $\int_{-2}^2 \int_0^{4-y^2} dx\, dy$

Think About It In Exercises 57 and 58, give a geometric argument for the given equality. Verify the equality analytically.

57. $\int_0^5 \int_x^{\sqrt{50-x^2}} x^2 y^2\, dy\, dx =$
$\int_0^5 \int_0^y x^2 y^2\, dx\, dy + \int_5^{5\sqrt{2}} \int_0^{\sqrt{50-y^2}} x^2 y^2\, dx\, dy$

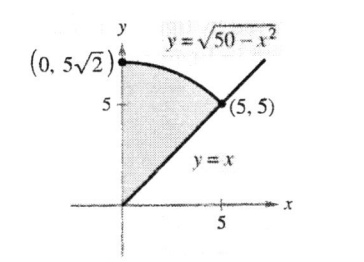

Figure for 57

Figure for 58

58. $\int_0^2 \int_{x^2}^{2x} x \sin y\, dy\, dx = \int_0^4 \int_{y/2}^{\sqrt{y}} x \sin y\, dx\, dy$

In Exercises 59–62, evaluate the iterated integral. (Note that it is necessary to switch the order of integration.)

59. $\int_0^2 \int_x^2 x\sqrt{1+y^3}\, dy\, dx$

60. $\int_0^2 \int_x^2 e^{-y^2}\, dy\, dx$

61. $\int_0^1 \int_y^1 \sin x^2\, dx\, dy$

62. $\int_0^2 \int_{y^2}^4 \sqrt{x} \sin x\, dx\, dy$

In Exercises 63–66, use a computer algebra system to evaluate the iterated integral.

63. $\int_0^2 \int_{x^2}^{2x} (x^3 + 3y^2)\, dy\, dx$

64. $\int_0^1 \int_y^{2y} \sin(x+y)\, dx\, dy$

65. $\int_0^4 \int_0^y \frac{2}{(x+1)(y+1)}\, dx\, dy$

66. $\int_0^a \int_0^{a-x} (x^2 + y^2)\, dy\, dx$

In Exercises 67 and 68, (a) sketch the region of integration, (b) switch the order of integration, and (c) use a computer algebra system to show that both orders yield the same value.

67. $\int_0^2 \int_{y^3}^{4\sqrt{2y}} (x^2 y - xy^2)\, dx\, dy$

68. $\int_0^2 \int_{\sqrt{4-x^2}}^{4-x^2/4} \frac{xy}{x^2+y^2+1}\, dy\, dx$

In Exercises 69–72, use a computer algebra system to approximate the iterated integral.

69. $\int_0^2 \int_0^{4-x^2} e^{xy}\, dy\, dx$

70. $\int_0^2 \int_x^2 \sqrt{16-x^3-y^3}\, dy\, dx$

71. $\int_0^{2\pi} \int_0^{1+\cos\theta} 6r^2 \cos\theta\, dr\, d\theta$

72. $\int_0^{\pi/2} \int_0^{1+\sin\theta} 15\theta r\, dr\, d\theta$

Getting at the Concept

73. Explain what is meant by an iterated integral. How is it evaluated?

74. Describe regions that are vertically simple and regions that are horizontally simple.

75. Give a geometric description of the region of integration if the inside and outside limits of integration are constants.

76. Why is it sometimes an advantage to change the order of integration?

True or False? In Exercises 77 and 78, determine whether the statement is true or false. If it is false, explain why or give an example that shows it is false.

77. $\int_a^b \int_c^d f(x,y)\, dy\, dx = \int_c^d \int_a^b f(x,y)\, dx\, dy$

78. $\int_0^1 \int_0^x f(x,y)\, dy\, dx = \int_0^1 \int_0^y f(x,y)\, dx\, dy$

Section 13.2 Double Integrals and Volume

- Use a double integral to represent the volume of a solid region.
- Use properties of double integrals.
- Evaluate a double integral as an iterated integral.

Double Integrals and Volume of a Solid Region

You already know that a definite integral over an *interval* uses a limit process to assign measure to quantities such as area, volume, arc length, and mass. In this section, we use a similar process to define the **double integral** of a function of two variables over a *region in the plane*.

Consider a continuous function f such that $f(x, y) \geq 0$ for all (x, y) in a region R in the xy-plane. The goal is to find the volume of the solid region lying between the surface given by

$$z = f(x, y) \qquad \text{Surface lying above the } xy\text{-plane}$$

and the xy-plane, as shown in Figure 13.8. You can begin by superimposing a rectangular grid over the region, as shown in Figure 13.9. The rectangles lying entirely within R form an **inner partition** Δ, whose **norm** $\|\Delta\|$ is defined as the length of the longest diagonal of the n rectangles. Next, choose a point (x_i, y_i) in each rectangle and form the rectangular prism whose height is $f(x_i, y_i)$, as shown in Figure 13.10. Because the area of the ith rectangle is

$$\Delta A_i \qquad \text{Area of } i\text{th rectangle}$$

it follows that the volume of the ith prism is

$$f(x_i, y_i) \, \Delta A_i \qquad \text{Volume of } i\text{th prism}$$

and you can approximate the volume of the solid region by the Riemann sum of the volumes of all n prisms,

$$\sum_{i=1}^{n} f(x_i, y_i) \, \Delta A_i \qquad \text{Riemann sum}$$

as shown in Figure 13.11. This approximation can be improved by tightening the mesh of the grid to form smaller and smaller rectangles, as shown in Example 1.

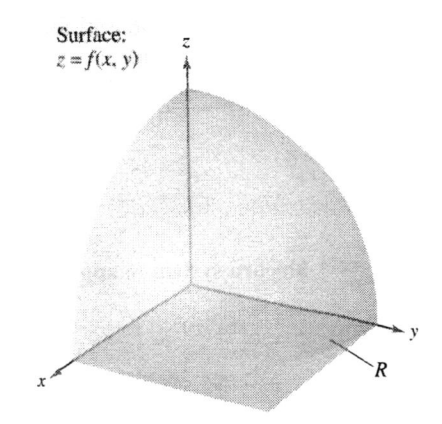

Surface:
$z = f(x, y)$

Figure 13.8

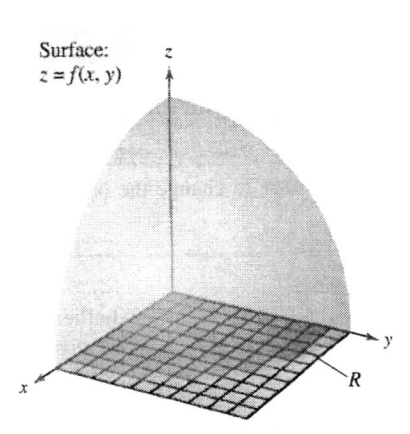

The rectangles lying within R form an inner partition of R.
Figure 13.9

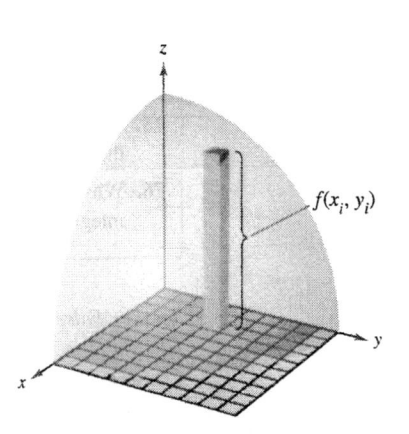

Rectangular prism whose base has an area of ΔA_i and whose height is $f(x_i, y_i)$
Figure 13.10

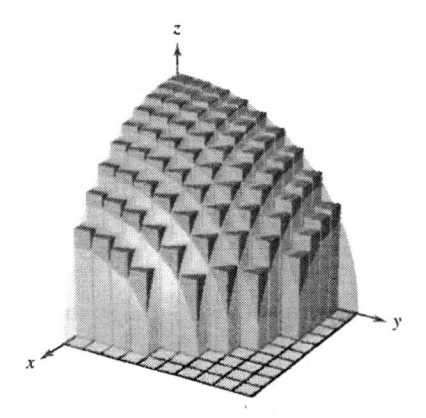

Volume approximated by rectangular prisms
Figure 13.11

Example 1 Approximating the Volume of a Solid

Approximate the volume of the solid lying between the paraboloid

$$f(x, y) = 1 - \frac{1}{2}x^2 - \frac{1}{2}y^2$$

and the square region R given by $0 \le x \le 1$, $0 \le y \le 1$. Use a partition made up of squares whose edges have a length of $\frac{1}{4}$.

Solution Begin by forming the specified partition of R. For this partition, it is convenient to choose the centers of the subregions as the points at which to evaluate $f(x, y)$.

$\left(\frac{1}{8}, \frac{1}{8}\right)$ $\left(\frac{1}{8}, \frac{3}{8}\right)$ $\left(\frac{1}{8}, \frac{5}{8}\right)$ $\left(\frac{1}{8}, \frac{7}{8}\right)$
$\left(\frac{3}{8}, \frac{1}{8}\right)$ $\left(\frac{3}{8}, \frac{3}{8}\right)$ $\left(\frac{3}{8}, \frac{5}{8}\right)$ $\left(\frac{3}{8}, \frac{7}{8}\right)$
$\left(\frac{5}{8}, \frac{1}{8}\right)$ $\left(\frac{5}{8}, \frac{3}{8}\right)$ $\left(\frac{5}{8}, \frac{5}{8}\right)$ $\left(\frac{5}{8}, \frac{7}{8}\right)$
$\left(\frac{7}{8}, \frac{1}{8}\right)$ $\left(\frac{7}{8}, \frac{3}{8}\right)$ $\left(\frac{7}{8}, \frac{5}{8}\right)$ $\left(\frac{7}{8}, \frac{7}{8}\right)$

Because the area of each square is $\Delta A_i = \frac{1}{16}$, you can approximate the volume by the sum

$$\sum_{i=1}^{16} f(x_i, y_i) \Delta A_i = \sum_{i=1}^{16} \left(1 - \frac{1}{2}x_i^2 - \frac{1}{2}y_i^2\right)\left(\frac{1}{16}\right)$$
$$\approx 0.672.$$

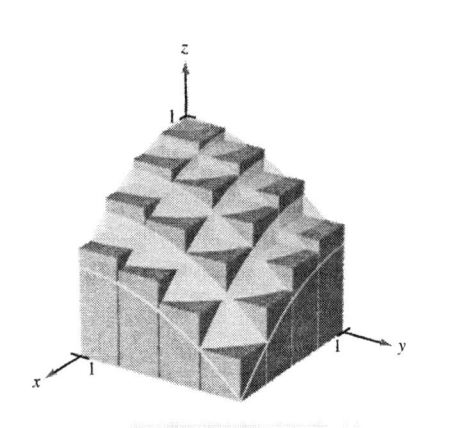

This approximation is shown graphically in Figure 13.12. The exact volume of the solid is $\frac{2}{3}$ (see Example 2). You can obtain a better approximation by using a finer partition. For example, with a partition of squares with sides of length $\frac{1}{10}$, the approximation is 0.668.

TECHNOLOGY Some three-dimensional graphing utilities are capable of sketching figures such as that shown in Figure 13.12. For instance, the sketch shown in Figure 13.13 was drawn with a computer program. In this sketch, note that each of the rectangular prisms lies within the solid region.

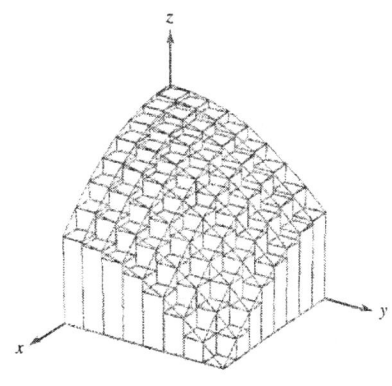

Surface:
$f(x, y) = 1 - \frac{1}{2}x^2 - \frac{1}{2}y^2$

Figure 13.12

In Example 1, note that by using finer partitions, you can obtain better approximations of the volume. This observation suggests that you could obtain the exact volume by taking a limit. That is,

$$\text{Volume} = \lim_{\|\Delta\| \to 0} \sum_{i=1}^{n} f(x_i, y_i) \Delta A_i.$$

The precise meaning of this limit is that the limit is equal to L if for every $\varepsilon > 0$ there exists a $\delta > 0$ such that

$$\left| L - \sum_{i=1}^{n} f(x_i, y_i) \Delta A_i \right| < \varepsilon$$

for all partitions Δ of the plane region R (that satisfy $\|\Delta\| < \delta$) and for all possible choices of x_i and y_i in the ith region.

Using the limit of a Riemann sum to define volume is a special case of using the limit to define a **double integral.** The general case, however, does not require that the function be positive or continuous.

Figure 13.13

EXPLORATION

The entries in the table represent the depth (in 10-yard units) of earth at the center of each square in the figure below.

x \ y	1	2	3
1	10	9	7
2	7	7	4
3	5	5	4
4	4	5	3

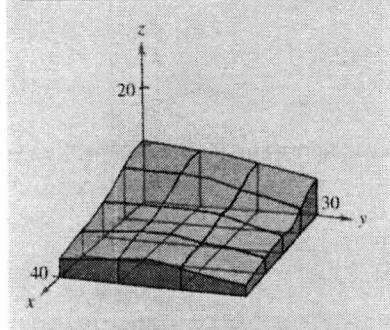

Approximate the number of cubic yards of earth in the first octant. This exploration was submitted by Robert Vojack, Ridgewood High School, Ridgewood, NJ.

Definition of Double Integral

If f is defined on a closed, bounded region R in the xy-plane, then the **double integral of f over R** is given by

$$\iint_R f(x, y)\, dA = \lim_{\|\Delta\| \to 0} \sum_{i=1}^{n} f(x_i, y_i)\, \Delta A_i$$

provided the limit exists. If the limit exists, then f is **integrable** over R.

NOTE Having defined a double integral, we will occasionally refer to a definite integral as a **single integral**.

Sufficient conditions for the double integral of f on the region R to exist are that R can be written as a union of a finite number of nonoverlapping (see Figure 13.14) subregions that are vertically or horizontally simple *and* that f is continuous on the region R.

A double integral can be used to find the volume of a solid region that lies between the xy-plane and the surface given by $z = f(x, y)$.

Volume of a Solid Region

If f is integrable over a plane region R and $f(x, y) \geq 0$ for all (x, y) in R, then the volume of the solid region that lies above R and below the graph of f is defined as

$$V = \iint_R f(x, y)\, dA.$$

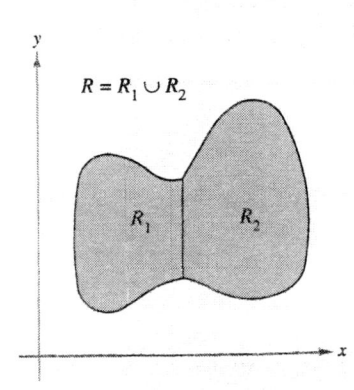

Two regions are nonoverlapping if their intersection is a set that has an area of 0. In this figure, the area of the line segment that is common to R_1 and R_2 is 0.
Figure 13.14

Properties of Double Integrals

Double integrals share many properties of single integrals.

THEOREM 13.1 Properties of Double Integrals

Let f and g be continuous over a closed, bounded plane region R, and let c be a constant.

1. $\iint_R cf(x, y)\, dA = c \iint_R f(x, y)\, dA$

2. $\iint_R [f(x, y) \pm g(x, y)]\, dA = \iint_R f(x, y)\, dA \pm \iint_R g(x, y)\, dA$

3. $\iint_R f(x, y)\, dA \geq 0$, if $f(x, y) \geq 0$.

4. $\iint_R f(x, y)\, dA \geq \iint_R g(x, y)\, dA$, if $f(x, y) \geq g(x, y)$.

5. $\iint_R f(x, y)\, dA = \iint_{R_1} f(x, y)\, dA + \iint_{R_2} f(x, y)\, dA$

where R is the union of two nonoverlapping subregions R_1 and R_2.

Evaluation of Double Integrals

Normally, the first step in evaluating a double integral is to rewrite it as an iterated integral. To show how this is done, we use a geometric model of a double integral as the volume of a solid.

Consider the solid region bounded by the plane $z = f(x, y) = 2 - x - 2y$ and the three coordinate planes, as shown in Figure 13.15. Each vertical cross section taken parallel to the yz-plane is a triangular region whose base has a length of $y = (2 - x)/2$ and whose height is $z = 2 - x$. This implies that for a fixed value of x, the area of the triangular cross section is

$$A(x) = \frac{1}{2}(\text{base})(\text{height}) = \frac{1}{2}\left(\frac{2-x}{2}\right)(2-x) = \frac{(2-x)^2}{4}.$$

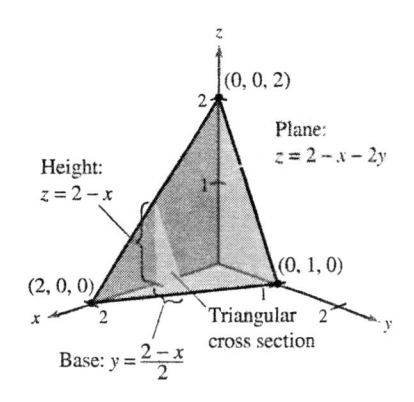

By the formula for the volume of a solid with known cross sections (Section 6.2), the volume of the solid is

$$\begin{aligned}\text{Volume} &= \int_a^b A(x)\,dx \\ &= \int_0^2 \frac{(2-x)^2}{4}\,dx \\ &= -\frac{(2-x)^3}{12}\Big]_0^2 = \frac{2}{3}.\end{aligned}$$

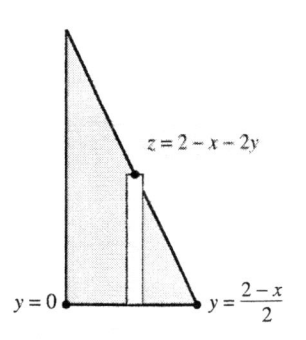

This procedure works no matter how $A(x)$ is obtained. In particular, you can find $A(x)$ by integration, as indicated in Figure 13.16. That is, you consider x to be constant, and integrate $z = 2 - x - 2y$ from 0 to $(2 - x)/2$ to obtain

$$\begin{aligned}A(x) &= \int_0^{(2-x)/2} (2 - x - 2y)\,dy \\ &= \Big[(2-x)y - y^2\Big]_0^{(2-x)/2} \\ &= \frac{(2-x)^2}{4}.\end{aligned}$$

Combining these results, you have the *iterated integral*

$$\text{Volume} = \iint_R f(x, y)\,dA = \int_0^2 \int_0^{(2-x)/2} (2 - x - 2y)\,dy\,dx.$$

To better understand this procedure, it helps to imagine the integration as two sweeping motions. For the inner integration, a vertical line sweeps out the area of a cross section. For the outer integration, the triangular cross section sweeps out the volume, as shown in Figure 13.17.

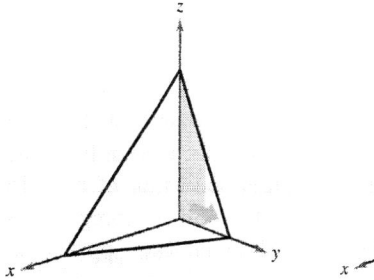

Volume: $\int_0^2 A(x)\,dx$

Figure 13.15

Triangular cross section
Figure 13.16

Integrate with respect to y to obtain the area of the cross section.
Figure 13.17

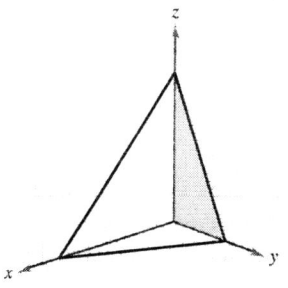

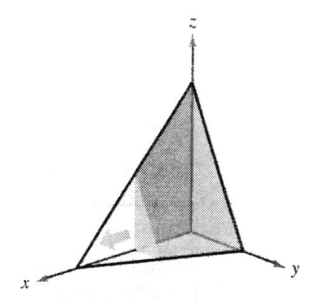

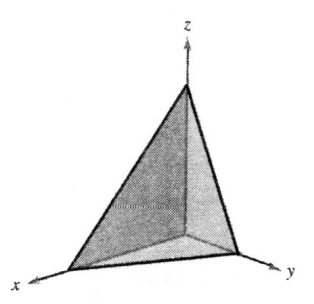

Integrate with respect to x to obtain the volume of the solid.

The following theorem was proved by the Italian mathematician Guido Fubini (1879–1943). The theorem states that if R is a vertically or horizontally simple region and f is continuous on R, the double integral of f on R is equal to an iterated integral.

THEOREM 13.2 Fubini's Theorem

Let f be continuous on a plane region R.

1. If R is defined by $a \leq x \leq b$ and $g_1(x) \leq y \leq g_2(x)$, where g_1 and g_2 are continuous on $[a, b]$, then

$$\int\int_R f(x, y)\, dA = \int_a^b \int_{g_1(x)}^{g_2(x)} f(x, y)\, dy\, dx.$$

2. If R is defined by $c \leq y \leq d$ and $h_1(y) \leq x \leq h_2(y)$, where h_1 and h_2 are continuous on $[c, d]$, then

$$\int\int_R f(x, y)\, dA = \int_c^d \int_{h_1(y)}^{h_2(y)} f(x, y)\, dx\, dy.$$

Example 2 **Evaluating a Double Integral as an Iterated Integral**

Evaluate

$$\int\int_R \left(1 - \tfrac{1}{2}x^2 - \tfrac{1}{2}y^2\right) dA$$

where R is the region given by $0 \leq x \leq 1$, $0 \leq y \leq 1$.

Solution Because the region R is a simple square, it is both vertically and horizontally simple, and you can use either order of integration. Suppose you choose $dy\, dx$ by placing a vertical representative rectangle in the region, as shown in Figure 13.18. This produces the following.

$$\int\int_R \left(1 - \tfrac{1}{2}x^2 - \tfrac{1}{2}y^2\right) dA = \int_0^1 \int_0^1 \left(1 - \tfrac{1}{2}x^2 - \tfrac{1}{2}y^2\right) dy\, dx$$

$$= \int_0^1 \left[\left(1 - \tfrac{1}{2}x^2\right)y - \tfrac{y^3}{6}\right]_0^1 dx$$

$$= \int_0^1 \left(\tfrac{5}{6} - \tfrac{1}{2}x^2\right) dx$$

$$= \left[\tfrac{5}{6}x - \tfrac{x^3}{6}\right]_0^1$$

$$= \tfrac{2}{3}$$

$R: 0 \leq x \leq 1$
$\quad\;\, 0 \leq y \leq 1$

$$\int\int_R f(x, y)\, dA = \int_0^1 \int_0^1 f(x, y)\, dy\, dx$$

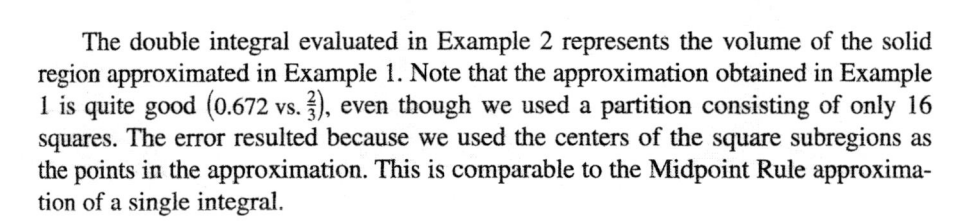

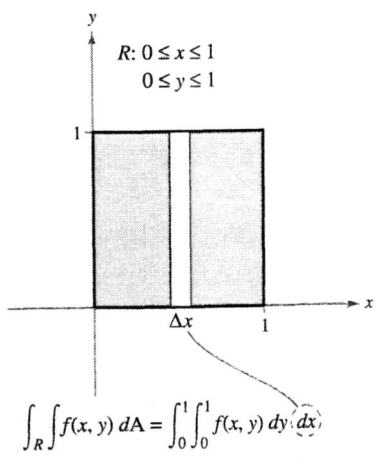

The volume of the solid region is $\tfrac{2}{3}$.
Figure 13.18

The double integral evaluated in Example 2 represents the volume of the solid region approximated in Example 1. Note that the approximation obtained in Example 1 is quite good $(0.672 \text{ vs. } \tfrac{2}{3})$, even though we used a partition consisting of only 16 squares. The error resulted because we used the centers of the square subregions as the points in the approximation. This is comparable to the Midpoint Rule approximation of a single integral.

EXPLORATION

Volume of a Paraboloid Sector
The solid in Example 3 has an elliptical (not a circular) base. Consider the region bounded by the circular paraboloid

$$z = a^2 - x^2 - y^2, \quad a > 0$$

and the *xy*-plane. How many ways do you now know for finding the volume of this solid? For instance, you could use the disk method to find the volume as a solid of revolution. Does each method involve integration?

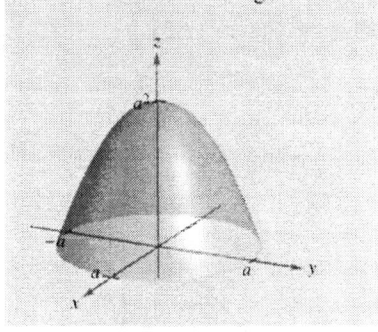

The difficulty of evaluating a single integral $\int_a^b f(x)\,dx$ usually depends on the function f, and not on the interval $[a, b]$. This is a major difference between single and double integrals. In the next example, we integrate a function similar to that in Examples 1 and 2. Notice that a change in the region R produces a much more difficult integration problem.

Example 3 Finding Volume by a Double Integral

Find the volume of the solid region bounded by the paraboloid $z = 4 - x^2 - 2y^2$ and the *xy*-plane.

Solution By letting $z = 0$, you can see that the base of the region in the *xy*-plane is the ellipse $x^2 + 2y^2 = 4$, as shown in Figure 13.19. This plane region is both vertically and horizontally simple, so the order $dy\,dx$ is appropriate.

Variable bounds for y: $\quad -\sqrt{\dfrac{(4-x^2)}{2}} \le y \le \sqrt{\dfrac{(4-x^2)}{2}}$

Constant bounds for x: $\quad -2 \le x \le 2$

The volume is given by

$$V = \int_{-2}^{2} \int_{-\sqrt{(4-x^2)/2}}^{\sqrt{(4-x^2)/2}} (4 - x^2 - 2y^2)\,dy\,dx$$

$$= \int_{-2}^{2} \left[(4 - x^2)y - \frac{2y^3}{3} \right]_{-\sqrt{(4-x^2)/2}}^{\sqrt{(4-x^2)/2}} dx$$

$$= \frac{4}{3\sqrt{2}} \int_{-2}^{2} (4 - x^2)^{3/2}\,dx$$

$$= \frac{4}{3\sqrt{2}} \int_{-\pi/2}^{\pi/2} 16 \cos^4 \theta\,d\theta \qquad x = 2\sin\theta$$

$$= \frac{64}{3\sqrt{2}}(2) \int_0^{\pi/2} \cos^4 \theta\,d\theta$$

$$= \frac{128}{3\sqrt{2}} \left(\frac{3\pi}{16} \right) \qquad \text{Wallis's Formula}$$

$$= 4\sqrt{2}\,\pi.$$

NOTE In Example 3, note the usefulness of Wallis's Formula to evaluate $\int_0^{\pi/2} \cos^n \theta\,d\theta$. You may want to review this formula in Section 7.3.

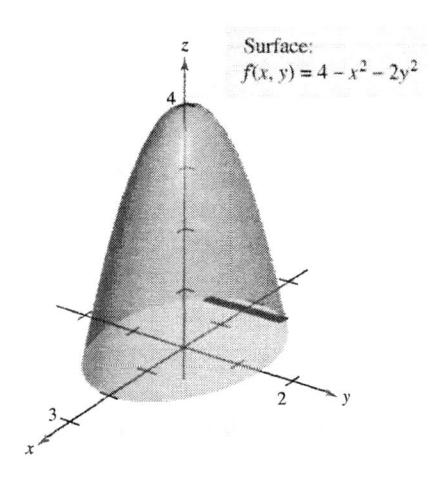

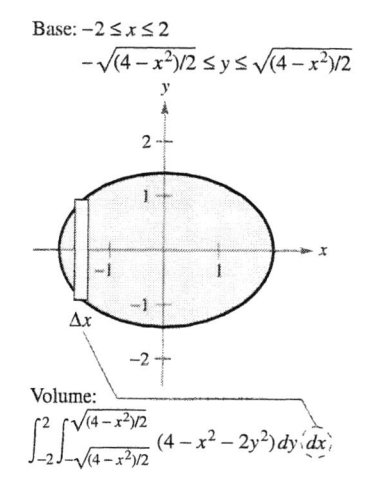

Figure 13.19

In Examples 2 and 3, the order of integration was optional, because the regions were both vertically and horizontally simple. Moreover, had you used the order $dx\,dy$, you would have obtained integrals of comparable difficulty. There are, however, some occasions in which one order of integration is much more convenient than the other. Example 4 shows such a case.

Example 4 Comparing Different Orders of Integration

Find the volume of the solid region R bounded by the surface

$$f(x, y) = e^{-x^2} \qquad \text{Surface}$$

and the planes $y = 0$, $y = x$, and $x = 1$, as shown in Figure 13.20.

Solution The base of R in the xy-plane is bounded by the lines $y = 0$, $x = 1$, and $y = x$. The two possible orders of integration are given in Figure 13.21.

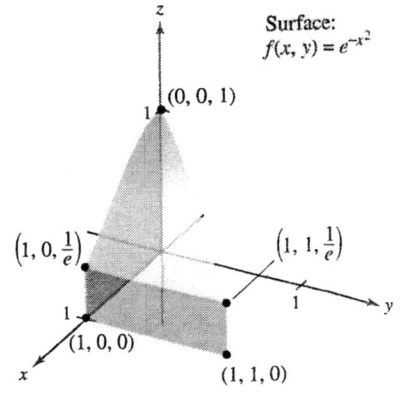

Base is bounded by $y = 0$, $y = x$, and $x = 1$.
Figure 13.20

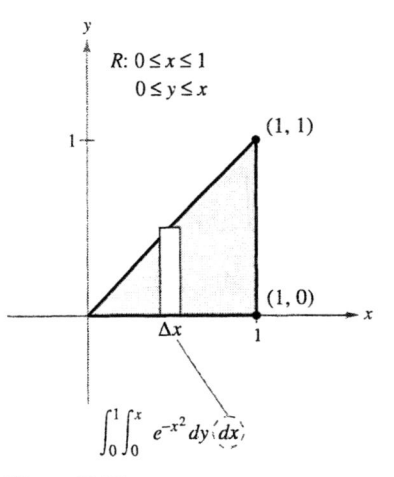

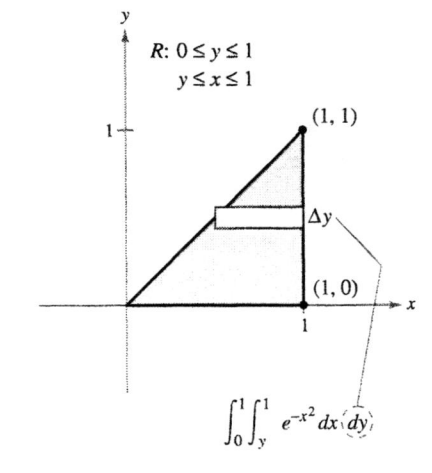

Figure 13.21

By setting up the corresponding iterated integrals, you can see that the order $dx\,dy$ requires the antiderivative $\int e^{-x^2}\,dx$, which is not an elementary function. On the other hand, the order $dy\,dx$ produces the integral

$$\int_0^1 \int_0^x e^{-x^2}\,dy\,dx = \int_0^1 e^{-x^2} y \Big]_0^x dx$$
$$= \int_0^1 xe^{-x^2}\,dx$$
$$= -\frac{1}{2}e^{-x^2}\Big]_0^1$$
$$= -\frac{1}{2}\left(\frac{1}{e} - 1\right)$$
$$= \frac{e-1}{2e}$$
$$\approx 0.316.$$

NOTE Try using a symbolic integration utility to evaluate the integral in Example 4.

Example 5 Volume of a Region Bounded by Two Surfaces

Find the volume of the solid region R bounded above by the paraboloid $z = 1 - x^2 - y^2$ and below by the plane $z = 1 - y$, as shown in Figure 13.22.

Solution Equating z-values, you can determine that the intersection of the two surfaces occurs on the right circular cylinder given by

$$1 - y = 1 - x^2 - y^2 \implies x^2 = y - y^2.$$

Because the volume of R is the difference between the volume under the paraboloid and the volume under the plane, you have

$$\text{Volume} = \int_0^1 \int_{-\sqrt{y-y^2}}^{\sqrt{y-y^2}} (1 - x^2 - y^2) \, dx \, dy - \int_0^1 \int_{-\sqrt{y-y^2}}^{\sqrt{y-y^2}} (1 - y) \, dx \, dy$$

$$= \int_0^1 \int_{-\sqrt{y-y^2}}^{\sqrt{y-y^2}} (y - y^2 - x^2) \, dx \, dy$$

$$= \int_0^1 \left[(y - y^2)x - \frac{x^3}{3} \right]_{-\sqrt{y-y^2}}^{\sqrt{y-y^2}} dy$$

$$= \frac{4}{3} \int_0^1 (y - y^2)^{3/2} \, dy$$

$$= \left(\frac{4}{3}\right)\left(\frac{1}{8}\right) \int_0^1 [1 - (2y - 1)^2]^{3/2} \, dy$$

$$= \frac{1}{6} \int_{-\pi/2}^{\pi/2} \frac{\cos^4 \theta}{2} \, d\theta \qquad 2y - 1 = \sin \theta$$

$$= \frac{1}{6} \int_0^{\pi/2} \cos^4 \theta \, d\theta$$

$$= \left(\frac{1}{6}\right)\left(\frac{3\pi}{16}\right) = \frac{\pi}{32}. \qquad \text{Wallis's Formula}$$

Paraboloid: $z = 1 - x^2 - y^2$
Plane: $z = 1 - y$

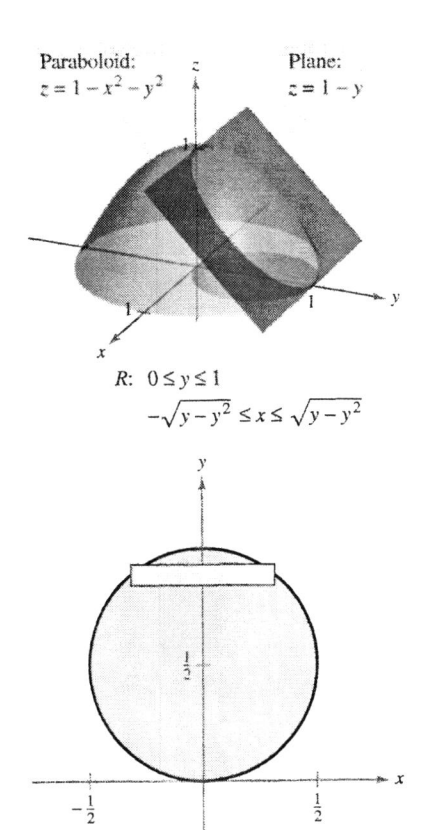

R: $0 \le y \le 1$
$-\sqrt{y - y^2} \le x \le \sqrt{y - y^2}$

The volume of the solid region is $\pi/32$.
Figure 13.22

EXERCISES FOR SECTION 13.2

Approximation In Exercises 1–4, approximate the integral $\int_R \int f(x, y) \, dA$ by dividing the rectangle R with vertices $(0, 0)$, $(4, 0)$, $(4, 2)$, and $(0, 2)$ into eight equal squares and finding the sum

$$\sum_{i=1}^{8} f(x_i, y_i) \, \Delta A_i$$

where (x_i, y_i) is the center of the ith square. Evaluate the iterated integral and compare it with the approximation.

1. $\int_0^4 \int_0^2 (x + y) \, dy \, dx$
2. $\frac{1}{2} \int_0^4 \int_0^2 x^2 y \, dy \, dx$
3. $\int_0^4 \int_0^2 (x^2 + y^2) \, dy \, dx$
4. $\int_0^4 \int_0^2 \frac{1}{(x + 1)(y + 1)} \, dy \, dx$

5. *Approximation* The table shows values of a function f over a square region R. Divide the region into 16 equal squares and select (x_i, y_i) to be the point in the ith square closest to the origin. Compare this approximation with that obtained by using the point in the ith square farthest from the origin.

$$\int_0^4 \int_0^4 f(x, y) \, dy \, dx.$$

x \ y	0	1	2	3	4
0	32	31	28	23	16
1	31	30	27	22	15
2	28	27	24	19	12
3	23	22	19	14	7
4	16	15	12	7	0

6. *Approximation* The figure shows the level curves for a function f over a square region R. Approximate the integral using four squares, selecting the midpoint of each square as (x_i, y_i).

$$\int_0^2 \int_0^2 f(x, y) \, dy \, dx.$$

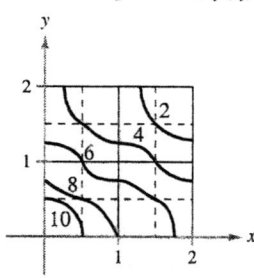

In Exercises 7–12, sketch the region R and evaluate the iterated integral $\int_R \int f(x, y) \, dA$.

7. $\displaystyle\int_0^2 \int_0^1 (1 + 2x + 2y) \, dy \, dx$
8. $\displaystyle\int_0^\pi \int_0^{\pi/2} \sin^2 x \cos^2 y \, dy \, dx$

9. $\displaystyle\int_0^6 \int_{y/2}^3 (x + y) \, dx \, dy$
10. $\displaystyle\int_0^4 \int_{\frac{1}{2}y}^{\sqrt{y}} x^2 y^2 \, dx \, dy$

11. $\displaystyle\int_{-a}^a \int_{-\sqrt{a^2-x^2}}^{\sqrt{a^2-x^2}} (x + y) \, dy \, dx$

12. $\displaystyle\int_0^1 \int_{y-1}^0 e^{x+y} \, dx \, dy + \int_0^1 \int_0^{1-y} e^{x+y} \, dx \, dy$

In Exercises 13–20, set up an integral for both orders of integration, and use the more convenient order to evaluate the integral over the region R.

13. $\displaystyle\int_R \int xy \, dA$

    R: rectangle with vertices $(0, 0), (0, 5), (3, 5), (3, 0)$

14. $\displaystyle\int_R \int \sin x \sin y \, dA$

    R: rectangle with vertices $(-\pi, 0), (\pi, 0), (\pi, \pi/2), (-\pi, \pi/2)$

15. $\displaystyle\int_R \int \frac{y}{x^2 + y^2} \, dA$

    R: triangle bounded by $y = x$, $y = 2x$, $x = 2$

16. $\displaystyle\int_R \int xe^y \, dA$

    R: triangle bounded by $y = 4 - x$, $y = 0$, $x = 0$

17. $\displaystyle\int_R \int -2y \ln x \, dA$

    R: region bounded by $y = 4 - x^2$, $y = 4 - x$

18. $\displaystyle\int_R \int \frac{y}{1 + x^2} \, dA$

    R: region bounded by $y = 0$, $y = \sqrt{x}$, $x = 4$

19. $\displaystyle\int_R \int x \, dA$

    R: sector of a circle in the first quadrant bounded by $y = \sqrt{25 - x^2}$, $3x - 4y = 0$, $y = 0$

20. $\displaystyle\int_R \int (x^2 + y^2) \, dA$

    R: semicircle bounded by $y = \sqrt{4 - x^2}$, $y = 0$

In Exercises 21–30, use a double integral to find the volume of the indicated solid.

21. $z = \dfrac{y}{2}$, $0 \le x \le 4$, $0 \le y \le 2$

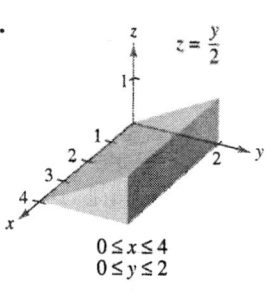

22. $z = 6 - 2y$, $0 \le x \le 4$, $0 \le y \le 2$

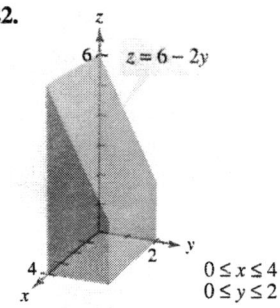

23. $z = 4 - x - y$, $y = x$, $y = 2$

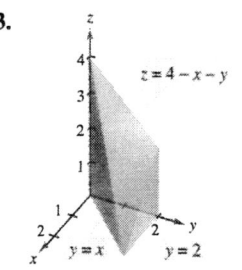

24. $z = 4$, $y = x$, $x = 2$

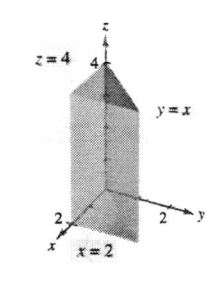

25. $2x + 3y + 4z = 12$

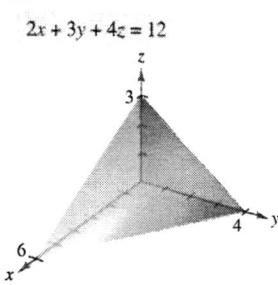

26. $x + y + z = 2$

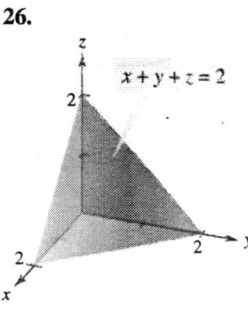

27. $z = 1 - xy$, $y = x$, $y = 1$

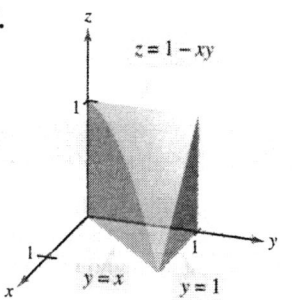

28. $z = 4 - y^2$, $y = x$, $y = 2$

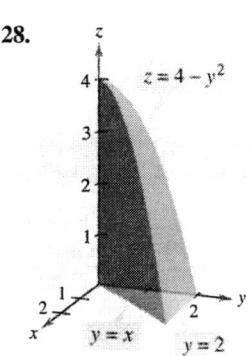

29. Improper integral
    $z = \dfrac{1}{(x+1)^2(y+1)^2}$, $0 \le x < \infty$, $0 \le y < \infty$

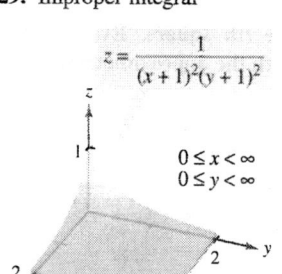

30. Improper integral
    $z = e^{-(x+y)/2}$, $0 \le x < \infty$, $0 \le y < \infty$

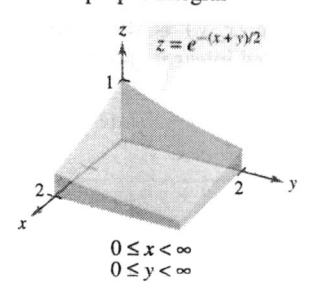

In Exercises 31 and 32, use a computer algebra system to find the volume of the solid.

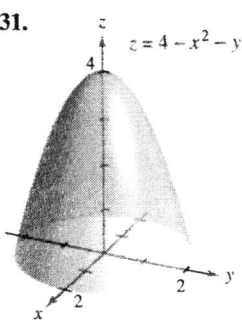

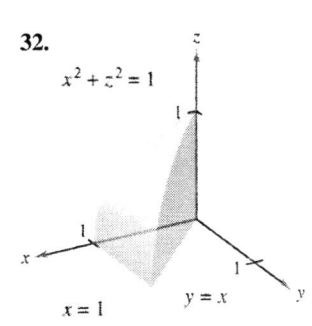

31. $z = 4 - x^2 - y^2$

32. $x^2 + z^2 = 1$, $x = 1$, $y = x$

In Exercises 33–40, set up a double integral to find the volume of the solid bounded by the graphs of the equations.

33. $z = xy$, $z = 0$, $y = x$, $x = 1$, first octant
34. $y = 0$, $z = 0$, $y = x$, $z = x$, $x = 0$, $x = 5$
35. $z = 0$, $z = x^2$, $x = 0$, $x = 2$, $y = 0$, $y = 4$
36. $x^2 + y^2 + z^2 = r^2$
37. $x^2 + z^2 = 1$, $y^2 + z^2 = 1$, first octant
38. $y = 4 - x^2$, $z = 4 - x^2$, first octant
39. $z = x + y$, $x^2 + y^2 = 4$, first octant
40. $z = \dfrac{1}{1 + y^2}$, $x = 0$, $x = 2$, $y \geq 0$

In Exercises 41 and 42, use Wallis's Formula to find the volume of the solid bounded by the graphs of the equations.

41. $z = x^2 + y^2$, $x^2 + y^2 = 4$, $z = 0$
42. $z = \sin^2 x$, $z = 0$, $0 \leq x \leq \pi$, $0 \leq y \leq 5$

In Exercises 43–46, use a computer algebra system to find the volume of the solid bounded by the graphs of the equations.

43. $z = 4 - x^2 - y^2$, $z = 0$
44. $x^2 = 9 - y$, $z^2 = 9 - y$, first octant
45. $z = \dfrac{2}{1 + x^2 + y^2}$, $z = 0$, $y = 0$, $x = 0$, $y = -0.5x + 1$
46. $z = \ln(1 + x + y)$, $z = 0$, $y = 0$, $x = 0$, $x = 4 - \sqrt{y}$

47. If f is a continuous function such that $0 \leq f(x, y) \leq 1$ over a region R of area 1, prove that
$$0 \leq \iint_R f(x, y)\, dA \leq 1.$$

48. Find the volume of the solid in the first octant bounded by the coordinate planes and the plane $(x/a) + (y/b) + (z/c) = 1$, where $a > 0$, $b > 0$, and $c > 0$.

In Exercises 49–52, evaluate the iterated integral. (Note that it is necessary to switch the order of integration.)

49. $\displaystyle \int_0^1 \int_{y/2}^{1/2} e^{-x^2}\, dx\, dy$

50. $\displaystyle \int_0^{\ln 10} \int_{e^x}^{10} \frac{1}{\ln y}\, dy\, dx$

51. $\displaystyle \int_0^1 \int_0^{\arccos y} \sin x \sqrt{1 + \sin^2 x}\, dx\, dy$

52. $\displaystyle \int_0^2 \int_{\frac{1}{2}x^2}^2 \sqrt{y} \cos y\, dy\, dx$

In Exercises 53–56, find the average value of $f(x, y)$ over the region R where

$$\text{Average} = \frac{1}{A} \iint_R f(x, y)\, dA$$

and where A is the area of R.

53. $f(x, y) = x$
    R: rectangle with vertices $(0, 0)$, $(4, 0)$, $(4, 2)$, $(0, 2)$
54. $f(x, y) = xy$
    R: rectangle with vertices $(0, 0)$, $(4, 0)$, $(4, 2)$, $(0, 2)$
55. $f(x, y) = x^2 + y^2$
    R: square with vertices $(0, 0)$, $(2, 0)$, $(2, 2)$, $(0, 2)$
56. $f(x, y) = e^{x+y}$
    R: triangle with vertices $(0, 0)$, $(0, 1)$, $(1, 1)$

Getting at the Concept

57. State the definition of a double integral. If the integrand is a nonnegative function over the region of integration, give the geometric interpretation of a double integral.

58. The following iterated integrals represent the solution to the same problem. Which iterated integral is easier to evaluate and why?
$$\int_0^4 \int_{x/2}^2 \sin y^2\, dy\, dx = \int_0^2 \int_0^{2y} \sin y^2\, dx\, dy$$

59. Let R be a region in the xy-plane whose area is B. If $f(x, y) = k$ for every point (x, y) in R, what is the value of $\iint_R f(x, y)\, dA$?

60. Let R represent a county in the northern part of the United States, and let $f(x, y)$ represent the total annual snowfall at the point (x, y) in R. Give an interpretation of each of the following.

(a) $\displaystyle \iint_R f(x, y)\, dA$
(b) $\displaystyle \frac{\iint_R f(x, y)\, dA}{\iint_R dA}$

61. Average Production The Cobb-Douglas production function for a company is

$$f(x, y) = 100x^{0.6}y^{0.4}$$

where x the number of units of labor and y is the number of units of capital. Estimate the average production level if the number of units of labor x varies between 200 and 250 and the number of units of capital y varies between 300 and 325.

62. Average Profit A firm's profit in marketing two products is

$$P = 192x + 576y - x^2 - 5y^2 - 2xy - 5000$$

where x and y represent the numbers of units of the two products. Use a computer algebra system to evaluate the double integral yielding the average weekly profit if x varies between 40 and 50 units and y varies between 45 and 60 units.

Probability A joint density function of the continuous random variables x and y is a function $f(x, y)$ satisfying the following properties.

(a) $f(x, y) \geq 0$ for all (x, y) (b) $\int_{-\infty}^{\infty} \int_{-\infty}^{\infty} f(x, y) \, dA = 1$

(c) $P[(x, y) \in R] = \int\int_R f(x, y) \, dA$

In Exercises 63–66, show that that the function is a joint density function and find the required probability.

63. $f(x, y) = \begin{cases} \frac{1}{10}, & 0 \leq x \leq 5, \ 0 \leq y \leq 2 \\ 0, & \text{elsewhere} \end{cases}$

$P(0 \leq x \leq 2, \ 1 \leq y \leq 2)$

64. $f(x, y) = \begin{cases} \frac{1}{4}xy, & 0 \leq x \leq 2, \ 0 \leq y \leq 2 \\ 0, & \text{elsewhere} \end{cases}$

$P(0 \leq x \leq 1, \ 1 \leq y \leq 2)$

65. $f(x, y) = \begin{cases} \frac{1}{27}(9 - x - y), & 0 \leq x \leq 3, \ 3 \leq y \leq 6 \\ 0, & \text{elsewhere} \end{cases}$

$P(0 \leq x \leq 1, \ 4 \leq y \leq 6)$

66. $f(x, y) = \begin{cases} e^{-x-y}, & x \geq 0, \ y \geq 0 \\ 0, & \text{elsewhere} \end{cases}$

$P(0 \leq x \leq 1, \ x \leq y \leq 1)$

67. Approximation The base of a pile of sand at a cement plant is rectangular with approximate dimensions of 20 meters by 30 meters. If the base is placed on the xy-plane with one vertex at the origin, the coordinates on the surface of the pile are $(5, 5, 3)$, $(15, 5, 6)$, $(25, 5, 4)$, $(5, 15, 2)$, $(15, 15, 7)$, and $(25, 15, 3)$. Approximate the volume of sand in the pile.

68. Programming Consider a continuous function $f(x, y)$ over the rectangular region R with vertices (a, c), (b, c), (a, d), and (b, d) where $a < b$ and $c < d$. Partition the intervals $[a, b]$ and $[c, d]$ into m and n subintervals, so that the subintervals in a given direction are of equal length. Write a program for a graphing utility to compute the sum

$$\sum_{i=1}^{n} \sum_{j=1}^{m} f(x_i, y_j) \Delta A_i \approx \int_a^b \int_c^d f(x, y) \, dA$$

where (x_i, y_j) is the center of a representative rectangle in R.

Approximation In Exercises 69–72, (a) use a computer algebra system to approximate the iterated integral, and (b) use the program in Exercise 68 to approximate the iterated integral for the indicated values of m and n.

69. $\int_0^1 \int_0^2 \sin \sqrt{x + y} \, dy \, dx$

$m = 4, \ n = 8$

70. $\int_0^2 \int_0^4 20e^{-x^3/8} \, dy \, dx$

$m = 10, \ n = 20$

71. $\int_4^6 \int_0^2 y \cos \sqrt{x} \, dx \, dy$

$m = 4, \ n = 8$

72. $\int_1^4 \int_1^2 \sqrt{x^3 + y^3} \, dx \, dy$

$m = 6, \ n = 4$

Approximation In Exercises 73 and 74, determine which value best approximates the volume of the solid between the xy-plane and the function over the region. (Make your selection on the basis of a sketch of the solid and *not* by performing any calculations.)

73. $f(x, y) = 4x$

R: square with vertices $(0, 0)$, $(4, 0)$, $(4, 4)$, $(0, 4)$

(a) -200 (b) 600 (c) 50 (d) 125 (e) 1000

74. $f(x, y) = \sqrt{x^2 + y^2}$

R: circle bounded by $x^2 + y^2 = 9$

(a) 50 (b) 500 (c) -500 (d) 5 (e) 5000

True or False? In Exercises 75 and 76, determine whether the statement is true or false. If it is false, explain why or give an example that shows it is false.

75. The volume of the sphere $x^2 + y^2 + z^2 = 1$ is given by the integral

$$V = 8 \int_0^1 \int_0^1 \sqrt{1 - x^2 - y^2} \, dx \, dy.$$

76. If $f(x, y) \leq g(x, y)$ for all (x, y) in R, and both f and g are continuous over R, then

$$\int\int_R f(x, y) \, dA \leq \int\int_R g(x, y) \, dA.$$

77. Let $f(x) = \int_1^x e^{t^2} \, dt$. Find the average value of f on the interval $[0, 1]$.

78. Find $\int_0^\infty \frac{e^{-x} - e^{-2x}}{x} \, dx$. $\left(\text{Hint: Evaluate } \int_1^2 e^{-xy} \, dy.\right)$

Section 13.3 Change of Variables: Polar Coordinates

- Write and evaluate double integrals in polar coordinates.

Double Integrals in Polar Coordinates

Some double integrals are *much* easier to evaluate in polar form than in rectangular form. This is especially true for regions such as circles, cardioids, and petal curves, and for integrands that involve $x^2 + y^2$.

In Section 9.4, you learned that the polar coordinates (r, θ) of a point are related to the rectangular coordinates (x, y) of the point as follows.

$$x = r \cos \theta \quad \text{and} \quad y = r \sin \theta$$

$$r^2 = x^2 + y^2 \quad \text{and} \quad \tan \theta = \frac{y}{x}$$

Example 1 Using Polar Coordinates to Describe a Region

Use polar coordinates to describe the regions shown in Figure 13.23.

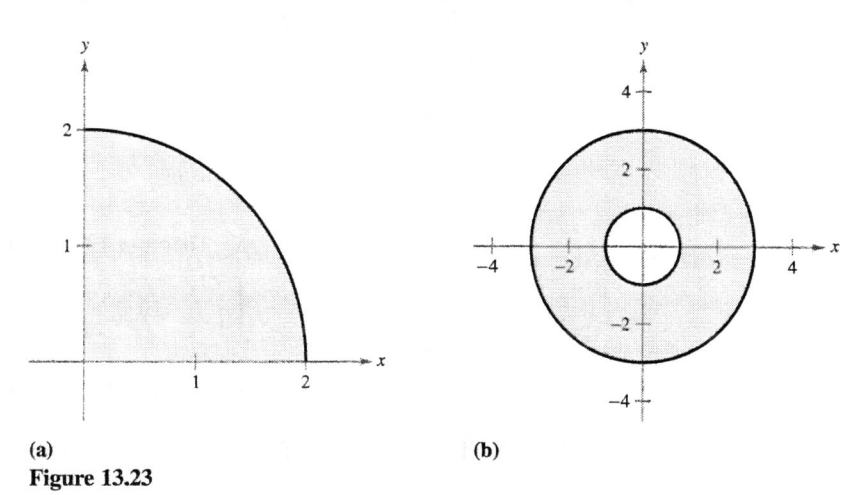

(a) (b)

Figure 13.23

Solution

a. The region R is a quarter circle of radius 2. It can be described in polar coordinates as

$$R = \{(r, \theta): 0 \leq r \leq 2, \ 0 \leq \theta \leq \pi/2\}.$$

b. The region R consists of all points between the concentric circles of radii 1 and 3. It can be described in polar coordinates as

$$R = \{(r, \theta): 1 \leq r \leq 3, \ 0 \leq \theta \leq 2\pi\}.$$

The regions in Example 1 are special cases of **polar sectors**

$$R = \{(r, \theta): r_1 \leq r \leq r_2, \ \theta_1 \leq \theta \leq \theta_2\} \quad \text{Polar sector}$$

as shown in Figure 13.24.

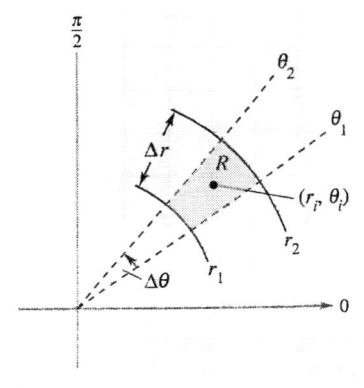

Polar sector
Figure 13.24

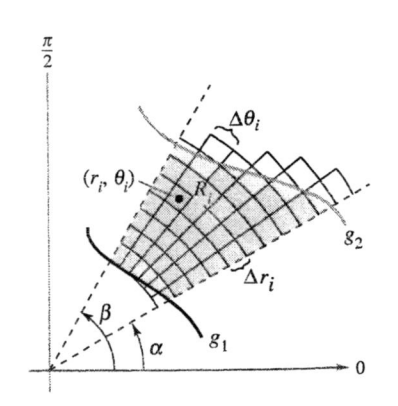

To define a double integral of a continuous function $z = f(x, y)$ in polar coordinates, consider a region R bounded by the graphs of $r = g_1(\theta)$ and $r = g_2(\theta)$ and the lines $\theta = \alpha$ and $\theta = \beta$. Instead of partitioning R into small rectangles, use a partition of small polar sectors. On R, superimpose a polar grid made of rays and circular arcs, as shown in Figure 13.25. The polar sectors R_i lying entirely within R form an **inner polar partition** Δ, whose **norm** $\|\Delta\|$ is the length of the longest diagonal of the n polar sectors.

Consider a specific polar sector R_i, as shown in Figure 13.26. It can be shown (see Exercise 61) that the area of R_i is

$$\Delta A_i = r_i \Delta r_i \Delta \theta_i \qquad \text{Area of } R_i$$

where $\Delta r_i = r_2 - r_1$ and $\Delta \theta_i = \theta_2 - \theta_1$. This implies that the volume of the solid of height $f(r_i \cos \theta_i, r_i \sin \theta_i)$ above R_i is approximately

$$f(r_i \cos \theta_i, r_i \sin \theta_i) r_i \Delta r_i \Delta \theta_i$$

and you have

$$\iint_R f(x, y) \, dA \approx \sum_{i=1}^{n} f(r_i \cos \theta_i, r_i \sin \theta_i) r_i \Delta r_i \Delta \theta_i.$$

The sum on the right can be interpreted as a Riemann sum for $f(r \cos \theta, r \sin \theta)r$. The region R corresponds to a *horizontally simple* region S in the $r\theta$-plane, as indicated in Figure 13.27. The polar sectors R_i correspond to rectangles S_i, and the area ΔA_i of S_i is $\Delta r_i \Delta \theta_i$. So, the right-hand side of the equation corresponds to the double integral

$$\iint_S f(r \cos \theta, r \sin \theta) r \, dA.$$

From this, you can apply Theorem 13.2 to write

$$\iint_R f(x, y) \, dA = \iint_S f(r \cos \theta, r \sin \theta) r \, dA$$
$$= \int_\alpha^\beta \int_{g_1(\theta)}^{g_2(\theta)} f(r \cos \theta, r \sin \theta) r \, dr \, d\theta.$$

This suggests the following theorem, the proof of which is discussed in Section 13.8.

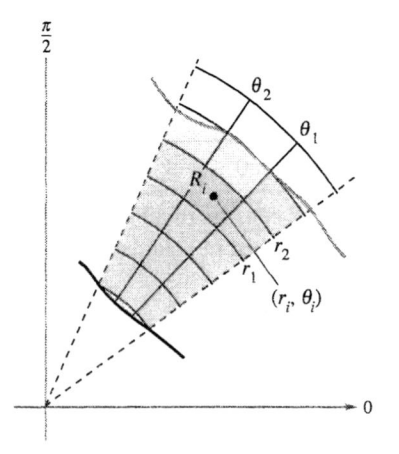

Polar grid is superimposed over region R.
Figure 13.25

The polar sector R_i is the set of all points (r, θ) such that $r_1 \leq r \leq r_2$ and $\theta_1 \leq \theta \leq \theta_2$.
Figure 13.26

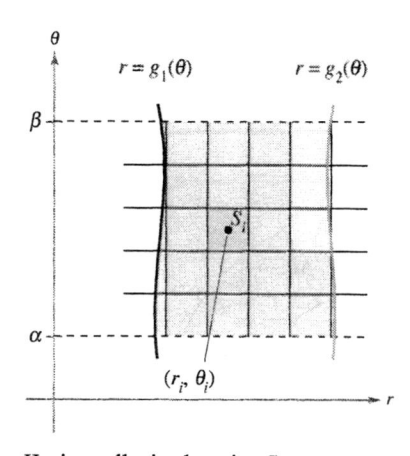

Horizontally simple region S
Figure 13.27

THEOREM 13.3 Change of Variables to Polar Form

Let R be a plane region consisting of all points $(x, y) = (r \cos \theta, r \sin \theta)$ satisfying the conditions $0 \leq g_1(\theta) \leq r \leq g_2(\theta)$, $\alpha \leq \theta \leq \beta$, where $0 \leq (\beta - \alpha) \leq 2\pi$. If g_1 and g_2 are continuous on $[\alpha, \beta]$ and f is continuous on R, then

$$\iint_R f(x, y) \, dA = \int_\alpha^\beta \int_{g_1(\theta)}^{g_2(\theta)} f(r \cos \theta, r \sin \theta) r \, dr \, d\theta.$$

NOTE If $z = f(x, y)$ is nonnegative on R, then the integral in Theorem 13.3 can be interpreted as the volume of the solid region between the graph of f and the region R.

The region R is restricted to two basic types, **r-simple** regions and **θ-simple** regions, as shown in Figure 13.28.

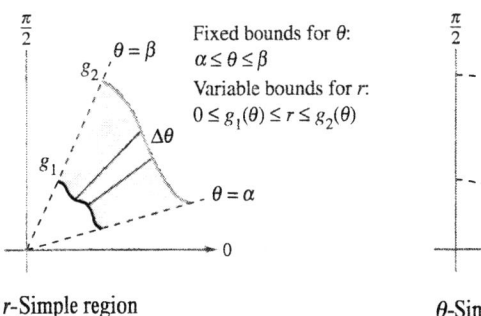

r-Simple region   θ-Simple region
Figure 13.28

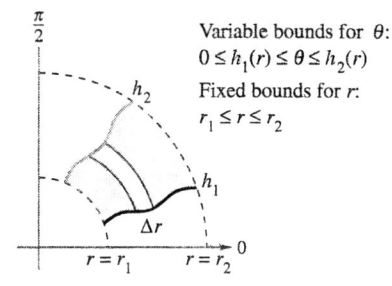

EXPLORATION

Volume of a Paraboloid In the Exploration feature on page 949, you were asked to summarize the different ways you know for finding the volume of the solid bounded by the paraboloid

$$z = a^2 - x^2 - y^2, \quad a > 0$$

and the *xy*-plane. You now know another way. Use it to find the volume of the solid.

R: $1 \leq r \leq \sqrt{5}$
$0 \leq \theta \leq 2\pi$

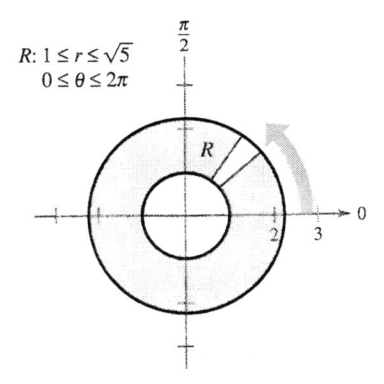

r-Simple region
Figure 13.29

Example 2   Evaluating a Double Polar Integral

Let R be the annular region lying between the two circles $x^2 + y^2 = 1$ and $x^2 + y^2 = 5$, as shown in Figure 13.29. Evaluate the integral $\iint_R (x^2 + y) \, dA$.

Solution The polar boundaries are $1 \leq r \leq \sqrt{5}$ and $0 \leq \theta \leq 2\pi$. Furthermore, $x^2 = (r \cos \theta)^2$ and $y = r \sin \theta$. So, you have

$$\iint_R (x^2 + y) \, dA = \int_0^{2\pi} \int_1^{\sqrt{5}} (r^2 \cos^2 \theta + r \sin \theta) r \, dr \, d\theta$$

$$= \int_0^{2\pi} \int_1^{\sqrt{5}} (r^3 \cos^2 \theta + r^2 \sin \theta) \, dr \, d\theta$$

$$= \int_0^{2\pi} \left(\frac{r^4}{4} \cos^2 \theta + \frac{r^3}{3} \sin \theta \right) \Big]_1^{\sqrt{5}} d\theta$$

$$= \int_0^{2\pi} \left(6 \cos^2 \theta + \frac{5\sqrt{5} - 1}{3} \sin \theta \right) d\theta$$

$$= \int_0^{2\pi} \left(3 + 3 \cos 2\theta + \frac{5\sqrt{5} - 1}{3} \sin \theta \right) d\theta$$

$$= \left(3\theta + 3 \frac{\sin 2\theta}{2} - \frac{5\sqrt{5} - 1}{3} \cos \theta \right) \Big]_0^{2\pi}$$

$$= 6\pi.$$

Surface: $z = \sqrt{16 - x^2 - y^2}$

R: $x^2 + y^2 \le 4$

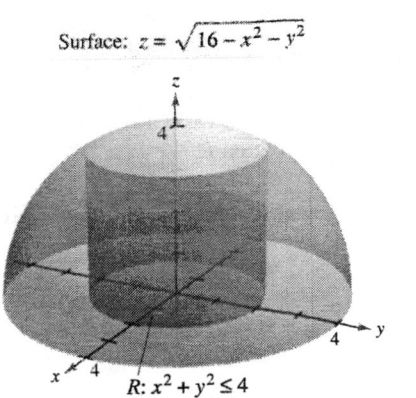

Figure 13.30

In Example 2, be sure to notice the extra factor of r in the integrand. This comes from the formula for the area of a polar sector. In differential notation, you can write

$$dA = r\, dr\, d\theta$$

which indicates that the area of a polar sector increases as you move away from the origin.

Example 3 Change of Variables to Polar Coordinates

Use polar coordinates to find the volume of the solid region bounded above by the hemisphere

$$z = \sqrt{16 - x^2 - y^2} \qquad \text{Hemisphere forms upper surface.}$$

and below by the circular region R given by

$$x^2 + y^2 \le 4 \qquad \text{Circular region forms lower surface.}$$

as shown in Figure 13.30.

Solution In Figure 13.30, you can see that R has the bounds

$$-\sqrt{4 - y^2} \le x \le \sqrt{4 - y^2}, \quad -2 \le y \le 2$$

and that $0 \le z \le \sqrt{16 - x^2 - y^2}$. In polar coordinates, the bounds are

$$0 \le r \le 2 \quad \text{and} \quad 0 \le \theta \le 2\pi$$

with height $z = \sqrt{16 - x^2 - y^2} = \sqrt{16 - r^2}$. Consequently, the volume V is given by

$$V = \iint_R f(x, y)\, dA = \int_0^{2\pi} \int_0^2 \sqrt{16 - r^2}\, r\, dr\, d\theta$$

$$= -\frac{1}{3} \int_0^{2\pi} \left[(16 - r^2)^{3/2}\right]_0^2 d\theta$$

$$= -\frac{1}{3} \int_0^{2\pi} \left(24\sqrt{3} - 64\right) d\theta$$

$$= -\frac{8}{3}\left(3\sqrt{3} - 8\right)\theta \Big]_0^{2\pi}$$

$$= \frac{16\pi}{3}\left(8 - 3\sqrt{3}\right) \approx 46.98.$$

NOTE To see the benefit of polar coordinates in Example 3, you should try to evaluate the corresponding rectangular iterated integral

$$\int_{-2}^{2} \int_{-\sqrt{4-y^2}}^{\sqrt{4-y^2}} \sqrt{16 - x^2 - y^2}\, dx\, dy.$$

TECHNOLOGY Any computer algebra system that can handle double integrals in rectangular coordinates can also handle double integrals in polar coordinates. The reason this is true is that once you have formed the iterated integral, its value is not changed by using different variables. In other words, if you use a computer algebra system to evaluate

$$\int_0^{2\pi} \int_0^2 \sqrt{16 - x^2}\, x\, dx\, dy$$

you should obtain the same value as that obtained in Example 3.

Just as with rectangular coordinates, the double integral

$$\iint_R dA$$

can be used to find the area of a region in the plane.

Example 4 Finding Areas of Polar Regions

Use a double integral to find the area enclosed by the graph of $r = 3\cos 3\theta$.

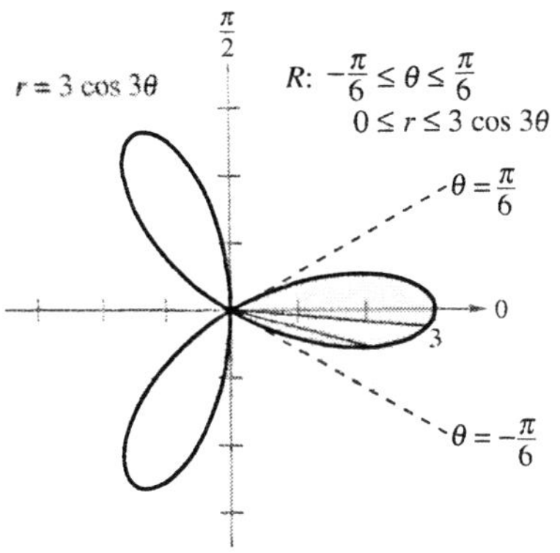

Solution Let R be one petal of the curve shown in Figure 13.31. This region is r-simple, and the boundaries are as follows.

$$-\frac{\pi}{6} \leq \theta \leq \frac{\pi}{6} \qquad \text{Fixed bounds on } \theta$$

$$0 \leq r \leq 3\cos 3\theta \qquad \text{Variable bounds on } r$$

So, the area of one petal is

$$\frac{1}{3}A = \iint_R dA = \int_{-\pi/6}^{\pi/6} \int_0^{3\cos 3\theta} r\, dr\, d\theta$$

$$= \int_{-\pi/6}^{\pi/6} \frac{r^2}{2}\bigg]_0^{3\cos 3\theta} d\theta$$

$$= \frac{9}{2}\int_{-\pi/6}^{\pi/6} \cos^2 3\theta\, d\theta$$

$$= \frac{9}{4}\int_{-\pi/6}^{\pi/6} (1 + \cos 6\theta)\, d\theta$$

$$= \frac{9}{4}\left[\theta + \frac{1}{6}\sin 6\theta\right]_{-\pi/6}^{\pi/6}$$

$$= \frac{3\pi}{4}.$$

Therefore, the total area is $A = 9\pi/4$.

The area of R is $3\pi/4$, and the total area is $9\pi/4$.
Figure 13.31

As illustrated in Example 4, the area of a region in the plane can be represented by

$$A = \int_\alpha^\beta \int_{g_1(\theta)}^{g_2(\theta)} r\, dr\, d\theta.$$

If $g_1(\theta) = 0$, you obtain

$$A = \int_\alpha^\beta \int_0^{g_2(\theta)} r\, dr\, d\theta = \int_\alpha^\beta \frac{r^2}{2}\bigg]_0^{g_2(\theta)} d\theta = \int_\alpha^\beta \frac{1}{2}(g_2(\theta))^2\, d\theta$$

which agrees with Theorem 9.13.

So far in this section, all of the examples of iterated integrals in polar form have been of the form

$$\int_\alpha^\beta \int_{g_1(\theta)}^{g_2(\theta)} f(r\cos\theta, r\sin\theta) r\, dr\, d\theta$$

in which the order of integration is with respect to r first. Sometimes you can obtain a simpler integration problem by switching the order of integration, as illustrated in the next example.

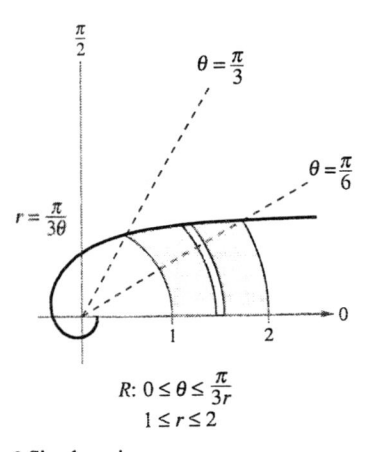

θ-Simple region
Figure 13.32

Example 5 Changing the Order of Integration

Find the area of the region bounded above by the spiral

$$r = \frac{\pi}{3\theta}$$

and below by the polar axis, between $r = 1$ and $r = 2$.

Solution The region is shown in Figure 13.32. The polar boundaries for the region are

$$1 \leq r \leq 2 \quad \text{and} \quad 0 \leq \theta \leq \frac{\pi}{3r}.$$

Hence, the area of the region can be evaluated as follows.

$$A = \int_1^2 \int_0^{\pi/3r} r \, d\theta \, dr = \int_1^2 r\theta \Big]_0^{\pi/3r} dr = \int_1^2 \frac{\pi}{3} dr = \frac{\pi r}{3}\Big]_1^2 = \frac{\pi}{3}$$

EXERCISES FOR SECTION 13.3

In Exercises 1–4, the region R for the integral $\int_R \int f(x, y) \, dA$ is shown. State whether you would use rectangular or polar coordinates to evaluate the integral.

1.

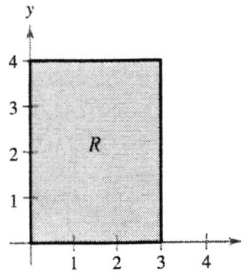

2.

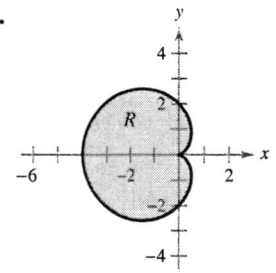

3.

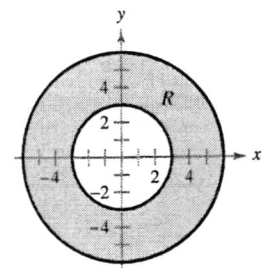

4.

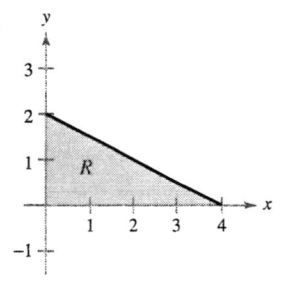

In Exercises 5–8, use polar coordinates to describe the region shown.

5.

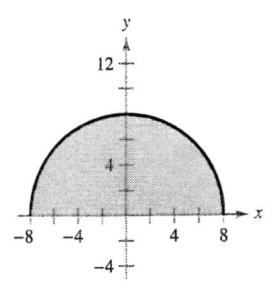

6.

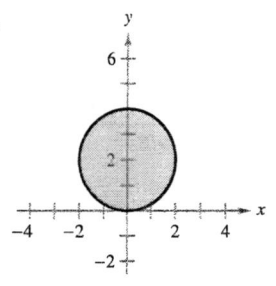

7.

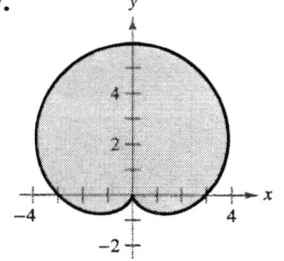

8.

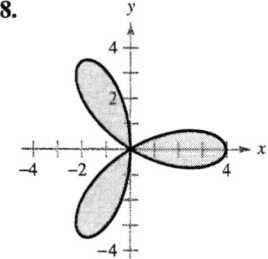

In Exercises 9–14, evaluate the double integral $\int_R \int f(r, \theta) \, dA$, and sketch the region R.

9. $\displaystyle\int_0^{2\pi} \int_0^6 3r^2 \sin \theta \, dr \, d\theta$

10. $\displaystyle\int_0^{\pi/4} \int_0^4 r^2 \sin \theta \cos \theta \, dr \, d\theta$

11. $\displaystyle\int_0^{\pi/2} \int_2^3 \sqrt{9 - r^2} \, r \, dr \, d\theta$

12. $\displaystyle\int_0^{\pi/2} \int_0^3 re^{-r^2} \, dr \, d\theta$

13. $\displaystyle\int_0^{\pi/2} \int_0^{1+\sin\theta} \theta r \, dr \, d\theta$

14. $\displaystyle\int_0^{\pi/2} \int_0^{1-\cos\theta} (\sin \theta) r \, dr \, d\theta$

In Exercises 15–20, evaluate the iterated integral by converting to polar coordinates.

15. $\displaystyle\int_0^a \int_0^{\sqrt{a^2-y^2}} y \, dx \, dy$

16. $\displaystyle\int_0^a \int_0^{\sqrt{a^2-x^2}} x \, dy \, dx$

17. $\displaystyle\int_0^3 \int_0^{\sqrt{9-x^2}} (x^2 + y^2)^{3/2} \, dy \, dx$

18. $\displaystyle\int_0^2 \int_y^{\sqrt{8-y^2}} \sqrt{x^2 + y^2} \, dx \, dy$

19. $\displaystyle\int_0^2 \int_0^{\sqrt{2x-x^2}} xy \, dy \, dx$

20. $\displaystyle\int_0^4 \int_0^{\sqrt{4y-y^2}} x^2 \, dx \, dy$

In Exercises 21 and 22, combine the sum of the two iterated integrals into a single iterated integral by converting to polar coordinates. Evaluate the resulting iterated integral.

21. $\displaystyle\int_0^2 \int_0^x \sqrt{x^2+y^2} \, dy \, dx + \int_2^{2\sqrt{2}} \int_0^{\sqrt{8-x^2}} \sqrt{x^2+y^2} \, dy \, dx$

22. $\int_0^{5\sqrt{2}/2} \int_0^x xy\, dy\, dx + \int_{5\sqrt{2}/2}^5 \int_0^{\sqrt{25-x^2}} xy\, dy\, dx$

In Exercises 23–26, use polar coordinates to set up and evaluate the double integral $\int_R \int f(x, y)\, dA$.

23. $f(x, y) = x + y$, $R: x^2 + y^2 \leq 4, x \geq 0, y \geq 0$
24. $f(x, y) = e^{-(x^2+y^2)/2}$, $R: x^2 + y^2 \leq 25, x \geq 0$
25. $f(x, y) = \arctan \frac{y}{x}$, $R: x^2 + y^2 \geq 1, x^2 + y^2 \leq 4, 0 \leq y \leq x$
26. $f(x, y) = 9 - x^2 - y^2$, $R: x^2 + y^2 \leq 9, x \geq 0, y \geq 0$

Volume **In Exercises 27–32, use a double integral in polar coordinates to find the volume of the solid bounded by the graphs of the equations.**

27. $z = xy$, $x^2 + y^2 = 1$, first octant
28. $z = x^2 + y^2 + 3$, $z = 0$, $x^2 + y^2 = 1$
29. $z = \sqrt{x^2 + y^2}$, $z = 0$, $x^2 + y^2 = 25$
30. $z = \ln(x^2 + y^2)$, $z = 0$, $x^2 + y^2 \geq 1$, $x^2 + y^2 \leq 4$
31. Inside the hemisphere $z = \sqrt{16 - x^2 - y^2}$ and inside the cylinder $x^2 + y^2 - 4x = 0$
32. Inside the hemisphere $z = \sqrt{16 - x^2 - y^2}$ and outside the cylinder $x^2 + y^2 = 1$

33. ***Volume*** Find a such that the volume inside the hemisphere $z = \sqrt{16 - x^2 - y^2}$ and outside the cylinder $x^2 + y^2 = a^2$ is one-half the volume of the hemisphere.

34. ***Volume*** Use a double integral in polar coordinates to find the volume of a sphere of radius a.

35. ***Volume*** Determine the diameter of a hole that is drilled vertically through the center of the solid bounded by the graphs of the equations
$$z = 25e^{-(x^2+y^2)/4}, \quad z = 0, \quad \text{and} \quad x^2 + y^2 = 16$$
if one-tenth of the volume of the solid is removed.

36. ***Machine Design*** The surfaces of a double-lobed cam are modeled by the inequalities $\frac{1}{4} \leq r \leq \frac{1}{2}(1 + \cos^2 \theta)$ and
$$\frac{-9}{4(x^2 + y^2 + 9)} \leq z \leq \frac{9}{4(x^2 + y^2 + 9)}$$
where all measurements are in inches.
(a) Use a computer algebra system to graph the cam.
(b) Use a computer algebra system to approximate the perimeter of the polar curve
$$r = \tfrac{1}{2}(1 + \cos^2 \theta).$$
This is the distance a roller must travel as it runs against the cam through one revolution of the cam.
(c) Use a computer algebra system to find the volume of steel in the cam.

In Exercises 37–42, use a double integral to find the area of the shaded region.

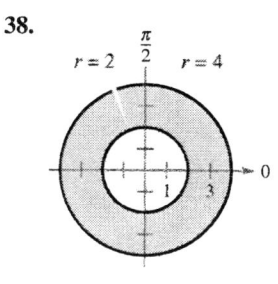

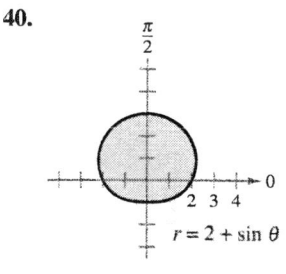

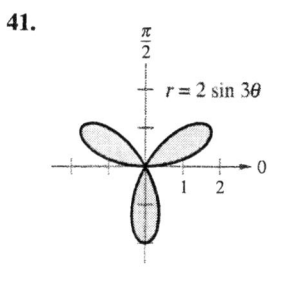

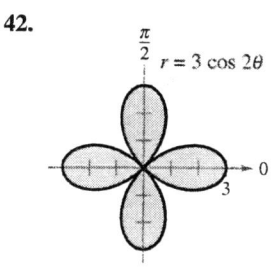

37. $r = 6 \cos \theta$
38. $r = 2$, $r = 4$
39. $r = 1 + \cos \theta$
40. $r = 2 + \sin \theta$
41. $r = 2 \sin 3\theta$
42. $r = 3 \cos 2\theta$

Getting at the Concept

43. Describe the partition of the region R of integration in the xy-plane when using polar coordinates to evaluate a double integral.

44. Explain how to change from rectangular coordinates to polar coordinates in a double integral.

45. In your own words, describe r-simple regions and θ-simple regions.

46. Each figure shows a region of integration for the double integral $\int_R \int f(x, y)\, dA$. For each region, state whether horizontal representative elements, vertical representative elements, or polar sectors would yield the easiest method for obtaining the limits of integration.

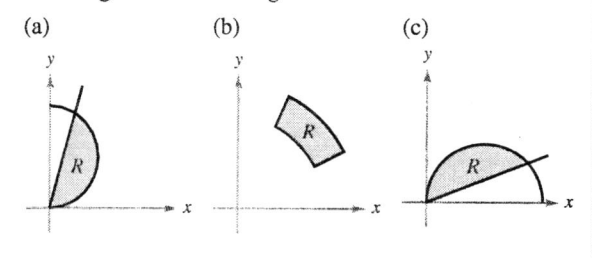

(a)   (b)   (c)

47. Think About It Consider the program you wrote to approximate double integrals in rectangular coordinates (Exercise 68, Section 13.2). If the program is used to approximate the double integral

$$\iint_R f(r, \theta) \, dA$$

in polar coordinates, how will you modify f when it is entered into the program? Because the limits of integration are constants, describe the plane region of integration.

48. Approximation Horizontal cross sections of a piece of ice that broke from a glacier are in the shape of a quarter of a circle with a radius of approximately 50 feet. The base is divided into 20 subregions, as shown in the figure. At the center of each subregion, the height of the ice is measured, yielding the following points in cylindrical coordinates.

$\left(5, \frac{\pi}{16}, 7\right), \left(15, \frac{\pi}{16}, 8\right), \left(25, \frac{\pi}{16}, 10\right), \left(35, \frac{\pi}{16}, 12\right), \left(45, \frac{\pi}{16}, 9\right),$

$\left(5, \frac{3\pi}{16}, 9\right), \left(15, \frac{3\pi}{16}, 10\right), \left(25, \frac{3\pi}{16}, 14\right), \left(35, \frac{3\pi}{16}, 15\right), \left(45, \frac{3\pi}{16}, 10\right),$

$\left(5, \frac{5\pi}{16}, 9\right), \left(15, \frac{5\pi}{16}, 11\right), \left(25, \frac{5\pi}{16}, 15\right), \left(35, \frac{5\pi}{16}, 18\right), \left(45, \frac{5\pi}{16}, 14\right),$

$\left(5, \frac{7\pi}{16}, 5\right), \left(15, \frac{7\pi}{16}, 8\right), \left(25, \frac{7\pi}{16}, 11\right), \left(35, \frac{7\pi}{16}, 16\right), \left(45, \frac{7\pi}{16}, 12\right)$

(a) Approximate the volume of the solid.

(b) Approximate the weight of the solid if ice weighs approximately 56 pounds per cubic foot.

(c) Approximate the number of gallons of water in the solid if there are 7.48 gallons of water per cubic foot.

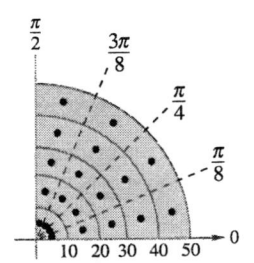

Approximation In Exercises 49 and 50, use a computer algebra system to approximate the iterated integral.

49. $\displaystyle\int_{\pi/4}^{\pi/2} \int_0^5 r\sqrt{1 + r^3} \sin \sqrt{\theta} \, dr \, d\theta$

50. $\displaystyle\int_0^{\pi/4} \int_0^4 5re^{\sqrt{r\theta}} \, dr \, d\theta$

Approximation In Exercises 51 and 52, determine which value best approximates the volume of the solid between the xy-plane and the function over the region. (Make your selection on the basis of a sketch of the solid and *not* by performing any calculations.)

51. $f(x, y) = 15 - 2y$; R: semicircle: $x^2 + y^2 = 16$, $y \geq 0$
(a) 100 (b) 200 (c) 300 (d) -200 (e) 800

52. $f(x, y) = xy + 2$; R: quarter circle: $x^2 + y^2 = 9$, $x \geq 0$, $y \geq 0$
(a) 25 (b) 8 (c) 100 (d) 50 (e) -30

True or False? In Exercises 53 and 54, determine whether the statement is true or false. If it is false, explain why or give an example that shows it is false.

53. If $\int_R\int f(r, \theta) \, dA > 0$, then $f(r, \theta) > 0$ for all (r, θ) in R.

54. If $f(r, \theta)$ is a constant function and the area of the region S is twice that of the region R, then $2 \int_R\int f(r, \theta) \, dA = \int_S\int f(r, \theta) \, dA$.

55. Probability The value of the integral $I = \displaystyle\int_{-\infty}^{\infty} e^{-x^2/2} \, dx$ is required in the development of the normal probability density function.

(a) Use polar coordinates to evaluate the improper integral.

$$I^2 = \left(\int_{-\infty}^{\infty} e^{-x^2/2} \, dx\right)\left(\int_{-\infty}^{\infty} e^{-y^2/2} \, dy\right)$$

$$= \int_{-\infty}^{\infty} \int_{-\infty}^{\infty} e^{-(x^2+y^2)/2} \, dA$$

(b) Use the result in part (a) to determine I.

FOR FURTHER INFORMATION For more information on this problem, see the article "Integrating e^{-x^2} Without Polar Coordinates" by William Dunham in *Mathematics Teacher*. To view this article, go to the website *www.matharticles.com*.

56. Use the result in Exercise 55 and a change of variables to evaluate each integral. No integration is required.

(a) $\displaystyle\int_{-\infty}^{\infty} e^{-x^2} \, dx$ (b) $\displaystyle\int_{-\infty}^{\infty} e^{-4x^2} \, dx$

57. Population The population density of a city is approximated by the model $f(x, y) = 4000e^{-0.01(x^2+y^2)}$, $x^2 + y^2 \leq 49$, where x and y are measured in miles. Integrate the density function over the indicated circular region to approximate the population of the city.

58. Probability Find k such that the function

$$f(x, y) = \begin{cases} ke^{-(x^2+y^2)}, & x \geq 0, y \geq 0 \\ 0, & \text{elsewhere} \end{cases}$$

is a probability density function.

59. Think About It Consider the region bounded by the graphs of $y = 2$, $y = 4$, $y = x$, and $y = \sqrt{3}x$ and the double integral $\int_R\int f \, dA$. Determine the limits of integration if the region R is divided into (a) horizontal representative elements, (b) vertical representative elements, and (c) polar sectors.

60. Repeat Exercise 59 for a region R bounded by the graph of the equation $(x - 2)^2 + y^2 = 4$.

61. Show that the area A of the polar sector R (see figure) is $A = r\Delta r\Delta\theta$, where $r = (r_1 + r_2)/2$ is the average radius of R.

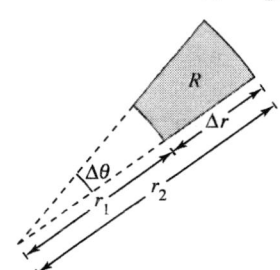

Section 13.4 Center of Mass and Moments of Inertia

- Find the mass of a planar lamina using a double integral.
- Find the center of mass of a planar lamina using double integrals.
- Find moments of inertia using double integrals.

Mass

In Section 6.6, we discussed several applications of integration involving a lamina of *constant* density ρ. For example, if the lamina corresponding to the region R, as shown in Figure 13.33, has a constant density ρ, then the mass of the lamina is given by

$$\text{Mass} = \rho A = \rho \iint_R dA = \iint_R \rho \, dA. \quad \text{Constant density}$$

If not otherwise stated, a lamina is assumed to have a constant density. In this section, however, we extend the definition of the term *lamina* to include thin plates of *variable* density. Double integrals can be used to find the mass of a lamina of *variable* density, where the density at (x, y) is given by the **density function** ρ.

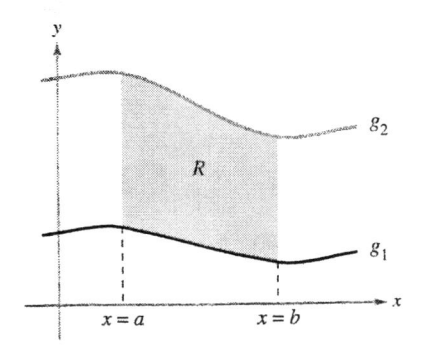

Lamina of constant density ρ
Figure 13.33

Definition of Mass of a Planar Lamina of Variable Density

If ρ is a continuous density function on the lamina corresponding to a plane region R, then the mass m of the lamina is given by

$$m = \iint_R \rho(x, y) \, dA. \quad \text{Variable density}$$

NOTE Density is normally expressed as mass per unit volume. For a planar lamina, however, density is mass per unit surface area.

Example 1 Finding the Mass of a Planar Lamina

Find the mass of the triangular lamina with vertices $(0, 0)$, $(0, 3)$, and $(2, 3)$, given that the density at (x, y) is $\rho(x, y) = 2x + y$.

Solution As shown in Figure 13.34, region R has the boundaries $x = 0$, $y = 3$, and $y = 3x/2$ (or $x = 2y/3$). Therefore, the mass of the lamina is

$$\begin{aligned}
m &= \iint_R (2x + y) \, dA = \int_0^3 \int_0^{2y/3} (2x + y) \, dx \, dy \\
&= \int_0^3 \left[x^2 + xy \right]_0^{2y/3} dy \\
&= \frac{10}{9} \int_0^3 y^2 \, dy \\
&= \frac{10}{9} \left[\frac{y^3}{3} \right]_0^3 \\
&= 10.
\end{aligned}$$

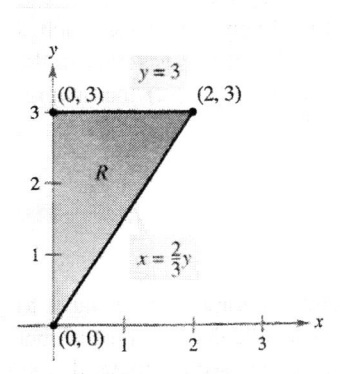

Lamina of variable density
$\rho(x, y) = 2x + y$
Figure 13.34

NOTE In Figure 13.34, note that the planar lamina is shaded so that the darkest shading corresponds to the densest part.

Example 2 Finding Mass by Polar Coordinates

Find the mass of the lamina corresponding to the first-quadrant portion of the circle

$$x^2 + y^2 = 4$$

where the density at the point (x, y) is proportional to the distance between the point and the origin, as shown in Figure 13.35.

Solution At any point (x, y), the density of the lamina is

$$\rho(x, y) = k\sqrt{(x - 0)^2 + (y - 0)^2}$$
$$= k\sqrt{x^2 + y^2}.$$

Because $0 \leq x \leq 2$ and $0 \leq y \leq \sqrt{4 - x^2}$, the mass is given by

$$m = \int\!\!\!\int_R k\sqrt{x^2 + y^2}\, dA$$
$$= \int_0^2 \int_0^{\sqrt{4-x^2}} k\sqrt{x^2 + y^2}\, dy\, dx.$$

To simplify the integration, you can change to polar coordinates, using the bounds $0 \leq \theta \leq \pi/2$ and $0 \leq r \leq 2$. So, the mass is

$$m = \int\!\!\!\int_R k\sqrt{x^2 + y^2}\, dA = \int_0^{\pi/2} \int_0^2 k\sqrt{r^2}\, r\, dr\, d\theta$$
$$= \int_0^{\pi/2} \int_0^2 kr^2\, dr\, d\theta$$
$$= \int_0^{\pi/2} \frac{kr^3}{3}\bigg]_0^2 d\theta$$
$$= \frac{8k}{3} \int_0^{\pi/2} d\theta$$
$$= \frac{8k}{3}\bigg[\theta\bigg]_0^{\pi/2}$$
$$= \frac{4\pi k}{3}.$$

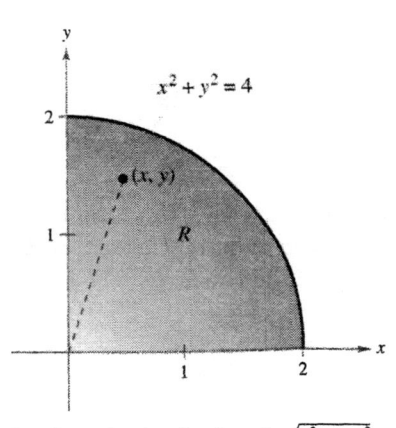

Density at (x, y): $\rho(x, y) = k\sqrt{x^2 + y^2}$
Figure 13.35

TECHNOLOGY On many occasions in this text, we have mentioned the benefits of computer programs that perform symbolic integration. Even if you use such a program regularly, you should remember that its greatest benefit comes only in the hands of a knowledgeable user. For instance, notice how much simpler the integral in Example 2 becomes when it is converted to polar form.

Rectangular Form	Polar Form
$\int_0^2 \int_0^{\sqrt{4-x^2}} k\sqrt{x^2 + y^2}\, dy\, dx$	$\int_0^{\pi/2} \int_0^2 kr^2\, dr\, d\theta$

If you have access to software that performs symbolic integration, try using it to evaluate both integrals. Some software programs cannot handle the first integral, but any program that can handle double integrals can evaluate the second integral.

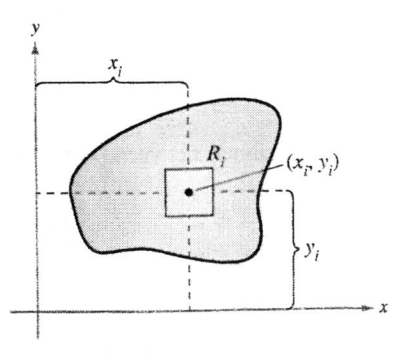

$M_x = (\text{mass})(y_i)$
$M_y = (\text{mass})(x_i)$
Figure 13.36

Moments and Center of Mass

For a lamina of variable density, moments of mass are defined in a manner similar to that used for the uniform density case. For a partition Δ of a lamina corresponding to a plane region R, consider the ith rectangle R_i of one area ΔA_i, as shown in Figure 13.36. Assume that the mass of R_i is concentrated at one of its interior points (x_i, y_i). The moment of mass of R_i with respect to the x-axis can be approximated by

$$(\text{Mass})(y_i) \approx [\rho(x_i, y_i) \Delta A_i](y_i).$$

Similarly, the moment of mass with respect to the y-axis can be approximated by

$$(\text{Mass})(x_i) \approx [\rho(x_i, y_i) \Delta A_i](x_i).$$

By forming the Riemann sum of all such products and taking the limits as the norm of Δ approaches 0, you obtain the following definitions of moments of mass with respect to the x- and y-axes.

Moments and Center of Mass of a Variable Density Planar Lamina

Let ρ be a continuous density function on the planar lamina R. The **moments of mass** with respect to the x- and y-axes are

$$M_x = \iint_R y\rho(x, y)\, dA \quad \text{and} \quad M_y = \iint_R x\rho(x, y)\, dA.$$

If m is the mass of the lamina, then the **center of mass** is

$$(\bar{x}, \bar{y}) = \left(\frac{M_y}{m}, \frac{M_x}{m}\right).$$

If R represents a simple plane region rather than a lamina, the point $(\bar{x}, \bar{y})$ is called the **centroid** of the region.

For some planar laminas, you can determine the center of mass (or one of its coordinates) using symmetry rather than using integration. For instance, consider the laminas shown in Figure 13.37. Using symmetry, you can see that $\bar{y} = 0$ for the first lamina and $\bar{x} = 0$ for the second lamina.

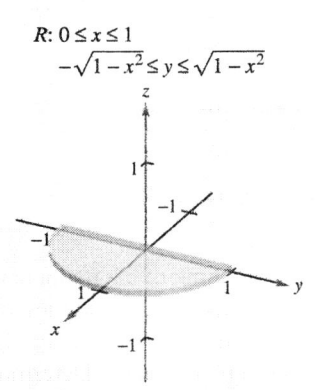

$R: 0 \leq x \leq 1$
$-\sqrt{1-x^2} \leq y \leq \sqrt{1-x^2}$

$R: -\sqrt{1-y^2} \leq x \leq \sqrt{1-y^2}$
$0 \leq y \leq 1$

Symmetric with respect to the x-axis   Symmetric with respect to the y-axis
Figure 13.37

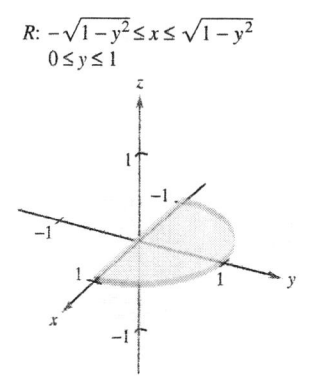

Variable density:
$\rho(x, y) = ky$

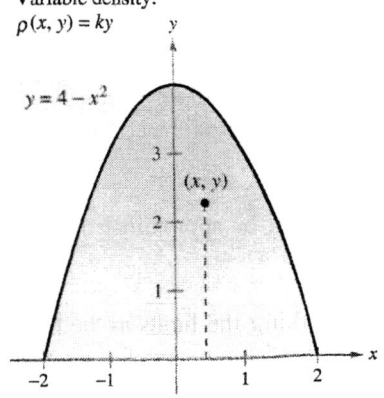

The parabolic region of variable density
Figure 13.38

Example 3   Finding the Center of Mass

Find the center of mass of the lamina corresponding to the parabolic region

$$0 \leq y \leq 4 - x^2 \qquad \text{Parabolic region}$$

where the density at the point (x, y) is proportional to the distance between (x, y) and the x-axis, as shown in Figure 13.38.

Solution   Because the lamina is symmetric with respect to the y-axis and

$$\rho(x, y) = ky$$

the center of mass lies on the y-axis. So, $\bar{x} = 0$. To find $\bar{y}$, first find the mass of the lamina.

$$\begin{aligned}
\text{Mass} &= \int_{-2}^{2} \int_{0}^{4-x^2} ky \, dy \, dx = \frac{k}{2} \int_{-2}^{2} y^2 \Big]_{0}^{4-x^2} dx \\
&= \frac{k}{2} \int_{-2}^{2} (16 - 8x^2 + x^4) \, dx \\
&= \frac{k}{2} \left[16x - \frac{8x^3}{3} + \frac{x^5}{5} \right]_{-2}^{2} \\
&= k \left(32 - \frac{64}{3} + \frac{32}{5} \right) \\
&= \frac{256k}{15}
\end{aligned}$$

Next, find the moment about the x-axis.

$$\begin{aligned}
M_x &= \int_{-2}^{2} \int_{0}^{4-x^2} (y)(ky) \, dy \, dx = \frac{k}{3} \int_{-2}^{2} y^3 \Big]_{0}^{4-x^2} dx \\
&= \frac{k}{3} \int_{-2}^{2} (64 - 48x^2 + 12x^4 - x^6) \, dx \\
&= \frac{k}{3} \left[64x - 16x^3 + \frac{12x^5}{5} - \frac{x^7}{7} \right]_{-2}^{2} \\
&= \frac{4096k}{105}
\end{aligned}$$

So,

$$\bar{y} = \frac{M_x}{m} = \frac{4096k/105}{256k/15} = \frac{16}{7}$$

and the center of mass is $\left(0, \frac{16}{7}\right)$.

Although you can think of the moments M_x and M_y as measuring the tendency to rotate about the x- or y-axis, the calculation of moments is usually an intermediate step toward a more tangible goal. The use of the moments M_x and M_y in Example 3 is typical—to find the center of mass. Determination of the center of mass is useful in a variety of applications that allow you to treat a lamina as if its mass were concentrated at just one point. Intuitively, you can think of the center of mass as the balancing point of the lamina. For instance, the lamina in Example 3 should balance on the point of a pencil placed at $\left(0, \frac{16}{7}\right)$, as shown in Figure 13.39.

Variable density:
$\rho(x, y) = ky$

$R: -2 \leq x \leq 2$
$0 \leq y \leq 4 - x^2$

Center of mass:
$\left(0, \frac{16}{7}\right)$

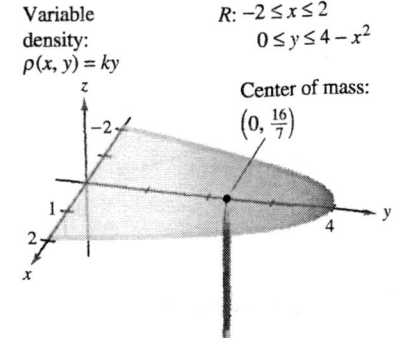

Figure 13.39

Moments of Inertia

The moments of M_x and M_y used in determining the center of mass of a lamina are sometimes called the **first moments** about the x- and y-axes. In each case, the moment is the product of a mass times a distance.

$$M_x = \int\!\!\int_R (y)\rho(x, y)\, dA \qquad M_y = \int\!\!\int_R (x)\rho(x, y)\, dA$$

$\qquad\qquad\qquad\;$ Distance $\;\;$ Mass $\qquad\qquad\quad\;$ Distance $\;\;$ Mass
$\qquad\qquad\qquad\;$ to x-axis $\qquad\qquad\qquad\qquad\;$ to y-axis

We now look at another type of moment—the **second moment,** or the **moment of inertia** of a lamina about a line. In the same way that mass is a measure of the tendency of matter to resist a change in straight-line motion, the moment of inertia about a line is a *measure of the tendency of matter to resist a change in rotational motion*. For example, if a particle of mass m is a distance d from a fixed line, its moment of inertia about the line is defined as

$$I = md^2 = (\text{mass})(\text{distance})^2.$$

As with moments of mass, you can generalize this concept to obtain the moments of inertia about the x- and y-axes of a lamina of variable density. These second moments are denoted by I_x and I_y, and in each case the moment is the product of a mass times the square of a distance.

$$I_x = \int\!\!\int_R (y^2)\rho(x, y)\, dA \qquad I_y = \int\!\!\int_R (x^2)\rho(x, y)\, dA$$

$\;\;$ Square of distance $\;\;$ Mass $\qquad\;$ Square of distance $\;\;$ Mass
$\;\;$ to x-axis $\qquad\qquad\qquad\qquad\;\;$ to y-axis

The sum of the moments I_x and I_y is called the **polar moment of inertia** and is denoted by I_0.

NOTE For a lamina in the xy-plane, I_0 represents the moment of inertia of the lamina about the z-axis. The term "polar moment of inertia" stems from the fact that the square of the polar distance r is used in the calculation.

$$I_0 = \int\!\!\int_R (x^2 + y^2)\rho(x, y)\, dA$$
$$= \int\!\!\int_R r^2 \rho(x, y)\, dA$$

Example 4 Finding the Moment of Inertia

Find the moment of inertia about the x-axis of the lamina in Example 3.

Solution From the definition of moment of inertia, you have

$$I_x = \int_{-2}^{2}\int_{0}^{4-x^2} y^2(ky)\, dy\, dx$$
$$= \frac{k}{4}\int_{-2}^{2} y^4 \Big]_0^{4-x^2} dx$$
$$= \frac{k}{4}\int_{-2}^{2} (256 - 256x^2 + 96x^4 - 16x^6 + x^8)\, dx$$
$$= \frac{k}{4}\left[256x - \frac{256x^3}{3} + \frac{96x^5}{5} - \frac{16x^7}{7} + \frac{x^9}{9}\right]_{-2}^{2}$$
$$= \frac{32{,}768k}{315}.$$

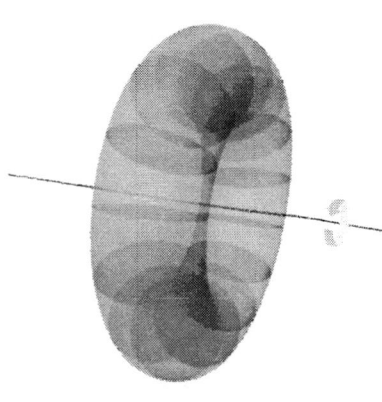

Planar lamina revolving at ω radians per second
Figure 13.40

The moment of inertia I of a revolving lamina can be used to measure its kinetic energy. For example, suppose a planar lamina is revolving about a line with an **angular speed** of ω radians per second, as shown in Figure 13.40. The kinetic energy of the revolving lamina is

$$E = \frac{1}{2}I\omega^2. \quad \text{Kinetic energy for rotational motion}$$

On the other hand, the kinetic energy of a mass m moving in a straight line at a velocity v is

$$E = \frac{1}{2}mv^2. \quad \text{Kinetic energy for linear motion}$$

So, the kinetic energy of a mass moving in a straight line is proportional to its mass, but the kinetic energy of a mass revolving about an axis is proportional to its moment of inertia.

The **radius of gyration** $\bar{\bar{r}}$ of a revolving mass m with moment of inertia I is defined to be

$$\bar{\bar{r}} = \sqrt{\frac{I}{m}}. \quad \text{Radius of gyration}$$

If the entire mass were located at a distance $\bar{\bar{r}}$ from its axis of revolution, it would have the same moment of inertia and, consequently, the same kinetic energy. For instance, the radius of gyration of the lamina in Example 4 about the x-axis is given by

$$\bar{\bar{y}} = \sqrt{\frac{I_x}{m}} = \sqrt{\frac{32{,}768k/315}{256k/15}} = \sqrt{\frac{128}{21}} \approx 2.47.$$

Example 5   Finding the Radius of Gyration

Find the radius of gyration about the y-axis for the lamina corresponding to the region $R\colon 0 \le y \le \sin x$, $0 \le x \le \pi$, where the density at (x, y) is given by $\rho(x, y) = x$.

Solution   The region R is shown in Figure 13.41. By integrating $\rho(x, y) = x$ over the region R, you can determine that the mass of the region is π. The moment of inertia about the y-axis is

$$\begin{aligned}
I_y &= \int_0^\pi \int_0^{\sin x} x^3 \, dy \, dx \\
&= \int_0^\pi x^3 y \Big]_0^{\sin x} dx \\
&= \int_0^\pi x^3 \sin x \, dx \\
&= \Big[(3x^2 - 6)(\sin x) - (x^3 - 6x)(\cos x)\Big]_0^\pi \\
&= \pi^3 - 6\pi.
\end{aligned}$$

So, the radius of gyration about the y-axis is

$$\bar{\bar{x}} = \sqrt{\frac{I_y}{m}} = \sqrt{\frac{\pi^3 - 6\pi}{\pi}} = \sqrt{\pi^2 - 6} \approx 1.97.$$

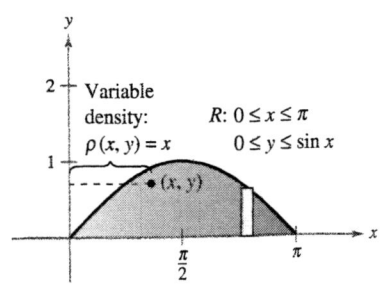

The radius of gyration about the y-axis is approximately 1.97.
Figure 13.41

EXERCISES FOR SECTION 13.4

In Exercises 1–4, find the mass of the lamina described by the inequalities, given that its density is $\rho(x, y) = xy$. (*Hint:* Some of the integrals are simpler in polar coordinates.)

1. $0 \leq x \leq 4,\ 0 \leq y \leq 3$
2. $x \geq 0,\ 0 \leq y \leq 9 - x^2$
3. $x \geq 0,\ 0 \leq y \leq \sqrt{4 - x^2}$
4. $x \geq 0,\ 3 \leq y \leq 3 + \sqrt{9 - x^2}$

In Exercises 5–8, find the mass and center of mass of the lamina for the indicated density.

5. R: rectangle with vertices $(0, 0),\ (a, 0),\ (0, b),\ (a, b)$
   (a) $\rho = k$ (b) $\rho = ky$ (c) $\rho = kx$
6. R: rectangle with vertices $(0, 0),\ (a, 0),\ (0, b),\ (a, b)$
   (a) $\rho = kxy$ (b) $\rho = k(x^2 + y^2)$
7. R: triangle with vertices $(0, 0),\ (b/2, h),\ (b, 0)$
   (a) $\rho = k$ (b) $\rho = ky$ (c) $\rho = kx$
8. R: triangle with vertices $(0, 0),\ (0, a),\ (a, 0)$
   (a) $\rho = k$ (b) $\rho = x^2 + y^2$

9. *Translations in the Plane* Translate the lamina in Exercise 5 to the right 5 units and determine the resulting center of mass.

10. *Conjecture* Use the result in Exercise 9 to make a conjecture about the change in the center of mass when a lamina of constant density is translated h units horizontally or k units vertically. Is the conjecture true if the density is not constant? Explain.

In Exercises 11–22, find the mass and center of mass of the lamina bounded by the graphs of the equations for the indicated density or densities. (*Hint:* Some of the integrals are simpler in polar coordinates.)

11. $y = \sqrt{a^2 - x^2},\ y = 0$
    (a) $\rho = k$ (b) $\rho = k(a - y)y$
12. $x^2 + y^2 = a^2,\ 0 \leq x,\ 0 \leq y$
    (a) $\rho = k$ (b) $\rho = k(x^2 + y^2)$
13. $y = \sqrt{x},\ y = 0,\ x = 4,\ \rho = kxy$
14. $y = x^3,\ y = 0,\ x = 2,\ \rho = kx$
15. $y = \dfrac{1}{1 + x^2},\ y = 0,\ x = -1,\ x = 1,\ \rho = k$
16. $xy = 4,\ x = 1,\ x = 4,\ \rho = kx^2$
17. $x = 16 - y^2,\ x = 0,\ \rho = kx$
18. $y = 9 - x^2,\ y = 0,\ \rho = ky^2$
19. $y = \sin \dfrac{\pi x}{L},\ y = 0,\ x = 0,\ x = L,\ \rho = ky$
20. $y = \cos \dfrac{\pi x}{L},\ y = 0,\ x = 0,\ x = \dfrac{L}{2},\ \rho = k$
21. $y = \sqrt{a^2 - x^2},\ 0 \leq y \leq x,\ \rho = k$
22. $y = \sqrt{a^2 - x^2},\ y = 0,\ y = x,\ \rho = k\sqrt{x^2 + y^2}$

In Exercises 23–26, use a computer algebra system to find the mass and center of mass of the lamina bounded by the graphs of the equations for the indicated density.

23. $y = e^{-x},\ y = 0,\ x = 0,\ x = 2,\ \rho = ky$
24. $y = \ln x,\ y = 0,\ x = 1,\ x = e,\ \rho = k/x$
25. $r = 2 \cos 3\theta,\ -\dfrac{\pi}{6} \leq \theta \leq \dfrac{\pi}{6},\ \rho = k$
26. $r = 1 + \cos \theta,\ \rho = k$

In Exercises 27–32, verify the given moment(s) of inertia and find $\bar{x}$ and $\bar{y}$. Assume that each lamina has a density of $\rho = 1$. (These regions are common shapes used in engineering.)

27. Rectangle
$I_x = \tfrac{1}{3} bh^3$
$I_y = \tfrac{1}{3} b^3 h$

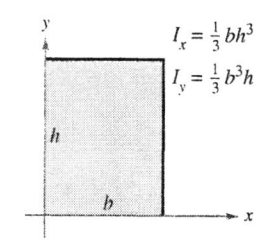

28. Right triangle
$I_x = \tfrac{1}{12} bh^3$
$I_y = \tfrac{1}{12} b^3 h$

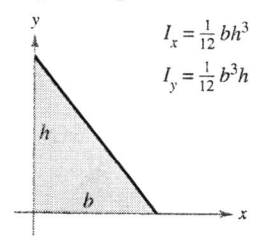

29. Circle
$I_0 = \tfrac{1}{2} \pi a^4$

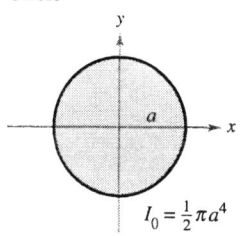

30. Semicircle
$I_0 = \tfrac{1}{4} \pi a^4$

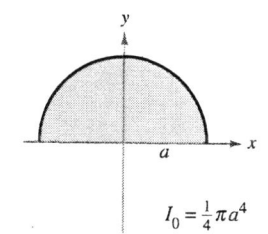

31. Quarter circle
$I_0 = \tfrac{1}{8} \pi a^4$

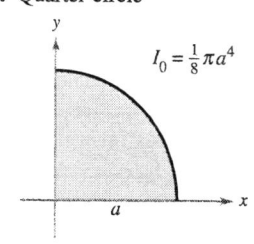

32. Ellipse
$I_0 = \tfrac{1}{4} \pi ab(a^2 + b^2)$

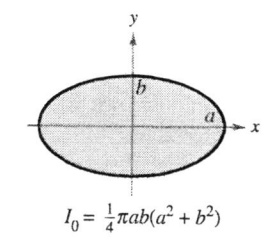

In Exercises 33–40, find I_x, I_y, I_0, $\bar{x}$ and $\bar{y}$ for the lamina bounded by the graphs of the equations. Use a computer algebra system to evaluate the double integrals.

33. $y = 0,\ y = b,\ x = 0,\ x = a,\ \rho = ky$
34. $y = \sqrt{a^2 - x^2},\ y = 0,\ \rho = ky$
35. $y = 4 - x^2,\ y = 0,\ x > 0,\ \rho = kx$
36. $y = x,\ y = x^2,\ \rho = kxy$
37. $y = \sqrt{x},\ y = 0,\ x = 4,\ \rho = kxy$
38. $y = x^2,\ y^2 = x,\ \rho = x^2 + y^2$
39. $y = x^2,\ y^2 = x,\ \rho = kx$
40. $y = x^3,\ y = 4x,\ \rho = k|y|$

In Exercises 41–46, set up the double integral required to find the moment of inertia I, about the indicated line, of the lamina bounded by the graphs of the equations. Use a computer algebra system to evaluate the double integral.

41. $x^2 + y^2 = b^2$, $\rho = k$, line: $x = a$ $(a > b)$
42. $y = 0$, $y = 2$, $x = 0$, $x = 4$, $\rho = k$, line: $x = 6$
43. $y = \sqrt{x}$, $y = 0$, $x = 4$, $\rho = kx$, line: $x = 6$
44. $y = \sqrt{a^2 - x^2}$, $y = 0$, $\rho = ky$, line: $y = a$
45. $y = \sqrt{a^2 - x^2}$, $y = 0$, $x \geq 0$, $\rho = k(a - y)$, line: $y = a$
46. $y = 4 - x^2$, $y = 0$, $\rho = k$, line: $y = 2$

Getting at the Concept

The center of mass of the lamina of constant density shown in the figure is $(2, \frac{8}{5})$. In Exercises 47–50, make a conjecture about how the center of mass $(\bar{x}, \bar{y})$ will change for the nonconstant density $\rho(x, y)$. Explain. (Make your conjecture *without* performing any calculations.)

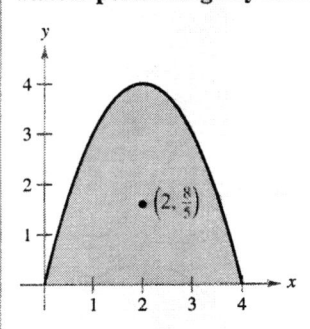

47. $\rho(x, y) = ky$
48. $\rho(x, y) = k|2 - x|$
49. $\rho(x, y) = kxy$
50. $\rho(x, y) = k(4 - x)(4 - y)$

51. Give the formulas for finding the moments and center of mass of a variable density planar lamina.
52. Give the formulas for finding the moments of inertia about the x- and y-axes for a variable density planar lamina.
53. In your own words, describe what the radius of gyration measures.

54. Prove the following Theorem of Pappus: Let R be a region in a plane and let L be a line in the same plane such that L does not intersect the interior of R. If r is the distance between the centroid of R and the line, then the volume V of the solid of revolution formed by revolving R about the line is given by $V = 2\pi rA$, where A is the area of R.

Hydraulics In Exercises 55–58, determine the location of the horizontal axis y_a at which a vertical gate in a dam is to be hinged so that there is no moment causing rotation under the indicated loading (see figure). The model for y_a is

$$y_a = \bar{y} - \frac{I_{\bar{y}}}{hA}$$

where $\bar{y}$ is the y-coordinate of the centroid of the gate, $I_{\bar{y}}$ is the moment of inertia of the gate about the line $y = \bar{y}$, h is the depth of the centroid below the surface, and A is the area of the gate.

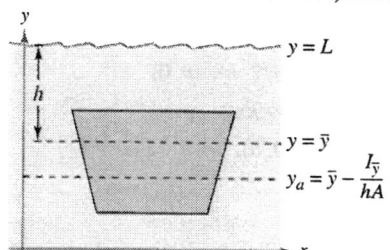

55. 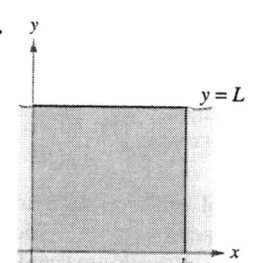
56. 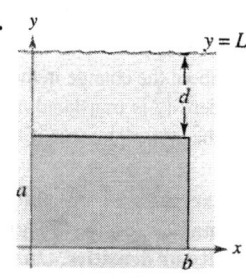
57. 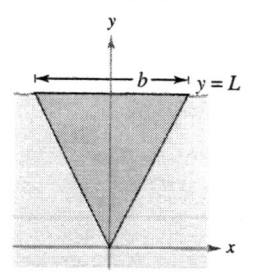
58.

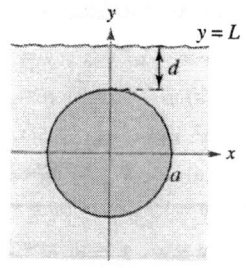

SECTION PROJECT: CENTER OF PRESSURE ON A SAIL

The center of pressure on a sail is that point (x_p, y_p) at which the total aerodynamic force may be assumed to act. If the sail is represented by a plane region R, the center of pressure is

$$x_p = \frac{\int_R \int xy\, dA}{\int_R \int y\, dA} \quad \text{and} \quad y_p = \frac{\int_R \int y^2\, dA}{\int_R \int y\, dA}.$$

Consider a triangular sail with vertices at $(0, 0)$, $(2, 1)$, and $(0, 5)$. Verify the values of the following three integrals.

$$\int_R \int y\, dA = 10$$

$$\int_R \int xy\, dA = \frac{35}{6}$$

$$\int_R \int y^2\, dA = \frac{155}{6}$$

Calculate the coordinates (x_p, y_p) of the center of pressure. Sketch a graph of the sail and indicate the location of the center of pressure.

Section 13.5 Surface Area

• Use a double integral to find the area of a surface.

Surface Area

At this point you know a great deal about the solid region lying between a surface and a closed and bounded region R in the xy-plane, as shown in Figure 13.42. For example, you know how to find the extrema of f on R (Section 12.8), the area of the base R of the solid (Section 13.1), the volume of the solid (Section 13.2), and the centroid of the base R (Section 13.4).

In this section, you will learn how to find the upper **surface area** of the solid. Later, you will learn how to find the centroid of the solid (Section 13.6) and the lateral surface area (Section 14.2).

To begin, consider a surface S given by

$$z = f(x, y) \qquad \text{Surface defined over a region } R$$

defined over a region R. Assume that R is closed and bounded and that f has continuous first partial derivatives. To find the surface area, construct an inner partition of R consisting of n rectangles, where the area of the ith rectangle R_i is $\Delta A_i = \Delta x_i \Delta y_i$, as shown in Figure 13.43. In each R_i let (x_i, y_i) be the point that is closest to the origin. At the point $(x_i, y_i, z_i) = (x_i, y_i, f(x_i, y_i))$ on the surface S, construct a tangent plane T_i. The area of the portion of the tangent plane that lies directly above R_i is approximately equal to the area of the surface lying directly above R_i. That is, $\Delta T_i \approx \Delta S_i$. Hence, the surface area of S is given by

$$\sum_{i=1}^{n} \Delta S_i \approx \sum_{i=1}^{n} \Delta T_i.$$

To find the area of the parallelogram ΔT_i, note that its sides are given by the vectors

$$\mathbf{u} = \Delta x_i \mathbf{i} + f_x(x_i, y_i)\, \Delta x_i \mathbf{k}$$

and

$$\mathbf{v} = \Delta y_i \mathbf{j} + f_y(x_i, y_i)\, \Delta y_i \mathbf{k}.$$

From Theorem 10.8, the area of ΔT_i is given by $\|\mathbf{u} \times \mathbf{v}\|$, where

$$\mathbf{u} \times \mathbf{v} = \begin{vmatrix} \mathbf{i} & \mathbf{j} & \mathbf{k} \\ \Delta x_i & 0 & f_x(x_i, y_i)\, \Delta x_i \\ 0 & \Delta y_i & f_y(x_i, y_i)\, \Delta y_i \end{vmatrix}$$

$$= -f_x(x_i, y_i)\, \Delta x_i \Delta y_i \mathbf{i} - f_y(x_i, y_i)\, \Delta x_i \Delta y_i \mathbf{j} + \Delta x_i \Delta y_i \mathbf{k}$$

$$= (-f_x(x_i, y_i)\mathbf{i} - f_y(x_i, y_i)\mathbf{j} + \mathbf{k})\, \Delta A_i.$$

So, the area of ΔT_i is $\|\mathbf{u} \times \mathbf{v}\| = \sqrt{[f_x(x_i, y_i)]^2 + [f_y(x_i, y_i)]^2 + 1}\; \Delta A_i$, and

$$\text{Surface area of } S \approx \sum_{i=1}^{n} \Delta S_i$$

$$\approx \sum_{i=1}^{n} \sqrt{1 + [f_x(x_i, y_i)]^2 + [f_y(x_i, y_i)]^2}\; \Delta A_i.$$

This suggests the following definition of surface area.

Surface:
$z = f(x, y)$

Region R in xy-plane

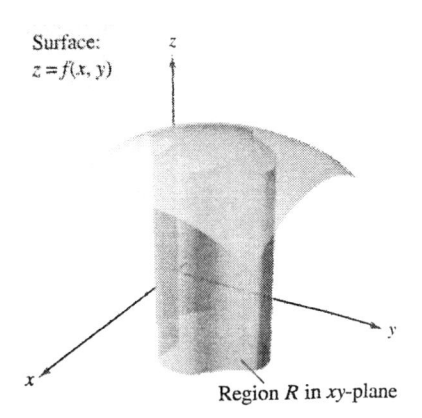

Figure 13.42

Surface:
$z = f(x, y)$

ΔT_i

$\Delta S_i \approx \Delta T_i$

R

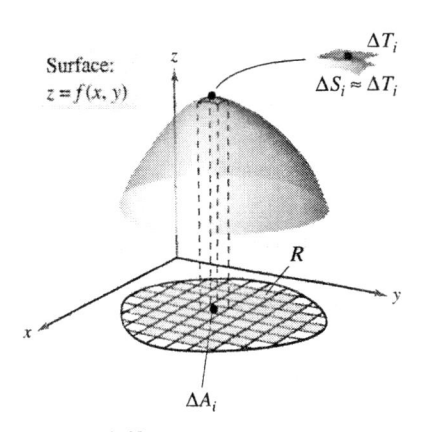

ΔA_i

Figure 13.43

Definition of Surface Area

If f and its first partial derivatives are continuous on the closed region R in the xy-plane, then the **area of the surface S** given by $z = f(x, y)$ over R is given by

$$\text{Surface area} = \iint_R dS$$
$$= \iint_R \sqrt{1 + [f_x(x, y)]^2 + [f_y(x, y)]^2} \, dA.$$

As an aid to remembering the double integral for surface area, it is helpful to note its similarity to the integral for arc length.

Length on x-axis: $\quad \displaystyle\int_a^b dx$

Arc length in xy-plane: $\quad \displaystyle\int_a^b ds = \int_a^b \sqrt{1 + [f'(x)]^2} \, dx$

Area in xy-plane: $\quad \displaystyle\iint_R dA$

Surface area in space: $\quad \displaystyle\iint_R dS = \iint_R \sqrt{1 + [f_x(x, y)]^2 + [f_y(x, y)]^2} \, dA$

Like integrals for arc length, integrals for surface area are often very difficult to evaluate. However, one type that is easily evaluated is demonstrated in the next example.

Example 1 The Surface Area of a Plane Region

Find the surface area of the portion of the plane

$$z = 2 - x - y$$

that lies above the circle $x^2 + y^2 \leq 1$ in the first quadrant, as shown in Figure 13.44.

Solution Because $f_x(x, y) = -1$ and $f_y(x, y) = -1$, the surface area is given by

$$S = \iint_R \sqrt{1 + [f_x(x, y)]^2 + [f_y(x, y)]^2} \, dA \quad \text{Formula for surface area}$$
$$= \iint_R \sqrt{1 + (-1)^2 + (-1)^2} \, dA \quad \text{Substitute.}$$
$$= \iint_R \sqrt{3} \, dA$$
$$= \sqrt{3} \iint_R dA.$$

Note that the last integral is simply $\sqrt{3}$ times the area of the region R. R is a quarter circle of radius 1, with an area of $\frac{1}{4}\pi(1^2)$ or $\pi/4$. So, the area of S is

$$S = \sqrt{3} \, (\text{area of } R)$$
$$= \sqrt{3}\left(\frac{\pi}{4}\right)$$
$$= \frac{\sqrt{3}\,\pi}{4}.$$

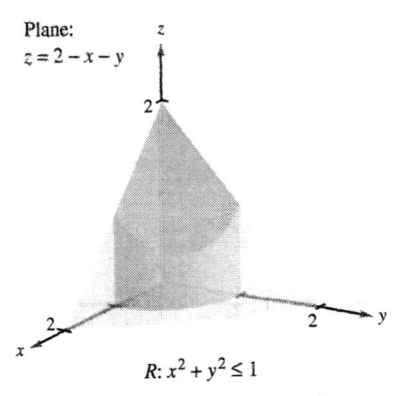

Plane: $z = 2 - x - y$

$R: x^2 + y^2 \leq 1$

The surface area of the portion of the plane that lies above the quarter circle is $\sqrt{3}\,\pi/4$.
Figure 13.44

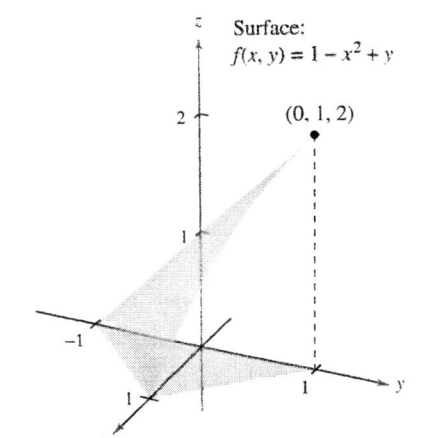

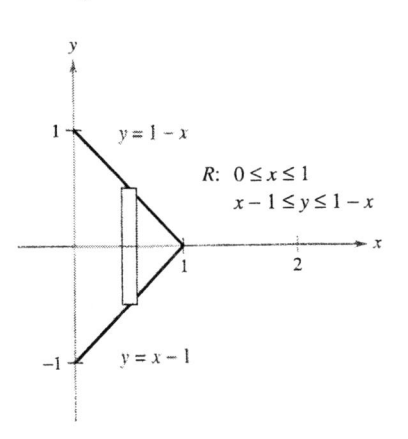

Example 2 Finding Surface Area

Find the area of the portion of the surface

$$f(x, y) = 1 - x^2 + y$$

that lies above the triangular region with vertices $(1, 0, 0)$, $(0, -1, 0)$, and $(0, 1, 0)$, as shown in Figure 13.45.

Solution Because $f_x(x, y) = -2x$ and $f_y(x, y) = 1$, you have

$$S = \iint_R \sqrt{1 + [f_x(x, y)]^2 + [f_y(x, y)]^2}\, dA = \iint_R \sqrt{1 + 4x^2 + 1}\, dA.$$

In Figure 13.45, you can see that the bounds for R are $0 \le x \le 1$ and $x - 1 \le y \le 1 - x$. So, the integral becomes

$$S = \int_0^1 \int_{x-1}^{1-x} \sqrt{2 + 4x^2}\, dy\, dx$$

$$= \int_0^1 y\sqrt{2 + 4x^2} \Big]_{x-1}^{1-x} dx$$

$$= \int_0^1 \left(2\sqrt{2 + 4x^2} - 2x\sqrt{2 + 4x^2}\right) dx \quad \text{Integration tables (Appendix C), Formula 26 and Power Rule}$$

$$= \left[x\sqrt{2 + 4x^2} + \ln\left(2x + \sqrt{2 + 4x^2}\right) - \frac{(2 + 4x^2)^{3/2}}{6}\right]_0^1$$

$$= \sqrt{6} + \ln(2 + \sqrt{6}) - \sqrt{6} - \ln\sqrt{2} + \frac{1}{3}\sqrt{2}$$

$$\approx 1.618.$$

Figure 13.45

Example 3 Change of Variables to Polar Coordinates

Find the surface area of the paraboloid $z = 1 + x^2 + y^2$ that lies above the unit circle, as shown in Figure 13.46.

Solution Because $f_x(x, y) = 2x$ and $f_y(x, y) = 2y$, you have

$$S = \iint_R \sqrt{1 + [f_x(x, y)]^2 + [f_y(x, y)]^2}\, dA = \iint_R \sqrt{1 + 4x^2 + 4y^2}\, dA.$$

You can convert to polar coordinates by letting $x = r\cos\theta$ and $y = r\sin\theta$. Then, because the region R is bounded by $0 \le r \le 1$ and $0 \le \theta \le 2\pi$, you have

$$S = \int_0^{2\pi}\int_0^1 \sqrt{1 + 4r^2}\, r\, dr\, d\theta$$

$$= \int_0^{2\pi} \frac{1}{12}(1 + 4r^2)^{3/2}\Big]_0^1 d\theta$$

$$= \int_0^{2\pi} \frac{5\sqrt{5} - 1}{12}\, d\theta$$

$$= \frac{5\sqrt{5} - 1}{12}\theta\Big]_0^{2\pi}$$

$$= \frac{\pi(5\sqrt{5} - 1)}{6}$$

$$\approx 5.33.$$

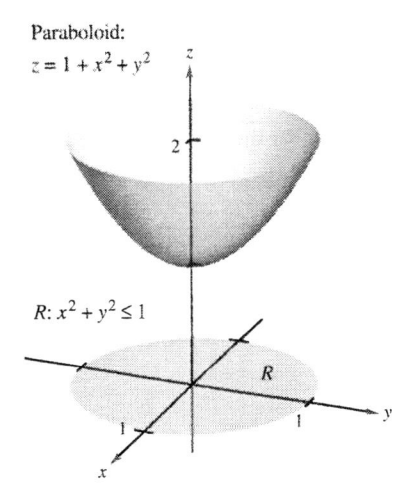

The surface area of the portion of the paraboloid that lies above the unit circle is approximately 5.33.
Figure 13.46

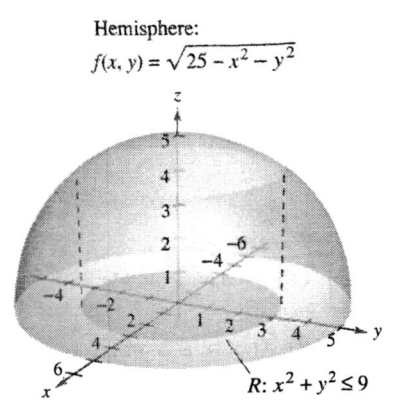

Hemisphere:
$f(x, y) = \sqrt{25 - x^2 - y^2}$

The surface area of the portion of the hemisphere that lies above the circle is 10π.
Figure 13.47

Example 4 Finding Surface Area

Find the surface area S of the portion of the hemisphere

$$f(x, y) = \sqrt{25 - x^2 - y^2} \qquad \text{Hemisphere}$$

that lies above the region R bounded by the circle $x^2 + y^2 \leq 9$, as shown in Figure 13.47.

Solution The first partial derivatives of f are

$$f_x(x, y) = \frac{-x}{\sqrt{25 - x^2 - y^2}} \quad \text{and} \quad f_y(x, y) = \frac{-y}{\sqrt{25 - x^2 - y^2}}$$

and, from the formula for surface area, you have

$$dS = \sqrt{1 + [f_x(x, y)]^2 + [f_y(x, y)]^2} \, dA$$

$$= \sqrt{1 + \left(\frac{-x}{\sqrt{25 - x^2 - y^2}}\right)^2 + \left(\frac{-y}{\sqrt{25 - x^2 - y^2}}\right)^2} \, dA$$

$$= \frac{5}{\sqrt{25 - x^2 - y^2}} \, dA.$$

Therefore, the surface area is

$$S = \iint_R \frac{5}{\sqrt{25 - x^2 - y^2}} \, dA.$$

You can convert to polar coordinates by letting $x = r \cos \theta$ and $y = r \sin \theta$. Then, because the region R is bounded by $0 \leq r \leq 3$ and $0 \leq \theta \leq 2\pi$, you obtain

$$S = \int_0^{2\pi} \int_0^3 \frac{5}{\sqrt{25 - r^2}} r \, dr \, d\theta$$

$$= 5 \int_0^{2\pi} \left[-\sqrt{25 - r^2} \right]_0^3 d\theta$$

$$= 5 \int_0^{2\pi} d\theta$$

$$= 10\pi.$$

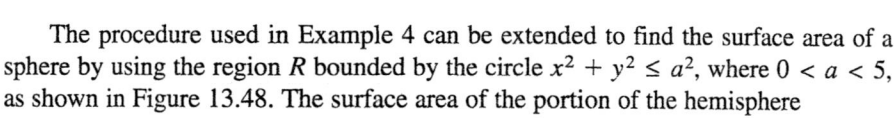

Hemisphere:
$f(x, y) = \sqrt{25 - x^2 - y^2}$

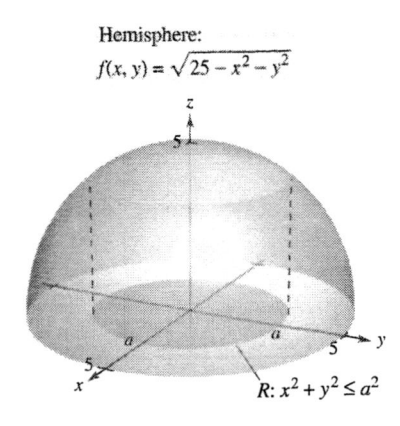

Figure 13.48

The procedure used in Example 4 can be extended to find the surface area of a sphere by using the region R bounded by the circle $x^2 + y^2 \leq a^2$, where $0 < a < 5$, as shown in Figure 13.48. The surface area of the portion of the hemisphere

$$f(x, y) = \sqrt{25 - x^2 - y^2}$$

lying above the circular region can be shown to be

$$S = \iint_R \frac{5}{\sqrt{25 - x^2 - y^2}} \, dA$$

$$= \int_0^{2\pi} \int_0^a \frac{5}{\sqrt{25 - r^2}} r \, dr \, d\theta$$

$$= 10\pi \left(5 - \sqrt{25 - a^2}\right).$$

By taking the limit as a approaches 5 and doubling the result, you obtain a total area of 100π. (The surface area of a sphere of radius r is $S = 4\pi r^2$.)

You can use Simpson's Rule or the Trapezoidal Rule to approximate the value of a double integral, *provided* you can get through the first integration. This is demonstrated in the next example.

Example 5 Approximating Surface Area by Simpson's Rule

Find the area of the surface of the paraboloid

$$f(x, y) = 2 - x^2 - y^2 \qquad \text{Paraboloid}$$

that lies above the square region bounded by $-1 \leq x \leq 1$ and $-1 \leq y \leq 1$, as shown in Figure 13.49.

Solution Using the partial derivatives

$$f_x(x, y) = -2x \quad \text{and} \quad f_y(x, y) = -2y$$

you have a surface area of

$$S = \iint_R \sqrt{1 + [f_x(x, y)]^2 + [f_y(x, y)]^2}\, dA$$
$$= \iint_R \sqrt{1 + (-2x)^2 + (-2y)^2}\, dA$$
$$= \iint_R \sqrt{1 + 4x^2 + 4y^2}\, dA.$$

In polar coordinates, the line $x = 1$ is given by $r \cos\theta = 1$ or $r = \sec\theta$, and you can determine from Figure 13.50 that one-fourth of the region R is bounded by

$$0 \leq r \leq \sec\theta \quad \text{and} \quad -\frac{\pi}{4} \leq \theta \leq \frac{\pi}{4}.$$

Letting $x = r\cos\theta$ and $y = r\sin\theta$ produces

$$\frac{1}{4}S = \frac{1}{4}\iint_R \sqrt{1 + 4x^2 + 4y^2}\, dA$$
$$= \int_{-\pi/4}^{\pi/4}\int_0^{\sec\theta} \sqrt{1 + 4r^2}\, r\, dr\, d\theta$$
$$= \int_{-\pi/4}^{\pi/4} \frac{1}{12}(1 + 4r^2)^{3/2}\Big]_0^{\sec\theta} d\theta$$
$$= \frac{1}{12}\int_{-\pi/4}^{\pi/4} \left[(1 + 4\sec^2\theta)^{3/2} - 1\right] d\theta.$$

Finally, using Simpson's Rule with $n = 10$, you can approximate this *single* integral to be

$$S = \frac{1}{3}\int_{-\pi/4}^{\pi/4} \left[(1 + 4\sec^2\theta)^{3/2} - 1\right] d\theta$$
$$\approx 7.45.$$

Paraboloid:
$f(x, y) = 2 - x^2 - y^2$

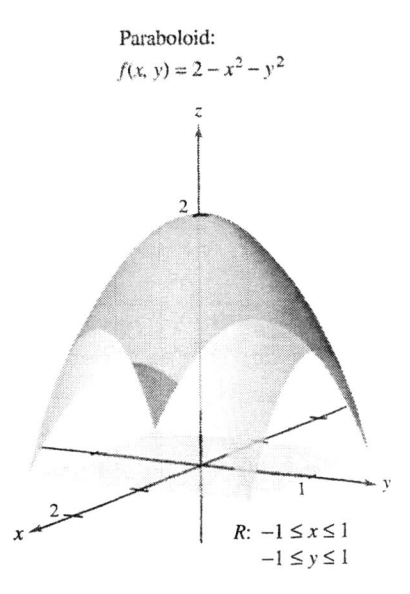

$R: -1 \leq x \leq 1$
$-1 \leq y \leq 1$

Figure 13.49

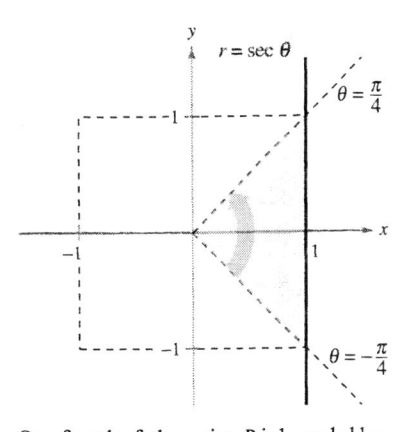

$r = \sec\theta$

One-fourth of the region R is bounded by
$0 \leq r \leq \sec\theta$ and $-\frac{\pi}{4} \leq \theta \leq \frac{\pi}{4}$.
Figure 13.50

TECHNOLOGY Most computer programs that are capable of performing symbolic integration for multiple integrals are also capable of performing numerical approximation techniques. If you have access to such software, try using it to approximate the value of the integral in Example 5.

EXERCISES FOR SECTION 13.5

In Exercises 1–14, find the area of the surface given by $z = f(x, y)$ over the region R. (*Hint:* Some of the integrals are simpler in polar coordinates.)

1. $f(x, y) = 2x + 2y$
   R: triangle with vertices $(0, 0), (2, 0), (0, 2)$

2. $f(x, y) = 15 + 2x - 3y$
   R: square with vertices $(0, 0), (3, 0), (0, 3), (3, 3)$

3. $f(x, y) = 8 + 2x + 2y$
   $R = \{(x, y): x^2 + y^2 \leq 4\}$

4. $f(x, y) = 10 + 2x - 3y$
   $R = \{(x, y): x^2 + y^2 \leq 9\}$

5. $f(x, y) = 9 - x^2$
   R: square with vertices $(0, 0), (3, 0), (0, 3), (3, 3)$

6. $f(x, y) = y^2$
   R: square with vertices $(0, 0), (3, 0), (0, 3), (3, 3)$

7. $f(x, y) = 2 + x^{3/2}$
   R: rectangle with vertices $(0, 0), (0, 4), (3, 4), (3, 0)$

8. $f(x, y) = 2 + \frac{2}{3}y^{3/2}$
   $R = \{(x, y): 0 \leq x \leq 2, 0 \leq y \leq 2 - x\}$

9. $f(x, y) = \ln|\sec x|$
   $R = \left\{(x, y): 0 \leq x \leq \frac{\pi}{4}, 0 \leq y \leq \tan x\right\}$

10. $f(x, y) = 9 + x^2 - y^2$
    $R = \{(x, y): x^2 + y^2 \leq 4\}$

11. $f(x, y) = \sqrt{x^2 + y^2}$
    $R = \{(x, y): 0 \leq f(x, y) \leq 1\}$

12. $f(x, y) = xy$
    $R = \{(x, y): x^2 + y^2 \leq 16\}$

13. $f(x, y) = \sqrt{a^2 - x^2 - y^2}$
    $R = \{(x, y): x^2 + y^2 \leq b^2, 0 < b < a\}$

14. $f(x, y) = \sqrt{a^2 - x^2 - y^2}$
    $R = \{(x, y): x^2 + y^2 \leq a^2\}$

In Exercises 15–18, find the area of the surface.

15. The portion of the plane $z = 24 - 3x - 2y$ in the first octant

16. The portion of the paraboloid $z = 16 - x^2 - y^2$ in the first octant

17. The portion of the sphere $x^2 + y^2 + z^2 = 25$ inside the cylinder $x^2 + y^2 = 9$

18. The portion of the cone $z = 2\sqrt{x^2 + y^2}$ inside the cylinder $x^2 + y^2 = 4$

In Exercises 19–24, write a double integral that represents the surface area of $z = f(x, y)$ over the region R. Use a computer algebra system to evaluate the double integral.

19. $f(x, y) = 2y + x^2$
    R: triangle with vertices $(0, 0), (1, 0), (1, 1)$

20. $f(x, y) = 2x + y^2$
    R: triangle with vertices $(0, 0), (2, 0), (2, 2)$

21. $f(x, y) = 4 - x^2 - y^2$
    $R = \{(x, y): 0 \leq f(x, y)\}$

22. $f(x, y) = x^2 + y^2$
    $R = \{(x, y): 0 \leq f(x, y) \leq 16\}$

23. $f(x, y) = 4 - x^2 - y^2$
    $R = \{(x, y): 0 \leq x \leq 1, 0 \leq y \leq 1\}$

24. $f(x, y) = \frac{2}{3}x^{3/2} + \cos x$
    $R = \{(x, y): 0 \leq x \leq 1, 0 \leq y \leq 1\}$

Approximation In Exercises 25 and 26, determine which value best approximates the surface area of $z = f(x, y)$ over the region R. (Make your selection on the basis of a sketch of the surface and *not* by performing any calculations.)

25. $f(x, y) = 10 - \frac{1}{2}y^2$
    R: square with vertices $(0, 0), (4, 0), (4, 4), (0, 4)$
    (a) 16   (b) 200   (c) -100 (d) 72   (e) 36

26. $f(x, y) = \frac{1}{4}\sqrt{x^2 + y^2}$
    R: circle bounded by $x^2 + y^2 = 9$
    (a) -100 (b) 150   (c) 9π (d) 55   (e) 500

In Exercises 27 and 28, use a computer algebra system to approximate the double integral that gives the surface area of the graph of f over the region $R = \{(x, y): 0 \leq x \leq 1, 0 \leq y \leq 1\}$.

27. $f(x, y) = e^x$ 28. $f(x, y) = \frac{2}{5}y^{5/2}$

In Exercises 29–34, set up a double integral that gives the area of the surface on the graph of f over the region R.

29. $f(x, y) = x^3 - 3xy + y^3$
    R: square with vertices $(1, 1), (-1, 1), (-1, -1), (1, -1)$

30. $f(x, y) = x^2 - 3xy - y^2$
    $R = \{(x, y): 0 \leq x \leq 4, 0 \leq y \leq x\}$

31. $f(x, y) = e^{-x} \sin y$
    $R = \{(x, y): x^2 + y^2 \leq 4\}$

32. $f(x, y) = \cos(x^2 + y^2)$
    $R = \left\{(x, y): x^2 + y^2 \leq \frac{\pi}{2}\right\}$

33. $f(x, y) = e^{xy}$
    $R = \{(x, y): 0 \leq x \leq 4, 0 \leq y \leq 10\}$

34. $f(x, y) = e^{-x} \sin y$
    $R = \{(x, y): 0 \leq x \leq 4, 0 \leq y \leq x\}$

Getting at the Concept

35. State the double integral definition of the area of a surface S given by $z = f(x, y)$ over the region R in the xy-plane.

36. Answer the following about the surface area S on a surface given by a positive function $z = f(x, y)$ over a region R in the xy-plane. Explain each answer.
    (a) Is it possible for S to equal the area of R?
    (b) Can S be greater than the area of R?
    (c) Can S be less than the area of R?

37. Find the surface area of the solid of intersection of the cylinders $x^2 + z^2 = 1$ and $y^2 + z^2 = 1$ (see figure).

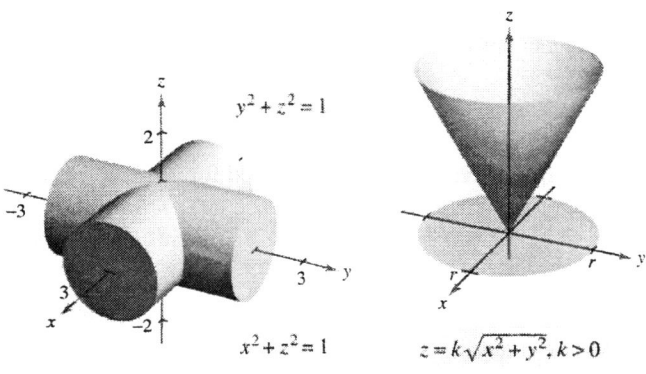

Figure for 37

Figure for 38

38. Show that the surface area of the cone $z = k\sqrt{x^2 + y^2}$, $k > 0$ over the circular region $x^2 + y^2 \leq r^2$ in the xy-plane is $\pi r^2 \sqrt{k^2 + 1}$ (see figure).

39. Building Design A new auditorium is built with a foundation in the shape of $\frac{1}{4}$ of a circle of radius 50 feet. Therefore, it forms a region R bounded by the graph of $x^2 + y^2 = 50^2$ with $x \geq 0$ and $y \geq 0$. The following equations are models for the floor and ceiling.

Floor: $z = \dfrac{x + y}{5}$

Ceiling: $z = 20 + \dfrac{xy}{100}$

(a) Calculate the volume of the room, which is needed to determine the heating and cooling requirements.

(b) Find the surface area of the ceiling.

40. Modeling Data A rancher builds a barn with dimensions 30 feet by 50 feet. The symmetrical shape and selected heights of the roof are shown in the figure.

(a) Use the regression capabilities of a graphing utility to fit the model $z = ay^3 + by^2 + cy + d$ to the roof line.

(b) Use the numerical integration capabilities of a graphing utility and the model in part (a) to approximate the volume of storage space in the barn.

(c) Use the numerical integration capabilities of a graphing utility and the model in part (a) to approximate the surface area of the roof.

(d) Approximate the arc length of the roof line and find the surface area of the roof by multiplying the arc length by the length of the barn. Compare the results and the integrations in parts (c) and (d).

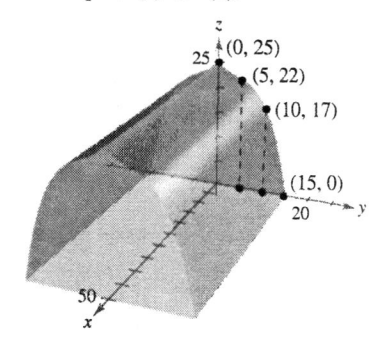

41. Product Design A company produces a spherical object of radius 25 centimeters. A hole of radius 4 centimeters is drilled through the center of the object. Find (a) the volume of the object and (b) the outer surface area of the object.

42. True or False? The surface area of the graph of a function $z = f(x, y)$ over a region R will increase if the graph is shifted k units vertically. Explain.

SECTION PROJECT CAPILLARY ACTION

A well-known property of liquids is that they will rise in narrow vertical channels—this property is called "capillary action." The figure shows two plates, which form a narrow wedge, in a container of liquid. The upper surface of the liquid follows a hyperbolic shape given by

$$z = \frac{k}{\sqrt{x^2 + y^2}}$$

where x, y, and z are measured in inches. The constant k depends on the angle of the wedge, the type of liquid, and the material that comprises the flat plates.

(a) Find the volume of the liquid that has risen in the wedge. (Assume $k = 1$.)

(b) Find the horizontal surface area of the liquid that has risen in the wedge.

Adaptation of capillary action problem from "Capillary Phenomena" by Thomas Greenslade, Jr., *Physics Teacher*, May 1992. Used by permission of the author.

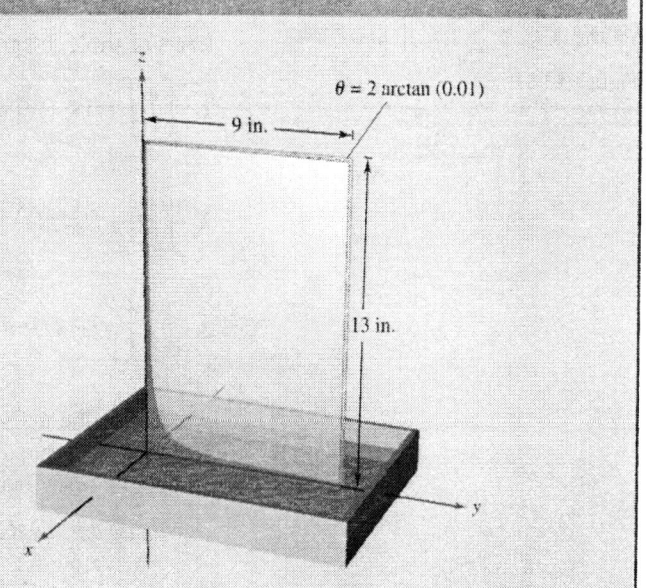

Section 13.6 Triple Integrals and Applications

- Use a triple integral to find the volume of a solid region.
- Find the center of mass and moments of inertia of a solid region.

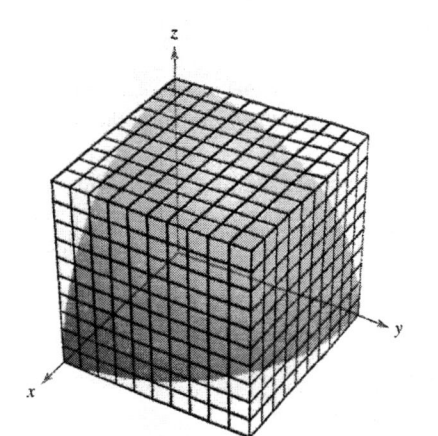

Triple Integrals

The procedure used to define a **triple integral** follows that used for double integrals. Consider a function f of three variables that is continuous over a bounded solid region Q. Then, encompass Q with a network of boxes and form the **inner partition** consisting of all boxes lying entirely within Q, as shown in Figure 13.51. The volume of the ith box is

$$\Delta V_i = \Delta x_i \Delta y_i \Delta z_i. \quad \text{Volume of } i\text{th box}$$

The **norm** $\|\Delta\|$ of the partition is the length of the longest diagonal of the n boxes in the partition. Choose a point (x_i, y_i, z_i) in each box and form the Riemann sum

$$\sum_{i=1}^{n} f(x_i, y_i, z_i) \Delta V_i.$$

Taking the limit as $\|\Delta\| \to 0$ leads to the following definition.

Definition of Triple Integral

If f is continuous over a bounded solid region Q, then the **triple integral of f over Q** is defined as

$$\iiint_Q f(x, y, z)\, dV = \lim_{\|\Delta\| \to 0} \sum_{i=1}^{n} f(x_i, y_i, z_i) \Delta V_i$$

provided the limit exists. The **volume** of the solid region Q is given by

$$\text{Volume of } Q = \iiint_Q dV.$$

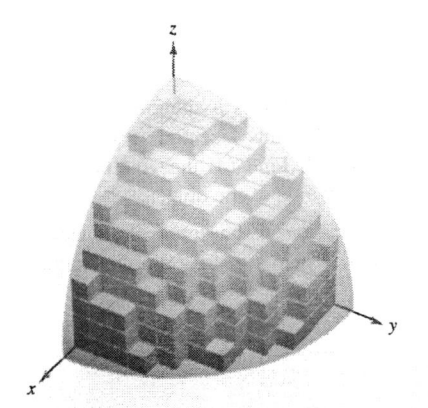

Solid region Q

Volume of $Q \approx \sum_{i=1}^{n} f(x_i, y_i, z_i)\, \Delta V_i$

Figure 13.51

Some of the properties of double integrals in Theorem 13.1 can be restated in terms of triple integrals

1. $\displaystyle\iiint_Q cf(x, y, z)\, dV = c\iiint_Q f(x, y, z)\, dV$

2. $\displaystyle\iiint_Q [f(x, y, z) \pm g(x, y, z)]\, dV = \iiint_Q f(x, y, z)\, dV \pm \iiint_Q g(x, y, z)\, dV$

3. $\displaystyle\iiint_Q f(x, y, z)\, dV = \iiint_{Q_1} f(x, y, z)\, dV + \iiint_{Q_2} f(x, y, z)\, dV$

where Q is the union of two nonoverlapping solid subregions Q_1 and Q_2. If the solid region Q is simple, the triple integral $\iiint f(x, y, z)\, dV$ can be evaluated with an iterated integral using one of the six possible orders of integration:

$$dx\, dy\, dz \quad dy\, dx\, dz \quad dz\, dx\, dy \quad dx\, dz\, dy \quad dy\, dz\, dx \quad dz\, dy\, dx.$$

EXPLORATION

Volume of a Paraboloid Sector On pages 949 and 957, you were asked to summarize the ways you know for finding the volume of the solid bounded by the paraboloid

$$z = a^2 - x^2 - y^2, \quad a > 0$$

and the xy-plane. You now know one more way. Use it to find the volume of the solid.

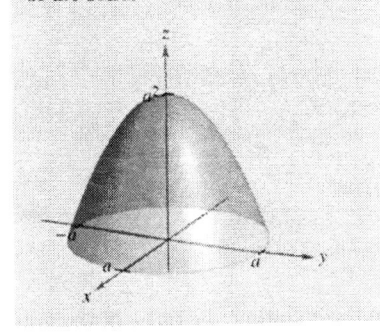

The following version of Fubini's Theorem describes a region that is considered simple with respect to the order $dz\, dy\, dx$. Similar descriptions can be given for the other five orders.

> **THEOREM 13.4 Evaluation by Iterated Integrals**
>
> Let f be continuous on a solid region Q defined by
>
> $$a \leq x \leq b, \quad h_1(x) \leq y \leq h_2(x), \quad g_1(x, y) \leq z \leq g_2(x, y)$$
>
> where $h_1, h_2, g_1,$ and g_2 are continuous functions. Then,
>
> $$\iiint_Q f(x, y, z)\, dV = \int_a^b \int_{h_1(x)}^{h_2(x)} \int_{g_1(x,y)}^{g_2(x,y)} f(x, y, z)\, dz\, dy\, dx.$$

To evaluate a triple iterated integral in the order $dz\, dy\, dx$, hold *both* x and y constant for the innermost integration. Then, hold x constant for the second integration.

Example 1 Evaluating a Triple Iterated Integral

Evaluate the triple iterated integral

$$\int_0^2 \int_0^x \int_0^{x+y} e^x(y + 2z)\, dz\, dy\, dx.$$

Solution For the first integration, hold x and y constant and integrate with respect to z.

$$\int_0^2 \int_0^x \int_0^{x+y} e^x(y + 2z)\, dz\, dy\, dx = \int_0^2 \int_0^x e^x(yz + z^2)\Big]_0^{x+y} dy\, dx$$

$$= \int_0^2 \int_0^x e^x(x^2 + 3xy + 2y^2)\, dy\, dx$$

For the second integration, hold x constant and integrate with respect to y.

$$\int_0^2 \int_0^x e^x(x^2 + 3xy + 2y^2)\, dy\, dx = \int_0^2 \left[e^x\left(x^2 y + \frac{3xy^2}{2} + \frac{2y^3}{3} \right) \right]_0^x dx$$

$$= \frac{19}{6} \int_0^2 x^3 e^x\, dx$$

$$= \frac{19}{6}\left[e^x(x^3 - 3x^2 + 6x - 6) \right]_0^2$$

$$= 19\left(\frac{e^2}{3} + 1 \right)$$

Example 1 demonstrates the integration order $dz\, dy\, dx$. For other orders, you can follow a similar procedure. For instance, to evaluate a triple iterated integral in the order $dx\, dy\, dz$, hold both y and z constant for the innermost integration and integrate with respect to x. Then, for the second integration, hold z constant and integrate with respect to y. Finally, for the third integration, integrate with respect to z.

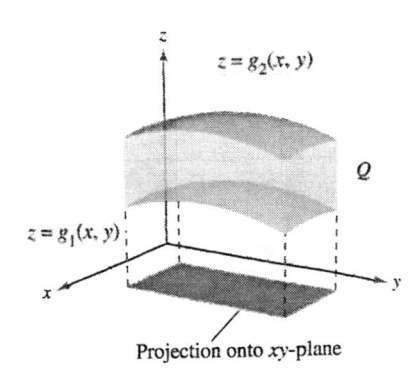

980 CHAPTER 13 Multiple Integration

To find the limits for a particular order of integration, it is generally advisable to first determine the innermost limits, which may be functions of the outer two variables. Then, by projecting the solid Q onto the coordinate plane of the outer two variables, you can determine their limits of integration by the methods used for double integrals. For instance, to evaluate

$$\iiint_Q f(x, y, z)\, dz\, dy\, dx$$

first determine the limits for z, and then the integral has the form

$$\iint \left[\int_{g_1(x, y)}^{g_2(x, y)} f(x, y, z)\, dz \right] dy\, dx.$$

By projecting the solid Q onto the xy-plane, you can determine the limits for x and y as you did for double integrals, as shown in Figure 13.52.

Solid region Q lies between two surfaces.
Figure 13.52

Example 2 Using a Triple Integral to Find Volume

Find the volume of the ellipsoidal solid given by $4x^2 + 4y^2 + z^2 = 16$.

Solution Because x, y, and z play similar roles in the equation, the order of integration is probably immaterial, and you arbitrarily choose $dz\, dy\, dx$. Moreover, you can simplify the calculation by considering only the portion of the ellipsoid lying in the first octant, as shown in Figure 13.53. From the order $dz\, dy\, dx$, you first determine the bounds for z.

$$0 \le z \le 2\sqrt{4 - x^2 - y^2}$$

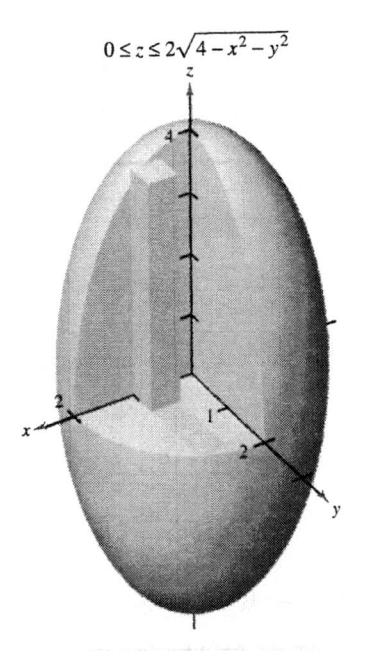

In Figure 13.54, you can see that the boundaries for x and y are $0 \le x \le 2$ and $0 \le y \le \sqrt{4 - x^2}$, so the volume of the ellipsoid is

$$V = \iiint_Q dV$$

$$= 8 \int_0^2 \int_0^{\sqrt{4-x^2}} \int_0^{2\sqrt{4-x^2-y^2}} dz\, dy\, dx$$

$$= 8 \int_0^2 \int_0^{\sqrt{4-x^2}} z \Big]_0^{2\sqrt{4-x^2-y^2}} dy\, dx$$

$$= 16 \int_0^2 \int_0^{\sqrt{4-x^2}} \sqrt{(4 - x^2) - y^2}\, dy\, dx \qquad \text{Integration tables (Appendix C), Formula 37}$$

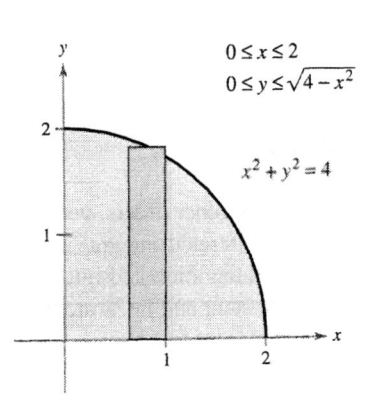

$$= 8 \int_0^2 \left[y\sqrt{4 - x^2 - y^2} + (4 - x^2) \arcsin\left(\frac{y}{\sqrt{4 - x^2}}\right) \right]_0^{\sqrt{4-x^2}} dx$$

$$= 8 \int_0^2 [0 + (4 - x^2)\arcsin(1) - 0 - 0]\, dx$$

$$= 8 \int_0^2 (4 - x^2)\left(\frac{\pi}{2}\right) dx$$

$$= 4\pi \left[4x - \frac{x^3}{3} \right]_0^2$$

$$= \frac{64\pi}{3}.$$

Ellipsoid: $4x^2 + 4y^2 + z^2 = 16$

The volume of the ellipsoid is $64\pi/3$.
Figure 13.53

Figure 13.54

Example 2 is unusual in that all six possible orders of integration produce integrals of comparable difficulty. Try setting up some other possible orders of integration to find the volume of the ellipsoid. For instance, the order $dx\,dy\,dz$ yields the integral

$$V = 8\int_0^4 \int_0^{\sqrt{16-z^2}/2} \int_0^{\sqrt{16-4y^2-z^2}/2} dx\,dy\,dz.$$

If you solve this integral, you will obtain the same volume obtained in Example 2. This is always the case—the order of integration does not affect the value of the integral. However, the order of integration often does affect the complexity of the integral. In Example 3, the given order of integration is not convenient, so you can change the order to simplify the problem.

Example 3 Changing the Order of Integration

Evaluate $\displaystyle\int_0^{\sqrt{\pi/2}} \int_x^{\sqrt{\pi/2}} \int_1^3 \sin y^2\,dz\,dy\,dx$.

Solution Note that after one integration in the given order, you would encounter the integral $2\int \sin(y^2)\,dy$, which is not an elementary function. To avoid this problem, change the order of integration to $dz\,dx\,dy$, so that y is the outer variable. The solid region Q is given by

$$0 \le x \le \sqrt{\frac{\pi}{2}}, \quad x \le y \le \sqrt{\frac{\pi}{2}}, \quad 1 \le z \le 3$$

as shown in Figure 13.55, and the projection of Q in the xy-plane yields the bounds

$$0 \le y \le \sqrt{\frac{\pi}{2}} \quad \text{and} \quad 0 \le x \le y.$$

Therefore, you have

$$V = \iiint_Q dV$$

$$= \int_0^{\sqrt{\pi/2}} \int_0^y \int_1^3 \sin(y^2)\,dz\,dx\,dy$$

$$= \int_0^{\sqrt{\pi/2}} \int_0^y z\sin(y^2)\Big]_1^3 dx\,dy$$

$$= 2\int_0^{\sqrt{\pi/2}} \int_0^y \sin(y^2)\,dx\,dy$$

$$= 2\int_0^{\sqrt{\pi/2}} x\sin(y^2)\Big]_0^y dy$$

$$= 2\int_0^{\sqrt{\pi/2}} y\sin(y^2)\,dy$$

$$= -\cos(y^2)\Big]_0^{\sqrt{\pi/2}}$$

$$= 1.$$

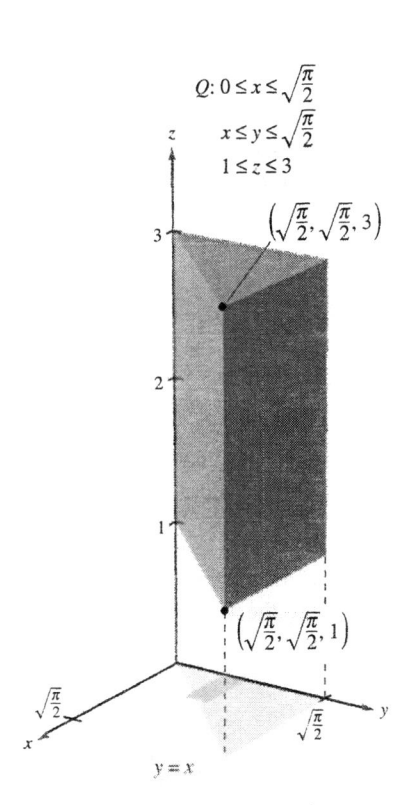

The volume of the solid region Q is 1.
Figure 13.55

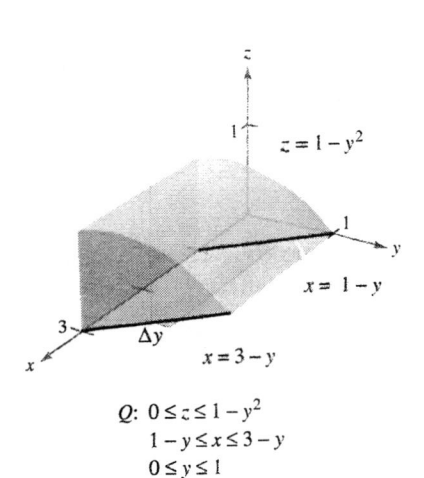

Figure 13.56

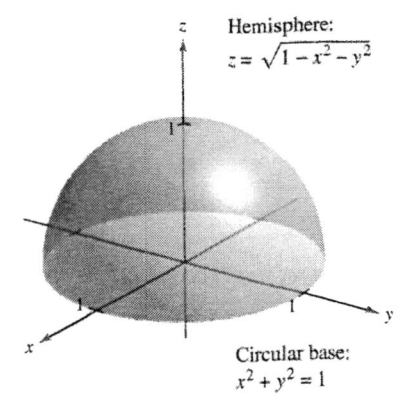

Q: $0 \le z \le 1 - y^2$
$1 - y \le x \le 3 - y$
$0 \le y \le 1$

Figure 13.57

Q: $-1 \le y \le 1$
$-\sqrt{1-y^2} \le x \le \sqrt{1-y^2}$
$0 \le z \le \sqrt{1-x^2-y^2}$

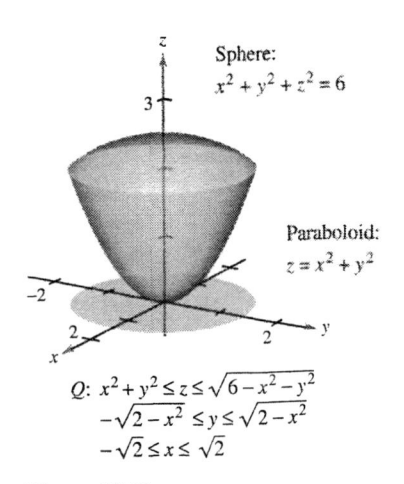

Figure 13.58

Q: $x^2 + y^2 \le z \le \sqrt{6 - x^2 - y^2}$
$-\sqrt{2-x^2} \le y \le \sqrt{2-x^2}$
$-\sqrt{2} \le x \le \sqrt{2}$

Example 4 Determining the Limits of Integration

Set up a triple integral for the volume of each solid region.

a. The region in the first octant bounded above by the cylinder $z = 1 - y^2$ and lying between the vertical planes $x + y = 1$ and $x + y = 3$

b. The upper hemisphere given by $z = \sqrt{1 - x^2 - y^2}$

c. The region bounded below by the paraboloid $z = x^2 + y^2$ and above by the sphere $x^2 + y^2 + z^2 = 6$

Solution

a. In Figure 13.56, note that the solid is bounded below by the xy-plane ($z = 0$) and above by the cylinder $z = 1 - y^2$. Therefore,

$$0 \le z \le 1 - y^2. \qquad \text{Bounds for } z$$

Projecting the region onto the xy-plane produces a parallelogram. Because two sides of the parallelogram are parallel to the x-axis, you have the following bounds:

$$1 - y \le x \le 3 - y \quad \text{and} \quad 0 \le y \le 1.$$

So, the volume of the region is given by

$$V = \iiint_Q dV = \int_0^1 \int_{1-y}^{3-y} \int_0^{1-y^2} dz\, dx\, dy.$$

b. For the upper hemisphere given by $z = \sqrt{1 - x^2 - y^2}$, you have

$$0 \le z \le \sqrt{1 - x^2 - y^2}. \qquad \text{Bounds for } z$$

In Figure 13.57, note that the projection of the hemisphere onto the xy-plane is the circle given by $x^2 + y^2 = 1$, and you can use either order $dx\, dy$ or $dy\, dx$. Choosing the first produces

$$-\sqrt{1 - y^2} \le x \le \sqrt{1 - y^2} \quad \text{and} \quad -1 \le y \le 1$$

which implies that the volume of the region is given by

$$V = \iiint_Q dV = \int_{-1}^{1} \int_{-\sqrt{1-y^2}}^{\sqrt{1-y^2}} \int_0^{\sqrt{1-x^2-y^2}} dz\, dx\, dy.$$

c. For the region bounded below by the paraboloid $z = x^2 + y^2$ and above by the sphere $x^2 + y^2 + z^2 = 6$, you have

$$x^2 + y^2 \le z \le \sqrt{6 - x^2 - y^2}. \qquad \text{Bounds for } z$$

The sphere and the paraboloid intersect when $z = 2$. Moreover, you can see in Figure 13.58 that the projection of the solid region onto the xy-plane is the circle given by $x^2 + y^2 = 2$. Using the order $dy\, dx$ produces

$$-\sqrt{2 - x^2} \le y \le \sqrt{2 - x^2} \quad \text{and} \quad -\sqrt{2} \le x \le \sqrt{2}$$

which implies that the volume of the region is given by

$$V = \iiint_Q dV = \int_{-\sqrt{2}}^{\sqrt{2}} \int_{-\sqrt{2-x^2}}^{\sqrt{2-x^2}} \int_{x^2+y^2}^{\sqrt{6-x^2-y^2}} dz\, dy\, dx.$$

Center of Mass and Moments of Inertia

In the remainder of this section, we discuss two applications of triple integrals that are important in engineering. Consider a solid region Q whose density at (x, y, z) is given by the **density function** ρ. The **center of mass** of a solid region Q of mass m is given by $(\bar{x}, \bar{y}, \bar{z})$, where

$$m = \iiint_Q \rho(x, y, z)\, dV \qquad \text{Mass of the solid}$$

$$M_{yz} = \iiint_Q x\rho(x, y, z)\, dV \qquad \text{First moment about } yz\text{-plane}$$

$$M_{xz} = \iiint_Q y\rho(x, y, z)\, dV \qquad \text{First moment about } xz\text{-plane}$$

$$M_{xy} = \iiint_Q z\rho(x, y, z)\, dV \qquad \text{First moment about } xy\text{-plane}$$

and

$$\bar{x} = \frac{M_{yz}}{m}, \quad \bar{y} = \frac{M_{xz}}{m}, \quad \bar{z} = \frac{M_{xy}}{m}.$$

The quantities M_{yz}, M_{xz}, and M_{xy} are called the **first moments** of the region Q about the yz-, xz-, and xy-planes.

The first moments for solid regions are taken about a plane, whereas the second moments for solids are taken about a line. The **second moments** (or **moments of inertia**) about the x-, y-, and z-axes are as follows.

$$I_x = \iiint_Q (y^2 + z^2)\rho(x, y, z)\, dV \qquad \text{Moment of inertia about } x\text{-axis}$$

$$I_y = \iiint_Q (x^2 + z^2)\rho(x, y, z)\, dV \qquad \text{Moment of inertia about } y\text{-axis}$$

$$I_z = \iiint_Q (x^2 + y^2)\rho(x, y, z)\, dV \qquad \text{Moment of inertia about } z\text{-axis}$$

For problems requiring the calculation of all three moments, considerable effort can be saved by applying the additive property of triple integrals and writing

$$I_x = I_{xz} + I_{xy}, \quad I_y = I_{yz} + I_{xy}, \quad \text{and} \quad I_z = I_{yz} + I_{xz}$$

where I_{xy}, I_{xz}, and I_{yz} are as follows.

$$I_{xy} = \iiint_Q z^2\rho(x, y, z)\, dV$$

$$I_{xz} = \iiint_Q y^2\rho(x, y, z)\, dV$$

$$I_{yz} = \iiint_Q x^2\rho(x, y, z)\, dV$$

EXPLORATION

Sketch the solid (of uniform density) bounded by $z = 0$ and $z = 1/(1 + x^2 + y^2)$, where $x^2 + y^2 \leq 1$. From your sketch, estimate the coordinates of the center of mass of the solid. Now use a computer algebra system to verify your estimate. What do you observe?

NOTE In engineering and physics, the moment of inertia of a mass is used to find the time required for a mass to reach a given speed of rotation about an axis, as indicated in Figure 13.59. The greater the moment of inertia, the longer a force must be applied for the mass to reach the given speed.

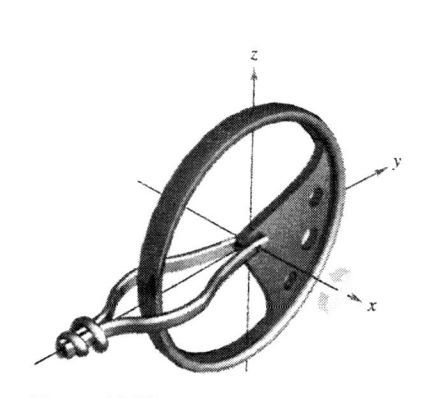

Figure 13.59

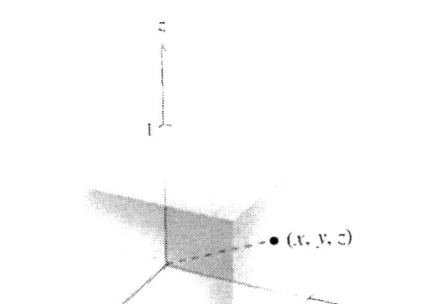

Variable density:
$\rho(x, y, z) = k(x^2 + y^2 + z^2)$
Figure 13.60

Find the center of mass of the unit cube shown in Figure 13.60, given that the density at the point (x, y, z) is proportional to the square of its distance from the origin.

Solution Because the density at (x, y, z) is proportional to the square of the distance between $(0, 0, 0)$ and (x, y, z), you have

$$\rho(x, y, z) = k(x^2 + y^2 + z^2).$$

You can use this density function to find the mass of the cube. Because of the symmetry of the region, any order of integration will produce an integral of comparable difficulty.

$$m = \int_0^1 \int_0^1 \int_0^1 k(x^2 + y^2 + z^2) \, dz \, dy \, dx$$

$$= k \int_0^1 \int_0^1 \left[(x^2 + y^2)z + \frac{z^3}{3} \right]_0^1 dy \, dx$$

$$= k \int_0^1 \int_0^1 \left(x^2 + y^2 + \frac{1}{3} \right) dy \, dx$$

$$= k \int_0^1 \left[\left(x^2 + \frac{1}{3} \right) y + \frac{y^3}{3} \right]_0^1 dx$$

$$= k \int_0^1 \left(x^2 + \frac{2}{3} \right) dx$$

$$= k \left[\frac{x^3}{3} + \frac{2x}{3} \right]_0^1 = k$$

The first moment about the yz-plane is

$$M_{yz} = k \int_0^1 \int_0^1 \int_0^1 x(x^2 + y^2 + z^2) \, dz \, dy \, dx$$

$$= k \int_0^1 x \left[\int_0^1 \int_0^1 (x^2 + y^2 + z^2) \, dz \, dy \right] dx.$$

Note that x can be factored out of the two inner integrals, because it is constant with respect to y and z. After factoring, the two inner integrals are the same as for the mass m. Hence, you have

$$M_{yz} = k \int_0^1 x \left(x^2 + \frac{2}{3} \right) dx$$

$$= k \left[\frac{x^4}{4} + \frac{x^2}{3} \right]_0^1$$

$$= \frac{7k}{12}.$$

So,

$$\bar{x} = \frac{M_{yz}}{m} = \frac{7k/12}{k} = \frac{7}{12}.$$

Finally, from the nature of ρ and the symmetry of x, y, and z in this solid region, you have $\bar{x} = \bar{y} = \bar{z}$, and the center of mass is $\left(\frac{7}{12}, \frac{7}{12}, \frac{7}{12} \right)$.

Example 6 Moments of Inertia for a Solid Region

Find the moments of inertia about the x- and y-axes for the solid region lying between the hemisphere

$$z = \sqrt{4 - x^2 - y^2}$$

and the xy-plane, given that the density at (x, y, z) is proportional to the distance between (x, y, z) and the xy-plane.

Solution The density of the region is given by $\rho(x, y, z) = kz$. Considering the symmetry of this problem, you know that $I_x = I_y$, and you need to compute only one moment, say I_x. From Figure 13.61, choose the order $dz\,dy\,dx$ and write

$$I_x = \iiint_Q (y^2 + z^2)\rho(x, y, z)\,dV$$

$$= \int_{-2}^{2}\int_{-\sqrt{4-x^2}}^{\sqrt{4-x^2}}\int_{0}^{\sqrt{4-x^2-y^2}} (y^2 + z^2)(kz)\,dz\,dy\,dx$$

$$= k\int_{-2}^{2}\int_{-\sqrt{4-x^2}}^{\sqrt{4-x^2}}\left[\frac{y^2 z^2}{2} + \frac{z^4}{4}\right]_0^{\sqrt{4-x^2-y^2}}\,dy\,dx$$

$$= k\int_{-2}^{2}\int_{-\sqrt{4-x^2}}^{\sqrt{4-x^2}}\left[\frac{y^2(4 - x^2 - y^2)}{2} + \frac{(4 - x^2 - y^2)^2}{4}\right]\,dy\,dx$$

$$= \frac{k}{4}\int_{-2}^{2}\int_{-\sqrt{4-x^2}}^{\sqrt{4-x^2}} [(4 - x^2)^2 - y^4]\,dy\,dx$$

$$= \frac{k}{4}\int_{-2}^{2}\left[(4 - x^2)^2 y - \frac{y^5}{5}\right]_{-\sqrt{4-x^2}}^{\sqrt{4-x^2}}\,dx$$

$$= \frac{k}{4}\int_{-2}^{2}\frac{8}{5}(4 - x^2)^{5/2}\,dx$$

$$= \frac{4k}{5}\int_{0}^{2}(4 - x^2)^{5/2}\,dx \qquad\qquad x = 2\sin\theta$$

$$= \frac{4k}{5}\int_{0}^{\pi/2} 64\cos^6\theta\,d\theta$$

$$= \left(\frac{256k}{5}\right)\left(\frac{5\pi}{32}\right) \qquad\qquad \text{Wallis's Formula}$$

$$= 8k\pi.$$

So, $I_x = 8k\pi = I_y$.

In Example 6, notice that the moments of inertia about the x- and y-axes are equal to each other. The moment about the z-axis, however, is different. Does it seem that the moment of inertia about the z-axis should be less than or greater than the moments calculated in Example 6? By performing the calculations, you can determine that

$$I_z = \tfrac{16}{3}k\pi.$$

This tells you that the solid shown in Figure 13.61 has a greater resistance to rotation about the x- or y-axis than about the z-axis.

$-2 \le x \le 2$
$-\sqrt{4 - x^2} \le y \le \sqrt{4 - x^2}$
$0 \le z \le \sqrt{4 - x^2 - y^2}$

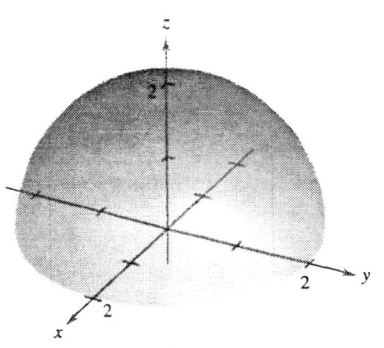

Hemisphere:
$z = \sqrt{4 - x^2 - y^2}$

Circular base:
$x^2 + y^2 = 4$

Variable density: $\rho(x, y, z) = kz$
Figure 13.61

EXERCISES FOR SECTION 13.6

In Exercises 1–8, evaluate the iterated integral.

1. $\int_0^3 \int_0^2 \int_0^1 (x+y+z)\,dx\,dy\,dz$

2. $\int_{-1}^1 \int_{-1}^1 \int_{-1}^1 x^2 y^2 z^2\,dx\,dy\,dz$

3. $\int_0^1 \int_0^x \int_0^{xy} x\,dz\,dy\,dx$

4. $\int_0^9 \int_0^{y/3} \int_0^{\sqrt{y^2-9x^2}} z\,dz\,dx\,dy$

5. $\int_1^4 \int_0^1 \int_0^x 2ze^{-x^2}\,dy\,dx\,dz$

6. $\int_1^4 \int_1^{e^2} \int_0^{1/xz} \ln z\,dy\,dz\,dx$

7. $\int_0^4 \int_0^{\pi/2} \int_0^{1-x} x\cos y\,dz\,dy\,dx$

8. $\int_0^{\pi/2} \int_0^{y/2} \int_0^{1/y} \sin y\,dz\,dx\,dy$

In Exercises 9 and 10, use a computer algebra system to evaluate the iterated integral.

9. $\int_0^2 \int_{-\sqrt{4-x^2}}^{\sqrt{4-x^2}} \int_0^{x^2} x\,dz\,dy\,dx$

10. $\int_0^{\sqrt{2}} \int_0^{\sqrt{2-x^2}} \int_{2x^2+y^2}^{4-y^2} y\,dz\,dy\,dx$

In Exercises 11 and 12, use a computer algebra system to approximate the iterated integral.

11. $\int_0^2 \int_0^{\sqrt{4-x^2}} \int_1^4 \frac{x^2 \sin y}{z}\,dz\,dy\,dx$

12. $\int_0^3 \int_0^{2-(2y/3)} \int_0^{6-2y-3z} ze^{-x^2 y^2}\,dx\,dz\,dy$

In Exercises 13–16, set up a triple integral for the volume of each solid.

13. The solid in the first octant bounded by the coordinate planes and the plane $z = 4 - x - y$
14. The solid bounded by $z = 9 - x^2$, $z = 0$, $x = 0$, and $y = 2x$
15. The solid bounded by the paraboloid $z = 9 - x^2 - y^2$ and the plane $z = 0$
16. The solid that is the common interior below the sphere $x^2 + y^2 + z^2 = 80$ and above the paraboloid $z = \frac{1}{2}(x^2 + y^2)$

Volume In Exercises 17–22, use a triple integral to find the volume of the solid bounded by the graphs of the equations.

17. $x = 4 - y^2, z = 0, z = x$
18. $z = xy, z = 0, x = 0, x = 1, y = 0, y = 1$
19. $x^2 + y^2 + z^2 = a^2$
20. $z = 36 - x^2 - y^2, z = 0$
21. $z = 4 - x^2, y = 4 - x^2$, first octant
22. $z = 9 - x^2, y = -x + 2, y = 0, z = 0, x \geq 0$

In Exercises 23–26, sketch the solid whose volume is given by the iterated integral and rewrite the integral using the indicated order of integration.

23. $\int_0^4 \int_0^{(4-x)/2} \int_0^{(12-3x-6y)/4} dz\,dy\,dx$

Rewrite using the order $dy\,dx\,dz$.

24. $\int_0^3 \int_0^{\sqrt{9-x^2}} \int_0^{6-x-y} dz\,dy\,dx$

Rewrite using the order $dz\,dx\,dy$.

25. $\int_0^1 \int_y^1 \int_0^{\sqrt{1-y^2}} dz\,dx\,dy$

Rewrite using the order $dz\,dy\,dx$.

26. $\int_0^2 \int_{2x}^4 \int_0^{\sqrt{y^2-4x^2}} dz\,dy\,dx$

Rewrite using the order $dx\,dy\,dz$.

In Exercises 27–30, list the six possible orders of integration for the triple integral over the solid Q

$$\iiint_Q xyz\,dV.$$

27. $Q = \{(x, y, z): 0 \leq x \leq 1, 0 \leq y \leq x, 0 \leq z \leq 3\}$
28. $Q = \{(x, y, z): 0 \leq x \leq 2, x^2 \leq y \leq 4, 0 \leq z \leq 2 - x\}$
29. $Q = \{(x, y, z): x^2 + y^2 \leq 9, 0 \leq z \leq 4\}$
30. $Q = \{(x, y, z): 0 \leq x \leq 1, y \leq 1 - x^2, 0 \leq z \leq 6\}$

Mass and Center of Mass In Exercises 31–34, find the mass and the indicated coordinates of the center of mass of the solid of indicated density bounded by the graphs of the equations.

31. Find $\bar{x}$ using $\rho(x, y, z) = k$.
    $Q: 2x + 3y + 6z = 12, x = 0, y = 0, z = 0$
32. Find $\bar{y}$ using $\rho(x, y, z) = ky$.
    $Q: 3x + 3y + 5z = 15, x = 0, y = 0, z = 0$
33. Find $\bar{z}$ using $\rho(x, y, z) = kx$.
    $Q: z = 4 - x, z = 0, y = 0, y = 4, x = 0$
34. Find $\bar{y}$ using $\rho(x, y, z) = k$.
    $Q: \frac{x}{a} + \frac{y}{b} + \frac{z}{c} = 1\ (a, b, c > 0), x = 0, y = 0, z = 0$

Mass and Center of Mass In Exercises 35 and 36, set up the triple integrals for finding the mass and the center of mass of the solid bounded by the graphs of the equations.

35. $x = 0, x = b, y = 0, y = b, z = 0, z = b$
    $\rho(x, y, z) = kxy$
36. $x = 0, x = a, y = 0, y = b, z = 0, z = c$
    $\rho(x, y, z) = kz$

Think About It The center of mass of a solid of constant density is shown in the figure. In Exercises 37–40, make a conjecture about how the center of mass $(\bar{x}, \bar{y}, \bar{z})$ will change for the nonconstant density $\rho(x, y, z)$. Explain.

37. $\rho(x, y, z) = kx$
38. $\rho(x, y, z) = kz$
39. $\rho(x, y, z) = k(y + 2)$
40. $\rho(x, y, z) = kxz^2(y + 2)^2$

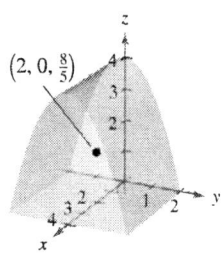

$(2, 0, \frac{8}{5})$

Centroid In Exercises 41–46, find the centroid of the solid region bounded by the graphs of the equations or described by the figure. Use a computer algebra system to evaluate the triple integrals. (Assume uniform density and find the center of mass.)

41. $z = \frac{h}{r}\sqrt{x^2 + y^2}, z = h$
42. $y = \sqrt{4 - x^2}, z = y, z = 0$
43. $z = \sqrt{4^2 - x^2 - y^2}, z = 0$
44. $z = \frac{1}{y^2 + 1}, z = 0, x = -2, x = 2, y = 0, y = 1$

45.

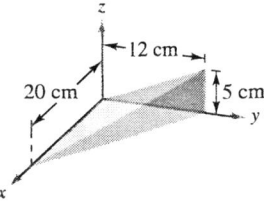

46. $(0, 0, 4)$, $(5, 0, 0)$, $(0, 3, 0)$

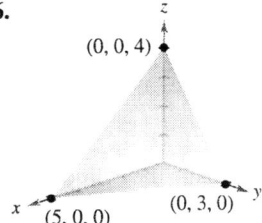

Moments of Inertia In Exercises 47–50, find I_x, I_y, and I_z for the solid of indicated density. Use a computer algebra system to evaluate the triple integrals.

47. (a) $\rho = k$
    (b) $\rho = kxyz$

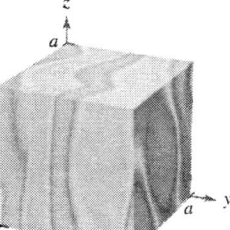

48. (a) $\rho(x, y, z) = k$
    (b) $\rho(x, y, z) = k(x^2 + y^2)$

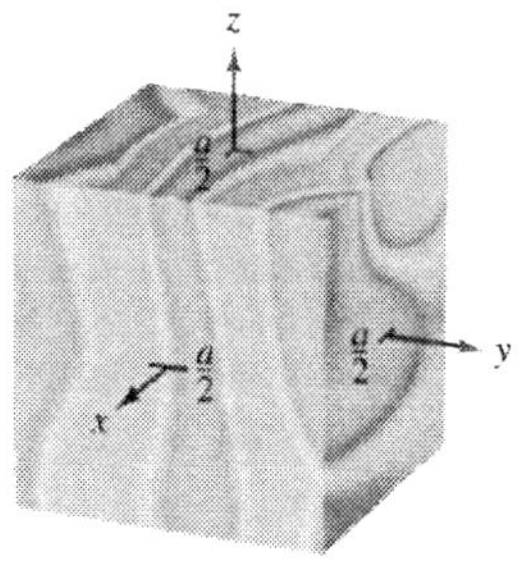

49. (a) $\rho(x, y, z) = k$
    (b) $\rho = ky$
    $z = 4 - x$

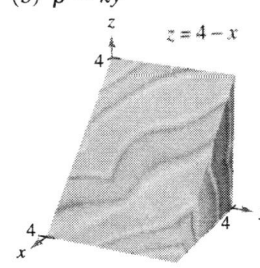

50. (a) $\rho = kz$
    (b) $\rho = k(4 - z)$
    $z = 4 - y^2$

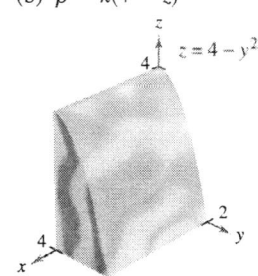

Moments of Inertia In Exercises 51 and 52, verify the moments of inertia for the solid of uniform density. Use a computer algebra system to evaluate the triple integrals.

51. $I_x = \frac{1}{12}m(3a^2 + L^2)$
    $I_y = \frac{1}{2}ma^2$
    $I_z = \frac{1}{12}m(3a^2 + L^2)$

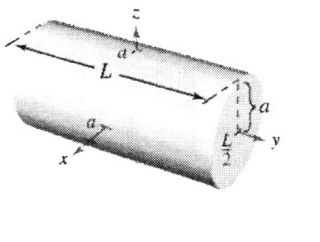

52. $I_x = \frac{1}{12}m(a^2 + b^2)$
    $I_y = \frac{1}{12}m(b^2 + c^2)$
    $I_z = \frac{1}{12}m(a^2 + c^2)$

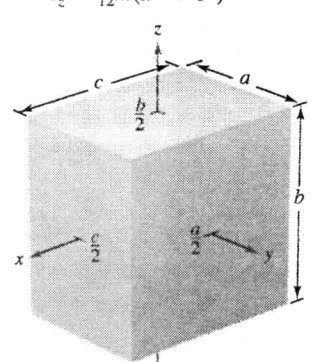

Moments of Inertia In Exercises 53 and 54, set up a triple integral that gives the moment of inertia about the z-axis of the solid Q of density ρ.

53. $Q = \{(x, y, z): -1 \le x \le 1, -1 \le y \le 1, 0 \le z \le 1 - x\}$
    $\rho = \sqrt{x^2 + y^2 + z^2}$
54. $Q = \{(x, y, z): x^2 + y^2 \le 1, 0 \le z \le 4 - x^2 - y^2\}$
    $\rho = kx^2$

Getting at the Concept

55. Define a triple integral and describe a method of evaluating a triple integral.

56. Give the number of possible orders of integration when evaluating a triple integral.

57. Consider solid A and solid B of equal weight shown in the figure.
    (a) Since the solids have the same weight, which has the greater density?
    (b) Which solid has the greater moment of inertia? Explain.
    (c) The solids are rolled down an inclined plane. If they are started at the same time and at the same height, which will reach the bottom first? Explain.

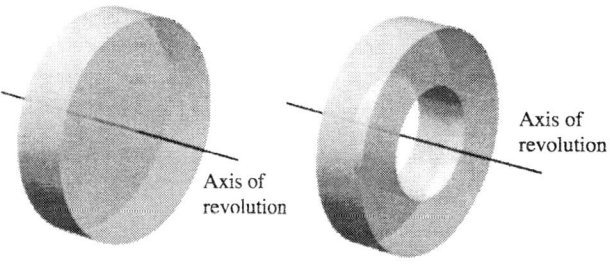

58. Determine whether the moment of inertia about the y-axis of the cylinder in Exercise 51 will increase or decrease for the nonconstant density $\rho(x, y, z) = \sqrt{x^2 + z^2}$ and $a = 4$.

Section 13.7 Triple Integrals in Cylindrical and Spherical Coordinates

- Write and evaluate a triple integral in cylindrical coordinates.
- Write and evaluate a triple integral in spherical coordinates.

Triple Integrals in Cylindrical Coordinates

Many common solid regions such as spheres, ellipsoids, cones, and paraboloids can yield difficult triple integrals in rectangular coordinates. In fact, it is precisely this difficulty that led to the introduction of nonrectangular coordinate systems. In this section, you will learn how to use *cylindrical* and *spherical* coordinates to evaluate triple integrals.

Recall from Section 10.7 that the rectangular conversion equations for cylindrical coordinates are

$$x = r \cos \theta$$
$$y = r \sin \theta$$
$$z = z.$$

NOTE An easy way to remember these conversions is to note that the equations for x and y are the same as in polar coordinates and z is unchanged.

In this coordinate system, the simplest solid region is a cylindrical block determined by

$$r_1 \leq r \leq r_2, \quad \theta_1 \leq \theta \leq \theta_2, \quad z_1 \leq z \leq z_2$$

as shown in Figure 13.62. To obtain the cylindrical coordinate form of a triple integral, suppose that Q is a solid region whose projection R onto the xy-plane can be described in polar coordinates. That is,

$$Q = \{(x, y, z): (x, y) \text{ is in } R, \quad h_1(x, y) \leq z \leq h_2(x, y)\}$$

and

$$R = \{(r, \theta): \theta_1 \leq \theta \leq \theta_2, \quad g_1(\theta) \leq r \leq g_2(\theta)\}.$$

If f is a continuous function on the solid Q, you can write the triple integral of f over Q as

$$\iiint_Q f(x, y, z) \, dV = \iint_R \left[\int_{h_1(x,y)}^{h_2(x,y)} f(x, y, z) \, dz \right] dA$$

where the double integral over R is evaluated in polar coordinates. That is, R is a plane region that is either r-simple or θ-simple. If R is r-simple, the iterated form of the triple integral in cylindrical form is

$$\iiint_Q f(x, y, z) \, dV = \int_{\theta_1}^{\theta_2} \int_{g_1(\theta)}^{g_2(\theta)} \int_{h_1(r\cos\theta, r\sin\theta)}^{h_2(r\cos\theta, r\sin\theta)} f(r \cos \theta, r \sin \theta, z) r \, dz \, dr \, d\theta.$$

NOTE This is only one of six possible orders of integration. The other five are $dz \, d\theta \, dr$, $dr \, dz \, d\theta$, $dr \, d\theta \, dz$, $d\theta \, dz \, dr$, and $d\theta \, dr \, dz$.

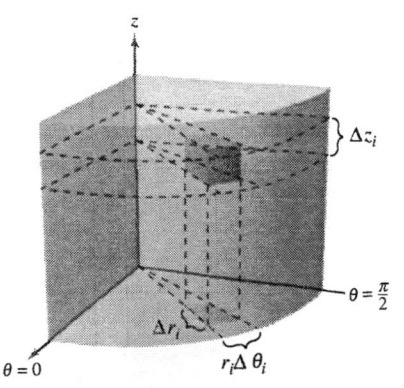

PIERRE SIMON DE LAPLACE (1749–1827)

One of the first to use a cylindrical coordinate system was the French mathematician Pierre Simon de Laplace. Laplace has been called the "Newton of France," and he published many important works in mechanics, differential equations, and probability.

Volume of cylindrical block:
$\Delta V_i = r_i \Delta r_i \Delta \theta_i \Delta z_i$
Figure 13.62

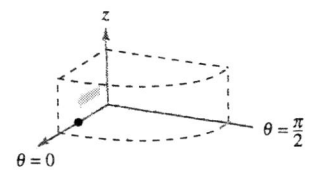

Integrate with respect to r.

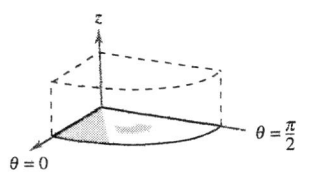

Integrate with respect to θ.

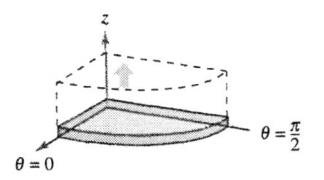

Integrate with respect to z.
Figure 13.63

To visualize a particular order of integration, it helps to view the iterated integral in terms of three sweeping motions—each adding another dimension to the solid. For instance, in the order $dr\, d\theta\, dz$, the first integration occurs in the r-direction as a point sweeps out a ray. Then, as θ increases, the line sweeps out a sector. Finally, as z increases, the sector sweeps out a solid wedge, as shown in Figure 13.63.

EXPLORATION

Volume of a Paraboloid Sector On pages 949, 957, and 979, you were asked to summarize the ways you know for finding the volume of the solid bounded by the paraboloid

$$z = a^2 - x^2 - y^2,\ a > 0$$

and the xy-plane. You now know one more way. Use it to find the volume of the solid. Compare the different methods. What are the advantages and disadvantages of each?

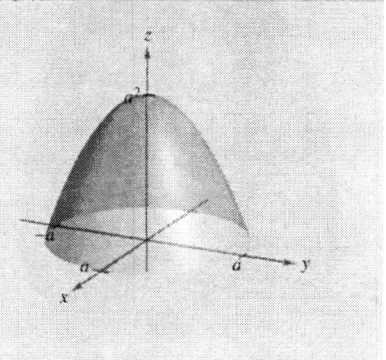

Example 1 Finding Volume by Cylindrical Coordinates

Find the volume of the solid region Q cut from the sphere

$$x^2 + y^2 + z^2 = 4 \quad \text{Sphere}$$

by the cylinder $r = 2\sin\theta$, as shown in Figure 13.64.

Solution Because $x^2 + y^2 + z^2 = r^2 + z^2 = 4$, the bounds on z are

$$-\sqrt{4 - r^2} \le z \le \sqrt{4 - r^2}.$$

Let R be the circular projection of the solid onto the $r\theta$-plane. Then the bounds on R are $0 \le r \le 2\sin\theta$ and $0 \le \theta \le \pi$. So, the volume of Q is

$$\begin{aligned}
V &= \int_0^\pi \int_0^{2\sin\theta} \int_{-\sqrt{4-r^2}}^{\sqrt{4-r^2}} r\, dz\, dr\, d\theta \\
&= 2\int_0^{\pi/2} \int_0^{2\sin\theta} 2r\sqrt{4-r^2}\, dr\, d\theta \\
&= 2\int_0^{\pi/2} \left[-\frac{2}{3}(4-r^2)^{3/2}\right]_0^{2\sin\theta} d\theta \\
&= \frac{4}{3}\int_0^{\pi/2} (8 - 8\cos^3\theta)\, d\theta \\
&= \frac{32}{3}\int_0^{\pi/2} [1 - (\cos\theta)(1 - \sin^2\theta)]\, d\theta \\
&= \frac{32}{3}\left[\theta - \sin\theta + \frac{\sin^3\theta}{3}\right]_0^{\pi/2} \\
&= \frac{16}{9}(3\pi - 4).
\end{aligned}$$

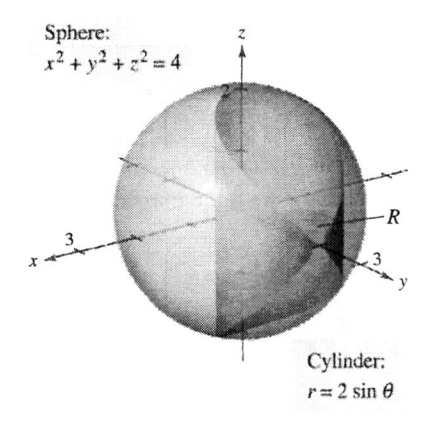

The volume of the solid region Q is $\frac{16}{9}(3\pi - 4)$.
Figure 13.64

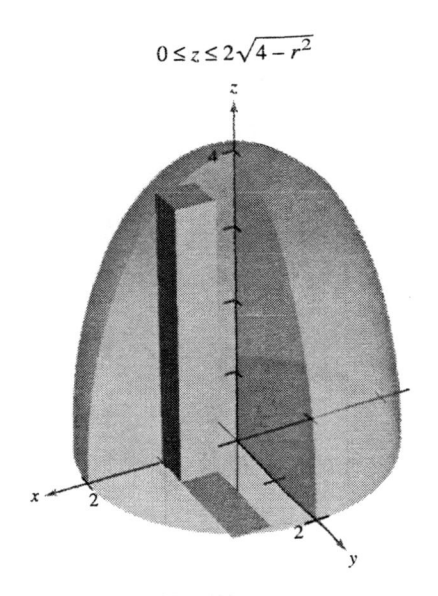

Example 2 Finding Mass by Cylindrical Coordinates

Find the mass of the ellipsoidal solid Q given by $4x^2 + 4y^2 + z^2 = 16$, lying above the xy-plane. The density at a point in the solid is proportional to the distance between the point and the xy-plane.

Solution The density function $\rho(r, \theta, z) = kz$. The bounds on z are

$$0 \leq z \leq \sqrt{16 - 4x^2 - 4y^2} = \sqrt{16 - 4r^2}$$

where $0 \leq r \leq 2$ and $0 \leq \theta \leq 2\pi$, as shown in Figure 13.65. The mass of the solid is

$$m = \int_0^{2\pi} \int_0^2 \int_0^{\sqrt{16-4r^2}} kzr \, dz \, dr \, d\theta$$

$$= \frac{k}{2} \int_0^{2\pi} \int_0^2 z^2 r \Big]_0^{\sqrt{16-4r^2}} dr \, d\theta$$

$$= \frac{k}{2} \int_0^{2\pi} \int_0^2 (16r - 4r^3) \, dr \, d\theta$$

$$= \frac{k}{2} \int_0^{2\pi} \left[8r^2 - r^4 \right]_0^2 d\theta$$

$$= 8k \int_0^{2\pi} d\theta = 16\pi k.$$

$0 \leq z \leq 2\sqrt{4-r^2}$

Ellipsoid: $4x^2 + 4y^2 + z^2 = 16$

The mass of the ellipsoidal solid is $16\pi k$.
Figure 13.65

Integration in cylindrical coordinates is useful when factors involving $x^2 + y^2$ appear in the integrand, as illustrated in Example 3.

Example 3 Finding a Moment of Inertia

Find the moment of inertia about the axis of symmetry of the solid bounded by the paraboloid $z = x^2 + y^2$ and the plane $z = 4$, as shown in Figure 13.66. The density at each point is proportional to the distance between the point and the z-axis.

Solution Because the z-axis is the axis of symmetry, and $\rho(x, y, z) = k\sqrt{x^2 + y^2}$, it follows that

$$I_z = \iiint_Q k(x^2 + y^2)\sqrt{x^2 + y^2} \, dV.$$

In cylindrical coordinates, $0 \leq r \leq \sqrt{x^2 + y^2} = \sqrt{z}$. Therefore, you have

$$I_z = k \int_0^4 \int_0^{2\pi} \int_0^{\sqrt{z}} r^2(r) r \, dr \, d\theta \, dz$$

$$= k \int_0^4 \int_0^{2\pi} \frac{r^5}{5} \Big]_0^{\sqrt{z}} d\theta \, dz$$

$$= k \int_0^4 \int_0^{2\pi} \frac{z^{5/2}}{5} d\theta \, dz$$

$$= k \int_0^4 \frac{z^{5/2}}{5} (2\pi) \, dz$$

$$= k \left[\left(\frac{2\pi}{5}\right)\left(\frac{2}{7}\right) z^{7/2} \right]_0^4 = \frac{512k\pi}{35}.$$

Q: Bounded by
$z = x^2 + y^2$
$z = 4$

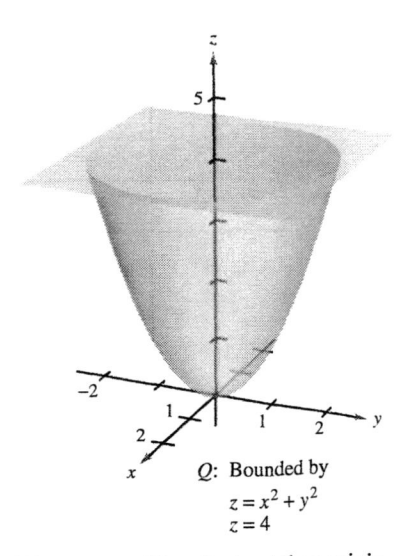

The moment of inertia about the z-axis is $512k\pi/35$.
Figure 13.66

Triple Integrals in Spherical Coordinates

Triple integrals involving spheres or cones are often easier to evaluate by converting to spherical coordinates. Recall from Section 10.7 that the rectangular conversion equations for spherical coordinates are

$$x = \rho \sin \phi \cos \theta$$
$$y = \rho \sin \phi \sin \theta$$
$$z = \rho \cos \phi.$$

In this coordinate system, the simplest region is a spherical block determined by

$$\{(\rho, \theta, \phi): \rho_1 \le \rho \le \rho_2, \ \theta_1 \le \theta \le \theta_2, \ \phi_1 \le \phi \le \phi_2\}$$

where $\rho_1 0$, $\theta_2 - \theta_1 \le 2\pi$, and $0 \le \phi_1 \le \phi_2 \le \pi$, as shown in Figure 13.67. If (ρ, θ, ϕ) is a point in the interior of such a block, then the volume of the block can be approximated by $\Delta V \approx \rho^2 \sin \phi \, \Delta \rho \, \Delta \phi \, \Delta \theta$ (see Exercise 46).

Using the usual process involving an inner partition, summation, and a limit, you can develop the following version of a triple integral in spherical coordinates for a continuous function f defined on the solid Q.

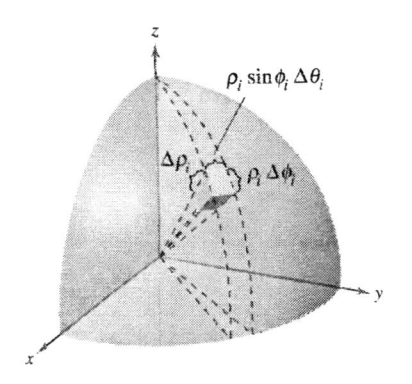

Spherical block:
$\Delta V_i \approx \rho_i^2 \sin \phi_i \, \Delta \rho_i \, \Delta \theta_i \, \Delta \phi_i$
Figure 13.67

$$\iiint_Q f(x, y, z) \, dV = \int_{\theta_1}^{\theta_2} \int_{\phi_1}^{\phi_2} \int_{\rho_1}^{\rho_2} f(\rho \sin \phi \cos \theta, \rho \sin \phi \sin \theta, \rho \cos \phi) \rho^2 \sin \phi \, d\rho \, d\phi \, d\theta.$$

This formula can be modified for different orders of integration and generalized to include regions with variable boundaries.

Like triple integrals in cylindrical coordinates, triple integrals in spherical coordinates are evaluated with iterated integrals. As with cylindrical coordinates, you can visualize a particular order of integration by viewing the iterated integral in terms of three sweeping motions—each adding another dimension to the solid. For instance, the iterated integral

$$\int_0^{2\pi} \int_0^{\pi/4} \int_0^3 \rho^2 \sin \phi \, d\rho \, d\phi \, d\theta$$

(which is used in Example 4) is illustrated in Figure 13.68.

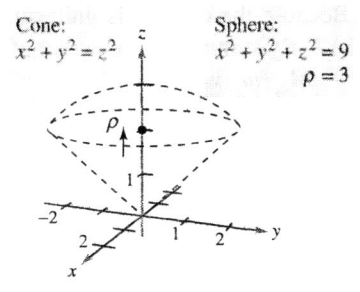

ρ varies from 0 to 3 with ϕ and θ held constant.
ϕ varies from 0 to $\pi/4$ with θ held constant.
θ varies from 0 to 2π.
Figure 13.68

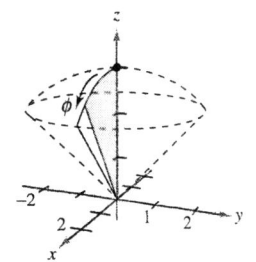

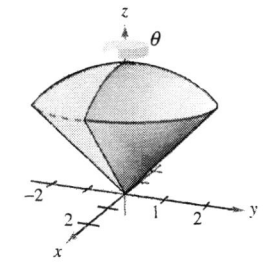

NOTE The Greek letter ρ used in spherical coordinates is not related to density. Rather, it is the three-dimensional analog of the r used in polar coordinates. For problems involving spherical coordinates and a density function, we will use a different symbol to denote density.

Upper nappe of cone:
$z^2 = x^2 + y^2$

Sphere:
$x^2 + y^2 + z^2 = 9$

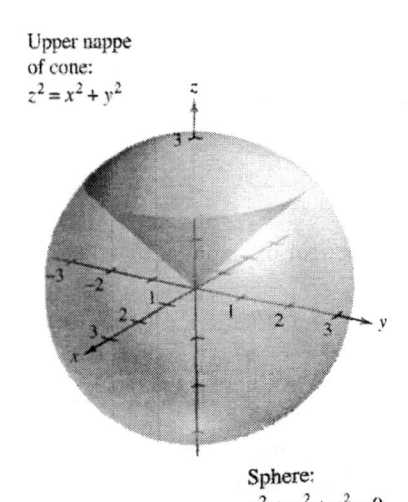

Figure 13.69

Example 4 Finding Volume in Spherical Coordinates

Find the volume of the solid region Q bounded below by the upper nappe of the cone $z^2 = x^2 + y^2$ and above by the sphere $x^2 + y^2 + z^2 = 9$, as shown in Figure 13.69.

Solution In spherical coordinates, the equation of the sphere is

$$\rho^2 = x^2 + y^2 + z^2 = 9 \implies \rho = 3.$$

Furthermore, the sphere and cone intersect when

$$(x^2 + y^2) + z^2 = (z^2) + z^2 = 9 \implies z = \frac{3}{\sqrt{2}}$$

and, because $z = \rho \cos \phi$, it follows that

$$\left(\frac{3}{\sqrt{2}}\right)\left(\frac{1}{3}\right) = \cos \phi \implies \phi = \frac{\pi}{4}.$$

Consequently, you can use the integration order $d\rho\, d\phi\, d\theta$, where $0 \leq \rho \leq 3$, $0 \leq \phi \leq \pi/4$, and $0 \leq \theta \leq 2\pi$. The volume is

$$V = \iiint_Q dV = \int_0^{2\pi} \int_0^{\pi/4} \int_0^3 \rho^2 \sin \phi \, d\rho\, d\phi\, d\theta$$

$$= \int_0^{2\pi} \int_0^{\pi/4} 9 \sin \phi \, d\phi\, d\theta$$

$$= 9 \int_0^{2\pi} \left[-\cos \phi \right]_0^{\pi/4} d\theta$$

$$= 9 \int_0^{2\pi} \left(1 - \frac{\sqrt{2}}{2}\right) d\theta = 9\pi(2 - \sqrt{2}) \approx 16.56.$$

Example 5 Finding the Center of Mass of a Solid Region

Find the center of mass of the solid region Q of uniform density, bounded below by the upper nappe of the cone $z^2 = x^2 + y^2$ and above by the sphere $x^2 + y^2 + z^2 = 9$.

Solution Because the density is uniform, you can consider the density at the point (x, y, z) to be k. By symmetry, the center of mass lies on the z-axis, and you need only calculate $\bar{z} = M_{xy}/m$, where $m = kV = 9k\pi(2 - \sqrt{2})$ from Example 4. Because $z = \rho \cos \phi$, it follows that

$$M_{xy} = \iiint_Q kz \, dV = k \int_0^3 \int_0^{2\pi} \int_0^{\pi/4} (\rho \cos \phi) \rho^2 \sin \phi \, d\phi\, d\theta\, d\rho$$

$$= k \int_0^3 \int_0^{2\pi} \rho^3 \left[\frac{\sin^2 \phi}{2}\right]_0^{\pi/4} d\theta\, d\rho$$

$$= \frac{k}{4} \int_0^3 \int_0^{2\pi} \rho^3 \, d\theta\, d\rho = \frac{k\pi}{2} \int_0^3 \rho^3 \, d\rho = \frac{81 k\pi}{8}.$$

So,

$$\bar{z} = \frac{M_{xy}}{m} = \frac{81 k\pi/8}{9k\pi(2 - \sqrt{2})} = \frac{9(2 + \sqrt{2})}{16} \approx 1.92$$

and the center of mass is approximately $(0, 0, 1.92)$.

EXERCISES FOR SECTION 13.7

In Exercises 1–6, evaluate the iterated integral.

1. $\int_0^4 \int_0^{\pi/2} \int_0^2 r \cos\theta \, dr \, d\theta \, dz$

2. $\int_0^{\pi/4} \int_0^2 \int_0^{2-r} rz \, dz \, dr \, d\theta$

3. $\int_0^{\pi/2} \int_0^{2\cos^2\theta} \int_0^{4-r^2} r \sin\theta \, dz \, dr \, d\theta$

4. $\int_0^{\pi/2} \int_0^{\pi} \int_0^2 e^{-\rho^3} \rho^2 \, d\rho \, d\theta \, d\phi$

5. $\int_0^{2\pi} \int_0^{\pi/4} \int_0^{\cos\phi} \rho^2 \sin\phi \, d\rho \, d\phi \, d\theta$

6. $\int_0^{\pi/4} \int_0^{\pi/4} \int_0^{\cos\theta} \rho^2 \sin\phi \cos\phi \, d\rho \, d\theta \, d\phi$

In Exercises 7 and 8, use a computer algebra system to evaluate the iterated integral.

7. $\int_0^4 \int_0^z \int_0^{\pi/2} re^r \, d\theta \, dr \, dz$

8. $\int_0^{\pi/2} \int_0^{\pi} \int_0^{\sin\theta} (2\cos\phi)\rho^2 \, d\rho \, d\theta \, d\phi$

In Exercises 9–12, sketch the solid region whose volume is given by the iterated integral, and evaluate the iterated integral.

9. $\int_0^{\pi/2} \int_0^3 \int_0^{e^{-r^2}} r \, dz \, dr \, d\theta$

10. $\int_0^{2\pi} \int_0^{\sqrt{3}} \int_0^{3-r^2} r \, dz \, dr \, d\theta$

11. $\int_0^{2\pi} \int_{\pi/6}^{\pi/2} \int_0^4 \rho^2 \sin\phi \, d\rho \, d\phi \, d\theta$

12. $\int_0^{2\pi} \int_0^{\pi} \int_2^5 \rho^2 \sin\phi \, d\rho \, d\phi \, d\theta$

In Exercises 13–16, convert the integral from rectangular coordinates to both cylindrical and spherical coordinates, and evaluate the simplest iterated integral.

13. $\int_{-2}^2 \int_{-\sqrt{4-x^2}}^{\sqrt{4-x^2}} \int_{x^2+y^2}^4 x \, dz \, dy \, dx$

14. $\int_0^2 \int_0^{\sqrt{4-x^2}} \int_0^{\sqrt{16-x^2-y^2}} \sqrt{x^2+y^2} \, dz \, dy \, dx$

15. $\int_{-a}^a \int_{-\sqrt{a^2-x^2}}^{\sqrt{a^2-x^2}} \int_a^{a+\sqrt{a^2-x^2-y^2}} x \, dz \, dy \, dx$

16. $\int_0^1 \int_0^{\sqrt{1-x^2}} \int_0^{\sqrt{1-x^2-y^2}} \sqrt{x^2+y^2+z^2} \, dz \, dy \, dx$

Volume In Exercises 17–20, use cylindrical coordinates to find the volume of the solid.

17. Solid inside both $x^2 + y^2 + z^2 = a^2$ and $\left(x - \frac{a}{2}\right)^2 + y^2 = \left(\frac{a}{2}\right)^2$

18. Solid inside $x^2 + y^2 + z^2 = 16$ and outside $z = \sqrt{x^2+y^2}$

19. Solid bounded by the graphs of the sphere $r^2 + z^2 = a^2$ and the cylinder $r = a\cos\theta$

20. Solid inside the sphere $x^2 + y^2 + z^2 = 4$ and above the cone $z^2 = x^2 + y^2$

Mass In Exercises 21 and 22, use cylindrical coordinates to find the mass of the solid Q.

21. $Q = \{(x, y, z): 0 \leq z \leq 9 - x - 2y, x^2 + y^2 \leq 4\}$
    $\rho(x, y, z) = k\sqrt{x^2 + y^2}$

22. $Q = \{(x, y, z): 0 \leq z \leq 12e^{-(x^2+y^2)}, x^2 + y^2 \leq 4, x \geq 0, y \geq 0\}$
    $\rho(x, y, z) = k$

In Exercises 23–28, use cylindrical coordinates to find the indicated characteristic of the cone (see figure).

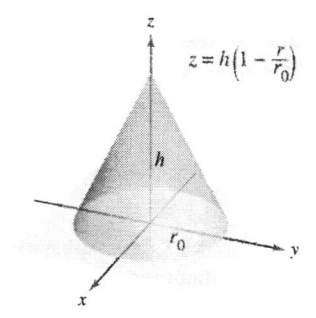

23. **Volume** Find the volume of the cone.

24. **Centroid** Find the centroid of the cone.

25. **Center of Mass** Find the center of mass of the cone assuming that its density at any point is proportional to the distance between the point and the axis of the cone. Use a computer algebra system to evaluate the triple integral.

26. **Center of Mass** Find the center of mass of the cone assuming that its density at any point is proportional to the distance between the point and the base. Use a computer algebra system to evaluate the triple integral.

27. **Moment of Inertia** Assume that the cone has uniform density and show that the moment of inertia about the z-axis is
$$I_z = \frac{3}{10}mr_0^2.$$

28. **Moment of Inertia** Assume that the density of the cone is
$$\rho(x, y, z) = k\sqrt{x^2 + y^2}$$
and find the moment of inertia about the z-axis.

Moment of Inertia In Exercises 29 and 30, use cylindrical coordinates to verify the given formula for the moment of inertia of the solid of uniform density.

29. Cylindrical shell: $I_z = \frac{1}{2}m(a^2 + b^2)$
    $0 < a \leq r \leq b, \ 0 \leq z \leq h$

30. Right circular cylinder: $I_z = \frac{3}{2}ma^2$
    $r = 2a\sin\theta, \ 0 \leq z \leq h$
    Use a computer algebra system to evaluate the triple integral.

Volume In Exercises 31 and 32, use spherical coordinates to find the volume of the solid.

31. The torus given by $\rho = 4 \sin \phi$ (Use a computer algebra system to evaluate the triple integral.)

32. The solid between the spheres $x^2 + y^2 + z^2 = a^2$ and $x^2 + y^2 + z^2 = b^2$, $b > a$, and inside the cone $z^2 = x^2 + y^2$

Mass In Exercises 33 and 34, use spherical coordinates to find the mass of the sphere $x^2 + y^2 + z^2 = a^2$ with the indicated density.

33. The density at any point is proportional to the distance between the point and the origin.

34. The density at any point is proportional to the distance of the point from the z-axis.

Center of Mass In Exercises 35 and 36, use spherical coordinates to find the center of mass of the solid of uniform density.

35. Hemispherical solid of radius r

36. Solid lying between two concentric hemispheres of radii r and R, where $r < R$

Moment of Inertia In Exercises 37 and 38, use spherical coordinates to find the moment of inertia about the z-axis of the solid of uniform density.

37. Solid bounded by the hemisphere $\rho = \cos \phi$, $\pi/4 \le \phi \le \pi/2$, and the cone $\phi = \pi/4$

38. Solid lying between two concentric hemispheres of radii r and R, where $r < R$

Getting at the Concept

39. Give the equations for the coordinate conversion from rectangular to cylindrical coordinates and vice versa.

40. Give the equations for the coordinate conversion from rectangular to spherical coordinates and vice versa.

41. Give the iterated form of the triple integral $\iiint_Q f(x, y, z)\, dV$ in cylindrical form.

42. Give the iterated form of the triple integral $\iiint_Q f(x, y, z)\, dV$ in spherical form.

43. Describe the surface whose equation is a coordinate equal to a constant for each of the coordinates in (a) the cylindrical coordinate system and (b) the spherical coordinate system.

44. When evaluating a triple integral with constant limits of integration in the cylindrical coordinate system, you are integrating over a part of what solid? What is the solid when you are in spherical coordinates?

45. Find the "volume" of the "four-dimensional sphere"
$$x^2 + y^2 + z^2 + w^2 = a^2$$
by evaluating
$$16 \int_0^a \int_0^{\sqrt{a^2-x^2}} \int_0^{\sqrt{a^2-x^2-y^2}} \int_0^{\sqrt{a^2-x^2-y^2-z^2}} dw\, dz\, dy\, dx.$$

46. Show that the volume of a spherical block can be approximated by
$$\Delta V \approx \rho^2 \sin \phi\, \Delta \rho\, \Delta \phi\, \Delta \theta.$$

SECTION PROJECT WRINKLED AND BUMPY SPHERES

In parts (a) and (b), find the volume of the wrinkled sphere or bumpy sphere. These solids are used as models for tumors.

(a) Wrinkled sphere

$\rho = 1 + 0.2 \sin 8\theta \sin \phi$

$0 \le \theta \le 2\pi, 0 \le \phi \le \pi$

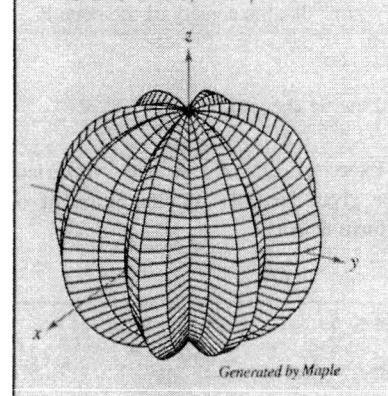

(b) Bumpy sphere

$\rho = 1 + 0.2 \sin 8\theta \sin 4\phi$

$0 \le \theta \le 2\pi, 0 \le \phi \le \pi$

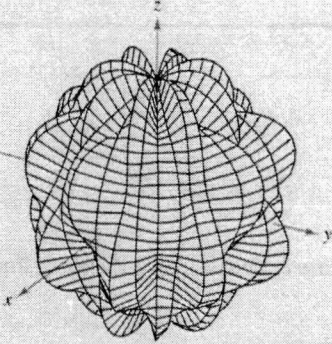

Generated by Maple

FOR FURTHER INFORMATION For more information, see page 934 or see the article "Heat Therapy for Tumors" by Leah Edelstein-Keshet in *The UMAP Journal*. To view this article, go to the website www.matharticles.com.

Section 13.8 Change of Variables: Jacobians

- Understand the concept of a Jacobian.
- Use a Jacobian to change variables in a double integral.

Jacobians

For the single integral
$$\int_a^b f(x)\, dx$$
you can change variables by letting $x = g(u)$, so that $dx = g'(u)\, du$, and obtain
$$\int_a^b f(x)\, dx = \int_c^d f(g(u)) g'(u)\, du$$
where $a = g(c)$ and $b = g(d)$. Note that the change-of-variables process introduces an additional factor $g'(u)$ into the integrand. This also occurs in the case of double integrals
$$\iint_R f(x, y)\, dA = \iint_S f(g(u, v), h(u, v)) \underbrace{\left| \frac{\partial x}{\partial u} \frac{\partial y}{\partial v} - \frac{\partial y}{\partial u} \frac{\partial x}{\partial v} \right|}_{\text{Jacobian}} du\, dv$$
where the change of variables $x = g(u, v)$ and $y = h(u, v)$ introduces a factor called the **Jacobian** of x and y with respect to u and v. In defining the Jacobian, it is convenient to use the following determinant notation.

Definition of the Jacobian

If $x = g(u, v)$ and $y = h(u, v)$, then the **Jacobian** of x and y with respect to u and v, denoted by $\partial(x, y)/\partial(u, v)$, is

$$\frac{\partial(x, y)}{\partial(u, v)} = \begin{vmatrix} \frac{\partial x}{\partial u} & \frac{\partial x}{\partial v} \\ \frac{\partial y}{\partial u} & \frac{\partial y}{\partial v} \end{vmatrix} = \frac{\partial x}{\partial u} \frac{\partial y}{\partial v} - \frac{\partial y}{\partial u} \frac{\partial x}{\partial v}.$$

Example 1 The Jacobian for Rectangular-to-Polar Conversion

Find the Jacobian for the change of variables defined by

$x = r \cos \theta$ and $y = r \sin \theta$.

Solution From the definition of a Jacobian, you obtain

$$\frac{\partial(x, y)}{\partial(r, \theta)} = \begin{vmatrix} \frac{\partial x}{\partial r} & \frac{\partial x}{\partial \theta} \\ \frac{\partial y}{\partial r} & \frac{\partial y}{\partial \theta} \end{vmatrix}$$
$$= \begin{vmatrix} \cos \theta & -r \sin \theta \\ \sin \theta & r \cos \theta \end{vmatrix}$$
$$= r \cos^2 \theta + r \sin^2 \theta$$
$$= r.$$

CARL GUSTAV JACOBI (1804–1851)

The Jacobian is named after the German mathematician Carl Gustav Jacobi. Jacobi is known for his work in many areas of mathematics, but his interest in integration stemmed from the problem of finding the circumference of an ellipse.

$T(r, \theta) = (r\cos\theta, r\sin\theta)$

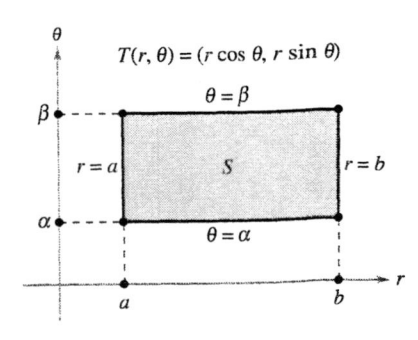

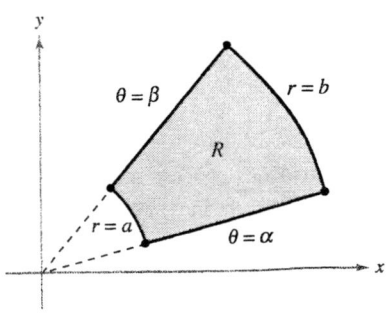

S is the region in the $r\theta$-plane that corresponds to R in the xy-plane.
Figure 13.70

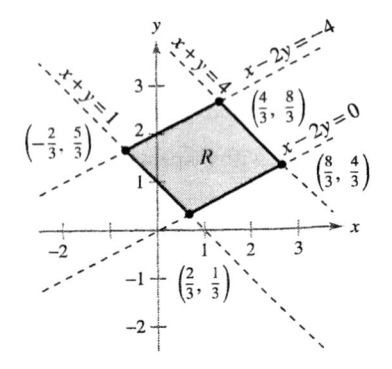

Region R in the xy-plane
Figure 13.71

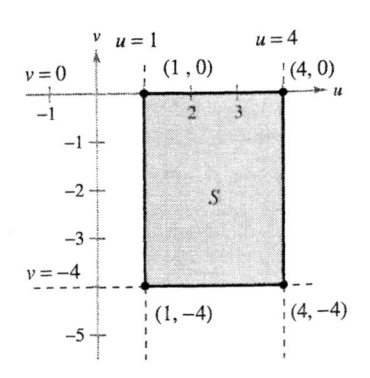

Region S in the uv-plane
Figure 13.72

Example 1 points out that the change of variables from rectangular to polar coordinates for a double integral can be written as

$$\iint_R f(x, y)\, dA = \iint_S f(r\cos\theta, r\sin\theta) r\, dr\, d\theta, \quad r > 0$$

$$= \iint_S f(r\cos\theta, r\sin\theta) \left| \frac{\partial(x, y)}{\partial(r, \theta)} \right| dr\, d\theta$$

where S is the region in the $r\theta$-plane that corresponds to the region R in the xy-plane, as shown in Figure 13.70. This formula is similar to that found on page 957.

In general, a change of variables is given by a one-to-one **transformation** T from a region S in the uv-plane to a region R in the xy-plane, to be given by

$$T(u, v) = (x, y) = (g(u, v), h(u, v))$$

where g and h have continuous first partial derivatives in the region S. Note that the point (u, v) lies in S and the point (x, y) lies in R. In most cases, you are hunting for a transformation for which the region S is simpler than the region R.

Example 2 Finding a Change of Variables to Simplify a Region

Let R be the region bounded by the lines

$$x - 2y = 0, \quad x - 2y = -4, \quad x + y = 4, \quad \text{and} \quad x + y = 1$$

as shown in Figure 13.71. Find a transformation T from a region S to R such that S is a rectangular region (with sides parallel to the u- or v-axis).

Solution To begin, let $u = x + y$ and $v = x - 2y$. Solving this system of equations for x and y produces $T(u, v) = (x, y)$, where

$$x = \frac{1}{3}(2u + v) \quad \text{and} \quad y = \frac{1}{3}(u - v).$$

The four boundaries for R in the xy-plane give rise to the following bounds for S in the uv-plane.

Bounds in the xy-Plane		Bounds in the uv-Plane
$x + y = 1$	⟹	$u = 1$
$x + y = 4$	⟹	$u = 4$
$x - 2y = 0$	⟹	$v = 0$
$x - 2y = -4$	⟹	$v = -4$

The region S is shown in Figure 13.72. Note that the transformation T maps the vertices of the region S onto the vertices of the region R. For instance,

$$T(1, 0) = \left(\tfrac{1}{3}[2(1) + 0], \tfrac{1}{3}[1 - 0]\right)$$
$$= \left(\tfrac{2}{3}, \tfrac{1}{3}\right)$$
$$T(4, 0) = \left(\tfrac{1}{3}[2(4) + 0], \tfrac{1}{3}[4 - 0]\right)$$
$$= \left(\tfrac{8}{3}, \tfrac{4}{3}\right)$$
$$T(4, -4) = \left(\tfrac{1}{3}[2(4) - 4], \tfrac{1}{3}[4 - (-4)]\right)$$
$$= \left(\tfrac{4}{3}, \tfrac{8}{3}\right)$$
$$T(1, -4) = \left(\tfrac{1}{3}[2(1) - 4], \tfrac{1}{3}[1 - (-4)]\right)$$
$$= \left(-\tfrac{2}{3}, \tfrac{5}{3}\right).$$

Change of Variables for Double Integrals

THEOREM 13.5 Change of Variables for Double Integrals

Let R and S be regions in the xy- and uv-planes that are related by the equations $x = g(u, v)$ and $y = h(u, v)$ such that each point in R is the image of a unique point in S. If f is continuous on R, g and h have continuous partial derivatives on S, and $\partial(x, y)/\partial(u, v)$ is nonzero on S, then

$$\int\!\!\int_R f(x, y)\, dx\, dy = \int\!\!\int_S f(g(u, v), h(u, v)) \left| \frac{\partial(x, y)}{\partial(u, v)} \right| du\, dv.$$

Proof Consider the case in which S is a rectangular region in the uv-plane with vertices (u, v), $(u + \Delta u, v)$, $(u + \Delta u, v + \Delta v)$, and $(u, v + \Delta v)$, as shown in Figure 13.73. The images of these vertices in the xy-plane are shown in Figure 13.74. If Δu and Δv are small, the continuity of g and h implies that R is approximately a parallelogram determined by the vectors $\overrightarrow{MN}$ and $\overrightarrow{MQ}$. So, the area of R is

$$\Delta A \approx \|\overrightarrow{MN} \times \overrightarrow{MQ}\|.$$

Moreover, for small Δu and Δv, the partial derivatives of g and h with respect to u can be approximated by

$$g_u(u, v) \approx \frac{g(u + \Delta u, v) - g(u, v)}{\Delta u}$$

and

$$h_u(u, v) \approx \frac{h(u + \Delta u, v) - h(u, v)}{\Delta u}.$$

Consequently,

$$\overrightarrow{MN} = [g(u + \Delta u, v) - g(u, v)]\mathbf{i} + [h(u + \Delta u, v) - h(u, v)]\mathbf{j}$$
$$\approx [g_u(u, v)\, \Delta u]\mathbf{i} + [h_u(u, v)\, \Delta u]\mathbf{j}$$
$$= \frac{\partial x}{\partial u} \Delta u \mathbf{i} + \frac{\partial y}{\partial u} \Delta u \mathbf{j}.$$

Similarly, you can approximate $\overrightarrow{MQ}$ by $\frac{\partial x}{\partial v} \Delta v \mathbf{i} + \frac{\partial y}{\partial v} \Delta v \mathbf{j}$, which implies that

$$\overrightarrow{MN} \times \overrightarrow{MQ} \approx \begin{vmatrix} \mathbf{i} & \mathbf{j} & \mathbf{k} \\ \frac{\partial x}{\partial u} \Delta u & \frac{\partial y}{\partial u} \Delta u & 0 \\ \frac{\partial x}{\partial v} \Delta v & \frac{\partial y}{\partial v} \Delta v & 0 \end{vmatrix} = \begin{vmatrix} \frac{\partial x}{\partial u} & \frac{\partial y}{\partial u} \\ \frac{\partial x}{\partial v} & \frac{\partial y}{\partial v} \end{vmatrix} \Delta u\, \Delta v \mathbf{k}.$$

It follows that, in Jacobian notation,

$$\Delta A \approx \|\overrightarrow{MN} \times \overrightarrow{MQ}\| \approx \left| \frac{\partial(x, y)}{\partial(u, v)} \right| \Delta u \Delta v.$$

Because this approximation improves as Δu and Δv approach 0, the limiting case can be written as

$$dA \approx \|\overrightarrow{MN} \times \overrightarrow{MQ}\| \approx \left| \frac{\partial(x, y)}{\partial(u, v)} \right| du\, dv.$$

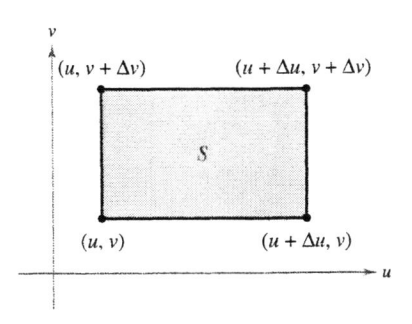

Area of $S = \Delta u\, \Delta v$
$\Delta u > 0$, $\Delta v > 0$
Figure 13.73

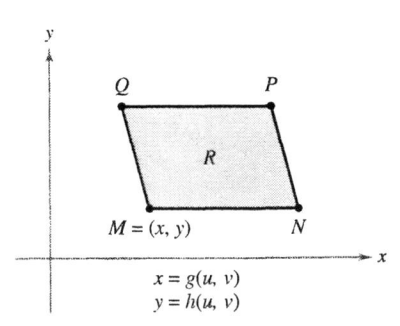

The vertices in the xy-plane are
$M(g(u, v), h(u, v))$, $N(g(u + \Delta u, v), h(u + \Delta u, v))$, $P(g(u + \Delta u, v + \Delta v), h(u + \Delta u, v + \Delta v))$, and $Q(g(u, v + \Delta v), h(u, v + \Delta v))$.
Figure 13.74

The next two examples show how a change of variables can simplify the integration process. The simplification can occur in various ways. You can make a change of variables to simplify either the *region R* or the *integrand f(x, y)*, or both.

Example 3   Using a Change of Variables to Simplify a Region

Let R be the region bounded by the lines

$$x - 2y = 0, \quad x - 2y = -4, \quad x + y = 4, \quad \text{and} \quad x + y = 1$$

as shown in Figure 13.75. Evaluate the double integral

$$\int_R\!\!\int 3xy\, dA.$$

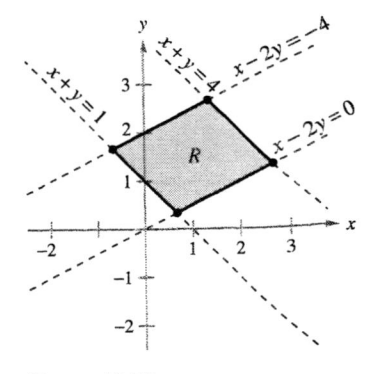

Figure 13.75

Solution From Example 2, you can use the following change of variables.

$$x = \frac{1}{3}(2u + v) \quad \text{and} \quad y = \frac{1}{3}(u - v)$$

The partial derivatives of x and y are

$$\frac{\partial x}{\partial u} = \frac{2}{3}, \quad \frac{\partial x}{\partial v} = \frac{1}{3}, \quad \frac{\partial y}{\partial u} = \frac{1}{3}, \quad \text{and} \quad \frac{\partial y}{\partial v} = -\frac{1}{3}$$

which implies that the Jacobian is

$$\frac{\partial(x, y)}{\partial(u, v)} = \begin{vmatrix} \frac{\partial x}{\partial u} & \frac{\partial x}{\partial v} \\ \frac{\partial y}{\partial u} & \frac{\partial y}{\partial v} \end{vmatrix}$$

$$= \begin{vmatrix} \frac{2}{3} & \frac{1}{3} \\ \frac{1}{3} & -\frac{1}{3} \end{vmatrix}$$

$$= -\frac{2}{9} - \frac{1}{9}$$

$$= -\frac{1}{3}.$$

Therefore, by Theorem 13.5, you obtain

$$\int_R\!\!\int 3xy\, dA = \int_S\!\!\int 3\left[\frac{1}{3}(2u + v)\frac{1}{3}(u - v)\right]\left|\frac{\partial(x, y)}{\partial(u, v)}\right| dv\, du$$

$$= \int_1^4\!\!\int_{-4}^0 \frac{1}{9}(2u^2 - uv - v^2)\, dv\, du$$

$$= \frac{1}{9}\int_1^4 \left[2u^2v - \frac{uv^2}{2} - \frac{v^3}{3}\right]_{-4}^0 du$$

$$= \frac{1}{9}\int_1^4 \left(8u^2 + 8u - \frac{64}{3}\right) du$$

$$= \frac{1}{9}\left[\frac{8u^3}{3} + 4u^2 - \frac{64}{3}u\right]_1^4$$

$$= \frac{164}{9}.$$

Example 4 Using a Change of Variables to Simplify an Integrand

Let R be the region bounded by the square with vertices $(0, 1)$, $(1, 2)$, $(2, 1)$, and $(1, 0)$. Evaluate the integral

$$\iint_R (x + y)^2 \sin^2(x - y)\, dA.$$

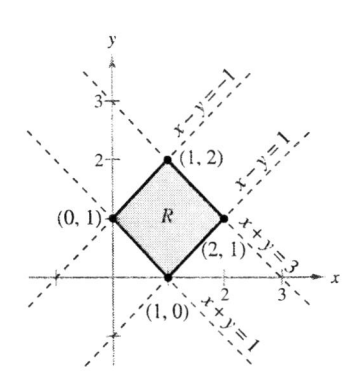

Solution Note that the sides of R lie on the lines $x + y = 1$, $x - y = 1$, $x + y = 3$, and $x - y = -1$, as shown in Figure 13.76. Letting $u = x + y$ and $v = x - y$, you can determine the bounds for region S in the uv-plane to be

$$1 \le u \le 3 \quad \text{and} \quad -1 \le v \le 1$$

as shown in Figure 13.77. Solving for x and y in terms of u and v produces

$$x = \frac{1}{2}(u + v) \quad \text{and} \quad y = \frac{1}{2}(u - v).$$

The partial derivatives of x and y are

$$\frac{\partial x}{\partial u} = \frac{1}{2}, \quad \frac{\partial x}{\partial v} = \frac{1}{2}, \quad \frac{\partial y}{\partial u} = \frac{1}{2}, \quad \text{and} \quad \frac{\partial y}{\partial v} = -\frac{1}{2}$$

which implies that the Jacobian is

$$\frac{\partial(x, y)}{\partial(u, v)} = \begin{vmatrix} \frac{\partial x}{\partial u} & \frac{\partial x}{\partial v} \\ \frac{\partial y}{\partial u} & \frac{\partial y}{\partial v} \end{vmatrix} = \begin{vmatrix} \frac{1}{2} & \frac{1}{2} \\ \frac{1}{2} & -\frac{1}{2} \end{vmatrix} = -\frac{1}{4} - \frac{1}{4} = -\frac{1}{2}.$$

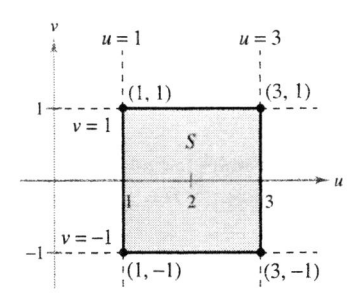

By Theorem 13.5, it follows that

$$\iint_R (x + y)^2 \sin^2(x - y)\, dA = \int_{-1}^{1} \int_{1}^{3} u^2 \sin^2 v \left(\frac{1}{2}\right) du\, dv$$

$$= \frac{1}{2} \int_{-1}^{1} (\sin^2 v) \frac{u^3}{3} \bigg]_1^3 dv$$

$$= \frac{13}{3} \int_{-1}^{1} \sin^2 v\, dv$$

$$= \frac{13}{6} \int_{-1}^{1} (1 - \cos 2v)\, dv$$

$$= \frac{13}{6} \left[v - \frac{1}{2} \sin 2v \right]_{-1}^{1}$$

$$= \frac{13}{6} \left[2 - \frac{1}{2} \sin 2 + \frac{1}{2} \sin(-2) \right]$$

$$= \frac{13}{6}(2 - \sin 2)$$

$$\approx 2.363.$$

Region R in the xy-plane
Figure 13.76

Region S in the uv-plane
Figure 13.77

In each of the change-of-variables examples in this section, the region S has been a rectangle with sides parallel to the u- or v-axis. Occasionally, a change of variables can be used for other types of regions. For instance, letting $T(u, v) = \left(x, \frac{1}{2}y\right)$ changes the circular region $u^2 + v^2 = 1$ to the elliptical region $x^2 + (y^2/4) = 1$.

EXERCISES FOR SECTION 13.8

In Exercises 1–8, find the Jacobian $\partial(x, y)/\partial(u, v)$ for the indicated change of variables.

1. $x = -\frac{1}{2}(u - v), y = \frac{1}{2}(u + v)$
2. $x = au + bv, y = cu + dv$
3. $x = u - v^2, y = u + v$
4. $x = uv - 2u, y = uv$
5. $x = u\cos\theta - v\sin\theta, y = u\sin\theta + v\cos\theta$
6. $x = u + a, y = v + a$
7. $x = e^u \sin v, y = e^u \cos v$
8. $x = \dfrac{u}{v}, y = u + v$

In Exercises 9 and 10, sketch the image S in the uv-plane of the region R in the xy-plane using the given transformations.

9. $x = 3u + 2v$
   $y = 3v$

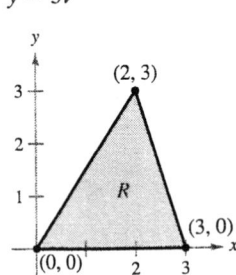

10. $x = \frac{1}{3}(4u - v)$
    $y = \frac{1}{3}(u - v)$

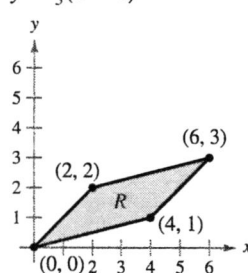

In Exercises 11–16, use the indicated change of variables to evaluate the double integral.

11. $\displaystyle\iint_R 4(x^2 + y^2)\, dA$
    $x = \frac{1}{2}(u + v)$
    $y = \frac{1}{2}(u - v)$

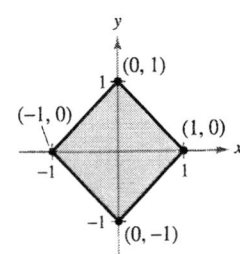

12. $\displaystyle\iint_R 60xy\, dA$
    $x = \frac{1}{2}(u + v)$
    $y = -\frac{1}{2}(u - v)$

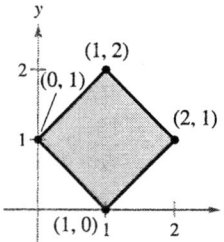

13. $\displaystyle\iint_R y(x - y)\, dA$
    $x = u + v$
    $y = u$

14. $\displaystyle\iint_R 4(x + y)e^{x-y}\, dA$
    $x = \frac{1}{2}(u + v)$
    $y = \frac{1}{2}(u - v)$

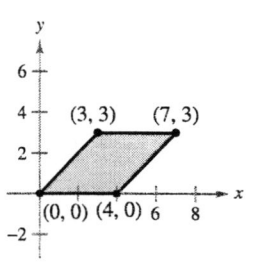

Figure for 13

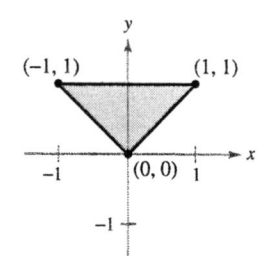

Figure for 14

15. $\displaystyle\iint_R e^{-xy/2}\, dA$
    $x = \sqrt{\dfrac{v}{u}},\ y = \sqrt{uv}$
    R: first quadrant region lying between the graphs of $y = \frac{1}{4}x$, $y = 2x$, $y = \dfrac{1}{x}$, $y = \dfrac{4}{x}$

16. $\displaystyle\iint_R y \sin xy\, dA$
    $x = \dfrac{u}{v}, y = v$
    R: region lying between the graphs of $xy = 1$, $xy = 4$, $y = 1$, $y = 4$

In Exercises 17–22, use a change of variables to find the volume of the solid region lying below the surface $z = f(x, y)$ and above the plane region R.

17. $f(x, y) = (x + y)e^{x-y}$
    R: region bounded by the square with vertices $(4, 0)$, $(6, 2)$, $(4, 4)$, $(2, 2)$

18. $f(x, y) = (x + y)^2 \sin^2(x - y)$
    R: region bounded by the square with vertices $(\pi, 0)$, $(3\pi/2, \pi/2)$, (π, π), $(\pi/2, \pi/2)$

19. $f(x, y) = \sqrt{(x - y)(x + 4y)}$
    R: region bounded by the parallelogram with vertices $(0, 0)$, $(1, 1)$, $(5, 0)$, $(4, -1)$

20. $f(x, y) = (3x + 2y)(2y - x)^{3/2}$
    R: region bounded by the parallelogram with vertices $(0, 0)$, $(-2, 3)$, $(2, 5)$, $(4, 2)$

21. $f(x, y) = \sqrt{x + y}$
    R: region bounded by the triangle with vertices $(0, 0)$, $(a, 0)$, $(0, a)$, where $a > 0$

22. $f(x, y) = \dfrac{xy}{1 + x^2 y^2}$
    R: region bounded by the graphs of $xy = 1$, $xy = 4$, $x = 1$, $x = 4$ (*Hint:* Let $x = u$, $y = v/u$.)

23. Consider the region R in the xy-plane bounded by the ellipse
$$\frac{x^2}{a^2} + \frac{y^2}{b^2} = 1$$
and the transformations $x = au$ and $y = bv$.
   (a) Sketch the graph of the region R and its image S under the indicated transformation.
   (b) Find $\partial(x, y)/\partial(u, v)$.
   (c) Find the area of the ellipse.

24. Use the result in Exercise 23 to find the volume of each dome-shaped solid lying below the surface $z = f(x, y)$ and above the elliptical region R. (*Hint:* After making the change of variables indicated by the results in Exercise 23, make a second change of variables to polar coordinates.)
   (a) $f(x, y) = 16 - x^2 - y^2$
   $R: \dfrac{x^2}{16} + \dfrac{y^2}{9} \leq 1$
   (b) $f(x, y) = A \cos\left(\dfrac{\pi}{2}\sqrt{\dfrac{x^2}{a^2} + \dfrac{y^2}{b^2}}\right)$
   $R: \dfrac{x^2}{a^2} + \dfrac{y^2}{b^2} \leq 1$

Getting at the Concept

25. State the definition of the Jacobian.
26. Describe how to use the Jacobian to change variables in double integrals.

In Exercises 27–30, find the Jacobian $\partial(x, y, z)/\partial(u, v, w)$ for the indicated change of variables. If $x = f(u, v, w), y = g(u, v, w)$, and $z = h(u, v, w)$, then the Jacobian of x, y, and z with respect to u, v, and w is

$$\frac{\partial(x, y, z)}{\partial(u, v, w)} = \begin{vmatrix} \dfrac{\partial x}{\partial u} & \dfrac{\partial x}{\partial v} & \dfrac{\partial x}{\partial w} \\ \dfrac{\partial y}{\partial u} & \dfrac{\partial y}{\partial v} & \dfrac{\partial y}{\partial w} \\ \dfrac{\partial z}{\partial u} & \dfrac{\partial z}{\partial v} & \dfrac{\partial z}{\partial w} \end{vmatrix}.$$

27. $x = u(1-v), y = uv(1-w), z = uvw$
28. $x = 4u - v, y = 4v - w, z = u + w$
29. **Spherical Coordinates**
    $x = \rho \sin\phi \cos\theta, y = \rho \sin\phi \sin\theta, z = \rho \cos\phi$
30. **Cylindrical Coordinates**
    $x = r\cos\theta, y = r\sin\theta, z = z$

REVIEW EXERCISES FOR CHAPTER 13

13.1 In Exercises 1 and 2, evaluate the integral.

1. $\displaystyle\int_1^{x^2} x \ln y \, dy$
2. $\displaystyle\int_y^{2y} (x^2 + y^2) \, dx$

In Exercises 3–6, evaluate the iterated integral. Change the coordinate system when convenient.

3. $\displaystyle\int_0^1 \int_0^{1+x} (3x + 2y) \, dy \, dx$
4. $\displaystyle\int_0^2 \int_{x^2}^{2x} (x^2 + 2y) \, dy \, dx$
5. $\displaystyle\int_0^3 \int_0^{\sqrt{9-x^2}} 4x \, dy \, dx$
6. $\displaystyle\int_0^{\sqrt{3}} \int_{2-\sqrt{4-y^2}}^{2+\sqrt{4-y^2}} dx \, dy$

Area In Exercises 7–14, write the limits for the double integral
$$\int_R\int f(x, y) \, dA$$
for both orders of integration. Compute the area of R by letting $f(x, y) = 1$ and integrating.

7. Triangle: vertices $(0, 0), (3, 0), (0, 1)$
8. Triangle: vertices $(0, 0), (3, 0), (2, 2)$
9. The larger area between the graphs of $x^2 + y^2 = 25$ and $x = 3$
10. Region bounded by the graphs of $y = 6x - x^2$ and $y = x^2 - 2x$
11. Region enclosed by the graph of $y^2 = x^2 - x^4$
12. Region bounded by the graphs of $x = y^2 + 1, x = 0, y = 0$, and $y = 2$
13. Region bounded by the graphs of $x = y + 3$ and $x = y^2 + 1$
14. Region bounded by the graphs of $x = -y$ and $x = 2y - y^2$

Think About It In Exercises 15 and 16, give a geometric argument for the given equality. Verify the equality analytically.

15. $\displaystyle\int_0^1 \int_{2y}^{2\sqrt{2-y^2}} (x+y) \, dx \, dy = \int_0^2 \int_0^{x/2} (x+y) \, dy \, dx +$
$\displaystyle\int_2^{2\sqrt{2}} \int_0^{\sqrt{8-x^2}/2} (x+y) \, dy \, dx$

16. $\displaystyle\int_0^2 \int_{3y/2}^{5-y} e^{x+y} \, dx \, dy = \int_0^3 \int_0^{2x/3} e^{x+y} \, dy \, dx + \int_3^5 \int_0^{5-x} e^{x+y} \, dy \, dx$

13.2 *Volume* In Exercises 17 and 18, use a multiple integral and a convenient coordinate system to find the volume of the solid.

17. Solid bounded by the graphs of $z = x^2 - y + 4, z = 0, y = 0, x = 0$, and $x = 4$
18. Solid bounded by the graphs of $z = x + y, z = 0, x = 0, x = 3$, and $y = x$

Think About It In Exercises 19 and 20, determine which value best approximates the volume of the solid between the xy-plane and the function over the region. (Make your selection on the basis of a sketch of the solid and *not* by performing any calculations.)

19. $f(x, y) = x + y$

    R: triangle with vertices $(0, 0), (3, 0), (3, 3)$

    (a) $\frac{9}{2}$ (b) 5   (c) 13   (d) 100   (e) -100

20. $f(x, y) = 10x^2y^2$

    R: circle bounded by $x^2 + y^2 = 1$

    (a) π (b) -15 (c) $\frac{2}{3}$ (d) 3   (e) 15

Probability In Exercises 21 and 22, find k such that the function is a joint density function and find the required probability, where

$$P(a \le x \le b, c \le y \le d) = \int_c^d \int_a^b f(x, y)\, dx\, dy.$$

21. $f(x, y) = \begin{cases} kxye^{-(x+y)}, & x \ge 0, y \ge 0 \\ 0, & \text{elsewhere} \end{cases}$

    $P(0 \le x \le 1, 0 \le y \le 1)$

22. $f(x, y) = \begin{cases} kxy, & 0 \le x \le 1, 0 \le y \le x \\ 0, & \text{elsewhere} \end{cases}$

    $P(0 \le x \le 0.5, 0 \le y \le 0.25)$

True or False? In Exercises 23–26, determine whether the statement is true or false. If it is false, explain why or give an example that shows it is false.

23. $\int_a^b \int_c^d f(x)g(y)\, dy\, dx = \left[\int_a^b f(x)\, dx\right]\left[\int_c^d g(y)\, dy\right]$

24. If f is continuous over R_1 and R_2, and

    $$\int\int_{R_1} dA = \int\int_{R_2} dA$$

    then

    $$\int\int_{R_1} f(x, y)\, dA = \int\int_{R_2} f(x, y)\, dA.$$

25. $\int_{-1}^1 \int_{-1}^1 \cos(x^2 + y^2)\, dx\, dy = 4\int_0^1 \int_0^1 \cos(x^2 + y^2)\, dx\, dy$

26. $\int_0^1 \int_0^1 \frac{1}{1 + x^2 + y^2}\, dx\, dy < \frac{\pi}{4}$

13.3 In Exercises 27 and 28, evaluate the iterated integral using polar coordinates.

27. $\int_0^h \int_0^x \sqrt{x^2 + y^2}\, dy\, dx$

28. $\int_0^4 \int_0^{\sqrt{16-y^2}} (x^2 + y^2)\, dx\, dy$

Volume In Exercises 29 and 30, use a multiple integral and a convenient coordinate system to find the volume of the solid.

29. Solid bounded by the graphs of $z = 0$ and $z = h$, outside the cylinder

    $$x^2 + y^2 = 1$$

    and inside the hyperboloid

    $$x^2 + y^2 - z^2 = 1$$

30. Solid that remains after drilling a hole of radius b through the center of a sphere of radius R $(b < R)$

31. Consider the region R in the xy-plane bounded by the graph of the equation

    $$(x^2 + y^2)^2 = 9(x^2 - y^2).$$

    (a) Convert the equation to polar coordinates. Use a graphing utility to graph the equation.

    (b) Use a double integral to find the area of the region R.

    (c) Use a computer algebra system to determine the volume of the solid over the region R and beneath the hemisphere $z = \sqrt{9 - x^2 - y^2}$.

32. Combine the sum of the two iterated integrals into a single iterated integral by converting to polar coordinates. Evaluate the resulting iterated integral.

    $$\int_0^{8/\sqrt{13}} \int_0^{3x/2} xy\, dy\, dx + \int_{8/\sqrt{13}}^4 \int_0^{\sqrt{16-x^2}} xy\, dy\, dx$$

13.4 **Mass and Center of Mass** In Exercises 33 and 34, find the mass and center of mass of the lamina of indicated density bounded by the graphs of the equations. Use a computer algebra system to evaluate the multiple integrals.

33. $y = 2x, y = 2x^3$, first quadrant

    (a) $\rho = kxy$ (b) $\rho = k(x^2 + y^2)$

34. $y = \frac{h}{2}\left(2 - \frac{x}{L} - \frac{x^2}{L^2}\right)$, $\rho = k$, first quadrant

In Exercises 35 and 36, find I_x, I_y, I_0, $\bar{\bar{x}}$, and $\bar{\bar{y}}$ for the lamina bounded by the graphs of the equations. Use a computer algebra system to evaluate the double integrals.

35. $y = 0, y = b, x = 0, x = a, \rho = kx$

36. $y = 4 - x^2, y = 0, x > 0, \rho = ky$

13.5 **Surface Area** In Exercises 37 and 38, find the area of the surface on the graph of the function $f(x, y)$ over the region R.

37. $f(x, y) = 16 - x^2 - y^2$

    $R = \{(x, y): x^2 + y^2 \le 16\}$

38. $f(x, y) = 16 - x - y^2$

    $R = \{(x, y): 0 \le x \le 2, 0 \le y \le x\}$

    Use a computer algebra system to evaluate the integral.

39. Surface Area Find the area of the surface of the cylinder $f(x, y) = 9 - y^2$ that lies above the triangle bounded by the graphs of the equations $y = x$, $y = -x$, and $y = 3$.

40. Surface Area The roof over the stage of an open air theater at a theme park is modeled by

$$f(x, y) = 25\left[1 + e^{-(x^2+y^2)/1000}\cos^2\left(\frac{x^2 + y^2}{1000}\right)\right]$$

where the stage is a semicircle bounded by the graphs of $y = \sqrt{50^2 - x^2}$ and $y = 0$.

(a) Use a computer algebra system to graph the surface.

(b) Use a computer algebra system to approximate the number of square feet of roofing required to cover the surface.

13.6 In Exercises 41–44, evaluate the iterated integral.

41. $\displaystyle\int_{-3}^{3}\int_{-\sqrt{9-x^2}}^{\sqrt{9-x^2}}\int_{x^2+y^2}^{9}\sqrt{x^2 + y^2}\,dz\,dy\,dx$

42. $\displaystyle\int_{-2}^{2}\int_{-\sqrt{4-x^2}}^{\sqrt{4-x^2}}\int_{0}^{(x^2+y^2)/2}(x^2 + y^2)\,dz\,dy\,dx$

43. $\displaystyle\int_{0}^{a}\int_{0}^{b}\int_{0}^{c}(x^2 + y^2 + z^2)\,dx\,dy\,dz$

44. $\displaystyle\int_{0}^{5}\int_{0}^{\sqrt{25-x^2}}\int_{0}^{\sqrt{25-x^2-y^2}}\frac{1}{1 + x^2 + y^2 + z^2}\,dz\,dy\,dx$

In Exercises 45 and 46, use a computer algebra system to evaluate the iterated integral.

45. $\displaystyle\int_{-1}^{1}\int_{-\sqrt{1-x^2}}^{\sqrt{1-x^2}}\int_{-\sqrt{1-x^2-y^2}}^{\sqrt{1-x^2-y^2}}(x^2 + y^2)\,dz\,dy\,dx$

46. $\displaystyle\int_{0}^{2}\int_{0}^{\sqrt{4-x^2}}\int_{0}^{\sqrt{4-x^2-y^2}}xyz\,dz\,dy\,dx$

13.7 Volume In Exercises 47 and 48, use a multiple integral to find the volume of the solid.

47. Solid inside the graphs of $r = 2\cos\theta$ and $r^2 + z^2 = 4$

48. Solid inside the graphs of $r^2 + z = 16$, $z = 0$, and $r = 2\sin\theta$

Center of Mass In Exercises 49–52, find the center of mass of the solid of uniform density bounded by the graphs of the equations.

49. Solid inside the hemisphere $\rho = \cos\phi$, $\pi/4 \leq \phi \leq \pi/2$, and outside the cone $\phi = \pi/4$

50. Wedge: $x^2 + y^2 = a^2$, $z = cy(c > 0)$, $y \geq 0$, $z \geq 0$

51. $x^2 + y^2 + z^2 = a^2$, first octant

52. $x^2 + y^2 + z^2 = 25$, $z = 4$ (the larger solid)

Moment of Inertia In Exercises 53 and 54, find the moment of inertia I_z of the solid of indicated density.

53. The solid of uniform density inside the paraboloid

$$z = 16 - x^2 - y^2$$

and outside the cylinder $x^2 + y^2 = 9$, $z \geq 0$

54. $x^2 + y^2 + z^2 = a^2$, density is proportional to the distance from the center

55. Investigation Consider a spherical segment of height h from a sphere of radius a, where $h \leq a$, and constant density $\rho(x, y, z) = k$ (see figure).

(a) Find the volume of the solid.

(b) Find the centroid of the solid.

(c) Use the result in part (b) to find the centroid of a hemisphere of radius a.

(d) Find $\displaystyle\lim_{h \to 0} \bar{z}$.

(e) Find I_z.

(f) Use the result in part (e) to find I_z for a hemisphere.

56. Moment of Inertia Find the moment of inertia about the z-axis of the ellipsoid

$$x^2 + y^2 + \frac{z^2}{a^2} = 1$$

where $a > 0$.

In Exercises 57 and 58, give a geometrical interpretation of the iterated integral.

57. $\displaystyle\int_{0}^{2\pi}\int_{0}^{\pi}\int_{0}^{6\sin\phi}\rho^2\sin\phi\,d\rho\,d\phi\,d\theta$

58. $\displaystyle\int_{0}^{\pi}\int_{0}^{2}\int_{0}^{1+r^2}r\,dz\,dr\,d\theta$

13.8 In Exercises 59 and 60, find the Jacobian for the change of variables.

59. $x = u + 3v$, $y = 2u - 3v$

60. $x = u^2 + v^2$, $y = u^2 - v^2$

In Exercises 61 and 62, use the indicated change of variables to evaluate the double integral.

61. $\displaystyle\iint_{R}\ln(x + y)\,dA$

$x = \frac{1}{2}(u + v)$, $y = \frac{1}{2}(u - v)$

62. $\displaystyle\iint_{R}\frac{x}{1 + x^2y^2}\,dA$

$x = u$, $y = \frac{v}{u}$

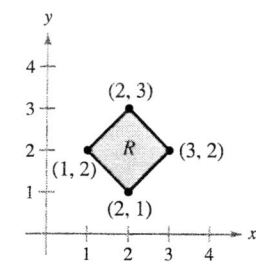

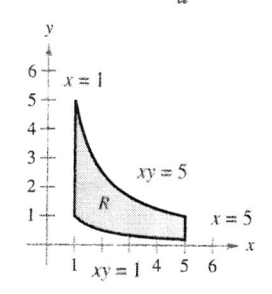

P.S. Problem Solving

1. (a) Find volume of the solid of intersection of the three cylinders $x^2 + z^2 = 1$, $y^2 + z^2 = 1$, and $x^2 + y^2 = 1$.

   (b) Use the Monte Carlo Method (see the exercises in Section 4.2) to confirm the answer in part (a). (*Hint:* Generate random points inside the cube of volume 8 centered at the origin.)

2. Let a, b, c, and d be positive real numbers. The first octant of the plane $ax + by + cz = d$ is shown in the figure. Show that the surface area of this portion of the plane is equal to

   $$\frac{A(R)}{c}\sqrt{a^2 + b^2 + c^2}$$

   where $A(R)$ is the area of the triangular region R in the xy-plane, as indicated in the figure.

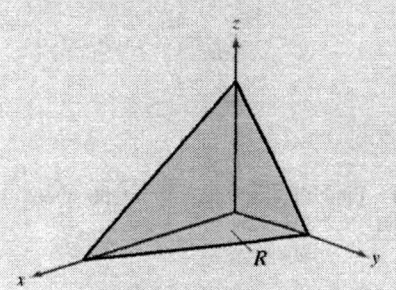

3. Derive Euler's famous result that was mentioned in Section 8.3, $\sum_{n=1}^{\infty} \frac{1}{n^2} = \frac{\pi^2}{6}$, by completing each step.

   (a) Prove that $\displaystyle\int \frac{dv}{2 - u^2 + v^2} = \frac{1}{\sqrt{2 - u^2}} \arctan \frac{v}{\sqrt{2 - u^2}} + C$.

   (b) Prove that $\displaystyle I_1 = \int_0^{\sqrt{2}/2} \int_{-u}^{u} \frac{2}{2 - u^2 + v^2} \, dv \, du = \frac{\pi^2}{18}$ by using the substitution $u = \sqrt{2} \sin \theta$.

   (c) Prove that

   $$I_2 = \int_{\sqrt{2}/2}^{\sqrt{2}} \int_{u - \sqrt{2}}^{-u + \sqrt{2}} \frac{2}{2 - u^2 + v^2} \, dv \, du$$

   $$= 4 \int_{\pi/6}^{\pi/2} \arctan \frac{1 - \sin \theta}{\cos \theta} \, d\theta$$

   by using the substitution $u = \sqrt{2} \sin \theta$.

   (d) Prove the trigonometric identity $\displaystyle\frac{1 - \sin \theta}{\cos \theta} = \tan\left(\frac{(\pi/2) - \theta}{2}\right)$.

   (e) Prove that $\displaystyle I_2 = \int_{\sqrt{2}/2}^{\sqrt{2}} \int_{u - \sqrt{2}}^{-u + \sqrt{2}} \frac{2}{2 - u^2 + v^2} \, dv \, du = \frac{\pi^2}{9}$.

   (f) Use the formula for the sum of an infinite geometric series to verify that $\displaystyle\sum_{n=1}^{\infty} \frac{1}{n^2} = \int_0^1 \int_0^1 \frac{1}{1 - xy} \, dx \, dy$.

   (g) Use the change of variables $\displaystyle u = \frac{x + y}{\sqrt{2}}$ and $\displaystyle v = \frac{y - x}{\sqrt{2}}$ to prove that $\displaystyle\sum_{n=1}^{\infty} \frac{1}{n^2} = \int_0^1 \int_0^1 \frac{1}{1 - xy} \, dx \, dy = I_1 + I_2 = \frac{\pi^2}{6}$.

4. Consider a circular lawn with a radius of 10 feet, as indicated in the figure. Assume that a sprinkler distributes water in a radial fashion according to the formula $f(r) = \dfrac{r}{16} - \dfrac{r^2}{160}$ (measured in cubic feet of water per hour per square foot of lawn), where r is the distance in feet from the sprinkler. Find the amount of water that is distributed in one hour in the following two annular regions.

   $A = \{(r, \theta) : 4 \leq r \leq 5, 0 \leq \theta \leq 2\pi\}$

   $B = \{(r, \theta) : 9 \leq r \leq 10, 0 \leq \theta \leq 2\pi\}$

   Is the distribution of water uniform? Determine the amount of water the entire lawn receives in one hour.

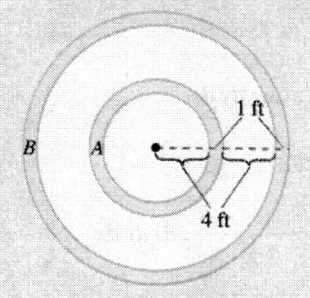

5. The figure below shows the region R bounded by the curves $y = \sqrt{x}$, $y = \sqrt{2x}$, $y = \dfrac{x^2}{3}$, and $y = \dfrac{x^2}{4}$. Use the change of variables $x = u^{1/3} v^{2/3}$ and $y = u^{2/3} v^{1/3}$ to find the area of the region R.

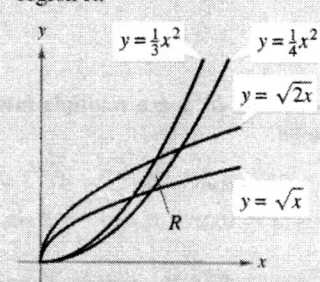

6. The figure shows a solid bounded below by the plane $z = 2$ and above by the sphere $x^2 + y^2 + z^2 = 8$.

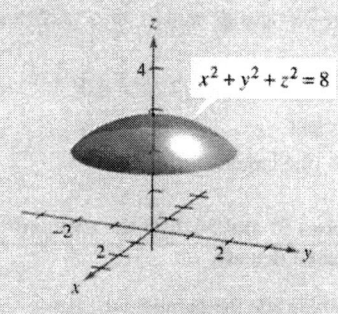

   (a) Find the volume of the solid using cylindrical coordinates.

   (b) Find the volume of the solid using spherical coordinates.

7. Sketch the solid whose volume is given by the sum of the iterated integrals $\int_0^6 \int_{z/2}^3 \int_{z/2}^y dx\,dy\,dz + \int_0^6 \int_3^{(12-z)/2} \int_{z/2}^{6-y} dx\,dy\,dz$. Then express the volume as a single iterated integral in the order $dy\,dz\,dx$.

8. Consider a continuous function $f(x, y)$ defined on the rectangular region R given by $a \leq x \leq b$ and $c \leq y \leq d$. Partition the interval $[a, b]$ into m subintervals of equal length Δx and partition the interval $[c, d]$ into n subintervals of equal length Δy, as indicated in the figure. The **Midpoint Rule** uses a double Riemann sum to approximate the double integral of f over R, as follows.

$$\int_a^b \int_c^d f(x, y)\,dA \approx \sum_{i=1}^m \sum_{j=1}^n f(x_i, y_j)\,\Delta y\,\Delta x$$

where (x_i, y_j) is the center of a representative rectangle in R. The base of a pile of rocks is a square measuring 4 meters on a side. The height of the pile at 1 meter intervals is given in the following table. Use the Midpoint Rule with $m = n = 2$ to estimate the volume of the pile of rocks.

	0	1	2	3	4
0	4	4.5	5	5	4
1	4	5	6	6	5
2	4.5	6	7	6	4
3	5	5	6	5	4
4	4	4	5	4	3.5

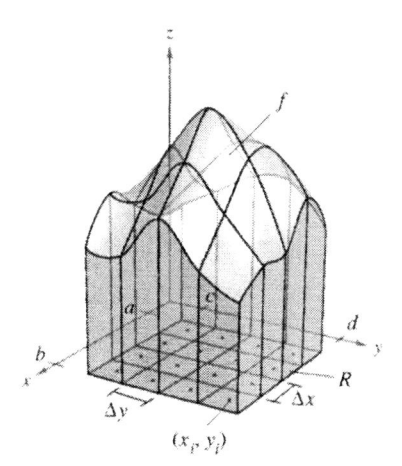

9. Evaluate the integral $\int_0^\infty x^2 e^{-x^2}\,dx$. (*Hint*: See Exercise 55, Section 13.3).

10. Evaluate the integral $\int_0^1 \sqrt{\ln(1/x)}\,dx$. (*Hint*: See Exercise 55, Section 13.3.)

11. Consider the function

$$f(x, y) = \begin{cases} ke^{-(x+y)/a}, & x \geq 0,\ y \geq 0 \\ 0, & \text{elsewhere}. \end{cases}$$

Find the relationship between the positive constants a and k such that f is a joint density function of the continuous random variables x and y.

12. From 1963 to 1986, the volume of the Great Salt Lake approximately tripled while its top surface area approximately doubled. Read the article "Relations between Surface Area and Volume in Lakes" by Daniel Cass and Gerald Wildenberg in *The College Mathematics Journal*. To view this article, go to the website *www.matharticles.com*. Then give examples of solids that have "water levels" a and b such that $V(b) = 3V(a)$ and $A(b) = 2A(a)$ (see figure), where V is volume and A is area.

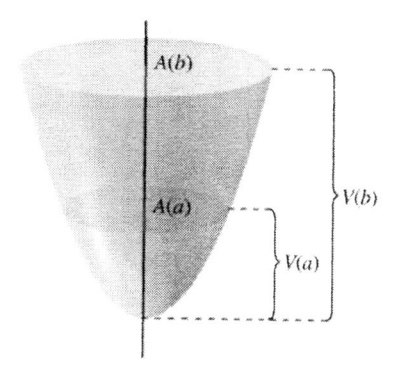

13. The angle between a plane P and the xy-plane is θ, where $0 \leq \theta < \pi/2$. The projection of a rectangular region in P onto the xy-plane is a rectangle whose sides have lengths Δx and Δy, as shown in the figure. Prove that the area of the rectangular region in P is $\sec\theta\,\Delta x\,\Delta y$.

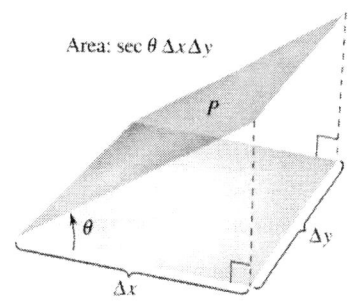

14. Use the result in Exercise 13 to order the planes in ascending order of their surface areas for a fixed region R in the xy-plane. Explain your ordering without doing any calculations.
   (a) $z_1 = 2 + x$
   (b) $z_2 = 5$
   (c) $z_3 = 10 - 5x + 9y$
   (d) $z_4 = 3 + x - 2y$

Mathematical Sculpture

Whether mathematics is seen as a science or as an art depends on one's perspective. One mathematician-sculptor, Helaman Ferguson, combines both viewpoints in a unique way. Ferguson's sculptures, which bear such names as *Cosine Wild Sphere* and *Esker Trefoil Torus*, are the concrete embodiments of mathematical concepts that incorporate ideas such as series expansions and vector fields into their creation. Some of the basic images of his work are tori and double tori, Möbius strips, and trefoil knots. One of his techniques is to use three-dimensional computer graphics to model his intended sculptures. The coordinates on the computer screen can then be used to direct the sculpting.

One example of Helaman Ferguson's work, *Umbilic Torus NC*, is shown below. This form can be written as a parametric surface using the following set of parametric equations.

$$x = \sin u \left[7 + \cos\left(\frac{u}{3} - 2v\right) + 2\cos\left(\frac{u}{3} + v\right) \right]$$

$$y = \cos u \left[7 + \cos\left(\frac{u}{3} - 2v\right) + 2\cos\left(\frac{u}{3} + v\right) \right]$$

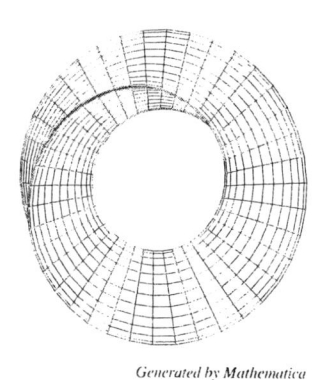

$$z = \sin\left(\frac{u}{3} - 2v\right) + 2\sin\left(\frac{u}{3} + v\right)$$

$$-\pi \leq u \leq \pi, \qquad -\pi \leq v \leq \pi$$

Generated by Mathematica

Mathematica rendition of *Umbilic Torus NC*

Photograph of *Umbilic Torus NC*

QUESTIONS

1. Explain how the umbilic torus shown above is similar to a Möbius strip. (A Möbius strip is shown in the photograph on the facing page.)

2. Use a three-dimensional computer algebra system to graph the torus. You will have to use the *parametric surface* graphing mode.

3. When viewed head-on (along the z-axis), the torus appears nearly circular. Is it possible to alter this appearance so that the shape of the torus is more like that of an elongated ellipse? If so, use a computer algebra system to sketch an "elliptical" torus.

The concepts presented here will be explored further in this chapter. For an extension of this application, see Lab 19 in the lab series that accompanies this text at college.hmco.com.

Vector Analysis 14

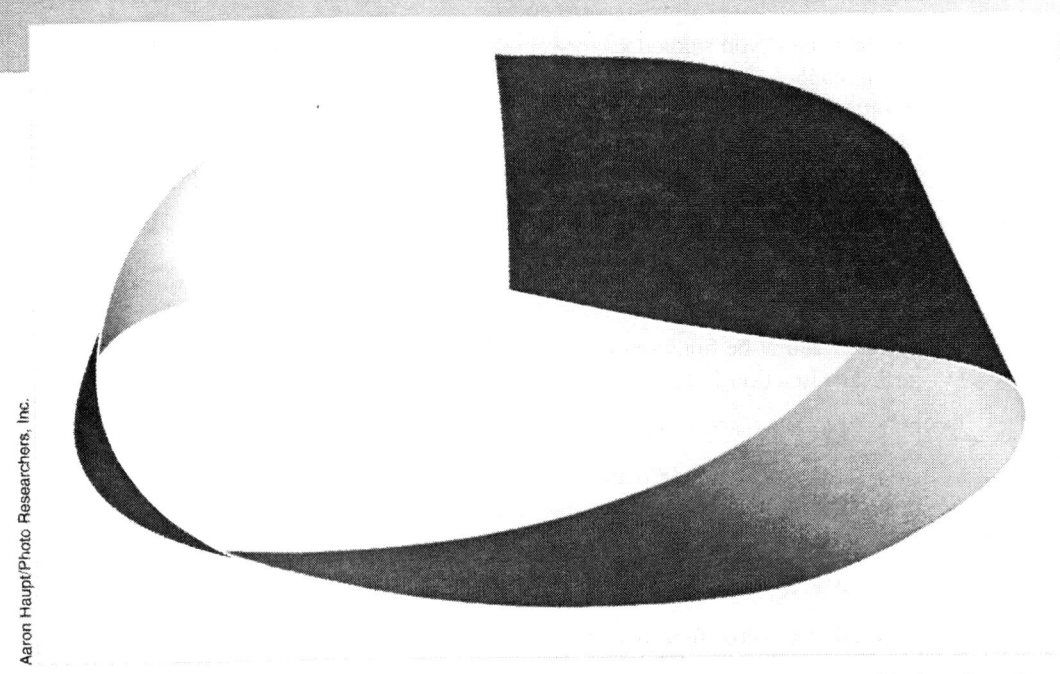

The Möbius strip, discovered by A.F. Möbius in 1858, was the first known one-sided surface. It can be constructed by twisting a strip of paper 180 degrees and attaching the ends together.

Cosine Wild Sphere

Esker Trefoil Torus

Helaman Ferguson holds a Ph.D. in mathematics and works out of his studio in Laurel, Maryland. In addition to exhibiting his works worldwide, he designs algorithms for operating machinery and for scientific visualization.

Section 14.1 Vector Fields

- Understand the concept of a vector field.
- Determine whether a vector field is conservative.
- Find the curl of a vector field.
- Find the divergence of a vector field.

Vector Fields

In Chapter 11, you studied vector-valued functions—functions that assign a vector to a *real number*. There you saw that vector-valued functions of real numbers are useful in representing curves and motion along a curve. In this chapter, you will study two other types of vector-valued functions—functions that assign a vector to a *point in the plane* or a *point in space*. Such functions are called **vector fields,** and they are useful in representing various types of **force fields** and **velocity fields.**

Definition of a Vector Field

Let M and N be functions of two variables x and y, defined on a plane region R. The function $\mathbf{F}$ defined by

$$\mathbf{F}(x, y) = M\mathbf{i} + N\mathbf{j} \qquad \text{Plane}$$

is called a **vector field over R.**

Let M, N, and P be functions of three variables x, y, and z, defined on a solid region Q in space. The function $\mathbf{F}$ defied by

$$\mathbf{F}(x, y, z) = M\mathbf{i} + N\mathbf{j} + P\mathbf{k} \qquad \text{Space}$$

is called a **vector field over Q.**

NOTE Although a vector field consists of infinitely many vectors, you can get a good idea of what the vector field looks like by sketching several representative vectors $\mathbf{F}(x, y)$ whose initial points are (x, y).

From this definition you can see that the *gradient* is one example of a vector field. For example, if

$$f(x, y) = x^2 + y^2$$

then the gradient of f

$$\nabla f(x, y) = f_x(x, y)\mathbf{i} + f_y(x, y)\mathbf{j} = 2x\mathbf{i} + 2y\mathbf{j} \qquad \text{Vector field in the plane}$$

is a vector field in the plane. From Chapter 12, the graphical interpretation of this field is a family of vectors, each of which points in the direction of maximum increase along the surface given by $z = f(x, y)$. For this particular function, the surface is a paraboloid and the gradient tells you that the direction of maximum increase along the surface is the direction given by the ray from the origin through the point (x, y).

Similarly, if

$$f(x, y, z) = x^2 + y^2 + z^2$$

then the gradient of f

$$\nabla f(x, y, z) = f_x(x, y, z)\mathbf{i} + f_y(x, y, z)\mathbf{j} + f_z(x, y, z)\mathbf{k}$$
$$= 2x\mathbf{i} + 2y\mathbf{j} + 2z\mathbf{k} \qquad \text{Vector field in space}$$

is a vector field in space.

A vector field is **continuous** at a point if each of its component functions M, N, and P is continuous at that point.

Some common *physical* examples of vector fields are **velocity fields, gravitational fields,** and **electric force fields.**

1. *Velocity fields* describe the motions of systems of particles in the plane or in space. For instance, Figure 14.1 shows the vector field determined by a wheel rotating on an axle. Notice that the velocity vectors are determined by the locations of their initial points—the farther a point is from the axle, the greater its velocity. Velocity fields are also determined by the flow of liquids through a container or by the flow of air currents around a moving object, as shown in Figure 14.2.

2. *Gravitational fields* are defined by **Newton's Law of Gravitation,** which states that the force of attraction exerted on a particle of mass m_1 located at (x, y, z) by a particle of mass m_2 located at $(0, 0, 0)$ is given by

$$\mathbf{F}(x, y, z) = \frac{-Gm_1m_2}{x^2 + y^2 + z^2}\mathbf{u}$$

where G is the gravitational constant and $\mathbf{u}$ is the unit vector in the direction from the origin to (x, y, z). In Figure 14.3, you can see that the gravitational field $\mathbf{F}$ has the properties that $\mathbf{F}(x, y, z)$ always points toward the origin, and that the magnitude of $\mathbf{F}(x, y, z)$ is the same at all points equidistant from the origin. A vector field with these two properties is called a **central force field.** Using the position vector

$$\mathbf{r} = x\mathbf{i} + y\mathbf{j} + z\mathbf{k}$$

for the point (x, y, z), you can express the gravitational field $\mathbf{F}$ as

$$\mathbf{F}(x, y, z) = \frac{-Gm_1m_2}{\|\mathbf{r}\|^2}\left(\frac{\mathbf{r}}{\|\mathbf{r}\|}\right)$$
$$= \frac{-Gm_1m_2}{\|\mathbf{r}\|^2}\mathbf{u}.$$

3. *Electric force fields* are defined by **Coulomb's Law,** which states that the force exerted on a particle with electric charge q_1 located at (x, y, z) by a particle with electric charge q_2 located at $(0, 0, 0)$ is given by

$$\mathbf{F}(x, y, z) = \frac{cq_1q_2}{\|\mathbf{r}\|^2}\mathbf{u}$$

where $\mathbf{r} = x\mathbf{i} + y\mathbf{j} + z\mathbf{k}$, $\mathbf{u} = \mathbf{r}/\|\mathbf{r}\|$, and c is a constant that depends on the choice of units for $\|\mathbf{r}\|$, q_1, and q_2.

Note that an electric force field has the same form as a gravitational field. That is,

$$\mathbf{F}(x, y, z) = \frac{k}{\|\mathbf{r}\|^2}\mathbf{u}.$$

Such a force field is called an **inverse square field.**

Definition of Inverse Square Field

Let $\mathbf{r}(t) = x(t)\mathbf{i} + y(t)\mathbf{j} + z(t)\mathbf{k}$ be a position vector. The vector field $\mathbf{F}$ is an **inverse square field** if

$$\mathbf{F}(x, y, z) = \frac{k}{\|\mathbf{r}\|^2}\mathbf{u}$$

where k is a real number and $\mathbf{u} = \mathbf{r}/\|\mathbf{r}\|$ is a unit vector in the direction of $\mathbf{r}$.

Velocity field

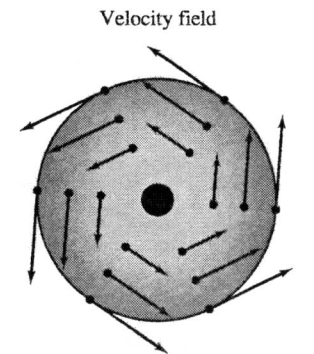

Rotating wheel
Figure 14.1

Air flow vector field
Figure 14.2

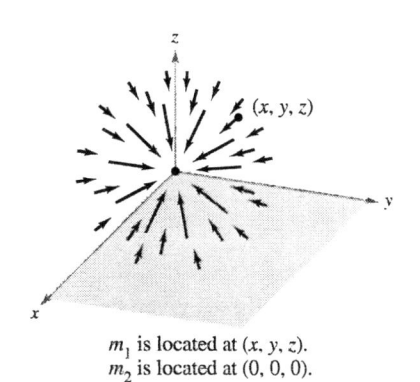

m_1 is located at (x, y, z).
m_2 is located at $(0, 0, 0)$.

Gravitational force field
Figure 14.3

Because vector fields consist of infinitely many vectors, it is not possible to actually create a sketch of the field. Instead, when you sketch a vector field, your goal is to sketch representative vectors that help you visualize the field.

Example 1 Sketching a Vector Field

Sketch some vectors in the vector field given by

$$\mathbf{F}(x, y) = -y\mathbf{i} + x\mathbf{j}.$$

Solution You could plot vectors at several random points in the plane. However, it is more enlightening to plot vectors of equal magnitude. This corresponds to finding level curves in scalar fields. In this case, vectors of equal magnitude lie on circles.

$$\|\mathbf{F}\| = c \qquad \text{Vectors of length } c$$
$$\sqrt{x^2 + y^2} = c$$
$$x^2 + y^2 = c^2 \qquad \text{Equation of circle}$$

To begin making the sketch, choose a value for c and plot several vectors on the resulting circle. For instance, the following vectors occur on the unit circle.

Point	Vector
$(1, 0)$	$\mathbf{F}(1, 0) = \mathbf{j}$
$(0, 1)$	$\mathbf{F}(0, 1) = -\mathbf{i}$
$(-1, 0)$	$\mathbf{F}(-1, 0) = -\mathbf{j}$
$(0, -1)$	$\mathbf{F}(0, -1) = \mathbf{i}$

These and several other vectors in the vector field are shown in Figure 14.4. Note in the figure that this vector field is similar to that given by the rotating wheel shown in Figure 14.1.

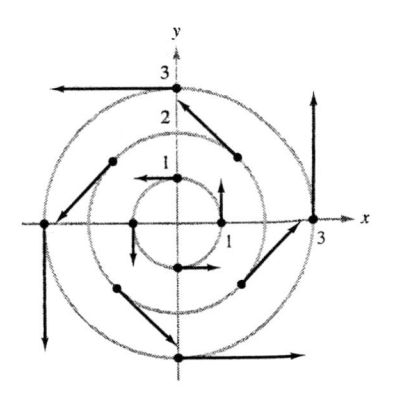

Vector field:
$\mathbf{F}(x, y) = -y\mathbf{i} + x\mathbf{j}$

Figure 14.4

Example 2 Sketching a Vector Field

Sketch some vectors in the vector field given by

$$\mathbf{F}(x, y) = 2x\mathbf{i} + y\mathbf{j}.$$

Solution For this vector field, vectors of equal length lie on ellipses given by

$$\|\mathbf{F}\| = \sqrt{(2x)^2 + (y)^2} = c$$

which implies that

$$4x^2 + y^2 = c^2.$$

For $c = 1$, sketch several vectors $2x\mathbf{i} + y\mathbf{j}$ of magnitude 1 at points on the ellipse given by

$$4x^2 + y^2 = 1.$$

For $c = 2$, sketch several vectors $2x\mathbf{i} + y\mathbf{j}$ of magnitude 2 at points on the ellipse given by

$$4x^2 + y^2 = 4.$$

These vectors are shown in Figure 14.5.

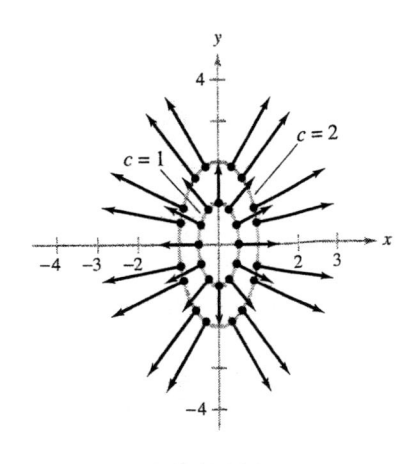

Vector field:
$\mathbf{F}(x, y) = 2x\mathbf{i} + y\mathbf{j}$

Figure 14.5

Example 3 Sketching a Velocity Field

Sketch some vectors in the velocity field given by
$$\mathbf{v}(x, y, z) = (16 - x^2 - y^2)\mathbf{k}$$
where $x^2 + y^2 \leq 16$.

Solution You can imagine that $\mathbf{v}$ describes the velocity of a liquid flowing through a tube of radius 4. Vectors near the z-axis are longer than those near the edge of the tube. For instance, at the point $(0, 0, 0)$, the velocity vector is $\mathbf{v}(0, 0, 0) = 16\mathbf{k}$, whereas at the point $(0, 3, 0)$, the velocity vector is $\mathbf{v}(0, 3, 0) = 7\mathbf{k}$. Figure 14.6 shows these and several other vectors for the velocity field. From the figure, you can see that the speed of the liquid is greater near the center of the tube than near the edges of the tube.

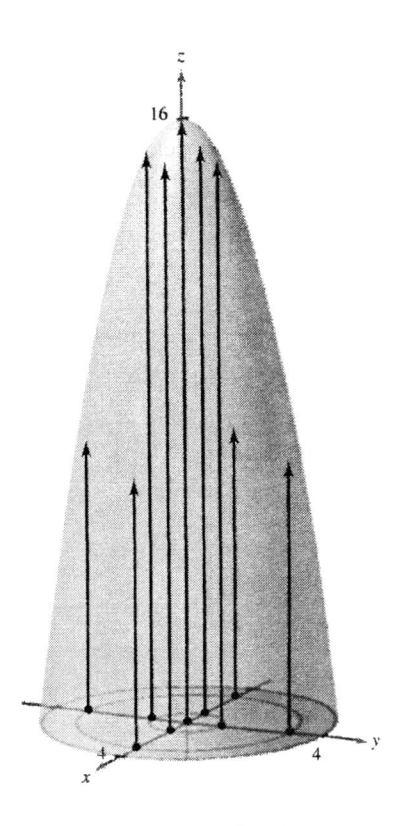

Velocity field:
$\mathbf{v}(x, y, z) = (16 - x^2 - y^2)\mathbf{k}$

Figure 14.6

Conservative Vector Fields

Notice in Figure 14.5 that all the vectors appear to be normal to the level curve from which they emanate. Because this is a property of gradients, it is natural to ask whether the vector field given by $\mathbf{F}(x, y) = 2x\mathbf{i} + y\mathbf{j}$ is the *gradient* for some differentiable function f. The answer is that some vector fields can be represented as the gradients of differentiable functions and some cannot—those that can are called **conservative** vector fields.

Definition of Conservative Vector Field

A vector field $\mathbf{F}$ is called **conservative** if there exists a differentiable function f such that $\mathbf{F} = \nabla f$. The function f is called the **potential function** for $\mathbf{F}$.

Example 4 Conservative Vector Fields

a. The vector field given by $\mathbf{F}(x, y) = 2x\mathbf{i} + y\mathbf{j}$ is conservative. To see this, consider the potential function $f(x, y) = x^2 + \frac{1}{2}y^2$. Because
$$\nabla f = 2x\mathbf{i} + y\mathbf{j} = \mathbf{F}$$
it follows that $\mathbf{F}$ is conservative.

b. Every inverse square field is conservative. To see this, let
$$\mathbf{F}(x, y, z) = \frac{k}{\|\mathbf{r}\|^2}\mathbf{u} \quad \text{and} \quad f(x, y, z) = \frac{-k}{\sqrt{x^2 + y^2 + z^2}}$$
where $\mathbf{u} = \mathbf{r}/\|\mathbf{r}\|$. Because
$$\nabla f = \frac{kx}{(x^2 + y^2 + z^2)^{3/2}}\mathbf{i} + \frac{ky}{(x^2 + y^2 + z^2)^{3/2}}\mathbf{j} + \frac{kz}{(x^2 + y^2 + z^2)^{3/2}}\mathbf{k}$$
$$= \frac{k}{x^2 + y^2 + z^2}\left(\frac{x\mathbf{i} + y\mathbf{j} + z\mathbf{k}}{\sqrt{x^2 + y^2 + z^2}}\right)$$
$$= \frac{k}{\|\mathbf{r}\|^2}\frac{\mathbf{r}}{\|\mathbf{r}\|}$$
$$= \frac{k}{\|\mathbf{r}\|^2}\mathbf{u}$$
it follows that $\mathbf{F}$ is conservative.

As can be seen in Example 4b, many important vector fields, including gravitational fields, magnetic fields, and electric force fields, are conservative. Most of the terminology in this chapter comes from physics. For example, the term "conservative" is derived from the classic physical law regarding the conservation of energy. This law states that the sum of the kinetic energy and the potential energy of a particle moving in a conservative force field is constant. (The kinetic energy of a particle is the energy due to its motion, and the potential energy is the energy due to its position in the force field.)

The following important theorem gives a necessary and sufficient condition for a vector field *in the plane* to be conservative.

THEOREM 14.1   Test for Conservative Vector Field in the Plane

Let M and N have continuous first partial derivatives on an open disk R. The vector field given by $\mathbf{F}(x, y) = M\mathbf{i} + N\mathbf{j}$ is conservative if and only if

$$\frac{\partial N}{\partial x} = \frac{\partial M}{\partial y}.$$

Proof   To prove that the given condition is necessary for $\mathbf{F}$ to be conservative, suppose there exists a potential function f such that

$$\mathbf{F}(x, y) = \nabla f(x, y) = M\mathbf{i} + N\mathbf{j}.$$

Then you have

$$f_x(x, y) = M \quad \Longrightarrow \quad f_{xy}(x, y) = \frac{\partial M}{\partial y}$$

$$f_y(x, y) = N \quad \Longrightarrow \quad f_{yx}(x, y) = \frac{\partial N}{\partial x}$$

and, by the equivalence of the mixed partials f_{xy} and f_{yx}, you can conclude that $\partial N/\partial x = \partial M/\partial y$ for all (x, y) in R. The sufficiency of the condition is proved in Section 14.4.

NOTE   Theorem 14.1 requires that the domain of $\mathbf{F}$ be an open disk. If R is simply an open region, the given condition is necessary but not sufficient to produce a conservative vector field.

Example 5   Testing for Conservative Vector Fields in the Plane

Decide whether the vector field given by $\mathbf{F}$ is conservative.

a. $\mathbf{F}(x, y) = x^2 y\mathbf{i} + xy\mathbf{j}$ **b.** $\mathbf{F}(x, y) = 2x\mathbf{i} + y\mathbf{j}$

Solution

a. The vector field given by $\mathbf{F}(x, y) = x^2 y\mathbf{i} + xy\mathbf{j}$ *is not* conservative because

$$\frac{\partial M}{\partial y} = \frac{\partial}{\partial y}[x^2 y] = x^2 \quad \text{and} \quad \frac{\partial N}{\partial x} = \frac{\partial}{\partial x}[xy] = y.$$

b. The vector field given by $\mathbf{F}(x, y) = 2x\mathbf{i} + y\mathbf{j}$ *is* conservative because

$$\frac{\partial M}{\partial y} = \frac{\partial}{\partial y}[2x] = 0 \quad \text{and} \quad \frac{\partial N}{\partial x} = \frac{\partial}{\partial x}[y] = 0.$$

Theorem 14.1 tells you whether a vector field is conservative. It does not tell you how to find a potential function of **F**. The problem is comparable to antidifferentiation. Sometimes you will be able to find a potential function by simple inspection. For instance, in Example 4 you observed that

$$f(x, y) = x^2 + \frac{1}{2}y^2$$

has the property that $\nabla f(x, y) = 2x\mathbf{i} + y\mathbf{j}$.

Example 6 Finding a Potential Function for F(x, y)

Find a potential function for

$$\mathbf{F}(x, y) = 2xy\mathbf{i} + (x^2 - y)\mathbf{j}.$$

Solution From Theorem 14.1 it follows that **F** is conservative because

$$\frac{\partial}{\partial y}[2xy] = 2x \quad \text{and} \quad \frac{\partial}{\partial x}[x^2 - y] = 2x.$$

If f is a function whose gradient is equal to $\mathbf{F}(x, y)$, then

$$\nabla f(x, y) = 2xy\mathbf{i} + (x^2 - y)\mathbf{j}$$

which implies that

$$f_x(x, y) = 2xy$$

and

$$f_y(x, y) = x^2 - y.$$

To reconstruct the function f from these two partial derivatives, integrate $f_x(x, y)$ with respect to x and $f_y(x, y)$ with respect to y, as follows.

$$f(x, y) = \int f_x(x, y)\, dx = \int 2xy\, dx = x^2y + g(y)$$

$$f(x, y) = \int f_y(x, y)\, dy = \int (x^2 - y)\, dy = x^2y - \frac{y^2}{2} + h(x)$$

Notice that $g(y)$ is constant with respect to x and $h(x)$ is constant with respect to y. To find a single expression that represents $f(x, y)$, let

$$g(y) = -\frac{1}{2}y^2 \quad \text{and} \quad h(x) = K.$$

Then, you can write

$$f(x, y) = x^2y + g(y) + K$$

$$= x^2y - \frac{y^2}{2} + K.$$

You can check this result by forming the gradient of f—it should be equal to the original function **F**.

NOTE Notice that the solution in Example 6 is comparable to that given by an indefinite integral. That is, the solution represents a family of potential functions, any two of which differ by a constant. To find a unique solution, you would have to be given an initial condition satisfied by the potential function.

Curl of a Vector Field

Theorem 14.1 has a counterpart for vector fields in space. Before stating that result, we define the **curl of a vector field** in space.

Definition of Curl of a Vector Field

The **curl** of $\mathbf{F}(x, y, z) = M\mathbf{i} + N\mathbf{j} + P\mathbf{k}$ is

$$\operatorname{curl} \mathbf{F}(x, y, z) = \nabla \times \mathbf{F}(x, y, z)$$
$$= \left(\frac{\partial P}{\partial y} - \frac{\partial N}{\partial z}\right)\mathbf{i} - \left(\frac{\partial P}{\partial x} - \frac{\partial M}{\partial z}\right)\mathbf{j} + \left(\frac{\partial N}{\partial x} - \frac{\partial M}{\partial y}\right)\mathbf{k}.$$

NOTE If $\operatorname{curl} \mathbf{F} = \mathbf{0}$, we say that $\mathbf{F}$ is **irrotational**.

The cross product notation used for curl comes from viewing the gradient ∇f as the result of the **differential operator** ∇ acting on the function f. In this context, you can use the following determinant form as an aid in remembering the formula for curl.

$$\operatorname{curl} \mathbf{F}(x, y, z) = \nabla \times \mathbf{F}(x, y, z)$$
$$= \begin{vmatrix} \mathbf{i} & \mathbf{j} & \mathbf{k} \\ \dfrac{\partial}{\partial x} & \dfrac{\partial}{\partial y} & \dfrac{\partial}{\partial z} \\ M & N & P \end{vmatrix}$$
$$= \left(\frac{\partial P}{\partial y} - \frac{\partial N}{\partial z}\right)\mathbf{i} - \left(\frac{\partial P}{\partial x} - \frac{\partial M}{\partial z}\right)\mathbf{j} + \left(\frac{\partial N}{\partial x} - \frac{\partial M}{\partial y}\right)\mathbf{k}$$

Example 7 Finding the Curl of a Vector Field

Find $\operatorname{curl} \mathbf{F}$ for the vector field given by

$$\mathbf{F}(x, y, z) = 2xy\mathbf{i} + (x^2 + z^2)\mathbf{j} + 2zy\mathbf{k}.$$

Is $\mathbf{F}$ irrotational?

Solution The curl of $\mathbf{F}$ is given by

$$\operatorname{curl} \mathbf{F}(x, y, z) = \nabla \times \mathbf{F}(x, y, z)$$
$$= \begin{vmatrix} \mathbf{i} & \mathbf{j} & \mathbf{k} \\ \dfrac{\partial}{\partial x} & \dfrac{\partial}{\partial y} & \dfrac{\partial}{\partial z} \\ 2xy & x^2 + z^2 & 2zy \end{vmatrix}$$
$$= \begin{vmatrix} \dfrac{\partial}{\partial y} & \dfrac{\partial}{\partial z} \\ x^2 + z^2 & 2zy \end{vmatrix}\mathbf{i} - \begin{vmatrix} \dfrac{\partial}{\partial x} & \dfrac{\partial}{\partial z} \\ 2xy & 2zy \end{vmatrix}\mathbf{j} + \begin{vmatrix} \dfrac{\partial}{\partial x} & \dfrac{\partial}{\partial y} \\ 2xy & x^2 + z^2 \end{vmatrix}\mathbf{k}$$
$$= (2z - 2z)\mathbf{i} - (0 - 0)\mathbf{j} + (2x - 2x)\mathbf{k}$$
$$= \mathbf{0}.$$

Because $\operatorname{curl} \mathbf{F} = \mathbf{0}$, $\mathbf{F}$ is irrotational.

---

indicates that in the Interactive 3.0 *CD-ROM and* Internet 3.0 *versions of this text (available at* college.hmco.com) *you will find an Open Exploration, which further explores this example using the computer algebra systems* Maple, Mathcad, Mathematica, *and* Derive.

Later in this chapter, we will assign a physical interpretation to the curl of a vector field. But for now, the primary use we make of curl is in the following test for conservative vector fields in space. The test states that for a vector field whose domain is all of three-dimensional space (or an open sphere), the curl is **0** at every point in the domain if and only if **F** is conservative. The proof is similar to that given for Theorem 14.1.

THEOREM 14.2 Test for Conservative Vector Field in Space

Suppose that M, N, and P have continuous first partial derivatives in an open sphere Q in space. The vector field given by $\mathbf{F}(x, y, z) = M\mathbf{i} + N\mathbf{j} + P\mathbf{k}$ is conservative if and only if

curl $\mathbf{F}(x, y, z) = \mathbf{0}$.

That is, **F** is conservative if and only if

$$\frac{\partial P}{\partial y} = \frac{\partial N}{\partial z}, \quad \frac{\partial P}{\partial x} = \frac{\partial M}{\partial z}, \quad \text{and} \quad \frac{\partial N}{\partial x} = \frac{\partial M}{\partial y}.$$

From Theorem 14.2, you can see that the vector field given in Example 7 is conservative because **curl** $\mathbf{F}(x, y, z) = \mathbf{0}$. Try showing that the vector field

$$\mathbf{F}(x, y, z) = x^3 y^2 z \mathbf{i} + x^2 z \mathbf{j} + x^2 y \mathbf{k}$$

is not conservative—you can do this by showing that its curl is

curl $\mathbf{F}(x, y, z) = (x^3 y^2 - 2xy)\mathbf{j} + (2xz - 2x^3 yz)\mathbf{k} \neq \mathbf{0}.$

For vector fields in space that pass the test for being conservative, you can find a potential function by following the same pattern used in the plane (as demonstrated in Example 6).

Example 8 Finding a Potential Function for $\mathbf{F}(x, y, z)$

NOTE Examples 6 and 8 are illustrations of a type of problem called *recovering a function from its gradient*. If you go on to take a course in differential equations, you will study other methods for solving this type of problem. One popular method gives an interplay between successive "partial integrations" and partial differentiations.

Find a potential function for $\mathbf{F}(x, y, z) = 2xy\mathbf{i} + (x^2 + z^2)\mathbf{j} + 2zy\mathbf{k}$.

Solution From Example 7, you know that the vector field given by **F** is conservative. If f is a function such that $\mathbf{F}(x, y, z) = \nabla f(x, y, z)$, then

$$f_x(x, y, z) = 2xy, \quad f_y(x, y, z) = x^2 + z^2, \quad \text{and} \quad f_z(x, y, z) = 2zy$$

and integrating with respect to x, y, and z separately produces

$$f(x, y, z) = \int M\, dx = \int 2xy\, dx = x^2 y + g(y, z)$$

$$f(x, y, z) = \int N\, dy = \int (x^2 + z^2)\, dy = x^2 y + z^2 y + h(x, z)$$

$$f(x, y, z) = \int P\, dz = \int 2zy\, dz = z^2 y + k(x, y).$$

Comparing these three versions of $f(x, y, z)$, you can conclude that

$$g(y, z) = z^2 y + K, \quad h(x, z) = K, \quad \text{and} \quad k(x, y) = x^2 y + K.$$

Therefore, $f(x, y, z)$ is given by

$$f(x, y, z) = x^2 y + z^2 y + K.$$

Divergence of a Vector Field

You have seen that the curl of a vector field $\mathbf{F}$ is itself a vector field. Another important function defined on a vector field is **divergence,** which is a scalar function.

NOTE Divergence can be viewed as a type of derivative of $\mathbf{F}$ in that, for vector fields representing velocities of moving particles, the divergence measures the rate of particle flow per unit volume at a point. In hydrodynamics (the study of fluid motion), a velocity field that is divergence free is called **incompressible.** In the study of electricity and magnetism, a vector field that is divergence free is called **solenoidal.**

Definition of Divergence of a Vector Field

The **divergence** of $\mathbf{F}(x, y) = M\mathbf{i} + N\mathbf{j}$ is

$$\text{div } \mathbf{F}(x, y) = \nabla \cdot \mathbf{F}(x, y) = \frac{\partial M}{\partial x} + \frac{\partial N}{\partial y}. \qquad \text{Plane}$$

The **divergence** of $\mathbf{F}(x, y, z) = M\mathbf{i} + N\mathbf{j} + P\mathbf{k}$ is

$$\text{div } \mathbf{F}(x, y, z) = \nabla \cdot \mathbf{F}(x, y, z) = \frac{\partial M}{\partial x} + \frac{\partial N}{\partial y} + \frac{\partial P}{\partial z}. \qquad \text{Space}$$

If div $\mathbf{F} = 0$, then $\mathbf{F}$ is said to be **divergence free.**

The dot product notation used for divergence comes from considering ∇ as a **differential operator,** as follows.

$$\nabla \cdot \mathbf{F}(x, y, z) = \left[\left(\frac{\partial}{\partial x}\right)\mathbf{i} + \left(\frac{\partial}{\partial y}\right)\mathbf{j} + \left(\frac{\partial}{\partial z}\right)\mathbf{k}\right] \cdot (M\mathbf{i} + N\mathbf{j} + P\mathbf{k})$$

$$= \frac{\partial M}{\partial x} + \frac{\partial N}{\partial y} + \frac{\partial P}{\partial z}$$

Example 9 **Finding the Divergence of a Vector Field**

Find the divergence at $(2, 1, -1)$ for the vector field

$$\mathbf{F}(x, y, z) = x^3 y^2 z \mathbf{i} + x^2 z \mathbf{j} + x^2 y \mathbf{k}.$$

Solution The divergence of $\mathbf{F}$ is

$$\text{div } \mathbf{F}(x, y, z) = \frac{\partial}{\partial x}[x^3 y^2 z] + \frac{\partial}{\partial y}[x^2 z] + \frac{\partial}{\partial z}[x^2 y] = 3x^2 y^2 z.$$

At the point $(2, 1, -1)$, the divergence is

$$\text{div } \mathbf{F}(2, 1, -1) = 3(2^2)(1^2)(-1) = -12.$$

There are many important properties of the divergence and curl of a vector field $\mathbf{F}$ (see Exercises 77–83). One that is used often is described in Theorem 14.3. You are asked to prove this theorem in Exercise 84.

THEOREM 14.3   Relationship Between Divergence and Curl

If $\mathbf{F}(x, y, z) = M\mathbf{i} + N\mathbf{j} + P\mathbf{k}$ is a vector field and M, N, and P have continuous second partial derivatives, then

$$\text{div }(\text{\bf curl } \mathbf{F}) = 0.$$

EXERCISES FOR SECTION 14.1

In Exercises 1–6, match the vector field with its graph. [The graphs are labeled (a), (b), (c), (d), (e), and (f).]

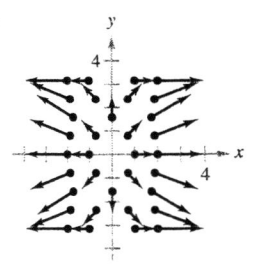

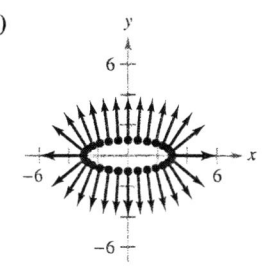

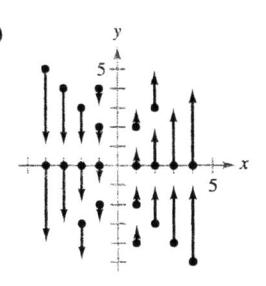

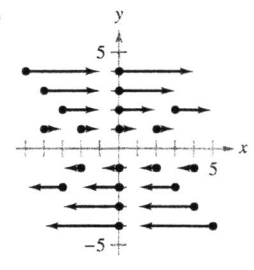

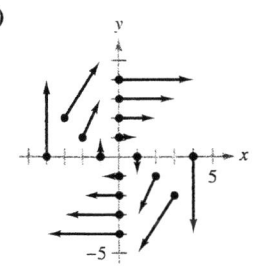

(a)

(b)

(c)

(d)

(e)

(f)

1. $\mathbf{F}(x, y) = x\mathbf{j}$
2. $\mathbf{F}(x, y) = y\mathbf{i}$
3. $\mathbf{F}(x, y) = x\mathbf{i} + 3y\mathbf{j}$
4. $\mathbf{F}(x, y) = y\mathbf{i} - x\mathbf{j}$
5. $\mathbf{F}(x, y) = \langle x, \sin y \rangle$
6. $\mathbf{F}(x, y) = \langle \frac{1}{2}xy, \frac{1}{4}x^2 \rangle$

In Exercises 7–16, sketch several representative vectors in the vector field.

7. $\mathbf{F}(x, y) = \mathbf{i} + \mathbf{j}$
8. $\mathbf{F}(x, y) = 2\mathbf{i}$
9. $\mathbf{F}(x, y) = x\mathbf{i} + y\mathbf{j}$
10. $\mathbf{F}(x, y) = x\mathbf{i} - y\mathbf{j}$
11. $\mathbf{F}(x, y, z) = 3y\mathbf{j}$
12. $\mathbf{F}(x, y) = x\mathbf{i}$
13. $\mathbf{F}(x, y) = 4x\mathbf{i} + y\mathbf{j}$
14. $\mathbf{F}(x, y) = (x^2 + y^2)\mathbf{i} + \mathbf{j}$
15. $\mathbf{F}(x, y, z) = \mathbf{i} + \mathbf{j} + \mathbf{k}$
16. $\mathbf{F}(x, y, z) = x\mathbf{i} + y\mathbf{j} + z\mathbf{k}$

In Exercises 17–20, use a computer algebra system to graph several representative vectors in the vector field.

17. $\mathbf{F}(x, y) = \frac{1}{8}(2xy\mathbf{i} + y^2\mathbf{j})$
18. $\mathbf{F}(x, y) = (2y - 3x)\mathbf{i} + (2y + 3x)\mathbf{j}$
19. $\mathbf{F}(x, y, z) = \dfrac{x\mathbf{i} + y\mathbf{j} + z\mathbf{k}}{\sqrt{x^2 + y^2 + z^2}}$
20. $\mathbf{F}(x, y, z) = x\mathbf{i} - y\mathbf{j} + z\mathbf{k}$

In Exercises 21–26, find the gradient vector field for the scalar function. (That is, find the conservative vector field for the potential function.)

21. $f(x, y) = 5x^2 + 3xy + 10y^2$
22. $f(x, y) = \sin 3x \cos 4y$
23. $f(x, y, z) = z - ye^{x^2}$
24. $f(x, y, z) = \dfrac{y}{z} + \dfrac{z}{x} - \dfrac{xz}{y}$
25. $g(x, y, z) = xy \ln(x + y)$
26. $g(x, y, z) = x \arcsin yz$

In Exercises 27–30, verify that the vector field is conservative.

27. $\mathbf{F}(x, y) = 12xy\mathbf{i} + 6(x^2 + y)\mathbf{j}$
28. $\mathbf{F}(x, y) = \dfrac{1}{x^2}(y\mathbf{i} - x\mathbf{j})$
29. $\mathbf{F}(x, y) = (\sin y)\mathbf{i} + x(\cos y)\mathbf{j}$
30. $\mathbf{F}(x, y) = \dfrac{1}{xy}(y\mathbf{i} - x\mathbf{j})$

In Exercises 31–34, determine if the vector field is conservative.

31. $\mathbf{F}(x, y) = 5y^2(3y\mathbf{i} - x\mathbf{j})$
32. $\mathbf{F}(x, y) = \dfrac{1}{\sqrt{x^2 + y^2}}(x\mathbf{i} + y\mathbf{j})$
33. $\mathbf{F}(x, y) = \dfrac{2}{y^2}e^{2x/y}(y\mathbf{i} - x\mathbf{j})$
34. $\mathbf{F}(x, y) = \dfrac{1}{\sqrt{1 - x^2y^2}}(y\mathbf{i} - x\mathbf{j})$

In Exercises 35–42, determine whether the vector field is conservative. If it is, find a potential function for the vector field.

35. $\mathbf{F}(x, y) = 2xy\mathbf{i} + x^2\mathbf{j}$
36. $\mathbf{F}(x, y) = \dfrac{1}{y^2}(y\mathbf{i} - 2x\mathbf{j})$
37. $\mathbf{F}(x, y) = xe^{x^2y}(2y\mathbf{i} + x\mathbf{j})$
38. $\mathbf{F}(x, y) = 3x^2y^2\mathbf{i} + 2x^3y\mathbf{j}$
39. $\mathbf{F}(x, y) = \dfrac{x\mathbf{i} + y\mathbf{j}}{x^2 + y^2}$
40. $\mathbf{F}(x, y) = \dfrac{2y}{x}\mathbf{i} - \dfrac{x^2}{y^2}\mathbf{j}$
41. $\mathbf{F}(x, y) = e^x(\cos y\mathbf{i} + \sin y\mathbf{j})$
42. $\mathbf{F}(x, y) = \dfrac{2x\mathbf{i} + 2y\mathbf{j}}{(x^2 + y^2)^2}$

In Exercises 43–46, find the curl of F at the indicated point.

Vector Field	Point
43. $\mathbf{F}(x, y, z) = xyz\mathbf{i} + y\mathbf{j} + z\mathbf{k}$	$(1, 2, 1)$
44. $\mathbf{F}(x, y, z) = x^2z\mathbf{i} - 2xz\mathbf{j} + yz\mathbf{k}$	$(2, -1, 3)$
45. $\mathbf{F}(x, y, z) = e^x \sin y\mathbf{i} - e^x \cos y\mathbf{j}$	$(0, 0, 3)$
46. $\mathbf{F}(x, y, z) = e^{-xyz}(\mathbf{i} + \mathbf{j} + \mathbf{k})$	$(3, 2, 0)$

In Exercises 47–50, use a computer algebra system to find the curl of the vector field F.

47. $\mathbf{F}(x, y, z) = \arctan \dfrac{x}{y}\mathbf{i} + \ln\sqrt{x^2 + y^2}\mathbf{j} + \mathbf{k}$
48. $\mathbf{F}(x, y, z) = \dfrac{yz}{y - z}\mathbf{i} + \dfrac{xz}{x - z}\mathbf{j} + \dfrac{xy}{x - y}\mathbf{k}$
49. $\mathbf{F}(x, y, z) = \sin(x - y)\mathbf{i} + \sin(y - z)\mathbf{j} + \sin(z - x)\mathbf{k}$
50. $\mathbf{F}(x, y, z) = \sqrt{x^2 + y^2 + z^2}(\mathbf{i} + \mathbf{j} + \mathbf{k})$

In Exercises 51–56, determine whether the vector field F is conservative. If it is, find a potential function for the vector field.

51. $\mathbf{F}(x, y, z) = \sin y \mathbf{i} - x \cos y \mathbf{j} + \mathbf{k}$
52. $\mathbf{F}(x, y, z) = e^z(y\mathbf{i} + x\mathbf{j} + \mathbf{k})$
53. $\mathbf{F}(x, y, z) = e^z(y\mathbf{i} + x\mathbf{j} + xy\mathbf{k})$
54. $\mathbf{F}(x, y, z) = y^2z^3\mathbf{i} + 2xyz^3\mathbf{j} + 3xy^2z^2\mathbf{k}$
55. $\mathbf{F}(x, y, z) = \dfrac{1}{y}\mathbf{i} - \dfrac{x}{y^2}\mathbf{j} + (2z - 1)\mathbf{k}$
56. $\mathbf{F}(x, y, z) = \dfrac{x}{x^2 + y^2}\mathbf{i} + \dfrac{y}{x^2 + y^2}\mathbf{j} + \mathbf{k}$

In Exercises 57–60, find the divergence of the vector field F.

57. $\mathbf{F}(x, y, z) = 6x^2\mathbf{i} - xy^2\mathbf{j}$
58. $\mathbf{F}(x, y, z) = xe^x\mathbf{i} + ye^y\mathbf{j}$
59. $\mathbf{F}(x, y, z) = \sin x \mathbf{i} + \cos y \mathbf{j} + z^2\mathbf{k}$
60. $\mathbf{F}(x, y, z) = \ln(x^2 + y^2)\mathbf{i} + xy\mathbf{j} + \ln(y^2 + z^2)\mathbf{k}$

In Exercises 61–64, find the divergence of the vector field F at the indicated point.

Vector Field	Point
61. $\mathbf{F}(x, y, z) = xyz\mathbf{i} + y\mathbf{j} + z\mathbf{k}$	$(1, 2, 1)$
62. $\mathbf{F}(x, y, z) = x^2z\mathbf{i} - 2xz\mathbf{j} + yz\mathbf{k}$	$(2, -1, 3)$
63. $\mathbf{F}(x, y, z) = e^x \sin y \mathbf{i} - e^x \cos y \mathbf{j}$	$(0, 0, 3)$
64. $\mathbf{F}(x, y, z) = \ln(xyz)(\mathbf{i} + \mathbf{j} + \mathbf{k})$	$(3, 2, 1)$

Getting at the Concept

65. Define a vector field in the plane and in space. Give some physical examples of vector fields.
66. What is a conservative vector field and how do you test for it in the plane and in space?
67. Define the curl of a vector field.
68. Define the divergence of a vector field in the plane and in space.

In Exercises 69 and 70, find curl $(\mathbf{F} \times \mathbf{G})$.

69. $\mathbf{F}(x, y, z) = \mathbf{i} + 2x\mathbf{j} + 3y\mathbf{k}$
    $\mathbf{G}(x, y, z) = x\mathbf{i} - y\mathbf{j} + z\mathbf{k}$

70. $\mathbf{F}(x, y, z) = x\mathbf{i} - z\mathbf{k}$
    $\mathbf{G}(x, y, z) = x^2\mathbf{i} + y\mathbf{j} + z^2\mathbf{k}$

In Exercises 71 and 72, find curl (curl F) = $\nabla \times (\nabla \times \mathbf{F})$.

71. $\mathbf{F}(x, y, z) = xyz\mathbf{i} + y\mathbf{j} + z\mathbf{k}$
72. $\mathbf{F}(x, y, z) = x^2z\mathbf{i} - 2xz\mathbf{j} + yz\mathbf{k}$

In Exercises 73 and 74, find div $(\mathbf{F} \times \mathbf{G})$.

73. $\mathbf{F}(x, y, z) = \mathbf{i} + 2x\mathbf{j} + 3y\mathbf{k}$
    $\mathbf{G}(x, y, z) = x\mathbf{i} - y\mathbf{j} + z\mathbf{k}$

74. $\mathbf{F}(x, y, z) = x\mathbf{i} - z\mathbf{k}$
    $\mathbf{G}(x, y, z) = x^2\mathbf{i} + y\mathbf{j} + z^2\mathbf{k}$

In Exercises 75 and 76, find div (curl F) = $\nabla \cdot (\nabla \times \mathbf{F})$.

75. $\mathbf{F}(x, y, z) = xyz\mathbf{i} + y\mathbf{j} + z\mathbf{k}$
76. $\mathbf{F}(x, y, z) = x^2z\mathbf{i} - 2xz\mathbf{j} + yz\mathbf{k}$

In Exercises 77–84, prove the property for vector fields F and G and scalar function f. (Assume that the required partial derivatives are continuous.)

77. curl $(\mathbf{F} + \mathbf{G}) = $ curl F + curl G
78. curl$(\nabla f) = \nabla \times (\nabla f) = \mathbf{0}$
79. div$(\mathbf{F} + \mathbf{G}) = $ div F + div G
80. div$(\mathbf{F} \times \mathbf{G}) = $ (curl F) $\cdot$ G $-$ F $\cdot$ (curl G)
81. $\nabla \times [\nabla f + (\nabla \times \mathbf{F})] = \nabla \times (\nabla \times \mathbf{F})$
82. $\nabla \times (f\mathbf{F}) = f(\nabla \times \mathbf{F}) + (\nabla f) \times \mathbf{F}$
83. div$(f\mathbf{F}) = f$ div F + $\nabla f \cdot \mathbf{F}$
84. div(curl F) = 0   (Theorem 14.3)

In Exercises 85–88, let $\mathbf{F}(x, y, z) = x\mathbf{i} + y\mathbf{j} + z\mathbf{k}$, and let $f(x, y, z) = \|\mathbf{F}(x, y, z)\|$.

85. Show that $\nabla(\ln f) = \dfrac{\mathbf{F}}{f^2}$.
86. Show that $\nabla\left(\dfrac{1}{f}\right) = -\dfrac{\mathbf{F}}{f^3}$.
87. Show that $\nabla f^n = nf^{n-2}\mathbf{F}$.
88. The **Laplacian** is the differential operator

$$\nabla^2 = \nabla \cdot \nabla = \dfrac{\partial^2}{\partial x^2} + \dfrac{\partial^2}{\partial y^2} + \dfrac{\partial^2}{\partial z^2}$$

and **Laplace's equation** is

$$\nabla^2 w = \dfrac{\partial^2 w}{\partial x^2} + \dfrac{\partial^2 w}{\partial y^2} + \dfrac{\partial^2 w}{\partial z^2} = 0.$$

Any function that satisfies this equation is called **harmonic**. Show that the function $1/f$ is harmonic.

89. *Wind Speed and Direction* The vector field in the figure gives the upper air wind speeds and directions over the United States on February 24, 2000. If the vectors of greater magnitude depict upper air winds of greater speed, compare the wind speeds and directions over the cities of Phoenix, Arizona and Atlanta, Georgia.

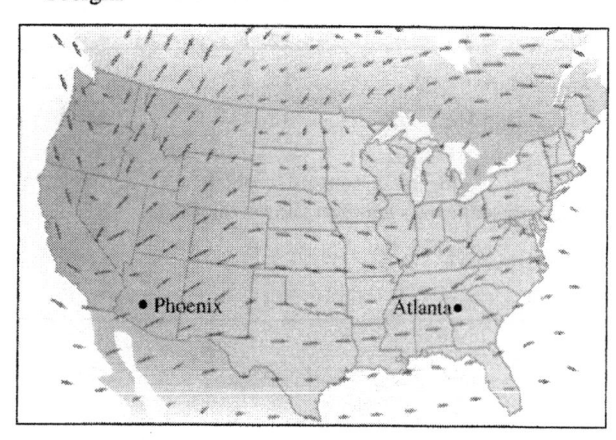

Section 14.2 Line Integrals

- Understand and use the concept of a piecewise smooth curve.
- Write and evaluate a line integral.
- Write and evaluate a line integral of a vector field.
- Write and evaluate a line integral in differential form.

Piecewise Smooth Curves

A classic property of gravitational fields is that, subject to certain physical constraints, the work done by gravity on an object moving between two points in the field is independent of the path taken by the object. One of the constraints is that the **path** must be a piecewise smooth curve. Recall that a plane curve C given by

$$\mathbf{r}(t) = x(t)\mathbf{i} + y(t)\mathbf{j}, \quad a \leq t \leq b$$

is **smooth** if

$$\frac{dx}{dt} \quad \text{and} \quad \frac{dy}{dt}$$

are continuous on $[a, b]$ and not simultaneously 0 on (a, b). Similarly, a space curve C given by

$$\mathbf{r}(t) = x(t)\mathbf{i} + y(t)\mathbf{j} + z(t)\mathbf{k}, \quad a \leq t \leq b$$

is **smooth** if

$$\frac{dx}{dt}, \quad \frac{dy}{dt}, \quad \text{and} \quad \frac{dz}{dt}$$

are continuous on $[a, b]$ and not simultaneously 0 on (a, b). A curve C is **piecewise smooth** if the interval $[a, b]$ can be partitioned into a finite number of subintervals, on each of which C is smooth.

Example 1 Finding a Piecewise Smooth Parametrization

Find a piecewise smooth parametrization of the graph of C shown in Figure 14.7.

Solution Because C consists of three line segments C_1, C_2, and C_3, you can construct a smooth parametrization for each segment and piece them together by making the last t-value in C_i correspond to the first t-value in C_{i+1}, as follows.

$$C_1: x(t) = 0, \qquad y(t) = 2t, \qquad z(t) = 0, \qquad 0 \leq t \leq 1$$
$$C_2: x(t) = t - 1, \quad y(t) = 2, \qquad z(t) = 0, \qquad 1 \leq t \leq 2$$
$$C_3: x(t) = 1, \qquad y(t) = 2, \qquad z(t) = t - 2, \quad 2 \leq t \leq 3$$

Therefore, C is given by

$$\mathbf{r}(t) = \begin{cases} 2t\mathbf{j}, & 0 \leq t \leq 1 \\ (t-1)\mathbf{i} + 2\mathbf{j}, & 1 \leq t \leq 2 \\ \mathbf{i} + 2\mathbf{j} + (t-2)\mathbf{k}, & 2 \leq t \leq 3 \end{cases}.$$

Because C_1, C_2, and C_3 are smooth, it follows that C is piecewise smooth.

Recall that parametrization of a curve induces an **orientation** to the curve. For instance, in Example 1, the curve is oriented such that the positive direction is from $(0, 0, 0)$, following the curve to $(1, 2, 1)$. Try finding a parametrization that induces the opposite orientation.

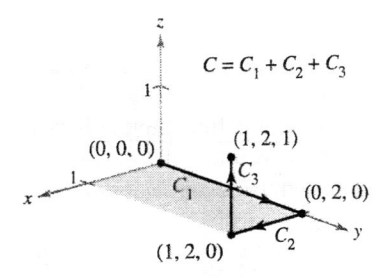

Figure 14.7

JOSIAH WILLARD GIBBS (1839–1903)

Many physicists and mathematicians have contributed to the theory and applications described in this chapter—Newton, Gauss, Laplace, Hamilton, and Maxwell, among others. However, the use of vector analysis to describe these results is attributed primarily to the American mathematical physicist Josiah Willard Gibbs.

Line Integrals

Up to this point in the text, you have studied various types of integrals. For a single integral

$$\int_a^b f(x)\, dx \qquad \text{Integrate over interval } [a, b].$$

you integrated over the interval $[a, b]$. Similarly, for a double integral

$$\iint_R f(x, y)\, dA \qquad \text{Integrate over region } R.$$

you integrated over the region R in the plane. In this section, you will study a new type of integral called a **line integral**

$$\int_C f(x, y)\, ds \qquad \text{Integrate over curve } C.$$

for which you integrate over a piecewise smooth curve C. (The terminology is somewhat unfortunate—this type of integral might be better described as a "curve integral.")

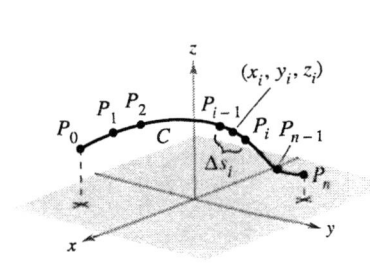

To introduce the concept of a line integral, consider the mass of a wire of finite length, given by a curve C in space. The density (mass per unit length) of the wire at the point (x, y, z) is given by $f(x, y, z)$. Partition the curve C by the points

$$P_0, P_1, \ldots, P_n$$

producing n subarcs, as shown in Figure 14.8. The length of the ith subarc is given by Δs_i. Next, choose a point (x_i, y_i, z_i) in each subarc. If the length of each subarc is small, the total mass of the wire can be approximated by the sum

$$\text{Mass of wire} \approx \sum_{i=1}^n f(x_i, y_i, z_i)\, \Delta s_i.$$

If you let $\|\Delta\|$ denote the length of the longest subarc and let $\|\Delta\|$ approach 0, it seems reasonable that the limit of this sum approaches the mass of the wire. This leads to the following definition.

Partitioning of curve C
Figure 14.8

Definition of Line Integral

If f is defined in a region containing a smooth curve C of finite length, then the **line integral of f along C** is given by

$$\int_C f(x, y)\, ds = \lim_{\|\Delta\| \to 0} \sum_{i=1}^n f(x_i, y_i)\, \Delta s_i \qquad \text{Plane}$$

or

$$\int_C f(x, y, z)\, ds = \lim_{\|\Delta\| \to 0} \sum_{i=1}^n f(x_i, y_i, z_i)\, \Delta s_i \qquad \text{Space}$$

provided this limit exists.

As with the integrals discussed in Chapter 13, evaluation of a line integral is best accomplished by converting to a definite integral. It can be shown that if f is *continuous*, the limit given above exists and is the same for all smooth parametrizations of C.

To evaluate a line integral over a plane curve C given by $\mathbf{r}(t) = x(t)\mathbf{i} + y(t)\mathbf{j}$, use the fact that

$$ds = \|\mathbf{r}'(t)\| \, dt = \sqrt{[x'(t)]^2 + [y'(t)]^2} \, dt.$$

A similar formula holds for a space curve, as indicated in the following theorem.

THEOREM 14.4 **Evaluation of a Line Integral as a Definite Integral**

Let f be continuous in a region containing a smooth curve C. If C is given by $\mathbf{r}(t) = x(t)\mathbf{i} + y(t)\mathbf{j}$, where $a \le t \le b$, then

$$\int_C f(x, y) \, ds = \int_a^b f(x(t), y(t)) \sqrt{[x'(t)]^2 + [y'(t)]^2} \, dt.$$

If C is given by $\mathbf{r}(t) = x(t)\mathbf{i} + y(t)\mathbf{j} + z(t)\mathbf{k}$, where $a \le t \le b$, then

$$\int_C f(x, y, z) \, ds = \int_a^b f(x(t), y(t), z(t)) \sqrt{[x'(t)]^2 + [y'(t)]^2 + [z'(t)]^2} \, dt.$$

Note that if $f(x, y, z) = 1$, the line integral gives the arc length of the curve C, as defined in Section 11.5. That is,

$$\int_C 1 \, ds = \int_a^b \|\mathbf{r}'(t)\| \, dt = \text{length of curve } C.$$

Example 2 **Evaluating a Line Integral**

NOTE The value of the line integral in Example 2 does not depend on the parametrization of the line segment C (any smooth parametrization will produce the same value). To convince yourself of this, try some other parametrizations, such as $x = 1 + 2t$, $y = 2 + 4t$, $z = 1 + 2t$, $-\frac{1}{2} \le t \le 0$, or $x = -t$, $y = -2t$, $z = -t$, $-1 \le t \le 0$.

Evaluate

$$\int_C (x^2 - y + 3z) \, ds$$

where C is the line segment shown in Figure 14.9.

Solution Begin by writing a parametric form of the equation of a line:

$$x = t, \quad y = 2t, \quad \text{and} \quad z = t, \quad 0 \le t \le 1.$$

Hence, $x'(t) = 1$, $y'(t) = 2$, and $z'(t) = 1$, which implies that

$$\sqrt{[x'(t)]^2 + [y'(t)]^2 + [z'(t)]^2} = \sqrt{1^2 + 2^2 + 1^2} = \sqrt{6}.$$

So, the line integral takes the following form.

$$\int_C (x^2 - y + 3z) \, ds = \int_0^1 (t^2 - 2t + 3t) \sqrt{6} \, dt$$

$$= \sqrt{6} \int_0^1 (t^2 + t) \, dt$$

$$= \sqrt{6} \left[\frac{t^3}{3} + \frac{t^2}{2} \right]_0^1$$

$$= \frac{5\sqrt{6}}{6}$$

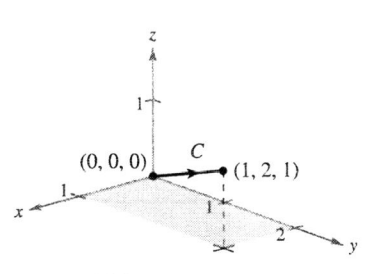

Figure 14.9

Suppose C is a path composed of smooth curves $C_1, C_2, \ldots, C_n$. If f is continuous on C, it can be shown that

$$\int_C f(x, y)\, ds = \int_{C_1} f(x, y)\, ds + \int_{C_2} f(x, y)\, ds + \cdots + \int_{C_n} f(x, y)\, ds.$$

This property is used in Example 3.

Example 3 Evaluating a Line Integral Over a Path

Evaluate $\int_C x\, ds$, where C is the piecewise smooth curve shown in Figure 14.10.

Solution Begin by integrating up the line $y = x$, using the following parametrization.

$$C_1:\ x = t,\quad y = t,\quad 0 \le t \le 1$$

For this curve, $\mathbf{r}(t) = t\mathbf{i} + t\mathbf{j}$, which implies that $x'(t) = 1$ and $y'(t) = 1$. So,

$$\sqrt{[x'(t)]^2 + [y'(t)]^2} = \sqrt{2}$$

and you have

$$\int_{C_1} x\, ds = \int_0^1 t\sqrt{2}\, dt = \frac{\sqrt{2}}{2} t^2 \Big|_0^1 = \frac{\sqrt{2}}{2}.$$

Next, integrate down the parabola $y = x^2$, using the parametrization

$$C_2:\ x = 1 - t,\quad y = (1 - t)^2,\quad 0 \le t \le 1.$$

For this curve, $\mathbf{r}(t) = (1 - t)\mathbf{i} + (1 - t)^2\mathbf{j}$, which implies that $x'(t) = -1$ and $y'(t) = -2(1 - t)$. So,

$$\sqrt{[x'(t)]^2 + [y'(t)]^2} = \sqrt{1 + 4(1 - t)^2}$$

and you have

$$\int_{C_2} x\, ds = \int_0^1 (1 - t)\sqrt{1 + 4(1 - t)^2}\, dt$$

$$= -\frac{1}{8}\left[\frac{2}{3}[1 + 4(1 - t)^2]^{3/2}\right]_0^1$$

$$= \frac{1}{12}(5^{3/2} - 1).$$

Consequently,

$$\int_C x\, ds = \int_{C_1} x\, ds + \int_{C_2} x\, ds = \frac{\sqrt{2}}{2} + \frac{1}{12}(5^{3/2} - 1) \approx 1.56.$$

For parametrizations given by $\mathbf{r}(t) = x(t)\mathbf{i} + y(t)\mathbf{j} + z(t)\mathbf{k}$, it is helpful to remember the form of ds as

$$ds = \|\mathbf{r}'(t)\|\, dt = \sqrt{[x'(t)]^2 + [y'(t)]^2 + [z'(t)]^2}\, dt.$$

This is demonstrated in Example 4.

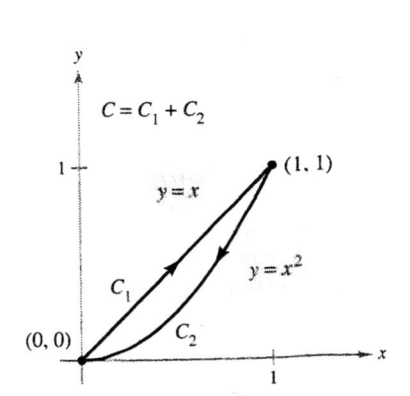

Figure 14.10

Example 4 Evaluating a Line Integral

Evaluate $\int_C (x + 2)\, ds$, where C is the curve represented by

$$\mathbf{r}(t) = t\mathbf{i} + \frac{4}{3}t^{3/2}\mathbf{j} + \frac{1}{2}t^2\mathbf{k}, \qquad 0 \le t \le 2.$$

Solution Because $\mathbf{r}'(t) = \mathbf{i} + 2t^{1/2}\mathbf{j} + t\mathbf{k}$, and

$$\|\mathbf{r}'(t)\| = \sqrt{[x'(t)]^2 + [y'(t)]^2 + [z'(t)]^2} = \sqrt{1 + 4t + t^2}$$

it follows that

$$\begin{aligned}
\int_C (x + 2)\, ds &= \int_0^2 (t + 2)\sqrt{1 + 4t + t^2}\, dt \\
&= \frac{1}{2}\int_0^2 2(t + 2)(1 + 4t + t^2)^{1/2}\, dt \\
&= \frac{1}{3}\Big[(1 + 4t + t^2)^{3/2}\Big]_0^2 \\
&= \frac{1}{3}(13\sqrt{13} - 1) \\
&\approx 15.29.
\end{aligned}$$

The next example shows how a line integral can be used to find the mass of a spring whose density varies. In Figure 14.11, note that the density of this spring increases as the spring spirals up the z-axis.

Example 5 Finding the Mass of a Spring

Find the mass of a spring in the shape of the circular helix

$$\mathbf{r}(t) = \frac{1}{\sqrt{2}}(\cos t\,\mathbf{i} + \sin t\,\mathbf{j} + t\mathbf{k}), \qquad 0 \le t \le 6\pi$$

where the density of the wire is $\rho(x, y, z) = 1 + z$. (See Figure 14.11.)

Solution Because

$$\|\mathbf{r}'(t)\| = \frac{1}{\sqrt{2}}\sqrt{(-\sin t)^2 + (\cos t)^2 + (1)^2} = 1$$

it follows that the mass of the spring is

$$\begin{aligned}
\text{Mass} &= \int_C (1 + z)\, ds = \int_0^{6\pi}\left(1 + \frac{t}{\sqrt{2}}\right) dt \\
&= \left[t + \frac{t^2}{2\sqrt{2}}\right]_0^{6\pi} \\
&= 6\pi\left(1 + \frac{3\pi}{\sqrt{2}}\right) \\
&\approx 144.47.
\end{aligned}$$

Density:
$\rho(x, y, z) = 1 + z$

$\mathbf{r}(t) = \frac{1}{\sqrt{2}}(\cos t\,\mathbf{i} + \sin t\,\mathbf{j} + t\mathbf{k})$

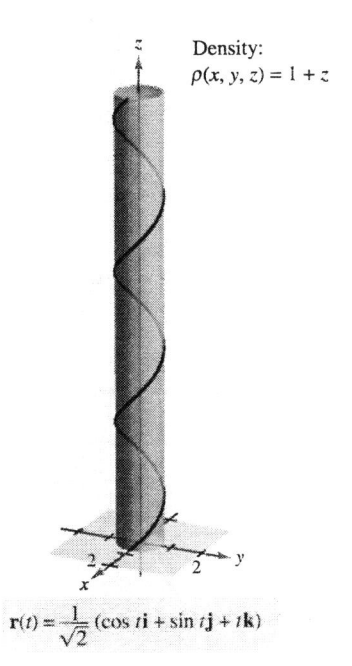

The mass of the spring is approximately 144.47.
Figure 14.11

Line Integrals of Vector Fields

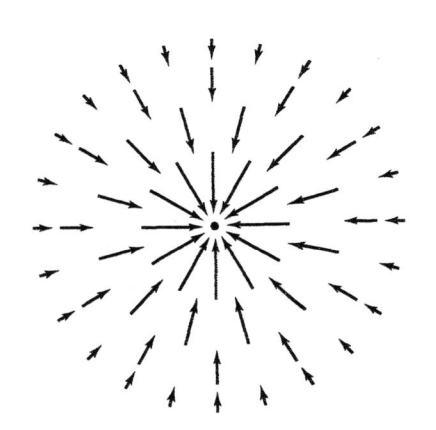

One of the most important physical applications of line integrals is that of finding the **work** done on an object moving in a force field. For example, Figure 14.12 shows an inverse square force field similar to the gravitational field of the sun. Note that the magnitude of the force along a circular path about the center is constant, whereas the magnitude of the force along a parabolic path varies from point to point.

To see how a line integral can be used to find work done in a force field **F**, consider an object moving along a path C in the field, as shown in Figure 14.13. To determine the work done by the force, you need consider only that part of the force that is acting in the same direction as that in which the object is moving (or the opposite direction). This means that at each point on C, you can consider the projection **F** · **T** of the force vector **F** onto the unit tangent vector **T**. On a small subarc of length Δs_i, the increment of work is

$$\Delta W_i = (\text{force})(\text{distance})$$
$$\approx [\mathbf{F}(x_i, y_i, z_i) \cdot \mathbf{T}(x_i, y_i, z_i)] \Delta s_i$$

where (x_i, y_i, z_i) is a point in the ith subarc. Consequently, the total work done is given by the following integral.

$$W = \int_C \mathbf{F}(x, y, z) \cdot \mathbf{T}(x, y, z) \, ds$$

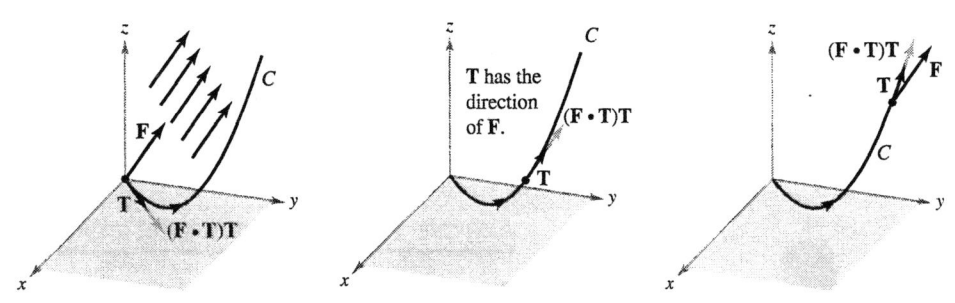

Inverse square force field **F**

Vectors along a parabolic path in the force field **F**
Figure 14.12

At each point on C, the force in the direction of motion is $(\mathbf{F} \cdot \mathbf{T})\mathbf{T}$.
Figure 14.13

This line integral appears in other contexts and is the basis of the following definition of the **line integral of a vector field.** Note in the definition that

$$\mathbf{F} \cdot \mathbf{T} \, ds = \mathbf{F} \cdot \frac{\mathbf{r}'(t)}{\|\mathbf{r}'(t)\|} \|\mathbf{r}'(t)\| \, dt$$
$$= \mathbf{F} \cdot \mathbf{r}'(t) \, dt$$
$$= \mathbf{F} \cdot d\mathbf{r}.$$

Definition of Line Integral of a Vector Field

Let **F** be a continuous vector field defined on a smooth curve C given by $\mathbf{r}(t)$, $a \le t \le b$. The **line integral** of **F** on C is given by

$$\int_C \mathbf{F} \cdot d\mathbf{r} = \int_C \mathbf{F} \cdot \mathbf{T} \, ds = \int_a^b \mathbf{F}(x(t), y(t), z(t)) \cdot \mathbf{r}'(t) \, dt.$$

Example 6 Work Done by a Force

Find the work done by the force field

$$\mathbf{F}(x, y, z) = -\frac{1}{2} x\mathbf{i} - \frac{1}{2} y\mathbf{j} + \frac{1}{4}\mathbf{k} \qquad \text{Force field } \mathbf{F}$$

on a particle as it moves along the helix given by

$$\mathbf{r}(t) = \cos t\,\mathbf{i} + \sin t\,\mathbf{j} + t\mathbf{k} \qquad \text{Space curve } C$$

from the point $(1, 0, 0)$ to $(-1, 0, 3\pi)$, as shown in Figure 14.14.

Solution Because

$$\mathbf{r}(t) = x(t)\mathbf{i} + y(t)\mathbf{j} + z(t)\mathbf{k}$$
$$= \cos t\,\mathbf{i} + \sin t\,\mathbf{j} + t\mathbf{k}$$

it follows that $x(t) = \cos t$, $y(t) = \sin t$, and $z(t) = t$. So, the force field can be written as

$$\mathbf{F}(x(t), y(t), z(t)) = -\frac{1}{2} \cos t\,\mathbf{i} - \frac{1}{2} \sin t\,\mathbf{j} + \frac{1}{4}\mathbf{k}.$$

To find the work done by the force field in moving a particle along the curve C, use the fact that

$$\mathbf{r}'(t) = -\sin t\,\mathbf{i} + \cos t\,\mathbf{j} + \mathbf{k}$$

and write the following.

$$W = \int_C \mathbf{F} \cdot d\mathbf{r}$$
$$= \int_a^b \mathbf{F}(x(t), y(t), z(t)) \cdot \mathbf{r}'(t)\, dt$$
$$= \int_0^{3\pi} \left(-\frac{1}{2} \cos t\,\mathbf{i} - \frac{1}{2} \sin t\,\mathbf{j} + \frac{1}{4}\mathbf{k}\right) \cdot (-\sin t\,\mathbf{i} + \cos t\,\mathbf{j} + \mathbf{k})\, dt$$
$$= \int_0^{3\pi} \left(\frac{1}{2} \sin t \cos t - \frac{1}{2} \sin t \cos t + \frac{1}{4}\right) dt$$
$$= \int_0^{3\pi} \frac{1}{4}\, dt$$
$$= \frac{1}{4} t \Big]_0^{3\pi}$$
$$= \frac{3\pi}{4}$$

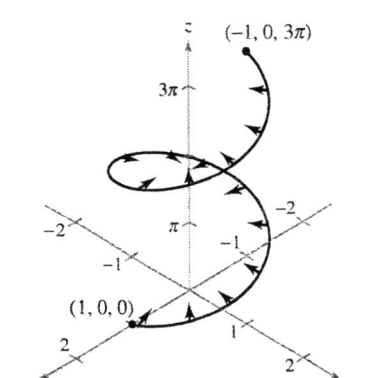

Figure 14.14

NOTE In Example 6, note that the x- and y-components of the force field end up contributing nothing to the total work. This occurs because *in this particular example* the z-component of the force field is the only portion of the force that is acting in the same (or opposite) direction in which the particle is moving (see Figure 14.15).

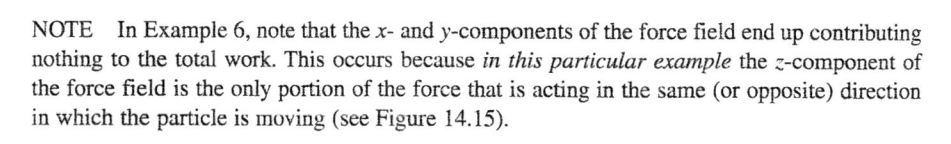

TECHNOLOGY The computer-generated view of the force field in Example 6 shown in Figure 14.15 indicates that each vector in the force field points toward the z-axis.

Generated by Mathematica

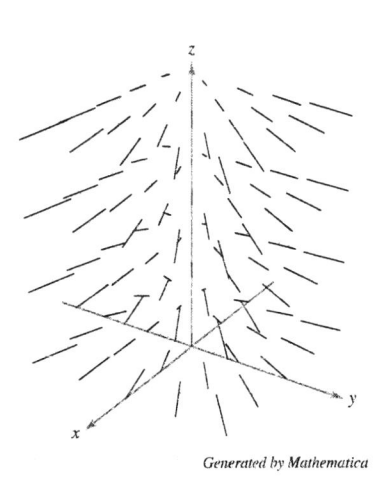

Figure 14.15

For line integrals of vector functions, the orientation of the curve C is important. If the orientation of the curve is reversed, the unit tangent vector $\mathbf{T}(t)$ is changed to $-\mathbf{T}(t)$, and you obtain

$$\int_{-C} \mathbf{F} \cdot d\mathbf{r} = -\int_{C} \mathbf{F} \cdot d\mathbf{r}.$$

Example 7 Orientation and Parametrization of a Curve

Let $\mathbf{F}(x, y) = y\mathbf{i} + x^2\mathbf{j}$ and evaluate the line integral $\int_C \mathbf{F} \cdot d\mathbf{r}$ for each parabolic curve (see Figure 14.16).

a. C_1: $\mathbf{r}_1(t) = (4 - t)\mathbf{i} + (4t - t^2)\mathbf{j}$, $\quad 0 \leq t \leq 3$
b. C_2: $\mathbf{r}_2(t) = t\mathbf{i} + (4t - t^2)\mathbf{j}$, $\quad 1 \leq t \leq 4$

Solution

a. Because $\mathbf{r}_1'(t) = -\mathbf{i} + (4 - 2t)\mathbf{j}$ and

$$\mathbf{F}(x(t), y(t)) = (4t - t^2)\mathbf{i} + (4 - t)^2\mathbf{j}$$

the line integral is

$$\int_{C_1} \mathbf{F} \cdot d\mathbf{r} = \int_0^3 [(4t - t^2)\mathbf{i} + (4 - t)^2\mathbf{j}] \cdot [-\mathbf{i} + (4 - 2t)\mathbf{j}] \, dt$$

$$= \int_0^3 (-4t + t^2 + 64 - 64t + 20t^2 - 2t^3) \, dt$$

$$= \int_0^3 (-2t^3 + 21t^2 - 68t + 64) \, dt$$

$$= \left[-\frac{t^4}{2} + 7t^3 - 34t^2 + 64t \right]_0^3$$

$$= \frac{69}{2}.$$

b. Because $\mathbf{r}_2'(t) = \mathbf{i} + (4 - 2t)\mathbf{j}$ and

$$\mathbf{F}(x(t), y(t)) = (4t - t^2)\mathbf{i} + t^2\mathbf{j}$$

the line integral is

$$\int_{C_2} \mathbf{F} \cdot d\mathbf{r} = \int_1^4 [(4t - t^2)\mathbf{i} + t^2\mathbf{j}] \cdot [\mathbf{i} + (4 - 2t)\mathbf{j}] \, dt$$

$$= \int_1^4 (4t - t^2 + 4t^2 - 2t^3) \, dt$$

$$= \int_1^4 (-2t^3 + 3t^2 + 4t) \, dt$$

$$= \left[-\frac{t^4}{2} + t^3 + 2t^2 \right]_1^4$$

$$= -\frac{69}{2}.$$

The answer in part (b) is the negative of that in part (a) because C_1 and C_2 represent opposite orientations of the same parabolic segment.

C_1: $\mathbf{r}_1(t) = (4 - t)\mathbf{i} + (4t - t^2)\mathbf{j}$
C_2: $\mathbf{r}_2(t) = t\mathbf{i} + (4t - t^2)\mathbf{j}$

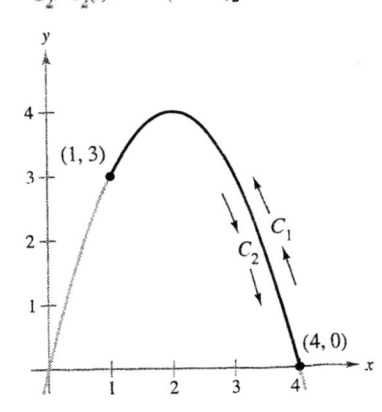

Figure 14.16

NOTE Although the value of the line integral in Example 7 depends on the orientation of C, it does not depend on the parametrization of C. To see this, let C_3 be represented by

$$\mathbf{r}_3 = (t + 2)\mathbf{i} + (4 - t^2)\mathbf{j}$$

where $-1 \leq t \leq 2$. The graph of this curve is the same parabolic segment shown in Figure 14.16. Does the value of the line integral over C_3 agree with the value over C_1 or C_2? Why or why not?

Line Integrals in Differential Form

A second commonly used form of line integrals is derived from the vector field notation used in the preceding section. If $\mathbf{F}$ is a vector field of the form $\mathbf{F}(x, y) = M\mathbf{i} + N\mathbf{j}$, and C is given by $\mathbf{r}(t) = x(t)\mathbf{i} + y(t)\mathbf{j}$, then $\mathbf{F} \cdot d\mathbf{r}$ is often written as $M\,dx + N\,dy$.

$$\int_C \mathbf{F} \cdot d\mathbf{r} = \int_C \mathbf{F} \cdot \frac{d\mathbf{r}}{dt}\,dt$$

$$= \int_a^b (M\mathbf{i} + N\mathbf{j}) \cdot (x'(t)\mathbf{i} + y'(t)\mathbf{j})\,dt$$

$$= \int_a^b \left(M\frac{dx}{dt} + N\frac{dy}{dt}\right)dt$$

$$= \int_C (M\,dx + N\,dy)$$

This **differential form** can be extended to three variables. The parentheses are often omitted, as follows.

$$\int_C M\,dx + N\,dy \quad \text{and} \quad \int_C M\,dx + N\,dy + P\,dz$$

Notice how this differential notation is used in Example 8.

NOTE The orientation of C affects the value of the differential form of a line integral. Specifically, if $-C$ has the orientation opposite to that of C, then

$$\int_{-C} M\,dx + N\,dy =$$
$$-\int_C M\,dx + N\,dy.$$

So, of the three line integral forms presented in this section, the orientation of C does not affect the form $\int_C f(x, y)\,ds$, but it does affect the vector form and the differential form.

Example 8 Evaluating a Line Integral in Differential Form

Let C be the circle of radius 3 given by

$$\mathbf{r}(t) = 3\cos t\,\mathbf{i} + 3\sin t\,\mathbf{j}, \quad 0 \le t \le 2\pi$$

and evaluate the line integral

$$\int_C y^3\,dx + (x^3 + 3xy^2)\,dy.$$

(See Figure 14.17.)

Solution Because $x = 3\cos t$ and $y = 3\sin t$, you have $dx = -3\sin t\,dt$ and $dy = 3\cos t\,dt$. So, the line integral is

$$\int_C M\,dx + N\,dy$$

$$= \int_C y^3\,dx + (x^3 + 3xy^2)\,dy$$

$$= \int_0^{2\pi} \left[(27\sin^3 t)(-3\sin t) + (27\cos^3 t + 81\cos t\sin^2 t)(3\cos t)\right]dt$$

$$= 81\int_0^{2\pi} (\cos^4 t - \sin^4 t + 3\cos^2 t\sin^2 t)\,dt$$

$$= 81\int_0^{2\pi} \left(\cos^2 t - \sin^2 t + \frac{3}{4}\sin^2 2t\right)dt$$

$$= 81\int_0^{2\pi} \left[\cos 2t + \frac{3}{4}\left(\frac{1 - \cos 4t}{2}\right)\right]dt$$

$$= 81\left[\frac{\sin 2t}{2} + \frac{3}{8}t - \frac{3\sin 4t}{32}\right]_0^{2\pi} = \frac{243\pi}{4}.$$

$\mathbf{r}(t) = 3\cos t\,\mathbf{i} + 3\sin t\,\mathbf{j}$

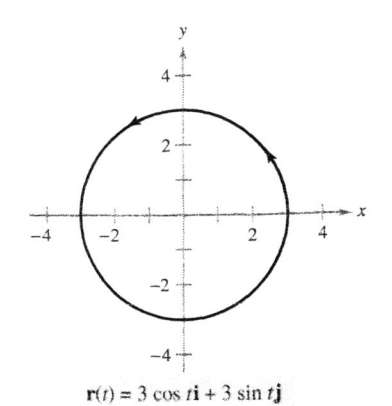

Figure 14.17

For curves represented by $y = g(x)$, $a \leq x \leq b$, you can let $x = t$ and obtain the parametric form

$$x = t \quad \text{and} \quad y = g(t), \quad a \leq t \leq b.$$

Because $dx = dt$ for this form, you have the option of evaluating the line integral in the variable x or t. This is demonstrated in Example 9.

Example 9 Evaluating a Line Integral in Differential Form

Evaluate

$$\int_C y \, dx + x^2 \, dy$$

where C is the parabolic arc given by $y = 4x - x^2$ from $(4, 0)$ to $(1, 3)$, as shown in Figure 14.18.

Solution Rather than converting to the parameter t, you can simply retain the variable x and write

$$y = 4x - x^2 \quad \Longrightarrow \quad dy = (4 - 2x) \, dx.$$

Then, in the direction from $(4, 0)$ to $(1, 3)$, the line integral is

$$\int_C y \, dx + x^2 \, dy = \int_4^1 [(4x - x^2) \, dx + x^2(4 - 2x) \, dx]$$

$$= \int_4^1 (4x + 3x^2 - 2x^3) \, dx$$

$$= \left[2x^2 + x^3 - \frac{x^4}{2} \right]_4^1 = \frac{69}{2}. \quad \text{See Example 7.}$$

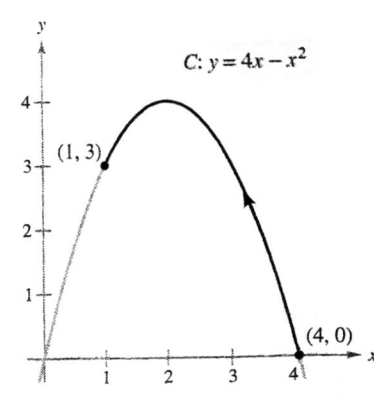

The line integral over C from $(4, 0)$ to $(1, 3)$ is $\frac{69}{2}$.
Figure 14.18

EXPLORATION

Finding Lateral Surface Area The figure below shows a piece of tin that has been cut from a circular cylinder. The base of the circular cylinder can be modeled by $x^2 + y^2 = 9$. At any point (x, y) on the base, the height of the object is given by

$$f(x, y) = 1 + \cos \frac{\pi x}{4}.$$

Explain how to use a line integral to find the surface area of the piece of tin.

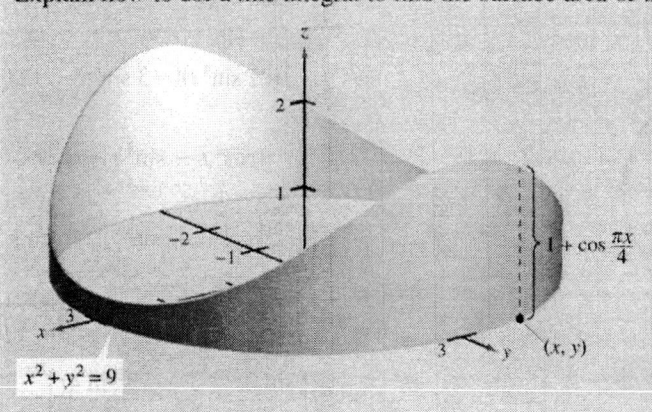

EXERCISES FOR SECTION 14.2

In Exercises 1–6, find a piecewise smooth parametrization of the path C.

1. 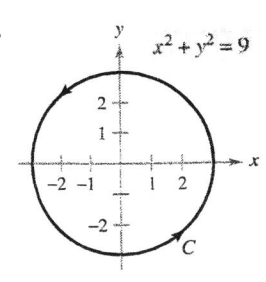
2. 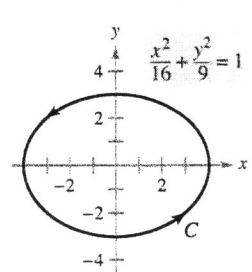
3. 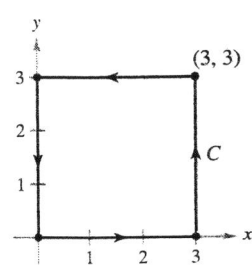
4. 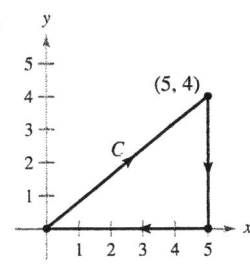
5. 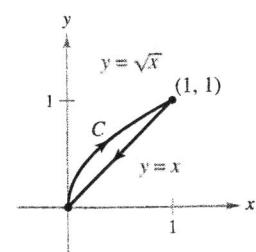
6.

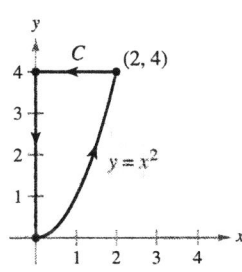

In Exercises 7–10, evaluate the line integral over the indicated path.

7. $\int_C (x - y)\, ds$
   C: $\mathbf{r}(t) = 4t\mathbf{i} + 3t\mathbf{j}$
   $0 \leq t \leq 2$

8. $\int_C 4xy\, ds$
   C: $\mathbf{r}(t) = t\mathbf{i} + (2 - t)\mathbf{j}$
   $0 \leq t \leq 2$

9. $\int_C (x^2 + y^2 + z^2)\, ds$
   C: $\mathbf{r}(t) = \sin t\mathbf{i} + \cos t\mathbf{j} + 8t\mathbf{k}$
   $0 \leq t \leq \pi/2$

10. $\int_C 8xyz\, ds$
    C: $\mathbf{r}(t) = 12t\mathbf{i} + 5t\mathbf{j} + 3\mathbf{k}$
    $0 \leq t \leq 2$

In Exercises 11–14, evaluate

$$\int_C (x^2 + y^2)\, ds$$

along the given path.

11. C: x-axis from $x = 0$ to $x = 3$
12. C: y-axis from $y = 1$ to $y = 10$
13. C: counterclockwise around the circle $x^2 + y^2 = 1$ from $(1, 0)$ to $(0, 1)$
14. C: counterclockwise around the circle $x^2 + y^2 = 4$ from $(2, 0)$ to $(0, 2)$

In Exercises 15–18, evaluate

$$\int_C (x + 4\sqrt{y})\, ds$$

along the given path.

15. C: line from $(0, 0)$ to $(1, 1)$
16. C: line from $(0, 0)$ to $(3, 9)$
17. C: counterclockwise around the triangle with vertices $(0, 0)$, $(1, 0)$ and $(0, 1)$
18. C: counterclockwise around the square with vertices $(0, 0)$, $(2, 0)$, $(2, 2)$, and $(0, 2)$

Mass In Exercises 19 and 20, find the total mass of two turns of a spring with density ρ in the shape of the circular helix

$\mathbf{r}(t) = 3\cos t\mathbf{i} + 3\sin t\mathbf{j} + 2t\mathbf{k}$.

19. $\rho(x, y, z) = \frac{1}{2}(x^2 + y^2 + z^2)$
20. $\rho(x, y, z) = z$

In Exercises 21–26, evaluate

$$\int_C \mathbf{F} \cdot d\mathbf{r}$$

where C is represented by $\mathbf{r}(t)$.

21. $\mathbf{F}(x, y) = xy\mathbf{i} + y\mathbf{j}$
    C: $\mathbf{r}(t) = 4t\mathbf{i} + t\mathbf{j}$, $0 \leq t \leq 1$

22. $\mathbf{F}(x, y) = xy\mathbf{i} + y\mathbf{j}$
    C: $\mathbf{r}(t) = 4\cos t\mathbf{i} + 4\sin t\mathbf{j}$, $0 \leq t \leq \pi/2$

23. $\mathbf{F}(x, y) = 3x\mathbf{i} + 4y\mathbf{j}$
    C: $\mathbf{r}(t) = 2\cos t\mathbf{i} + 2\sin t\mathbf{j}$, $0 \leq t \leq \pi/2$

24. $\mathbf{F}(x, y) = 3x\mathbf{i} + 4y\mathbf{j}$
    C: $\mathbf{r}(t) = t\mathbf{i} + \sqrt{4 - t^2}\mathbf{j}$, $-2 \leq t \leq 2$

25. $\mathbf{F}(x, y, z) = x^2y\mathbf{i} + (x - z)\mathbf{j} + xyz\mathbf{k}$
    C: $\mathbf{r}(t) = t\mathbf{i} + t^2\mathbf{j} + 2\mathbf{k}$, $0 \leq t \leq 1$

26. $\mathbf{F}(x, y, z) = x^2\mathbf{i} + y^2\mathbf{j} + z^2\mathbf{k}$
    C: $\mathbf{r}(t) = 2\sin t\mathbf{i} + 2\cos t\mathbf{j} + \frac{1}{2}t^2\mathbf{k}$, $0 \leq t \leq \pi$

In Exercises 27 and 28, use a computer algebra system to evaluate the integral

$$\int_C \mathbf{F} \cdot d\mathbf{r}$$

where C is represented by $\mathbf{r}(t)$.

27. $\mathbf{F}(x, y, z) = x^2 z \mathbf{i} + 6y \mathbf{j} + yz^2 \mathbf{k}$
    $C: \mathbf{r}(t) = t\mathbf{i} + t^2\mathbf{j} + \ln t\mathbf{k}, \quad 1 \le t \le 3$

28. $\mathbf{F}(x, y, z) = \dfrac{x\mathbf{i} + y\mathbf{j} + z\mathbf{k}}{\sqrt{x^2 + y^2 + z^2}}$
    $C: \mathbf{r}(t) = t\mathbf{i} + t\mathbf{j} + e^t\mathbf{k}, \quad 0 \le t \le 2$

Work In Exercises 29–34, find the work done by the force field $\mathbf{F}$ on an object moving along the indicated path.

29. $\mathbf{F}(x, y) = -x\mathbf{i} - 2y\mathbf{j}$
    $C: y = x^3$ from $(0, 0)$ to $(2, 8)$

30. $\mathbf{F}(x, y) = x^2\mathbf{i} - xy\mathbf{j}$
    $C: x = \cos^3 t, y = \sin^3 t$ from $(1, 0)$ to $(0, 1)$

31. $\mathbf{F}(x, y) = 2x\mathbf{i} + y\mathbf{j}$
    C: counterclockwise around the triangle with vertices $(0, 0)$, $(1, 0)$, and $(1, 1)$

32. $\mathbf{F}(x, y) = -y\mathbf{i} - x\mathbf{j}$
    C: counterclockwise along the semicircle $y = \sqrt{4 - x^2}$ from $(2, 0)$ to $(-2, 0)$

33. $\mathbf{F}(x, y, z) = x\mathbf{i} + y\mathbf{j} - 5z\mathbf{k}$
    $C: \mathbf{r}(t) = 2\cos t\mathbf{i} + 2\sin t\mathbf{j} + t\mathbf{k}, \quad 0 \le t \le 2\pi$

34. $\mathbf{F}(x, y, z) = yz\mathbf{i} + xz\mathbf{j} + xy\mathbf{k}$
    C: line from $(0, 0, 0)$ to $(5, 3, 2)$

35. **Work** Find the work done by a person weighing 150 pounds walking exactly one revolution up a circular helical staircase of radius 3 feet if the person rises 10 feet.

36. **Work** A particle moves along the path $y = x^2$ from the point $(0, 0)$ to the point $(1, 1)$. The force field $\mathbf{F}$ is measured at five points along the path and the results are shown in the table. Use Simpson's Rule or a graphing utility to approximate the work done by the force field.

(x, y)	$(0, 0)$	$\left(\frac{1}{4}, \frac{1}{16}\right)$	$\left(\frac{1}{2}, \frac{1}{4}\right)$	$\left(\frac{3}{4}, \frac{9}{16}\right)$	$(1, 1)$
$\mathbf{F}(x, y)$	$\langle 5, 0 \rangle$	$\langle 3.5, 1 \rangle$	$\langle 2, 2 \rangle$	$\langle 1.5, 3 \rangle$	$\langle 1, 5 \rangle$

In Exercises 37 and 38, evaluate $\int_C \mathbf{F} \cdot d\mathbf{r}$ for each curve. Discuss the orientation of the curve and its effect on the value of the integral.

37. $\mathbf{F}(x, y) = x^2\mathbf{i} + xy\mathbf{j}$
    (a) $\mathbf{r}_1(t) = 2t\mathbf{i} + (t - 1)\mathbf{j}, \quad 1 \le t \le 3$
    (b) $\mathbf{r}_2(t) = 2(3 - t)\mathbf{i} + (2 - t)\mathbf{j}, \quad 0 \le t \le 2$

38. $\mathbf{F}(x, y) = x^2 y\mathbf{i} + xy^{3/2}\mathbf{j}$
    (a) $\mathbf{r}_1(t) = (t + 1)\mathbf{i} + t^2\mathbf{j}, \quad 0 \le t \le 2$
    (b) $\mathbf{r}_2(t) = (1 + 2\cos t)\mathbf{i} + (4\cos^2 t)\mathbf{j}, \quad 0 \le t \le \pi/2$

In Exercises 39–42, demonstrate the property that

$$\int_C \mathbf{F} \cdot d\mathbf{r} = 0$$

regardless of the initial and terminal points of C, if the tangent vector $\mathbf{r}'(t)$ is orthogonal to the force field $\mathbf{F}$.

39. $\mathbf{F}(x, y) = y\mathbf{i} - x\mathbf{j}$
    $C: \mathbf{r}(t) = t\mathbf{i} - 2t\mathbf{j}$

40. $\mathbf{F}(x, y) = -3y\mathbf{i} + x\mathbf{j}$
    $C: \mathbf{r}(t) = t\mathbf{i} - t^3\mathbf{j}$

41. $\mathbf{F}(x, y) = (x^3 - 2x^2)\mathbf{i} + \left(x - \dfrac{y}{2}\right)\mathbf{j}$
    $C: \mathbf{r}(t) = t\mathbf{i} + t^2\mathbf{j}$

42. $\mathbf{F}(x, y) = x\mathbf{i} + y\mathbf{j}$
    $C: \mathbf{r}(t) = 3\sin t\mathbf{i} + 3\cos t\mathbf{j}$

In Exercises 43–46, evaluate the line integral over the path C given by $x = 2t$, $y = 10t$, where $0 \le t \le 1$.

43. $\displaystyle\int_C (x + 3y^2)\, dy$

44. $\displaystyle\int_C (x + 3y^2)\, dx$

45. $\displaystyle\int_C xy\, dx + y\, dy$

46. $\displaystyle\int_C (3y - x)\, dx + y^2\, dy$

In Exercises 47–54, evaluate the integral

$$\int_C (2x - y)\, dx + (x + 3y)\, dy$$

along the path.

47. C: x-axis from $x = 0$ to $x = 5$
48. C: y-axis from $y = 0$ to $y = 2$
49. C: line segments from $(0, 0)$ to $(3, 0)$ and $(3, 0)$ to $(3, 3)$
50. C: line segments from $(0, 0)$ to $(0, -3)$ and $(0, -3)$ to $(2, -3)$
51. C: arc on $y = 1 - x^2$ from $(0, 1)$ to $(1, 0)$
52. C: arc on $y = x^{3/2}$ from $(0, 0)$ to $(4, 8)$
53. C: parabolic path $x = t$, $y = 2t^2$, from $(0, 0)$ to $(2, 8)$
54. C: elliptic path $x = 4\sin t$, $y = 3\cos t$, from $(0, 3)$ to $(4, 0)$

Lateral Surface Area In Exercises 55–62, find the area of the lateral surface (see figure) over the curve C in the xy-plane and under the surface $z = f(x, y)$, where

Lateral surface area $= \displaystyle\int_C f(x, y)\, ds$.

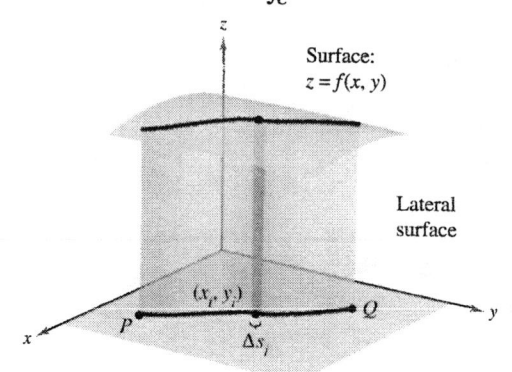

55. $f(x, y) = h$, C: line from $(0, 0)$ to $(3, 4)$
56. $f(x, y) = y$, C: line from $(0, 0)$ to $(4, 4)$
57. $f(x, y) = xy$, C: $x^2 + y^2 = 1$ from $(1, 0)$ to $(0, 1)$
58. $f(x, y) = x + y$, C: $x^2 + y^2 = 1$ from $(1, 0)$ to $(0, 1)$
59. $f(x, y) = h$, C: $y = 1 - x^2$ from $(1, 0)$ to $(0, 1)$
60. $f(x, y) = y + 1$, C: $1 - x^2$ from $(1, 0)$ to $(0, 1)$
61. $f(x, y) = xy$, C: $y = 1 - x^2$ from $(1, 0)$ to $(0, 1)$
62. $f(x, y) = x^2 - y^2 + 4$, C: $x^2 + y^2 = 4$

63. *Engine Design* A tractor engine has a steel component with a circular base modeled by the vector-valued function $\mathbf{r}(t) = 2 \cos t\,\mathbf{i} + 2 \sin t\,\mathbf{j}$. Its height is given by $z = 1 + y^2$. All measurements of the component are given in centimeters.
    (a) Find the lateral surface area of the component.
    (b) If the component is in the form of a shell of thickness 0.2 centimeter, use the result in part (a) to approximate the amount of steel used in its manufacture.
    (c) Make a sketch of the component.

64. *Building Design* The ceiling of a building has a height above the floor given by $z = 20 + \frac{1}{4}x$, and one of the walls follows a path modeled by $y = x^{3/2}$. Find the surface area of the wall if $0 \le x \le 40$. (All measurements are given in feet.)

Approximation In Exercises 65 and 66, determine which value best approximates the lateral surface area over the curve C in the xy-plane and under the surface $z = f(x, y)$. (Make your selection on the basis of a sketch of the surface and *not* by performing any calculations.)

65. $f(x, y) = e^{xy}$
    C: line from $(0, 0)$ to $(2, 2)$
    (a) 54   (b) 25   (c) -250 (d) 75   (e) 100

66. $f(x, y) = y$
    C: $y = x^2$ from $(0, 0)$ to $(2, 4)$
    (a) 2   (b) 4   (c) 8   (d) 16

67. *Investigation* The top outer edge of a solid with vertical sides and resting on the xy-plane is modeled by
    $$\mathbf{r}(t) = 3 \cos t\,\mathbf{i} + 3 \sin t\,\mathbf{j} + (1 + \sin^2 2t)\mathbf{k}$$
    where all measurements are in centimeters. The intersection of the plane $y = b\,(-3 < b < 3)$ with the top of the solid is a horizontal line.
    (a) Use a computer algebra system to graph the solid.
    (b) Use a computer algebra system to approximate the lateral surface area of the solid.
    (c) Find (if possible) the volume of the solid.

68. *Investigation* Determine the value of c such that the work done by the force field
    $$\mathbf{F}(x, y) = 15[(4 - x^2y)\mathbf{i} - xy\mathbf{j}]$$
    on an object moving along the parabolic path $y = c(1 - x^2)$ between the points $(-1, 0)$ and $(1, 0)$ is a minimum. Compare the result with the work required to move the object along the straight-line path connecting the points.

Getting at the Concept

69. Define a line integral of a function f along a smooth curve C in the plane and in space. How do you evaluate the line integral as a definite integral?

70. Define a line integral of a continuous vector field $\mathbf{F}$ on a smooth curve C. How do you evaluate the line integral as a definite integral?

71. Order the surfaces in ascending order of the lateral surface area under the surface and over the curve $y = \sqrt{x}$ from $(0, 0)$ to $(4, 2)$ in the xy-plane. Explain your ordering without doing any calculations.
    (a) $z_1 = 2 + x$ (b) $z_2 = 5 + x$
    (c) $z_3 = 2$ (d) $z_4 = 10 + x + 2y$

72. For each of the following, determine whether the work done in moving an object from the first to the second point through the force field shown in the figure is positive, negative, or zero. Explain your answer.
    (a) From $(-3, -3)$ to $(3, 3)$
    (b) From $(-3, 0)$ to $(0, 3)$
    (c) From $(5, 0)$ to $(0, 3)$

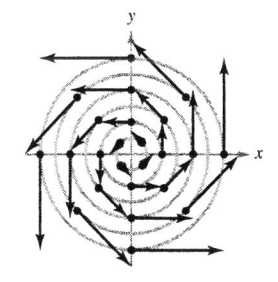

True or False? In Exercises 73–76, determine whether the statement is true or false. If it is false, explain why or give an example that shows it is false.

73. If C is given by $x(t) = t$, $y(t) = t$, $0 \le t \le 1$, then
    $$\int_C xy\,ds = \int_0^1 t^2\,dt.$$

74. If $C_2 = -C_1$, then $\int_{C_1} f(x, y)\,ds + \int_{C_2} f(x, y)\,ds = 0$.

75. The vector functions $\mathbf{r}_1 = t\mathbf{i} + t^2\mathbf{j}$, $0 \le t \le 1$, and $\mathbf{r}_2 = (1 - t)\mathbf{i} + (1 - t)^2\mathbf{j}$, $0 \le t \le 1$, define the same curve.

76. If $\int_C \mathbf{F} \cdot \mathbf{T}\,ds = 0$, then $\mathbf{F}$ and $\mathbf{T}$ are orthogonal.

Section 14.3 Conservative Vector Fields and Independence of Path

- Understand and use the Fundamental Theorem of Line Integrals.
- Understand the concept of independence of path.
- Understand the concept of conservation of energy.

Fundamental Theorem of Line Integrals

In the preceding section we pointed out that in a gravitational field the work done by gravity on an object moving between two points in the field is independent of the path taken by the object. In this section, you will study an important generalization of this result—it is called the **Fundamental Theorem of Line Integrals**.

We begin with an example in which the line integral of a *conservative vector field* is evaluated over three different paths.

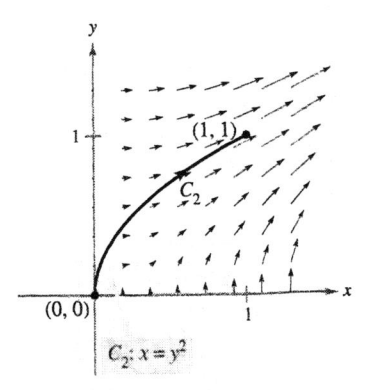

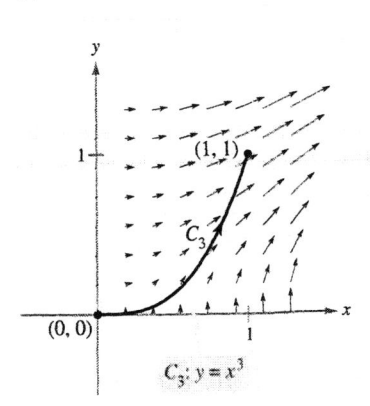

Example 1 Line Integral of a Conservative Vector Field

Find the work done by the force field

$$\mathbf{F}(x, y) = \frac{1}{2}xy\mathbf{i} + \frac{1}{4}x^2\mathbf{j}$$

on a particle that moves from (0, 0) to (1, 1) along each path.

a. $C_1: y = x$ **b.** $C_2: x = y^2$ **c.** $C_3: y = x^3$

Solution (See Figure 14.19.)

a. Let $\mathbf{r}(t) = t\mathbf{i} + t\mathbf{j}$ for $0 \leq t \leq 1$, so that

$$d\mathbf{r} = (\mathbf{i} + \mathbf{j})\, dt \quad \text{and} \quad \mathbf{F}(x, y) = \frac{1}{2}t^2\mathbf{i} + \frac{1}{4}t^2\mathbf{j}.$$

Then, the work done is

$$W = \int_{C_1} \mathbf{F} \cdot d\mathbf{r} = \int_0^1 \frac{3}{4}t^2\, dt = \frac{1}{4}t^3 \Big]_0^1 = \frac{1}{4}.$$

b. Let $\mathbf{r}(t) = t\mathbf{i} + \sqrt{t}\mathbf{j}$ for $0 \leq t \leq 1$, so that

$$d\mathbf{r} = \left(\mathbf{i} + \frac{1}{2\sqrt{t}}\mathbf{j}\right) dt \quad \text{and} \quad \mathbf{F}(x, y) = \frac{1}{2}t^{3/2}\mathbf{i} + \frac{1}{4}t^2\mathbf{j}.$$

Then, the work done is

$$W = \int_{C_2} \mathbf{F} \cdot d\mathbf{r} = \int_0^1 \frac{5}{8}t^{3/2}\, dt = \frac{1}{4}t^{5/2} \Big]_0^1 = \frac{1}{4}.$$

c. Let $\mathbf{r}(t) = \frac{1}{2}t\mathbf{i} + \frac{1}{8}t^3\mathbf{j}$ for $0 \leq t \leq 2$, so that

$$d\mathbf{r} = \left(\frac{1}{2}\mathbf{i} + \frac{3}{8}t^2\mathbf{j}\right) dt \quad \text{and} \quad \mathbf{F}(x, y) = \frac{1}{32}t^4\mathbf{i} + \frac{1}{16}t^2\mathbf{j}.$$

Then, the work done is

$$W = \int_{C_3} \mathbf{F} \cdot d\mathbf{r} = \int_0^2 \frac{5}{128}t^4\, dt = \frac{1}{128}t^5 \Big]_0^2 = \frac{1}{4}.$$

The work done by a conservative vector field is the same for all paths.
Figure 14.19

In Example 1, note that the vector field $\mathbf{F}(x, y) = \frac{1}{2}xy\mathbf{i} + \frac{1}{4}x^2\mathbf{j}$ is conservative because $\mathbf{F}(x, y) = \nabla f(x, y)$, where $f(x, y) = \frac{1}{4}x^2y$. In such cases, the following theorem states that the value of $\int_C \mathbf{F} \cdot d\mathbf{r}$ is given by

$$\int_C \mathbf{F} \cdot d\mathbf{r} = f(x(1), y(1)) - f(x(0), y(0)) = \frac{1}{4} - 0 = \frac{1}{4}.$$

NOTE Notice how the Fundamental Theorem of Line Integrals is similar to the Fundamental Theorem of Calculus (Section 4.4), which states that

$$\int_a^b f(x)\, dx = F(b) - F(a)$$

where $F'(x) = f(x)$.

THEOREM 14.5 Fundamental Theorem of Line Integrals

Let C be a piecewise smooth curve lying in an open region R and given by

$$\mathbf{r}(t) = x(t)\mathbf{i} + y(t)\mathbf{j}, \quad a \leq t \leq b.$$

If $\mathbf{F}(x, y) = M\mathbf{i} + N\mathbf{j}$ is conservative in R, and M and N are continuous in R, then

$$\int_C \mathbf{F} \cdot d\mathbf{r} = \int_C \nabla f \cdot d\mathbf{r} = f(x(b), y(b)) - f(x(a), y(a))$$

where f is a potential function of $\mathbf{F}$. That is, $\mathbf{F}(x, y) = \nabla f(x, y)$.

Proof We provide a proof only for a smooth curve. For piecewise smooth curves, the procedure is carried out separately on each smooth portion. Because $\mathbf{F}(x, y) = \nabla f(x, y) = f_x(x, y)\mathbf{i} + f_y(x, y)\mathbf{j}$, it follows that

$$\int_C \mathbf{F} \cdot d\mathbf{r} = \int_a^b \mathbf{F} \cdot \frac{d\mathbf{r}}{dt}\, dt$$

$$= \int_a^b \left[f_x(x, y)\frac{dx}{dt} + f_y(x, y)\frac{dy}{dt} \right] dt$$

and, by the Chain Rule (Theorem 12.6), you have

$$\int_C \mathbf{F} \cdot d\mathbf{r} = \int_a^b \frac{d}{dt}[f(x(t), y(t))]\, dt$$

$$= f(x(b), y(b)) - f(x(a), y(a)).$$

The last step is an application of the Fundamental Theorem of Calculus.

In space, the Fundamental Theorem of Line Integrals takes the following form. Let C be a piecewise smooth curve lying in an open region Q and given by

$$\mathbf{r}(t) = x(t)\mathbf{i} + y(t)\mathbf{j} + z(t)\mathbf{k}, \quad a \leq t \leq b.$$

If $\mathbf{F}(x, y, z) = M\mathbf{i} + N\mathbf{j} + P\mathbf{k}$ is conservative and M, N, and P are continuous, then

$$\int_C \mathbf{F} \cdot d\mathbf{r} = \int_C \nabla f \cdot d\mathbf{r}$$
$$= f(x(b), y(b), z(b)) - f(x(a), y(a), z(a))$$

where $\mathbf{F}(x, y, z) = \nabla f(x, y, z)$.

The Fundamental Theorem of Line Integrals states that if the vector field $\mathbf{F}$ is conservative, then the line integral between any two points is simply the difference in the values of the *potential* function f at these points.

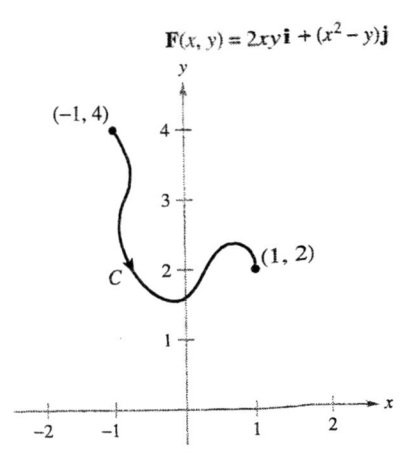

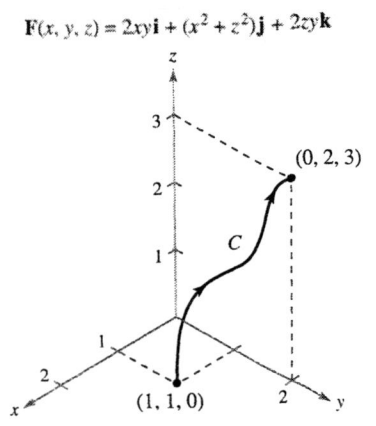

Figure 14.20
Using the Fundamental Theorem of Line Integrals, $\int_C \mathbf{F} \cdot d\mathbf{r} = 4$.

$\mathbf{F}(x, y) = 2xy\mathbf{i} + (x^2 - y)\mathbf{j}$

Example 2 Using the Fundamental Theorem of Line Integrals

Evaluate $\int_C \mathbf{F} \cdot d\mathbf{r}$, where C is a piecewise smooth curve from $(-1, 4)$ to $(1, 2)$ and

$$\mathbf{F}(x, y) = 2xy\mathbf{i} + (x^2 - y)\mathbf{j}.$$

(See Figure 14.20.)

Solution From Example 6 in Section 14.1, you know that $\mathbf{F}$ is the gradient of f where

$$f(x, y) = x^2y - \frac{1}{2}y^2 + K.$$

Consequently, $\mathbf{F}$ is conservative, and by the Fundamental Theorem of Line Integrals, it follows that

$$\int_C \mathbf{F} \cdot d\mathbf{r} = f(1, 2) - f(-1, 4)$$
$$= \left[1^2(2) - \frac{1}{2}(2^2)\right] - \left[(-1)^2(4) - \frac{1}{2}(4^2)\right]$$
$$= 4.$$

Note that it is unnecessary to include a constant K as part of f, because it is canceled by subtraction.

Example 3 Using the Fundamental Theorem of Line Integrals

Evaluate $\int_C \mathbf{F} \cdot d\mathbf{r}$, where C is a piecewise smooth curve from $(1, 1, 0)$ to $(0, 2, 3)$ and

$$\mathbf{F}(x, y, z) = 2xy\mathbf{i} + (x^2 + z^2)\mathbf{j} + 2zy\mathbf{k}.$$

(See Figure 14.21.)

Solution From Example 8 in Section 14.1, you know that $\mathbf{F}$ is the gradient of f where $f(x, y, z) = x^2y + z^2y + K$. Consequently, $\mathbf{F}$ is conservative, and by the Fundamental Theorem of Line Integrals, it follows that

$$\int_C \mathbf{F} \cdot d\mathbf{r} = f(0, 2, 3) - f(1, 1, 0)$$
$$= [(0)^2(2) + (3)^2(2)] - [(1)^2(1) + (0)^2(1)]$$
$$= 17.$$

Figure 14.21
Using the Fundamental Theorem of Line Integrals, $\int_C \mathbf{F} \cdot d\mathbf{r} = 17$.

$\mathbf{F}(x, y, z) = 2xy\mathbf{i} + (x^2 + z^2)\mathbf{j} + 2zy\mathbf{k}$

In Examples 2 and 3, be sure you see that the value of the line integral is the same for any smooth curve C that has the given initial and terminal points. For instance, in Example 3, try evaluating the line integral for the curve given by

$$\mathbf{r}(t) = (1 - t)\mathbf{i} + (1 + t)\mathbf{j} + 3t\mathbf{k}.$$

You should obtain

$$\int_C \mathbf{F} \cdot d\mathbf{r} = \int_0^1 (30t^2 + 16t - 1)\, dt$$
$$= 17.$$

Independence of Path

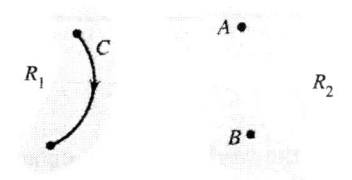

From the Fundamental Theorem of Line Integrals it is clear that if **F** is continuous and conservative in an open region R, the value of $\int_C \mathbf{F} \cdot d\mathbf{r}$ is the same for every piecewise smooth curve C from one fixed point in R to another fixed point in R. This result is described by saying that the line integral $\int_C \mathbf{F} \cdot d\mathbf{r}$ is **independent of path** in the region R.

A region in the plane (or in space) is **connected** if any two points in the region can be joined by a piecewise smooth curve lying entirely within the region, as shown in Figure 14.22. In open regions that are *connected*, the path independence of $\int_C \mathbf{F} \cdot d\mathbf{r}$ is equivalent to the condition that **F** is conservative.

R_1 is connected. R_2 is not connected.

Figure 14.22

THEOREM 14.6 Independence of Path and Conservative Vector Fields

If **F** is continuous on an open connected region, then the line integral
$$\int_C \mathbf{F} \cdot d\mathbf{r}$$
is independent of path if and only if **F** is conservative.

Proof If **F** is conservative, then, by the Fundamental Theorem of Line Integrals, the line integral is independent of path. We establish the converse for a plane region R. Let $\mathbf{F}(x, y) = M\mathbf{i} + N\mathbf{j}$, and let (x_0, y_0) be a fixed point in R. If (x, y) is any point in R, choose a piecewise smooth curve C running from (x_0, y_0) to (x, y), and define f by

$$f(x, y) = \int_C \mathbf{F} \cdot d\mathbf{r}$$
$$= \int_C M\, dx + N\, dy.$$

The existence of C in R is guaranteed by the fact that R is connected. You can show that f is a potential function of **F** by considering two different paths between (x_0, y_0) and (x, y). For the *first* path, choose (x_1, y) in R such that $x \neq x_1$. This is possible because R is open. Then choose C_1 and C_2, as shown in Figure 14.23. Using the independence of path, it follows that

$$f(x, y) = \int_C M\, dx + N\, dy$$
$$= \int_{C_1} M\, dx + N\, dy + \int_{C_2} M\, dx + N\, dy.$$

Because the first integral does not depend on x, and because $dy = 0$ in the second integral, you have

$$f(x, y) = g(y) + \int_{C_2} M\, dx$$

and it follows that the partial derivative of f with respect to x is $f_x(x, y) = M$. For the *second* path, choose a point (x, y_1). Using reasoning similar to that used for the first path, you can conclude that $f_y(x, y) = N$. Therefore,

$$\nabla f(x, y) = f_x(x, y)\mathbf{i} + f_y(x, y)\mathbf{j}$$
$$= M\mathbf{i} + N\mathbf{j}$$
$$= \mathbf{F}(x, y)$$

and it follows that **F** is conservative.

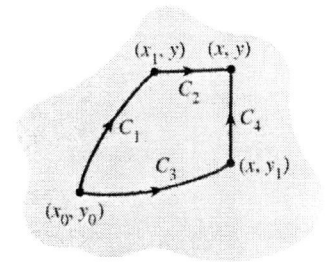

Figure 14.23

Example 4 **Finding Work in a Conservative Force Field**

For the force field given by

$$\mathbf{F}(x, y, z) = e^x \cos y \, \mathbf{i} - e^x \sin y \, \mathbf{j} + 2\mathbf{k}$$

show that $\int_C \mathbf{F} \cdot d\mathbf{r}$ is independent of path, and calculate the work done by $\mathbf{F}$ on an object moving along a curve C from $(0, \pi/2, 1)$ to $(1, \pi, 3)$.

Solution Writing the force field in the form $\mathbf{F}(x, y, z) = M\mathbf{i} + N\mathbf{j} + P\mathbf{k}$, you have $M = e^x \cos y$, $N = -e^x \sin y$, and $P = 2$, and it follows that

$$\frac{\partial P}{\partial y} = 0 = \frac{\partial N}{\partial z}$$

$$\frac{\partial P}{\partial x} = 0 = \frac{\partial M}{\partial z}$$

$$\frac{\partial N}{\partial x} = -e^x \sin y = \frac{\partial M}{\partial y}.$$

Hence, $\mathbf{F}$ is conservative. If f is a potential function of $\mathbf{F}$, then

$$f_x(x, y, z) = e^x \cos y$$
$$f_y(x, y, z) = -e^x \sin y$$
$$f_z(x, y, z) = 2.$$

By integrating with respect to x, y, and z separately, you obtain

$$f(x, y, z) = \int f_x(x, y, z) \, dx = \int e^x \cos y \, dx = e^x \cos y + g(y, z)$$

$$f(x, y, z) = \int f_y(x, y, z) \, dy = \int -e^x \sin y \, dy = e^x \cos y + h(x, z)$$

$$f(x, y, z) = \int f_z(x, y, z) \, dz = \int 2 \, dz = 2z + k(x, y).$$

By comparing these three versions of $f(x, y, z)$, you can conclude that

$$f(x, y, z) = e^x \cos y + 2z + K.$$

Therefore, the work done by $\mathbf{F}$ along *any* curve C from $(0, \pi/2, 1)$ to $(1, \pi, 3)$ is

$$W = \int_C \mathbf{F} \cdot d\mathbf{r}$$
$$= \left[e^x \cos y + 2z \right]_{(0, \pi/2, 1)}^{(1, \pi, 3)}$$
$$= (-e + 6) - (0 + 2)$$
$$= 4 - e.$$

How much work would be done if the object in Example 4 moved from the point $(0, \pi/2, 1)$ to $(1, \pi, 3)$ and then back to the starting point $(0, \pi/2, 1)$? The Fundamental Theorem of Line Integrals states that there is zero work done. Remember that, by definition, work can be negative. Hence, by the time the object gets back to its starting point, the amount of work that registers positively is canceled out by the amount of work that registers negatively.

SECTION 14.3 Conservative Vector Fields and Independence of Path

A curve C given by $\mathbf{r}(t)$ for $a \leq t \leq b$ is **closed** if $\mathbf{r}(a) = \mathbf{r}(b)$. By the Fundamental Theorem of Line Integrals, you can conclude that if $\mathbf{F}$ is continuous and conservative on an open region R, then the line integral over every closed curve C is 0.

THEOREM 14.7 Equivalent Conditions

Let $\mathbf{F}(x, y, z) = M\mathbf{i} + N\mathbf{j} + P\mathbf{k}$ have continuous first partial derivatives in an open connected region R, and let C be a piecewise smooth curve in R. The following conditions are equivalent.

1. $\mathbf{F}$ is conservative. That is, $\mathbf{F} = \nabla f$ for some function f.
2. $\int_C \mathbf{F} \cdot d\mathbf{r}$ is independent of path.
3. $\int_C \mathbf{F} \cdot d\mathbf{r} = 0$ for every *closed* curve C in R.

NOTE Theorem 14.7 gives you options for evaluating a line integral involving a conservative vector field. You can use a potential function, or it might be more convenient to choose a particularly simple path, such as a straight line.

Example 5 Evaluating a Line Integral

Evaluate $\int_{C_1} \mathbf{F} \cdot d\mathbf{r}$, where

$$\mathbf{F}(x, y) = (y^3 + 1)\mathbf{i} + (3xy^2 + 1)\mathbf{j}$$

and C_1 is the semicircular path from $(0, 0)$ to $(2, 0)$, as shown in Figure 14.24.

Solution You have the following three options.

a. You can use the method presented in the preceding section to evaluate the line integral along the *given curve*. To do this, you can use the parametrization $\mathbf{r}(t) = (1 - \cos t)\mathbf{i} + \sin t\mathbf{j}$, where $0 \leq t \leq \pi$. For this parametrization, it follows that $d\mathbf{r} = \mathbf{r}'(t)\,dt = (\sin t\mathbf{i} + \cos t\mathbf{j})\,dt$, and

$$\int_{C_1} \mathbf{F} \cdot d\mathbf{r} = \int_0^\pi (\sin t + \sin^4 t + \cos t + 3\sin^2 t \cos t - 3\cos^2 t \sin^2 t)\,dt.$$

This integral should dampen your enthusiasm for this option.

b. You can try to find a *potential function* and evaluate the line integral by the Fundamental Theorem of Line Integrals. Using the technique demonstrated in Example 4, you can find the potential function to be $f(x, y) = xy^3 + x + y + K$, and, by the Fundamental Theorem,

$$W = \int_{C_1} \mathbf{F} \cdot d\mathbf{r} = f(2, 0) - f(0, 0) = 2.$$

c. Knowing that $\mathbf{F}$ is conservative, you have a third option. Because the value of the line integral is independent of path, you can replace the semicircular path with a *simpler path*. Suppose you choose the straight-line path C_2 from $(0, 0)$ to $(2, 0)$. Then, $\mathbf{r}(t) = t\mathbf{i}$, where $0 \leq t \leq 2$. So, $d\mathbf{r} = \mathbf{i}\,dt$ and $\mathbf{F}(x, y) = (y^3 + 1)\mathbf{i} + (3xy^2 + 1)\mathbf{j} = \mathbf{i} + \mathbf{j}$, so that

$$\int_{C_1} \mathbf{F} \cdot d\mathbf{r} = \int_{C_2} \mathbf{F} \cdot d\mathbf{r} = \int_0^2 1\,dt = t\Big]_0^2 = 2.$$

Of the three options, obviously the third one is the easiest.

$C_1: \mathbf{r}(t) = (1 - \cos t)\mathbf{i} + \sin t\mathbf{j}$

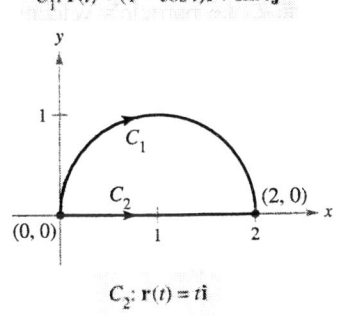

$C_2: \mathbf{r}(t) = t\mathbf{i}$

Figure 14.24

Conservation of Energy

In 1840, the English physicist Michael Faraday wrote, "Nowhere is there a pure creation or production of power without a corresponding exhaustion of something to supply it." This statement represents the first formulation of one of the most important laws of physics—the **Law of Conservation of Energy**. In modern terminology, the law is stated as follows: *In a conservative force field, the sum of the potential and kinetic energies of an object remains constant from point to point.*

You can use the Fundamental Theorem of Line Integrals to derive this law. From physics, the **kinetic energy** of a particle of mass m and speed v is $k = \frac{1}{2}mv^2$. The **potential energy** p of a particle at point (x, y, z) in a conservative vector field $\mathbf{F}$ is defined as $p(x, y, z) = -f(x, y, z)$, where f is the potential function for $\mathbf{F}$. Consequently, the work done by $\mathbf{F}$ along a smooth curve C from A to B is

$$W = \int_C \mathbf{F} \cdot d\mathbf{r} = f(x, y, z)\Big]_A^B$$
$$= -p(x, y, z)\Big]_A^B$$
$$= p(A) - p(B)$$

as indicated in Figure 14.25. In other words, work W is equal to the difference in the potential energies of A and B. Now, suppose that $\mathbf{r}(t)$ is the position vector for a particle moving along C from $A = \mathbf{r}(a)$ to $B = \mathbf{r}(b)$. At any time t, the particle's velocity, acceleration, and speed are $\mathbf{v}(t) = \mathbf{r}'(t)$, $\mathbf{a}(t) = \mathbf{r}''(t)$, and $v(t) = \|\mathbf{v}(t)\|$. So, by Newton's Second Law of Motion, $\mathbf{F} = m\mathbf{a}(t) = m(\mathbf{v}'(t))$, and the work done by $\mathbf{F}$ is

$$W = \int_C \mathbf{F} \cdot d\mathbf{r} = \int_a^b \mathbf{F} \cdot \mathbf{r}'(t)\, dt$$
$$= \int_a^b \mathbf{F} \cdot \mathbf{v}(t)\, dt = \int_a^b [m\mathbf{v}'(t)] \cdot \mathbf{v}(t)\, dt$$
$$= \int_a^b m[\mathbf{v}'(t) \cdot \mathbf{v}(t)]\, dt$$
$$= \frac{m}{2} \int_a^b \frac{d}{dt}[\mathbf{v}(t) \cdot \mathbf{v}(t)]\, dt$$
$$= \frac{m}{2} \int_a^b \frac{d}{dt}[\|\mathbf{v}(t)\|^2]\, dt$$
$$= \frac{m}{2}\Big[\|\mathbf{v}(t)\|^2\Big]_a^b$$
$$= \frac{m}{2}\Big[[v(t)]^2\Big]_a^b$$
$$= \frac{1}{2}m[v(b)]^2 - \frac{1}{2}m[v(a)]^2$$
$$= k(B) - k(A).$$

Equating these two results for W produces

$$p(A) - p(B) = k(B) - k(A)$$
$$p(A) + k(A) = p(B) + k(B)$$

which implies that the sum of the potential and kinetic energies remains constant from point to point.

MICHAEL FARADAY (1791–1867)

Several philosophers of science have considered Faraday's Law of Conservation of Energy to be the greatest generalization ever conceived by humankind. Many physicists have contributed to our knowledge of this law. Two early and influential ones were James Prescott Joule (1818–1889) and Hermann Ludwig Helmholtz (1821–1894).

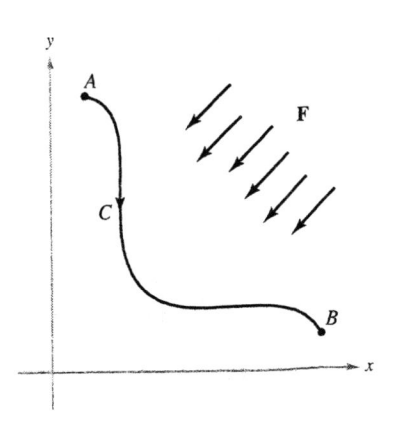

The work done by $\mathbf{F}$ along C is
$$W = \int_C \mathbf{F} \cdot d\mathbf{r} = p(A) - p(B).$$
Figure 14.25

EXERCISES FOR SECTION 14.3

In Exercises 1–4, show that the value of $\int_C \mathbf{F} \cdot d\mathbf{r}$ is the same for both parametric representations of C.

1. $\mathbf{F}(x, y) = x^2\mathbf{i} + xy\mathbf{j}$
   (a) $\mathbf{r}_1(t) = t\mathbf{i} + t^2\mathbf{j}, \quad 0 \le t \le 1$
   (b) $\mathbf{r}_2(\theta) = \sin\theta\mathbf{i} + \sin^2\theta\mathbf{j}, \quad 0 \le \theta \le \frac{\pi}{2}$

2. $\mathbf{F}(x, y) = (x^2 + y^2)\mathbf{i} - x\mathbf{j}$
   (a) $\mathbf{r}_1(t) = t\mathbf{i} + \sqrt{t}\mathbf{j}, \quad 0 \le t \le 4$
   (b) $\mathbf{r}_2(w) = w^2\mathbf{i} + w\mathbf{j}, \quad 0 \le w \le 2$

3. $\mathbf{F}(x, y) = y\mathbf{i} - x\mathbf{j}$
   (a) $\mathbf{r}_1(\theta) = \sec\theta\mathbf{i} + \tan\theta\mathbf{j}, \quad 0 \le \theta \le \frac{\pi}{3}$
   (b) $\mathbf{r}_2(t) = \sqrt{t+1}\mathbf{i} + \sqrt{t}\mathbf{j}, \quad 0 \le t \le 3$

4. $\mathbf{F}(x, y) = y\mathbf{i} + x^2\mathbf{j}$
   (a) $\mathbf{r}_1(t) = (2 + t)\mathbf{i} + (3 - t)\mathbf{j}, \quad 0 \le t \le 3$
   (b) $\mathbf{r}_2(w) = (2 + \ln w)\mathbf{i} + (3 - \ln w)\mathbf{j}, \quad 1 \le w \le e^3$

In Exercises 5–10, determine whether or not the vector field is conservative.

5. $\mathbf{F}(x, y) = e^x(\sin y\mathbf{i} + \cos y\mathbf{j})$
6. $\mathbf{F}(x, y) = 15x^2y^2\mathbf{i} + 10x^3y\mathbf{j}$
7. $\mathbf{F}(x, y) = \frac{1}{y^2}(y\mathbf{i} + x\mathbf{j})$
8. $\mathbf{F}(x, y, z) = y\ln z\mathbf{i} - x\ln z\mathbf{j} + \frac{xy}{z}\mathbf{k}$
9. $\mathbf{F}(x, y, z) = y^2z\mathbf{i} + 2xyz\mathbf{j} + xy^2\mathbf{k}$
10. $\mathbf{F}(x, y, z) = \sin yz\mathbf{i} + xz\cos yz\mathbf{j} + xy\sin yz\mathbf{k}$

In Exercises 11–24, find the value of the line integral

$$\int_C \mathbf{F} \cdot d\mathbf{r}.$$

(*Hint:* If **F** is conservative, the integration may be easier on an alternative path.)

11. $\mathbf{F}(x, y) = 2xy\mathbf{i} + x^2\mathbf{j}$
    (a) $\mathbf{r}_1(t) = t\mathbf{i} + t^2\mathbf{j}, \quad 0 \le t \le 1$
    (b) $\mathbf{r}_2(t) = t\mathbf{i} + t^3\mathbf{j}, \quad 0 \le t \le 1$

12. $\mathbf{F}(x, y) = ye^{xy}\mathbf{i} + xe^{xy}\mathbf{j}$
    (a) $\mathbf{r}_1(t) = t\mathbf{i} - (t - 3)\mathbf{j}, \quad 0 \le t \le 3$
    (b) The closed path consisting of line segments from (0, 3) to (0, 0), and then from (0, 0) to (3, 0)

13. $\mathbf{F}(x, y) = y\mathbf{i} - x\mathbf{j}$
    (a) $\mathbf{r}_1(t) = t\mathbf{i} + t\mathbf{j}, \quad 0 \le t \le 1$
    (b) $\mathbf{r}_2(t) = t\mathbf{i} + t^2\mathbf{j}, \quad 0 \le t \le 1$
    (c) $\mathbf{r}_3(t) = t\mathbf{i} + t^3\mathbf{j}, \quad 0 \le t \le 1$

14. $\mathbf{F}(x, y) = xy^2\mathbf{i} + 2x^2y\mathbf{j}$
    (a) $\mathbf{r}_1(t) = t\mathbf{i} + \frac{1}{t}\mathbf{j}, \quad 1 \le t \le 3$
    (b) $\mathbf{r}_2(t) = (t + 1)\mathbf{i} - \frac{1}{3}(t - 3)\mathbf{j}, \quad 0 \le t \le 2$

15. $\int_C y^2\, dx + 2xy\, dy$

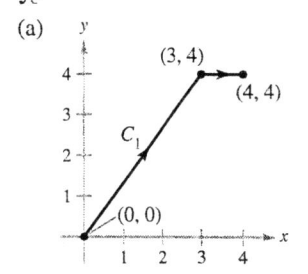

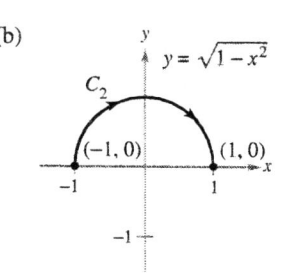

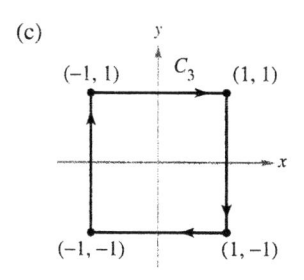

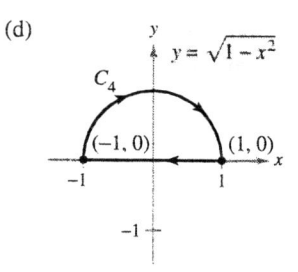

(a) C_1: line segments from (0, 0) to (3, 4) to (4, 4)
(b) C_2: upper semicircle $y = \sqrt{1 - x^2}$ from (1, 0) to (−1, 0)
(c) C_3: square with vertices (−1, 1), (1, 1), (1, −1), (−1, −1)
(d) C_4: semicircle $y = \sqrt{1 - x^2}$ from (−1, 0) to (1, 0)

16. $\int_C (2x - 3y + 1)\, dx - (3x + y - 5)\, dy$

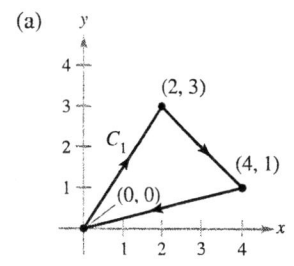

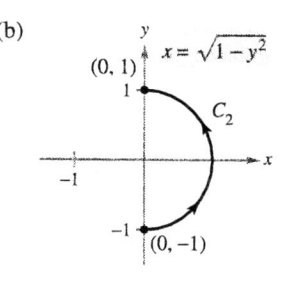

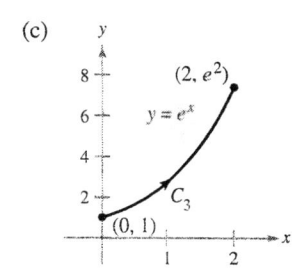

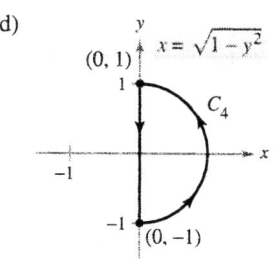

(a) C_1: triangle with vertices (0, 0), (2, 3), (4, 1)
(b) C_2: semicircle $x = \sqrt{1 - y^2}$ from (0, 1) to (0, −1)
(c) C_3: curve $y = e^x$ from (0, 1) to $(2, e^2)$
(d) C_4: semicircle $x = \sqrt{1 - y^2}$ from (0, −1) to (0, 1)

17. $\int_C 2xy\, dx + (x^2 + y^2)\, dy$

(a) C: ellipse $\dfrac{x^2}{25} + \dfrac{y^2}{16} = 1$ from (5, 0) to (0, 4)
(b) C: parabola $y = 4 - x^2$ from (2, 0) to (0, 4)

18. $\int_C (x^2 + y^2)\, dx + 2xy\, dy$

(a) $\mathbf{r}_1(t) = t^3\mathbf{i} + t^2\mathbf{j}, \quad 0 \le t \le 2$

(b) $\mathbf{r}_2(t) = 2\cos t\,\mathbf{i} + 2\sin t\,\mathbf{j}, \quad 0 \le t \le \dfrac{\pi}{2}$

19. $\mathbf{F}(x, y, z) = yz\mathbf{i} + xz\mathbf{j} + xy\mathbf{k}$

(a) $\mathbf{r}_1(t) = t\mathbf{i} + 2\mathbf{j} + t\mathbf{k}, \quad 0 \le t \le 4$

(b) $\mathbf{r}_2(t) = t^2\mathbf{i} + t\mathbf{j} + t^2\mathbf{k}, \quad 0 \le t \le 2$

20. $\mathbf{F}(x, y, z) = \mathbf{i} + z\mathbf{j} + y\mathbf{k}$

(a) $\mathbf{r}_1(t) = \cos t\,\mathbf{i} + \sin t\,\mathbf{j} + t^2\mathbf{k}, \quad 0 \le t \le \pi$

(b) $\mathbf{r}_2(t) = (1 - 2t)\mathbf{i} + \pi^2 t\mathbf{k}, \quad 0 \le t \le 1$

21. $\mathbf{F}(x, y, z) = (2y + x)\mathbf{i} + (x^2 - z)\mathbf{j} + (2y - 4z)\mathbf{k}$

(a) $\mathbf{r}_1(t) = t\mathbf{i} + t^2\mathbf{j} + \mathbf{k}, \quad 0 \le t \le 1$

(b) $\mathbf{r}_2(t) = t\mathbf{i} + t\mathbf{j} + (2t - 1)^2\mathbf{k}, \quad 0 \le t \le 1$

22. $\mathbf{F}(x, y, z) = -y\mathbf{i} + x\mathbf{j} + 3xz^2\mathbf{k}$

(a) $\mathbf{r}_1(t) = \cos t\,\mathbf{i} + \sin t\,\mathbf{j} + t\mathbf{k}, \quad 0 \le t \le \pi$

(b) $\mathbf{r}_2(t) = (1 - 2t)\mathbf{i} + \pi t\mathbf{k}, \quad 0 \le t \le 1$

23. $\mathbf{F}(x, y, z) = e^z(y\mathbf{i} + x\mathbf{j} + xy\mathbf{k})$

(a) $\mathbf{r}_1(t) = 4\cos t\,\mathbf{i} + 4\sin t\,\mathbf{j} + 3\mathbf{k}, \quad 0 \le t \le \pi$

(b) $\mathbf{r}_2(t) = (4 - 8t)\mathbf{i} + 3\mathbf{k}, \quad 0 \le t \le 1$

24. $\mathbf{F}(x, y, z) = y\sin z\,\mathbf{i} + x\sin z\,\mathbf{j} + xy\cos x\,\mathbf{k}$

(a) $\mathbf{r}_1(t) = t^2\mathbf{i} + t^2\mathbf{j}, \quad 0 \le t \le 2$

(b) $\mathbf{r}_2(t) = 4t\mathbf{i} + 4t\mathbf{j}, \quad 0 \le t \le 1$

In Exercises 25–34, evaluate the line integral using the Fundamental Theorem of Line Integrals. Use a computer algebra system to verify your results.

25. $\int_C (y\mathbf{i} + x\mathbf{j}) \cdot d\mathbf{r}$

C: smooth curve from $(0, 0)$ to $(3, 8)$

26. $\int_C [2(x + y)\mathbf{i} + 2(x + y)\mathbf{j}] \cdot d\mathbf{r}$

C: smooth curve from $(-2, 2)$ to $(4, 3)$

27. $\int_C \cos x \sin y\, dx + \sin x \cos y\, dy$

C: smooth curve from $(0, -\pi)$ to $\left(\dfrac{3\pi}{2}, \dfrac{\pi}{2}\right)$

28. $\int_C \dfrac{y\,dx - x\,dy}{x^2 + y^2}$

C: smooth curve from $(1, 1)$ to $(2\sqrt{3}, 2)$

29. $\int_C e^x \sin y\, dx + e^x \cos y\, dy$

C: cycloid $x = \theta - \sin\theta,\ y = 1 - \cos\theta$ from $(0, 0)$ to $(2\pi, 0)$

30. $\int_C \dfrac{2x}{(x^2 + y^2)^2}\, dx + \dfrac{2y}{(x^2 + y^2)^2}\, dy$

C: circle $(x - 4)^2 + (y - 5)^2 = 9$ clockwise from $(7, 5)$ to $(1, 5)$

31. $\int_C (y + 2z)\, dx + (x - 3z)\, dy + (2x - 3y)\, dz$

(a) C: line segment from $(0, 0, 0)$ to $(1, 1, 1)$

(b) C: line segments from $(0, 0, 0)$ to $(0, 0, 1)$ to $(1, 1, 1)$

(c) C: line segments from $(0, 0, 0)$ to $(1, 0, 0)$ to $(1, 1, 0)$ to $(1, 1, 1)$

32. Repeat Exercise 31 using the integral

$$\int_C zy\, dx + xz\, dy + xy\, dz.$$

33. $\int_C -\sin x\, dx + z\, dy + y\, dz$

C: smooth curve from $(0, 0, 0)$ to $\left(\dfrac{\pi}{2}, 3, 4\right)$

34. $\int_C 6x\, dx - 4z\, dy - (4y - 20z)\, dz$

C: smooth curve from $(0, 0, 0)$ to $(4, 3, 1)$

Work In Exercises 35 and 36, find the work done by the force field **F** in moving an object from P to Q.

35. $\mathbf{F}(x, y) = 9x^2y^2\mathbf{i} + (6x^3y - 1)\mathbf{j}$

$P(0, 0),\ Q(5, 9)$

36. $\mathbf{F}(x, y) = \dfrac{2x}{y}\mathbf{i} - \dfrac{x^2}{y^2}\mathbf{j}$

$P(-3, 2),\ Q(1, 4)$

37. Work A stone weighing 1 pound is attached to the end of a 2-foot string and is whirled horizontally with one end held fixed. It makes 1 revolution per second. Find the work done by the force **F** that keeps the stone moving in a circular path. [*Hint:* Use Force = (mass)(centripetal acceleration).]

38. Work If $\mathbf{F}(x, y, z) = a_1\mathbf{i} + a_2\mathbf{j} + a_3\mathbf{k}$ is a constant force vector field, show that the work done in moving a particle along any path from P to Q is

$$W = \mathbf{F} \cdot \overrightarrow{PQ}.$$

39. Kinetic and Potential Energy The kinetic energy of an object moving through a conservative force field is decreasing at a rate of 10 units per minute. At what rate is the potential energy changing?

40. Work To allow a means of escape for workers in a hazardous job 50 meters above ground level, a slide wire has been installed. It runs from their position to a point on the ground 50 meters from the base of the installation where they are located. Show that the work done by the gravitational force field for a 150-pound man moving the length of the slide wire is the same for the following two paths.

(a) $\mathbf{r}(t) = t\mathbf{i} + (50 - t)\mathbf{j}$

(b) $\mathbf{r}(t) = t\mathbf{i} + \tfrac{1}{50}(50 - t)^2\mathbf{j}$

41. Work Can you find a path for the slide wire in Exercise 40 such that the work done by the gravitational force field would differ from the amounts of work done for the two paths given? Explain why or why not.

42. Let $\mathbf{F}(x, y) = \dfrac{y}{x^2 + y^2}\mathbf{i} - \dfrac{x}{x^2 + y^2}\mathbf{j}$.

   (a) Show that
   $$\dfrac{\partial N}{\partial x} = \dfrac{\partial M}{\partial y}$$
   where
   $$M = \dfrac{y}{x^2 + y^2} \text{ and } N = \dfrac{-x}{x^2 + y^2}.$$

   (b) If $\mathbf{r}(t) = \cos t\,\mathbf{i} + \sin t\,\mathbf{j}$ for $0 \le t \le \pi$, find $\int_C \mathbf{F} \cdot d\mathbf{r}$.

   (c) If $\mathbf{r}(t) = \cos t\,\mathbf{i} - \sin t\,\mathbf{j}$ for $0 \le t \le \pi$, find $\int_C \mathbf{F} \cdot d\mathbf{r}$.

   (d) If $\mathbf{r}(t) = \cos t\,\mathbf{i} + \sin t\,\mathbf{j}$ for $0 \le t \le 2\pi$, find $\int_C \mathbf{F} \cdot d\mathbf{r}$. Why doesn't this contradict Theorem 14.7?

   (e) Show that
   $$\nabla\left(\arctan \dfrac{x}{y}\right) = \mathbf{F}.$$

Getting at the Concept

43. State the Fundamental Theorem of Line Integrals.

44. What does it mean that a line integral is independent of path? State the method for determining if a line integral is independent of path.

45. Consider the force field shown in the figure.

   (a) Give a verbal argument that the force field is not conservative because you can identify two paths that require different amounts of work to move an object from $(-4, 0)$ to $(3, 4)$. Identify two paths and state which requires the greater amount of work. To print an enlarged copy of the graph, go to the website *www.mathgraphs.com*.

   (b) Give a verbal argument that the force field is not conservative because you can find a closed curve C such that
   $$\int_C \mathbf{F} \cdot d\mathbf{r} \neq 0.$$

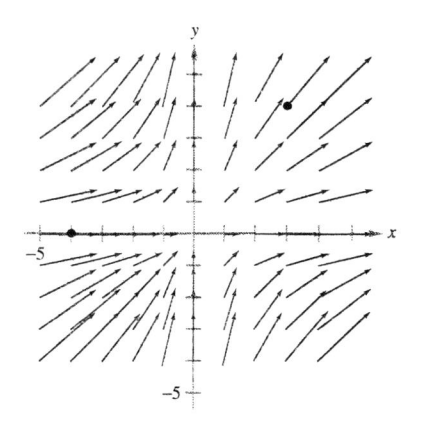

46. *Wind Speed and Direction* The map shows the jet stream wind speed vectors over the United States for February 24, 2000. In planning a flight from Dallas to Atlanta in a small plane at an altitude of 5000 feet, is the amount of fuel required independent of the flight path? Is the vector field conservative? Explain.

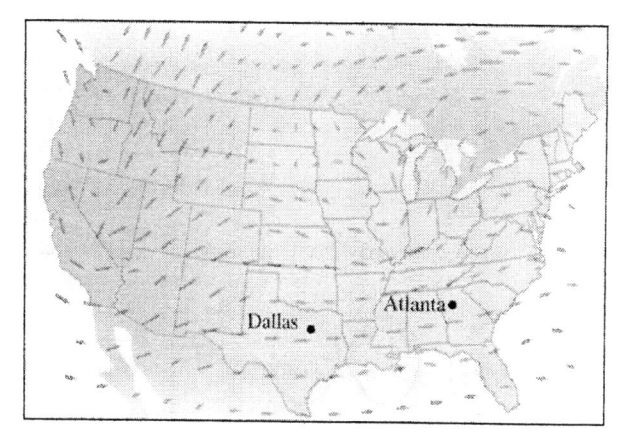

True or False? In Exercises 47–50, determine whether the statement is true or false. If it is false, explain why or give an example that shows it is false.

47. If C_1, C_2, and C_3 have the same initial and terminal points and $\int_{C_1} \mathbf{F} \cdot d\mathbf{r}_1 = \int_{C_2} \mathbf{F} \cdot d\mathbf{r}_2$, then $\int_{C_1} \mathbf{F} \cdot d\mathbf{r}_1 = \int_{C_3} \mathbf{F} \cdot d\mathbf{r}_3$.

48. If $\mathbf{F} = y\mathbf{i} + x\mathbf{j}$ and C is given by $\mathbf{r}(t) = (4\sin t)\mathbf{i} + (3\cos t)\mathbf{j}$, $0 \le t \le \pi$, then $\int_C \mathbf{F} \cdot d\mathbf{r} = 0$.

49. If $\mathbf{F}$ is conservative in a region R bounded by a simple closed path and C lies within R, then $\int_C \mathbf{F} \cdot d\mathbf{r}$ is independent of path.

50. If $\mathbf{F} = M\mathbf{i} + N\mathbf{j}$ and $\partial M/\partial x = \partial N/\partial y$, then $\mathbf{F}$ is conservative.

51. f is called *harmonic* if $\dfrac{\partial^2 f}{\partial x^2} + \dfrac{\partial^2 f}{\partial y^2} = 0$.

   Prove that if f is harmonic, then
   $$\int_C \left(\dfrac{\partial f}{\partial y} dx - \dfrac{\partial f}{\partial x} dy\right) = 0$$
   where C is a smooth closed curve in the plane.

Section 14.4 Green's Theorem

- Use Green's Theorem to evaluate a line integral.
- Use alternative forms of Green's Theorem.

Green's Theorem

In this section, you will study **Green's Theorem,** named after the English mathematician George Green (1793–1841). This theorem states that the value of a double integral over a *simply connected* plane region R is determined by the value of a line integral around the boundary of R.

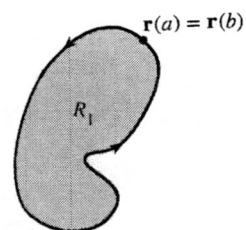

A curve C given by $\mathbf{r}(t) = x(t)\mathbf{i} + y(t)\mathbf{j}$, where $a \leq t \leq b$, is **simple** if it does not cross itself—that is, $\mathbf{r}(c) \neq \mathbf{r}(d)$ for all c and d in the open interval (a, b). A plane region R is **simply connected** if its boundary consists of *one* simple closed curve, as shown in Figure 14.26.

Figure 14.26

Simply connected

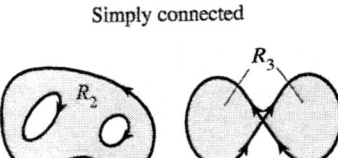

Not simply connected

THEOREM 14.8 Green's Theorem

Let R be a simply connected region with a piecewise smooth boundary C, oriented counterclockwise (that is, C is traversed *once* so that the region R always lies to the *left*). If M and N have continuous partial derivatives in an open region containing R, then

$$\int_C M\,dx + N\,dy = \iint_R \left(\frac{\partial N}{\partial x} - \frac{\partial M}{\partial y}\right) dA.$$

Proof We give a proof only for a region that is both vertically simple and horizontally simple, as shown in Figure 14.27.

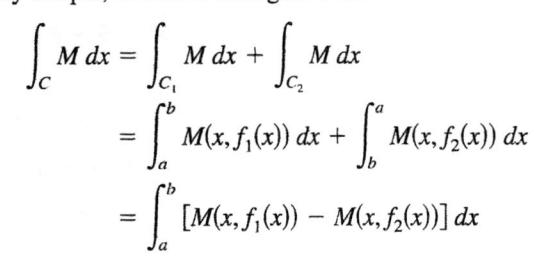

$$\int_C M\,dx = \int_{C_1} M\,dx + \int_{C_2} M\,dx$$

$$= \int_a^b M(x, f_1(x))\,dx + \int_b^a M(x, f_2(x))\,dx$$

$$= \int_a^b [M(x, f_1(x)) - M(x, f_2(x))]\,dx$$

On the other hand,

$$\iint_R \frac{\partial M}{\partial y}\,dA = \int_a^b \int_{f_1(x)}^{f_2(x)} \frac{\partial M}{\partial y}\,dy\,dx$$

$$= \int_a^b M(x, y)\Big]_{f_1(x)}^{f_2(x)}\,dx$$

$$= \int_a^b [M(x, f_2(x)) - M(x, f_1(x))]\,dx.$$

Consequently,

$$\int_C M\,dx = -\iint_R \frac{\partial M}{\partial y}\,dA.$$

Similarly, you can use $g_1(y)$ and $g_2(y)$ to show that $\int_C N\,dy = \iint_R \partial N/\partial x\,dA$. By adding the integrals $\int_C M\,dx$ and $\int_C N\,dy$, you obtain the conclusion stated in the theorem.

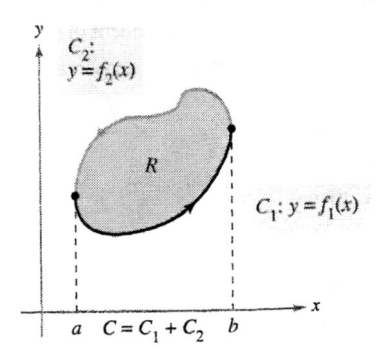

R is vertically simple.

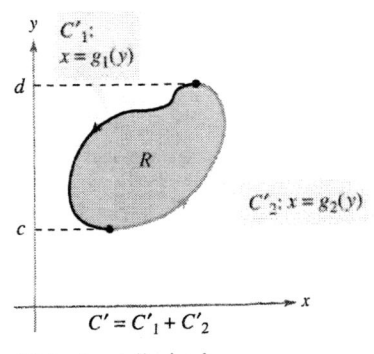

R is horizontally simple.

Figure 14.27

Example 1 Using Green's Theorem

Use Green's Theorem to evaluate the line integral

$$\int_C y^3 \, dx + (x^3 + 3xy^2) \, dy$$

where C is the path from $(0, 0)$ to $(1, 1)$ along the graph of $y = x^3$ and from $(1, 1)$ to $(0, 0)$ along the graph of $y = x$, as shown in Figure 14.28.

Solution Because $M = y^3$ and $N = x^3 + 3xy^2$, it follows that

$$\frac{\partial N}{\partial x} = 3x^2 + 3y^2 \quad \text{and} \quad \frac{\partial M}{\partial y} = 3y^2.$$

Applying Green's Theorem, you then have

$$\int_C y^3 \, dx + (x^3 + 3xy^2) \, dy = \int\int_R \left(\frac{\partial N}{\partial x} - \frac{\partial M}{\partial y}\right) dA$$

$$= \int_0^1 \int_{x^3}^{x} [(3x^2 + 3y^2) - 3y^2] \, dy \, dx$$

$$= \int_0^1 \int_{x^3}^{x} 3x^2 \, dy \, dx$$

$$= \int_0^1 3x^2 y \Big]_{x^3}^{x} dx$$

$$= \int_0^1 (3x^3 - 3x^5) \, dx$$

$$= \left[\frac{3x^4}{4} - \frac{x^6}{2}\right]_0^1$$

$$= \frac{1}{4}.$$

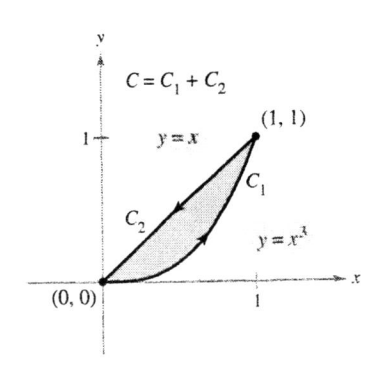

C is simple and closed, and the region R always lies to the left of C.
Figure 14.28

GEORGE GREEN (1793–1841)

Green, a self-educated miller's son, first published the theorem that bears his name in 1828 in an essay on electricity and magnetism. At that time there was almost no mathematical theory to explain electrical phenomena. "Considering how desirable it was that a power of universal agency, like electricity, should, as far as possible, be submitted to calculation, . . . I was induced to try whether it would be possible to discover any general relations existing between this function and the quantities of electricity in the bodies producing it."

Green's Theorem cannot be applied to every line integral. Among other restrictions stated in Theorem 14.8, the curve C must be simple and closed. When Green's Theorem does apply, however, it can save time. To see this, try using the techniques described in Section 14.2 to evaluate the line integral in Example 1. To do this, you would need to write the line integral as

$$\int_C y^3 \, dx + (x^3 + 3xy^2) \, dy =$$

$$\int_{C_1} y^3 \, dx + (x^3 + 3xy^2) \, dy + \int_{C_2} y^3 \, dx + (x^3 + 3xy^2) \, dy$$

where C_1 is the cubic path given by

$$\mathbf{r}(t) = t\mathbf{i} + t^3\mathbf{j}$$

from $t = 0$ to $t = 1$, and C_2 is the line segment given by

$$\mathbf{r}(t) = (1 - t)\mathbf{i} + (1 - t)\mathbf{j}$$

from $t = 0$ to $t = 1$.

1044   CHAPTER 14   Vector Analysis

$\mathbf{F}(x, y) = y^3\mathbf{i} + (x^3 + 3xy^2)\mathbf{j}$

The work done by **F** as a particle travels once around the circle is $\dfrac{243\pi}{4}$.
Figure 14.29

Example 2   Using Green's Theorem to Calculate Work

While subject to the force

$$\mathbf{F}(x, y) = y^3\mathbf{i} + (x^3 + 3xy^2)\mathbf{j}$$

a particle travels once around the circle of radius 3 shown in Figure 14.29. Use Green's Theorem to find the work done by **F**.

Solution   From Example 1, you know by Green's Theorem that

$$\int_C y^3\, dx + (x^3 + 3xy^2)\, dy = \iint_R 3x^2\, dA.$$

In polar coordinates, using $x = r\cos\theta$ and $dA = r\, dr\, d\theta$, the work done is

$$\begin{aligned}
W &= \iint_R 3x^2\, dA = \int_0^{2\pi}\!\int_0^3 3(r\cos\theta)^2\, r\, dr\, d\theta \\
&= 3\int_0^{2\pi}\!\int_0^3 r^3\cos^2\theta\, dr\, d\theta \\
&= 3\int_0^{2\pi} \frac{r^4}{4}\cos^2\theta\Big]_0^3 d\theta \\
&= 3\int_0^{2\pi} \frac{81}{4}\cos^2\theta\, d\theta \\
&= \frac{243}{8}\int_0^{2\pi} (1 + \cos 2\theta)\, d\theta \\
&= \frac{243}{8}\left[\theta + \frac{\sin 2\theta}{2}\right]_0^{2\pi} \\
&= \frac{243\pi}{4}.
\end{aligned}$$

When evaluating line integrals over closed curves, remember that for conservative vector fields (those for which $\partial N/\partial x = \partial M/\partial y$), the value of the line integral is 0. This is easily seen from the statement of Green's Theorem:

$$\int_C M\, dx + N\, dy = \iint_R \left(\frac{\partial N}{\partial x} - \frac{\partial M}{\partial y}\right) dA = 0.$$

Example 3   Green's Theorem and Conservative Vector Fields

Evaluate the line integral

$$\int_C y^3\, dx + 3xy^2\, dy$$

where C is the path shown in Figure 14.30.

Solution   From this line integral, $M = y^3$ and $N = 3xy^2$. So, $\partial N/\partial x = 3y^2$ and $\partial M/\partial y = 3y^2$. This implies that the vector field $\mathbf{F} = M\mathbf{i} + N\mathbf{j}$ is conservative, and because C is closed, you can conclude that

$$\int_C y^3\, dx + 3xy^2\, dy = 0.$$

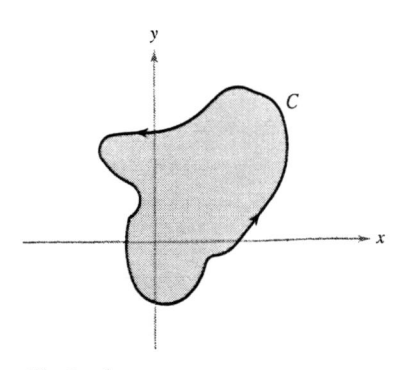

C is closed.
Figure 14.30

Example 4 Using Green's Theorem for a Piecewise Smooth Curve

Evaluate

$$\int_C (\arctan x + y^2)\, dx + (e^y - x^2)\, dy$$

where C is the path enclosing the annular region shown in Figure 14.31.

Solution In polar coordinates, R is given by $1 \leq r \leq 3$ for $0 \leq \theta \leq \pi$. Moreover,

$$\frac{\partial N}{\partial x} - \frac{\partial M}{\partial y} = -2x - 2y = -2(r\cos\theta + r\sin\theta).$$

So, by Green's Theorem,

$$\int_C (\arctan x + y^2)\, dx + (e^y - x^2)\, dy = \int\!\!\int_R -2(x+y)\, dA$$

$$= \int_0^\pi \int_1^3 -2r(\cos\theta + \sin\theta) r\, dr\, d\theta$$

$$= \int_0^\pi -2(\cos\theta + \sin\theta) \frac{r^3}{3}\bigg]_1^3 d\theta$$

$$= \int_0^\pi \left(-\frac{52}{3}\right)(\cos\theta + \sin\theta)\, d\theta$$

$$= -\frac{52}{3}\bigg[\sin\theta - \cos\theta\bigg]_0^\pi$$

$$= -\frac{104}{3}.$$

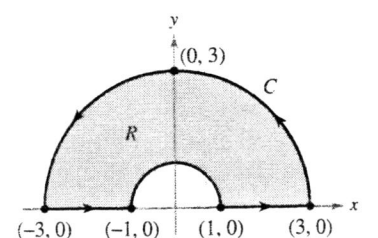

C is piecewise smooth.
Figure 14.31

In Examples 1, 2, and 4, Green's Theorem was used to evaluate line integrals as double integrals. You can also use the theorem to evaluate double integrals as line integrals. One useful application occurs when $\partial N/\partial x - \partial M/\partial y = 1$.

$$\int_C M\, dx + N\, dy = \int\!\!\int_R \left(\frac{\partial N}{\partial x} - \frac{\partial M}{\partial y}\right) dA$$

$$= \int\!\!\int_R 1\, dA \qquad \frac{\partial N}{\partial x} - \frac{\partial M}{\partial y} = 1$$

$$= \text{area of region } R$$

Among the many choices for M and N satisfying the stated condition, the choice of $M = -y/2$ and $N = x/2$ produces the following line integral for the area of region R.

THEOREM 14.9 Line Integral for Area

If R is a plane region bounded by a piecewise smooth simple closed curve C, oriented counterclockwise, then the area of R is given by

$$A = \frac{1}{2}\int_C x\, dy - y\, dx.$$

Example 5 Finding Area by a Line Integral

Use a line integral to find the area of the ellipse

$$\frac{x^2}{a^2} + \frac{y^2}{b^2} = 1.$$

Solution Using Figure 14.32, you can induce a counterclockwise orientation to the elliptical path by letting

$$x = a \cos t \quad \text{and} \quad y = b \sin t, \quad 0 \le t \le 2\pi.$$

Therefore, the area is

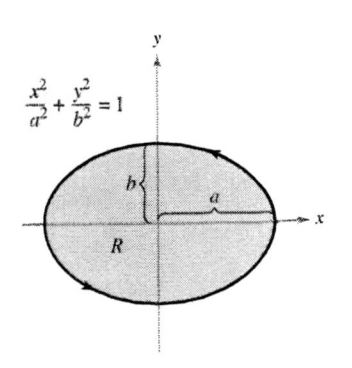

$$\begin{aligned}
A &= \frac{1}{2} \int_C x \, dy - y \, dx = \frac{1}{2} \int_0^{2\pi} [(a \cos t)(b \cos t) \, dt - (b \sin t)(-a \sin t) \, dt] \\
&= \frac{ab}{2} \int_0^{2\pi} (\cos^2 t + \sin^2 t) \, dt \\
&= \frac{ab}{2} \Big[t \Big]_0^{2\pi} \\
&= \pi ab.
\end{aligned}$$

The area of the ellipse is πab.
Figure 14.32

Green's Theorem can be extended to cover some regions that are not simply connected. This is demonstrated in the next example.

Example 6 Green's Theorem Extended to a Region with a Hole

Let R be the region inside the ellipse $(x^2/9) + (y^2/4) = 1$ and outside the circle $x^2 + y^2 = 1$. Evaluate the line integral

$$\int_C 2xy \, dx + (x^2 + 2x) \, dy$$

where $C = C_1 + C_2$ is the boundary of R, as shown in Figure 14.33.

Solution To begin, we introduce the line segments C_3 and C_4, as shown in Figure 14.33. Note that because the curves C_3 and C_4 have opposite orientations, the line integrals over them cancel. Furthermore, you can apply Green's Theorem to the region R using the boundary $C_1 + C_4 + C_2 + C_3$ to obtain

$$\begin{aligned}
\int_C 2xy \, dx + (x^2 + 2x) \, dy &= \int\!\!\!\int_R \left(\frac{\partial N}{\partial x} - \frac{\partial M}{\partial y} \right) dA \\
&= \int\!\!\!\int_R (2x + 2 - 2x) \, dA \\
&= 2 \int\!\!\!\int_R dA \\
&= 2(\text{area of } R) \\
&= 2(\pi ab - \pi r^2) \\
&= 2[\pi(3)(2) - \pi(1^2)] \\
&= 10\pi.
\end{aligned}$$

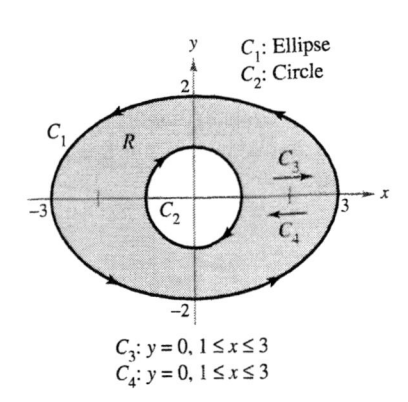

C_1: Ellipse
C_2: Circle
C_3: $y = 0, 1 \le x \le 3$
C_4: $y = 0, 1 \le x \le 3$

Figure 14.33

In Section 14.1, we listed a necessary and sufficient condition for conservative vector fields. There, we proved only one direction of the proof. We now outline the other direction, using Green's Theorem. Let $\mathbf{F}(x, y) = M\mathbf{i} + N\mathbf{j}$ be defined on an open disk R. We want to show that if M and N have continuous first partial derivatives and

$$\frac{\partial M}{\partial y} = \frac{\partial N}{\partial x}$$

then $\mathbf{F}$ is conservative. Suppose that C is a closed path forming the boundary of a connected region lying in R. Then, using the fact that $\partial M/\partial y = \partial N/\partial x$, you can apply Green's Theorem to conclude that

$$\begin{aligned}
\int_C \mathbf{F} \cdot d\mathbf{r} &= \int_C M\, dx + N\, dy \\
&= \int\int_R \left(\frac{\partial N}{\partial x} - \frac{\partial M}{\partial y}\right) dA \\
&= 0.
\end{aligned}$$

This, in turn, is equivalent to showing that $\mathbf{F}$ is conservative (see Theorem 14.7).

Alternative Forms of Green's Theorem

We conclude this section with the derivation of two vector forms of Green's Theorem for regions in the plane. The extension of these vector forms to three dimensions is the basis for the discussion in the remaining sections of this chapter. If $\mathbf{F}$ is a vector field in the plane, you can write

$$\mathbf{F}(x, y, z) = M\mathbf{i} + N\mathbf{j} + 0\mathbf{k}$$

so that the curl of $\mathbf{F}$, as described in Section 14.1, is given by

$$\begin{aligned}
\mathbf{curl}\ \mathbf{F} = \nabla \times \mathbf{F} &= \begin{vmatrix} \mathbf{i} & \mathbf{j} & \mathbf{k} \\ \frac{\partial}{\partial x} & \frac{\partial}{\partial y} & \frac{\partial}{\partial z} \\ M & N & 0 \end{vmatrix} \\
&= -\frac{\partial N}{\partial z}\mathbf{i} + \frac{\partial M}{\partial z}\mathbf{j} + \left(\frac{\partial N}{\partial x} - \frac{\partial M}{\partial y}\right)\mathbf{k}.
\end{aligned}$$

Consequently,

$$\begin{aligned}
(\mathbf{curl}\ \mathbf{F}) \cdot \mathbf{k} &= \left[-\frac{\partial N}{\partial z}\mathbf{i} + \frac{\partial M}{\partial z}\mathbf{j} + \left(\frac{\partial N}{\partial x} - \frac{\partial M}{\partial y}\right)\mathbf{k}\right] \cdot \mathbf{k} \\
&= \frac{\partial N}{\partial x} - \frac{\partial M}{\partial y}.
\end{aligned}$$

With appropriate conditions on $\mathbf{F}$, C, and R, you can write Green's Theorem in the vector form

$$\begin{aligned}
\int_C \mathbf{F} \cdot d\mathbf{r} &= \int\int_R \left(\frac{\partial N}{\partial x} - \frac{\partial M}{\partial y}\right) dA \\
&= \int\int_R (\mathbf{curl}\ \mathbf{F}) \cdot \mathbf{k}\, dA. \qquad \text{First alternative form}
\end{aligned}$$

The extension of this vector form of Green's Theorem to surfaces in space produces **Stokes's Theorem,** discussed in Section 14.8.

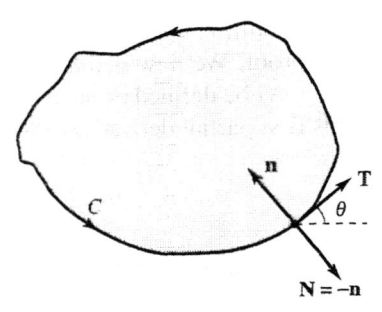

$T = \cos\theta \mathbf{i} + \sin\theta \mathbf{j}$

$\mathbf{n} = \cos\left(\theta + \dfrac{\pi}{2}\right)\mathbf{i} + \sin\left(\theta + \dfrac{\pi}{2}\right)\mathbf{j}$
$= -\sin\theta \mathbf{i} + \cos\theta \mathbf{j}$
$\mathbf{N} = \sin\theta \mathbf{i} - \cos\theta \mathbf{j}$

Figure 14.34

For the second vector form of Green's Theorem, assume the same conditions for $\mathbf{F}$, C, and R. Using the arc length parameter s for C, you have $\mathbf{r}(s) = x(s)\mathbf{i} + y(s)\mathbf{j}$. So, a unit tangent vector $\mathbf{T}$ to curve C is given by

$$\mathbf{r}'(s) = \mathbf{T} = x'(s)\mathbf{i} + y'(s)\mathbf{j}.$$

From Figure 14.34 you can see that the *outward* unit normal vector $\mathbf{N}$ can then be written as

$$\mathbf{N} = y'(s)\mathbf{i} - x'(s)\mathbf{j}.$$

Consequently, for $\mathbf{F}(x, y) = M\mathbf{i} + N\mathbf{j}$, you can apply Green's Theorem to obtain

$$\int_C \mathbf{F} \cdot \mathbf{N}\, ds = \int_a^b (M\mathbf{i} + N\mathbf{j}) \cdot (y'(s)\mathbf{i} - x'(s)\mathbf{j})\, ds$$

$$= \int_a^b \left(M\frac{dy}{ds} - N\frac{dx}{ds} \right) ds$$

$$= \int_C M\, dy - N\, dx$$

$$= \int_C -N\, dx + M\, dy$$

$$= \iint_R \left(\frac{\partial M}{\partial x} + \frac{\partial N}{\partial y} \right) dA \qquad \text{Green's Theorem}$$

$$= \iint_R \text{div } \mathbf{F}\, dA.$$

Therefore,

$$\int_C \mathbf{F} \cdot \mathbf{N}\, ds = \iint_R \text{div } \mathbf{F}\, dA. \qquad \text{Second alternative form}$$

The extension of this form to three dimensions is called the **Divergence Theorem**, discussed in Section 14.7. The physical interpretations of divergence and curl will be discussed in Sections 14.7 and 14.8.

EXERCISES FOR SECTION 14.4

In Exercises 1–4, verify Green's Theorem by evaluating both integrals

$$\int_C y^2\, dx + x^2\, dy = \iint_R \left(\frac{\partial N}{\partial x} - \frac{\partial M}{\partial y} \right) dA$$

for the indicated path.

1. C: square with vertices $(0, 0)$, $(4, 0)$, $(4, 4)$, $(0, 4)$
2. C: triangle with vertices $(0, 0)$, $(4, 0)$, $(4, 4)$
3. C: boundary of the region lying between the graphs of $y = x$ and $y = x^2/4$
4. C: circle given by $x^2 + y^2 = 1$

In Exercises 5 and 6, verify Green's Theorem by using a computer algebra system to evaluate both integrals

$$\int_C xe^y\, dx + e^x\, dy = \iint_R \left(\frac{\partial N}{\partial x} - \frac{\partial M}{\partial y} \right) dA$$

for the indicated path.

5. C: circle given by $x^2 + y^2 = 4$
6. C: boundary of the region lying between the graphs of $y = x$ and $y = x^3$

In Exercises 7–10, use Green's Theorem to evaluate the integral

$$\int_C (y - x)\, dx + (2x - y)\, dy$$

for the indicated path.

7. C: boundary of the region lying between the graphs of $y = x$ and $y = x^2 - x$
8. C: $x = 2\cos\theta$, $y = \sin\theta$
9. C: boundary of the region lying inside the rectangle bounded by $x = -5$, $x = 5$, $y = -3$, and $y = 3$, and outside the square bounded by $x = -1$, $x = 1$, $y = -1$, and $y = 1$
10. C: boundary of the region lying inside the semicircle $y = \sqrt{25 - x^2}$ and outside the semicircle $y = \sqrt{9 - x^2}$

In Exercises 11–20, use Green's Theorem to evaluate the line integral.

11. $\int_C 2xy\, dx + (x + y)\, dy$

    C: boundary of the region lying between the graphs of $y = 0$ and $y = 4 - x^2$

12. $\int_C y^2\, dx + xy\, dy$

    C: boundary of the region lying between the graphs of $y = 0$, $y = \sqrt{x}$, and $x = 9$

13. $\int_C (x^2 - y^2)\, dx + 2xy\, dy$

    C: $x^2 + y^2 = a^2$

14. $\int_C (x^2 - y^2)\, dx + 2xy\, dy$

    C: $r = 1 + \cos\theta$

15. $\int_C 2\arctan\frac{y}{x}\, dx + \ln(x^2 + y^2)\, dy$

    C: $x = 4 + 2\cos\theta$, $y = 4 + \sin\theta$

16. $\int_C e^x \cos 2y\, dx - 2e^x \sin 2y\, dy$

    C: $x^2 + y^2 = a^2$

17. $\int_C \sin x \cos y\, dx + (xy + \cos x \sin y)\, dy$

    C: boundary of the region lying between the graphs of $y = x$, and $y = \sqrt{x}$

18. $\int_C (e^{-x^2/2} - y)\, dx + (e^{-y^2/2} + x)\, dy$

    C: boundary of the region lying between the graphs of the circle $x = 6\cos\theta$, $y = 6\sin\theta$ and the ellipse $x = 3\cos\theta$, $y = 2\sin\theta$

19. $\int_C xy\, dx + (x + y)\, dy$

    C: boundary of the region lying between the graphs of $x^2 + y^2 = 1$ and $x^2 + y^2 = 9$

20. $\int_C 3x^2 e^y\, dx + e^y\, dy$

    C: boundary of the region lying between the squares with vertices $(1, 1)$, $(-1, 1)$, $(-1, -1)$, and $(1, -1)$, and $(2, 2)$, $(-2, 2)$, $(-2, -2)$, and $(2, -2)$

Work In Exercises 21–24, use Green's Theorem to calculate the work done by the force F on a particle that is moving counterclockwise around the closed path C.

21. $\mathbf{F}(x, y) = xy\mathbf{i} + (x + y)\mathbf{j}$

    C: $x^2 + y^2 = 4$

22. $\mathbf{F}(x, y) = (e^x - 3y)\mathbf{i} + (e^y + 6x)\mathbf{j}$

    C: $r = 2\cos\theta$

23. $\mathbf{F}(x, y) = (x^{3/2} - 3y)\mathbf{i} + (6x + 5\sqrt{y})\mathbf{j}$

    C: boundary of the triangle with vertices $(0, 0)$, $(5, 0)$, and $(0, 5)$

24. $\mathbf{F}(x, y) = (3x^2 + y)\mathbf{i} + 4xy^2\mathbf{j}$

    C: boundary of the region lying between the graphs of $y = \sqrt{x}$, $y = 0$, and $x = 9$

Area In Exercises 25–28, use a line integral to find the area of the region R.

25. R: region bounded by the graph of $x^2 + y^2 = a^2$
26. R: triangle bounded by the graphs of $x = 0$, $3x - 2y = 0$, and $x + 2y = 8$
27. R: region bounded by the graphs of $y = 2x + 1$ and $y = 4 - x^2$
28. R: region inside the loop of the folium of Descartes bounded by the graph of

    $x = \dfrac{3t}{t^3 + 1}$, $y = \dfrac{3t^2}{t^3 + 1}$

Getting at the Concept

29. State Green's Theorem.
30. Give the line integral for the area of a region R bounded by a piecewise smooth simple curve C.
31. Write three force fields in the plane such that the line integral of each field on the unit circle centered at the origin is 0.

32. Use Green's Theorem to verify the line integral formulas.

    (a) The centroid of the region having area A bounded by the simple closed path C is

    $\bar{x} = \dfrac{1}{2A}\int_C x^2\, dy$, $\bar{y} = -\dfrac{1}{2A}\int_C y^2\, dx$.

    (b) The area of a plane region bounded by the simple closed path C given in polar coordinates is

    $A = \dfrac{1}{2}\int_C r^2\, d\theta$.

Centroid In Exercises 33–36, use a computer algebra system and the results in Exercise 32 to find the centroid of the region.

33. R: region bounded by the graphs of $y = 0$ and $y = 4 - x^2$
34. R: region bounded by the graphs of $y = \sqrt{a^2 - x^2}$ and $y = 0$
35. R: region bounded by the graphs of $y = x^3$ and $y = x$, $0 \le x \le 1$
36. R: triangle with vertices $(-a, 0)$, $(a, 0)$, and (b, c), where $-a \le b \le a$

Area In Exercises 37–40, use a computer algebra system and the results in Exercise 32 to find the area of the region bounded by the graph of the polar equation.

37. $r = a(1 - \cos\theta)$
38. $r = a\cos 3\theta$
39. $r = 1 + 2\cos\theta$ (inner loop)
40. $r = \dfrac{3}{2 - \cos\theta}$

41. **Think About It** Let

    $I = \int_C \dfrac{y\, dx - x\, dy}{x^2 + y^2}$

    where C is a circle orientated counterclockwise. Show that $I = 0$ if C does not contain the origin. What is I if C contains the origin?

42. (a) Let C be the line segment joining (x_1, y_1) and (x_2, y_2). Show that
$$\int_C -y\, dx + x\, dy = x_1 y_2 - x_2 y_1.$$

(b) Let (x_1, y_1), (x_2, y_2), ..., (x_n, y_n) be the vertices of a polygon. Prove that the area enclosed is
$$\tfrac{1}{2}[(x_1 y_2 - x_2 y_1) + (x_2 y_3 - x_3 y_2) + \cdots + (x_{n-1} y_n - x_n y_{n-1}) + (x_n y_1 - x_1 y_n)].$$

Area In Exercises 43 and 44, use the result of Exercise 42(b) to find the area enclosed by the polygon with the given vertices.

43. Pentagon: $(0, 0), (2, 0), (3, 2), (1, 4), (-1, 1)$
44. Hexagon: $(0, 0), (2, 0), (3, 2), (2, 4), (0, 3), (-1, 1)$

45. *Investigation* Consider the line integral
$$\int_C y^n\, dx + x^n\, dy$$
where C is the boundary of the region lying between the graphs of $y = \sqrt{a^2 - x^2}$ $(a > 0)$ and $y = 0$.

(a) Use a computer algebra system to verify Green's Theorem for n, an odd integer from 1 through 7.

(b) Use a computer algebra system to verify Green's Theorem for n, an even integer from 2 through 8.

(c) For n an odd integer, make a conjecture about the value of the integral.

In Exercises 46 and 47, prove the identity where R is a simply connected region with boundary C. Assume that the required partial derivatives of the scalar functions f and g are continuous. The expressions $D_N f$ and $D_N g$ are the derivatives in the direction of the outward normal vector N of C, and are defined by
$$D_N f = \nabla f \cdot \mathbf{N}, \quad D_N g = \nabla g \cdot \mathbf{N}.$$

46. Green's first identity:
$$\iint_R (f\nabla^2 g + \nabla f \cdot \nabla g)\, dA = \int_C f D_N g\, ds$$

[*Hint:* Use the second alternative form of Green's Theorem and the property div $(f\mathbf{G}) = f$ div $\mathbf{G} + \nabla f \cdot \mathbf{G}$.]

47. Green's second identity:
$$\iint_R (f\nabla^2 g - g\nabla^2 f)\, dA = \int_C (f D_N g - g D_N f)\, ds$$

(*Hint:* Use Exercise 46 twice.)

48. Use Green's Theorem to prove that
$$\int_C f(x)\, dx + g(y)\, dy = 0$$
if f and g are differentiable functions and C is a piecewise smooth simple closed path.

49. Let $\mathbf{F} = M\mathbf{i} + N\mathbf{j}$, where M and N have continuous first partial derivatives in a simply connected region R. Prove that if C is simple, smooth, and closed, and $N_x = M_y$, then
$$\int_C \mathbf{F} \cdot d\mathbf{r} = 0.$$

SECTION PROJECT **HYPERBOLIC AND TRIGONOMETRIC FUNCTIONS**

(a) Sketch the plane curve represented by the vector-valued function $\mathbf{r}(t) = \cosh t\mathbf{i} + \sinh t\mathbf{j}$ on the interval $0 \leq t \leq 5$. Show that the rectangular equation corresponding to $\mathbf{r}(t)$ is the hyperbola $x^2 - y^2 = 1$. Verify your sketch by using a graphing utility to graph the hyperbola.

(b) Let $P = (\cosh \phi, \sinh \phi)$ be the point on the hyperbola corresponding to $\mathbf{r}(\phi)$ for $\phi > 0$. Use the formula for area
$$A = \frac{1}{2} \int_C x\, dy - y\, dx$$
to show that the area of the region indicated in the figure is $\tfrac{1}{2}\phi$.

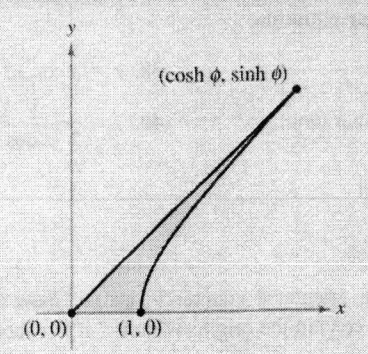

(c) Show that the area of the indicated region is also given by the integral
$$A = \int_0^{\sinh \phi} \left[\sqrt{1 + y^2} - (\coth \phi) y\right] dy.$$

Confirm your answer in part (b) by numerically approximating this integral for $\phi = 1, 2, 4$, and 10.

(d) Consider the unit circle given by $x^2 + y^2 = 1$. Let θ be the angle formed by the x-axis and the radius to (x, y). The area of the corresponding sector is $\tfrac{1}{2}\theta$. That is, the trigonometric functions $f(\theta) = \cos \theta$ and $g(\theta) = \sin \theta$ could have been defined to be the coordinates of that point $(\cos \theta, \sin \theta)$ on the unit circle that determines a sector of area $\tfrac{1}{2}\theta$. Write a short paragraph explaining how you could define the hyperbolic functions in a similar manner, using the "unit hyperbola" $x^2 - y^2 = 1$.

Section 14.5 Parametric Surfaces

- Understand the definition of a parametric surface.
- Find a set of parametric equations to represent a surface.
- Find a normal vector and a tangent plane to a parametric surface.
- Find the area of a parametric surface.

Parametric Surfaces

You already know how to represent a curve in the plane or in space by a set of parametric equations—or, equivalently, by a vector-valued function.

$$\mathbf{r}(t) = x(t)\mathbf{i} + y(t)\mathbf{j} \qquad \text{Plane curve}$$
$$\mathbf{r}(t) = x(t)\mathbf{i} + y(t)\mathbf{j} + z(t)\mathbf{k} \qquad \text{Space curve}$$

In this section, you will learn how to represent a surface in space by a set of parametric equations—or by a vector-valued function. For curves, note that the vector-valued function $\mathbf{r}$ is a function of a *single* parameter t. For surfaces, the vector-valued function is a function of *two* parameters u and v.

Definition of Parametric Surface

Let x, y, and z be functions of u and v that are continuous on a domain D in the uv-plane. The set of points (x, y, z) given by

$$\mathbf{r}(u, v) = x(u, v)\mathbf{i} + y(u, v)\mathbf{j} + z(u, v)\mathbf{k} \qquad \text{Parametric surface}$$

is called a **parametric surface**. The equations

$$x = x(u, v), \quad y = y(u, v), \quad \text{and} \quad z = z(u, v)$$

are the **parametric equations** for the surface.

If S is a parametric surface given by the vector-valued function $\mathbf{r}$, then S is traced out by the position vector $\mathbf{r}(u, v)$ as the point (u, v) moves throughout the domain D, as shown in Figure 14.35.

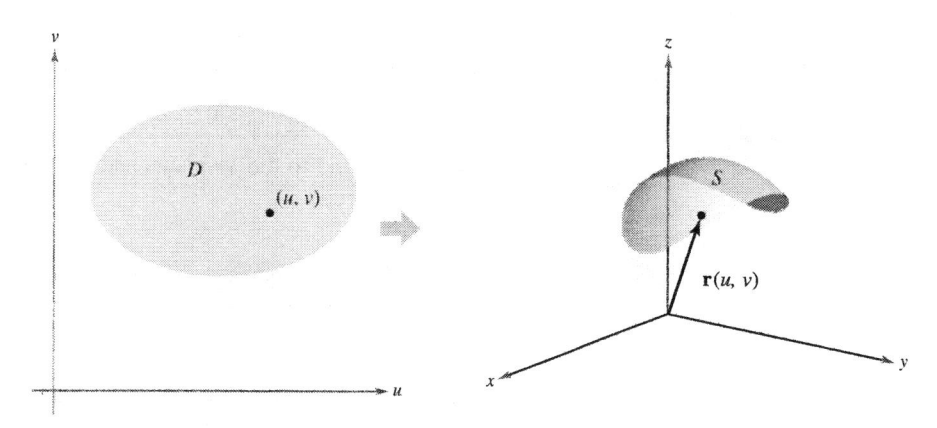

Figure 14.35

TECHNOLOGY Some computer algebra systems are capable of sketching surfaces that are represented parametrically. If you have access to such software, try using it to sketch some of the surfaces in the examples and exercises in this section.

Example 1 Sketching a Parametric Surface

Identify and sketch the parametric surface S given by

$$\mathbf{r}(u, v) = 3\cos u\mathbf{i} + 3\sin u\mathbf{j} + v\mathbf{k}$$

where $0 \leq u \leq 2\pi$ and $0 \leq v \leq 4$.

Solution Because $x = 3\cos u$ and $y = 3\sin u$, you know that for each point (x, y, z) on the surface, x and y are related by the equation $x^2 + y^2 = 3^2$. In other words, each cross section of S taken parallel to the xy-plane is a circle of radius 3, centered on the z-axis. Because $z = v$, where $0 \leq v \leq 4$, you can see that the surface is a right circular cylinder of height 4. The radius of the cylinder is 3, and the z-axis forms the axis of the cylinder, as shown in Figure 14.36.

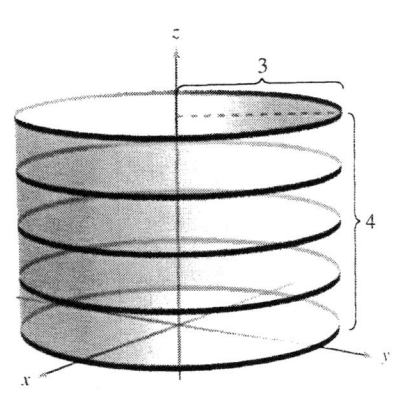

Figure 14.36

As with parametric representations of curves, parametric representations of surfaces are not unique. That is, there are many other sets of parametric equations that could be used to represent the surface shown in Figure 14.36.

Example 2 Sketching a Parametric Surface

Identify and sketch the parametric surface S given by

$$\mathbf{r}(u, v) = \sin u \cos v\mathbf{i} + \sin u \sin v\mathbf{j} + \cos u\mathbf{k}$$

where $0 \leq u \leq \pi$ and $0 \leq v \leq 2\pi$.

Solution To identify the surface, you can try to use trigonometric identities to eliminate the parameters. After some experimentation, you can discover that

$$\begin{aligned} x^2 + y^2 + z^2 &= (\sin u \cos v)^2 + (\sin u \sin v)^2 + (\cos u)^2 \\ &= \sin^2 u \cos^2 v + \sin^2 u \sin^2 v + \cos^2 u \\ &= \sin^2 u(\cos^2 v + \sin^2 v) + \cos^2 u \\ &= \sin^2 u + \cos^2 u \\ &= 1. \end{aligned}$$

So, each point on S lies on the unit sphere, centered at the origin, as shown in Figure 14.37. For fixed $u = d_i$, $\mathbf{r}(u, v)$ traces out latitude circles

$$x^2 + y^2 = \sin^2 d_i, \quad 0 \leq d_i \leq \pi$$

that are parallel to the xy-plane, and for fixed $v = c_i$, $\mathbf{r}(u, v)$ traces out longitude (or meridian) circles.

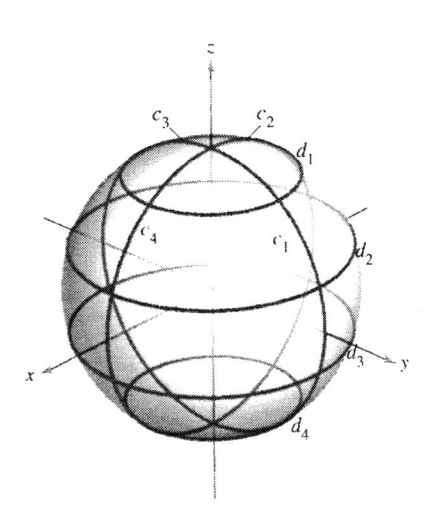

Figure 14.37

NOTE To further convince yourself that the vector-valued function in Example 2 traces out the entire unit sphere, recall that the parametric equations

$$x = \rho \sin \phi \cos \theta, \quad y = \rho \sin \phi \sin \theta, \quad \text{and} \quad z = \rho \cos \phi$$

where $0 \leq \theta \leq 2\pi$ and $0 \leq \phi \leq \pi$, describe the conversion from rectangular to spherical coordinates, as discussed in Section 10.7.

Finding Parametric Equations for Surfaces

In Examples 1 and 2, you were asked to identify the surface described by a given set of parametric equations. The reverse problem—that of writing a set of parametric equations for a given surface—is generally more difficult. One type of surface for which this problem is straightforward, however, is a surface that is given by $z = f(x, y)$. You can parametrize such a surface as

$$\mathbf{r}(x, y) = x\mathbf{i} + y\mathbf{j} + f(x, y)\mathbf{k}.$$

Example 3 Representing a Surface Parametrically

Write a set of parametric equations for the cone given by

$$z = \sqrt{x^2 + y^2}$$

as shown in Figure 14.38.

Solution Because this surface is given in the form $z = f(x, y)$, you can let x and y be the parameters. Then the cone is represented by the vector-valued function

$$\mathbf{r}(x, y) = x\mathbf{i} + y\mathbf{j} + \sqrt{x^2 + y^2}\,\mathbf{k}$$

where (x, y) varies over the entire xy-plane.

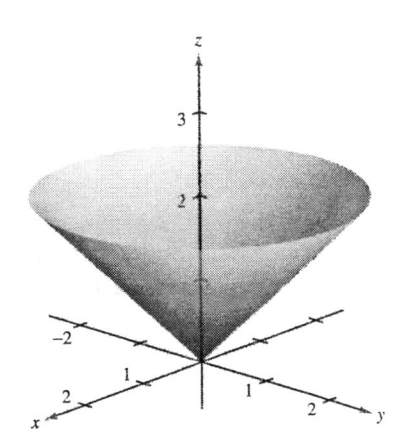

Figure 14.38

A second type of surface that is easily represented parametrically is a surface of revolution. For instance, to represent the surface formed by revolving the graph of $y = f(x)$, $a \leq x \leq b$, about the x-axis, use

$$x = u, \quad y = f(u) \cos v, \quad \text{and} \quad z = f(u) \sin v$$

where $a \leq u \leq b$ and $0 \leq v \leq 2\pi$.

Example 4 Representing a Surface of Revolution Parametrically

Write a set of parametric equations for the surface of revolution obtained by revolving

$$f(x) = \frac{1}{x}, \quad 1 \leq x \leq 10$$

about the x-axis.

Solution Use the parameters u and v as described above to write

$$x = u, \quad y = f(u) \cos v = \frac{1}{u} \cos v, \quad \text{and} \quad z = f(u) \sin v = \frac{1}{u} \sin v$$

where $1 \leq u \leq 10$ and $0 \leq v \leq 2\pi$. The resulting surface is a portion of *Gabriel's Horn*, as shown in Figure 14.39.

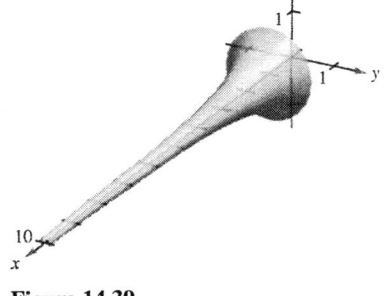

Figure 14.39

The surface of revolution in Example 4 is formed by revolving the graph of $y = f(x)$ about the x-axis. For other types of surfaces of revolution, a similar parametrization can be used. For instance, to parametrize the surface formed by revolving the graph of $x = f(z)$ about the z-axis, you can use

$$z = u, \quad x = f(u) \cos v, \quad \text{and} \quad y = f(u) \sin v.$$

Normal Vectors and Tangent Planes

Let S be a parametric surface given by

$$\mathbf{r}(u, v) = x(u, v)\mathbf{i} + y(u, v)\mathbf{j} + z(u, v)\mathbf{k}$$

over an open region D such that x, y, and z have continuous partial derivatives on D. The **partial derivatives of r** with respect to u and v are defined as

$$\mathbf{r}_u = \frac{\partial x}{\partial u}(u, v)\mathbf{i} + \frac{\partial y}{\partial u}(u, v)\mathbf{j} + \frac{\partial z}{\partial u}(u, v)\mathbf{k}$$

and

$$\mathbf{r}_v = \frac{\partial x}{\partial v}(u, v)\mathbf{i} + \frac{\partial y}{\partial v}(u, v)\mathbf{j} + \frac{\partial z}{\partial v}(u, v)\mathbf{k}.$$

Each of these partial derivatives is a vector-valued function that can be interpreted geometrically in terms of tangent vectors. For instance, if $v = v_0$ is held constant, then $\mathbf{r}(u, v_0)$ is a vector-valued function of a single parameter and defines a curve C_1 that lies on the surface S. The tangent vector to C_1 at the point $(x(u_0, v_0), y(u_0, v_0), z(u_0, v_0))$ is given by

$$\mathbf{r}_u(u_0, v_0) = \frac{\partial x}{\partial u}(u_0, v_0)\mathbf{i} + \frac{\partial y}{\partial u}(u_0, v_0)\mathbf{j} + \frac{\partial z}{\partial u}(u_0, v_0)\mathbf{k}$$

as shown in Figure 14.40. In a similar way, if $u = u_0$ is held constant, then $\mathbf{r}(u_0, v_0)$ is a vector-valued function of a single parameter and defines a curve C_2 that lies on the surface S. The tangent vector to C_2 at the point $(x(u_0, v_0), y(u_0, v_0), z(u_0, v_0))$ is given by

$$\mathbf{r}_v(u_0, v_0) = \frac{\partial x}{\partial v}(u_0, v_0)\mathbf{i} + \frac{\partial y}{\partial v}(u_0, v_0)\mathbf{j} + \frac{\partial z}{\partial v}(u_0, v_0)\mathbf{k}.$$

If the normal vector $\mathbf{r}_u \times \mathbf{r}_v$ is not $\mathbf{0}$ for any (u, v) in D, the surface S is called **smooth** and will have a tangent plane. Informally, a smooth surface is one that has no sharp points or cusps. For instance, spheres, ellipsoids, and paraboloids are smooth, whereas the cone given in Example 3 is not smooth.

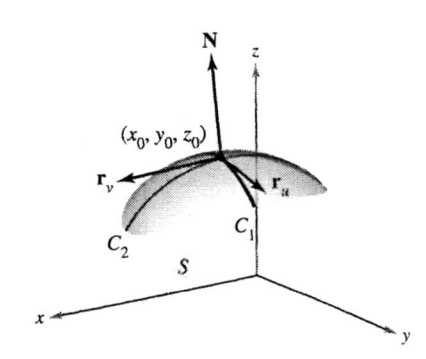

Figure 14.40

Normal Vector to a Smooth Parametric Surface

Let S be a smooth parametric surface

$$\mathbf{r}(u, v) = x(u, v)\mathbf{i} + y(u, v)\mathbf{j} + z(u, v)\mathbf{k}$$

defined over an open region D in the uv-plane. Let (u_0, v_0) be a point in D. A normal vector at the point

$$(x_0, y_0, z_0) = (x(u_0, v_0), y(u_0, v_0), z(u_0, v_0))$$

is given by

$$\mathbf{N} = \mathbf{r}_u(u_0, v_0) \times \mathbf{r}_v(u_0, v_0) = \begin{vmatrix} \mathbf{i} & \mathbf{j} & \mathbf{k} \\ \dfrac{\partial x}{\partial u} & \dfrac{\partial y}{\partial u} & \dfrac{\partial z}{\partial u} \\ \dfrac{\partial x}{\partial v} & \dfrac{\partial y}{\partial v} & \dfrac{\partial z}{\partial v} \end{vmatrix}.$$

NOTE Figure 14.40 shows the normal vector $\mathbf{r}_u \times \mathbf{r}_v$. The vector $\mathbf{r}_v \times \mathbf{r}_u$ is also normal to S and points in the opposite direction.

Example 5 **Finding a Tangent Plane to a Parametric Surface**

Find an equation of the tangent plane to the paraboloid given by

$$\mathbf{r}(u, v) = u\mathbf{i} + v\mathbf{j} + (u^2 + v^2)\mathbf{k}$$

at the point $(1, 2, 5)$.

Solution The point in the uv-plane that is mapped to the point $(x, y, z) = (1, 2, 5)$ is $(u, v) = (1, 2)$. The partial derivatives of $\mathbf{r}$ are

$$\mathbf{r}_u = \mathbf{i} + 2u\mathbf{k} \quad \text{and} \quad \mathbf{r}_v = \mathbf{j} + 2v\mathbf{k}.$$

The normal vector is given by

$$\mathbf{r}_u \times \mathbf{r}_v = \begin{vmatrix} \mathbf{i} & \mathbf{j} & \mathbf{k} \\ 1 & 0 & 2u \\ 0 & 1 & 2v \end{vmatrix} = -2u\mathbf{i} - 2v\mathbf{j} + \mathbf{k}$$

which implies that the normal vector at $(1, 2, 5)$ is $\mathbf{r}_u \times \mathbf{r}_v = -2\mathbf{i} - 4\mathbf{j} + \mathbf{k}$. So, an equation of the tangent plane at $(1, 2, 5)$ is

$$-2(x - 1) - 4(y - 2) + (z - 5) = 0$$
$$-2x - 4y + z = -5.$$

The tangent plane is shown in Figure 14.41.

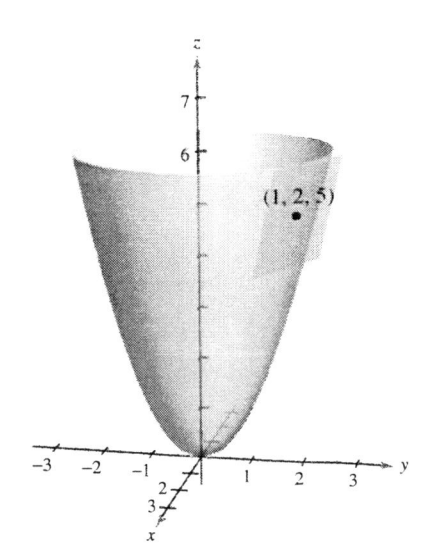

Figure 14.41

Area of a Parametric Surface

To define the area of a parametric surface, you can use a development that is similar to that given in Section 13.5. Begin by constructing an inner partition of D consisting of n rectangles, where the area of the ith rectangle D_i is $\Delta A_i = \Delta u_i \Delta v_i$, as shown in Figure 14.42. In each D_i let (u_i, v_i) be the point that is closest to the origin. At the point $(x_i, y_i, z_i) = (x(u_i, v_i), y(u_i, v_i), z(u_i, v_i))$ on the surface S, construct a tangent plane T_i. The area of the portion of S that corresponds to D_i, ΔT_i, can be approximated by a parallelogram in the tangent plane. That is, $\Delta T_i \approx \Delta S_i$. Hence the surface of S is given by $\Sigma \Delta S_i \approx \Sigma \Delta T_i$. The area of the parallelogram in the tangent plane is

$$\|\Delta u_i \mathbf{r}_u \times \Delta v_i \mathbf{r}_v\| = \|\mathbf{r}_u \times \mathbf{r}_v\| \Delta u_i \Delta v_i$$

which leads to the following definition.

Area of a Parametric Surface

Let S be a smooth parametric surface

$$\mathbf{r}(u, v) = x(u, v)\mathbf{i} + y(u, v)\mathbf{j} + z(u, v)\mathbf{k}$$

defined over an open region D in the uv-plane. If each point on the surface S corresponds to exactly one point in the domain D, then the **surface area** of S is given by

$$\text{Surface area} = \iint_S dS = \iint_D \|\mathbf{r}_u \times \mathbf{r}_v\| \, dA$$

where $\mathbf{r}_u = \dfrac{\partial x}{\partial u}\mathbf{i} + \dfrac{\partial y}{\partial u}\mathbf{j} + \dfrac{\partial z}{\partial u}\mathbf{k}$ and $\mathbf{r}_v = \dfrac{\partial x}{\partial v}\mathbf{i} + \dfrac{\partial y}{\partial v}\mathbf{j} + \dfrac{\partial z}{\partial v}\mathbf{k}$.

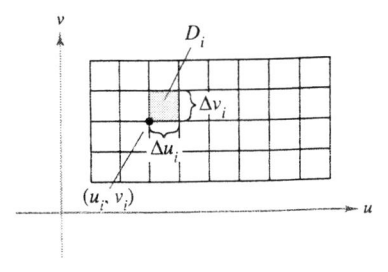

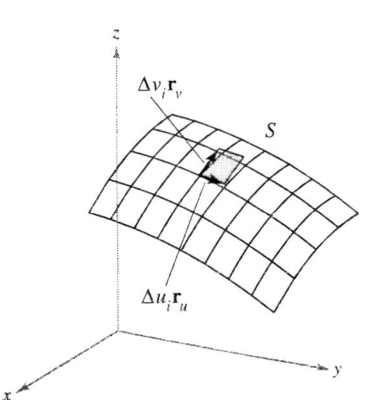

Figure 14.42

For a surface S given by $z = f(x, y)$, this formula for surface area corresponds to that given in Section 13.5. To see this, you can parametrize the surface using the vector-valued function

$$\mathbf{r}(x, y) = x\mathbf{i} + y\mathbf{j} + f(x, y)\mathbf{k}$$

defined over the region R in the xy-plane. Using

$$\mathbf{r}_x = \mathbf{i} + f_x(x, y)\mathbf{k} \quad \text{and} \quad \mathbf{r}_y = \mathbf{j} + f_y(x, y)\mathbf{k}$$

you have

$$\mathbf{r}_x \times \mathbf{r}_y = \begin{vmatrix} \mathbf{i} & \mathbf{j} & \mathbf{k} \\ 1 & 0 & f_x(x, y) \\ 0 & 1 & f_y(x, y) \end{vmatrix} = -f_x(x, y)\mathbf{i} - f_y(x, y)\mathbf{j} + \mathbf{k}$$

and $\|\mathbf{r}_x \times \mathbf{r}_y\| = \sqrt{[f_x(x, y)]^2 + [f_y(x, y)]^2 + 1}$. This implies that the surface area of S is

$$\text{Surface area} = \iint_R \|\mathbf{r}_x \times \mathbf{r}_y\|\, dA$$

$$= \iint_R \sqrt{1 + [f_x(x, y)]^2 + [f_y(x, y)]^2}\, dA.$$

Example 6 **Finding Surface Area**

NOTE The surface in Example 6 does not quite fulfill the hypothesis that each point on the surface corresponds to exactly one point in D. For this surface, $\mathbf{r}(u, 0) = \mathbf{r}(u, 2\pi)$ for any fixed value of u. However, because the overlap consists of only a semicircle (which has no area), you can still apply the formula for the area of a parametric surface.

Find the surface area of the unit sphere given by

$$\mathbf{r}(u, v) = \sin u \cos v\, \mathbf{i} + \sin u \sin v\, \mathbf{j} + \cos u\, \mathbf{k}$$

where the domain D is given by $0 \le u \le \pi$ and $0 \le v \le 2\pi$.

Solution Begin by calculating $\mathbf{r}_u$ and $\mathbf{r}_v$.

$$\mathbf{r}_u = \cos u \cos v\, \mathbf{i} + \cos u \sin v\, \mathbf{j} - \sin u\, \mathbf{k}$$
$$\mathbf{r}_v = -\sin u \sin v\, \mathbf{i} + \sin u \cos v\, \mathbf{j}$$

The cross product of these two vectors is

$$\mathbf{r}_u \times \mathbf{r}_v = \begin{vmatrix} \mathbf{i} & \mathbf{j} & \mathbf{k} \\ \cos u \cos v & \cos u \sin v & -\sin u \\ -\sin u \sin v & \sin u \cos v & 0 \end{vmatrix}$$

$$= \sin^2 u \cos v\, \mathbf{i} + \sin^2 u \sin v\, \mathbf{j} + \sin u \cos u\, \mathbf{k}$$

which implies that

$$\|\mathbf{r}_u \times \mathbf{r}_v\| = \sqrt{(\sin^2 u \cos v)^2 + (\sin^2 u \sin v)^2 + (\sin u \cos u)^2}$$
$$= \sqrt{\sin^4 u + \sin^2 u \cos^2 u}$$
$$= \sqrt{\sin^2 u}$$
$$= \sin u. \qquad \sin u > 0 \text{ for } 0 \le u \le \pi$$

Finally, the surface area of the sphere is

$$A = \iint_D \|\mathbf{r}_u \times \mathbf{r}_v\|\, dA = \int_0^{2\pi} \int_0^{\pi} \sin u\, du\, dv$$
$$= \int_0^{2\pi} 2\, dv$$
$$= 4\pi.$$

EXPLORATION

For the torus in Example 7, describe the function $\mathbf{r}(u, v)$ for fixed u. Then describe the function $\mathbf{r}(u, v)$ for fixed v.

Example 7 Finding Surface Area

Find the surface area of the torus given by

$$\mathbf{r}(u, v) = (2 + \cos u) \cos v\mathbf{i} + (2 + \cos u) \sin v\mathbf{j} + \sin u\mathbf{k}$$

where the domain D is given by $0 \le u \le 2\pi$ and $0 \le v \le 2\pi$. (See Figure 14.43.)

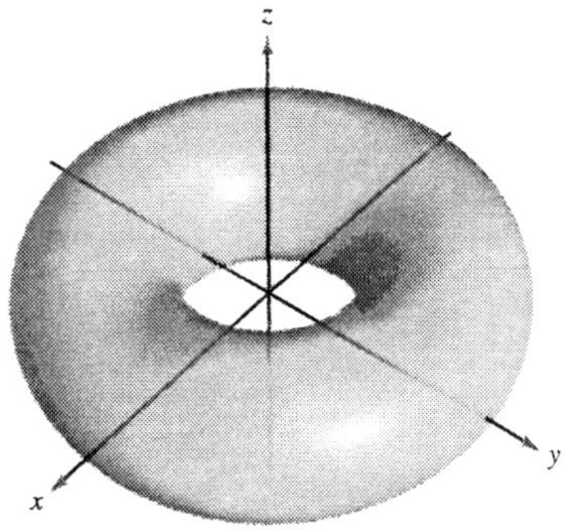

Figure 14.43

Solution Begin by calculating $\mathbf{r}_u$ and $\mathbf{r}_v$.

$$\mathbf{r}_u = -\sin u \cos v\mathbf{i} - \sin u \sin v\mathbf{j} + \cos u\mathbf{k}$$
$$\mathbf{r}_v = -(2 + \cos u) \sin v\mathbf{i} + (2 + \cos u) \cos v\mathbf{j}$$

The cross product of these two vectors is

$$\begin{aligned} \mathbf{r}_u \times \mathbf{r}_v &= \begin{vmatrix} \mathbf{i} & \mathbf{j} & \mathbf{k} \\ -\sin u \cos v & -\sin u \sin v & \cos u \\ -(2 + \cos u) \sin v & (2 + \cos u) \cos v & 0 \end{vmatrix} \\ &= -(2 + \cos u)(\cos v \cos u\mathbf{i} + \sin v \cos u\mathbf{j} + \sin u\mathbf{k}) \end{aligned}$$

which implies that

$$\begin{aligned} \|\mathbf{r}_u \times \mathbf{r}_v\| &= (2 + \cos u)\sqrt{(\cos v \cos u)^2 + (\sin v \cos u)^2 + \sin^2 u} \\ &= (2 + \cos u)\sqrt{\cos^2 u(\cos^2 v + \sin^2 v) + \sin^2 u} \\ &= (2 + \cos u)\sqrt{\cos^2 u + \sin^2 u} \\ &= 2 + \cos u. \end{aligned}$$

Finally, the surface area of the torus is

$$\begin{aligned} A = \int\!\!\int_D \|\mathbf{r}_u \times \mathbf{r}_v\|\, dA &= \int_0^{2\pi}\!\int_0^{2\pi} (2 + \cos u)\, du\, dv \\ &= \int_0^{2\pi} 4\pi\, dv \\ &= 8\pi^2. \end{aligned}$$

If the surface S is a surface of revolution, you can show that the formula for surface area given in Section 6.4 is equivalent to the formula given in this section. For instance, suppose f is a nonnegative function such that f' is continuous over the interval $[a, b]$. Let S be the surface of revolution formed by revolving the graph of f, where $a \le x \le b$, about the x-axis. From Section 6.4, you know that the surface area is given by

$$\text{Surface area} = 2\pi \int_a^b f(x)\sqrt{1 + [f'(x)]^2}\, dx.$$

To represent S parametrically, let $x = u$, $y = f(u) \cos v$, and $z = f(u) \sin v$, where $a \le u \le b$ and $0 \le v \le 2\pi$. Then,

$$\mathbf{r}(u, v) = u\mathbf{i} + f(u) \cos v\mathbf{j} + f(u) \sin v\mathbf{k}.$$

Try showing that the formula

$$\text{Surface area} = \int\!\!\int_D \|\mathbf{r}_u \times \mathbf{r}_v\|\, dA$$

is equivalent to the formula given above.

EXERCISES FOR SECTION 14.5

In Exercises 1–4, match the vector-valued function with its graph. [The graphs are labeled (a), (b), (c), and (d).]

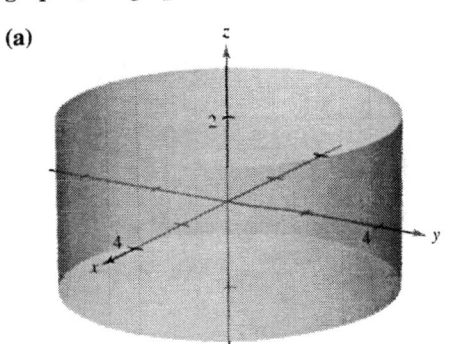

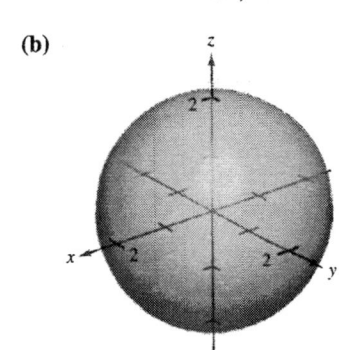

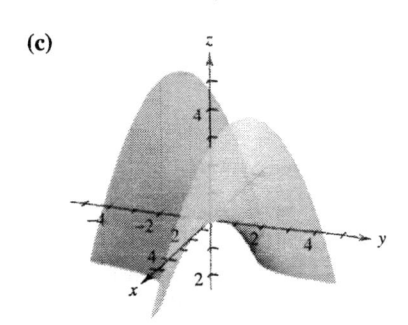

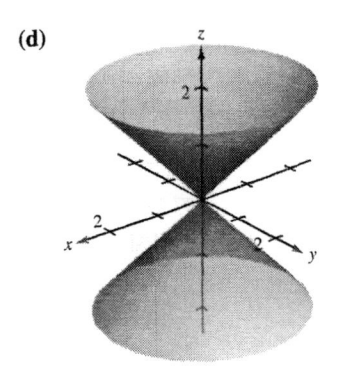

(a)

(b)

(c)

(d)

1. $\mathbf{r}(u, v) = u\mathbf{i} + v\mathbf{j} + uv\mathbf{k}$
2. $\mathbf{r}(u, v) = u \cos v\mathbf{i} + u \sin v\mathbf{j} + u\mathbf{k}$
3. $\mathbf{r}(u, v) = 2 \cos v \cos u\mathbf{i} + 2 \cos v \sin u\mathbf{j} + 2 \sin v\mathbf{k}$
4. $\mathbf{r}(u, v) = 4 \cos u\mathbf{i} + 4 \sin u\mathbf{j} + v\mathbf{k}$

In Exercises 5–8, find the rectangular equation for the surface by eliminating the parameters from the vector-valued function. Identify the surface and sketch its graph.

5. $\mathbf{r}(u, v) = u\mathbf{i} + v\mathbf{j} + \frac{v}{2}\mathbf{k}$
6. $\mathbf{r}(u, v) = 2u \cos v\mathbf{i} + 2u \sin v\mathbf{j} + \frac{1}{2}u^2\mathbf{k}$
7. $\mathbf{r}(u, v) = 2 \cos u\mathbf{i} + v\mathbf{j} + 2 \sin u\mathbf{k}$
8. $\mathbf{r}(u, v) = 3 \cos v \cos u\mathbf{i} + 3 \cos v \sin u\mathbf{j} + 5 \sin v\mathbf{k}$

Think About It In Exercises 9–12, determine how the graph of the surface $\mathbf{s}(u, v)$ differs from the graph of $\mathbf{r}(u, v) = u \cos v\mathbf{i} + u \sin v\mathbf{j} + u^2\mathbf{k}$ (see figure) where $0 \leq u \leq 2$ and $0 \leq v \leq 2\pi$. (It is not necessary to graph s.)

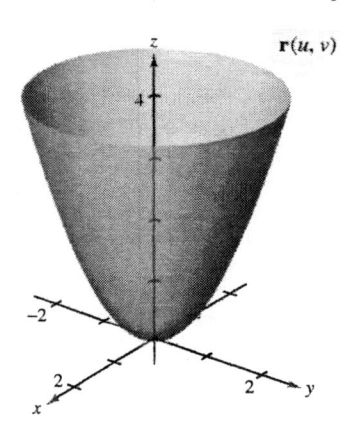

9. $\mathbf{s}(u, v) = u \cos v\mathbf{i} + u \sin v\mathbf{j} - u^2\mathbf{k}$
   $0 \leq u \leq 2, \quad 0 \leq v \leq 2\pi$
10. $\mathbf{s}(u, v) = u \cos v\mathbf{i} + u^2\mathbf{j} + u \sin v\mathbf{k}$
    $0 \leq u \leq 2, \quad 0 \leq v \leq 2\pi$
11. $\mathbf{s}(u, v) = u \cos v\mathbf{i} + u \sin v\mathbf{j} + u^2\mathbf{k}$
    $0 \leq u \leq 3, \quad 0 \leq v \leq 2\pi$
12. $\mathbf{s}(u, v) = 4u \cos v\mathbf{i} + 4u \sin v\mathbf{j} + u^2\mathbf{k}$
    $0 \leq u \leq 2, \quad 0 \leq v \leq 2\pi$

In Exercises 13–18, use a computer algebra system to graph the surface represented by the vector-valued function.

13. $\mathbf{r}(u, v) = 2u \cos v\mathbf{i} + 2u \sin v\mathbf{j} + u^4\mathbf{k}$
    $0 \leq u \leq 1, \quad 0 \leq v \leq 2\pi$
14. $\mathbf{r}(u, v) = 2 \cos v \cos u\mathbf{i} + 4 \cos v \sin u\mathbf{j} + \sin v\mathbf{k}$
    $0 \leq u \leq 2\pi, \quad 0 \leq v \leq 2\pi$
15. $\mathbf{r}(u, v) = 2 \sinh u \cos v\mathbf{i} + \sinh u \sin v\mathbf{j} + \cosh u\mathbf{k}$
    $0 \leq u \leq 2, \quad 0 \leq v \leq 2\pi$
16. $\mathbf{r}(u, v) = 2u \cos v\mathbf{i} + 2u \sin v\mathbf{j} + v\mathbf{k}$
    $0 \leq u \leq 1, \quad 0 \leq v \leq 3\pi$
17. $\mathbf{r}(u, v) = (u - \sin u)\cos v\mathbf{i} + (1 - \cos u)\sin v\mathbf{j} + u\mathbf{k}$
    $0 \leq u \leq \pi, \quad 0 \leq v \leq 2\pi$

18. $\mathbf{r}(u, v) = \cos^3 u \cos v \mathbf{i} + \sin^3 u \sin v \mathbf{j} + u\mathbf{k}$
$0 \le u \le \dfrac{\pi}{2}, \quad 0 \le v \le 2\pi$

In Exercises 19–26, find a vector-valued function whose graph is the indicated surface.

19. The plane $z = y$
20. The plane $x + y + z = 6$
21. The cylinder $x^2 + y^2 = 16$
22. The cylinder $4x^2 + y^2 = 16$
23. The cylinder $z = x^2$
24. The ellipsoid $\dfrac{x^2}{9} + \dfrac{y^2}{4} + \dfrac{z^2}{1} = 1$
25. The part of the plane $z = 4$ that lies inside the cylinder $x^2 + y^2 = 9$
26. The part of the paraboloid $z = x^2 + y^2$ that lies inside the cylinder $x^2 + y^2 = 9$

Surface of Revolution **In Exercises 27–30, write a set of parametric equations for the surface of revolution obtained by revolving the graph of the function about the indicated axis.**

Function	Axis of Revolution
27. $y = \dfrac{x}{2}, \quad 0 \le x \le 6$	x-axis
28. $y = x^{3/2}, \quad 0 \le x \le 4$	x-axis
29. $x = \sin z, \quad 0 \le z \le \pi$	z-axis
30. $z = 4 - y^2, \quad 0 \le y \le 2$	y-axis

Tangent Plane **In Exercises 31–34, find an equation of the tangent plane to the surface given by the vector-valued function at the indicated point.**

31. $\mathbf{r}(u, v) = (u + v)\mathbf{i} + (u - v)\mathbf{j} + v\mathbf{k}, \quad (1, -1, 1)$

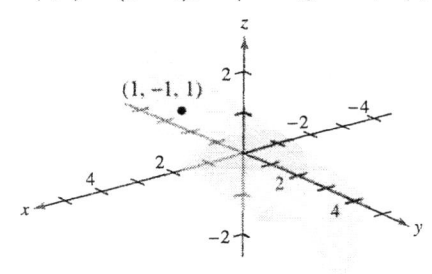

32. $\mathbf{r}(u, v) = u\mathbf{i} + v\mathbf{j} + \sqrt{uv}\,\mathbf{k}, \quad (1, 1, 1)$

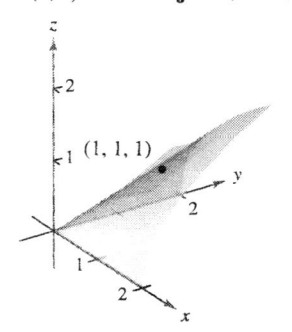

33. $\mathbf{r}(u, v) = 2u \cos v \mathbf{i} + 3u \sin v \mathbf{j} + u^2 \mathbf{k}, \quad (0, 6, 4)$

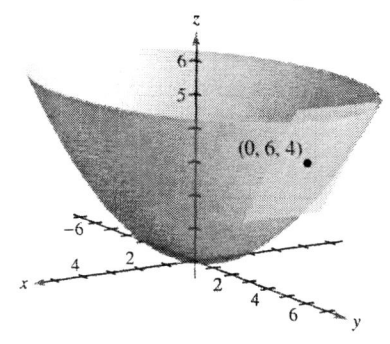

34. $\mathbf{r}(u, v) = 2u \cosh v \mathbf{i} + 2u \sinh v \mathbf{j} + \tfrac{1}{2}u^2 \mathbf{k}, \quad (-4, 0, 2)$

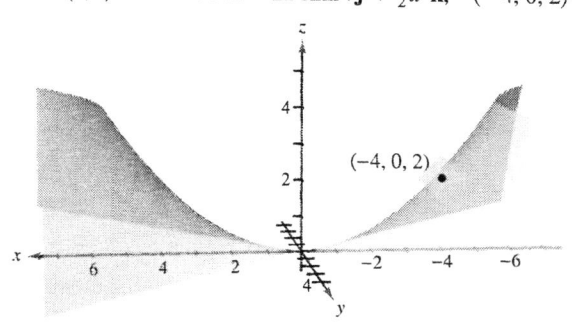

Area **In Exercises 35–42, find the area of the surface over the indicated region. Use a computer algebra system to verify your results.**

35. The part of the plane
$$\mathbf{r}(u, v) = 2u\mathbf{i} - \dfrac{v}{2}\mathbf{j} + \dfrac{v}{2}\mathbf{k}$$
where $0 \le u \le 2$ and $0 \le v \le 1$

36. The part of the paraboloid
$$\mathbf{r}(u, v) = 4u \cos v \mathbf{i} + 4u \sin v \mathbf{j} + u^2 \mathbf{k}$$
where $0 \le u \le 2$ and $0 \le v \le 2\pi$

37. The part of the cylinder
$$\mathbf{r}(u, v) = a \cos u \mathbf{i} + a \sin u \mathbf{j} + v\mathbf{k}$$
where $0 \le u \le 2\pi$ and $0 \le v \le b$

38. The sphere
$$\mathbf{r}(u, v) = a \sin u \cos v \mathbf{i} + a \sin u \sin v \mathbf{j} + a \cos u \mathbf{k}$$
where $0 \le u \le \pi$ and $0 \le v \le 2\pi$

39. The part of the cone
$$\mathbf{r}(u, v) = au \cos v \mathbf{i} + au \sin v \mathbf{j} + u\mathbf{k}$$
where $0 \le u \le b$ and $0 \le v \le 2\pi$

40. The torus $\mathbf{r}(u, v) = (a + b \cos v)\cos u \mathbf{i} + (a + b \cos v)\sin u \mathbf{j} + b \sin v \mathbf{k}$, where $a > b$, $0 \le u \le 2\pi$, and $0 \le v \le 2\pi$

41. The surface of revolution $\mathbf{r}(u, v) = \sqrt{u} \cos v \mathbf{i} + \sqrt{u} \sin v \mathbf{j} + u\mathbf{k}$, where $0 \le u \le 4$ and $0 \le v \le 2\pi$

42. The surface of revolution
$$\mathbf{r}(u, v) = \sin u \cos v \mathbf{i} + u\mathbf{j} + \sin u \sin v \mathbf{k}$$
where $0 \le u \le \pi$ and $0 \le v \le 2\pi$

Getting at the Concept

43. Define a parametric surface.

44. Give the double integral that yields the surface area of a parametric surface over an open region D.

45. The four figures are graphs of the surface
$$\mathbf{r}(u, v) = u\mathbf{i} + \sin u \cos v\mathbf{j} + \sin u \sin v\mathbf{k},$$
$$0 \leq u \leq \frac{\pi}{2}, \quad 0 \leq v \leq 2\pi.$$

Match each of the four graphs with the point in space from which the surface is viewed. The four points are $(10, 0, 0)$, $(-10, 10, 0)$, $(0, 10, 0)$, and $(10, 10, 10)$.

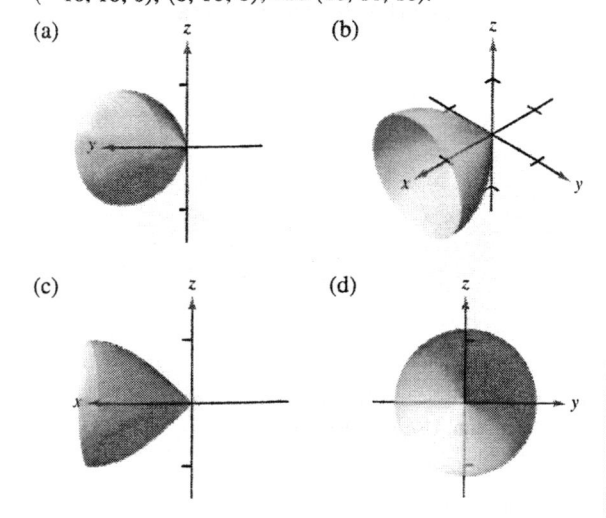

(a) (b) (c) (d)

46. Use a computer algebra system to sketch three views of the graph of the vector-valued function
$$\mathbf{r}(u, v) = u \cos v\mathbf{i} + u \sin v\mathbf{j} + v\mathbf{k}, \quad 0 \leq u \leq \pi, \quad 0 \leq v \leq \pi$$
from the points $(10, 0, 0)$, $(0, 0, 10)$, and $(10, 10, 10)$.

47. Investigation Use a computer algebra system to graph the torus
$$\mathbf{r}(u, v) = (a + b \cos v) \cos u\mathbf{i} + (a + b \cos v) \sin u\mathbf{j} + b \sin v\mathbf{k}$$
for each set of values of a and b, where $0 \leq u \leq 2\pi$ and $0 \leq v \leq 2\pi$. Use the results to describe the effect of a and b on the shape of the torus.

(a) $a = 4, \quad b = 1$ (b) $a = 4, \quad b = 2$
(c) $a = 8, \quad b = 1$ (d) $a = 8, \quad b = 3$

48. Investigation Consider the function in Exercise 16.

(a) Sketch a graph of the function where u is held constant at $u = 1$. Identify the graph.

(b) Sketch a graph of the function where v is held constant at $v = 2\pi/3$. Identify the graph.

(c) Assume that a surface is represented by the vector-valued function $\mathbf{r} = \mathbf{r}(u, v)$. What generalization can you make about the graph of the function if one of the parameters is held constant?

49. The surface of the dome on a new museum is given by
$$\mathbf{r}(u, v) = 20 \sin u \cos v\mathbf{i} + 20 \sin u \sin v\mathbf{j} + 20 \cos u\mathbf{k}$$
where $0 \leq u \leq \pi/3$ and $0 \leq v \leq 2\pi$ and $\mathbf{r}$ is in meters. Find the surface area of the dome.

50. Find a vector-valued function for the hyperboloid
$$x^2 + y^2 - z^2 = 1$$
and determine the tangent plane at $(1, 0, 0)$.

51. Graph and find the area of one turn of the spiral ramp
$$\mathbf{r}(u, v) = u \cos v\mathbf{i} + u \sin v\mathbf{j} + 2v\mathbf{k}$$
where $0 \leq u \leq 3, \, 0 \leq v \leq 2\pi$.

52. Let f be a nonnegative function such that f' is continuous over the interval $[a, b]$. Let S be the surface of revolution formed by revolving the graph of f, where $a \leq x \leq b$, about the x-axis. Let $x = u$, $y = f(u) \cos v$, and $z = f(u) \sin v$, where $a \leq u \leq b$ and $0 \leq v \leq 2\pi$. Then, S is represented parametrically by
$$\mathbf{r}(u, v) = u\mathbf{i} + f(u) \cos v\mathbf{j} + f(u) \sin v\mathbf{k}.$$

Show that the following formulas are equivalent.

$$\text{Surface area} = 2\pi \int_a^b f(x)\sqrt{1 + [f'(x)]^2} \, dx$$

$$\text{Surface area} = \iint_D \|\mathbf{r}_u \times \mathbf{r}_v\| \, dA$$

53. Open-Ended Project The parametric equations
$$x = 3 + \sin u[7 - \cos(3u - 2v) - 2\cos(3u + v)]$$
$$y = 3 + \cos u[7 - \cos(3u - 2v) - 2\cos(3u + v)]$$
$$z = \sin(3u - 2v) + 2\sin(3u + v)$$
where $-\pi \leq u \leq \pi$ and $-\pi \leq v \leq \pi$, represent the surface shown below. Try to create your own parametric surface using a computer algebra system.

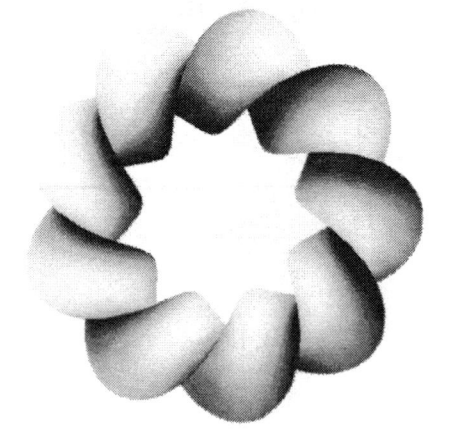

Section 14.6 Surface Integrals

- Evaluate a surface integral as a double integral.
- Evaluate a surface integral for a parametric surface.
- Determine the orientation of a surface.
- Understand the concept of a flux integral.

Surface Integrals

The remainder of this chapter deals primarily with **surface integrals**. We begin by considering surfaces given by $z = g(x, y)$. Later in this section we will consider more general surfaces given in parametric form.

Let S be a surface given by $z = g(x, y)$ and let R be its projection onto the xy-plane, as shown in Figure 14.44. Suppose that g, g_x, and g_y are continuous at all points in R and that f is defined on S. Employing the procedure used to find surface area in Section 13.5, evaluate f at (x_i, y_i, z_i) and form the sum

$$\sum_{i=1}^{n} f(x_i, y_i, z_i) \, \Delta S_i$$

where $\Delta S_i \approx \sqrt{1 + [g_x(x_i, y_i)]^2 + [g_y(x_i, y_i)]^2} \, \Delta A_i$. Provided the limit as $\|\Delta\|$ approaches 0 exists, the **surface integral of f over S** is defined as

$$\iint_S f(x, y, z) \, dS = \lim_{\|\Delta\| \to 0} \sum_{i=1}^{n} f(x_i, y_i, z_i) \, \Delta S_i.$$

This integral can be evaluated by a double integral.

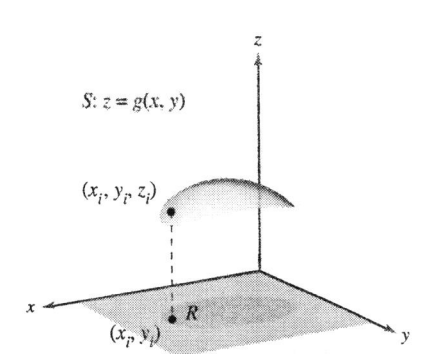

Scalar function f assigns a number to each point of S.
Figure 14.44

THEOREM 14.10 Evaluating a Surface Integral

Let S be a surface with equation $z = g(x, y)$ and let R be its projection onto the xy-plane. If g, g_x, and g_y are continuous on R and f is continuous on S, then the surface integral of f over S is

$$\iint_S f(x, y, z) \, dS = \iint_R f(x, y, g(x, y)) \sqrt{1 + [g_x(x, y)]^2 + [g_y(x, y)]^2} \, dA.$$

For surfaces described by functions of x and z (or y and z), you can make the following adjustments to Theorem 14.10. If S is the graph of $y = g(x, z)$ and R is its projection onto the xz-plane, then

$$\iint_S f(x, y, z) \, dS = \iint_R f(x, g(x, z), z) \sqrt{1 + [g_x(x, z)]^2 + [g_z(x, z)]^2} \, dA.$$

If S is the graph of $x = g(y, z)$ and R is its projection onto the yz-plane, then

$$\iint_S f(x, y, z) \, dS = \iint_R f(g(y, z), y, z) \sqrt{1 + [g_y(y, z)]^2 + [g_z(y, z)]^2} \, dA.$$

If $f(x, y, z) = 1$, the surface integral over S yields the surface area of S. For instance, suppose the surface S is the plane given by $z = x$, where $0 \leq x \leq 1$ and $0 \leq y \leq 1$. The surface area of S is $\sqrt{2}$ square units. Try verifying that $\iint_S f(x, y, z) \, dS = \sqrt{2}$.

Example 1 Evaluating a Surface Integral

Evaluate the surface integral

$$\iint_S (y^2 + 2yz)\, dS$$

where S is the first-octant portion of the plane $2x + y + 2z = 6$.

Solution Begin by writing S as

$$z = \frac{1}{2}(6 - 2x - y)$$

$$g(x, y) = \frac{1}{2}(6 - 2x - y).$$

Using the partial derivatives $g_x(x, y) = -1$ and $g_y(x, y) = -\frac{1}{2}$, you can write

$$\sqrt{1 + [g_x(x, y)]^2 + [g_y(x, y)]^2} = \sqrt{1 + 1 + \frac{1}{4}} = \frac{3}{2}.$$

Using Figure 14.45 and Theorem 14.10, you obtain

$$\iint_S (y^2 + 2yz)\, dS = \iint_R f(x, y, g(x, y))\sqrt{1 + [g_x(x, y)]^2 + [g_y(x, y)]^2}\, dA$$

$$= \iint_R \left[y^2 + 2y\left(\frac{1}{2}\right)(6 - 2x - y)\right]\left(\frac{3}{2}\right) dA$$

$$= 3\int_0^3 \int_0^{2(3-x)} y(3 - x)\, dy\, dx$$

$$= 6\int_0^3 (3 - x)^3\, dx$$

$$= -\frac{3}{2}(3 - x)^4 \Big]_0^3$$

$$= \frac{243}{2}.$$

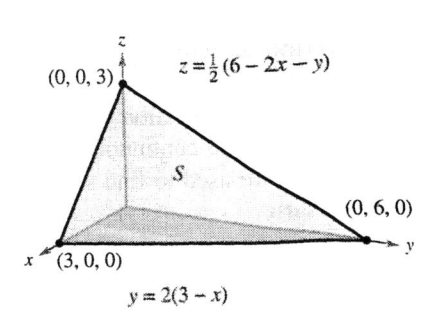

Figure 14.45

An alternative solution to Example 1 would be to project S onto the yz-plane, as shown in Figure 14.46. Then, $x = \frac{1}{2}(6 - y - 2z)$, and

$$\sqrt{1 + [g_y(y, z)]^2 + [g_z(y, z)]^2} = \sqrt{1 + \frac{1}{4} + 1} = \frac{3}{2}.$$

So, the surface integral is

$$\iint_S (y^2 + 2yz)\, dS = \iint_R f(g(y, z), y, z)\sqrt{1 + [g_y(y, z)]^2 + [g_x(y, z)]^2}\, dA$$

$$= \int_0^6 \int_0^{(6-y)/2} (y^2 + 2yz)\left(\frac{3}{2}\right) dz\, dy$$

$$= \frac{3}{8}\int_0^6 (36y - y^3)\, dy$$

$$= \frac{243}{2}.$$

Try reworking Example 1 by projecting S onto the xz-plane.

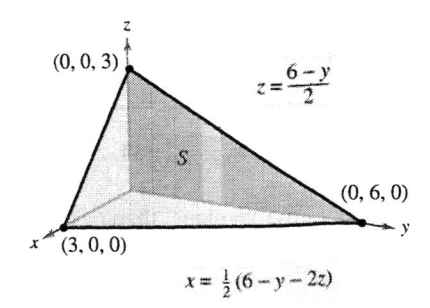

Figure 14.46

In Example 1, you could have projected the surface S onto any one of the three coordinate planes. In Example 2, S is a portion of a cylinder centered about the x-axis, and you can project it onto either the xz-plane or the xy-plane.

Example 2 Evaluating a Surface Integral

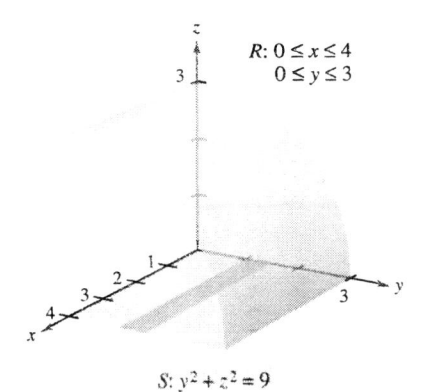

Evaluate the surface integral

$$\iint_S (x + z)\, dS$$

where S is the first-octant portion of the cylinder $y^2 + z^2 = 9$ between $x = 0$ and $x = 4$, as shown in Figure 14.47.

Solution Project S onto the xy-plane, so that $z = g(x, y) = \sqrt{9 - y^2}$, and obtain

$$\sqrt{1 + [g_x(x, y)]^2 + [g_y(x, y)]^2} = \sqrt{1 + \left(\frac{-y}{\sqrt{9 - y^2}}\right)^2}$$

$$= \frac{3}{\sqrt{9 - y^2}}.$$

Theorem 14.10 does not apply directly because g_y is not continuous when $y = 3$. However, you can apply the theorem for $0 \le b < 3$ and then take the limit as b approaches 3, as follows.

$$\iint_S (x + z)\, dS = \lim_{b \to 3^-} \int_0^b \int_0^4 \left(x + \sqrt{9 - y^2}\right) \frac{3}{\sqrt{9 - y^2}}\, dx\, dy$$

$$= \lim_{b \to 3^-} 3 \int_0^b \int_0^4 \left(\frac{x}{\sqrt{9 - y^2}} + 1\right) dx\, dy$$

$$= \lim_{b \to 3^-} 3 \int_0^b \left[\frac{x^2}{2\sqrt{9 - y^2}} + x\right]_0^4 dy$$

$$= \lim_{b \to 3^-} 3 \int_0^b \left(\frac{8}{\sqrt{9 - y^2}} + 4\right) dy$$

$$= \lim_{b \to 3^-} 3 \left[4y + 8 \arcsin \frac{y}{3}\right]_0^b$$

$$= \lim_{b \to 3^-} 3 \left(4b + 8 \arcsin \frac{b}{3}\right)$$

$$= 36 + 24\left(\frac{\pi}{2}\right)$$

$$= 36 + 12\pi$$

Figure 14.47
$R: 0 \le x \le 4$, $0 \le y \le 3$
$S: y^2 + z^2 = 9$

TECHNOLOGY Some computer algebra systems are capable of evaluating improper integrals. If you have access to such computer software, try using it to evaluate the improper integral

$$\int_0^3 \int_0^4 \left(x + \sqrt{9 - y^2}\right) \frac{3}{\sqrt{9 - y^2}}\, dx\, dy.$$

Do you obtain the same result as in Example 2?

You have already seen that if the function f defined on the surface S is simply $f(x, y, z) = 1$, the surface integral yields the *surface area* of S.

$$\text{Area of surface} = \iint_S 1 \, dS$$

On the other hand, if S is a lamina of variable density and $\rho(x, y, z)$ is the density at the point (x, y, z), then the *mass* of the lamina is given by

$$\text{Mass of lamina} = \iint_S \rho(x, y, z) \, dS.$$

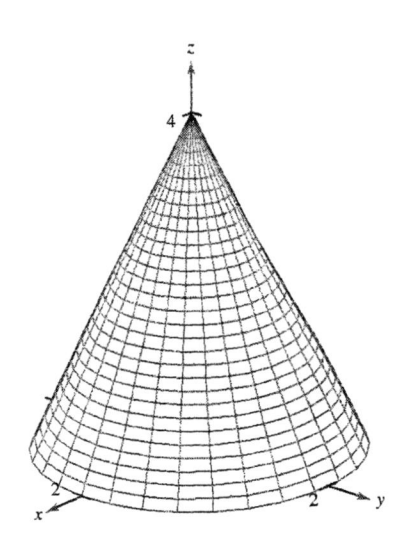

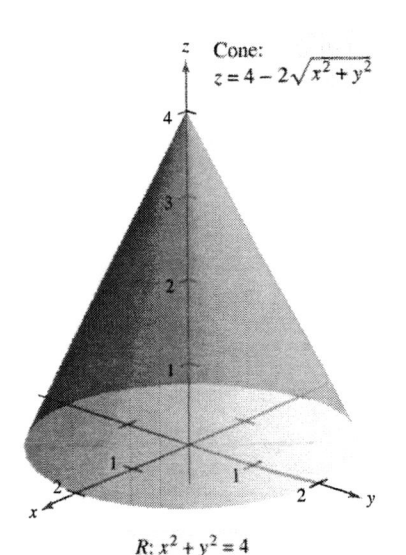

Example 3 Finding the Mass of a Surface Lamina

A cone-shaped surface lamina S is given by

$$z = 4 - 2\sqrt{x^2 + y^2}, \quad 0 \le z \le 4$$

as shown in Figure 14.48. At each point on S, the density is proportional to the distance between the point and the z-axis. Find the mass m of the lamina.

Solution Projecting S onto the xy-plane produces

$$S: z = 4 - 2\sqrt{x^2 + y^2} = g(x, y), \quad 0 \le z \le 4$$
$$R: x^2 + y^2 \le 4$$

with a density of $\rho(x, y, z) = k\sqrt{x^2 + y^2}$. Using a surface integral, you can find the mass to be

$$\begin{aligned}
m &= \iint_S \rho(x, y, z) \, dS \\
&= \iint_R k\sqrt{x^2 + y^2} \sqrt{1 + [g_x(x, y)]^2 + [g_y(x, y)]^2} \, dA \\
&= k \iint_R \sqrt{x^2 + y^2} \sqrt{1 + \frac{4x^2}{x^2 + y^2} + \frac{4y^2}{x^2 + y^2}} \, dA \\
&= k \iint_R \sqrt{5} \sqrt{x^2 + y^2} \, dA \\
&= k \int_0^{2\pi} \int_0^2 (\sqrt{5} r) r \, dr \, d\theta \quad \text{Polar coordinates} \\
&= \frac{\sqrt{5} k}{3} \int_0^{2\pi} r^3 \Big]_0^2 \, d\theta \\
&= \frac{8\sqrt{5} k}{3} \int_0^{2\pi} d\theta \\
&= \frac{8\sqrt{5} k}{3} \Big[\theta\Big]_0^{2\pi} = \frac{16\sqrt{5} k\pi}{3}.
\end{aligned}$$

TECHNOLOGY Try using a computer algebra system to confirm the result shown in Example 3. When we did using *Derive*, we obtained

$$k \int_{-2}^{2} \int_{-\sqrt{4-y^2}}^{\sqrt{4-y^2}} \sqrt{5} \sqrt{x^2 + y^2} \, dx \, dy = \frac{16\sqrt{5} k\pi}{3}.$$

Cone: $z = 4 - 2\sqrt{x^2 + y^2}$

$R: x^2 + y^2 = 4$

Density: $\rho(x, y, z) = k\sqrt{x^2 + y^2}$
Figure 14.48

Parametric Surfaces and Surface Integrals

For a surface S given by the vector-valued function

$$\mathbf{r}(u, v) = x(u, v)\mathbf{i} + y(u, v)\mathbf{j} + z(u, v)\mathbf{k} \qquad \text{Parametric surface}$$

defined over a region D in the uv-plane, you can show that the surface integral of $f(x, y, z)$ over S is given by

$$\iint_S f(x, y, z)\, dS = \iint_D f(x(u, v), y(u, v), z(u, v)) \|\mathbf{r}_u(u, v) \times \mathbf{r}_v(u, v)\|\, dA.$$

Note the similarity to a line integral over a space curve C.

$$\int_C f(x, y, z)\, dS = \int_a^b f(x(t), y(t), z(t)) \|\mathbf{r}'(t)\|\, dt \qquad \text{Line integral}$$

NOTE Notice that ds and dS can be written as $ds = \|\mathbf{r}'(t)\|\, dt$ and $dS = \|\mathbf{r}_u(u, v) \times \mathbf{r}_v(u, v)\|\, dA$.

Example 4 Evaluating a Surface Integral

Example 2 demonstrated an evaluation of the surface integral

$$\iint_S (x + z)\, dS$$

where S is the first-octant portion of the cylinder $y^2 + z^2 = 9$ between $x = 0$ and $x = 4$ (see Figure 14.49). Reevaluate this integral in parametric form.

Solution In parametric form, the surface is given by

$$\mathbf{r}(x, \theta) = x\mathbf{i} + 3\cos\theta\mathbf{j} + 3\sin\theta\mathbf{k}$$

where $0 \le x \le 4$ and $0 \le \theta \le \pi/2$. To evaluate the surface integral in parametric form, begin by calculating the following.

$$\mathbf{r}_x = \mathbf{i}$$
$$\mathbf{r}_\theta = -3\sin\theta\mathbf{j} + 3\cos\theta\mathbf{k}$$
$$\mathbf{r}_x \times \mathbf{r}_\theta = \begin{vmatrix} \mathbf{i} & \mathbf{j} & \mathbf{k} \\ 1 & 0 & 0 \\ 0 & -3\sin\theta & 3\cos\theta \end{vmatrix} = -3\cos\theta\mathbf{j} - 3\sin\theta\mathbf{k}$$
$$\|\mathbf{r}_x \times \mathbf{r}_\theta\| = \sqrt{9\cos^2\theta + 9\sin^2\theta} = 3$$

Therefore, the surface integral can be evaluated as follows.

$$\iint_D (x + 3\sin\theta)3\, dA = \int_0^4 \int_0^{\pi/2} (3x + 9\sin\theta)\, d\theta\, dx$$
$$= \int_0^4 \left[3x\theta - 9\cos\theta \right]_0^{\pi/2} dx$$
$$= \int_0^4 \left(\frac{3\pi}{2}x + 9 \right) dx$$
$$= \left[\frac{3\pi}{4}x^2 + 9x \right]_0^4$$
$$= 12\pi + 36$$

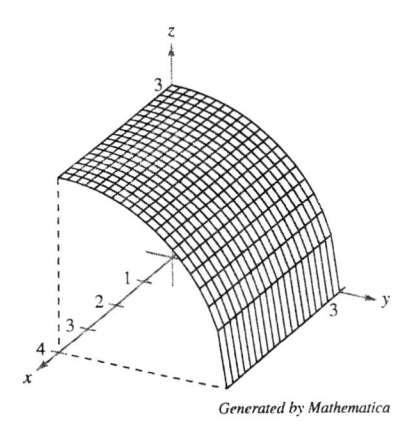

Figure 14.49

Orientation of a Surface

Unit normal vectors are used to induce an orientation to a surface S in space. A surface is called **orientable** if a unit normal vector $\mathbf{N}$ can be defined at every nonboundary point of S in such a way that the normal vectors vary continuously over the surface S. If this is possible, S is called an **oriented surface.**

An orientable surface S has two distinct sides. So, when you orient a surface, you are selecting one of the two possible unit normal vectors. If S is a closed surface such as a sphere, it is customary to choose the unit normal vector $\mathbf{N}$ to be the one that points outward from the sphere.

Most common surfaces, such as spheres, paraboloids, ellipses, and planes, are orientable. (See Exercise 35 for an example of a surface that is *not* orientable.) Moreover, for an orientable surface, the gradient vector provides a convenient way to find a unit normal vector. That is, for an orientable surface S given by

$$z = g(x, y) \qquad \text{Orientable surface}$$

let

$$G(x, y, z) = z - g(x, y).$$

Then, S can be oriented by either the unit normal vector

$$\mathbf{N} = \frac{\nabla G(x, y, z)}{\|\nabla G(x, y, z)\|}$$

$$= \frac{-g_x(x, y)\mathbf{i} - g_y(x, y)\mathbf{j} + \mathbf{k}}{\sqrt{1 + [g_x(x, y)]^2 + [g_y(x, y)]^2}} \qquad \text{Upward unit normal}$$

or the unit normal vector

$$\mathbf{N} = \frac{-\nabla G(x, y, z)}{\|\nabla G(x, y, z)\|}$$

$$= \frac{g_x(x, y)\mathbf{i} + g_y(x, y)\mathbf{j} - \mathbf{k}}{\sqrt{1 + [g_x(x, y)]^2 + [g_y(x, y)]^2}} \qquad \text{Downward unit normal}$$

as shown in Figure 14.50. If the smooth orientable surface S is given in parametric form by

$$\mathbf{r}(u, v) = x(u, v)\mathbf{i} + y(u, v)\mathbf{j} + z(u, v)\mathbf{k} \qquad \text{Parametric surface}$$

the unit normal vectors are given by

$$\mathbf{N} = \frac{\mathbf{r}_u \times \mathbf{r}_v}{\|\mathbf{r}_u \times \mathbf{r}_v\|}$$

and

$$\mathbf{N} = \frac{\mathbf{r}_v \times \mathbf{r}_u}{\|\mathbf{r}_v \times \mathbf{r}_u\|}.$$

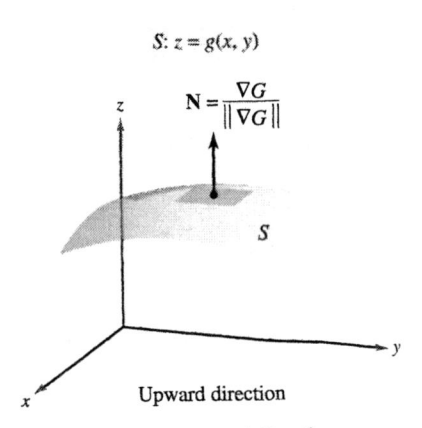

S is oriented in an upward direction.

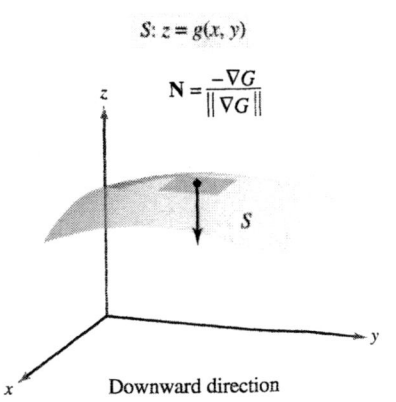

S is oriented in an downward direction.
Figure 14.50

NOTE Suppose that the orientable surface is given by $y = g(x, z)$ or $x = g(y, z)$. Then you can use the gradient vector

$$\nabla G(x, y, z) = -g_x(x, z)\mathbf{i} + \mathbf{j} - g_z(x, z)\mathbf{k} \qquad G(x, y, z) = y - g(x, z)$$

or

$$\nabla G(x, y, z) = \mathbf{i} - g_y(y, z)\mathbf{j} - g_z(y, z)\mathbf{k} \qquad G(x, y, z) = x - g(y, z)$$

to orient the surface.

Flux Integrals

One of the principal applications involving the vector form of a surface integral relates to the flow of a fluid through a surface S. Suppose an oriented surface S is submerged in a fluid having a continuous velocity field $\mathbf{F}$. Let ΔS be the area of a small patch of the surface S over which $\mathbf{F}$ is nearly constant. Then the amount of fluid crossing this region per unit of time is approximated by the volume of the column of height $\mathbf{F} \cdot \mathbf{N}$, as shown in Figure 14.51. That is,

$$\Delta V = (\text{height})(\text{area of base}) = (\mathbf{F} \cdot \mathbf{N})\Delta S.$$

Consequently, the volume of fluid crossing the surface S per unit of time (called the **flux of F across S**) is given by the surface integral in the following definition.

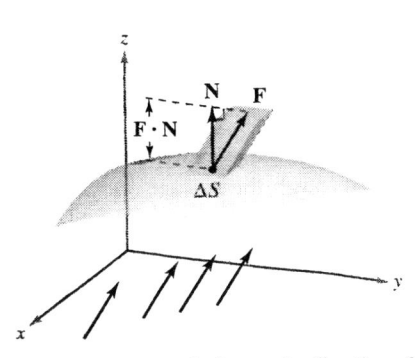

The velocity field $\mathbf{F}$ indicates the direction of the fluid flow.
Figure 14.51

Definition of Flux Integral

Let $\mathbf{F}(x, y, z) = M\mathbf{i} + N\mathbf{j} + P\mathbf{k}$, where M, N, and P have continuous first partial derivatives on the surface S oriented by a unit normal vector $\mathbf{N}$. The **flux integral of F across S** is given by

$$\iint_S \mathbf{F} \cdot \mathbf{N}\, dS.$$

Geometrically, a flux integral is the surface integral over S of the *normal component* of $\mathbf{F}$. If $\rho(x, y, z)$ is the density of the fluid at (x, y, z), the flux integral

$$\iint_S \rho \mathbf{F} \cdot \mathbf{N}\, dS$$

represents the *mass* of the fluid flowing across S per unit of time.

To evaluate a flux integral for a surface given by $z = g(x, y)$, let $G(x, y, z) = z - g(x, y)$. Then, $\mathbf{N}\, dS$ can be written as follows.

$$\begin{aligned}
\mathbf{N}\, dS &= \frac{\nabla G(x, y, z)}{\|\nabla G(x, y, z)\|}\, dS \\
&= \frac{\nabla G(x, y, z)}{\sqrt{(g_x)^2 + (g_y)^2 + 1}} \sqrt{(g_x)^2 + (g_y)^2 + 1}\, dA \\
&= \nabla G(x, y, z)\, dA
\end{aligned}$$

THEOREM 14.11 Evaluating a Flux Integral

Let S be an oriented surface given by $z = g(x, y)$ and let R be its projection onto the xy-plane.

$$\iint_S \mathbf{F} \cdot \mathbf{N}\, dS = \iint_R \mathbf{F} \cdot [-g_x(x, y)\mathbf{i} - g_y(x, y)\mathbf{j} + \mathbf{k}]\, dA \qquad \text{Oriented upward}$$

$$\iint_S \mathbf{F} \cdot \mathbf{N}\, dS = \iint_R \mathbf{F} \cdot [g_x(x, y)\mathbf{i} + g_y(x, y)\mathbf{j} - \mathbf{k}]\, dA \qquad \text{Oriented downward}$$

For the first integral, the surface is oriented upward, and for the second integral, the surface is oriented downward.

Example 5 Using a Flux Integral to Find the Rate of Mass Flow

Let S be the portion of the paraboloid

$$z = g(x, y) = 4 - x^2 - y^2$$

lying above the xy-plane, oriented by an upward unit normal vector, as shown in Figure 14.52. A fluid of constant density ρ is flowing through the surface S according to the vector field

$$\mathbf{F}(x, y, z) = x\mathbf{i} + y\mathbf{j} + z\mathbf{k}.$$

Find the rate of mass flow through S.

Solution Begin by computing the partial derivatives of g.

$$g_x(x, y) = -2x$$

and

$$g_y(x, y) = -2y$$

The rate of mass flow through the surface S is

$$\iint_S \rho \mathbf{F} \cdot \mathbf{N}\, dS = \rho \iint_R \mathbf{F} \cdot [-g_x(x, y)\mathbf{i} - g_y(x, y)\mathbf{j} + \mathbf{k}]\, dA$$

$$= \rho \iint_R [x\mathbf{i} + y\mathbf{j} + (4 - x^2 - y^2)\mathbf{k}] \cdot (2x\mathbf{i} + 2y\mathbf{j} + \mathbf{k})\, dA$$

$$= \rho \iint_R [2x^2 + 2y^2 + (4 - x^2 - y^2)]\, dA$$

$$= \rho \iint_R (4 + x^2 + y^2)\, dA$$

$$= \rho \int_0^{2\pi} \int_0^2 (4 + r^2) r\, dr\, d\theta \qquad \text{Polar coordinates}$$

$$= \rho \int_0^{2\pi} 12\, d\theta$$

$$= 24\pi\rho.$$

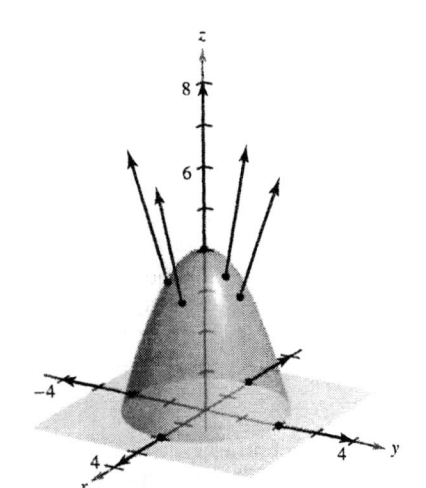

Figure 14.52

For an oriented surface S given by the vector-valued function

$$\mathbf{r}(u, v) = x(u, v)\mathbf{i} + y(u, v)\mathbf{j} + z(u, v)\mathbf{k} \qquad \text{Parametric surface}$$

defined over a region D in the uv-plane, you can define the flux integral of $\mathbf{F}$ across S as

$$\iint_S \mathbf{F} \cdot \mathbf{N}\, dS = \iint_D \mathbf{F} \cdot \left(\frac{\mathbf{r}_u \times \mathbf{r}_v}{\|\mathbf{r}_u \times \mathbf{r}_v\|}\right) \|\mathbf{r}_u \times \mathbf{r}_v\|\, dA$$

$$= \iint_D \mathbf{F} \cdot (\mathbf{r}_u \times \mathbf{r}_v)\, dA.$$

Note the similarity of this integral to the line integral

$$\int_C \mathbf{F} \cdot d\mathbf{r} = \int_C \mathbf{F} \cdot \mathbf{T}\, ds.$$

A summary of formulas for line and surface integrals is presented on page 1070.

Example 6   Finding the Flux of an Inverse Square Field

Find the flux over the sphere S given by

$$x^2 + y^2 + z^2 = a^2 \qquad \text{Sphere } S$$

where $\mathbf{F}$ is an inverse square field given by

$$\mathbf{F}(x, y, z) = \frac{q}{\|\mathbf{r}\|^2}\frac{\mathbf{r}}{\|\mathbf{r}\|} = \frac{q\mathbf{r}}{\|\mathbf{r}\|^3} \qquad \text{Inverse square field } \mathbf{F}$$

and $\mathbf{r} = x\mathbf{i} + y\mathbf{j} + z\mathbf{k}$. Assume S is oriented outward, as shown in Figure 14.53.

Solution   The sphere is given by

$$\begin{aligned}\mathbf{r}(u, v) &= x(u, v)\mathbf{i} + y(u, v)\mathbf{j} + z(u, v)\mathbf{k} \\ &= a \sin u \cos v\,\mathbf{i} + a \sin u \sin v\,\mathbf{j} + a \cos u\,\mathbf{k}\end{aligned}$$

where $0 \le u \le \pi$ and $0 \le v \le 2\pi$. The partial derivatives of $\mathbf{r}$ are

$$\mathbf{r}_u(u, v) = a \cos u \cos v\,\mathbf{i} + a \cos u \sin v\,\mathbf{j} - a \sin u\,\mathbf{k}$$

and

$$\mathbf{r}_v(u, v) = -a \sin u \sin v\,\mathbf{i} + a \sin u \cos v\,\mathbf{j}$$

which implies that the normal vector $\mathbf{r}_u \times \mathbf{r}_v$ is

$$\begin{aligned}\mathbf{r}_u \times \mathbf{r}_v &= \begin{vmatrix} \mathbf{i} & \mathbf{j} & \mathbf{k} \\ a\cos u \cos v & a\cos u \sin v & -a\sin u \\ -a\sin u \sin v & a\sin u \cos v & 0 \end{vmatrix} \\ &= a^2(\sin^2 u \cos v\,\mathbf{i} + \sin^2 u \sin v\,\mathbf{j} + \sin u \cos u\,\mathbf{k}).\end{aligned}$$

Now, using

$$\begin{aligned}\mathbf{F}(x, y, z) &= \frac{q\mathbf{r}}{\|\mathbf{r}\|^3} \\ &= q\frac{x\mathbf{i} + y\mathbf{j} + z\mathbf{k}}{\|x\mathbf{i} + y\mathbf{j} + z\mathbf{k}\|^3} \\ &= \frac{q}{a^3}(a\sin u \cos v\,\mathbf{i} + a\sin u \sin v\,\mathbf{j} + a\cos u\,\mathbf{k})\end{aligned}$$

it follows that

$$\begin{aligned}\mathbf{F}\cdot(\mathbf{r}_u \times \mathbf{r}_v) &= \frac{q}{a^3}[(a\sin u \cos v\,\mathbf{i} + a\sin u \sin v\,\mathbf{j} + a\cos u\,\mathbf{k}) \cdot \\ &\qquad a^2(\sin^2 u \cos v\,\mathbf{i} + \sin^2 u \sin v\,\mathbf{j} + \sin u \cos u\,\mathbf{k})] \\ &= q(\sin^3 u \cos^2 v + \sin^3 u \sin^2 v + \sin u \cos^2 u) \\ &= q \sin u.\end{aligned}$$

Finally, the flux over the sphere S is given by

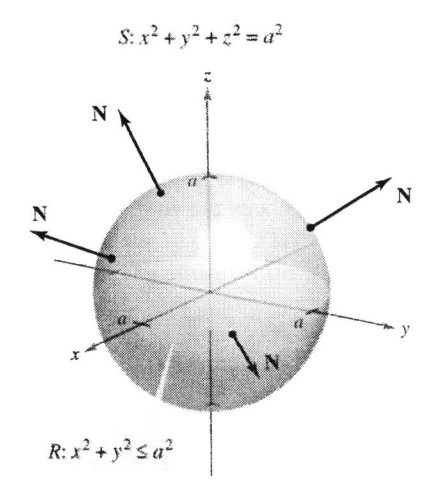

$$\begin{aligned}\iint_S \mathbf{F}\cdot\mathbf{N}\,dS &= \iint_D (q\sin u)\,dA \\ &= \int_0^{2\pi}\!\!\int_0^{\pi} q\sin u\,du\,dv \\ &= 4\pi q.\end{aligned}$$

Figure 14.53

The result in Example 6 shows that the flux across a sphere S in an inverse square field is independent of the radius of S. In particular, if $\mathbf{E}$ is an electric field, the result in Example 6, along with Coulomb's Law, yields one of the basic laws of electrostatics, known as **Gauss's Law:**

$$\iint_S \mathbf{E} \cdot \mathbf{N}\, dS = 4\pi q \qquad \text{Gauss's Law}$$

where q is a point charge located at the center of the sphere. Gauss's Law is valid for more general closed surfaces that enclose the origin, and relates the flux out of the surface to the total charge q inside the surface.

We conclude this section with a summary of different forms of line integrals and surface integrals.

Summary of Line and Surface Integrals

Line Integrals

$$ds = \|\mathbf{r}'(t)\|\, dt = \sqrt{[x'(t)]^2 + [y'(t)]^2 + [z'(t)]^2}\, dt$$

$$\int_C f(x, y, z)\, ds = \int_a^b f(x(t), y(t), z(t))\, ds \qquad \text{Scalar form}$$

$$\int_C \mathbf{F} \cdot d\mathbf{r} = \int_C \mathbf{F} \cdot \mathbf{T}\, ds$$
$$= \int_a^b \mathbf{F}(x(t), y(t), z(t)) \cdot \mathbf{r}'(t)\, dt \qquad \text{Vector form}$$

Surface Integrals $[z = g(x, y)]$

$$dS = \sqrt{1 + [g_x(x, y)]^2 + [g_y(x, y)]^2}\, dA$$

$$\iint_S f(x, y, z)\, dS = \iint_R f(x, y, g(x, y))\sqrt{1 + [g_x(x, y)]^2 + [g_y(x, y)]^2}\, dA \qquad \text{Scalar form}$$

$$\iint_S \mathbf{F} \cdot \mathbf{N}\, dS = \iint_R \mathbf{F} \cdot [-g_x(x, y)\mathbf{i} - g_y(x, y)\mathbf{j} + \mathbf{k}]\, dA \qquad \text{Vector form (upward normal)}$$

Surface Integrals (*parametric form*)

$$dS = \|\mathbf{r}_u(u, v) \times \mathbf{r}_v(u, v)\|\, dA$$

$$\iint_S f(x, y, z)\, dS = \iint_D f(x(u, v), y(u, v), z(u, v))\, dS \qquad \text{Scalar form}$$

$$\iint_S \mathbf{F} \cdot \mathbf{N}\, dS = \iint_D \mathbf{F} \cdot (\mathbf{r}_u \times \mathbf{r}_v)\, dA \qquad \text{Vector form}$$

EXERCISES FOR SECTION 14.6

In Exercises 1–4, evaluate $\iint_S (x - 2y + z)\, dS$.

1. $S: z = 4 - x, \quad 0 \le x \le 4, \quad 0 \le y \le 4$
2. $S: z = 15 - 2x + 3y, \quad 0 \le x \le 2, \quad 0 \le y \le 4$
3. $S: z = 10, \quad x^2 + y^2 \le 1$
4. $S: z = \frac{2}{3}x^{3/2}, \quad 0 \le x \le 1, \quad 0 \le y \le x$

In Exercises 5 and 6, evaluate $\iint_S xy\, dS$.

5. $S: z = 6 - x - 2y$, first octant
6. $S: z = h, \quad 0 \le x \le 2, \quad 0 \le y \le \sqrt{4 - x^2}$

In Exercises 7 and 8, use a computer algebra system to evaluate
$\iint_S xy\, dS$.

7. $S: z = 9 - x^2, \quad 0 \le x \le 2, \quad 0 \le y \le x$
8. $S: z = \frac{1}{2}xy, \quad 0 \le x \le 4, \quad 0 \le y \le 4$

In Exercises 9 and 10, use a computer algebra system to evaluate
$\iint_S (x^2 - 2xy)\, dS$.

9. $S: z = 10 - x^2 - y^2, \quad 0 \le x \le 2, \quad 0 \le y \le 2$
10. $S: z = \cos x, \quad 0 \le x \le \frac{\pi}{2}, \quad 0 \le y \le \frac{1}{2}x$

Mass In Exercises 11 and 12, find the mass of the surface lamina S of density ρ.

11. $S: 2x + 3y + 6z = 12$, first octant, $\rho(x, y, z) = x^2 + y^2$
12. $S: z = \sqrt{a^2 - x^2 - y^2}, \quad \rho(x, y, z) = kz$

In Exercises 13–16, evaluate $\iint_S f(x, y)\, dS$.

13. $f(x, y) = y + 5$
    $S: \mathbf{r}(u, v) = u\mathbf{i} + v\mathbf{j} + \frac{v}{2}\mathbf{k}, \quad 0 \le u \le 1, \quad 0 \le v \le 2$

14. $f(x, y) = x + y$
    $S: \mathbf{r}(u, v) = 2\cos u\, \mathbf{i} + 2\sin u\, \mathbf{j} + v\mathbf{k}$
    $0 \le u \le \frac{\pi}{2}, \quad 0 \le v \le 2$

15. $f(x, y) = xy$
    $S: \mathbf{r}(u, v) = 2\cos u\, \mathbf{i} + 2\sin u\, \mathbf{j} + v\mathbf{k}$
    $0 \le u \le \frac{\pi}{2}, \quad 0 \le v \le 2$

16. $f(x, y) = x + y$
    $S: \mathbf{r}(u, v) = 4u\cos v\, \mathbf{i} + 4u\sin v\, \mathbf{j} + 3u\mathbf{k}$
    $0 \le u \le 4, \quad 0 \le v \le \pi$

In Exercises 17–22, evaluate $\iint_S f(x, y, z)\, dS$.

17. $f(x, y, z) = x^2 + y^2 + z^2$
    $S: z = x + 2, \quad x^2 + y^2 \le 1$

18. $f(x, y, z) = \frac{xy}{z}$
    $S: z = x^2 + y^2, \quad 4 \le x^2 + y^2 \le 16$

19. $f(x, y, z) = \sqrt{x^2 + y^2 + z^2}$
    $S: z = \sqrt{x^2 + y^2}, \quad x^2 + y^2 \le 4$

20. $f(x, y, z) = \sqrt{x^2 + y^2 + z^2}$
    $S: z = \sqrt{x^2 + y^2}, \quad (x - 1)^2 + y^2 \le 1$

21. $f(x, y, z) = x^2 + y^2 + z^2$
    $S: x^2 + y^2 = 9, \quad 0 \le x \le 3, \quad 0 \le y \le 3, \quad 0 \le z \le 9$

22. $f(x, y, z) = x^2 + y^2 + z^2$
    $S: x^2 + y^2 = 9, \quad 0 \le x \le 3, \quad 0 \le z \le x$

In Exercises 23–28, find the flux of $\mathbf{F}$ through S,
$$\iint_S \mathbf{F} \cdot \mathbf{N}\, dS$$
where $\mathbf{N}$ is the upward unit normal vector to S.

23. $\mathbf{F}(x, y, z) = 3z\mathbf{i} - 4\mathbf{j} + y\mathbf{k}$
    $S: x + y + z = 1$, first octant

24. $\mathbf{F}(x, y, z) = x\mathbf{i} + y\mathbf{j}$
    $S: 2x + 3y + z = 6$, first octant

25. $\mathbf{F}(x, y, z) = x\mathbf{i} + y\mathbf{j} + z\mathbf{k}$
    $S: z = 9 - x^2 - y^2, \quad z \ge 0$

26. $\mathbf{F}(x, y, z) = x\mathbf{i} + y\mathbf{j} + z\mathbf{k}$
    $S: x^2 + y^2 + z^2 = 36$, first octant

27. $\mathbf{F}(x, y, z) = 4\mathbf{i} - 3\mathbf{j} + 5\mathbf{k}$
    $S: z = x^2 + y^2, \quad x^2 + y^2 \le 4$

28. $\mathbf{F}(x, y, z) = x\mathbf{i} + y\mathbf{j} - 2z\mathbf{k}$
    $S: z = \sqrt{a^2 - x^2 - y^2}$

In Exercises 29 and 30, find the flux of $\mathbf{F}$ over the closed surface. (Let $\mathbf{N}$ be the outward unit normal vector of the surface.)

29. $\mathbf{F}(x, y, z) = 4xy\mathbf{i} + z^2\mathbf{j} + yz\mathbf{k}$
    S: unit cube bounded by $x = 0, x = 1, y = 0, y = 1, z = 0, z = 1$

30. $\mathbf{F}(x, y, z) = (x + y)\mathbf{i} + y\mathbf{j} + z\mathbf{k}$
    $S: z = 1 - x^2 - y^2, \quad z = 0$

Getting at the Concept

31. Define a surface integral of the scalar function f over a surface $z = g(x, y)$. Explain how to evaluate the surface integral.

32. Describe an orientable surface.

33. Define a flux integral and explain how it is evaluated.

34. Is the surface shown in the figure orientable?

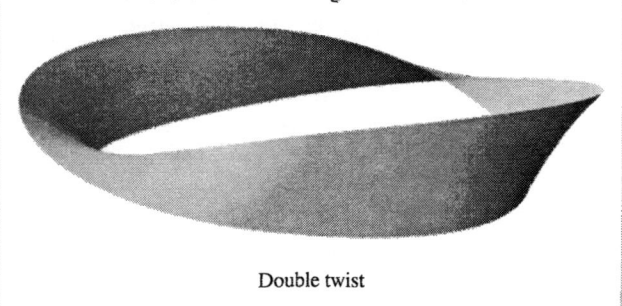

Double twist

35. *Investigation*

(a) Use a computer algebra system to graph the vector-valued function

$$\mathbf{r}(u, v) = (4 - v \sin u) \cos(2u)\mathbf{i} + (4 - v \sin u) \sin(2u)\mathbf{j} + v \cos u \mathbf{k}, \quad 0 \leq u \leq \pi, \quad -1 \leq v \leq 1.$$

This surface is called a Möbius strip.

(b) Explain why this surface is not orientable.

(c) Use a computer algebra system to graph the space curve represented by $\mathbf{r}(u, 0)$. Identify the curve.

(d) Construct a Möbius strip by cutting a strip of paper, making a single twist, and pasting the ends together.

(e) Cut the Möbius strip along the space curve graphed in part (c), and describe the result.

36. *Electrical Charge* Let $\mathbf{E} = yz\mathbf{i} + xz\mathbf{j} + xy\mathbf{k}$ be an electrostatic field. Use Gauss's Law to find the total charge enclosed by the closed surface consisting of the hemisphere $z = \sqrt{1 - x^2 - y^2}$ and its circular base in the xy-plane.

Moment of Inertia In Exercises 37 and 38, use the following formulas for the moments of inertia about the coordinate axes of a surface lamina of density ρ.

$$I_x = \iint_S (y^2 + z^2)\rho(x, y, z) \, dS$$

$$I_y = \iint_S (x^2 + z^2)\rho(x, y, z) \, dS$$

$$I_z = \iint_S (x^2 + y^2)\rho(x, y, z) \, dS$$

37. Verify that the moment of inertia of a conical shell of uniform density about its axis is $\frac{1}{2}ma^2$, where m is the mass and a is the radius.

38. Verify that the moment of inertia of a spherical shell of uniform density about its diameter is $\frac{2}{3}ma^2$, where m is the mass and a is the radius.

Moment of Inertia In Exercises 39 and 40, find I_z for the given lamina with uniform density of 1. Use a computer algebra system to verify your results.

39. $x^2 + y^2 = a^2, \quad 0 \leq z \leq h$

40. $z = x^2 + y^2, \quad 0 \leq z \leq h$

Flow Rate In Exercises 41 and 42, use a computer algebra system to find the rate of mass flow of a fluid of density ρ through the surface S oriented upward if the velocity field is given by $\mathbf{F}(x, y, z) = 0.5z\mathbf{k}$.

41. $S: z = 16 - x^2 - y^2, \quad z \geq 0$

42. $S: z = \sqrt{16 - x^2 - y^2}$

SECTION PROJECT: HYPERBOLOID OF ONE SHEET

Consider the parametric surface given by the function

$$\mathbf{r}(u, v) = a \cosh u \cos v \mathbf{i} + a \cosh u \sin v \mathbf{j} + b \sinh u \mathbf{k}.$$

(a) Use a graphing utility to graph $\mathbf{r}$ for various values of the constants a and b. Describe the effect of the constants on the shape of the surface.

(b) Show that the surface is a hyperboloid of one sheet given by

$$\frac{x^2}{a^2} + \frac{y^2}{a^2} - \frac{z^2}{b^2} = 1.$$

(c) For fixed values $u = u_0$, describe the curves given by

$$\mathbf{r}(u_0, v) = a \cosh u_0 \cos v \mathbf{i} + a \cosh u_0 \sin v \mathbf{j} + b \sinh u_0 \mathbf{k}.$$

(d) For fixed values $v = v_0$, describe the curves given by

$$\mathbf{r}(u, v_0) = a \cosh u \cos v_0 \mathbf{i} + a \cosh u \sin v_0 \mathbf{j} + b \sinh u \mathbf{k}.$$

(e) Find a normal vector to the surface at $(u, v) = (0, 0)$.

Section 14.7 Divergence Theorem

- Understand and use the Divergence Theorem.
- Use the Divergence Theorem to calculate flux.

Divergence Theorem

Recall from Section 14.4 that an alternative form of Green's Theorem is

$$\int_C \mathbf{F} \cdot \mathbf{N}\, ds = \int\!\!\int_R \left(\frac{\partial M}{\partial x} + \frac{\partial N}{\partial y}\right) dA$$

$$= \int\!\!\int_R \operatorname{div} \mathbf{F}\, dA.$$

In an analogous way, the **Divergence Theorem** gives the relationship between a triple integral over a solid region Q and a surface integral over the surface of Q. In the statement of the theorem, the surface S is **closed** in the sense that it forms the complete boundary of the solid Q. Regions bounded by spheres, ellipsoids, cubes, tetrahedrons, or combinations of these surfaces are typical examples of closed surfaces. Assume that Q is a solid region on which a triple integral can be evaluated, and that the closed surface S is oriented by *outward* unit normal vectors, as shown in Figure 14.54. With these restrictions on S and Q, we state the following theorem.

S_1: Oriented by upward unit normal vector

S_2: Oriented by downward unit normal vector

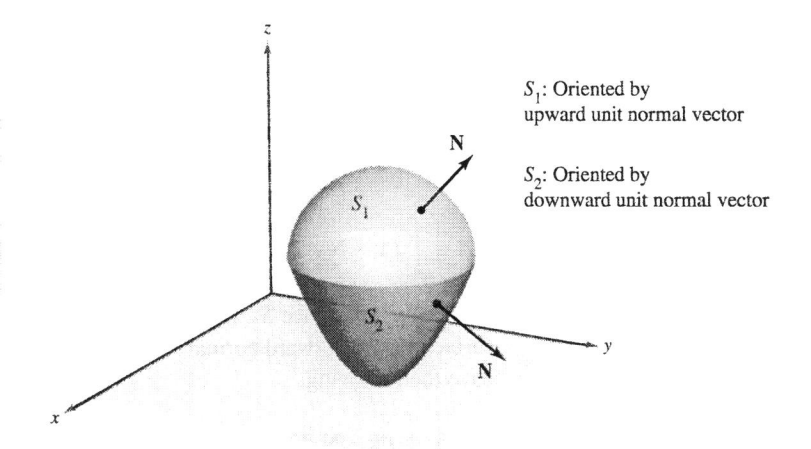

Figure 14.54

CARL FRIEDRICH GAUSS (1777–1855)

The **Divergence Theorem** is also called **Gauss's Theorem**, after the famous German mathematician Carl Friedrich Gauss. Gauss is recognized, with Newton and Archimedes, as one of the three greatest mathematicians in history. One of his many contributions to mathematics was made at the age of 22, when, as part of his doctoral dissertation, he proved the *Fundamental Theorem of Algebra*.

THEOREM 14.12 The Divergence Theorem

Let Q be a solid region bounded by a closed surface S oriented by a unit normal vector directed outward from Q. If $\mathbf{F}$ is a vector field whose component functions have continuous partial derivatives in Q, then

$$\int\!\!\int_S \mathbf{F} \cdot \mathbf{N}\, dS = \int\!\!\int\!\!\int_Q \operatorname{div} \mathbf{F}\, dV.$$

NOTE As noted at the left above, the Divergence Theorem is sometimes called Gauss's Theorem. It is also sometimes called Ostrogradsky's Theorem, after the Russian mathematician Michel Ostrogradsky (1801–1861).

Proof If you let $\mathbf{F}(x, y, z) = M\mathbf{i} + N\mathbf{j} + P\mathbf{k}$, the theorem takes the form

$$\iint_S \mathbf{F} \cdot \mathbf{N}\, dS = \iint_S (M\mathbf{i} \cdot \mathbf{N} + N\mathbf{j} \cdot \mathbf{N} + P\mathbf{k} \cdot \mathbf{N})\, dS$$

$$= \iiint_Q \left(\frac{\partial M}{\partial x} + \frac{\partial N}{\partial y} + \frac{\partial P}{\partial z} \right) dV.$$

You can prove this by verifying that the following three equations are valid.

$$\iint_S M\mathbf{i} \cdot \mathbf{N}\, dS = \iiint_Q \frac{\partial M}{\partial x}\, dV$$

$$\iint_S N\mathbf{j} \cdot \mathbf{N}\, dS = \iiint_Q \frac{\partial N}{\partial y}\, dV$$

$$\iint_S P\mathbf{k} \cdot \mathbf{N}\, dS = \iiint_Q \frac{\partial P}{\partial z}\, dV$$

Because the verifications of the three equations are similar, we will discuss only the third. We restrict the proof to a **simple solid** region with upper surface

$$z = g_2(x, y) \qquad \text{Upper surface}$$

and lower surface

$$z = g_1(x, y) \qquad \text{Lower surface}$$

whose projections onto the xy-plane coincide and form region R. If Q has a lateral surface like S_3 in Figure 14.55, then a normal vector is horizontal, which implies that $P\mathbf{k} \cdot \mathbf{N} = 0$. Consequently, you have

$$\iint_S P\mathbf{k} \cdot \mathbf{N}\, dS = \iint_{S_1} P\mathbf{k} \cdot \mathbf{N}\, dS + \iint_{S_2} P\mathbf{k} \cdot \mathbf{N}\, dS + 0.$$

On the upper surface S_2, the outward normal vector is upward, whereas on the lower surface S_1, the outward normal vector is downward. Therefore, by Theorem 14.11, you have the following.

$$\iint_{S_1} P\mathbf{k} \cdot \mathbf{N}\, dS = \iint_R P(x, y, g_1(x, y))\mathbf{k} \cdot \left(\frac{\partial g_1}{\partial x}\mathbf{i} + \frac{\partial g_1}{\partial y}\mathbf{j} - \mathbf{k} \right) dA$$

$$= -\iint_R P(x, y, g_1(x, y))\, dA$$

$$\iint_{S_2} P\mathbf{k} \cdot \mathbf{N}\, dS = \iint_R P(x, y, g_2(x, y))\mathbf{k} \cdot \left(-\frac{\partial g_2}{\partial x}\mathbf{i} - \frac{\partial g_2}{\partial y}\mathbf{j} + \mathbf{k} \right) dA$$

$$= \iint_R P(x, y, g_2(x, y))\, dA$$

Adding these results, you obtain

$$\iint_S P\mathbf{k} \cdot \mathbf{N}\, dS = \iint_R [P(x, y, g_2(x, y)) - P(x, y, g_1(x, y))]\, dA$$

$$= \iint_R \left[\int_{g_1(x, y)}^{g_2(x, y)} \frac{\partial P}{\partial z}\, dz \right] dA$$

$$= \iiint_Q \frac{\partial P}{\partial z}\, dV.$$

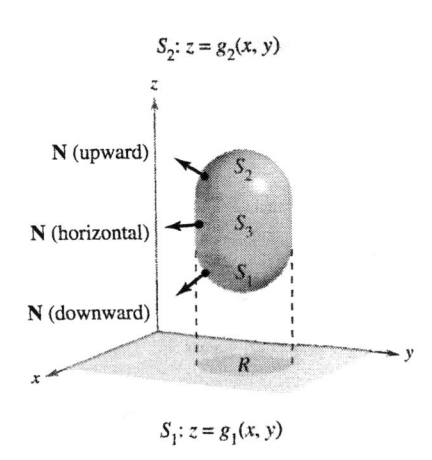

Figure 14.55

Example 1 Using the Divergence Theorem

Let Q be the solid region bounded by the coordinate planes and the plane $2x + 2y + z = 6$, and let $\mathbf{F} = x\mathbf{i} + y^2\mathbf{j} + z\mathbf{k}$. Find

$$\iint_S \mathbf{F} \cdot \mathbf{N}\, dS$$

where S is the surface of Q.

Solution From Figure 14.56, you can see that Q is bounded by four subsurfaces. So, you would need four *surface integrals* to evaluate

$$\iint_S \mathbf{F} \cdot \mathbf{N}\, dS.$$

However, by the Divergence Theorem, you need only one triple integral. Because

$$\text{div } \mathbf{F} = \frac{\partial M}{\partial x} + \frac{\partial N}{\partial y} + \frac{\partial P}{\partial z}$$
$$= 1 + 2y + 1$$
$$= 2 + 2y$$

you have

$$\iint_S \mathbf{F} \cdot \mathbf{N}\, dS = \iiint_Q \text{div } \mathbf{F}\, dV$$
$$= \int_0^3 \int_0^{3-y} \int_0^{6-2x-2y} (2 + 2y)\, dz\, dx\, dy$$
$$= \int_0^3 \int_0^{3-y} (2z + 2yz)\Big]_0^{6-2x-2y} dx\, dy$$
$$= \int_0^3 \int_0^{3-y} (12 - 4x + 8y - 4xy - 4y^2)\, dx\, dy$$
$$= \int_0^3 \Big[12x - 2x^2 + 8xy - 2x^2y - 4xy^2\Big]_0^{3-y} dy$$
$$= \int_0^3 (18 + 6y - 10y^2 + 2y^3)\, dy$$
$$= \Big[18y + 3y^2 - \frac{10y^3}{3} + \frac{y^4}{2}\Big]_0^3$$
$$= \frac{63}{2}.$$

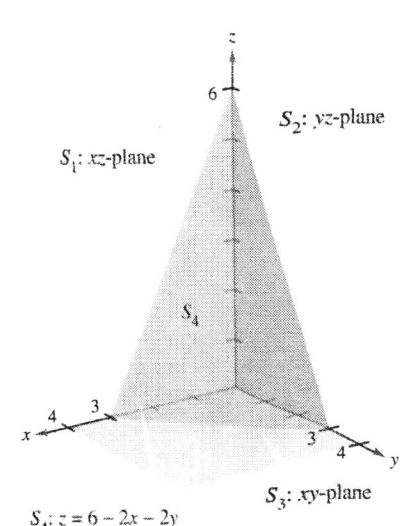

Figure 14.56

S_1: xz-plane
S_2: yz-plane
S_3: xy-plane
S_4: $z = 6 - 2x - 2y$

TECHNOLOGY If you have access to a computer algebra system that can evaluate triple-iterated integrals, try using it to verify the result in Example 1. When you are using such a utility, note that the first step is to convert the triple integral to an iterated integral—this step must be done by hand. To give yourself some practice with this important step, try finding the limits of integration for the following iterated integrals. Then use a computer to verify that the value is the same as that obtained in Example 1.

$$\int_?^? \int_?^? \int_?^? (2 + 2y)\, dy\, dz\, dx, \quad \int_?^? \int_?^? \int_?^? (2 + 2y)\, dx\, dy\, dz$$

Example 2 **Verifying the Divergence Theorem**

Let Q be the solid region between the paraboloid
$$z = 4 - x^2 - y^2$$
and the xy-plane. Verify the Divergence Theorem for
$$\mathbf{F}(x, y, z) = 2z\mathbf{i} + x\mathbf{j} + y^2\mathbf{k}.$$

Solution From Figure 14.57 you can see that the outward normal vector for the surface S_1 is $\mathbf{N}_1 = -\mathbf{k}$, whereas the outward normal vector for the surface S_2 is
$$\mathbf{N}_2 = \frac{2x\mathbf{i} + 2y\mathbf{j} + \mathbf{k}}{\sqrt{4x^2 + 4y^2 + 1}}.$$

So, by Theorem 14.11, you have

$$\iint_S \mathbf{F} \cdot \mathbf{N}\, dS$$
$$= \iint_{S_1} \mathbf{F} \cdot \mathbf{N}_1\, dS + \iint_{S_2} \mathbf{F} \cdot \mathbf{N}_2\, dS$$
$$= \iint_{S_1} \mathbf{F} \cdot (-\mathbf{k})\, dS + \iint_{S_2} \mathbf{F} \cdot (2x\mathbf{i} + 2y\mathbf{j} + \mathbf{k})\, dS$$
$$= \iint_R -y^2\, dA + \iint_R (4xz + 2xy + y^2)\, dA$$
$$= -\int_{-2}^{2}\int_{-\sqrt{4-y^2}}^{\sqrt{4-y^2}} y^2\, dx\, dy + \int_{-2}^{2}\int_{-\sqrt{4-y^2}}^{\sqrt{4-y^2}} (4xz + 2xy + y^2)\, dx\, dy$$
$$= \int_{-2}^{2}\int_{-\sqrt{4-y^2}}^{\sqrt{4-y^2}} (4xz + 2xy)\, dx\, dy$$
$$= \int_{-2}^{2}\int_{-\sqrt{4-y^2}}^{\sqrt{4-y^2}} [4x(4 - x^2 - y^2) + 2xy]\, dx\, dy$$
$$= \int_{-2}^{2}\int_{-\sqrt{4-y^2}}^{\sqrt{4-y^2}} (16x - 4x^3 - 4xy^2 + 2xy)\, dx\, dy$$
$$= \int_{-2}^{2} \left[8x^2 - x^4 - 2x^2y^2 + x^2y \right]_{-\sqrt{4-y^2}}^{\sqrt{4-y^2}} dy$$
$$= \int_{-2}^{2} 0\, dy$$
$$= 0.$$

On the other hand, because
$$\text{div } \mathbf{F} = \frac{\partial}{\partial x}[2z] + \frac{\partial}{\partial y}[x] + \frac{\partial}{\partial z}[y^2] = 0 + 0 + 0 = 0$$

you can apply the Divergence Theorem to obtain the equivalent result
$$\iint_S \mathbf{F} \cdot \mathbf{N}\, dS = \iiint_Q \text{div } \mathbf{F}\, dV$$
$$= \iiint_Q 0\, dV = 0.$$

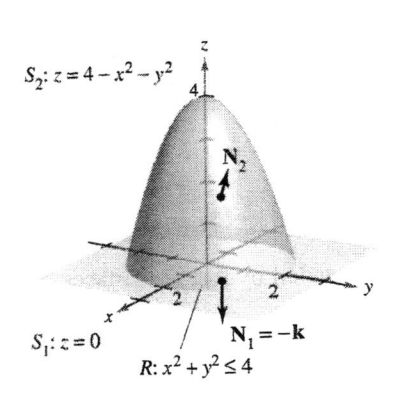

$S_2: z = 4 - x^2 - y^2$
$S_1: z = 0$
$R: x^2 + y^2 \leq 4$
$\mathbf{N}_1 = -\mathbf{k}$

Figure 14.57

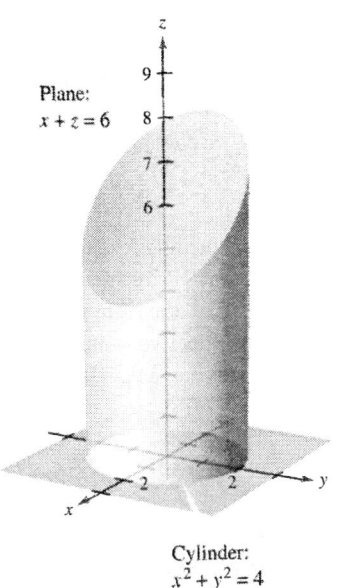

Example 3 Using the Divergence Theorem

Let Q be the solid bounded by the cylinder $x^2 + y^2 = 4$, the plane $x + z = 6$, and the xy-plane, as shown in Figure 14.58. Find

$$\iint_S \mathbf{F} \cdot \mathbf{N} \, dS$$

where S is the surface of Q and

$$\mathbf{F}(x, y, z) = (x^2 + \sin z)\mathbf{i} + (xy + \cos z)\mathbf{j} + e^y\mathbf{k}.$$

Solution Direct evaluation of this surface integral would be difficult. However, by the Divergence Theorem, you can evaluate the integral as follows.

$$\iint_S \mathbf{F} \cdot \mathbf{N} \, dS = \iiint_Q \text{div } \mathbf{F} \, dV$$

$$= \iiint_Q (2x + x + 0) \, dV$$

$$= \iiint_Q 3x \, dV$$

$$= \int_0^{2\pi} \int_0^2 \int_0^{6 - r\cos\theta} (3r \cos \theta) r \, dz \, dr \, d\theta$$

$$= \int_0^{2\pi} \int_0^2 (18r^2 \cos \theta - 3r^3 \cos^2 \theta) \, dr \, d\theta$$

$$= \int_0^{2\pi} (48 \cos \theta - 12 \cos^2 \theta) d\theta$$

$$= \left[48 \sin \theta - 6\left(\theta + \frac{1}{2} \sin 2\theta\right) \right]_0^{2\pi}$$

$$= -12\pi$$

Notice that cylindrical coordinates with $x = r \cos \theta$ and $dV = r \, dz \, dr \, d\theta$ were used to evaluate the triple integral.

Even though we stated the Divergence Theorem for a simple solid region Q bounded by a closed surface, the theorem is also valid for regions that are the finite unions of simple solid regions. For example, let Q be the solid bounded by the closed surfaces S_1 and S_2, as shown in Figure 14.59. To apply the Divergence Theorem to this solid, let $S = S_1 \cup S_2$. The normal vector $\mathbf{N}$ to S is given by $-\mathbf{N}_1$ on S_1 and by $\mathbf{N}_2$ on S_2. So, you can write

$$\iiint_Q \text{div } \mathbf{F} \, dV = \iint_S \mathbf{F} \cdot \mathbf{N} \, dS$$

$$= \iint_{S_1} \mathbf{F} \cdot (-\mathbf{N}_1) \, dS + \iint_{S_2} \mathbf{F} \cdot \mathbf{N}_2 \, dS$$

$$= -\iint_{S_1} \mathbf{F} \cdot \mathbf{N}_1 \, dS + \iint_{S_2} \mathbf{F} \cdot \mathbf{N}_2 \, dS.$$

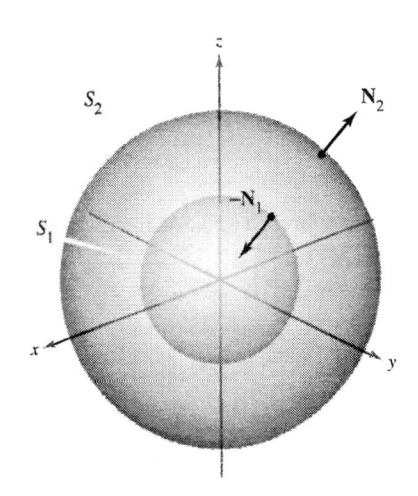

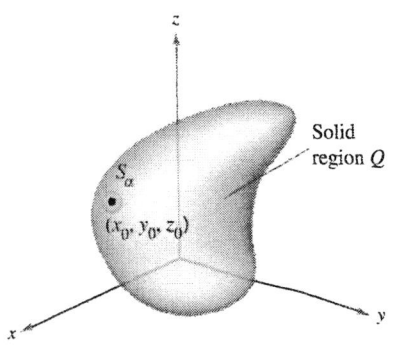

Figure 14.60

Flux and the Divergence Theorem

To help understand the Divergence Theorem, consider the two sides of the equation

$$\iint_S \mathbf{F} \cdot \mathbf{N}\, dS = \iiint_Q \operatorname{div} \mathbf{F}\, dV.$$

You know from Section 14.6 that the flux integral on the left determines the total fluid flow across the surface S per unit of time. This can be approximated by summing the fluid flow across small patches of the surface. The triple integral on the right measures this same fluid flow across S, but from a very different perspective—namely, by calculating the flow of fluid into (or out of) small *cubes* of volume ΔV_i. The flux of the ith cube is approximately

$$\text{Flux of } i\text{th cube} \approx \operatorname{div} \mathbf{F}(x_i, y_i, z_i)\, \Delta V_i$$

for some point (x_i, y_i, z_i) in the ith cube. Note that for a cube in the interior of Q, the gain (or loss) of fluid through any one of its six sides is offset by a corresponding loss (or gain) through one of the sides of an adjacent cube. After summing over all the cubes in Q, the only fluid flow that is not canceled by adjoining cubes is that on the outside edges of the cubes on the boundary. So, the sum

$$\sum_{i=1}^{n} \operatorname{div} \mathbf{F}(x_i, y_i, z_i)\, \Delta V_i$$

approximates the total flux into (or out of) Q, and therefore through the surface S.

To see what is meant by the divergence of $\mathbf{F}$ at a point, we consider ΔV_α to be the volume of a small sphere S_α of radius α and center (x_0, y_0, z_0), contained in region Q, as shown in Figure 14.60. Applying the Divergence Theorem to S_α produces

$$\text{Flux of } \mathbf{F} \text{ across } S_\alpha = \iiint_{Q_\alpha} \operatorname{div} \mathbf{F}\, dV$$
$$\approx \operatorname{div} \mathbf{F}(x_0, y_0, z_0)\, \Delta V_\alpha$$

where Q_α is the interior of S_α. Consequently, you have

$$\operatorname{div} \mathbf{F}(x_0, y_0, z_0) \approx \frac{\text{flux of } \mathbf{F} \text{ across } S_\alpha}{\Delta V_\alpha}$$

and, by taking the limit as $\alpha \to 0$, you obtain the divergence of $\mathbf{F}$ at the point (x_0, y_0, z_0).

$$\operatorname{div} \mathbf{F}(x_0, y_0, z_0) = \lim_{\alpha \to 0} \frac{\text{flux of } \mathbf{F} \text{ across } S_\alpha}{\Delta V_\alpha}$$
$$= \text{flux per unit volume at } (x_0, y_0, z_0)$$

The point (x_0, y_0, z_0) in a vector field is classified as a source, a sink, or incompressible, as follows.

1. **Source,** if $\operatorname{div} \mathbf{F} > 0$ (Figure 14.61a)
2. **Sink,** if $\operatorname{div} \mathbf{F} < 0$ (Figure 14.61b)
3. **Incompressible,** if $\operatorname{div} \mathbf{F} = 0$ (Figure 14.61c)

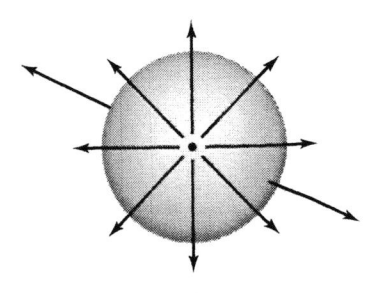

NOTE In hydrodynamics, a *source* is a point at which additional fluid is considered as being introduced to the region occupied by the fluid. A *sink* is a point at which fluid is considered as being removed.

(a) Source: div $\mathbf{F} > 0$

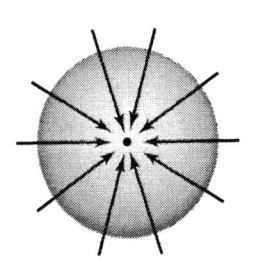

(b) Sink: div $\mathbf{F} < 0$

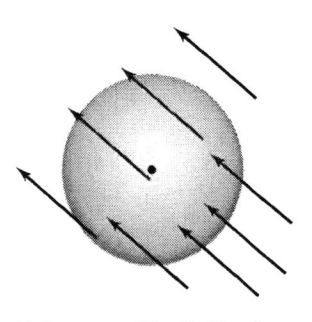

(c) Incompressible: div $\mathbf{F} = 0$

Figure 14.61

Example 4 Calculating Flux by the Divergence Theorem

Let Q be the region bounded by the sphere $x^2 + y^2 + z^2 = 4$. Find the outward flux of the vector field $\mathbf{F}(x, y, z) = 2x^3\mathbf{i} + 2y^3\mathbf{j} + 2z^3\mathbf{k}$ through the sphere.

Solution By the Divergence Theorem, you have

$$\text{Flux across } S = \iint_S \mathbf{F} \cdot \mathbf{N} \, dS = \iiint_Q \text{div } \mathbf{F} \, dV$$

$$= \iiint_Q 6(x^2 + y^2 + z^2) \, dV$$

$$= 6 \int_0^2 \int_0^\pi \int_0^{2\pi} \rho^4 \sin\phi \, d\theta \, d\phi \, d\rho \quad \text{Spherical coordinates}$$

$$= 6 \int_0^2 \int_0^\pi 2\pi \rho^4 \sin\phi \, d\phi \, d\rho$$

$$= 12\pi \int_0^2 2\rho^4 \, d\rho$$

$$= 24\pi \left(\frac{32}{5}\right)$$

$$= \frac{768\pi}{5}.$$

EXERCISES FOR SECTION 14.7

In Exercises 1–4, verify the Divergence Theorem by evaluating

$$\iint_S \mathbf{F} \cdot \mathbf{N} \, dS$$

as a surface integral and as a triple integral.

1. $\mathbf{F}(x, y, z) = 2x\mathbf{i} - 2y\mathbf{j} + z^2\mathbf{k}$

   S: cube bounded by the planes $x = 0$, $x = a$, $y = 0$, $y = a$, $z = 0$, $z = a$

2. $\mathbf{F}(x, y, z) = 2x\mathbf{i} - 2y\mathbf{j} + z^2\mathbf{k}$

   S: cylinder $x^2 + y^2 = 4$, $0 \le z \le h$

3. $\mathbf{F}(x, y, z) = (2x - y)\mathbf{i} - (2y - z)\mathbf{j} + z\mathbf{k}$

   S: surface bounded by the plane $2x + 4y + 2z = 12$ and the coordinate planes

4. $\mathbf{F}(x, y, z) = xy\mathbf{i} + z\mathbf{j} + (x + y)\mathbf{k}$

   S: surface bounded by the planes $y = 4$ and $z = 4 - x$ and the coordinate planes

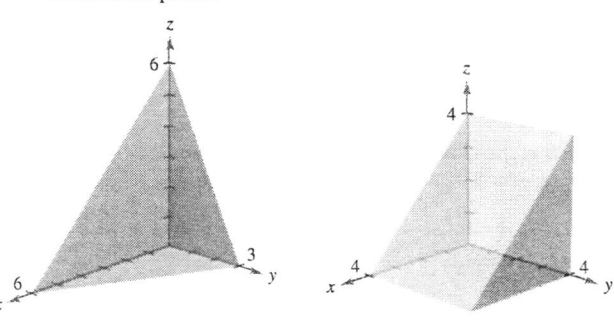

Figure for 3

Figure for 4

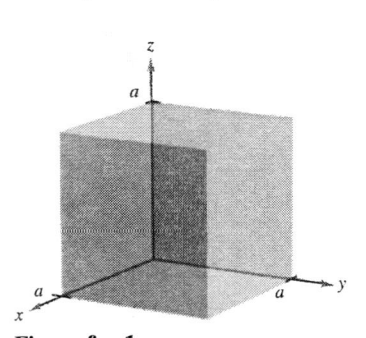

Figure for 1

Figure for 2

In Exercises 5–16, use the Divergence Theorem to evaluate

$$\iint_S \mathbf{F} \cdot \mathbf{N}\, dS$$

and find the outward flux of **F** through the surface of the solid bounded by the graphs of the equations. Use a computer algebra system to verify your results.

5. $\mathbf{F}(x, y, z) = x^2\mathbf{i} + y^2\mathbf{j} + z^2\mathbf{k}$
   S: $x = 0, x = a, y = 0, y = a, z = 0, z = a$

6. $\mathbf{F}(x, y, z) = x^2 z^2 \mathbf{i} - 2y\mathbf{j} + 3xyz\mathbf{k}$
   S: $x = 0, x = a, y = 0, y = a, z = 0, z = a$

7. $\mathbf{F}(x, y, z) = x^2\mathbf{i} - 2xy\mathbf{j} + xyz^2\mathbf{k}$
   S: $z = \sqrt{a^2 - x^2 - y^2}, z = 0$

8. $\mathbf{F}(x, y, z) = xy\mathbf{i} + yz\mathbf{j} - yz\mathbf{k}$
   S: $z = \sqrt{a^2 - x^2 - y^2}, z = 0$

9. $\mathbf{F}(x, y, z) = x\mathbf{i} + y\mathbf{j} + z\mathbf{k}$
   S: $x^2 + y^2 + z^2 = 4$

10. $\mathbf{F}(x, y, z) = xyz\mathbf{j}$
    S: $x^2 + y^2 = 9, z = 0, z = 4$

11. $\mathbf{F}(x, y, z) = x\mathbf{i} + y^2\mathbf{j} - z\mathbf{k}$
    S: $x^2 + y^2 = 9, z = 0, z = 4$

12. $\mathbf{F}(x, y, z) = (xy^2 + \cos z)\mathbf{i} + (x^2 y + \sin z)\mathbf{j} + e^z\mathbf{k}$
    S: $z = \frac{1}{2}\sqrt{x^2 + y^2}, z = 8$

13. $\mathbf{F}(x, y, z) = x^3\mathbf{i} + x^2 y\mathbf{j} + x^2 e^y\mathbf{k}$
    S: $z = 4 - y, z = 0, x = 0, x = 6, y = 0$

14. $\mathbf{F}(x, y, z) = xe^z\mathbf{i} + ye^z\mathbf{j} + e^z\mathbf{k}$
    S: $z = 4 - y, z = 0, x = 0, x = 6, y = 0$

15. $\mathbf{F}(x, y, z) = xy\mathbf{i} + 4y\mathbf{j} + xz\mathbf{k}$
    S: $x^2 + y^2 + z^2 = 9$

16. $\mathbf{F}(x, y, z) = 2(x\mathbf{i} + y\mathbf{j} + z\mathbf{k})$
    S: $z = \sqrt{4 - x^2 - y^2}, z = 0$

In Exercises 17 and 18, evaluate

$$\iint_S \operatorname{curl} \mathbf{F} \cdot \mathbf{N}\, dS$$

where S is the closed surface of the solid bounded by the graphs of $x = 4$, $z = 9 - y^2$, and the coordinate planes.

17. $\mathbf{F}(x, y, z) = (4xy + z^2)\mathbf{i} + (2x^2 + 6yz)\mathbf{j} + 2xz\mathbf{k}$

18. $\mathbf{F}(x, y, z) = xy \cos z\mathbf{i} + yz \sin x\mathbf{j} + xyz\mathbf{k}$

19. State the Divergence Theorem.

20. How do you determine if a point (x_0, y_0, z_0) in a vector field is a source, a sink, or incompressible?

21. Use the Divergence Theorem to verify that the volume of the solid bounded by a surface S is

$$\iint_S x\, dy\, dz = \iint_S y\, dz\, dx = \iint_S z\, dx\, dy.$$

22. Verify the result in Exercise 21 for the cube bounded by $x = 0$, $x = a$, $y = 0$, $y = a$, $z = 0$, and $z = a$.

23. Verify that

$$\iint_S \operatorname{curl} \mathbf{F} \cdot \mathbf{N}\, dS = 0$$

for any closed surface S.

24. For the constant vector field given by

$$\mathbf{F}(x, y, z) = a_1\mathbf{i} + a_2\mathbf{j} + a_3\mathbf{k}$$

verify that

$$\iint_S \mathbf{F} \cdot \mathbf{N}\, dS = 0$$

where V is the volume of the solid bounded by the closed surface S.

25. Given the vector field

$$\mathbf{F}(x, y, z) = x\mathbf{i} + y\mathbf{j} + z\mathbf{k}$$

verify that

$$\iint_S \mathbf{F} \cdot \mathbf{N}\, dS = 3V$$

where V is the volume of the solid bounded by the closed surface S.

26. Given the vector field

$$\mathbf{F}(x, y, z) = x\mathbf{i} + y\mathbf{j} + z\mathbf{k}$$

verify that

$$\frac{1}{\|\mathbf{F}\|}\iint_S \mathbf{F} \cdot \mathbf{N}\, dS = \frac{3}{\|\mathbf{F}\|}\iiint_Q dV.$$

In Exercises 27 and 28, prove the identity, assuming that Q, S, and $\mathbf{N}$ meet the conditions of the Divergence Theorem and that the required partial derivatives of the scalar functions f and g are continuous. The expressions $D_\mathbf{N} f$ and $D_\mathbf{N} g$ are the derivatives in the direction of the vector $\mathbf{N}$ and are defined by

$$D_\mathbf{N} f = \nabla f \cdot \mathbf{N}, \quad D_\mathbf{N} g = \nabla g \cdot \mathbf{N}.$$

27. $\iiint_Q (f\nabla^2 g + \nabla f \cdot \nabla g)\, dV = \iint_S f D_\mathbf{N} g\, dS$

    [Hint: Use div $(f\mathbf{G}) = f\operatorname{div} \mathbf{G} + \nabla f \cdot \mathbf{G}$.]

28. $\iiint_Q (f\nabla^2 g - g\nabla^2 f)\, dV = \iint_S (f D_\mathbf{N} g - g D_\mathbf{N} f)\, dS$

    (Hint: Use Exercise 27 twice.)

Section 14.8
Stokes's Theorem

- Understand and use Stokes's Theorem.
- Use curl to analyze the motion of a rotating liquid.

Stokes's Theorem

A second higher-dimension analog of Green's Theorem is called **Stokes's Theorem**, after the English mathematical physicist George Gabriel Stokes. Stokes was part of a group of English mathematical physicists referred to as the Cambridge School, which included William Thomson (Lord Kelvin) and James Clerk Maxwell. In addition to making contributions to physics, Stokes worked with infinite series and differential equations, as well as with the integration results presented in this section.

Stokes's Theorem gives the relationship between a surface integral over an oriented surface S and a line integral along a closed space curve C forming the boundary of S, as shown in Figure 14.62. The positive direction along C is counterclockwise relative to the normal vector **N**. That is, if you imagine grasping the normal vector **N** with your right hand, with your thumb pointing in the direction of **N**, your fingers will point in the positive direction C, as shown in Figure 14.63.

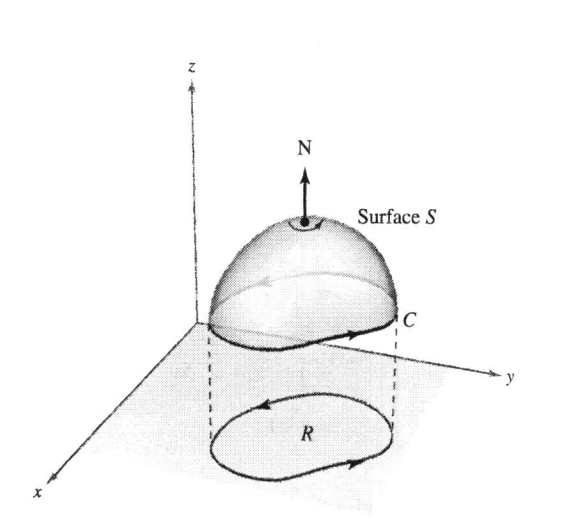

GEORGE GABRIEL STOKES (1819–1903)

Stokes became a Lucasian professor of mathematics at Cambridge in 1849. Five years later, he published the theorem that bears his name as a prize examination question there.

Figure 14.62

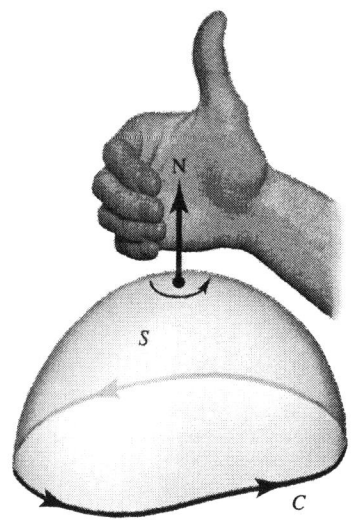

Direction along C is counterclockwise relative to N.
Figure 14.63

THEOREM 14.13 Stokes's Theorem

Let S be an oriented surface with unit normal vector **N**, bounded by a piecewise smooth simple closed curve C. If **F** is a vector field whose component functions have continuous partial derivatives on an open region containing S and C, then

$$\int_C \mathbf{F} \cdot d\mathbf{r} = \iint_S (\text{curl } \mathbf{F}) \cdot \mathbf{N} \, dS.$$

NOTE The line integral may be expressed in the differential form $\int_C M\,dx + N\,dy + P\,dz$ or in the vector form $\int_C \mathbf{F} \cdot \mathbf{T}\,ds$.

Example 1 Using Stokes's Theorem

Let C be the oriented triangle lying in the plane $2x + 2y + z = 6$, as shown in Figure 14.64. Evaluate

$$\int_C \mathbf{F} \cdot d\mathbf{r}$$

where $\mathbf{F}(x, y, z) = -y^2\mathbf{i} + z\mathbf{j} + x\mathbf{k}$.

Solution Using Stokes's Theorem, begin by finding the curl of $\mathbf{F}$.

$$\text{curl } \mathbf{F} = \begin{vmatrix} \mathbf{i} & \mathbf{j} & \mathbf{k} \\ \frac{\partial}{\partial x} & \frac{\partial}{\partial y} & \frac{\partial}{\partial z} \\ -y^2 & z & x \end{vmatrix} = -\mathbf{i} - \mathbf{j} + 2y\mathbf{k}$$

Considering $z = 6 - 2x - 2y = g(x, y)$, you can use Theorem 14.11 for an upward normal vector to obtain

$$\int_C \mathbf{F} \cdot d\mathbf{r} = \int\int_S (\text{curl } \mathbf{F}) \cdot \mathbf{N} \, dS$$

$$= \int\int_R (-\mathbf{i} - \mathbf{j} + 2y\mathbf{k}) \cdot [-g_x(x, y)\mathbf{i} - g_y(x, y)\mathbf{j} + \mathbf{k}] \, dA$$

$$= \int\int_R (-\mathbf{i} - \mathbf{j} + 2y\mathbf{k}) \cdot (2\mathbf{i} + 2\mathbf{j} + \mathbf{k}) \, dA$$

$$= \int_0^3 \int_0^{3-y} (2y - 4) \, dx \, dy$$

$$= \int_0^3 (-2y^2 + 10y - 12) \, dy$$

$$= \left[-\frac{2y^3}{3} + 5y^2 - 12y \right]_0^3$$

$$= -9.$$

Try evaluating the line integral in Example 1 directly, *without* using Stokes's Theorem. One way to do this would be to consider C as the union of C_1, C_2, and C_3, as follows.

C_1: $\mathbf{r}_1(t) = (3 - t)\mathbf{i} + t\mathbf{j}$, $\quad 0 \le t \le 3$
C_2: $\mathbf{r}_2(t) = (6 - t)\mathbf{j} + (2t - 6)\mathbf{k}$, $\quad 3 \le t \le 6$
C_3: $\mathbf{r}_3(t) = (t - 6)\mathbf{i} + (18 - 2t)\mathbf{k}$, $\quad 6 \le t \le 9$

The value of the line integral is

$$\int_C \mathbf{F} \cdot d\mathbf{r} = \int_{C_1} \mathbf{F} \cdot \mathbf{r}_1'(t) \, dt + \int_{C_2} \mathbf{F} \cdot \mathbf{r}_2'(t) \, dt + \int_{C_3} \mathbf{F} \cdot \mathbf{r}_3'(t) \, dt$$

$$= \int_0^3 t^2 \, dt + \int_3^6 (-2t + 6) \, dt + \int_6^9 (-2t + 12) \, dt$$

$$= 9 - 9 - 9$$

$$= -9.$$

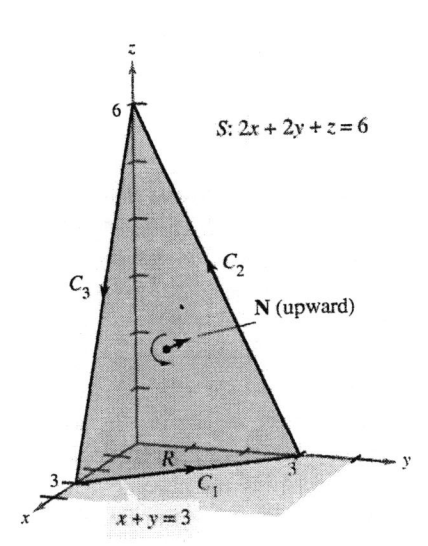

Figure 14.64

Example 2 Verifying Stokes's Theorem

Verify Stokes's Theorem for $\mathbf{F}(x, y, z) = 2z\mathbf{i} + x\mathbf{j} + y^2\mathbf{k}$, where S is the surface of the paraboloid $z = 4 - x^2 - y^2$ and C is the trace of S in the xy-plane, as shown in Figure 14.65.

Solution As a *surface integral*, you have $z = g(x, y) = 4 - x^2 - y^2$ and

$$\mathbf{curl\ F} = \begin{vmatrix} \mathbf{i} & \mathbf{j} & \mathbf{k} \\ \dfrac{\partial}{\partial x} & \dfrac{\partial}{\partial y} & \dfrac{\partial}{\partial z} \\ 2z & x & y^2 \end{vmatrix} = 2y\mathbf{i} + 2\mathbf{j} + \mathbf{k}.$$

By Theorem 14.11 for an upward normal vector $\mathbf{N}$, you obtain

$$\iint_S (\mathbf{curl\ F}) \cdot \mathbf{N}\, dS = \iint_R (2y\mathbf{i} + 2\mathbf{j} + \mathbf{k}) \cdot (2x\mathbf{i} + 2y\mathbf{j} + \mathbf{k})\, dA$$

$$= \int_{-2}^{2} \int_{-\sqrt{4-y^2}}^{\sqrt{4-y^2}} (4xy + 4y + 1)\, dx\, dy$$

$$= \int_{-2}^{2} \left[2x^2 y + (4y + 1)x \right]_{-\sqrt{4-y^2}}^{\sqrt{4-y^2}} dy$$

$$= \int_{-2}^{2} 2(4y + 1)\sqrt{4 - y^2}\, dy$$

$$= \int_{-2}^{2} \left(8y\sqrt{4 - y^2} + 2\sqrt{4 - y^2} \right) dy$$

$$= \left[-\frac{8}{3}(4 - y^2)^{3/2} + y\sqrt{4 - y^2} + 4 \arcsin \frac{y}{2} \right]_{-2}^{2}$$

$$= 4\pi.$$

As a *line integral*, you can parametrize C by

$$\mathbf{r}(t) = 2 \cos t\, \mathbf{i} + 2 \sin t\, \mathbf{j} + 0\mathbf{k}, \quad 0 \le t \le 2\pi.$$

For $\mathbf{F}(x, y, z) = 2z\mathbf{i} + x\mathbf{j} + y^2\mathbf{k}$, you obtain

$$\int_C \mathbf{F} \cdot d\mathbf{r} = \int_C M\, dx + N\, dy + P\, dz$$

$$= \int_C 2z\, dx + x\, dy + y^2\, dz$$

$$= \int_0^{2\pi} [0 + 2 \cos t (2 \cos t) + 0]\, dt$$

$$= \int_0^{2\pi} 4 \cos^2 t\, dt$$

$$= 2 \int_0^{2\pi} (1 + \cos 2t)\, dt$$

$$= 2 \left[t + \frac{1}{2} \sin 2t \right]_0^{2\pi}$$

$$= 4\pi.$$

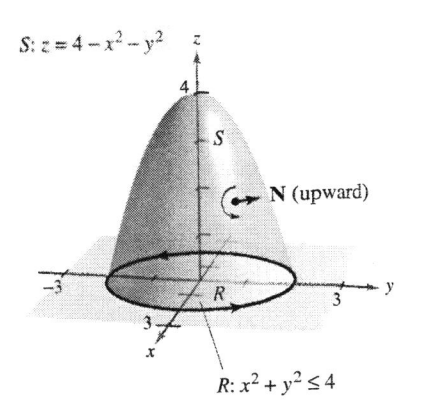

Figure 14.65

$S: z = 4 - x^2 - y^2$
$\mathbf{N}$ (upward)
$R: x^2 + y^2 \le 4$

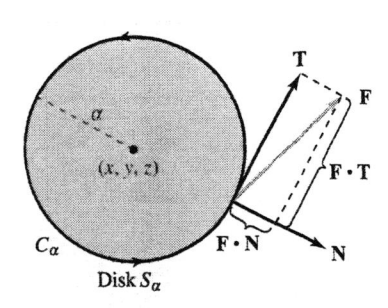

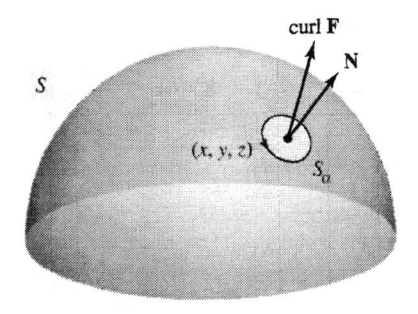

Physical Interpretation of Curl

Stokes's Theorem provides insight into a physical interpretation of curl. In a vector field **F**, let S_α be a *small* circular disk of radius α, centered at (x, y, z) and with boundary C_α, as shown in Figure 14.66. At each point on the circle C_α, **F** has a normal component **F** · **N** and a tangential component **F** · **T**. The more closely **F** and **T** are aligned, the greater the value of **F** · **T**. So, a fluid tends to move along the circle rather than across it. Consequently, you say that the line integral around C_α measures the **circulation of F around C_α**. That is,

$$\int_{C_\alpha} \mathbf{F} \cdot \mathbf{T}\, ds = \text{circulation of } \mathbf{F} \text{ around } C_\alpha.$$

Now consider a small disk S_α to be centered at some point (x, y, z) on the surface S, as shown in Figure 14.67. On such a small disk, **curl F** is nearly constant, because it varies little from its value at (x, y, z). Moreover, **curl F** · **N** is also nearly constant on S_α, because all unit normals to S_α are about the same. Consequently, Stokes's Theorem yields

$$\int_{C_\alpha} \mathbf{F} \cdot \mathbf{T}\, ds = \iint_{S_\alpha} (\text{curl } \mathbf{F}) \cdot \mathbf{N}\, dS$$

$$\approx (\text{curl } \mathbf{F}) \cdot \mathbf{N} \iint_{S_\alpha} dS$$

$$\approx (\text{curl } \mathbf{F}) \cdot \mathbf{N}(\pi \alpha^2).$$

Therefore,

$$(\text{curl } \mathbf{F}) \cdot \mathbf{N} \approx \frac{\int_{C_\alpha} \mathbf{F} \cdot \mathbf{T}\, ds}{\pi \alpha^2}$$

$$= \frac{\text{circulation of } \mathbf{F} \text{ around } C_\alpha}{\text{area of disk } S_\alpha}$$

$$= \text{rate of circulation.}$$

Assuming conditions are such that the approximation improves for smaller and smaller disks ($\alpha \to 0$), it follows that

$$(\text{curl } \mathbf{F}) \cdot \mathbf{N} = \lim_{\alpha \to 0} \frac{1}{\pi \alpha^2} \int_{C_\alpha} \mathbf{F} \cdot \mathbf{T}\, ds$$

which is referred to as the **rotation of F about N**. That is,

curl $\mathbf{F}(x, y, z) \cdot \mathbf{N}$ = rotation of **F** about **N** at (x, y, z).

In this case, the rotation of **F** is maximum when **curl F** and **N** have the same direction. Normally, this tendency to rotate will vary from point to point on the surface S, and Stokes's Theorem

$$\underbrace{\iint_S (\text{curl } \mathbf{F}) \cdot \mathbf{N}\, dS}_{\text{Surface integral}} = \underbrace{\int_C \mathbf{F} \cdot d\mathbf{r}}_{\text{Line integral}}$$

says that the collective measure of this *rotational* tendency taken over the entire surface S (surface integral) is equal to the tendency of a fluid to *circulate* around the boundary C (line integral).

Example 3 An Application of Curl

A liquid is swirling around in a cylindrical container of radius 2, so that its motion is described by the velocity field

$$\mathbf{F}(x, y, z) = -y\sqrt{x^2 + y^2}\,\mathbf{i} + x\sqrt{x^2 + y^2}\,\mathbf{j}$$

as shown in Figure 14.68. Find

$$\iint_S (\text{curl } \mathbf{F}) \cdot \mathbf{N}\, dS$$

where S is the upper surface of the cylindrical container.

Solution The curl of $\mathbf{F}$ is given by

$$\text{curl } \mathbf{F} = \begin{vmatrix} \mathbf{i} & \mathbf{j} & \mathbf{k} \\ \dfrac{\partial}{\partial x} & \dfrac{\partial}{\partial y} & \dfrac{\partial}{\partial z} \\ -y\sqrt{x^2+y^2} & x\sqrt{x^2+y^2} & 0 \end{vmatrix} = 3\sqrt{x^2 + y^2}\,\mathbf{k}.$$

Letting $\mathbf{N} = \mathbf{k}$, you have

$$\iint_S (\text{curl } \mathbf{F}) \cdot \mathbf{N}\, dS = \iint_R 3\sqrt{x^2 + y^2}\, dA$$
$$= \int_0^{2\pi} \int_0^2 (3r) r\, dr\, d\theta$$
$$= \int_0^{2\pi} r^3 \Big]_0^2 d\theta$$
$$= \int_0^{2\pi} 8\, d\theta$$
$$= 16\pi.$$

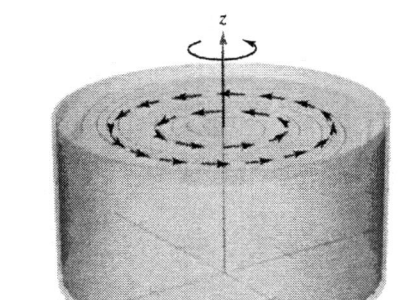

Figure 14.68

NOTE If **curl F** = **0** throughout region Q, the rotation of **F** about each unit normal **N** is 0. That is, **F** is irrotational. From earlier work, you know that this is a characteristic of conservative vector fields.

Summary of Integration Formulas

Fundamental Theorem of Calculus:

$$\int_a^b F'(x)\, dx = F(b) - F(a)$$

Fundamental Theorem of Line Integrals:

$$\int_C \mathbf{F} \cdot d\mathbf{r} = \int_C \nabla f \cdot d\mathbf{r} = f(x(b), y(b)) - f(x(a), y(a))$$

Green's Theorem:

$$\int_C M\, dx + N\, dy = \iint_R \left(\frac{\partial N}{\partial x} - \frac{\partial M}{\partial y}\right) dA = \int_C \mathbf{F} \cdot \mathbf{T}\, ds = \int_C \mathbf{F} \cdot d\mathbf{r} = \iint_R (\text{curl } \mathbf{F}) \cdot \mathbf{k}\, dA$$

$$\int_C \mathbf{F} \cdot \mathbf{N}\, ds = \iint_R \text{div } \mathbf{F}\, dA$$

Divergence Theorem:

$$\iint_S \mathbf{F} \cdot \mathbf{N}\, dS = \iiint_Q \text{div } \mathbf{F}\, dV$$

Stokes's Theorem:

$$\int_C \mathbf{F} \cdot d\mathbf{r} = \iint_S (\text{curl } \mathbf{F}) \cdot \mathbf{N}\, dS$$

EXERCISES FOR SECTION 14.8

In Exercises 1–6, find the curl of the vector field F.

1. $F(x, y, z) = (2y - z)\mathbf{i} + xyz\mathbf{j} + e^z\mathbf{k}$
2. $F(x, y, z) = x^2\mathbf{i} + y^2\mathbf{j} + x^2\mathbf{k}$
3. $F(x, y, z) = 2z\mathbf{i} - 4x^2\mathbf{j} + \arctan x\mathbf{k}$
4. $F(x, y, z) = x\sin y\mathbf{i} - y\cos x\mathbf{j} + yz^2\mathbf{k}$
5. $F(x, y, z) = e^{x^2+y^2}\mathbf{i} + e^{y^2+z^2}\mathbf{j} + xyz\mathbf{k}$
6. $F(x, y, z) = \arcsin y\mathbf{i} + \sqrt{1-x^2}\mathbf{j} + y^2\mathbf{k}$

In Exercises 7–10, verify Stokes's Theorem by evaluating

$$\int_C \mathbf{F} \cdot \mathbf{T}\, dS$$

as a line integral and as a double integral.

7. $F(x, y, z) = (-y + z)\mathbf{i} + (x - z)\mathbf{j} + (x - y)\mathbf{k}$
   S: $z = \sqrt{1 - x^2 - y^2}$
8. $F(x, y, z) = (-y + z)\mathbf{i} + (x - z)\mathbf{j} + (x - y)\mathbf{k}$
   S: $z = 4 - x^2 - y^2$, $z \geq 0$
9. $F(x, y, z) = xyz\mathbf{i} + y\mathbf{j} + z\mathbf{k}$
   S: $3x + 4y + 2z = 12$, first octant
10. $F(x, y, z) = z^2\mathbf{i} + x^2\mathbf{j} + y^2\mathbf{k}$
    S: $z = y^2$, $0 \leq x \leq a$, $0 \leq y \leq a$

In Exercises 11–20, use Stokes's Theorem to evaluate

$$\int_C \mathbf{F} \cdot d\mathbf{r}.$$

Use a computer algebra system to verify your results.

11. $F(x, y, z) = 2y\mathbf{i} + 3z\mathbf{j} + x\mathbf{k}$
    C: triangle with vertices $(0, 0, 0)$, $(0, 2, 0)$, $(1, 1, 1)$
12. $F(x, y, z) = \arctan\dfrac{x}{y}\mathbf{i} + \ln\sqrt{x^2 + y^2}\mathbf{j} + \mathbf{k}$
    C: triangle with vertices $(0, 0, 0)$, $(1, 1, 1)$, $(0, 0, 2)$
13. $F(x, y, z) = z^2\mathbf{i} + x^2\mathbf{j} + y^2\mathbf{k}$
    S: $z = 4 - x^2 - y^2$, $z \geq 0$
14. $F(x, y, z) = 4xz\mathbf{i} + y\mathbf{j} + 4xy\mathbf{k}$
    S: $z = 9 - x^2 - y^2$, $z \geq 0$
15. $F(x, y, z) = z^2\mathbf{i} + y\mathbf{j} + xz\mathbf{k}$
    S: $z = \sqrt{4 - x^2 - y^2}$
16. $F(x, y, z) = x^2\mathbf{i} + z^2\mathbf{j} - xyz\mathbf{k}$
    S: $z = \sqrt{4 - x^2 - y^2}$
17. $F(x, y, z) = -\ln\sqrt{x^2 + y^2}\mathbf{i} + \arctan\dfrac{x}{y}\mathbf{j} + \mathbf{k}$
    S: $z = 9 - 2x - 3y$ over one petal of $r = 2\sin 2\theta$ in the first octant
18. $F(x, y, z) = yz\mathbf{i} + (2 - 3y)\mathbf{j} + (x^2 + y^2)\mathbf{k}$
    S: the first-octant portion of $x^2 + z^2 = 16$ over $x^2 + y^2 = 16$
19. $F(x, y, z) = xyz\mathbf{i} + y\mathbf{j} + z\mathbf{k}$
    S: $z = x^2$, $0 \leq x \leq a$, $0 \leq y \leq a$
    N is the downward unit normal to the surface.
20. $F(x, y, z) = xyz\mathbf{i} + y\mathbf{j} + z\mathbf{k}$
    S: the first-octant portion of $z = x^2$ over $x^2 + y^2 = a^2$

Motion of a Liquid In Exercises 21 and 22, the motion of a liquid in a cylindrical container of radius 1 is described by the velocity field $F(x, y, z)$. Find

$$\iint_S (\operatorname{curl} \mathbf{F}) \cdot \mathbf{N}\, dS$$

where S is the upper surface of the cylindrical container.

21. $F(x, y, z) = \mathbf{i} + \mathbf{j} - 2\mathbf{k}$
22. $F(x, y, z) = -z\mathbf{i} + y\mathbf{k}$

Getting at the Concept

23. State Stokes's Theorem.
24. Give a physical interpretation of curl.

25. Let f and g be scalar functions with continuous partial derivatives, and let C and S satisfy the conditions of Stokes's Theorem. Verify each identity.

    (a) $\displaystyle\int_C (f\nabla g) \cdot d\mathbf{r} = \iint_S (\nabla f \times \nabla g) \cdot \mathbf{N}\, dS$
    (b) $\displaystyle\int_C (f\nabla f) \cdot d\mathbf{r} = 0$
    (c) $\displaystyle\int_C (f\nabla g + g\nabla f) \cdot d\mathbf{r} = 0$

26. Demonstrate the results in Exercise 25 for the functions $f(x, y, z) = xyz$ and $g(x, y, z) = z$. Let S be the hemisphere $z = \sqrt{4 - x^2 - y^2}$.

27. Let C be a constant vector. Let S be an oriented surface with a unit normal vector N, bounded by a smooth curve C. Prove that

    $$\iint_S \mathbf{C} \cdot \mathbf{N}\, dS = \frac{1}{2}\int_C (\mathbf{C} \times \mathbf{r}) \cdot d\mathbf{r}.$$

REVIEW EXERCISES FOR CHAPTER 14

14.1 In Exercises 1 and 2, sketch several representative vectors in the vector field. Use a computer algebra system to verify your results.

1. $\mathbf{F}(x, y, z) = x\mathbf{i} + \mathbf{j} + 2\mathbf{k}$
2. $\mathbf{F}(x, y) = \mathbf{i} - 2y\mathbf{j}$

In Exercises 3 and 4, find the gradient vector field for the scalar function.

3. $f(x, y, z) = 8x^2 + xy + z^2$
4. $f(x, y, z) = x^2 e^{yz}$

In Exercises 5–12, determine if $\mathbf{F}$ is conservative. If it is, find the potential function f.

5. $\mathbf{F}(x, y) = \dfrac{1}{y}\mathbf{i} - \dfrac{y}{x^2}\mathbf{j}$
6. $\mathbf{F}(x, y) = -\dfrac{y}{x^2}\mathbf{i} + \dfrac{1}{x}\mathbf{j}$
7. $\mathbf{F}(x, y) = (6xy^2 - 3x^2)\mathbf{i} + (6x^2y + 3y^2 - 7)\mathbf{j}$
8. $\mathbf{F}(x, y) = (-2y^3 \sin 2x)\mathbf{i} + 3y^2(1 + \cos 2x)\mathbf{j}$
9. $\mathbf{F}(x, y, z) = (4xy + z)\mathbf{i} + (2x^2 + 6y)\mathbf{j} + 2z\mathbf{k}$
10. $\mathbf{F}(x, y, z) = (4xy + z^2)\mathbf{i} + (2x^2 + 6yz)\mathbf{j} + 2xz\mathbf{k}$
11. $\mathbf{F}(x, y, z) = \dfrac{yz\mathbf{i} - xz\mathbf{j} - xy\mathbf{k}}{y^2 z^2}$
12. $\mathbf{F}(x, y, z) = \sin z(y\mathbf{i} + x\mathbf{j} + \mathbf{k})$

In Exercises 13–20, find (a) the divergence of the vector field $\mathbf{F}$ and (b) the curl of the vector field $\mathbf{F}$.

13. $\mathbf{F}(x, y, z) = x^2\mathbf{i} + y^2\mathbf{j} + z^2\mathbf{k}$
14. $\mathbf{F}(x, y, z) = xy^2\mathbf{j} - zx^2\mathbf{k}$
15. $\mathbf{F}(x, y, z) = (\cos y + y \cos x)\mathbf{i} + (\sin x - x \sin y)\mathbf{j} + xyz\mathbf{k}$
16. $\mathbf{F}(x, y, z) = (3x - y)\mathbf{i} + (y - 2z)\mathbf{j} + (z - 3x)\mathbf{k}$
17. $\mathbf{F}(x, y, z) = \arcsin x\,\mathbf{i} + xy^2\mathbf{j} + yz^2\mathbf{k}$
18. $\mathbf{F}(x, y, z) = (x^2 - y)\mathbf{i} - (x + \sin^2 y)\mathbf{j}$
19. $\mathbf{F}(x, y, z) = \ln(x^2 + y^2)\mathbf{i} + \ln(x^2 + y^2)\mathbf{j} + z\mathbf{k}$
20. $\mathbf{F}(x, y, z) = \dfrac{z}{x}\mathbf{i} + \dfrac{z}{y}\mathbf{j} + z^2\mathbf{k}$

14.2 In Exercises 21–26, evaluate the line integral over the indicated path(s).

21. $\displaystyle\int_C (x^2 + y^2)\,ds$

   (a) C: line segment from $(-1, -1)$ to $(2, 2)$
   (b) C: $x^2 + y^2 = 16$, one revolution counterclockwise, starting at $(4, 0)$

22. $\displaystyle\int_C xy\,ds$

   (a) C: line segment from $(0, 0)$ to $(5, 4)$
   (b) C: counterclockwise around the triangle with vertices $(0, 0), (4, 0), (0, 2)$

23. $\displaystyle\int_C (x^2 + y^2)\,ds$

   C: $\mathbf{r}(t) = (\cos t + t \sin t)\mathbf{i} + (\sin t - t \cos t)\mathbf{j}, \quad 0 \le t \le 2\pi$

24. $\displaystyle\int_C x\,ds$

   C: $\mathbf{r}(t) = (t - \sin t)\mathbf{i} + (1 - \cos t)\mathbf{j}, \quad 0 \le t \le 2\pi$

25. $\displaystyle\int_C (2x - y)\,dx + (x + 3y)\,dy$

   (a) C: line segment from $(0, 0)$ to $(2, -3)$
   (b) C: counterclockwise around the circle $x = 3\cos t$, $y = 3 \sin t$

26. $\displaystyle\int_C (2x - y)\,dx + (x + 3y)\,dy$

   C: $\mathbf{r}(t) = (\cos t + t \sin t)\mathbf{i} + (\sin t - t \sin t)\mathbf{j}, \quad 0 \le t \le \pi/2$

In Exercises 27 and 28, use a computer algebra system to evaluate the line integral over the indicated path.

27. $\displaystyle\int_C (2x + y)\,ds$
    $\mathbf{r}(t) = a\cos^3 t\,\mathbf{i} + a\sin^3 t\,\mathbf{j}$,
    $0 \le t \le \pi/2$

28. $\displaystyle\int_C (x^2 + y^2 + z^2)\,ds$
    $\mathbf{r}(t) = t\mathbf{i} + t^2\mathbf{j} + t^{3/2}\mathbf{k}$,
    $0 \le t \le 4$

In Exercises 29 and 30, find the lateral surface area over the curve C in the xy-plane and under the surface $z = f(x, y)$.

29. $f(x, y) = 5 + \sin(x + y)$
    C: $y = 3x$ from $(0, 0)$ to $(2, 6)$
30. $f(x, y) = 12 - x - y$
    C: $y = x^2$ from $(0, 0)$ to $(2, 4)$

14.3 In Exercises 31–36, evaluate $\displaystyle\int_C \mathbf{F} \cdot d\mathbf{r}$.

31. $\mathbf{F}(x, y) = xy\mathbf{i} + x^2\mathbf{j}$
    C: $\mathbf{r}(t) = t^2\mathbf{i} + t^3\mathbf{j}, \quad 0 \le t \le 1$
32. $\mathbf{F}(x, y) = (x - y)\mathbf{i} + (x + y)\mathbf{j}$
    C: $\mathbf{r}(t) = 4\cos t\,\mathbf{i} + 3\sin t\,\mathbf{j}, \quad 0 \le t \le 2\pi$
33. $\mathbf{F}(x, y, z) = x\mathbf{i} + y\mathbf{j} + z\mathbf{k}$
    C: $\mathbf{r}(t) = 2\cos t\,\mathbf{i} + 2\sin t\,\mathbf{j} + t\mathbf{k}, \quad 0 \le t \le 2\pi$
34. $\mathbf{F}(x, y, z) = (2y - z)\mathbf{i} + (z - x)\mathbf{j} + (x - y)\mathbf{k}$
    C: curve of intersection of $x^2 + z^2 = 4$ and $y^2 + z^2 = 4$ from $(2, 2, 0)$ to $(0, 0, 2)$
35. $\mathbf{F}(x, y, z) = (y - z)\mathbf{i} + (z - x)\mathbf{j} + (x - y)\mathbf{k}$
    C: curve of intersection of $z = x^2 + y^2$ and $x + y = 0$ from $(-2, 2, 8)$ to $(2, -2, 8)$
36. $\mathbf{F}(x, y, z) = (x^2 - z)\mathbf{i} + (y^2 + z)\mathbf{j} + x\mathbf{k}$
    C: curve of intersection of $z = x^2$ and $x^2 + y^2 = 4$ from $(0, -2, 0)$ to $(0, 2, 0)$

In Exercises 37 and 38, use a computer algebra system to evaluate the line integral.

37. $\int_C xy\, dx + (x^2 + y^2)\, dy$

    C: $y = x^2$ from $(0, 0)$ to $(2, 4)$ and $y = 2x$ from $(2, 4)$ to $(0, 0)$

38. $\int_C \mathbf{F} \cdot d\mathbf{r}$

    $\mathbf{F}(x, y) = (2x - y)\mathbf{i} + (2y - x)\mathbf{j}$

    C: $\mathbf{r}(t) = (2\cos t + 2t \sin t)\mathbf{i} + (2\sin t - 2t\cos t)\mathbf{j}$, $0 \le t \le \pi$

39. **Work** Find the work done by the force field $\mathbf{F} = x\mathbf{i} - \sqrt{y}\mathbf{j}$ along the path $y = x^{3/2}$ from $(0, 0)$ to $(4, 8)$.

40. **Work** Find the work done by the engines of a 20-ton aircraft if it climbs 2000 feet while making a 90° turn in a circular arc of radius 10 miles.

In Exercises 41 and 42, use the Fundamental Theorem of Line Integrals to evaluate the integral.

41. $\int_C 2xyz\, dx + x^2 z\, dy + x^2 y\, dz$

    C: smooth curve from $(0, 0, 0)$ to $(1, 4, 3)$

42. $\int_C y\, dx + x\, dy + \frac{1}{z}\, dz$

    C: smooth curve from $(0, 0, 1)$ to $(4, 4, 4)$

43. Evaluate the line integral $\int_C y^2\, dx + 2xy\, dy$.

    (a) C: $\mathbf{r}(t) = (1 + 3t)\mathbf{i} + (1 + t)\mathbf{j}$, $0 \le t \le 1$

    (b) C: $\mathbf{r}(t) = t\mathbf{i} + \sqrt{t}\mathbf{j}$, $1 \le t \le 4$

    (c) Use the Fundamental Theorem of Line Integrals, where C is a smooth curve from $(1, 1)$ to $(4, 2)$.

14.4

44. **Area and Centroid** Consider the region bounded by the x-axis and one arch of the cycloid with parametric equations $x = a(\theta - \sin\theta)$ and $y = a(1 - \cos\theta)$. Use line integrals to find (a) the area of the region and (b) the centroid of the region.

In Exercises 45–50, use Green's Theorem to evaluate the line integral.

45. $\int_C y\, dx + 2x\, dy$

    C: boundary of the square with vertices $(0, 0)$, $(0, 2)$, $(2, 0)$, $(2, 2)$

46. $\int_C xy\, dx + (x^2 + y^2)\, dy$

    C: boundary of the square with vertices $(0, 0)$, $(0, 2)$, $(2, 0)$, $(2, 2)$

47. $\int_C xy^2\, dx + x^2 y\, dy$

    C: $x = 4\cos t$, $y = 2\sin t$

48. $\int_C (x^2 - y^2)\, dx + 2xy\, dy$

    C: $x^2 + y^2 = a^2$

49. $\int_C xy\, dx + x^2\, dy$

    C: boundary of the region between the graphs of $y = x^2$ and $y = x$

50. $\int_C y^2\, dx + x^{4/3}\, dy$

    C: $x^{2/3} + y^{2/3} = 1$

14.5 **In Exercises 51 and 52, use a computer algebra system to graph the surface represented by the vector-valued function.**

51. $\mathbf{r}(u, v) = \sec u \cos v\mathbf{i} + (1 + 2\tan u)\sin v\mathbf{j} + 2u\mathbf{k}$

    $0 \le u \le \frac{\pi}{3}$, $0 \le v \le 2\pi$

52. $\mathbf{r}(u, v) = e^{-u/4}\cos v\mathbf{i} + e^{-u/4}\sin v\mathbf{j} + \frac{u}{6}\mathbf{k}$

    $0 \le u \le 4$, $0 \le v \le 2\pi$

53. **Investigation** Consider the surface represented by the vector-valued function

    $\mathbf{r}(u, v) = 3\cos v \cos u\mathbf{i} + 3\cos v \sin u\mathbf{j} + \sin v\mathbf{k}$.

    Use a computer algebra system when a graph is required.

    (a) Graph the surface for $0 \le u \le 2\pi$ and $-\frac{\pi}{2} \le v \le \frac{\pi}{2}$.

    (b) Graph the surface for $0 \le u \le 2\pi$ and $\frac{\pi}{4} \le v \le \frac{\pi}{2}$.

    (c) Graph the surface for $0 \le u \le \frac{\pi}{4}$ and $0 \le v \le \frac{\pi}{2}$.

    (d) Graph and identify the space curve for $0 \le u \le 2\pi$ and $v = \frac{\pi}{4}$.

    (e) Use a computer algebra system to approximate the area of the surface graphed in part (b).

    (f) Use a computer algebra system to approximate the area of the surface graphed in part (c).

14.6

54. Evaluate the surface integral $\iint_S z\, dS$ over the surface S:

    $\mathbf{r}(u, v) = (u + v)\mathbf{i} + (u - v)\mathbf{j} + \sin v\mathbf{k}$

    where $0 \le u \le 2$ and $0 \le v \le \pi$.

55. Use a computer algebra system to graph the surface S and approximate the surface integral $\iint_S (x + y)\, dS$, where S is the surface S: $\mathbf{r}(u, v) = u\cos v\mathbf{i} + u\sin v\mathbf{j} + (u - 1)(2 - u)\mathbf{k}$ over $0 \le u \le 2$ and $0 \le v \le 2\pi$.

56. *Mass* A cone-shaped surface lamina S is given by

$$z = a(a - \sqrt{x^2 + y^2}), \quad 0 \leq z \leq a^2.$$

At each point on S, the density is proportional to the distance between the point and the z-axis.

(a) Sketch the cone-shaped surface.

(b) Find the mass m of the lamina.

14.7 **In Exercises 57 and 58, verify the Divergence Theorem by evaluating** $\iint_S \mathbf{F} \cdot \mathbf{N}\, dS$ **as a surface integral and as a triple integral.**

57. $\mathbf{F}(x, y, z) = x^2\mathbf{i} + xy\mathbf{j} + z\mathbf{k}$

Q: solid region bounded by the coordinate planes and the plane $2x + 3y + 4z = 12$

58. $\mathbf{F}(x, y, z) = x\mathbf{i} + y\mathbf{j} + z\mathbf{k}$

Q: solid region bounded by the coordinate planes and the plane $2x + 3y + 4z = 12$

14.8 **In Exercises 59 and 60, verify Stokes's Theorem by evaluating**

$\int_C \mathbf{F} \cdot d\mathbf{r}$ **as a line integral and as a double integral.**

59. $\mathbf{F}(x, y, z) = (\cos y + y \cos x)\mathbf{i} + (\sin x - x \sin y)\mathbf{j} + xyz\mathbf{k}$

S: portion of $z = y^2$ over the square in the xy-plane with vertices $(0, 0)$, $(a, 0)$, (a, a), $(0, a)$

$\mathbf{N}$ is the upward unit normal to the surface.

60. $\mathbf{F}(x, y, z) = (x - z)\mathbf{i} + (y - z)\mathbf{j} + x^2\mathbf{k}$

S: first-octant portion of the plane $3x + y + 2z = 12$

SECTION PROJECT: THE PLANIMETER

You have learned many calculus techniques for finding the area of a planar region. Engineers use a mechanical device called a *planimeter* for measuring planar areas, which is based on the area formula given in Theorem 14.9 (page 1045). As you can see in the figure, the planimeter is fixed at point O (but free to pivot) and has a hinge at A. The end of the tracer arm AB moves counterclockwise around the region R. A small wheel at B is perpendicular to $\overline{AB}$ and is marked with a scale to measure how much it rolls as B traces out the boundary of region R. In this project you will show that the area of R is given by the length L of the tracer arm $\overline{AB}$ multiplied by the distance D that the wheel rolls.

Assume that point B traces out the boundary of R for $a \leq t \leq b$. Point A will move back and forth along a circular arc around the origin O. Let $\theta(t)$ denote the angle in the figure and let $(x(t), y(t))$ denote the coordinates of A.

(a) Show that the vector $\overrightarrow{OB}$ is given by the vector-valued function

$$\mathbf{r}(t) = [x(t) + L \cos \theta(t)]\mathbf{i} + [y(t) + L \sin \theta(t)]\mathbf{j}.$$

(b) Show that the following two integrals are equal to zero.

$$I_1 = \int_a^b \frac{1}{2} L^2 \frac{d\theta}{dt}\, dt$$

$$I_2 = \int_a^b \frac{1}{2}\left(x \frac{dy}{dt} - y \frac{dx}{dt}\right) dt$$

(c) Use the integral $\int_a^b [x(t) \sin \theta(t) - y(t) \cos \theta(t)]'\, dt$ to show that the following two integrals are equal.

$$I_3 = \int_a^b \frac{1}{2} L \left(y \sin \theta \frac{d\theta}{dt} + x \cos \theta \frac{d\theta}{dt}\right) dt$$

$$I_4 = \int_a^b \frac{1}{2} L \left(-\sin \theta \frac{dx}{dt} + \cos \theta \frac{dy}{dt}\right) dt$$

(d) Let $\mathbf{N} = -\sin \theta \mathbf{i} + \cos \theta \mathbf{j}$. Explain why the distance D that the wheel rolls is given by

$$D = \int_C \mathbf{N} \cdot \mathbf{T}\, ds.$$

(e) Show that the area of region R is given by $I_1 + I_2 + I_3 + I_4 = DL$.

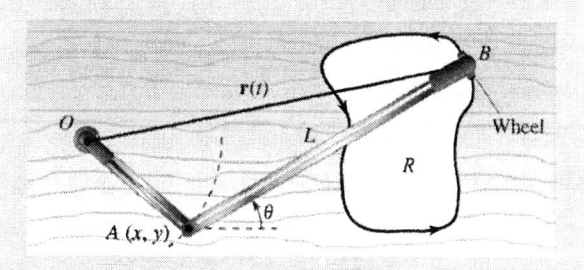

FOR FURTHER INFORMATION For more information about using calculus to find irregular areas, see "The Amateur Scientist" by C. L. Strong in the August 1958 issue of *Scientific American*.

P.S. Problem Solving

1. Heat flows from areas of higher temperature to areas of lower temperature in the direction of greatest change. As a result, measuring heat flux involves the gradient of the temperature. The flux depends on the area of the surface. It is the normal direction to the surface that is important, because heat that flows in directions tangential to the surface will give no heat loss. So, assume that the heat flux across a portion of the surface of area ΔS is given by $\Delta H \approx -k\nabla T \cdot \mathbf{n}\, dS$, where T is the temperature, $\mathbf{n}$ is the unit normal to the surface in the direction of the heat flow, and k is the thermal diffusivity of the material. The heat flux across the surface S is given by

$$H = \iint_S -k\nabla T \cdot \mathbf{n}\, dS.$$

Consider a single heat source located at the origin with temperature

$$T(x, y, z) = \frac{25}{\sqrt{x^2 + y^2 + z^2}}.$$

(a) Calculate the heat flux across the surface

$$S = \left\{(x, y, z): z = \sqrt{1 - x^2},\ -\frac{1}{2} \leq x \leq \frac{1}{2},\ 0 \leq y \leq 1\right\}$$

as indicated in the figure.

(b) Repeat the calculation in part (a) using the parametrization

$$x = \cos u,\ y = v,\ z = \sin u,\ \frac{\pi}{3} \leq u \leq \frac{2\pi}{3},\ 0 \leq v \leq 1.$$

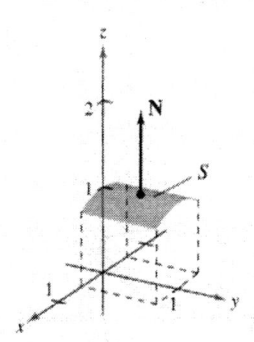

2. Consider a single heat source located at the origin with temperature

$$T(x, y, z) = \frac{25}{\sqrt{x^2 + y^2 + z^2}}.$$

(a) Calculate the heat flux across the surface

$$S = \left\{(x, y, z): z = \sqrt{1 - x^2 - y^2},\ x^2 + y^2 \leq 1\right\}$$

as indicated in the figure.

(b) Repeat the calculation in part (a) using the parametrization

$$x = \sin u \cos v,\ y = \sin u \sin v,\ z = \cos u,\ 0 \leq u \leq \frac{\pi}{2},\ 0 \leq v \leq 2\pi.$$

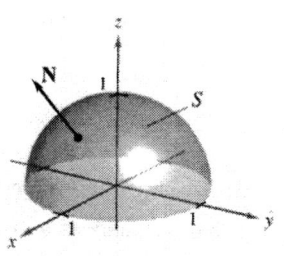

Figure for 2

3. Consider a wire of density $\rho(x, y, z)$ given by the space curve

$$C:\ \mathbf{r}(t) = x(t)\mathbf{i} + y(t)\mathbf{j} + z(t)\mathbf{k},\ a \leq t \leq b.$$

The **moments of inertia** about the x-, y-, and z-axes are given by

$$I_x = \int_C (y^2 + z^2)\rho(x, y, z)\, ds$$

$$I_y = \int_C (x^2 + z^2)\rho(x, y, z)\, ds$$

$$I_z = \int_C (x^2 + y^2)\rho(x, y, z)\, ds.$$

Find the moments of inertia for a wire of uniform density $\rho = 1$ in the shape of the helix

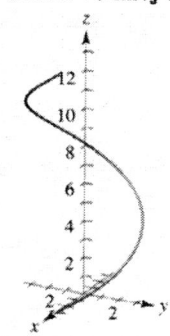

$$\mathbf{r}(t) = 3\cos t\,\mathbf{i} + 3\sin t\,\mathbf{j} + 2t\,\mathbf{k},\ 0 \leq t \leq 2\pi.$$

4. Find the moments of inertia for the wire of density $\rho = \dfrac{1}{1+t}$ given by the curve

$$C:\ \mathbf{r}(t) = \frac{t^2}{2}\mathbf{i} + t\mathbf{j} + \frac{2\sqrt{2}\,t^{3/2}}{3}\mathbf{k},\ 0 \leq t \leq 1.$$

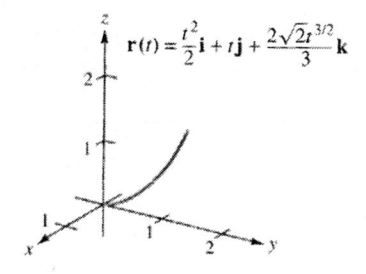

5. Use a line integral to find the area bounded by one arch of the cycloid
$$x(\theta) = a(\theta - \sin\theta), \quad y(\theta) = a(1 - \cos\theta), \quad 0 \le \theta \le 2\pi$$
as indicated in the figure.

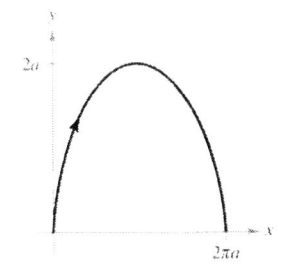

6. Use a line integral to find the area bounded by the two loops of the eight-curve
$$x(t) = \frac{1}{2}\sin 2t, \quad y(t) = \sin t, \quad 0 \le t \le 2\pi$$
as indicated in the figure.

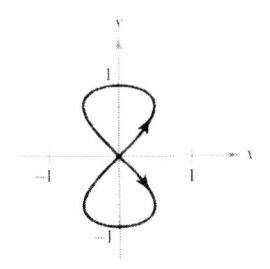

7. The force field $\mathbf{F}(x, y) = (x + y)\mathbf{i} + (x^2 + 1)\mathbf{j}$ acts on an object moving from the point $(0, 0)$ to the point $(0, 1)$.
   (a) Find the work done if the object moves along the path $x = 0$, $0 \le y \le 1$.
   (b) Find the work done if the object moves along the path $x = y - y^2$, $0 \le y \le 1$.
   (c) Suppose the object moves along the path $x = c(y - y^2)$, $0 \le y \le 1$, $c > 0$. Find the value of the constant c that minimizes the work.

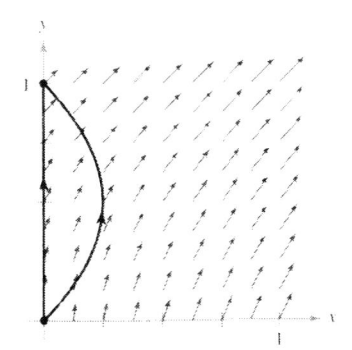

8. The force field $\mathbf{F}(x, y) = (3x^2y^2)\mathbf{i} + (2x^3y)\mathbf{j}$ is shown in the figure below. Three particles move from the point $(1, 1)$ to the point $(2, 4)$ along different paths. Explain why the work done is the same for each particle, and find the value of the work.

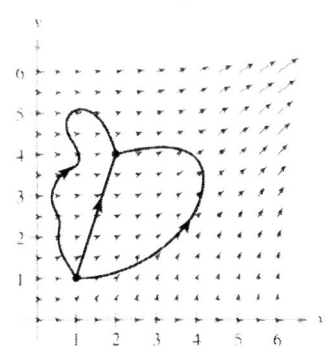

9. Let S be a smooth oriented surface with normal vector $\mathbf{N}$, bounded by a smooth simple closed curve C. Let $\mathbf{v}$ be a constant vector, and prove that
$$\iint_S (2\mathbf{v} \cdot \mathbf{N})\, dS = \int_C (\mathbf{v} \times \mathbf{r}) \cdot d\mathbf{r}.$$

10. How does the area of the ellipse $\dfrac{x^2}{a^2} + \dfrac{y^2}{b^2} = 1$ compare with the magnitude of the work done by the force field
$$\mathbf{F}(x, y) = -\frac{1}{2}y\mathbf{i} + \frac{1}{2}x\mathbf{j}$$
on a particle that moves once around the ellipse?

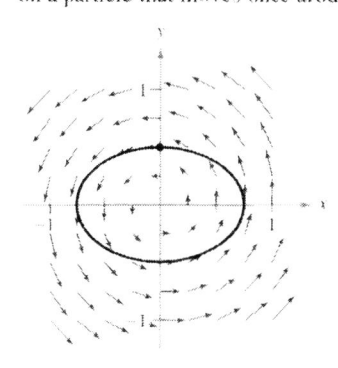

11. A cross section of earth's magnetic field can be represented as a vector field in which the center of earth is located at the origin and the positive y-axis points in the direction of the magnetic north pole. The equation for this field is
$$\mathbf{F}(x, y) = M(x, y)\mathbf{i} + N(x, y)\mathbf{j}$$
$$= \frac{m}{(x^2 + y^2)^{5/2}}[3xy\mathbf{i} + (2y^2 - x^2)\mathbf{j}]$$
where m is the magnetic moment of earth. Show that this vector field is conservative.

Appendices

Appendix A *Additional Topics in Differential Equations*   A2
Appendix B *Proofs of Selected Theorems*   A9
Appendix C *Integration Tables*   A27

The remaining appendices are located on the website that accompanies this text at *college.hmco.com*.

Appendix D *Precalculus Review*
Appendix E *Rotation and the General Second-Degree Equation*
Appendix F *Complex Numbers*
Appendix G *Business and Economic Applications*

Additional Topics in Differential Equations

A

- Use a slope field to sketch solutions of a differential equation.
- Use Euler's Method to approximate a solution of a differential equation.
- Solve a first-order linear differential equation.

Slope Fields

In this appendix, you will study two techniques for approximating solutions of differential equations of the form $y' = F(x, y)$. The first technique is a graphical approach that uses **slope fields,** or *direction fields*. The second technique is a numerical approach and is called *Euler's method.*

Consider a differential equation of the form

$$y' = F(x, y). \quad \text{Differential equation}$$

You can interpret this differential equation graphically to mean that the slope of the graph of each solution at the point (x, y) is y'. You can use a slope field to visualize the family of solutions. To sketch a slope field, pick several points (x, y) and draw short line segments with slope $F(x, y)$. The slope field shows the general shape of all the solutions. An initial condition is needed to sketch a particular solution, as shown in Example 1.

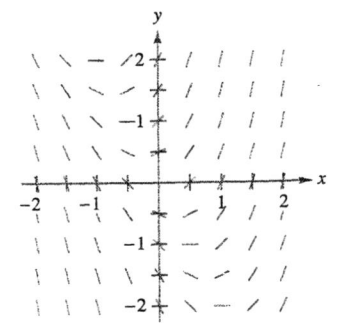

Example 1 Sketching a Solution Using a Slope Field

Sketch a slope field for the differential equation

$$y' = 2x + y.$$

Use the slope field to sketch the solution that passes through the point $(1, 1)$.

Solution Make a table showing the slope at several points. The table shown is a small sample. The slope at many other points should be calculated to get a representative slope field. Next draw line segments at the points with their respective slopes, as shown in Figure A.1.

x	-2	-2	-1	-1	0	0	1	1	2	2
y	-1	1	-1	1	-1	1	-1	1	-1	1
$y' = 2x + y$	-5	-3	-3	-1	-1	1	1	3	3	5

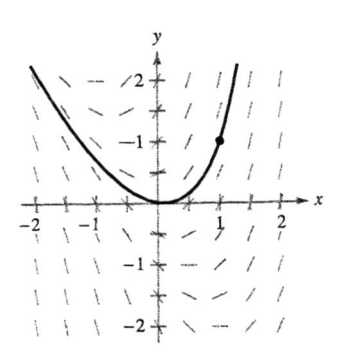

After the slope field is drawn, start at the initial point $(1, 1)$ and move to the right in the direction of the line segment. Continue to draw the solution curve so that it moves parallel to the line segments. Do the same to the left of $(1, 1)$. The resulting solution is shown in Figure A.2.

Slope field for $y' = 2x + y$
Figure A.1

Particular solution for $y' = 2x + y$ passing through $(1, 1)$
Figure A.2

NOTE Drawing a slope field by hand is tedious. In practice, slope fields are usually drawn using a graphing utility.

Euler's Method

Euler's Method is a numerical approach to approximate the particular solution of the differential equation $y' = F(x, y)$ that passes through the point (x_0, y_0). From the given information, you know that the graph of the solution passes through the point (x_0, y_0) and has a slope of $F(x_0, y_0)$ at this point. This gives you a "starting point" for approximating the solution.

From this starting point, you can proceed in the direction indicated by the slope. Using a small step h, move along the tangent line until you arrive at the point (x_1, y_1), where

$$x_1 = x_0 + h \quad \text{and} \quad y_1 = y_0 + hF(x_0, y_0)$$

as shown in Figure A.3. If you think of (x_1, y_1) as a new starting point, you can repeat the process to obtain a second point (x_2, y_2). The values of x_i and y_i are as follows.

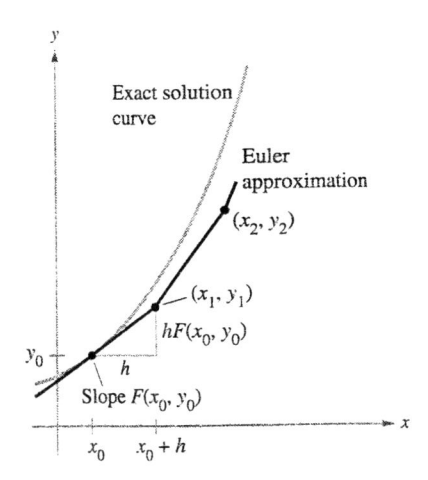

$$x_1 = x_0 + h \qquad y_1 = y_0 + hF(x_0, y_0)$$
$$x_2 = x_1 + h \qquad y_2 = y_1 + hF(x_1, y_1)$$
$$\vdots \qquad\qquad \vdots$$
$$x_n = x_{n-1} + h \qquad y_n = y_{n-1} + hF(x_{n-1}, y_{n-1})$$

Figure A.3

NOTE You can obtain better approximations to the exact solution by choosing smaller and smaller step sizes.

Example 2 Approximating a Solution Using Euler's Method

Use Euler's Method to approximate the particular solution of the differential equation

$$y' = x - y$$

passing through $(0, 1)$. Use a step of $h = 0.1$.

Solution Using $h = 0.1$, $x_0 = 0$, $y_0 = 1$, and $F(x, y) = x - y$, you have $x_0 = 0$, $x_1 = 0.1$, $x_2 = 0.2$, $x_3 = 0.3$, . . . , and

$$y_1 = y_0 + hF(x_0, y_0) = 1 + (0.1)(0 - 1) = 0.9$$
$$y_2 = y_1 + hF(x_1, y_1) = 0.9 + (0.1)(0.1 - 0.9) = 0.82$$
$$y_3 = y_2 + hF(x_2, y_2) = 0.82 + (0.1)(0.2 - 0.82) = 0.758.$$

The first ten approximations are shown in the table. You can plot these values to see a graph of the approximate solution, as shown in Figure A.4.

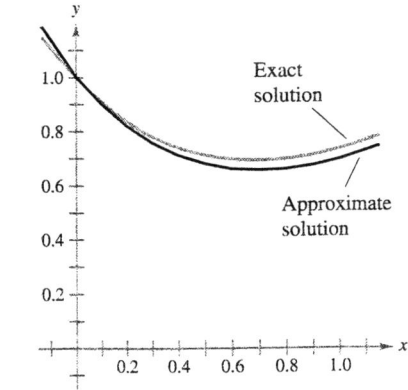

Figure A.4

n	0	1	2	3	4	5	6	7	8	9	10
x_n	0	0.1	0.2	0.3	0.4	0.5	0.6	0.7	0.8	0.9	1.0
y_n	1	0.900	0.820	0.758	0.712	0.681	0.663	0.657	0.661	0.675	0.697

NOTE For the differential equation in Example 2, you can find the exact solution to be $y = x - 1 + 2e^{-x}$. Figure A.4 compares this exact solution with the approximate solution obtained in Example 2.

First-Order Linear Differential Equations

As a final topic in this appendix, you will learn how to solve a very important class of first-order differential equations—first-order *linear* differential equations.

Definition of a First-Order Linear Differential Equation

A first-order linear differential equation is an equation of the form

$$\frac{dy}{dx} + P(x)y = Q(x)$$

where P and Q are continuous functions of x. This first-order linear differential equation is said to be in **standard form.**

To solve a first-order linear differential equation, you can use an *integrating factor* $u(x)$, which converts the left side into the derivative of the product $u(x)y$. That is, you need a factor $u(x)$ such that

$$u(x)\frac{dy}{dx} + u(x)P(x)y = \frac{d[u(x)y]}{dx}$$

$$u(x)y' + u(x)P(x)y = u(x)y' + yu'(x)$$

$$u(x)P(x)y = yu'(x)$$

$$P(x) = \frac{u'(x)}{u(x)}$$

$$\ln|u(x)| = \int P(x)\,dx + C_1$$

$$u(x) = Ce^{\int P(x)\,dx}.$$

Because you don't need the most general integrating factor, let $C = 1$. Multiplying the original equation $y' + P(x)y = Q(x)$ by $u(x) = e^{\int P(x)dx}$ produces

$$y'e^{\int P(x)\,dx} + yP(x)e^{\int P(x)\,dx} = Q(x)e^{\int P(x)\,dx}$$

$$\frac{d}{dx}\left[ye^{\int P(x)\,dx}\right] = Q(x)e^{\int P(x)\,dx}.$$

The general solution is given by

$$ye^{\int P(x)\,dx} = \int Q(x)e^{\int P(x)\,dx}\,dx + C.$$

THEOREM A.1 Solution of a First-Order Linear Differential Equation

An integrating factor for the first-order linear differential equation

$$y' + P(x)y = Q(x)$$

is $u(x) = e^{\int P(x)\,dx}$. The solution of the differential equation is

$$ye^{\int P(x)\,dx} = \int Q(x)e^{\int P(x)\,dx}\,dx + C.$$

STUDY TIP Rather than memorizing this formula, just remember that multiplication by the integrating factor $e^{\int P(x)\,dx}$ converts the left side of the differential equation into the derivative of the product $ye^{\int P(x)\,dx}$.

Example 3 Solving a First-Order Linear Differential Equation

Find the general solution of $xy' - 2y = x^2$.

Solution The *standard form* of the given equation is

$$y' + P(x)y = Q(x)$$
$$y' - \left(\frac{2}{x}\right)y = x. \qquad \text{Standard form}$$

So, $P(x) = -2/x$, and you have

$$\int P(x)\,dx = -\int \frac{2}{x}\,dx = -\ln x^2$$

$$e^{\int P(x)\,dx} = e^{-\ln x^2} = \frac{1}{x^2}. \qquad \text{Integrating factor}$$

Therefore, multiplying both sides of the standard form by $1/x^2$ yields

$$\frac{y'}{x^2} - \frac{2y}{x^3} = \frac{1}{x}$$

$$\frac{d}{dx}\left[\frac{y}{x^2}\right] = \frac{1}{x}$$

$$\frac{y}{x^2} = \int \frac{1}{x}\,dx$$

$$\frac{y}{x^2} = \ln|x| + C$$

$$y = x^2(\ln|x| + C). \qquad \text{General solution}$$

Several solution curves (for $C = -2, -1, 0, 1, 2, 3$, and 4) are shown in Figure A.5.

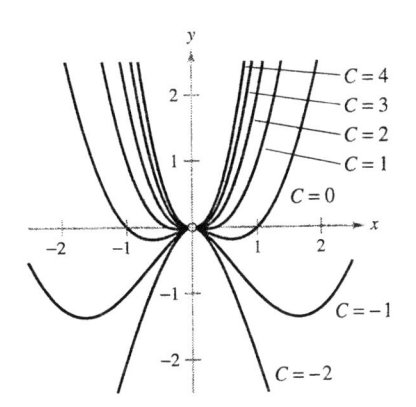

Figure A.5

Example 4 Solving a First-Order Linear Differential Equation

Find the general solution of $y' - y\tan t = 1$, $-\frac{\pi}{2} < t < \frac{\pi}{2}$.

Solution The equation is already in the standard form $y' + P(t)y = Q(t)$. So,

$$\int P(t)\,dt = -\int \tan t\,dt = \ln|\cos t|$$

which implies that the integrating factor is $e^{\int P(t)\,dt} = e^{\ln|\cos t|} = |\cos t|$.

A quick check shows that $\cos t$ is also an integrating factor. So, multiplying $y' - y\tan t = 1$ by $\cos t$ produces

$$\frac{d}{dt}[y\cos t] = \cos t$$

$$y\cos t = \int \cos t\,dt$$

$$y\cos t = \sin t + C$$

$$y = \tan t + C\sec t. \qquad \text{General solution}$$

Several solution curves are shown in Figure A.6.

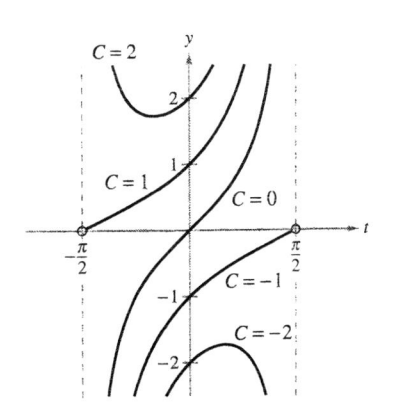

Figure A.6

Application

A simple electrical circuit consists of electric current I (in amperes), a resistance R (in ohms), an inductance L (in henrys), and a constant electromotive force E (in volts), as shown in Figure A.7. According to Kirchhoff's Second Law, if the switch S is closed when $t = 0$, the applied electromotive force (voltage) is equal to the sum of the voltage drops in the rest of the circuit. This in turn means that the current I satisfies the differential equation

$$L\frac{dI}{dt} + RI = E.$$

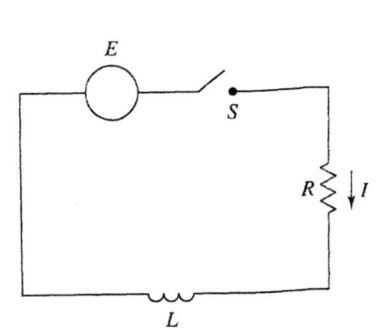

Figure A.7

Example 5 An Electric Circuit Problem

Find the current I as a function of time t (in seconds), given that I satisfies the differential equation $L(dI/dt) + RI = \sin 2t$, where R and L are nonzero constants.

Solution In standard form, the given linear equation is

$$\frac{dI}{dt} + \frac{R}{L}I = \frac{1}{L}\sin 2t.$$

Let $P(t) = R/L$, so that $e^{\int P(t)\,dt} = e^{(R/L)t}$, and, by Theorem A.1,

$$Ie^{(R/L)t} = \frac{1}{L}\int e^{(R/L)t}\sin 2t\,dt = \frac{1}{4L^2 + R^2}e^{(R/L)t}(R\sin 2t - 2L\cos 2t) + C.$$

So, the general solution is

$$I = e^{-(R/L)t}\left[\frac{1}{4L^2 + R^2}e^{(R/L)t}(R\sin 2t - 2L\cos 2t) + C\right]$$

$$I = \frac{1}{4L^2 + R^2}(R\sin 2t - 2L\cos 2t) + Ce^{-(R/L)t}.$$

EXERCISES FOR APPENDIX A

In Exercises 1 and 2, a differential equation and its slope field are given. Determine the slope (if possible) in the slope field at the points given in the table.

x	-4	-2	0	2	4	8
y	2	0	4	4	6	8
dy/dx						

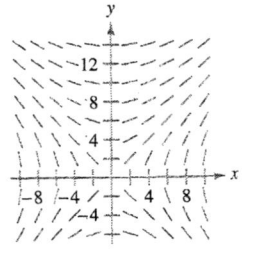

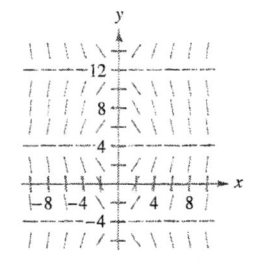

1. $\dfrac{dy}{dx} = \dfrac{x}{y}$

2. $\dfrac{dy}{dx} = x\cos\dfrac{\pi y}{8}$

In Exercises 3–6, (a) sketch an approximate solution of the differential equation satisfying the initial condition by hand on the slope field, (b) find the particular solution that satisfies the initial condition, and (c) use a graphing utility to graph the particular solution. Compare the graph with the hand-drawn graph of part (a).

Differential Equation	Initial Condition
3. $\dfrac{dy}{dx} = e^x - y$	$(0, 1)$
4. $y' + 2y = \sin x$	$(0, 4)$

Figure for 3

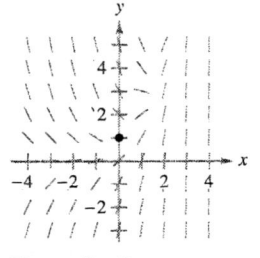

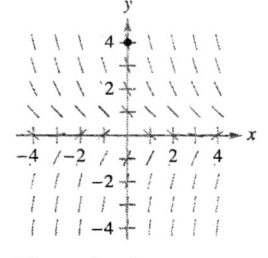

Figure for 4

Differential Equation	Initial Condition
5. $y' = \csc x + y \cot x$	$(1, 1)$
6. $y' = \csc x - y \cot x$	$(1, 2)$

Figure for 5

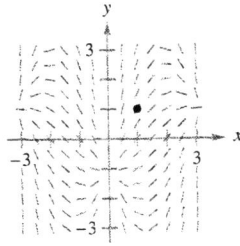

Figure for 6

In Exercises 7 and 8, use a computer algebra system to sketch the slope field for the differential equation and graph the solution satisfying the specified initial condition.

7. $\dfrac{dy}{dx} = 0.4y(3 - x)$, $y(0) = 1$

8. $\dfrac{dy}{dx} = \dfrac{1}{2}e^{-x/8} \sin \dfrac{\pi y}{4}$, $y(0) = 2$

Euler's Method In Exercise 9–14, use Euler's method to make a table of values for the approximate solution of the differential equation with the specified initial value. Use n steps of size h.

9. $y' = x + y$, $y(0) = 2$, $n = 10$, $h = 0.1$
10. $y' = x + y$, $y(0) = 2$, $n = 20$, $h = 0.05$
11. $y' = 3x - 2y$, $y(0) = 3$, $n = 10$, $h = 0.05$
12. $y' = 0.5x(3 - y)$, $y(0) = 1$, $n = 5$, $h = 0.4$
13. $y' = e^{xy}$, $y(0) = 1$, $n = 10$, $h = 0.1$
14. $y' = \cos x + \sin y$, $y(0) = 5$, $n = 10$, $h = 0.1$

True or False? In Exercises 15 and 16, determine whether the statement is true or false. If it is false, explain why or give an example that shows it is false.

15. $y' + x\sqrt{y} = x^2$ is a first-order linear differential equation.
16. $y' + xy = e^x y$ is a first-order linear differential equation.

In Exercises 17–32, solve the first-order linear differential equation.

17. $\dfrac{dy}{dx} + \left(\dfrac{1}{x}\right)y = 3x + 4$
18. $\dfrac{dy}{dx} + \left(\dfrac{2}{x}\right)y = 3x + 1$
19. $\dfrac{dy}{dx} - 3x^2 y = e^{x^3}$
20. $\dfrac{dy}{dx} - \dfrac{3y}{x^2} = \dfrac{1}{x^2}$
21. $y' - y = \cos x$
22. $y' + 2xy = 2x$
23. $(x + y) dx - x dy = 0$
24. $(2y - e^x) dx + x dy = 0$
25. $(3y + \sin 2x) dx - dy = 0$
26. $(y - 1)\sin x dx - dy = 0$
27. $(x - 1)y' + y = x^2 - 1$
28. $y' + 5y = e^{5x}$
29. $dy = (y \tan x + 2e^x) dx$
30. $xy' + y = \sin x$
31. $xy' - ay = bx^4$
32. $y' = y + 2x(y - e^x)$

In Exercises 33–40, find the particular solution of the differential equation that satisfies the boundary condition.

Differential Equation	Boundary Condition
33. $y' \cos^2 x + y - 1 = 0$	$y(0) = 5$
34. $x^3 y' + 2y = e^{1/x^2}$	$y(1) = e$
35. $y' + y \tan x = \sec x + \cos x$	$y(0) = 1$
36. $y' + y \sec x = \sec x$	$y(0) = 4$
37. $y' + \left(\dfrac{1}{x}\right)y = 0$	$y(2) = 2$
38. $y' + (2x - 1)y = 0$	$y(1) = 2$
39. $x dy = (x + y + 2) dx$	$y(1) = 10$
40. $2xy' - y = x^3 - x$	$y(4) = 2$

In Exercises 41 and 42, (a) use a graphing utility to graph the slope field for the differential equation, (b) find the particular solutions of the differential equation passing through the specified points, and (c) use a graphing utility to graph the particular solutions on the slope field.

Differential Equation	Points
41. $\dfrac{dy}{dx} - \dfrac{1}{x}y = x^2$	$(-2, 4)$, $(2, 8)$
42. $\dfrac{dy}{dx} + (\cot x)y = x$	$(1, 1)$, $(3, -1)$

Electrical Circuits In Exercises 43–46, use the differential equation for electrical circuits given by

$$L \dfrac{dI}{dt} + RI = E.$$

In this equation, I is the current, R is the resistance, L is the inductance, and E is the electromotive force (voltage).

43. Solve the differential equation given a constant voltage E_0.

44. Use the result of Exercise 43 to find the equation for the current if $I(0) = 0$, $E_0 = 110$ volts, $R = 550$ ohms, and $L = 4$ henrys. When does the current reach 90% of its limiting value?

45. Solve the differential equation given a periodic electromotive force $E_0 \sin \omega t$.

46. Verify that the solution of Exercise 45 can be written in the form

$$I = ce^{-(R/L)t} + \frac{E_0}{\sqrt{R^2 + \omega^2 L^2}} \sin(\omega t + \phi)$$

where ϕ, the phase angle, is given by $\arctan(-\omega L/R)$. (Note that the exponential term approaches 0 as $t \to \infty$. This implies that the current approaches a periodic function.)

47. *Population Growth* When predicting population growth, demographers must consider birth and death rates as well as the net change caused by the difference between the rates of immigration and emigration. Let P be the population at time t and let N be the net increase per unit time resulting from the difference between immigration and emigration. So, the rate of growth of the population is given by

$$\frac{dP}{dt} = kP + N, \quad N \text{ is constant.}$$

Solve this differential equation to find P as a function of time if at time $t = 0$ the size of the population is P_0.

48. *Investment Growth* A large corporation starts at time $t = 0$ to continuously invest part of its receipts at a rate of P dollars per year in a fund for future corporate expansion. Assume that the fund earns r percent interest per year compounded continuously. So, the rate of growth of the amount A in the fund is given by

$$\frac{dA}{dt} = rA + P$$

where $A = 0$ when $t = 0$. Solve this differential equation for A as a function of t.

Investment Growth In Exercises 49 and 50, use the result of Exercise 48.

49. Find A for the following.

(a) $P = \$100,000$, $r = 6\%$, and $t = 5$ years

(b) $P = \$250,000$, $r = 5\%$, and $t = 10$ years

50. Find t if the corporation needs $\$800,000$ and it can invest $\$75,000$ per year in a fund earning 8% interest compounded continuously.

51. *Investment* Let $A(t)$ be the amount in a fund earning interest at an annual rate r compounded continuously. If a continuous cash flow of P dollars per year is withdrawn from the fund, the rate of change of A is given by the differential equation

$$\frac{dA}{dt} = rA - P$$

where $A = A_0$ when $t = 0$. Solve this differential equation for A as a function of t.

52. *Investment* A retired couple plans to withdraw P dollars per year from a retirement account of $\$500,000$ earning 10% compounded continuously. Use the result of Exercise 51 and a graphing utility to graph the function A for each of the following continuous annual cash flows. Use the graphs to describe what happens to the balance in the fund for each of the cases.

(a) $P = \$40,000$

(b) $P = \$50,000$

(c) $P = \$60,000$

53. *Intravenous Feeding* Glucose is added intravenously to the bloodstream at the rate of q units per minute, and the body removes glucose from the bloodstream at a rate proportional to the amount present. Assume $Q(t)$ is the amount of glucose in the bloodstream at time t.

(a) Determine the differential equation describing the rate of change with respect to time of glucose in the bloodstream.

(b) Solve the differential equation from part (a), letting $Q = Q_0$ when $t = 0$.

(c) Find the limit of $Q(t)$ as $t \to \infty$.

54. *Learning Curve* The management at a certain factory has found that the maximum number of units a worker can produce in a day is 30. The rate of increase in the number of units N produced with respect to time t in days by a new employee is proportional to $30 - N$.

(a) Determine the differential equation describing the rate of change of performance with respect to time.

(b) Solve the differential equation from part (a).

(c) Find the particular solution for a new employee who produced ten units on the first day at the factory and 19 units on the twentieth day.

In Exercises 55–58, match the differential equation with its solution.

Differential Equation	Solution
55. $y' - 2x = 0$	(a) $y = Ce^{x^2}$
56. $y' - 2y = 0$	(b) $y = -\frac{1}{2} + Ce^{x^2}$
57. $y' - 2xy = 0$	(c) $y = x^2 + C$
58. $y' - 2xy = x$	(d) $y = Ce^{2x}$

Proofs of Selected Theorems

> **THEOREM 1.2** Properties of Limits (Properties 2, 3, 4, and 5) (page 57)
>
> Let b and c be real numbers, let n be a positive integer, and let f and g be functions with the following limits.
>
> $$\lim_{x \to c} f(x) = L \quad \text{and} \quad \lim_{x \to c} g(x) = K$$
>
> **2.** Sum or difference: $\lim_{x \to c} [f(x) \pm g(x)] = L \pm K$
>
> **3.** Product: $\lim_{x \to c} [f(x)g(x)] = LK$
>
> **4.** Quotient: $\lim_{x \to c} \dfrac{f(x)}{g(x)} = \dfrac{L}{K}$, provided $K \neq 0$
>
> **5.** Power: $\lim_{x \to c} [f(x)]^n = L^n$

Proof To prove Property 2, choose $\varepsilon > 0$. Because $\varepsilon/2 > 0$, you know that there exists $\delta_1 > 0$ such that $0 < |x - c| < \delta_1$ implies $|f(x) - L| < \varepsilon/2$. You also know that there exists $\delta_2 > 0$ such that $0 < |x - c| < \delta_2$ implies $|g(x) - K| < \varepsilon/2$. Let δ be the smaller of δ_1 and δ_2; then $0 < |x - c| < \delta$ implies that

$$|f(x) - L| < \frac{\varepsilon}{2} \quad \text{and} \quad |g(x) - K| < \frac{\varepsilon}{2}.$$

So, you can apply the Triangle Inequality to conclude that

$$|[f(x) + g(x)] - (L + K)| \leq |f(x) - L| + |g(x) - K| < \frac{\varepsilon}{2} + \frac{\varepsilon}{2} = \varepsilon$$

which implies that

$$\lim_{x \to c} [f(x) + g(x)] = L + K = \lim_{x \to c} f(x) + \lim_{x \to c} g(x).$$

The proof that

$$\lim_{x \to c} [f(x) - g(x)] = L - K$$

is similar.

To prove Property 3, given that

$$\lim_{x \to c} f(x) = L \quad \text{and} \quad \lim_{x \to c} g(x) = K$$

you can write

$$f(x)g(x) = [f(x) - L][g(x) - K] + [Lg(x) + Kf(x)] - LK.$$

Because the limit of $f(x)$ is L, and the limit of $g(x)$ is K, you have

$$\lim_{x \to c} [f(x) - L] = 0 \quad \text{and} \quad \lim_{x \to c} [g(x) - K] = 0.$$

Let $0 < \varepsilon < 1$. Then there exists $\delta > 0$ such that if $0 < |x - c| < \delta$, then
$$|f(x) - L - 0| < \varepsilon \quad \text{and} \quad |g(x) - K - 0| < \varepsilon$$
which implies that
$$|[f(x) - L][g(x) - K] - 0| = |f(x) - L||g(x) - K| < \varepsilon\varepsilon < \varepsilon.$$
Hence,
$$\lim_{x \to c} [f(x) - L][g(x) - K] = 0.$$
Furthermore, by Property 1, you have
$$\lim_{x \to c} Lg(x) = LK \quad \text{and} \quad \lim_{x \to c} Kf(x) = KL.$$
Finally, by Property 2, you obtain
$$\lim_{x \to c} f(x)g(x) = \lim_{x \to c} [f(x) - L][g(x) - K] + \lim_{x \to c} Lg(x) + \lim_{x \to c} Kf(x) - \lim_{x \to c} LK$$
$$= 0 + LK + KL - LK$$
$$= LK.$$

To prove Property 4, note that it is sufficient to prove that
$$\lim_{x \to c} \frac{1}{g(x)} = \frac{1}{K}.$$
Then you can use Property 3 to write
$$\lim_{x \to c} \frac{f(x)}{g(x)} = \lim_{x \to c} f(x) \frac{1}{g(x)} = \lim_{x \to c} f(x) \cdot \lim_{x \to c} \frac{1}{g(x)} = \frac{L}{K}.$$
Let $\varepsilon > 0$. Because $\lim_{x \to c} g(x) = K$, there exists $\delta_1 > 0$ such that if
$$0 < |x - c| < \delta_1, \text{ then } |g(x) - K| < \frac{|K|}{2}$$
which implies that
$$|K| = |g(x) + [|K| - g(x)]| \leq |g(x)| + ||K| - g(x)| < |g(x)| + \frac{|K|}{2}.$$
That is, for $0 < |x - c| < \delta_1$,
$$\frac{|K|}{2} < |g(x)| \quad \text{or} \quad \frac{1}{|g(x)|} < \frac{2}{|K|}.$$
Similarly, there exists a $\delta_2 > 0$ such that if $0 < |x - c| < \delta_2$, then
$$|g(x) - K| < \frac{|K|^2}{2} \varepsilon.$$
Let δ be the smaller of δ_1 and δ_2. For $0 < |x - c| < \delta$, you have
$$\left| \frac{1}{g(x)} - \frac{1}{K} \right| = \left| \frac{K - g(x)}{g(x)K} \right| = \frac{1}{|K|} \cdot \frac{1}{|g(x)|} |K - g(x)| < \frac{1}{|K|} \cdot \frac{2}{|K|} \cdot \frac{|K|^2}{2} \varepsilon = \varepsilon.$$
So, $\lim_{x \to c} \dfrac{1}{g(x)} = \dfrac{1}{K}.$

Finally, the proof of Property 5 can be obtained by a straightforward application of mathematical induction coupled with Property 3.

> **THEOREM 1.4 The Limit of a Function Involving a Radical (page 58)**
>
> Let n be a positive integer. The following limit is valid for all c if n is odd, and is valid for $c > 0$ if n is even.
>
> $$\lim_{x \to c} \sqrt[n]{x} = \sqrt[n]{c}.$$

Proof Consider the case for which $c > 0$ and n is any positive integer. For a given $\varepsilon > 0$, you need to find $\delta > 0$ such that

$$\left| \sqrt[n]{x} - \sqrt[n]{c} \right| < \varepsilon \quad \text{whenever} \quad 0 < |x - c| < \delta$$

which is the same as saying

$$-\varepsilon < \sqrt[n]{x} - \sqrt[n]{c} < \varepsilon \quad \text{whenever} \quad -\delta < x - c < \delta.$$

Assume $\varepsilon < \sqrt[n]{c}$, which implies that $0 < \sqrt[n]{c} - \varepsilon < \sqrt[n]{c}$. Now, let δ be the smaller of the two numbers.

$$c - \left(\sqrt[n]{c} - \varepsilon \right)^n \quad \text{and} \quad \left(\sqrt[n]{c} + \varepsilon \right)^n - c$$

Then you have

$$-\delta < x - c < \delta$$
$$-\left[c - \left(\sqrt[n]{c} - \varepsilon \right)^n \right] < x - c < \left(\sqrt[n]{c} + \varepsilon \right)^n - c$$
$$\left(\sqrt[n]{c} - \varepsilon \right)^n - c < x - c < \left(\sqrt[n]{c} + \varepsilon \right)^n - c$$
$$\left(\sqrt[n]{c} - \varepsilon \right)^n < x < \left(\sqrt[n]{c} + \varepsilon \right)^n$$
$$\sqrt[n]{c} - \varepsilon < \sqrt[n]{x} < \sqrt[n]{c} + \varepsilon$$
$$-\varepsilon < \sqrt[n]{x} - \sqrt[n]{c} < \varepsilon.$$

> **THEOREM 1.5 The Limit of a Composite Function (page 59)**
>
> If f and g are functions such that $\lim_{x \to c} g(x) = L$ and $\lim_{x \to L} f(x) = f(L)$, then
>
> $$\lim_{x \to c} f(g(x)) = f\left(\lim_{x \to c} g(x) \right) = f(L).$$

Proof For a given $\varepsilon > 0$, you must find $\delta > 0$ such that

$$|f(g(x)) - f(L)| < \varepsilon \quad \text{whenever} \quad 0 < |x - c| < \delta.$$

Because the limit of $f(x)$ as $x \to L$ is $f(L)$, you know there exists $\delta_1 > 0$ such that

$$|f(u) - f(L)| < \varepsilon \quad \text{whenever} \quad |u - L| < \delta_1.$$

Moreover, because the limit of $g(x)$ as $x \to c$ is L, you know there exists $\delta > 0$ such that

$$|g(x) - L| < \delta_1 \quad \text{whenever} \quad 0 < |x - c| < \delta.$$

Finally, letting $u = g(x)$, you have

$$|f(g(x)) - f(L)| < \varepsilon \quad \text{whenever} \quad 0 < |x - c| < \delta.$$

THEOREM 1.7 Functions That Agree at All But One Point (page 60)

Let c be a real number and let $f(x) = g(x)$ for all $x \neq c$ in an open interval containing c. If the limit of $g(x)$ as x approaches c exists, then the limit of $f(x)$ also exists and

$$\lim_{x \to c} f(x) = \lim_{x \to c} g(x).$$

Proof Let L be the limit of $g(x)$ as $x \to c$. Then, for each $\varepsilon > 0$ there exists a $\delta > 0$ such that $f(x) = g(x)$ in the open intervals $(c - \delta, c)$ and $(c, c + \delta)$, and

$$|g(x) - L| < \varepsilon \quad \text{whenever} \quad 0 < |x - c| < \delta.$$

Because $f(x) = g(x)$ for all x in the open interval other than $x = c$, it follows that

$$|f(x) - L| < \varepsilon \quad \text{whenever} \quad 0 < |x - c| < \delta.$$

So, the limit of $f(x)$ as $x \to c$ is also L.

THEOREM 1.8 The Squeeze Theorem (page 63)

If $h(x) \leq f(x) \leq g(x)$ for all x in an open interval containing c, except possibly at c itself, and if

$$\lim_{x \to c} h(x) = L = \lim_{x \to c} g(x)$$

then $\lim_{x \to c} f(x)$ exists and is equal to L.

Proof For $\varepsilon > 0$ there exist δ_1 and δ_2 such that

$$|h(x) - L| < \varepsilon \quad \text{whenever} \quad 0 < |x - c| < \delta_1$$

and

$$|g(x) - L| < \varepsilon \quad \text{whenever} \quad 0 < |x - c| < \delta_2.$$

Because $h(x) \leq f(x) \leq g(x)$ for all x in an open interval containing c, except possibly at c itself, there exists $\delta_3 > 0$ such that $h(x) \leq f(x) \leq g(x)$ for $0 < |x - c| < \delta_3$. Let δ be the smallest of δ_1, δ_2, and δ_3. Then, if $0 < |x - c| < \delta$, it follows that $|h(x) - L| < \varepsilon$ and $|g(x) - L| < \varepsilon$, which implies that

$$-\varepsilon < h(x) - L < \varepsilon \quad \text{and} \quad -\varepsilon < g(x) - L < \varepsilon$$

$$L - \varepsilon < h(x) \quad \text{and} \quad g(x) < L + \varepsilon.$$

Now, because $h(x) \leq f(x) \leq g(x)$, it follows that $L - \varepsilon < f(x) < L + \varepsilon$, which implies that $|f(x) - L| < \varepsilon$. Therefore,

$$\lim_{x \to c} f(x) = L.$$

> **THEOREM 1.14 Vertical Asymptotes (page 82)**
>
> Let f and g be continuous on an open interval containing c. If $f(c) \neq 0$, $g(c) = 0$, and there exists an open interval containing c such that $g(x) \neq 0$ for all $x \neq c$ in the interval, then the graph of the function given by
>
> $$h(x) = \frac{f(x)}{g(x)}$$
>
> has a vertical asymptote at $x = c$.

Proof Consider the case for which $f(c) > 0$, and there exists $b > c$ such that $c < x < b$ implies $g(x) > 0$. Then for $M > 0$, choose δ_1 such that

$$0 < x - c < \delta_1 \quad \text{implies that} \quad \frac{f(c)}{2} < f(x) < \frac{3f(c)}{2}$$

and δ_2 such that

$$0 < x - c < \delta_2 \quad \text{implies that} \quad 0 < g(x) < \frac{f(c)}{2M}.$$

Now let δ be the smaller of δ_1 and δ_2. Then it follows that

$$0 < x - c < \delta \quad \text{implies that} \quad \frac{f(x)}{g(x)} > \frac{f(c)}{2}\left[\frac{2M}{f(c)}\right] = M.$$

Therefore, it follows that

$$\lim_{x \to c^+} \frac{f(x)}{g(x)} = \infty$$

and the line $x = c$ is a vertical asymptote of the graph of h.

> **Alternative Form of the Derivative (page 99)**
>
> The derivative of f at c is given by
>
> $$f'(c) = \lim_{x \to c} \frac{f(x) - f(c)}{x - c}$$
>
> provided this limit exists.

Proof The derivative of f at c is given by

$$f'(c) = \lim_{\Delta x \to 0} \frac{f(c + \Delta x) - f(c)}{\Delta x}.$$

Let $x = c + \Delta x$. Then $x \to c$ as $\Delta x \to 0$. So, replacing $c + \Delta x$ by x, you have

$$f'(c) = \lim_{\Delta x \to 0} \frac{f(c + \Delta x) - f(c)}{\Delta x} = \lim_{x \to c} \frac{f(x) - f(c)}{x - c}.$$

> **THEOREM 2.10 The Chain Rule (page 128)**
>
> If $y = f(u)$ is a differentiable function of u, and $u = g(x)$ is a differentiable function of x, then $y = f(g(x))$ is a differentiable function of x and
>
> $$\frac{dy}{dx} = \frac{dy}{du} \cdot \frac{du}{dx} \quad \text{or, equivalently,} \quad \frac{d}{dx}[f(g(x))] = f'(g(x))g'(x).$$

Proof In Section 2.4, we let $h(x) = f(g(x))$ and used the alternative form of the derivative to show that $h'(c) = f'(g(c))g'(c)$, provided $g(x) \neq g(c)$ for values of x other than c. Now consider a more general proof. Begin by considering the derivative of f.

$$f'(x) = \lim_{\Delta x \to 0} \frac{f(x + \Delta x) - f(x)}{\Delta x} = \lim_{\Delta x \to 0} \frac{\Delta y}{\Delta x}$$

For a fixed value of x, define a function η such that

$$\eta(\Delta x) = \begin{cases} 0, & \Delta x = 0 \\ \dfrac{\Delta y}{\Delta x} - f'(x), & \Delta x \neq 0. \end{cases}$$

Because the limit of $\eta(\Delta x)$ as $\Delta x \to 0$ doesn't depend on the value of $\eta(0)$, you have

$$\lim_{\Delta x \to 0} \eta(\Delta x) = \lim_{\Delta x \to 0} \left[\frac{\Delta y}{\Delta x} - f'(x) \right] = 0$$

and you can conclude that η is continuous at 0. Moreover, because $\Delta y = 0$ when $\Delta x = 0$, the equation

$$\Delta y = \Delta x \eta(\Delta x) + \Delta x f'(x)$$

is valid whether Δx is zero or not. Now, by letting $\Delta u = g(x + \Delta x) - g(x)$, you can use the continuity of g to conclude that

$$\lim_{\Delta x \to 0} \Delta u = \lim_{\Delta x \to 0} [g(x + \Delta x) - g(x)] = 0$$

which implies that

$$\lim_{\Delta x \to 0} \eta(\Delta u) = 0.$$

Finally,

$$\Delta y = \Delta u \eta(\Delta u) + \Delta u f'(u) \to \frac{\Delta y}{\Delta x} = \frac{\Delta u}{\Delta x} \eta(\Delta u) + \frac{\Delta u}{\Delta x} f'(u), \quad \Delta x \neq 0$$

and taking the limit as $\Delta x \to 0$, you have

$$\frac{dy}{dx} = \frac{du}{dx} \left[\lim_{\Delta x \to 0} \eta(\Delta u) \right] + \frac{du}{dx} f'(u) = \frac{du}{dx}(0) + \frac{du}{dx} f'(u)$$

$$= \frac{du}{dx} f'(u) = \frac{du}{dx} \cdot \frac{dy}{du}.$$

Concavity Interpretation (page 184)

1. Let f be differentiable on an open interval I. If the graph of f is concave *upward* on I, then the graph of f lies *above* all of its tangent lines on I.
2. Let f be differentiable on an open interval I. If the graph of f is concave *downward* on I, then the graph of f lies *below* all of its tangent lines on I.

Proof Assume that f is concave upward on $I = (a, b)$. Then, f' is increasing on (a, b). Let c be a point in the interval $I = (a, b)$. The equation of the tangent line to the graph of f at c is given by

$$g(x) = f(c) + f'(c)(x - c).$$

If x is in the open interval (c, b), then the directed distance from point $(x, f(x))$ (on the graph of f) to the point $(x, g(x))$ (on the tangent line) is given by

$$\begin{aligned}d &= f(x) - [f(c) + f'(c)(x - c)]\\ &= f(x) - f(c) - f'(c)(x - c).\end{aligned}$$

Moreover, by the Mean Value Theorem there exists a number z in (c, x) such that

$$f'(z) = \frac{f(x) - f(c)}{x - c}.$$

So, you have

$$\begin{aligned}d &= f(x) - f(c) - f'(c)(x - c)\\ &= f'(z)(x - c) - f'(c)(x - c)\\ &= [f'(z) - f'(c)](x - c).\end{aligned}$$

The second factor $(x - c)$ is positive because $c < x$. Moreover, because f' is increasing, it follows that the first factor $[f'(z) - f'(c)]$ is also positive. Therefore, $d > 0$ and you can conclude that the graph of f lies above the tangent line at x. If x is in the open interval (a, c), a similar argument can be given. This proves the first statement. The proof of the second statement is similar.

THEOREM 3.10 Limits at Infinity (page 193)

If r is a positive rational number, and c is any real number, then

$$\lim_{x \to \infty} \frac{c}{x^r} = 0.$$

Furthermore, if x^r is defined when $x < 0$, then $\lim_{x \to -\infty} \frac{c}{x^r} = 0$.

Proof Begin by proving that

$$\lim_{x \to \infty} \frac{1}{x} = 0.$$

For $\varepsilon > 0$, let $M = 1/\varepsilon$. Then, for $x > M$, you have

$$x > M = \frac{1}{\varepsilon} \quad \Rightarrow \quad \frac{1}{x} < \varepsilon \quad \Rightarrow \quad \left|\frac{1}{x} - 0\right| < \varepsilon.$$

Therefore, by the definition of a limit at infinity, you can conclude that the limit of $1/x$ as $x \to \infty$ is 0. Now, using this result, and letting $r = m/n$, you can write the following.

$$\lim_{x \to \infty} \frac{c}{x^r} = \lim_{x \to \infty} \frac{c}{x^{m/n}}$$
$$= c \left[\lim_{x \to \infty} \left(\frac{1}{\sqrt[n]{x}} \right)^m \right]$$
$$= c \left(\lim_{x \to \infty} \sqrt[n]{\frac{1}{x}} \right)^m$$
$$= c \left(\sqrt[n]{\lim_{x \to \infty} \frac{1}{x}} \right)^m$$
$$= c \left(\sqrt[n]{0} \right)^m$$
$$= 0$$

The proof of the second part of the theorem is similar.

THEOREM 4.2   Summation Formulas (page 254)

1. $\sum_{i=1}^{n} c = cn$
2. $\sum_{i=1}^{n} i = \frac{n(n+1)}{2}$
3. $\sum_{i=1}^{n} i^2 = \frac{n(n+1)(2n+1)}{6}$
4. $\sum_{i=1}^{n} i^3 = \frac{n^2(n+1)^2}{4}$

Proof   The proof of Property 1 is straightforward. By adding c to itself n times, you obtain a sum of cn.

To prove Property 2, write the sum in increasing and decreasing order and add corresponding terms as follows.

$$\sum_{i=1}^{n} i = 1 + 2 + 3 + \cdots + (n-1) + n$$

$$\sum_{i=1}^{n} i = n + (n-1) + (n-2) + \cdots + 2 + 1$$

$$2\sum_{i=1}^{n} i = \underbrace{(n+1) + (n+1) + (n+1) + \cdots + (n+1) + (n+1)}_{n \text{ terms}}$$

Therefore,

$$\sum_{i=1}^{n} i = \frac{n(n+1)}{2}.$$

To prove Property 3, use mathematical induction. First, if $n = 1$, the result is true because

$$\sum_{i=1}^{1} i^2 = 1^2 = 1 = \frac{1(1+1)(2+1)}{6}.$$

Now, assuming the result is true for $n = k$, you can show that it is true for $n = k + 1$, as follows.

$$\sum_{i=1}^{k+1} i^2 = \sum_{i=1}^{k} i^2 + (k + 1)^2$$

$$= \frac{k(k + 1)(2k + 1)}{6} + (k + 1)^2$$

$$= \frac{k + 1}{6}(2k^2 + k + 6k + 6)$$

$$= \frac{k + 1}{6}[(2k + 3)(k + 2)]$$

$$= \frac{(k + 1)(k + 2)[2(k + 1) + 1]}{6}$$

Property 4 can be proved using a similar argument with mathematical induction.

THEOREM 4.8 Preservation of Inequality (page 272)

1. If f is integrable and nonnegative on the closed interval $[a, b]$, then

$$0 \le \int_a^b f(x)\, dx.$$

2. If f and g are integrable on the closed interval $[a, b]$, and $f(x) \le g(x)$ for every x in $[a, b]$, then

$$\int_a^b f(x)\, dx \le \int_a^b g(x)\, dx.$$

Proof To prove Property 1, suppose, on the contrary, that

$$\int_a^b f(x)\, dx = I < 0.$$

Then, let $a = x_0 < x_1 < x_2 < \cdots < x_n = b$ be a partition of $[a, b]$, and let

$$R = \sum_{i=1}^{n} f(c_i)\, \Delta x_i$$

be a Riemann sum. Because $f(x) \ge 0$, it follows that $R \ge 0$. Now, for $\|\Delta\|$ sufficiently small, you have $|R - I| < -I/2$, which implies that

$$\sum_{i=1}^{n} f(c_i)\, \Delta x_i = R < I - \frac{I}{2} < 0$$

which is not possible. From this contradiction, you can conclude that

$$0 \le \int_a^b f(x)\, dx.$$

To prove Property 2 of the theorem, note that $f(x) \le g(x)$ implies that $g(x) - f(x) \ge 0$. Hence, you can apply the result of Property 1 to conclude that

$$0 \le \int_a^b [g(x) - f(x)]\, dx$$

$$0 \le \int_a^b g(x)\, dx - \int_a^b f(x)\, dx$$

$$\int_a^b f(x)\, dx \le \int_a^b g(x)\, dx.$$

Properties of the Natural Logarithmic Function (page 315)

$$\lim_{x \to 0^+} \ln x = -\infty \quad \text{and} \quad \lim_{x \to \infty} \ln x = \infty$$

Proof To begin, show that $\ln 2 \ge \frac{1}{2}$. From the Mean Value Theorem for Integrals, you can write

$$\ln 2 = \int_1^2 \frac{1}{x}\, dx = (2 - 1)\frac{1}{c} = \frac{1}{c}$$

where c is in $[1, 2]$. This implies that

$$1 \le c \le 2$$

$$1 \ge \frac{1}{c} \ge \frac{1}{2}$$

$$1 \ge \ln 2 \ge \frac{1}{2}.$$

Now, let N be any positive (large) number. Because $\ln x$ is increasing, it follows that if $x > 2^{2N}$, then

$$\ln x > \ln 2^{2N} = 2N \ln 2.$$

However, because $\ln 2 \ge \frac{1}{2}$, it follows that

$$\ln x > 2N \ln 2 \ge 2N\left(\frac{1}{2}\right) = N.$$

This verifies the second limit. To verify the first limit, let $z = 1/x$. Then, $z \to \infty$ as $x \to 0^+$, and you can write

$$\lim_{x \to 0^+} \ln x = \lim_{x \to 0^+} \left(-\ln \frac{1}{x}\right)$$

$$= \lim_{z \to \infty} (-\ln z)$$

$$= -\lim_{z \to \infty} \ln z$$

$$= -\infty$$

> **THEOREM 5.8 Continuity and Differentiability of Inverse Functions**
> **(page 336)**
>
> Let f be a function whose domain is an interval I. If f has an inverse function, then the following statements are true.
>
> 1. If f is continuous on its domain, then f^{-1} is continuous on its domain.
> 2. If f is increasing on its domain, then f^{-1} is increasing on its domain.
> 3. If f is decreasing on its domain, then f^{-1} is decreasing on its domain.
> 4. If f is differentiable at c and $f'(c) \neq 0$, then f^{-1} is differentiable at $f(c)$.

Proof To prove Property 1, first show that if f is continuous on I and has an inverse function, then f is strictly monotonic on I. Suppose that f were not strictly monotonic. Then there would exist numbers x_1, x_2, x_3 in I such that $x_1 < x_2 < x_3$, but $f(x_2)$ is not between $f(x_1)$ and $f(x_3)$. Without loss of generality, assume $f(x_1) < f(x_3) < f(x_2)$. By the Intermediate Value Theorem, there exists a number x_0 between x_1 and x_2 such that $f(x_0) = f(x_3)$. So, f is not one-to-one and cannot have an inverse function. Hence, f must be strictly monotonic.

Because f is continuous, the Intermediate Value Theorem implies that the set of values of f,

$$\{f(x): x \in I\},$$

forms an interval J. Assume that a is an interior point of J. From the previous argument, $f^{-1}(a)$ is an interior point of I. Let $\varepsilon > 0$. There exists $0 < \varepsilon_1 < \varepsilon$ such that

$$I_1 = (f^{-1}(a) - \varepsilon_1, f^{-1}(a) + \varepsilon_1) \subseteq I.$$

Because f is strictly monotonic on I_1, the set of values $\{f(x): x \in I_1\}$ forms an interval $J_1 \subseteq J$. Let $\delta > 0$ such that $(a - \delta, a + \delta) \subseteq J_1$. Finally, if

$$|y - a| < \delta, \text{ then } |f^{-1}(y) - f^{-1}(a)| < \varepsilon_1 < \varepsilon.$$

Hence, f^{-1} is continuous at a. A similar proof can be given if a is an endpoint.

To prove Property 2, let y_1 and y_2 be in the domain of f^{-1}, with $y_1 < y_2$. Then, there exist x_1 and x_2 in the domain of f such that

$$f(x_1) = y_1 < y_2 = f(x_2).$$

Because f is increasing, $f(x_1) < f(x_2)$ holds precisely when $x_1 < x_2$. Therefore,

$$f^{-1}(y_1) = x_1 < x_2 = f^{-1}(y_2),$$

which implies that f^{-1} is increasing. (Property 3 can be proved in a similar way.)

Finally, to prove Property 4, consider the limit

$$(f^{-1})'(a) = \lim_{y \to a} \frac{f^{-1}(y) - f^{-1}(a)}{y - a}$$

where a is in the domain of f^{-1} and $f^{-1}(a) = c$. Because f is differentiable at c, f is continuous at c, and so is f^{-1} at a. So, $y \to a$ implies that $x \to c$, and you have

$$(f^{-1})'(a) = \lim_{x \to c} \frac{x - c}{f(x) - f(c)}$$

$$= \lim_{x \to c} \frac{1}{\left(\dfrac{f(x) - f(c)}{x - c}\right)}$$

$$= \frac{1}{\lim_{x \to c} \dfrac{f(x) - f(c)}{x - c}}$$

$$= \frac{1}{f'(c)}.$$

Hence, $(f^{-1})'(a)$ exists, and f^{-1} is differentiable at $f(c)$.

THEOREM 5.9 The Derivative of an Inverse Function (page 336)

Let f be a function that is differentiable on an interval I. If f has an inverse function g, then g is differentiable at any x for which $f'(g(x)) \neq 0$. Moreover,

$$g'(x) = \frac{1}{f'(g(x))}, \quad f'(g(x)) \neq 0.$$

Proof From the proof of Theorem 5.8, letting $a = x$, you know that g is differentiable. Using the Chain Rule, differentiate both sides of the equation $x = f(g(x))$ to obtain

$$1 = f'(g(x)) \frac{d}{dx}[g(x)].$$

Because $f'(g(x)) \neq 0$, you can divide by this quantity to obtain

$$\frac{d}{dx}[g(x)] = \frac{1}{f'(g(x))}.$$

THEOREM 5.15 A Limit Involving e (page 355)

$$\lim_{x \to \infty} \left(1 + \frac{1}{x}\right)^x = \lim_{x \to \infty} \left(\frac{x + 1}{x}\right)^x = e$$

Proof Let $y = \lim_{x \to \infty} \left(1 + \dfrac{1}{x}\right)^x$. Taking the natural logs of both sides, you have

$$\ln y = \ln\left[\lim_{x \to \infty} \left(1 + \frac{1}{x}\right)^x\right].$$

Because the natural logarithmic function is continuous, you can write

$$\ln y = \lim_{x \to \infty} \left[x \ln\left(1 + \frac{1}{x}\right) \right] = \lim_{x \to \infty} \left\{ \frac{\ln[1 + (1/x)]}{1/x} \right\}.$$

Letting $x = \frac{1}{t}$, you have

$$\ln y = \lim_{t \to 0^+} \frac{\ln(1 + t)}{t}$$

$$= \lim_{t \to 0^+} \frac{\ln(1 + t) - \ln 1}{t}$$

$$= \frac{d}{dx} \ln x \text{ at } x = 1$$

$$= \frac{1}{x} \text{ at } x = 1$$

$$= 1.$$

Finally, because $\ln y = 1$, you know that $y = e$, and you can conclude that

$$\lim_{x \to \infty} \left(1 + \frac{1}{x}\right)^x = e.$$

THEOREM 7.3 The Extended Mean Value Theorem (page 531)

If f and g are differentiable on an open interval (a, b) and continuous on $[a, b]$ such that $g'(x) \neq 0$ for any x in (a, b), then there exists a point c in (a, b) such that

$$\frac{f'(c)}{g'(c)} = \frac{f(b) - f(a)}{g(b) - g(a)}.$$

Proof You can assume that $g(a) \neq g(b)$, because otherwise, by Rolle's Theorem, it would follow that $g'(x) = 0$ for some x in (a, b). Now, define $h(x)$ to be

$$h(x) = f(x) - \left[\frac{f(b) - f(a)}{g(b) - g(a)}\right] g(x).$$

Then

$$h(a) = f(a) - \left[\frac{f(b) - f(a)}{g(b) - g(a)}\right] g(a) = \frac{f(a)g(b) - f(b)g(a)}{g(b) - g(a)}$$

and

$$h(b) = f(b) - \left[\frac{f(b) - f(a)}{g(b) - g(a)}\right] g(b) = \frac{f(a)g(b) - f(b)g(a)}{g(b) - g(a)}$$

and by Rolle's Theorem there exists a point c in (a, b) such that

$$h'(c) = f'(c) - \frac{f(b) - f(a)}{g(b) - g(a)} g'(c) = 0$$

which implies that

$$\frac{f'(c)}{g'(c)} = \frac{f(b) - f(a)}{g(b) - g(a)}.$$

> **THEOREM 7.4 L'Hôpital's Rule (page 531)**
>
> Let f and g be functions that are differentiable on an open interval (a, b) containing c, except possibly at c itself. Assume that $g'(x) \neq 0$ for all x in (a, b), except possibly at c itself. If the limit of $f(x)/g(x)$ as x approaches c produces the indeterminate form $0/0$, then
>
> $$\lim_{x \to c} \frac{f(x)}{g(x)} = \lim_{x \to c} \frac{f'(x)}{g'(x)}$$
>
> provided the limit on the right exists (or is infinite). This result also applies if the limit of $f(x)/g(x)$ as x approaches c produces any one of the indeterminate forms ∞/∞, $(-\infty)/\infty$, $\infty/(-\infty)$, or $(-\infty)/(-\infty)$.

You can use the Extended Mean Value Theorem to prove L'Hôpital's Rule. Of the several different cases of this rule, the proof of only one case is illustrated. The remaining cases where $x \to c^-$ and $x \to c$ are left for you to prove.

Proof Consider the case for which

$$\lim_{x \to c^+} f(x) = 0 \quad \text{and} \quad \lim_{x \to c^+} g(x) = 0.$$

Define the following new functions:

$$F(x) = \begin{cases} f(x), & x \neq c \\ 0, & x = c \end{cases} \quad \text{and} \quad G(x) = \begin{cases} g(x), & x \neq c \\ 0, & x = c \end{cases}.$$

For any x, $c < x < b$, F and G are differentiable on $(c, x]$ and continuous on $[c, x]$. You can apply the Extended Mean Value Theorem to conclude that there exists a number z in (c, x) such that

$$\frac{F'(z)}{G'(z)} = \frac{F(x) - F(c)}{G(x) - G(c)} = \frac{F(x)}{G(x)} = \frac{f'(z)}{g'(z)} = \frac{f(x)}{g(x)}.$$

Finally, by letting x approach c from the right, $x \to c^+$, we have $z \to c^+$ because $c < z < x$, and

$$\lim_{x \to c^+} \frac{f(x)}{g(x)} = \lim_{x \to c^+} \frac{f'(z)}{g'(z)} = \lim_{z \to c^+} \frac{f'(z)}{g'(z)} = \lim_{x \to c^+} \frac{f'(x)}{g'(x)}.$$

> **THEOREM 8.16 Absolute Convergence (page 593)**
>
> If the series $\Sigma |a_n|$ converges, then the series Σa_n also converges.

Proof Because $0 \leq a_n + |a_n| \leq 2|a_n|$ for all n, the series

$$\sum_{n=1}^{\infty} (a_n + |a_n|)$$

converges by comparison with the convergent series

$$\sum_{n=1}^{\infty} 2|a_n|.$$

Furthermore, because $a_n = (a_n + |a_n|) - |a_n|$, you can write

$$\sum_{n=1}^{\infty} a_n = \sum_{n=1}^{\infty} (a_n + |a_n|) - \sum_{n=1}^{\infty} |a_n|$$

where both series on the right converge. Hence it follows that Σa_n converges.

> **THEOREM 8.19** Taylor's Theorem (page 611)
>
> If a function f is differentiable through order $n + 1$ in an interval I containing c, then, for each x in I, there exists z between x and c such that
>
> $$f(x) = f(c) + f'(c)(x - c) + \frac{f''(c)}{2!}(x - c)^2 + \cdots + \frac{f^{(n)}(c)}{n!}(x - c)^n + R_n(x)$$
>
> where
>
> $$R_n(x) = \frac{f^{(n+1)}(z)}{(n + 1)!}(x - c)^{n+1}.$$

Proof To find $R_n(x)$, fix x in I ($x \neq c$) and write

$$R_n(x) = f(x) - P_n(x)$$

where $P_n(x)$ is the nth Taylor polynomial for $f(x)$. Then let g be a function of t defined by

$$g(t) = f(x) - f(t) - f'(t)(x - t) - \cdots - \frac{f^{(n)}(t)}{n!}(x - t)^n - R_n(x)\frac{(x - t)^{n+1}}{(x - c)^{n+1}}.$$

The reason for defining g in this way is that differentiation with respect to t has a telescoping effect. For example, you have

$$\frac{d}{dt}[-f(t) - f'(t)(x - t)] = -f'(t) + f'(t) - f''(t)(x - t)$$
$$= -f''(t)(x - t).$$

The result is that the derivative $g'(t)$ simplifies to

$$g'(t) = -\frac{f^{(n+1)}(t)}{n!}(x - t)^n + (n + 1)R_n(x)\frac{(x - t)^n}{(x - c)^{n+1}}$$

for all t between c and x. Moreover, for a fixed x,

$$g(c) = f(x) - [P_n(x) + R_n(x)] = f(x) - f(x) = 0$$

and

$$g(x) = f(x) - f(x) - 0 - \cdots - 0 = f(x) - f(x) = 0.$$

Therefore, g satisfies the conditions of Rolle's Theorem, and it follows that there is a number z between c and x such that $g'(z) = 0$. Substituting z for t in the equation for $g'(t)$ and then solving for $R_n(x)$, you obtain

$$g'(z) = -\frac{f^{(n+1)}(z)}{n!}(x - z)^n + (n + 1)R_n(x)\frac{(x - z)^n}{(x - c)^{n+1}} = 0$$

$$R_n(x) = \frac{f^{(n+1)}(z)}{(n + 1)!}(x - c)^{n+1}.$$

Finally, because $g(c) = 0$, you have

$$0 = f(x) - f(c) - f'(c)(x - c) - \cdots - \frac{f^{(n)}(c)}{n!}(x - c)^n - R_n(x)$$

$$f(x) = f(c) + f'(c)(x - c) + \cdots + \frac{f^{(n)}(c)}{n!}(x - c)^n + R_n(x).$$

> **THEOREM 8.20 Convergence of a Power Series (page 617)**
>
> For a power series centered at c, precisely one of the following is true.
>
> 1. The series converges only at c.
> 2. There exists a real number $R > 0$ such that the series converges absolutely for $|x - c| < R$, and diverges for $|x - c| > R$.
> 3. The series converges absolutely for all x.
>
> The number R is the **radius of convergence** of the power series. If the series converges only at c, the radius of convergence is $R = 0$, and if the series converges for all x, the radius of convergence is $R = \infty$. The set of all values of x for which the power series converges is the **interval of convergence** of the power series.

Proof In order to simplify the notation, we will prove the theorem for the power series $\Sigma\, a_n x^n$ centered at $x = 0$. The proof for a power series centered at $x = c$ follows easily. A key step in this proof uses the Completeness Property of the set of real numbers: If a nonempty set S of real numbers has an upper bound, then it must have a least upper bound (see page 563).

We must show that if a power series $\Sigma\, a_n x^n$ converges at $x = d$, $d \neq 0$ then it converges for all b satisfying $|b| < |d|$. Because $\Sigma\, a_n x^n$ converges, $\lim_{x \to \infty} a_n d^n = 0$. Hence, there exists $N > 0$ such that $a_n d^n < 1$ for all $n \geq N$. Then for $n \geq N$,

$$|a_n b^n| = \left|a_n b^n \frac{d^n}{d^n}\right| = |a_n d^n| \left|\frac{b^n}{d^n}\right| < \left|\frac{b^n}{d^n}\right|.$$

So, for $|b| < |d|$, $\left|\frac{b}{d}\right| < 1$ which implies that

$$\Sigma \left|\frac{b^n}{d^n}\right|$$

is a convergent geometric series. By the Comparison Test, the series $\Sigma\, a_n b^n$ converges.

Similarly, if the power series $\Sigma\, a_n x^n$ diverges at $x = b$, where $b \neq 0$, then it diverges for all d satisfying $|d| > |b|$. If $\Sigma\, a_n d^n$ converged, then the above argument would imply that $\Sigma\, a_n b^n$ converged as well.

Finally, to prove the theorem, suppose that neither case 1 nor case 3 is true. Then there exist points b and d such that $\Sigma\, a_n x^n$ converges to b and diverges at d. Let $S = \{x: \Sigma\, a_n x^n \text{ converges}\}$. S is nonempty because $b \in S$. If $b \in S$ then $|x| \leq |d|$, which shows that $|d|$ is an upper bound for the nonempty set S. By the Completeness Property, S has a least upper bound, R.

Now, if $|x| > R$, then $x \notin S$ so $\Sigma\, a_n x^n$ diverges. And if $|x| < R$, then $|x|$ is not an upper bound for S, so there exists b in S satisfying $|b| > |x|$. Since $b \in S$, $\Sigma\, a_n b^n$ converges, which implies that $\Sigma\, a_n x^n$ converges.

APPENDIX B   Proofs of Selected Theorems   **A25**

> **THEOREM 9.16   Classification of Conics by Eccentricity (page 702)**
>
> Let F be the fixed point (*focus*) and D be a fixed line (*directrix*) in the plane. Let P be another point in the plane and let e (*eccentricity*) be the ratio of the distance between P and F to the distance between P and D. The collection of all points P with a given eccentricity is a conic.
>
> 1. The conic is an ellipse if $0 < e < 1$.
> 2. The conic is a parabola if $e = 1$.
> 3. The conic is a hyperbola if $e > 1$.

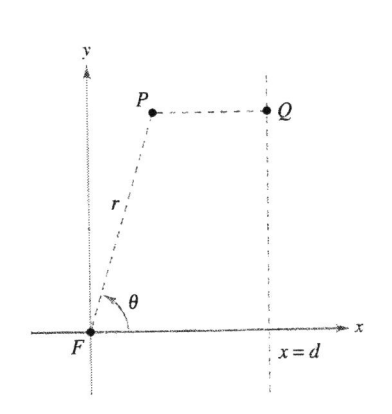

Figure B.1

Proof   If $e = 1$, then, by definition, the conic must be a parabola. If $e \neq 1$, then you can consider the focus F to lie at the origin and the directrix $x = d$ to lie to the right of the origin, as shown in Figure B.1. For the point $P = (r, \theta) = (x, y)$, you have $|PF| = r$ and $|PQ| = d - r\cos\theta$. Given that $e = |PF|/|PQ|$, it follows that

$$|PF| = |PQ|e \implies r = e(d - r\cos\theta).$$

By converting to rectangular coordinates and squaring both sides, you obtain

$$x^2 + y^2 = e^2(d - x)^2 = e^2(d^2 - 2dx + x^2).$$

Completing the square produces

$$\left(x + \frac{e^2 d}{1 - e^2}\right)^2 + \frac{y^2}{1 - e^2} = \frac{e^2 d^2}{(1 - e^2)^2}.$$

If $e < 1$, this equation represents an ellipse. If $e > 1$, then $1 - e^2 < 0$, and the equation represents a hyperbola.   ∎

> **THEOREM 12.4   Sufficient Condition for Differentiability (page 870)**
>
> If f is a function of x and y, where f_x and f_y are continuous in an open region R, then f is differentiable on R.

$\Delta z = f(x + \Delta x, y + \Delta y) - f(x, y)$
Figure B.2

Proof   Let S be the surface defined by $z = f(x, y)$, where $f, f_x,$ and f_y are continuous at (x, y). Let A, B, and C be points on surface S, as shown in Figure B.2. From this figure, you can see that the change in f from point A to point C is given by

$$\begin{aligned}\Delta z &= f(x + \Delta x, y + \Delta y) - f(x, y) \\ &= [f(x + \Delta x, y) - f(x, y)] + [f(x + \Delta x, y + \Delta y) - f(x + \Delta x, y)] \\ &= \Delta z_1 + \Delta z_2.\end{aligned}$$

Between A and B, y is fixed and x changes. Hence, by the Mean Value Theorem, there is a value x_1 between x and $x + \Delta x$ such that

$$\Delta z_1 = f(x + \Delta x, y) - f(x, y) = f_x(x_1, y)\,\Delta x.$$

Similarly, between B and C, x is fixed and y changes, and there is a value y_1 between y and $y + \Delta y$ such that

$$\Delta z_2 = f(x + \Delta x, y + \Delta y) - f(x + \Delta x, y) = f_y(x + \Delta x, y_1)\,\Delta y.$$

By combining these two results, you can write

$$\Delta z = \Delta z_1 + \Delta z_2 = f_x(x_1, y)\Delta x + f_y(x + \Delta x, y_1)\,\Delta y.$$

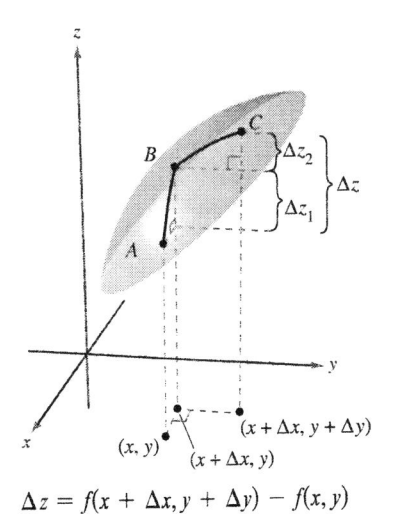

If you define ε_1 and ε_2 as

$$\varepsilon_1 = f_x(x_1, y) - f_x(x, y) \quad \text{and} \quad \varepsilon_2 = f_y(x + \Delta x, y_1) - f_y(x, y)$$

it follows that

$$\Delta z = \Delta z_1 + \Delta z_2 = [\varepsilon_1 + f_x(x, y)]\Delta x + [\varepsilon_2 + f_y(x, y)]\Delta y$$
$$= [f_x(x, y)\Delta x + f_y(x, y)\Delta y] + \varepsilon_1 \Delta x + \varepsilon_2 \Delta y.$$

By the continuity of f_x and f_y and the fact that $x \leq x_1 \leq x + \Delta x$ and $y \leq y_1 \leq y + \Delta y$, it follows that $\varepsilon_1 \to 0$ and $\varepsilon_2 \to 0$ as $\Delta x \to 0$ and $\Delta y \to 0$. Therefore, by definition, f is differentiable.

THEOREM 12.6 Chain Rule: One Independent Variable (page 876)

Let $w = f(x, y)$, where f is a differentiable function of x and y. If $x = g(t)$ and $y = h(t)$, where g and h are differentiable functions of t, then w is a differentiable function of t, and

$$\frac{dw}{dt} = \frac{\partial w}{\partial x}\frac{dx}{dt} + \frac{\partial w}{\partial y}\frac{dy}{dt}.$$

Proof Because g and h are differentiable functions of t, you know that both Δx and Δy approach zero as Δt approaches zero. Moreover, because f is a differentiable function of x and y, you know that

$$\Delta w = \frac{\partial w}{\partial x}\Delta x + \frac{\partial w}{\partial y}\Delta y + \varepsilon_1 \Delta x + \varepsilon_2 \Delta y$$

where both ε_1 and $\varepsilon_2 \to 0$ as $(\Delta x, \Delta y) \to (0, 0)$. So, for $\Delta t \neq 0$, we have

$$\frac{\Delta w}{\Delta t} = \frac{\partial w}{\partial x}\frac{\Delta x}{\Delta t} + \frac{\partial w}{\partial y}\frac{\Delta y}{\Delta t} + \varepsilon_1 \frac{\Delta x}{\Delta t} + \varepsilon_2 \frac{\Delta y}{\Delta t}$$

from which it follows that

$$\frac{dw}{dt} = \lim_{\Delta t \to 0} \frac{\Delta w}{\Delta t} = \frac{\partial w}{\partial x}\frac{dx}{dt} + \frac{\partial w}{\partial y}\frac{dy}{dt} + 0\left(\frac{dx}{dt}\right) + 0\left(\frac{dy}{dt}\right)$$
$$= \frac{\partial w}{\partial x}\frac{dx}{dt} + \frac{\partial w}{\partial y}\frac{dy}{dt}.$$

Integration Tables

Forms Involving u^n

1. $\int u^n \, du = \dfrac{u^{n+1}}{n+1} + C, \ n \neq -1$

2. $\int \dfrac{1}{u} \, du = \ln|u| + C$

Forms Involving $a + bu$

3. $\int \dfrac{u}{a+bu} \, du = \dfrac{1}{b^2}\big(bu - a\ln|a+bu|\big) + C$

4. $\int \dfrac{u}{(a+bu)^2} \, du = \dfrac{1}{b^2}\left(\dfrac{a}{a+bu} + \ln|a+bu|\right) + C$

5. $\int \dfrac{u}{(a+bu)^n} \, du = \dfrac{1}{b^2}\left[\dfrac{-1}{(n-2)(a+bu)^{n-2}} + \dfrac{a}{(n-1)(a+bu)^{n-1}}\right] + C, \ n \neq 1, 2$

6. $\int \dfrac{u^2}{a+bu} \, du = \dfrac{1}{b^3}\left[-\dfrac{bu}{2}(2a - bu) + a^2 \ln|a+bu|\right] + C$

7. $\int \dfrac{u^2}{(a+bu)^2} \, du = \dfrac{1}{b^3}\left(bu - \dfrac{a^2}{a+bu} - 2a\ln|a+bu|\right) + C$

8. $\int \dfrac{u^2}{(a+bu)^3} \, du = \dfrac{1}{b^3}\left[\dfrac{2a}{a+bu} - \dfrac{a^2}{2(a+bu)^2} + \ln|a+bu|\right] + C$

9. $\int \dfrac{u^2}{(a+bu)^n} \, du = \dfrac{1}{b^3}\left[\dfrac{-1}{(n-3)(a+bu)^{n-3}} + \dfrac{2a}{(n-2)(a+bu)^{n-2}} - \dfrac{a^2}{(n-1)(a+bu)^{n-1}}\right] + C, \ n \neq 1, 2, 3$

10. $\int \dfrac{1}{u(a+bu)} \, du = \dfrac{1}{a}\ln\left|\dfrac{u}{a+bu}\right| + C$

11. $\int \dfrac{1}{u(a+bu)^2} \, du = \dfrac{1}{a}\left(\dfrac{1}{a+bu} + \dfrac{1}{a}\ln\left|\dfrac{u}{a+bu}\right|\right) + C$

12. $\int \dfrac{1}{u^2(a+bu)} \, du = -\dfrac{1}{a}\left(\dfrac{1}{u} + \dfrac{b}{a}\ln\left|\dfrac{u}{a+bu}\right|\right) + C$

13. $\int \dfrac{1}{u^2(a+bu)^2} \, du = -\dfrac{1}{a^2}\left[\dfrac{a+2bu}{u(a+bu)} + \dfrac{2b}{a}\ln\left|\dfrac{u}{a+bu}\right|\right] + C$

APPENDIX C   Integration Tables

Forms Involving $a + bu + cu^2$, $b^2 \neq 4ac$

14. $\displaystyle\int \frac{1}{a + bu + cu^2}\, du = \begin{cases} \dfrac{2}{\sqrt{4ac - b^2}} \arctan \dfrac{2cu + b}{\sqrt{4ac - b^2}} + C, & b^2 < 4ac \\[2mm] \dfrac{1}{\sqrt{b^2 - 4ac}} \ln \left| \dfrac{2cu + b - \sqrt{b^2 - 4ac}}{2cu + b + \sqrt{b^2 - 4ac}} \right| + C, & b^2 > 4ac \end{cases}$

15. $\displaystyle\int \frac{u}{a + bu + cu^2}\, du = \frac{1}{2c}\left(\ln|a + bu + cu^2| - b \int \frac{1}{a + bu + cu^2}\, du \right)$

Forms Involving $\sqrt{a + bu}$

16. $\displaystyle\int u^n \sqrt{a + bu}\, du = \frac{2}{b(2n + 3)}\left[u^n(a + bu)^{3/2} - na \int u^{n-1}\sqrt{a + bu}\, du \right]$

17. $\displaystyle\int \frac{1}{u\sqrt{a + bu}}\, du = \begin{cases} \dfrac{1}{\sqrt{a}} \ln \left| \dfrac{\sqrt{a + bu} - \sqrt{a}}{\sqrt{a + bu} + \sqrt{a}} \right| + C, & a > 0 \\[2mm] \dfrac{2}{\sqrt{-a}} \arctan \sqrt{\dfrac{a + bu}{-a}} + C, & a < 0 \end{cases}$

18. $\displaystyle\int \frac{1}{u^n\sqrt{a + bu}}\, du = \frac{-1}{a(n-1)}\left[\frac{\sqrt{a + bu}}{u^{n-1}} + \frac{(2n - 3)b}{2} \int \frac{1}{u^{n-1}\sqrt{a + bu}}\, du \right], \ n \neq 1$

19. $\displaystyle\int \frac{\sqrt{a + bu}}{u}\, du = 2\sqrt{a + bu} + a \int \frac{1}{u\sqrt{a + bu}}\, du$

20. $\displaystyle\int \frac{\sqrt{a + bu}}{u^n}\, du = \frac{-1}{a(n-1)}\left[\frac{(a + bu)^{3/2}}{u^{n-1}} + \frac{(2n - 5)b}{2} \int \frac{\sqrt{a + bu}}{u^{n-1}}\, du \right], \ n \neq 1$

21. $\displaystyle\int \frac{u}{\sqrt{a + bu}}\, du = \frac{-2(2a - bu)}{3b^2}\sqrt{a + bu} + C$

22. $\displaystyle\int \frac{u^n}{\sqrt{a + bu}}\, du = \frac{2}{(2n + 1)b}\left(u^n \sqrt{a + bu} - na \int \frac{u^{n-1}}{\sqrt{a + bu}}\, du \right)$

Forms Involving $a^2 \pm u^2$, $a > 0$

23. $\displaystyle\int \frac{1}{a^2 + u^2}\, du = \frac{1}{a} \arctan \frac{u}{a} + C$

24. $\displaystyle\int \frac{1}{u^2 - a^2}\, du = -\int \frac{1}{a^2 - u^2}\, du = \frac{1}{2a} \ln \left| \frac{u - a}{u + a} \right| + C$

25. $\displaystyle\int \frac{1}{(a^2 \pm u^2)^n}\, du = \frac{1}{2a^2(n - 1)}\left[\frac{u}{(a^2 \pm u^2)^{n-1}} + (2n - 3) \int \frac{1}{(a^2 \pm u^2)^{n-1}}\, du \right], \ n \neq 1$

Forms Involving $\sqrt{u^2 \pm a^2}$, $a > 0$

26. $\displaystyle\int \sqrt{u^2 \pm a^2}\, du = \frac{1}{2}\left(u\sqrt{u^2 \pm a^2} \pm a^2 \ln|u + \sqrt{u^2 \pm a^2}| \right) + C$

27. $\displaystyle\int u^2\sqrt{u^2 \pm a^2}\, du = \frac{1}{8}\left[u(2u^2 \pm a^2)\sqrt{u^2 \pm a^2} - a^4 \ln|u + \sqrt{u^2 \pm a^2}| \right] + C$

28. $\displaystyle\int \frac{\sqrt{u^2 + a^2}}{u}\, du = \sqrt{u^2 + a^2} - a \ln \left| \frac{a + \sqrt{u^2 + a^2}}{u} \right| + C$

29. $\int \dfrac{\sqrt{u^2 - a^2}}{u}\,du = \sqrt{u^2 - a^2} - a\,\text{arcsec}\,\dfrac{|u|}{a} + C$

30. $\int \dfrac{\sqrt{u^2 \pm a^2}}{u^2}\,du = \dfrac{-\sqrt{u^2 \pm a^2}}{u} + \ln|u + \sqrt{u^2 \pm a^2}| + C$

31. $\int \dfrac{1}{\sqrt{u^2 \pm a^2}}\,du = \ln|u + \sqrt{u^2 \pm a^2}| + C$

32. $\int \dfrac{1}{u\sqrt{u^2 + a^2}}\,du = \dfrac{-1}{a} \ln\left|\dfrac{a + \sqrt{u^2 + a^2}}{u}\right| + C$

33. $\int \dfrac{1}{u\sqrt{u^2 - a^2}}\,du = \dfrac{1}{a}\,\text{arcsec}\,\dfrac{|u|}{a} + C$

34. $\int \dfrac{u^2}{\sqrt{u^2 \pm a^2}}\,du = \dfrac{1}{2}\left(u\sqrt{u^2 \pm a^2} \mp a^2 \ln|u + \sqrt{u^2 \pm a^2}|\right) + C$

35. $\int \dfrac{1}{u^2\sqrt{u^2 \pm a^2}}\,du = \mp \dfrac{\sqrt{u^2 \pm a^2}}{a^2 u} + C$

36. $\int \dfrac{1}{(u^2 \pm a^2)^{3/2}}\,du = \dfrac{\pm u}{a^2\sqrt{u^2 \pm a^2}} + C$

Forms Involving $\sqrt{a^2 - u^2},\ a > 0$

37. $\int \sqrt{a^2 - u^2}\,du = \dfrac{1}{2}\left(u\sqrt{a^2 - u^2} + a^2 \arcsin \dfrac{u}{a}\right) + C$

38. $\int u^2\sqrt{a^2 - u^2}\,du = \dfrac{1}{8}\left[u(2u^2 - a^2)\sqrt{a^2 - u^2} + a^4 \arcsin \dfrac{u}{a}\right] + C$

39. $\int \dfrac{\sqrt{a^2 - u^2}}{u}\,du = \sqrt{a^2 - u^2} - a \ln\left|\dfrac{a + \sqrt{a^2 - u^2}}{u}\right| + C$

40. $\int \dfrac{\sqrt{a^2 - u^2}}{u^2}\,du = \dfrac{-\sqrt{a^2 - u^2}}{u} - \arcsin \dfrac{u}{a} + C$

41. $\int \dfrac{1}{\sqrt{a^2 - u^2}}\,du = \arcsin \dfrac{u}{a} + C$

42. $\int \dfrac{1}{u\sqrt{a^2 - u^2}}\,du = \dfrac{-1}{a} \ln\left|\dfrac{a + \sqrt{a^2 - u^2}}{u}\right| + C$

43. $\int \dfrac{u^2}{\sqrt{a^2 - u^2}}\,du = \dfrac{1}{2}\left(-u\sqrt{a^2 - u^2} + a^2 \arcsin \dfrac{u}{a}\right) + C$

44. $\int \dfrac{1}{u^2\sqrt{a^2 - u^2}}\,du = \dfrac{-\sqrt{a^2 - u^2}}{a^2 u} + C$

45. $\int \dfrac{1}{(a^2 - u^2)^{3/2}}\,du = \dfrac{u}{a^2\sqrt{a^2 - u^2}} + C$

Forms Involving sin u or cos u

46. $\displaystyle\int \sin u \, du = -\cos u + C$

47. $\displaystyle\int \cos u \, du = \sin u + C$

48. $\displaystyle\int \sin^2 u \, du = \frac{1}{2}(u - \sin u \cos u) + C$

49. $\displaystyle\int \cos^2 u \, du = \frac{1}{2}(u + \sin u \cos u) + C$

50. $\displaystyle\int \sin^n u \, du = -\frac{\sin^{n-1} u \cos u}{n} + \frac{n-1}{n}\int \sin^{n-2} u \, du$

51. $\displaystyle\int \cos^n u \, du = \frac{\cos^{n-1} u \sin u}{n} + \frac{n-1}{n}\int \cos^{n-2} u \, du$

52. $\displaystyle\int u \sin u \, du = \sin u - u \cos u + C$

53. $\displaystyle\int u \cos u \, du = \cos u + u \sin u + C$

54. $\displaystyle\int u^n \sin u \, du = -u^n \cos u + n\int u^{n-1} \cos u \, du$

55. $\displaystyle\int u^n \cos u \, du = u^n \sin u - n\int u^{n-1} \sin u \, du$

56. $\displaystyle\int \frac{1}{1 \pm \sin u} \, du = \tan u \mp \sec u + C$

57. $\displaystyle\int \frac{1}{1 \pm \cos u} \, du = -\cot u \pm \csc u + C$

58. $\displaystyle\int \frac{1}{\sin u \cos u} \, du = \ln|\tan u| + C$

Forms Involving tan u, cot u, sec u, csc u

59. $\displaystyle\int \tan u \, du = -\ln|\cos u| + C$

60. $\displaystyle\int \cot u \, du = \ln|\sin u| + C$

61. $\displaystyle\int \sec u \, du = \ln|\sec u + \tan u| + C$

62. $\displaystyle\int \csc u \, du = \ln|\csc u - \cot u| + C$

63. $\int \tan^2 u \, du = -u + \tan u + C$

64. $\int \cot^2 u \, du = -u - \cot u + C$

65. $\int \sec^2 u \, du = \tan u + C$

66. $\int \csc^2 u \, du = -\cot u + C$

67. $\int \tan^n u \, du = \dfrac{\tan^{n-1} u}{n-1} - \int \tan^{n-2} u \, du, \; n \neq 1$

68. $\int \cot^n u \, du = -\dfrac{\cot^{n-1} u}{n-1} - \int (\cot^{n-2} u) \, du, \; n \neq 1$

69. $\int \sec^n u \, du = \dfrac{\sec^{n-2} u \tan u}{n-1} + \dfrac{n-2}{n-1} \int \sec^{n-2} u \, du, \; n \neq 1$

70. $\int \csc^n u \, du = -\dfrac{\csc^{n-2} u \cot u}{n-1} + \dfrac{n-2}{n-1} \int \csc^{n-2} u \, du, \; n \neq 1$

71. $\int \dfrac{1}{1 \pm \tan u} \, du = \dfrac{1}{2}(u \pm \ln|\cos u \pm \sin u|) + C$

72. $\int \dfrac{1}{1 \pm \cot u} \, du = \dfrac{1}{2}(u \mp \ln|\sin u \pm \cos u|) + C$

73. $\int \dfrac{1}{1 \pm \sec u} \, du = u + \cot u \mp \csc u + C$

74. $\int \dfrac{1}{1 \pm \csc u} \, du = u - \tan u \pm \sec u + C$

Forms Involving Inverse Trigonometric Functions

75. $\int \arcsin u \, du = u \arcsin u + \sqrt{1 - u^2} + C$

76. $\int \arccos u \, du = u \arccos u - \sqrt{1 - u^2} + C$

77. $\int \arctan u \, du = u \arctan u - \ln\sqrt{1 + u^2} + C$

78. $\int \text{arccot}\, u \, du = u \, \text{arccot}\, u + \ln\sqrt{1 + u^2} + C$

79. $\int \text{arcsec}\, u \, du = u \, \text{arcsec}\, u - \ln\left|u + \sqrt{u^2 - 1}\right| + C$

80. $\int \text{arccsc}\, u \, du = u \, \text{arccsc}\, u + \ln\left|u + \sqrt{u^2 - 1}\right| + C$

Forms Involving e^u

81. $\displaystyle\int e^u \, du = e^u + C$

82. $\displaystyle\int u e^u \, du = (u - 1)e^u + C$

83. $\displaystyle\int u^n e^u \, du = u^n e^u - n \int u^{n-1} e^u \, du$

84. $\displaystyle\int \frac{1}{1 + e^u} \, du = u - \ln(1 + e^u) + C$

85. $\displaystyle\int e^{au} \sin bu \, du = \frac{e^{au}}{a^2 + b^2}(a \sin bu - b \cos bu) + C$

86. $\displaystyle\int e^{au} \cos bu \, du = \frac{e^{au}}{a^2 + b^2}(a \cos bu + b \sin bu) + C$

Forms Involving $\ln u$

87. $\displaystyle\int \ln u \, du = u(-1 + \ln u) + C$

88. $\displaystyle\int u \ln u \, du = \frac{u^2}{4}(-1 + 2 \ln u) + C$

89. $\displaystyle\int u^n \ln u \, du = \frac{u^{n+1}}{(n+1)^2}[-1 + (n+1) \ln u] + C, \ n \neq -1$

90. $\displaystyle\int (\ln u)^2 \, du = u\left[2 - 2 \ln u + (\ln u)^2\right] + C$

91. $\displaystyle\int (\ln u)^n \, du = u(\ln u)^n - n \int (\ln u)^{n-1} \, du$

A114   Answers to Odd-Numbered Exercises

Chapter 9
Section 9.1 (page 660)

1. h 2. a 3. e 4. b
5. f 6. g 7. c 8. d

9. Vertex: $(0, 0)$
   Focus: $\left(-\frac{3}{2}, 0\right)$
   Directrix: $x = \frac{3}{2}$

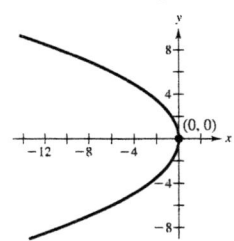

11. Vertex: $(-3, 2)$
    Focus: $\left(-\frac{13}{4}, 2\right)$
    Directrix: $x = -\frac{11}{4}$

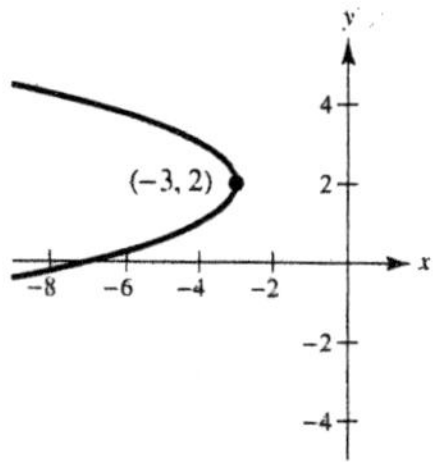

13. Vertex: $(-1, 2)$
    Focus: $(0, 2)$
    Directrix: $x = -2$

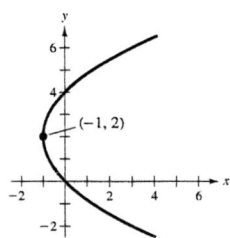

15. Vertex: $(-2, 2)$
    Focus: $(-2, 1)$
    Directrix: $y = 3$

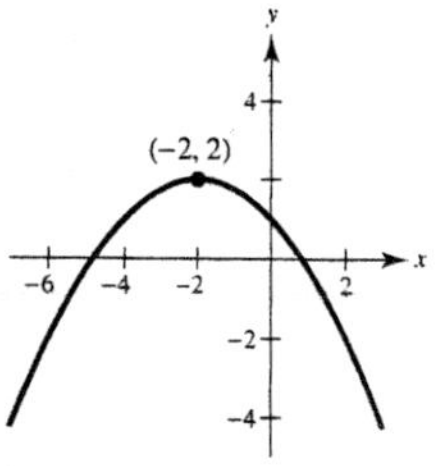

17. Vertex: $\left(\frac{1}{4}, -\frac{1}{2}\right)$
    Focus: $\left(0, -\frac{1}{2}\right)$
    Directrix: $x = \frac{1}{2}$

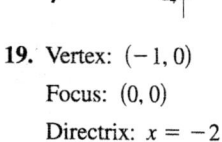

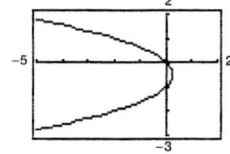

19. Vertex: $(-1, 0)$
    Focus: $(0, 0)$
    Directrix: $x = -2$

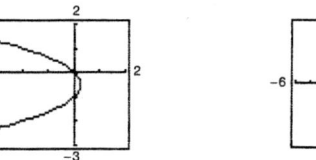

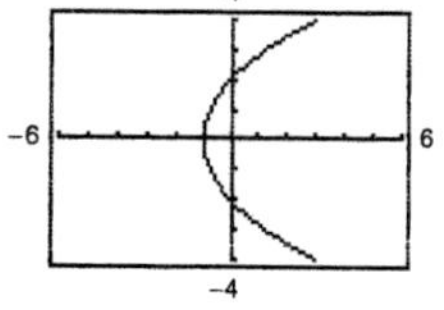

21. $y^2 - 4y + 8x - 20 = 0$
23. $x^2 - 24y + 96 = 0$
25. $x^2 + y - 4 = 0$
27. $5x^2 - 14x - 3y + 9 = 0$

29. Center: $(0, 0)$
    Foci: $(\pm\sqrt{3}, 0)$
    Vertices: $(\pm 2, 0)$
    $e = \dfrac{\sqrt{3}}{2}$

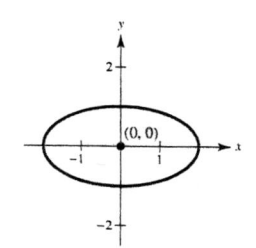

31. Center: $(1, 5)$
    Foci: $(1, 9), (1, 1)$
    Vertices: $(1, 10), (1, 0)$
    $e = \dfrac{4}{5}$

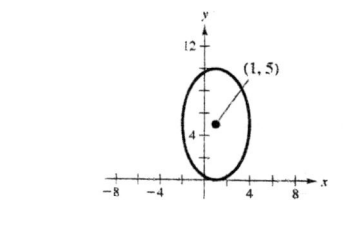

33. Center: $(-2, 3)$
    Foci: $(-2, 3 \pm \sqrt{5})$
    Vertices: $(-2, 6), (-2, 0)$
    $e = \dfrac{\sqrt{5}}{3}$

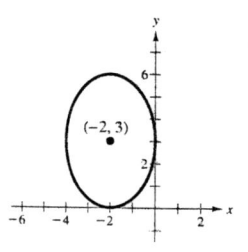

35. Center: $\left(\dfrac{1}{2}, -1\right)$
    Foci: $\left(\dfrac{1}{2} \pm \sqrt{2}, -1\right)$
    Vertices: $\left(\dfrac{1}{2} \pm \sqrt{5}, -1\right)$

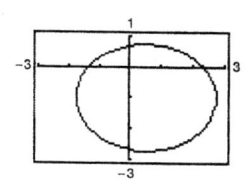

    To obtain the graph, solve for y and get
    $y_1 = -1 + \sqrt{\dfrac{57 + 12x - 12x^2}{20}}$ and
    $y_2 = -1 - \sqrt{\dfrac{57 + 12x - 12x^2}{20}}$.
    Graph these equations in the same viewing window.

37. Center: $\left(\dfrac{3}{2}, -1\right)$
    Foci: $\left(\dfrac{3}{2} - \sqrt{2}, -1\right), \left(\dfrac{3}{2} + \sqrt{2}, -1\right)$
    Vertices: $\left(-\dfrac{1}{2}, -1\right), \left(\dfrac{7}{2}, -1\right)$

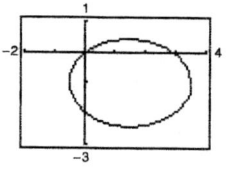

    To obtain the graph, solve for y and get
    $y_1 = -1 + \sqrt{\dfrac{7 + 12x - 4x^2}{8}}$ and
    $y_2 = -1 - \sqrt{\dfrac{7 + 12x - 4x^2}{8}}$.
    Graph these equations in the same viewing window.

39. $\dfrac{x^2}{9} + \dfrac{y^2}{5} = 1$ **41.** $\dfrac{(x-3)^2}{9} + \dfrac{(y-5)^2}{16} = 1$

43. $\dfrac{x^2}{16} + \dfrac{7y^2}{16} = 1$

45. Center: $(0, 0)$
Foci: $(0, \pm\sqrt{5})$
Vertices: $(0, \pm 1)$

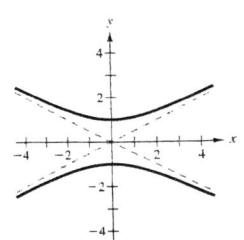

47. Center: $(1, -2)$
Foci: $(1 \pm \sqrt{5}, -2)$
Vertices: $(-1, -2), (3, -2)$

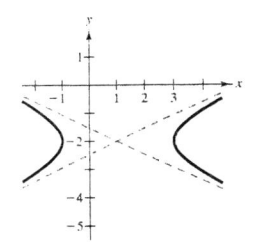

49. Center: $(2, -3)$
Foci: $(2 \pm \sqrt{10}, -3)$
Vertices: $(1, -3), (3, -3)$

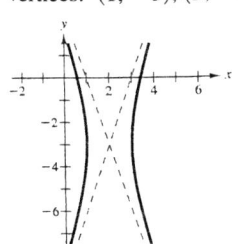

51. Degenerate hyperbola
Graph is two lines
$y = -3 \pm \frac{1}{3}(x+1)$
intersecting at $(-1, -3)$.

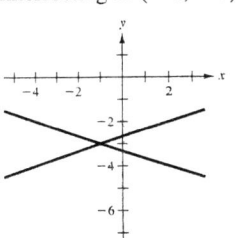

53. Center: $(1, -3)$
Foci: $(1, -3 \pm 2\sqrt{5})$
Vertices: $(1, -3 \pm \sqrt{2})$

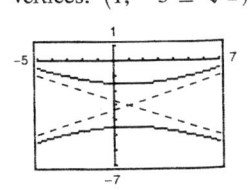

55. Center: $(1, -3)$
Foci: $(1 \pm \sqrt{10}, -3)$
Vertices: $(-1, -3), (3, -3)$

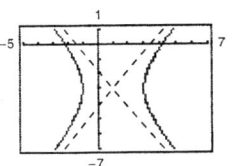

57. $\dfrac{x^2}{1} - \dfrac{y^2}{9} = 1$ **59.** $\dfrac{y^2}{9} - \dfrac{(x-2)^2}{9/4} = 1$

61. $\dfrac{y^2}{4} - \dfrac{x^2}{12} = 1$ **63.** $\dfrac{(x-3)^2}{9} - \dfrac{(y-2)^2}{4} = 1$

65. (a) $(6, \sqrt{3})$: $2x - 3\sqrt{3}y - 3 = 0$
$(6, -\sqrt{3})$: $2x + 3\sqrt{3}y - 3 = 0$
(b) $(6, \sqrt{3})$: $9x + 2\sqrt{3}y - 60 = 0$
$(6, -\sqrt{3})$: $9x - 2\sqrt{3}y - 60 = 0$

67. Ellipse   **69.** Parabola   **71.** Circle
73. Circle   **75.** Hyperbola

77. (a) A parabola is the set of all points (x, y) that are equidistant from a fixed line and a fixed point not on the line.
(b) For directrix $y = k - p$: $(x - h)^2 = 4p(y - k)$
For directrix $x = h - p$: $(y - k)^2 = 4p(x - h)$
(c) If P is a point on a parabola, then the tangent line to the parabola at P makes equal angles with the line passing through P and the focus, and with the line passing through P parallel to the axis of the parabola.

79. (a) A hyperbola is the set of all points (x, y) for which the absolute value of the difference between the distances from two distinct fixed points is constant.
(b) Transverse axis is horizontal: $\dfrac{(x-h)^2}{a^2} - \dfrac{(y-k)^2}{b^2} = 1$
Transverse axis is vertical: $\dfrac{(y-k)^2}{a^2} - \dfrac{(x-h)^2}{b^2} = 1$
(c) Transverse axis is horizontal:
$y = k + \dfrac{b}{a}(x-h)$ and $y = k - \dfrac{b}{a}(x-h)$
Transverse axis is vertical:
$y = k + \dfrac{a}{b}(x-h)$ and $y = k - \dfrac{a}{b}(x-h)$

81. $\frac{9}{4}$ meters   **83.** $y = 2ax_0 x - ax_0^2$

85. (a) Proof   (b) Proof

87. $x_0 = \dfrac{2\sqrt{3}}{3}$; Distance from hill: $\dfrac{2\sqrt{3}}{3} - 1$

89. $\dfrac{16(4 + 3\sqrt{3} - 2\pi)}{3} \approx 15.536$ square feet

91. (a) $y = \dfrac{1}{180}x^2$
(b) $10\left[2\sqrt{13} + 9\ln\left(\dfrac{2 + \sqrt{13}}{3}\right)\right] \approx 128.4$ meters

93.   **95.**

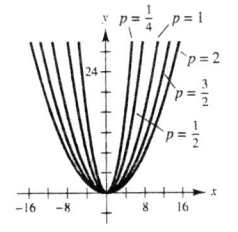

As p increases, the graph of $x^2 = 4py$ gets wider.

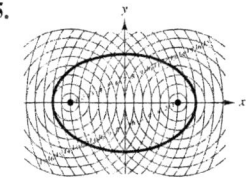

97. The tacks should be placed 1.5 feet from the center. The string should be $2a = 5$ feet long.

99.
$e = \dfrac{c}{a}$
$A + P = 2a$
$a = \dfrac{A+P}{2}$
$c = a - P = \dfrac{A+P}{2} - P = \dfrac{A-P}{2}$
$e = \dfrac{c}{a} = \dfrac{\frac{(A-P)}{2}}{\frac{(A+P)}{2}} = \dfrac{A-P}{A+P}$

101. $e \approx 0.9672$ **103.** $\left(0, \frac{25}{3}\right)$

105. Minor-axis endpoints: $(-6, -2), (0, -2)$
Major-axis endpoints: $(-3, -6), (-3, 2)$

107. (a) Area = 2π
(b) Volume = $\dfrac{8\pi}{3}$
Surface area = $\dfrac{2\pi(9 + 4\sqrt{3}\pi)}{9} \approx 21.48$
(c) Volume = $\dfrac{16\pi}{3}$
Surface area = $\dfrac{4\pi[6 + \sqrt{3}\ln(2 + \sqrt{3})]}{3} \approx 34.69$

109. 37.96   **111.** 40   **113.** $\dfrac{(x-6)^2}{9} - \dfrac{(y-2)^2}{7} = 1$

115.

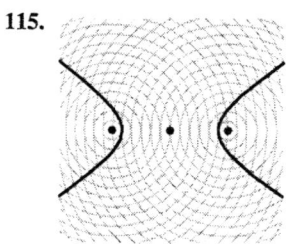

117. Proof

119. $x = \dfrac{-90 + 96\sqrt{2}}{7} \approx 6.538$

$y = \dfrac{160 - 96\sqrt{2}}{7} \approx 3.462$

121. There are four points of intersection.
At $\left(\dfrac{\sqrt{2}ac}{\sqrt{2a^2 - b^2}}, \dfrac{b^2}{\sqrt{2}\sqrt{2a^2 - b^2}}\right)$, the slopes of the tangent lines are $y'_e = -\dfrac{c}{a}$ and $y'_h = \dfrac{a}{c}$.

Since the slopes are negative reciprocals, the tangent lines are perpendicular. Similarly, the curves are perpendicular at the other three points of intersection.

123. False. See the definition of a parabola.   **125.** True
127. False. $y^2 - x^2 + 2x + 2y = 0$ yields two intersecting lines.
129. True

Section 9.2 (page 672)

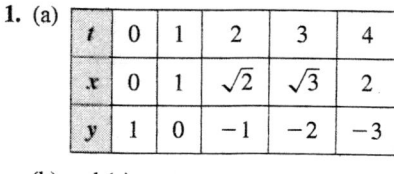

1. (a)

t	0	1	2	3	4
x	0	1	$\sqrt{2}$	$\sqrt{3}$	2
y	1	0	-1	-2	-3

(b) and (c)

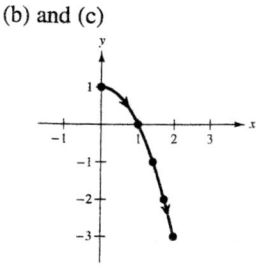

(d) $y = 1 - x^2,\ x \geq 0$

3. $2x - 3y + 5 = 0$

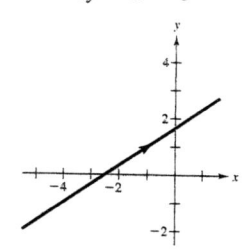

5. $y = (x - 1)^2$

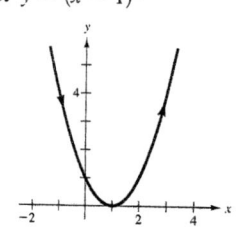

7. $y = \tfrac{1}{2}x^{2/3}$

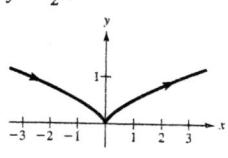

9. $y = x^2 - 2,\ x \geq 0$

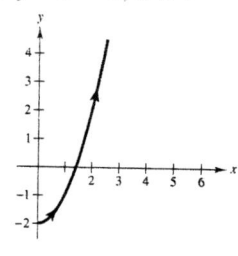

11. $y = \dfrac{x + 1}{x}$

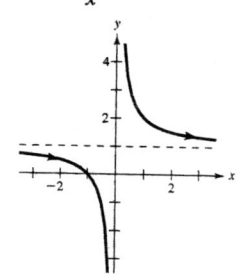

13. $y = \dfrac{|x - 4|}{2}$

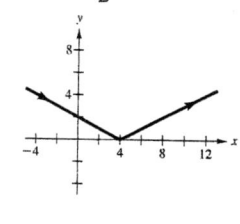

15. $y = x^3 + 1,\ x > 0$

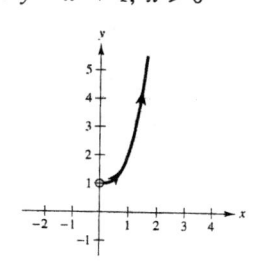

17. $y = \dfrac{1}{x},\ |x| \geq 1$

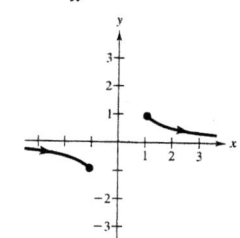

19. $x^2 + y^2 = 9$

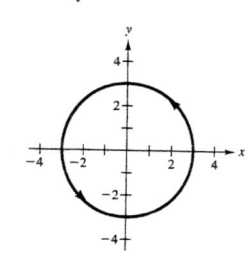

21. $\dfrac{x^2}{16} + \dfrac{y^2}{4} = 1$

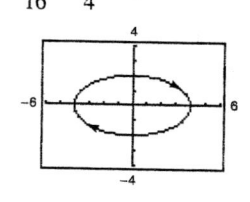

23. $\dfrac{(x - 4)^2}{4} + \dfrac{(y + 1)^2}{1} = 1$

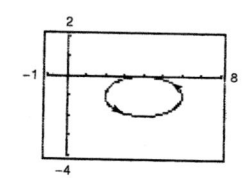

25. $\dfrac{(x-4)^2}{4} + \dfrac{(y+1)^2}{16} = 1$ **27.** $\dfrac{x^2}{16} - \dfrac{y^2}{9} = 1$

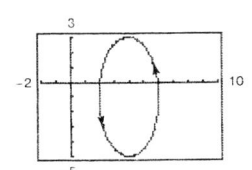

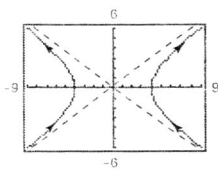

29. $y = \ln x$ 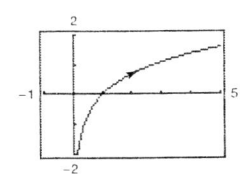 **31.** $y = \dfrac{1}{x^3}$, $x > 0$

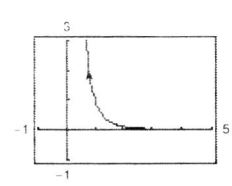

33. Each curve represents a portion of the line $y = 2x + 1$.

	Domain	Orientation	Smooth
(a)	$-\infty < x < \infty$	Up	Yes
(b)	$-1 \le x \le 1$	Oscillates	No, $\dfrac{dx}{d\theta} = \dfrac{dy}{d\theta} = 0$ when $\theta = 0, \pm\pi, \pm 2\pi, \ldots$
(c)	$0 < x < \infty$	Down	Yes
(d)	$0 < x < \infty$	Up	Yes

35. (a) and (b) represent the parabola $y = 2(1 - x^2)$ for $-1 \le x \le 1$. The curve is smooth. The orientation is from right to left in part (a) and in part (b).

37. (a)

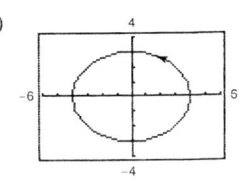

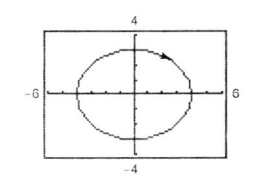

(b) The orientation is reversed.
(c) The orientation is reversed.
(d) Answers will vary. For example,
$x = 2 \sec t \quad x = 2 \sec(-t)$
$y = 5 \sin t \quad y = 5 \sin(-t)$
have the same graphs, but their orientation is reversed.

39. $y - y_1 = \dfrac{y_2 - y_1}{x_2 - x_1}(x - x_1)$ **41.** $\dfrac{(x-h)^2}{a^2} + \dfrac{(y-k)^2}{b^2} = 1$

43. $x = 5t$
$y = -2t$
(Solution is not unique.)

45. $x = 2 + 4\cos\theta$
$y = 1 + 4\sin\theta$
(Solution is not unique.)

47. $x = 5\cos\theta$
$y = 3\sin\theta$
(Solution is not unique.)

49. $x = 4\sec\theta$
$y = 3\tan\theta$
(Solution is not unique.)

51. $x = t$
$y = 3t - 2$;
$x = t - 3$
$y = 3t - 11$
(Solution is not unique.)

53. $x = t$
$y = t^3$;
$x = \tan t$
$y = \tan^3 t$
(Solution is not unique.)

55.

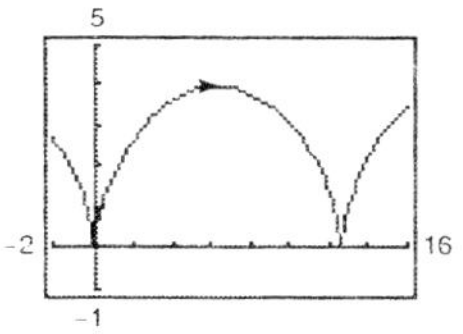

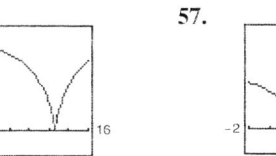

57.

Not smooth when $\theta = 2n\pi$

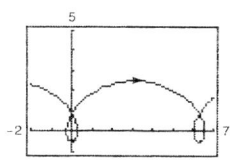

59.

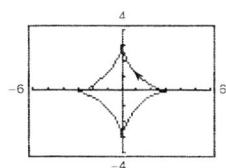

61.

Not smooth when $\theta = \tfrac{1}{2}n\pi$

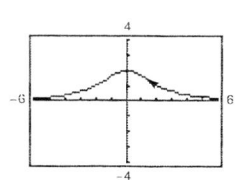

63. See page 665. **65.** See page 670.
67. d **68.** a **69.** b **70.** c
71. $x = a\theta - b\sin\theta$; $y = a - b\cos\theta$
73. False. The graph of the parametric equations is the portion of the line $y = x$ when $x \ge 0$.
75. (a) $x = \left(\dfrac{440}{3}\cos\theta\right)t$; $y = 3 + \left(\dfrac{440}{3}\sin\theta\right)t - 16t^2$

(b)

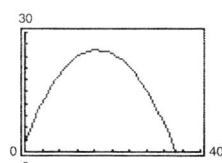

Not a home run

(c)

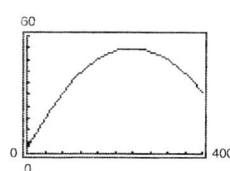

Home run

(d) $19.4°$

Section 9.3 (page 681)

1. $-\dfrac{2}{t}$ **3.** -1

5. $\dfrac{dy}{dx} = \dfrac{3}{2}$, $\dfrac{d^2y}{dx^2} = 0$; neither concave upward nor concave downward

7. $\dfrac{dy}{dx} = 2t + 3$, $\dfrac{d^2y}{dx^2} = 2$

At $t = -1$, $\dfrac{dy}{dx} = 1$, $\dfrac{d^2y}{dx^2} = 2$; concave upward

9. $\dfrac{dy}{dx} = -\cot\theta$, $\dfrac{d^2y}{dx^2} = -\dfrac{\csc^3\theta}{2}$

At $\theta = \dfrac{\pi}{4}$, $\dfrac{dy}{dx} = -1$, $\dfrac{d^2y}{dx^2} = -\sqrt{2}$; concave downward

11. $\dfrac{dy}{dx} = 2\csc\theta$, $\dfrac{d^2y}{dx^2} = -2\cot^3\theta$

At $\theta = \dfrac{\pi}{6}$, $\dfrac{dy}{dx} = 4$, $\dfrac{d^2y}{dx^2} = -6\sqrt{3}$; concave downward

13. $\dfrac{dy}{dx} = -\tan\theta$, $\dfrac{d^2y}{dx^2} = \dfrac{\sec^4\theta \csc\theta}{3}$

At $\theta = \dfrac{\pi}{4}$, $\dfrac{dy}{dx} = -1$, $\dfrac{d^2y}{dx^2} = \dfrac{4\sqrt{2}}{3}$; concave upward

15. $\left(-\dfrac{2}{\sqrt{3}}, \dfrac{3}{2}\right)$: $3\sqrt{3}x - 8y + 18 = 0$

$(0, 2)$: $y - 2 = 0$

$\left(2\sqrt{3}, \dfrac{1}{2}\right)$: $\sqrt{3}x + 8y - 10 = 0$

17. (a) and (d)

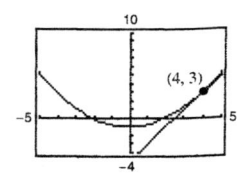

(b) At $t = 2$, $\dfrac{dx}{dt} = 2$, $\dfrac{dy}{dt} = 4$, and $\dfrac{dy}{dx} = 2$.

(c) $y = 2x - 5$

19. (a) and (d)

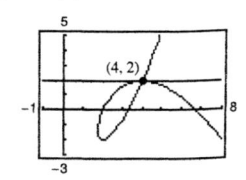

(b) At $t = -1$, $\dfrac{dx}{dt} = -3$, $\dfrac{dy}{dt} = 0$, and $\dfrac{dy}{dx} = 0$.

(c) $y = 2$

21. $y = \pm\dfrac{3}{4}x$

23. Horizontal: $(1, 0), (-1, \pi), (1, -2\pi)$

Vertical: $\left(\dfrac{\pi}{2}, 1\right), \left(-\dfrac{3\pi}{2}, -1\right), \left(\dfrac{5\pi}{2}, 1\right)$

25. Horizontal: $(1, 0)$
Vertical: none

27. Horizontal: $(0, -2), (2, 2)$
Vertical: none

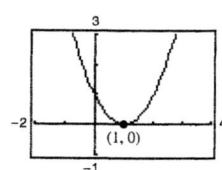

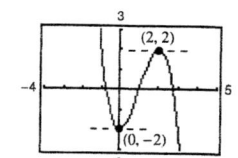

29. Horizontal: $(0, 3), (0, -3)$
Vertical: $(3, 0), (-3, 0)$

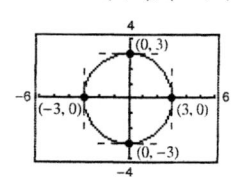

31. Horizontal: $(4, 0), (4, -2)$
Vertical: $(2, -1), (6, -1)$

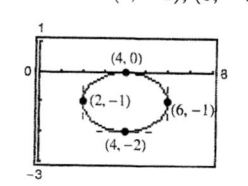

33. Horizontal: none
Vertical: $(1, 0), (-1, 0)$

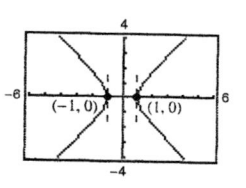

35. $2\sqrt{5} + \ln(2 + \sqrt{5}) \approx 5.916$ **37.** $\sqrt{2}(1 - e^{-\pi/2}) \approx 1.12$

39. $\dfrac{1}{12}\left[\ln(\sqrt{37} + 6) + 6\sqrt{37}\right] \approx 3.249$ **41.** $6a$ **43.** $8a$

45. (a)

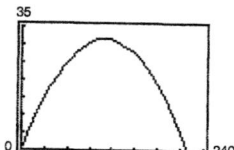

(b) 219.2 feet
(c) 230.8 feet

(d) The range is maximized when $\theta = 45°$; the arc length is maximized when $\theta = 90°$.

47. (a)

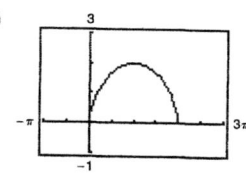

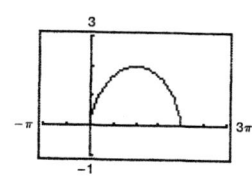

(b) The average speed of the particle on the second path is twice the average speed of the particle on the first path.

(c) 4π

49. (a) $32\pi\sqrt{5}$ (b) $16\pi\sqrt{5}$ **51.** 32π **53.** $\dfrac{12\pi a^2}{5}$

55. See Theorem 9.7, Parametric Form of the Derivative, on page 675.

57. Answers will vary. Example:

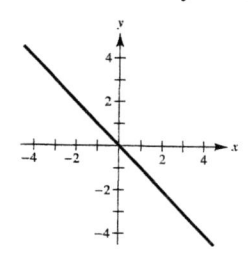

59. See Theorem 9.8, Arc Length in Parametric Form, on page 678.

61. $2\pi r^2(1 - \cos\theta)$ **63.** $\left(\dfrac{3}{4}, \dfrac{8}{5}\right)$ **65.** 36π **67.** $\dfrac{3\pi}{2}$

69. d **70.** b **71.** f **72.** c **73.** a **74.** e

75.

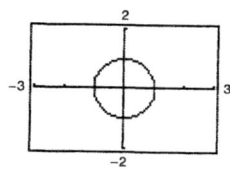

(a) Circle of radius 1 and center at $(0, 0)$ except the point $(-1, 0)$

(b) As t increases from -20 to 0, the speed increases, and as t increases from 0 to 20, the speed decreases.

77. False: $\dfrac{d^2y}{dx^2} = \dfrac{\dfrac{d}{dt}\left[\dfrac{g'(t)}{f'(t)}\right]}{f'(t)} = \dfrac{f'(t)g''(t) - g'(t)f''(t)}{[f'(t)]^3}.$

Section 9.4 (page 691)

1. (0, 4)

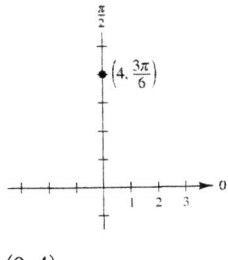

3. $(-2, 2\sqrt{3}) \approx (-2, 3.464)$

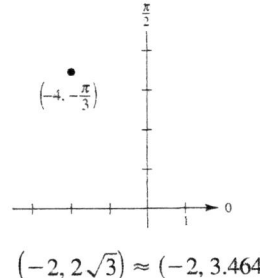

5. (−1.004, 0.996)

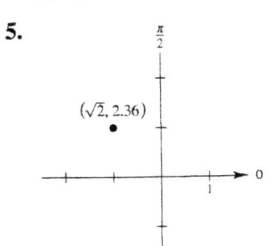

7.

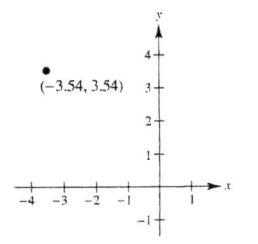

9.

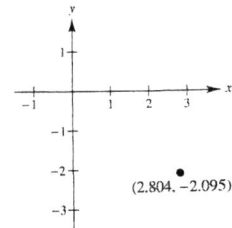

11. $\left(\sqrt{2}, \dfrac{\pi}{4}\right), \left(-\sqrt{2}, \dfrac{5\pi}{4}\right)$

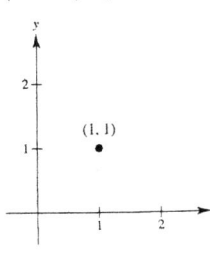

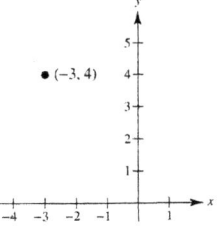

13. (5, 2.214), (−5, 5.356)

15. (3.606, −0.588)

17. (2.833, 0.490)

19. (a) (b)

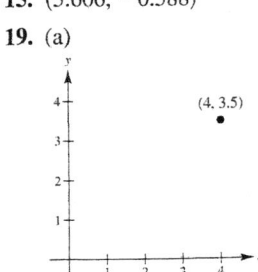

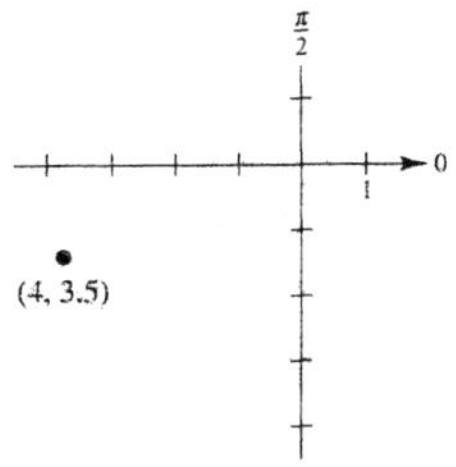

21. $r = a$

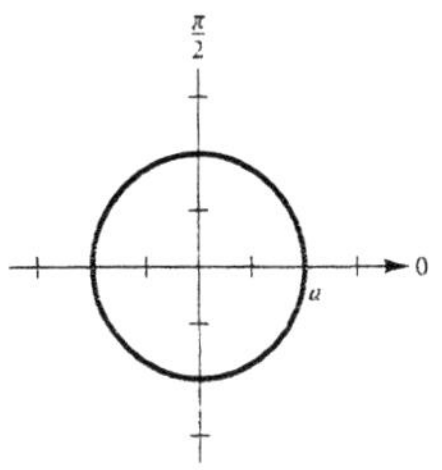

23. $r = 4\csc\theta$

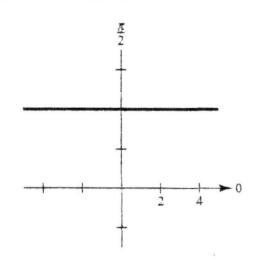

25. $r = \dfrac{-2}{3\cos\theta - \sin\theta}$

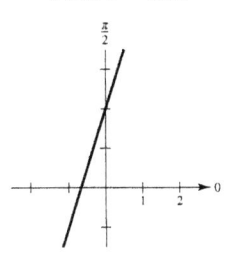

27. $r = 9\csc^2\theta\cos\theta$

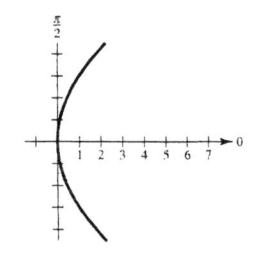

29. $x^2 + y^2 = 9$

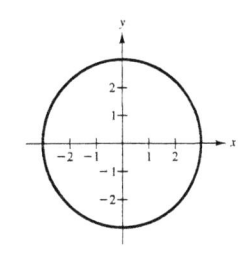

31. $x^2 + y^2 - y = 0$

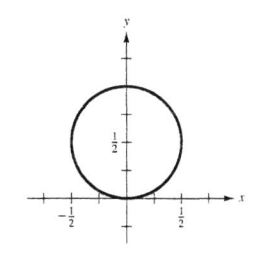

33. $\sqrt{x^2 + y^2} = \arctan\dfrac{y}{x}$

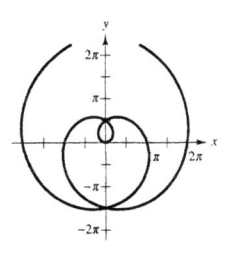

35. $x - 3 = 0$

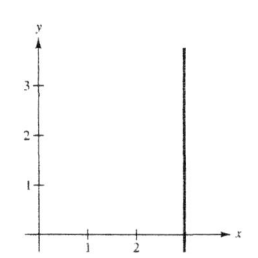

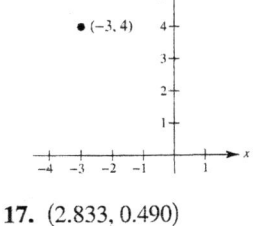

37. $0 \le \theta < 2\pi$

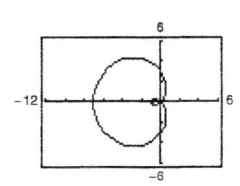

39. $0 \le \theta < 2\pi$

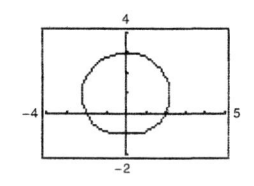

41. $-\pi < \theta < \pi$

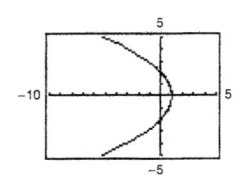

43. $0 \le \theta < 4\pi$

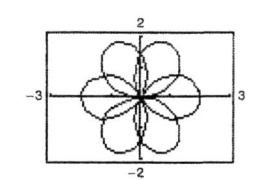

45. $0 \le \theta < \pi/2$

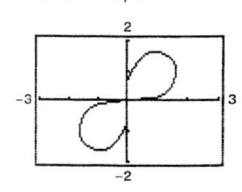

47. $(x - h)^2 + (y - k)^2 = h^2 + k^2$
Center: (h, k)
Radius: $\sqrt{h^2 + k^2}$

49. $2\sqrt{5}$ **51.** 5.6

53. $\dfrac{dy}{dx} = \dfrac{2 \cos \theta (3 \sin \theta + 1)}{6 \cos^2 \theta - 2 \sin \theta - 3}$

$\left(5, \dfrac{\pi}{2}\right): \dfrac{dy}{dx} = 0$

$(2, \pi): \dfrac{dy}{dx} = -\dfrac{2}{3}$

$\left(-1, \dfrac{3\pi}{2}\right): \dfrac{dy}{dx} = 0$

55. (a) and (b)

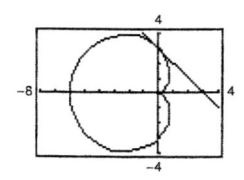

(c) -1

57. (a) and (b)

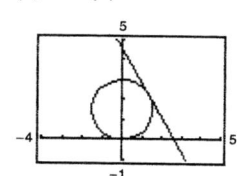

(c) $-\sqrt{3}$

59. Horizontal: $\left(2, \dfrac{3\pi}{2}\right), \left(\dfrac{1}{2}, \dfrac{\pi}{6}\right), \left(\dfrac{1}{2}, \dfrac{5\pi}{6}\right)$

Vertical: $\left(\dfrac{3}{2}, \dfrac{7\pi}{6}\right), \left(\dfrac{3}{2}, \dfrac{11\pi}{6}\right)$

61. $\left(5, \dfrac{\pi}{2}\right), \left(1, \dfrac{3\pi}{2}\right)$

63. $(0, 0), (1.4142, 0.7854),$
$(1.4142, 2.3562)$

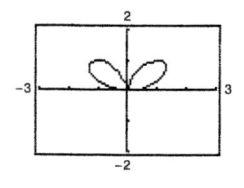

65. $(7, 1.5708), (3, 4.7124)$

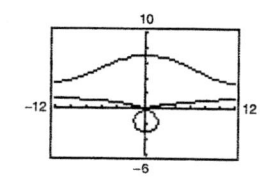

67. $\theta = 0$

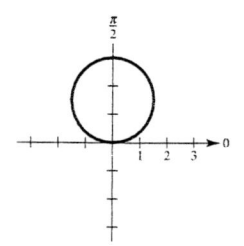

69. $\theta = \dfrac{\pi}{2}$

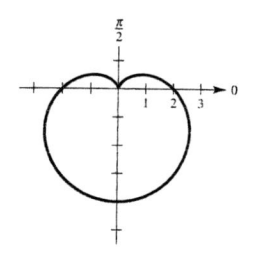

71. $\theta = \dfrac{\pi}{6}, \dfrac{\pi}{2}, \dfrac{5\pi}{6}$

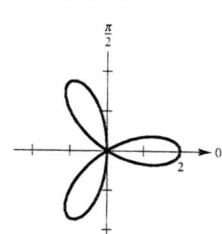

73. $\theta = 0, \dfrac{\pi}{2}$

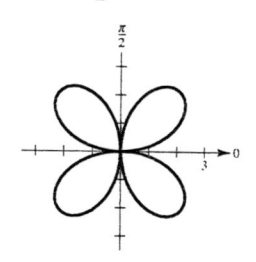

75.

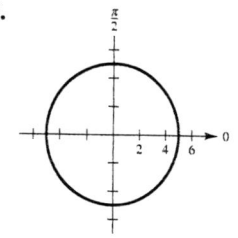

77.

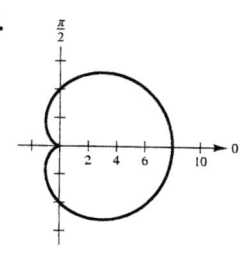

79.

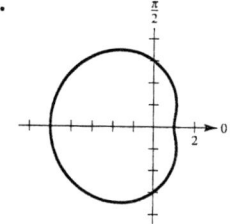

81.

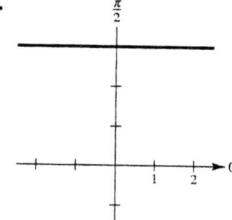

83.

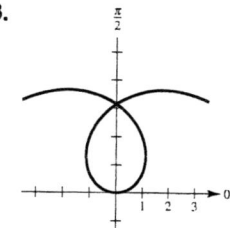

85.

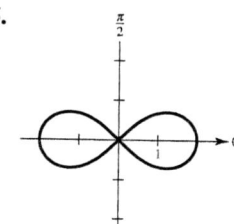

87.

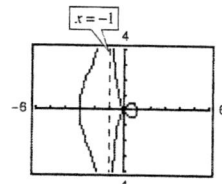

89.

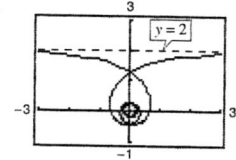

91. The rectangular coordinate system is a collection of points of the form (x, y), where x is the directed distance from the y-axis to the point and y is the directed distance from the x-axis to the point. Every point has a unique representation.

The polar coordinate system is a collection of points of the form (r, θ), where r is the directed distance from the origin O to a point P and θ is the directed angle, measured counterclockwise, from the polar axis to the segment $\overline{OP}$. Polar coordinates do not have unique representations.

93. $r = a$: Circle of radius a centered at the pole
$\theta = b$: Line passing through the pole

95. c **96.** b **97.** a **98.** d

99. (a) [graph] (b) [graph] (c) [graph]

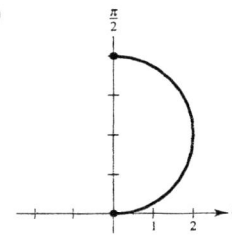

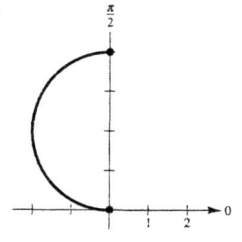

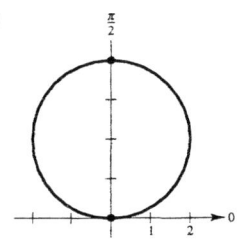

101. Proof

103. (a) $r = 2 - \sin\left(\theta - \dfrac{\pi}{4}\right)$
$= 2 - \dfrac{\sqrt{2}(\sin\theta - \cos\theta)}{2}$ 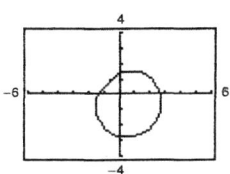
(b) $r = 2 + \cos\theta$ 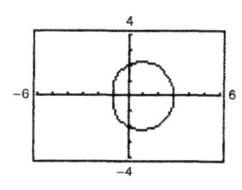
(c) $r = 2 + \sin\theta$ 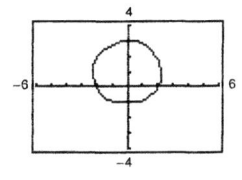
(d) $r = 2 - \cos\theta$

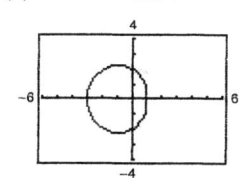

105. (a) [graph] (b) [graph]

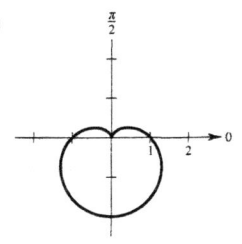

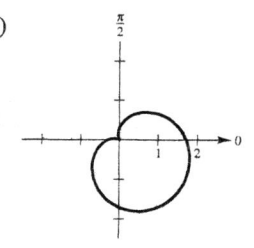

107. $\psi = \dfrac{\pi}{2}$ 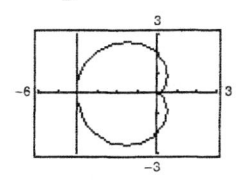 **109.** $\psi = 0$

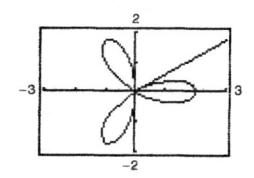

111. $\psi = \dfrac{\pi}{3}, 60°$ 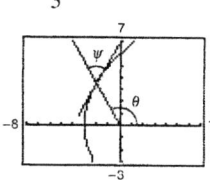 **113.** True **115.** True

Section 9.5 (page 700)

1. 16π **3.** $\dfrac{\pi}{3}$ **5.** $\dfrac{\pi}{8}$ **7.** $\dfrac{3\pi}{2}$

9. $\dfrac{2\pi - 3\sqrt{3}}{2}$ **11.** $\pi + 3\sqrt{3}$

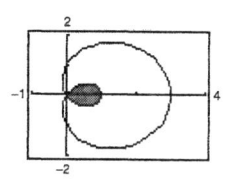

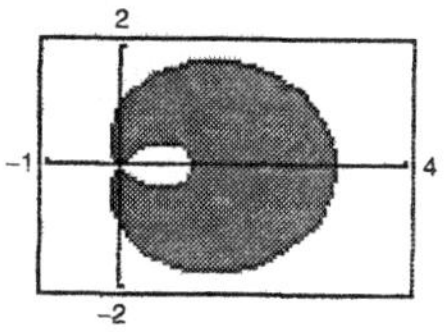

13. $\left(1, \dfrac{\pi}{2}\right), \left(1, \dfrac{3\pi}{2}\right), (0,0)$

15. $\left(\dfrac{2-\sqrt{2}}{2}, \dfrac{3\pi}{4}\right), \left(\dfrac{2+\sqrt{2}}{2}, \dfrac{7\pi}{4}\right), (0,0)$

17. $\left(\dfrac{3}{2}, \dfrac{\pi}{6}\right), \left(\dfrac{3}{2}, \dfrac{5\pi}{6}\right), (0,0)$ **19.** $(2,4), (-2,-4)$

21. $\left(2, \dfrac{\pi}{12}\right), \left(2, \dfrac{5\pi}{12}\right), \left(2, \dfrac{7\pi}{12}\right), \left(2, \dfrac{11\pi}{12}\right)$
$\left(2, \dfrac{13\pi}{12}\right), \left(2, \dfrac{17\pi}{12}\right), \left(2, \dfrac{19\pi}{12}\right), \left(2, \dfrac{23\pi}{12}\right)$

23. $(-0.581, \pm 2.607), (2.581, \pm 1.376)$

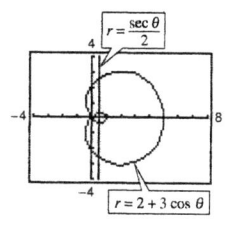

25. $(0,0), (0.935, 0.363), (0.535, -1.006)$
The graphs reach the pole at different times (θ-values).

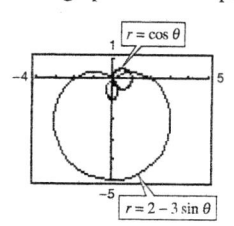

27. $\dfrac{4}{3}(4\pi - 3\sqrt{3})$

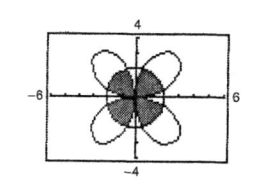

29. $11\pi - 24$

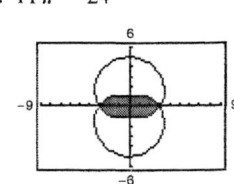

31. $\frac{2}{3}(4\pi - 3\sqrt{3})$

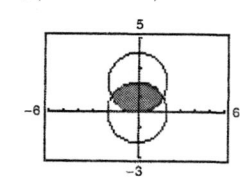

33. $\frac{5\pi a^2}{4}$ **35.** $\frac{a^2}{2}(\pi - 2)$

37. (a) $(x^2 + y^2)^{3/2} = ax^2$

(b)

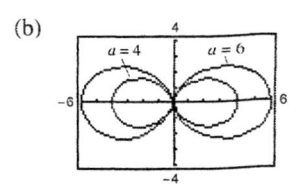

(c) $\frac{15\pi}{2}$

39. The area enclosed by the function is $\frac{\pi a^2}{4}$ if n is odd and is $\frac{\pi a^2}{2}$ if n is even.

41. $2\pi a$ **43.** 8

45.

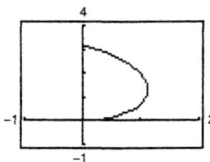

≈ 4.16

47.

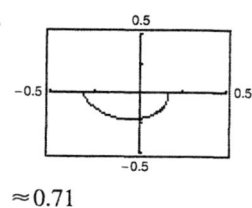

≈ 0.71

49.

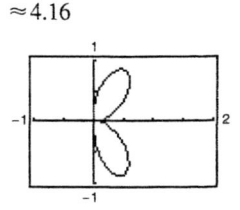

≈ 4.39

51. 36π **53.** $\frac{2\pi\sqrt{1+a^2}}{1+4a^2}(e^{\pi a} - 2a)$ **55.** 21.87

57. Area $= \frac{1}{2}\int_\alpha^\beta r^2\, d\theta$; Arc length $= \int_\alpha^\beta \sqrt{r^2 + \left(\frac{dr}{d\theta}\right)^2}\, d\theta$

59. The integral (a) yields the correct arc length.

61. $4\pi^2 ab$

63. False. The graphs of $f(\theta) = 1$ and $g(\theta) = -1$ coincide.

65. In parametric form,

$$s = \int_a^b \sqrt{\left(\frac{dx}{dt}\right)^2 + \left(\frac{dy}{dt}\right)^2}\, dt.$$

Using θ instead of t gives $x = r\cos\theta$ and $y = r\sin\theta$. Let $r = f(\theta)$. Now we have $x = f(\theta)\cos\theta$ and $y = f(\theta)\sin\theta$.

So, $\frac{dx}{d\theta} = f'(\theta)\cos\theta - f(\theta)\sin\theta$ and

$\frac{dy}{d\theta} = f'(\theta)\sin\theta + f(\theta)\cos\theta$.

(continued)

It follows that

$$\left(\frac{dx}{d\theta}\right)^2 + \left(\frac{dy}{d\theta}\right)^2 = [f'(\theta)\cos\theta - f(\theta)\sin\theta]^2$$
$$+ [f'(\theta)\sin\theta + f(\theta)\cos\theta]^2$$
$$= [f(\theta)]^2 + [f'(\theta)]^2.$$

Therefore, $s = \int_\alpha^\beta \sqrt{[f(\theta)]^2 + [f'(\theta)]^2}\, d\theta$.

Section 9.6 (page 707)

1.

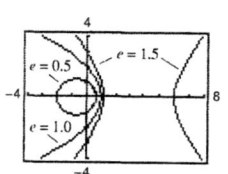

(a) Parabola
(b) Ellipse
(c) Hyperbola

3.

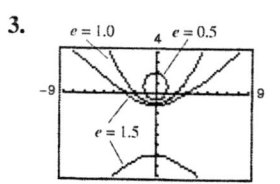

(a) Parabola
(b) Ellipse
(c) Hyperbola

5. (a) Ellipse

As $e \to 1^-$, the ellipse becomes more elliptical, and as $e \to 0^+$, it becomes more circular.

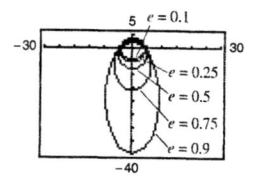

(b) Parabola

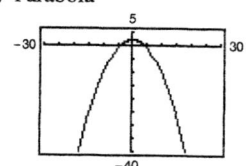

(c) Hyperbola

As $e \to 1^+$, the hyperbola opens more slowly, and as $e \to \infty$, it opens more rapidly.

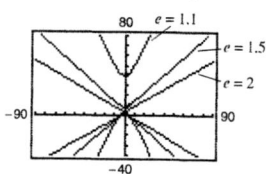

7. c   **8.** f   **9.** a   **10.** e   **11.** b   **12.** d

13. Parabola

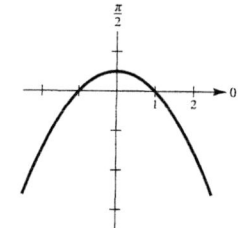

15. Ellipse

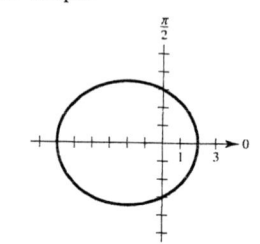

17. Ellipse

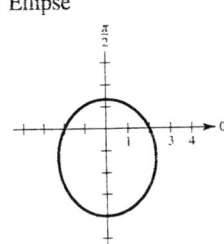

19. Hyperbola

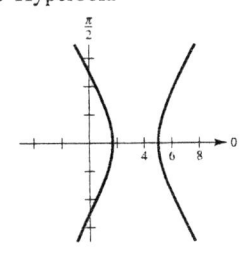

21. Hyperbola

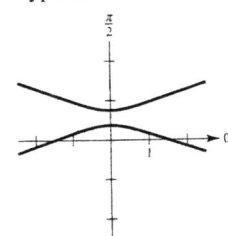

23. Ellipse

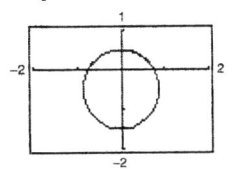

25. Parabola

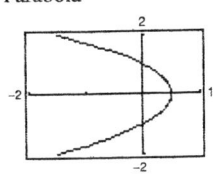

27. Rotated $\pi/4$ radians counterclockwise

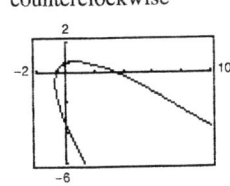

29. Rotated $\pi/6$ radians clockwise

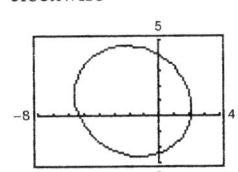

31. $r = \dfrac{5}{5 + 3\cos\left(\theta + \dfrac{\pi}{4}\right)}$

33. $r = \dfrac{1}{1 - \cos\theta}$

35. $r = \dfrac{1}{2 + \sin\theta}$

37. $r = \dfrac{2}{1 + 2\cos\theta}$

39. $r = \dfrac{2}{1 - \sin\theta}$

41. $r = \dfrac{16}{5 + 3\cos\theta}$

43. $r = \dfrac{9}{4 - 5\sin\theta}$

45. If $0 < e < 1$, the conic is an ellipse.
    If $e = 1$, the conic is a parabola.
    If $e > 1$, the conic is a hyperbola.

47. (a) Hyperbola (b) Ellipse
    (c) Parabola   (d) Hyperbola

49. $r^2 = \dfrac{9}{1 - (16/25)\cos^2\theta}$

51. $r^2 = \dfrac{-16}{1 - (25/9)\cos^2\theta}$

53. 10.88

55. $r = \dfrac{345{,}996{,}000}{43{,}373 - 40{,}627\cos\theta}$; 11,004 miles

57. $r = \dfrac{92{,}931{,}075.2223}{1 - 0.0167\cos\theta}$
    Perihelion: 91,404,618 miles
    Aphelion: 94,509,382 miles

59. $r = \dfrac{5.537 \times 10^9}{1 - 0.2481\cos\theta}$
    Perihelion: 4.436×10^9 kilometers
    Aphelion: 7.364×10^9 kilometers

61. (a) 9.341×10^{18} square kilometers; 21.867 years
    (b) 0.8995 radians; Larger angle with the smaller ray to generate an equal area
    (c) Part (a): 2.559×10^9 kilometers; 1.17×10^8 kilometers per year
        Part (b): 4.119×10^9 kilometers; 1.88×10^8 kilometers per year

63. Let $r_1 = \dfrac{ed}{1 + \sin\theta}$ and $r_2 = \dfrac{ed}{1 - \sin\theta}$.

    The points of intersection of r_1 and r_2 are $(ed, 0)$ and (ed, π). The slope of the tangent line to r_1 at $(ed, 0)$ is -1 and at (ed, π) is 1. The slope of the tangent line to r_2 at $(ed, 0)$ is 1 and at (ed, π) is -1. Therefore, at $(ed, 0)$, $m_1 m_2 = -1$ and at (ed, π), $m_1 m_2 = -1$ and the curves intersect at right angles.

Review Exercises for Chapter 9 (page 709)

1. d   2. b   3. a   4. c

5. Circle
   Center: $\left(\dfrac{1}{2}, -\dfrac{3}{4}\right)$
   Radius: 1

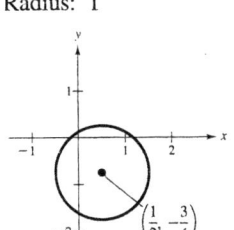

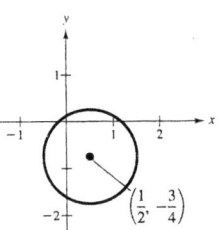

7. Hyperbola
   Center: $(-4, 3)$
   Vertices: $\left(-4 \pm \sqrt{2}, 3\right)$

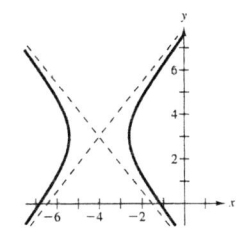

9. Ellipse
   Center: $(2, -3)$
   Vertices: $\left(2, -3 \pm \dfrac{\sqrt{2}}{2}\right)$

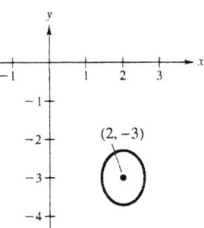

11. $y^2 - 4y - 12x + 4 = 0$

13. $\dfrac{(x-2)^2}{25} + \dfrac{y^2}{21} = 1$

15. $\dfrac{x^2}{16} - \dfrac{y^2}{20} = 1$

17. 15.87   19. $4x + 4y - 7 = 0$

21. (a) 192π cubic feet   (b) 7057.3 pounds
    (c) 4.212 feet   (d) 429.105 square feet

23. $4y + 3x - 11 = 0$

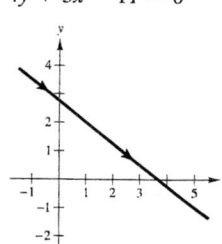

25. $x^2 + y^2 = 36$

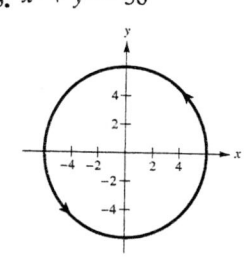

27. $(x-2)^2 - (y-3)^2 = 1$

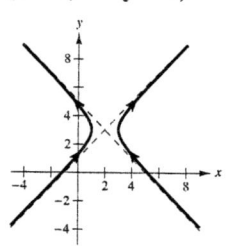

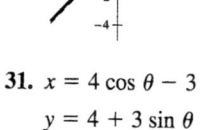

29. $x = 5t - 2$
$y = 6 - 4t$

31. $x = 4\cos\theta - 3$
$y = 4 + 3\sin\theta$

33.

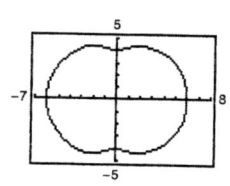

35. (a)

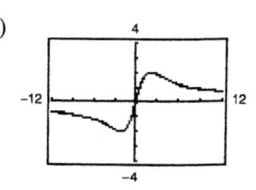

(b) From $x = 2\cot\theta$, it follows that $\cot\theta = \dfrac{x}{2}$. Substituting into $y = 4\sin\theta\cos\theta$ results in
$$y = 4\left(\dfrac{x}{\sqrt{x^2+4}}\right)\left(\dfrac{2}{\sqrt{x^2+4}}\right).$$
This simplifies to $y = \dfrac{8x}{x^2+4}$ or $8x = (4+x^2)y$.

37. (a) $\dfrac{dy}{dx} = -\dfrac{3}{4}$; Horizontal tangents: none

(b) $y = \dfrac{-3x + 11}{4}$ (c)

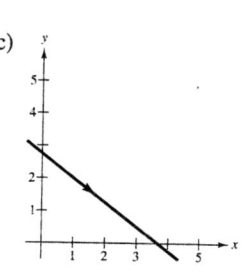

39. (a) $\dfrac{dy}{dx} = -2t^2$; Horizontal tangents: none

(b) $y = 3 + \dfrac{2}{x}$

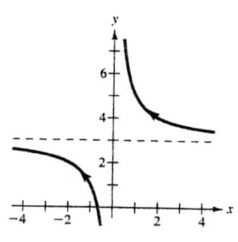

41. (a) $\dfrac{dy}{dx} = \dfrac{(t-1)(2t+1)^2}{t^2(t-2)^2}$; Horizontal tangents: $\left(\dfrac{1}{3}, -1\right)$

(b) $y = \dfrac{4x^2}{(5x-1)(x-1)}$ (c)

43. (a) $\dfrac{dy}{dx} = -\dfrac{5}{2}\cot\theta$; Horizontal tangents: $(3, 7), (3, -3)$

(b) $\dfrac{(x-3)^2}{4} + \dfrac{(y-2)^2}{25} = 1$ (c)

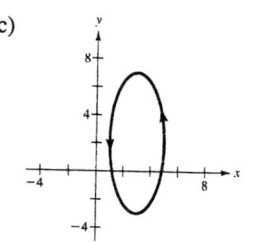

45. (a) $\dfrac{dy}{dx} = -4\tan\theta$; Horizontal tangents: none

(b) $x^{2/3} + (y/4)^{2/3} = 1$ (c)

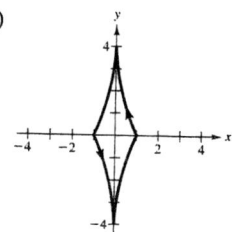

47. (a) and (c)

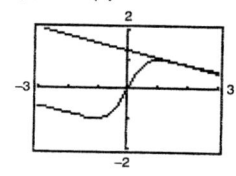

(b) $\dfrac{dx}{d\theta} = -4, \dfrac{dy}{d\theta} = 1, \dfrac{dy}{dx} = -\dfrac{1}{4}$

49. $\dfrac{\pi^2 r}{2}$

51.

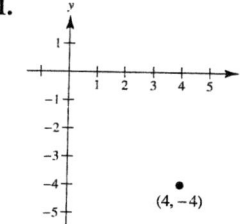

$\left(4\sqrt{2}, \dfrac{7\pi}{4}\right), \left(-4\sqrt{2}, \dfrac{3\pi}{4}\right)$

53. $x^2 + y^2 - 3x = 0$ **55.** $(x^2 + y^2 + 2x)^2 = 4(x^2 + y^2)$
57. $(x^2 + y^2)^2 = x^2 - y^2$ **59.** $y^2 = x^2\left(\dfrac{4-x}{4+x}\right)$
61. $r = a\cos^2\theta \sin\theta$ **63.** $r^2 = a^2\theta^2$
65. Circle   **67.** Line 
69. Cardioid 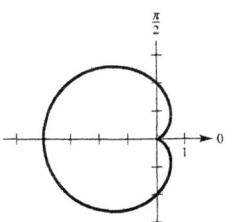   **71.** Limaçon 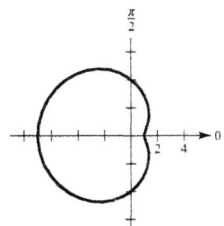
73. Rose curve 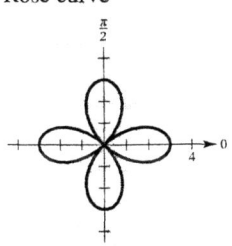   **75.** Rose curve 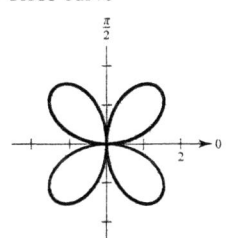
77. 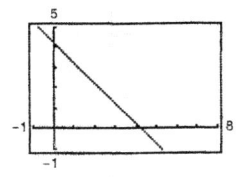   **79.** 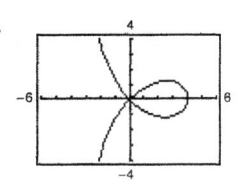
81. (a) $\pm\dfrac{\pi}{3}$
(b) Vertical: $(-1, 0)$, $(3, \pi)$, $(\tfrac{1}{2}, \pm 1.318)$
Horizontal: $(-0.686, \pm 0.568)$, $(2.186, \pm 2.206)$
(c) 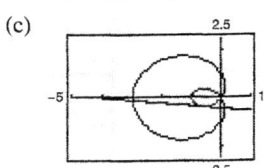
83. $\arctan\left(\dfrac{2\sqrt{3}}{3}\right) \approx 49.1°$

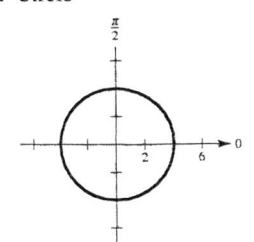

85. $r_1 = 1 + \cos\theta$; $r_2 = 1 - \cos\theta$
The points of intersection are $\left(1, \dfrac{\pi}{2}\right), \left(1, \dfrac{3\pi}{2}\right)$.
$m_{r_1} = \dfrac{-\sin^2\theta + \cos\theta(1 + \cos\theta)}{-\sin\theta\cos\theta - \sin\theta(1 + \cos\theta)}$
m_{r_1} at $\left(1, \dfrac{\pi}{2}\right) = 1$; m_{r_1} at $\left(1, \dfrac{3\pi}{2}\right) = -1$
$m_{r_2} = \dfrac{\sin^2\theta + \cos\theta(1 - \cos\theta)}{\sin\theta\cos\theta - \sin\theta(1 - \cos\theta)}$
m_{r_2} at $\left(1, \dfrac{\pi}{2}\right) = -1$; m_{r_2} at $\left(1, \dfrac{3\pi}{2}\right) = 1$
So, $m_{r_1} = -\dfrac{1}{m_{r_2}}$ and the graphs are orthogonal.

87. $A = 2\left(\dfrac{1}{2}\right)\displaystyle\int_0^\pi (2 + \cos\theta)^2\, d\theta \approx 14.14$

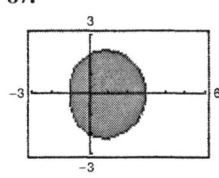

89. $A = 2\left(\dfrac{1}{2}\right)\displaystyle\int_0^{\pi/2} \sin^2\theta \cos^4\theta\, d\theta \approx 0.10$

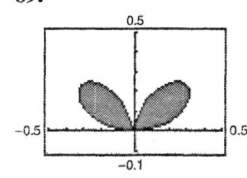

91. $A = 2\left(\dfrac{1}{2}\right)\displaystyle\int_0^{\pi/2} 4\sin 2\theta\, d\theta \approx 4.00$

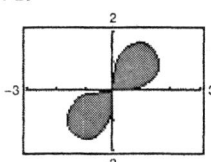

93. $A = 2\left(\dfrac{1}{2}\displaystyle\int_0^{\pi/3} 4\, d\theta + \dfrac{1}{2}\int_{\pi/3}^{\pi/2} 16\cos^2\theta\, d\theta\right) \approx 4.91$

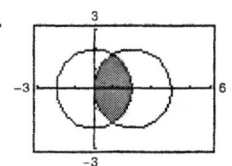

95. $8a$
97. Parabola   **99.** Ellipse

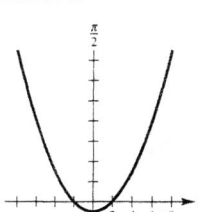

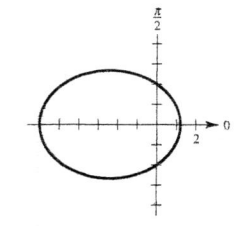

101. Hyperbola **103.** $r = 10 \sin \theta$

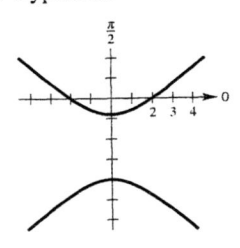

105. $r = \dfrac{4}{1 - \cos \theta}$ **107.** $r = \dfrac{5}{3 - 2\cos \theta}$

P.S. Problem Solving (page 712)

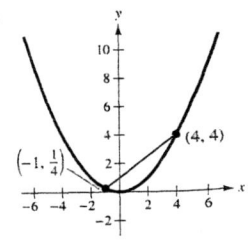

1. (a)

(b) The slope of the tangent line to the parabola at $\left(-1, \tfrac{1}{4}\right)$ is $-\tfrac{1}{2}$. The slope of the tangent line to the parabola at $(4, 4)$ is 2. The product of the two slopes is -1 and therefore the tangent lines are perpendicular.

(c) The directrix of the parabola is $y = -1$. The equations of the two tangent lines are $y = -\tfrac{1}{2}x - \tfrac{1}{4}$ and $y = 2x - 4$. They intersect at the point $\left(\tfrac{3}{2}, -1\right)$, which lies on the directrix.

3. Proof

5. (a) $r = 2a \tan \theta \sin \theta$

(b) $x = \dfrac{2at^2}{1 + t^2}$

$y = \dfrac{2at^3}{1 + t^2}$

(c) $y^2 = \dfrac{x^3}{2a - x}$

7. $x = a \arccos\left(\dfrac{a - y}{a}\right) - \sqrt{2ay - y^2},\ 0 \le y \le 2a$

9. ∞

11. (a) Area of triangle $= \tfrac{1}{2} \times$ base $\times$ height
$= \tfrac{1}{2}(1)(\tan \alpha)$
$= \tfrac{1}{2} \tan \alpha$

and $A(\alpha) = \tfrac{1}{2} \int_0^\alpha \sec^2 \theta \, d\theta$
$= \tfrac{1}{2}\left[\tan \theta\right]_0^\alpha$
$= \tfrac{1}{2} \tan \alpha$

(b) $\int_0^\alpha \sec^2 \theta \, d\theta = \left[\tan \theta\right]_0^\alpha$
$= \tan \alpha$

(c) $\dfrac{d}{d\alpha}(\tan \alpha) = \sec^2 \alpha$

13. $r = \dfrac{1}{\sqrt{2}} de^{(\pi/4 - \theta)}$

15. (a) First plane: $x_1 = \cos 70(150 - 375t)$
$y_1 = \sin 70(150 - 375t)$
Second plane: $x_2 = \cos 45(450t - 190)$
$y_2 = \sin 45(190 - 450t)$

(b) $\{[\cos 45(450t - 190) - \cos 70(150 - 375t)]^2$
$+ [\sin 45(190 - 450t) - \sin 70(150 - 375t)]^2\}^{1/2}$

(c)

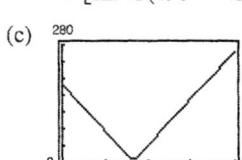

0.4145 hours; Yes

17. $r = \cos 5\theta + n \cos \theta$

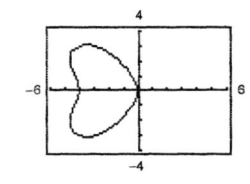

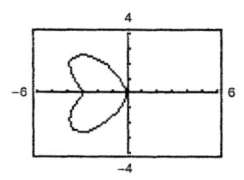

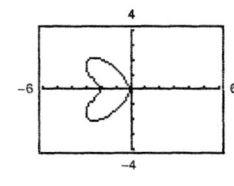

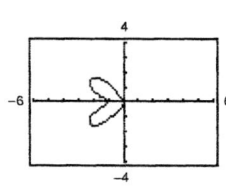

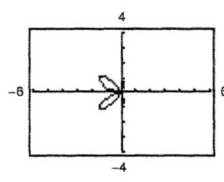

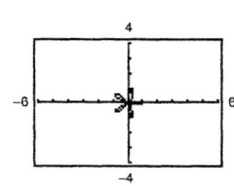

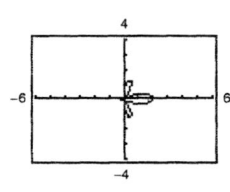

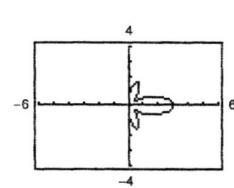

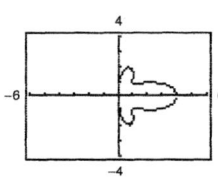

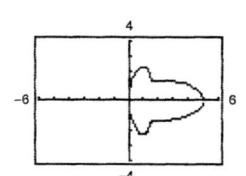

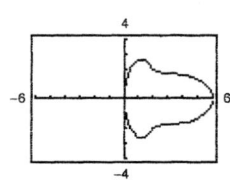

$n = 1, 2, 3, 4, 5$ produce "bells"; $n = -1, -2, -3, -4, -5$ produce "hearts."

Chapter 10
Section 10.1 (page 723)

1. (a) $\langle 4, 2 \rangle$
(b) [graph showing vector **v** from origin to (4, 2)]

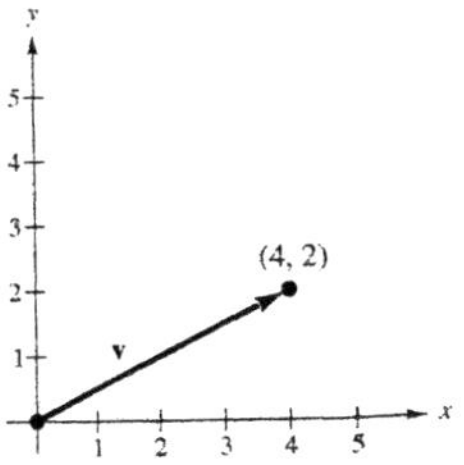

3. (a) $\langle -7, 0 \rangle$
(b) [graph showing vector **v** from origin to (−7, 0)]

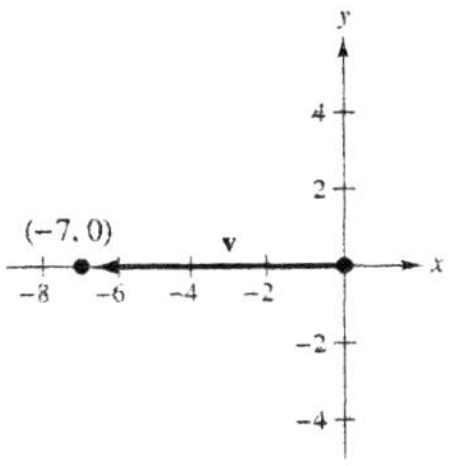

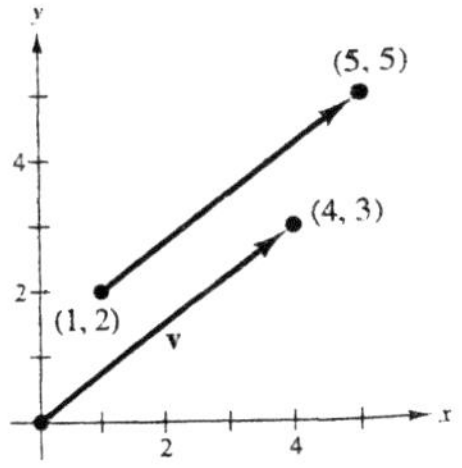

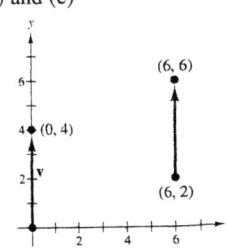

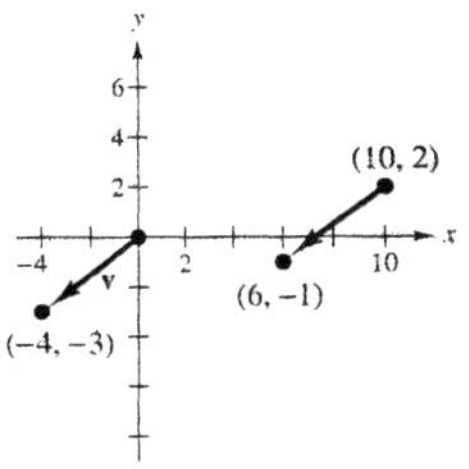

5. $\mathbf{u} = \mathbf{v} = \langle 2, 4 \rangle$ **7.** $\mathbf{u} = \mathbf{v} = \langle 6, -5 \rangle$

9. (a) and (c) [graph with points (1, 2), (4, 3), (5, 5)]
(b) $\langle 4, 3 \rangle$

11. (a) and (c) [graph with points (−4, −3), (6, −1), (10, 2)]
(b) $\langle -4, -3 \rangle$

13. (a) and (c) [graph with points (0, 4), (6, 2), (6, 6)]
(b) $\langle 0, 4 \rangle$

15. (a) and (c) [graph with points $(-1, \tfrac{5}{3})$, $(\tfrac{1}{2}, 3)$, $(\tfrac{3}{2}, \tfrac{4}{3})$]
(b) $\langle -1, \tfrac{5}{3} \rangle$

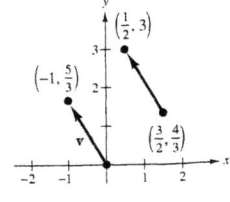

17. (a) $\langle 4, 6 \rangle$ [graph showing 2**v** from (2, 3) to (4, 6)]
(b) $\langle -6, -9 \rangle$ [graph showing −3**v** to (−6, −9)]
(c) $\langle 7, \tfrac{21}{2} \rangle$ [graph showing $\tfrac{7}{2}\mathbf{v}$ from (2, 3) to $(7, \tfrac{21}{2})$] 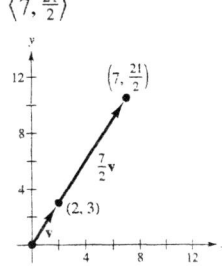
(d) $\langle \tfrac{4}{3}, 2 \rangle$ [graph showing $\tfrac{2}{3}\mathbf{v}$ to $(\tfrac{4}{3}, 2)$]

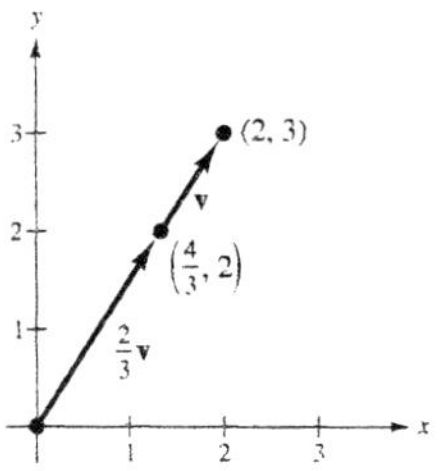

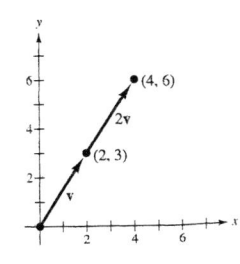

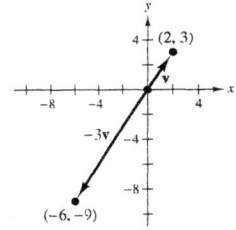

19. [graph showing vector −**u**]

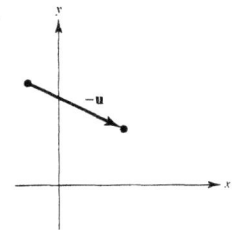

21. [graph showing vectors **u**, −**v**, **u** − **v**]

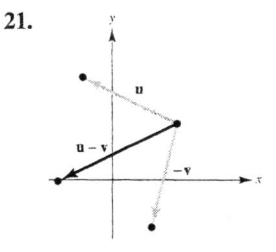

23. (a) $\langle \tfrac{8}{3}, 6 \rangle$ (b) $\langle -2, -14 \rangle$ (c) $\langle 18, -7 \rangle$

25. $\langle 3, -\tfrac{3}{2} \rangle$ [graph showing **u** and $\mathbf{v} = \tfrac{3}{2}\mathbf{u}$]

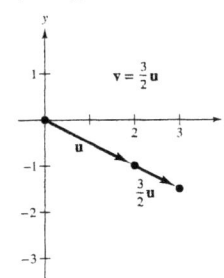

27. $\langle 4, 3 \rangle$ [graph showing **u**, 2**w**, and $\mathbf{v} = \mathbf{u} + 2\mathbf{w}$]

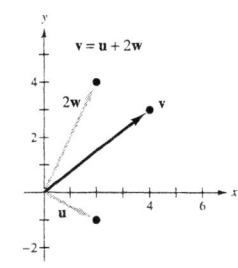

29. (3, 5)   **31.** 5   **33.** $\sqrt{61}$ **35.** 4

37. $\left\langle \dfrac{\sqrt{17}}{17}, \dfrac{4\sqrt{17}}{17} \right\rangle$ **39.** $\left\langle \dfrac{3\sqrt{34}}{34}, \dfrac{5\sqrt{34}}{34} \right\rangle$

41. (a) $\sqrt{2}$ (b) $\sqrt{5}$ (c) 1   (d) 1   (e) 1   (f) 1

43. (a) $\sqrt{5}/2$ (b) $\sqrt{13}$ (c) $\sqrt{85}/2$
(d) 1   (e) 1   (f) 1

45. $\|\mathbf{u}\| + \|\mathbf{v}\| = \sqrt{5} + \sqrt{41}$ and $\|\mathbf{u} + \mathbf{v}\| = \sqrt{74}$
$\sqrt{74} < \sqrt{5} + \sqrt{41}$

47. $\langle 2\sqrt{2}, 2\sqrt{2} \rangle$ **49.** $\langle 1, \sqrt{3} \rangle$ **51.** $\langle 3, 0 \rangle$

53. $\langle -\sqrt{3}, 1 \rangle$ **55.** $\left\langle \dfrac{2 + 3\sqrt{2}}{2}, \dfrac{3\sqrt{2}}{2} \right\rangle$

57. $\langle 2\cos 4 + \cos 2, 2\sin 4 + \sin 2 \rangle$

59. Answers will vary. Example: A scalar is a single real number such as 2. A vector is a line segment having both direction and magnitude. The vector $\langle \sqrt{3}, 1 \rangle$, given in component form, has a direction of $\pi/6$ and a magnitude of 2.

61. The process of dividing a vector by its magnitude is called normalization of a vector.

63. $a = 1, b = 1$ **65.** $a = 1, b = 2$ **67.** $a = \tfrac{2}{3}, b = \tfrac{1}{3}$

69. (a) $\pm \dfrac{1}{\sqrt{10}} \langle 1, 3 \rangle$ (b) $\pm \dfrac{1}{\sqrt{10}} \langle 3, -1 \rangle$

71. (a) $\pm \dfrac{1}{5} \langle -4, 3 \rangle$ (b) $\pm \dfrac{1}{5} \langle 3, 4 \rangle$ **73.** $\left\langle -\dfrac{\sqrt{2}}{2}, \dfrac{\sqrt{2}}{2} \right\rangle$

75. (a)–(c) Answers will vary.   77. 1.33, 132.5°
79. (a) Direction: $\alpha = 11.8°$
    Magnitude: 440.2 N
    (b) $M = \sqrt{(275 + 180\cos\theta)^2 + (180\sin\theta)^2}$
    $\alpha = \arctan\left(\dfrac{180\sin\theta}{275 + 180\cos\theta}\right)$

(c)

θ	0°	30°	60°	90°	120°
M	455.0	440.2	396.9	328.7	241.9
α	0°	11.8°	23.1°	33.2°	40.1°

θ	150°	180°
M	149.3	95.0
α	37.1°	0°

(d) [graphs of M and α vs θ]

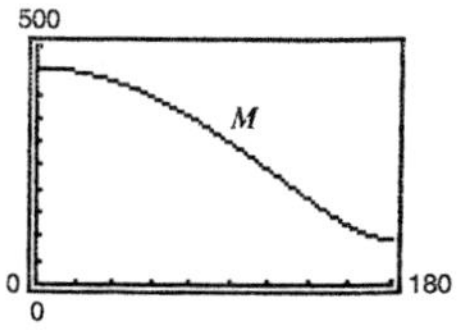

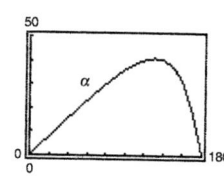

(e) M decreases because the forces change from acting in the same direction to acting in opposite directions as θ increases from 0° to 180°.

81. 71.3°, 228.5 pounds
83. (a) $\theta = 0°$ (b) $\theta = 180°$
    (c) No, the resultant can only be less than or equal to the sum.
85. $(-4, -1), (6, 5), (10, 3)$
87. Tension in cable $AC \approx 1758.8$ pounds
    Tension in cable $BC \approx 1305.4$ pounds
89. Horizontal: 1193.43 feet per second
    Vertical: 125.43 feet per second
91. 38.3° north of west   93. $T_2 = 157.316$
    882.9 kilometers per hour $T_3 = 3692.482$
95. Proof   97. Proof
99. True   101. True   103. False. $\|a\mathbf{i} + b\mathbf{j}\| = \sqrt{2}|a|$

Section 10.2 (page 732)

1. [3D plot with points $(2, 1, 3)$ and $(-1, 2, 1)$]
3. [3D plot with points $(5, -2, 2)$ and $(5, -2, -2)$]
5. $A(2, 3, 4)$ 7. $(-3, 4, 5)$ 9. $(10, 0, 0)$ 11. 0
   $B(-1, -2, 2)$
13. 6 units above the xy-plane
15. 4 units in front of the yz-plane
17. To the left of the xz-plane and either above, below, or on the xy-plane and either in front of, behind, or on the yz-plane
19. Within 3 units of the xz-plane
21. 3 units below the xy-plane, to the right of the xz-plane and in front of the yz-plane, *or* 3 units below the xy-plane, to the left of the xz-plane and behind the yz-plane
23. 1. Above the xy-plane and (a) to the right of the xz-plane and behind the yz-plane or (b) to the left of the xz-plane and in front of the yz-plane or,
    2. Below the xy-plane and (a) to the right of the xz-plane and in front of the yz-plane or (b) to the left of the xz-plane and behind the yz-plane
25. $\sqrt{65}$ 27. $\sqrt{61}$
29. $3, 3\sqrt{5}, 6$ 31. $6, 6, 2\sqrt{10}$
    Right triangle   Isosceles triangle
33. $(0, 0, 5), (2, 2, 6), (2, -4, 9)$
35. $\left(\dfrac{3}{2}, -3, 5\right)$ 37. $(x - 0)^2 + (y - 2)^2 + (z - 5)^2 = 4$
39. $(x - 1)^2 + (y - 3)^2 + (z - 0)^2 = 10$
41. $(x - 1)^2 + (y + 3)^2 + (z + 4)^2 = 25$
    Center: $(1, -3, -4)$
    Radius: 5
43. $\left(x - \dfrac{1}{3}\right)^2 + (y + 1)^2 + z^2 = 1$
    Center: $\left(\dfrac{1}{3}, -1, 0\right)$
    Radius: 1
45. A solid sphere with center $(0, 0, 0)$ and radius 6
47. (a) $\langle -2, 2, 2 \rangle$ 49. (a) $\langle -3, 0, 3 \rangle$
    (b) [3D vector plot $(-2, 2, 2)$]   (b) [3D vector plot $(-3, 0, 3)$]
51. $\mathbf{u} = \langle 1, -1, 6 \rangle$ 53. $\mathbf{u} = \langle -1, 0, -1 \rangle$
    $\|\mathbf{u}\| = \sqrt{38}$ $\|\mathbf{u}\| = \sqrt{2}$
    $\dfrac{\mathbf{u}}{\|\mathbf{u}\|} = \dfrac{1}{\sqrt{38}}\langle 1, -1, 6 \rangle$ $\dfrac{\mathbf{u}}{\|\mathbf{u}\|} = \dfrac{1}{\sqrt{2}}\langle -1, 0, -1 \rangle$
55. (a) and (c)   57. $(3, 1, 8)$
    [3D plot with points $(3, 3, 4)$, $(-1, 2, 3)$, $(0, 0, 0)$, $(4, 1, 1)$]
    (b) $\langle 4, 1, 1 \rangle$

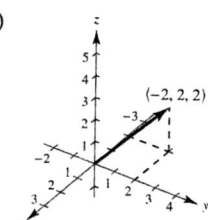

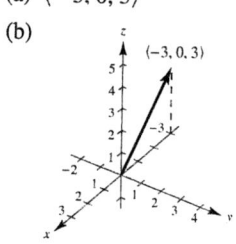

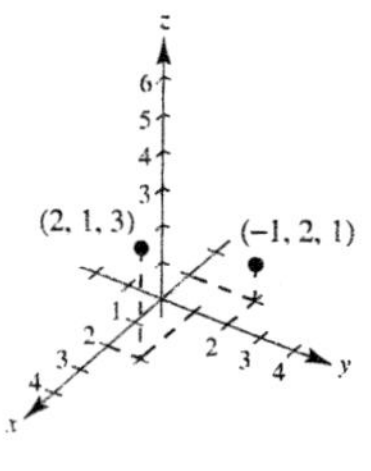

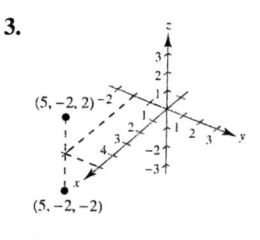

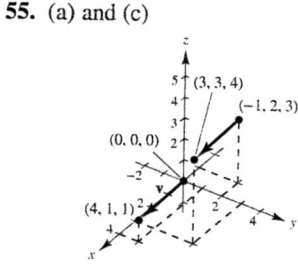

59. (a) [graph showing (2, 4, 4)] (b) [graph showing (−1, −2, −2)]
(c) [graph showing ⟨3/2, 3, 3⟩] (d) [graph showing (0, 0, 0)]

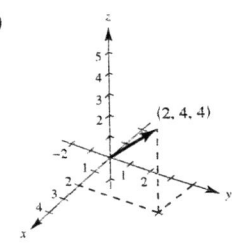

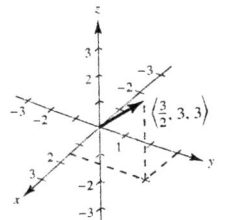

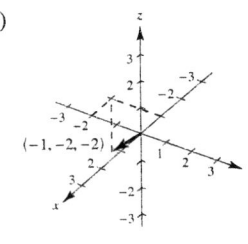

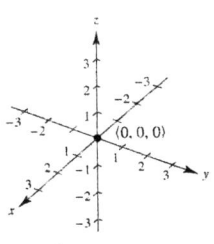

61. ⟨−1, 0, 4⟩   **63.** ⟨6, 12, 6⟩   **65.** ⟨7/2, 3, 5/2⟩

67. a and b   **69.** a   **71.** Collinear   **73.** Not collinear

75. $\overrightarrow{AB} = \langle 1, 2, 3 \rangle$
$\overrightarrow{CD} = \langle 1, 2, 3 \rangle$
$\overrightarrow{BD} = \langle -2, 1, 1 \rangle$
$\overrightarrow{AC} = \langle -2, 1, 1 \rangle$
Since $\overrightarrow{AB} = \overrightarrow{CD}$ and $\overrightarrow{BD} = \overrightarrow{AC}$, the given points form the vertices of a parallelogram.

77. 0   **79.** $\sqrt{14}$ **81.** $\sqrt{34}$

83. (a) $\frac{1}{3}\langle 2, -1, 2 \rangle$ (b) $-\frac{1}{3}\langle 2, -1, 2 \rangle$

85. (a) $\frac{1}{\sqrt{38}}\langle 3, 2, -5 \rangle$ (b) $-\frac{1}{\sqrt{38}}\langle 3, 2, -5 \rangle$

87. (a)–(d) Answers will vary.   **89.** $\pm\frac{5}{3}$ **91.** $\langle 0, \frac{10}{\sqrt{2}}, \frac{10}{\sqrt{2}} \rangle$

93. $\langle 1, -1, \frac{1}{2} \rangle$

95. $\langle 0, \sqrt{3}, \pm 1 \rangle$ **97.** (2, −1, 2)

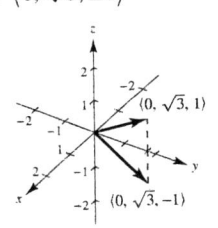

[graph showing $(0, \sqrt{3}, 1)$ and $(0, \sqrt{3}, -1)$]

99. (a) [graph with v and u]

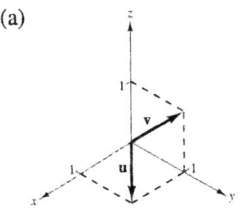

(b) $\mathbf{w} = a\mathbf{u} + b\mathbf{v}$
$= a\mathbf{i} + (a + b)\mathbf{j} + b\mathbf{k}$
$= \mathbf{0}$
$a = 0, a + b = 0, b = 0$

(c) $\mathbf{w} = a\mathbf{u} + b\mathbf{v}$
$= a\mathbf{i} + (a + b)\mathbf{j} + b\mathbf{k}$
$= \mathbf{i} + 2\mathbf{j} + \mathbf{k}$
$a = 1, a + b = 2, b = 1$

(d) $\mathbf{w} = a\mathbf{u} + b\mathbf{v}$
$= a\mathbf{i} + (a + b)\mathbf{j} + b\mathbf{k}$
$= \mathbf{i} + 2\mathbf{j} + 3\mathbf{k}$
$a = 1, a + b = 2, b = 3$

So, $b = 1$ and $b = 3$. This is not possible.

101. $d = \sqrt{(x_2 - x_1)^2 + (y_2 - y_1)^2 + (z_2 - z_1)^2}$

103. Two nonzero vectors **u** and **v** are parallel if there is some scalar c such that $\mathbf{u} = c\mathbf{v}$.

105. (a) $T = \frac{8L}{\sqrt{L^2 - 18^2}}, L > 18$

(b)
L	20	25	30	35	40	45	50
T	18.4	11.5	10	9.3	9.0	8.7	8.6

(c) [graph with $L = 18$ and $T = 8$ asymptotes]

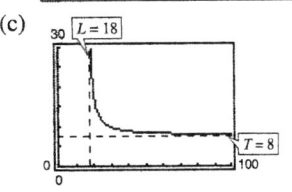

(d) Proof   (e) 30 inches

107. $\frac{\sqrt{3}}{3}\langle 1, 1, 1 \rangle$

109. Tension in cable AB: 202.919 N
Tension in cable AC: 157.909 N
Tension in cable AD: 226.521 N

111. $(x - \frac{4}{3})^2 + (y - 3)^2 + (z + \frac{1}{3})^2 = \frac{44}{9}$

Section 10.3 (page 741)

1. (a) −6   (b) 25   (c) 25   (d) ⟨−12, 18⟩   (e) −12

3. (a) 2   (b) 29   (c) 29   (d) ⟨0, 12, 10⟩   (e) 4

5. (a) 1   (b) 6   (c) 6   (d) $\mathbf{i} - \mathbf{k}$ (e) 2

7. $17,139.05; Total revenue

9. 20   **11.** $\frac{\pi}{2}$ **13.** $\arccos\left(-\frac{1}{5\sqrt{2}}\right) \approx 98.1°$

15. $\arccos\left(\frac{\sqrt{2}}{3}\right) \approx 61.9°$ **17.** $\arccos\left(-\frac{8\sqrt{13}}{65}\right) \approx 116.3°$

19. Neither   **21.** Orthogonal

23. Neither   **25.** Orthogonal

27. $\cos\alpha = \frac{1}{3}$ **29.** $\cos\alpha = 0$

$\cos\beta = \frac{2}{3}$ $\cos\beta = \frac{3}{\sqrt{13}}$

$\cos\gamma = \frac{2}{3}$ $\cos\gamma = -\frac{2}{\sqrt{13}}$

31. $\alpha \approx 43.3°, \beta \approx 61.0°, \gamma \approx 119.0°$

33. $\alpha \approx 100.5°, \beta \approx 24.1°, \gamma \approx 68.6°$

35. Magnitude: 124.310 pounds
$\alpha = 29.48°, \beta = 61.39°, \gamma = 96.53°$

37. $\arccos\left(\dfrac{1}{\sqrt{3}}\right) \approx 54.7°$

39. $\alpha = 90°$, $\beta = 45°$, $\gamma = 45°$

41. $\langle 4, -1 \rangle$ 43. $\langle 2, 1, 1 \rangle$

45. (a) $\left\langle \dfrac{5}{2}, \dfrac{1}{2} \right\rangle$ 47. (a) $\left\langle 0, \dfrac{33}{25}, \dfrac{44}{25} \right\rangle$
    (b) $\left\langle -\dfrac{1}{2}, \dfrac{5}{2} \right\rangle$ (b) $\left\langle 2, -\dfrac{8}{25}, \dfrac{6}{25} \right\rangle$

49. See "Definition of Dot Product," page 735.

51. (a) $\theta = \dfrac{\pi}{2}$ (b) $0 < \theta < \dfrac{\pi}{2}$ (c) $\dfrac{\pi}{2} < \theta < \pi$

53. In space, direction is measured in terms of the angles between a nonzero vector **v** and the three unit vectors **i**, **j**, and **k**. The angles α, β, and γ are the direction angles of **v**, where α is the angle between **v** and **i**, β is the angle between **v** and **j**, and γ is the angle between **v** and **k**. The direction cosines of **v** are $\cos \alpha$, $\cos \beta$, and $\cos \gamma$.

55. (a) The vectors are parallel.   (b) The vectors are orthogonal.

57. (a)–(c) Answers will vary.   59. Answers will vary.

61. $\langle 0, 0 \rangle$

63. Answers will vary. Example: $\langle 4, 3 \rangle$ and $\langle -4, -3 \rangle$

65. Answers will vary. Example: $\langle 2, 0, 3 \rangle$ and $\langle -2, 0, -3 \rangle$

67. (a) 8335.1 pounds   (b) 47,270.8 pounds

69. 425 foot-pounds   71. 72

73. False. For example, $\langle 1, 1 \rangle \cdot \langle 2, 3 \rangle = 5$ and $\langle 1, 1 \rangle \cdot \langle 1, 4 \rangle = 5$, but $\langle 2, 3 \rangle \neq \langle 1, 4 \rangle$.

75. Proof   77. $\mathbf{u} \cdot \mathbf{v} = \cos \alpha \cos \beta + \sin \alpha \sin \beta$; Proof

79. Proof   81. Proof

Section 10.4 (page 750)

1. $-\mathbf{k}$ 3. $\mathbf{i}$

5. $-\mathbf{j}$

7. (a) $-22\mathbf{i} + 16\mathbf{j} - 23\mathbf{k}$ 9. (a) $17\mathbf{i} - 33\mathbf{j} - 10\mathbf{k}$
   (b) $22\mathbf{i} - 16\mathbf{j} + 23\mathbf{k}$ (b) $-17\mathbf{i} + 33\mathbf{j} + 10\mathbf{k}$
   (c) $\mathbf{0}$ (c) $\mathbf{0}$

11. $\langle -1, -1, -1 \rangle$ 13. $\langle 0, 0, 54 \rangle$ 15. $\langle -2, 3, -1 \rangle$

17.

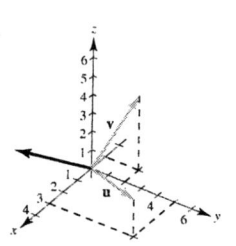

19.

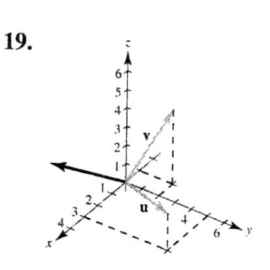

21. $\left\langle -70, -23, \dfrac{57}{2} \right\rangle$

$\left\langle \dfrac{-140}{\sqrt{24{,}965}}, \dfrac{-46}{\sqrt{24{,}965}}, \dfrac{57}{\sqrt{24{,}965}} \right\rangle$

23. $\left\langle -\dfrac{71}{20}, -\dfrac{11}{5}, \dfrac{5}{4} \right\rangle$

$\left\langle \dfrac{-71}{\sqrt{7602}}, \dfrac{-44}{\sqrt{7602}}, \dfrac{25}{\sqrt{7602}} \right\rangle$

25. Answers will vary.   27. 1   29. $6\sqrt{5}$

31. $2\sqrt{83}$ 33. $\dfrac{3\sqrt{13}}{2}$ 35. $\dfrac{\sqrt{16{,}742}}{2}$

37. $10 \cos 40 \approx 7.66$ foot-pounds

39. (a) $90 \sin \theta$

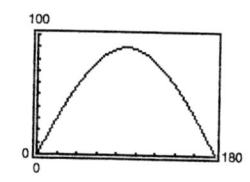

(b) $45\sqrt{2} \approx 63.64$

(c) $\theta = 90$. This is what should be expected. When $\theta = 90$ the pipe wrench is horizontal.

41. 1   43. 6   45. 2   47. 75

49. See "Definition of Cross Product of Two Vectors in Space," page 744.

51. The magnitude of the cross product will increase by a factor of 4.

53. False. The cross product of two vectors is not defined in a two-dimensional coordinate system.

55. True   57. Proof   59. Proof   61. Proof   63. Proof

Section 10.5 (page 759)

1. (a)

(b) $P = (1, 2, 2)$, $Q = (10, -1, 17)$, $\overrightarrow{PQ} = \langle 9, -3, 15 \rangle$
(There are many correct answers.) The components of the vector and the coefficients of t are proportional because the line is parallel to $\overrightarrow{PQ}$.

(c) $\left(-\dfrac{1}{5}, \dfrac{12}{5}, 0\right)$, $(7, 0, 12)$, $\left(0, \dfrac{7}{3}, \dfrac{1}{3}\right)$

	Parametric Equations	Symmetric Equations	Direction Numbers
3.	$x = t$ $y = 2t$ $z = 3t$	$x = \dfrac{y}{2} = \dfrac{z}{3}$	1, 2, 3
5.	$x = -2 + 2t$ $y = 4t$ $z = 3 - 2t$	$\dfrac{x+2}{2} = \dfrac{y}{4} = \dfrac{z-3}{-2}$	2, 4, −2
7.	$x = 1 + 3t$ $y = -2t$ $z = 1 + t$	$\dfrac{x-1}{3} = \dfrac{y}{-2} = \dfrac{z-1}{1}$	3, −2, 1
9.	$x = 5 + 17t$ $y = -3 - 11t$ $z = -2 - 9t$	$\dfrac{x-5}{17} = \dfrac{y+3}{-11} = \dfrac{z+2}{-9}$	17, −11, −9
11.	$x = 2 + 8t$ $y = 3 + 5t$ $z = 12t$	$\dfrac{x-2}{8} = \dfrac{y-3}{5} = \dfrac{z}{12}$	8, 5, 12

13. $x = 2$
$y = 3$
$z = 4 + t$

15. a and b   17. $L_1 = L_2$ and is parallel to L_3.

19. $(2, 3, 1)$, $\cos\theta = \dfrac{7\sqrt{17}}{51}$

21. Nonintersecting

23. $(7, 8, -1)$

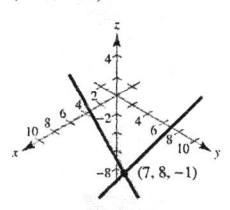

25. (a) $P = (0, 0, -1)$, $Q = (0, -2, 0)$, $R = (3, 4, -1)$
$\overrightarrow{PQ} = \langle 0, -2, 1\rangle$, $\overrightarrow{PR} = \langle 3, 4, 0\rangle$
(There are many correct answers.)
(b) $\overrightarrow{PQ} \times \overrightarrow{PR} = \langle -4, 3, 6\rangle$
The components of the cross product are proportional to the coefficients of the variables in the equation. The cross product is parallel to the normal vector.

27. $x - 2 = 0$ 29. $2x + 3y - z = 10$
31. $x - y + 2z = 12$ 33. $3x + 9y - 7z = 0$
35. $4x - 3y + 4z = 10$ 37. $z = 3$ 39. $x + y + z = 5$
41. $7x + y - 11z = 5$ 43. $y - z = -1$
45. Orthogonal   47. Neither; 83.5°   49. Parallel

51.

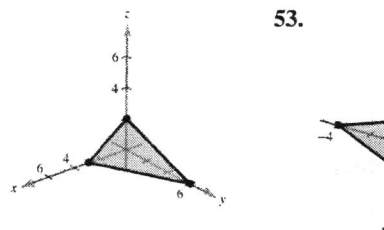

53.

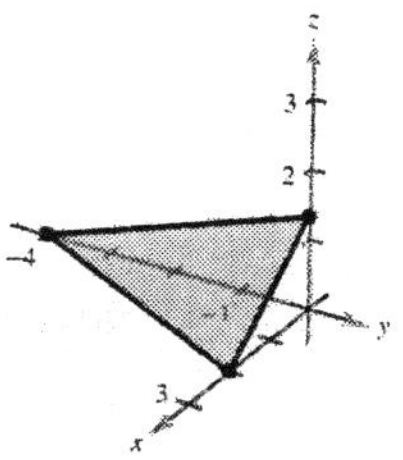

55.

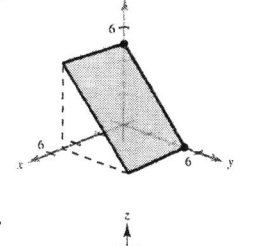

57.

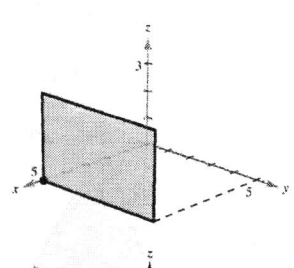

59.

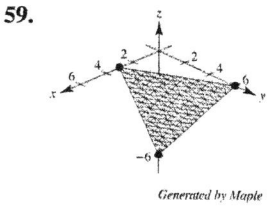

61.

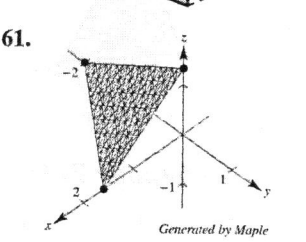

63. $P_1 = P_4$ and is parallel to P_2.
65. The planes have intercepts at $(c, 0, 0)$, $(0, c, 0)$, and $(0, 0, c)$ for each value of c.
67. $x = 2$
$y = 1 + t$
$z = 1 + 2t$
69. $(2, -3, 2)$ The line does not lie in the plane.
71. Nonintersecting
73. $\dfrac{6\sqrt{14}}{7}$ 75. $\dfrac{11\sqrt{6}}{6}$ 77. $\dfrac{2\sqrt{26}}{13}$
79. $\dfrac{27\sqrt{94}}{188}$ 81. $\dfrac{\sqrt{2533}}{17}$
83. Parametric equations: $x = x_1 + at$, $y = y_1 + bt$, and $z = z_1 + ct$
Symmetric equations: $\dfrac{x - x_1}{a} = \dfrac{y - y_1}{b} = \dfrac{z - z_1}{c}$
You need a vector $\mathbf{v} = \langle a, b, c\rangle$ parallel to the line and a point $P(x_1, y_1, z_1)$ on the line.
85. Simultaneously solve the two linear equations representing the planes and substitute the values back into one of the original equations. Then, choose a value for t and form the corresponding parametric equations for the line of intersection.
87. (a) Sphere
$x^2 + y^2 + z^2 - 6x + 4y - 10z + 22 = 0$
(b) Planes
$4x - 3y + z = 10 \pm 4\sqrt{26}$

89. (a)

Year	1980	1985	1990	1994	1995	1996	1997
x	3.1	3.2	4.9	5.8	6.2	6.4	6.6
y	6.3	7.9	9.1	8.7	8.2	8.0	7.7
z	16.5	13.9	10.2	8.8	8.4	8.4	8.2
z'	16.2	14.2	9.8	8.6	8.4	8.3	8.2

(b) Consumption of the third type decreases.
(c)

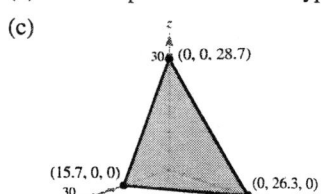

91. True

Section 10.6 (page 771)

1. c **2.** e **3.** f **4.** b **5.** d **6.** a
7. Plane   **9.** Right circular cylinder 
11. Parabolic cylinder 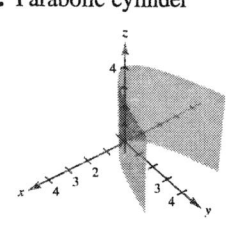   **13.** Elliptic cylinder 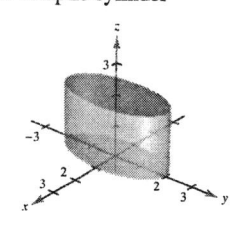
15. Cylinder 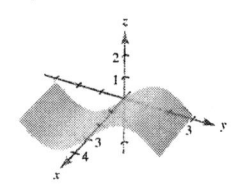
17. (a) (20, 0, 0)
(b) (10, 10, 20)
(c) (0, 0, 20)
(d) (0, 20, 0)
19. Ellipsoid   **21.** Hyperboloid of one sheet 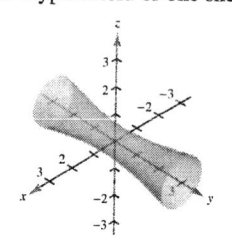
23. Elliptic paraboloid 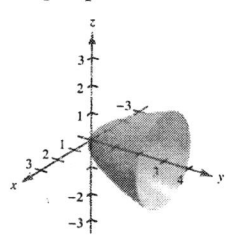 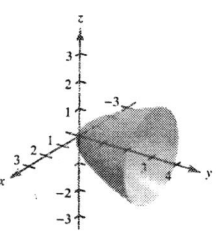   **25.** Hyperbolic paraboloid 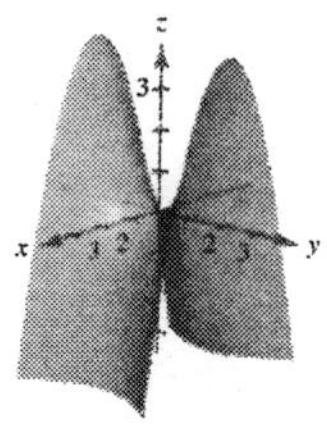
27. Elliptic cone 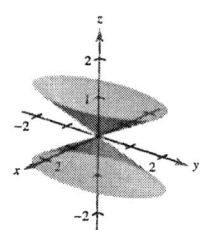 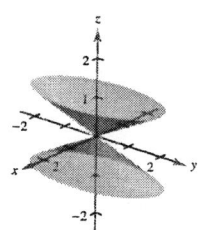   **29.** Ellipsoid 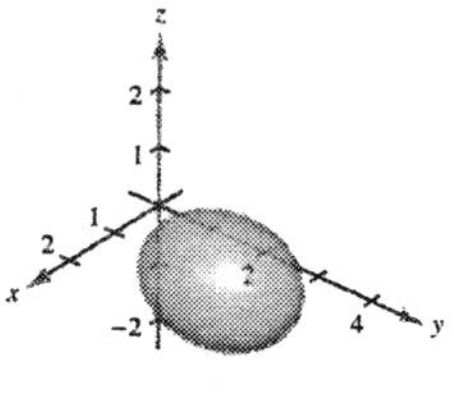
31. 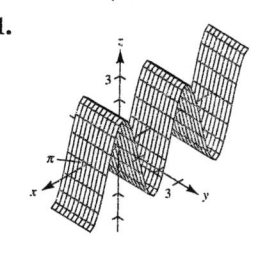   **33.** 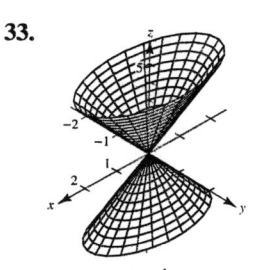
35. 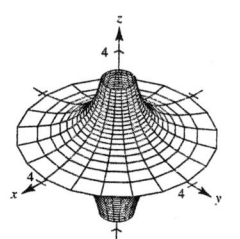   **37.** 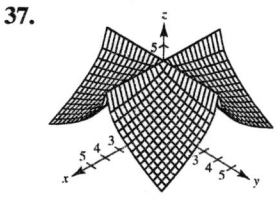
39. 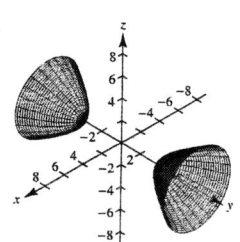   **41.** 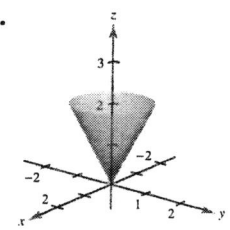
43. 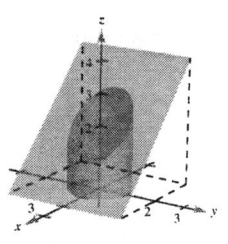
45. $x^2 + z^2 = 4y$ **47.** $4x^2 + 4y^2 = z^2$ **49.** $y^2 + z^2 = \dfrac{4}{x^2}$
51. $y = \sqrt{2z} \ (\text{or } x = \sqrt{2z})$

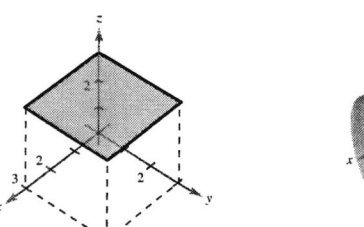

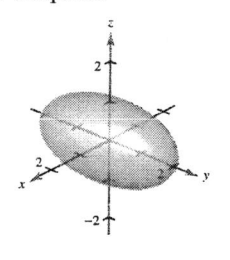

53. Let C be a curve in a plane and let L be a line not in a parallel plane. The set of all lines parallel to L and intersecting C is called a cylinder. C is called the generating curve of the cylinder, and the parallel lines are called rulings.
55. See pages 765 and 766.
57. $\dfrac{128\pi}{3}$
59. (a) Major axis: $4\sqrt{2}$ (b) Major axis: $8\sqrt{2}$
    Minor axis: 4   Minor axis: 8
    Foci: $(0, \pm 2, 2)$ Foci: $(0, \pm 4, 8)$
61. $x^2 + z^2 = 8y$; Elliptic paraboloid
63. $\dfrac{x^2}{3963^2} + \dfrac{y^2}{3963^2} + \dfrac{z^2}{3942^2} = 1$
65. $x = at, y = -bt, z = 0$;
    $x = at, y = bt + ab^2, z = 2abt + a^2b^2$
67. The Klein bottle does not have both an "inside" and an "outside." It is formed by inserting the small open end through the side of the bottle and making it contiguous with the top of the bottle.

Section 10.7 (page 778)

1. $(5, 0, 2)$ 3. $(1, \sqrt{3}, 2)$ 5. $(-2\sqrt{3}, -2, 3)$
7. $\left(5, \dfrac{\pi}{2}, 1\right)$ 9. $\left(2, \dfrac{\pi}{3}, 4\right)$ 11. $\left(2\sqrt{2}, -\dfrac{\pi}{4}, -4\right)$
13. $r^2 + z^2 = 10$ 15. $r = \sec\theta\tan\theta$
17. $x^2 + y^2 = 4$ 19. $x - \sqrt{3}y = 0$

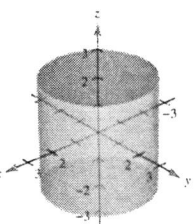

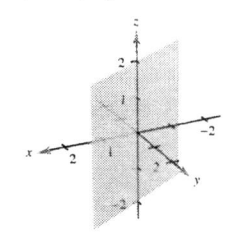

21. $x^2 + y^2 - 2y = 0$ 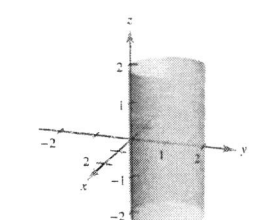   23. $x^2 + y^2 + z^2 = 4$

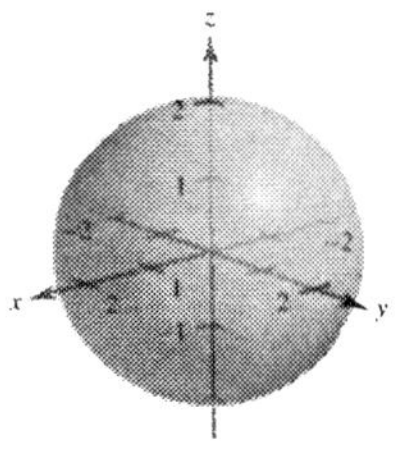

25. $\left(4, 0, \dfrac{\pi}{2}\right)$ 27. $\left(4\sqrt{2}, \dfrac{2\pi}{3}, \dfrac{\pi}{4}\right)$ 29. $\left(4, \dfrac{\pi}{6}, \dfrac{\pi}{6}\right)$
31. $(\sqrt{6}, \sqrt{2}, 2\sqrt{2})$ 33. $(0, 0, 12)$ 35. $\left(\dfrac{5}{2}, \dfrac{5}{2}, -\dfrac{5\sqrt{2}}{2}\right)$
37. (a) Answers will vary.
    (b) $(5.385, -0.927, 1.190)$
39. $\rho = 6$ 41. $\rho = 3\csc\phi$

43. $x^2 + y^2 + z^2 = 4$ 45. $3x^2 + 3y^2 - z^2 = 0$

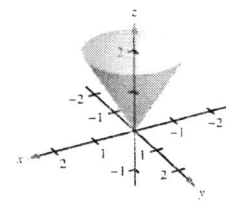

47. $x^2 + y^2 + (z-2)^2 = 4$ 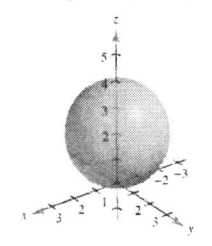   49. $x^2 + y^2 = 1$

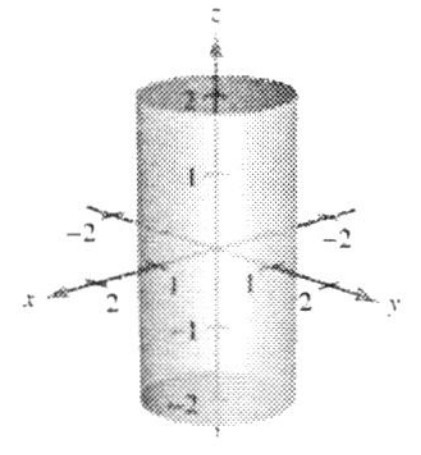

51. $\left(4, \dfrac{\pi}{4}, \dfrac{\pi}{2}\right)$ 53. $\left(4\sqrt{2}, \dfrac{\pi}{2}, \dfrac{\pi}{4}\right)$
55. $\left(2\sqrt{13}, -\dfrac{\pi}{6}, \arccos\left[\dfrac{3}{\sqrt{13}}\right]\right)$ 57. $\left(13, \pi, \arccos\left[\dfrac{5}{13}\right]\right)$
59. $\left(10, \dfrac{\pi}{6}, 0\right)$ 61. $(36, \pi, 0)$
63. $\left(3\sqrt{3}, -\dfrac{\pi}{6}, 3\right)$ 65. $\left(4, \dfrac{7\pi}{6}, 4\sqrt{3}\right)$

	Rectangular	Cylindrical	Spherical
67.	$(4, 6, 3)$	$(7.211, 0.983, 3)$	$(7.810, 0.983, 1.177)$
69.	$(4.698, 1.710, 8)$	$\left(5, \dfrac{\pi}{9}, 8\right)$	$(9.434, 0.349, 0.559)$
71.	$(-7.071, 12.247, 14.142)$	$(14.142, 2.094, 14.142)$	$\left(20, \dfrac{2\pi}{3}, \dfrac{\pi}{4}\right)$
73.	$(3, -2, 2)$	$(3.606, -0.588, 2)$	$(4.123, -0.588, 1.064)$
75.	$\left(\dfrac{5}{2}, \dfrac{4}{3}, -\dfrac{3}{2}\right)$	$(2.833, 0.490, -1.5)$	$(3.206, 0.490, 2.058)$
77.	$(-3.536, 3.536, -5)$	$\left(5, \dfrac{3\pi}{4}, -5\right)$	$(7.071, 2.356, 2.356)$
79.	$(2.804, -2.095, 6)$	$(-3.5, 2.5, 6)$	$(6.946, 5.641, 0.528)$

81. d   82. e   83. c   84. a   85. f   86. b
87. Rectangular to cylindrical:
$r^2 = x^2 + y^2, \tan\theta = \dfrac{y}{x}, z = z$
Cylindrical to rectangular:
$x = r\cos\theta, y = r\sin\theta, z = z$
89. Rectangular to spherical:
$\rho^2 = x^2 + y^2 + z^2, \tan\theta = \dfrac{y}{x}, \phi = \arccos\left(\dfrac{z}{\sqrt{x^2+y^2+z^2}}\right)$
Spherical to rectangular:
$x = \rho\sin\phi\cos\theta, y = \rho\sin\phi\sin\theta, z = \rho\cos\phi$

91. (a) $r^2 + z^2 = 16$ (b) $\rho = 4$
93. (a) $r^2 + (z-1)^2 = 1$ (b) $\rho = 2\cos\phi$
95. (a) $r = 4\sin\theta$ (b) $\rho = \dfrac{4\sin\theta}{\sin\phi} = 4\sin\theta\csc\phi$
97. (a) $r^2 = \dfrac{9}{\cos^2\theta - \sin^2\theta}$ (b) $\rho^2 = \dfrac{9\csc^2\phi}{\cos^2\theta - \sin^2\theta}$

99.

101.

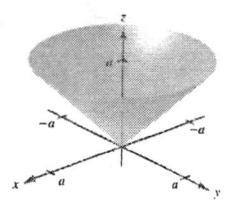

103.

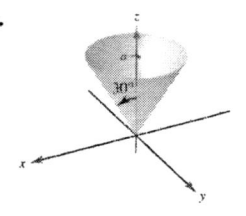

105. Rectangular: $0 \le x \le 10$
$0 \le y \le 10$
$0 \le z \le 10$

107. Spherical: $4 \le \rho \le 6$ 109. Ellipse

Review Exercises for Chapter 10 (page 780)

1. (a) $\mathbf{u} = 3\mathbf{i} - \mathbf{j}$
   $\mathbf{v} = 4\mathbf{i} + 2\mathbf{j}$
   (b) $2\sqrt{5}$ (c) $10\mathbf{i}$
3. $\mathbf{v} = \langle -4, 4\sqrt{3} \rangle$ 5. $\dfrac{10}{\sqrt{11}} \approx 3.015$ feet
7. $(-5, 4, 0)$
9. Above the xy-plane and to the right of the xz-plane *or* below the xy-plane and to the left of the xz-plane
11. $(x-3)^2 + (y+2)^2 + (z-6)^2 = \dfrac{225}{4}$
13. Center: $(2, 3, 0)$
    Radius: 3

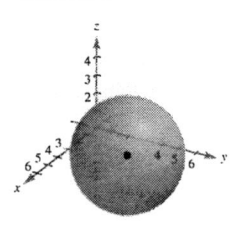

15. $\mathbf{u} = \langle 2, 5, -10 \rangle$

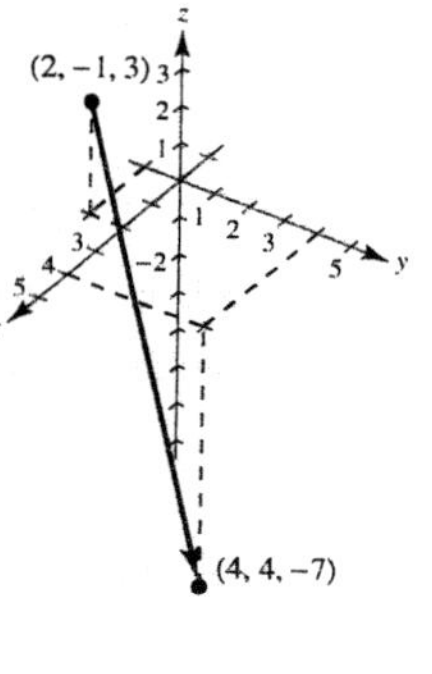

17. Collinear   19. $\dfrac{1}{\sqrt{38}}\langle 2, 3, 5 \rangle$
21. (a) $\mathbf{u} = \langle -1, 4, 0 \rangle$, $\mathbf{v} = \langle -3, 0, 6 \rangle$ 23. Orthogonal
    (b) 3   (c) 45
25. $\theta = \arccos\left(\dfrac{\sqrt{2}+\sqrt{6}}{4}\right) = 15°$ 27. π

29. Answers will vary. Example: $\langle -6, 5, 0 \rangle$, $\langle 6, -5, 0 \rangle$
31. $\mathbf{u} \cdot \mathbf{u} = 14 = \|\mathbf{u}\|^2$ 33. $\left\langle -\dfrac{15}{14}, \dfrac{5}{7}, -\dfrac{5}{14} \right\rangle$
35. $\dfrac{1}{\sqrt{5}}(-2\mathbf{i} - \mathbf{j})$ or $\dfrac{1}{\sqrt{5}}(2\mathbf{i} + 2\mathbf{j})$
37. 4   39. $\sqrt{285}$ 41. $100\sec 20° \approx 106.4$ pounds
43. (a) $x = 1, y = 2 + t, z = 3$ (b) None
45. (a) $x = t, y = -1 + t, z = 1$ (b) $x = y + 1, z = 1$
47. $x + 2y = 1$ 49. $\dfrac{8}{7}$ 51. $\dfrac{\sqrt{30}}{3}$

53.

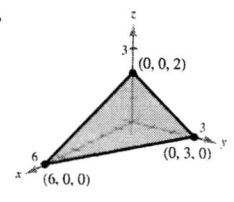

55.

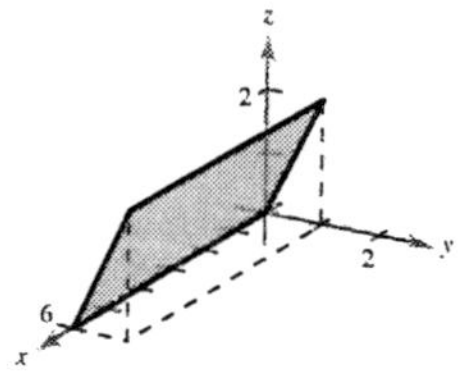

57.

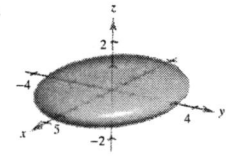

59.

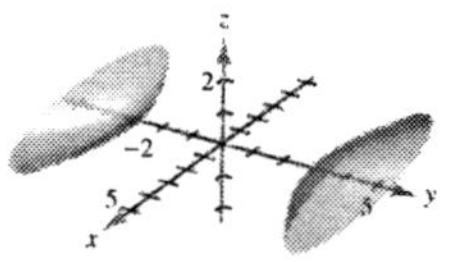

61. (a) $x^2 + y^2 - 2z + 2 = 0$
    (b) $4\pi \approx 12.6$ cubic centimeters
    (c) $\dfrac{225\pi}{64} \approx 11.0$ cubic centimeters

63. (a) $\left(4, \dfrac{3\pi}{4}, 2\right)$ (b) $\left(2\sqrt{5}, \dfrac{3\pi}{4}, \arccos\left[\dfrac{\sqrt{5}}{5}\right]\right)$
65. $\left(50\sqrt{5}, -\dfrac{\pi}{6}, \arccos\left[\dfrac{1}{\sqrt{5}}\right]\right)$ 67. $\left(\dfrac{25\sqrt{2}}{2}, -\dfrac{\pi}{4}, -\dfrac{25\sqrt{2}}{2}\right)$
69. (a) $r^2\cos 2\theta = 2z$ (b) $\rho = 2\sec 2\theta \cos\phi \csc^2\phi$

P.S. Problem Solving (page 782)

1. Proof   3. Proof
5. (a) $\dfrac{3\sqrt{2}}{2} \approx 2.12$ (b) $\sqrt{5} \approx 2.24$
7. (a) $\dfrac{\pi}{2}$ (b) $\dfrac{1}{2}(\pi ab)k^2$
   (c) $V = \dfrac{1}{2}(\pi ab)k^2$
   $V = \dfrac{1}{2}(\text{area of base})\text{height}$

9. (a)   (b)

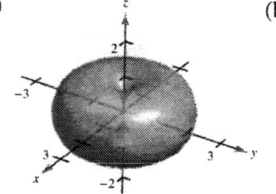

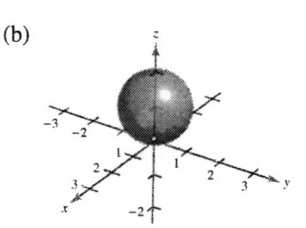

11. Proof

13. (a) Tension: $\dfrac{2\sqrt{3}}{3} \approx 1.1547$ pounds

   Magnitude of **u**: $\dfrac{\sqrt{3}}{3} \approx 0.5774$ pounds

   (b) $\|\mathbf{u}\| = \tan \theta$; $T = \sec \theta$; Domain: $0 \le \theta \le 90°$

   (c)
θ	0°	10°	20°	30°	40°
T	1	1.0154	1.0642	1.1547	1.3054
$\|\mathbf{u}\|$	0	0.1763	0.3640	0.5774	0.8391

θ	50°	60°
T	1.5557	2
$\|\mathbf{u}\|$	1.1918	1.7321

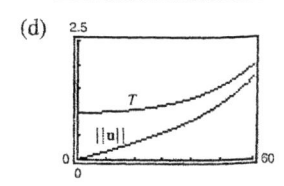

   (d) [graph]

   (e) Both are increasing functions.

   (f) $\lim\limits_{\theta \to \pi/2^-} T = \infty$ and $\lim\limits_{\theta \to \pi/2^-} \|\mathbf{u}\| = \infty$

   Yes. As θ increases, both T and $\|\mathbf{u}\|$ increase.

15. $\langle 0, 0, \cos\alpha \sin\beta - \cos\beta \sin\alpha \rangle$; Proof

17. $D = \dfrac{|\overrightarrow{PQ} \cdot \mathbf{n}|}{\|\mathbf{n}\|}$

   $= \dfrac{|\mathbf{w} \cdot (\mathbf{u} \times \mathbf{v})|}{\|\mathbf{u} \times \mathbf{v}\|} = \dfrac{|(\mathbf{u} \times \mathbf{v}) \cdot \mathbf{w}|}{\|\mathbf{u} \times \mathbf{v}\|} = \dfrac{|\mathbf{u} \cdot (\mathbf{v} \times \mathbf{w})|}{\|\mathbf{u} \times \mathbf{v}\|}$

19. $a_1, b_1, c_1,$ and a_2, b_2, c_2 are two sets of direction numbers for the same line. The line is parallel to both $\mathbf{u} = a_1\mathbf{i} + b_1\mathbf{j} + c_1\mathbf{k}$ and $\mathbf{v} = a_2\mathbf{i} + b_2\mathbf{j} + c_2\mathbf{k}$. Therefore, **u** and **v** are parallel, and there exists a scalar d such that $\mathbf{u} = d\mathbf{v}$, $a_1\mathbf{i} + b_1\mathbf{j} + c_1\mathbf{k} = d(a_2\mathbf{i} + b_2\mathbf{j} + c_2\mathbf{k})$, $a_1 = a_2d$, $b_1 = b_2d$, $c_1 = c_2d$.

Chapter 11

Section 11.1 (page 791)

1. $(-\infty, 0) \cup (0, \infty)$ 3. $(0, \infty)$
5. $[0, \infty)$ 7. $(-\infty, \infty)$
9. (a) $\tfrac{1}{2}\mathbf{i}$ (b) $\mathbf{j}$ (c) $\tfrac{1}{2}(s+1)^2\mathbf{i} - s\mathbf{j}$
   (d) $\tfrac{1}{2}\Delta t(\Delta t + 4)\mathbf{i} - \Delta t\mathbf{j}$

11. (a) $\ln 2\mathbf{i} + \dfrac{1}{2}\mathbf{j} + 6\mathbf{k}$ (b) Not possible

    (c) $\ln(t-4)\mathbf{i} + \dfrac{1}{t-4}\mathbf{j} + 3(t-4)\mathbf{k}$

    (d) $\ln(1+\Delta t)\mathbf{i} - \dfrac{\Delta t}{1+\Delta t}\mathbf{j} + 3\Delta t\mathbf{k}$

13. $\sqrt{1+t^2}$ 15. $t^2(5t-1)$; The dot product is a scalar.
17. b   18. c   19. d   20. a

21. (a) $(-20, 0, 0)$ (b) $(10, 20, 10)$
    (c) $(0, 0, 20)$ (d) $(20, 0, 0)$

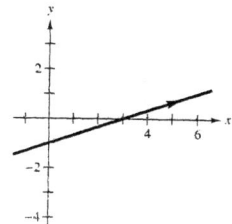

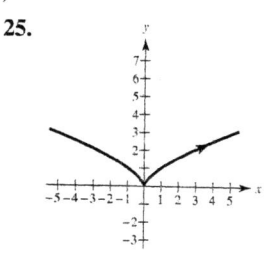

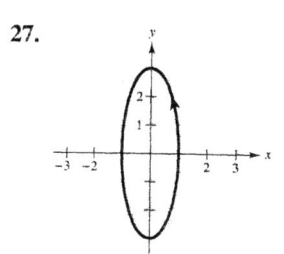

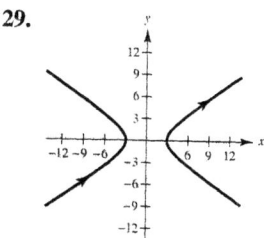

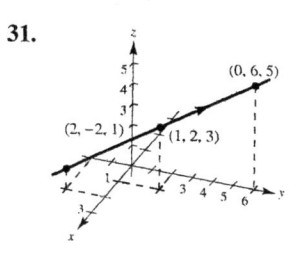

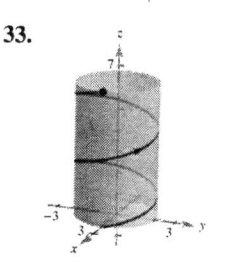

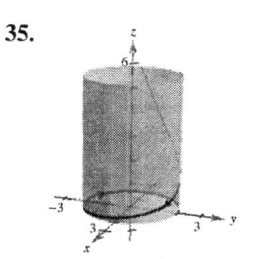

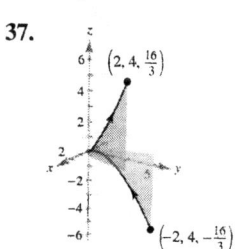

23. [graph]   25. [graph]

27. [graph]   29. [graph]

31. [graph with points $(2, -2, 1)$, $(1, 2, 3)$, $(0, 6, 5)$]   33. [graph]

35. [graph]   37. [graph with points $(2, 4, \tfrac{16}{3})$ and $(-2, 4, -\tfrac{16}{3})$]

39. Parabola   41. Helix

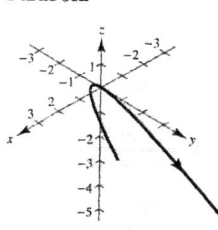

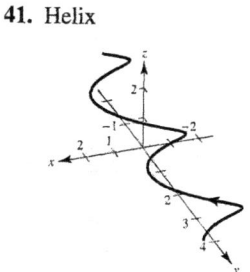

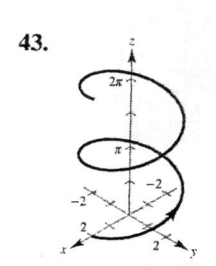

[graph]   [graph]

43. [graph]   (a) The helix is translated 2 units back on the x-axis.

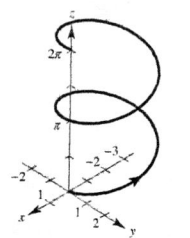

    [graph]

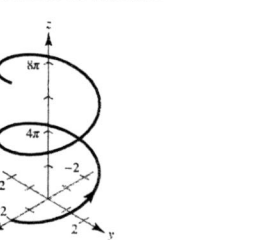

(b) The height of the helix increases at a faster rate.
(c) The orientation of the graph is reversed. 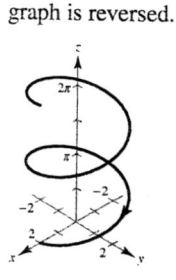
(d) The axis of the helix is the x-axis. 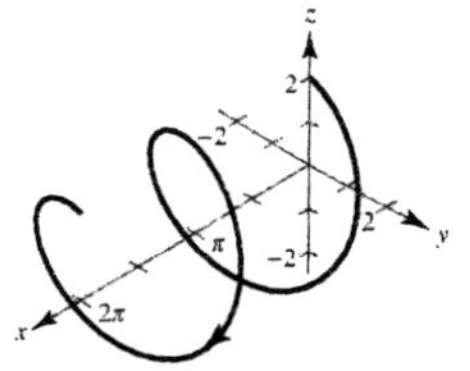
(e) The radius of the helix is increased from 2 to 6.

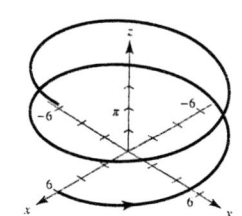

45.–51. Answers are not unique.
45. $\mathbf{r}(t) = t\mathbf{i} + (4 - t)\mathbf{j}$ **47.** $\mathbf{r}(t) = t\mathbf{i} + (t - 2)^2\mathbf{j}$
49. $\mathbf{r}(t) = 5\cos t\mathbf{i} + 5\sin t\mathbf{j}$ **51.** $\mathbf{r}(t) = 4\sec t\mathbf{i} + 2\tan t\mathbf{j}$
53. $\mathbf{r}(t) = \langle 2 - 2t, 3 + 5t, 8t \rangle$

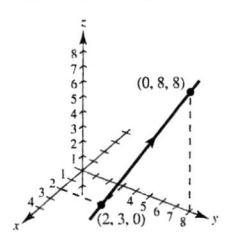

55. $\mathbf{r}_1(t) = t\mathbf{i}, \quad 0 \le t \le 4$
$\mathbf{r}_2(t) = (4 - 4t)\mathbf{i} + 6t\mathbf{j}, \quad 0 \le t \le 1$
$\mathbf{r}_3(t) = (6 - t)\mathbf{j}, \quad 0 \le t \le 6$

57. $\mathbf{r}_1(t) = t\mathbf{i} + t^2\mathbf{j}, \quad 0 \le t \le 2$
$\mathbf{r}_2(t) = (2 - t)\mathbf{i}, \quad 0 \le t \le 2$
$\mathbf{r}_3(t) = (4 - t)\mathbf{j}, \quad 0 \le t \le 4$

59. $\mathbf{r}(t) = t\mathbf{i} - t\mathbf{j} + 2t^2\mathbf{k}$ **61.** $\mathbf{r}(t) = 2\sin t\mathbf{i} + 2\cos t\mathbf{j} + 4\sin^2 t\mathbf{k}$

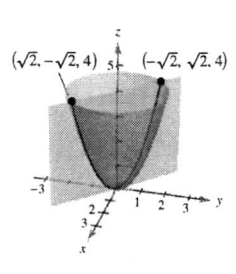

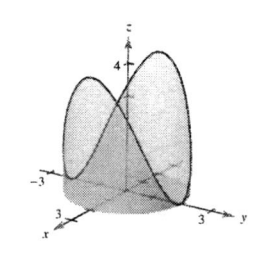

63. $\mathbf{r}(t) = (1 + \sin t)\mathbf{i} + \sqrt{2}\cos t\mathbf{j} + (1 - \sin t)\mathbf{k}$ and
$\mathbf{r}(t) = (1 + \sin t)\mathbf{i} - \sqrt{2}\cos t\mathbf{j} + (1 - \sin t)\mathbf{k}$

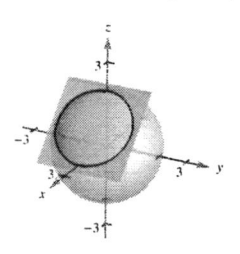

65. $\mathbf{r}(t) = t\mathbf{i} + t\mathbf{j} + \sqrt{4 - t^2}\mathbf{k}$

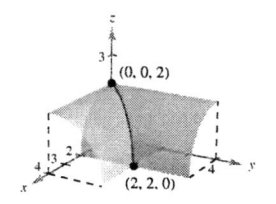

67. Let $x = t$, $y = 2t\cos t$, and $z = 2t\sin t$. Then
$y^2 + z^2 = (2t\cos t)^2 + (2t\sin t)^2$
$= 4t^2\cos^2 t + 4t^2\sin^2 t$
$= 4t^2(\cos^2 t + \sin^2 t)$
$= 4t^2.$
Since $x = t$, $y^2 + z^2 = 4x^2$.

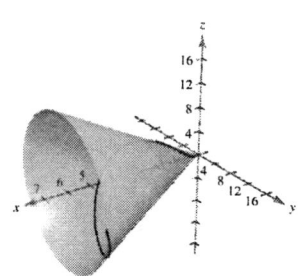

69. $2\mathbf{i} + 2\mathbf{j} + \tfrac{1}{2}\mathbf{k}$ **71.** 0   **73.** Limit does not exist.
75. $(-\infty, 0), (0, \infty)$ **77.** $[-1, 1]$
79. $\left(-\dfrac{\pi}{2} + n\pi, \dfrac{\pi}{2} + n\pi\right)$, n is an integer.
81. A function of the form $\mathbf{r}(t) = f(t)\mathbf{i} + g(t)\mathbf{j}$ (plane) or $\mathbf{r}(t) = f(t)\mathbf{i} + g(t)\mathbf{j} + h(t)\mathbf{k}$ (space) is a vector-valued function, where the component functions f, g, and h are real-valued functions of the parameter t.
83. (a) $\mathbf{s}(t) = t^2\mathbf{i} + (t - 3)\mathbf{j} + (t + 3)\mathbf{k}$
(b) $\mathbf{s}(t) = (t^2 - 2)\mathbf{i} + (t - 3)\mathbf{j} + t\mathbf{k}$
(c) $\mathbf{s}(t) = t^2\mathbf{i} + (t + 2)\mathbf{j} + t\mathbf{k}$
85. Proof   **87.** Proof   **89.** True

Section 11.2 (page 800)

1. $\mathbf{r}(2) = 4\mathbf{i} + 2\mathbf{j}$
   $\mathbf{r}'(2) = 4\mathbf{i} + \mathbf{j}$

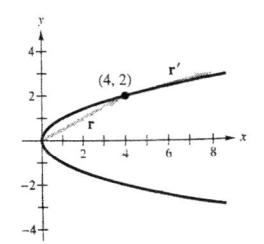

3. $\mathbf{r}\left(\dfrac{\pi}{2}\right) = \mathbf{j}$
   $\mathbf{r}'\left(\dfrac{\pi}{2}\right) = -\mathbf{i}$

$\mathbf{r}'(t_0)$ is tangent to the curve at t_0.

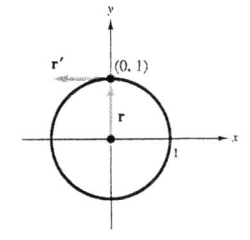

$\mathbf{r}'(t_0)$ is tangent to the curve at t_0.

5. (a) and (b)

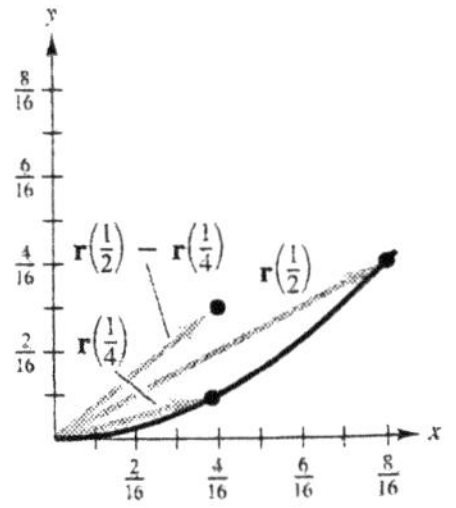

   (c) The vector
   $$\dfrac{\mathbf{r}(1/2) - \mathbf{r}(1/4)}{(1/2) - (1/4)}$$
   approximates the tangent vector $\mathbf{r}'(1/4)$.

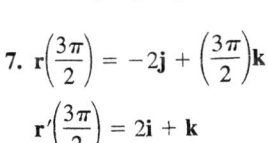

7. $\mathbf{r}\left(\dfrac{3\pi}{2}\right) = -2\mathbf{j} + \left(\dfrac{3\pi}{2}\right)\mathbf{k}$
   $\mathbf{r}'\left(\dfrac{3\pi}{2}\right) = 2\mathbf{i} + \mathbf{k}$

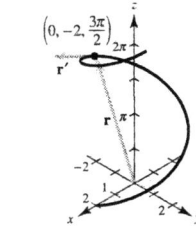

9. $6\mathbf{i} - 14t\mathbf{j} + 3t^2\mathbf{k}$
11. $-3a \sin t \cos^2 t\, \mathbf{i} + 3a \sin^2 t \cos t\, \mathbf{j}$
13. $-e^{-t}\mathbf{i}$
15. $\langle \sin t + t \cos t, \cos t - t \sin t, 1\rangle$
17. (a) $6t\mathbf{i} + \mathbf{j}$ (b) $18t^3 + t$
19. (a) $-4\cos t\,\mathbf{i} - 4\sin t\,\mathbf{j}$ (b) 0
21. (a) $\mathbf{i} + t\mathbf{k}$ (b) $\dfrac{t^3}{2} + t$
23. (a) $\langle \cos t - t \sin t, \sin t + t \cos t, 0\rangle$ (b) t
25. $\dfrac{\mathbf{r}'(-1/4)}{\|\mathbf{r}'(-1/4)\|} = \dfrac{1}{\sqrt{4\pi^2+1}}(\sqrt{2}\pi\mathbf{i} + \sqrt{2}\pi\mathbf{j} - \mathbf{k})$

$\dfrac{\mathbf{r}''(-1/4)}{\|\mathbf{r}''(-1/4)\|} = \dfrac{1}{2\sqrt{\pi^4+4}}(-\sqrt{2}\pi^2\mathbf{i} + \sqrt{2}\pi^2\mathbf{j} + 4\mathbf{k})$

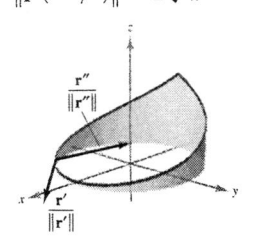

27. $(-\infty, 0), (0, \infty)$
29. $\left(\dfrac{n\pi}{2}, \dfrac{(n+1)\pi}{2}\right)$
31. $(-\infty, \infty)$
33. $(-\infty, 0), (0, \infty)$
35. $\left(-\dfrac{\pi}{2} + n\pi, \dfrac{\pi}{2} + n\pi\right)$, n is an integer.
37. (a) $\mathbf{i} + 3\mathbf{j} + 2t\mathbf{k}$ (b) $2\mathbf{k}$ (c) $8t + 9t^2 + 5t^4$
   (d) $-\mathbf{i} + (9 - 2t)\mathbf{j} + (6t - 3t^2)\mathbf{k}$
   (e) $8t^3\mathbf{i} + (12t^2 - 4t^3)\mathbf{j} + (3t^2 - 24t)\mathbf{k}$
   (f) $\dfrac{10 + 2t^2}{\sqrt{10+t^2}}$

39. $\theta(t) = \arccos\left(\dfrac{-7 \sin t \cos t}{\sqrt{9\sin^2 t + 16\cos^2 t}\,\sqrt{9\cos^2 t + 16\sin^2 t}}\right)$

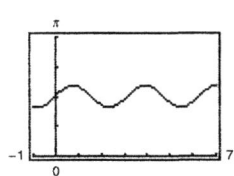

Maximum: $\theta\left(\dfrac{\pi}{4}\right) = \theta\left(\dfrac{5\pi}{4}\right) \approx 1.855$

Minimum: $\theta\left(\dfrac{3\pi}{4}\right) = \theta\left(\dfrac{7\pi}{4}\right) \approx 1.287$

Orthogonal: $\dfrac{n\pi}{2}$, n is an integer

41. $\mathbf{r}'(t) = 3\mathbf{i} - 2t\mathbf{j}$
43. $t^2\mathbf{i} + t\mathbf{j} + t\mathbf{k} + \mathbf{C}$
45. $\ln t\,\mathbf{i} + t\mathbf{j} - \tfrac{2}{5}t^{5/2}\mathbf{k} + \mathbf{C}$
47. $(t^2 - t)\mathbf{i} + t^4\mathbf{j} + 2t^{3/2}\mathbf{k} + \mathbf{C}$
49. $\tan t\,\mathbf{i} + \arctan t\,\mathbf{j} + \mathbf{C}$
51. $4\mathbf{i} + \tfrac{1}{2}\mathbf{j} - \mathbf{k}$
53. $a\mathbf{i} + a\mathbf{j} + \dfrac{\pi}{2}\mathbf{k}$
55. $2e^{2t}\mathbf{i} + 3(e^t - 1)\mathbf{j}$
57. $600\sqrt{3}t\mathbf{i} + (-16t^2 + 600t)\mathbf{j}$
59. $\left(\dfrac{2 - e^{-t^2}}{2}\right)\mathbf{i} + (e^{-t} - 2)\mathbf{j} + (t+1)\mathbf{k}$

61. See "Definition of the Derivative of a Vector-Valued Function" and Figure 11.8 on page 794.
63. The three components of $\mathbf{u}$ are increasing functions of t at $t = t_0$.
65. Proof 67. Proof 69. Proof 71. Proof
73. False: Let $\mathbf{r}(t) = \cos t\,\mathbf{i} + \sin t\,\mathbf{j} + \mathbf{k}$, then $\dfrac{d}{dt}[\|\mathbf{r}(t)\|] = 0$, but $\|\mathbf{r}'(t)\| = 1$.

Section 11.3 (page 808)

1. $\mathbf{v}(1) = 3\mathbf{i} + \mathbf{j}$
   $\mathbf{a}(1) = \mathbf{0}$

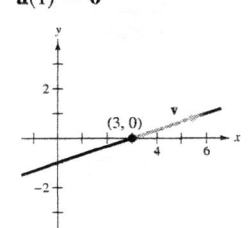

3. $\mathbf{v}(2) = 4\mathbf{i} + \mathbf{j}$
   $\mathbf{a}(2) = 2\mathbf{i}$

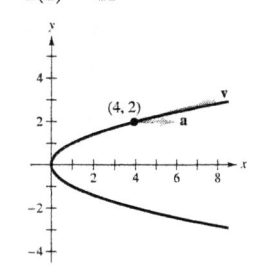

5. $\mathbf{v}\left(\dfrac{\pi}{4}\right) = -\sqrt{2}\mathbf{i} + \sqrt{2}\mathbf{j}$

$\mathbf{a}\left(\dfrac{\pi}{4}\right) = -\sqrt{2}\mathbf{i} - \sqrt{2}\mathbf{j}$

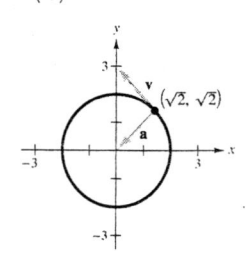

7. $\mathbf{v}(\pi) = 2\mathbf{i}$

$\mathbf{a}(\pi) = -\mathbf{j}$

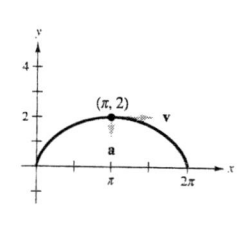

9. $\mathbf{v}(t) = \mathbf{i} + 2\mathbf{j} + 3\mathbf{k}$
$s(t) = \sqrt{14}$
$\mathbf{a}(t) = \mathbf{0}$

11. $\mathbf{v}(t) = \mathbf{i} + 2t\mathbf{j} + t\mathbf{k}$
$s(t) = \sqrt{1 + 5t^2}$
$\mathbf{a}(t) = 2\mathbf{j} + \mathbf{k}$

13. $\mathbf{v}(t) = \mathbf{i} + \mathbf{j} - \dfrac{t}{\sqrt{9 - t^2}}\mathbf{k}$

$s(t) = \sqrt{\dfrac{18 - t^2}{9 - t^2}}$

$\mathbf{a}(t) = \dfrac{-9}{(9 - t^2)^{3/2}}\mathbf{k}$

15. $\mathbf{v}(t) = 4\mathbf{i} - 3\sin t\mathbf{j} + 3\cos t\mathbf{k}$
$s(t) = 5$
$\mathbf{a}(t) = -3\cos t\mathbf{j} - 3\sin t\mathbf{k}$

17. (a) $x = 1 + t$ (b) $(1.100, -1.200, 0.325)$
$y = -1 - 2t$
$z = \tfrac{1}{4} + \tfrac{3}{4}t$

19. $\mathbf{v}(t) = t(\mathbf{i} + \mathbf{j} + \mathbf{k})$

$\mathbf{r}(t) = \dfrac{t^2}{2}(\mathbf{i} + \mathbf{j} + \mathbf{k})$

$\mathbf{r}(2) = 2(\mathbf{i} + \mathbf{j} + \mathbf{k})$

21. $\mathbf{v}(t) = \left(\dfrac{t^2}{2} + \dfrac{9}{2}\right)\mathbf{j} + \left(\dfrac{t^2}{2} - \dfrac{1}{2}\right)\mathbf{k}$

$\mathbf{r}(t) = \left(\dfrac{t^3}{6} + \dfrac{9}{2}t - \dfrac{14}{3}\right)\mathbf{j} + \left(\dfrac{t^3}{6} - \dfrac{1}{2}t + \dfrac{1}{3}\right)\mathbf{k}$

$\mathbf{r}(2) = \dfrac{17}{3}\mathbf{j} + \dfrac{2}{3}\mathbf{k}$

23. The velocity of an object involves both magnitude and direction of motion, whereas speed involves only magnitude.

25. $\mathbf{r}(t) = 44\sqrt{3}t\mathbf{i} + (10 + 44t - 16t^2)\mathbf{j}$

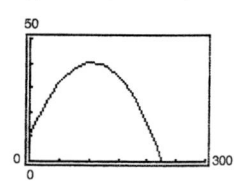

27. $v_0 = 40\sqrt{6}$ feet per second; 78 feet

29. $x(t) = t(v_0 \cos\theta)$
$y(t) = t(v_0 \sin\theta) - 16t^2 + h$

$y = \dfrac{x}{v_0 \cos\theta}(v_0 \sin\theta) - 16\left(\dfrac{x^2}{v_0^2 \cos^2\theta}\right) + h$

$= (\tan\theta)x - \left(\dfrac{16}{v_0^2}\sec^2\theta\right)x^2 + h$

31. (a) $y = -0.004x^2 + 0.37x + 6$
$\mathbf{r}(t) = t\mathbf{i} + (-0.004t^2 + 0.37t + 6)\mathbf{j}$

(b)

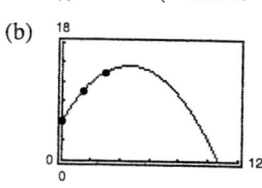

(c) 14.56 feet

(d) Initial velocity: 67.4 feet per second; $\theta \approx 20.14°$

33. (a) $\mathbf{r}(t) = \left(\dfrac{440}{3}\cos\theta_0\right)t\mathbf{i} + \left[3 + \left(\dfrac{440}{3}\sin\theta_0\right)t - 16t^2\right]\mathbf{j}$

(b)

The minimum angle appears to be $\theta_0 = 20°$.

(c) $\theta_0 \approx 19.38°$

35. (a) $v_0 = 28.78$ feet per second; $\theta = 58.28°$

(b) $v_0 \approx 32$ feet per second

37. $1.91°$

39. (a)

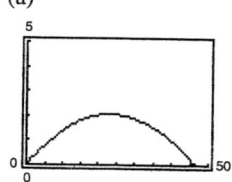

(b)

Maximum height: 2.1 feet
Range: 46.6 feet

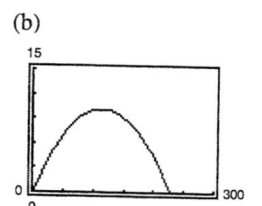

Maximum height: 10.0 feet
Range: 227.8 feet

(c)

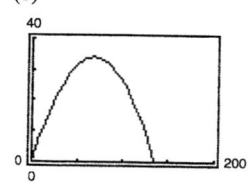

(d)

Maximum height: 34.0 feet
Range: 136.1 feet

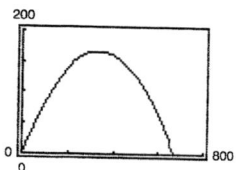

Maximum height: 166.5 feet
Range: 666.1 feet

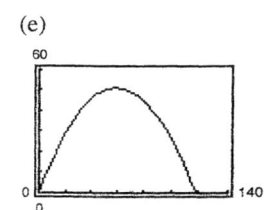

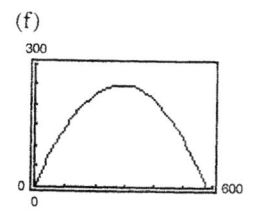

(e) Maximum height: 51.0 feet
Range: 117.9 feet

(f) Maximum height: 249.8 feet
Range: 576.9 feet

41. Maximum height: 129.1 meters
Range: 886.3 meters

43. $\mathbf{v}(t) = b\omega[(1 - \cos \omega t)\mathbf{i} + \sin \omega t \mathbf{j}]$
$\mathbf{a}(t) = b\omega^2(\sin \omega t \mathbf{i} + \cos \omega t \mathbf{j})$
(a) $\|\mathbf{v}(t)\| = 0$ when $\omega t = 0, 2\pi, 4\pi, \ldots$
(b) $\|\mathbf{v}(t)\|$ is maximum when $\omega t = \pi, 3\pi, \ldots$

45. $\mathbf{v}(t) = -b\omega \sin \omega t \mathbf{i} + b\omega \cos \omega t \mathbf{j}$
$\mathbf{v}(t) \cdot \mathbf{r}(t) = 0$

47. $\mathbf{a}(t) = -b\omega^2(\cos \omega t \mathbf{i} + \sin \omega t \mathbf{j}) = -\omega^2 \mathbf{r}(t)$

49. $8\sqrt{10}$ feet per second   51. Proof   53. Proof

55. (a) $\mathbf{v}(t) = -6 \sin t \mathbf{i} + 3 \cos t \mathbf{j}$
$\|\mathbf{v}(t)\| = 3\sqrt{3 \sin^2 t + 1}$
$\mathbf{a}(t) = -6 \cos t \mathbf{i} - 3 \sin t \mathbf{j}$

(b)
t	0	$\pi/4$	$\pi/2$	$2\pi/3$	π
Speed	3	$3\sqrt{10}/2$	6	$3\sqrt{13}/2$	3

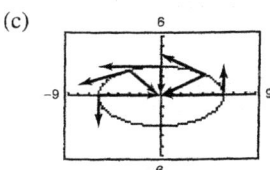

(c)

(d) The speed is increasing when the angle between $\mathbf{v}$ and $\mathbf{a}$ is in the interval $[0, \pi/2)$, and decreasing when the angle is in the interval $(\pi/2, \pi]$.

Section 11.4 (page 817)

1. $\mathbf{T}(1) = \frac{\sqrt{2}}{2}(\mathbf{i} + \mathbf{j})$ 3. $\mathbf{T}\left(\frac{\pi}{4}\right) = \frac{\sqrt{2}}{2}(-\mathbf{i} + \mathbf{j})$

5. $\mathbf{T}(0) = \frac{\sqrt{2}}{2}(\mathbf{i} + \mathbf{k})$ 7. $\mathbf{T}(0) = \frac{\sqrt{5}}{5}(2\mathbf{j} + \mathbf{k})$
$x = t$ $x = 2$
$y = 0$ $y = 2t$
$z = t$ $z = t$

9. $\mathbf{T}\left(\frac{\pi}{4}\right) = \frac{1}{2}\langle -\sqrt{2}, \sqrt{2}, 0 \rangle$
$x = \sqrt{2} - \sqrt{2}t$
$y = \sqrt{2} + \sqrt{2}t$
$z = 4$

11. $\mathbf{T}(3) = \frac{1}{19}\langle 1, 6, 18 \rangle$
$x = 3 + t$
$y = 9 + 6t$
$z = 18 + 18t$

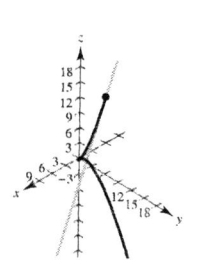

13. Tangent line: $x = 1 + t$
$y = t$
$z = 1 + \frac{1}{2}t$
$\mathbf{r}(1.1) \approx \langle 1.1, 0.1, 1.05 \rangle$

15. $1.2°$

17. $\mathbf{N}(2) = \frac{\sqrt{5}}{5}(-2\mathbf{i} + \mathbf{j})$ 19. $\mathbf{N}\left(\frac{3\pi}{4}\right) = \frac{\sqrt{2}}{2}(\mathbf{i} - \mathbf{j})$

21. $\mathbf{v}(t) = 4\mathbf{i}$ 23. $\mathbf{v}(t) = 8t\mathbf{i}$
$\mathbf{a}(t) = 0$ $\mathbf{a}(t) = 8\mathbf{i}$
$\mathbf{T}(t) = \mathbf{i}$ $\mathbf{T}(t) = \mathbf{i}$
$\mathbf{N}(t)$ is undefined. The path $\mathbf{N}(t)$ is undefined. The path
is a line and the speed is   is a line and the speed is
constant.   variable.

25. $\mathbf{T}(1) = \frac{\sqrt{2}}{2}(\mathbf{i} - \mathbf{j})$ 27. $\mathbf{T}\left(\frac{\pi}{2}\right) = \frac{\sqrt{2}}{2}(-\mathbf{i} + \mathbf{j})$
$\mathbf{N}(1) = \frac{\sqrt{2}}{2}(\mathbf{i} + \mathbf{j})$ $\mathbf{N}\left(\frac{\pi}{2}\right) = -\frac{\sqrt{2}}{2}(\mathbf{i} + \mathbf{j})$
$a_\mathbf{T} = -\sqrt{2}$ $a_\mathbf{T} = \sqrt{2}e^{\pi/2}$
$a_\mathbf{N} = \sqrt{2}$ $a_\mathbf{N} = \sqrt{2}e^{\pi/2}$

29. $\mathbf{T}(t_0) = (\cos \omega t_0)\mathbf{i} + (\sin \omega t_0)\mathbf{j}$
$\mathbf{N}(t_0) = (-\sin \omega t_0)\mathbf{i} + (\cos \omega t_0)\mathbf{j}$
$a_\mathbf{T} = \omega^2$
$a_\mathbf{N} = \omega^3 t_0$

31. $\mathbf{T}(t) = -\sin(\omega t)\mathbf{i} + \cos(\omega t)\mathbf{j}$
$\mathbf{N}(t) = -\cos(\omega t)\mathbf{i} - \sin(\omega t)\mathbf{j}$
$a_\mathbf{T} = 0$
$a_\mathbf{N} = a\omega^2$

33. $\|\mathbf{v}(t)\| = a\omega$; The speed is constant because $a_\mathbf{T} = 0$.

35. $\mathbf{r}(2) = 2\mathbf{i} + \frac{1}{2}\mathbf{j}$
$\mathbf{T}(2) = \frac{\sqrt{17}}{17}(4\mathbf{i} - \mathbf{j})$
$\mathbf{N}(2) = \frac{\sqrt{17}}{17}(\mathbf{i} + 4\mathbf{j})$

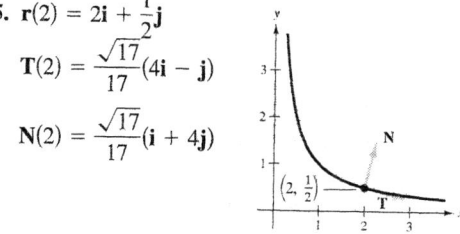

37. $\mathbf{T}(1) = \frac{\sqrt{14}}{14}(\mathbf{i} + 2\mathbf{j} - 3\mathbf{k})$
$\mathbf{N}(1)$ is undefined.
$a_\mathbf{T}$ is undefined.
$a_\mathbf{N}$ is undefined.

39. $T(1) = \frac{\sqrt{6}}{6}(i + 2j + k)$

$N(1) = \frac{\sqrt{30}}{30}(-5i + 2j + k)$

$a_T = \frac{5\sqrt{6}}{6}$

$a_N = \frac{\sqrt{30}}{6}$

41. $T\left(\frac{\pi}{2}\right) = \frac{1}{5}(4i - 3j)$

$N\left(\frac{\pi}{2}\right) = -k$

$a_T = 0$

$a_N = 3$

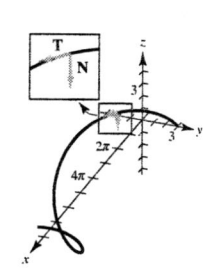

43. Let C be a smooth curve represented by r on an open interval I. The unit tangent vector $T(t)$ at t is defined as

$T(t) = \frac{r'(t)}{\|r'(t)\|}$, $r'(t) \neq 0$.

The principal unit normal vector $N(t)$ at t is defined as

$N(t) = \frac{T'(t)}{\|T'(t)\|}$, $T'(t) \neq 0$.

The tangential and normal components of acceleration are defined as follows

$a(t) = a_T T(t) + a_N N(t)$.

45. The particle's motion is in a straight line.

47. (a) $t = \frac{1}{2}$: $a_T = \frac{\sqrt{2}\pi^2}{2}$, $a_N = \frac{\sqrt{2}\pi^2}{2}$

$t = 1$: $a_T = 0$, $a_N = \pi^2$

$t = \frac{3}{2}$: $a_T = -\frac{\sqrt{2}\pi^2}{2}$, $a_N = \frac{\sqrt{2}\pi^2}{2}$

(b) $t = \frac{1}{2}$: Increasing since $a_T > 0$.

$t = 1$: Maximum since $a_T = 0$.

$t = \frac{3}{2}$: Decreasing since $a_T < 0$.

49. $T\left(\frac{\pi}{2}\right) = \frac{\sqrt{17}}{17}(-4i + k)$

$N\left(\frac{\pi}{2}\right) = -j$

$B\left(\frac{\pi}{2}\right) = \frac{\sqrt{17}}{17}(i + 4k)$

51. $a_T = \frac{-32(v_0 \sin\theta - 32t)}{\sqrt{v_0^2 \cos^2\theta + (v_0 \sin\theta - 32t)^2}}$

$a_N = \frac{32 v_0 \cos\theta}{\sqrt{v_0^2 \cos^2\theta + (v_0 \sin\theta - 32t)^2}}$

At maximum height, $a_T = 0$ and $a_N = 32$.

53. (a) $4\sqrt{625\pi^2 + 1} \approx 314$ miles per hour

(b) $a_T = 0$, $a_N = 1000\pi^2$

$a_T = 0$ because the speed is constant.

55. (a) The centripetal component is quadrupled.

(b) The centripetal component is halved.

57. 4.83 miles per second **59.** 4.67 miles per second

61. Proof **63.** Proof

Section 11.5 (page 828)

1. $4\sqrt{10}$ **3.** $6a$

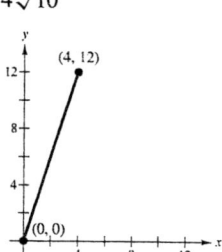

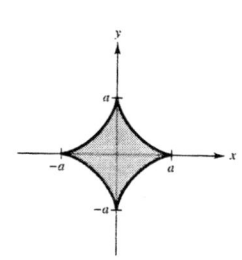

5. (a) $r(t) = (50t\sqrt{2})i + (3 + 50t\sqrt{2} - 16t^2)j$

(b) $\frac{649}{8} \approx 81$ feet (c) 315.5 feet (d) 362.9 feet

7. $2\sqrt{14}$ **9.** $2\pi\sqrt{a^2 + b^2}$

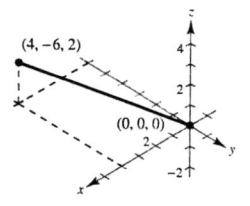

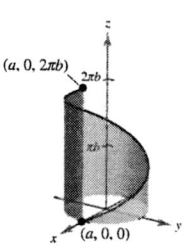

11. 8.37

13. (a) $2\sqrt{21} \approx 9.165$ (b) 9.529

(c) Increase the number of line segments. (d) 9.571

15. (a) $s = \sqrt{5}t$

(b) $r(s) = 2\cos\frac{s}{\sqrt{5}}i + 2\sin\frac{s}{\sqrt{5}}j + \frac{s}{\sqrt{5}}k$

(c) $s = \sqrt{5}$: $(1.081, 1.683, 1.000)$

$s = 4$: $(-0.433, 1.953, 1.789)$

(d) Proof

17. 0 **19.** $\frac{2}{5}$ **21.** 0 **23.** $\frac{\sqrt{2}}{2}$ **25.** $\frac{1}{4}$ **27.** $\frac{1}{a}$

29. $\frac{\sqrt{2}}{2}e^{-t}$ **31.** $\frac{1}{\omega t}$ **33.** $\frac{\sqrt{5}}{(1+5t^2)^{3/2}}$ **35.** $\frac{3}{25}$

37. $K = 0$, $1/K$ is undefined.

39. $K = \frac{4}{17^{3/2}}$, $\frac{1}{K} = \frac{17^{3/2}}{4}$ **41.** $K = \frac{1}{a}$, $\frac{1}{K} = a$

43. (a) $\left(x - \frac{\pi}{2}\right)^2 + y^2 = 1$

(b) Because the curvature is not as great, the radius of the curvature is greater.

45. $(x-1)^2 + \left(y - \frac{5}{2}\right)^2 = \left(\frac{1}{2}\right)^2$ **47.** $(x+2)^2 + (y-3)^2 = 8$

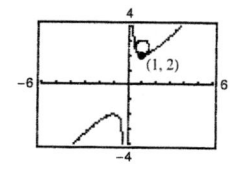

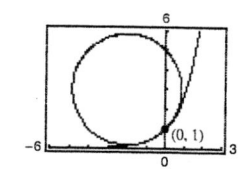

49.

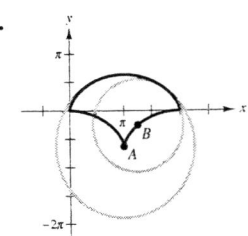

51. (a) (1, 3) (b) 0

53. (a) $K \to \infty$ as $x \to 0$ (b) 0   **55.** (1, 3)

57. $K = \dfrac{|y''|}{[1 + (y')^2]^{3/2}}$

The curvature is 0 at every point at which $y'' = 0$.

59. If C is a smooth curve given by $\mathbf{r}(t) = x(t)\mathbf{i} + y(t)\mathbf{j} + z(t)\mathbf{k}$ on an interval $[a, b]$, then the arc length of C on the interval is
$$s = \int_a^b \sqrt{[x'(t)]^2 + [y'(t)]^2 + [z'(t)]^2}\, dt = \int_a^b \|\mathbf{r}'(t)\|\, dt.$$

61. The graph is linear.

63. $x^2 + 4y^2 = 4$

Endpoints of major axis: $(\pm 2, 0)$
Endpoints of minor axis: $(0, \pm 1)$

$2x + 8yy' = 0$

$y' = -\dfrac{x}{4y}$

$y'' = -\dfrac{1}{4y^3}$

$K = \dfrac{16}{(16y^2 + x^2)^{3/2}}$

$K = \dfrac{16}{(16 - 3x^2)^{3/2}}$; Since $-2 \le x \le 2$, K is greatest when $x = \pm 2, y = 0$.

$K = \dfrac{16}{(12y^2 + 4)^{3/2}}$; Since $-1 \le y \le 1$, K is smallest when $y = \pm 1, x = 0$.

65. (a) $K = \dfrac{2|6x^2 - 1|}{(16x^6 - 16x^4 + 4x^2 + 1)^{3/2}}$

(b) $x = 0$: $x^2 + \left(y + \dfrac{1}{2}\right)^2 = \dfrac{1}{4}$

$x = 1$: $x^2 + \left(y - \dfrac{1}{2}\right)^2 = \dfrac{5}{4}$

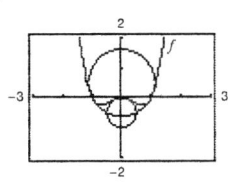

(c)

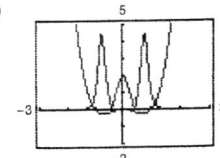

The curvature tends to be greatest near the extrema of the function and decreases as $x \to \pm \infty$. However, f and K do not have the same critical numbers.

Critical numbers of f: $x = 0, \pm \dfrac{\sqrt{2}}{2} \approx \pm 0.7071$

Critical numbers of K: $x = 0, \pm 0.7647, \pm 0.4082$

67. (a) 12.25 units (b) $\tfrac{1}{2}$

69. Given $y = f(x)$: $K = \dfrac{|y''|}{(1 + [y']^2)^{3/2}}$

$R = \dfrac{1}{K}$

The center of the circle is on the normal line at a distance of R from (x, y).

Equation of normal line: $y - y_0 = -\dfrac{1}{y'}(x - x_0)$

$\sqrt{(x - x_0)^2 + \left[-\dfrac{1}{y'}(x - x_0)\right]^2} = \dfrac{(1 + [y']^2)^{3/2}}{|y''|}$

$(x - x_0)^2 \left[1 + \dfrac{1}{(y')^2}\right] = \dfrac{(1 + [y']^2)^3}{(y'')^2}$

$(x - x_0)^2 = \dfrac{(y')^2 (1 + [y']^2)^2}{(y'')^2}$

$x - x_0 = \dfrac{y'(1 + [y']^2)}{y''} = y'z$

$x_0 = x - y'z$

$y - y_0 = -\dfrac{1}{y'}[x - (x - y'z)] = -z$

$y_0 = y + z$

Thus, $(x_0, y_0) = (x - y'z, y + z)$.

Center: $(-2, 3)$

71. $\dfrac{3}{2\sqrt{2(1 + \sin\theta)}}$ **73.** $\dfrac{2}{|a|}$

75. (a) 0 (b) 0   **77.** $\tfrac{1}{4}$ **79.** Proof

81. $K = \dfrac{1}{4a}\left|\csc \dfrac{\theta}{2}\right|$

Minimum: $K = \dfrac{1}{4a}$

There is no maximum.

83. 3327.5 pounds

85. Proof   **87.** Proof   **89.** Proof   **91.** Proof

Review Exercises for Chapter 11 (page 832)

1. (a) All reals except $n\pi$, n is an integer
(b) Continuous except at $t = n\pi$, n is an integer

3. (a) $(0, \infty)$ (b) Continuous for all $t > 0$

5. (a) $\mathbf{i}$
(b) $-3\mathbf{i} + 4\mathbf{j} + \tfrac{8}{3}\mathbf{k}$
(c) $(2c - 1)\mathbf{i} + (c - 1)^2\mathbf{j} + \tfrac{1}{3}(1 - c)^3\mathbf{k}$
(d) $2\Delta t\mathbf{i} + \Delta t(\Delta t + 2)\mathbf{j} - \tfrac{1}{3}\Delta t[(\Delta t)^2 + 3\Delta t + 3]\mathbf{k}$

7.   **9.**

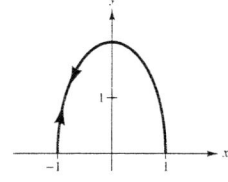

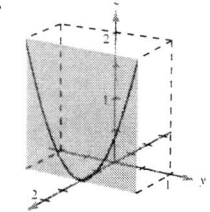

11.

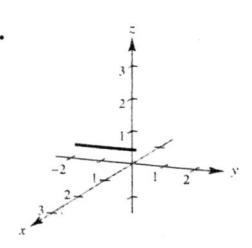

13.

15. $\mathbf{r}_1(t) = 4t\mathbf{i} + 3t\mathbf{j}, \quad 0 \le t \le 1$
$\mathbf{r}_2(t) = 4\mathbf{i} + (3-t)\mathbf{j}, \quad 0 \le t \le 3$
$\mathbf{r}_3(t) = (4-t)\mathbf{i}, \quad 0 \le t \le 4$

17. $\mathbf{r}(t) = \langle -2 + 7t, -3 + 4t, 8 - 10t \rangle$
(Answer is not unique.)

19. $x = t, y = -t, z = 2t^2$

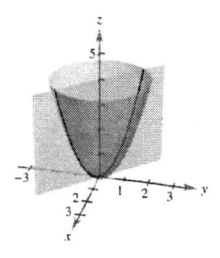

21. $4\mathbf{i} + \mathbf{k}$

23. (a) $3\mathbf{i} + \mathbf{j}$ (b) $\mathbf{0}$ (c) $4t + 3t^2$
(d) $-5\mathbf{i} + (2t - 2)\mathbf{j} + 2t^2\mathbf{k}$
(e) $\dfrac{10t - 1}{\sqrt{10t^2 - 2t + 1}}$
(f) $\left(\dfrac{8}{3}t^3 - 2t^2\right)\mathbf{i} - 8t^3\mathbf{j} + (9t^2 - 2t + 1)\mathbf{k}$

25. $x(t)$ and $y(t)$ are increasing functions at $t = t_0$, and $z(t)$ is a decreasing function at $t = t_0$.

27. $\sin t\mathbf{i} + (t \sin t + \cos t)\mathbf{j} + \mathbf{C}$

29. $\tfrac{1}{2}(t\sqrt{1+t^2} + \ln|t + \sqrt{1+t^2}|) + \mathbf{C}$

31. $\mathbf{r}(t) = (t^2 + 1)\mathbf{i} + (e^t + 2)\mathbf{j} - (e^{-t} + 4)\mathbf{k}$

33. $\tfrac{32}{3}\mathbf{j}$ 35. $2(e-1)\mathbf{i} - 8\mathbf{j} - 2\mathbf{k}$

37. $\mathbf{v}(t) = \langle -3\cos^2 t \sin t, 3\sin^2 t \cos t, 3 \rangle$
$\|\mathbf{v}(t)\| = 3\sqrt{\sin^2 t \cos^2 t + 1}$
$\mathbf{a}(t) = \langle 3\cos t(3\sin^2 t - 1), 3\sin t(2\cos^2 t - \sin^2 t), 0 \rangle$

39. $x(t) = t, y(t) = 16 + 8t, z(t) = 2 + \tfrac{1}{2}t$
$\mathbf{r}(4.1) \approx (0.1, 16.8, 2.05)$

41. 152 feet

43. 34.9 meters per second

45. $\mathbf{v} = 5\mathbf{i}$
$\|\mathbf{v}\| = 5$
$\mathbf{a} = 0$
$\mathbf{a} \cdot \mathbf{T} = 0$
$\mathbf{a} \cdot \mathbf{N}$ does not exist.

47. $\mathbf{v} = \mathbf{i} + \dfrac{1}{2\sqrt{t}}\mathbf{j}$
$\|\mathbf{v}\| = \dfrac{\sqrt{4t+1}}{2\sqrt{t}}$
$\mathbf{a} = -\dfrac{1}{4t\sqrt{t}}\mathbf{j}$
$\mathbf{a} \cdot \mathbf{T} = \dfrac{-1}{4t\sqrt{t}\sqrt{4t+1}}$
$\mathbf{a} \cdot \mathbf{N} = \dfrac{1}{2t\sqrt{4t+1}}$

49. $\mathbf{v} = e^t\mathbf{i} - e^{-t}\mathbf{j}$
$\|\mathbf{v}\| = \sqrt{e^{2t} + e^{-2t}}$
$\mathbf{a} = e^t\mathbf{i} + e^{-t}\mathbf{j}$
$\mathbf{a} \cdot \mathbf{T} = \dfrac{e^{2t} - e^{-2t}}{\sqrt{e^{2t} + e^{-2t}}}$
$\mathbf{a} \cdot \mathbf{N} = \dfrac{2}{\sqrt{e^{2t} + e^{-2t}}}$

51. $\mathbf{v} = \mathbf{i} + 2t\mathbf{j} + t\mathbf{k}$
$\|\mathbf{v}\| = \sqrt{1 + 5t^2}$
$\mathbf{a} = 2\mathbf{j} + \mathbf{k}$
$\mathbf{a} \cdot \mathbf{T} = \dfrac{5t}{\sqrt{1 + 5t^2}}$
$\mathbf{a} \cdot \mathbf{N} = \dfrac{\sqrt{5}}{\sqrt{1 + 5t^2}}$

53. $x = -\sqrt{2} - \sqrt{2}t$
$y = \sqrt{2} - \sqrt{2}t$
$z = \dfrac{3\pi}{4} + t$

55. 4.56 miles per second

57. $5\sqrt{13}$

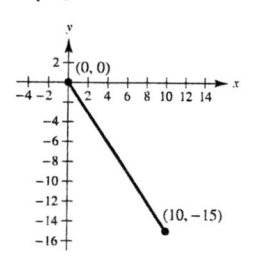

59. 60

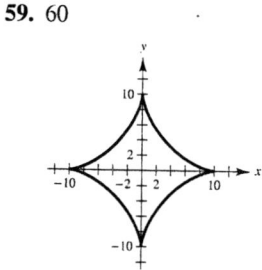

61. $3\sqrt{29}$

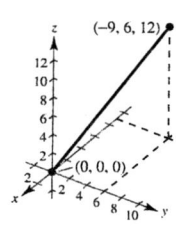

63. $\dfrac{\sqrt{65}\pi}{2}$

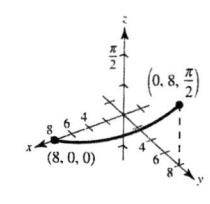

65. $\dfrac{\sqrt{5}\pi}{2}$ 67. 0   69. $\dfrac{2\sqrt{5}}{(4 + 5t^2)^{3/2}}$

71. $K = \dfrac{1}{17\sqrt{17}}; r = 17\sqrt{17}$

73. $K = \dfrac{1}{2\sqrt{2}}; r = 2\sqrt{2}$

75. The curvature changes abruptly from zero to a nonzero constant.

P.S. Problem Solving (page 834)

1. (a) a (b) πa (c) $k = \pi a$
3. Initial speed: 447.21 feet per second; $\theta \approx 63.43°$
5. By the definition of arc length,
$$s(t) = \int_\pi^t \sqrt{(1-\cos\theta)^2 + \sin^2\theta}\, d\theta.$$ This equals
$$\frac{-2\sqrt{2}\sin t}{\sqrt{1-\cos t}}, \pi < t < 2\pi.$$ Also, $K(t) = \frac{1}{2\sqrt{2}\sqrt{1-\cos t}}$.
So $\rho(t) = 2\sqrt{2}\sqrt{1-\cos t}$. Therefore,
$$s^2 + \rho^2 = \left(\frac{-2\sqrt{2}\sin t}{\sqrt{1-\cos t}}\right)^2 + \left(2\sqrt{2}\sqrt{1-\cos t}\right)^2 = 16.$$
7. Proof
9. Unit tangent: $\langle -\frac{4}{5}, 0, \frac{3}{5}\rangle$
   Unit normal: $\langle 0, -1, 0\rangle$
   Binormal: $\langle \frac{3}{5}, 0, \frac{4}{5}\rangle$

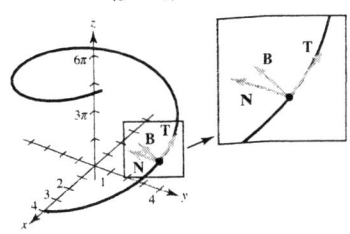

11. (a) Proof   (b) Proof
13. (a)   (b) 6.766

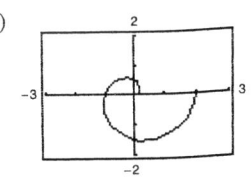

(c) $K = \dfrac{\pi(\pi^2 t^2 + 2)}{(\pi^2 t^2 + 1)^{3/2}}$
   $K(0) = 2\pi$
   $K(1) = \dfrac{\pi(\pi^2 + 2)}{(\pi^2 + 1)^{3/2}} \approx 1.04$
   $K(2) \approx 0.51$

(d)

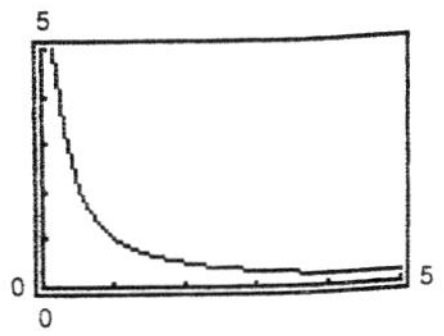

(e) $\lim\limits_{t\to\infty} K = 0$
(f) As $t \to \infty$, the graph spirals outward and the curvature decreases.

Chapter 12

Section 12.1 (page 846)

1. z is a function of x and y.   3. z is not a function of x and y.
5. (a) $\frac{3}{2}$ (b) $-\frac{1}{4}$ (c) 6 (d) $\frac{5}{y}$ (e) $\frac{x}{2}$ (f) $\frac{5}{t}$
7. (a) 5 (b) $3e^2$ (c) $\frac{2}{e}$ (d) $5e^x$ (e) xe^2 (f) te^t
9. (a) $\frac{2}{3}$ (b) 0 11. (a) $\sqrt{2}$ (b) $3\sin 1$
13. (a) 4 (b) 6
15. (a) $2x + \Delta x, \Delta x \neq 0$ (b) $-2, \Delta y \neq 0$
17. Domain: $\{(x, y): x^2 + y^2 \leq 4\}$
    Range: $0 \leq z \leq 2$
19. Domain: $\{(x, y): -1 \leq x + y \leq 1\}$
    Range: $-\frac{\pi}{2} \leq z \leq \frac{\pi}{2}$
21. Domain: $\{(x, y): y < -x + 4\}$
    Range: all real numbers
23. Domain: $\{(x, y): x \neq 0, y \neq 0\}$
    Range: all real numbers
25. Domain: $\{(x, y): y \neq 0\}$
    Range: $z > 0$
27. Domain: $\{(x, y): x \neq 0, y \neq 0\}$
    Range: $|z| > 0$
29. (a) $(20, 0, 0)$ (b) $(-15, 10, 20)$
    (c) $(20, 15, 25)$ (d) $(20, 20, 0)$

31.   33.

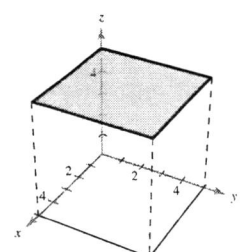

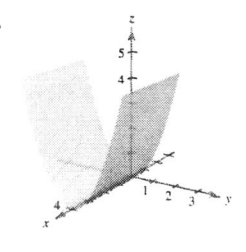

35.    37.

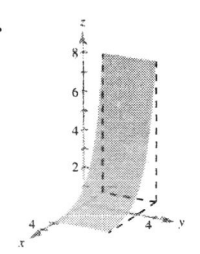

39. 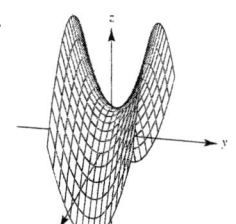   41.

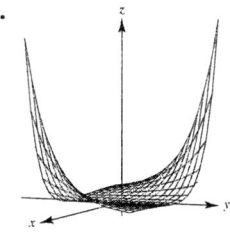

43. (a) [graph]

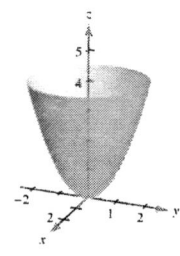

(b) g is a vertical translation of f 2 units upward.
(c) g is a horizontal translation of f 2 units to the right.
(d) g is a reflection of f in the xy-plane followed by a vertical translation 4 units upward.
(e) [graphs of $z = f(1, y)$ and $z = f(x, 1)$]

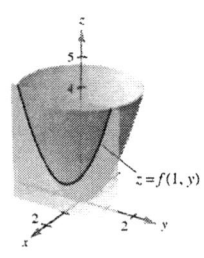

45. c **46.** d **47.** b **48.** a

49. Lines: $x + y = c$

51. Circles centered at $(0, 0)$
Radius ≤ 5

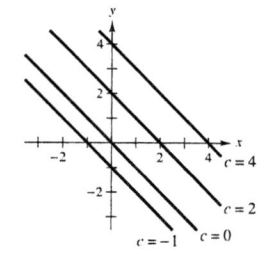

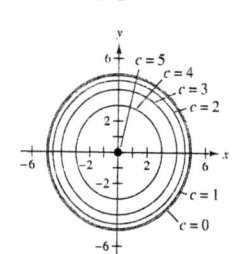

53. Hyperbolas: $xy = c$

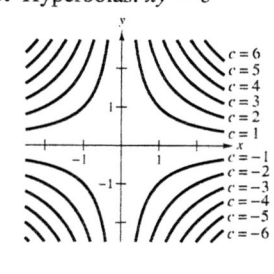

55. Circles passing through $(0, 0)$
Centered at $\left(\dfrac{1}{2c}, 0\right)$

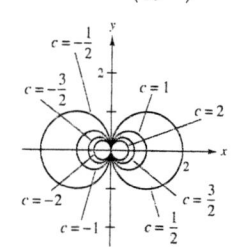

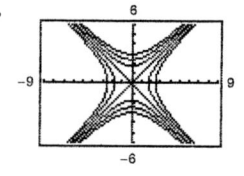

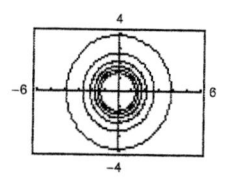

57. [graph] **59.** [graph]

61. Let D be a set of ordered pairs of real numbers. If to each ordered pair (x, y) in D there corresponds a unique real number $f(x, y)$, then f is called a function of x and y.

63. No. Example: $z = e^{-(x^2+y^2)}$

65. The surface may be shaped like a saddle. For example, let $f(x, y) = xy$. The graph is not unique; any vertical translation will produce the same level curves.

67.

Tax Rate	Inflation Rate		
	0	0.03	0.05
0	$2593.74	$1929.99	$1592.33
0.28	$2004.23	$1491.34	$1230.42
0.35	$1877.14	$1396.77	$1152.40

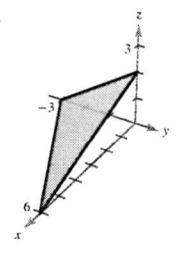

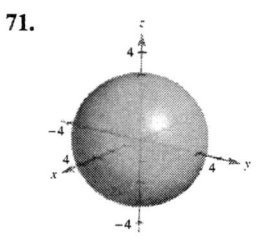

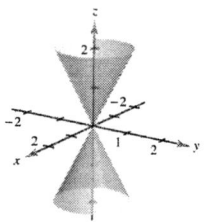

69. [graph] **71.** [sphere graph]

73. [graph]

75. (a) 243 board-feet
(b) 507 board-feet

77. [level curves graph] **79.** $C = 0.75xy + 0.80(xz + yz)$

81. (a) $k = \dfrac{520}{3}$
(b) $P = \dfrac{520T}{3V}$

The level curves are lines.

83. (a) C (b) A (c) B

85. (a) The boundaries between colors represent level curves.
(b) No: the colors represent intervals of different lengths.
(c) Use more colors.

87. False: let $f(x, y) = 4$.

89. False: let $f(x, y) = xy^2$. Then $f(ax, ay) = a^3 f(x, y)$.

Section 12.2 (page 856)

1. Proof **3.** 2 **5.** 15 **7.** 5, continuous

9. -3, continuous for $x \neq y$

11. 0, continuous for $xy \neq -1$, $y \neq 0$, $\left|\dfrac{x}{y}\right| \leq 1$

13. $\frac{1}{e^2}$, continuous   15. $2\sqrt{2}$, continuous for $x + y + z \geq 0$

17. 1, continuous

19. Continuous except at $(0, 0)$; the limit does not exist.

21.

(x, y)	$(1, 0)$	$(0.5, 0)$	$(0.1, 0)$	$(0.01, 0)$	$(0.001, 0)$
$f(x, y)$	0	0	0	0	0

$y = 0$: 0

(x, y)	$(1, 1)$	$(0.5, 0.5)$	$(0.1, 0.1)$
$f(x, y)$	$\frac{1}{2}$	$\frac{1}{2}$	$\frac{1}{2}$

(x, y)	$(0.01, 0.01)$	$(0.001, 0.001)$
$f(x, y)$	$\frac{1}{2}$	$\frac{1}{2}$

$y = x$: $\frac{1}{2}$

Limit does not exist.

Continuous except at $(0, 0)$

23.

(x, y)	$(1, 1)$	$(0.25, 0.5)$	$(0.01, 0.1)$
$f(x, y)$	$-\frac{1}{2}$	$-\frac{1}{2}$	$-\frac{1}{2}$

(x, y)	$(0.0001, 0.01)$	$(0.000001, 0.001)$
$f(x, y)$	$-\frac{1}{2}$	$-\frac{1}{2}$

$x = y^2$: $-\frac{1}{2}$

(x, y)	$(-1, 1)$	$(-0.25, 0.5)$	$(-0.01, 0.1)$
$f(x, y)$	$\frac{1}{2}$	$\frac{1}{2}$	$\frac{1}{2}$

(x, y)	$(-0.0001, 0.01)$	$(-0.000001, 0.001)$
$f(x, y)$	$\frac{1}{2}$	$\frac{1}{2}$

$x = -y^2$: $\frac{1}{2}$

Limit does not exist.

Continuous except at $(0, 0)$

25. f is continuous except at $(0, 0)$.

g is continuous.

f has a removable discontinuity at $(0, 0)$.

27. 0   29. Limit does not exist.

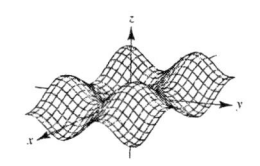

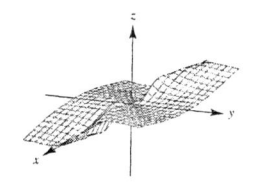

31. Limit does not exist

33. 1   35. 0

37. Continuous except at $(0, 0, 0)$

39. Continuous   41. Continuous

43. Continuous for $y \neq \frac{3x}{2}$ 45. (a) $2x$ (b) -4

47. (a) $2 + y$ (b) $x - 3$

49. See "Definition of the Limit of a Function of Two Variables," on page 851; show that the value of $\lim_{(x, y) \to (x_0, y_0)} f(x, y)$ is not the same for two different paths to (x_0, y_0).

51. No: the existence of $f(2, 3)$ has no bearing on the existence of the limit as $(x, y) \to (2, 3)$.

53. Proof   55. True

57. False: let $f(x, y) = \begin{cases} \ln(x^2 + y^2), & x \neq 0, y \neq 0 \\ 0, & x = 0, y = 0 \end{cases}$.

Section 12.3 (page 865)

1. $f_x = (4, 1) < 0$ 3. $f_y = (4, 1) > 0$

5. $f_x(x, y) = 2$
   $f_y(x, y) = -3$

7. $\frac{\partial z}{\partial x} = \sqrt{y}$
   $\frac{\partial z}{\partial y} = \frac{x}{2\sqrt{y}}$

9. $\frac{\partial z}{\partial x} = 2x - 5y$
   $\frac{\partial z}{\partial y} = -5x + 6y$

11. $\frac{\partial z}{\partial x} = 2xe^{2y}$
    $\frac{\partial z}{\partial y} = 2x^2 e^{2y}$

13. $\frac{\partial z}{\partial x} = \frac{2x}{x^2 + y^2}$
    $\frac{\partial z}{\partial y} = \frac{2y}{x^2 + y^2}$

15. $\frac{\partial z}{\partial x} = \frac{-2y}{x^2 - y^2}$
    $\frac{\partial z}{\partial y} = \frac{2x}{x^2 - y^2}$

17. $\frac{\partial z}{\partial x} = \frac{x^3 - 4y^3}{x^2 y}$
    $\frac{\partial z}{\partial y} = \frac{-x^3 + 16y^3}{2xy^2}$

19. $h_x(x, y) = -2xe^{-(x^2+y^2)}$
    $h_y(x, y) = -2ye^{-(x^2+y^2)}$

21. $f_x(x, y) = \frac{x}{\sqrt{x^2 + y^2}}$
    $f_y(x, y) = \frac{y}{\sqrt{x^2 + y^2}}$

23. $\frac{\partial z}{\partial x} = 2\sec^2(2x - y)$
    $\frac{\partial z}{\partial y} = -\sec^2(2x - y)$

25. $\frac{\partial z}{\partial x} = ye^y \cos xy$
    $\frac{\partial z}{\partial y} = e^y(x \cos xy + \sin xy)$

27. $f_x(x, y) = 1 - x^2$
$f_y(x, y) = y^2 - 1$

29. $f_x(x, y) = 2$
$f_y(x, y) = 3$

31. $f_x(x, y) = \dfrac{1}{2\sqrt{x+y}}$
$f_y(x, y) = \dfrac{1}{2\sqrt{x+y}}$

33. $g_x(1, 1) = -2$
$g_y(1, 1) = -2$

35. $\dfrac{\partial z}{\partial x} = -1$
$\dfrac{\partial z}{\partial y} = 0$

37. $\dfrac{\partial z}{\partial x} = \dfrac{1}{4}$
$\dfrac{\partial z}{\partial y} = \dfrac{1}{4}$

39. $\dfrac{\partial z}{\partial x} = -\dfrac{1}{4}$
$\dfrac{\partial z}{\partial y} = \dfrac{1}{4}$

41. $-\dfrac{1}{2}$

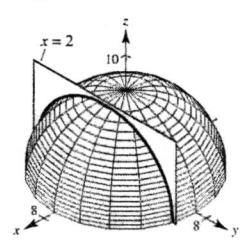

43. 18

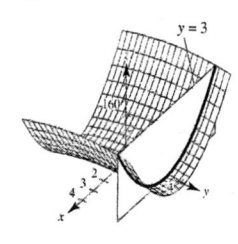

45. $x = -6, y = 4$ **47.** $x = 1, y = 1$

49. (a) f_y (b) f_x

f_x represents the slope in the x-direction, and f_y represents the slope in the y-direction.

51. $\dfrac{\partial w}{\partial x} = \dfrac{x}{\sqrt{x^2+y^2+z^2}}$
$\dfrac{\partial w}{\partial y} = \dfrac{y}{\sqrt{x^2+y^2+z^2}}$
$\dfrac{\partial w}{\partial z} = \dfrac{z}{\sqrt{x^2+y^2+z^2}}$

53. $F_x(x, y, z) = \dfrac{x}{x^2+y^2+z^2}$
$F_y(x, y, z) = \dfrac{y}{x^2+y^2+z^2}$
$F_z(x, y, z) = \dfrac{z}{x^2+y^2+z^2}$

55. $H_x(x, y, z) = \cos(x + 2y + 3z)$
$H_y(x, y, z) = 2\cos(x + 2y + 3z)$
$H_z(x, y, z) = 3\cos(x + 2y + 3z)$

57. $\dfrac{\partial^2 z}{\partial x^2} = 2$
$\dfrac{\partial^2 z}{\partial y^2} = 6$
$\dfrac{\partial^2 z}{\partial y \partial x} = \dfrac{\partial^2 z}{\partial x \partial y} = -2$

59. $\dfrac{\partial^2 z}{\partial x^2} = \dfrac{y^2}{(x^2+y^2)^{3/2}}$
$\dfrac{\partial^2 z}{\partial y^2} = \dfrac{x^2}{(x^2+y^2)^{3/2}}$
$\dfrac{\partial^2 z}{\partial y \partial x} = \dfrac{\partial^2 z}{\partial x \partial y} = \dfrac{-xy}{(x^2+y^2)^{3/2}}$

61. $\dfrac{\partial^2 z}{\partial x^2} = e^x \tan y$
$\dfrac{\partial^2 z}{\partial y^2} = 2e^x \sec^2 y \tan y$
$\dfrac{\partial^2 z}{\partial y \partial x} = \dfrac{\partial^2 z}{\partial x \partial y} = e^x \sec^2 y$

63. $\dfrac{\partial^2 z}{\partial x^2} = \dfrac{2xy}{(x^2+y^2)^2}$
$\dfrac{\partial^2 z}{\partial y^2} = \dfrac{-2xy}{(x^2+y^2)^2}$
$\dfrac{\partial^2 z}{\partial y \partial x} = \dfrac{\partial^2 z}{\partial x \partial y} = \dfrac{y^2-x^2}{(x^2+y^2)^2}$

65. $\dfrac{\partial z}{\partial x} = \sec y$
$\dfrac{\partial z}{\partial y} = x \sec y \tan y$
$\dfrac{\partial^2 z}{\partial x^2} = 0$
$\dfrac{\partial^2 z}{\partial y^2} = x \sec y (\sec^2 y + \tan^2 y)$
$\dfrac{\partial^2 z}{\partial y \partial x} = \dfrac{\partial^2 z}{\partial x \partial y} = \sec y \tan y$
No values of x and y exist such that $f_x(x, y) = f_y(x, y) = 0$.

67. $\dfrac{\partial z}{\partial x} = \dfrac{y^2 - x^2}{x(x^2+y^2)}$
$\dfrac{\partial z}{\partial y} = \dfrac{-2y}{x^2+y^2}$
$\dfrac{\partial^2 z}{\partial x^2} = \dfrac{x^4 - 4x^2y^2 - y^4}{x^2(x^2+y^2)^2}$
$\dfrac{\partial^2 z}{\partial y^2} = \dfrac{2(y^2-x^2)}{(x^2+y^2)^2}$
$\dfrac{\partial^2 z}{\partial y \partial x} = \dfrac{\partial^2 z}{\partial x \partial y} = \dfrac{4xy}{(x^2+y^2)^2}$
No values of x and y exist such that $f_x(x, y) = f_y(x, y) = 0$.

69. $f_{xyy}(x, y, z) = f_{yxy}(x, y, z) = f_{yyx}(x, y, z) = 0$

71. $f_{xyy}(x, y, z) = f_{yxy}(x, y, z) = f_{yyx}(x, y, z) = z^2 e^{-x} \sin yz$

73. $\dfrac{\partial^2 z}{\partial x^2} + \dfrac{\partial^2 z}{\partial y^2} = 0 + 0 = 0$

75. $\dfrac{\partial^2 z}{\partial x^2} + \dfrac{\partial^2 z}{\partial y^2} = e^x \sin y - e^x \sin y = 0$

77. $\dfrac{\partial^2 z}{\partial t^2} = -c^2 \sin(x - ct) = c^2 \dfrac{\partial^2 z}{\partial x^2}$

79. $\dfrac{\partial z}{\partial t} = -e^{-t} \cos \dfrac{x}{c} = c^2 \dfrac{\partial^2 z}{\partial x^2}$

81. See "Definition of Partial Derivatives of a Function of Two Variables," on page 859.

83.

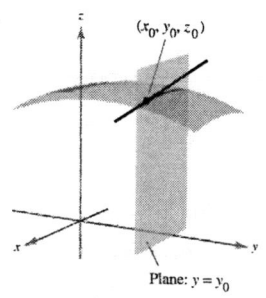

$\partial f/\partial x$ represents the slope of the curve formed by the intersection of the surface $z = f(x, y)$ and the plane $y = y_0$ at any point on the curve.

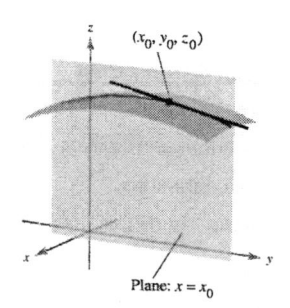

$\partial f/\partial y$ represents the slope of the curve formed by the intersection of the surface $z = f(x, y)$ and the plane $x = x_0$ at any point on the curve.

85.

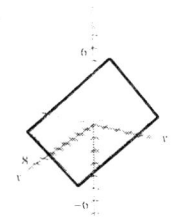

87. (a) $\dfrac{\partial C}{\partial x} = 183, \dfrac{\partial C}{\partial y} = 237$

(b) The fireplace-insert stove results in the cost increasing at a faster rate because the coefficient of y is greater in magnitude than the coefficient of x.

89. An increase in either charge for food and housing or tuition will cause a decrease in the number of applicants.

91. $\dfrac{\partial T}{\partial x} = -2.4°$ per meter, $\dfrac{\partial T}{\partial y} = -9°$ per meter

93.
$$T = \dfrac{PV}{nR} \Rightarrow \dfrac{\partial T}{\partial P} = \dfrac{V}{nR}$$
$$P = \dfrac{nRT}{V} \Rightarrow \dfrac{\partial P}{\partial V} = -\dfrac{nRT}{V^2}$$
$$V = \dfrac{nRT}{P} \Rightarrow \dfrac{\partial V}{\partial T} = \dfrac{nR}{P}$$
$$\dfrac{\partial T}{\partial P} \cdot \dfrac{\partial P}{\partial V} \cdot \dfrac{\partial V}{\partial T} = -\dfrac{nRT}{VP} = -\dfrac{nRT}{nRT} = -1$$

95. (a) $\dfrac{\partial z}{\partial x} = -1.83, \dfrac{\partial z}{\partial y} = -1.09$

(b) For every decrease of 1.83 gallons of whole milk there is an increase of one gallon of skim milk. For every decrease of 1.09 gallons of whole milk there is an increase of one gallon of reduced-fat milk.

97. (a) $f_x(x, y) = \dfrac{y(x^4 + 4x^2y^2 - y^4)}{(x^2 + y^2)^2}$

$f_y(x, y) = \dfrac{x(x^4 - 4x^2y^2 - y^4)}{(x^2 + y^2)^2}$

(b) $f_x(0, 0) = 0, f_y(0, 0) = 0$

(c) $f_{xy}(0, 0) = -1, f_{yx}(0, 0) = 1$

(d) f_{xy} or f_{yx} or both are not continuous at $(0, 0)$.

99. True   **101.** True

Section 12.4 (page 874)

1. $dz = 6xy^3\,dx + 9x^2y^2\,dy$ **3.** $dz = \dfrac{2}{(x^2 + y^2)^2}(x\,dx + y\,dy)$

5. $dz = (\cos y + y\sin x)\,dx - (x\sin y + \cos x)\,dy$

7. $dz = (e^x \sin y)\,dx + (e^x \cos y)\,dy$

9. $dw = 2z^3y\cos x\,dx + 2z^3\sin x\,dy + 6z^2y\sin x\,dz$

11. (a) $f(1, 2) = 4, f(1.05, 2.1) = 3.4875, \Delta z = -0.5125$

(b) $dz = -0.5$

13. (a) $f(1, 2) \approx 0.90930, f(1.05, 2.1) \approx 0.90637, \Delta z \approx -0.00293$

(b) $dz \approx 0.00385$

15. (a) $f(1, 2) = -5, f(1.05, 2.1) = -5.25, \Delta z = -0.25$

(b) $dz = -0.25$

17. 0.094   **19.** -0.012

21. If $z = f(x, y)$ and Δx and Δy are increments of x and y, and x and y are independent variables, then the total differential of the dependent variable z is

$$dz = \dfrac{\partial z}{\partial x}dx + \dfrac{\partial z}{\partial y}dy = f_x(x, y)\,\Delta x + f_y(x, y)\,\Delta y.$$

23. The approximation of Δz by dz is called a linear approximation, where dz represents the change in height of a plane that is tangent to the surface at the point $P(x_0, y_0)$.

25. $dA = h\,dl + l\,dh$

27.

Δr	Δh	dV	ΔV	$\Delta V - dV$
0.1	0.1	4.7124	4.8391	0.1267
0.1	-0.1	2.8274	2.8264	-0.0010
0.001	0.002	0.0565	0.0565	0.0001
-0.0001	0.0002	-0.0019	-0.0019	0.0000

29. (a) $dz = -1.83\,dx - 1.09\,dy$

(b) $dz = \pm 0.73; \dfrac{dz}{z} \approx 11.67\%$

31. 10%   **33.** ± 0.24 square inch

35. (a) $V = 18 \sin \theta$ cubic feet; $\theta = \pi/2$

(b) 1.047 cubic feet

37. 7%   **39.** $L \approx 8.096 \times 10^{-4} \pm 6.6 \times 10^{-6}$ microhenrys

41. Answers will vary.   **43.** Answers will vary.
Example: Example:
$\varepsilon_1 = \Delta x$ $\varepsilon_1 = y\,\Delta x$
$\varepsilon_2 = 0$ $\varepsilon_2 = 2x\,\Delta x + (\Delta x)^2$

45. Proof

47. Answers will vary. For example, we can use the equation $F = ma$.

Then $dF = \dfrac{\partial F}{\partial m}dm + \dfrac{\partial F}{\partial a}da$

$= a\,dm + m\,da.$

We can estimate the possible propagated errors when given the error in measurement.

Section 12.5 (page 882)

1. $2(e^{2t} - e^{-2t})$ **3.** $e^t \sec(\pi - t)[1 - \tan(\pi - t)]$

5. $2\cos 2t$ **7.** $4e^{2t}$ **9.** $3(2t^2 - 1)$ **11.** -2.04

13. $\dfrac{-8\cos t \sin t(1 + 2\sin^4 t + 2\cos^4 t)}{(1 + 4\cos^2 t \sin^2 t)^2}$

15. $\dfrac{\partial w}{\partial s} = 4s, 8$
$\dfrac{\partial w}{\partial t} = 4t, -4$

17. $\dfrac{\partial w}{\partial s} = 2s \cos 2t, 0$
$\dfrac{\partial w}{\partial t} = -2s^2 \sin 2t, -18$

19. $\dfrac{\partial w}{\partial r} = 0$
$\dfrac{\partial w}{\partial \theta} = 8\theta$

21. $\dfrac{\partial w}{\partial r} = 0$
$\dfrac{\partial w}{\partial \theta} = 1$

23. $\dfrac{\partial w}{\partial s} = t^2(3s^2 - t^2)$
$\dfrac{\partial w}{\partial t} = 2st(s^2 - 2t^2)$

25. $\dfrac{\partial w}{\partial s} = \dfrac{te^{(s-t)/(s+t)}(s^2 + 4st + t^2)}{(s+t)^2}$
$\dfrac{\partial w}{\partial t} = \dfrac{se^{(s-t)/(s+t)}(s^2 + t^2)}{(s+t)^2}$

27. $\dfrac{3y - 2x + 2}{2y - 3x + 1}$

29. $-\dfrac{x + y(x^2 + y^2)}{y + x(x^2 + y^2)}$

31. $\dfrac{\partial z}{\partial x} = \dfrac{-x}{z}$
$\dfrac{\partial z}{\partial y} = \dfrac{-y}{z}$

33. $\dfrac{\partial z}{\partial x} = \dfrac{-\sec^2(x+y)}{\sec^2(y+z)}$
$\dfrac{\partial z}{\partial y} = -1 - \dfrac{\sec^2(x+y)}{\sec^2(y+z)}$

35. $\dfrac{\partial z}{\partial x} = \dfrac{-x}{y+z}$
$\dfrac{\partial z}{\partial y} = \dfrac{-z}{y+z}$

37. $\dfrac{\partial z}{\partial x} = -\dfrac{ze^{xz} + y}{xe^{xz}}$
$\dfrac{\partial z}{\partial y} = -e^{-xz}$

39. $\dfrac{\partial w}{\partial x} = \dfrac{-yz - zw}{xz - yz + 2w}$
$\dfrac{\partial w}{\partial y} = \dfrac{-xz + zw}{xz - yz + 2w}$
$\dfrac{\partial w}{\partial z} = \dfrac{yw - xy - xw}{xz - yz + 2w}$

41. $\dfrac{\partial w}{\partial x} = \dfrac{y \sin xy}{z}$
$\dfrac{\partial w}{\partial y} = \dfrac{x \sin xy - z \cos yz}{z}$
$\dfrac{\partial w}{\partial z} = -\dfrac{y \cos yz + w}{z}$

43. $1; xf_x(x,y) + yf_y(x,y) = \dfrac{xy}{\sqrt{x^2+y^2}} = 1 f(x,y)$.

45. $0; xf_x(x,y) + yf_y(x,y) = \dfrac{xe^{x/y}}{y} - \dfrac{xe^{x/y}}{y} = 0$.

47. $\dfrac{dw}{dt} = \dfrac{\partial w}{\partial x} \cdot \dfrac{dx}{dt} + \dfrac{\partial w}{\partial y} \cdot \dfrac{dy}{dt}$

49. The explicit form of a function of two variables is of the form $z = f(x, y)$, as in $z = x^2 + y^2$. The implicit form of a function of two variables is of the form $F(x, y, z) = 0$, as in $z - x^2 - y^2 = 0$.

51. $\dfrac{\sqrt{2}}{10}(15 + \pi)$ square meters per hour

53. $\dfrac{dV}{dt} = 1536\pi$ cubic inches per minute

$\dfrac{dS}{dt} = \dfrac{36\pi}{5}(20 + 9\sqrt{10})$ square inches per minute
(Surface area includes base.)

55. $28m$ square centimeters per second

57. (a) $\tan(\theta + \phi) = \dfrac{4}{x} = \dfrac{\tan \theta + \tan \phi}{1 - \tan \theta \tan \phi}$

Also, $\tan \phi = \dfrac{2}{x}$. Therefore,

$4\left(1 - \tan \theta \cdot \dfrac{2}{x}\right) = x\left(\tan \theta + \dfrac{2}{x}\right)$

$4 - \dfrac{8}{x} \tan \theta = x \tan \theta + 2$

$x^2 \tan \theta - 2x + 8 \tan \theta = 0$

(b) $\dfrac{d\theta}{dx} = \dfrac{2 \cos^2 \theta - 2x \cos \theta \sin \theta}{x^2 + 8}$

(c) $x = 2\sqrt{2}$

59. $\dfrac{\partial w}{\partial u} = \dfrac{\partial w}{\partial x}\dfrac{dx}{du} + \dfrac{\partial w}{\partial y}\dfrac{dy}{du} = \dfrac{\partial w}{\partial x} - \dfrac{\partial w}{\partial y}$

$\dfrac{\partial w}{\partial v} = \dfrac{\partial w}{\partial x}\dfrac{dx}{dv} + \dfrac{\partial w}{\partial y}\dfrac{dy}{dv} = -\dfrac{\partial w}{\partial x} + \dfrac{\partial w}{\partial y}$

$\dfrac{\partial w}{\partial u} + \dfrac{\partial w}{\partial v} = 0$.

61. (a) Proof   (b) Proof   63. Proof

Section 12.6 (page 893)

1. $\dfrac{\sqrt{3} - 5}{2}$ 3. $\dfrac{5\sqrt{2}}{2}$ 5. $-\dfrac{7}{25}$ 7. $-e$ 9. $\dfrac{2\sqrt{6}}{3}$

11. $\dfrac{(8 + \pi)\sqrt{6}}{24}$ 13. $\sqrt{2}(x + y)$ 15. $\left(\dfrac{2 + \sqrt{3}}{2}\right)\cos(2x - y)$

17. $-7\sqrt{2}$ 19. $\dfrac{7\sqrt{19}}{19}$ 21. $3\mathbf{i} - 10\mathbf{j}$

23. $-6 \sin 25 \mathbf{i} + 8 \sin 25 \mathbf{j} \approx 0.7941\mathbf{i} - 1.0588\mathbf{j}$

25. $6\mathbf{i} + 13\mathbf{j} - 9\mathbf{k}$ 27. $2\sqrt{5}$ 29. $-\dfrac{2\sqrt{5}}{5}$

31. $\tan y \mathbf{i} + x \sec^2 y \mathbf{j}, \sqrt{17}$

33. $\dfrac{2}{3(x^2 + y^2)}(x\mathbf{i} + y\mathbf{j}), \dfrac{2\sqrt{5}}{15}$ 35. $\dfrac{x\mathbf{i} + y\mathbf{j} + z\mathbf{k}}{\sqrt{x^2 + y^2 + z^2}}, 1$

37. $e^{yz}\mathbf{i} + xze^{yz}\mathbf{j} + xye^{yz}\mathbf{k}; \sqrt{65}$

39.

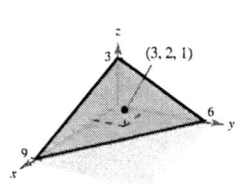

41. (a) $\dfrac{2 + 3\sqrt{3}}{12}$
(b) $\dfrac{3 - 2\sqrt{3}}{12}$

43. (a) $-\dfrac{1}{5}$ (b) $-\dfrac{11\sqrt{10}}{60}$ 45. $\dfrac{\sqrt{13}}{6}$

47.

49. $-2\mathbf{i} - 4\mathbf{j}, 2\sqrt{5}$

51. (a) Answers will vary. Example: $-4\mathbf{i} + \mathbf{j}$
(b) $-\frac{2}{5}\mathbf{i} + \frac{1}{10}\mathbf{j}$ (c) $\frac{2}{5}\mathbf{i} - \frac{1}{10}\mathbf{j}$
The direction opposite that of the gradient

53. (a)

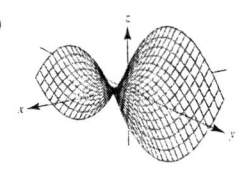

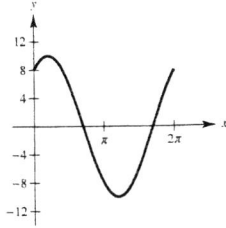

(b) $D_\mathbf{u} f(4, -3) = 8\cos\theta + 6\sin\theta$

(c) $\theta \approx 2.21, \theta \approx 5.36$
Directions in which there is no change in f
(d) $\theta \approx 0.64, \theta \approx 3.79$
Directions of greatest rate of change in f
(e) 10; Magnitude of the greatest rate of change
(f)

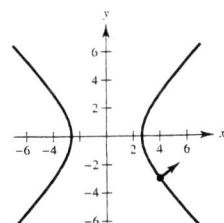

Orthogonal to the level curve

55. $6\mathbf{i} + 8\mathbf{j}$ **57.** $-\frac{1}{2}\mathbf{j}$
59. $\frac{\sqrt{257}}{257}(16\mathbf{i} - \mathbf{j})$ **61.** $\frac{\sqrt{85}}{85}(9\mathbf{i} - 2\mathbf{j})$

63. $\frac{1}{625}(7\mathbf{i} - 24\mathbf{j})$
65. The directional derivative of $z = f(x, y)$ in the direction of $\mathbf{u} = \cos t\mathbf{i} + \sin t\mathbf{j}$ is
$$D_\mathbf{u} f(x, y) = \lim_{t \to 0} \frac{f(x + t\cos\theta, y + t\sin\theta) - f(x, y)}{t}$$
if the limit exists.

67. Let $f(x, y)$ be a function of two variables and let $\mathbf{u} = \cos\theta \mathbf{i} + \sin\theta \mathbf{j}$ be a unit vector.
(a) If $\theta = 0°$, then $D_\mathbf{u} f = \frac{\partial f}{\partial x}$.
(b) If $\theta = 90°$, then $D_\mathbf{u} f = \frac{\partial f}{\partial y}$.

69. Answers will vary. Example:

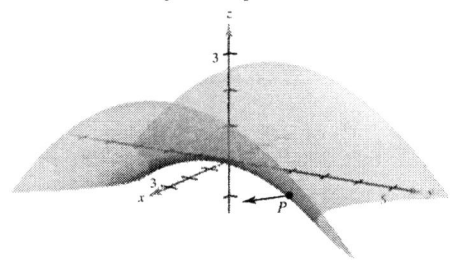

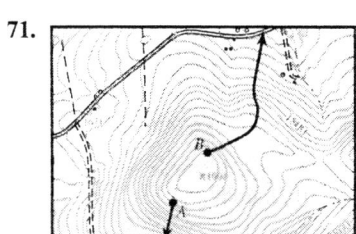

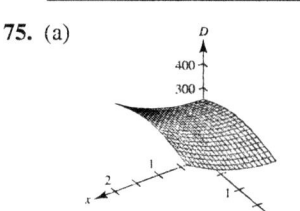

71. **73.** $y^2 = 10x$

75. (a) (b) Graph $-D$.
(c) 315 meters
(d) 60.0
(e) 55.5
(f) $60.0\mathbf{i} + 55.5\mathbf{j}$

77. True **79.** True **81.** $f(x, y, z) = e^x \cos y + \frac{1}{2}z^2 + C$

Section 12.7 (page 902)

1. The level surface can be written as $3x - 5y + 3z = 15$, which is an equation of a plane in space.
3. The level surface can be written as $4x^2 + 9y^2 - 4z^2 = 0$, which is an elliptic cone that lies on the z-axis.
5. $\frac{\sqrt{3}}{3}(\mathbf{i} + \mathbf{j} + \mathbf{k})$ **7.** $\frac{\sqrt{2}}{10}(3\mathbf{i} + 4\mathbf{j} - 5\mathbf{k})$
9. $\frac{\sqrt{2049}}{2049}(32\mathbf{i} + 32\mathbf{j} - \mathbf{k})$ **11.** $\frac{\sqrt{3}}{3}(\mathbf{i} - \mathbf{j} + \mathbf{k})$
13. $\frac{\sqrt{113}}{113}(-\mathbf{i} - 6\sqrt{3}\mathbf{j} + 2\mathbf{k})$ **15.** $6x + 2y + z = 35$
17. $3x + 4y - 5z = 0$ **19.** $10x - 8y - z = 9$
21. $2x - z = -2$ **23.** $3x + 4y - 25z = 25(1 - \ln 5)$
25. $x - 4y + 2z = 18$ **27.** $x + y + z = 1$
29. $2x + 4y + z = 14$ **31.** $3x + 2y + z = -6$
$\frac{x-1}{2} = \frac{y-2}{4} = \frac{z-4}{1}$ $\frac{x+2}{3} = \frac{y+3}{2} = \frac{z-6}{1}$
33. $x - y + 2z = \frac{\pi}{2}$
$\frac{x-1}{1} = \frac{y-1}{-1} = \frac{z - (\pi/4)}{2}$

35. (a) Line: $x = 1, y = 1, z = 1 - t$
Plane: $z = 1$
(b) Line: $x = -1, y = 2 + \frac{6}{25}t, z = -\frac{4}{5} - t$
Plane: $6y - 25z - 32 = 0$
(c)

(d) At $(1, 1, 1)$ the tangent plane is parallel to the xy-plane, implying that the surface is level. At $\left(-1, 2, -\frac{4}{5}\right)$, the function does not change in the x-direction.

37. $F_x(x_0, y_0, z_0)(x - x_0) + F_y(x_0, y_0, z_0)(y - y_0) + F_z(x_0, y_0, z_0)(z - z_0) = 0$

39. (a) $\frac{x - 2}{1} = \frac{y - 1}{-2} = \frac{z - 2}{1}$ (b) $\frac{\sqrt{10}}{5}$, not orthogonal

41. (a) $\frac{x - 3}{4} = \frac{y - 3}{4} = \frac{z - 4}{-3}$ (b) $\frac{16}{25}$, not orthogonal

43. (a) $\frac{y - 1}{1} = \frac{z - 1}{-1}, x = 2$ (b) 0, orthogonal

45. (a) $x = 1 + t$
$y = 2 - 2t$
$z = 4$
$\theta \approx 48.2°$
(b)

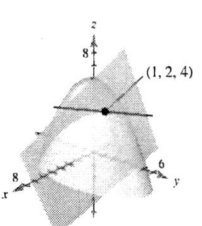

47. 86.0° **49.** 77.4°
51. $(0, 3, 12)$

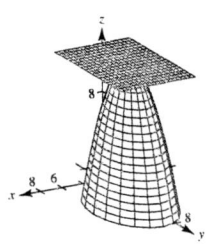

The function is maximum.

53. $x = 4e^{-4kt}, y = 3e^{-2kt}, z = 10e^{-8kt}$

55. $F(x, y, z) = \frac{x^2}{a^2} + \frac{y^2}{b^2} + \frac{z^2}{c^2} - 1$
$F_x(x, y, z) = \frac{2x}{a^2}$
$F_y(x, y, z) = \frac{2y}{b^2}$
$F_z(x, y, z) = \frac{2z}{c^2}$
Plane: $\frac{2x_0}{a^2}(x - x_0) + \frac{2y_0}{b^2}(y - y_0) + \frac{2z_0}{c^2}(z - z_0) = 0$
$\frac{x_0 x}{a^2} + \frac{y_0 y}{b^2} + \frac{z_0 z}{c^2} = 1$

57. $F(x, y, z) = a^2x^2 + b^2y^2 - z^2$
$F_x(x, y, z) = 2a^2x$
$F_y(x, y, z) = 2b^2y$
$F_z(x, y, z) = -2z$
Plane: $2a^2x_0(x - x_0) + 2b^2y_0(y - y_0) - 2z_0(z - z_0) = 0$
$a^2x_0x + b^2y_0y - z_0z = 0$
Hence, the plane passes through the origin.

59. (a) $P_1(x, y) = 1 + x - y$
(b) $P_2(x, y) = 1 + x - y + \frac{1}{2}x^2 - xy + \frac{1}{2}y^2$
(c) If $x = 0, P_2(0, y) = 1 - y + \frac{1}{2}y^2$.
This is the second-degree Taylor polynomial for e^{-y}.
If $y = 0, P_2(x, 0) = 1 + x + \frac{1}{2}x^2$.
This is the second-degree Taylor polynomial for e^x.

(d)
x	y	$f(x, y)$	$P_1(x, y)$	$P_2(x, y)$
0	0	1	1	1
0	0.1	0.9048	0.9000	0.9050
0.2	0.1	1.1052	1.1000	1.1050
0.2	0.5	0.7408	0.7000	0.7450
1	0.5	1.6487	1.5000	1.6250

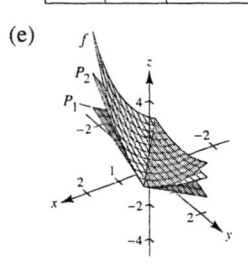

(e)

61. Proof

Section 12.8 (page 911)

1. Relative minimum: $(1, 3, 0)$ **3.** Relative minimum: $(0, 0, 1)$

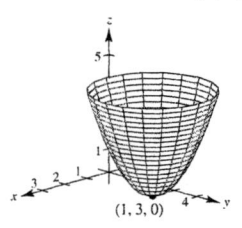

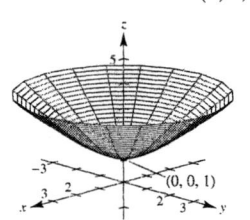

5. Relative minimum: $(-1, 3, -4)$

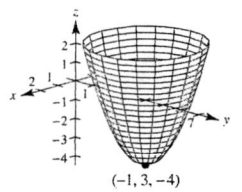

7. Relative minimum: $(-1, 1, -4)$
9. Relative maximum: $(8, 16, 74)$

11. Relative minimum: $(1, 2, -1)$
13. Relative minimum: $(0, 0, 3)$
15. Relative maximum: $(0, 0, 4)$
17. Relative maximum: $(-1, 0, 2)$
    Relative minimum: $(1, 0, -2)$ 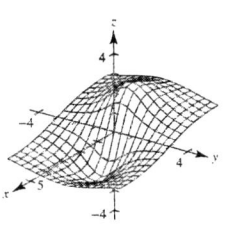
19. Relative minimum: $(0, 0, 0)$
    Relative maxima: $(0, \pm 1, 4)$
    Saddle points: $(\pm 1, 0, 1)$ 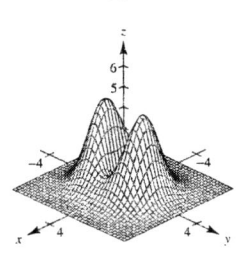
21. Saddle point: $(1, -2, -1)$ 23. Saddle point: $(0, 0, 0)$
25. Saddle point: $(0, 0, 0)$
    Relative minimum: $(1, 1, -1)$
27. There are no critical numbers.
29. z is never negative. Minimum: $z = 0$ when $x = y \neq 0$. 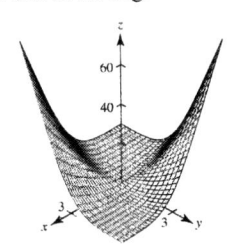
31. Insufficient information   33. Saddle point
35. (a) The function f defined on a region R containing (x_0, y_0) has a relative minimum at (x_0, y_0) if $f(x, y) \geq f(x_0, y_0)$.
    (b) The function f defined on a region R containing (x_0, y_0) has a relative maximum at (x_0, y_0) if $f(x, y) \leq f(x_0, y_0)$.
    (c) A saddle point is a critical point (x_0, y_0) that is not an extrema.
    (d) Let f be defined on an open region R containing (x_0, y_0). The point (x_0, y_0) is a critical point of f if one of the following is true:
    1. $f_x(x_0, y_0) = 0$ and $f_y(x_0, y_0) = 0$
    2. $f_x(x_0, y_0)$ or $f_y(x_0, y_0)$ does not exist
37. Answers will vary.   39. Answers will vary.
    Example: Example:
    No extrema   Saddle point

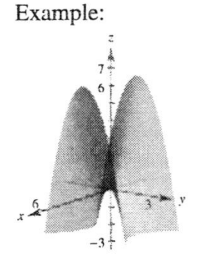

41. Point A is a saddle point.   43. $-4 < f_{xy}(3, 7) < 4$
45. Saddle point: $(0, 0, 0)$
    Test fails
47. Absolute minima: $(1, a, 0), (b, -4, 0)$
    Test fails
49. Absolute minimum: $(0, 0, 0)$
    Test fails
51. Relative minimum: $(0, 3, -1)$
53. Absolute maximum: $(0, 1, 10)$
    Absolute minimum: $(1, 2, 5)$
55. Absolute maxima: $(\pm 2, 4, 28)$
    Absolute minimum: $(0, 1, -2)$
57. Absolute maxima: $(2, 1, 6), (-2, -1, 6)$
    Absolute minima: $\left(-\frac{1}{2}, 1, -\frac{1}{4}\right), \left(\frac{1}{2}, -1, -\frac{1}{4}\right)$
59. Absolute maxima: $(2, 2, 16), (-2, -2, 16)$
    Absolute minima: $(x, -x, 0), |x| \leq 2$
61. Absolute maximum: $(1, 1, 1)$
    Absolute minimum: $(0, 0, 0)$
63. False. Let $f(x, y) = |1 - x - y|$ at the point $(0, 0, 1)$.

Section 12.9 (page 917)

1. $\dfrac{6\sqrt{14}}{7}$ 3. 6   5. 10, 10, 10
7. 10, 10, 10   9. $36 \times 18 \times 18$ inches
11. Let $a + b + c = k$.
$$V = \frac{4\pi abc}{3} = \frac{4}{3}\pi ab(k - a - b) = \frac{4}{3}(kab - a^2b - ab^2)$$
$$\left.\begin{array}{l} V_a = \frac{4}{3}\pi(kb - 2ab - b^2) = 0 \\ V_b = \frac{4}{3}\pi(ka - a^2 - 2ab) = 0 \end{array}\right\} \begin{array}{l} kb - 2ab - b^2 = 0 \\ ka - a^2 - 2ab = 0 \end{array}$$
So, $a = b$ and $b = k/3$. Thus, $a = b = c = k/3$.
13. Let x, y, and z be the length, width, and height, respectively, and let V_0 be the given volume. Then $V_0 = xyz$ and $z = V_0/xy$. The surface area is
$$S = 2xy + 2yz + 2xz = 2\left(xy + \frac{V_0}{x} + \frac{V_0}{y}\right).$$
$$\left.\begin{array}{l} S_x = 2\left(y - \dfrac{V_0}{x^2}\right) = 0 \\ S_y = 2\left(x - \dfrac{V_0}{y^2}\right) = 0 \end{array}\right\} \begin{array}{l} x^2 y - V_0 = 0 \\ xy^2 - V_0 = 0 \end{array}$$
So, $x = \sqrt[3]{V_0}$, $y = \sqrt[3]{V_0}$, and $z = \sqrt[3]{V_0}$.
15. $x = \dfrac{\sqrt{2}}{2} \approx 0.707$ kilometer
    $y = \dfrac{3\sqrt{2} + 2\sqrt{3}}{6} \approx 1.284$ kilometers
17. Each edge of $w/3$ inches is turned up 60° from the horizontal.
19. $x_1 = 3$; $x_2 = 6$ 21. $x_1 = 275$; $x_2 = 110$

23. (a) $S = \sqrt{x^2 + y^2} + \sqrt{(x+2)^2 + (y-2)^2} + \sqrt{(x-4)^2 + (y-2)^2}$

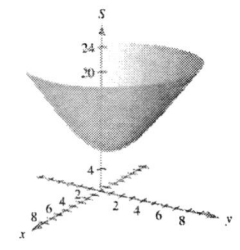

The surface has a minimum.

(b) $S_x = \dfrac{x}{\sqrt{x^2+y^2}} + \dfrac{x+2}{\sqrt{(x+2)^2+(y-2)^2}} + \dfrac{x-4}{\sqrt{(x-4)^2+(y-2)^2}}$

$S_y = \dfrac{y}{\sqrt{x^2+y^2}} + \dfrac{y-2}{\sqrt{(x+2)^2+(y-2)^2}} + \dfrac{y-2}{\sqrt{(x-4)^2+(y-2)^2}}$

(c) $-\dfrac{1}{\sqrt{2}}\mathbf{i} - \left(\dfrac{1}{\sqrt{2}} - \dfrac{2}{\sqrt{10}}\right)\mathbf{j}$

$\theta \approx 186.0°$

(d) $t = 1.344$; $(x_2, y_2) \approx (0.05, 0.90)$

(e) $(x_4, y_4) \approx (0.06, 0.45)$; $S = 7.266$

(f) $-\nabla S(x, y)$ gives the direction of greatest rate of decrease of S. Use $\nabla S(x, y)$ when finding a maximum.

25. Write the equation to be maximized or minimized as a function of two variables. Take the partial derivatives and set them equal to zero or undefined to obtain the critical points. Use the Second Partials Test to test for relative extrema using the critical points. Check the boundary points.

27. (a) $y = \frac{3}{4}x + \frac{4}{3}$ (b) $\frac{1}{6}$

29. (a) $y = -2x + 4$ (b) 2

31. $y = \frac{37}{43}x + \frac{7}{43}$

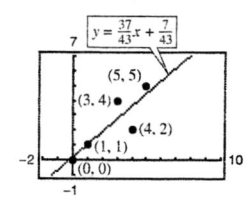

33. $y = -\frac{175}{148}x + \frac{945}{148}$

35. (a) $y = 1.724x + 79.733$

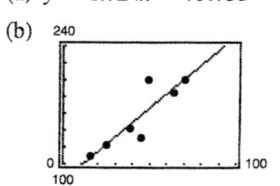

(b)

(c) 1.724

37. $y = 14x + 19$

41.4 bushels per acre

39. $a\sum_{i=1}^{n} x_i^4 + b\sum_{i=1}^{n} x_i^3 + c\sum_{i=1}^{n} x_i^2 = \sum_{i=1}^{n} x_i^2 y_i$

$a\sum_{i=1}^{n} x_i^3 + b\sum_{i=1}^{n} x_i^2 + c\sum_{i=1}^{n} x_i = \sum_{i=1}^{n} x_i y_i$

$a\sum_{i=1}^{n} x_i^2 + b\sum_{i=1}^{n} x_i + cn = \sum_{i=1}^{n} y_i$

41. $y = \frac{3}{7}x^2 + \frac{6}{5}x + \frac{26}{35}$ **43.** $y = x^2 - x$

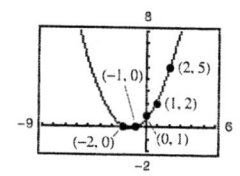

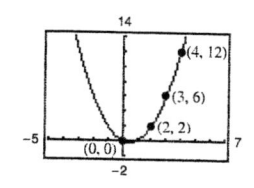

45. (a) $y = -0.22x^2 + 9.66x - 1.79$

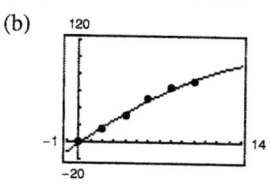

(b)

47. (a) $\ln P = -0.1499h + 9.3018$

(b) $P = 10{,}957.7e^{-0.1499h}$

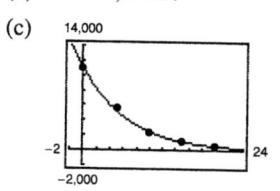

(c)

(d) Proof

Section 12.10 (page 927)

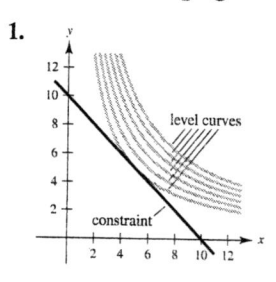

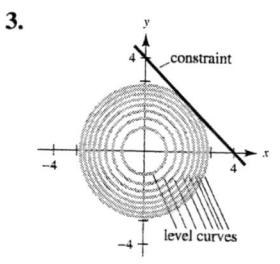

1. $f(5, 5) = 25$ **3.** $f(2, 2) = 8$

5. $f(2, 4) = -12$ **7.** $f(25, 50) = 2600$

9. $f(1, 1) = 2$ **11.** $f(2, 2) = e^4$

13. Maxima: $f\left(\dfrac{\sqrt{2}}{2}, \dfrac{\sqrt{2}}{2}\right) = \dfrac{5}{2}$

$f\left(-\dfrac{\sqrt{2}}{2}, -\dfrac{\sqrt{2}}{2}\right) = \dfrac{5}{2}$

Minima: $f\left(-\dfrac{\sqrt{2}}{2}, \dfrac{\sqrt{2}}{2}\right) = -\dfrac{1}{2}$

$f\left(\dfrac{\sqrt{2}}{2}, -\dfrac{\sqrt{2}}{2}\right) = -\dfrac{1}{2}$

15. $f(2, 2, 2) = 12$ **17.** $f\left(\frac{1}{3}, \frac{1}{3}, \frac{1}{3}\right) = \frac{1}{3}$

19. $f(8, 16, 8) = 1024$ **21.** $f(3, \frac{3}{2}, 1) = 6$

23. $\dfrac{\sqrt{13}}{13}$ **25.** $\sqrt{3}$

27. $x = \dfrac{10 + 2\sqrt{265}}{15}$

$y = \dfrac{5 + \sqrt{265}}{15}$

$z = \dfrac{-1 + \sqrt{265}}{3}$

29. Optimization problems that have restrictions or constraints on the values that can be used to produce the optimal solution are called constrained optimization problems.

31. $36 \times 18 \times 18$ inches   **33.** $\sqrt[3]{360} \times \sqrt[3]{360} \times \frac{4}{3}\sqrt[3]{360}$ feet

35. $\dfrac{2\sqrt{3}a}{3} \times \dfrac{2\sqrt{3}b}{3} \times \dfrac{2\sqrt{3}c}{3}$ **37.** Proof   **39.** $\dfrac{2}{3}$

41. $P\left(\dfrac{3125}{6}, \dfrac{6250}{3}\right) = 147{,}314$

43. $x = 50\sqrt{2}$

$y = 200\sqrt{2}$

Cost = \$13,576.45

45. (a) $g\left(\dfrac{\pi}{3}, \dfrac{\pi}{3}, \dfrac{\pi}{3}\right) = \dfrac{1}{8}$

(b)

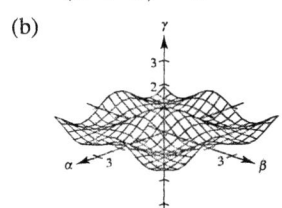

Maximum values occur when $\alpha = \beta$.

Review Exercises for Chapter 12 (page 929)

1. Not a function because every $f(x, y)$ does not have a unique z-value.

3.   **5.**

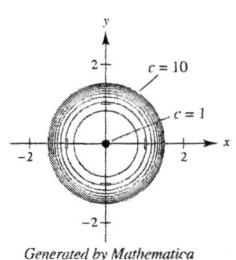

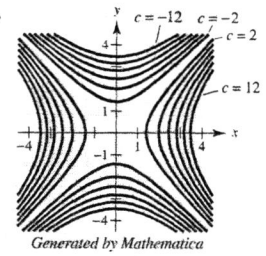

7.   **9.**

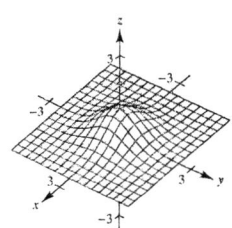

11. Continuous except at $(0, 0)$

Limit: $\dfrac{1}{2}$

13. Continuous except at $(0, 0)$

Limit does not exist.

15. $f_x(x, y) = e^x \cos y$

$f_y(x, y) = -e^x \sin y$

17. $\dfrac{\partial z}{\partial x} = e^y + ye^x$ **19.** $g_x(x, y) = \dfrac{y(y^2 - x^2)}{(x^2 + y^2)^2}$

$\dfrac{\partial z}{\partial y} = xe^y + e^x$ $g_y(x, y) = \dfrac{x(x^2 - y^2)}{(x^2 + y^2)^2}$

21. $f_x(x, y, z) = \dfrac{-yz}{x^2 + y^2}$

$f_y(x, y, z) = \dfrac{xz}{x^2 + y^2}$

$f_z(x, y, z) = \arctan \dfrac{y}{x}$

23. $u_x(x, t) = cne^{-n^2t} \cos nx$

$u_t(x, t) = -cn^2 e^{-n^2t} \sin nx$

25. Answers will vary. Example:

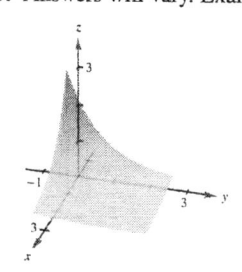

27. $f_{xx}(x, y) = 6$

$f_{yy}(x, y) = 12y$

$f_{xy}(x, y) = f_{yx}(x, y) = -1$

29. $h_{xx}(x, y) = -y \cos x$

$h_{yy}(x, y) = -x \sin y$

$h_{xy}(x, y) = h_{yx}(x, y) = \cos y - \sin x$

31. $\dfrac{\partial^2 z}{\partial x^2} + \dfrac{\partial^2 z}{\partial y^2} = 2 + (-2) = 0$

33. $\dfrac{\partial^2 z}{\partial x^2} + \dfrac{\partial^2 z}{\partial y^2} = \dfrac{6x^2y - 2y^3}{(x^2 + y^2)^3} + \dfrac{-6x^2y + 2y^3}{(x^2 + y^2)^3} = 0$

35. $\left(\sin\dfrac{y}{x} - \dfrac{y}{x}\cos\dfrac{y}{x}\right)dx + \left(\cos\dfrac{y}{x}\right)dy$

37. 0.6538 centimeter, 5.03%   **39.** $\pm \pi$ cubic inches

41. $\dfrac{dw}{dt} = \dfrac{10t + 4}{5t^2 + 4t + 25}$

43. $\dfrac{\partial u}{\partial r} = 2r$ **45.** $\dfrac{\partial z}{\partial x} = \dfrac{2xy - z}{x + 2y + 2z}$

$\dfrac{\partial u}{\partial t} = 2t$ $\dfrac{\partial z}{\partial y} = \dfrac{x^2 - 2z}{x + 2y + 2z}$

47. 0   **49.** $\dfrac{2}{3}$ **51.** $\left\langle -\dfrac{1}{2}, 0 \right\rangle, \dfrac{1}{2}$

53. $\left\langle -\dfrac{\sqrt{2}}{2}, -\dfrac{\sqrt{2}}{2} \right\rangle, 1$ **55.** $\dfrac{27}{\sqrt{793}}\mathbf{i} - \dfrac{8}{\sqrt{793}}\mathbf{j}$

57. Tangent plane: $4x + 4y - z = 8$
Normal line: $x = 2 + 4t, y = 1 + 4t, z = 4 - t$

59. Tangent plane: $z = 4$
Normal line: $x = 2, y = -3, z = 4 + t$

61. $\dfrac{x-2}{1} = \dfrac{y-1}{2}, z = 3$ **63.** $\theta \approx 36.7°$

65. Relative minimum: $\left(\dfrac{3}{2}, \dfrac{9}{4}, -\dfrac{27}{16}\right)$
Saddle point: $(0, 0, 0)$

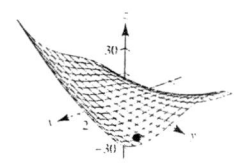

67. Relative minimum: $(1, 1, 3)$

69. The level curves are hyperbolas. The critical point $(0, 0)$ may be a saddle point or an extrema.

71. $x_1 = 94, x_2 = 157$ **73.** $f(49.4, 253) = 13{,}201.8$

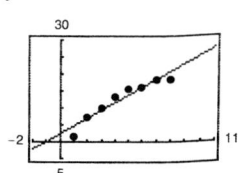

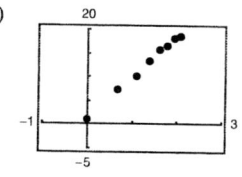

75. (a) $y = 2.29t + 2.34$ (b) More closely linear

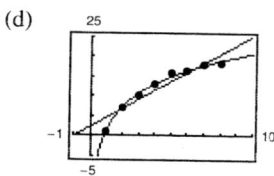

(c) $y = 1.54 + 8.37 \ln t$ (d) Logarithmic model is a better fit.

77. Maximum: $f\left(\dfrac{1}{3}, \dfrac{1}{3}, \dfrac{1}{3}\right) = \dfrac{1}{3}$

79. $x = \dfrac{\sqrt{2}}{2} \approx 0.707$ kilometer; $y = \dfrac{\sqrt{3}}{3} \approx 0.577$ kilometer;

$z = \dfrac{60 - 3\sqrt{2} - 2\sqrt{3}}{6} \approx 8.716$ kilometers

P.S. Problem Solving (page 932)

1. (a) 12 square units

(b) Let $f(a, b, c) = s(s - a)(s - b)(s - c)$ with $a + b + c =$ constant.
Then, $f_a(a, b, c) = -s(s - b)(s - c)$
$f_b(a, b, c) = -s(s - a)(s - c)$
$f_c(a, b, c) = -s(s - a)(s - b)$.
Maximum occurs when $a = b = c$. Therefore, the triangle is equilateral.

(c) Let $f(a, b, c) = a + b + c$ with
$A^2 = s(s - a)(s - b)(s - c) =$ constant
and $f_a = \lambda g_a, f_b = \lambda g_b$, and $f_c = \lambda g_c$.
Then, $-\lambda s(s - b)(s - c) = 1$
$-\lambda s(s - a)(s - c) = 1$
$-\lambda s(s - a)(s - b) = 1$.
Minimum occurs when $a = b = c$. Therefore, the triangle is equilateral.

3. (a) $y_0 z_0(x - x_0) + x_0 z_0(y - y_0) + x_0 y_0(z - z_0) = 0$

(b) $x_0 y_0 z_0 = 1$
$z_0 = \dfrac{1}{x_0 y_0}$
Then the tangent plane is
$y_0\left(\dfrac{1}{x_0 y_0}\right)(x - x_0) + x_0\left(\dfrac{1}{x_0 y_0}\right)(y - y_0) + x_0 y_0\left(z - \dfrac{1}{x_0 y_0}\right) = 0.$
Intercepts: $(3x_0, 0, 0), (0, 3y_0, 0), \left(0, 0, \dfrac{3}{x_0 y_0}\right)$
$V = \dfrac{1}{3} bh = \dfrac{9}{2}$

5. No; Yes

7. $2\sqrt[3]{150} \times 2\sqrt[3]{150} \times \dfrac{5\sqrt[3]{150}}{3}$

9. (a) $x\dfrac{\partial f}{\partial x} + y\dfrac{\partial f}{\partial y} = xCy^{1-a}ax^{a-1} + yCx^a(1-a)y^{-a}$
$= ax^a Cy^{1-a} + (1-a)x^a C(y^{1-a})$
$= Cx^a y^{1-a}[a + (1-a)]$
$= Cx^a y^{1-a}$
$= f(x, y)$

(b) $f(tx, ty) = C(tx)^a(ty)^{1-a}$
$= Ctx^a y^{1-a}$
$= tCx^a y^{1-a}$
$= tf(x, y)$

11. (a) $x = 32\sqrt{2}t$
$y = 32\sqrt{2}t - 16t^2$

(b) $\alpha = \arctan\left(\dfrac{y}{x + 50}\right)$
$= \arctan\left(\dfrac{32\sqrt{2}t - 16t^2}{32\sqrt{2}t + 50}\right)$

(c) $\dfrac{d\alpha}{dt} = \dfrac{-16(8\sqrt{2}t^2 + 25t - 25\sqrt{2})}{64t^4 - 256\sqrt{2}t^3 + 1024t^2 + 800\sqrt{2}t + 625}$

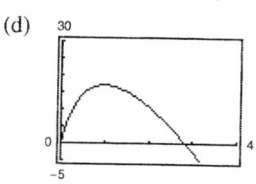

(d) No; The rate of change of α is greatest when the projectile is closest to the camera.

(e) α is maximum when $t = 0.98$ seconds.

No; the projectile is at its maximum height when $t = \sqrt{2} \approx 1.41$ seconds.

13. (a)

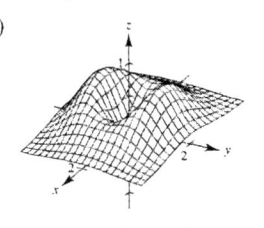

Minimum: $(0, 0, 0)$
Maxima: $(0, \pm 1, 2e^{-1})$
Saddle points: $(\pm 1, 0, e^{-1})$

(b)

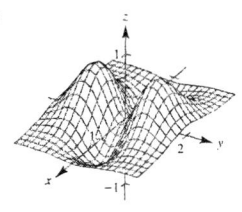

Minima: $(\pm 1, 0, -e^{-1})$
Maxima: $(0, \pm 1, 2e^{-1})$
Saddle point: $(0, 0, 0)$

(c) $\alpha > 0$
Minimum: $(0, 0, 0)$
Maxima: $(0, \pm 1, \beta e^{-1})$
Saddle points: $(\pm 1, 0, \alpha e^{-1})$

$\alpha < 0$
Minima: $(\pm 1, 0, \alpha e^{-1})$
Maxima: $(0, \pm 1, \beta e^{-1})$
Saddle point: $(0, 0, 0)$

15. (a) 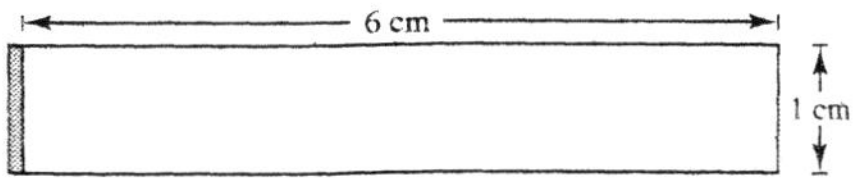 6 cm × 1 cm

(b) 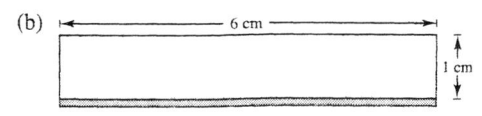 6 cm × 1 cm

(c) Height

(d) $dl = 0.01, dh = 0: dA = 0.01$
$dl = 0, dh = 0.01: dA = 0.06$

17. Answers will vary. For example: A minor variance in measurement could create a major variance in results. Also, close examination of the variable(s) having the greatest effect in an applied formula greatly affects the results of that application.

19. $u(x, t) = \frac{1}{2}[f(x - ct) + f(x + ct)]$

Let $r = x - ct$ and $s = x + ct$. Then $u(r, s) = \frac{1}{2}[f(r) + f(s)]$.

$\frac{\partial u}{\partial t} = \frac{1}{2}\frac{df}{dr}(-c) + \frac{1}{2}\frac{df}{ds}(c)$

$\frac{\partial^2 u}{\partial t^2} = \frac{c^2}{2}\left[\frac{d^2f}{dr^2} + \frac{d^2f}{ds^2}\right]$

$\frac{\partial u}{\partial x} = \frac{1}{2}\frac{df}{dr}(1) + \frac{1}{2}\frac{df}{ds}(1)$

$\frac{\partial^2 u}{\partial x^2} = \frac{1}{2}\left[\frac{d^2f}{dr^2} + \frac{d^2f}{ds^2}\right]$

Thus, $\frac{\partial^2 u}{\partial t^2} = c^2 \frac{\partial^2 u}{\partial x^2}$.

Chapter 13

Section 13.1 (page 942)

1. $\frac{3x^2}{2}$ **3.** $y \ln(2y)$ **5.** $\frac{4x^2 - x^4}{2}$ **7.** $\frac{y}{2}[(\ln y)^2 - y^2]$

9. $x^2(1 - e^{-x^2} - x^2 e^{-x^2})$ **11.** 3   **13.** $\frac{1}{3}$ **15.** $\frac{20}{3}$

17. $\frac{2}{3}$ **19.** 4   **21.** $\frac{\pi^2}{32} + \frac{1}{8}$ **23.** $\frac{1}{2}$ **25.** Diverges

27. 24   **29.** $\frac{16}{3}$ **31.** $\frac{9}{2}$ **33.** $\frac{8}{3}$ **35.** 5   **37.** πab

39. $\int_0^4 \int_0^y f(x, y)\, dx\, dy = \int_0^4 \int_x^4 f(x, y)\, dy\, dx$

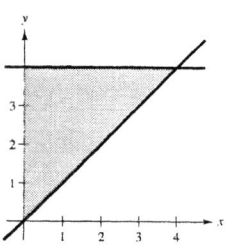

41. $\int_{-2}^2 \int_0^{\sqrt{4-x^2}} f(x, y)\, dy\, dx = \int_0^2 \int_{-\sqrt{4-y^2}}^{\sqrt{4-y^2}} f(x, y)\, dx\, dy$

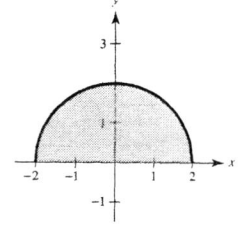

43. $\int_1^{10} \int_0^{\ln y} f(x, y)\, dx\, dy = \int_0^{\ln 10} \int_{e^x}^{10} f(x, y)\, dy\, dx$

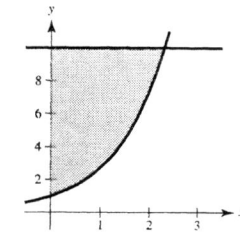

45. $\int_{-1}^1 \int_{x^2}^1 f(x, y)\, dy\, dx = \int_0^1 \int_{-\sqrt{y}}^{\sqrt{y}} f(x, y)\, dx\, dy$

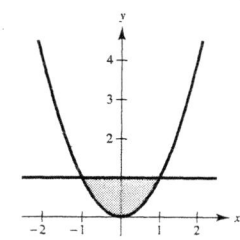

47. $\int_0^1 \int_0^2 dy\, dx = \int_0^2 \int_0^1 dx\, dy = 2$

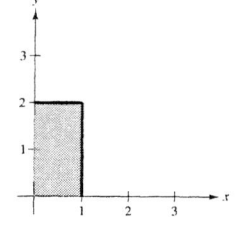

A156 Answers to Odd-Numbered Exercises

49. $\int_0^1 \int_{-\sqrt{1-y^2}}^{\sqrt{1-y^2}} dx\, dy = \int_{-1}^1 \int_0^{\sqrt{1-x^2}} dy\, dx = \dfrac{\pi}{2}$

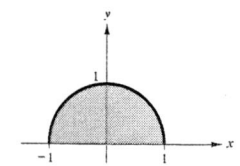

51. $\int_0^2 \int_0^x dy\, dx + \int_2^4 \int_0^{4-x} dy\, dx = \int_0^2 \int_y^{4-y} dx\, dy = 4$

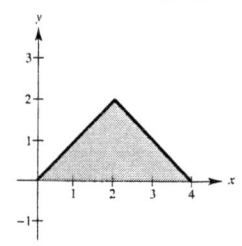

53. $\int_0^2 \int_{x/2}^1 dy\, dx = \int_0^1 \int_0^{2y} dx\, dy = 1$

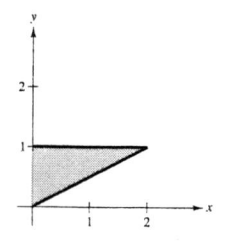

55. $\int_0^1 \int_{y^2}^{\sqrt[3]{y}} dx\, dy = \int_0^1 \int_{x^3}^{\sqrt{x}} dy\, dx = \dfrac{5}{12}$

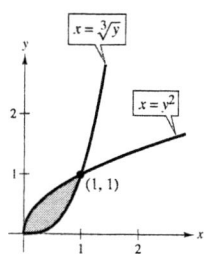

57. Common region of integration is the sector of the circle shown in the figure.

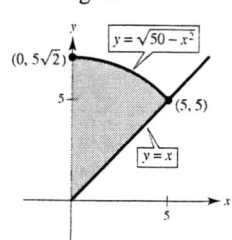

Value of the integrals: $\dfrac{15{,}625\pi}{24}$

59. $\dfrac{26}{9}$ **61.** $\tfrac{1}{2}(1 - \cos 1) \approx 0.230$

63. $\dfrac{1664}{105}$ **65.** $(\ln 5)^2$

67. (a)

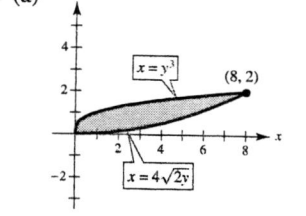

(b) $\int_0^2 \int_{y^3}^{4\sqrt{2y}} (x^2 y - xy^2)\, dx\, dy = \int_0^8 \int_{x^2/32}^{\sqrt[3]{x}} (x^2 y - xy^2)\, dy\, dx$

(c) $\dfrac{67{,}520}{693}$

69. 20.5648 **71.** $\dfrac{15\pi}{2}$

73. Integration of a function of several variables. Integrate with respect to one variable while holding the other variables constant.

75. If all four limits of integration are constant, the region of integration is rectangular.

77. True

Section 13.2 (page 951)

1. 24 (approximation is exact)

3. Approximation: 52; Exact: $\dfrac{160}{3}$

5. 400; 272

7. 8

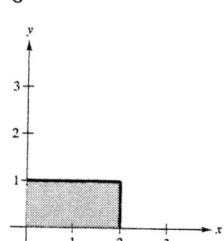

9. 36

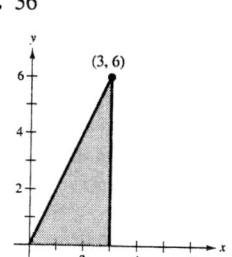

11. 0

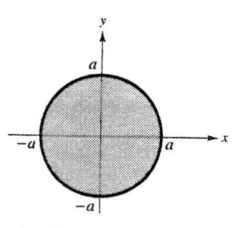

13. $\int_0^3 \int_0^5 xy\, dy\, dx = \dfrac{225}{4}$

$\int_0^5 \int_0^3 xy\, dx\, dy = \dfrac{225}{4}$

15. $\int_0^2 \int_x^{2x} \dfrac{y}{x^2 + y^2}\, dy\, dx = \ln \dfrac{5}{2}$

$\int_0^2 \int_{y/2}^y \dfrac{y}{x^2 + y^2}\, dx\, dy + \int_2^4 \int_{y/2}^2 \dfrac{y}{x^2 + y^2}\, dx\, dy = \ln \dfrac{5}{2}$

17. $\int_0^1 \int_{4-x}^{4-x^2} -2y \ln x \, dy \, dx = \frac{26}{25}$

$\int_3^4 \int_{4-y}^{\sqrt{4-y}} -2y \ln x \, dx \, dy = \frac{26}{25}$

19. $\int_0^3 \int_{4y/3}^{\sqrt{25-y^2}} x \, dx \, dy = 25$

$\int_0^4 \int_0^{3x/4} x \, dy \, dx + \int_4^5 \int_0^{\sqrt{25-x^2}} x \, dy \, dx = 25$

21. 4 **23.** 4 **25.** 12 **27.** $\frac{3}{8}$ **29.** 1 **31.** 8π

33. $\int_0^1 \int_0^x xy \, dy \, dx = \frac{1}{8}$ **35.** $\int_0^2 \int_0^4 x^2 \, dy \, dx = \frac{32}{3}$

37. $2\int_0^1 \int_0^x \sqrt{1-x^2} \, dy \, dx = \frac{2}{3}$

39. $\int_0^2 \int_0^{\sqrt{4-x^2}} (x+y) \, dy \, dx = \frac{16}{3}$

41. 8π **43.** 8π **45.** 1.2315 **47.** Proof

49. $1 - e^{-1/4} \approx 0.221$ **51.** $\frac{1}{3}[2\sqrt{2}-1]$ **53.** 2 **55.** $\frac{2}{3}$

57. See "Definition of Double Integral" on page 946. The double integral of a function $f(x, y) \geq 0$ over the region of integration yields the volume of that region.

59. kB **61.** 25,645.24

63. $f(x, y) \geq 0$ for all (x, y) and

$\int_{-\infty}^{\infty} \int_{-\infty}^{\infty} f(x, y) \, dA = \int_0^5 \int_0^2 \frac{1}{10} \, dy \, dx = \int_0^5 \frac{1}{5} \, dx = 1$

$P(0 \leq x \leq 2, 1 \leq y \leq 2) = \int_0^2 \int_1^2 \frac{1}{10} \, dy \, dx = \int_0^2 \frac{1}{10} \, dx = \frac{1}{5}$

65. $f(x, y) \geq 0$ for all (x, y) and

$\int_{-\infty}^{\infty} \int_{-\infty}^{\infty} f(x, y) \, dA = \int_0^3 \int_3^6 \frac{1}{27}(9 - x - y) \, dy \, dx$

$= \int_0^3 \left(\frac{1}{2} - \frac{1}{9}x\right) dx = 1$

$P(0 \leq x \leq 1, 4 \leq y \leq 6) = \int_0^1 \int_4^6 \frac{1}{27}(9 - x - y) \, dy \, dx$

$= \int_0^1 \frac{2}{27}(4 - x) \, dx = \frac{7}{27}$

67. 2500 cubic meters **69.** (a) 1.784 (b) 1.788

71. (a) 11.057 (b) 11.041

73. d **75.** False: $V = 8\int_0^1 \int_0^{\sqrt{1-y^2}} \sqrt{1-x^2-y^2} \, dx \, dy$.

77. $\frac{1}{2}(1-e)$

Section 13.3 (page 960)

1. Rectangular **3.** Polar

5. The region R is a half circle of radius 8. It can be described in polar coordinates as

$R = \{(r, \theta): 0 \leq r \leq 8, 0 \leq \theta \leq \pi\}$.

7. The region R is a cardioid with $a = b = 3$. It can be described in polar coordinates as

$R = \{(r, \theta): 0 \leq r \leq 3 + 3\sin\theta, 0 \leq \theta \leq 2\pi\}$.

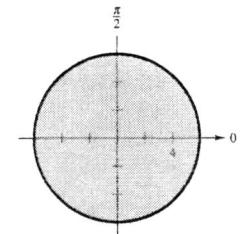

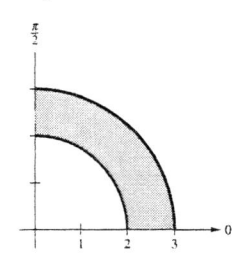

9. 0 **11.** $\frac{5\sqrt{5}\pi}{6}$

13. $\frac{9}{8} + \frac{3\pi^2}{32}$

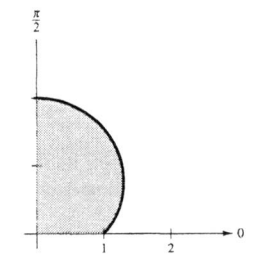

15. $\frac{a^3}{3}$ **17.** $\frac{243\pi}{10}$ **19.** $\frac{2}{3}$

21. $\int_0^{\pi/4} \int_0^{2\sqrt{2}} r^2 \, dr \, d\theta = \frac{4\sqrt{2}\pi}{3}$

23. $\int_0^{\pi/2} \int_0^2 r^2(\cos\theta + \sin\theta) \, dr \, d\theta = \frac{16}{3}$

25. $\int_0^{\pi/4} \int_1^2 r\theta \, dr \, d\theta = \frac{3\pi^2}{64}$ **27.** $\frac{1}{8}$ **29.** $\frac{250\pi}{3}$

31. $\frac{64}{9}(3\pi - 4)$ **33.** $2\sqrt{4 - 2\sqrt[3]{2}}$ **35.** 1.2858

37. 9π **39.** $\frac{3\pi}{2}$ **41.** π

43. Let R be a region bounded by the graphs of $r = g_1(\theta)$ and $r = g_2(\theta)$ and the lines $\theta = a$ and $\theta = b$. When using polar coordinates to evaluate a double integral over R, R can be partitioned into small polar sectors.

45. r-simple regions have fixed bounds for θ and variable bounds for r.

θ-simple regions have variable bounds for θ and fixed bounds for r.

47. Insert a factor of r; Sector of a circle

49. 56.051 **51.** c

53. False: let $f(r, \theta) = r - 1$ and let R be a sector where $0 \leq r \leq 6$ and $0 \leq \theta \leq \pi$.

55. (a) 2π (b) $\sqrt{2\pi}$ **57.** 486,788

59. (a) $\int_2^4 \int_{y/\sqrt{3}}^y f\,dx\,dy$

(b) $\int_{2/\sqrt{3}}^2 \int_2^{\sqrt{3}x} f\,dy\,dx + \int_2^{4/\sqrt{3}} \int_x^{\sqrt{3}x} f\,dy\,dx + \int_{4/\sqrt{3}}^4 \int_x^4 f\,dy\,dx$

(c) $\int_{\pi/4}^{\pi/3} \int_{2\csc\theta}^{4\csc\theta} f r\,dr\,d\theta$

61. $A = \dfrac{\Delta\theta r_2^2}{2} - \dfrac{\Delta\theta r_1^2}{2} = \Delta\theta\left(\dfrac{r_1+r_2}{2}\right)(r_2-r_1) = r\Delta r\Delta\theta$

Section 13.4 (page 969)

1. $m = 36$ 3. $m = 2$

5. (a) $m = kab$, $\left(\dfrac{a}{2},\dfrac{b}{2}\right)$ (b) $m = \dfrac{kab^2}{2}$, $\left(\dfrac{a}{2},\dfrac{2b}{3}\right)$

(c) $m = \dfrac{ka^2 b}{2}$, $\left(\dfrac{2a}{3},\dfrac{b}{2}\right)$

7. (a) $m = \dfrac{kbh}{2}$, $\left(\dfrac{b}{2},\dfrac{h}{3}\right)$ (b) $m = \dfrac{kh^2 b}{6}$, $\left(\dfrac{b}{2},\dfrac{h}{2}\right)$

(c) $m = \dfrac{khb^2}{4}$, $\left(\dfrac{7b}{12},\dfrac{h}{3}\right)$

9. (a) $\left(\dfrac{a}{2}+5,\dfrac{b}{2}\right)$ (b) $\left(\dfrac{a}{2}+5,\dfrac{2b}{3}\right)$

(c) $\left(\dfrac{2(a^2+15a+75)}{3(a+10)},\dfrac{b}{2}\right)$

11. (a) $m = \dfrac{k\pi a^2}{2}$, $\left(0,\dfrac{4a}{3\pi}\right)$

(b) $m = \dfrac{ka^4}{24}(16-3\pi)$, $\left(0,\dfrac{a}{5}\left[\dfrac{15\pi-32}{16-3\pi}\right]\right)$

13. $m = \dfrac{32k}{3}$, $\left(3,\dfrac{8}{7}\right)$ 15. $m = \dfrac{k\pi}{2}$, $\left(0,\dfrac{\pi+2}{4\pi}\right)$

17. $m = \dfrac{8192k}{15}$, $\left(\dfrac{64}{7},0\right)$ 19. $m = \dfrac{kL}{4}$, $\left(\dfrac{L}{2},\dfrac{16}{9\pi}\right)$

21. $m = \dfrac{k\pi a^2}{8}$, $\left(\dfrac{4\sqrt{2}a}{3\pi},\dfrac{4a(2-\sqrt{2})}{3\pi}\right)$

23. $m = \dfrac{k}{4}(1-e^{-4})$, $\left(\dfrac{e^4-5}{2(e^4-1)},\dfrac{4}{9}\left[\dfrac{e^6-1}{e^6-e^2}\right]\right)$

25. $m = \dfrac{k\pi}{3}$, $\left(\dfrac{81\sqrt{3}}{40\pi},0\right)$

27. $\bar{x} = \dfrac{\sqrt{3}b}{3}$ 29. $\bar{x} = \dfrac{a}{2}$ 31. $\bar{x} = \dfrac{a}{2}$

$\bar{y} = \dfrac{\sqrt{3}h}{3}$ $\bar{y} = \dfrac{a}{2}$ $\bar{y} = \dfrac{a}{2}$

33. $I_x = \dfrac{kab^4}{4}$ 35. $I_x = \dfrac{32k}{3}$

$I_y = \dfrac{kb^2 a^3}{6}$ $I_y = \dfrac{16k}{3}$

$I_0 = \dfrac{3kab^4 + 2ka^3 b^2}{12}$ $I_0 = 16k$

$\bar{\bar{x}} = \dfrac{\sqrt{3}a}{3}$ $\bar{\bar{x}} = \dfrac{2\sqrt{3}}{3}$

$\bar{\bar{y}} = \dfrac{\sqrt{2}b}{2}$ $\bar{\bar{y}} = \dfrac{2\sqrt{6}}{3}$

37. $I_x = 16k$

$I_y = \dfrac{512k}{5}$

$I_0 = \dfrac{592k}{5}$

$\bar{\bar{x}} = \dfrac{4\sqrt{15}}{5}$

$\bar{\bar{y}} = \dfrac{\sqrt{6}}{2}$

39. $I_x = \dfrac{3k}{56}$

$I_y = \dfrac{k}{18}$

$I_0 = \dfrac{55k}{504}$

$\bar{\bar{x}} = \dfrac{\sqrt{30}}{9}$

$\bar{\bar{y}} = \dfrac{\sqrt{70}}{14}$

41. $2k\int_{-b}^{b}\int_0^{\sqrt{b^2-x^2}}(x-a)^2\,dy\,dx = \dfrac{k\pi b^2}{4}(b^2+4a^2)$

43. $\int_0^4 \int_0^{\sqrt{x}} kx(x-6)^2\,dy\,dx = \dfrac{42{,}752k}{315}$

45. $\int_0^a \int_0^{\sqrt{a^2-x^2}} k(a-y)(y-a)^2\,dy\,dx = ka^5\left(\dfrac{7\pi}{16}-\dfrac{17}{15}\right)$

47. $\bar{y}$ will increase.   49. $\bar{x}$ and $\bar{y}$ will both increase.

51. Let ρ be a continuous density function on the planar lamina R. The moments of mass with respect to the x- and y-axes are

$M_x = \iint_R y\rho(x,y)\,dA$ and $M_y = \iint_R x\rho(x,y)\,dA$.

If m is the mass of the lamina, then the center of mass is $(\bar{x},\bar{y}) = \left(\dfrac{M_y}{m},\dfrac{M_x}{m}\right)$.

53. See definition on page 968.

55. $\dfrac{L}{3}$ 57. $\dfrac{L}{2}$

Section 13.5 (page 976)

1. 6   3. 12π 5. $\dfrac{3}{4}\left[6\sqrt{37}+\ln(\sqrt{37}+6)\right]$

7. $\dfrac{4}{27}(31\sqrt{31}-8)$ 9. $\sqrt{2}-1$ 11. $\sqrt{2}\pi$

13. $2\pi a(a-\sqrt{a^2-b^2})$ 15. $48\sqrt{14}$ 17. 20π

19. $\int_0^1 \int_0^x \sqrt{5+4x^2}\,dy\,dx = \dfrac{27-5\sqrt{5}}{12}$

21. $\int_{-2}^{2}\int_{-\sqrt{4-x^2}}^{\sqrt{4-x^2}} \sqrt{1+4x^2+4y^2}\,dy\,dx$

$= \int_0^{2\pi}\int_0^2 \sqrt{1+4r^2}\,r\,dr\,d\theta$

$= \dfrac{\pi}{6}(17\sqrt{17}-1)$

23. $\int_0^1\int_0^1 \sqrt{1+4x^2+4y^2}\,dy\,dx = 1.8616$

25. e   27. 2.0035

29. $\int_{-1}^{1}\int_{-1}^{1} \sqrt{1+9(x^2-y)^2+9(y^2-x)^2}\,dy\,dx$

31. $\int_{-2}^{2}\int_{-\sqrt{4-x^2}}^{\sqrt{4-x^2}} \sqrt{1+e^{-2x}}\,dy\,dx$

33. $\int_0^4\int_0^{10}\sqrt{1+e^{2xy}(x^2+y^2)}\,dy\,dx$

35. If f and its first partial derivatives are continuous on the closed region R in the xy-plane, then the area of the surface S given by $z = f(x, y)$ over R is
$$\iint_R \sqrt{1 + [f_x(x, y)]^2 + [f_y(x, y)]^2} \, dA.$$
37. 16   **39.** (a) 30,415.74 cubic feet   (b) 2081.53 square feet
41. (a) $812\pi\sqrt{609}$ cubic centimeters
   (b) $100\pi\sqrt{609}$ square centimeters

Section 13.6 (page 986)

1. 18   **3.** $\dfrac{1}{10}$ **5.** $\dfrac{15}{2}\left(1 - \dfrac{1}{e}\right)$ **7.** $-\dfrac{40}{3}$ **9.** $\dfrac{128}{15}$

11. 2.44167   **13.** $V = \int_0^4 \int_0^{4-x} \int_0^{4-x-y} dz \, dy \, dx$

15. $V = \int_{-3}^3 \int_{-\sqrt{9-y^2}}^{\sqrt{9-y^2}} \int_0^{9-x^2-y^2} dz \, dx \, dy$

17. $\dfrac{256}{15}$ **19.** $\dfrac{4\pi a^3}{3}$ **21.** $\dfrac{256}{15}$

23. $\int_0^3 \int_0^{(12-4z)/3} \int_0^{(12-4z-3x)/6} dy \, dx \, dz$

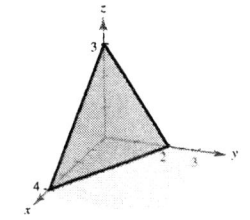

25. $\int_0^1 \int_0^x \int_0^{\sqrt{1-y^2}} dz \, dy \, dx$

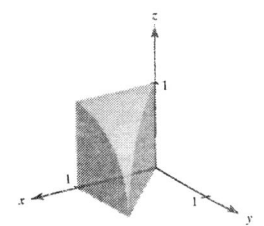

27. $\int_0^1 \int_0^x \int_0^3 xyz \, dz \, dy \, dx,$ $\int_0^1 \int_y^1 \int_0^3 xyz \, dz \, dx \, dy,$

$\int_0^1 \int_0^3 \int_0^x xyz \, dy \, dz \, dx,$ $\int_0^3 \int_0^1 \int_0^x xyz \, dy \, dx \, dz,$

$\int_0^3 \int_0^1 \int_y^1 xyz \, dx \, dy \, dz,$ $\int_0^1 \int_0^3 \int_y^1 xyz \, dx \, dz \, dy$

29. $\int_{-3}^3 \int_{-\sqrt{9-x^2}}^{\sqrt{9-x^2}} \int_0^4 xyz \, dz \, dy \, dx,$ $\int_{-3}^3 \int_{-\sqrt{9-y^2}}^{\sqrt{9-y^2}} \int_0^4 xyz \, dz \, dx \, dy$

$\int_{-3}^3 \int_0^4 \int_{-\sqrt{9-x^2}}^{\sqrt{9-x^2}} xyz \, dy \, dz \, dx,$ $\int_0^4 \int_{-3}^3 \int_{-\sqrt{9-x^2}}^{\sqrt{9-x^2}} xyz \, dy \, dx \, dz$

$\int_0^4 \int_{-3}^3 \int_{-\sqrt{9-y^2}}^{\sqrt{9-y^2}} xyz \, dx \, dy \, dz,$ $\int_{-3}^3 \int_0^4 \int_{-\sqrt{9-y^2}}^{\sqrt{9-y^2}} xyz \, dx \, dz \, dy$

31. $m = 8k$ **33.** $m = \dfrac{128k}{3}$

$\bar{x} = \dfrac{3}{2}$ $\bar{z} = 1$

35. $m = k\int_0^b \int_0^b \int_0^b xy \, dz \, dy \, dx$

$M_{yz} = k\int_0^b \int_0^b \int_0^b x^2 y \, dz \, dy \, dx$

$M_{xz} = k\int_0^b \int_0^b \int_0^b xy^2 \, dz \, dy \, dx$

$M_{xy} = k\int_0^b \int_0^b \int_0^b xyz \, dz \, dy \, dx$

37. $\bar{x}$ will be greater than 2, and $\bar{y}$ and $\bar{z}$ will be unchanged.
39. $\bar{x}$ and $\bar{z}$ will be unchanged, and $\bar{y}$ will be greater than 0.

41. $\left(0, 0, \dfrac{3h}{4}\right)$ **43.** $\left(0, 0, \dfrac{3}{2}\right)$ **45.** $\left(5, 6, \dfrac{5}{4}\right)$

47. (a) $I_x = \dfrac{2ka^5}{3}$ (b) $I_x = \dfrac{ka^8}{8}$

$I_y = \dfrac{2ka^5}{3}$ $I_y = \dfrac{ka^8}{8}$

$I_z = \dfrac{2ka^5}{3}$ $I_z = \dfrac{ka^8}{8}$

49. (a) $I_x = 256k$ (b) $I_x = \dfrac{2048k}{3}$

$I_y = \dfrac{512k}{3}$ $I_y = \dfrac{1024k}{3}$

$I_z = 256k$ $I_z = \dfrac{2048k}{3}$

51. Proof   **53.** $\int_{-1}^1 \int_{-1}^1 \int_0^{1-x} (x^2 + y^2)\sqrt{x^2 + y^2 + z^2} \, dz \, dy \, dx$

55. See "Definition of Triple Integral" on page 978 and "Theorem 13.4: Evaluation by Iterated Integrals" on page 979.

57. (a) Solid B
   (b) Solid B has the greater moment of inertia because it is more dense.
   (c) Solid A will reach the bottom first. Since Solid B has a greater moment of inertia, it has a greater resistance to rotational motion.

Section 13.7 (page 993)

1. 8   **3.** $\dfrac{52}{45}$ **5.** $\dfrac{\pi}{8}$ **7.** $\pi(e^4 + 3)$

9. $\dfrac{\pi}{4}(1 - e^{-9})$ **11.** $\dfrac{64\sqrt{3}\pi}{3}$

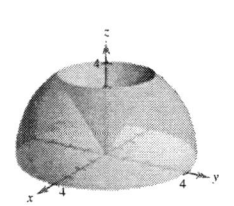

13. Cylindrical: $\int_0^{2\pi}\int_0^2\int_{r^2}^4 r^2 \cos\theta \, dz \, dr \, d\theta = 0$

Spherical: $\int_0^{2\pi}\int_0^{\arctan(1/2)}\int_0^{4\sec\phi} \rho^3 \sin^2\phi \cos\theta \, d\rho \, d\phi \, d\theta$

$+ \int_0^{2\pi}\int_{\arctan(1/2)}^{\pi/2}\int_0^{\cot\phi\csc\phi} \rho^3 \sin^2\phi \cos\phi \, d\rho \, d\phi \, d\theta = 0$

15. Cylindrical: $\int_0^{2\pi}\int_0^a\int_a^{a+\sqrt{a^2-r^2}} r^2 \cos\theta \, dz \, dr \, d\theta = 0$

Spherical: $\int_0^{\pi/4}\int_0^{2\pi}\int_{a\sec\phi}^{2a\cos\phi} \rho^3 \sin^2\phi \cos\theta \, d\rho \, d\theta \, d\phi = 0$

17. $\dfrac{2a^3}{9}(3\pi - 4)$ 19. $\dfrac{2a^3}{9}(3\pi - 4)$ 21. $48k\pi$

23. $\dfrac{\pi r_0^2 h}{3}$ 25. $\left(0, 0, \dfrac{h}{5}\right)$

27. $I_z = 4k \int_0^{\pi/2}\int_0^{r_0}\int_0^{h(r_0-r)/r_0} r^3 \, dz \, dr \, d\theta$

$= \dfrac{3}{10} m r_0^2$

29. Proof   31. $16\pi^2$ 33. $k\pi a^4$

35. $\left(0, 0, \dfrac{3r}{8}\right)$ 37. $\dfrac{k\pi}{192}$

39. Rectangular to cylindrical: $r^2 = x^2 + y^2$

$\tan\theta = \dfrac{y}{x}$

$z = z$

Cylindrical to rectangular: $x = r\cos\theta$

$y = r\sin\theta$

$z = z$

41. $\int_{\theta_1}^{\theta_2}\int_{g_1(\theta)}^{g_2(\theta)}\int_{h_1(r\cos\theta, r\sin\theta)}^{h_2(r\cos\theta, r\sin\theta)} f(r\cos\theta, r\sin\theta, z) r \, dz \, dr \, d\theta$

43. (a) r constant: right circular cylinder about z-axis

θ constant: plane parallel to z-axis

z constant: plane parallel to xy-plane

(b) ρ constant: sphere

θ constant: plane parallel to z-axis

ϕ constant: cone

45. $\dfrac{1}{2}\pi^2 a^4$

Section 13.8 (page 1000)

1. $-\dfrac{1}{2}$ 3. $1 + 2v$ 5. 1 7. $-e^{2u}$

9. [graph showing triangle with vertices (0,1) and (1,0)]

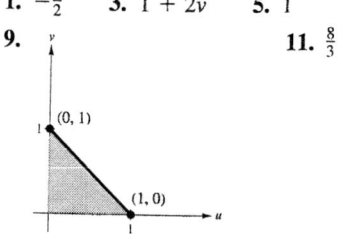

11. $\dfrac{8}{3}$ 13. 36

15. $(e^{-1/2} - e^{-2}) \ln 8 \approx 0.9798$

17. $12(e^4 - 1)$ 19. $\dfrac{100}{9}$ 21. $\dfrac{2}{5}a^{5/2}$

23. (a) [graphs of ellipse R and circle S]

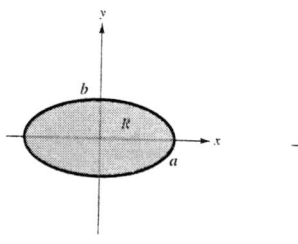

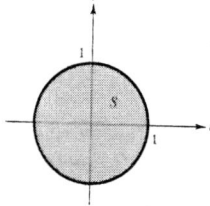

(b) ab (c) πab

25. See "Definition of the Jacobian," on page 995.

27. $u^2 v$ 29. $-\rho^2 \sin\phi$

Review Exercises for Chapter 13 (page 1001)

1. $x - x^3 + x^3 \ln x^2$ 3. $\dfrac{29}{6}$ 5. 36

7. $\int_0^3\int_0^{(3-x)/3} dy \, dx = \int_0^1\int_0^{3-3y} dx \, dy = \dfrac{3}{2}$

9. $\int_{-5}^3\int_{-\sqrt{25-x^2}}^{\sqrt{25-x^2}} dy \, dx$

$= \int_{-5}^{-4}\int_{-\sqrt{25-y^2}}^{\sqrt{25-y^2}} dx \, dy + \int_{-4}^4\int_{-\sqrt{25-y^2}}^3 dx \, dy$

$+ \int_4^5\int_{-\sqrt{25-y^2}}^{\sqrt{25-y^2}} dx \, dy$

$= \dfrac{25\pi}{2} + 12 + 25\arcsin\dfrac{3}{5} \approx 67.36$

11. $4\int_0^1\int_0^{x\sqrt{1-x^2}} dy \, dx = 4\int_0^{1/2}\int_{\sqrt{(1-\sqrt{1-4y^2})/2}}^{\sqrt{(1+\sqrt{1-4y^2})/2}} dx \, dy = \dfrac{4}{3}$

13. $\int_2^5\int_{x-3}^{\sqrt{x-1}} dy \, dx + 2\int_1^2\int_0^{\sqrt{x-1}} dy \, dx = \int_{-1}^2\int_{y^2+1}^{y+3} dx \, dy = \dfrac{9}{2}$

15. Integration over the common region R shown in the figure

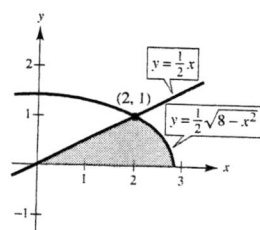

[graph showing curves $y = \tfrac{1}{2}x$ and $y = \tfrac{1}{2}\sqrt{8-x^2}$ intersecting at (2, 1)]

Both integrals yield $\dfrac{4}{3} + \dfrac{4}{3}\sqrt{2}$

17. $\dfrac{3296}{15}$ 19. c   21. $k = 1$, 0.070 23. True

25. True   27. $\dfrac{h^3}{6}[\ln(\sqrt{2} + 1) + \sqrt{2}]$ 29. $\dfrac{\pi h^3}{3}$

31. (a) $r = 3\sqrt{\cos 2\theta}$

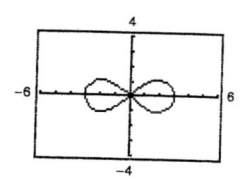

[graph of lemniscate in window $[-6, 6] \times [-4, 4]$]

(b) 9 (c) $3(3\pi - 16\sqrt{2} + 20) \approx 20.392$

33. (a) $m = \dfrac{k}{4}, \left(\dfrac{32}{45}, \dfrac{64}{55}\right)$ (b) $m = \dfrac{17k}{30}, \left(\dfrac{936}{1309}, \dfrac{784}{663}\right)$

35. $I_x = \dfrac{ka^2b^3}{6}$

$I_y = \dfrac{ka^4b}{4}$

$I_0 = \dfrac{2ka^2b^3 + 3ka^4b}{12}$

$\bar{\bar{x}} = \dfrac{a}{\sqrt{2}}$

$\bar{\bar{y}} = \dfrac{b}{\sqrt{3}}$

37. $\dfrac{\pi}{6}(65\sqrt{65} - 1)$ **39.** $\dfrac{1}{6}(37\sqrt{37} - 1)$ **41.** $\dfrac{324\pi}{5}$

43. $\dfrac{abc}{3}(a^2 + b^2 + c^2)$ **45.** $\dfrac{8\pi}{15}$ **47.** $\dfrac{32}{3}\left(\dfrac{\pi}{2} - \dfrac{2}{3}\right)$

49. $\left(0, 0, \dfrac{1}{4}\right)$ **51.** $\left(\dfrac{3a}{8}, \dfrac{3a}{8}, \dfrac{3a}{8}\right)$ **53.** $\dfrac{833k\pi}{3}$

55. (a) $\dfrac{1}{3}\pi h^2(3a - h)$ (b) $\left(0, 0, \dfrac{3(2a - h)^2}{4(3a - h)}\right)$

(c) $\left(0, 0, \dfrac{3}{8}a\right)$ (d) a (e) $\dfrac{\pi}{30}h^3(20a^2 - 15ah + 3h^2)$

(f) $\dfrac{4}{15}\pi a^5$

57. Volume of a torus formed by a circle of radius 3, centered at $(0, 3, 0)$ and revolved about the z-axis

59. -9 **61.** $5\ln 5 - 3\ln 3 - 2 \approx 2.751$

P.S. Problem Solving (page 1004)

1. (a) $8(2 - \sqrt{2})$ (b) Programs will vary.
3. (a) Proof (b) Proof (c) Proof (d) Proof
  (e) Proof (f) Proof (g) Proof
5. $\dfrac{1}{3}$
7.

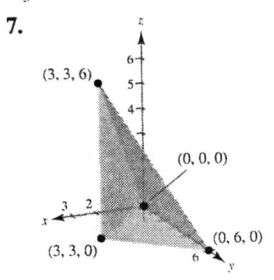

$\int_0^3 \int_0^{2x} \int_x^{6-x} dy\, dz\, dx = 18$

9. $\dfrac{\sqrt{\pi}}{4}$

11. If $a, k > 0$, then $1 = ka^2$ or $a = \dfrac{1}{\sqrt{k}}$.

13. Proof

Chapter 14

Section 14.1 (page 1017)

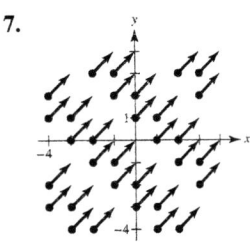

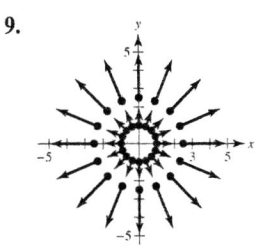

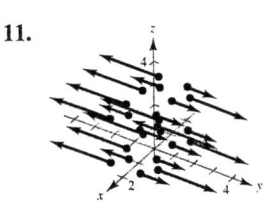

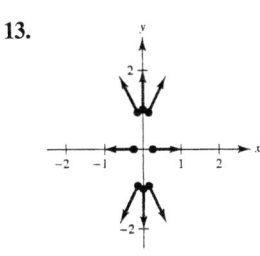

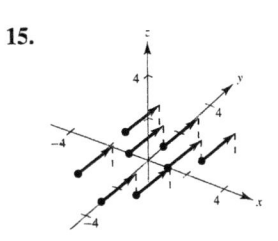

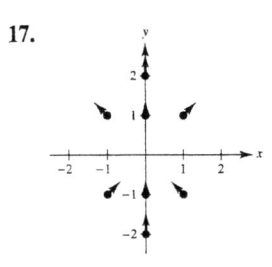

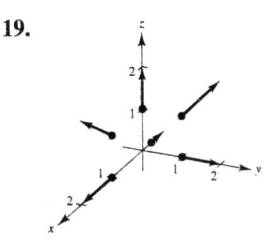

1. c   **2.** d   **3.** b   **4.** e   **5.** a   **6.** f
7.   **9.**
11.   **13.**
15.   **17.**
19.

21. $(10x + 3y)\mathbf{i} + (3x + 20y)\mathbf{j}$ **23.** $-2xye^{x^2}\mathbf{i} - e^{x^2}\mathbf{j} + \mathbf{k}$

25. $\left[\dfrac{xy}{x + y} + y\ln(x + y)\right]\mathbf{i} + \left[\dfrac{xy}{x + y} + x\ln(x + y)\right]\mathbf{j}$

27. Proof   **29.** Proof
31. Not conservative   **33.** Conservative
35. Conservative: $f(x, y) = x^2y + K$
37. Conservative: $f(x, y) = e^{x^2y} + K$
39. Conservative: $f(x, y) = \dfrac{1}{2}\ln(x^2 + y^2) + K$
41. Not conservative   **43.** $2\mathbf{j} - \mathbf{k}$ **45.** $-2\mathbf{k}$
47. $\dfrac{2x}{x^2 + y^2}\mathbf{k}$ **49.** $\cos(y - z)\mathbf{i} + \cos(z - x)\mathbf{j} + \cos(x - y)\mathbf{k}$
51. Not conservative   **53.** Conservative: $f(x, y, z) = xye^z + K$
55. Conservative: $f(x, y, z) = \dfrac{x}{y} + z^2 - z + K$
57. $12x - 2xy$ **59.** $\cos x - \sin y + 2z$ **61.** 4 **63.** 0

65. See "Definition of a Vector Field" on page 1008. Some physical examples of vector fields include velocity fields, gravitational fields, and electric force fields.

67. See "Definition of Curl of a Vector Field" on page 1014.

69. $6x\mathbf{j} - 3y\mathbf{k}$ 71. $z\mathbf{j} + y\mathbf{k}$ 73. $2z + 3x$ 75. 0

77.–83. Proof

85. $f(x, y, z) = \|\mathbf{F}(x, y, z)\| = \sqrt{x^2 + y^2 + z^2}$

$\ln f = \frac{1}{2}\ln(x^2 + y^2 + z^2)$

$\nabla \ln f = \frac{x}{x^2 + y^2 + z^2}\mathbf{i} + \frac{y}{x^2 + y^2 + z^2}\mathbf{j} + \frac{z}{x^2 + y^2 + z^2}\mathbf{k}$

$= \frac{\mathbf{F}}{f^2}$

87. $f^n = \|\mathbf{F}(x, y, z)\|^n = \left(\sqrt{x^2 + y^2 + z^2}\right)^n$

$\nabla f^n = n\left(\sqrt{x^2 + y^2 + z^2}\right)^{n-1}\left(\frac{x\mathbf{i} + y\mathbf{i} + z\mathbf{k}}{\sqrt{x^2 + y^2 + z^2}}\right)$

$= nf^{n-2}\mathbf{F}$

89. The winds are stronger over Phoenix than over Atlanta. Also, although the winds over both cities are northeasterly, they are more toward the north over Phoenix and more toward the east over Atlanta.

Section 14.2 (page 1029)

1. $\mathbf{r}(t) = 3\cos t\mathbf{i} + 3\sin t\mathbf{j}, \quad 0 \leq t \leq 2\pi$

3. $\mathbf{r}(t) = \begin{cases} t\mathbf{i}, & 0 \leq t \leq 3 \\ 3\mathbf{i} + (t-3)\mathbf{j}, & 3 \leq t \leq 6 \\ (9-t)\mathbf{i} + 3\mathbf{j}, & 6 \leq t \leq 9 \\ (12-t)\mathbf{j}, & 9 \leq t \leq 12 \end{cases}$

5. $\mathbf{r}(t) = \begin{cases} t\mathbf{i} + \sqrt{t}\mathbf{j}, & 0 \leq t \leq 1 \\ (2-t)\mathbf{i} + (2-t)\mathbf{j}, & 1 \leq t \leq 2 \end{cases}$

7. 10 9. $\frac{\sqrt{65}\pi}{6}(3 + 16\pi^2)$ 11. 9

13. $\frac{\pi}{2}$ 15. $\frac{19\sqrt{2}}{6}$ 17. $\frac{19}{6}(1 + \sqrt{2})$

19. $\frac{2\sqrt{13}\pi}{3}(27 + 64\pi^2) \approx 4973.8$ 21. $\frac{35}{6}$ 23. 2

25. $-\frac{17}{15}$ 27. 249.49 29. -66 31. 0 33. $-10\pi^2$

35. 1500 foot-pounds

37. (a) $\frac{236}{3}$; Orientation is from left to right so the value is positive.
    (b) $-\frac{236}{3}$; Orientation is from right to left so the value is negative.

39. $\mathbf{F}(t) = -2t\mathbf{i} - t\mathbf{j}$

$\mathbf{r}'(t) = \mathbf{i} - 2\mathbf{j}$

$\mathbf{F}(t) \cdot \mathbf{r}'(t) = -2t + 2t = 0$

$\int_C \mathbf{F} \cdot d\mathbf{r} = 0$

41. $\mathbf{F}(t) = (t^3 - 2t^2)\mathbf{i} + \left(t - \frac{t^2}{2}\right)\mathbf{j}$

$\mathbf{r}'(t) = \mathbf{i} + 2t\mathbf{j}$

$\mathbf{F}(t) \cdot \mathbf{r}'(t) = t^3 - 2t^2 + 2t^2 - t^3 = 0$

$\int_C \mathbf{F} \cdot d\mathbf{r} = 0$

43. 1010 45. $\frac{190}{3}$ 47. 25 49. $\frac{63}{2}$

51. $-\frac{11}{6}$ 53. $\frac{316}{3}$ 55. $5h$ 57. $\frac{1}{2}$

59. $\frac{h}{4}\left[2\sqrt{5} + \ln(2 + \sqrt{5})\right]$ 61. $\frac{1}{120}(25\sqrt{5} - 11)$

63. (a) $12\pi \approx 37.70$ square centimeters
    (b) $\frac{12\pi}{5} \approx 7.54$ cubic centimeters
    (c)

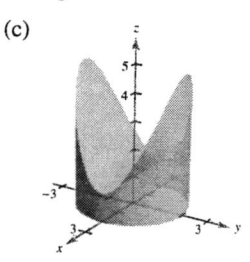

65. b

67. (a)

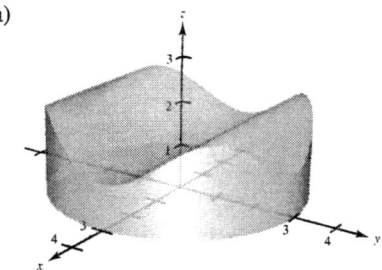

(b) 9π square centimeters

(c) Volume $= 2\int_0^3 2\sqrt{9 - y^2}\left[1 + 4\frac{y^2}{9}\left(1 - \frac{y^2}{9}\right)\right] dy$

$= \frac{27\pi}{2} \approx 42.412$ cubic centimeters

69. See "Definition of Line Integral" on page 1020 and "Evaluation of a Line Integral as a Definite Integral" on page 1021.

71. z_3, z_1, z_2, z_4; The greater the height of the surface over the curve $y = \sqrt{x}$, the greater the lateral surface area.

73. False: $\int_C xy\,ds = \sqrt{2}\int_0^1 t^2\,dt$.

75. False: the orientations are different.

Section 14.3 (page 1039)

1. (a) $\int_0^1 (t^2 + 2t^4)\,dt = \frac{11}{15}$
   (b) $\int_0^{\pi/2} (\sin^2\theta\cos\theta + 2\sin^4\theta\cos\theta)\,d\theta = \frac{11}{15}$

3. (a) $\int_0^{\pi/3} (\sec\theta \tan^2\theta - \sec^3\theta)\, d\theta \approx -1.317$

(b) $\int_0^3 \left[\frac{\sqrt{t}}{2\sqrt{t+1}} - \frac{\sqrt{t-1}}{2\sqrt{t}}\right] dt \approx -1.317$

5. Conservative **7.** Not conservative **9.** Conservative

11. (a) 1 (b) 1 **13.** (a) 0 (b) $-\frac{1}{3}$ (c) $-\frac{1}{2}$

15. (a) 64 (b) 0 (c) 0 (d) 0

17. (a) $\frac{64}{3}$ (b) $\frac{64}{3}$ **19.** (a) 32 (b) 32

21. (a) $\frac{2}{3}$ (b) $\frac{17}{6}$ **23.** (a) 0 (b) 0

25. 24 **27.** -1 **29.** 0

31. (a) 0 (b) 0 (c) 0

33. 11 **35.** 30,366 **37.** 0

39. Increasing at the rate of 10 units per minute

41. No, the force field is conservative.

43. See Theorem 14.5 on page 1033.

45. (a) The direct path along the line segment joining $(-4, 0)$ to $(3, 4)$ requires less work than the path going from $(-4, 0)$ to $(-4, 4)$ and then to $(3, 4)$.

(b) The closed curve given by the line segments joining $(-4, 0)$, $(-4, 4)$, $(3, 4)$, and $(-4, 0)$ satisfies $\int_C \mathbf{F} \cdot d\mathbf{r} \neq 0$.

47. False. It would be true if $\mathbf{F}$ were conservative.

49. True **51.** Proof

Section 14.4 (page 1048)

1. 0 **3.** $\frac{32}{15}$ **5.** 19.99 **7.** $\frac{4}{3}$ **9.** 56 **11.** $\frac{32}{3}$

13. 0 **15.** 0 **17.** $\frac{1}{12}$ **19.** 8π **21.** 4π **23.** $\frac{225}{2}$

25. πa^2 **27.** $\frac{32}{3}$

29. See Theorem 14.8 on page 1042.

31. Answers will vary. Example:
$\mathbf{F}_1(x, y) = y\mathbf{i} + x\mathbf{j}$
$\mathbf{F}_2(x, y) = x^2\mathbf{i} + y^2\mathbf{j}$
$\mathbf{F}_3(x, y) = 2xy\mathbf{i} + x^2\mathbf{j}$

33. $\left(0, \frac{8}{5}\right)$ **35.** $\left(\frac{8}{15}, \frac{8}{21}\right)$

37. $\frac{3\pi a^2}{2}$ **39.** $\pi - \frac{3\sqrt{3}}{2}$

41. $\int_C \mathbf{F} \cdot d\mathbf{r} = \int_C M\, dx + N\, dy = \iint_R \left(\frac{\partial N}{\partial x} - \frac{\partial M}{\partial y}\right) dA = 0$; $I = -2\pi$ when C is a circle that contains the origin.

43. $\frac{19}{2}$

45. (a) $n = 1$: 0 (b) $n = 2$: $-\frac{4}{3}a^3$ (c) 0
$n = 3$: 0 $n = 4$: $-\frac{16}{15}a^5$
$n = 5$: 0 $n = 6$: $-\frac{32}{35}a^7$
$n = 7$: 0 $n = 8$: $-\frac{256}{315}a^9$

47. Proof **49.** Proof

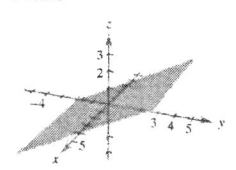

Section 14.5 (page 1058)

1. c **2.** d **3.** b **4.** a

5. $y - 2z = 0$ **7.** $x^2 + z^2 = 4$
Plane Cylinder

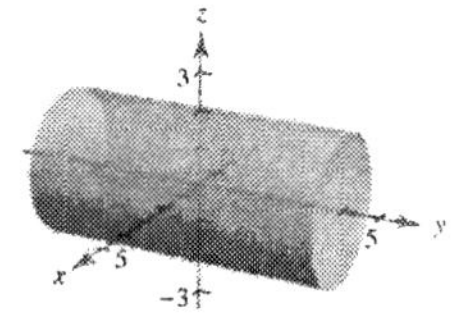

9. The paraboloid is reflected (inverted) through the xy-plane.

11. The height of the paraboloid is increased from 4 to 9.

13. **15.**

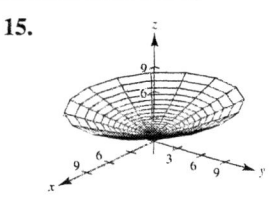

17.

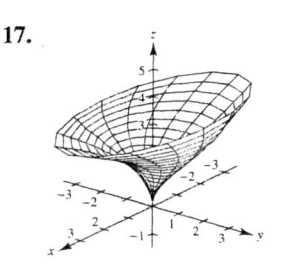

19. $\mathbf{r}(u, v) = u\mathbf{i} + v\mathbf{j} + v\mathbf{k}$

21. $\mathbf{r}(u, v) = 4\cos u\mathbf{i} + 4\sin u\mathbf{j} + v\mathbf{k}$

23. $\mathbf{r}(u, v) = u\mathbf{i} + v\mathbf{j} + u^2\mathbf{k}$

25. $\mathbf{r}(u, v) = v\cos u\mathbf{i} + v\sin u\mathbf{j} + 4\mathbf{k}, \quad 0 \leq v \leq 3$

27. $x = u, y = \frac{u}{2}\cos v, z = \frac{u}{2}\sin v, \quad 0 \leq u \leq 6, 0 \leq v \leq 2\pi$

29. $x = \sin u \cos v, y = \sin u \sin v, z = u$
$0 \leq u \leq \pi, 0 \leq v \leq 2\pi$

31. $x - y - 2z = 0$ **33.** $4y - 3z = 12$ **35.** $2\sqrt{2}$

37. $2\pi ab$ **39.** $\pi ab^2\sqrt{a^2 + 1}$

41. $\frac{\pi}{6}(17\sqrt{17} - 1) \approx 36.177$

43. See "Definition of Parametric Surface" on page 1051.

45. (a) $(-10, 10, 0)$ (b) $(10, 10, 10)$
(c) $(0, 10, 0)$ (d) $(10, 0, 0)$

47. (a) (b) (c) (d)

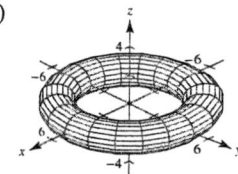

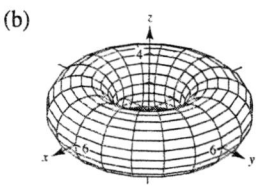

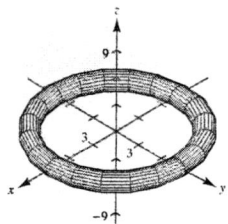

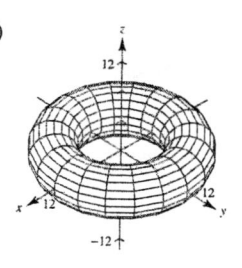

The radius of the generating circle that is revolved about the z-axis is b, and its center is a units from the axis of revolution.

49. 400π square meters

51. $2\pi\left[\frac{3}{2}\sqrt{13} + 2\ln(3 + \sqrt{13}) - 2\ln 2\right]$

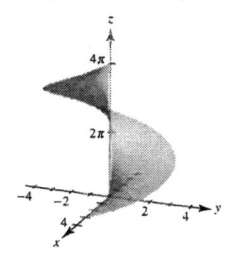

53. Answers will vary. Example: Let
$x = (2 - u)(5 + \cos v) \cos 3\pi u$
$y = (2 - u)(5 + \cos v) \sin 3\pi u$
$z = 5u + (2 - u) \sin v$
where $-\pi \le u \le \pi$ and $-\pi \le v \le \pi$.

Section 14.6 (page 1071)

1. 0 **3.** 10π **5.** $\frac{27\sqrt{6}}{2}$ **7.** $\frac{391\sqrt{17} + 1}{240}$

9. -11.47 **11.** $\frac{364}{3}$ **13.** $6\sqrt{5}$ **15.** 8

17. $\frac{19\sqrt{2}\pi}{4}$ **19.** $\frac{32\pi}{3}$ **21.** 486π **23.** $-\frac{4}{3}$

25. $\frac{243\pi}{2}$ **27.** 20π **29.** $\frac{5}{2}$

31. The surface integral of f over a surface S, where S is given by $z = g(x, y)$, is defined as
$$\iint_S f(x, y, z)\, dS = \lim_{\|\Delta\| \to 0} \sum_{i=1}^{n} f(x_i, y_i, z_i)\, \Delta S_i;$$
see Theorem 14.10 on page 1061.

33. See "Definition of Flux Integral," on page 1067; see Theorem 14.11 on page 1067.

35. (a)

(b) If a normal vector at a point P on the surface is moved around the Möbius strip once, it will point in the opposite direction.

(c)

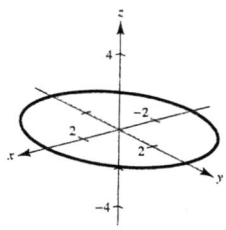

Circle

(d) Construction

(e) A strip with a double twist that is twice as long as the Möbius strip.

37. Proof

39. $2\pi a^3 h$ **41.** $64\pi\rho$

Section 14.7 (page 1079)

1. a^4 **3.** 18 **5.** $3a^4$ **7.** 0 **9.** 32π
11. 0 **13.** 2304 **15.** 144π **17.** 0
19. See Theorem 14.12 on page 1073.
21. Proof **23.** Proof **25.** Proof **27.** Proof

Section 14.8 (page 1086)

1. $-xy\mathbf{i} - \mathbf{j} + (yz - 2)\mathbf{k}$ **3.** $\left(2 - \frac{1}{1 + x^2}\right)\mathbf{j} - 8x\mathbf{k}$

5. $z(x - 2e^{y^2 + z^2})\mathbf{i} - yz\mathbf{j} - 2ye^{x^2 + y^2}\mathbf{k}$ **7.** 2π **9.** 0

11. 1 **13.** 0 **15.** 0 **17.** $\frac{8}{3}$ **19.** $\frac{a^5}{4}$ **21.** 0

23. See Theorem 14.13 on page 1081.
25. Proof **27.** Proof

Review Exercises for Chapter 14 (page 1087)

1.

3. $(16x + y)\mathbf{i} + x\mathbf{j} + 2z\mathbf{k}$

5. Not conservative

7. Conservative: $f(x, y) = 3x^2y^2 - x^3 + y^3 - 7y + K$

Answers to Odd-Numbered Exercises **A165**

9. Not conservative
11. Conservative: $f(x, y, z) = \dfrac{x}{yz} + K$
13. (a) div **F** $= 2x + 2y + 2z$ (b) curl **F** $= 0$
15. (a) div **F** $= -y \sin x - x \cos y + xy$
    (b) curl **F** $= xz\mathbf{i} - yz\mathbf{j}$
17. (a) div **F** $= \dfrac{1}{\sqrt{1-x^2}} + 2xy + 2yz$
    (b) curl **F** $= z^2\mathbf{i} + y^2\mathbf{k}$
19. (a) div **F** $= \dfrac{2x + 2y}{x^2 + y^2} + 1$ (b) curl **F** $= \dfrac{2x - 2y}{x^2 + y^2}\mathbf{k}$
21. (a) $6\sqrt{2}$ (b) 128π 23. $2\pi^2(1 + 2\pi^2)$
25. (a) $\dfrac{35}{2}$ (b) 18π 27. $\dfrac{9a^2}{5}$
29. $\dfrac{\sqrt{10}}{4}(41 - \cos 8) \approx 32.528$ 31. $\dfrac{5}{7}$ 33. $2\pi^2$
35. $\dfrac{64}{3}$ 37. $\dfrac{4}{3}$ 39. $\dfrac{8}{3}(3 - 4\sqrt{2}) \approx -7.085$
41. 12   43. (a) 15 (b) 15 (c) 15
45. 4   47. 0   49. $\dfrac{1}{12}$
51.

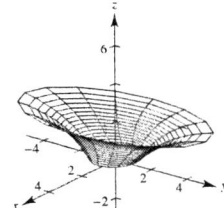

53. (a) 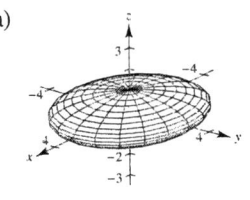   (b) 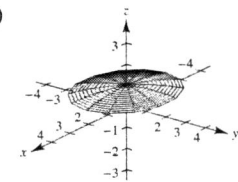
    (c) 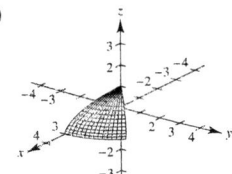   (d)

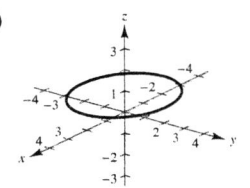

    (e) 14.436   (f) 4.269   Circle
55.

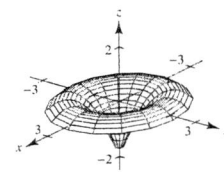

    0
57. 66   59. $\dfrac{2a^6}{5}$

P.S. Problem Solving (page 1090)

1. (a) $\dfrac{25\sqrt{2}}{6}k\pi$ (b) $\dfrac{25\sqrt{2}}{6}k\pi$
3. $I_x = \dfrac{\sqrt{13}\pi}{3}(27 + 32\pi^2)$
   $I_y = \dfrac{\sqrt{13}\pi}{3}(27 + 32\pi^2)$
   $I_z = 18\sqrt{13}\pi$
5. $3a^2\pi$ 7. (a) 1 (b) $\dfrac{13}{15}$ (c) $\dfrac{5}{2}$ 9. Proof
11. $M = 3mxy(x^2 + y^2)^{-5/2}$
    $\dfrac{\partial M}{\partial y} = \dfrac{3mx(x^2 - 4y^2)}{(x^2 + y^2)^{7/2}}$
    $N = m(2y^2 - x^2)(x^2 + y^2)^{-5/2}$
    $\dfrac{\partial N}{\partial x} = \dfrac{3mx(x^2 - 4y^2)}{(x^2 + y^2)^{7/2}}$
    Therefore, $\dfrac{\partial N}{\partial x} = \dfrac{\partial M}{\partial y}$ and **F** is conservative.

Appendix

Appendix A (page A6)

1.

x	-4	-2	0	2	4	8
y	2	0	4	4	6	8
dy/dx	-2	Undef.	0	$\dfrac{1}{2}$	$\dfrac{2}{3}$	1

3. (a) Answers will vary.   (b) $y = \dfrac{1}{2}(e^x + e^{-x})$

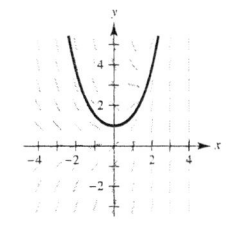

   (c)

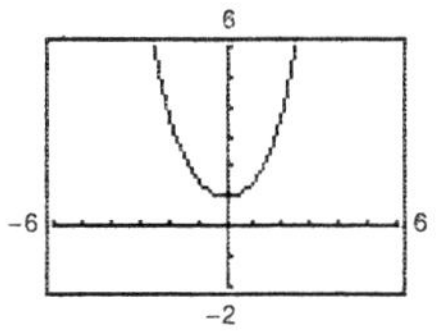

5. (a) Answers will vary.

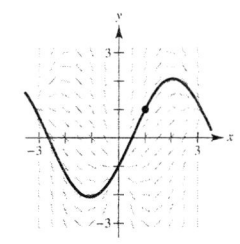

(b) $y = -\cos x + 1.8305 \sin x$
(c)

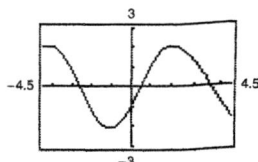

7.

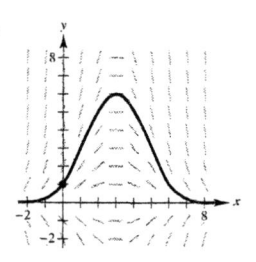

9.

n	0	1	2	3	4	5
x_n	0	0.1	0.2	0.3	0.4	0.5
y_n	2	2.2	2.43	2.693	2.9923	3.3315

n	6	7	8	9	10
x_n	0.6	0.7	0.8	0.9	1.0
y_n	3.7147	4.1462	4.6308	5.1738	5.7812

11.

n	0	1	2	3	4	5
x_n	0	0.05	0.10	0.15	0.20	0.25
y_n	3	2.7	2.4375	2.2088	2.0104	1.8393

n	6	7	8	9	10
x_n	0.30	0.35	0.40	0.45	0.50
y_n	1.6929	1.5686	1.4643	1.3778	1.3075

13.

n	0	1	2	3	4	5
x_n	0	0.1	0.2	0.3	0.4	0.5
y_n	1	1.1	1.2116	1.3390	1.4885	1.6699

n	6	7	8	9	10
x_n	0.6	0.7	0.8	0.9	1.0
y_n	1.9003	2.2131	2.6838	3.5398	5.9584

15. False. $y' + xy = x^2$ is linear.

17. $y = x^2 + 2x + \dfrac{C}{x}$ 19. $y = e^{x^3}(x + C)$

21. $y = \tfrac{1}{2}(\sin x - \cos x) + Ce^x$ 23. $y = x(\ln|x| + C)$

25. $y = -\tfrac{1}{13}(3 \sin 2x + 2 \cos 2x) + Ce^{3x}$

27. $y = \dfrac{x^3 - 3x + C}{3(x - 1)}$

29. $y = e^x(1 + \tan x) + C \sec x$

31. $y = \dfrac{bx^4}{4 - a} + Cx^a$

33. $y = 1 + 4e^{-\tan x}$

35. $y = \sin x + (x + 1) \cos x$ 37. $y = \dfrac{4}{x}$

39. $y = x \ln|x| + 12x - 2$

41. (a)

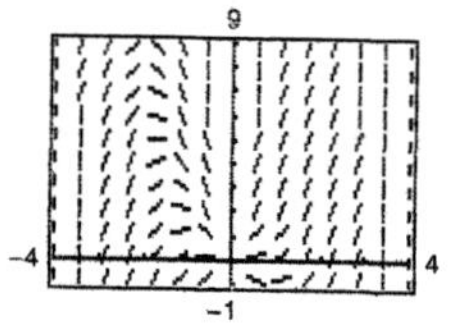

(b) $(-2, 4)$: $y = \tfrac{1}{2}x(x^2 - 8)$
$(2, 8)$: $y = \tfrac{1}{2}x(x^2 + 4)$
(c)

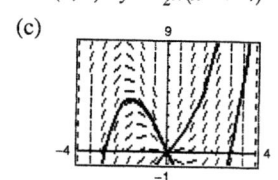

43. $I = \dfrac{E_0}{R} + Ce^{-Rt/L}$

45. $I = Ce^{-Rt/L} + \dfrac{E_0}{R^2 + \omega^2 L^2}(R \sin \omega t - \omega L \cos \omega t)$

47. $P = -\dfrac{N}{k} + \left(\dfrac{N}{k} + P_0\right)e^{kt}$

49. (a) \$583,098.01 (b) \$3,243,606.35

51. $A = \dfrac{P}{r} + \left(A_0 - \dfrac{P}{r}\right)e^{rt}$

53. (a) $\dfrac{dQ}{dt} = q - kQ$

(b) $Q = \dfrac{q}{k} + \left(Q_0 - \dfrac{q}{k}\right)e^{-kt}$

(c) $\dfrac{q}{k}$

55. c   56. d   57. a   58. b

Index of Applications

Engineering and Physical Sciences

Absolute zero, 72
Acceleration, 126, 152, 252, 307
Acceleration due to gravity, 31, 123
Acceleration on the moon, 157
Acid rain, 849
Adiabatic expansion, 151
Air pressure, 407
Air traffic control, 150, 713, 819
Aircraft glide rate, 190
Aircraft separation, 252
Altitude of a plane, 152
Anamorphic art, 693
Angle of elevation, 151, 152
Angle of elevation of a camera, 147
Angular rate of change, 387
Annual snowfall, 953
Annual temperature range, 849
Antenna radiation, 700
Apparent temperature, 867
Architecture, 662
Area, 37, 116, 125, 149, 216, 217, 218
Area of end of a log, 233
Area of a lot, 264, 306
Asteroid Apollo, 706
Atmospheric pressure, 323, 349, 367, 920
Auditorium lights, 734
Auditorium noise level, 368
Autocatalytic chemical reaction, 216
Automobile aerodynamics, 29
Automobile engine, 920
Average displacement, 496
Average field strength, 514
Average speed, 86
Average velocity of a falling object, 111
Barge towing, 783
Beam deflection, 191, 661
Beam strength, 34, 218
Bessel function, 624
Boiling temperature, 35
Bouncing ball, 572, 644
Boyle's Law, 86, 125, 458
Braking load, 743
Breaking strength of steel, 359
Bridge design, 662
Brinell hardness, 34
Building construction, 734
Building design, 420, 529, 977, 1031
Building a pipeline, 920
Bulb design, 448
Buoyant force, 474, 479

Cable tension, 726
Camera surveillance design, 152
Capillary action, 977
Car performance, 34, 35
Carbon dating, 367
Catenary, 403
Cavalieri's Theorem, 431
Center of mass of a conversion van window, 468
Center of mass of a section of a boat's hull, 469
Center of pressure on a sail, 970
Centripetal acceleration, 819
Centripetal force, 819, 833
Centroid, 477
Changing shadow length, 152
Charles's Law, 72
Chemical reaction, 378, 404, 523, 931
Chemical release from a storage tank, 350
Circular motion, 810, 818, 835
Cissoid of Diocles, 712
Climb rate for an airplane, 406
Comparing fluid forces, 511
Constant flow rate, 28
Constructing an arch dam, 410
Construction cost, 848
Construction of the Gateway Arch in St. Louis, 405
Construction of a semielliptical arch, 663
Conveyer design, 16
Cornu spiral, 712, 834
Cross section of a canal, 220
Curtate cycloid, 673
Cycloidal motion, 809, 818
Cycloids, 674
Daily temperature, 35, 135
Deceleration, 252
Depth, 149, 150, 155, 190
Dimensions of a barn, 977
Dimensions of a box, 872, 928
Distance between cities, 783
Distance traveled, 286, 574
Doppler effect, 135
Drag force, 931
Earthquake intensity, 368
Electric motor, 87
Electric power in a circuit, 182
Electrical charge, 1072
Electrical circuits, A6, A7
Electrical power, 875
Electrical resistance, 151, 183
Electricity, 299
Electromagnetic theory, 549

Electronically controlled thermostat, 28
Enclosing a maximum area, 215
Engine design, 239, 1031
Engine efficiency, 201
Engine power, 227
Epicycloid, 674
Equilibrium, 780
Eruption of Old Faithful, 1
Escape velocity, 91, 251
Evaporation rate, 151
Evolute, 830
Explorer 663, 708
Exploring new planets, 648
Falling object, 33
Ferris wheel, 835
Fluid flow, 155
Fluid force on a circular plate, 475, 477
Fluid force on a circular window, 473, 514
Fluid force of gasoline, 474, 475
Fluid force on a rectangular plate, 475
Fluid force on a submarine porthole, 475
Fluid force on a submerged metal sheet, 471, 474
Fluid force on a tank wall, 474
Fluid force on a vertical gate in a dam, 472
Fluid force against a vertical irrigation canal gate, 475
Fluid force against a vertical region, 477
Fluid force on a vertical stern of a boat, 475
Fluid force on the vertical walls of a swimming pool, 477, 479
Fluid force of water, 474
Force, 285, 740
Force on a concrete form, 474
Free-falling object, 67, 88
Frictional force, 827, 831
Gabriel's Horn, 546
Gravity, 92
Halley's comet, 663, 705
Hanging power cables, 398
Harmonic motion, 37, 135, 235, 349, 407
Heat flow, 1090
Heat transfer, 331
Height of a tower, 929
Highway design, 167, 190, 833
Hooke's Law, 33, 456
Horizontal motion, 154
Hours of daylight, 32
Hydraulics, 970
Hyperbolic detection system, 659
Hyperbolic mirror, 664
Ideal Gas Law, 849, 867, 883

Illumination from a light source, 219, 238
Inductance, 875
Inflating a balloon, 146
Instantaneous velocity, 112, 171
Investigating the ocean floor, 895
Kepler's Law, 831
Kinetic and potential energy, 1040
Koch snowflake, 554, 589
Lawn sprinkler construction, 167
Length, 29
Length of a catenary, 448, 477
Length of an electric cable, 443
Length of the Gateway Arch, 448
Length of a hypotenuse, 217
Length of pursuit, 447
Length of a recording tape, 679
Linear vs. angular speed, 152
Linear and angular velocity, 155
Load supports, 734
Load-supporting cables, 742, 743
Lunar gravity, 251
Machine design, 151, 781, 961
Machine part, 438, 492
Magnetic field, 1091
Making a Mercator map, 480
Manufacturing, 426, 431
Mass of a spring, 1023, 1029
Mass on the surface of the earth, 459
Mathematical sculpting, 1006
Maximum angle, 883
Maximum angle subtended by a camera lens, 384
Maximum area of an aluminum sheet, 918
Maximum area of a fitness room, 217
Maximum volume of a box, 211, 217
Maximum volume of a package, 218
Maximum volume of a rectangular box, 913, 917, 918
Maximum volume of a rectangular package, 917, 928
Measuring force, 731
Mechanical design, 420, 513, 762
Meteorology, 849, 895
Minimum area, 213
Minimum distance, 213, 220, 237
Minimum force, 220
Minimum heat loss, 932
Minimum length, 214, 237, 780
Minimum length of a power line, 218
Minimum material for a tank, 932
Minimum surface area of a cylinder, 218
Minimum travel time, 219, 227
Moment of a force, 748
Motion of a liquid, 1085, 1086
Motion of a particle, 792, 819, 832
Moving ladder, 150
Navigation, 664, 726

Newton's Law of Cooling, 116, 365, 368
Oblateness of Saturn, 439
Ohm's Law, 234
Optical illusion, 143
Optimization, 750, 751, 926
Orbital speed, 819
Orbit of the earth, 663
Orbit of the moon, 654
Oxygen level in a pond, 197
Packaging, 158
Path of a car, 830
Path of a heat-seeking particle, 890, 895
Path of an object, 877
Path of a projectile, 180, 674, 682
Pendulum, 135, 234, 875
Planetary motion, 708, 709
Planimeter, 1089
Plastics and cooling, 312
Playground slide, 792
Position of a pipe, 150
Power lines, 505
Power output of a battery, 167
Probability of iron in ore samples, 309
Producing a machine part, 430
Projectile motion, 154, 234, 514, 642, 674, 726, 806, 807, 808, 809, 817, 819, 828, 833, 834, 933
Projectile range, 219
Race car cornering, 784
Radioactive decay, 363, 367, 378, 407
Radioactive half-life, 351
Radio reception, 379
Radio and television reception, 662
Rainbows, 183
Rate of change, 86
Rate of change of the level of the Connecticut River, 221
Rate of mass flow of a fluid, 1068, 1072
Rectilinear motion, 251, 287
Refraction of light, 928
Refrigeration, 155
Resistance, 875
Resistance of copper wire, 9
Resultant force, 722, 725
Resultant speed and direction of an airplane, 723
Ripples, 29, 145
Roadway design, 151
Rocket velocity, 553
Rolling a ball bearing, 182
Roof area, 447
Rotary engine, 710
Satellite antenna, 709
Satellite orbit, 663, 833, 835
Satellite receiving dish, 836
Screw efficiency, 238
Shape of the earth, 772

Shared load, 726
Slope of a water-ski ramp, 12
Snell's Law of Refraction, 928
Solar collector, 661
Sound intensity, 39, 323, 368
Sound location, 664
Specific gravity, 191
Speed of an airplane, 147
Speed of an automobile, 154
Speed of sound, 280
Sphereflake, 575
Spiral staircase, 832
Statics, 467
Stopping distance of an automobile, 116, 126
Stress test, 37
Strophoid, 712
Surface area, 149, 155
Surface area of a honeycomb, 167
Surface area of a pond, 479
Surface area of a roof, 1003
Surface area of a satellite dish, 662
Surveying, 234
Suspension bridges, 714
Swimming speed, 40
Temperature, 323, 505, 895
Temperature conversion, 17
Temperature distribution, 848, 867, 894, 928
Temperature in a heat exchanger, 201
Temperature of a metal plate, 889
Tension in the rope of a tetherball, 783
Theory of relativity, 86
Thermometer reading, 409
Throwing a shot-put, 810
Tidal energy, 458
Topographic map, 143
Topography, 894, 895
Torque, 750, 781
Tower guy wire, 734
Trachea contraction, 182
Tractrix, 323, 401, 402, 404, 514, 683
Velocity, 116, 252, 307
Velocity and acceleration, 803, 804, 805, 808, 810
Velocity of a baseball player, 151
Velocity of a boat, 150
Velocity of a car, 310
Velocity of a piston, 148
Velocity of a plane, 150
Velocity in a resisting medium, 538
Velocity of a shadow, 151, 155, 157
Velocity of a sliding plank, 150
Vertical motion, 115, 153, 172, 173, 248, 251, 394, 404, 407
Vibrating spring, 153, 496
Volume, 29, 79, 116, 125, 149, 216

Volume of a ball bearing, 230
Volume of concrete in a ramp, 430
Volume of a fuel tank, 429
Volume of a gasoline tank, 477
Volume of a goblet, 830
Volume of the Great Salt Lake, 1005
Volume of a lab glass, 430
Volume of a piece of ice, 962
Volume of a pond, 439
Volume of a pontoon, 436
Volume of a propane tank, 848
Volume of sand, 954
Volume of a storage shed, 438
Volume of a storage tank, 513
Volume of a trough, 875
Volume of a vase, 449
Volume of a water tank, 430, 663, 709
Wankel rotary engine, 240
Water supply, 299
Wave equation, 933
Wave motion, 135
Wind chill, 874
Wind speed, 1018, 1041
Witch of Agnesi, 793
Work, 306
Work done in closing a door, 741
Work done in compressing a spring, 452, 456
Work done by a constant force, 456
Work done by an electric force, 458
Work done in emptying a tank of oil, 454
Work done by the engines of an aircraft, 1088
Work done by an expanding gas, 455
Work done by a force, 477
Work done by a force field, 1025, 1030, 1032, 1036, 1040, 1044, 1049, 1088, 1091
Work done by a hydraulic cylinder, 529
Work done by a hydraulic press, 458
Work done in lifting a chain, 455, 457, 477
Work done in lifting an object, 450
Work done in moving a particle, 743
Work done in moving a rocket in outer space, 548
Work done in moving a space module into orbit, 453, 543
Work done by a person, 1030
Work done in propulsion, 456
Work done in pulling an object, 743, 780
Work done in pulling a wagon, 743
Work done in pumping diesel fuel, 457
Work done in pumping gasoline, 457
Work done in pumping water, 456, 457
Work done in pumping a well, 477
Work done in stretching a spring, 456, 477
Work done in using a demolition crane, 457
Work done in winding up a cable, 477
Work done in wood splitting, 458

Business and Economics

Advertising costs, 227
Air conditioning costs, 311
Annuities, 575
Apartment rental, 18
Automobile costs, 34
Average cost, 191, 201
Average price, 331
Average production level, 954
Average profit, 285, 954
Average sales, 285
Bankruptcy, 183
Break-even analysis, 9, 36
Budget deficit, 419
Capitalized cost, 549
Cobb-Douglas production function, 843, 848, 924, 932
Compound interest, 356, 358, 359, 360, 367, 406, 538, 565, 643, 644
Consumer price index, 9
Consumer surplus, 479
Cost, 136, 339
Cost of a cargo container, 928
Cost of equipment, 287
Cost of an industrial tank, 218
Cost of overnight delivery, 89
Cost of a telephone call, 55
Declining sales, 364
Demand, 237, 919
Depreciation, 36, 284, 299, 348, 357, 358, 406, 574, 643
Diminishing returns, 220
Dollar value of a product, 17
Fertilizer sales, 311
Fuel cost, 116, 309
Fuel economy, 407
Government expenditures, 566
Home mortgage, 323, 409
Hospital room costs, 566
Income, 575
Inflation, 358, 566
Inventory cost, 167, 191, 236
Inventory management, 79, 116
Inventory replenishment, 125
Investment, 565, 848, 867, A8
Investment growth, A8
Locating a distribution center, 918
Marginal cost, 867
Marginal productivity, 867
Marginal utility, 867
Marketing, 574
Maximum profit, 220, 914, 918, 931
Medical expenditures, 867
Minimum cost, 219, 220, 237, 918, 928, 931
National debt, 368
National defense outlays, 236
National deficit, 236
Present value, 496, 551, 575
Probability of selling a product, 574
Producer surplus, 479
Product design, 977
Production level, 928, 931
Profit, 182, 234, 420
Rate of disbursement, 299
Receipts and expenditures, 420
Reimbursed expenses, 18
Reorder costs, 172
Revenue, 367, 420, 741, 918
Sales, 173, 236, 299, 331, 368
Sales growth, 191
Sales for H. J. Heinz Company, 566, 575
Sales increase, 409
Sales for Wal-Mart, 849
Service revenue for cellular telephone industry, 476
Straight-line depreciation, 18
Tourist spending, 574
Value of a car, 349

Social and Behavioral Sciences

Air conditioner use, 849
Amount of money given to philanthropy, 359
Automobile costs, 34
Carbon dioxide concentration, 7
Cost of clean air, 89
Cost of removing pollutants, 523
Energy consumption, 33
Health maintenance organizations, 34
Illegal drugs, 86
Learning curve, 367, A8
Learning theory, 359
Mean height of American men, 549
Medical expenditures, 772
Memory model, 496
Number of MDs in the United States, 344
Number of motor homes, 125
Population, 16, 367
Population density of a city, 962
Population growth, 125, A8
Probability of recall, 309
Salary increase, 419, 476, 575
Total compensation, 644
Traffic control, 216
Traffic flow, 239
University applicants, 867
Waiting in line, 350, 848

Women in the work force, 152, 919
World population, 920

Life Sciences

Average population size, 529
Average rate of change of population, 12
Bacterial culture growth, 210, 356, 367
Blood flow, 285
Blood types, 918
Carcinogens, 33
Career choice, 18
Circulatory system, 135
Commuting, 28
Concentration of a chemical in the bloodstream, 182
Epidemic model, 523
Farm size, 9, 28
Forest defoliation, 359
Forestry, 848
Height vs. arm span, 30
Hyperthermia treatments for tumors, 934
Intravenous feeding, A8
Length of warblers, 551
Life insurance policies, 514
Medicine, 227
Points of vision, 920
Population growth, 250, 331, 359, 364, 407, 647, A8
Property, 449
Rainfall, 299
Respiratory cycle, 285, 309
Running, 79, 201
Systolic blood pressure, 919
Timber yield, 359
Tree growth, 250
Weight gain, 379
Wheat yield, 919
Wildflower diversity, 904
Wildlife population, 375

General

Average scores, 18
Baseball, 875
Boating, 38
Buffon's needle experiment, 284
Cantor set, 646
Cantor's disappearing table, 576
Dog's path, 713
Estimating the number of customers, 286
Fruit consumption, 662
Lawn sprinkler, 1004
Milk consumption, 762, 867, 874
Natural gas usage, 286
Pasture fencing, 38
Probability of range for battery life, 350
Probability of tossing a coin, 631
Quiz scores, 33
Sailing, 379
Solera method, 589
Swimming pool, 79
Telephone charges, 79
Typing speed, 191, 201

Index

A

Abel, Niels Henrik (1802–1829), 225
Absolute convergence, 593
Absolute extrema, 905
Absolute maximum, 905
Absolute maximum value on an interval, 160
Absolute minimum, 905
Absolute minimum value on an interval, 160
Absolute value, derivative of, 320
Absolute value function, 22
Absolute Value Theorem, 560
Acceleration, 803, 826
    centripetal component of, 815
    due to gravity, 123
    normal component of, 815
    tangential component of, 815
    vector, 803, 814
Accumulation function, 281
Additive interval property, 270
Agnesi, Maria (1718–1799), 195
d'Alembert, Jean Le Rond (1717–1783), 859
Algebraic function, 25
    limit of, 57
Algebraic properties of the cross product, 745
Alternating series, 590
    harmonic, 593
Alternating series remainder, 592
Alternating Series Test, 590
Alternative form of the directional derivative, 887
Alternative forms of Green's Theorem, 1047, 1048
Angle
    between two nonzero vectors, 736
    between two planes, 754
    of incidence, 652
    of inclination of a plane, 900
    of reflection, 652
Angular speed, 968
Antiderivative, 242
    of a composite function, 288
    notation for, 243
Antidifferentiation (or indefinite integration), 243
Appolonius (262–190 B.C.), 650
Arc length, 440, 441
    function, 821
    parameter, 821, 822
    in parametric form, 678
    of a polar curve, 698
    of a space curve, 820

Arccosecant function, 380
Arccosine function, 380
Arccotangent function, 380
Archimedes (287–212 B.C.), 255
    spiral of, 679, 686
Arcsecant function, 380
Arcsine function, 380
    series for, 638
Arctangent function, 380
    series for, 638
Area, 255
    by an iterated integral, 938
    given by a line integral, 1045
    of a parametric surface, 1055
    of a plane region, 256, 259, 938
    in polar coordinates, 694
    of a rectangle, 255
    of a region between intersecting curves, 414
    of a region between two curves, 412, 413
    of a surface, 971, 972, 1055
    of a surface of revolution, 444, 445, 680, 699
Asymptote
    horizontal, 193
    of a hyperbola, 657
    slant, 204
    vertical, 81
Average rate of change, 12
Average value of a function on an interval, 279
Average velocity, 111
Axis
    conjugate, of a hyperbola, 657
    major, of an ellipse, 653
    minor, of an ellipse, 653
    of a parabola, 651
    of revolution, 421
    transverse, of a hyperbola, 657

B

Barrow, Isaac (1630–1677), 141
Base, 317
    of an exponential function, 351
    of a logarithmic function, 352
    of a natural logarithm, 317
Basic differentiation rules for elementary functions, 385
Basic equation for partial fractions, 517
    guidelines for solving, 521
Basic integration rules, 244, 391
Basic types of transformations, 23
Bernoulli, James (1654–1705), 671, 684
Bernoulli, John (1667–1748), 515

Binomial series, 637
Bisection method, 76
Boundary point, 850
Bounded
    above, 563
    below, 563
    monotonic sequence, 563
    region, 905
    sequence, 563
Brachistochrone problem, 671
Brahe, Tycho, 705
Breteuil, Emilie de (1706–1749), 451
Buoyant force, 474

C

Cancellation, 61
Cantor, Georg (1845–1918), 646
Capitalized cost, 549
Cardioid, 690
Catenary, 398
Cauchy, Augustine-Louis (1789–1857), 73
Cauchy-Riemann differential equations, 883
Cauchy-Schwarz Inequality, 743
Center
    of curvature, 825
    of an ellipse, 653
    of gravity, 461, 462
    of a hyperbola, 657
    of mass, 460, 461, 462, 463, 965, 983
        in a one-dimensional system, 460, 461
        of a planar lamina, 463
        of a planar lamina of variable density, 965
        in a two-dimensional system, 462
    of a power series, 616
Centered at c, 605
Central force field, 1009
Centripetal component of acceleration, 815
Centroid, 464, 965
Chain Rule, 127
    functions of several variables, 876
    implicit differentiation, 881
    one independent variable, 876
    and trigonometric functions, 132
    two independent variables, 878
Change of variables, 291
    for definite integrals, 294
    for double integrals, 997
        guidelines for making, 292
    for homogeneous equations, 374
    for an indefinite integral, 291
    to polar form, 957
    using a Jacobian, 995

A171

INDEX

Change in x, 95
Change in y, 95
Circulation, 1084
Circumscribed rectangle, 257
Cissoid of Diocles, 712
Classification of conics by eccentricity, 702
Closed curve, 1037
Closed disk, 850
Closed region, 850
Closed surface, 1073
Cobb-Douglas production function, 843
Coefficient
  leading, 24
  of a polynomial function, 24
Common logarithmic function, 352
Comparing gradients, 901
Comparison of disk and shell methods, 434
Comparison Test
  Direct, 583
  Limit, 585
Completeness of real numbers, 563
Completing the square, 389
Component form of a vector in the plane, 716, 717
Component functions, 786
Component of acceleration
  centripetal, 815
  normal, 815
  tangential, 815
Composite function, 25, 839
  antiderivative of, 288
  continuity of, 73, 855
  limit of, 59
Composition of two functions, 25
Compound interest formulas, 355
Computer graphics, 844
Concavity, 184
  test for, 185
Conditional convergence, 593
Conic(s), 650
  classification of, 702
  degenerate, 650
  in polar form, 703
Conjugate axis of a hyperbola, 657
Connected region, 1035
Conservative vector fields, 1011
  independence of path, 1035
  test for, 1012, 1015
Constant function, 24
Constant of integration, 243
Constant Multiple Rule for differentiation, 108
Constant of proportionality, 362
Constant rule for differentiation, 105
Constant term, 24
Constraint, 921
Continuity
  on a closed interval, 71

of a composite function, 73, 855
and differentiability, 872
of a function of three variables, 856
of a function of two variables, 854
and integrability, 267
of an inverse function, 336
from the left, 71
on an open interval, 68
at a point, 57, 68
of a polynomial function, 73
properties of, 73
of a radical function, 73
of a rational function, 73
from the right, 71
of a trigonometric function, 73
of a vector-valued function, 790
Continuous
  everywhere, 68
  in an open region, 854, 856
  at a point, 854, 856
Continuous compounding, 355
Continuously differentiable, 440
Contour line, 841
Convergence, 557, 567
  of an improper integral, 540, 543
  of an infinite series, 567
    absolute, 593
    Alternating Series Test, 590
    conditional, 593
    Direct Comparison Test, 583
    geometric series, 569
    Integral Test, 577
    Limit Comparison Test, 585
    power series, 617
    of a p-series, 579
    Ratio Test, 597
    Root Test, 600
    summary of tests, 602
  interval of, 617
  of Newton's Method, 224
  radius of, 617
  of a sequence, 557
  of Taylor series, 634
Convergent sequence, 224
Convergent series, 567
Conversion
  coordinate, 685
  cylindrical to rectangular, 773
  polar to rectangular, 685
  rectangular to cylindrical, 773
  rectangular to polar, 685
  rectangular to spherical, 776
  spherical to rectangular, 776
Coordinate conversion, 685
Coordinate system
  cylindrical, 773
  polar, 684
  spherical, 776

three-dimensional, 727
Coordinates of a point in space, 727
Copernicus, Nicholas (1473–1543), 653
Cornu spiral, 712, 834
Cosecant function
  derivative of, 121
  integral of, 244, 329
  inverse of, 380
Cosine function
  derivative of, 110
  integral of, 244, 329
  integrals involving, 497
    guidelines for evaluating, 497
  inverse of, 380
  rational function of sine and, 527
  series for, 638
Cotangent function
  derivative of, 121
  integral of, 329
  inverse of, 380
Coulomb, Charles (1736–1806), 452
Coulomb's Law, 452, 1009
Critical numbers, 162
  and relative extrema, 162
Critical point, 906
  and relative extrema, 906
Cross product
  properties of, 745, 746
  of two vectors, 744
Cubic function, 24
Curl of a vector field, 1014
  and divergence, 1016
Curtate cycloid, 673
Curvature, 823
  center of, 825
  circle of, 825
  formulas for, 824
  radius of, 825
  in rectangular coordinates, 825
  related to acceleration and speed, 826
  in space, 824
Curve
  closed, 1037
  graph of, 665
  level, 841
  orientation of, 666, 1019
  piecewise smooth, 670, 1019
  plane, 665
  simple, 1042
  smooth, 440, 670, 796, 1019
  in space, 786
Curve sketching, summary of, 202
Cusps, 796
Cycloid, 670
  curtate, 673
  prolate, 677
Cylinder, 763
Cylindrical coordinate system, 773

conversion to rectangular, 773
conversion of rectangular to, 773
integration in, 988
Cylindrical surface, 763

D

Decay, 362
Decreasing function, 174
   test for, 174
Definite integral, 267
   change of variables for, 294
   evaluation of, 276
   properties of, 270, 271
   of a vector-valued function, 798
Degenerate conic, 650
Degree of a polynomial function, 24
Delta, δ, 850
   neighborhood, 850
Density, 463
Density function, 963, 983
Dependent variable, 19, 838
Derivative(s)
   of an absolute value function, 320
   alternate form of, 99
   of an arc length function, 821
   for bases other than e, 353
   Chain Rule, 127
   Constant Multiple Rule, 108
   Constant Rule, 105
   of a cosecant function, 121
   of a cosine function, 110
   of a cotangent function, 121
   Difference Rule, 109
   directional, 884, 885, 887, 892
   of an exponential function, base a, 353
   of a function, 97
   General Power Rule, 129
   higher-order, 123
   of a hyperbolic function, 397
   of an inverse cosecant function, 383
   of an inverse cosine function, 383
   of an inverse cotangent function, 383
   of an inverse function, 336
   of an inverse hyperbolic function, 401
   of an inverse secant function, 383
   of an inverse sine function, 383
   of an inverse tangent function, 383
   of an inverse trigonometric function, 383
   from the left, 99
   of a logarithmic function, base a, 353
   of a natural exponential function, 343
   of the natural logarithmic function, 318
   notation, 97
   in parametric form, 675
   partial, 859, 1054
   Power Rule (real exponents), 106
   of power series, 621

Product Rule, 117
Quotient Rule, 119
   from the right, 99
   of a secant function, 121
   second, 123
   of sine function, 110
   Sum Rule, 109
   summary of rules, 133
   of a tangent function, 121
   third, 123
   of trigonometric functions, 121
   of a vector-valued function, 794
Descartes, René (1596–1650), 2, 95
Determinant form of a cross product, 744
Determinate forms of a limit, 536
Difference quotient, 95
Difference Rule for differentiation, 109
Differentiability, 870
Differentiability and continuity, 99, 101, 872
Differentiable function, 97
   on a closed interval, 99
   on an open interval, 97
   in a region, 870
   in three variables, 871
   of two variables, 870
   vector-valued, 794
Differential, 869
   form, 231
   formulas, 231
   operator, 1014, 1016
   total, 869
   of x, 229
   of y, 229
Differential equation, 243, 361
   general solution of, 243, 369
   homogeneous, 373
   linear first-order, A4
      solution of, A4
   linear homogeneous, 373
   logistics, 238
   particular solution of, 370
   separation of variables, 371
   singular solution of, 369
   solution of, 369
Differential form of a line integral, 1027
Differentiation, 97
   implicit, 137
   partial, 859
   rules for elementary functions, 385
   of vector-valued functions, 795
Direct Comparison Test, 583
Direct substitution, 57
Directed line segment, 716
Direction angles, 738
Direction cosines, 738
Direction field (slope field), 315, A2

Direction numbers, 752
Direction vector, 752
Directional derivative, 884, 885, 887, 892
   alternative form, 887
   of a function of three variables, 892
Directrix
   of a cylinder, 763
   of a parabola, 651
Dirichlet, Peter Gustav (1850–1859), 51
Dirichlet function, 51
Discontinuity, 69
   infinite, 540
   nonremovable, 69
   removable, 69
Disk, 421, 850
   closed, 850
   method, 421, 422
      compared to shell method, 434
   open, 850
Distance
   between a point and a line in space, 758
   between a point and a plane in space, 757
   between two points in space, 728
Distance Formula
   in space, 728
Divergence, 540, 543, 557, 567
   of an improper integral, 540, 543
   of infinite series, 567
      Direct Comparison Test, 583
      geometric, 569
      Integral Test, 577
      Limit Comparison Test, 585
      nth-Term Test for, 571
      power, 617
      p-series, 579
      Ratio Test, 597
      Root Test, 600
      summary of tests, 602
   of a sequence, 557
   series, 567
   of a vector field, 1016
   and curl, 1016
Divergence free, 1016
Divergence Theorem, 1049, 1073
   and flux, 1078
Domain
   of a function, 19
   of a function of two variables, 838
   of a vector-valued function, 787
Doomsday equation, 409
Dot product, 735
   properties of, 735
Double integral, 944, 945, 946
   in polar coordinates, 955
   properties of, 946
Doyle Log Rule, 848
Dyne, 450

E

e, the number, 317
Eccentricity
   of a conic, 702
   of an ellipse, 655
   of a hyperbola, 658
Eight curve, 156
Electric force fields, 1009
Elementary function, 24, 385
   basic differentiation rules for, 385
   power series for, 638
Eliminating the parameter, 667
Ellipse, 653
   center of, 653
   eccentricity of, 655
   foci of, 653
   major axis of, 653
   minor axis of, 653
   polar form of equation of, 702
   reflective property of, 655
   standard equation of, 653
   vertices of, 653
Ellipsoid, 764
Elliptic cone, 764
Elliptic paraboloid, 764
Endpoint convergence, 619
Endpoint extrema, 160
Energy
   conservation of, 1038
   kinetic, 1038
   potential, 1038
Epicycloid, 674, 678
Epsilon, ε, 52
Equal vectors, 717
Equation
   of cylinders, 763
   doomsday, 409
   graph of, 2
   of a horizontal line, 14
   of a line, 11
   logistics, 409
   parametric, 665, 752
   solution point of, 2
   of a tangent plane, 897
   of a vertical line, 14
Equilibrium, 460
Equipotential lines, 841
Equivalent
   conditions, 1037
   directed line segments, 716
Error
   percent, 230
   relative, 230
   in Taylor's Theorem, 611
   in Trapezoidal and Simpson's Rule, 304
Error propagation, 230
Euler, Leonhard (1707–1783), 19, 24, 341, 515, 859

Euler's method, A3, A7
Evaluation
   of double integrals, 947
   of a function, 19
   of a flux integral, 1067
   of iterated integrals, 979
   of a limit, 57, 60
   of a surface integral, 1061
Even function, 26
   integration of, 296
Everywhere continuous, 68
Existence of an inverse function, 334
Existence of a limit, 52, 71
Existence theorem, 75
Expanded about *c* (or centered at *c*), 605
Expected value, 548
Explicit form of a function, 19, 137
Exponential decay, 362
Exponential function
   base *a*, 351
      derivative of, 353
   derivative of, 343
   integration of, 345
   inverse of, 341
   operations with, 342
   properties of, 342
   series for, 638
Exponential growth, 362
Extended Mean Value Theorem, 531
Extrema
   absolute, 160, 905
   applications of, 913
   on a closed interval, 163
      guidelines for finding, 163
   endpoint, 160
   relative, 161, 905
Extreme value of a function on an interval, 160
Extreme Value Theorem, 160, 905

F

Factorial, 559
Faraday, Michael (1791–1867), 1038
Fermat, Pierre de (1601–1665), 162
First Derivative Test, 176
First moment, 967, 983
First partial derivative, 859
   notation, 860
First-order differential equation, A4
First-order linear differential equation, A4
Fluid force, 470
Fluid pressure, 470
Flux and the Divergence Theorem, 1078
Flux integral, 1067
Focal chord, 651
Focus of an ellipse, 653
Focus of a parabola, 651
Force, 450

   buoyant, 474, 479
   constant, 450
   exerted by a fluid, 471
   of friction, 827
   variable, 451
   as a vector, 722
Force fields, 1008
   central, 1009
   conservative, 1011
Formulas for curvature, 824
Fourier, Joseph (1768–1830), 625
Fourier Sine Series, 496
Free-falling object, 67, 88
Fresnel function, 310
Friction, 827
Fubini, Guido (1879–1943), 948
Fubini's Theorem, 948
   for a triple integral, 979
Function, 6, 19
   accumulation, 281
   addition of, 25
   algebraic, 25
   antiderivative of, 242
   average value on an interval, 279
   component, 786
   composite, 25, 839
   concave downward, 184
   concave upward, 184
   constant, 24
   continuity, 68
   cubic, 24
   decreasing, 174
   derivative of, 97
   difference of, 25
   differentiable, 97
   domain of, 19
   elementary, 24
   even, 26
   explicit form of, 19, 137
   exponential, 341
   extreme value of, 160
   gamma, 548
   graph of, 22
      guidelines for analyzing, 202
   greatest integer, 70
   Heaviside, 38
   homogeneous, 373
   hyperbolic, 395
   hyperbolic cosecant, 395
   hyperbolic cosine, 395
   hyperbolic cotangent, 395
   hyperbolic secant, 395
   hyperbolic sine, 395
   hyperbolic tangent, 395
   implicit form of, 19, 137
   increasing, 174
   integrable, 946
   inverse, 332

inverse cosecant, 380
inverse cosine, 380
inverse cotangent, 380
inverse hyperbolic, 399
inverse hyperbolic cosecant, 399
inverse hyperbolic cosine, 399
inverse hyperbolic cotangent, 399
inverse hyperbolic secant, 399
inverse hyperbolic sine, 399
inverse hyperbolic tangent, 399
inverse secant, 380
inverse sine, 380
inverse tangent, 380
limit of, 48
linear, 24
natural logarithmic, 314
notation, 19
odd, 26
one-to-one, 21
onto, 21
orthogonal, 505
polynomial, 24, 839
position, 111
potential, 1011
product of, 25
quadratic, 24
range of, 19
rational, 25, 839
relative maximum of, 176
relative minimum of, 176
of several variables, 838
step, 70
strictly monotonic, 175
of three variables, 892
   continuity of, 856
   directional derivative, 892
   gradient of, 892
   partial derivative of, 862
transcendental, 25
transformation of graph of, 23
of two variables, 838
   continuity of, 854
   domain of, 838
   gradient of, 887
   graph of, 840
   homogeneous, 373
   limit of, 851
   partial derivative of, 859
   range of, 838
vector-valued, 786
velocity, 112
zero of, 26, 222
Fundamental Theorem of Calculus, 275
  guidelines for using, 276
Fundamental Theorem of Calculus, Second, 282
Fundamental Theorem of Line Integrals, 1032, 1033

G

Gabriel's Horn, 546
Galilei, Galileo (1564–1642), 385
Galios, Evariste (1811–1832), 225
Gamma function, 548
Gauss, Carl Friedrich (1777–1855), 254, 1073
Gauss's Law, 1070
Gauss's Theorem, 1073
General antiderivative, 243
General form
  of an equation of a line, 14
  of an equation of a plane in space, 753
  of a second-degree equation, 650
General harmonic series, 579
General Power Rule
  of differentiation, 129
  for integration, 293
General second-degree equation, 650
General solution of a differential equation, 243, 369
Generating curve (or directrix), 763
Geometric power series, 625
Geometric properties of the cross product, 746
Geometric property of the triple scalar product, 749
Geometric series, 569
  convergence of, 569
  divergence of, 569
Gibbs, Josiah Willard (1839–1903), 745, 1019
Goldbach, Christian, 341
Golden ratio, 566
Grad, 887
Gradient
  of a function of three variables, 892
  of a function of two variables, 887
  normal to level curves, 891
  normal to level surfaces, 901
  properties of, 888
Graph(s)
  of common functions, 22
  of a curve, 665
  of an equation, 2
  of a function, 22
    sketching, 202
    tangent line to, 95
  of a function of two variables, 840
  intercept of, 4
  of a natural logarithmic function, 314
  of a parametric equation, 665
  symmetry of, 5
Gravitation, Newton's Law of Universal, 452
Gravitational fields, 1009
Gravity
  acceleration due to, 123
  center of, 461, 462
  force due to, 806
Greatest integer function, 70
Green, George (1793–1841), 1043
Green's Theorem, 1042
  alternative forms, 1047, 1048
Gregory, James (1638–1675), 621, 629
Growth and decay, 362
Gyration, radius of, 968

H

Half-life, 363
Halley, Edmund (1656–1742), 705
Halley's Comet, 705
Hamilton, William Rowan (1805–1865), 718
Harmonic series, 579
  alternating, 593
  general, 579
Heaviside, Oliver (1850–1925), 38
Heaviside function, 38
Helix, 787
Helmholtz, Hermann Ludwig (1821–1894), 1038
Herschel, Caroline (1750–1848), 659
Higher-order derivative, 123
  partial, 863
Homogeneous differential equation, 373
Homogeneous function, 373
Hooke, Robert (1635–1703), 452
Hooke's Law, 452
Horizontal asymptote, 193
Horizontal component of a vector, 721
Horizontal line, equation of, 14
Horizontal line test, 334
Horizontal shift of a graph of a function, 23
Horizontally simple region, 938
Huygens, Christian (1629–1695), 440
Hypatia (370–415), 650
Hyperbola, 657
  asymptote of, 657
  center of, 657
  conjugate axis of, 657
  eccentricity of, 658
  polar form of equation of, 702
  standard form of equation of, 657
  transverse axis of, 657
  vertex of, 657
Hyperbolic cosecant function, 395
  derivative of, 397
  graph of, 396
  identities for, 397
  integration of, 397
  inverse of, 399
Hyperbolic cosine function, 395
  derivative of, 397
  graph of, 396
  identities for, 397

integration of, 397
inverse of, 399
Hyperbolic cotangent function, 395
   derivative of, 397
   graph of, 396
   identities for, 397
   integration of, 397
   inverse of, 399
Hyperbolic functions, 395
Hyperbolic identities, 397
Hyperbolic paraboloid, 764
Hyperbolic secant function, 395
   derivative of, 397
   graph of, 396
   identities for, 397
   integration of, 397
   inverse of, 399
Hyperbolic sine function, 395
   derivative of, 397
   graph of, 396
   identities for, 397
   integration of, 397
   inverse of, 399
Hyperbolic tangent function, 395
   derivative of, 397
   graph of, 396
   identities for, 397
   integration of, 397
   inverse of, 399
Hyperboloid
   of one sheet, 764
   of two sheets, 764
Hypocycloid, 674

I

Identities
   hyperbolic, 397
Identity function, 22
Image of x under f, 19
Implicit differentiation, 137
   guidelines for, 138
Implicit form of a function, 19, 137
Implicit partial differentiation, 880
Improper integral, 540
   convergence of, 540, 543
   divergence of, 540, 543
   with infinite discontinuities, 543
   with infinite limits of integration, 540
   special type of, 546
Incidence, angle of, 652
Inclination, angle of, 900
Incompressible, 1016, 1078
Increasing function, 174
   test for, 174
Indefinite integral, 243
Indefinite integral (or antiderivative) of a vector-valued function, 798

Independence of path, 1035
Independent variable, 19, 838
Indeterminate form, 61, 530
Index of summation, 253
Inertia, moment of, 967, 983
   polar, 967
Infinite discontinuity, 540
Infinite limit, 80
   at infinity, 198
   from the left, 80
   properties of, 84
   from the right, 80
Infinite series (or series), 567
   alternating, 590
   convergence of, 567
   divergence of, 567
   geometric, 569
   harmonic, 579
   limit at, 192
   nth partial sum of, 567
   properties of, 571
   p-series, 579
   sequence of partial sums of, 567
   sum of, 567
   telescoping, 568
   term of, 567
Inflection point, 186
Initial condition, 247, 370
Initial point, 716
Initial value of exponential growth and decay models, 362
Inner partition, 944, 978
   polar, 956
Inner product, 505, 735
Inner radius, 424
Inscribed rectangle, 257
Inside limits of integration, 937
Instantaneous velocity, 112
Integrability and continuity, 267
Integrable function, 946
Integral
   definite, 267
   definition of, 267
   double, 944, 945, 946
   flux, 1067
   improper, 540
   indefinite, 243
   iterated, 937
   line, 1020
   single, 946
   surface, 1061
   triple, 978
Integral Test, 577
Integration, 243, 540
   constant of, 243
   completing the square, 389
   of even and odd functions, 296
   General Power Rule, 293

   guidelines for, 327
   of a hyperbolic function, 397
   involving inverse hyperbolic function, 401
   involving inverse trigonometric functions, 388
   involving logarithmic functions, 324
   involving secant and tangent, 500
   involving sine and cosine, 497, 502
   limits of, 937
   Log Rule, 324
   lower limit of, 267
   by partial fraction, 515
   by parts, 488
      guidelines for, 488
      summary, 493
      tabular method, 493
   of power series, 621
   region of, 937
   rules, 244, 391
   rules for exponential functions, 345
   by substitution, 288
   summary of formulas, 1085
   by tables, 524
   of trigonometric functions, 328, 329
   by trigonometric substitution, 506
   upper limit of, 267
   of a vector-valued function, 798
Integration formulas, special, 510
Intercept of a graph, 4
Interest, compound, 355
Interior point, 850, 856
Intermediate Value Theorem, 75
Interval
   of convergence, 617
   partition of, 266
Inverse cosecant function, 380
   derivative of, 383
   graph of, 381
Inverse cosine function, 380
   derivative of, 383
   graph of, 381
Inverse cotangent function, 380
   derivative of, 383
   graph of, 381
Inverse function, 332
   continuity of, 336
   derivative of, 336
   existence of, 334
   graph of, 333
      guidelines for finding, 335
   horizontal line test, 334
   properties of, 352
   reflective property of, 333
Inverse hyperbolic cosecant function, 399
   derivative of, 401
   graph of, 400
   integrals involving, 401

Inverse hyperbolic cosine function, 399
  derivative of, 401
  graph of, 400
  integrals involving, 401
Inverse hyperbolic cotangent function, 399
  derivative of, 401
  graph of, 400
  integrals involving, 401
Inverse hyperbolic functions, 399
Inverse hyperbolic secant function, 399
  derivative of, 401
  graph of, 400
  integrals involving, 401
Inverse hyperbolic sine function, 399
  derivative of, 401
  graph of, 400
  integrals involving, 401
Inverse hyperbolic tangent function, 399
  derivative of, 401
  graph of, 400
  integrals involving, 401
Inverse secant function, 380
  derivative of, 383
  graph of, 381
Inverse sine function, 380
  derivative of, 383
  graph of, 381
Inverse square field, 1009
Inverse tangent function, 380
  derivative of, 383
  graph of, 381
Inverse trigonometric function(s), 380
  derivative of, 383
  graph of, 381
  integrals involving, 388
  properties of, 382
Involute of a circle, 710
Irrotational, 1014
Isobars, 841
Isotherm, 841
Isothermal surfaces, 844
Iterated integral, 937
Iteration, 222
ith term of a sum, 253

J

Jacobi, Carl Gustav (1804–1851), 995
Jacobian, 995
Joule, 450
Joule, James Prescott (1818–1889), 1038

K

Kappa curve, 141
Kepler, Johannes (1571–1630), 656, 705
Kepler's Laws, 705
Kinetic energy, 1038

Koch snowflake, 554, 589
Kovalevsky, Sonya (1850–1891), 850

L

Lagrange, Joseph Louis (1736–1813), 170, 922
Lagrange form of the remainder, 611
Lagrange multiplier, 922
  method of, 921, 922
    with two constraints, 926
Lagrange's Theorem, 922
Lambert, Johann Heinrich (1728–1777), 395
Lamina, planar, 463
Laplace, Pierre Simon de (1749–1827), 988
Laplace's equation, 866, 929
Laplacian, 1018
Latus rectum, 651
Law of Conservation of Energy, 1038
Law of refraction, 928
Leading coefficient, 24
Leading coefficient test, 24
Least squares
  method of, 915
  regression line, 915, 916
  regression quadratic, 919
Least upper bound of a sequence, 563
Left-handed orientation, 727
Legendre, Adrien-Marie (1752–1833), 916
Leibniz, Gottfried Wilhelm (1646–1716), 19, 231
Leibniz notation for derivatives or differentials, 231
Lemniscate, 39, 140, 690
Length
  of an arc, 440, 441
  of the moment arm, 460
  of a scalar multiple, 720
  of a vector, 716, 717
Level curves (or contour lines), 841
Level surfaces, 843
L'Hôpital, Guillaume François Antoine de (1661–1704), 531
L'Hôpital's Rule, 531
Limaçon, 690
Limit
  of an algebraic function, 57
  of a composite function, 59
  definition of, 52
  determinate form, 536
  evaluation of, 57
  existence of, 52, 71
  of a function, 48
  of a function of two variables, 851
  indeterminate form, 61, 530
  infinite, 80
  at infinity, 192

  of integration
    inside, 937
    outside, 937
  involving e, 355
  from the left, 70
  of lower sum, 259
  nonexistence of, 50
  of nth term of a convergent series, 571
  one-sided, 70
  of a polynomial function, 58
  properties of, 57
  of a radical function, 58
  of a rational function, 58
  from the right, 70
  of a sequence, 557
  of the slope of a line, 45
  of a series, 567
  strategy for finding, 60
  trigonometric, 63
  of a trigonometric function, 59
  of upper sum, 259
  of a vector-valued function, 789
Limit Comparison Test, 585
Line(s)
  general form of equation, 14
  parallel, 14
  perpendicular, 14
  point-slope equation of, 11
  sketching the graph of, 13
  slope of, 10
  slope-intercept equation, 13
  in space, 752
    direction number of, 752
    direction vector for, 752
    parametric equations of, 752
    symmetric equations of, 752
Line of impact, 896
Line integral, 1020
  for area, 1045
  as a definite integral, 1021
  differential form of, 1027
  Fundamental Theorem of, 1032, 1033
  independence of path, 1035
  of a vector field, 1024
Linear approximation, 871
Linear combination, 721
Linear factors, 517
Linear function, 24
Linear regression, 7, 30
Locus, 650
Log Rule for Integration, 324
Logarithmic differentiation, 319
Logarithmic function
  base a, 352
  common, 352
  integral involving, 324
  natural, 314
Logarithmic properties, 315

Logistics curve, 523
Logistics differential equation, 238
Logistics equation, 409
Lower bound of a sequence, 563
Lower bound of summation, 253
Lower limit of integration, 267
Lower sum, 257
 limit of, 259

M

Macintyre, Sheila Scott (1910–1960), 497
Maclaurin, Colin (1698–1746), 632
Maclaurin polynomials, 607
Maclaurin series, 632
Magnitude, 716, 717
Major axis of an ellipse, 653
Marginal productivity of money, 924
Mass, 459, 963
 center of, 460, 461, 462, 463, 965, 983
 moments of, 463, 965
 of a planar lamina, 963
Mathematical model, 7, 915
Maximum, 905
 absolute, 160, 905
 of a function on an interval, 160
 relative, 161, 905, 908
Maximum problems (applied), 212
Maxwell, James (1831–1879), 718
Mean Value Theorem, 170
 extended, 238, 531
 for integrals, 278
Measurement, system of, 459
Method of Lagrange Multipliers, 921, 922
 with two constraints, 926
Method of least squares, 915
Method of partial fractions, 515
Midpoint between two points in space, 728
Midpoint Rule, 263, 1005
Minimum, 905
 absolute, 160, 905
 of a function on an interval, 160
 relative, 161, 905, 908
Minimum problems (applied), 212
Minor axis of an ellipse, 653
Mixed partial derivatives, 863
 equality of, 864
Model, mathematical, 7, 915
Moment
 first, 967, 983
 of a force about a point, 748
 of inertia, 967, 983, 1090
  polar, 967
  second, 967, 983
 about a line, 460
 of m about the point P, 460, 748
 of mass, 965
 of mass of a planar lamina of variable density, 965
 about the origin, 460, 461
 about a point, 460
 about the x-axis, 462, 463
 about the y-axis, 462, 463
Moment arm, length of, 460
Monotonic, 562
Monotonic function, 175
Monotonic sequence, 562
Motion of a projectile, 806
Multiple integral
 iterated, 937
 in nonrectangular coordinates, 955, 988
 triple, 978
Mutually orthogonal, 376

N

n factorial, 559
Napier, John (1550–1617), 314
Natural exponential function, 341
Natural logarithmic base, 317
Natural logarithmic function, 314
 base of, 317
 derivative of, 318
 graph of, 314
 properties of, 315
 series for, 638
Negative of a vector, 718
Neighborhood in the plane, 850
Newton, Isaac (1642–1727), 94, 222, 684
Newton's Law of Cooling, 365
Newton's Law of Universal Gravitation, 452
Newton's Method, 222
 convergence of, 224
Newton's Second Law of Motion, 806
Node, 796
Noether, Emmy (1882–1935), 720
Nonremovable discontinuity, 69, 854
Norm
 of a partition, 266, 944, 956, 978
 of a vector, 717
Normal component of acceleration, 815
Normal line to a surface, 896, 897
Normal probability density function, 548
Normal vector, 812, 1054
 to a smooth parametric surface, 1054
 unit, 812
Normalization of a vector, 720
nth Maclaurin polynomial, 607
nth partial sum, 567
nth Taylor polynomial, 607
nth term of a sequence, 556
nth-Term Test for Divergence, 571
Number e, 317
Numerical integration, 300

O

Octants, 727
Odd function, 26
 integration of, 296
Ohm's law, 234
One-sided limits, 70
One-to-one function, 21
Onto function, 21
Open disk, 850
Open interval
 continuity on, 68
 differentiable on, 97
Open region, 850, 856
Open sphere, 856
Optimization problems, 750, 751, 926
 guidelines for, 212
Order of a differential equation, 369
Order of integration, 940, 950
Orientation of a curve in the plane, 666
Orientation of a curve in space, 1019
Oriented surface, 1066
Origin, 684
 symmetry, 5
Orthogonal, 505
 functions, 505
 projections, 739
 trajectory, 376
 vectors, 737
Ostrogradsky, Michel (1801–1861), 1073
Outer radius, 424
Outside limits of integration, 937

P

Pappus
 Second Theorem of, 469
 Theorem of, 466
Pappus of Alexandria (ca. 300), 466
Parabola, 2, 651
 axis of, 651
 directrix of, 651
 focal chord of, 651
 focus of, 651
 latus rectum of, 651
 polar form of equation of, 702
 reflective property of, 652
 standard form of equation of, 651
 vertex of, 651
Parabolic spandrel, 468
Paraboloid, 766
Parallel lines, 14
Parallel planes, 754
Parallel vectors, 730
Parallelepiped, volume of, 749
Parameter, 665
Parametric equations, 665, 1051
 and arc length, 678

area of surface of revolution, 680
of a line in space, 752
for surfaces, 1053
Parametric form of the derivative, 675
Parametric surface, 1051
area of, 1055
equation for, 1051
and surface integrals, 1065
Partial derivative, 1054
first, 859
of a function of three or more variables, 862
of a function of two variables, 859
higher-order, 863
implicit, 880
mixed, 863
notation for, 860
Partial differentiation, 859
Partial fraction(s), 515
Partial sums, sequence of, 567
Particular solution, 247
of a differential equation, 370
Partition
inner, 944, 978
polar, 956
of an interval, 266
norm of, 266, 944, 956
regular, 266
Pascal, Blaise (1623–1662), 470
Pascal's Principle, 470
Path, 851, 1019
Pattern recognition, 288
for sequences, 560
Pear-shaped cuartic, 156
Percent error, 230
Perpendicular lines, 14
Physical interpretation of curl, 1084
Piecewise smooth curve, 670, 1019
Planar lamina, 463
Plane(s)
distance between a point and, 757
general equation of, 753
parallel, 754
perpendicular, 754
in space, 753
standard equation of, 753
tangent to a surface, 897
Plane curve, 665
Plane region, area of, 256, 259, 938
Point of inflection, 186
Point of intersection of polar graphs, 696
Point of intersection of two graphs, 6
Point-plotting method, 2
Point-slope form of an equation of a line, 11
Point-slope method, 2
Polar axis, 684
Polar coordinate system, 684

Polar coordinates, 684
and arc length, 698
and area, 694
area of surface of revolution, 699
conversion of rectangular to, 685
double integral, 955
equation of conics in, 703
graphing techniques for, 686
Polar graphs, 686
graphs, special, 690
and points of intersection, 696
Polar moment of inertia, 967
Polar sectors, 955
Pole (or origin), 684, 773
Polynomial approximation, 605
Polynomial function, 24
continuity of, 73
degree of, 24
limit of, 58
of two variables, 839
Position function, 111
for a projectile, 807
Position vector, 806
Potential energy, 1038
Potential function, 1011
Pound mass, 459
Power Rule
for differentiation, 106
General, 129
for real exponents, 354
Power series, 616
centered at c, 616
convergence of, 617
derivative of, 621
divergence of, 617
for elementary functions, 638
geometric, 625
integration, 621
interval of convergence, 619
operations with, 627
properties of, 621
Preservation of inequality, 272
Pressure, 470
Primary equation, 211, 212
Principal unit normal vector, 812
Probability density function, 548
Procedures for fitting integrands to basic rules, 485
Product Rule for differentiation, 117
Projection, 739
using the dot product, 740
Prolate cycloid, 677
Propagated error, 230
Properties
of continuity, 73
of continuous function of two variables, 854

of cross product, 745, 746
of definite integrals, 270, 271
of the derivative, 796
of the dot product, 735
of double integrals, 946
of functions defined by power series, 621
of the gradient, 888
of infinite limits, 84
of infinite series, 571
of inverse functions, 352
of inverse trigonometric functions, 382
of limits, 57
of limits of sequences, 558
of the natural exponential function, 342
of the natural logarithmic function, 315
of vector operations, 719
Proportionality constant of exponential growth and decay models, 362
p-series, 579
convergence of, 579
divergence of, 579

Q

Quadratic factors, 519
Quadratic function, 24
Quadric surface, 764
Quotient rule for differentiation, 119
Quotient of two polynomials, 25

R

Radial lines, 684
Radical function
continuity of, 73
limit of, 58
of convergence, 617
of curvature, 825
of gyration, 968
Radius of convergence, 617
Radius function, 769
Ramanujan, Srinivasa (1887–1920), 629
Range
of a function, 19
of a function of two variables, 838
Raphson, Joseph (1648–1715), 222
Rate of change, 12
average, 12
instantaneous, 171
Ratio, 12
Ratio Test, 597
Rational function, 25
continuity of, 73
limit of, 58
of sine and cosine, 527
of two variables, 839
Rationalization, 61

Real-valued function f of a real variable, 19
Rearrangement of series, 594
Rectangle, area of, 255
Rectangular coordinates
    conversion to cylindrical, 773
    conversion to polar, 685
    conversion to spherical, 776
Rectifiable, 440
Reduction formulas, 526
Reflection of graph of a function, 23, 333
Reflective property
    of an ellipse, 655
    of inverse functions, 333
    of a parabola, 652
Reflective surface, 652
Refraction, Snell's Law of, 928
Region in the plane
    area of, 938
    centroid of, 965
    closed, 850
    connected, 1035
    horizontally simple, 938
    of integration, 937
    open, 850, 856
    simple solid, 1074
    simply connected, 1042
    between two curves, area of, 412, 413
    vertically simple, 938
Regular partition, 266
Related rates, 144
    guidelines for problem solving with, 145
Relation, 19
Relative error, 230
Relative extrema, 161, 905
    and critical numbers, 162
    and critical points, 906
    First Derivative Test for, 176
    Second Derivative Test for, 188
Relative maximum, 161, 905, 908
    First Derivative Test for, 176
    of a function, 176
    Second Derivative Test for, 188
    Second Partials Test for, 908
Relative minimum, 161, 905, 908
    First Derivative Test for, 176
    of a function, 176
    Second Derivative Test for, 188
    Second Partials Test for, 908
Removable discontinuity, 69, 854
Representative rectangle, 412
Resultant force, 722
Resultant vector, 718
Riemann, Georg Friedrich Bernhard (1826–1866), 266
Riemann sum, 266
Riemann zeta function, 581
Right-handed orientation, 727
Rolle, Michel (1652–1719), 168

Rolle's Theorem, 168
Root Test, 600
Rose curve, 687, 690
Rotation, 1084
r-simple region, 957
Rulings, 763

S

Saddle point, 908
Scalar, 716
Scalar field, 841
Scalar multiple, 718
Scalar multiplication, 718
Scalar product, 735
Scalar quantities, 716
Secant function
    derivative of, 121
    integral of, 329
    inverse of, 380
    and tangent, integrals involving, 500
Secant line, 45, 95
Second derivative, 123
Second Derivative Test, 188
Second Fundamental Theorem of Calculus, 282
Second moment, 967, 983
Second Partials Test, 908
Second Theorem of Pappus, 469
Secondary equation, 212
Separable differential equations, 371
Separation of variables, 371
Sequence, 556
    bounded, 563
    convergence of, 224, 557
    divergence of, 557
    least upper bound, 563
    limit of, 557
    monotonic, 562
    nth term, 556
    of partial sums, 567
    pattern recognition for, 560
    properties of, 558
    Squeeze Theorem, 559
    term of, 556
    upper bound of, 563
Series, 567
    absolute convergence, 593
    alternating, 590
    binomial, 637
    convergence of, 567
        guidelines for testing, 601
    divergence of, 567
        guidelines for testing, 601
    geometric, 569
    harmonic, 579
    infinite, 567
    Maclaurin, 632, 633
    nth partial sum, 567

    power, 616
    properties of, 571
    p-series, 579
    sequence of partial sums of, 567
    sum of, 567
    summary of tests for, 602
    Taylor, 632, 633
    telescoping, 568
    term of, 567
Shell method, 432, 433
    compared to the disk method, 434
Shift of a graph, 23
Sigma notation, 253
Signum function, 79
Simple curve, 1042
Simple solid region, 1074
Simply connected region, 1042
Simpson, Thomas (1710–1761), 302
Simpson's Rule, 302, 303
    error in, 304
Sine function
    derivative of, 110
    integral of, 329
    integrals involving, 497
        guidelines for evaluating, 497
    inverse of, 380
    series for, 638
Singular solutions of a differential equation, 369
Sink, 1078
Sketching planes in space, 756
Slant asymptote, 204
Slope
    field, 315, A2
    of a graph of f at $x = c$, 95
    of a horizontal line, 10
    of a line, 10
    of parallel lines, 14
    of perpendicular lines, 14
    in polar form, 688
    of a surface in the x-direction, 860
    of a surface in the y-direction, 860
Slope-intercept equation of a line, 13
Slug, 459
Smooth curve, 440, 670, 796, 1019
    on an open interval, 796
Smooth surface, 1054
Snell's Law of Refraction, 928
Solenoidal, 1016
Solid of revolution, 421
    volume of, 422, 424, 433
Solids with known cross sections, 426
Solution curves of a differential equation, 370
Solution point of an equation, 2
Somerville, Mary Fairfax (1780–1872), 838
Source, 1078
Space curve, 786

arc length of, 820
curvature of, 824
Special integration formulas, 510
Special polar graphs, 690
Speed, 112, 802, 803, 826
Sphere, 728
    equation of, 728
Spherical coordinate system, 776
Spherical coordinates, 776
    conversion to cylindrical, 776
    conversion to rectangular, 776
    triple integrals in, 991
Spiral of Archimedes, 679, 686
Square root function, 22
Square root symbol, 58
Squared errors, sum of, 915
Squeeze Theorem, 63
    for sequences, 559
Standard form
    of an equation of an ellipse, 653
    of an equation of a hyperbola, 657
    of an equation of a parabola, 651
    of an equation of a plane in space, 753
    of an equation of a sphere, 728
    of the equations of quadric surfaces, 764, 765, 766
    of a first-order linear differential equation, A4
Standard position of a vector, 717
Standard unit vector, 721
    notation, 729
    in the plane, 721
    in space, 729
Step functions, 70
Stokes, George Gabriel (1819–1903), 1081
Stokes's Theorem, 1047, 1081
Strategy for finding limits, 60
Strictly monotonic function, 175
Strophoid, 712
Substitution
    integration by, 288
    for rational functions of sine and cosine, 527
Sufficient condition for differentiability, 870
Suiseth, Richard, 567
Sum
    of infinite series, 567
    of two functions, 25
Sum Rule for differentiation, 109
Sum of the squared errors, 915
Summary of common integrals using integration by parts, 493
Summary of compound interest formulas, 355
Summary of differentiation rules, 133
Summary of equations of lines, 14
Summary of integration formulas, 1085
Summary of line and surface integrals, 1070

Summary of tests for series, 602
Summary of velocity, acceleration, and curvature, 828
Summation formulas, 254
Surface, 763
    closed, 1073
    isothermal, 844
    level, 843
    orientable, 1066
    orientation of, 1066
    in space, 763
Surface area, 971, 972, 1055
Surface integrals, 1061
Surface of revolution, 444, 769
    area of, 445, 680
    area in parametric form, 680
    area in polar form, 699
Symmetric equations of a line in space, 752
Symmetric with respect to (a, b), 408
Symmetry
    of a graph, 5
    with respect to the origin, 5
    with respect to the x-axis, 5
    with respect to the y-axis, 5
    tests for, 5

T

Table of values, 2
Tables, integration by, 524
Tabular method, 493
Tangent function
    derivative of, 121
    integral of, 329
    inverse of, 380
Tangent line, 45, 95, 675, 688
    to a curve, 812
    to the graph of a function, 95
    at the pole, 689
    vertical, 97
Tangent line approximation, 228
Tangent plane to a surface, 896, 897, 1054
    equation of, 897
Tangent vector, 802, 811
Tangential component of acceleration, 815
Tautochrone problem, 671
Taylor, Brook (1685–1731), 607, 621, 632
Taylor polynomials, 156, 607
    remainder of, 611
Taylor series, 632, 633
    convergence of, 634
    guidelines for finding, 636
Taylor's Theorem, 611
Telescoping series, 568
Term of a sequence, 556
Term of a series, 567
Terminal point, 716
    of a vector, 716

Test for concavity, 185
Test for conservative vector field, 1011
    in the plane, 1012
    in space, 1015
Test for decreasing function, 174
Test for even and odd functions, 26
Test for increasing function, 174
Test for symmetry, 5
Tests for convergence
    Alternating Series Test, 590
    Direct Comparison Test, 583
    Integral Test, 577
    Limit Comparison Test, 585
    Ratio Test, 597
    Root Test, 600
    summary of, 602
Theorem of Pappus, 466
θ-simple region, 957
Third derivative, 123
Thomson, William (1824–1907), 1081
Three-dimensional coordinate system, 727
Topographic map, 841
Torque, 461, 748
Torus, 466
Total differential, 869
Total mass, 461, 462
Trace, 756
    of a surface, 764
Tractrix, 401
Transcendental functions, 25
Transformation, 996
    of graph of function, 23
Transverse axis of a hyperbola, 657
Trapezoidal Rule, 300, 301
    error in, 304
Triangle Inequality, 721
Trigonometric functions(s)
    continuity of, 73
    derivative of, 121
    integration of, 329
    inverse, 380
    limit of, 59, 63
Trigonometric substitution, 506
Triple integral, 978
    in cylindrical coordinates, 988
    in spherical coordinates, 991
Triple scalar product, 748
    properties of, 749
Two-point gaussian quadrature approximation, 310

U

Unit normal vector, 812
Unit tangent vector, 811
Unit vector, 717, 720
    standard, 721
Upper bound of a sequence, 563
Upper bound of summation, 253

Upper limit of integration, 267
Upper sum, 257
    limit of, 259
u-substitution, 288

V

Variable
    dependent, 19, 838
    independent, 19, 838
    of integration, 243
Vector(s)
    acceleration, 803, 814
    addition of, 718
    angle between, 736
    components, 717, 721, 739
    cross product of, 744
    difference of, 718
    dot product of, 735
    equality of, 717, 729
    initial point, 716
    length of, 716, 717
    linear combination of, 721
    magnitude, 716, 717
    negative of, 718
    norm of, 717
    normalization of, 720
    orthogonal, 737
    parallel, 730
    in the plane, 716
    position, 806
    projection of, 739, 740
    properties of, 719
    resultant, 718
    scalar multiple, 718
    scalar multiplication, 718
    in space, 729
    standard position, 717
    subtraction, 718
    tangent, 802
    terminal point of, 716
    triple scalar product, 748, 749
    unit, 717, 720
    velocity, 802, 803
    zero, 717, 729
Vector addition in space, 729
Vector field, 1008
    circulation of, 1084
    conservative, 1011
    curl of, 1014
    divergence of, 1016
    divergence free, 1016
    incompressible, 1016, 1078
    irrotational, 1014
    line integral, of, 1024
    rotation of, 1084
    sink, 1078
    solenoidal, 1016
    source, 1078
Vector operations, 718
Vector product, 744
Vector space, 720
Vector-valued functions, 786
    continuous on an open interval, 790
    continuous at a point, 790
    derivative of, 794
    differentiation of, 795
    domain, 787
    integration of, 798
    limit of, 789
    properties of derivative, 796
    summary of properties, 828
Velocity, 802, 803
    average, 111
    of a free-falling object, 67
    function, 112
    instantaneous, 112
    summary of acceleration and curvature, 828
Velocity field, 1008, 1009
Velocity vector, 802, 803
Vertex
    of an ellipse, 653
    of a hyperbola, 657
    of a parabola, 651
Vertical asymptote, 81
Vertical component of a vector, 721
Vertical line, equation of, 14
Vertical line test, 22
Vertical shift of a graph of a function, 23
Vertical tangent line, 97
Vertically simple region, 938
Volume
    by disk method, 422
    by double integration, 946
    by shell method, 433
    of solid region, 944, 946
    of solid of revolution, 422, 424, 433
    of solids with known cross sections, 426
    by triple integration, 978

W

Wallis, John (1616–1703), 499
Wallis's Formulas, 499
Washer, 424
Washer method, 424
Wave equation, 933
Weierstrass, Karl (1815–1897), 850, 906
Weight, 459
Witch of Agnesi, 793
Work, vector form, 741
Work done by a constant force, 450
Work done by a variable force, 451
Work given by a line integral, 1024
Wren, Christopher, 678

X

x-axis symmetry, 5
x-coordinate, 727
x-intercept, 4
xy-plane, 727
xz-plane, 727

Y

y-axis symmetry, 5
y-coordinate, 727
y-intercept, 4
Young, Grace Chisholm (1868–1944), 42
yz-plane, 727

Z

z-axis, 727
z-coordinate, 727
Zero factorial, 559
Zero vector, 717, 729
Zeros of function, 26, 222

ALGEBRA

Factors and Zeros of Polynomials
Let $p(x) = a_n x^n + a_{n-1} x^{n-1} + \cdots + a_1 x + a_0$ be a polynomial. If $p(a) = 0$, then a is a *zero* of the polynomial and a solution of the equation $p(x) = 0$. Furthermore, $(x - a)$ is a *factor* of the polynomial.

Fundamental Theorem of Algebra
An nth degree polynomial has n (not necessarily distinct) zeros. Although all of these zeros may be imaginary, a real polynomial of odd degree must have at least one real zero.

Quadratic Formula
If $p(x) = ax^2 + bx + c$, and $0 \leq b^2 - 4ac$, then the real zeros of p are $x = \left(-b \pm \sqrt{b^2 - 4ac}\right)/2a$.

Special Factors
$x^2 - a^2 = (x - a)(x + a)$ $\qquad\qquad\qquad\qquad x^3 - a^3 = (x - a)(x^2 + ax + a^2)$

$x^3 + a^3 = (x + a)(x^2 - ax + a^2)$ $\qquad\qquad\quad x^4 - a^4 = (x^2 - a^2)(x^2 + a^2)$

Binomial Theorem
$(x + y)^2 = x^2 + 2xy + y^2$ $\qquad\qquad\qquad\qquad (x - y)^2 = x^2 - 2xy + y^2$

$(x + y)^3 = x^3 + 3x^2 y + 3xy^2 + y^3$ $\qquad\qquad\quad (x - y)^3 = x^3 - 3x^2 y + 3xy^2 - y^3$

$(x + y)^4 = x^4 + 4x^3 y + 6x^2 y^2 + 4xy^3 + y^4$ $\qquad (x - y)^4 = x^4 - 4x^3 y + 6x^2 y^2 - 4xy^3 + y^4$

$(x + y)^n = x^n + nx^{n-1} y + \dfrac{n(n-1)}{2!} x^{n-2} y^2 + \cdots + nxy^{n-1} + y^n$

$(x - y)^n = x^n - nx^{n-1} y + \dfrac{n(n-1)}{2!} x^{n-2} y^2 - \cdots \pm nxy^{n-1} \mp y^n$

Rational Zero Theorem
If $p(x) = a_n x^n + a_{n-1} x^{n-1} + \cdots + a_1 x + a_0$ has integer coefficients, then every *rational zero* of p is of the form $x = r/s$, where r is a factor of a_0 and s is a factor of a_n.

Factoring by Grouping
$acx^3 + adx^2 + bcx + bd = ax^2(cx + d) + b(cx + d) = (ax^2 + b)(cx + d)$

Arithmetic Operations
$ab + ac = a(b + c) \qquad\qquad \dfrac{a}{b} + \dfrac{c}{d} = \dfrac{ad + bc}{bd} \qquad\qquad \dfrac{a+b}{c} = \dfrac{a}{c} + \dfrac{b}{c}$

$\dfrac{\left(\dfrac{a}{b}\right)}{\left(\dfrac{c}{d}\right)} = \left(\dfrac{a}{b}\right)\left(\dfrac{d}{c}\right) = \dfrac{ad}{bc} \qquad\qquad \dfrac{\left(\dfrac{a}{b}\right)}{c} = \dfrac{a}{bc} \qquad\qquad \dfrac{a}{\left(\dfrac{b}{c}\right)} = \dfrac{ac}{b}$

$a\left(\dfrac{b}{c}\right) = \dfrac{ab}{c} \qquad\qquad \dfrac{a-b}{c-d} = \dfrac{b-a}{d-c} \qquad\qquad \dfrac{ab + ac}{a} = b + c$

Exponents and Radicals
$a^0 = 1, \quad a \neq 0 \qquad (ab)^x = a^x b^x \qquad a^x a^y = a^{x+y} \qquad \sqrt{a} = a^{1/2} \qquad \dfrac{a^x}{a^y} = a^{x-y} \qquad \sqrt[n]{a} = a^{1/n}$

$\left(\dfrac{a}{b}\right)^x = \dfrac{a^x}{b^x} \qquad\qquad \sqrt[n]{a^m} = a^{m/n} \qquad a^{-x} = \dfrac{1}{a^x} \qquad \sqrt[n]{ab} = \sqrt[n]{a}\sqrt[n]{b} \qquad (a^x)^y = a^{xy} \qquad \sqrt[n]{\dfrac{a}{b}} = \dfrac{\sqrt[n]{a}}{\sqrt[n]{b}}$